Günter Obe
Vijayalaxmi (Editors)

Chromosomal Alterations

Methods, Results and Importance in Human Health

With 99 Figures, 11 in Color, and 47 Tables

Springer

Professor Dr. Günter Obe
Gershwinstrasse 33
14513 Teltow
Germany

Professor Dr. Vijayalaxmi
Department of Radiation Oncology
The University of Texas
Health Science Center
7703 Floyd Curl Drive
San Antonio, TX 78229
USA

Cover:

In Remembrance of Sarah

Chromosomal aberrations in a metaphase of cultured fibroblasts exposed to mitomycin C (50 ng/ml for 24 hrs). Fibroblasts are from a girl with Fanconi anemia, complementation group FA-N, caused by mutations in the PALB2 gene (Reid S et al., Nat Genet 39, 162-164, 2007). The checkpoint defect in FA-N also leads to aneuploidy due to non-disjunction.
Sarah died from childhood cancer at the age of 12 months due to severe chromosome instability.

ISBN: 978-3-642-09059-2 eISBN: 978-3-540-71414-9

Springer-Verlag is a part of Springer Science+Business Media

springer.com

Editor: Dr. Dieter Czeschlik, Heidelberg
Desk Editor: Anette Lindqvist, Heidelberg

Cover Design: Design & Production, Heidelberg

Printed on acid-free paper 31/3152-HM 5 4 3 2 1 0

Günter Obe • Vijayalaxmi (Editors)

Chromosomal Alterations

Methods, Results and Importance in Human Health

Preface

This book deals mainly with chromosomal alterations which represent changes in the structure (chromosome mutations) or number (genome mutations) of chromosomes and with sister chromatid exchanges (SCE).

Double-strand breaks (DSB) in DNA are the ultimate lesions for the formation of chromosome mutation and SCE evaluated in the light microscope. Typical chromosomal alterations are chromosomal aberrations (CA) and micronuclei (MN). SCE result from repair of lesions during S phase ("true" SCE), or from chromosome-type CA induced during G1 phase ("false" SCE). Numerical alterations of chromosomes are discussed with respect to their importance in clinical cytogenetics, to their ability to give rise to MN containing whole chromosomes and to their origin during divisions of CA containing cells.

Evaluation of chromosomal alterations, especially CA, MN and SCE, has long been used in basic and applied research. Investigations into the mechanism(s) of the origin of chromosomal alterations lead to insights into the structure and function of chromosomes. Increased indices of chromosomal alterations in cells, in vitro and in vivo, are used as indicators of exposure to biological, chemical and physical genotoxins in our environment. Furthermore, elevated genotoxicity has been correlated with increased carcinogenicity. This fits with the observation of specific types of CA in cancer cells and with a positive correlation between elevated frequencies of CA and cancer risk in human populations.

DSB are induced directly by powerful DNA-damaging agents such as ionizing radiation and restriction endonucleases which lead to the formation of CA in the same cell cycle stage in which the DSB are induced: in G0/G1 phase as chromosome-type CA, in S phase as both chromosome-type and chromatid-type CA (depending on whether DSB occur in unreplicated or in replicated DNA, respectively) and in G2 phase as chromatid-type CA. Most genotoxic agents induce DNA lesions other than DSB which during S phase lead to chromatid-type CA and to "true" SCE.

CA, MN and SCE are analyzed in the light microscope after appropriate staining (generally Giemsa stain). Special pretreatment of fixed metaphase chromosomes on slides followed by staining with Giemsa stain leads to specific banding patterns that allow in-depth analyses of specific types of CA such as translocations and intra chromosomal and interchromosomal distributions of CA. Application of fluorescence in situ hybridization (FISH) opened new insights into unexpected complexities of CA. Following differential substitution of chromosomal DNA with specific agents such as bromodeoxyuridine, SCE can be made visible by staining with Giemsa stain. MN result from CA or from whole chromosomes not distributed to the cell poles. Analyses of MN are usually carried out in binucleate second-division cells after exposure to a mutagen. FISH techniques are extremely useful to investigate mechanistic aspects of the formation of MN.

The book starts with a chapter on structural and functional aspects of human chromosomes. Chapters 2–8 describe DNA lesions leading to chromosomal alterations (including gene mutations), and the ability of the cells to repair such lesions. Chapters 9 and 10 explain how DNA damage can be measured by means of γ-H2AX foci and by the comet assay. Chapters 11–14 explore mechanisms of the origin of CA (Chap. 11), the influence of nuclear and chromatin structure on CA formation (Chaps. 12, 13) and CA and SCE in telomeric regions (Chap. 14). Chapters 15–17 are devoted to MN and SCE. FISH methods and their applications for the analysis of chromosomal alterations are presented in Chaps. 18 and 19. In the following chapters topics such as changing patterns of chromosomal alterations in ongoing cell cycles (Chap. 20), human-population monitoring (Chap. 21), biological dosimetry (Chaps. 22–24) and CA in peripheral lymphocytes of astronauts (Chap. 25) are included. Chapters 26–29 are devoted to the question of possible carcinogenic and chromosome-damaging effects of low- and high-frequency electromagnetic fields. The importance of chromosomal alterations in cancer cells and as indicators of cancer risk in human populations is highlighted in Chaps. 30 and 31.

We would like to thank our colleagues who generously and willingly contributed to this book.

The editors hope that the book will serve as a textbook for graduate students in biological sciences, residents in radiology and radiation oncology, as well as researchers interested in occupational exposures, ionizing and non-ionizing radiation, environmental mutagenesis and carcinogenesis.

January 2007
Günter Obe
Vijayalaxmi

Contents

List of Contributors

SIDDHARTH G. ADHVARYU
Department of Pathology, University of Texas Health Science Center,
7703 Floyd Curl Drive, San Antonio, TX 78229-3900, USA

PASCALE V. BELLIER
Consumer and Clinical Radiation Protection Bureau, Health Canada,
775 Brookfield Road, Ottawa, ON K1A 1C1, Canada

FERDINANDO BERSANI
Department of Physics, University of Bologna, Viale Berti Pichat 6/2, 40127
Bologna, Italy

STEFANO BONASSI
Unit of Molecular Epidemiology, National Cancer Research Institute, Largo
Rosanna Benzi 10, 16132 Genoa, Italy

PETER E. BRYANT
Bute Medical School, University of St. Andrews, St. Andrews,
Fife KY16 9TS, UK

MARKUS CHRISTMANN
Department of Toxicology, University of Mainz, Obere Zahlbacher Str.
67, 55131 Mainz, Germany

ILSE CHUDOBA
MetaSystems, Robert-Bosch-Str. 6, 68804 Altlussheim, Germany

CHRISTOPH CREMER
Kirchhoff Institute of Physics, INF 227, 69120 Heidelberg, Germany

RINNE DE BONT
Study Centre Carcinogenesis and Primary Prevention of Cancer,
Department of Radiotherapy, Nuclear Medicine and Experimental
Cancerology, Ghent University, University Hospital 3K3,
De Pintelaan 185, 9000 Ghent, Belgium

María Vittoria Di Tomaso
Department of Genetic Toxicology and Chromosome Pathology, Instituto de Investigaciones Biológicas Clemente Estable, Ministry of Education and Culture, Avenida Italia 3318, CP 11600, Montevideo, Uruguay

Máximo E. Drets
Laboratory of Quantitative Microscopy, Instituto de Investigaciones Biológicas Clemente Estable, Av. Italia 3318, 11600 Montevideo, Uruguay

Sarah Driessen
femu – Research Center for Bioelectromagnetic Interaction, RWTH Aachen University, 30 Pauwelsstrasse, 52074 Aachen, Germany

Marco Durante
Department of Physics, University Federico II, Monte S. Angelo, Via Cintia, 80126 Naples, Italy

Alan A. Edwards
Radiation Protection Division, Health Protection Agency, Chilton, Didcot OX11 0RQ, UK

Michael Fenech
CSIRO Human Nutrition, Genome Health Nutrigenomics Project, PO Box 10041, Adelaide BC, Adelaide, SA 5000, Australia

Kerry George
Wyle Laboratories, 1290 Hercules Drive, Houston, TX 77058-2759, USA

Wolfgang Goedecke
Department Biology and Geography, University of Duisburg-Essen, Universitätsstraße 5, 45117 Essen, Germany

Makoto Hayashi
Division of Genetics and Mutagenesis, National Institute of Health Sciences, 1-18-1 Kamiyoga, Setagaya-ku, Tokyo 158-8501, Japan

George Iliakis
Institute of Medical Radiation Biology, University of Duisburg-Essen Medical School, 45122 Essen, Germany

Christian Johannes
Department of Genetics, Universität Duisburg-Essen, Universitätsstraße 5, 45117 Essen, Germany

Bernd Kaina
Department of Toxicology, University of Mainz, Obere Zahlbacher Str. 67, 55131 Mainz, Germany

Jürgen Kiefer
Am Dornacker 4, 35435 Wettenberg, Germany

Frank Klubertz
femu – Research Center for Bioelectromagnetic Interaction, RWTH Aachen University, 30 Pauwelsstrasse, 52074 Aachen, Germany

Gregor Kreth
Kirchhoff Institute of Physics, INF 227, 69120 Heidelberg, Germany

Tracy Lightfoot
Epidemiology and Genetics Unit, Department of Health Sciences, University of York, York, UK

David C. Lloyd
Radiation Protection Division, Health Protection Agency, Chilton, Didcot OX11 0RQ, UK

Wilner Martínez-López
Department of Genetic Toxicology and Chromosome Pathology, Instituto de Investigaciones Biológicas Clemente Estable, Ministry of Education and Culture, Avenida Italia 3318, CP 11600, Montevideo, Uruguay, and Institute of Biology, Faculty of Sciences, Montevideo, Uruguay

James P. McNamee
Consumer and Clinical Radiation Protection Bureau, Health Canada, 775 Brookfield Road, Ottawa, ON K1A 1C1, Canada

Leticia Méndez-Acuña
Department of Genetic Toxicology and Chromosome Pathology, Instituto de Investigaciones Biológicas Clemente Estable, Ministry of Education and Culture, Avenida Italia 3318, CP 11.600, Montevideo, Uruguay

William F. Morgan
Radiation Oncology Research Laboratory, and Marlene and Stewart Greenebaum Cancer Center, BRB 7-011, University of Maryland, 655W. Baltimore St., Baltimore, MD 21201-1551, USA

María Mühlmann
Molecular Cytogenetic Laboratory, Radiobiology Department, National Commission of Atomic Energy, Buenos Aires, Argentina

Wolfgang-Ulrich Müller
Institut für Medizinische Strahlenbiologie, Universität Duisburg-Essen, Universitätsklinikum Essen, Hufelandstraße 55, 45122 Essen, Germany

HEIDEMARIE NEITZEL
Institut für Humangenetik, Charité – Universitätsmedizin Berlin, Campus Virchow-Klinikum, Augustenburger Platz 1, 13353 Berlin, Germany

GÜNTER OBE
Gershwinstrasse 33, 14513 Teltow, Germany

URSULA OESTREICHER
Division for Radiation Protection and Health, Federal Office for Radiation Protection, Ingolstaedter Landstrasse 1, 85764 Oberschleissheim, Germany

GABRIEL E. PANTELIAS
Institute of Nuclear Technology and Radiation Protection, National Centre for Scientific Research "Demokritos", 15310 Ag. Paraskevi Attikis, Athens, Greece

LOTHAR RIEDEL
Bundesamt für Verbraucherschutz und Lebensmittelsicherheit, Mauerstraße 39-42, 10117 Berlin, Germany

HORST ROMM
Division for Radiation Protection and Health, Federal Office for Radiation Protection, Ingolstaedter Landstrasse 1, 85764 Oberschleissheim, Germany

WYNAND P. ROOS
Department of Toxicology, University of Mainz, Obere Zahlbacher Str. 67, 55131 Mainz, Germany

PAVEL RÖSSNER
Laboratory of Genetic Ecotoxicology, Institute of Experimental Medicine AS CR, Vídeňská 1083, 142 20 Prague 4, Czech Republic

FEDERICO F. SANTIÑAQUE
Laboratory of Quantitative Microscopy, Instituto de Investigaciones Biológicas Clemente Estable, Av. Italia 3318, 11600 Montevideo, Uruguay

JOHN R.K. SAVAGE
34 City Road, Tilehurst, Reading RG31 5HB, UK

MARIA ROSARIA SCARFÌ
CNR Institute for Electromagnetic Sensing of Environment, Via Diocleziano 328, 80124 Naples, Italy

JOACHIM SCHÜZ
Institute of Cancer Epidemiology, Danish Cancer Society, Copenhagen, Denmark

Jiri Silny
Femu – Research Center for Bioelectromagnetic Interaction, RWTH Aachen University, 30 Pauwelsstrasse, 52074 Aachen, Germany

Marianne B. Sowa
Chemical Structure & Dynamics, Pacific Northwest National Laboratory, Richland, WA 99354, USA

Radim J. Sram
Laboratory of Genetic Ecotoxicology, Institute of Experimental Medicine AS CR, Vídeňská 1083, 142 20 Prague 4, Czech Republic

Günter Stephan
Division for Radiation Protection and Health, Federal Office for Radiation Protection, Ingolstaedter Landstrasse 1, 85764 Oberschleissheim, Germany

Marta Szłuińska
Radiation Protection Division, Health Protection Agency, Chilton, Didcot OX11 0RQ, UK

Irena Szumiel
Department of Radiobiology and Health Protection, Institute of Nuclear Chemistry & Technology, ul Dorodna 16, 03-195 Warsaw, Poland

Georgia I. Terzoudi
Institute of Nuclear Technology and Radiation Protection, National Centre for Scientific Research "Demokritos", 15310 Ag. Paraskevi Attikis, Athens, Greece

Marc Trimborn
Institut für Medizinische Genetik, Charité – Universitätsmedizin Berlin, Campus Virchow-Klinikum, Augustenburger Platz 1, 13353 Berlin, Germany

Nik van Larebeke
Study Centre Carcinogenesis and Primary Prevention of Cancer, Department of Radiotherapy, Nuclear Medicine and Experimental Cancerology, Ghent University, University Hospital 3K3, De Pintelaan 185, 9000 Ghent, Belgium

Ursula von Recklinghausen
Institute of Hygiene and Occupational Health, Universitätsklinikum Essen, Hufelandstraße 55, 45122 Essen, Germany

Clemens von Sonntag
Max-Planck-Institut für Bioanorganische Chemie, Stiftstr. 34-36, 45413 Mülheim an der Ruhr, Germany, and Leibniz-Institut für Oberflächenmodifizierung (IOM), Permoserstr. 15, 04303 Leipzig, Germany

Minli Wang
Institute of Medical Radiation Biology, University of Duisburg-Essen Medical School, 45122 Essen, Germany

Roman Wienert
femu – Research Center for Bioelectromagnetic Interaction, RWTH Aachen University, 30 Pauwelsstrasse, 52074 Aachen, Germany

Andrzej Wojcik
Department of Radiation Biology and Health Protection, Institute of Nuclear Chemistry and Technology, ul. Dorodna 16, 03-195 Warsaw, Poland, and Department of Radiobiology and Immunology, Institute of Biology, Swietokrzyska Academy, ul. Swietokrzyska 15, 25-406 Kielce, Poland

Wenqi Wu
Institute of Medical Radiation Biology, University of Duisburg-Essen Medical School, 45122 Essen, Germany

1 Human Chromosomes: Structural and Functional Aspects

Heidemarie Neitzel and Marc Trimborn

Abstract

Understanding the structural and molecular basis of the mitotic chromosome remains a basic challenge in cell biology and cytogenetics. The chromosomal behavior during cell division was first described in 1882. At the beginning of the last century, the chromosome theory of inheritance combined the cytological observations with the principles of Mendelian inheritance deduced from breeding experiments. This fundamental theory explained for the first time that the chromosomes are the units of heredity which are arrayed linearly on chromosomes as well as genetic linkage, chromosomal recombination, and the independent assortment of alleles localized on different chromosomes. The chromosome theory of inheritance was the prerequisite for the important improvements in the fields of experimental and clinical cytogenetics. After the definition of the exact number of human chromosomes $2n = 46$ in 1955, it was demonstrated that different human aneuploidies are the leading cause of fetal loss, birth defects, and mental retardation and that duplications or deletions of even smaller chromosomal segments have profound consequences on normal gene expression, resulting in severe developmental and physiologic abnormalities. In the 1980s and 1990s, the field of conventional cytogenetics was revolutionized by the introduction of molecular cytogenetic techniques resulting in the recognition of the importance of subtle cytogenetic aberrations, such as microdeletions and microduplications. Our insights into cell cycle progression, which is coordinated by a complex network of checkpoints to monitor chromosome structure, with DNA repair and spindle formation led to the identification of Mendelian disorders affecting chromosome integrity, some of them associated with high genomic instability and a markedly increased cancer risk. It was the thorough observation of chromosomal changes by many researchers during the last four decades which paved the way for our understanding of the underlying mechanisms of many congenital disorders, as well as of chromosome surveillance, DNA repair, and cancer susceptibility.

Günter Obe and Vijayalaxmi (Eds.)
Chromosomal Alterations: Methods, Results and Importance in Human Health

1.1 History of Chromosome Research

"Chromosomes have attracted many microscopists not only because these sausage-like bodies represent vehicles of genetic material (and hence, are biologically important) but also because they are hypnotically beautiful objects" (Hsu 1979).

The first cytologist who described chromosome behavior during cell division and how chromosomes move during mitosis was Walter Flemming (1882) in 1882. His terms "prophase," "metaphase," and "anaphase" are still used to describe the different steps of mitosis. In 1888 the structures were termed "chromosomes" (Greek *chroma* meaning "color" and *soma* meaning "body") by the German anatomist Heinrich Waldeyer, because they were particularly well stained with a certain nuclear dye. Flemming's work and the rediscovery of Mendel's laws were the basis of the chromosome theory of inheritance. Using the fruit fly *Drosophila melanogaster* as a model organism, Thomas Hunt Morgan and his students at Columbia University, who included such important geneticists as Alfred Sturtevant, Hermann Muller, and Calvin Bridges, made many important contributions to genetics. They showed that genes, strung on chromosomes, are the units of heredity which are arrayed linearly on chromosomes. They described the independent assortment of alleles localized on different chromosomes, X-linked inheritance, genetic linkage, and chromosomal recombination.

The chromosome theory of inheritance combined the cytological observations with the principles of Mendelian inheritance deduced from breeding experiments (Fig. 1.1). The two homologous chromosomes in somatic cells correspond to two alleles, one of each inherited from the mother and the father. The chiasmata observed in meiosis I corresponds to the recombination or crossing over events. The segregation of homologous chromosomes during meiosis correlates with the segregation of alleles into separate gametes.

In 1953 Francis Crick and James Watson described the double-helix structure of DNA and concluded in their paper, published in *Nature*, "It has not escaped our notice that the specific pairing we have postulated immediately suggests a possible copying mechanism for the genetic material" (Watson and Crick 1953). Subsequently, Matthew Meselson and Franklin Stahl (1958) demonstrated that DNA replicates semiconservatively, with each strand in a DNA molecule from the parent generation pairing with a new strand in the daughter generation.

In 1955 Tjio and Levan (1956) defined the exact number of human chromosomes $2n = 46$ and in 1959 Lejeune et al. (1959) described the first aneuploidy in humans, demonstrating that children with Down syndrome have three instead of two copies of chromosome 21. Although more than 100 years has passed since the first observation of the chromosome (Flemming 1882) a number of researchers have continued to study chromosomes and their behavior during the cell cycle. Important improvements were made during

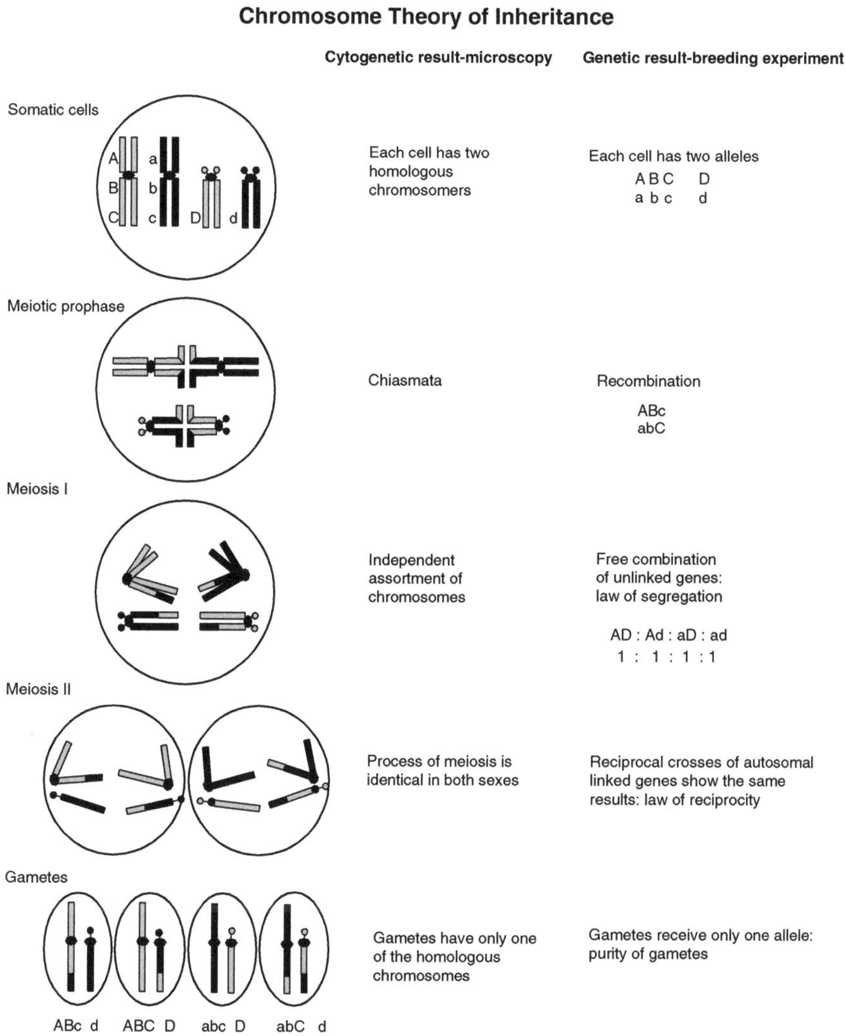

Fig. 1.1 Chromosome theory of inheritance (for details, see the text)

the 1960s and 1970s. The establishment of lymphocyte cultures as an easily accessible source for chromosome preparations (Nowell 1960) and the development of the differential banding methods in the 1970s led to considerable improvement in the fields of experimental and clinical cytogenetics (Caspersson et al. 1969; Drets and Shaw 1971; Dutrillaux and Lejeune 1971; Patil et al. 1971; Sumner et al. 1971). In the 1980s and 1990s, the field of conventional cytogenetics was again revolutionized by the introduction of molecular cytogenetic techniques such as fluorescence in situ hybridization (FISH), spectral karyotyping, and comparative genomic hybridization (CGH), leading to considerable progress especially in clinical cytogenetics (Cremer et al. 1986; Pinkel et al. 1986; Kallioniemi et al. 1992; Schrock et al. 1996). As a result, there was an increased appreciation of the importance of "subtle" constitutional cytogenetic aberrations, such as microdeletions and imprinting disorders.

1.2 Composition and Compartmentalization of Human Chromosomes

Human DNA is composed of 60% single-copy DNA sequences and 40% repetitive DNAs. The characteristic of human and all other mammalian genomes is its compartmentalization, which finds its expression under the light microscope as G-, R-, T-, and C-bands (Korenberg and Rykowski 1988; Holmquist 1989, 1992). The different chromatin domains differ not only in their AT/GC content but also in their gene content, their replication timing, and their repetitive elements. G-, R-, and T-bands are defined as euchromatin containing most of the protein coding sequences which make up approximately 2% of the total DNA. The highest gene density is found in the telomeric T-bands, followed by the Giemsa-light bands or R-bands. These two compartments harbor predominantly the housekeeping genes which are essential for the metabolism of each single cell. R- and T-bands are accompanied by short repetitive nonviral retroposons, so-called short interspersed nucleotide elements (SINEs). These repeats are propagated in the genome by retroposition and can be classified as either autonomous or nonautonomous elements. The most prominent nonautonomous SINE member is the Alu family in humans and the B1 family in mice. The most abundant members of the autonomous retroposons are the long interspersed nucleotide elements (LINEs), which are several kilobases long and preferentially located in the dark G-bands where the gene density is much lower (Chen and Manuelidis 1989; Ostertag 2001). The C-bands equate to the constitutive heterochromatin, are devoid of protein coding sequences, and are thought to be genetically inert. Constitutive heterochromatin is mainly composed of satellite DNA containing simple repetitive elements, organized as complex tandem arrays (Waye and Willard 1989; Willard 1989; Fig. 1.2).

The different compartments are characterized by differences in their replication timing during S phase. Incorporation of the synthetic thymidine analog bromodeoxyuridine during S phase leads to a characteristic banding pattern which shows that the T- and R-bands replicate in the first half of S phase, while the dark G-bands replicate in the second half. At the very end of S phase the constitutive heterochromatin is replicated as well as the inactivated X chromosome in mammalian females designated as facultative heterochromatin.

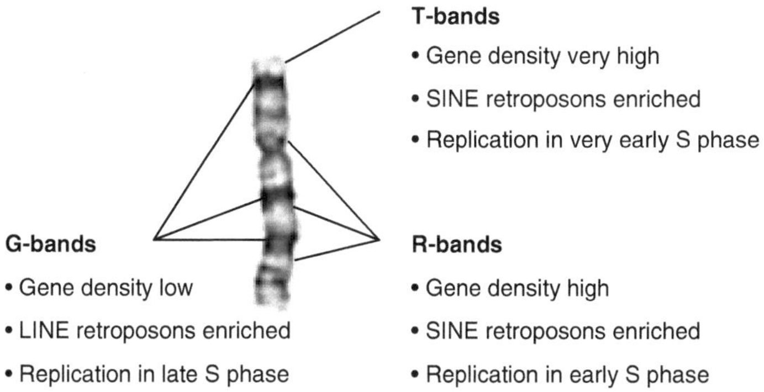

Fig. 1.2 Compartmentalization of human chromosomes. *LINE* long interspersed nucleotide element, *SINE* short interspersed nucleotide element

The differences in gene content of the different compartments can be impressively illustrated by the gene mapping data of the Human Genome Project (Fig. 1.4). The smallest human autosome, chromosome 21, which has a broad Giemsa-dark band comprises 46.9 Mbp of DNA and accommodates 352 genes. In contrast, chromosome 22 consists mainly of R-band material, it is 50 Mbp in physical length, only slightly bigger than chromosome 21, and harbors 742 genes (National Center for Biotechnology Information, Map Viewer, build 36).

1.3 The Human Karyotype and Clinical Cytogenetics

In humans, the normal diploid number of chromosomes is 46, consisting of 22 pairs of autosomal chromosomes numbered 1–22 and one pair of sex chromosomes (XX in females and XY in males) (Fig. 1.3). The genome is estimated to contain approximately 25,000 genes which are distributed along 23 chromosome pairs. As mentioned already, even the smallest autosome, chromosome 21, contains 352 genes. Thus, it is not surprising that duplications or deletions of chromosomes, or even small chromosome segments, have profound consequences on normal gene expression, leading to severe developmental and physiologic abnormalities.

Deviations in the number or structure of the 46 human chromosomes are astonishingly common, despite their severe deleterious consequences. Chromosomal disorders occur in an estimated 10–25% of all pregnancies. They are the leading cause of fetal loss and, among pregnancies continuing to term, are the leading known cause of birth defects and mental retardation. The most common chromosomal aberrations are trisomies for various chromosomes, indicating that chromosome segregation at meiosis is an extremely error-prone process in

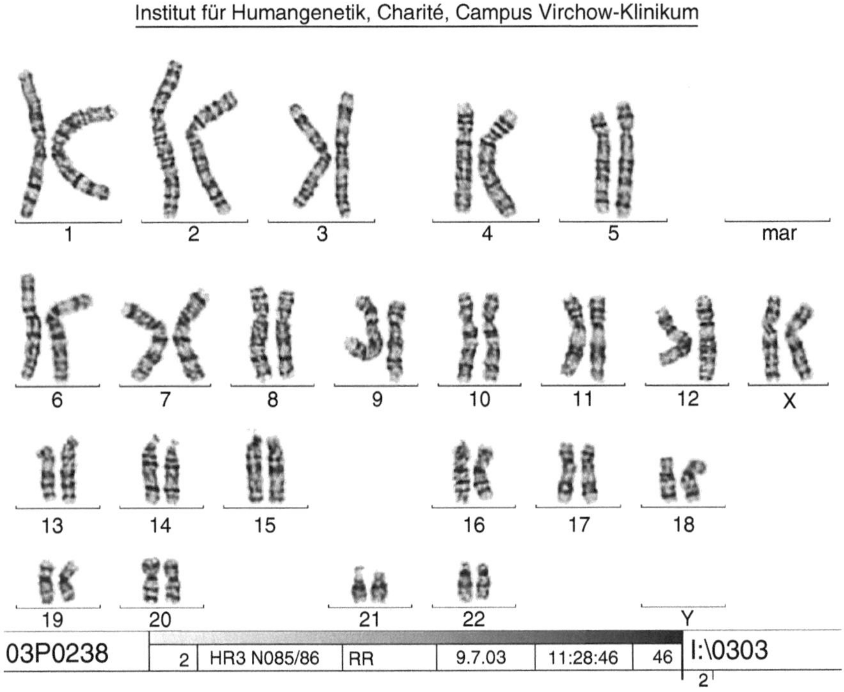

Fig. 1.3 Human female karyotype 46,XX after G-banding

humans. Most chromosomal aneuploidies originate from female meiosis and contribute significantly to pregnancy failures, particularly in women of advanced maternal age. The most common autosomal trisomy is trisomy 21, with a frequency of approximately 1 in 700 live births. In contrast, trisomy 22 results in spontaneous miscarriage due to the extent of imbalance since chromosome 22 carries twice as many genes (742 genes) as chromosome 21. In addition to trisomy 21, only two other autosomal trisomies with low gene density, trisomy 13 (551 genes) and trisomy 18 (432 genes), occur in live births with a prevalence of 1 in 10,000 and 1 in 20,000, respectively. However, most conceptions with trisomy 13 and trisomy 18 result in fetal loss during pregnancy or are associated with death in infancy, typically occurring during the first year of life. Autosomal trisomies of all other chromosomes are not compatible with survival to term, indicating that dose imbalance of the number of autosomal genes has a severe effect on developmental processes and survival (Fig. 1.4). This is further illustrated by the fact that all full autosomal monosomies do not survive to term. Karyotype–phenotype correlation studies on the basis of clinical findings demonstrate that monosomies for an autosomal segment cause more, and more severe alterations to the phenotype and restrict survival more than do trisomies for the same segment.

In contrast to autosomal imbalances, sex chromosome trisomies such as 47,XXX and 47,XXY have few phenotypic complications owing to the

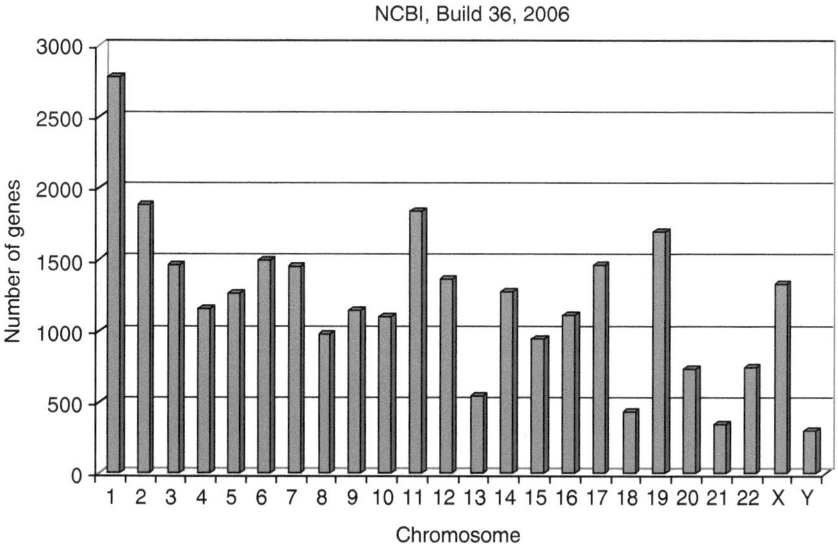

Fig. 1.4 Number of genes mapped to human chromosomes (National Center for Biotechnology Information, *NCBI*, Map Viewer, build 36)

mechanism of X chromosome inactivation, or as in 47,XYY owing the low number of Y-linked genes, most of which are involved in testicular development or spermatogenesis.

The introduction of molecular cytogenetic techniques such as FISH revolutionized the field of conventional cytogenetics, leading to considerable progress especially in clinical cytogenetics (Cremer et al. 1986; Pinkel et al. 1986; Kallioniemi et al. 1992; Schrock et al. 1996; Fig. 1.5). Subsequently, numerous deletion and duplication syndromes have been described that are too small to be detected under the microscope using conventional cytogenetic methods, such as G-banding. The molecular cytogenetic methods have expanded the possibilities for precise genetic diagnoses, which are extremely important for clinical management of patients and appropriate counseling of their families. Depending on the size of the deletion or duplication, specific FISH probes are employed to identify the aberration (Tonnies 2002).

Since most microdeletion/microduplication syndromes are defined by a common deleted/duplicated region, the abnormal dose of genes located within these regions can explain the phenotypic similarities among individuals with a specific syndrome. Consequently, detailed genotype–phenotype correlations provide a unique resource towards the genetic dissection of complex phenotypes such as congenital heart defects, mental and growth retardation, and specific cognitive and behavioral components of humans. For example, the Williams–Beuren syndrome (WBS), a neurodevelopmental disorder caused by a microdeletion at 7q11.23, provides one of the most convincing models of a relationship that links genes with human cognition and behavior. Detailed

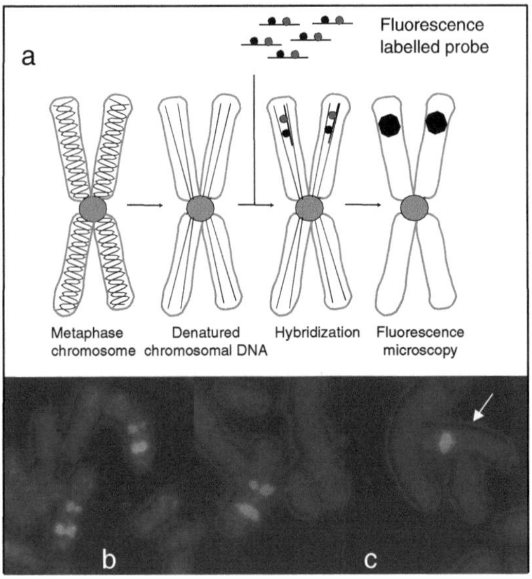

Fig. 1.5 **a** Principle of fluorescence in situ hybridization (FISH), **b** FISH with the Elastin probe for Williams–Beuren syndrome (WBS) and a control region in a normal individual, and **c** in a patient with WBS deletion showing the missing signal in one of the homologous chromosomes (*arrow*)

molecular characterization of the deletion alongside well-defined cognitive profiling in WBS provides a unique opportunity to investigate the neuromolecular basis of complex cognitive behavior and develop integrated approaches to study gene function and genotype–phenotype correlations (Osborne 1999; Mervis and Klein-Tasman 2000).

Furthermore, the molecular analysis of these chromosomal aberrations has led to a growing understanding of their mechanisms of origin, indicating that certain regions of the human genome are especially prone to structural rearrangements due to the presence of repetitive sequence elements. Interaction between chromosome-specific repetitive DNA leads to gain, loss, or inversion of the intervening sequence by nonallelic homologous recombination between misaligned repetitive elements. Where a particular region contains dose-sensitive or imprinted genes, this can lead to a specific genetic disease: loss of 7q11.23 results in Williams syndrome (Osborne 1999), loss of 22q11 results in Di George syndrome/velocardiofacial syndrome (Cuneo 2001), and loss of 15q11–q13 results in either Prader–Willi syndrome or Angelman syndrome (Cassidy et al. 2000; Horsthemke and Buiting 2006; Thomas et al. 2006). The available data demonstrate that the majority of rearrangements, approximately three or four, are interchromosomal. Therefore, they are likely to have arisen as the result of unequal meiotic crossing over between repetitive elements on different chromosome homologues. The remaining intrachromosomal rearrangements are also likely to be meiotic, although for these cases a postzygotic error during mitosis cannot be excluded (Thomas et al. 2006).

1.4 Cell Cycle and Chromosome Cycle

The cell cycle consists of four distinct phases: G1 phase, S phase, G2 phase, and M phase, or mitosis. G1, S, and G2 comprise together the interphase. G1 phase is a growth phase with high metabolic activity increasing the amount of cytoplasm and important organelles for preparing the cell for duplicating its DNA in S phase. In G2 phase, the cell continues with growth and metabolism in preparation for undergoing mitosis. During M phase the replicated chromatids segregate to the two daughter cells.

The chromosomes at G1, S, and G2 phases can be directly visualized after premature chromosome condensation (PCC) (Rao et al. 1982; Fig. 1.6). Fusion between mitotic and interphase cells results in a rapid chromosome condensation, with dissolution of the nuclear membrane due to the activity of the mitosis-promoting factor (MPF) in the interphase cells. The morphology of PCC chromosomes varies according to the stage of the interphase cell at the time of fusion. Thus, the PCC at G1 phase are very long with single chromatids and those at G2 phase are elongated with slender double chromatids. PCC chromosomes at S phase are characterized by their fragmented, pulverized appearance. The gaps of S-PCC represent the sites of DNA replication.

In all phases of the cell cycle, the surveillance of the chromosomal integrity is crucial for the genetic processes. Therefore, a complex network of checkpoints has evolved to monitor chromosome structure and coordinate cell cycle progression with DNA repair and spindle formation. Checkpoints are coordinated series of responses that delay progress through the cell cycle at a particular phase or transition in response to the lack of appropriate conditions

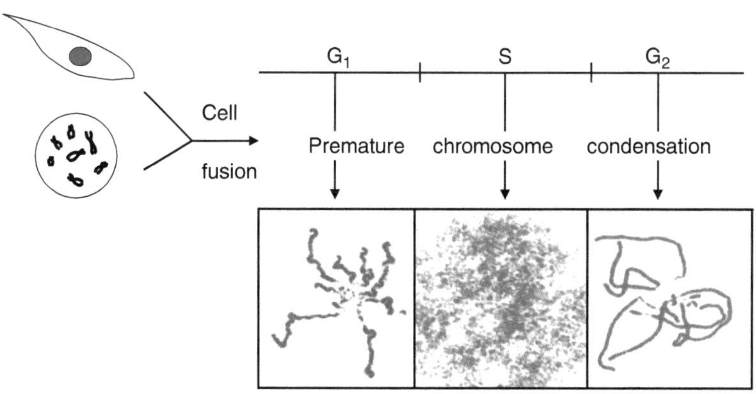

Fig. 1.6 Visualization of G1-, S-, and G2-phase chromosomes after premature chromosome condensation

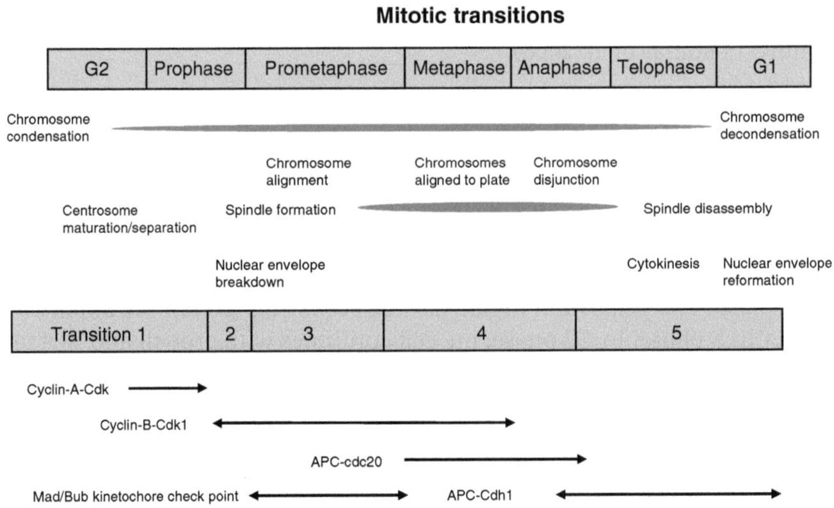

Fig. 1.7 Mitotic transitions according to the classification of Pines and Rieder (2001)

for progression. Delay allows enough time for crucial processes such as DNA repair and spindle attachment to be completed before continuing the cell-division cycle or initiating apoptosis (Hartwell and Weinert 1989; Hartwell and Kastan 1994; Nurse 1997; Clarke and Gimenez-Abian 2000; Morrison and Rieder 2004).

Especially, cell entry into mitosis is under the control of a tightly regulated network of protein kinases, cyclins, and protein phosphatases. According to Pines and Rieder (Pines and Rieder 2001) G2 phase and mitosis can be subdivided into five transitional phases which are characterized not only by the structural and behavioral changes of the chromosomes and the spindle, but also, at the molecular level, by the activation and inactivation of cell cycle regulators such as the cyclin-dependent kinases (Cdks) and the anaphase-promoting complex (APC) (Fig. 1.7). In vertebrates, the G2–M transition is initiated by the increase of cyclin-A-Cdk2 throughout the G2 phase of the cell cycle, resulting in chromosome condensation in the absence of significant cyclin-B1-Cdk1 activity. Subsequently, the cyclin-B1-Cdk1 complex, also known as mitosis-promoting factor (MPF), is activated as a result of its dephosphorylation by Cdc25 and rapidly accumulates in the nucleus, followed by the breakdown of the nuclear envelope and the entry of the cell into metaphase.

1.5 Shaping the Metaphase Chromosome

Chromosomes are complex and highly dynamic structures containing DNA, histones, and non-histone proteins. Understanding the structural and molecular basis of mitotic chromosome condensation remains a basic challenge in cell biology. During cell cycle progress towards mitosis, the chromosome undergoes progressive morphological conversion. From prophase to metaphase, apparently amorphous interphase chromatin is reorganized into individual chromosomes, with a pair of separate sister chromatids. This process, referred to as chromosome assembly, implying chromosome condensation and sister-chromatid resolution, is an essential prerequisite for the faithful segregation of duplicated genetic information into two daughter cells.

The human genome contains 6×10^9 bp of DNA per diploid cell, corresponding to approximately 1.7 m of DNA which is organized in chromatin fibers. The basic unit of chromatin is the nucleosome, which consists of 146 bp of DNA wound around an octamer of histone proteins, two each of histones H2A, H2B, H3, and H4 accounting for the first sixfold to sevenfold linear compaction of the DNA (Bednar et al. 1998).

The binding of linker histone H1 to linker DNA sequences, so-called scaffold-associated regions (SARs), localized between nucleosomes leads to further chromatin compaction – the 30-nm fiber – generating another sixfold to sevenfold compaction. It has been proposed that the 30-nm chromatin fiber is accomplished by a unique three-dimensional zigzag folding pattern rather than by supercoiling. In mitosis, the 30-nm fiber must compact another 200- to 500-fold to achieve the final 10,000- to 20,000-fold linear compaction of the mitotic chromosome (Swedlow and Hirano 2003; Fig. 1.8). Early studies suggested the presence of a chromosome scaffold composed of non-histone proteins that serve as the backbone of the mitotic chromosome. Two major scaffold proteins, topoisomerase II and SC2, a structural maintenance of chromosomes (SMC) family member, have been shown to be part of the scaffold and involved in condensation. *cis* sites for condensation are expected to lie along the chromosomal axis in vivo. Scaffold models propose that the SARs are the *cis*-acting DNA sequences that serve as binding sites for DNA topoisomerase II, a major component of the chromosome scaffold, resulting in radial DNA loops of approximately 50–100 kb in length (Earnshaw et al. 1985; Earnshaw and Heck 1985). Yet the radial loop model has remained highly controversial.

The highly organized chromatin can be modified further by various mechanisms, such as posttranslational modifications of histones, ATP-dependent chromatin remodeling, and the exchange of histone proteins.

1.6 Cohesion and Condensins

The discovery of SMC proteins led to rapid progress in our understanding of chromosome organization and behavior. The identification and characterization of cohesion and the condensins, SMC-containing complexes, demonstrated them to be key regulators that function in chromosome assembly and segregation during mitosis (Hirano 2005, 2006; Losada and Hirano 2005; Nasmyth and Haering 2005). SMC proteins are ubiquitous in organisms from

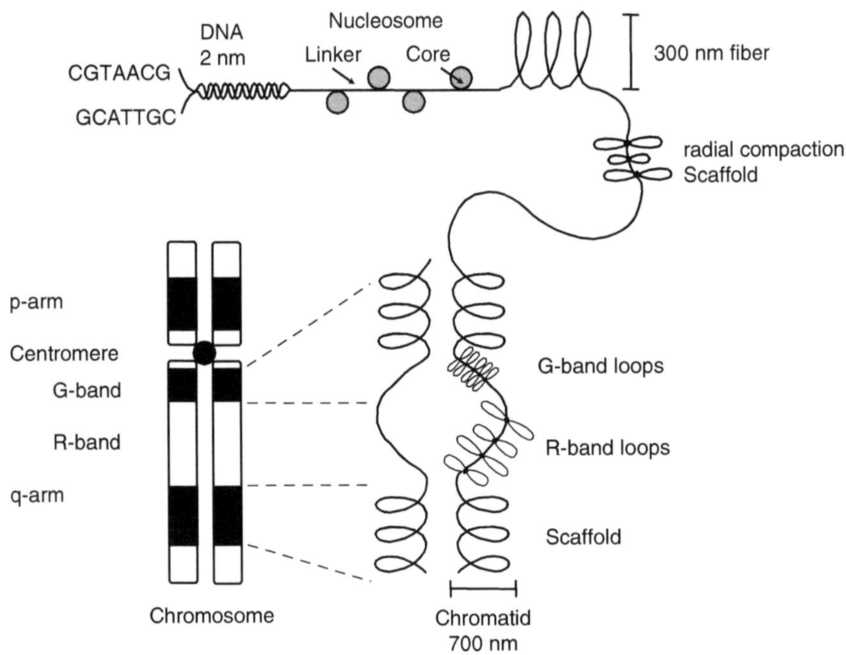

Fig. 1.8 Model of chromatin compaction from DNA double helix to metaphase chromosome

bacteria to humans. The SMC proteins SMC1 and SMC3 constitute the core of the cohesin complex and bind the non-SMC proteins Scc1 and Scc3 to form a ringlike structure that mediates sister-chromatid cohesion. A model of the interaction of cohesion predicts that the establishment of sister-chromatid cohesion is accomplished when a replication fork passes through the cohesin ring that is preloaded during the G1 phase of the cell cycle. The sister chromatids generated by DNA replication become aligned along the entire length of their arms and at the kinetochore. In vertebrates, most cohesin dissociates from chromatin at prophase, and only a small population, enriched in the pericentromeric region, remains on the chromosomes until metaphase. Cohesion is essential for the congression and alignment of chromosomes from prometaphase to metaphase. At the onset of anaphase, loss of cohesion triggers the separation of sister chromatids, allowing them to be pulled apart to opposite poles of the cell.

The core component of the two condensin complexes is the monomers SMC2 and SMC4 (Hirano 2005). The SMC cores are bound by different sets of regulatory subunits forming the functional complexes. Vertebrates have two different condensin complexes, condensin I and condensin II, each containing a unique set of regulatory subunits. CAP-G, CAP-D2, and CAP-H bind to the condensin core to form condensin I (Hirano et al. 1997), whereas condensin II is defined by its regulatory subunits CAP-G2, CAP-D3, and CAP-H2 (Ono et al. 2003). Condensin I (and possibly condensin II as well) has the ability to introduce positive helical tension into double-stranded DNA in vitro (Kimura and Hirano 1997; Kimura et al. 1999; Bazett-Jones et al. 2002; Hagstrom et al. 2002). While the two complexes cooperate to assemble metaphase chromosomes (Ono et al. 2003), their behaviors are regulated differently during the cell cycle (Hirota et al. 2004; Ono et al. 2004; Trimborn et al. 2006). Condensin II is nuclear throughout the cell cycle and participates in an early stage of prophase chromosome condensation within the nucleus, whereas condensin I gains access to chromosomes only after the breakdown of the nuclear envelope. Both complexes finally bind to the central chromatid axes in an alternate pattern. The molecular mechanism underlying the differential regulation of the two condensin complexes remains to be determined, but it was proposed that sequential activation of cyclin A/Cdk and cyclin B/Cdk could be responsible for the successive loading of condensin I and condensin II (Hirano 2005). The loading of condensins is a prerequisite for the proper assembly and segregation of metaphase chromosomes. The diverse functions of the SMC complexes, however, range far beyond chromosome segregation and may include gene regulation and DNA repair.

1.7 DNA Repair

The DNA damage checkpoints cause cell cycle delay before or during the decisive cell cycle transitions of replication and mitosis (G1/S, intra-S, G2/M checkpoints) (Sancar et al. 2004). This involves a number of highly conserved proteins that sense the damage and signal the cell cycle machinery. Central to this network are two protein kinases, ataxia telangiectasia mutated (ATM) and ATM-Rad3 related (ATR). These kinases sense the DNA damage and start signaling cascades which finally result in cell cycle arrest or induce apopoptic pathways. ATM kinase is primarily activated by DNA double-strand breaks (DSBs) induced by ionizing irradiation, whereas ATR kinase responds to UV-induced and replication-specific DNA damage. ATR kinase activation demands the association with the protein ATRIP and two additional complexes RAD17 and 9-1-1. The MRE11-NBS-RAD50 (MNR) complex plays a crucial role in the ATM kinase mediated answer on DNA DSBs. It was generally thought that ATM kinase and ATR kinase work independently. Recent reports, however, indicate that ATM kinase and nuclease activity of meiotic recombination 11 (MRE11) are required for the processing of DNA DSBs to generate the replication protein A (RPA) coated single-stranded DNA that is needed for ATR kinase recruitment (Watson and Crick 1953; Adams et al. 2006; Jazayeri et al. 2006). Critical for the signal transduction following the damage detection are a vast number of mediators, many of them (e.g., TOPBP1, 53BP1, and BRCA1) containing BRCT domains. By means of these transducers and the checkpoint kinases CHK1 and CHK2, the signal is transferred to the target/effector proteins like CDC25 phophatases, p53, or SMC1 (Lee 2002; Kastan and Bartek 2004; Sancar et al. 2004; Li and Zou 2005). DNA DSBs are repaired by two distinct but connected pathways: nonhomologous end-joining (NHEJ) and homologous repair (HR). NHEJ rejoins the two ends of a DSB by simple ligation in an error-prone process, while HR uses a homologous template to copy and restore the information disrupted by the break. This promotes error-free repair. Since the information is usually copied from an intact sister chromatid, HR is the preferred pathway in S and G2 phases, while NHEJ is the predominant pathway in the G1 phase (reviewed in Lisby and Rothstein 2004).

1.8 Mendelian Disorders Affecting Chromosome Integrity

1.8.1 Chromosome Instability Disorders

Our insights into DNA repair processes in normal cells have been considerably improved by the identification of the underlying genetic defects of chromosome instability syndromes. Chromosome instability syndromes are a group of inherited disorders associated with high genomic instability and a markedly increased cancer risk. In the following, we focus on two instability syndromes, one implicated in the repair of DSBs and one involved in the repair of DNA cross-links.

Patients affected by Nijmegen breakage syndrome (NBS) have biallelic mutations in the *NBS1* gene, mapped on chromosome 8q21. It encodes a 95-kDa protein called nibrin, a member of the hMre11hRad50 protein complex, involved in the ATM-dependent DNA damage signaling pathway of cellular response to DSBs. The affected patients present with microcephaly, a distinct facial appearance, growth retardation, immunodeficiency, cytogenetic abnormalities, radiosensitivity, and high susceptibility to lymphoid malignancy. In 40% of patients, a malignancy occurs before the age of 21 (Varon et al. 1998, 2001, Varon et al. 2003; Kitagawa and Kastan 2005; Kruger et al. 2007).

Fanconi anemia is a genetically heterogeneous, autosomal recessive or X-recessive chromosome instability disorder with increased hypersensitivity to cross-linking agents. At least 12 genetic complementation groups have been described (FA-A, FA-B, FA-C, FA-D1, FA-D2, FA-E, FA-F, FA-G, FA-I, FA-J, FA-L, FA-M) and all except FA-I have been linked to a distinct gene. All Fanconi anemia proteins act in a single pathway involved in DNA cross-link repair. Most Fanconi anemia proteins form a complex that activates the FANCD2 protein via monoubiquitination, which is prerequisite for the activation of *BRCA2*, a gene which was originally identified in families with increased breast and ovarian cancer susceptibility. Disruption of any of the Fanconi anemia proteins results in an increased chromosomal instability (Fig. 1.9a). Fanconi anemia patients have a high risk for bone marrow failure, aplastic anemia, myelodysplastic syndrome, acute myeloid leukemia, and, later in life, epithelial malignancies. The most life-threatening early event in most complementation groups is bone marrow failure, which occurs typically during the first decade of life (Wang and D'Andrea 2004; Kennedy and D'Andrea 2005; Bagby and Alter 2006; Lyakhovich and Surralles 2006).

These two examples might be sufficient to demonstrate that any mechanism impairing chromosome surveillance and chromosomal integrity has profound effects on cancer formation and progression.

1.8.2 PCC Syndrome

The first description of a disorder in humans affecting the fundamental process of chromosome condensation was reported in 2002 (Neitzel et al. 2002). The patients' chromosomes display PCC (PCC syndrome) in early G2 phase commencing as soon as 1 h after completion of S phase and also delayed decondensation after mitosis (Fig. 1.9b).

In 2004, it was demonstrated that PCC syndrome is caused by mutations in the *MCPH1* gene encoding microcephalin (Trimborn et al. 2004). Microcephalin encompasses 835 amino acids and contains one N-terminal und two C-terminal BRCT domains (BRCA1 C-terminus) linking its function to DNA checkpoint control and/or DNA repair. The clinical phenotype is characterized by microcephaly, growth retardation, and mental retardation. These findings implicated microcephalin as a novel regulator of chromosome condensation and linked the apparently disparate fields of neurogenesis and chromosome biology.

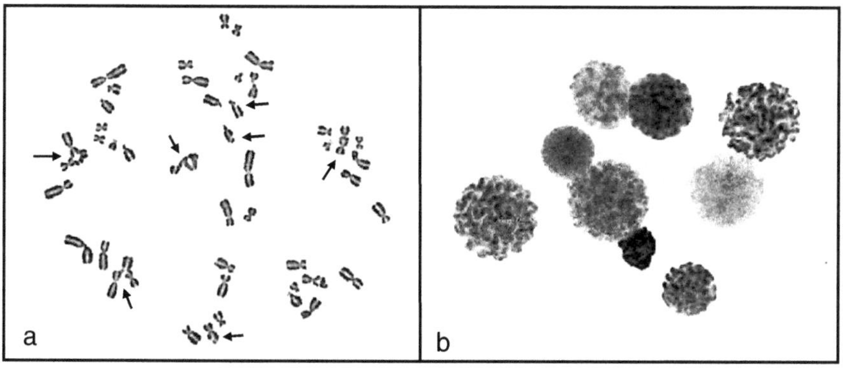

Fig. 1.9 Chromosomal instability syndromes: **a** increased chromosomal breakage in lymphocytes of a Fanconi anemia patient; **b** aberrant chromosome condensation in a patient with *MCPH1* autosomal recessive primary microcephaly

The misregulation of chromosome condensation in *MCPH1* deficiency is mediated by the SMC protein condensin II (Trimborn et al. 2006).

In patient cells with *MCPH1* deficiency, knockdown of condensin II subunits leads to a pronounced reduction of cells with the condensation defects in both G1 and G2 phases of the cell cycle. In contrast, knockdown of condensin I subunits does not reverse the cellular phenotype. Consistently, condensin I stays in the cytoplasm in the prophaselike cells of *MCPH1* patients. These results offer a molecular explanation for the aberrant chromosome condensation in *MCPH1* deficiency. In normal cells, microcephalin acts as a negative regulator of condensin II which prevents PCC until the onset of prophase and allows timely decondensation after mitosis.

1.8.3 Further Syndromes Affecting Structural Maintenance of the Chromosome

In the last few years a growing number of genes that regulate genome surveillance and cell cycle progression have been linked to developmental and progressive neurological diseases. Some of these are involved in chromosome dynamics, spindle formation, and the centrosome cycle. Mutations in the gene *NIPBL*, the human counterpart of *Scc2*, were shown to cause Brachmann/de Lange syndrome (Krantz et al. 2004; Tonkin et al. 2004), which is associated with growth retardation, microcephaly, and limb malformations. Scc2 is crucial for sister-chromatid cohesion and replication licensing (Furuya et al. 1998; Ciosk et al. 2000; Gillespie and Hirano 2004). Expression of *NIPBL* in developing limbs of human embryos was shown by in situ hybridization (Tonkin et al. 2004). A dual role for Nipped-B, the *Drosophila* homologue of *NIPBL*, in sister-chromatid cohesion and developmental regulation connected with limb formation was recently confirmed (Rollins et al. 2004). Mutations in yet another cohesin factor gene ESCO2 cause Roberts syndrome (Vega et al. 2005), characterized by premature centromere separation. Brachmann/de Lange syndrome and Roberts syndrome share clinical symptoms, such as growth retardation, microcephaly, and intriguingly limb malformations.

BUB1B is mutated in mosaic variegated aneuploidy (MVA), an autosomal recessive disorder characterized by mosaic aneuploidies, predominantly trisomies and monosomies, involving multiple different chromosomes and tissues (Hanks et al. 2004; Hanks and Rahman 2005). *BUB1B* encodes a key protein in the mitotic spindle checkpoint (Sudakin et al. 2001) and is involved in sister-chromatid cohesion (Kitajima et al. 2005). Affected individuals present with severe intrauterine growth retardation, various congenital abnormalities, microcephaly, developmental delay, and a high risk of malignancy.

These few examples illustrate the relevance of genes involved in the surveillance of the chromosomal integrity, cell cycle progression, and DNA repair. Deficiencies of these genes due to mutation do not only result in multiple congenital abnormalities of almost all tissues but also in a high risk for the development of various malignancies. Furthermore, these examples demonstrate impressively the importance of the field of human cytogenetics: in all cases the cytogenetic observation of distinctive chromosomal features, such as increased chromosomal instability, PCC, increased somatic nondisjunction, or premature centromere division, preceded the identification of the underlying gene defects.

Hsu's annotation in 1979 that chromosomes are "hypnotically beautiful objects" still holds true. Beyond it, the thorough observation of chromosomal changes by many researchers during the last four decades paved the way for our understanding of the underlying mechanisms of many congenital disorders, as well as of chromosome surveillance, DNA repair, and cancer susceptibility.

References

Adams KE, Medhurst AL, Dart DA, Lakin ND (2006) Recruitment of ATR to sites of ionising radiation-induced DNA damage requires ATM and components of the MRN protein complex. Oncogene 25:3894–3904

Bagby GC, Alter BP (2006) Fanconi anemia. Semin Hematol 43:147–156

Bazett-Jones DP, Kimura K, Hirano T (2002) Efficient supercoiling of DNA by a single condensin complex as revealed by electron spectroscopic imaging. Mol Cell 9:1183–1190

Bednar J, Horowitz RA, Grigoryev SA, Carruthers LM, Hansen JC, Koster AJ, Woodcock CL (1998) Nucleosomes, linker DNA, and linker histone form a unique structural motif that directs the higher-order folding and compaction of chromatin. Proc Natl Acad Sci USA 95:14173–14178

Caspersson T, Zech L, Modest EJ, Foley GE, Wagh U, Simonsson E (1969) Chemical differentiation with fluorescent alkylating agents in Vicia faba metaphase chromosomes. Exp Cell Res 58:128–140

Cassidy SB, Dykens E, Williams CA (2000) Prader-Willi and Angelman syndromes: sister imprinted disorders. Am J Med Genet 97:136–146

Chen TL, Manuelidis L (1989) SINEs and LINEs cluster in distinct DNA fragments of Giemsa band size. Chromosoma 98:309–316

Ciosk R, Shirayama M, Shevchenko A, Tanaka T, Toth A, Nasmyth K (2000) Cohesin's binding to chromosomes depends on a separate complex consisting of Scc2 and Scc4 proteins. Mol Cell 5:243–254

Clarke DJ, Gimenez-Abian JF (2000) Checkpoints controlling mitosis. BioEssays 22:351–363

Cremer T, Landegent J, Bruckner A, Scholl HP, Schardin M, Hager HD, Devilee P, Pearson P, van der Ploeg M (1986) Detection of chromosome aberrations in the human interphase nucleus by visualization of specific target DNAs with radioactive and non-radioactive in situ hybridization techniques: diagnosis of trisomy 18 with probe L1.84. Hum Genet 74:346–352

Cuneo BF (2001) 22q11.2 deletion syndrome: DiGeorge, velocardiofacial, and conotruncal anomaly face syndromes. Curr Opin Pediatr 13:465–472

Drets ME, Shaw MW (1971) Specific banding patterns of human chromosomes. Proc Natl Acad Sci USA 68:2073–2077

Dutrillaux B, Lejeune J (1971) (A new technique of analysis of the human karyotype). C R Acad Sci Hebd Seances Acad Sci D 272:2638–2640

Earnshaw WC, Heck MM (1985) Localization of topoisomerase II in mitotic chromosomes. J Cell Biol 100:1716–1725

Earnshaw WC, Halligan B, Cooke CA, Heck MM, Liu LF (1985) Topoisomerase II is a structural component of mitotic chromosome scaffolds. J Cell Biol 100:1706–1715
Flemming W (1882) Zellsubstanz, Kern und Zelltheilung. Vogel, Leipzig
Furuya K, Takahashi K, Yanagida M (1998) Faithful anaphase is ensured by Mis4, a sister chromatid cohesion molecule required in S phase and not destroyed in G1 phase. Genes Dev 12:3408–3418
Gillespie PJ, Hirano T (2004) Scc2 couples replication licensing to sister chromatid cohesion in Xenopus egg extracts. Curr Biol 14:1598–1603
Hagstrom KA, Holmes VF, Cozzarelli NR, Meyer BJ (2002) C. elegans condensin promotes mitotic chromosome architecture, centromere organization, and sister chromatid segregation during mitosis and meiosis. Genes Dev 16:729–742
Hanks S, Rahman N (2005) Aneuploidy-cancer predisposition syndromes: a new link between the mitotic spindle checkpoint and cancer. Cell Cycle 4:225–227
Hanks S, Coleman K, Reid S, Plaja A, Firth H, Fitzpatrick D, Kidd A, Mehes K, Nash R, Robin N, Shannon N, Tolmie J, Swansbury J, Irrthum A, Douglas J, Rahman N (2004) Constitutional aneuploidy and cancer predisposition caused by biallelic mutations in BUB1B. Nat Genet 36:1159–1161
Hartwell LH, Kastan MB (1994) Cell cycle control and cancer. Science 266:1821–1828
Hartwell LH, Weinert TA (1989) Checkpoints: controls that ensure the order of cell cycle events. Science 246:629–634
Hirano T (2005) Condensins: organizing and segregating the genome. Curr Biol 15:R265–275
Hirano T (2006) At the heart of the chromosome: SMC proteins in action. Nat Rev Mol Cell Biol 7:311–322
Hirano T, Kobayashi R, Hirano M (1997) Condensins, chromosome condensation protein complexes containing XCAP-C, XCAP-E and a Xenopus homolog of the Drosophila Barren protein. Cell 89:511–521
Hirota T, Gerlich D, Koch B, Ellenberg J, Peters JM (2004) Distinct functions of condensin I and II in mitotic chromosome assembly. J Cell Sci 117:6435–6445
Holmquist GP (1989) Evolution of chromosome bands: molecular ecology of noncoding DNA. J Mol Evol 28:469–486
Holmquist GP (1992) Chromosome bands, their chromatin flavors, and their functional features. Am J Hum Genet 51:17–37
Horsthemke B, Buiting K (2006) Imprinting defects on human chromosome 15. Cytogenet Genome Res 113:292–299
Hsu TC (1979) Human and mammalian cytogenetics. An historical perspective. Springer, New York
Jazayeri A, Falck J, Lukas C, Bartek J, Smith GC, Lukas J, Jackson SP (2006) ATM- and cell cycle-dependent regulation of ATR in response to DNA double-strand breaks. Nat Cell Biol 8:37–45
Kallioniemi A, Kallioniemi OP, Sudar D, Rutovitz D, Gray JW, Waldman F, Pinkel D (1992) Comparative genomic hybridization for molecular cytogenetic analysis of solid tumors. Science 258:818–821
Kastan MB, Bartek J (2004) Cell-cycle checkpoints and cancer. Nature 432:316–323
Kennedy RD, D'Andrea AD (2005) The Fanconi anemia/BRCA pathway: new faces in the crowd. Genes Dev 19:2925–2940
Kimura K, Hirano T (1997) ATP-dependent positive supercoiling of DNA by 13S condensin: a biochemical implication for chromosome condensation. Cell 90:625–634
Kimura K, Rybenkov VV, Crisona NJ, Hirano T, Cozzarelli NR (1999) 13S condensin actively reconfigures DNA by introducing global positive writhe: implications for chromosome condensation. Cell 98:239–248
Kitagawa R, Kastan MB (2005) The ATM-dependent DNA damage signaling pathway. Cold Spring Harbor Symp Quant Biol 70:99–109
Kitajima TS, Hauf S, Ohsugi M, Yamamoto T, Watanabe Y (2005) Human Bub1 defines the persistent cohesion site along the mitotic chromosome by affecting Shugoshin localization. Curr Biol 15:353–359

Korenberg JR, Rykowski MC (1988) Human genome organization: Alu, lines, and the molecular structure of metaphase chromosome bands. Cell 53:391–400

Krantz ID, McCallum J, DeScipio C, Kaur M, Gillis LA, Yaeger D, Jukofsky L, Wasserman N, Bottani A, Morris CA, Nowaczyk MJ, Toriello H, Bamshad MJ, Carey JC, Rappaport E, Kawauchi S, Lander AD, Calof AL, Li HH, Devoto M, Jackson LG (2004) Cornelia de Lange syndrome is caused by mutations in NIPBL, the human homolog of Drosophila melanogaster Nipped-B. Nat Genet 36:631–635

Kruger L, Demuth I, Neitzel H, Varon R, Sperling K, Chrzanowska KH, Seemanova E, Digweed M (2007) Cancer incidence in Nijmegen breakage syndrome is modulated by the amount of a variant NBS protein. Carcinogenesis 28:107–111

Lee EY (2002) BRCA1 and Chk1 in G2/M checkpoint: a new order of regulation. Cell Cycle 1:178–180

Lejeune J, Turpin R, Gautier M (1959) (Chromosomic diagnosis of mongolism). Arch Fr Pediatr 16:962–963

Li L, Zou L (2005) Sensing, signaling, and responding to DNA damage: organization of the checkpoint pathways in mammalian cells. J Cell Biochem 94:298–306

Lisby M, Rothstein R (2004) DNA damage checkpoint and repair centers. Curr Opin Cell Biol 16:328–334

Losada A, Hirano T (2005) Dynamic molecular linkers of the genome: the first decade of SMC proteins. Genes Dev 19:1269–1287

Lyakhovich A, Surralles J (2006) Disruption of the Fanconi anemia/BRCA pathway in sporadic cancer. Cancer Lett 232:99–106

Mervis CB, Klein-Tasman BP (2000) Williams syndrome: cognition, personality, and adaptive behavior. Ment Retard Dev Disabil Res Rev 6:148–158

Meselson M, Stahl FW (1958) The replication of DNA. Cold Spring Harbor Symp Quant Biol 23:9–12

Morrison C, Rieder CL (2004) Chromosome damage and progression into and through mitosis in vertebrates. DNA Repair (Amst) 3:1133–1139

Nasmyth K, Haering CH (2005) The structure and function of SMC and kleisin complexes. Annu Rev Biochem 74:595–648

Neitzel H, Neumann LM, Schindler D, Wirges A, Tonnies H, Trimborn M, Krebsova A, Richter R, Sperling K (2002) Premature chromosome condensation in humans associated with microcephaly and mental retardation: a novel autosomal recessive condition. Am J Hum Genet 70:1015–1022

Nowell PC (1960) Phytohemagglutinin: an initiator of mitosis in cultures of normal human leukocytes. Cancer Res 20:462–466

Nurse P (1997) Checkpoint pathways come of age. Cell 91:865–867

Ono T, Losada A, Hirano M, Myers MP, Neuwald AF, Hirano T (2003) Differential contributions of condensin I and condensin II to mitotic chromosome architecture in vertebrate cells. Cell 115:109–121

Ono T, Fang Y, Spector DL, Hirano T (2004) Spatial and temporal regulation of Condensins I and II in mitotic chromosome assembly in human cells. Mol Biol Cell 15:3296–3308

Osborne LR (1999) Williams-Beuren syndrome: unraveling the mysteries of a microdeletion disorder. Mol Genet Metab 67:1–10

Ostertag EM, Kazazian HH Jr (2001) Biology of mammalian L1 retrotransposons. Annu Rev Genet 35:501–538

Patil SR, Merrick S, Lubs HA (1971) Identification of each human chromosome with a modified Giemsa stain. Science 173:821–822

Pines J, Rieder CL (2001) Re-staging mitosis: a contemporary view of mitotic progression. Nat Cell Biol 3:E3–6

Pinkel D, Straume T, Gray JW (1986) Cytogenetic analysis using quantitative, high-sensitivity, fluorescence hybridization. Proc Natl Acad Sci USA 83:2934–2938

Rao PN, Johnson RT, Sperling K (1982) Premature chromosome condensation. Academic, New York

Rollins RA, Korom M, Aulner N, Martens A, Dorsett D (2004) Drosophila nipped-B protein supports sister chromatid cohesion and opposes the stromalin/Scc3 cohesion factor to facilitate long-range activation of the cut gene. Mol Cell Biol 24:3100–3111
Sancar A, Lindsey-Boltz LA, Unsal-Kacmaz K, Linn S (2004) Molecular mechanisms of mammalian DNA repair and the DNA damage checkpoints. Annu Rev Biochem 73:39–85
Schrock E, du Manoir S, Veldman T, Schoell B, Wienberg J, Ferguson-Smith MA, Ning Y, Ledbetter DH, Bar-Am I, Soenksen D, Garini Y, Ried T (1996) Multicolor spectral karyotyping of human chromosomes. Science 273:494–497
Sudakin V, Chan GK, Yen TJ (2001) Checkpoint inhibition of the APC/C in HeLa cells is mediated by a complex of BUBR1, BUB3, CDC20, and MAD2. J Cell Biol 154:925–936
Sumner AT, Evans HJ, Buckland RA (1971) New technique for distinguishing between human chromosomes. Nat New Biol 232:31–32
Swedlow JR, Hirano T (2003) The making of the mitotic chromosome: modern insights into classical questions. Mol Cell 11:557–569
Thomas NS, Durkie M, Potts G, Sandford R, Van Zyl B, Youings S, Dennis NR, Jacobs PA (2006) Parental and chromosomal origins of microdeletion and duplication syndromes involving 7q11.23, 15q11-q13 and 22q11. Eur J Hum Genet 14:831–837
Tjio JH, Levan A (1956) The chromosome number of man. Hereditas 42:1–6
Tonkin ET, Wang TJ, Lisgo S, Bamshad MJ, Strachan T (2004) NIPBL, encoding a homolog of fungal Scc2-type sister chromatid cohesion proteins and fly Nipped-B, is mutated in Cornelia de Lange syndrome. Nat Genet 36:636–641
Tonnies H (2002) Modern molecular cytogenetic techniques in genetic diagnostics. Trends Mol Med 8:246–250
Trimborn M, Bell SM, Felix C, Rashid Y, Jafri H, Griffiths PD, Neumann LM, Krebs A, Reis A, Sperling K, Neitzel H, Jackson AP (2004) Mutations in microcephalin cause aberrant regulation of chromosome condensation. Am J Hum Genet 75:261–266
Trimborn M, Schindler D, Neitzel H, Hirano T (2006) Misregulated chromosome condensation in MCPH1 primary microcephaly is mediated by condensin II. Cell Cycle 5:322–326
Varon R, Vissinga C, Platzer M, Cerosaletti KM, Chrzanowska KH, Saar K, Beckmann G, Seemanova E, Cooper PR, Nowak NJ, Stumm M, Weemaes CM, Gatti RA, Wilson RK, Digweed M, Rosenthal A, Sperling K, Concannon P, Reis A (1998) Nibrin, a novel DNA double-strand break repair protein, is mutated in Nijmegen breakage syndrome. Cell 93:467–476
Varon R, Reis A, Henze G, von Einsiedel HG, Sperling K, Seeger K (2001) Mutations in the Nijmegen Breakage Syndrome gene (NBS1) in childhood acute lymphoblastic leukemia (ALL). Cancer Res 61:3570–3572
Varon R, Schoch C, Reis A, Hiddemann WC, Sperling K, Schnittger S (2003) Mutation analysis of the Nijmegen breakage syndrome gene (NBS1) in nineteen patients with acute myeloid leukemia with complex karyotypes. Leuk Lymphoma 44:1931–1934
Vega H, Waisfisz Q, Gordillo M, Sakai N, Yanagihara I, Yamada M, van Gosliga D, Kayserili H, Xu C, Ozono K, Jabs EW, Inui K, Joenje H (2005) Roberts syndrome is caused by mutations in ESCO2, a human homolog of yeast ECO1 that is essential for the establishment of sister chromatid cohesion. Nat Genet 37:468–470
Wang X, D'Andrea AD (2004) The interplay of Fanconi anemia proteins in the DNA damage response. DNA Repair (Amst) 3:1063–1069
Watson GD, Crick FHC (1953) Molecular structure of nucleic acids: a structure for deoxyribose nucleic acid. Nature 171:737–738
Waye JS, Willard HF (1989) Human beta satellite DNA: genomic organization and sequence definition of a class of highly repetitive tandem DNA. Proc Natl Acad Sci USA 86:6250–6254
Willard HF (1989) The genomics of long tandem arrays of satellite DNA in the human genome. Genome 31:737–744

2 DNA Lesions Induced by Ionizing Radiation

CLEMENS VON SONNTAG

Abstract

When ionizing radiation is absorbed in cells, a fraction of the absorbed energy is deposited in their DNA ("direct effect") and in the water surrounding it ("indirect effect," mainly owing to the formation of $^{\bullet}OH$ radicals). These ionization events induce a series of free-radical reactions that eventually show up in biological end points such as mutations and reproductive cell death. Precursors of these biological end points are various kinds of damaged DNA sites, predominantly damaged bases, single-stand breaks, double-strand breaks and DNA–protein cross-links. Molecular oxygen enhances and thiols reduce the severity of the DNA damage. The mechanisms leading to these lesions are discussed in the light of our present knowledge of the underlying free-radical chemistry.

2.1 Introduction

For biological effects such as point mutations and chromosomal mutations induced by ionizing radiation, DNA is the obvious target, but there is also a report that mutations are induced when the cytoplasm is hit by an α-particle (Wu et al. 1999). For other biological end points such as cell death, DNA is also considered to be the main target, but there is a phenomenon, the bystander effect (a cell is hit by a single particle of ionizing radiation – a neighboring cell dies by apoptosis; Shao et al. 2004), that cannot be accounted for by radiation-induced DNA damage.

The cell's repair enzymes cope with most of the radiation-induced DNA damage very effectively, and what we see as radiation-induced DNA damage on the cellular level is the damage that remains after the much greater DNA damage originally caused has been dealt with by the interplay of "chemical repair" (see later) and enzymatic repair. To give the reader an idea of the number of lesions and of the efficiency of the repair machinery, the DNA damage caused in a mammalian cell by a dose of 1 Gy is shown in Table 2.1.

At this dose, the lethality is only near 50%. Yet, as many as several thousand damaged bases, 1,000 single-strand breaks (SSBs), 150 DNA–protein cross-links

Günter Obe and Vijayalaxmi (Eds.)
Chromosomal Alterations: Methods, Results and Importance in Human Health

Table 2.1 Some of the damage in a mammalian cell nucleus from 1 Gy of low linear energy transfer radiation (Ward 1988; Goodhead 1994)

Initial physical damage	
Ionizations in the cell nucleus	~100,000
Ionizations directly in DNA	~2,000
Excitations directly in DNA	~2,000
Selected biochemical damage	
Single-strand breaks	1,000
8-Oxo-A (typical single-base damage)	700
Double-strand breaks	40
DNA–protein cross-links	150
Selected cellular effects	
Lethal events	~0.2–0.8
Chromosome aberrations	~1
Hprt mutations	$\sim 10^{-5}$

and 40 double-strand breaks (DSBs) have been caused. Obviously, most of this is repaired effectively. For example, a DSB as such is not yet a sufficiently severe type of DNA damage to cause cell death, and it has been concluded that there must be a much more complex type of damage responsible for cell death. There is evidence that clustered damage beyond a DSB is responsible for or is at least a major contributor to cell death by ionizing radiation. Yet, inadequately repaired DSBs are thought to be the origin of chromosomal mutations and induction of cancer; for a detailed review on the production and the fidelity of repair of endogenous and ionizing-radiation-induced DSBs see Vilenchik and

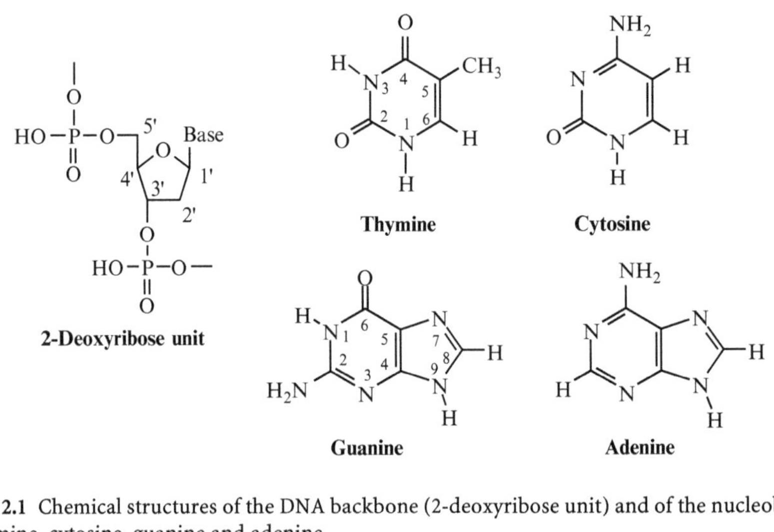

Fig. 2.1 Chemical structures of the DNA backbone (2-deoxyribose unit) and of the nucleobases thymine, cytosine, guanine and adenine

Knudson (2003) and for the details on endogenous and ionizing-radiation-induced DNA base damage see De Bont and van Larebeke (2004).

While UV radiation (at $\lambda > 200$ nm) practically only damages the pyrimidines thymine and cytosine, ionizing radiation also damages to a major extent the purines guanine and adenine as well as the 2-deoxyribose moiety (Fig. 2.1). Thus, SSBs and DSBs are also important lesions induced by ionizing radiation (Table 2.1).

The free-radical-induced DNA damage has been dealt with recently in a monograph (von Sonntag 2006) in much more detail then space allows here. Only some salient points can be addressed, and for further reading the reader is referred to that book and the original literature cited therein.

2.2 Energy Deposition by Ionizing Radiation

The energy of high-energy photons such as γ-rays and X-rays below 1 MeV is absorbed by the Compton effect and the photo effect. Above 1.01 MeV (twice the rest mass of the electron), pair formation, i.e., creation of an electron–positron pair, starts to play a role. Compton and photo effects generate high-energy electrons. The energy loss of charged particles is described by the Bethe equation which relates the energy loss (linear energy transfer) of the charged particle to its mass, charge and energy. Multiply charged and heavy particles (e.g., α-particles) lose their energy more effectively (densely ionizing radiation) than singly charged light particles such as high-energy electrons (β-particles). As their energy decreases, electrons also become more densely ionizing, i.e., they cause further ionizations in the very neighborhood of their formation. Considering these mechanisms of energy absorption, it is understood why high-energy electrons (1–10 MeV) and γ-rays (^{60}Co γ-rays 1.3 MeV, ^{137}Cs γ-rays 0.5 MeV) have practically the same linear energy transfer.

Upon energy loss of high-energy electrons, approximately 100 eV of energy is deposited per ionization event (on average), whereby a radical cation and an electron are generated. Reactions 1 and 2 show this for the two relevant targets that we are interested in here, the DNA and the water that surrounds it (the cell consists of about 70% water, and DNA is highly hydrated). The electron may still contain enough energy to cause further ionizations in the very near neighborhood. These areas, containing a number of ionization and occasionally also electronic-excitation events (reactions 3 and 4), are called spurs. In the case of sparsely ionizing radiation, these spurs do not overlap. In densely ionizing radiation, however, they form cylinders of spurs that are called tracks.

$$\mathrm{DNA} \xrightarrow[\text{radiation}]{\text{ionizing}} \mathrm{DNA}^{\bullet+} + e^- \tag{1}$$

$$\mathrm{H_2O} \xrightarrow[\text{radiation}]{\text{ionizing}} \mathrm{H_2O}^{\bullet+} + e^- \tag{2}$$

$$DNA \xrightarrow[\text{radiation}]{\text{ionizing}} DNA^* \quad (3)$$

$$H_2O \xrightarrow[\text{radiation}]{\text{ionizing}} H_2O^* \quad (4)$$

The water radical cation, produced in reaction 2, is a very strong acid and immediately loses a proton to neighboring water molecules thereby forming an $^{\bullet}OH$ radical (reaction 5). The electron becomes hydrated by water (reaction 6). Electronically excited water can decompose into $^{\bullet}OH$ and $H^{\bullet}$ (reaction 7). Thus, three kinds of free radicals are formed side by side in the spurs, $^{\bullet}OH$, e_{aq}^{-} and $H^{\bullet}$. To match the charge of the electrons, an equivalent amount of H^{+} is also present.

$$H_2O^{\bullet+} \rightarrow {}^{\bullet}OH + H^{+} \quad (5)$$

$$e^{-} + nH_2O \rightarrow e_{aq}^{-} \quad (6)$$

$$H_2O^* \rightarrow {}^{\bullet}OH + H^{\bullet} \quad (7)$$

A spur may contain more than one reactive intermediate, and they can interact with one another (e.g., $^{\bullet}OH + e_{aq}^{-} \rightarrow OH^{-}$; $H^{+} + e_{aq}^{-} \rightarrow H^{\bullet}$; $^{\bullet}OH + H^{\bullet} \rightarrow H_2O$; $2HO^{\bullet} \rightarrow H_2O_2$; $2e_{aq}^{-} + 2H^{+} \rightarrow H_2$; $2H^{\bullet} \rightarrow H_2$) in competition with a reaction with a neighboring solute such as DNA.

The radiation-chemical yield is expressed by the *G* value. Originally, it was given in the units of molecules formed (or destroyed) per 100 eV of absorbed energy. The electronvolt is not an SI unit and hence is usually no longer in use. These two units may be converted into one another taking into account that 1 molecule $(100\ eV)^{-1} = 1.036 \times 10^{-7}\ mol\ J^{-1}$.

2.3 Direct and Indirect Effect

As depicted in reactions 1 and 3, the energy of the ionizing radiation may be absorbed by DNA itself, and this part of the DNA damage by ionizing radiation is called the *direct effect*. When the ionizing radiation is absorbed by the water that surrounds DNA (reactions 2 and 4), reactive intermediates, notably $^{\bullet}OH$ (reaction 5), are generated that can also damage DNA. This contribution is called the *indirect effect*. There have been a number of attempts to disentangle the two contributions by the addition of $^{\bullet}OH$ scavengers. The results are not really clear-cut (von Sonntag 2006), but it is obvious that there must be these two contributions. It is hoped that microdosimetric approaches using Monte Carlo type calculations may sharpen our view on this topic; for a review on the contribution of quantum-mechanical and Monte Carlo calculations in understanding DNA damage, see von Sonntag (2007).

2.4 Cerenkov Radiation

High-energy electrons that pass through a dense medium such as a cell lose a small fraction of their energy as Cerenkov radiation. This UV radiation is also absorbed by DNA and causes the well-known UV lesions, notably pyrimidine dimers and 6,4-photoproducts. The studies that are concerned with this effect were discussed in detail by von Sonntag (1987), and the relevant literature is also mentioned in von Sonntag (2006).

2.5 Hole and Electron Transfer Through DNA

In reaction 1, a DNA radical cation is formed. There is a rapid long-range hole transfer through DNA. The radical cation ends up at G, the nucleobase with the lowest ionization potential. The ionization potentials of clustered Gs, GG and GGG, are even lower than that of a single G, and the hole ends up preferentially at a GGG site (Saito et al. 1998). In competition with this rapid hole transfer, there are other reactions such as the reaction of the hole with water or with traps other than GGG. For example, the DNA radiation product 8-oxo-G may serve as a trap on account of its very low ionization potential (Shafirovich et al. 2001). This hole transfer is presently a very hot topic in DNA free-radical research, and a large number of excellent experimental as well as theoretical studies have been reported over the last few years. For details, the reader is referred to some reviews (Giese 2000; Schuster 2000; O'Neill and Barton 2004; Nakatani and Saito 2004; Berlin et al. 2004; von Sonntag 2006).

Electron transfer can also occur over some distance, whereby T, the nucleobase that has the highest electron affinity, is the trapping site (von Sonntag 2006). Our understanding of electron transfer through DNA is not yet as detailed as that of the corresponding hole transfer (Wagenknecht 2003, 2006).

2.6 Formation of Base Radicals

Radicals (often referred to as free radicals) are species with unpaired electrons. These unpaired electrons give the radicals a pronounced reactivity. Yet, depending on the type of radical, the reactivity may vary considerably (many orders of magnitude). The $^{\bullet}OH$ radicals that are generated in reaction 5 are among the most reactive free radicals. With DNA, they react preferentially with the nucleobases, and only a minor fraction abstracts a hydrogen atom from the sugar moiety (see later). Despite its high reduction potential (1.9 V at pH 7; Wardman 1989), the $^{\bullet}OH$ radical undergoes electron transfer reactions only very reluctantly. The radicals rather react by addition and by hydrogen abstraction. Although the •OH radical reacts with the nucleobases at close to diffusion-controlled rates, there is a remarkable regioselectivity (Fujita and Steenken 1981). This reaction is controlled by kinetics rather than by thermodynamics; for a quantum-chemical approach to rationalize this, see von Sonntag (2006). For example, in thymine it adds preferentially to the C6 position (60%, reaction 8) rather than to the C5 position (30%, reaction 10), although the latter reaction is thermodynamically favored by 46 kJ mol^{-1} (S. Naumov and C. von Sonntag 2006, unpublished results). Only a small fraction abstracts a hydrogen atom from the methyl group (approximately 10%, reaction 9), despite the fact that this reaction seems to be thermodynamically the most favored one.

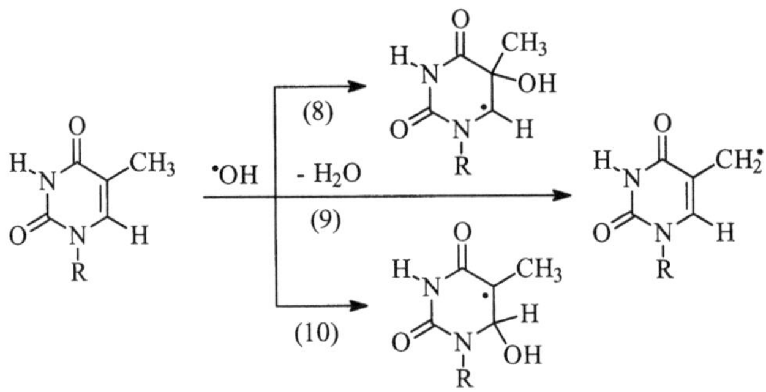

In the purines, C4 and C8 are the preferred sites for •OH radical attack, and formation of the thermodynamically strongly favored guanyl and adenyl radicals (G•; A•) only occurs upon release of water from the •OH adduct radicals (e.g., reaction 11).

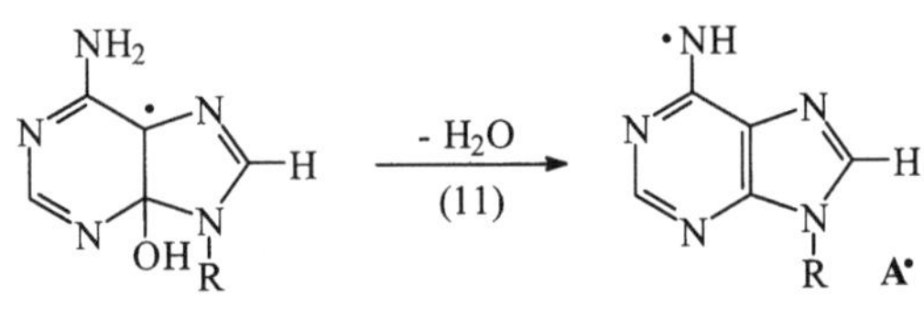

An addition to the C8 position gives rise to a radical that readily opens the ring (e.g., reaction 12), and the ring-open radicals are the precursors of the well-documented FAPY products (FAPY-A is 4,6-diamino-5-formamidopyrimidine from adenine; FAPY-G is 2,6-diamino-4-hydroxy-5-formamidopyrimidine from guanine; e.g., reaction 13). Mechanistic details of this rapid ring-opening reaction have not yet been fully elucidated, and reaction 12 shows only one of the various possible intermediates.

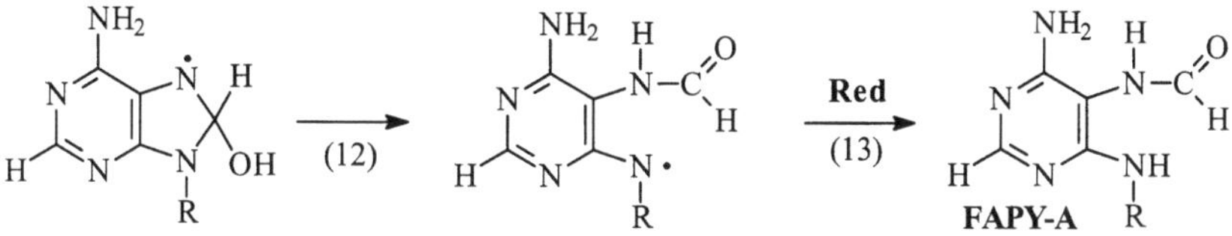

The nucleobase radical cations formed according to reaction 1 release a heteroatom-bound proton quite readily, and in the case of A and G this leads to the formation of the thermodynamically favored $G^{\bullet}$ and $A^{\bullet}$ radicals (e.g., reaction 14).

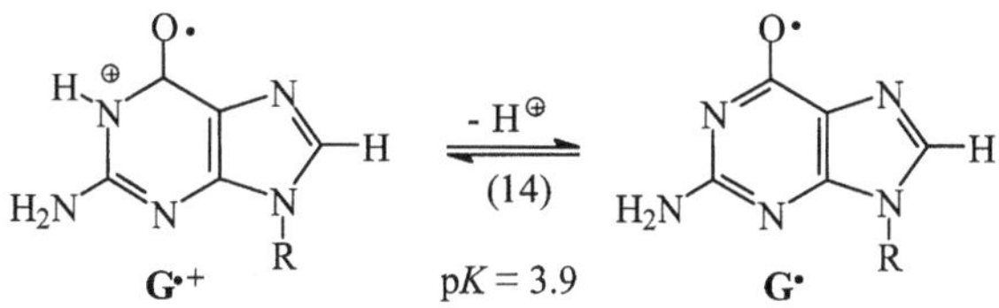

In thymidine, the N3 radical that is formed upon deprotonation of the radical cation (equilibrium 16) is not the thermodynamically favored species and eventually is converted into the more stable allylic radical (reaction 15).

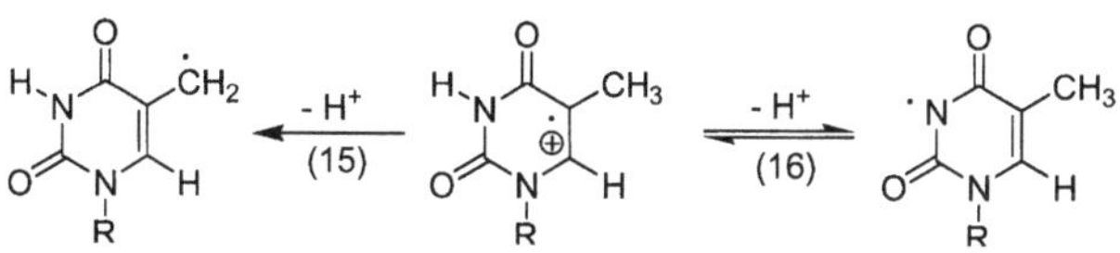

2.7 Base Damage

For the conversion of a radical into a final diamagnetic product, the radical must be either reduced or oxidized. In this respect, oxygen is not a good oxidant, since it usually converts the radical only into the corresponding peroxyl radical (reaction 17), i.e., the radical character is not lost.

$$R^{\bullet} + O_2 \rightarrow ROO^{\bullet} \tag{17}$$

In certain cases, transition metal ions such as Fe^{3+}/Fe^{2+} may act as oxidants/reductants and thiols such as glutathione (GSH) can undergo hydrogen-transfer reactions (reaction 18). GSH is the most abundant thiol in cells. Its concentration within the cell is not uniform but is typically 1–2 mM (Wardman and von Sonntag 1995).

$$R^{\bullet} + GSH \rightarrow RH + GS^{\bullet} \quad (18)$$

In DNA free-radical chemistry, reaction 18 has been termed "chemical repair," although it is obvious that a hydrogen donation to radicals such as those formed in reactions 8 and 10 do not cause these radicals to revert into thymine but rather gives rise to C5OH,C6H-dihydrothymine and C5H,C6OH-dihydrothymine, respectively. This is of importance when discussing reactions 17 and 18 in a cellular environment (see later).

In model systems on which our knowledge of the mechanism of free-radical-induced DNA damage is largely based, the radials decay bimolecularly at practically diffusion controlled rates, and these reactions lead to dimers (reaction 19) and disproportionation products (reaction 20).

$$2HR^{\bullet} \rightarrow HR - HR \quad (19)$$

$$2HR^{\bullet} \rightarrow RH_2 + R \quad (20)$$

It is a particular feature of ionizing radiation that with a high probability more than one radical is formed within a very small volume. If these radicals are sufficiently close to one another, they may decay bimolecularly. Much more problematic is the situation when only one radical is formed (or remains when the close-by ones have decayed) on the DNA without another nearby radical to react with. Such radicals may become very long lived. There may be some flexibility of the DNA macromolecule, although the negative charges of the phosphate groups at its surface tend to keep it largely rodlike by electrostatic repulsion. There are always counterions condensed at the DNA surface that reduce the repulsive forces and thus increase the flexibility. In man-made polymers that have been studied as model systems, bimolecular decay processes also take place in charged polymers, albeit with a slower rate than in uncharged coiled polymers. For DNA, we lack detailed information on the rates, but the DNA radical lifetimes are not very long also in the absence of any added external oxidant/reductant. The techniques that are available at present for the determination of lifetimes of DNA radicals, pulse radiolysis and electron paramagnetic resonance, require radical concentrations that generate many radicals per DNA macromolecule. Under such conditions, the decay kinetics do not follow simple second-order kinetics (Ulanski et al. 1994; von Sonntag 2006), and extrapolation to a few radicals per macromolecule is fraught with very large errors. An additional problem is the fact that in cells the DNA is wrapped up by the nucleohistones, and this situation will further lower the mobility of DNA radicals. In the absence of O_2, this increases the probability of reaction 18 with respect to a bimolecular decay of DNA radicals. In the presence of O_2, i.e., when DNA peroxyl radicals

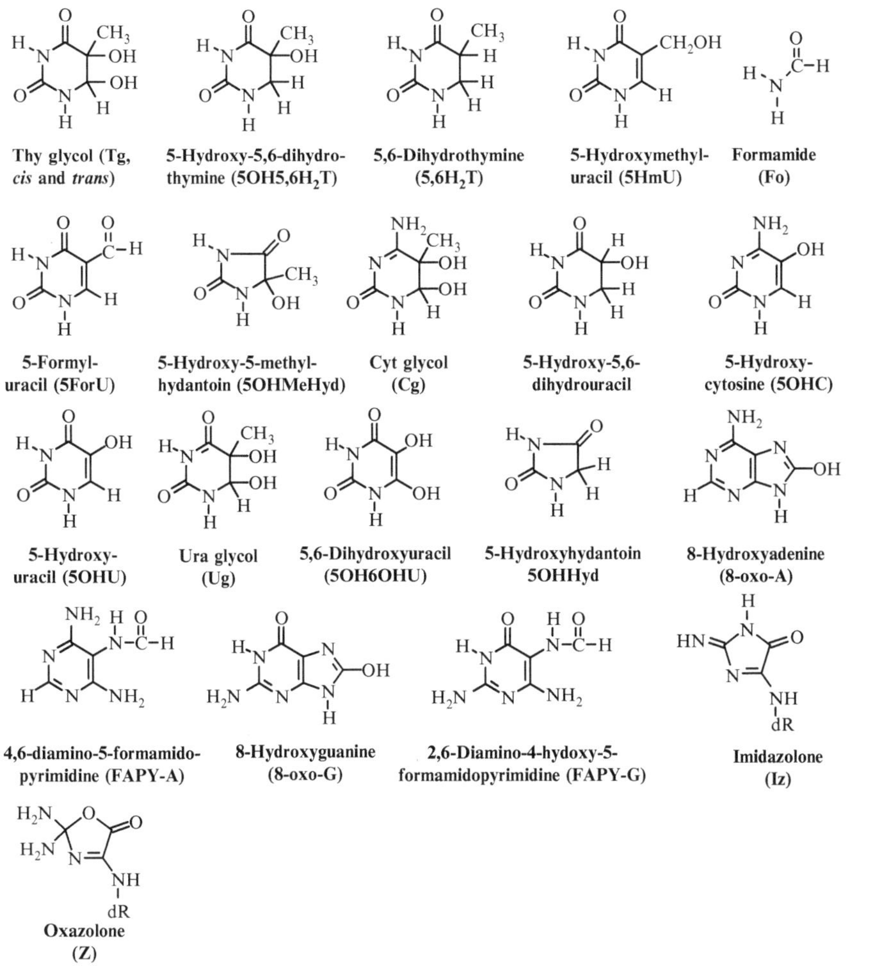

Fig. 2.2 DNA base lesions as known from model systems and from DNA γ-irradiated in aqueous solution

dominate, one is confronted with the same problem of the rate of the bimolecular decay of the DNA peroxyl radicals as only a few, at the most, can eliminate $HO_2^{\bullet}/O_2^{\bullet-}$. The rate for reaction of GSH with peroxyl radicals is slow. An originally reported high rate constant (Schulte-Frohlinde et al. 1986) has been revised to a much lower value (Hildenbrand and Schulte-Frohlinde 1997). The rate constant for reaction of $HO_2^{\bullet}$ with dithiothreitol has been measured at 120 M^{-1} s^{-1} (Lal et al. 1997), and as a first approximation one may assume that the rate constant for reaction of DNA peroxyl radicals with GSH is of the same order of magnitude. With a GSH concentration of 1 mM, this gives a rate of decay near 0.1 s^{-1} and a lifetime of the DNA peroxyl radicals with respect to their reaction with GSH on the order of 10 s. Such long lifetimes may allow DNA peroxyl radicals to undergo reactions that cannot be studied with nucleobase/nucleotide peroxyl radicals. The latter are too short-lived (fractions of a second at reasonable dose rates). Their long lifetimes may allow DNA peroxyl radicals to undergo damage-amplification reactions such as hydrogen abstraction and addition to double bonds of the nucleobases (see later) and thus increase the severity of radiation-induced damage.

In the presence of O_2, there is always some $O_2^{\bullet-}$ formed. This is a rather long-lived (on the basis of its bimolecular decay) freely diffusing radical that could play a role in the termination of DNA peroxyl radicals. Although hydroperoxides are likely products of this reaction, there is evidence that this cannot be the only pathway for how such species react with one another (von Sonntag 2006).

After having addressed some of the mechanistic uncertainties, it seems worthwhile to report the alteration of the nucleobases as far as we know them from model studies. A major part has also been found to be formed in DNA when γ-irradiated in aqueous solution and even in γ-irradiated cells. In Fig. 2.2, the products formed in the absence and presence of O_2 are compiled (see also Fig. 2.3).

In the "repair reaction" (reaction 18) a thiyl radical ($GS^\bullet$) is formed. These radicals are not as inert as commonly thought, and undergo also the reverse of reaction 18 (von Sonntag 2006). Even if they do not abstract a hydrogen from the sugar moiety or the thymine methyl group in DNA, in multiple radical sites (spurs) they are likely to interact with DNA radicals that are still around after one of the DNA radicals has been repaired.

2.8 Single-Strand Breaks

We have only little information as to the involvement of the direct effect on the formation of SSBs in cellular systems (it must be of considerable importance). However, there is a wealth of information on the reactions of the $^\bullet OH$ radical. To induce a SSB, the $^\bullet OH$ radical must abstract a hydrogen atom from the sugar moiety. In DNA in aqueous solution, SSBs occur with a marked preference at local structures where the minor groove is wide and the sugar moiety is more accessible to the $^\bullet OH$ radical. The situation has been successfully modeled by Monte Carlo type calculations to a perfection that even the protective effects of bound proteins are well reflected (Begusova et al. 2001); for further references, see von Sonntag (2006).

Mechanistically, strand breakage as a consequence of hydrogen abstraction at C4′ is understood best (Dizdaroglu et al. 1975). A neighboring phosphate group is eliminated heterolytically (e.g., reaction 21). The resulting radical cation may react with water (e.g., reactions 22 and 23; for a more

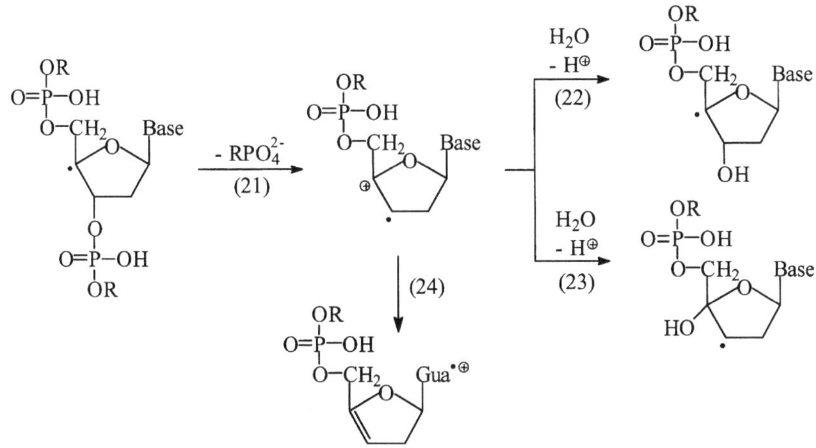

detailed description, see von Sonntag 2006) or oxidize a nearby guanine (reaction 24) with subsequent hole transfer through DNA (see earlier).

In the presence of O_2, the peroxyl radical at C4′ also gives rise to a strand break. This lesion is characterized by a 3′-phosphoglycolate end group (reactions 25–27; for details, see von Sonntag 2006).

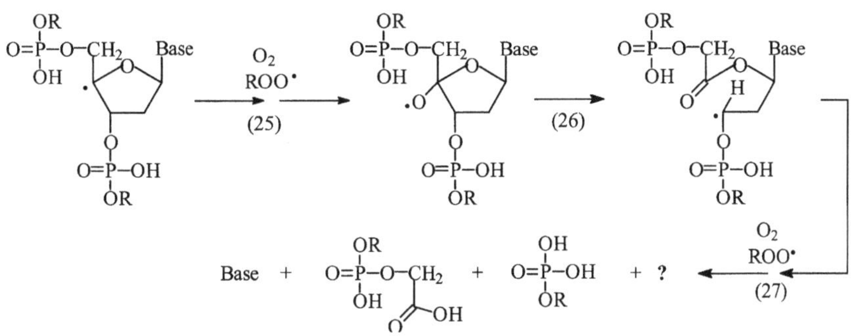

A similar sequence of events may be written for the C5′ peroxyl radical (von Sonntag 2006). In both cases, an interaction with a second peroxyl radical is required. The implication of this was discussed above.

2.9 Double Lesions

It has already been mentioned that the deposition of ionizing radiation occurs in packages of, on average, about 100 eV. Thus, several radicals may be generated in close proximity in these spurs. When the spurs overlap, such as in the tracks of densely ionizing radiation, the number of nearby radicals confined in a small volume increases. When the spurs/tracks hit the DNA in a cell, more than one DNA lesion is likely to be formed in close proximity. This also implies that a spur/track may touch both DNA and nucleohistone. In this case, DNA and nucleohistone radicals are formed in close proximity. When the spur/track only streaks at its edge the DNA, a single DNA lesion may occur, but the more the spur/track and DNA overlap, the more the probability of forming double and multiple lesions will increase. Base lesions are more abundant than sugar lesions, and since one of the prominent sugar lesions is a SSB, they are much more prominent than SSBs. As far double lesions are concerned, the most prominent ones must be two nearby base lesions (on the same and on the opposite strand). Next in frequency must be a double lesion consisting of a damaged base and a SSB (on the same and on the opposite strand). A DSB, formed by two nearby SSBs on opposite strands (two nearby SSBs on the same strand are counted as one SSB), is of even lower probability. With sparsely ionizing radiation, there are 40 DSBs versus 1,000 SSBs (Table 2.1). With densely ionizing radiation, this ratio is considerably higher, and clustered lesions beyond a double lesion are also more prominent (see later). These multiple lesions when processed by repair enzymes may lead to severer lesions than the processing of single lesions. For example, excision repair of a damaged base opposite a SSB will result in a DSB. There is increasing evidence that double lesions can also be formed as a consequence of the attack of a single radical, in contrast to the spur effect just discussed. Such reactions may have to be taken into account when discussing the formation of endogenous DSBs. Some aspects will be dealt with now.

It has been suggested that the sugar radical that remains after a SSB has been formed may abstract with low probability a hydrogen from the sugar moiety of the opposite strand thereby causing a DSB (Siddiqi and Bothe 1987). This flip-over concept has been used to explain the low DSB yields (3%) induced by photons whose energies are just sufficient to cause a single ionization (Prise et al. 2000). The flip-over mechanism has been challenged (Ward 1988), and as an alternative it has been suggested (von Sonntag 2006) that the formation of the second SSB could be due to an epithermal electron that is released in the photoionization process (reactions 28–30) (Li et al. 2003; Becker et al. 2003).

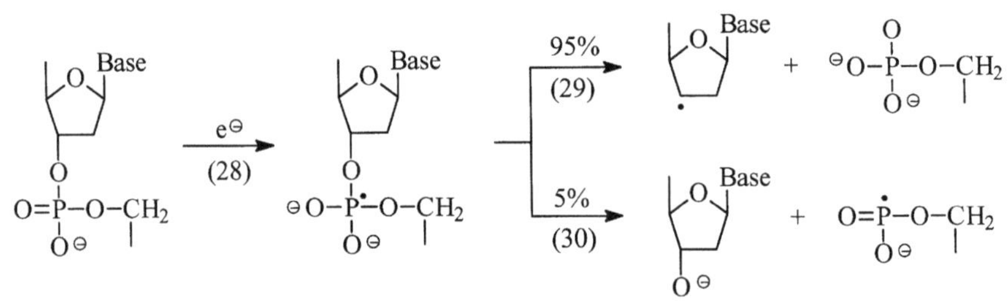

The flip-over mechanism is one example where two damaged sites may be generated from one single radical originally formed. There are other examples that are based on a much better footing. The well-documented 5′, 8-cyclonucleotide lesions cA and cG are another example. Although they may originate from a C5′ radical, there is evidence from model systems that base radicals are likely precursors, although the mechanistic details are still to be elucidated (von Sonntag 2006). There are a number of vicinal lesions that belong to this group (Fig. 2.3; for mechanistic aspects and further details, see von Sonntag 2006).

Further studies on such damage-amplification reactions will hopefully shed some light on the formation of endogenous DSBs and other higher-order lesions that may be responsible for chromosomal mutations.

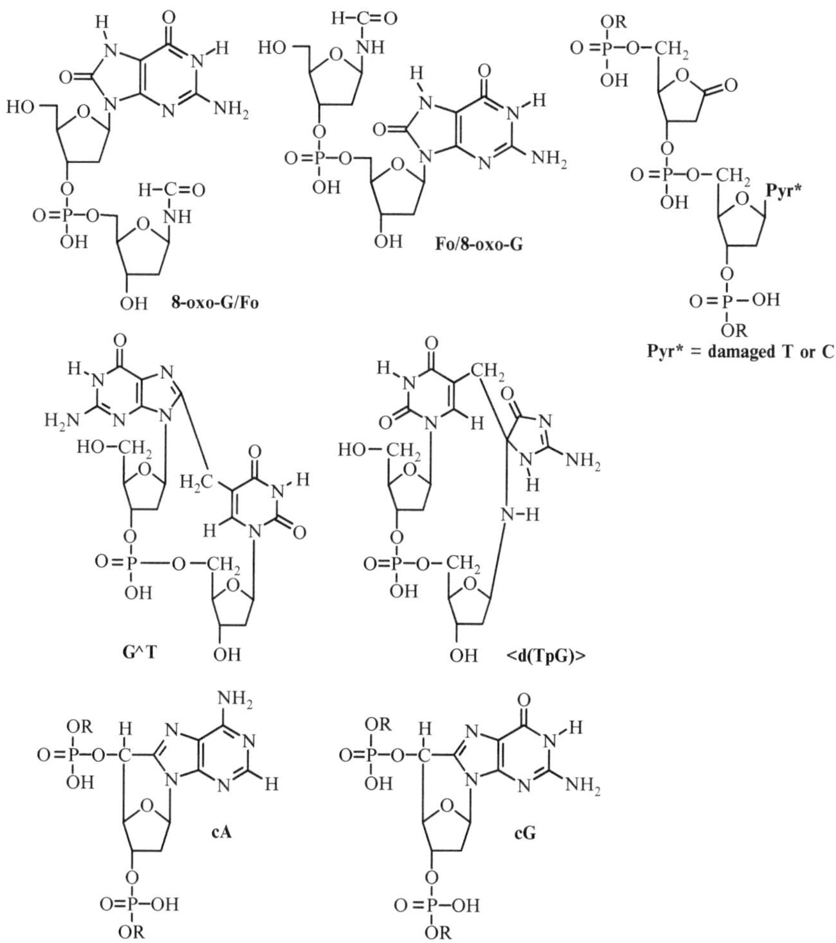

Fig. 2.3 Double lesions, partially vicinal lesions induced by a single •OH radical, as detected in DNA or relevant model systems

2.10 DNA–DNA Cross-Links

There must be a given likelihood that two radicals on opposite strands formed within a spur within a very close distance recombine. Such a process leads to an interstrand DNA–DNA cross-link. There are, however, also single-radical reactions that give rise to such a lesion (Hong and Greenberg 2005).

2.11 DNA–Protein Cross-Links

As seen from Table 2.1, DNA–protein cross-links are even more abundant cellular lesions than DSBs. Mechanistically trivial is the combination of two radicals formed in close proximity, a DNA radical and a protein radical. As DNA and DNA-bound proteins may be overlapped by the same spur, this mechanism is a very likely one for the major part of the DNA–protein cross-links. Yet, there seem to be other precursors, as it has been shown that peroxidized proteins can covalently link to DNA (Gebicki and Gebicki 1999). Mechanistic details are still missing.

2.12 Clustered Lesions Beyond Double Lesions

It was mentioned in Sect. 2.1 that clustered lesions beyond a double lesion may be very difficult to repair, and this type of damage has been considered to have a high probability of being lethal. Clustered lesions beyond double lesions may be formed in spurs and tracks by a trivial mechanism, i.e., the local formation of more than two damaged sites. Densely ionizing radiation must induce more of this type of damage than sparsely ionizing radiation, and it is believed that the higher relative biological efficiency of densely ionizing radiation is due to this effect. Yet, there is also a given possibility that after a double lesion has been formed by a single radical the damage-amplification reaction continues to proceed. Since this will be a slow process that stands a good chance of being intercepted by cellular thiols (see later), it will be difficult to find experimental evidence for its occurrence.

2.13 Oxygen Effect and Chemical Repair

In cells, radiation-induced DNA damage as expressed by SSB/DSB formation or by biological end points such as reproductive cell death are enhanced in the presence of oxygen. This phenomenon has been termed “oxygen effect” and been attributed to the competition of damage fixation by O_2 (reaction 17)) and "chemical repair" by the cell's thiol pool (reaction 18) (Alper 1983). At the time, it was not fully clear that the chemical repair rarely restores the former integrity of DNA. Thus, it is somewhat too simplistic to call reaction 18 a "repair" reaction. Yet, reduction by GSH may lower the severity of the damage. For example, let us take the situation of a multiply damaged DNA site that in the presence of O_2 would give rise to a DSB. The reduction of one of the two sugar radicals by GSH prevents the formation of one of the two close-by SSBs that would give rise to a DSB, i.e., the damage is lowered from the DSB to the SSB level. That DSBs largely originate from two close-by SSBs has been confirmed by measuring the oxygen enhancement ratio (Prise et al. 1993) and the repair rates for these two lesions (Prise et al. 1999). The prevention of a SSB by hydrogen donation is an atypical case, as here a proper repair is not unlikely. Thus, there must also be a benefit when a base radical is inadequately repaired. Earlier, it was mentioned that most damage-amplification reactions are induced by peroxyl radicals. As GSH "repair" prevents their formation, such damage-amplification reactions are intercepted. As a consequence, the level of damage is lowered and the subsequent enzymatic repair facilitated.

2.14 DNA Damage by UV-C and Ionizing Radiation – a Comparison

UVC radiation damages mainly the pyrimidines, and the major products by far are pyrimidine cyclobutane dimers and the 6,4-photoproducts (Cadet 1990). In the terminology used here, they are doubly damaged base lesions. There are only very low yields of singly damaged base sites such as FAPY-A and 8-oxo-G and of SSBs. The presence of radical scavengers such as O_2 or GSH has no or only a negligible effect. In contrast, ionizing radiation produces many singly damaged base sites with a large variety of chemical changes in pyrimidines and purines alike. SSBs are abundant lesions. Double base lesions are not confined to the same strand as those induced by UVC radiation, but also occur on opposite strands. DSBs, DNA–DNA interstrand and DNA–protein cross-links are other double lesions. Clustered lesions beyond a double lesion are another typical feature of ionizing radiation, as are the effects of O_2 and radiation quality.

In this comparison, it is worth mentioning that when in DNA T is substituted by 5BrU the photochemistry changes dramatically, and free-radical processes now dominate. Among others, SSBs and DNA–protein cross-links are observed (for references and mechanistic aspects, see von Sonntag 2006). The observation, that sister-chromatid exchanges are induced led to the conclusion that interstrand DNA cross-links may also be an important lesion (Wojcik et al. 2003) and this has now been substantiated experimentally (Wojcik et al. 2006).

The large variety of damaged sites induced by ionizing radiation is a different challenge for cellular repair systems from UV damage. It seems that highly specific enzymes such as the photoreactivating enzyme that repairs cyclobutane-type dimers effectively do not exist for ionizing-radiation damage. In many aspects, ionizing radiation appears to resemble the damage caused by "oxygen stress," which is considered to be free radical in nature as well. "Oxygen stress" and possibly also ionizing radiation have, like UV radiation, a long evolutionary carrier, and cells have adapted their repair system to this challenge. To a large extent, such repair enzymes have to be generalists and have to be able to cope with a large set of lesions. With this in mind, the large variation of products formed by ionizing radiation may not be that much of a problem, but the interesting questions remain at what rate such lesions are repaired and what are the reasons for and the products of a misrepair. These are the key questions in our attempt to understand the effects of ionizing radiation on biological end points such as chromosomal mutations and reproductive cell death.

References

Alper T (1983) Oxygen as radiosensitizer: methods of analysis. Int J Radiat Biol 44:313–314

Becker D, Bryant-Friedrich A, Trzasko C-A, Sevilla MD (2003) Electron spin resonance study of DNA irradiated with an argon-ion beam: evidence for formation of sugar phosphate backbone radicals. Radiat Res 160:174–185

Begusova M, Spotheim-Maurizot M, Sy D, Michalik V, Charlier M (2001) RADACK, a stochastic simulation of hydroxyl radical attack on DNA. J Biomol Struct Dyn 19:141–158

Berlin YA, Kurnikov IV, Beratan D, Ratner MA, Burin AL (2004) DNA electron transfer processes: some theoretical notions. Top Curr Chem 237:1–36

Cadet J (1990) The photochemistry of nucleic acids. In: Morrison H (ed) Photochemistry and the nucleic acids. Wiley-Interscience, New York, pp 1–272

De Bont R, van Larebeke N (2004) Endogeneous DNA damage in humans: a review of quantitative data. Mutagenesis 19:169–185

Dizdaroglu M, von Sonntag C, Schulte-Frohlinde D (1975) Strand breaks and sugar release by γ-irradiation of DNA in aqueous solution. J Am Chem Soc 97:2277–2278

Fujita S, Steenken S (1981) Pattern of OH radical addition to uracil and methyl- and carboxyl-substituted uracils. Electron transfer of OH adducts with *N,N,N′,N′*-tetramethyl-p-phenylenediamine and tetranitromethane. J Am Chem Soc 103:2540–2545

Gebicki S, Gebicki JM (1999) Crosslinking of DNA and proteins induced by protein hydroperoxides. Biochem J 338:629–636

Giese B (2000) Long-distance charge transport in DNA: the hopping mechanism. Acc Chem Res 33:631–636

Goodhead DT (1994) Initial events in the cellular effect of the ionizing radiations: clustered damage in DNA. Int J Radiat Biol 65:7–17

Hildenbrand K, Schulte-Frohlinde D (1997) Time-resolved EPR studies on the reaction rates of peroxyl radicals of poly(acrylic acid) and of calf thymus DNA with glutathione. Re-examination of a rate constant for DNA. Int J Radiat Biol 71:377–385

Hong IS, Greenberg MM (2005) Efficient DNA interstrand cross-link formation from a nucleotide radical. J Am Chem Soc 127:3692–3693

Lal M, Rao R, Fang X, Schuchmann H-P, von Sonntag C (1997) Radical-induced oxidation of dithiothreitol in acidic oxygenated aqueous solution: a chain reaction. J Am Chem Soc 119:5735–5739
Li X, Sevilla MD, Sanche L (2003) Density functional theory studies of electron interaction with DNA: can zero eV electrons induce strand breaks? J Am Chem Soc 125:13668–13699
Nakatani K, Saito I (2004) Charge transport in duplex DNA containing modified nucleotide bases. Top Curr Chem 236:163–186
O'Neill MA, Barton JK (2004) DNA-mediated charge transport chemistry and biology. Top Curr Chem 236:67–115
Prise KM, Davies S, Michael BD (1993) Evidence for induction of DNA double-strand breaks at paired radical sites. Radiat Res 134:102–106
Prise KM, Gillies NE, Michael BD (1999) Further evidence for double-strand breaks originating from a paired radical precursor from studies of oxygen fixation processes. Radiat Res 151:635–641
Prise KM, Folkard M, Michael BD, Hopkirk A, Munro IH (2000) Critical energies for SSB and DSB induction in plasmid DNA by low-energy photons: action spectra for strand-break induction in plasmid DNA irradiated in vacuum. Int J Radiat Biol 76:881–890
Saito I, Nakamura T, Nakatani K, Yoshioka Y, Yamaguchi K, Sugiyama H (1998) Mapping of the hot spots for DNA damage by one-electron oxidation: Efficacy of GG doublets and GGG triplets as a trap in long-range hole migration. J Am Chem Soc 120:12868–12687
Schulte-Frohlinde D, Behrens G, Önal A (1986) Lifetime of peroxyl radicals of poly(U), poly(A) and single- and double-stranded DNA and the rate of their reaction with thiols. Int J Radiat Biol 50:103–110
Schuster GB (2000) Long-range charge transfer in DNA: transient structural distortions control the distance dependence. Acc Chem Res 33:253–260
Shafirovich V, Cadet J, Gasparutto D, Dourandin A, Huang W, Geacintov NE (2001) Direct spectroscopic observation of 8-oxo-7,8-dihydro-2′-deoxyguanosine radicals in double-stranded DNA generated by one-electron oxidation at a distance by 2-aminopurine radicals. J Phys Chem B 105:586–592
Shao C, Folkard M, Michael BD, Prise KM (2004) Targeted cytoplasmic irradiation induces bystander responses. Proc Nat Acad Sci USA 101:13495–13500
Siddiqi MA, Bothe E (1987) Single- and double-strand break formation in DNA irradiated in aqueous solution: Dependence on dose and OH radical scavenger concentration. Radiat Res 112:449–463
Ulanski P, Bothe E, Rosiak JM, von Sonntag C (1994) OH-Radical-induced crosslinking and strand breakage of poly(vinyl alcohol) in aqueous solution in the absence and presence of oxygen. A pulse radiolysis and product study. Macromol Chem Phys 195:1443–1461
Vilenchik MM, Knudson AG (2003) Endogeneous DNA double-strand breaks: Production, fidelity of repair, and induction of cancer. Proc Nat Acad Sci USA 100:12871–12876
von Sonntag C (1987) The chemical basis of radiation biology. Taylor and Francis, London
von Sonntag C (2006) Free-radical-induced DNA damage and its repair. A chemical perspective. Springer, Berlin
von Sonntag C (2007) Free-radical-induced DNA damage as approached by quantum mechanics and Monte Carlo calculations: an overview from the standpoint of an experimentalist. Adv Quantum Chem 52:5–20
Wagenknecht H-A (2003) Reductive electron transfer and transport of excess electrons in DNA. Angew Chem Int Ed Engl 42:2454–2460
Wagenknecht H-A (2006) Electron transfer processes in DNA: mechanisms, biological relevance and applications in DNA analytics. Nat Prod Rep 23:973–1006
Ward JF (1988) DNA damage produced by ionizing radiation in mammalian cells: identities, mechanisms of formation, and repairability. Prog Nucleic Acid Res Mol Biol 35:95–125
Wardman P (1989) Reduction potentials of one-electron couples involving free radicals in aqueous solution. J Phys Chem Ref Data 18:1637–1755

Wardman P, von Sonntag C (1995) Kinetic factors that control the fate of thiyl radicals in cells. Methods Enzymol 251:31–45

Wojcik A, von Sonntag C, Obe G (2003) Application of the biotin-dUTP chromosome-labelling technique to study the role of 5-bromo-2′-deoxyuridine in the formation of UV-induced sister chromatid exchanges in CHO cells. J Photochem Photobiol B 69:139–144

Wojcik A, Bochenek A, Lankoff A, Lisowska H, Padjas A, Szumiel I, von Sonntag C, Obe G (2006) DNA interstrand crosslinks are induced in cells prelaballed with 5-bromo-2′-deoxyuridine and exposed to UVC radiation. J Photochem Photobiol B 84:15–20

Wu LJ, Randers-Pehrson G, Xu A, Waldren CA, Yu Z, Yu Z, Hei TK (1999) Targeted cytoplasmic irradiation with alpha particles induces mutations in mammalian cells. Proc Nat Acad Sci USA 96:4959–4964

3 Effects of Ultraviolet Radiation on DNA

JÜRGEN KIEFER

Abstract

UV radiation is selectively absorbed by DNA, mainly in the UV-B and the UV-C regions. Excitation of DNA leads to typical photoproducts, the most important being pyrimidine dimers and the so-called 6-4-photoproduct. The yield of strand breaks is very low directly after exposure, but they may be formed indirectly upon further incubation. Although DNA absorbs only weakly in the UV-A region, it may be damaged indirectly via endogenous photosensitisers. Photoproducts are subject to repair processes which are genetically determined. Photoreactivation splits selectively pyrimidine dimers but it is limited to lower eukaryotes and does not occur in mammalian cells. Nucleotide excision repair acts on various types of damage. UV-induced damage manifests itself also at the level of chromosomes. Chromatid-type aberrations are more frequent with UV irradiation than chromosomal aberrations. The action spectrum for their formation demonstrates that an indirect mechanism plays an important role in the UV-A region. Also micronuclei are found after broadband UV exposure which are presumably due to indirectly formed double-strand breaks. UV radiation is a strong inducer of sister-chromatid exchanges, in contrast to ionising radiation, where the yields are low. This suggests that strand breaks play only a minor role.

3.1 Introduction

Ultraviolet (UV) radiation is a part of optical radiation and covers the wavelength region between 100 and 480 nm, corresponding to quantum energies from about 2.5 to 12.5 eV. The UV region is commonly subdivided as given in Table 3.1.

The most energetic part is called the "Schumann UV" or "vacuum UV" because it is strongly absorbed by atmospheric gases. It plays little practical role on Earth but may have some significance in open space. The most important natural source of UV radiation is the sun, which emits UV radiation over the whole range. The ozone layer absorbs most of the UV-C radiation, so only UV-B and UV-A radiation seem to be of significance. The destruction of

Günter Obe and Vijayalaxmi (Eds.)
Chromosomal Alterations: Methods, Results and Importance in Human Health

Table 3.1 Wavelength regions of UV radiation and their common designations

Wavelength (nm)	Quantum energy(eV)	Name
100–200	12.5–6	Vacuum UV, Schumann UV
200–280	6–4.4	UV-C
280–315	4.4–4	UV-B
315–380	4–3.3	UV-A

stratospheric ozone reduces its shielding effect, so the short-wavelength edge of terrestrial solar radiation is shifted towards higher quantum energies, leading to increased hazards for humans, animals and plants.

It is clear from the second column of Table 3.1 that the quantum energies are not sufficient to cause ionisations. Interaction of UV radiation with matter proceeds via selective absorption and excitation. The absorption spectra of biological molecules are, therefore, of great importance for the understanding of action of UV radiation because they determine the probability of energy deposition. Strongly absorbing molecules are also called "chromophores". If radiation is absorbed in essential biomolecules (e.g. DNA) their excitation may lead to photochemical alterations causing biological effects like cell inactivation or mutations. But even if biomolecules are not directly excited because they do not absorb the impinging radiation, they may be indirectly affected via energy transfer from excited states of absorbing "photosensitisers". An important example of this process is the "photodynamic action".

The wavelength dependence of a biological effect can be used to identify the chromophore from where it initiated. This principle of "action spectroscopy" (see later) constitutes an important analytical tool and played some role in the history of molecular biology although this was not really appreciated at the time. As long ago as 1930, Gates (1930) showed that the spectral dependence of bactericidal action of UV radiation followed closely the absorption spectrum of nucleic acids and was distinctly different from that of proteins, which at that time were thought to be the building blocks of the genetic material. A few years later, similar results were obtained for mutagenesis (Knapp et al 1939; Emmons and Hollaender 1939), demonstrating the importance of nucleic acids before this was biochemically proven (and honoured by award of the Nobel Prize) by Avery et al. (1944).

Figure 3.1 displays the absorption spectrum of DNA. It has a maximum around 265 nm (in the UV-C region) but there is still also substantial absorption of UV-B radiation, while the molecule is virtually translucent for UV-A radiation.

Although frequently used, the term "dose", meaning "deposited energy per unit mass", is misleading in connection with UV radiation exposure. In most cases it stands for "energy fluence" (E), i.e. impinging energy per unit area, which is measured in joules per square metre. In many health-related publications also "weighted doses" are used. They are obtained by weighting all wavelengths according to their efficiency to cause a certain effect

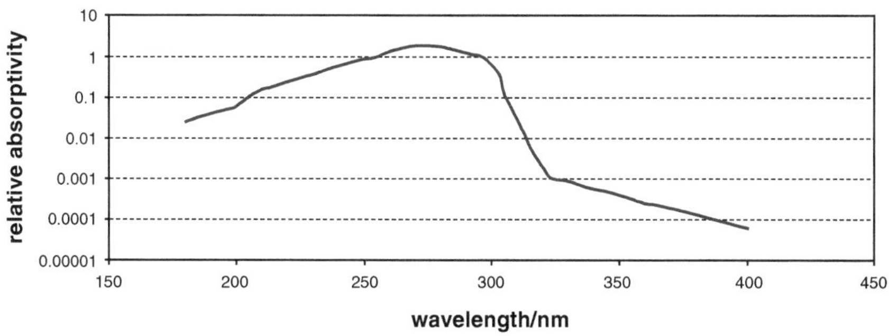

Fig. 3.1 Absorption spectrum of DNA. (Data from Sutherland and Griffin 1981)

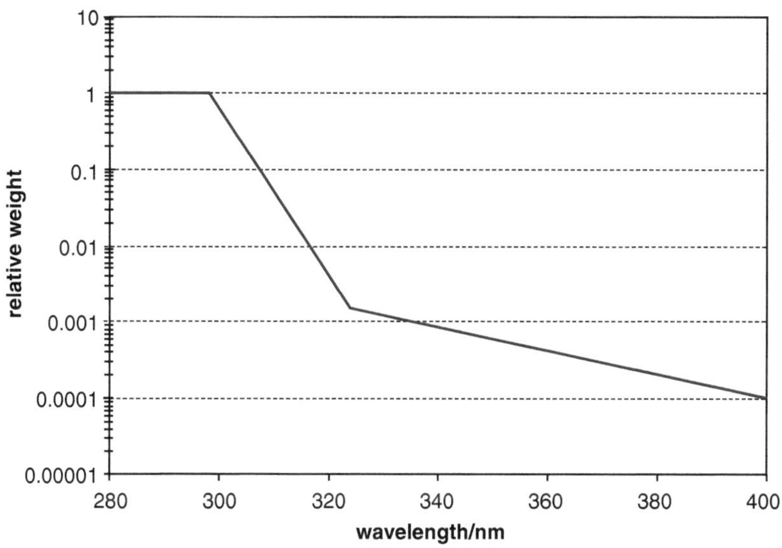

Fig. 3.2 The CIE-weighting function for erythema induction (McKinley and Diffey 1987)

(e.g. sunburn) and summing over the whole spectrum. These weighted doses are unfortunately given in the same units, mostly joules per square metre, which may lead to misunderstandings. The most important weighting function is that for erythema induction. Figure 3.2 shows it in a standardised form as defined by the Commission Internationale de l'Eclairage; International Commission on Illumination (CIE) (McKinley and Diffey 1987).

In photochemistry "photon fluence" (Φ), i.e. the number of photons per unit area, is more customary. The relationship between energy and photon fluence is simple:

$$E = h\nu\Phi,$$

where h is Planck's constant and ν the frequency.

3.2 Primary Photoproducts in DNA

As indicated already, one has to distinguish between direct UV action and energy transfer from photosensitisers. Types and yields of photoproducts may differ considerably between these two cases. As DNA absorbs most strongly around 265 nm, low-pressure mercury lamps emitting nearly monochromatic radiation at 254 nm are often used for assessing the direct action of UV radiation on cells and their constituents. More recently broadband sources for UV-B and UV-A radiation have also been employed. To determine product yields as a function of wavelength for the construction of action spectra, either broadband sources coupled to monochromators or tunable lasers are used.

The spectrum of DNA alterations induced by UV radiation is distinctly different from that seen with ionising radiation. The most important photoproducts are *cis-syn* cyclobutane dimers between adjacent pyrimidines (CPD) on the same DNA strand; thymine–thymine dimers are the most frequent. Another damage type is also characterised by a covalent linkage between two pyrimidine bases but with a different structure involving the 6-position and the 4-position of the ring. It is thus commonly termed the "6-4-photoproduct". Figure 3.3 shows the molecular structures. Other damage types, like DNA–protein cross-links and directly induced single-strand breaks, play only a minor role as seen from Table 3.2, where yields of DNA changes upon exposure to 254-nm radiation are listed.

Douki et al. (2000) using modern chromatographic techniques arrived at essentially the same results. The same group summarised UV-mediated damage to DNA with special emphasis on various wavelength regions and simulated sunlight (Cadet et al. 2005). UV-A radiation is of particular interest as it forms the largest contribution from terrestrial solar radiation but is only very weakly absorbed by DNA. This had sometimes prompted the conclusion that it is not hazardous to human health and that even extensive exposure (e.g. on sunbeds) does not carry any cancer risk. This view has now been revised. The yields of CPDs and the 6-4-photoproducts are much smaller in the UV-A region compared with those in the UV-B region but a new class of damage is seen with longer wavelengths, namely oxidative products and strand breaks. This situation is comparable to that with low

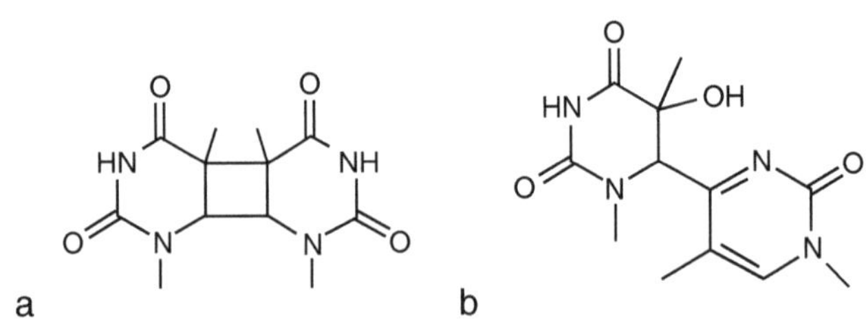

Fig. 3.3 UV photoproducts in DNA. **a** Cyclobutane-type pyrimidine dimer, **b** 6-4-photoproduct

Table 3.2 Yields of photoproducts in mammalian cells with exposure to 254-nm UV radiation

Product	Yield $(10^8\ u^a)^{-1}\ (J\ m^{-2})^{-1}$	Relative yield	Reference
CPD	2.0	1	Rosenstein and Mitchell (1987)
6-4-Product	0.4	0.2	Rosenstein and Mitchell (1987)
DNA–protein cross-links	1.3×10^{-4}	6.6×10^{-5}	Peak et al. (1985)
DNA single-strand breaks	4.1×10^{-4}	2×10^{-4}	Rosenstein and Ducore (1983)

CPD cis-syn cyclobutane dimers between adjacent pyrimidines
[a] u is the atomic mass unit

Table 3.3 DNA lesions in UV-A exposed and γ-irradiated human monocytes (Cadet et al. 2005)

Lesion	Yield per kJ m^{-2} UV-A radiation	Yield per Gy γ-rays
8-Oxo-7,8-dihydo-2′-deoxyguanosine	0.98	11
DNA strand breaks (single-strand breaks, double-strand breaks, alkali labile sites)	0.9	130
Endonuclease III sensitive sites[a]	0.3	53
FPG-sensitive sites[b]	1.9	48

FPG formamidopyrimidine DNA glycosylase
[a] Mostly oxidised pyrimidine bases
[b] Mostly modified purine bases

linear energy transfer ionising radiation and suggests a free-radical mechanism. Table 3.3 lists some of the products in cells exposed to UV-A radiation in comparison with those in cells after γ-irradiation. It is seen that there is a great similarity in the kind of products, but relative yields are different. While DNA breaks figure most prominently with γ-rays, modified purine bases are most important with UV-A radiation.

Nevertheless, the formation of strand breaks whose yield is negligible with short-wavelength UV is very remarkable as they are also held responsible for most of the hazardous effects of ionising radiation. This is strictly true, however, only for DNA double-strand breaks (DSBs). It is not clear at present whether they are also induced directly by UV-A radiation.

DSBs may, however, be formed indirectly when photoproducts remain unrepaired in cellular DNA. This was first suggested by experiments in a yeast mutant conditionally defective for DSB repair. (Kiefer 1987; Frankenberg-Schwager et al. 1987). It was previously assumed that the accumulation of excision repair related single-strand breaks may also lead to DSBs (Bradley 1981; Bradley and Taylor 1983), but it was later shown that UV-induced DSBs were also involved in excision-defective cells (Kiefer and Feige 1993). Experiments with transformed human fibroblasts defective in the ability to bypass pyrimidine dimers during DNA replication (XP variants) showed that DSBs originate at stalled replication forks (Limoli et al. 2002). DSBs can be specifically determined by an immunofluorescence assay for the phosphorylated histone γH2AX which is formed at DSB sites in DNA (Rogakou et al. 1998, 1999). A complex of recombination proteins (Mre11 complex) localises at stalled replication forks (Limoli et al. 2000). Simultaneous immunofluorescence staining both for γH2AX and for the Mre11 complex showed that both types of foci colocalise in UV-exposed cells. This is in line with an increased induction of sister-chromatid exchanges (SCEs) in these repair-deficient cells.

3.3 Action Spectra

The dependence of a given biological effect on the wavelength of the impinging radiation is often called an "action spectrum". The fundamental interest in action spectra lies in the possibility to identify the molecule responsible for a photobiological action by comparing the action spectrum with the absorption spectra of candidate cellular constituents. In order to reach this goal, a number of experimental precautions have to be taken but these cannot be discussed in detail here (see e.g. Kiefer 1990). A necessary condition, however, which must be mentioned in the context of this chapter is that the radiation must reach the intracellular target without being modulated by intervening material. This is easy to achieve with in vitro experiments with cell cultures (provided the cell suspension or the cell layer is not too dense), but is rather difficult in vivo. A prominent example is the so-called action spectrum for erythema induction by UV radiation. The cells responsible lie in the skin's stratum germinativum at a depth of some micrometres covered by the epidermis. This layer absorbs strongly in the UV-C region and partly also in the UV-B region (Parrish et al. 1978), which causes a strong distortion of the emission spectrum of an external light source. The resultant erythema action spectrum is thus a convolution of skin absorption and the action spectrum for cellular effects in the stratum germinativum.

Figures 3.4 and 3.5 show action spectra for the formation of some of the photoproducts mentioned already. The induction efficiency of CPDs and 6-4-photoproducts follows closely DNA absorption, which is not the case for single-strand breaks and DNA–protein cross-links, where a second peak is seen between 350 and 450 nm. This behaviour indicates that another chromophore is involved as discussed already.

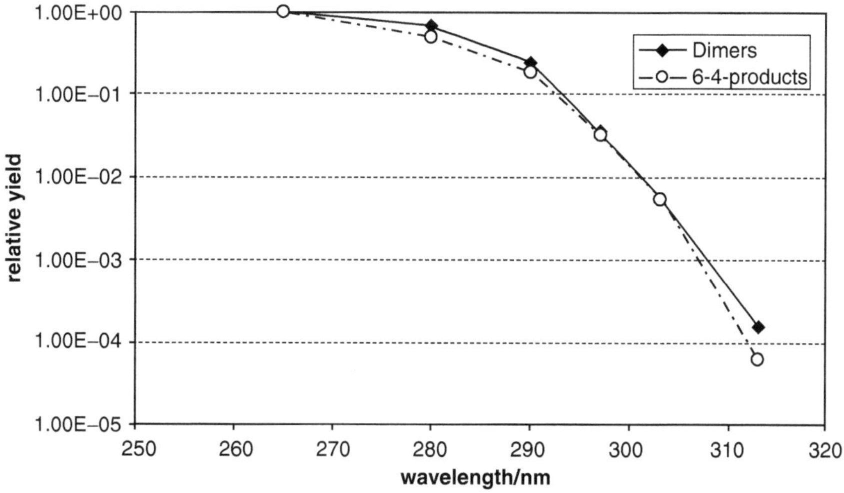

Fig. 3.4 Action spectra for the formation of thymine dimers and 6-4-photoproducts. (After Rosenstein and Mitchell 1987)

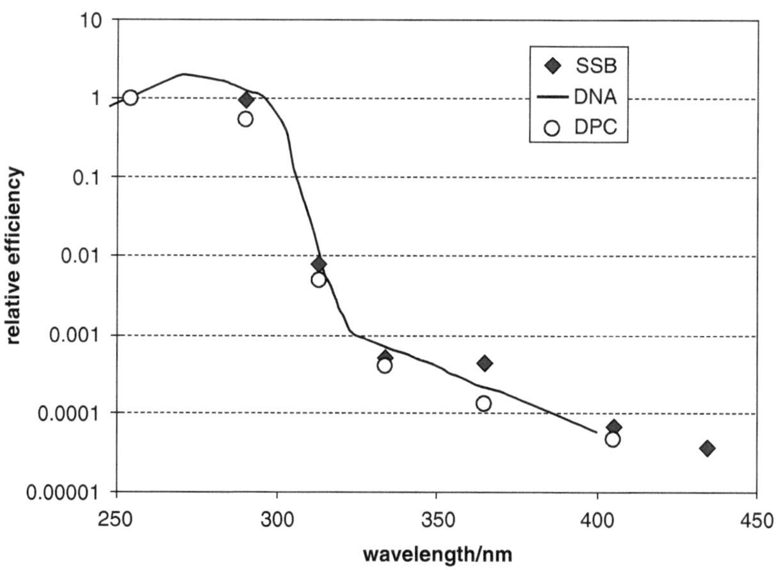

Fig. 3.5 Action spectra for the formation of single-strand breaks (*SSB*) and DNA–protein cross-links (*DPC*). The *line* shows DNA absorption for comparison. (After Peak and Peak 1991)

3.4 Reversal and Repair of Primary Photoproducts

The history of processes by which genetic damage is repaired or its effect circumvented is tightly linked to the study of the interaction of UV radiation on cells and subcellular systems. There is a wealth of literature on this topic, and even a short review is outside the scope of this chapter (see e.g. Walker et al. 1995). CPDs play here a leading role because of their specific properties. Their formation is a steady-state process: they are not only induced by UV radiation but also split at the same time. The efficiency for both pathways depends on the wavelength: the relative maximum for formation is at 280 nm, that for splitting is at 240 nm. This provides a "photochemical switch" between induction and reversal of these photoproducts. As a consequence of this behaviour, the CPD-induction dose–effect curve saturates at higher fluences, mirroring a steady state between product formation and reversal. It has to be pointed out that this is a merely photochemical process and is not related to biological repair phenomena.

There is, however, also a light-driven enzymatic reaction by which CPDs can be split. It is called "photoreactivation" and has an efficiency maximum at longer wavelengths, around 365 nm. The underlying molecular processes are now quite well understood. The enzyme now termed "photolyase" is not active in all cells, especially not in higher mammalian systems, but in marsupialia, birds and reptiles as well as in most microbes (reviewed in Weber 2005). Photoreactivation was assumed to be unique for CPDs, so one had a rather simple tool to judge their contribution to photobiological effects but, of course, only in those systems where the enzyme is present and active. More recently, however, also a photolyase activity for the 6-4-photoproduct has been reported (Todo et al. 1993). The gene has been cloned and its product used for analysing the role of specific photoproducts (Otoshi et al. 2000).

A more universal and probably the most important repair process is excision repair. It was also first found in UV-exposed bacteria (Setlow and Carrier 1964; Boyce and Howard-Flanders 1964) but subsequent research revealed its presence also in mammalian (including human) cells (reviewed in Walker et al. 1995). It is controlled by a large number of genetic loci. In the human disease *Xeroderma pigmentosum* (XP) (Cleaver 1968) some of them are altered, leading to an extreme sensitivity to UV radiation and also to proneness to skin tumours and also other types of cancer. This recessively inherited condition of which seven complementation groups are known (Cleaver et al. 1999) is very rare and affects only about one in one million people. In vitro cultured XP cells constitute a very valuable tool for the elucidation of repair processes.

Excision repair is very complex and a detailed description is outside the scope of this chapter. Briefly, the DNA strand with altered nucleotides is incised and the damaged part removed. The single-strand gap is then sealed by repair replication, with the remaining undamaged DNA strand serving as a template. Nucleotide excision repair (NER), as it is termed now, is thus essentially error-free (apart from the background error rate during replication) and its action is not restricted to UV-induced lesions but virtually to all types of altered nucleotides.

It was found later that the efficiency of NER is not the same over the genome and depends on transcriptional activity in the affected part. Transcription coupled repair (TCR), dealing exclusively with lesions on the actively transcribed strand, is considerably faster than global genome repair, which acts on damage in the whole cellular genome (Mellon et al. 1987; reviewed in van Zeeland et al. 2005). TCR is controlled by genes which are defective in the human disease Cockayne's syndrome (CS). Two complementation groups are known (Cleaver et al. 1999).

Although NER is the most important repair process in UV-exposed cells, other processes may also contribute significantly to lesion removal or circumvention, e.g. homologous recombination (HR) and lesion bypass. The latter plays a role when the density of lesions on both DNA strands is so high that excision gaps may overlap and thus repair replication can no longer work efficiently (Lehmann 2006).

As pointed out already, DNA DSBs are not directly induced at least by UV-B and UV-C radiation, but they may be formed indirectly. It is well known from studies with ionising radiation that they can be repaired by essentially two pathways, namely HR, which is essentially error-free, and nonhomologous end-joining (NHEJ), which may lead to chromosomal misarrangements. These pathways may also be important for DSBs formed indirectly upon UV exposure.

A recent review of all relevant repair pathways has been provided by Christmann et al. (2003) and should be consulted for further details.

3.5 Cytological Effects

As with other mutagenic or clastogenic agents, various cytological end points have been investigated with UV radiation, including structural chromosomal aberrations, micronuclei and SCEs.

3.5.1 Chromosomal Aberrations

Chromosomal aberrations in mammalian cells upon UV exposure were first studied in a systematic manner by Chu (1965a, b). He first used unsynchronised exponential cultures of Chinese hamster cells and found both chromatid as well as chromosome aberrations whose yields, however, depended on the time interval between exposure and scoring. Chromosome aberrations were seen only 42 h after exposure, and their number was always distinctly lower than those of chromatid-type aberrations. It can be assumed that chromosome-type aberrations are derived from chromatid-type aberrations during the subsequent cell cycle. The action spectrum (Fig. 3.6) follows in the short-wavelength region essentially DNA absorption, but there is a clear deviation above 320 nm, where breakage efficiency is much higher than expected on the basis of DNA absorption. This was not noted at the time but suggests a different action mechanism for the UV-A region.

In the work described in the second paper (Chu 1965b), synchronised populations were used. Although there was an indication that S phase is most sensitive in terms of the induction of aberrations by UV radiation, definite conclusions could not be drawn. When cells are exposed to short-wavelength UV radiation, mainly aberrations of the chromatid type are induced. Pyrimidine dimers and other photoproducts confined to a single DNA strand block replication in S phase and strand breaks are formed, leading to chromosome aberrations. UV radiation is thus considered as an S-phase-dependent agent (Bender et al. 1973; Natarajan et al. 1980).

There is general agreement that DNA DSBs are the principal lesions in the formation process of chromosomal aberrations (Obe et al. 2002). DSBs are, however, not directly formed by UV-B and UV-C radiation if the fluences are limited to physiological ranges. Nevertheless, all types of chromosomal aberrations are found in mammalian cells after UV exposure. This suggests that DSBs are created either by secondary mechanisms acting on primary photoproducts or as a result of stalled replication. DNA single-strand gaps

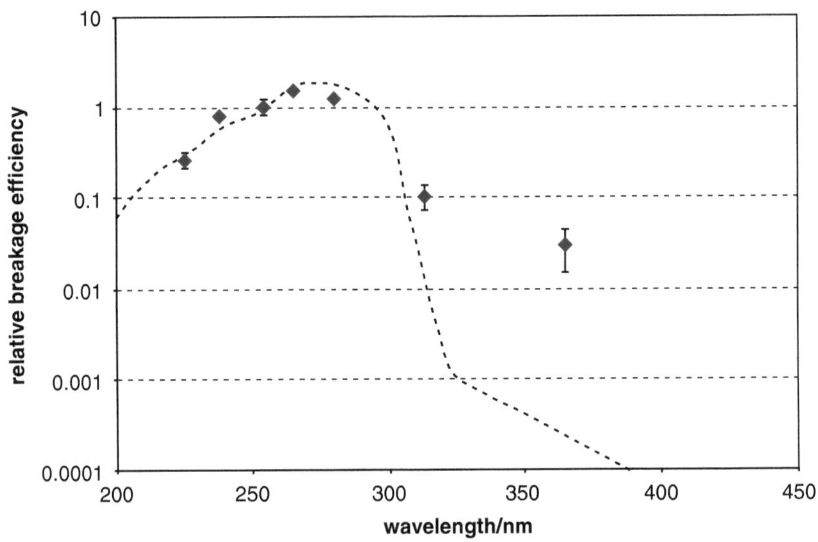

Fig. 3.6 Action spectrum for the formation of chromatid aberrations. The *dotted line* shows DNA absorption. (After Chu 1965a)

are part of NER. If their number is high, they may overlap and lead to DSBs. As the sequence of incision and subsequent repair replication is well coordinated, the probability for this to occur is low with not too high fluences. If gap filling is inhibited by specific chemicals, e.g. by ara-C or aphidicolin, single strands accumulate and DSBs are created (Bradley and Taylor 1983). Holmberg and Gumauskas (1990) demonstrated that in unstimulated human lymphocytes the yield of chromosomal aberrations after UV-C exposure is very low but is substantially increased when ara-C is added. The fluence–effect curve for dicentrics could be fitted to a purely quadratic function, indicating that they are formed via the interactions of overlapping single-strand gaps whose number is proportional to that of the primary photoproducts.

The process just described does not seem to play an important part in cells with normal and uninhibited repair capacities. In these cases DSBs are formed when stalled replication forks are disentangled. Their number depends thus on the number of photolesions still remaining in the DNA at the beginning of S phase and increases linearly with UV radiation fluence (Limoli et al. 2002).

The yield of chromosomal aberrations after UV exposure is inversely linked with the cell's capacity for the removal of photoproducts from the DNA before it enters S phase. Consequently deficiencies in NER lead to increased numbers of cells carrying chromosomal aberrations (Chipchase and Melton 2002). TCR seems to play an important part in this respect as shown in studies with various Chinese hamster cell lines homologous to human CS cells (Proietti de Santis et al. 2001). It is estimated that only 5–8% of the eukaryotic genome is actively transcribed. TCR is at first sight not expected to have a great impact on DNA lesions compared with the remainder of the genome. The experiments demonstrate, however, a drastic increase in UV sensitivity. This suggests that even relatively few remaining lesions in actively transcribed strands are critical for cell survival. This assumption is borne out by the fact that TCR-deficient cells undergo apoptosis in the second cell cycle after exposure. This raises the possibility that the accumulation of chromosomal aberrations may serve as a signal for apoptosis (Proietti de Santis et al. 2001).

3.5.2 Micronuclei

Micronuclei are assumed to be formed as a result of chromosome breakage. They can be assessed either by microscopical observation of interphase cells or by cytofluorometric measurements (Fenech 2000). The determination of micronuclei is simpler and less time consuming than the observation of structural chromosomal aberrations, but is also open to experimental pitfalls (Fenech 2000). So far it has been mainly employed to study the differential effects of various wavelength ranges, initially concentrating on the UV-B range. Keulers et al. (1998) showed in human peripheral blood lymphocytes that broadband UV-B radiation induces micronuclei in a fluence-dependent manner following a linear–quadratic relationship. More recently also UV-A radiation, which is a major component of sunlight, was investigated in more detail. As already pointed out, DNA absorption is very weak in this region and it has to be assumed that DNA is not damaged by direct photon absorption but via photosensitised processes. There are no studies on the whole range of structural chromosomal aberrations with UV-A radiation and most of the studies are on micronuclei induction. Rapp and Greulich (2004) showed that in human keratinocytes DSBs induced by UV-A radiation if not repaired lead to micronuclei. If these cells are exposed to 100 kJ m^{-2} of broadband UV-A radiation their survival is reduced to only about 70%, but the number of cells with micronuclei increases with further incubation until day 28, which is about the natural lifespan of these cells in culture (Phillipson et al. 2002). This suggests that UV-A radiation induces genomic instability similarly to ionising radiation (reviewed in Kronenberg 1994).

The micronuclei assay can also be used to assess the effect of UV-B radiation treatment of patients with skin diseases (Ferahbas et al. 2004). It could be shown that in lymphocytes isolated from their blood exposure to narrowband UV-B radiation in several treatment sessions leads to significant levels of damage.

3.5.3 Sister-Chromatid Exchanges

UV radiation is a potent inducer of SCEs. The classic way to evaluate SCEs is to label one DNA strand selectively with bromodeoxyuridine, which causes fainting of Giemsa staining so that the different strands can be differentiated in the microscopic picture. It was not realised initially that the bromine atom may be photochemically active and interfere with the SCE formation process. This problem was overcome by the introduction of another labelling method, namely the use of biotin-16-2′-deoxyuridine-5′-triphosphate, which is photochemically inert (Wojcik et al. 2004). Since this technique is comparatively new, older results have to be treated with some reservation. Wolff et al. (1974) observed that SCEs are only formed after UV exposure if the cells pass through S phase. They originate from long-lived lesions which may be induced in any of the cell cycle phases. The action spectrum below 320 nm coincides with the DNA absorption spectrum (Nussbaum 1995), but this does not give any information about the initiating lesion as the different photoproducts cannot be differentiated on the basis of short-wavelength action spectra (Figs. 3.4, 3.5). Wojcik et al. (2004) suggested that DNA interstrand cross-links may be the starting point of SCEs because chemicals like mitomycin C are also very potent SCE inducers (Kaina and Aurich 1985), but this, of course, is only indirect evidence and further clarification is required. Recent experiments by Wojcik et al. (2006) with a modified comet assay showed that cross-links may be formed as a result of radiation in 5-bromo-2′-deoxyuridine labelled cells, which supports the original assumption. It seems, however, that DNA DSBs do not play a major role as initiators because ionising radiation has comparatively low SCE yields (Wojcik et al. 2004).

3.6 Conclusions

Recent years have seen the development of powerful and sensitive analytical techniques to identify and quantify UV-induced photoproducts in DNA. The results demonstrate that not only the UV-B and UV-C regions, where DNA absorbs strongly, play a role for photobiological and photomedical effects but also the UV-A region, where DNA is virtually translucent. This points to the importance of indirect damage pathways presumably involving photosensitised reactions. Cytogenetic studies may be helpful for the understanding of the underlying basic mechanism of damage recognition and repair as well as for elucidation of the development of UV-related diseases like skin cancer. They may in future also aid the identification of genetically determined individual sensitivities.

References

Avery OT, MacLeod CM, McCarthy M (1944) Studies on the chemical nature of the substance inducing transformation of pneumococcal types. J Exp Med 79:137–158

Bender MA, Griggs HG, Walker PL (1973) Mechanisms of chromosomal aberration production I. Aberration induction by ultraviolet light. Mutat Res 20:387–402

Boyce R, Howard-Flanders P (1964) Release of ultraviolet light-induced thymine dimers from DNA in *E. coli* K-12. Proc Natl Acad Sci USA 51:293–300

Bradley MO (1981) Double-strand breaks in DNA caused by repair of damage due to ultraviolet light. J Supramol Struct Cell Biochem 16:337–343

Bradley MO, Taylor VI (1981) DNA double-strand breaks induced in normal human cells during the repair of ultraviolet light damage. Proc Natl Acad Sci USA 78:3619–3623

Bradley MO, Taylor VI (1983) Repair-induced DNA double strand breaks after ultraviolet-light and either aphidicolin or 1-β-D-abinofuranosylcytosine/hydroxyurea. Carcinogenesis 4:1533–1537

Cadet J, Sage E, Douki T (2005) Ultraviolet radiation-mediated damage to cellular DNA. Mutat Res 571:3–17

Chipchase MD, Melton DW (2002) The formation of UV-induced chromosome aberrations involves ERCC1 and XPF but not other nucleotide excision repair genes. DNA Repair (Amst) 1:335–340

Christmann M, Tomicic MT, Roos WP, Kaina B (2003) Mechanisms of human DNA repair: an update. Toxicology 193:3–34

Chu EHY (1965a) Effects of ultraviolet radiation on mammalian cells. I. Induction of chromosome aberrations. Mutat Res 2:75–94

Chu EHY (1965b) Effects of ultraviolet radiation on mammalian cells. II. Differential UV and x-ray sensitivity of chromosomes to breakage in 5-aminouracil synchronized cell populations. Genetics 52:1279-1294

Cleaver JE (1968) Defective repair replication of DNA in xeroderma pigmentosum. Nature 218:652–656
Cleaver JE, Thompson LH, Richardson AS, States JC (1999) A summary of mutations in the UV-sensitive disorders: xeroderma pigmentosum, Cockayne syndrome, and trichothiodystrophy. Hum Mutat 14:9–22
Douki T, Court M, Sauvaigo S, Odin F, Cadet J (2000) Formation of the main UV-induced thymine dimeric lesions within isolated and cellular DNA as measured by high performance liquid chromatography-tandem mass spectrometry. J Biol Chem 275: 11678–11685
Emmons W, Hollaender A (1939) The action of ultraviolet radiation on dermatophytes. 11. Mutations induced in cultures of dermatophytes by exposure of spores to monochromatic ultraviolet radiation. Am J Bot 26:467–475
Fenech M (2000) The in vitro micronucleus technique. Mutat Res 455:81–95
Ferahbas A, Donmez-Altuntas H, Hamurcu Z, Aktas E, Utas S (2004) Micronucleus evaluation in mitogen-stimulated lymphocytes of narrow-band (311 nm TL01) UVB-treated patients. Photodermatol Photoimmunol Photomed 20:81–85
Frankenberg-Schwager M, Frankenberg D, Harbich R (1987) Possible occurrence of DNA double-strand breaks during repair of u.v.-induced damage in yeast. Int J Radiat Biol 52:107–113
Gates FL (1930) A study of bactericidal action of ultraviolet light. II. The absorption of ultraviolet light by bacteria. J. Gen Physiol 14:31–42
Holmberg M, Gumauskas E (1990) Chromosome-type exchange aberrations are induced by inhibiting repair of UVC-induced DNA lesions in quiescent human lymphocytes. Mutat Res 232:261–266
Kaina B, Aurich O (1985) Dependency of the yield of sister-chromatid exchanges induced by alkylating agents on fixation time. Possible involvement of secondary lesions in sister-chromatid exchange induction. Mutat Res 149:451–461
Keulers RA, de Roon AR, de Roode S, Tates AD (1998) The induction and analysis of micronuclei and cell killing by ultraviolet-B radiation in human peripheral blood lymphocytes. Photochem Photobiol 67:426–432
Kiefer J (1987) UV response of the temperature-conditional rad 54 mutant of the yeast *Saccharomyces cerevisiae*. Mutat Res 191:9–12
Kiefer J (1990) Biological radiation action. Springer, Berlin
Kiefer J, Feige M (1993) The significance of DNA double-strand breaks in the UV inactivation of yeast cells. Mutat Res 299:219–224
Knapp E, Reuss,A, Risse A, Schreiber H (1939) Quantitative Analyse der mutationsauslosenden Wirkung monochromatischen u.v.-Lichtes. Naturwissenschaften 27:304
Kronenberg A (1994) Radiation-induced genomic instability. Int J Radiat Biol 66:603–609
Lehmann A (2006) Translesion synthesis in mammalian cells. Exp Cell Res 312:2673–2676
Limoli CL, Giedzinski E, Morgan WF, Cleaver JE (2000) Polymerase η deficiency in the xeroderma pigmentosum variant uncovers an overlap between the S phase checkpoint and double-strand break repair. Proc Natl Acad Sci USA 97:7939–7946
Limoli CL, Giedzinski E, Bonner WM, Cleaver JE (2002) UV-induced replication arrest in the Xeroderma variants leads to DNA double strand breaks, g-H2AX formation and Mre11 relocalisation. Proc Natl Acad Sci USA 99:233–238
McKinley AF, Diffey BL (1987) A reference spectrum for ultraviolet induced erythema in human skin. CIE J 6:17–22
Mellon I, Spivak G, Hanawalt PC (1987) Selective removal of transcription-blocking DNA damage from the transcribed strand of the mammalian DHFR gene. Cell 51:241–249
Natarajan AT, Kihlmann BA, Obe G (1980) Use of the 5-bromodeoxyuridine-labelling technique for exploring mechanisms involved in the formation of chromosomal aberrations. Mutat Res 73:307–317
Nussbaum A (1995) Aktionsspektroskopie über die Induktion von Schwesterchromatifaustäuschen (SCEs) an V79-Hamsterzellen und ein neues Modell der SCE-Entstehung. PhD thesis, University of Giessen

Obe G, Pfeiffer P, Savage JRK, Johannes C, Goedecke W, Jeppesen P, Natarajan AT, Martinez-Lopez W, Folle GA, Drets ME (2002) Chromosomal aberrations: formation, identification and distribution. Mutat Res 504:17–36
Otoshi E, Yagi T, Mori T, Matsunaga T, Nikaido O, Kim ST, Hitomi K, Ikenaga M, Todo T (2000) Respective roles of cyclobutane pyrimidine dimers, (6-4)photoproducts, and minor photoproducts in ultraviolet mutagenesis of repair-deficient xeroderma pigmentosum A cells. Cancer Res 60:1729–1735
Parrish JA, Anderson RR, Urbach J, Pitts D (1978) UV-A. Biological effects of ultraviolet radiation with emphasis on human responses to longwave ultraviolet. Plenum, New York
Peak JG, Peak, MJ (1991) Comparison of initial yields of DNA-to-protein crosslinks and single strand breaks induced in cultured human cells by far- and near-ultraviolet light, blue light and X-rays. Mutat Res 246:187–191
Peak JG, Peak MJ, Sikorski RS, Jones CA (1985) Induction of DNA-protein crosslinks in human cells by ultraviolet and visible radiations: action spectrum. Photochem. Photobiol 41:295–302
Phillipson RP, Tobi SE, Morris JA, McMillan TJ (2002) UV-A induces persistent genomic instability in human keratinocytes through an oxidative stress mechanism. Free Radic Biol Med 32:474–480
Proietti de Santis L, Garcia CL, Balajee AS, Latini P, Pichierri P, Nikaido O, Stefanini M, Palitti F (2002) Transcription coupled repair efficiency determines the cell cycle progression and apoptosis after UV exposure in hamster cells. DNA Repair (Amst) 28:209–223
Rapp A, Greulich KO (2004) After double-strand break induction by UV-A, homologous recombination and nonhomologous endjoining cooperate at the same DSB if both systems are available. J Cell Sci 117:4935–4945
Rogakou EP, Pilch DR, Ort AH, Ivanova VS, Bonner WM (1998) DNA double-stranded breaks induce histone H2AX phosphorylation on serine 139. J Biol Chem 273:5858–5868
Rogakou EP, Boon C, Redon C, Bonner WM (1999) Megabase chromatin domains involved in DNA double-strand breaks in vivo. J Cell Biol 146:905–916
Rosenstein BS, Ducore JM (1983) Induction of DNA strand breaks in normal human fibroblasts exposed to monochromatic ultraviolet and visible wavelengths in the 240-546 nm range. Photochem Photobiol 38:51–55
Rosenstein BS, Mitchell DL (1987) Action spectra for the induction of pyrimidine(6-4)pyrimidone photoproducts and cyclobutane pyrimidine dimers in normal human skin fibroblasts. Photochem Photobiol 45:775–780
Setlow RB, Carrier W (1964) The disappearance of thymine dimmers from DNA: an error correcting mechanisms. Proc Natl Acad Sci USA 51:226–231
Sutherland JC, Griffin KP (1981) Absorption spectrum of DNA for wavelengths longer than 320 nm. Radiat Res 86:399–409
Todo T, Takemori H, Ryo H, Ihara M, Matsunaga T, Nikaido O, Sato K, Nomura, T (1993) A new photoreactivating enzyme that specifically repairs ultraviolet light-induced (6-4)photoproducts. Nature 361:371–374
van Zeeland AA, Vreeswijk MPG, de Gruijl F, van Kranen HJ, Vrieling H, Mullenders LFH (2005) Transcription-coupled repair: Impact on UV-induced mutagenesis in cultures rodent cells and mouse skin tumours. Mutat Res 577:170–178
Walker GC, Siede W, Friedberg EC (1995) DNA repair and mutagenesis. American Society for Microbiology, Washington
Weber S (2005) Light-driven enzymatic catalysis of DNA repair: a review of recent biophysical studies on photolyase. Biochim Biophys Acta 1707:1–23
Wojcik A, Bruckmann E, Obe G (2004) Insights into the mechanisms of sister chromatid exchange formation. Cytogenet Genome Res 104:304–309
Wojcik A, Bochenek A, Lankoff H, Lisowska A, Padjas I, Szumiel C, von Sonntag, Obe G (2006) DNA interstrand crosslinks are induced in cells prelabelled with 5-bromo-2′-deoxyuridine and exposed to UVC radiation. J Photochem Photobiol B 84:15–20
Wolff S, Bodycote J, Painter RB (1974) Sister chromatid exchanges induced in Chinese hamster cells by UV irradiation of different stages of the cell cycle: the necessity for cells to pass through S. Mutat Res 25:73–81

4 Double Strand Break Repair Mechanisms in Mammalian Cells

WOLFGANG GOEDECKE

Abstract

Double-strand breaks (DSBs) represent major threats in chromosomal DNA. They arise either as intermediate structures during recombination, replication and repair events or as potentially lethal lesions introduced by ionising radiation or drugs. In any case, DSBs have to be eliminated immediately, because of the recombinogenic capacity of the DNA ends generated by the DSB that increases the risk for undesired chromosomal aberrations (CAs). In order to cope with DSBs, cells exhibit two different sets of repair activities, namely non-homlogous end joining (NHEJ) and homologous recombinational repair (HRR). Both pathways are regulated during the cell cycle with peak activities either in G1 phase (NHEJ) or in late S and G2 phases (HRR). While NHEJ has the capacity to join arbitrary DNA ends together, HRR depends on the presence of a second undamaged sequence in the genome providing a template for the reconstitution of the sequence at the DSB site. While most of the proteins identified so far participate in one of the two repair pathways, a few proteins are known to participate in both pathways. These last group of proteins is supposed to be involved in the selection process of the DNA repair pathway. Mutations in one DSB repair system result in the accumulation of DSBs and increased levels of CAs. As a result, genome instability is observed in cells with impaired DSB repair functions.

4.1 Introduction

Double-strand breaks (DSBs) in chromosomal DNA are considered to represent intermediate structures which exist transiently during replication, recombination and repair of the DNA. They occur either spontaneously based on the cellular DNA metabolism or by exogeneous damage created by ionising radiation or DNA-damaging drugs. Therefore it is obvious that threats generated by DSBs are hardly to be avoided and DSBs from exogenous sources increase previously existing levels. Since DSBs are the ultimate lesions for the generation of chromosomal aberrations (CAs), accumulation of DSBs will raise the probability of the creation of CAs. Generally about 50 DSBs are induced by

Günter Obe and Vijayalaxmi (Eds.)
Chromosomal Alterations: Methods, Results and Importance in Human Health

2 Gy of sparsely ionising irradiation and nearly the same number of DSBs occur spontaneously in each mammalian cell during a single cell division (Vilenchik and Knudson 2003).

The severe effects caused by DSBs result mainly from their abilities to generate DNA structures with free ends. These free DNA ends are potent inducers of various kinds of recombinational activities, such as rearrangements of the immunoglobulin genes, for example. Balanced translocations which are frequently seen in tumour cells are considered to be the result of uncontrolled or accidental recombination events; therefore, it is a great achievement for each cell to control the number of DSBs and to ensure genomic stability by preventing CAs. This is achieved by DSB repair mechanisms which are able to eliminate most of the DSBs immediately after they have been generated.

During evolution, two strategies of DSB repair were established, namely the non-homologous end-joining (NHEJ) and the homologous recombinational repair (HRR) systems. Both appeared early during evolution and are present in probacteria or bacteria, underlining their significance for survival of living cells (Bowater and Doherty 2006; Cavalier-Smith 2002). NHEJ is characterised by the ability to join free DNA ends, irrespective of their sequence and structure. The activity is a universal pathway for the elimination of DSBs. Since nucleolytic degradation and non-templated addition of nucleotides occur frequently at free DNA ends, this repair pathway comprises a high risk for the generation of de novo mutations at sites of former DSBs. This susceptibility to mutagenesis is a direct consequence of the NHEJ system to join any pair of free DNA ends, because sequence alterations that are attached to ends are incorporated into the junction of the joined molecule.

In contrast to NHEJ, HRR is able to restore the original sequence of the broken DNA molecule. This decreases the risk for the generation of small deletions and insertions at the site of the former DSB. HRR activity is non-mutagenic at sites of DSBs but only a small fraction of breaks is accessible for repair by this mechanism. In fact, HRR is limited to those DSBs which are located within sequences having at least one further and undamaged copy somewhere else in the genome (Obe et al. 2002).

4.2 NHEJ Reaction

4.2.1 NHEJ Mechanism and Junction Formation

Most of our knowledge about the mechanisms underlying the NHEJ reaction results from studies with cell-free extracts. Several organisms and cell types have been used for the preparation of such extracts (Pfeiffer et al. 2005). Since NHEJ is the dominant DSB repair activity in vertebrate cells, most of our knowledge is derived from mammalian cell cultures.

The NHEJ activity is assayed by either circle formation or concatenation of linear DNA substrate molecules. Making use of restriction endonucleases, we can easily generate them from plasmid molecules and they permit the generation of substrates with defined and non-compatible combinations of DNA ends. It should be emphasised that DNA ends created with restriction enzymes provide structures with either blunt or 3′ or 5′ protruding single strands (PSS). These end structures differ from those generated by ionising radiation. Nevertheless, it was shown that DNA ends generated by ionising radiation are also suitable substrates for NHEJ reactions (Odersky et al. 2002).

Depending on the nature of the DNA ends, two modes of junctions may be observed, as outlined in Fig. 4.1. The first is designated as the overlap type and utilises matches of a few base pairs to align DNA ends for end-joining reactions. Substrates preferentially processed by the overlap type are ends with PSS of the same polarity (3′-PSS vs. 3′-PSS and 5′-PSS vs. 5′-PSS). The second pathway is the fill-in type. This mode is used to join blunt ends, PSS of opposing polarity (3′-PSS vs. 5′-PSS) or combinations of blunt and PSS ends. Only a minor fraction of junctions are associated with a more pronounced loss of nucleotides during junction formation. Interestingly, the 3′-OH of one of the DNA ends can be used to prime DNA synthesis on a template of the other end (Thode et al. 1990). DNA ends with PSS longer than ten nucleotides are not used as substrates for the NHEJ mechanism; these structures may be converted into joinable DNA ends by successive degradation of the PSS tail (Beyert et al. 1994; Daley and Wilson 2005).

The mechanisms described above have been confirmed by in vivo data from different species and cell types. If linear substrate molecules are introduced into cells by either transfection or injection and the sequences of their

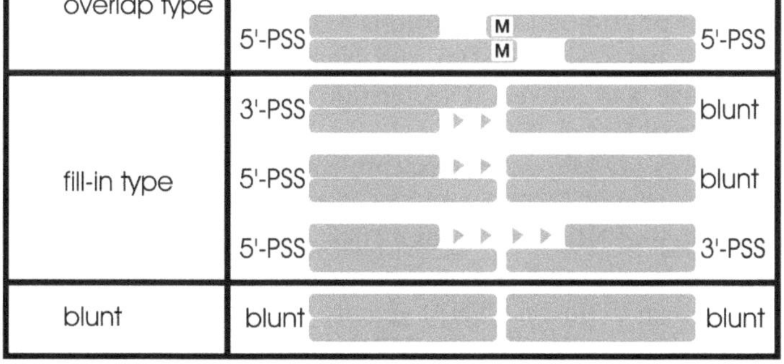

Fig. 4.1 The two modes of non-homologous end-joining (NHEJ) reactions depending on DNA end configurations. Ends of the same polarity are joined by overlap formation and ends of opposite polarity or with at least one blunt end involved are joined by the fill-in mode. White squares labelled with *M* indicate bases that match arbitrarily. *PSS* protruding single strand

junctions are reinvestigated after DSB repair, they look similar to those generated in vitro (Roth and Wilson 1986). Another type of evidence comes from investigating the junctions of DSBs generated by the rare cutting endonuclease *SceI*. This enzyme from *Saccharomyces cerevisiae*, where it functions in intron homing, has a long recognition sequence of 18 base pairs, which is absent in mammalian genomes. In genetically engineered cell lines, where this site is introduced into the mammalian genome, introduction of *SceI* allows site-specific cleavage at a single site and the junctions observed after NHEJ resemble those found by the joining in cell-free extracts (Dahm-Daphi et al. 2005).

4.2.2 Proteins Involved in NHEJ

Both types of the NHEJ pathway, the one using arbitrary matches between complementary base pairs (overlap type) and the one conserving PSS (fill-in type), tend to avoid formation of de novo mutations. They are therefore called the "high-fidelity" NHEJ pathways. The proteins involved in the high-fidelity end joining are called core proteins. They are DNA-PK, Artemis, XRCC4 and Ligase IV (O'Driscoll and Jeggo 2006). In addition to these core proteins, the polymerases Polμ and Polλ are both recruited for the gap filling by templated addition of nucleotides. It is assumed, that both polymerases play a redundant role during gap filling, but in contrast to Polλ, only Polμ is able to perform gap filling across the DSB (Nick McElhinny et al. 2005). Terminal desoxytransferase (TdT) is a further DNA polymerase acting at DSBs. The protein is involved in the non-templated addition of nucleotides, which is occasionally observed in cells of the immune system (Paull 2005).

All these proteins assemble at sites of DSBs. DNA-PK is a DNA-dependent phosphokinase, composed of three subunits, the DNA-PK catalytic subunit (DNA-PKcs) and a DNA-binding heterodimer formed by the two subunits Ku70 and Ku80. The DNA-PK complex is a DNA-dependent kinase, which is recruited under physiological conditions to DNA ends owing to its binding to the Ku70/Ku80 heterodimer (Dip and Naegeli 2005). The function of the Artemis nuclease is to convert DNA termini into ligatable structures (Drouet et al. 2006; Niewolik et al. 2006).

Other proteins are known that assemble at sites of DSBs. These are ataxia telangiectasia mutated (ATM) and the components of the MRN complex, MRE11, RAD50 and NBS1. ATM, like DNA-PKcs, is another member of the protein kinase family and is a component of the MRN complex, namely MRE11 was shown to possess NHEJ activity in vitro (Paull and Gellert 2000).

Some evidence suggests that the MRN complex is involved in the second pathway of NHEJ reactions. Since this pathway is characterised by extended deletions that occur as a result of loss of terminal nucleotides at the site of the DSBs, this pathway is called "low-fidelity" NHEJ. It is tempting to assume that junctions formed by low-fidelity end joining are processed by a

MRN-dependent activity, representing a second NHEJ pathway associated with a different set of proteins.

This model is supported by the fact that mutants defective in one of the core components are still able to perform NHEJ reactions, but junctions resulting from these activities are usually characterised by elevated levels of terminal nucleotide losses and an increased usage of patch homologies to align the juxtaposed strands. Data from MRE11-depleted extracts of the clawed frog *Xenopus laevis* and MRE11 knockout chicken DT40 cells show that in these systems the Ku-dependent high-fidelity NHEJ pathway remains unaffected. This indicates that in these cells the high-fidelity pathway is dominant over the low-fidelity pathway (Evans and Hearing 2005).

4.3 DSB Repair by HRR

HRR mechanisms have been extensively studied in procaryotes and lower eucaryotes such as the budding yeast *Saccharomyces cerevisiae*. A reason that might account for the overrepresentation of data from these organisms is the predominance of HRR compared with that in vertebrates. In order to explain the frequently observed association of gene conversions with crossover events especially during meiotic recombination, the DSB repair model was established as a standard model for HRR (Szostak et al. 1983). Nevertheless it should be noted that there are alternative models and that no single model is able to explain all aspects of HRR assays. Other models are the synthesis dependent strand annealing (SDSA) model and the break induced replication (BIR) model. For a comprehensive description of the different HRR mechanisms, see Paques and Haber (1999).

A common feature of DSB repair by HRR is the existence of an undamaged template sequence somewhere else in the genome that serves as a donor for the sequence information of the broken molecule. In somatic cells this donor sequence is usually provided by the sister chromatids. Compared with templates from other locations (ectopic sites) in mammalian cells, the sister chromatid is used several orders of magnitude more frequently. This is a clear difference between mammalian cells and *Saccharomyces cerevisae*, where the difference in recombination between sister chromatids and ectopic sites is much less pronounced (Johnson and Jasin 2001). The strong bias towards sister-chromatid preference suggests that HRR takes place predominantly during late S phase and probably early G2 phase. In contrast to this, NHEJ mainly occurs in prereplicative G1 cells (Fig. 4.2).

The homologous chromosome is efficiently used as a donor sequence during prophase I of meiosis, when both homologous chromosomes are aligned by the synaptonemal complex. In contrast to meiotic recombination, which is characterised by gene conversions associated with crossovers, HRR in somatic cells is generally not associated with crossovers (Johnson and Jasin 2001).

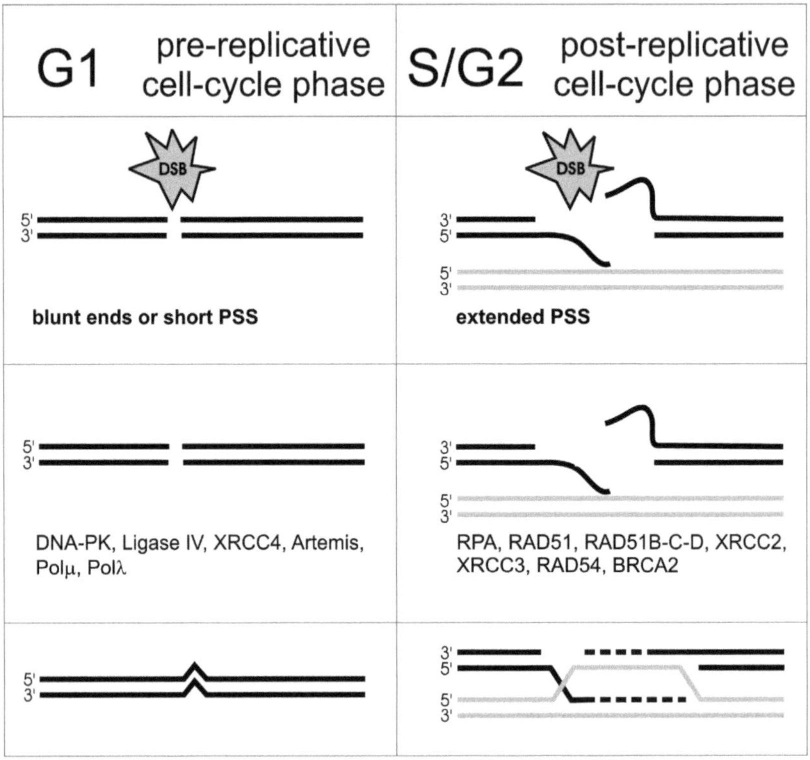

Fig. 4.2 NHEJ is the dominant pathway in prereplicative cells. The single-strand protrusions (*PSS*) should not be longer than ten nucleotides. In contrast, homologous recombinational repair is able to process longer PSS tails and is dominant in postreplicative cells. Enzymes involved in one of the two mechanisms are indicated and explained in the text. The *kink* symbolises the high risk of mutation formation during NHEJ, while the indicated double Holliday structure indicates the precise restoration of the break. *DSB* double-strand break

The underlying mechanisms for HRR in meiosis and somatic cells may differ. The DNA ends used in meiotic recombination differ from those breaks that are generated by irradiation or replication errors, since they are tagged by the spo11 protein, an enzyme which is generates the DSBs during prophase I of meiosis (Neale et al. 2005). It is therefore problematic to discuss meiotic DSBs as a model for DSB repair, because the DSBs used to initiate HRR during meiosis do not resemble the free ends of DNA, which are the major threats of DSBs in somatic cells. Meiotic recombination represents a highly specialised system to guarantee homologous exchange of DNA molecules, and it should not be taken as a model system for the repair of accidental DSBs.

4.3.1 HRR Mechanisms

The common scheme of HRR involves copying of sequence information from the unbroken template molecule to the site of the DSB. HRR events are carried out by a series of successive steps, namely (1) presynapsis, in which extended protruding 3′-tails are generated which are much longer than the PSS appearing during NHEJ, (2) synapsis, where strand invasion of these tails into the undamaged template sequence generates a structure called a displacement loop (D-loop) and (3) postsynapsis, which includes branch migration and resolution of the double Holliday junctions (dHJ).

The main function of DSB repair by HRR is the restoration of stalled replication forks. This is supported by a specific activation of HRR during S-phase, which is not the case for the NHEJ pathway (Rothkamm et al. 2003; Segurado et al. 2002). The tight correlation between HRR and replication is further indicated by genetic evidence. Recombinogenic ends are generated by either stalled or collapsed replication forks (Michel et al. 2001). Mutants like ATM and ataxia-telangiectasia related (ATR), which are defective in S-phase control, are also impaired in HRR, indicating the close relationship between cell cycle and DSB repair. The same holds for mutants deficient for sister-chromatid cohesion. Loss of sister-chromatid cohesion abolishes postreplicative DSB repair. This indicates that sister chromatids generated during replication are required as template molecules for HRR repair in somatic cells.

4.3.2 Proteins Involved in HRR

The components of the MRN complex and exonucleases like Exo1 are key components for the generation of long single-stranded 3′-tails needed for strand invasion. Immediately after generation of the 3′-tails, they become stabilised by the heterotrimeric replication factor A (RPA) complex. Strand transfer and D-loop formation require several further RAD52 proteins, namely RAD52 and Rad51B, Rad51C, Rad51D, Xrcc2, Xrcc3 and BRCA2. While the proteins involved in these early recombination steps are quite well known, the subsequent steps are less clear. Among other things, they require the function of the RAD54 protein and, in order to resolve the dHJ, the action of BLM, topo3α, sgs1 and others (Heyer et al. 2006).

4.4 Impaired DSB Repair and Chromosomal Aberrations

As outlined already, impaired DSB repair leads to elevated levels of DSBs and to the persistence of DSBs over longer periods of time (Iliakis et al. 2004). Since DSBs themselves have a recombinogenic potential, one would expect increased frequencies of CAs in cells with impaired DSB repair systems.

Inherited diseases are known which show the direct link between DSB repair and CAs. Around 125 genes are involved in overall DNA-repair activity, and those participating in DSB repair show elevated levels of CAs, if they are mutated (Wood et al. 2001; Ronen and Glickman 2001).

The effect of NHEJ proteins on CAs was shown from patient data as well as from mouse models. Screening of peripheral lymphocytes of patients suffering from untreated cancer showed elevated levels of CAs. The number of CAs correlated with the activity of DNA-PK in these cells. This result indicate that, in addition to impaired DSB repair on the DNA level, there is also inefficiency of DSB repair on the chromosomal level (Someya et al. 2006). Translocations, which are a hallmark of aberrant DSB repair, are frequently observed in transgenic mice, knocked out for NHEJ genes. Mouse embryonic fibroblasts from Ku70 and Ligase IV deficient mice show a dramatic spontaneous increase of chromosomal fragmentations and non-reciprocal translocations, even in the absence of genotoxic drugs or irradiation. This result is consistent with elevated levels of unprocessed DSBs, acting as a primary source for CAs (Ferguson et al. 2000).

Elevated frequencies of CAs are also observed in cells with disturbed functions in HRR. Cell lines derived from Chinese hamster ovary cells which are deficient in members of the RAD51 family (XRCC2 and XRCC3) accumulate CAs such as breaks, deletions and unbalanced translocations (Cui et al. 1999). Together with the previously mentioned results from mutants in the NHEJ system, these data clearly show that problems in one of the two DSB repair systems lead to CAs and genome instability. Although there is striking evidence that DSBs are the ultimate lesions that are converted into CAs by an incorrect action of the aforementioned repair pathways ("misrepair"), the exact mechanisms of how this happens are not quite clear.

4.5 Fragile Sites Are Associated with DNA DSBs

Common fragile sites (CFSs) are specific chromosomal loci that appear as constrictions or gaps on metaphase chromosomes. They are induced in cultured cells treated with drugs that partially block DNA replication or stimulate premature chromosome condensation (El Achkar et al. 2005). Examples of these two groups are aphidicolin and calyculin A, respectively. In contrast to the CFSs, which are present in all individuals, rare fragile sites (RFSs) only appear in in cells with extended amplifications of repetitive elements at a certain chromosomal locus and are therefore not present in all individuals. This rather depends on the allele of the repetitive element, which may change from one individual to the other owing to dynamic mutations that change the length of the allele (Pearson et al. 2005). CFSs and RFSs are both characterised by stretches of repetitive DNA of either AT-rich or GC-rich sequences underlining their common basis on the level of the DNA sequence (Zlotorynski et al. 2003).

The fact that replication problems as well as premature chromosome condensation lead to the induction of fragile sites indicates that the basis of this phenomenon is the disturbance of the chronology of chromosome replication and condensation during the cell cycle. Therefore proteins regulating the cell cycle, like the ATR protein, also affect fragile site stability (Tibbetts et al. 2000).

The observed correlation of chromosomal breakpoints of balanced translocations involved in cancer and the high frequency of sister-chromatid exchanges associated with CFSs lead to the suggestion that DSBs are involved in the establishment of fragile sites (Arlt et al. 2003). This proposal has recently been supported by the detection of DSB repair proteins at CFSs (Schwartz et al. 2005). Further evidence comes from downregulation of DSB repair proteins involved in HRR (RAD51) and NHEJ (DNA-PK and Ligase IV). In the absence of these proteins, a high number of constrictions are induced, in agreement with the model that proteins from both DSB repair pathways are involved in stabilising fragile sites (Schwartz et al. 2005). Obviously, DSBs generated in repetitive DNA are converted into fragile sites, if they fail to be repaired before the onset of the next cycle of chromosome condensation.

4.6 Conclusions

Free DNA ends generated by DSBs can be joined to other free ends present in the genome or they are able to invade intact DNA molecules, initiating recombination events. Since these activities bear the risk for the generation of CAs, DSBs, if possible, are eliminated immediately after they are created. Two major pathways are available, the universal NHEJ pathway and the HRR pathway. The latter one is restricted to DSBs with extended sequence similarities between the broken DNA molecule and other sequences in the genome. Usually the unbroken DNA molecule is provided by the sister chromatid. This is the reason why HRR takes place predominantly in postreplicative cells. Consistently NHEJ is detected in prereplcative cell cycle phases.

In cases of "misrepair", the repair mechanisms fail to eliminate DSBs by joining of the originally linked DNA ends. In these cases, DSB repair pathways create new junctions that are detected as translocations on the chromosomal level. Therefore, a consequence of unmeant DSB repair is the generation of chromosomal aberrations.

References

Arlt MF, Casper AM, Glover TW (2003) Common fragile sites. Cytogenet Genome Res 100:92–100

Beyert N, Reichenberger S, Peters M, Hartung M, Gottlich B, Goedecke W, Vielmetter W, Pfeiffer P (1994) Nonhomologous DNA end joining of synthetic hairpin substrates in Xenopus laevis egg extracts. Nucleic Acids Res 22:1643–1650

Bowater R, Doherty AJ (2006) Making ends meet: repairing breaks in bacterial DNA by non-homologous end-joining. PLoS Genet 2:e8
Cavalier-Smith T (2002) Origins of the machinery of recombination and sex. Heredity 88:125–141
Cui X, Brenneman M, Meyne J, Oshimura M, Goodwin EH, Chen DJ (1999) The XRCC2 and XRCC3 repair genes are required for chromosome stability in mammalian cells. Mutat Res 434:75–88
Dahm-Daphi J, Hubbe P, Horvath F, El Awady RA, Bouffard KE, Powell SN, Willers H (2005) Nonhomologous end-joining of site-specific but not of radiation-induced DNA double-strand breaks is reduced in the presence of wild-type p53. Oncogene 24:1663–1672
Daley JM, Wilson TE (2005) Rejoining of DNA double-strand breaks as a function of overhang length. Mol Cell Biol 25:896–906
Dip R, Naegeli H (2005) More than just strand breaks: the recognition of structural DNA discontinuities by DNA-dependent protein kinase catalytic subunit. FASEB J 19:704–715
Drouet J, Frit P, Delteil C, de Villartay JP, Salles B, Calsou P (2006) Interplay between Ku, Artemis, and the DNA-dependent protein kinase catalytic subunit at DNA ends. J Biol Chem 281:27784–27793
El Achkar E, Gerbault-Seureau M, Muleris M, Dutrillaux B, Debatisse M (2005) Premature condensation induces breaks at the interface of early and late replicating chromosome bands bearing common fragile sites. Proc Natl Acad Sci USA 102:18069–18074
Evans JD, Hearing P (2005) Relocalization of the Mre11-Rad50-Nbs1 complex by the adenovirus E4 ORF3 protein is required for viral replication. J Virol 79:6207–6215
Ferguson DO, Sekiguchi JM, Chang S, Frank KM, Gao Y, DePinho RA, Alt FW (2000) The non-homologous end-joining pathway of DNA repair is required for genomic stability and the suppression of translocations. Proc Natl Acad Sci USA 97:6630–6633
Heyer WD, Li X, Rolfsmeier M, Zhang XP (2006) Rad54: the Swiss Army knife of homologous recombination? Nucleic Acids Res 34:4115–4125
Iliakis G, Wang H, Perrault AR, Boecker W, Rosidi B, Windhofer F, Wu W, Guan J, Terzoudi G, Pantelias G (2004) Mechanisms of DNA double strand break repair and chromosome aberration formation. Cytogenet Genome Res 104:14–20
Johnson RD, Jasin M (2001) Double-strand-break-induced homologous recombination in mammalian cells. Biochem Soc Trans 29:196–201
Michel B, Flores MJ, Viguera E, Grompone G, Seigneur M, Bidnenko V (2001) Rescue of arrested replication forks by homologous recombination. Proc Natl Acad Sci USA 98:8181–8188
Neale MJ, Pan J, Keeney S (2005) Endonucleolytic processing of covalent protein-linked DNA double-strand breaks. Nature 436:1053–1057
Nick McElhinny SA, Havener JM, Garcia-Diaz M, Juarez R, Bebenek K, Kee BL, Blanco L, Kunkel TA, Ramsden DA (2005) A gradient of template dependence defines distinct biological roles for family X polymerases in nonhomologous end joining. Mol Cell 19:357–366
Niewolik D, Pannicke U, Lu H, Ma Y, Wang LC, Kulesza P, Zandi E, Lieber MR, Schwarz K (2006) DNA-PKcs dependence of artemis endonucleolytic activity: differences between hairpins and 5′ or 3′ overhangs. J Biol Chem 10:33900–33909
O'Driscoll M, Jeggo PA (2006) The role of double-strand break repair – insights from human genetics. Nat Rev Genet 7:45–54
Obe G, Pfeiffer P, Savage JR, Johannes C, Goedecke W, Jeppesen P, Natarajan AT, Martinez-Lopez W, Folle GA, Drets ME (2002) Chromosomal aberrations: formation, identification and distribution. Mutat Res 504:17–36
Odersky A, Panyutin IV, Panyutin IG, Schunck C, Feldmann E, Goedecke W, Neumann RD, Obe G, Pfeiffer P (2002) Repair of sequence-specific 125I-induced double-strand breaks by non-homologous DNA end joining in mammalian cell-free extracts. J Biol Chem 277:11756–11764
Paques F and Haber JE (1999) Multiple pathways of recombination induced by double-strand breaks in Saccharomyces cerevisiae. Microbiol Mol Biol Rev 63:349–404
Paull TT (2005) Saving the ends for last: the role of pol mu in DNA end joining. Mol Cell 19:294–296

Paull TT, Gellert M (2000) A mechanistic basis for Mre11-directed DNA joining at microhomologies. Proc Natl Acad Sci USA 97:6409–6414

Pearson CE, Edamura KN, Cleary JD (2005) Repeat instability: mechanisms of dynamic mutations. Nat Rev Genet 6:729–742

Pfeiffer P, Feldmann E, Odersky A, Kuhfittig-Kulle S, Goedecke W (2005) Analysis of DNA double-strand break repair by nonhomologous end joining in cell-free extracts from mammalian cells. Methods Mol Biol 291:351–371

Ronen A, Glickman BW (2001) Human DNA repair genes. Environ Mol Mutagen 37:241–283

Roth DB, Wilson JH (1986) Nonhomologous recombination in mammalian cells: role for short sequence homologies in the joining reaction. Mol Cell Biol 6:4295–4304

Rothkamm K, Kruger I, Thompson LH, Lobrich M (2003) Pathways of DNA double-strand break repair during the mammalian cell cycle. Mol Cell Biol 23:5706–5715

Schwartz M, Zlotorynski E, Goldberg M, Ozeri E, Rahat A, le Sage C, Chen BP, Chen DJ, Agami R, Kerem B (2005) Homologous recombination and nonhomologous end-joining repair pathways regulate fragile site stability. Genes Dev 19:2715–2726

Segurado M, Gomez M, Antequera F (2002) Increased recombination intermediates and homologous integration hot spots at DNA replication origins. Mol Cell 10:907–916

Someya M, Sakata K, Matsumoto Y, Yamamoto H, Monobe M, Ikeda H, Ando K, Hosoi Y, Suzuki N, Hareyama M (2006) The association of DNA-dependent protein kinase activity with chromosomal instability and risk of cancer. Carcinogenesis 27:117–122

Szostak JW, Orr-Weaver TL, Rothstein RJ, Stahl FW (1983) The double-strand-break repair model for recombination. Cell 33:25–35

Thode S, Schafer A, Pfeiffer P, Vielmetter W (1990) A novel pathway of DNA end-to-end joining. Cell 60:921–928

Tibbetts RS, Cortez D, Brumbaugh KM, Scully R, Livingston D, Elledge SJ, Abraham RT (2000) Functional interactions between BRCA1 and the checkpoint kinase ATR during genotoxic stress. Genes Dev 14:2989–3002

Vilenchik MM, Knudson AG (2003) Endogenous DNA double-strand breaks: production, fidelity of repair, and induction of cancer. Proc Natl Acad Sci USA 100:12871–12876

Wood RD, Mitchell M, Sgouros J, Lindahl T (2001) Human DNA repair genes. Science 291:1284–1289

Zlotorynski E, Rahat A, Skaug J, Ben Porat N, Ozeri E, Hershberg R, Levi A, Scherer SW, Margalit H, Kerem B (2003) Molecular basis for expression of common and rare fragile sites. Mol Cell Biol 23:7143–7151

5 Backup Pathways of Nonhomologous End Joining May Have a Dominant Role in the Formation of Chromosome Aberrations

George Iliakis, Wenqi Wu, Minli Wang, Georgia I. Terzoudi, and Gabriel E. Pantelias

Abstract

Chromosome aberrations are large-scale chromatin rearrangements produced in cells after exposure to ionizing radiation and other DNA-damaging agents. They have been directly implicated in cell killing, mutation induction and oncogenic transformation and have spearheaded a great number of investigations focused on their qualitative and quantitative analysis. The central aim of modern cytogenetics is the description at the molecular level of the observations at the chromosome level. There is consensus that the majority of chromosome aberrations are the ultimate consequence of errors in the repair of DNA double-strand breaks (DSBs). Here, we outline the molecular aspects of DSB repair pathways and discuss them vis-à-vis the two prominent theories of chromosome aberration formation. Although both homology-dependent and homology-independent pathways contribute to DSB repair, homology-independent pathways are predominantly responsible for the initial removal of the majority of DSBs from the genome. From the homology-independent pathways, both DNA-PK-dependent nonhomologous end joining (D-NHEJ) and backup nonhomologous end joining (B-NHEJ) have the potential of forming chromosome aberrations, but it is likely that most of the aberrations derive from a combination of low-frequency misjoining events during D-NHEJ, and from much higher frequency misjoining events during B-NHEJ. Further elucidation of the mechanism of DSB repair at the level of chromatin will allow a more complete molecular description of the process.

5.1 Introduction

Chromosome aberrations are large-scale chromatin rearrangements produced in cells after exposure to ionizing radiation (IR) and other DNA-damaging agents (Cornforth 2006; Cornforth and Bedford 1993; Evans 1962; Hlatky et al. 2002; Natarajan 2002; Obe et al. 2002; Revell 1974). Such rearrangements, which are directly observable at metaphase, have been directly implicated in cell killing, mutation induction and oncogenic transformation and have spearheaded a great number of investigations focused on their

Günter Obe and Vijayalaxmi (Eds.)
Chromosomal Alterations: Methods, Results and Importance in Human Health

qualitative and quantitative analysis (Savage 1975). The complex nature of chromosomal organization at the molecular level and the lack of information on the underlying repair mechanisms, together with the diversity in the mode of action between agents producing similar chromosomal changes, have led to the adoption in the field of a descriptive terminology that is still in use today. The central aim of modern cytogenetics is the description at the molecular level of the observations at the chromosome level.

Efforts to interpret cytogenetic results from early studies culminated in the formulation of two seemingly incompatible theories of aberration formation (Revell 1974; Savage 1998; Savage and Harvey 1991): The classical "breakage-and-reunion" theory and the "exchange" theory. Both theories are based on generalizations and abstractions put forward at a time when our understandings of DNA repair and chromatin structure were not nearly as well developed as they are today. Probably as a result of this limitation, and owing to their individual success in describing distinct sets of data, it has not been possible to conclusively refute one of these theories in favor of the other.

The breakage-and-reunion theory postulates that radiation induces a distinct physical break in the chromosome (chromonema) that leads to the generation of open ends. While the majority (over 90%) of these ends restitute (rejoin), some remain unrejoined and directly lead to terminal deletions. Others rejoin illegitimately with ends of a neighboring break(s), causing chromosome exchanges. The break in this model is at the chromosome level (not the DNA level), and the ends are assumed to be free to move around (flapping ends) and to participate in promiscuous rejoining events that lead to chromosome exchanges.

In the exchange theory, the primary lesion is not a chromosome break (i.e., not a breakage in the chromonema), but an unstable (repairable) recombinogenic lesion of undefined nature that has the potential of interacting pairwise with a similar lesion to cause chromosome exchanges. The mediation of such, presumably nonrandom, interactions in the absence of chromosome breaks is an essential feature of the model according to which chromatid (or chromosome) breaks arise only indirectly from incomplete exchanges. Exchanges in this model are the result of error-prone processing and explaining complex exchanges will require additional assumptions.

Although the breakage-and-reunion theory is clearly favored among researchers, the arguments for the exchange theory have not been conclusively refuted, and when challenged they are "rescued" by invoking scoring limitations (Revell 1974; Savage 1998; Savage and Harvey 1991); as a result, the debate continuous. In many ways the ongoing debate of these theories reflects the fact that despite large strides in our understanding of DNA metabolism, our knowledge of the chromosomal organization of DNA remains incomplete and the same holds for some aspects of DNA metabolism, including repair, in the context of chromatin. As a result, the formulation of a molecular theory of chromosome aberration formation that would settle the debate is lacking.

On the positive side, there is consensus that the majority of chromosome aberrations are the ultimate consequence of errors in the repair of DNA double-strand breaks (DSBs). By focusing on a concrete and measurable DNA lesion, a solid foundation is generated for reexamining the fundamental assumptions of the aforementioned theories and for comparing them with our present knowledge of DSB repair and chromatin structure. Since both theories require two lesions for the formation of an exchange aberration, the issues that need to be explained include:

1. Are DSBs breaking the chromosome and if not under what conditions can a DNA DSB lead to a chromosome break?
2. If DSBs do not break the chromosome directly, can they be considered as the undefined recombinogenic lesion of the exchange theory? Or, can a DSB that does not break the chromosome still lead to the formation of a chromosome aberration?
3. Are DSBs "flapping" around "searching" for exchange partners? Under what conditions could this happen and which repair pathways are then involved in the formation of chromosome aberrations?
4. Is processing required before a DSB can lead to chromosome aberrations and if so which pathway of DSB repair is involved?
5. Are certain DSB repair pathways more likely than others to cause formation of chromosome aberrations?

To begin addressing these questions, we briefly outline salient features of the repair pathways implicated in the removal of DSBs, and discuss their potential involvement in the formation of chromosome aberrations.

5.2 Pathways of DSB Repair

Here, we outline molecular aspects of the DSB repair pathways that are relevant to chromosome aberration formation but refrain from an exhaustive molecular description of the participating proteins and their properties. For a more detailed description of these pathways, the reader is referred to Chap. 4 by Goedecke.

It is well established that cells of higher eukaryotes possess efficient pathways for repairing DSBs (Jackson 2002; Lieber et al. 2003; Valerie and Povirk 2003). These include two homology-dependent pathways, homologous recombination repair (HRR) and single-strand annealing (SSA), and two homology-independent pathways, DNA-PK-dependent nonhomologous end joining (D-NHEJ) and backup nonhomologous end joining (B-NHEJ) (Fig. 5.1).

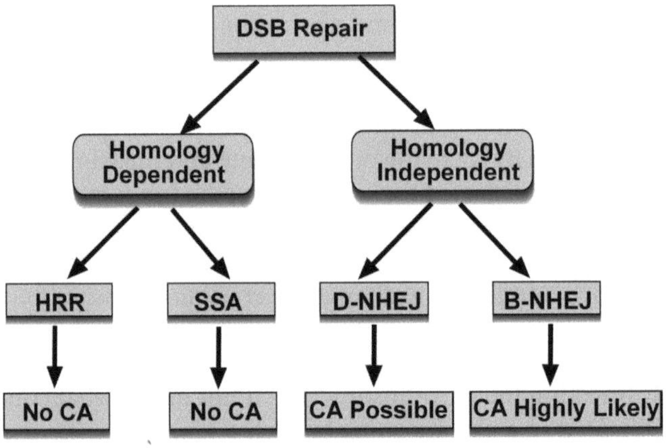

Fig. 5.1 In cells of higher eukaryotes double-strand breaks (*DSBs*) can in principle be removed by homology-dependent and homology-independent pathways. Homologous recombination repair (*HRR*) and single-strand annealing (*SSA*) are the main homology-dependent pathways and are unlikely to cause chromosome aberrations (*CA*). Homology-independent pathways are classified as DNA-PK-dependent nonhomologous end joining (*D-NHEJ*) and backup nonhomologous end joining (*B-NHEJ*) and have a varying potential regarding the formation of chromosome aberrations (see text)

5.2.1 Homology-Dependent Pathways of DSB Repair

In genetics homologous recombination describes the process through which reciprocal exchanges occur between homologous chromosomes during meiosis. This process is typically initiated by the programmed generation of DSBs and occurs without loss or gain of genetic material. In mitotic cells, homologous recombination can be employed to repair randomly induced DSBs in a process that utilizes the sister chromatid and is therefore thought to be active mainly in the late S phase and the G2 phase of the cell cycle. This repair pathway is well characterized at the molecular level and utilizes the genes of the Rad52 epistasis group of genes (Griffin and Thacker 2004; West 2003). By design HRR, if correctly initiated, is error-free and will not lead to chromosome aberrations. However, aberrant initiation of HRR is possible and has been documented in certain genetic diseases in humans. In a similar manner it is therefore possible that aberrant initiation of homologous recombination between repeat sequences in ectopic locations causes genomic rearrangements that manifest as chromosome aberrations (Griffin and Thacker 2004). Yet, chromosome aberration junctions fail to show the sequence homology that would have been expected had they been initiated by aberrant homologous recombination events. It is therefore unlikely that this repair pathway has a significant contribution to the formation of chromosome aberrations.

SSA is a variant of HRR (Griffin and Thacker 2004) and can in principle operate when a DSB is induced between two flanking homologous repeat sequences (Fig. 5.1). During SSA one of the two repeats and the intervening sequence are lost, thus making the process error-prone. While the molecular requirements of SSA are in general similar to those of HRR, distinct differences have been noted. As with HRR, canonical use of SSA will not cause genomic rearrangements that will manifest as chromosome aberrations. Aberrant use may of course cause genomic rearrangements, but here again the expected homology in the sequences around the break is not found.

Thus, homology-dependent pathways of DNA DSB repair are unlikely to be among the main players in the formation of chromosome aberrations and will not be considered further in the discussion that follows (Fig. 5.1).

5.2.2 Homology-Independent Pathways of DSB Repair

In their simplest manifestation, homology-independent pathways of DSB repair remove DSBs from the DNA by simply ligating the two ends (Lieber et al. 2003; Valerie and Povirk 2003). As the term implies, there are no homology requirements for the rejoining event, and because nucleotides are often removed from or added to the DNA ends (also to make them ligatable), sequence changes in the vicinity of the break are likely. Thus, homology-independent pathways satisfy only one aspect of DNA DSB repair, i.e., the restoration in the integrity of the DNA molecule, but do not ensure preservation of sequence; they should therefore be classified as error-prone at this level.

Although DSB repair by a homology-independent pathway was first characterized in higher eukaryotes, where it appears to dominate DSB removal, it is now clear that the pathway is evolutionarily conserved and active from bacteria to humans (Lieber et al. 2003; Valerie and Povirk 2003).

One characteristic of homology-independent pathways of DSB repair is that there is no "proofreading" mechanism incorporated to ensure the exclusive rejoining of the original ends – synonymous to the restoration of the DNA chain in each chromosome. DNA ends from different DSBs can be joined and although this can cause large-scale genomic rearrangements, there is no known cellular surveillance system registering and responding to this type of error. It appears that elimination of DNA ends, even through misrejoining, is the primary goal of these repair pathways, raising the question as to how the adverse consequences of misrejoining are minimized within the cell. As we discuss later, the short half life of the DSBs, the privilege of spatial proximity of the original ends within one DSB, and the possible restriction in motion imposed upon the DNA ends by the coordinated action of DNA repair proteins are parameters shifting the balance towards the joining of the original ends (correct rejoining) – at least for certain repair pathways. As a result of these intrinsic properties and limitations, it comes as no surprise that these repair pathways are considered as the major sources of chromosome aberrations (Iliakis et al. 2004).

In recent years the term nonhomologous end joining (NHEJ) has been predominantly used to describe homology-independent pathways of DSB repair and it will also be used here. While the term NHEJ was initially introduced to describe a DSB repair pathway utilizing, among other proteins, DNA-dependent protein kinase (DNA-PK; see later), evidence is being accumulated for the operation of a DNA-PK independent pathway of NHEJ that operates as a backup to the DNA-PK dependent pathway (see later). To distinguish between the two pathways of NHEJ, the term D-NHEJ was introduced to describe the DNA-PK dependent pathway and the term B-NHEJ to describe the backup DNA-PK independent pathway (Wang et al. 2001a, 2003, 2006); these terms will also be used here. In the following sections we describe the molecular characteristics of D-NHEJ and B-NHEJ and discuss their ramifications for chromosome aberration formation.

5.2.2.1 DNA-PK-Dependent NHEJ

This DSB repair pathway was the first homology-independent pathway to be characterized and a great deal of information regarding its molecular details is now known (Lieber et al. 2003; Valerie and Povirk 2003). Following exposure to IR and the induction of random DSBs in the genome, the first protein thought to interact with the generated DNA ends is Ku. Ku is a heterodimer of Ku70 and Ku80, with strong DNA binding properties and a toroidal structure that allows the DNA strand to be tightly threaded and stabilized (Fig. 5.2). Ku is very abundant, with nearly half a million molecules per cell, which is in vast excess to the 20–40 DSBs induced per gray of X-rays. The high equilibrium constant of Ku for DNA ends and its high abundance ensure that Ku, when available, will quickly capture the DNA ends. To avoid blocking and to enable the further end processing, Ku diffuses inwards along the DNA strand by an ATP-independent mechanism.

Binding of Ku to DNA ends is quickly followed by the recruitment of another protein, DNA-PK catalytic subunit – DNA-PKcs, with DNA binding properties and DNA-dependent serine/threonine kinase activity that are strongly stimulated by Ku. Although DNA-PKcs is thought to be approximately fivefold less abundant than Ku, ample amounts are present to support processing of a staggering number of DSBs. Binding of this large protein (molecular mass 469 kDa), together with Ku, to each of the DNA ends at a DSB is thought to generate a landing pad for the recruitment of components required for further processing. In addition, the reported association of the DNA-PKcs with the nuclear matrix (Mauldin et al. 2002) raises the possibility that the rejoining apparatus benefits from further stabilization provided by this nuclear structure. Although there is evidence that Ku and DNA-PKcs can mediate DNA end-to-end associations, the exact mechanism by which the synapsis of DNA ends is achieved in preparation for the rejoining step remains largely unknown. This is of course a critical question with direct ramifications for our mechanistic understanding of chromosome aberration formation.

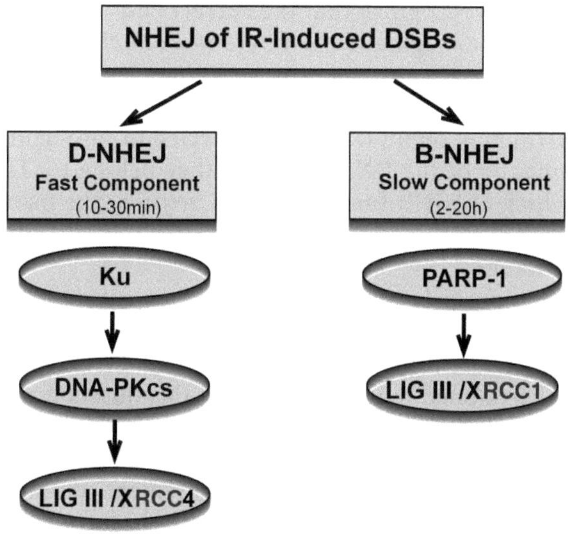

Fig. 5.2 One can distinguish between two pathways of NHEJ. D-NHEJ is fast and utilizes DNA-PK and DNA ligase IV, whereas B-NHEJ is slow and may utilize PARP-1 and DNA ligase III. See Fig. 5.1 for an outline of the contribution of each pathway to the formation of chromosome aberrations. *IR* ionizing radiation

Before final rejoining, and because IR also damages the sugar–phosphate backbone (Lieber et al. 2003; Valerie and Povirk 2003), processing is required to make ends ligatable. The nucleases and DNA polymerases suspected to operate at this step are discussed elsewhere in this volume. It is at this step that alterations in the original sequence around the DSB occur with the frequent loss or addition of nucleotides. This is the first level of error-prone operation for D-NHEJ. However, since sequence errors introduced at this stage are not causing chromosome aberrations, they will not be further discussed.

Ligation of the processed ends and restoration of DNA integrity is provided by DNA ligase IV, which functions as a complex with the stabilizing factor XRCC4 (Fig. 5.2). The XRCC4–DNA ligase IV complex does not bind stably to DNA, but is probably recruited to the DSB site by Ku–DNA interactions.

Studies in intact cells have shown that D-NHEJ is the dominant pathway utilized for the repair of DSBs in cells of higher eukaryotes (Wang et al. 2001a). One of its most salient features is the speed of operation. Nearly 90% of DSBs induced by sparsely IR are removed with half lives of 10–15 min, suggesting a highly efficient repair apparatus that captures, processes and rejoins DNA ends through precisely regulated functional coordination of its constituents (DiBiase et al. 2000). The biochemical properties of the participating proteins further suggest that naked DNA ends will not be present in the cell for more than a few seconds after their induction and that protein captured ends will be removed from the genome within minutes.

How does the cell recognize that the correct ends are being rejoined by D-NHEJ and that the original DNA fibers in each chromosome are restituted? As mentioned already, there is no known surveillance mechanism incorporated into the D-NHEJ apparatus to satisfy this requirement. In contrast, D-NHEJ may rejoin DNA ends detected in sufficient spatial proximity independently of the DNA molecule they belong to. This is the second level at which D-NHEJ functions in an error-prone manner, and it is evident that errors at this level will be directly relevant to chromosome aberration formation.

How is it then possible that in cells of higher eukaryotes a DSB repair system evolved, which not only introduces sequence alterations but which also allows for illegitimate recombination events that can lead to chromosome aberrations? Since an answer to this question remains elusive, it provides fertile ground for speculation. DSB rejoining takes place in the context of chromatin and may involve interactions of D-NHEJ constituents with the nuclear matrix (see before). This requirement confines topologically the rejoining event within a region of chromatin in the vicinity of the DSB. Since DSBs are rare events in the normal life cycle of a cell (when induced, e.g., by reactive oxygen species generated as by-products of metabolic processes), the probability that more than one DSB will be induced in the same region of chromatin is very small and therefore illegitimate recombination is an unlikely event. This fact together with the extremely high speed of D-NHEJ operation generate conditions for predominantly correct rejoining.

Can we speculate that D-NHEJ is a predominantly error-free pathway of DSB repair with insignificant contribution to chromosome aberration formation? Such a postulate may be plausible for the approximately 90% of DSBs that are repaired with the fast kinetics (if other DSBs are not present in the immediate vicinity; see later). However, it is well established that a small percentage (10–20%) of IR-induced DSBs are rejoined with slow kinetics, similar to those of interphase chromosome breaks (measured by the technique of premature chromosome condensation). Since this subset of DSBs appears to break chromosomes upon premature condensation (Cornforth and Bedford 1993) and is rejoined slowly, it may consist of DSBs whose rejoining by D-NHEJ is somehow compromised. To the extent that the same D-NHEJ proteins operate on these DSBs, the reduction in rejoining efficiency could either reflect a requirement for more extensive processing, or a failure in the synapsis of DNA ends (see before). Although extensive end processing will somewhat delay ultimate rejoining, it is unlikely to be responsible for the 5–10 times slower rejoining kinetics frequently observed, unless it interferes with the function of Ku. Therefore, the possibility of synapsis failure becomes particularly attractive as it also allows for illegitimate recombination events causing the observed chromosome aberrations. Circumstances leading to synapsis failure are discussed later.

Thus, D-NHEJ is an efficient pathway of DSB repair with a high potential of capturing, processing and ultimately rejoining DNA ends, restoring the original DNA molecules, albeit with possible sequence alterations in the vicinity of the DSB. Possible failure of this system at the synapsis step is accompanied by a drastic reduction in the rejoining kinetics and may be associated with illegitimate recombination events that lead to chromosome aberrations.

5.2.2.2 Backup NHEJ

Is D-NHEJ the only means a cell has to rejoin DSBs by homology-independent processes? Frequently in the literature, postulates to this effect are made, implicitly or explicitly, based on the limited DSB rejoining observed under certain (not always optimal) conditions in cells with defects in components of D-NHEJ. Our laboratory studied this question extensively as part of an effort to quantify the contribution of homology-directed processes in DSB repair and their coordination with homology-independent processes (Wang et al. 2001a, 2003, 2006).

Human tumor cells deficient in DNA-PKcs, studied under conditions optimal for repair (DiBiase et al. 2000), rejoin, like wild-type cells, IR-induced DSBs with kinetics that displays a fast and a slow component. While wild-type cells rejoin the majority of DSBs with fast kinetics, leaving only an estimated 10% for rejoining with slow kinetics, DNA-PKcs deficient cells rejoin a large proportion of DSBs with slow kinetics. Notably, the half lives of the fast and the slow component of rejoining as well as the total number of DSBs rejoined are similar in DNA-PKcs mutant and wild-type cells. Nearly complete rejoining of DSBs was also observed in cells deficient in Ku (Wang et al. 2001a) or DNA ligase IV (Wang et al. 2001b), as well as after treatment with DNA-PK inhibitors (Wang et al. 2001b). In addition, cells of different origins showed similar responses, suggesting that these observations are not species-specific.

These results strongly suggest that cells must be endowed with a DSB repair system operating with slow kinetics and handling a large proportion of IR-induced DSBs whenever D-NHEJ is genetically or chemically compromised. Although at first the slower kinetics of operation raised the possibility that homology-directed repair underlies this component of DSB repair, a study utilizing mutants with defects in diverse components of homologous recombination failed to generate evidence for a significant contribution by these pathways (Wang et al. 2001a). We postulated therefore that slow rejoining in cells with compromised D-NHEJ reflects the functions of an alternative pathway operating independently of homology and serving as a backup to D-NHEJ, thus termed backup NHEJ – B-NHEJ (Wang et al. 2001a). The observation that slow DSB rejoining is error-prone (Loebrich et al. 1995) is in line with the properties of homology-independent pathways of DSB repair but is incompatible with the mostly error-free operating HRR. Operation of such a NHEJ pathway in cells is also supported by the presence of end-joining activities in cellular extracts that do not rely on D-NHEJ components (Cheong et al. 1999; Perrault et al. 2004; Wang et al. 2003).

Genetic and biochemical studies have provided evidence that ligation during B-NHEJ is carried out by DNA ligase III (Wang et al. 2005) and that its operation is aided by PARP-1 (Wang et al. 2006). PARP-1 has DNA end binding properties similar to those of Ku and can potentially play a role in DSB repair (Fig. 5.2). However, in irradiated wild-type cells the higher affinity of Ku for DSBs and an excessive number of single-strand breaks and base damages that recruit PARP-1 limit its contribution to DSB repair. When essential components of D-NHEJ are absent, PARP-1 can be recruited for DSB repair, particularly in the absence of Ku. Thus, the repair module PARP-1/DNA ligase III/XRCC1 (PLX), hitherto regarded as central in repair of single-strand breaks, is implicated in the repair of DSBs (Wang et al. 2006). This backup function does not require novel activities, because a DSB can be seen as two single-strand breaks, and repair of single-strand breaks is the area of specialization of the PLX complex.

What makes B-NHEJ slow? The most plausible explanation is that this pathway of DSB repair lacks the efficient synapsis characteristics provided in D-NHEJ by the coordinated function of its constituents on the nuclear matrix. As a result, rejoining relies on synapsis by random encounters of the freely diffusing DNA ends, and relatively weak interactions of the participating proteins with the DSB (DNA ligase III, unlike DNA ligase IV, binds DNA) that make rejoining inefficient. There is evidence that junctions generated when D-NHEJ is compromised show more extensive deletions and increased incidence of microhomologies (Roth and Wilson 1986; Verkaik et al. 2002), suggesting end stabilization by hydrogen bonding. Inefficient rejoining increases the half life of DNA ends inside the nucleus and the risk of nuclease-mediated degradation and/or illegitimate recombination. Indeed both of these aspects have been experimentally documented (Lieber et al. 2003; Valerie and Povirk 2003).

The above postulate suggests similarities between slow rejoining observed for a few DSBs in wild-type cells and slow rejoining of a much larger proportion of DSBs in D-NHEJ deficient cells. Slow rejoining of DSBs in wild-type cells, to the extent that it remains DNA-PK dependent, must therefore be enzymatically distinct from B-NHEJ. The similarity in the kinetics suggests that synapsis problems can have similar consequences for the enzymatic apparatus of the D-NHEJ or the B-NHEJ. However, we speculate that in the case of B-NHEJ difficulty in DNA end synapsis is the consequence of suboptimal enzymatic processing, whereas in wild-type cells it derives from the characteristics of the lesion that abrogate the normal function of the pathway. From the perspective of chromosome aberration formation, both methods of synapsis abrogation may have similar consequences.

Cytogenetic data provide direct support for the operation of backup pathways of NHEJ and indicate their propensity for illegitimate recombination. Thus, cells deficient in DNA-PKcs show large increases in the incidence of exchanges, scored as dicentrics and complex exchanges (Ferguson et al. 2000; Martin et al. 2005; Virsik-Köpp et al. 2003, 2005). The presence of chromosome and chromatid exchanges in these cells is direct evidence for DNA-PKcs independent rejoining activity and their substantial increase in these cells is an indication that this repair pathway is prone to illegitimate recombination events (Fig. 5.1).

5.3 Homology-Independent End Joining and Chromosome Aberration Formation

Assuming that chromosome aberrations are the visual manifestation of errors during DSB repair, the following speculative ideas can be offered regarding the contribution of the already discussed repair pathways in their formation.

The fast component of D-NHEJ has the highest probability of error-free operation in the sense that illegitimate recombination events are rare. This is supported by studies evaluating DSB-rejoining fidelity using a combination of pulsed-field gel electrophoresis and Southern blotting (Loebrich et al. 1995). Furthermore, there is evidence that exchange-type aberrations are forming with half lives on the order of hours, different from what would be expected from a contribution of this repair component (Cornforth 2006; Cornforth and Bedford 1993; Evans 1962; Hlatky et al. 2002; Natarajan 2002; Obe et al. 2002; Revell 1974). DSBs removed by the slow component of D-NHEJ (assuming that D-NHEJ processes slowly repairing DSBs), which may also be responsible for chromosome breakage upon premature condensation (Cornforth and Bedford 1993), are the most likely contributors of chromosome aberrations in wild-type cells. The presumed deficiency in the synapsis of DNA ends owing to the nature of the underlying lesion delays rejoining and increases the probability of interaction with other DSBs, which may result in events of illegitimate recombination causing exchange-type aberrations. Some of these DSBs may also remain unrejoined, giving rise to acentric fragments.

As a potential contributor to chromosome aberrations in wild-type cells, and the major contributor of chromosome aberrations in D-NHEJ deficient cells, one should consider B-NHEJ. This backup repair pathway is normally suppressed by D-NHEJ (Perrault et al. 2004; Wang et al. 2003) but comes to the fore whenever D-NHEJ is compromised. Suboptimally regulated synapsis results in a low speed of rejoining and allows with high probability illegitimate recombination events. Despite the high probability for misjoining, B-NHEJ must be advantageous to the cell. This is indicated by the fact that the survival curves of D-NHEJ deficient cells have slopes suggesting accumulation of many (10–20) DSBs before registering a lethal event (DiBiase et al. 2000). In contrast, yeast deficient in homologous recombination dies upon registering approximately one DSB.

The operation of backup pathways of NHEJ is not limited to radiation-induced DSBs but it extends to other phenotypic manifestations where processing of DSBs is expected to occur. Thus, backup pathways of DSB repair bring together the *c-myc* and *Igh* locus and cause B-cell lymphomas in mice with defects in *Ku*, *LIG4* or *XRCC4* (Difilippantonio et al. 2000; Ferguson et al. 2000; Frank et al. 2000; Gao et al. 2000; Karanjawala et al. 1999; Ramiro et al. 2006; Zhu et al. 2002). Functionally equivalent pathways may generate the aberrant junctions manifesting chromosome instability in the same mutants. Similar pathways of end joining generate the few V(D)J junctions observed in cells with defects in D-NHEJ (Bogue et al. 1997; Lee et al. 2004; Li et al. 1995) and are implicated in antibody class switching occurring under the same conditions (Ramiro et al. 2006). Thus, B-NHEJ, when utilized, helps the cell to restore its genome and thus presumably to avert cell death. However, its error-prone nature causes genomic instability and cancer in the affected organism.

While prominent when the classic pathway is compromised, B-NHEJ may be utilized to a limited degree in cells with functional D-NHEJ as well. This can occur when a DSB is induced in a nuclear region where Ku or other key components of D-NHEJ are not immediately available; e.g., because they are already recruited to a neighboring DSB. Although this type of event is expected to be rare overall, its contribution to the total chromosome aberration load in a wild-type cell may be significant. B-NHEJ may also come to the fore when D-NHEJ becomes somehow compromised during processing of a DSB.

5.4 DSBs, NHEJ and Theories of Chromosome Aberration Formation

Here, we reconsider the fundamental postulates of the breakage-and-reunion and the exchange theories and discuss them under the assumption that DSBs, if processed incorrectly by homology-independent pathways of DSB repair, will initiate chromosome aberrations.

The fundamental assumption of the breakage-and-reunion theory is that the initiator lesion breaks the chromonema (Revell 1974; Savage 1998; Savage and Harvey 1991). Are DSBs capable of directly breaking the chromosome? The extremely complex folding patterns and levels of organization of DNA in the metaphase chromosome make it seem intuitively quite unlikely that a DSB will break the chromosome directly. And there is direct evidence to support this assumption. Irradiation of cells at metaphase, or of interphase cells after premature chromosome condensation, almost never breaks chromosomes (Cornforth and Bedford 1993). Thus, the fundamental postulate of the breakage-and-reunion theory that the underlying lesion directly breaks the chromonema is not tenable when the entirety of IR-induced DSBs are considered.

On the other hand, premature chromosome condensation clearly demonstrates that chromosome breakage can be a consequence of a DSB if chromatin condensation is initiated before repair can been completed (Cornforth and Bedford 1993). It can be postulated therefore that although DSBs by themselves must not break the chromosome, they nevertheless have the potential of being "translated" to a chromosome break through the action of chromatin-reorganizing processes. However, under these conditions, it is not the DSB that ultimately breaks the chromosome but the fortuitously concurrent, repair-independent process of chromatin condensation (decondensation may also have a similar outcome). This is a major departure from the direct chromosome breakage requirement of the breakage-and-reunion theory.

Can all IR-induced DSBs be "transformed" to chromosome breaks by the induction of chromatin conformation changes, such as premature chromosome condensation? Data suggest that only a relatively small fraction of IR-induced DSBs can be translated to interphase chromosome breaks (about 10–20%) even when precautions are taken to minimize repair during chromosome condensation (Okayasu and Iliakis 1993; Okayasu et al. 1993). Can we consider this fraction of DSBs as the only relevant fraction for chromosome aberration formation and apply to it the breakage-and-reunion theory? To do this we will have to assume that the remaining DSBs are incapable of participating in exchanges, but as we have discussed already this cannot be considered as generally true.

The above analysis imposes the status of chromatin and its conformational change as collateral for the consequences of a DSB on the integrity of the chromosome. The same parameters are also likely to have consequences for the repair outcome. For a DSB repair pathway operating with the limitations of D-NHEJ (inaccurate DNA sequence restoration at the DSB and more importantly the lack of a mechanism for correct rejoining-partner selection) the best means for minimizing interference is speed. Yet, the level of chromatin condensation at the site of the DSB as a result of ongoing metabolic processes (transcription or replication) and changes in that structure shortly thereafter are parameters that should be considered in the formulation of generalized theories of chromosome aberration formation.

The recognition that DSBs do not by themselves break the chromosome makes them good candidates for the undefined recombinogenic (potential of illegitimate recombination) lesions of the exchange theory. The exchange theory postulates that the initial damage is not a break, but an unstable lesion, termed the primary event of damage, which tends to decay towards a normal or undetectable state (Revell 1974; Savage 1998; Savage and Harvey 1991). If such events are brought together in pairs, they may take part in an exchange process, and so give rise to the aberrations seen at metaphase. Incomplete exchange will lead to a discontinuity that will be seen as a break. Breaks are therefore secondary.

Both pathways of NHEJ can in principle be recombinogenic if more than one DSB is induced in close proximity. But the interaction does not have to be limited to pairs and will not require active movement. Spatially proximate DSBs can be induced by random ionization events initiated by one or more electron traversals and will be located in the DNA of the same chromosome (highest probability event), or in the DNA of adjacently organized chromosomes.

The ends of each DSB are a priori spatially privileged in the sense that they have the highest possible initial proximity, and this will favor restitution. However, within an interaction radius, which will depend on the characteristics of the repair pathways involved, interaction with other DSBs is possible. On the basis of the high speed and the well-coordinated operation of the enzymatic apparatus, we anticipate small interaction radii (i.e., only very closely located DSBs will have a chance to interact) for DSBs handled by the fast component of D-NHEJ (Fig. 5.2). A significantly greater radius of interaction will be available for DSBs handled by the slow component of D-NHEJ. The maximum radius of interaction will be available for DSBs handled slowly by B-NHEJ. Increase of linear energy transfer will increase the probability of multiple DSBs within the sphere of interaction. Furthermore, changes in chromatin reorganization after DSB induction, taking place independently of repair, may change the proximity of the DNA ends and may bring close together "remotely" induced DSBs, thus modulating repair and the probability for illegitimate recombination.

In this model, all DSBs are potentially recombinogenic despite the fact that they are not capable of breaking the chromosome. The geometric size of the interaction radius is determined by the repair pathway active at the site. The domain organization of the chromosomes inside the nucleus makes it plausible that when two or more DSBs are induced within a given interaction radius, these will belong predominantly to the same chromosome and may even be closely located within the DNA molecule. This means that many illegitimate events may not be visible cytogenetically. This has direct consequences in the analysis of acentric fragments and is discussed below.

The recombinogenic potential of DSBs is a direct consequence of the repair pathways involved and does not require that DSBs break the chromosome directly. Since the ultimate goal of these repair pathways is the removal of DNA ends, the outcome when successful (albeit in an error-prone manner) will be complete exchanges. Incomplete exchanges indicate that interference after the determining initial misjoining event may prevent rejoining of remaining ends. It is noteworthy in this regard that activation of checkpoints in G1 and G2 phases of the cell cycle strives, among other things, the delay of cell cycle progression associated, scheduled alterations in chromatin conformation (Iliakis 1997; Iliakis et al. 2003).

If the initial chromosome break is only an unlikely outcome of the induction of a DSB, the idea of "flapping" ends "searching" for exchange partners (Savage 1998) should be considered unrealistic at the chromosome level. Can this idea be transferred to DSBs? On the basis of the above discussion of DSB repair pathways, freedom of movement for DNA ends can be considered as significant only for the small percentage of DSBs rejoined with slow kinetics by D-NHEJ, and probably for the majority of DSBs rejoined by B-NHEJ. (Note that when D-NHEJ is compromised, a small percentage of DSBs still rejoin with fast kinetics, suggesting efficient synapsis by alternative mechanisms, probably as a result of their location in chromatin.) This freedom of movement defines the spheres of interaction introduced above to discuss the formation of exchanges. Thus, "flapping" DNA, but not chromosome, ends may be generated under certain conditions and may be contributing to the formation of chromosome aberrations.

The key discrepancy between the breakage-and-reunion theory and the exchange theory concerns the origin of acentric fragments. While in the exchange theory acentric fragments are only produced by incomplete exchanges, in the breakage-and-reunion theory they are the direct product of the lack of chromosome end processing (Revell 1974; Savage 1998; Savage and Harvey 1991). A direct consequence of the underlying assumptions of acentric fragment formation by the two theories is a linear dose–effect relationship for the breakage-and-reunion theory and a linear–quadratic dose–effect relationship for the exchange theory. Interestingly, both functional dependencies have been observed in different organisms, making it difficult to settle the issue on this basis. This issue is further complicated by the often discussed difficulties in scoring acentric fragments (Revell 1974; Savage 1998; Savage and Harvey 1991) and limitations in the detection of recombination events associated with relatively small fragments of DNA (Savage 2004).

The breakage-and-reunion theory of course does not deny the existence of breaks arising from incomplete exchanges, but assumes that the contribution from this source is negligible when compared with those arising from failed restitution of single breaks. This raises the question as to the processes that could cause failed restitution of DSBs and the most likely source here again is changes in chromatin conformation shortly after induction of the DSB that may pull the DNA ends apart, thus compromising subsequent rejoining by NHEJ. Mechanistically, the dispute is translated to the question as to when the cell gives up looking for a partner for a DNA end given that even after chromatin reorganization the two ends will remain reasonably close together. A high degree of persistence will favor the exchange theory and will make the first exchange the "point of no return" for exchange aberration formation, whereas a low degree of persistence will support the breakage-and-reunion theory and will make the "pulling apart" of ends the point of no return.

It would be an omission to leave the subject without mentioning achromatic lesions (Savage 2004). This form of chromosomal change has generated much controversy and has introduced many errors in the analysis of acentric fragments. Yet, the question as to whether achromatic lesions contribute to the formation of chromosome aberrations remains open. The reduced staining in these regions of the chromosomes suggests changes in chromatin structure, and one wonders whether they reflect chromatin processing associated with DSB repair. Their frequent occurrence in cells irradiated in the G2 phase, but not in the G1 phase, and their complete disappearance upon repair possibly makes them intriguing manifestations of ongoing repair processes. Molecular characterization of achromatic lesions and connection with known molecular processes such as generation of γ-H2AX or homologous recombination will be mostly instructive.

The question whether IR induces qualitative different types of DSBs remains a central question not only in the field of cytogenetics but in radiation biology in general. Different "quality" may be dictated indirectly by the context, e.g., the state of condensation and chromatin organization in the vicinity of the break, or directly by the type of DSB.

The most common qualitative distinction between DSBs is that of a complex DSB, understood as a combination between a simple DSB with base damages and/or single-strand breaks a few nucleotides apart (Ward 1985, 1988). It is thought that complex DSBs impose additional requirements on the DSB repair pathways that reduce their efficiency or inhibit their function (Riballo et al. 2004). As mentioned above though, such proximal lesions must not always present great difficulty to the DNA end processing systems, and additional damaged nucleotides may be removed by mechanisms similar to those developed to process DNA ends.

A different level of complication may be generated when two or more DSBs are induced in close proximity in the DNA, with proximity here ranging from a few base pairs to several million base pairs (Johnston et al. 1998). The local destabilization of DNA and chromatin produced under these conditions may generate a major obstacle for the repair system and may complicate synapsis of the DNA ends. This form of "complex" breaks may have a higher probability of forming chromosome aberrations, both acentric fragments and exchange aberrations.

In summary, although both homology-dependent and homology-independent pathways contribute to the ultimate repair of DSBs, homology-independent pathways are responsible for the initial removal of the vast majority of DSBs from the genome. From the homology-independent pathways of DSB repair both D-NHEJ and B-NHEJ have the potential of forming chromosome aberrations, but it is likely that most of the aberrations derive from misjoining events associated with the slow repair by D-NHEJ of a subset of DSBs, and with the majority of DSBs rejoined by B-NHEJ. DSBs do not break the chromosomes directly, and as a consequence of the characteristics of NHEJ are highly recombinogenic. This satisfies the essential requirements of the exchange theory that becomes breakage-and-reunion theory at the DNA level. Further elucidation of the mechanism of DSB repair at the level of chromatin will allow the description of processes leading to chromosome aberrations at the molecular level and will settle a dispute that has challenged the field intellectually for many years. Down to the basic concepts, both theories appear to have made correct assumptions and like so often in life, the truth may lie somewhere in the middle.

Acknowledgements: This work was supported by grants from the German Research Foundation (DFG), the German Ministry of Education and Research (BMBF) and the EU.

References

Bogue MA, Wang C, Zhu C, Roth DB (1997) V(D)J Recombination in Ku86-deficient mice: distinct effects on coding, signal, and hybrid joint formation. Immunity 7:37–47

Cheong N, Perrault AR, Wang H, Wachsberger P, Mammen P, Jackson I, Iliakis G (1999) DNA-PK-independent rejoining of DNA double-strand breaks in human cell extracts *in vitro*. Int J Radiat Biol 75:67–81

Cornforth MN (2006) Perspectives on the formation of radiation-induced exchange aberrations. DNA Repair (Amst) 5:1182–1191

Cornforth MN, Bedford JS (1993) Ionizing radiation damage and its early development in chromosomes. Adv Radiat Biol 17:423–496

DiBiase SJ, Zeng Z-C, Chen R, Hyslop T, Curran WJ, Jr., Iliakis G (2000) DNA-dependent protein kinase stimulates an independently active, nonhomologous, end-joining apparatus. Cancer Res 60:1245–1253

Difilippantonio MJ, Zhu J, Chen HT, Meffre E, Nussenzweig NC, Max EE, Ried T, Nussenzweig A (2000) DNA repair protein Ku80 suppresses chromosomal aberrations and malignant transformation. Nature 404:510–514

Evans HJ (1962) Chromosome aberrations induced by ionizing radiations. Int Rev Cytol 13:221–321

Ferguson DO, Sekiguchi JM, Chang S, Frank KM, Gao Y, DePinho RA, Alt FW (2000) The nonhomologous end-joining pathway of DNA repair is required for genomic stability and the suppression of translocations. Proc Natl Acad Sci USA 97:6630–6633

Frank KM, Sharpless NE, Gao Y, Sekiguchi JM, Fergson DO, Zhu C, Manis JP, Horner J, DePinho RA, Alt FW (2000) DNA ligase IV deficiency in mice leads to defective neurogenesis and embryonic lethality via the p53 pathway. Mol Cell 5:993–1002

Gao Y, Ferguson DO, Xie W, Manis JP, Sekiguchi JA, Frank KM, Chaudhuri J, Horner J, DePinho RA, Alt FW (2000) Interplay of p53 and DNA-repair protein XRCC4 in tumorigenesis, genomic stability and development. Nature 404:897–900

Griffin CS, Thacker J (2004) The role of homologous recombination repair in the formation of chromosome aberrations. Cytogenet Genome Res 104:21–27

Hlatky L, Sachs RK, Vazquez M, Cornforth MN (2002) Radiation-induced chromosome aberrations: insights gained from biophysical modeling. BioEssays 24:714–723

Iliakis G (1997) Cell cycle regulation in irradiated and non-irradiated cells. Semin Oncol 24:602–615

Iliakis G, Wang Y, Guan J, Wang H (2003) DNA damage checkpoint control in cells exposed to ionizing radiation. Oncogene 22:5834–5847

Iliakis G, Wang H, Perrault AR, Boecker W, Rosidi B, Windhofer F, Wu W, Guan J, Terzoudi G, Pantelias G (2004) Mechanisms of DNA double strand break repair and chromosome aberration formation. Cytogenet Genome Res 104:14–20

Jackson SP (2002) Sensing and repairing DNA double-strand breaks. Carcinogenesis 23:687–696

Johnston PJ, MacPhail SH, Stamato TD, Kirchgessner CU, Olive PL (1998) Higher-order chromatin structure-dependent repair of DNA double-strand breaks: involvement of the V(D)J recombination double-strand break repair pathway. Radiat Res 149:455–462

Karanjawala ZE, Grawunder U, Hsieh C-L, Lieber MR (1999) The nonhomologous DNA end joining pathway is important for chromosome stability in primary fibroblasts. Curr Biol 9:1501–1504

Lee GS, Neiditch MB, Salus SS, Roth DB (2004) RAG proteins shepherd double-strand breaks to a specific pathway, suppressing error-prone repair, but RAG nicking initiates homologous recombination. Cell 117:171–184

Li Z, Otevrel T, Gao Y, Cheng W-L, Seed B, Stamato TD, Taccioli GE, Alt FW (1995) The XRCC4 gene encodes a novel protein involved in DNA double-strand break repair and V(D)J recombination. Cell 83:1079–1089

Lieber MR, Ma Y, Pannicke U, Schwarz K (2003) Mechanism and regulation of human non-homologous DNA end-joining. Nat Rev Mol Cell Biol 4:712–720
Loebrich M, Rydberg B, Cooper PK (1995) Repair of x-ray-induced DNA double-strand breaks in specific *Not* I restriction fragments in human fibroblasts: Joining of correct and incorrect ends. Proc Natl Acad Sci USA 92:12050–12054
Martin M, Genesca A, Latre L, Jaco I, Taccioli GE, Egozcue J, Blasco MA, Iliakis G, Tusell L (2005) Postreplicative joining of DNA double-strand breaks causes genomic instability in DNA-PKcs-deficient mouse embryonic fibroblasts. Cancer Res 65:10223–10232
Mauldin SK, Getts RC, Liu We, Stamato TD (2002) DNA-PK-dependent binding of DNA ends to plasmids containing nuclear matrix attachment region DNA sequences: evidence for assembly of a repair complex. Nucleic Acids Res 30:4075–4087
Natarajan AT (2002) Chromosome aberrations: past, present and future. Mutat Res 504:3–16
Obe G, Pfeiffer P, Savage JRK, Johannes C, Goedecke W, Jeppesen P, Natarajan AT, Martinez-Lopez W, Folle GA, Drets ME (2002) Chromosomal aberrations: formation, identification and distribution. Mutat Res 504:17–36
Okayasu R, Iliakis G (1993) Ionizing radiation induces two forms of interphase chromosome breaks in CHO cells that rejoin with different kinetics and show different sensitivity to treatment in hypertonic medium or b-araA. Radiat Res 136:262–270
Okayasu R, Varlotto J, Iliakis G (1993) Hypertonic treatment does not affect the radiation yield of interphase chromosome breaks in DNA double-strand break repair deficient *xrs-5* cells. Radiat Res 135: 171–177
Perrault R, Wang H, Wang M, Rosidi B, Iliakis G (2004) Backup pathways of NHEJ are suppressed by DNA-PK. J Cell Biochem 92:781–794
Ramiro AR, Jankovic M, Callen E, Difilippantonio S, Chen H-T, McBride KM, Eisenreich TR, Chen J, Dickins RA, Lowe SW, Nussenzweig A, Nussenzweig MC (2006) Role of genomic instability and p53 in AID-induced c-myc-Igh translocations. Nature 440:105–109
Revell SH (1974) The breakage-and-reunion theory and the exchange theory for chromosomal aberrations induced by ionizing radiation: a short history. Adv Radiat Biol 4:367–416
Riballo E, Kühne M, Rief N, Doherty A, Smith GCM, Recio M-J, Reis C, Dahm K, Fricke A, Krempler A, Parker AR, Jackson SP, Gennery A, Jeggo PA, Loebrich M (2004) A pathway of double-strand break rejoining dependent upon ATM, Artemis, and proteins locating to g-H2AX foci. Mol Cell 16:715–724
Roth DB, Wilson JH (1986) Nonhomologous recombination in mammalian cells: Role for short sequence homologies in the joining reaction. Mol Cell Biol 6:4295–4304
Savage JRK (1975) Classification and relationships of induced chromosomal structural changes. J Med Genet 12:103–122
Savage JRK (1998) A brief survey of aberration origin theories. Mutat Res 404:139–147
Savage JRK (2004) On the nature of visible chromosomal gaps and breaks Cytogenet Genome Res 104:46–55
Savage JRK, Harvey AN (1991) Revell revisited. Mutat Res 250:307-317
Valerie K, Povirk LF (2003) Regulation and mechanisms of mammalian double-strand break repair. Oncogene 22:5792–5812
Verkaik NS, Esveldt-van Lange REE, van Heemst D, Brüggenwirth HT, Hoeijmakers JHJ, Zdzienicka MZ, van Gent DC (2002) Different types of V(D)J recombination and end-joining defects in DNA double-strand break repair mutant mammalian cells. Eur J Immunol 32:701–709
Virsik-Köpp P, Rave-Fränk M, Hofman-Hüther H, Schmidberger H (2003) Role of DNA-PK in the process of aberration formation as studied in irradiated human glioblastoma cell lines M059K and M059J. Int J Radiat Biol 79:61–68
Virsik-Köpp P, Hofman-Hüther H, Rave-Fränk M, Schmidberger H (2005) The Effect of Wortmannin on Radiation-Induced Chromosome Aberration Formation in the Radioresistant Tumor Cell Line WiDr. Radiat Res 164:148–156
Wang H, Zeng Z-C, Bui T-A, Sonoda E, Takata M, Takeda S, Iliakis G (2001a) Efficient rejoining of radiation-induced DNA double-strand breaks in vertebrate cells deficient in genes of the RAD52 epistasis group. Oncogene 20:2212–2224

Wang H, Zhao-Chong Z, Perrault AR, Cheng X, Qin W, Iliakis G (2001b) Genetic evidence for the involvement of DNA ligase IV in the DNA-PK-dependent pathway of non-homologous end joining in mammalian cells. Nucleic Acids Res 29:1653–1660
Wang H, Perrault AR, Takeda Y, Qin W, Wang H, Iliakis G (2003) Biochemical evidence for Ku-independent backup pathways of NHEJ. Nucleic Acids Res 31:5377–5388
Wang H, Rosidi B, Perrault R, Wang M, Zhang L, Windhofer F, Iliakis G (2005) DNA Ligase III as a candidate component of backup pathways of nonhomologous end joining. Cancer Res 65:4020–4030
Wang M, Wu W, Wu W, Rosidi B, Zhang L, Wang H, Iliakis G (2006) PARP-1 and Ku compete for repair of DNA double strand breaks by distinct NHEJ pathways. Nucleic Acids Res 34:6170–6182
Ward JF (1985) Biochemistry of DNA lesions. Radiat Res 104:S103–S111
Ward JF (1988) DNA damage produced by ionizing radiation in mammalian cells: Identities, mechanisms of formation, and reparability. Prog Nucl Acid Res 35:95–125
West SC (2003) Molecular views of recombination proteins and their control. Nat Rev Mol Cell Biol 4:1–11
Zhu C, Mills KD, Ferguson DO, Lee C, Manis J, Fleming J, Gao Y, Morton CC, Alt FW (2002) Unrepaired DNA breaks in p53-deficient cells lead to oncogenic gene amplification subsequent to translocations. Cell 109:811–821

6 Targeted and Non-Targeted Induction of Chromosomal Rearrangements After Exposure to Ionizing Radiation

William F. Morgan and Marianne B. Sowa

Abstract

Exposure to ionizing radiation can result in chromosomal rearrangements in a cell cycle dependent manner. This is presumably due to the deposition of energy in the nucleus of the irradiated cell causing DNA double-strand cleavage and the failure of the DNA repair machinery to faithfully restore the integrity of the genetic material. There is also increasing evidence that an irradiated cell can signal to a non-irradiated "bystander" cell and elicit cellular responses, including chromosomal rearrangements, in these non-targeted cells. These signals can be communicated between targeted (irradiated) cells and non-targeted (non-irradiated) cells by both cell-to-cell gap junction communication mechanisms as well as soluble factors secreted into the culture medium. In this chapter we review the evidence for targeted and non-targeted induction of chromosomal rearrangements after exposure to ionizing radiation and speculate about potential mechanisms and their significance in radiation risk assessment.

6.1 Introduction

Analysis of chromosomal rearrangements has been used for many years to determine whether an individual has been exposed to ionizing radiation, and to provide a reliable estimate of the exposure dose (reviewed in Cornforth 1998). A rich and well-documented literature indicates that cytogenetic analysis can be used to determine exposures to radiation doses as low as 15–20 cGy, and thus provides the most sensitive and reliable assay for radiation exposure. Implicit in interpreting these types of analysis is that the induced cytogenetic alterations result from the deposition of energy by ionizing radiation and subsequent cellular responses. Specifically, radiation exposure leads to DNA damage, namely, single and double stranded DNA breaks, DNA–DNA and DNA–protein cross-links, and DNA base damages (Ward 1988). The cell then responds to this damage by initiating changes in gene expression, inducing cell cycle checkpoint control strategies and signal transduction pathways, and activating DNA repair processes.

Günter Obe and Vijayalaxmi (Eds.)
Chromosomal Alterations: Methods, Results and Importance in Human Health

The general consensus in the field has long been that chromosomal rearrangements result from the non-rejoining or misrejoining of radiation-induced DNA double-strand breaks (King et al. 1994). That is, deletion-type aberrations can result from the non-rejoining of an induced break or acentric rings induced by two double-strand breaks. Inversions result in misrejoining of breaks within a chromosome and exchange-type alterations result from the misrejoining of breaks on two or more different chromosomes, leading to insertions, rings, polycentric chromosomes, and/or reciprocal translocations. Given that in most radiation-exposure scenarios cells from irradiated individuals would be in G0/G1 phase, the dominant molecular process giving rise to misrejoining events would likely be non-homologous end joining.

In the present context, chromosomal rearrangements can be equated with the "breakage first hypothesis" (Sax 1938) that predicts breaks must be "open" at the same time and within close enough proximity to interact. While certain fundamental aspects of the formation of chromosomal rearrangements remain contentious, breakage followed by either non-rejoining of the break or misrejoining of induced breaks can satisfactorily describe most of the observed cytogenetic alterations in irradiated cells. In the past decade however, a host of different laboratories have described a number of non-targeted effects associated with exposure to ionizing radiation. In contrast to the traditional paradigm that a cell must be irradiated, or "targeted," to exhibit those detrimental effects associated with radiation exposure, these non-targeted effects occur in cells that were not subject to energy-deposition events because they were not irradiated. These non-targeted effects include radiation-induced chromosomal instability, bystander effects, the death-inducing effect, clastogenic factors, as well as abscopal effects described in the clinical literature as effects occurring outside the directly irradiated tissue volume (reviewed in Morgan 2003a,b). They occur in cells that were not irradiated but that either were neighbors of irradiated cells or received soluble signals from irradiated cells. A schematic summarizing the concept of targeted and non-targeted effects associated with observed chromosomal rearrangements is presented in Fig. 6.1. In this chapter we will review the evidence for non-targeted radiation-induced chromosomal instability, and discuss how factors produced by irradiated cells may communicate the legacy of radiation exposure to non-targeted cells resulting in DNA double strand break formation and ultimately chromosomal rearrangements.

6.2 Radiation-Induced Chromosomal Instability

First described by Kadhim et al. (1992), radiation-induced chromosomal instability refers to the induction of chromosomal rearrangements in the progeny of an irradiated cell. Instability was observed in cells that had undergone multiple cell division cycles after exposure to ionizing radiation. This

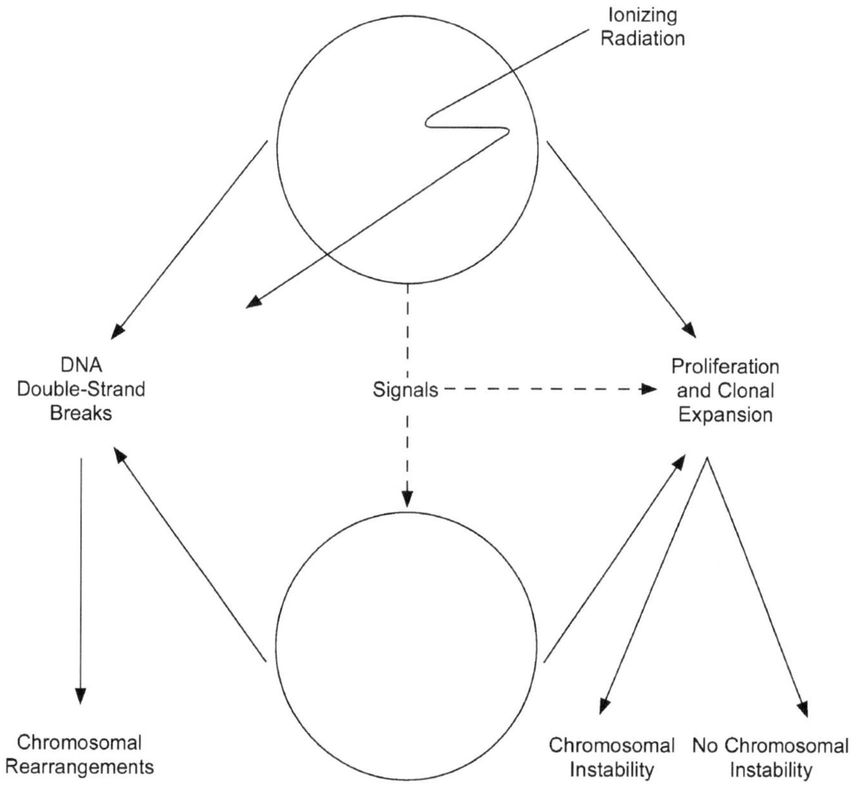

Fig. 6.1 A schematic summarizing the concept of targeted and non-targeted chromosomal aberration induction after ionizing radiation. A cell is irradiated (targeted) breaks are induced that can lead to cytogenetic damage. An irradiated cell can also send a signal to a non-irradiated (non-targeted) cell via cell-to-cell gap junction communication and/or soluble shed/secreted factors. These factors can elicit DNA cleavage leading to the induction of chromosomal rearrangements. Clonal expansion of both targeted and non-targeted cells can also result in chromosomal instability in the progeny multiple cell generations after the initial challenge and/or signal

demonstration of delayed cytogenetic damage has since been confirmed by many investigators (reviewed in Morgan 2003a), and suggests that ionizing radiation can elicit a memory in an irradiated cell that can manifest as cytogenetic changes weeks, even months, after cellular irradiation. At the time as these chromosomal alterations are observed, the radiation-induced DNA lesions are presumably long gone, either as a consequence of DNA repair processes or by dilution during successive cell division cycles. Thus, the memory of ionizing radiation must include the delayed induction of genetic lesions in the progeny of an irradiated cell. Radiation-induced chromosomal instability has been described in vivo in both animal studies (Plumb et al. 1998; Watson et al. 2000, 2001) and in exposed human populations

(Nakanishi et al. 1999; Salomaa et al. 1998); however, these data are somewhat controversial and currently the subject of intense debate (Tawn et al. 2000; Whitehouse and Tawn 2001).

This conceptual understanding of the mechanism(s) of radiation-induced chromosomal instability is further complicated by convincing evidence demonstrating that chromosomal instability can be observed in the progeny of cells that were not irradiated, i.e., not subject to energy-deposition events. Using low fluences of α-particles, Kadhim et al. (1992) showed chromosomal instability in a significantly greater number of clonogenic survivors than could possibly have been hit by radiation. Extending these initial observations, Lorimore et al. (1998) described chromosomal instability in cells that were protected from radiation exposure by a metal grid, while cells around them had been lethally irradiated and were thus non-viable. These non-targeted surviving cells were likely subject to signals from irradiated cells communicated via cell-to-cell gap junctions, secreted or shed factors produced by an irradiated cell and being available to interact with a non-targeted cell by exposure in the culture medium, or possibly a combination of both (Sowa Resat and Morgan 2004). It should be stressed at this stage that radiation-induced chromosomal instability in vivo also appears to have a significant non-targeted component (Watson et al. 2000, 2001) that might also involve the transmission of soluble signaling factors.

To explain the transfer of a signal from a targeted cell to a non-targeted cell that can induce chromosomal rearrangements, these signals must be soluble, relatively long lived, and capable of reaching the nucleus of the non-targeted cell and inducing a DNA double-strand break. A precedence for this has been described in vivo in humans occupationally (Gemignani et al. 1999), therapeutically (Hollowell and Littlefield 1968; Littlefield et al. 1969), or accidentally exposed to ionizing radiation (Goh and Sumner 1968; Pant and Kamada 1977). Blood plasma isolated from these irradiated individuals can induce chromosomal breakage when mixed and co-cultured with peripheral blood lymphocytes from non-irradiated individuals. These so-called clastogenic factors (Parsons et al. 1954) indicate that soluble factors can be induced in irradiated humans and these can cause cytogenetic damage in normal healthy lymphocytes.

6.3 The Death-Inducing Effect

We were interested in whether secreted soluble factors in the culture medium could provide a mechanism for perpetuating chromosomal instability in our GM10115 human hamster hybrid cell system. We knew that one characteristic of our unstable cell clones was that they demonstrated persistently elevated levels of reactive oxygen species (ROS; Limoli et al. 1998, 2003). Furthermore, addition of antioxidants could reduce the levels of induced chromosomal instability (Limoli et al. 2001b). While this could be explained in part by dysfunctional mitochondria (Kim et al. 2006), it was difficult to reconcile how a single radiation exposure could induce mitochondrial abnormalities that expressed themselves at delayed times after irradiation and persisted over time. We proposed that the initial radiation exposure and subsequent induction of ROS initiated cycles of cytokine secretion that in turn stimulated ROS production producing cytokine secretion – ROS production, etc. (Morgan et al. 2002). This would provide a source for the mitochondrial dysfunction (Kim et al. 2006) and persistently elevated levels of ROS (Limoli et al. 2003), as well as an extracellular mechanism driving the generation of the novel chromosome rearrangements over time that characterize our chromosomally unstable clones.

If this indeed were the case, then medium from our chromosomally unstable cell clones should be capable of generating chromosomal instability in cells that had never been exposed to ionizing radiation. This would be similar to the transmission of factors from irradiated to non-irradiated cells leading to chromosome destabilization as described by Kadhim et al. (1992) and Lorimore et al. (1998). To test this hypothesis Nagar et al (2003a) selected three chromosomally unstable cell clones derived from irradiated GM10115 cells, one irradiated but chromosomally stable derivative of GM10115 cells, and parental GM10115 cells for investigation. All clones were grown to confluence, the medium was replaced with fresh medium for 48 h, after which time it was decanted from the cells, filtered, and transferred to flasks containing 100 GM10115 cells plated 4 h earlier. The idea was to clonally expand surviving cells and investigate chromosomal instability in these clones, the logic being that if factors secreted by unstable cells do lead to cytogenetic rearrangements, then instability should be observed in colonies cultured in medium from unstable clones, but not in colonies expanded in medium from either parental GM10115 cells or the irradiated but chromosomally stable clone. However, rather than exhibiting chromosomal instability, Nagar et al. (2003a) found that medium from two (clones LS-12 and Fe-10-3) of the three unstable cells was almost completely cytotoxic to GM10115 cells. This cytotoxic effect after transfer of medium from these unstable clones was called the death-inducing effect (DIE). The DIE was not observed after medium transfer from parental cells or the irradiated but chromosomally stable clone.

Nagar and colleagues then investigated how the DIE killed non-irradiated GM10115 cells. They observed that after transfer of medium from unstable clones there was a rapid induction of γH2AX foci formation in recipient cells (Nagar et al. 2003b). Formation of double-strand breaks induces phosphorylation of histone H2AX and this phosphorylated form, γH2AX, forms foci at the sites of DNA cleavage (Rogakou et al. 1998). Presumably then, DIE medium contained soluble factors secreted by unstable cells that could induce DNA double-strand cleavage. This was a significant observation because it indicates a source for the DNA double-strand breakage that would drive the chromosomal instability observed in our unstable clones. Given the preceding discussion, this lends credence to the hypothesis that the DIE factor(s) are responsible for perpetuating chromosomal instability in the unstable clone by providing a source for the genetic damage that leads to chromosomal alterations in the progeny of the irradiated cell. In those naive cells cultured in DIE medium, the induced DNA cleavage led to micronuclei formation and ultimately cell death by apoptosis.

So why do chromosomally unstable clones not die when cultured in this medium? It was rapidly apparent when initially expanding these unstable clones that many showed a general failure to thrive. In fact, one well-described characteristic of radiation-induced genomic instability is delayed lethal mutations (Seymour and Mothersill 1997) or delayed reproductive cell death (Chang and Little 1991). This is observed as a reduced plating efficiency in some cell clones surviving radiation exposure. Many of our unstable GM10115 cells also show reduced clonogenic survival (Limoli et al. 1997; Marder and Morgan 1993), and this generally manifests itself as persistently high levels of apoptosis (Limoli et al. 1998; Nagar et al. 2005). We had originally thought that this high level of apoptosis might contribute to perpetuating chromosomal instability because lytic products from apoptosing cells might also cause DNA strand breakage. Analysis of the apoptotic profiles of two of our unstable clones Fe-10-3 and LS-12 that showed the DIE led us to conclude that because of the significantly different levels of apoptosis in these two clones, lytic products associated with apoptosis are unlikely to provide the driving force for the DIE (Nagar et al. 2005). This conclusion was supported by a series of mixing experiments: We took a flask of GM10115 cells grown to confluence and lethally irradiated it with 20 Gy of X-rays. The irradiated cells plus medium were mixed with 100 non-irradiated GM10115 cells and survival of the non-irradiated cells was determined by colony formation. Survival in the non-irradiated cells was normal and was not affected by filtering out the dead/dying cells and just transferring the medium, or irradiation of the medium alone. Consequently it is our hypothesis that the DIE results from soluble factor(s) released by unstable cells that is/are cytotoxic to non-irradiated cells.

Despite the observed delayed reproductive cell death in many of our unstable clones, the majority of cells are reproductively viable and unstable clones have been cultured off and on for more than 4 years following irradiation. So why is DIE medium not cytotoxic to these cells? It is our hypothesis that this is due to a combination of reasons. Firstly, it takes unstable cells some time in culture to secrete sufficient amounts of DIE factor(s) into the culture medium to produce an almost complete cytotoxic response (Nagar et al. 2003a). Therefore our standard protocol of subculturing 5×10^5 cells twice a week provides ample time for them to attach, replicate, and divide before they really begin to secrete/shed the DIE factor(s). By the time the DIE is fully expressed, there are a large number of target cells that will soon be subcultured again. Secondly, we hypothesize that the unstable clones have "adapted" to this medium. Nevertheless they are susceptible to breakage induced by DIE factors that leads to persistent chromosomal instability and elevated apoptosis. To this end we examined whether this "adaptation" to a potentially cytotoxic environment conferred a more radiation resistant phenotype to our unstable clones (Limoli et al. 2001a). Chromosomal instability and radiation sensitivity were compared in 30 independently isolated clones. We observed a broader distribution of radiosensitivity and a higher mean surviving fraction after 2 Gy of X-rays in chromosomally unstable clones. Cytogenetic and DNA double strand break rejoining assays suggested that sensitivity was a function of DNA repair efficiency. In the unstable population, the more radioresistant clones also had significantly lower plating efficiencies. We proposed that the process of chromosome breakage and recombination that accompanies chromosome instability might provide some selective pressure for more radioresistant variants (Limoli et al. 2001a).

6.4 Chromosomal Instability Induced by DIE Medium from Unstable Clones

Despite the surprisingly effective cytotoxic effect of DIE medium, occasionally single cell clones can survive and form colonies. Nagar and Morgan (2005) expanded 20 clonal isolates of GM10115 cells surviving in DIE medium from either LS-12 or Fe-10-3 unstable cell clones. Surviving colonies were expanded in a 1:1 mixture of DIE medium to fresh medium, a ratio that we have previously demonstrated to be cytotoxic to non-irradiated GM10115 cells (Nagar et al. 2003a). Cytogenetic analysis of these clones revealed that one of 20 clones surviving in medium from LS-12 cells and three of 20 clones cultured in medium from Fe-10-3 cells showed chromosomal instability. This indicates that DIE medium can elicit the unstable phenotype, but why the significant majority of surviving clones do not manifest chromosomal instability is not immediately apparent. However, with time in culture in DIE medium, those clones demonstrating instability demonstrated increasing levels and complexity of delayed chromosomal rearrangements (Nagar and Morgan 2005).

6.5 A Role for Non-Targeted Bystander Effects in Radiation-Induced Chromosomal Instability

Non-targeted bystander effects refer to those responses occurring in cells that were not subject to energy-deposition events following ionizing radiation. Instead, these bystander cells may have been neighbors of irradiated cells, or physically separated, but susceptible to secreted signals from irradiated cells. This signal produced by an irradiated cell can result in a number of bystander responses in non-targeted cells. These responses include increased sister-chromatid exchanges, chromosomal rearrangements, micronuclei, gene mutations, apoptosis, genomic instability, transformation, altered cellular proliferation, and a variety of damage-inducible stress responses (reviewed in Morgan 2003a, b). A role for both gap junction communication (Azzam et al. 2001) and secreted soluble factors (Mothersill and Seymour 1998) in communicating the bystander response has been described (Sowa Resat and Morgan 2004). Consequently the transmission of radiation-induced chromosomal instability might well be considered a bystander-like response (Lorimore et al. 2003; Morgan et al. 2002). Candidate molecules responsible for the bystander signal include ROS and reactive nitrogen species as well as proteins associated with cellular stress responses (Azzam et al. 2003; Lehnert and Goodwin 1997; Mothersill et al. 2000; Narayanan et al. 1997, 1999; Shao et al. 2002, 2003).

Recent studies indicate that the signal from an irradiated cell to the non-targeted bystander cell can induce DNA cleavage in the recipient cell as indicated by phosphorylation of histone H2AX (Sokolov et al. 2005; Yang et al. 2005). This suggests that DNA double-strand breakage may be responsible for the observed bystander effects, which in turn would explain why the well-documented bystander effects in non-targeted cells are the same as the deleterious effects described in directly irradiated cells. A role for DNA breaks in mediating the bystander effect is also supported by studies indicating that the bystander response is significantly enhanced in repair deficient xrs5 cell lines following exposure to low fluences of α-particles (Nagasawa and Little 2002; Nagasawa et al. 2003). Nagasawa and Little (2002) have proposed that the primary DNA damage in bystander Chinese hamster ovary cells is oxidative base damage. This leads to a relatively small bystander effect for gross chromosomal aberrations as compared with mutations or sister-chromatid exchanges in wild-type cells, but to a significant bystander effect in xrs-5 cells because of non-repaired DNA strand breaks caused by opposed oxidative lesions.

6.6 Implications of Non-Targeted Effects for Cytogenetic Biodosimetry

It is premature at present to speculate on the significance of non-targeted effects of radiation on the use of chromosome analysis following human exposure to ionizing radiation. Despite of a large body of in vitro information, there continues to be considerable uncertainty regarding the causal relationship between non-targeted effects and the observed effects occurring in exposed individuals that might be attributable to radiation. The available information provides some support for potential associations, but not causation, and further research focusing on mechanistic information regarding effects occurring at delayed times after radiation exposure should help clarify this issue.

6.7 Conclusions

It is well established that exposure to ionizing radiation results in chromosomal rearrangements in a cell cycle independent manner. It is generally agreed that potential cytogenetic effects in irradiated (targeted) cells should be analyzed at the first metaphase following exposure to avoid complications associated with the formation of derived aberrations. Over the last decade, reports of chromosomal instability occurring in cells multiple generations after irradiation have indicated that a legacy of radiation insult can be transmitted from an irradiated cell to a non-irradiated (non-targeted) bystander cell and/or the progeny of that irradiated or non-targeted cell. Whatever the mechanism, these data indicate that chromosomal rearrangements can occur in non-irradiated cells. These non-targeted cells were either bystanders at the time of irradiation, or the progeny of irradiated cells. It appears that a signal from an irradiated cell can elicit molecular changes and subsequent responses in non-irradiated cells that results in chromosomal rearrangements. Furthermore, once chromosomal instability is initiated it can be perpetuated over time by bridge–breakage–fusion cycles (Kaplan et al. 1997; Marder and Morgan 1993) and DIE-type factors secreted by unstable cells (Nagar and Morgan 2005). Clearly identification and characterization of the signal(s) responsible for the DIE and other non-targeted bystander effects is critical for understanding the molecular mechanism(s) leading to chromosomal rearrangements in cells not directly exposed to DNA-damaging agents.

Acknowledgement: This research was supported by the Office of Science (BER), US Department of Energy, grant no. DE-FG02-05ER64082.

References

Azzam EI, de Toledo SM, Little JB (2001) Direct evidence for the participation of gap junction-mediated intercellular communication in the transmission of damage signals from alpha - particle irradiated to nonirradiated cells. Proc Natl Acad Sci USA 98:473–478

Azzam EI, de Toledo SM, Little JB (2003) Oxidative metabolism, gap junctions and the ionizing radiation-induced bystander effect. Oncogene 22:7050–7057

Chang WP, Little JB (1991) Delayed reproductive cell death in X-irradiated Chinese hamster ovary cells. Int J Radiat Biol 60:483–496

Cornforth MN (1998) Radiation-induced damage and the formation of chromosomal aberrations. In: Hoekstra MF, Nickoloff JA (eds) DNA damage and repair: DNA repair in higher eukaryotes, vol 2. Humana, Totawa, pp 559–585

Gemignani F, Ballardin M, Maggiani F, Rossi AM, Antonelli A, Barale R (1999) Chromosome aberrations in lymphocytes and clastogenic factors in plasma detected in Belarus children 10 years after Chernobyl accident. Mutat Res 446:245–253

Goh K, Sumner H (1968) Breaks in normal human chromosomes: are they induced by a transferable substance in the plasma of persons exposed to total-body irradiation? Radiat Res 35:171–181

Hollowell JG Jr, Littlefield LG (1968) Chromosome damage induced by plasma of x-rayed patients: an indirect effect of x-ray. Proc Soc Exp Biol Med 129:240–244
Kadhim MA, Macdonald DA, Goodhead DT, Lorimore SA, Marsden SJ, Wright EG (1992) Transmission of chromosomal instability after plutonium alpha-particle irradiation. Nature 355:738–740
Kaplan MI, Limoli CL, Morgan WF (1997) Perpetuating radiation-induced chromosomal instability. Radiat Oncol Invest 5:124–128
Kim GK, Chandrasekaran K, Morgan WF (2006) Mitochondrial dysfunction, persistently elevated levels of reactive oxygen species, and radiation induced genomic instability: a review. Mutagenesis 21:361–367
King JS, Phillips JW, Morgan WF (1994) The role of DNA double strand break rejoining in chromosome damage and repair. In: Obe G, Natarajan AT (eds) Chromosomal alterations: origin and significance. Springer, Berlin, pp 64–75
Lehnert BE, Goodwin EH (1997) Extracellular factor(s) following exposure to alpha particles can cause sister chromatid exchanges in normal human cells. Cancer Res 57:2164–2171
Limoli CL, Kaplan MI, Corcoran J, Meyers M, Boothman DA, Morgan WF (1997) Chromosomal instability and its relationship to other end points of genomic instability. Cancer Res 57:5557–5563
Limoli CL, Hartmann A, Shephard L, Yang CR, Boothman DA, Bartholomew J, Morgan WF (1998) Apoptosis, reproductive failure, and oxidative stress in Chinese hamster ovary cells with compromised genomic integrity. Cancer Res 58:3712–3718
Limoli CL, Corcoran JJ, Jordan R, Morgan WF, Schwartz JL (2001a) A role for chromosomal instability in the development of and selection for radioresistant cell variants. Br J Cancer 84:489–492
Limoli CL, Kaplan MI, Giedzinski E, Morgan WF (2001b) Attenuation of radiation-induced genomic instability by free radical scavengers and cellular proliferation. Free Radic Biol Med 31:10–19
Limoli CL, Giedzinski E, Morgan WF, Swarts SG, Jones GD, Hyun W (2003) Persistent oxidative stress in chromosomally unstable cells. Cancer Res 63:3107–3111
Littlefield LG, Hollowell JG Jr, Pool WH Jr (1969) Chromosomal aberrations induced by plasma from irradiated patients: an indirect effect of X radiation. Further observations and studies of a control population. Radiology 93:879–886
Lorimore SA, Kadhim MA, Pocock DA, Papworth D, Stevens DL, Goodhead DT, Wright EG (1998) Chromosomal instability in the descendants of unirradiated surviving cells after alpha-particle irradiation. Proc Natl Acad Sci USA 95:5730–5733
Lorimore SA, Coates PJ, Wright EG (2003) Radiation-induced genomic instability and bystander effects: inter-related nontargeted effects of exposure to ionizing radiation. Oncogene 22:7058–7069
Marder BA, Morgan WF (1993) Delayed chromosomal instability induced by DNA damage. Mol Cell Biol 13: 6667–6677
Morgan WF (2003a) Non-targeted and delayed effects of exposure to ionizing radiation: I. Radiation-induced genomic instability and bystander effects in vitro. Radiat Res 159:567–580
Morgan WF (2003b) Non-targeted and delayed effects of exposure to ionizing radiation: II. Radiation-induced genomic instability and bystander effects in vivo, clastogenic factors and transgenerational effects. Radiat Res 159:581–596
Morgan WF, Hartmann A, Limoli CL, Nagar S, Ponnaiya B (2002) Bystander effects in radiation-induced genomic instability. Mutat Res 504:91–100
Mothersill C, Seymour CB (1998) Cell-cell contact during gamma irradiation is not required to induce a bystander effect in normal human keratinocytes: evidence for release during irradiation of a signal controlling survival into the medium. Radiat Res 149:256–262
Mothersill C, Stamato TD, Perez ML, Cummins R, Mooney R, Seymour CB (2000) Involvement of energy metabolism in the production of 'bystander effects' by radiation. Br J Cancer 82:1740–1746

Nagar S, Morgan WF (2005) The death-inducing effect and genomic instability. Radiat Res 163:316–323

Nagar S, Smith LE, Morgan WF (2003a) Characterization of a novel epigenetic effect of ionizing radiation: the death inducing effect. Cancer Res 63:324–328

Nagar S, Smith LE, Morgan WF (2003b) Mechanisms of cell death associated with death-inducing factors from genomically unstable cell lines. Mutagenesis 18:549–560

Nagar S, Smith LE, Morgan WF (2005) Variation in apoptosis profiles in radiation-induced genomically unstable cell lines. Radiat Res 163:324–331

Nagasawa H, Little J (2002) Bystander effect for chromosomal aberrations induced in wild-type and repair deficient CHO cells by low fluences of alpha particles. Mutat Res 508:121

Nagasawa H, Huo L, Little JB (2003) Increased bystander mutagenic effect in DNA double-strand break repair-deficient mammalian cells. Int J Radiat Biol 79:35–41

Nakanishi M, Tanaka K, Shintani T, Takahashi T, Kamada N (1999) Chromosomal instability in acute myelocytic leukemia and myelodysplastic syndrome patients among atomic bomb survivors. J Radiat Res 40:159–167

Narayanan PK, Goodwin EH, Lehnert BE (1997) Alpha particles initiate biological production of superoxide anions and hydrogen peroxide in human cells. Cancer Res 57:3963–3971

Narayanan PK, LaRue KE, Goodwin EH, Lehnert BE (1999) Alpha particles induce the production of interleukin-8 by human cells. Radiat Res 152:57–63

Pant GS, Kamada N (1977) Chromosome aberrations in normal leukocytes induced by the plasma of exposed individuals. Hiroshima J Med Sci 26: 149–154

Parsons WB, Watkins CH, Pease GL, Childs DS (1954) Changes in sternal bone marrow following roentgen-ray therapy to the spleen in chronic granulocytic leukaemia. Cancer 7:179–189

Plumb M, Cleary H, Wright E (1998) Genetic instability in radiation-induced leukaemias: mouse models. Int J Radiat Biol 74:711–720

Rogakou EP, Pilch DR, Orr AH, Ivanova VS, Bonner WM (1998) DNA double-stranded breaks induce histone H2AX phosphorylation on serine 139. J Biol Chem 273:5858–5868

Salomaa S, Holmberg K, Lindholm C, Mustonen R, Tekkel M, Veidebaum T, Lambert B (1998) Chromosomal instability in in vivo radiation exposed subjects. Int J Radiat Biol 74:771–779

Sax K (1938) Chromosome aberrations induced by X-rays. Genetics 23:494–516

Seymour CB, Mothersill C (1997) Delayed expression of lethal mutations and genomic instability in the progeny of human epithelial cells that survived in a bystander-killing environment. Radiat Oncol Invest 5:106–110

Shao C, Furusawa Y, Aoki M, Matsumoto H, Ando K (2002) Nitric oxide-mediated bystander effect induced by heavy-ions in human salivary gland tumour cells. Int J Radiat Biol 78:837–844

Shao C, Stewart V, Folkard M, Michael BD, Prise KM (2003) Nitric oxide-mediated signaling in the bystander response of individually targeted glioma cells. Cancer Res 63:8437–8442

Sokolov MV, Smilenov LB, Hall EJ, Panyutin IG, Bonner WM, Sedelnikova OA (2005) Ionizing radiation induces DNA double-strand breaks in bystander primary human fibroblasts. Oncogene 24:7257–7265

Sowa Resat MB, Morgan WF (2004) Radiation-induced genomic instability: a role for secreted soluble factors in communicating the radiation response to non-irradiated cells. J Cell Biochem 92:1013–1019

Tawn EJ, Whitehouse CA, Martin FA (2000) Sequential chromosome aberration analysis following radiotherapy – no evidence for enhanced genomic instability. Mutat Res 465:45–51

Ward JF (1988) DNA damage produced by ionizing radiation in mammalian cells: identities, mechanisms of formation, and reparability. Prog Nucleic Acid Res Mol Biol 35:95–125

Watson GE, Lorimore SA, Macdonald DA, Wright EG (2000) Chromosomal instability in unirradiated cells induced in vivo by a bystander effect of ionizing radiation. Cancer Res 60: 5608–5611

Watson GE, Pocock DA, Papworth D, Lorimore SA, Wright EG (2001) In vivo chromosomal instability and transmissible aberrations in the progeny of haemopoietic stem cells induced by high- and low-LET radiations. Int J Radiat Biol 77:409–417

Whitehouse CA, Tawn EJ (2001) No evidence for chromosomal instability in radiation workers with in vivo exposure to plutonium. Radiat Res 156:467–475
Yang H, Asaad N, Held KD (2005) Medium-mediated intercellular communication is involved in bystander responses of X-ray-irradiated normal human fibroblasts. Oncogene 24:2096–2103

7 DNA Methylation Damage: Formation, Repair and Biological Consequences

MARKUS CHRISTMANN, WYNAND P. ROOS, AND BERND KAINA

Abstract

DNA methylating agents are powerful mutagens and carcinogens. They are endogenously formed, are present in food, beverages and tobacco smoke and are being used in cancer therapy. Methylating agents attack DNA at 13 sites. The main mutagenic and carcinogenic DNA lesion is O^6-methylguanine, whereas both N-methylation lesions and O^6-methylguanine contribute to sister-chromatid exchanges, chromosomal aberrations and cell death. The contribution depends on the relative level of adducts induced (S_N1 versus S_N2 agents), the alkyltransferase (MGMT) and mismatch repair activity as well as base excision repair capacity. The repair of the lesions and the genotoxic pathways triggered by them are discussed.

7.1 Methylating Agents as Environmental Carcinogens

Many compounds found in the environment methylate DNA. This methylation occurs on different sites of the DNA. Depending on how these lesions are processed by the cell, the lesions can give rise either to mutation or to cell death. The carcinogenic properties of methylating agents are due to the fact that they induce mutations. The most abundant methylating agents found in the environment are *N*-nitroso compounds.

The sources of *N*-nitroso compounds are food and tobacco smoke. For food, humans are mainly exposed to *N*-nitroso compounds through the consumption of cured meats and different fish products (Fiddler 1975). These compounds can also be formed in the body when nitrite or nitrogen oxides are ingested along with a nitrosable substrate like a secondary amine (Wogan and Tannenbaum 1975). Food contains non-volatile and volatile *N*-nitroso compounds, but it is thought that only the volatile compounds in food are mutagenic and therefore carcinogenic. A large group of volatile *N*-nitroso compounds found in food is the carcinogenic *N*-nitrosamines. The most abundant *N*-nitrosamines, in order of decreasing carcinogenicity, are *N*-nitrosodiethylamine (NDEA), *N*-nitrosodimethylamine (NDMA), *N*-nitrosopyrrolidine (NPYR) and *N*-nitrosopiperidine (NPIP) (Jagerstad and Skog 2005). Although

Günter Obe and Vijayalaxmi (Eds.)
Chromosomal Alterations: Methods, Results and Importance in Human Health

the non-volatile food *N*-nitroso compounds are not thought to be carcinogenic, they can be converted to carcinogenic volatile *N*-nitrosamines by heat; an example is *N*-nitrosoproline that is formed while frying bacon owing to the nitrosation of the amino acid proline that then becomes decarboxylated to form NPYR (Walker 1990).

N-Nitrosamines found in tobacco smoke are normally divided into two groups, namely, non-specific volatile nitrosamines and non-volatile tobacco-specific nitrosamines (TSNAs). The non-specific *N*-nitrosamines include NDMA, NDEA, *N*-nitrosoethylmethylamine, *N*-nitrosodiethanolamine, NPYR, and *N*-nitroso-n-butylamine (Mitacek et al. 1999). The four most widely reported TSNAs are *N*-nitrosoanabasine, *N*-nitrosoanabatine, 4-(methylnitrosamino)-1-(3-pyridyl)-1-butanone (NNK), and *N*-nitrosonornicotine (NNN).

N-Nitrosamines require metabolic activation to become mutagenic and carcinogenic. The hydroxylation is catalysed mainly by CYP2E1 (Guengerich and Shimada 1991; Lai and Arcos 1980; Lin et al. 1999), and to a lesser extent by CYP2A6 (Guengerich and Shimada 1991; Kamataki et al. 1999). Because of the involvement of the cytochrome P450 monooxygenase system in the activation of these compounds, it mainly occurs in the liver, but also to a lesser extent in other tissue types.

The DNA lesion mainly responsible for the mutagenicity, and therefore the carcinogenicity, of methylating agents is O^6-methylguanine (O6MeG). O6MeG forms GC-AT point mutations in proliferating cells if it is not repaired by O6MeG–DNA methyltransferase (MGMT). For a more mechanistic discussion on the mutagenicity of methylating agents see Sect. 7.5. The carcinogenic properties of dietary *N*-nitrosamines are strongly supported by experiments in animal models (Bartsch and Montesano 1984). Also, the carcinogenic properties of the TSNAs NNN and NNK have been shown in animals (Hecht et al. 1980).

7.2 Anticancer Drugs

As stated already, depending on how the DNA lesions are processed by the cell, methylating agents can give rise to either mutation or cell death. The death-provoking properties (cytotoxicity) of methylating agents have found an application in cancer therapy. Thus, methylating agents are widely used as anticancer chemotherapeutics as they exhibit antitumour properties. The best known DNA methylating anticancer chemotherapeutics today are procarbazine (PCB), dacarbazine (DIC), streptozotocin (STZ) and temozolomide (TMZ).

PCB(PCZ, N-methyl hydrazine, Natulan®, Matulane®) is used for the treatment of Hodgkin's disease, non-Hodgkin's lymphoma and less frequently for the treatment of brain tumours. PCB requires metabolic activation in order to accomplish its antitumour function. It is metabolized by the cytochrome

P450 isoforms CYP1A and CYP2B to the DNA reactive methyl radicals (Goria-Gatti et al. 1992).

DIC (imidazole carboxamide, dimethyltriazenoimidazole carboxamide, DTIC®-Dome) is also used for the treatment of Hodgkin's disease. Apart from Hodgkin's disease, it is used for the treatment of malignant melanoma, neuroblastoma and soft-tissue sarcomas. DIC is inactive until it is metabolized in the liver by cytochromes P450 to form the reactive N-demethylated species 5-(3-hydroxymethyl-3-methyltriazen-1-yl)-imidazole-4-carboxamide and 5-(3-methyltriazen-1-yl)-imidazole-4-carboxamide (MTIC) (Hill 1975; Skibba et al. 1970). This reaction is predominantly catalysed by CYP1A2, by CYP2E1 at higher substrate concentrations, and CYP1A1 contributes only a little (Reid et al. 1999).

STZ (NSC 85998, Zanosar®) is primarily used for the treatment of pancreatic (islet cell) cancer. The STZ molecule is structurally similar enough to glucose that it is transported into the cell by the glucose transport protein GLUT2. As the islet cells are particularly rich in GLUT2, it is clear why STZ is used for the treatment of this cancer (Brentjens and Saltz 2001; Wang and Gleichmann 1998). It is less frequently used in the treatment of carcinoid tumours, colon cancer, hepatic cancer, Hodgkin's disease, pancreatic (ductal) cancer and prostate cancer. STZ does not require metabolic activation and is known to directly methylate DNA (Bolzan and Bianchi 2002).

TMZ (SCHS2.365, NSC 362856, Temodal®, Temodar®) is the newest of the methylating chemotherapeutics. As it can pass the blood–brain barrier, it is used in the treatment of astrocytoma, glioblastoma and brain metastasis from solid tumours (Newlands et al. 1997). TMZ is also used for the treatment of melanoma. After oral administration, TMZ is rapidly absorbed with almost 100% bioavailability and undergoes spontaneous hydrolysis at physiological pH into its active metabolite, MTIC (Ostermann et al. 2004), and does therefore not need activation.

The cytotoxic action of the methylating agent chemotherapeutics is due to the DNA lesion O6MeG. As MGMT repairs this lesion, tumours with high MGMT will be resistant to these agents. A second requirement for the cytotoxicity of these agents is mismatch repair (MMR). A more detailed explanation on the mechanism of cytotoxicity of methylating agents will be given later in this chapter.

7.3 DNA Lesions, Sites of Attack

N-Nitrosamines and *N*-nitrosamides are the methylating agents with the highest biological relevance. *N*-Nitrosamides are unstable, undergo spontaneous hydrolysis and form reactive electrophilic species (e.g. carbenium cations). In contrast, *N*-nitrosamines are stable and need metabolic activation in order to form reactive electrophilic species. These simple methylating agents reacts with 13 different sites on the DNA (Fig. 7.1) via

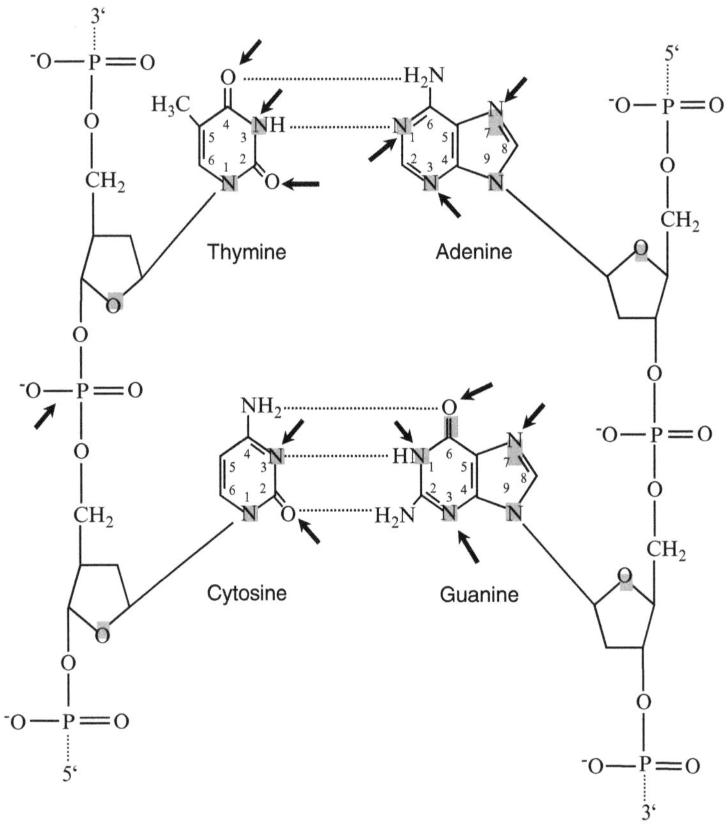

Fig. 7.1 Sites of attack of methylating agents

nucleophilic substitution reactions, leading to the formation of N-methylated and O-methylated purines and pyrimidines as well as phosphotriesters (reviewed in Beranek 1990).

Depending on the electrophilicity of the methylating agents, they react via either S_N1 or S_N2 mechanisms. S_N1 reactions follow first-order kinetics that is dependent on the formation of an electrophilic carbocation. This intermediate is covalently bound by the nucleophilic centre on the DNA. S_N2 reactions are bimolecular and dependent on steric accessibility, involving a transition state where the electrophile attacks the nucleophilic centre. This leads to formation of a transition complex, which thereafter forms the methylated product and a free leaving group. Therefore, the reactivity of different electrophilic methylating agents depends on the stability of the carbocation and/or the strength of the leaving group. Thus, nitrosoureas such as *N*-methyl-*N′*-nitro-*N*-nitrosoguanidine (MNNG) preferentially bind to oxygen via S_N1 reactions, whereas methyl methanesulfonate (MMS) binds to the most nucleophilic site (N7 position of guanine) and reacts via S_N2 reaction. This different reactivity explains the strong variation between different methylating agents concerning methylation of the O6 and N7 positions of guanine. Because of this, upon MNNG treatment 7–9% of all methylation products are O6MeG, whereas for MMS treatment only 0.3% are O6MeG (Beranek 1990).

Methyl DNA adducts have different stabilities. Therefore *N3*-methylpurines and *N7*-methylpurines are easily depurinated at neutral pH. Only some adducts, such as *N7*-methylguanine (N7MeG), *N3*-methyladenine (N3MeA) and O6MeG are stable for longer times ($T_{1/2}$ in vitro 40–80, 7–16 and 20 h, respectively). Besides representing the most stable lesions, N7MeG (80–85%) and N3MeA (8–20%) are also the most frequent methylation lesions. However, the biologically most critical lesion is O6MeG. Another premutagenic methylation lesion is *O*^{4}-methylthymine (O4MeT), which is induced in very small amounts (less than 0.4%). O6MeG and O4MeT are repaired by MGMT, N7MeG and N3MeA are repaired by the base excision repair (BER) system and *N1*-methyladenine (N1MeA) and *N3*-methylcytosine (N3MeC) are repaired by the three AlkB homologous (ABH1, ABH2, ABH3) proteins (Fig. 7.2).

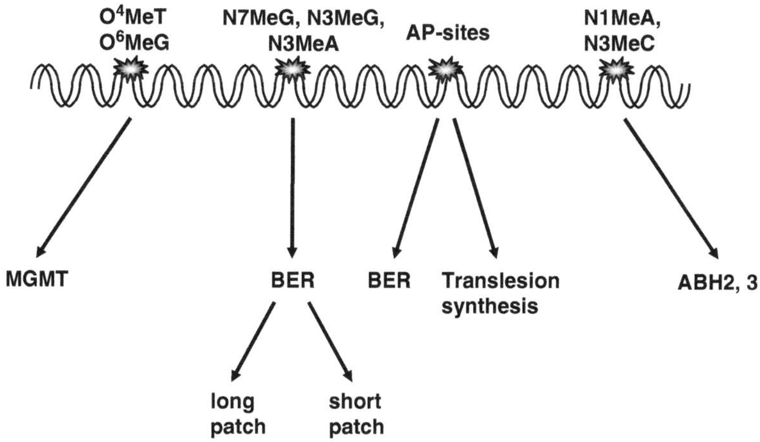

Fig. 7.2 Repair and tolerance of different methylation lesions via damage reversal (MGMT, ABH), base excision repair and translesion synthesis. *O^{6}MeG* O^6-methylguanine, *N7MeG* *N7*-methylguanine, *N3MeG* *N3*-methylguanine, *N3MeA* *N3*-methyladenine, *N1MeA* *N1*-methyladenine, *N3MeC* *N3*-methylcytosine, *MGMT* *O6*-methylguanine–DNA methyltransferase, *BER* base excision repair

7.4 Repair of DNA Methylation Damage

7.4.1 Reversal Repair

7.4.1.1 Single-Step Repair by MGMT

In *Escherichia coli*, resistance to methylating agents is mediated by the induction of the four genes *Ada, alkA, alkB* and *aidB* (reviewed in Sedgwick and Lindahl 2002). Ada repairs O6MeG, O^4-methylthymine and methyl-phosphotriester. In human cells, O^6-methylation lesions are repaired in a single-step reaction by MGMT (also known as ATase, AGT, AGAT), which is the mammalian homologue of the Ada gene product of *E. coli*. The human MGMT protein transfers the methyl group from the methylated guanine in a one-step reaction onto an internal cysteine residue in its active centre (reviewed in Pegg et al. 1995), leading to irreversible inactivation of the MGMT protein. This one-step reaction restores the guanine in the DNA to its unmethylated state. MGMT-deficient cells are unable to repair O6MeG and are therefore sensitive to methylating agents (Day et al. 1980a; Yarosh et al. 1983), while MGMT transfection provokes expression-dependent resistance to O^6-methylating agents (Kaina et al. 1991). MGMT knockout mice are viable, show a higher frequency of spontaneous tumours and are sensitive to treatment with methylating agents (Tsuzuki et al. 1996). Conversely, MGMT overexpressing mice show a reduced frequency of tumours in liver, thymus or skin (Becker et al. 1996; Dumenco et al. 1993; Nakatsuru et al. 1993; Zhou et al. 2001). MGMT overexpression specifically protects against tumour initiation without affecting tumour promotion in two-stage carcinogenesis experiments (Becker et al. 1996, 1997). It also protects against tumour progression (Becker et al. 2003) upon exposure to O^6-methylating agents. MGMT expression is highly regulated by methylation of both the gene and the promoter. Methylation of the promoter provokes inhibition of expression (Qian et al. 1995), which was shown to happen frequently during tumour formation (Hegi et al. 2004, 2005; Toyooka et al. 2001).

7.4.1.2 DNA Damage Reversal by AlkB Homologues

As already mentioned, in *E. coli* repair of DNA methylation damage is also performed by AlkB. Expression of AlkB in human cells increases their resistance to methylating agents (Chen et al. 1994). AlkB specifically repairs methylation damage in both single-stranded and double-stranded DNA and binds preferentially to single-stranded DNA in vitro (Dinglay et al. 2000). AlkB belongs to a superfamily of 2-oxoglutarate- and iron-dependent dioxygenases (Falnes et al. 2002; Trewick et al. 2002). AlkB repairs DNA methylation damage such as N1MeA and N3MeC (Fig. 7.1) in an oxygen-, ketoglutarate- and Fe(II)-dependent reaction, by coupling oxidative decarboxylation of ketoglutarate to hydroxylation of methylated bases (reviewed in Falnes et al. 2002). In humans, eight AlkB homologues were identified, and were designated as ABH1–ABH8 (Duncan et al. 2002; Kurowski et al. 2003). Whereas for ABH1 and ABH4–ABH8 the biological function is unknown, ABH2 and ABH3 belong, similarly to AlkB, to the superfamily of ketoglutarate- and Fe(II)-dependent dioxygenases and repair DNA methylation damage N1MeA and N3MeC (Aas et al. 2003; Duncan et al. 2002). During this repair, ABH2 and ABH3 differ in their template specificity; ABH3 repairs RNA and single-stranded DNA, whereas ABH2 preferentially repairs single-stranded and double-stranded DNA. It is thought that ABH2 and ABH3 remove the methylation in an α-ketoglutarate-dependent mechanism similar to the AlkB protein, involving oxidative demethylation leading to direct damage reversal and restoration of the undamaged base. By generation of ABH2- and ABH3-knockout mice, it was shown recently that mouse ABH2 acts as the primary oxidative demethylase for repairing N1MeA and N3MeC lesions in DNA (Ringvoll et al. 2006).

7.4.2 Mismatch Repair

The MMR system is responsible for removal of DNA base mismatches caused by spontaneous and induced base deamination, oxidation, methylation and replication errors (Modrich and Lahue 1996; Umar and Kunkel 1996). The main targets of MMR are base mismatches such as G/T, arising from deamination of 5-methylcytosine, G/G, A/C and C/C (Fang and Modrich 1993).

The recognition of mismatches is performed by the so-called MutSα complex, which binds to the lesions. MutSα is composed of the bacterial MutS homologous proteins MSH2 (Fishel et al. 1993; Leach et al. 1993) and MSH6 (Palombo et al. 1995). For the efficient binding to mismatches, phosphorylation of the MutSα complex is required (Christmann et al. 2002). MMR removes bases specifically from the newly synthesized DNA strand; however, the exact mechanism of the strand discrimination is unclear, but it involves the formation of an ADP/ATP-dependent sliding clamp (Gradia et al. 1999). In this translocation model, MutSα uses the energy gained by ATP hydrolysis to actively translocate along the DNA from the site of mismatch recognition to a site responsible for signalling the strand specificity (most likely a DNA single-strand break), where the assembly of the MutLα complex occurs (Blackwell et al. 2001). MutLα consists of the MutL homologous MMR proteins MLH1 and PMS2 (Li and Modrich 1995; Nicolaides et al. 1994; Papadopoulos et al. 1994). The excision of the DNA strand containing the mispaired base is performed by exonuclease I (Genschel et al. 2002) and new synthesis of DNA is performed by Polδ (Longley et al. 1997).

MMR does not play a direct role in the repair of methylated DNA; it rather counteracts the protective effect of MGMT by converting O6MeG into a cytotoxic secondary lesion. It was shown that cells which have a defect in MMR are resistant to methylating agents (Branch et al. 1993; Dosch et al. 1998). In MMR-deficient cells, the so-called tolerant phenotype, O6MeG induces mutations owing to the loss of its cytotoxic properties. As mentioned above, MMR is able to convert O6MeG into a cytotoxic lesion. This is initiated by binding of MutSα to O6MeG paired with cytosine or thymine (Duckett et al. 1996). During the subsequent repair of these faulty base pairs, MMR inserts thymine opposite O6MeG. During the following replication cycles, this leads to futile cycles of DNA repair (Karran and Bignami 1994) leading to the generation of double-strand breaks (DSBs), which represent the ultimate cytotoxic DNA lesions (Kaina et al. 1997).

In addition, the binding of MMR to O6MeG paired with thymine signals different mechanisms such as cell cycle regulation and apoptosis. It was shown that treatment with MNNG arrested only MMR-proficient cells in the G2-phase after treatment, which was based on activation of ATR and CHK1 (Stojic et al. 2004). ATR directly interacts with MSH2 (Wang and Qin 2003) and is recruited to O6MeG/T mismatches (Yoshioka et al. 2006). In addition, processing of O6MeG/T by MMR is also necessary for p38K activation, which can signal cell cycle arrest in glioma and colon cancer cell lines treated with TMZ (Hirose et al. 2003).

7.4.3 Base Excision Repair

While O6MeG is subject to direct reversal, lesions at the N-position (N7MeG, N3MeA and *N*3-methylguanine, N3MeG) are mainly repaired via the BER pathway. The BER mechanism, which is shown in Fig. 7.3, depends on the following steps: recognition, base removal and incision. The first step in BER is carried out by specific DNA glycosylases, which recognize and remove damaged bases by hydrolysing the N-glycosidic bond. In human cells, 12 different DNA glycosylases have been found which are characterized by different substrate specificities and their ability to cut the DNA backbone. Type I glycosylases remove modified bases leaving an AP site in DNA, whereas type II enzymes remove the base and subsequently cleave the AP site by an endogenous 3′ endonuclease activity giving rise to a single-strand break. For type I glycosylases, incision into the phosphodiester bond of the AP site occurs by AP endonuclease (APE1 alias APEX, Ref-1), resulting in 5′-deoxyribose-5-phosphate (5′-dRP) and 3′OH (Wilson and Barsky 2001). The glycosylase responsible for the removal of N-methylated bases is called methylpurine glycosylase [MPG, alias 3-alkyladenine DNA glycosylase (AAG) or alkylpurine DNA N-glycosylase (APNG)]. MPG is the homologue of *E. coli* AlkA and is a type I glycosylase (O'Connor 1993). It releases N3MeA, N7MeG, and N3MeG from DNA (Chakravarti et al. 1991) and imbalanced expression renders cells more sensitive to methylating agents (Coquerelle et al. 1995). MPG-knockout mice are viable and show an increased level of mutations following MMS treatment (Elder et al. 1998).

The insertion of the first nucleotide is not dependent on the chemical structure of the AP site and is performed by DNA polymerase β (Polβ), which

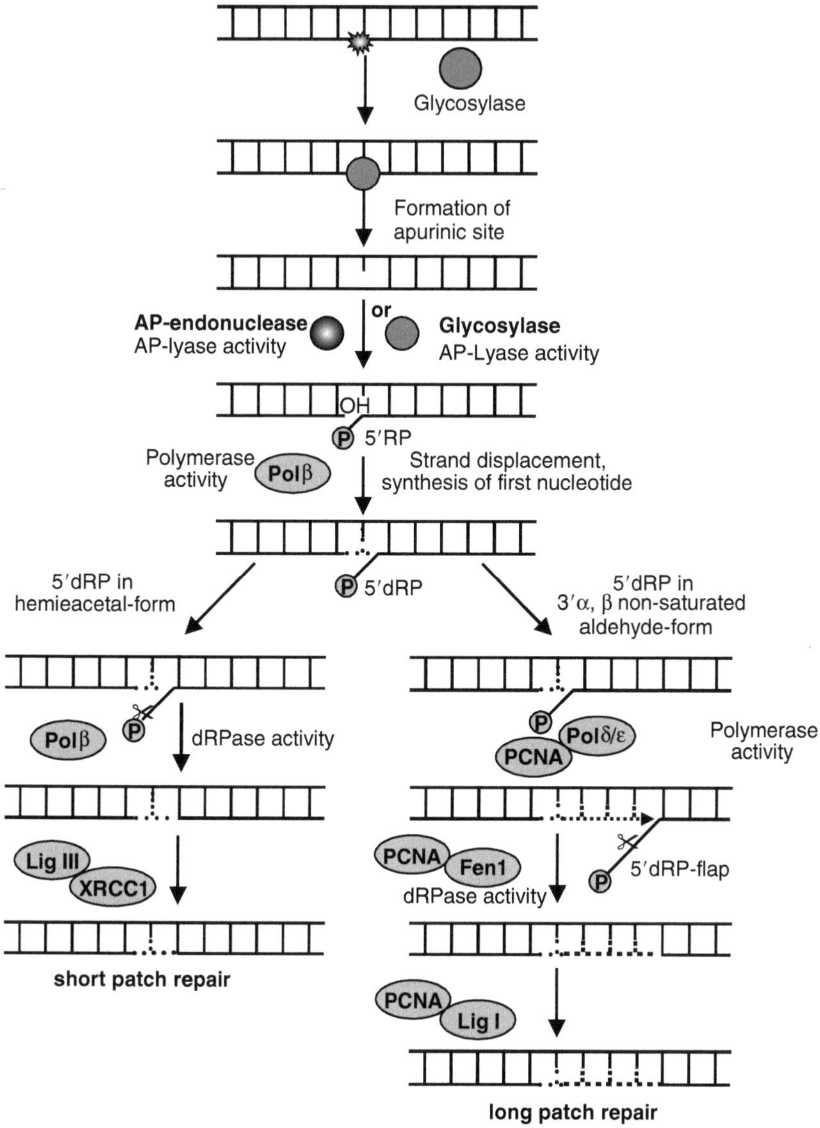

Fig. 7.3 Base excision repair: short-patch and long-patch pathway

inserts a single nucleotide (Dianov et al. 1999; Sobol and Wilson 2001). The critical step in the decision between short-patch and long-patch BER is the removal of 5′-dRP upon the insertion of the first nucleotide. Polβ has lyase activity and is able to catalyse the release of the hemiacetal form of 5′-dRP residues from incised AP sites by β-elimination (Matsumoto and Kim 1995;

Prasad et al. 1998; Sobol et al. 2000). Oxidized or reduced AP sites, 3′-unsaturated aldehydes or 3′-phosphates are resistant to β-elimination by Polβ (Nakamura et al. 2000) and have to be processed by proliferating cell nuclear antigen (PCNA) dependent long-patch repair (Frosina et al. 1996; Matsumoto et al. 1999). During long-patch repair further DNA synthesis is accomplished by Polε or Polδ together with PCNA (Stucki et al. 1998), resulting in repair patches of up to ten nucleotides. The deoxyribose phosphate flap structure is removed by PCNA-stimulated flap endonuclease (FEN1; Klungland and Lindahl 1997). The ligation step is performed by DNA ligase III in complex with XRCC1 (short patch) or DNA ligase I (long patch).

7.5 Mechanism of Mutagenicity and Carcinogenicity

As discussed in the previous sections, many carcinogens in the environment (Bartsch and Montesano 1984; Peto et al. 1984), in tobacco smoke (DeMarini 2004) and food (Jagerstad and Skog 2005) as well as endogenous metabolic products (Marnett and Burcham 1993) generate O6MeG. This lesion is highly mutagenic because of its mispairing properties. O6MeG mispairs with thymine and is, therefore, the major source of GC→AT point mutations following methylating agent exposure. It is a very important tumorigenic lesion (Becker et al. 1996) since point mutations arising from O6MeG contribute significantly to cancer induction (Gerson 2004) and cancer progression (Becker et al. 2003).

O6MeG does not cause DNA replication blockade, but is a mispairing lesion (Toorchen and Topal 1983). Consequently, it mispairs with thymine in the first S phase following exposure of DNA to methylating agents. This mispair only occurs if O6MeG is not repaired by MGMT before the cell enters the S phase. As not all cells express MGMT, [the MGMT expression levels vary 100-fold between individuals and tissue types (Margison et al. 2003)], and owing to the fact that MGMT can be depleted by chronic exposure of cells to methylating agents this can occur at a significant frequency. Once the O6MeG/thymine mismatch forms, it may not be recognized by MMR as the binding affinity of MutSα to O6MeG/thymine is much lower than the binding affinity of MutSα to guanine/thymine mismatches (Dosch et al. 1998). Therefore, it may remain in the DNA through the subsequent cell cycle phases and in the next S phase thymine will pair with adenine, while O6MeG will again mispair with thymine. Subsequent mispairs of O6MeG with thymine depend on the stability (half life) of O6MeG and the replication rate of the cells. MGMT is still able to remove the methyl group from guanine when it is mispaired with thymine. From the preceding discussion of the mechanism it is clear that O6MeG-mediated GC→AT point mutation formation can only occur in the absence of MGMT, and that replication is absolutely required for the formation of these mutations.

Mutations cause cancer owing to the inactivation of proteins that suppress cancer, or owing to hyperactivation of proteins that promote cancer. An example of the former is the cancer-suppressor p53 (Carson and Lois 1995). p53 plays important roles in cell cycle progression, DNA repair and apoptosis. By inactivation of this protein by mutation, the chance of a cell becoming cancerous increases significantly. An example where a mutation may lead to hyperactivation of a protein is when epidermal growth factor receptor (EGFR) is mutated and therefore signals continuously. This has been observed in many tumour types (Garcia de Palazzo et al. 1993; Ro et al. 1988; Xu et al. 1984).

Apart from the fact that humans are exposed to environmental and endogenous carcinogenic methylating compounds every day, methylating agents are also used in the clinic for the treatment of cancer. The cytotoxic properties of methylating agents will be discussed in the next section. The mutagenic properties of these chemotherapeutics may lead to therapy-based carcinogenesis by exactly the same mechanism described above. Not only may these methylating agents induce tumour formation in other sites of the body; they also induce mutations in the tumour itself (Lees et al. 2004).

7.6 Mechanism of Cytotoxicity of Methylating Agents by Apoptosis

PCB, DIC, STZ and TMZ are used in cancer therapy owing to the cytotoxicity of these methylating agents. The critical cytotoxic lesion is O6MeG (Day et al. 1980b; Glassner et al. 1999; Kaina et al. 1991; Qin et al. 1999; Sklar and Strauss 1981). Today, O6MeG provides perhaps the best understood example for a well-defined DNA lesion that triggers apoptosis. Evidence that O6MeG is not only a genotoxic lesion in cells and mice (Glassner et al. 1999; Kaina et al. 1991; Qin et al. 1999) but that it also triggers apoptosis was provided by comparing cells deficient and proficient for the DNA repair protein MGMT (Kaina et al. 1997; Meikrantz et al. 1998; Tominaga et al. 1997). Cells expressing MGMT show very low levels of apoptosis after treatment with O^6-methylating agents. This clearly shows that O6MeG induces apoptosis. However, the lesion does not directly induce apoptosis; it needs to be converted into a secondary apoptosis signalling lesion. Therefore, O6MeG is considered as a proapoptotic DNA lesion.

A requirement for apoptosis triggered by O6MeG is MMR (Hickman and Samson 1999; Pepponi et al. 2003). This is because O6MeG mispairs, during replication, with thymine. This gives rise to GC→AT point mutations in the absence of MMR. In the presence of MMR, the O6MeG/thymine mismatch is recognized by the heterodimeric protein complex MutSα, and MutSα binding has been shown to directly correlate with cell survival (Dosch et al. 1998). Thereafter, MutLα assembles to the complex and along with exonuclease I and Polδ removes the thymine opposite O6MeG. Because of the mispairing properties of O6MeG, thymine will be reinserted. At this stage it is reasonable to hypothesize that this results in a futile repair cycle (Karran and Bignami 1994) that ends up with the formation of severe secondary lesions (presumably gaped DNA) that significantly interfere with the subsequent DNA replication. Blocking of replication at this site may lead to collapse of replication forks, ending up with the formation of DSBs. This has indeed been shown to occur in MGMT-deficient MMR-proficient cells (Ochs and Kaina 2000). MMR-deficient cells are highly resistant to O6MeG DNA lesions because they do not undergo apoptosis. They survive (and thus tolerate O6MeG lesions) at the expense of a high level of GC→AT point mutations (Karran and Bignami 1994). Therefore, MMR-driven apoptosis is believed to eliminate premutated cells from the population, reducing the mutation pressure. It should be noted, however, that only a small amount of O6MeG (about 30 lesions) gives rise to detectable genotoxic effects (sister-chromatid exchanges, SCEs) without an apparent threshold dose, whereas apoptosis is induced at a higher dose level, displaying a threshold. This explains why a single lesion, O6MeG, is so powerful mutagenic and carcinogenic and also highly efficient in evoking the apoptotic pathway mediated by MMR.

The model of O6MeG-triggered apoptosis implies that DNA replication is essential. This has been shown experimentally. Using proliferating and non-proliferating human peripheral blood lymphocytes (PBLCs), Roos et al. (2004) showed that for clinically relevant doses of the methylating anticancer drug TMZ only the proliferating PBLCs underwent apoptosis. Thus, proliferation and MutSα-dependent MMR are essential components of O6MeG-triggered apoptosis. A wave of DSBs preceded apoptosis (Roos et al. 2004).

MutSα not only binds to O6MeG/thymine mispairs but also to UV-C (Hoffman et al. 2005), cisplatin (Claij and te Riele 2004), *N*-acetyl-2-aminofluorene and benzo[*a*]pyrene adducts in DNA (Li et al. 1996). On the basis of this finding it has been speculated that MutSα may directly signal downstream to the apoptotic machinery, thereby acting as a sensor for non-repaired DNA damage. Contradicting this "sensor" model are the findings that (1) there is no general cross-resistance of tolerant cells to UV-C, cisplatin, benzo[a]pyrene and other agents, (2) apoptosis is a late response occurring in the first to fourth post-treatment cell cycle and (3) DNA replication is required for O6MeG-triggered apoptosis. For isogenic embryonic stem cells expressing and not expressing MSH2, it has been shown that MMR does not impact on cisplatin resistance (Claij and te Riele 2004). Therefore, the model of futile MMR on O6MeG/thymine lesions causing DSBs at blocked replication forks that ultimately trigger apoptosis is favoured (Ochs and Kaina 2000). DSBs were indeed observed following O6MeG induction, and they preceded apoptosis in a replication-dependent manner (Ochs and Kaina 2000; Roos et al. 2004).

The apoptotic pathways employed by cells following the exposure to O^6-methylating agents differ depending on their p53 status. In cells mutated in the p53 gene, cells make use of the mitochondrial apoptotic pathway (Ochs and Kaina 2000). Consequently, following O6MeG induction, Chinese hamster ovary

(CHO) cells (mutated for p53) show a decline in Bcl-2, which increases the Bax to Bcl-2 ratio and allows for the release of cytochrome *c* from the mitochondria. This release of cytochrome *c* leads to the activation of the apoptosome consisting of Apaf-1, ATP, cytochrome *c* and procaspase-9. In turn, caspase-3 becomes activated and cleaves the inhibitor of caspase-activated DNase (ICAD), thereby freeing caspase-activated DNase (CAD). CAD finally degrades DNA, resulting in the characteristic apoptotic DNA fragmentation (Ochs and Kaina 2000). In human PBLCs and lymphoblastoid cell lines wild-type for p53, O6MeG was shown to activate the death receptor apoptotic pathway (Fas/CD95/Apo-1) (Dunkern et al. 2003; Roos et al. 2004). In line with this is the finding that execution of apoptosis can be attenuated by blocking Fas (Roos et al. 2004). Thus the data show that the same DNA lesion, O6MeG, is able to trigger both the mitochondrial and the death receptor (Fas, CD95, Apo-1) pathway, stressing the importance of the cellular background.

Since O^6-methylating agents (TMZ, DIC) are being used for the treatment of glioblastomas and malignant melanomas, the question arises as to what is the apoptotic mechanism evoked in these tumours. In gliomas, O6MeG was shown to trigger apoptosis by activating both the mitochondrial (in p53-mutated tumour cells) and the p53-controlled Fas-dependent pathway (Roos et al. 2007). MGMT and p53 are therefore main components of TMZ resistance of gliomas.

7.7 Mechanism of Formation of SCEs and Aberrations

7.7.1 Critical Lesions

The identification of methylation lesions responsible for recombinogenic and clastogenic effects of the agents was based on molecular dosimetry (van Zeeland et al. 1983; Vogel and Natarajan 1979) and later on the comparison of genetically engineered isogenic cell lines that are MGMT-deficient, MGMT- proficient and MGMT-tolerant (MGMT-deficient, MMR-impaired) (reviewed in Kaina et al. 1993). The data revealed that (1) agents like MMS and dimethyl sulfate (DMS) that predominantly induce N-methylations (S_N2 agents) exhibit high clastogenic but low point-mutagenic potency, indicating that N-methylation lesions are responsible for clastogenic effects; (2) agents like 1-methyl-1-nitrosourea (MNU) and MNNG that induce a higher level of O-methylations (S_N1 agents) are clastogenic and point-mutagenic, confirming that O-methylations are responsible for the induction of point-mutations; (3) cells deficient in MGMT are highly sensitive to the clastogenic and SCE-inducing effects of S_N1 agents (e.g. MNU and MNNG) compared with the same cells expressing MGMT. The same holds true for tolerant cells (Kaina 2004). From this it has been concluded that O6MeG (the minor lesion O4MeT is usually neglected because of its rarity) is a critical lesion leading to SCEs and chromatid-type aberrations. Additionally, it was shown that MGMT-deficient Chinese hamster cells treated with MNU or MNNG induced SCEs only in the second posttreatment cell cycle detectable in the second mitosis following treatment. In contrast, MMS is able to induce SCEs both in the treatment and in the posttreatment cell cycle (Kaina and Aurich 1985). Similar data were obtained for chromosomal aberrations for which the frequency after treatment with MNNG was low in the first mitosis and much higher in the second mitosis following treatment (Kaina 2004). Interestingly, if the same cells were transfected with a MGMT expression vector, aberrations in M2 cells were reduced to nearly background levels (Kaina 2004). These data indicated that O6MeG needs two replication cycles in order to be converted into SCEs and chromatid-type aberrations. The data also revealed that the methylation lesion O6MeG is converted during the first round of replication into a secondary DNA lesion that acts as the ultimate trigger for either SCEs or aberrations (Kaina 1985b; Kaina et al. 1993). In summary, the data showed that in cells deficient for MGMT, O6MeG is the predominant recombinogenic and clastogenic lesion, with the effect only being detectable in the second mitosis following treatment. In cells that are MGMT-proficient, N-methylation products (N7MeG, N3MeA, N3MeG) appear to be the predominant SCE and clastogenic lesions, and provoke the genotoxic effect already in the treatment cell cycle, visible as SCEs or aberrations in the first mitosis following treatment. It should be noted that the dose of the mutagenic agent is also important. If the dose of the methylating agent is high, N-methylations will also contribute to genotoxicity even in MGMT-expressing cells.

On the basis of these findings it becomes clear that in human fibroblasts and primary Chinese hamster cells, both expressing MGMT, the highest clastogenic response is observed in the first mitosis following treatment, which is due to non-repaired N-methylation lesions. In contrast, MGMT-deficient V79 and CHO cells (and many other established cell lines) display the highest clastogenic effect in the second mitosis following treatment, which is due to non-repaired O6MeG lesions (Kaina 2004). For S_N2 agents (MMS, DMS), the aberration maximum is expected to occur in the first mitosis following treatment because of the relatively high number of N-alkylations compared with the amount O6MeG the agents induce in DNA.

7.7.2 Involvement of MMR in Clastogenicity

MGMT-deficient cells defective in MMR display lower MNNG-induced aberration frequencies than MGMT-deficient cells not defective in MMR (Armstrong and Galloway 1996; Galloway et al. 1995; Kaina 2004; Kaina et al. 1995); therefore, it has been suggested that MMR is essential for the conversion of O6MeG lesions into SCEs and aberrations (Kaina et al. 1997). The process is similar to the one described above in explaining the killing effect of the lesion. The heterodimeric complex MutSα, composed of MSH2 and MSH6, binds to O6MeG/thymine mispairs, which are formed during replication of O6MeG lesions. MLH1 and PMS2 assemble to the lesion (Sibghat-Ullah et al. 2001) in order to repair the mismatch. This, however, is a futile process generating secondary lesions that interfere with the subsequent round of DNA replication, leading to DSBs (Kaina et al. 1997; Ochs and Kaina 2000). This model provides an explanation of why SCEs and aberrations are formed in response to O6MeG in a replication-dependent manner almost exclusively in the first posttreatment cell cycle (appearing in the second mitosis following treatment). Although this model still awaits further verification, it provides a firm basis for explaining most of the data obtained in this field.

7.7.3 Conversion Rates

The probability that O6MeG will be converted into SCEs has been determined in two independent studies. For MGMT-deficient CHO-9 cells, a correlation was found between 30 O6MeG adducts producing one SCE (30:1) (Kaina et al. 1993). For human glioblastoma cells, it was reported that 42 O6MeG adducts produce one SCE (42:1) (Rasouli-Nia et al. 1994). This is quite a high conversion frequency compared, for example, with that of cyclobutane dimers, for which a relation of 600:1 was estimated (De Weerd-Kastelein et al. 1977). It should be noted, however, that this relation might need to be reestimated in view of the fact that UV-C induced 5-bromo-2′-deoxyuridine (BrdU) photolysis may contribute to SCE formation (Wojcik et al. 2004). The conversion probabilities for chromosomal aberrations are much lower, and are estimated to be in the range of 22,000:1 and 147,000:1 for aberrations induced in the second and the first mitosis following treatment, respectively (Kaina et al. 1993). For lethal events, the values between 360:1 (Kaina et al. 1993) and 6,650:1 (Rasouli-Nia et al. 1994) have been calculated. The close correlation between O6MeG lesions and SCEs is striking. Taking into account that O6MeG mispairs in only a third of base pairings (provided equal amounts of deoxynucleotides are present) (Abbott and Saffhill 1979), the probability of the mispaired lesion O6MeG/thymine converting into SCEs would even be higher, i.e. about 10:1.

O6MeG/thymine mispaired lesions are obviously extremely potent in inducing SCEs. The reason for this may be found in the processing of O6MeG/thymine lesions by MMR. Thus, it is tempting to speculate that MMR-generated lesions in the DNA cause a severe block in DNA replication, which is circumvented by a recombination process that finally leads to a SCE (Fig. 7.4). There are many lesions that block DNA replication although they seem to induce SCEs at a much lower level, such as cyclobutane pyrimidine dimers. It may therefore be speculated that most of the presumed replication-blocking lesions, if they encounter a replication fork, are tolerated by translesion synthesis (*replication bypass*) accomplished by low-fidelity SOS polymerases (Lehman 2002), reading through non-instructive lesions without initiating the recombination process that leads to SCEs. The MMR-mediated lesion induced by O6MeG/thymine can presumably not be handled

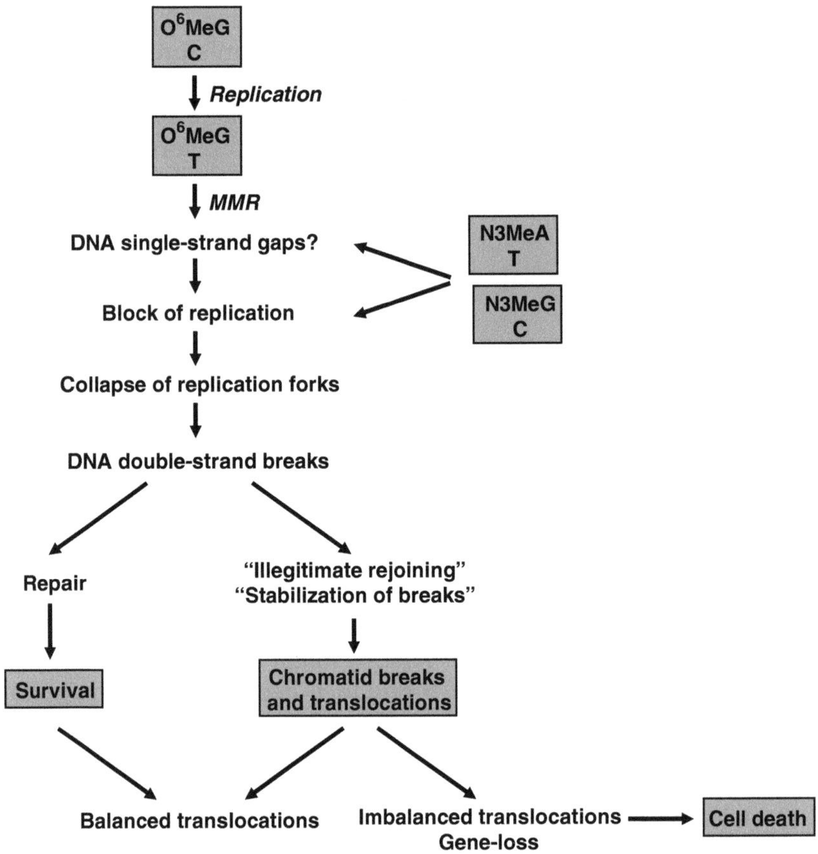

Fig. 7.4 Model of conversion of O^6-methylguanine lesions into DNA double-strand breaks and, further, into chromosomal aberrations and cell-killing effects. *MMR* mismatch repair

by SOS polymerases and, therefore, recombination bypass will be the preferred mode of lesion tolerance. It should be noted that the doses of MNNG and MNU inducing SCEs in the second mitosis following treatment do not cause any significant inhibition of DNA replication (as measured by thymidine incorporation and the delay in entry of BrdU-labelled cells into mitosis) (Kaina 1985a; Kaina and Aurich 1985), suggesting that recombination bypass leading to SCEs is a very efficient and fast process of lesion circumvention.

7.7.4 Critical Role of Inhibition of Replication Fork Movement

While SCEs are induced at very low dose levels, and are therefore a very sensitive indicator of DNA damage, higher doses of MNNG or MNU are required for the induction of chromosomal aberrations (Kaina 1985a). If we compare

SCEs and aberrations with DNA synthesis inhibition, it becomes clear that DNA synthesis is inhibited in the dose range that induces aberrations in the first mitosis following treatment (Kaina 1985a, 1998). On the basis of the overlap in dose–response of replication inhibition and aberrations in the first mitosis following treatment, it has been proposed that inhibition of replication fork movement causes nuclease attack at stalled replication forks that results in DSBs, which cause chromatid-type aberrations (Kaina 1998). Various N-methylation lesions such as N3MeA block DNA replication. Also, apurinic sites resulting from spontaneous hydrolysis of N-methylpurines such as N7MeG and N3MeA are critical candidates for blocking DNA replication, thus leading to DSBs. If they are not repaired, or are illegitimately recombined, these DSBs will be translated into chromatid-type aberrations in the first mitosis following treatment. A similar process may give rise to chromatid-type aberrations in the second mitosis following treatment if O6MeG/thymine lesions processed by MMR block DNA replication in the second cell cycle after treatment (Kaina et al. 1997). It will be an interesting topic of future research to figure out the molecular players and pathways that are involved in this highly vulnerable process.

Acknowledgement: This work was supported by the Deutsche Forschungsgemeinschaft, grants KA724/13-1 and 13-2.

References

Aas P.A, Otterlei M, Falnes PO, Vagbo CB, Skorpen F, Akbari M, Sundheim O, Bjoras M, Slupphaug G, Seeberg E et al (2003) Human and bacterial oxidative demethylases repair alkylation damage in both RNA and DNA. Nature 421:859–863

Abbott PJ, Saffhill R (1979) DNA synthesis with methylated poly(dC-dG) templates. Evidence for a competitive nature to miscoding by O6-methylguanine. Biochim Biophys Acta 562:51–61

Armstrong MJ, Galloway SM (1996) Mismatch repair provokes chromosome aberrations in hamster cells treated with methylating agents or 6-thioguanine, but not with ethylating agents. Mutat Res 373:167–178

Bartsch H, Montesano R (1984) Relevance of nitrosamines to human cancer. Carcinogenesis 5:1381–1393

Becker K, Dosch J, Gregel CM, Martin BA, Kaina B (1996) Targeted expression of human O(6)-methylguanine-DNA methyltransferase (MGMT) in transgenic mice protects against tumor initiation in two-stage skin carcinogenesis. Cancer Res 56:3244–3249

Becker K, Gregel CM, Kaina B (1997). The DNA repair protein O6-methylguanine-DNA methyltransferase protects against skin tumor formation induced by antineoplastic chloroethylnitrosourea. Cancer Res 57:3335–3338

Becker K, Gregel C, Fricke C, Komitowski D, Dosch J, Kaina B (2003) DNA repair protein MGMT protects against N-methyl-N-nitrosourea-induced conversion of benign into malignant tumors. Carcinogenesis 24:541–546

Beranek DT (1990) Distribution of methyl and ethyl adducts following alkylation with monofunctional alkylating agents. Mutat Res 231:11–30

Beranek DT, Weis CC, Swenson DH (1980) A comprehensive quantitative analysis of methylated and ethylated DNA using high pressure liquid chromatography. Carcinogenesis 1:595–606

Blackwell LJ, Bjornson KP, Allen DJ, Modrich P (2001) Distinct MutS DNA-binding modes that are differentially modulated by ATP binding and hydrolysis. J Biol Chem 276:34339–34347
Bolzan AD, Bianchi MS (2002) Genotoxicity of streptozotocin. Mutat Res 512:121–134
Branch P, Aquilina G, Bignami M, Karran P (1993) Defective mismatch binding and a mutator phenotype in cells tolerant to DNA damage. Nature 362:652–654
Brentjens R, Saltz L (2001) Islet cell tumors of the pancreas: the medical oncologist's perspective. Surg Clin North Am 81:527–542
Carson DA, Lois A (1995) Cancer progression and p53. Lancet 346:1009–1011
Chakravarti D, Ibeanu GC, Tano K, Mitra S (1991) Cloning and expression in Escherichia coli of a human cDNA encoding the DNA repair protein N-methylpurine-DNA glycosylase. J Biol Chem 266:15710–15715
Chen BJ, Carroll P, Samson L (1994) The Escherichia coli AlkB protein protects human cells against alkylation-induced toxicity. J Bacteriol 176:6255–6261
Christmann M, Tomicic MT, Kaina B (2002). Phosphorylation of mismatch repair proteins MSH2 and MSH6 affecting MutSalpha mismatch-binding activity. Nucleic Acids Res 30:1959–1966
Claij N, te Riele H (2004). Msh2 deficiency does not contribute to cisplatin resistance in mouse embryonic stem cells. Oncogene 23:260–266
Coquerelle T, Dosch J, Kaina B (1995) Overexpression of N-methylpurine-DNA glycosylase in Chinese hamster ovary cells renders them more sensitive to the production of chromosomal aberrations by methylating agents – a case of imbalanced DNA repair. Mutat Res 336:9–17
Day RS III, Ziolkowski CHJ, Scudiero DA, Meyer SA, Lubiniecki AS, Girardi AJ, Galloway SM, Bynum GD (1980a) Defective repair of alkylated DNA by human tumour and SV40-transformed human cell strains. Nature 288:724–727
Day RS III, Ziolkowski CHJ, Scudiero DA, Meyer SA, Mattern MR (1980b) Human tumor cell strains defective in the repair of alkylation damage. Carcinogenesis 1:21–32
DeMarini DM (2004) Genotoxicity of tobacco smoke and tobacco smoke condensate: a review. Mutat Res 567:447–474
De Weerd-Kastelein EA, Keijzer W, Rainaldi G, Bootsma D (1977) Induction of sister chromatid exchange in Xeroderma pigmentosum cells after exposure to ultraviolet light. Mutat Res 45:253–261
Dianov GL, Prasad R, Wilson SH, Bohr VA (1999) Role of DNA polymerase beta in the excision step of long patch mammalian base excision repair. J Biol Chem 274:13741–13743
Dinglay S, Trewick SC, Lindahl T, Sedgwick B (2000) Defective processing of methylated single-stranded DNA by E. coli AlkB mutants. Genes Dev 14:2097–2105
Dosch J, Christmann M, Kaina B (1998) Mismatch G-T binding activity and MSH2 expression is quantitatively related to sensitivity of cells to methylating agents. Carcinogenesis 19:567–573
Duckett DR, Drummond JT, Murchie AI, Reardon JT, Sancar A, Lilley DM, Modrich P (1996) Human MutSalpha recognizes damaged DNA base pairs containing O6-methylguanine, O4-methylthymine, or the cisplatin-d(GpG) adduct. Proc Natl Acad Sci USA 93:6443–6447
Dumenco LL, Allay E, Norton K, Gerson SL (1993) The prevention of thymic lymphomas in transgenic mice by human O6-alkylguanine-DNA alkyltransferase. Science 259:219–222
Duncan T, Trewick SC, Koivisto P, Bates PA, Lindahl T, Sedgwick B (2002) Reversal of DNA alkylation damage by two human dioxygenases. Proc Natl Acad Sci USA 99:16660–16665
Dunkern T, Roos W, Kaina B (2003) Apoptosis induced by MNNG in human TK6 lymphoblastoid cells is p53 and Fas/CD95/Apo-1 related. Mutat Res 544:167–172
Elder RH, Jansen JG, Weeks RJ, Willington MA, Deans B, Watson AJ, Mynett KJ, Bailey JA, Cooper DP, Rafferty JA et al (1998) Alkylpurine-DNA-N-glycosylase knockout mice show increased susceptibility to induction of mutations by methyl methanesulfonate. Mol Cell Biol 18:5828–5837
Falnes PO, Johansen RF, Seeberg E (2002) AlkB-mediated oxidative demethylation reverses DNA damage in Escherichia coli. Nature 419:178–182
Fang WH, Modrich P (1993) Human strand-specific mismatch repair occurs by a bidirectional mechanism similar to that of the bacterial reaction. J Biol Chem 268:11838–11844

Fiddler (1975) The occurrence and determination of N-nitroso compounds. Toxicol Appl Pharmacol 31:352–360
Fishel R, Lescoe MK, Rao MR, Copeland NG, Jenkins NA, Garber J, Kane M, Kolodner R (1993) The human mutator gene homolog MSH2 and its association with hereditary nonpolyposis colon cancer. Cell 75:1027–1038
Frosina G, Fortini P, Rossi O, Carrozzino F, Raspaglio G, Cox LS, Lane DP, Abbondandolo A, Dogliotti E (1996) Two pathways for base excision repair in mammalian cells. J Biol Chem 271:9573–9578
Galloway SM, Greenwood SK, Hill RB, Bradt CI, Bean CL (1995) A role for mismatch repair in production of chromosome aberrations by methylating agents in human cells. Mutat Res 346:231–245
Garcia de Palazzo IE, Adams GP, Sundareshan P, Wong AJ, Testa JR, Bigner DD, Weiner LM (1993) Expression of mutated epidermal growth factor receptor by non-small cell lung carcinomas. Cancer Res 53:3217–3220
Genschel J, Bazemore LR, Modrich P (2002) Human exonuclease I is required for 5′ and 3′ mismatch repair. J Biol Chem 277:13302–13311
Gerson SL (2004) MGMT: its role in cancer aetiology and cancer therapeutics. Nat Rev Cancer 4:296–307
Glassner BJ, Weeda G, Allan JM, Broekhof JL, Carls NH, Donker I, Engelward BP, Hampson RJ, Hersmus R, Hickman MJ (1999) DNA repair methyltransferase (Mgmt) knockout mice are sensitive to the lethal effects of chemotherapeutic alkylating agents. Mutagenesis 14:339–347
Goria-Gatti L, Iannone A, Tomasi A, Poli G, Albano E (1992) In vitro and in vivo evidence for the formation of methyl radical from procarbazine: a spin-trapping study. Carcinogenesis 13:799–805
Gradia S, Subramanian D, Wilson T, Acharya S, Makhov A, Griffith J, Fishel R (1999) hMSH2-hMSH6 forms a hydrolysis-independent sliding clamp on mismatched DNA. Mol Cell 3:255–261
Guengerich FP, Shimada T (1991) Oxidation of toxic and carcinogenic chemicals by human cytochrome P-450 enzymes. Chem Res Toxicol 4:391–407
Hecht SS, Chen CB, Ohmori T, Hoffmann D (1980) Comparative carcinogenicity in F344 rats of the tobacco-specific nitrosamines, N′-nitrosonornicotine and 4-(N-methyl-N-nitrosamino)-1-(3-pyridyl)-1-butanone. Cancer Res 40:298–302
Hegi ME, Diserens AC, Godard S, Dietrich PY, Regli L, Ostermann S, Otten P, Van Melle G, de Tribolet N, Stupp R (2004) Clinical trial substantiates the predictive value of O-6-methylguanine-DNA methyltransferase promoter methylation in glioblastoma patients treated with temozolomide. Clin Cancer Res 10:1871–1874
Hegi ME, Diserens AC, Gorlia T, Hamou MF, de Tribolet N, Weller M, Kros JM, Hainfellner JA, Mason W, Mariani L et al (2005) MGMT gene silencing and benefit from temozolomide in glioblastoma. N Engl J Med 352:997–1003
Hickman MJ, Samson LD (1999) Role of DNA mismatch repair and p53 in signaling induction of apoptosis by alkylating agents. Proc Natl Acad Sci USA 96:10764–10769
Hill DL (1975) Microsomal metabolism of triazenylimidazoles. Cancer Res 35:3106–3110
Hirose Y, Katayama M, Stokoe D, Haas-Kogan DA, Berger MS, Pieper RO (2003) The p38 mitogen-activated protein kinase pathway links the DNA mismatch repair system to the G2 checkpoint and to resistance to chemotherapeutic DNA-methylating agents. Mol Cell Biol 23:8306–8315
Hoffman PD, Wang H, Lawrence CW, Iwai S, Hanaoka F, Hays JB (2005) Binding of MutS mismatch repair protein to DNA containing UV photoproducts, "mismatched" opposite Watson–Crick and novel nucleotides, in different DNA sequence contexts. DNA Repair (Amst) 4:983–993
Jagerstad M, Skog K (2005) Genotoxicity of heat-processed foods. Mutat Res 574:156–172
Kaina B (1985a) The interrelationship between SCE induction, cell survival, mutagenesis, aberration formation and DNA synthesis inhibition in V79 cells treated with N-methyl-N-nitrosourea or N-methyl-N′-nitro-N-nitrosoguanidine. Mutat Res 142:49–54
Kaina B (1985b) Secondary lesions and the induction of SCEs and chromosomal aberrations. In: Proceedings of the 13th international congress on biochemistry, Amsterdam, p 31

Kaina B (1998) Critical steps in alkylation-induced aberration formation. Mutat Res 404:119–124
Kaina B (2004) Mechanisms and consequences of methylating agent-induced SCEs and chromosomal aberrations: a long road traveled and still a far way to go. Cytogenet Genome Res 104:77–86
Kaina B, Aurich O (1985) Dependency of the yield of sister-chromatid exchanges induced by alkylating agents on fixation time. Possible involvement of secondary lesions in sister-chromatid exchange induction. Mutat Res 149:451–461
Kaina B, Fritz G, Mitra S, Coquerelle T (1991) Transfection and expression of human O6-methylguanine-DNA methyltransferase (MGMT) cDNA in Chinese hamster cells: the role of MGMT in protection against the genotoxic effects of alkylating agents. Carcinogenesis 12:1857–1867
Kaina B, Fritz G, Coquerelle T (1993) Contribution of O6-alkylguanine and N-alkylpurines to the formation of sister chromatid exchanges, chromosomal aberrations, and gene mutations: new insights gained from studies of genetically engineered mammalian cell lines. Environ Mol Mutagen 22:283–292
Kaina B, Fritz G, Eschabch E, Preuss I (1995) Transfer of genes affecting resistance of mammalian cells to alkylating drugs – implication for cancer chemotherapy. In: Zeller WJ, Eisenbrand G, Hellmann K (eds) Reduction of anticancer drug toxicity. Pharmacological, biologic and gene therapeutic approaches. Karger, Basel, pp 125–141
Kaina B, Ziouta A, Ochs K, Coquerelle T (1997) Chromosomal instability, reproductive cell death and apoptosis induced by O6-methylguanine in Mex–, Mex+ and methylation-tolerant mismatch repair compromised cells: facts and models. Mutat Res 381:227–241
Kamataki T, Nunoya K, Sakai Y, Kushida H, Fujita K (1999) Genetic polymorphism of CYP2A6 in relation to cancer. Mutat Res 428:125–130
Karran P, Bignami M (1994) DNA damage tolerance, mismatch repair and genome instability. BioEssays 16:833–839
Klungland A, Lindahl T (1997) Second pathway for completion of human DNA base excision-repair: reconstitution with purified proteins and requirement for DNaseIV (FEN1). EMBO J 16:3341–3348
Kurowski MA, Bhagwat AS, Papaj G, Bujnicki JM (2003) Phylogenomic identification of five new human homologs of the DNA repair enzyme AlkB. BMC Genomics 4:48
Lai DY, Arcos JC (1980) Minireview: dialkylnitrosamine bioactivation and carcinogenesis. Life Sci 27:2149–2165
Leach FS, Nicolaides NC, Papadopoulos N, Liu B, Jen J, Parsons R, Peltomaki P, Sistonen P, Aaltonen LA, Nystrom-Lahti M et al (1993) Mutations of a mutS homolog in hereditary nonpolyposis colorectal cancer. Cell 75:1215–1225
Lees NP, Harrison KL, Hall CN, Margison GP, Povey AC (2004) Reduced MGMT activity in human colorectal adenomas is associated with K-ras GC→AT transition mutations in a population exposed to methylating agents. Carcinogenesis 25:1243–1247
Lehman AR (2002) Replication of damaged DNA in mammalian cells: new solutions to an old problem. Mutat Res 509:23–34
Li GM, Modrich P (1995) Restoration of mismatch repair to nuclear extracts of H6 colorectal tumor cells by a heterodimer of human MutL homologs. Proc Natl Acad Sci USA 92:1950–1954
Li GM, Wang H, Romano LJ (1996) Human MutSalpha specifically binds to DNA containing aminofluorene and acetylaminofluorene adducts. J Biol Chem 271:24084–24088
Lin HL, Parsels LA, Maybaum J, Hollenberg PF (1999) N-Nitrosodimethylamine-mediated cytotoxicity in a cell line expressing P450 2E1: evidence for apoptotic cell death. Toxicol Appl Pharmacol 157:117–124
Longley MJ, Pierce AJ, Modrich P (1997) DNA polymerase delta is required for human mismatch repair in vitro. J Biol Chem 272:10917–10921
Margison GP, Povey AC, Kaina B, Santibanez Koref MF (2003) Variability and regulation of O6-alkylguanine-DNA alkyltransferase. Carcinogenesis 24:625–635
Marnett LJ, Burcham PC (1993) Endogenous DNA adducts: potential and paradox. Chem Res Toxicol 6:771–785
Matsumoto Y, Kim K (1995) Excision of deoxyribose phosphate residues by DNA polymerase beta during DNA repair. Science 269:699–702

Matsumoto Y, Kim K, Hurwitz J, Gary R, Levin DS, Tomkinson AE, Park MS (1999) Reconstitution of proliferating cell nuclear antigen-dependent repair of apurinic/apyrimidinic sites with purified human proteins. J Biol Chem 274:33703–33708
Meikrantz W, Bergom MA, Memisoglu A, Samson L (1998) O6-Alkylguanine DNA lesions trigger apoptosis. Carcinogenesis 19:369–372
Mitacek EJ, Brunnemann KD, Hoffmann D, Limsila T, Suttajit M, Martin N, Caplan LS (1999) Volatile nitrosamines and tobacco-specific nitrosamines in the smoke of Thai cigarettes: a risk factor for lung cancer and a suspected risk factor for liver cancer in Thailand. Carcinogenesis 20:133–137
Modrich P, Lahue R (1996) Mismatch repair in replication fidelity, genetic recombination, and cancer biology. Annu Rev Biochem 65:101–133
Nakamura J, La DK, Swenberg JA (2000) 5′-nicked apurinic/apyrimidinic sites are resistant to beta-elimination by beta-polymerase and are persistent in human cultured cells after oxidative stress. J Biol Chem 275:5323–5328
Nakatsuru Y, Matsukuma S, Nemoto N, Sugano H, Sekiguchi M, Ishikawa T (1993) O^6-Methylguanine-DNA methyltransferase protects against nitrosamine-induced hepatocarcinogenesis. Proc Natl Acd Sci USA 90:6468–6472
Newlands ES, Stevens MF, Wedge SR, Wheelhouse RT, Brock C (1997) Temozolomide: a review of its discovery, chemical properties, pre-clinical development and clinical trials. Cancer Treat Rev 23:35–61
Nicolaides NC, Papadopoulos N, Liu B, Wei YF, Carter KC, Ruben SM, Rosen CA, Haseltine WA, Fleischmann RD, Fraser CM et al (1994) Mutations of two PMS homologues in hereditary nonpolyposis colon cancer. Nature 371:75–80
O'Connor TR (1993) Purification and characterization of human 3-methyladenine-DNA glycosylase. Nucleic Acids Res 21:5561–5569
Ochs K, Kaina, B. (2000) Apoptosis induced by DNA damage O6-methylguanine is Bcl-2 and caspase-9/3 regulated and Fas/caspase-8 independent. Cancer Res 60:5815–5824
Ostermann S, Csajka C, Buclin T, Leyvraz S, Lejeune F, Decosterd LA, Stupp R (2004) Plasma and cerebrospinal fluid population pharmacokinetics of temozolomide in malignant glioma patients. Clin Cancer Res 10:3728–3736
Palombo F, Gallinari P, Iaccarino I, Lettieri T, Hughes M, D'Arrigo A, Truong O, Hsuan JJ, Jiricny J (1995) GTBP, a 160-kilodalton protein essential for mismatch-binding activity in human cells. Science 268:1912–1914
Papadopoulos N, Nicolaides NC, Wei YF, Ruben SM, Carter KC, Rosen CA, Haseltine WA, Fleischmann RD, Fraser CM, Adams MD et al (1994) Mutation of a mutL homolog in hereditary colon cancer. Science 263:1625–1629
Pegg AE, Dolan ME, Moschel RC (1995) Structure, function, and inhibition of O6-alkylguanine-DNA alkyltransferase. Prog Nucleic Acid Res Mol Biol 51:167–223
Pepponi R, Marra G, Fuggetta MP, Falcinelli S, Pagani E, Bonmassar E, Jiricny J, D'Atri S (2003) The effect of O6-alkylguanine-DNA alkyltransferase and mismatch repair activities on the sensitivity of human melanoma cells to temozolomide, 1,3-bis(2-chloroethyl)1-nitrosourea, and cisplatin. J Pharmacol Exp Ther 304:661–668
Peto R, Gray R, Brantom P, Grasso P (1984) Nitrosamine carcinogenesis in 5120 rodents: chronic administration of sixteen different concentrations of NDEA, NDMA, NPYR and NPIP in the water of 4440 inbred rats, with parallel studies on NDEA alone of the effect of age of starting (3, 6 or 20 weeks) and of species (rats, mice or hamsters). IARC Sci Publ 627–665
Prasad R, Beard WA, Strauss PR, Wilson SH (1998) Human DNA polymerase beta deoxyribose phosphate lyase. Substrate specificity and catalytic mechanism. J Biol Chem 273:15263–15270
Qian X, von Wronski MA, Brent TP (1995) Localization of methylation sites in the human O^6-methylguanine-DNA methyltransferase promoter: correlation with gene suppression. Cacinogenesis 16:1385–1390
Qin X, Zhou H, Liu L, Gerson SL (1999) Transgenic expression of human MGMT blocks the hypersensitivity of PMS2-deficient mice to low dose MNU thymic lymphomagenesis. Carcinogenesis 20:1667–1673

Rasouli-Nia A, Sibghat-Ullah, Mirzayans R, Paterson MC, Day RS III (1994) On the quantitative relationship between O6-methylguanine residues in genomic DNA and production of sister-chromatid exchanges, mutations and lethal events in a Mer– human tumor cell line. Mutat Res 314:99–113

Reid JM, Kuffel MJ, Miller JK, Rios R, Ames MM (1999) Metabolic activation of dacarbazine by human cytochromes P450: the role of CYP1A1, CYP1A2, and CYP2E1. Clin Cancer Res 5:2192–2197

Ringvoll J, Nordstrand LM, Vagbo CB, Talstad V, Reite K, Aas PA, Lauritzen KH, Liabakk NB, Bjork A, Doughty RW et al (2006) Repair deficient mice reveal mABH2 as the primary oxidative demethylase for repairing 1meA and 3meC lesions in DNA. EMBO J 25:2189–2198

Ro J, North SM, Gallick GE, Hortobagyi GN, Gutterman JU, Blick M (1988) Amplified and overexpressed epidermal growth factor receptor gene in uncultured primary human breast carcinoma. Cancer Res 48:161–164

Roos W, Baumgartner M, Kaina B (2004) Apoptosis triggered by DNA damage O6-methylguanine in human lymphocytes requires DNA replication and is mediated by p53 and Fas/CD95/-Apo-1. Oncogene 23:359–367

Roos WP, Batista LF, Naumann SC, Wick W, Weller M, Menck CF, Kaina B (2007) Apoptosis in malignant glioma cells triggered by the temozolomide-induced DNA lesion O(6)-methylguanine. Oncogene 26:186–197

Sedgwick B, Lindahl T (2002) Recent progress on the Ada response for inducible repair of DNA alkylation damage. Oncogene 21:8886–8894

Sibghat-Ullah L, Day RS III, Dobler KD, Paterson MC (2001) Initiation of strand incision at G:T and O6-methylguanine:T base mismatches in DNA by human cell extracts. Nucleic Acids Res 29:2409–2417

Skibba JL, Beal DD, Ramirez G, Bryan GT (1970) N-demethylation the antineoplastic agent4(5)-(3,3-dimethyl-1-triazeno)imidazole-5(4)-carboxamide by rats and man. Cancer Res 30:147–150

Sklar RM, Strauss B (1981) Removal of O^6-methylguanine from DNA of normal and xeroderma pigmentosum-derived lymphoblastoid lines. Nature 289:417–420

Sobol RW, Wilson SH (2001) Mammalian DNA beta-polymerase in base excision repair of alkylation damage. Prog Nucleic Acid Res Mol Biol 68:57–74

Sobol RW, Prasad R, Evenski A, Baker A, Yang XP, Horton JK, Wilson SH (2000) The lyase activity of the DNA repair protein beta-polymerase protects from DNA-damage-induced cytotoxicity. Nature 405:807–810

Stojic L, Mojas N, Cejka P, Di Pietro M, Ferrari S, Marra G, Jiricny J (2004) Mismatch repair-dependent G2 checkpoint induced by low doses of SN1 type methylating agents requires the ATR kinase. Genes Dev 18:1331–1344

Stucki M, Pascucci B, Parlanti E, Fortini P, Wilson SH, Hubscher U, Dogliotti E (1998) Mammalian base excision repair by DNA polymerases delta and epsilon. Oncogene 17:835–843

Tominaga Y, Tsuzuki T, Shiraishi A, Kawate H, Sekiguchi M (1997) Alkylation-induced apoptosis of embryonic stem cells in which the gene for DNA-repair, methyltransferase, had been disrupted by gene targeting. Carcinogenesis 18:889–896

Toorchen D, Topal MD (1983) Mechanisms of chemical mutagenesis and carcinogenesis: effects on DNA replication of methylation at the O6-guanine position of dGTP. Carcinogenesis 4:1591–1597

Toyooka S, Toyooka KO, Maruyama R, Virmani AK, Girard L, Miyajima K, Harada K, Ariyoshi Y, Takahashi T. Sugio K et al (2001) DNA methylation profiles of lung tumors. Mol Cancer Ther 1:61–67

Trewick SC, Henshaw TF, Hausinger RP, Lindahl T, Sedgwick B (2002) Oxidative demethylation by Escherichia coli AlkB directly reverts DNA base damage. Nature 419:174–178

Tsuzuki T, Sakumi K, Shiraishi A, Kawate H, Igarashi H, Iwakuma T, Tominaga Y, Zhang S, Shimizu S, Ishikawa T et al. (1996) Targeted disruption of the DNA repair methyltransferase gene renders mice hypersensitive to alkylating agent. Carcinogenesis 17:1215–1220

Umar A, Kunkel TA (1996) DNA-replication fidelity, mismatch repair and genome instability in cancer cells. Eur J Biochem 238:297–307 van Zeeland AA, Mohn GR, Aaron CS, Glickman BW, Brendel M, de Serres FJ, Hung CY, Brockman HE (1983) Molecular dosimetry of the chemical mutagen ethyl methanesulfonate. Quantitative comparison of the mutagenic potency in Neurospora crassa and Saccharomyces cerevisiae. Mutat Res 119:45–54

Vogel E, Natarajan AT (1979) The relation between reaction kinetics and mutagenic action of mono-functional alkylating agents in higher eukaryotic systems. I. Recessive lethal mutations and translocations in Drosophila. Mutat Res 62:51–100

Walker R (1990) Nitrates, nitrites and N-nitrosocompounds: a review of the occurrence in food and diet and the toxicological implications. Food Addit Contam 7:717-768

Wang Y, Qin J (2003) MSH2 and ATR form a signaling module and regulate two branches of the damage response to DNA methylation. Proc Natl Acad Sci USA 100:15387–15392

Wang Z, Gleichmann H (1998) GLUT2 in pancreatic islets: crucial target molecule in diabetes induced with multiple low doses of streptozotocin in mice. Diabetes 47:50–56

Wilson DM, Barsky D (2001) The major human abasic endonuclease: formation, consequences and repair of abasic lesions in DNA. Mutat Res 485:283–307

Wogan GN, Tannenbaum SR (1975) Environmental N-nitroso compounds: implications for public health. Toxicol Appl Pharmacol 31:375–383

Wojcik A, Bruckmann E, Obe G (2004) Insights into the mechanisms of sister chromatid exchange formation. Cytogent Genome Res 104:301–309

Xu YH, Richert N, Ito S, Merlino GT, Pastan I (1984) Characterization of epidermal growth factor receptor gene expression in malignant and normal human cell lines. Proc Natl Acad Sci USA 81:7308–7312

Yarosh DB, Foote RS, Mitra S, Day RS 3rd (1983) Repair of O6-methylguanine in DNA by demethylation is lacking in Mer– human tumor cell strains. Carcinogenesis 4:199–205

Yoshioka K, Yoshioka Y, Hsieh P (2006) ATR kinase activation mediated by MutSalpha and MutLalpha in response to cytotoxic O6-methylguanine adducts. Mol Cell 22:501–510

Zhou ZQ, Manguino D, Kewitt K, Intano GW, McMahan CA, Herbert DC, Hanes M, Reddick R, Ikeno Y, Walter CA (2001) Spontaneous hepatocellular carcinoma is reduced in transgenic mice overexpressing human O6-methylguanine-DNA methyltransferase. Proc Natl Acad Sci USA 98:12566–12571

8 Adducts, Sister-Chromatid Exchanges and Mutations Following Benzo[*a*]pyrene Exposure: a Review of Quantitative Data Followed by Some Considerations Regarding Risk

RINNE DE BONT AND NIK VAN LAREBEKE

Abstract

We conducted a literature study (papers published until September 2004) concerning quantitative aspects of the formation of DNA adducts and the induction of sister-chromatid exchanges (SCE) and mutations following benzo[*a*]pyrene (BaP) exposure. Mean adduct levels per dose unit after exposure in vitro were higher in both human (11 studies) and animal (five studies) cells for doses below 5 μM (2.47 and 1.445 adducts per 10^6 nucleotides per μM, respectively) than for doses of 5 μM and above (0.104 and 0.106 adducts per 10^6 nucleotides per μM). In vivo in humans BaP adducts (0.054–0.918 adducts per 10^6 nucleotides per μM) occur in much lower concentrations than endogenous adducts. The total quantity of nine endogenous adducts in human cells in vivo amounts to between 4.97 and 69.05 adducts per 10^6 nucleotides. Mean SCE levels per dose unit after exposure in vitro were higher in both human and animal cells for doses below 5 μM BaP (1.48 and 13.32 SCEs per cell per μM, respectively) than for doses of 5 μM and above (0.18 and 0.97 SCEs per cell per μM). In human cells in vitro (three studies, all below 5 μM) a non-significant downward trend in mutation frequency per dose unit with increasing dose was reported, and in Chinese hamster ovary cells doses below 5 μM BaP resulted in a higher mean mutation frequency per micromole per litre (in the *HGPRT* or the *XPRT* gene) than doses of 5 μM and above (4.30 vs. 0.88 per 10^5 cells per micromole per litre). Calculations on cancer risk associated with a certain level of BaP adducts and on the relation between environmental exposures and a threshold level for induction of DNA repair are presented.

8.1 Introduction

The potent carcinogen benzo[*a*]pyrene (BaP), the most extensively studied of the polycyclic aromatic hydrocarbons (PAHs), is present in car exhaust fumes, coke oven emissions, cigarette smoke, environmental air and a variety of foods. Metabolic activation of BaP proceeds by two major mechanisms, monooxygenation to yield bay-region diolepoxides, the most important of which is BaP-7,8-dihydrodiol-9,10-epoxide (BPDE), giving rise to stable DNA

Günter Obe and Vijayalaxmi (Eds.)
Chromosomal Alterations: Methods, Results and Importance in Human Health

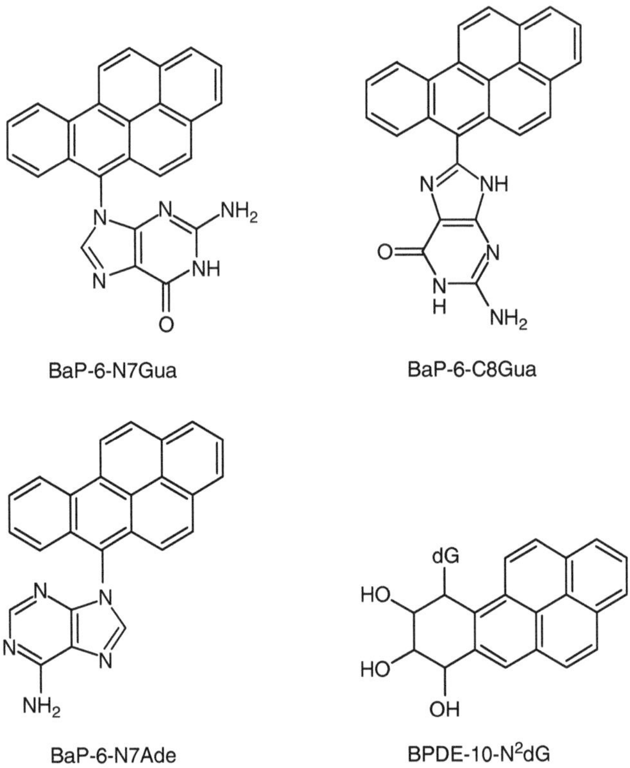

Fig. 8.1 Structures of some benzo[*a*]pyrene (*BaP*) adducts

adducts, and one-electron oxidation with formation of radical cations leading to mostly unstable depurinating DNA adducts. The most important stable DNA adduct is the BPDE adduct to deoxyguanosine, BPDE-10-N2dG, but other adducts, including BPDE-N6dA and, to a lesser extent, BPDE-N4dC are also formed. The major depurinating adducts following the one-electron pathway are BaP-6-N7dG, BaP-6-C8dG and BaP-6-N7dA (Banasiewicz et al. 2004). Figure 8.1 shows the structure of some BaP DNA adducts.

To investigate the dose–effect relationship following exposure to BaP, we conducted a literature study concerning quantitative aspects of the formation of DNA adducts and the induction of sister-chromatid exchanges (SCEs) and mutations.

8.2 Methodological Approach

We sought all potentially informative peer-reviewed publications reporting studies on DNA adducts, SCEs and mutations following BaP exposure. We included in vitro studies and in vivo human observations. Using Medline, we searched for articles published until September 2004. Studies that used dose units not comparable and not transformable to the dose units we used in our analysis were excluded.

For the sake of comparison, DNA adduct levels are expressed in a standardized way, including the number of adducts per 10^6 nucleotides.

Possible confounding factors that complicate the comparison of different studies are the method of adduct measurement, the cell or tissue types investigated, the use of external metabolic activation and the duration of exposure. To correct for these factors we performed an ANOVA analysis using the statistical programme Statview.

Published data suggest that BPDE adducts constitute the vast majority of BaP adducts. From Shugart (1985) and Li et al. (1996) we calculated that BPDE-dG adducts constituted about 71.1% of all the BPDE adducts and from several other studies (Ashurst and Cohen 1980, 1981; DiGiovanni et al. 1985; Bodell et al. 1989; Ross et al. 1990; Van Schooten et al. 1991; Rogan et al. 1993; Devanesan et al. 1992; Roggeband et al. 1994a, b; Chen et al. 1996) that BPDE-dG adducts constitute about 77.2 % of all BaP adducts, so it is probable that BPDE adducts comprise almost all BaP adducts For further calculations, we assumed that BPDE adducts represent 100% of all BaP adducts.

We list quantitative in vivo data concerning BaP adducts in humans and compare these data with levels of endogenous DNA adducts (De Bont and van Larebeke 2004).

8.3 Results

8.3.1 Adducts

8.3.1.1 BaP Exposure in Vitro: Dose–Effect Relationship for Adduct Formation

The entire set of adduct levels measured in human cells (Table 8.1; 11 studies totalling 23 measured adduct levels) were plotted against the exposure dose (Fig. 8.2). The data were divided into adducts following low-dose exposure (below 5 μM BaP) and adducts following exposure to BaP doses of 5 μM and above. BaP. Linear regression shows an R^2 value of 0.579 at doses below 5 μM and an R^2 value of 0.719 at doses of 5 μM and above. In linear regression the slope of the curve for doses below 5 μM (1.676 with significance of the t value of 0.002) is more than 10 times higher than that for doses of 5 μM and above (0.105 with a significance of the t value of 0.004). We also calculated the mean adduct level per dose unit to be higher for doses below 5 μM (2.47 adducts per 10^6 nucleotides per μM) than for doses of 5 μM and above (0.104 adducts per 10^6 nucleotides per μM). Possible confounding factors that could complicate the comparison of different studies are the method of adduct measurement

Table 8.1 Benzo[*a*]pyrene (BaP) adducts following exposure of human cells to BaP in vitro

Method	Cell or tissue type	Dose (μM)	Adducts/10^6 nucleotides	Reference
HPLC	Monocytes	0.30	0.680	Rüdiger et al. (1985)
32P	Peripheral lymphocytes	8.00	0.140	Canova et al. (1996)
		20.00	0.200	
Technique based on radioactivity	Skin fibroblasts	0.25	2.260	Rüdiger et al. (1979)
32P	Lung fibroblasts	0.01	0.018	Binková et al. (2000)
		1.00	2.370	
HPLC	Mammary epithelial cells	2.00	5.440	Moore et al. (1987)
32P	White blood cells	10.00	2.200	Roggeband et al. (1994a)
		25.00	4.200	
		50.00	9.300	
		100.00	11.800	
HPLC	Epidermal cells	0.40	1.500	Theall et al. (1981)
		4.00	6.000	
HPLC	Bladder bronchus	1.00	6.400	Stoner et al. (1982)
		1.00	3.100	
32P	White blood cells	100.00	9.300	Roggeband et al. (1993)
32P	Lymphocytes	0.01	0.004	Wiencke et al. (1990)
		0.05	0.013	
		0.10	0.027	
		0.50	0.199	
		1.00	0.317	
		5.00	0.556	
32P	Fibroblasts	50.00	0.675	Hanelt et al. (1997)

HPLC high performance liquid chromatography, *32P* ^{32}P-postlabelling

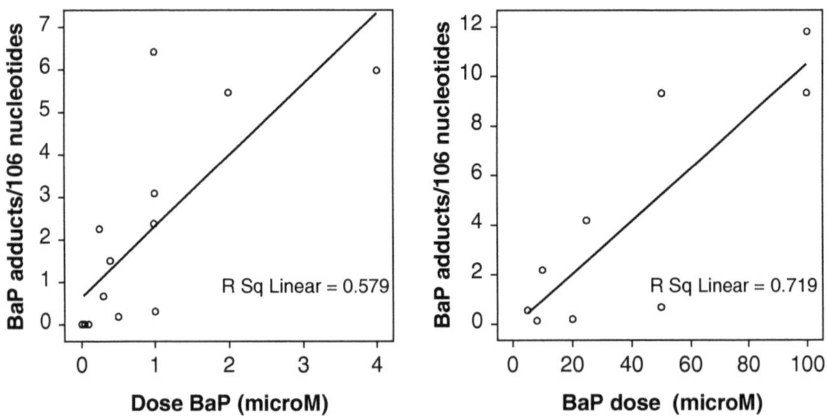

Fig. 8.2 BaP adducts in human tissues following BaP exposure in vitro: doses below 5 μM (*left*); doses of 5 μM and above (*right*)

(high-performance liquid chromatography or ^{32}P-postlabelling), the cell or tissue types investigated (blood, lung, skin, bladder, mammary epithelial tissue), the use of external metabolic activation (S9) and the duration of exposure (2–48 h). To correct for these factors we performed an ANOVA analysis after logarithmic transformation of the adducts-per-dose values using the statistical program Statview. The difference in the number of adducts per dose between the high-exposure group and the low-exposure group was statistically significant ($p = 0.0054$).

In vitro tests on animal cells (six studies totalling 11 measured adduct levels) showed that doses from 1 to 55 μM gave rise to adduct levels ranging from 0.1 to 11 adducts per 10^6 nucleotides (Table 8.2). The mean adduct level per dose unit was higher for doses below 5 μM (1.445 adducts per 10^6 nucleotides per μM) than for higher doses (0.10^6 adducts per 10^6 nucleotides) but there were too few data to test the significance of the difference in an ANOVA test taking into account the confounding factors.

8.3.1.2 *Adducts in Vivo*

Data on adduct levels after in vivo treatment of animals with BaP (seven mouse studies with 16 measured adduct levels and one rat study) are shown in Table 8.3. Owing to the limited number of data, it was not possible to compare the number of adducts per dose between high doses and low doses in an ANOVA analysis integrating all covariates.

Table 8.2 BaP adducts following exposure of animal cells to BaP in vitro

Method	Cell or tissue type	Dose (μM)	Adducts/10^6 nucleotides	Reference
32P	Mouse foetal liver	8.0	1.44	Canova et al. (1996)
		20.0	2.40	
		40.0	4.22	
HPLC	Rat dermal fibroblasts	1.2	3.15	Alexandrov et al. (1988)
	Rabbit dermal fibroblasts	1.2	1.68	
HPLC	Rat mammary epithelial cells	2.0	3.31	Moore et al. (1987)
HPLC	Rat hepatocytes	55.0	1.02	Ashurst et al. (1980)
		55.0	1.71	
Sephadex LH 20 chromatography	Rat lung	1.0	0.10	Deckers-Schmelzle et al. (1978)
32P	Rat trachea	19.8	3.70	Roggeband et al. (1994b)
	Hamster trachea	19.8	11.0	

Table 8.3 Animals: BaP adducts following BaP exposure in vivo

Method	Cell or tissue type	Dose (mg/kg)	Adducts/10^6 nucleotides	Reference
32P	Rat lung	10	0.031	Godschalk et al. (1997)
	Rat liver	10	0.029	
	Rat hart	10	0.063	
32P	Mouse skin	0.63[a]	0.125	Talaska et al. (1996)
ELISA	Mouse skin	1.6[a]	0.230	Albert et al. (1996)
		3.2[a]	0.393	
		6.4[a]	0.525	
[1,3-^{3}H]BPDE radiolabelling	Mouse liver	25[a]	0.864	Divi et al. (2002)
		25[a]	1.376	
32P		50[a]	0.114	
		50[a]	0.221	
DELFIA		25[a]	0.043	
		50[a]	0.079	
CIA		25[a]	0.110	
		50[a]	0.170	
HPLC	Mouse skin	10[a]	1.75	Bartley et al. (1982)
32P	Mouse skin	2.52[a]	1.07	Bodell et al. (1989)
32P	Mouse skin	2.52[a]	3.56	Rogan et al. (1993)
HPLC	Mouse skin	2.52[a]	3.15	DiGiovanni et al. (1985)

BDPE BaP-7,8-dihydrodiol-9,10-epoxide, *ELISA* enzyme-linked immunosorbent assay, *DELFIA* dissociation-enhanced lanthanide fluoroimmunoassay, *CIA* Chemiluminescence immunoassay
[a]Dose units transformed into milligrams per kilograms using a mean weight of mice of 20g

8.3.1.3 BPDE Adducts Following BPDE Exposure

We also considered five studies that quantify BPDE-dG adducts or total BPDE adducts following BPDE exposure (data not shown). On the basis of all available studies on BPDE adduct formation after in vitro exposure to BPDE or BaP, we calculated, by dividing the number of adducts by the dose, that 1 μM of BPDE forms the same number of BPDE adducts as 18.58 μM BaP. The number of adducts formed per microgram of BPDE at BPDE exposures equivalent to BaP exposures below 5 μM was lower than at BPDE exposures equivalent to BaP exposures of 5 μM and above. but taking the method of adduct measurement into account this difference was not significant

8.3.1.4 Human BaP Adducts in Vivo: Comparison with Endogenous Adducts

Table 8.4 provides quantitative in vivo data concerning BaP adducts in humans. BaP and BPDE adducts (0.054–0.918 and 0.013–0.787 adducts per 10^6

Table 8.4 BaP adducts in humans in vivo

Method	Cell or tissue type	Exposure type	Adducts/10^6 nucleotides	Reference
BaP adducts				
HPLC	Monocytes	Cancer patients	0.918	Nowak et al. (1992)
		Controls	0.698	
32P	Leukocytes	Controls: non-smokers	0.0054	Arnould et al. (1999)
		Smokers	0.0177	
		Factory workers	0.0057–0.389	
BPDE adducts				
IAC	Placenta		0.100	Manchester et al. (1988)
USERIA	Lymphocytes	Coke oven workers	0.558	Haugen et al. (1986)
USERIA	White blood cells	Factory workers	0.013–0.787	Shamsuddin et al. (1985)
ELISA	White blood cells	Factory workers	0.051	Van Schooten et al. (1990)
		Controls	0.027	
ELISA	Placenta	Smokers	0.066–0.393	Arnould et al. (1997)
		Non-smokers	0.066–0.087	
	Umbilical cord blood	Smokers	0.066–0.145	
		Non-smokers	0.066	
HPLC (1991)	Lung tissue		0.010–0.400	Weston and Bowman
HPLC	Lymphocytes	Coke oven workers	0.505	Weston et al. (1989)
32P	White blood cells	Controls	0.098–0.300	Rojas et al. (1994)
HPLC	Lymphocyte-monocyte fraction	Coke oven workers	0.044	Pavanello et al. (2005)
HPLC	Lung tissue	Cancer patients	0.006–0.099	Alexandrov et al. (1992)
HPLC	Lymphocyte-monocyte fraction	Controls	0.039	Pavanello et al. (1999)
		Anode aluminium workers	0.115	
		Chimney sweeps	0.055	
		Coke-oven workers	0.195	
		Psoriatic patients	0.023	

IAC immunoaffinity chromatography, *USERIA* ultrasensitive enzymatic radioimmunoassay

nucleotides, respectively) occur in much lower concentrations than several mainly endogenous adducts. DNA damage produced by reactive oxygen species (ROS) is the most frequently occurring damage. ROS are produced in part through exposure to certain agents, but are produced mainly through endogenous processes, including cellular respiration, biosynthetic and biodegrading processes of normal intermediary metabolism and activation of phagocytotic cells. Among the oxidatively modified bases, 8-oxo-dG is apparently the most abundant. Steady-state levels of this adduct range from 0.24 to 59.70 adducts per10^6 nucleotides in white blood cells (De Bont and van Larebeke 2004). Other abundant oxidatively modified bases comprise the formamido pyrimidine adducts of adenine and guanine and 5-hydroxyuracil and uracil glycol. Lipid peroxidation leads to propano adducts (up to 0.40 adducts per 10^6 nucleotides in humans), etheno adducts, including N^2-3-ethenodeoxyguanosine, whose levels were similar to those of propano adducts, and malondialdehyde adducts, the most abundant of which is the adduct to dG (M1G; pyrimido[1,2-α]purin-10(3*H*)-one) (up to 0.9 adducts per10^6 nucleotides in humans) (De Bont and van Larebeke 2004). Base propenals, arising from oxidative degradation of DNA, might even lead to more M1G adduct formation than malondialdehyde derived from lipid peroxidation (Dedon et al. 1998). Several endogenous oestrogen metabolites, particularly the catechol oestrogens, can be oxidized to quinones that lead to several types of DNA adducts (Liehr 2000). Endogenous alkylating agents can lead to alkyl adducts, the most frequent being 7-methylguanine (up to 1.35 adducts per 10^6 nucleotides in lymphocytes of non-smokers) (De Bont and van Larebeke 2004). Lipid peroxidation and glycation can lead to reactive carbonyl species such as glyoxal and methylglyoxal (Roberts et al. 2003). To get an idea of the total number of endogenous DNA adducts in white blood cells we considered the quantitative data regarding endogenous DNA adducts in white blood cells as summarized by De Bont and van Larebeke (2004). Only the following endogenous adducts were included: 5-hydroxy-2′-deoxycytidine, 5-hydroxy-2′-deoxyuridine,5,6-dihydroxy-5, 6-dihydro-2′-deoxyuridine, 8-oxo-7,8-dihydro-2′-deoxyguanosine, 7-methylguanine, O^6-methylguanine, 7-(2-hydroxyethyl)guanine, 7-alkylguanine and 7-ethyldeoxyguanosine. The total quantity of these endogenous adducts in human cells in vivo amounts to between 4.97 and 69.05 adducts per 10^6 nucleotides. The minimum level (4.97) was reached by adding up the lowest level observed for each of the endogenous adduct types included, while the maximum level (69.05) was based on the highest observed level.

8.3.2 Sister-Chromatid Exchanges

We found eight studies investigating SCEs following BaP exposure in vitro: five human studies and three animal studies (24 measurements) (Table 8.5). Seven of the eight studies determined more than two SCE levels and revealed a supralinear dose–response curve. Our analysis of the eight studies showed that, for humans, BaP doses below 5 μM induce a higher mean number of SCEs per micromole per litre of BaP than doses of 5 μM and above (1.48 vs. 0.18 SCEs per cell). The same can be concluded for animal studies (13.32 vs. 0.97 SCEs per cell). For the entire dataset this difference was statistically significant when corrected for species and the use of metabolic stimulation (after logarithmic transformation of the number of adducts per dose) ($p = 0.046$).

Table 8.5 Sister-chromatid exchanges (SCEs) following BaP exposure in human and animal cells in vitro

Cell or tissue type	Dose (μM)	SCEs/cell[a]	Reference
Humans			
Lymphocytes	50	3.40	Hartmann and Speit (1995)
	100	4.90	
	500	6.30	
Lymphocytes	1.00	0.08	Inoue et al. (1983)
	10.00	2.32	
	100.00	5.92	
Lymphocytes	39.60	3.44	Salama et al. (2001)
	198.00	3.94	
Lymphocytes	0.80	1.60	Jacono et al. (1992)
	20.00	8.80	
	40.00	12.20	
Lymphocytes	1.00	3.23	Dosaka et al. (1987)
	10.00	5.97	
	100.00	9.50	
Animals			
Rat liver epithelial cells	50	8.00	Cunningham and Ringrose (1983)
	200	23.50	
	400	27.70	
	2,000	35.60	
Chinese hamster ovary cells	0.25	7.37	Ray-Chaudhuri et al. (1980)
	2.50	13.67	
	25.00	14.07	
Rat liver epithelial cells	1.00	5.00	Tong et al. (1981)
	10.00	48.60	
	100.00	103.30	

[a]After subtraction of blank control value

8.3.3 Mutations

8.3.3.1 Mutations in Vivo in Animals Following BaP Exposure

We found only three in vivo mutation studies following BaP exposure and comparison between high and low doses was not possible (Table 8.6). Several other in vivo studies used dose units not comparable to the dose unit we used in our analysis (micromoles per litre) (Dean et al. 1998; Brooks et al. 1999). Brooks et al. (1999) showed a clear linear relationship between the dose and the mutation frequency. Hakura et al. (1998) measured the mutation frequency in 11 different tissues and showed a high variation in mutation frequencies (from 1.7×10^5 in brain to 267.6×10^5 in colon).

Table 8.6 Mutation frequency following BaP exposure in vivo

Cell or tissue type	Gene or locus	Dose (mg/kg)	Mutation frequency[a] ($\times 10^{-5}$)/cell	Reference
Mouse colon	*LacZ*	625	267.60	Hakura et al. (1998)
Mouse ilium		625	176.40	
Mouse forestomach		625	137.00	
Mouse bone marrow		625	87.00	
Mouse spleen		625	80.70	
Mouse glandular stomach		625	49.90	
Mouse liver		625	17.90	
Mouse lung		625	16.60	
Mouse kidney		625	11.70	
Mouse heart		625	8.90	
Mouse brain		625	1.70	
Mouse liver	*LacI*	120	6.40	Shane et al. (2000)
Mouse splenic T cells	*Hprt*	50	0.09	Monroe et al. (1998)
		150	0.91	
	LacI	50	4.40	
		150	13.80	
	CII	150	1.90	
	CI	50	1.70	
		150	0.50	

[a]After subtraction of blank control value

8.3.3.2 Mutations in Vitro Following BaP Exposure

We found three human studies (seven measuring points) and three animal studies (20 measuring points) measuring the mutation frequency in cells in vitro (Table 8.7). In human cells all experimental exposures were below 5 μM, demonstrating that mutations are already formed at low doses. The mutation frequency per dose unit showed a non-significant downward trend with increasing dose and the same trend was observed when only studies without metabolic activation were included. In Chinese hamster ovary cells doses below 5 μM BaP resulted in a higher mean mutation frequency per micromole per litre (in the *HGPRT* or the *XPRT* gene) than doses of 5 μM and above (4.30 vs. 0.88 per 10^5 cells per micromole per litre).

8.4 Discussion

PAHs present in ambient air, in tobacco smoke and in food contribute to the induction of cancer in the lung and in several other organs and to the induction of cardiovascular and airway disease (Ramos and Moorthy 2005;

Table 8.7 Mutation frequency following BaP exposure in vitro

Cell or tissue type	Gene or locus	Dose (μM)	Mutation frequency[a] ($\times 10^{-5}$)/cell	Reference
Humans				
Lymphoblastoid cells	*HGPRT*[b]	0.020	0.08	Danheiser et al. (1989)
		0.100	0.21	
		0.500	0.28	
		1.000	0.38	
Fibroblast without BER for BaP adducts	6-Thioguanine resistance[c]	0.100	6.50	Aust et al. (1980)
		0.200	9.90	
Epidermal keratinocytes	*HGPRT*[b]	3.960	3.08	Allen-Hoffmann and Rheinwald (1984)
Animals				
Chinese hamster ovary cells	*HGPRT*[b]	0.396	1.45	Machanoff et al. (1981)
		0.990	7.83	
		1.980	12.99	
		3.960	12.15	
		7.920	13.44	
		15.840	7.08	
Chinese hamster ovary cells	*XPRT*[d]	0.396	2.80	Oberly et al. (1993)
		1.980	2.80	
		3.960	2.10	
		7.920	6.20	
		15.840	13.50	
		31.680	18.90	
Chinese hamster V79 cells	Azaguanine resistance	0.100	5.00	Huberman et al. (1976)
		0.400	45.00	
		1.200	70.00	
		4.000	312.00	
	Ouabian resistance	0.200	0.30	
		0.400	6.80	
		1.200	18.80	
		4.00	106.80	

[a]After subtraction of blank control value
[b]Hypoxanthine-guanine phosphoribosyl transferase gene
[c]Mutations inactivating the *HGPRT* gene lead to resistance to 6-thioguanine
[d]Xanthine-guanine phosphoribosyl transferase gene

Miller et al. 2004). There are many different PAHs and each parental PAH can, for example through atmospheric reactions, give rise to several different derivatives. DNA adducts of BaP, the guide substance most often used to assess exposure to PAHs, have been found in human peripheral blood cells and in human tissues on most occasions when appropriate techniques have been applied. All humans are probably exposed to PAHs.

8.4.1 Cancer Risk Associated with a Certain Level of BaP Adducts

The data on the occurrence of BaP adducts and on the cancer incidence among workers in coke ovens and in other industries with exposure to PAHs can be used to estimate the cancer risk associated with a certain level of BaP adducts.

At least two considerations limit the meaning of such estimations. First, internal exposure can also originate from food sources and to a lesser extent from exposure through the skin. This might explain the fact that non-smoking controls, not exposed to coke oven emissions, show relatively high values of BaP adducts compared with coke oven workers or smokers (values for coke oven workers are a maximum 10 times higher). Exposures through food or skin have, however, a much lower risk for lung cancer than exposure through the airways (Culp et al. 2000) and although they might have the same risk for cancers not related to the airways, the total cancer risk will be substantially lower. So for people exposed through food or skin, the level of BaP adducts will lead to an overestimation of cancer risk. Second, both experimental and epidemiological studies have indicated that, for the induction of a certain lung tumour incidence, inhaled cigarette smoke contains about 100 times less BaP and inhaled diesel exhaust fumes contains about 1,000 times less BaP than the exhaust from coke ovens or heated tar pitch associated with the same risk of lung cancer (Pot and Heinrich 1990). So for smokers or people exposed to diesel exhaust fumes the level of BaP adducts will lead to an underestimation of cancer risk. Notwithstanding these and also other considerations, risk estimations based on data from workers in coke ovens and in other industries with exposure to PAHs can have some relevance as many people are exposed to similar complex mixtures and as clearly superior approaches are not available.

We calculated the mean BaP adduct level per dose unit (micrograms per cubic metre) in vivo in coke oven and other factory workers to estimate the cancer risk associated with the presence of BaP adducts based on the cancer risk experienced by these workers. Table 8.8 shows studies that measured BPDE or BaP adducts in white blood cells of coke oven or factory workers. We calculated the mean adduct level per dose unit to be 0.418 adducts per /10^6 nucleotides per microgram per cubic metre.

The unit risk for lung cancer from BaP (after 70 years of exposure) is 1.1–3.3 × 10^{-3}/μg BaP/m^3 (OEHHA 1994) so 0.418 adducts per 10^6 nucleotides could result in a unit lung cancer risk of 1.1–3.3 × 10^{-3} after exposure to pure BaP. The number of endogenous DNA adducts in human DNA, which is on the order of 4.97– 69.05 adducts per 10^6 nucleotides, is far greater than the number of BaP adducts. This load of endogenous DNA adducts might underlie the spontaneous incidence of lung cancer, which is probably on the order of 6.6 × 10^{-3}, the cumulative incidence of lung cancer up to age 75 in Chennai, India, in the period 1978–1982 (Muir et al. 1987). This suggests that the pathogenic potency of exogenous high molecular weight adducts such as those

Table 8.8 BPDE or BaP adducts in white blood cells of coke oven or factory workers

Reference	Dose (μg/m^3)	Adduct level (BPDE adducts/10^6 nucleotides)	Adducts per dose unit
Haugen et al. (1986)	2.5	0.210[b]	0.084
	8	0.255	0.032
Arnould et al. (1999)	0.64	0.034	0.053
Pavanello et al. (1999)	0.0289[a]	0.039	1.349
	0.0867[a]	0.115	1.326
	0.202[a]	0.055	0.272
	0.347[a]	0.195	0.562
Pavanello et al. (2005)	2.69[a]	0.044	0.016
Van Schooten et al. (1990)	0.8	0.051	0.064
			Mean 0.418

[a]Exposure doses were originally expressed in micromoles of 1-pyrenol per mole of creatine; 1 μmol 1-pyrenol/mol creatine = 0.289 μg BaP/m^3
[b]Excluding the women in the study who had a very large number of adducts

formed by BaP is higher than that of most endogenous adducts. That this might be the case is supported by the finding that in male F344 rats, 0.85 aflatoxine adducts per 10^6 nucleotides are associated with a liver tumour incidence of 50% (Buss et al. 1990) at 2 years. The reported spontaneous liver tumour incidence in male F344 rats is only 1–2%. The level of endogenous DNA adducts as measured by Nath and Chung (1994), Chung et al. (1999, 2000), Wu et al. (1999), Walker et al. (1993), Föst et al. (1989), Nair et al. (1995), Loureiro et al. (2002), Schuler et al. (1997), Fraga et al. (1990) and Biancini et al. (1993), only including acrolein-dG, crotonaldehyde-dG, N^7-(2-hydroxyethyl)guanine, HNE-dG, etheno-dA, etheno-dC, 1,N^2-etheno-2′-dG, 8-oxo-dG and 7-methyldeoxyguanosine, amounts to 2.18–137.3 adducts per 10^6 nucleotides. The minimum level (2.18) was reached by adding up the lowest level observed for each of the endogenous adduct types considered, while the maximum level (137.3) was based on the highest level observed.

Occupational exposures involve complex mixtures of PAHs other than BaP. So BaP adducts can be considered as a measure of total PAH exposure, the impact of which on cancer incidence is much greater than that of the component BaP alone. We based the following calculations of the lung cancer risk associated with PAH exposure in industries measured as BaP on the meta-analysis of Armstrong et al. (2004). Using data from 39 studies, they calculated the unit relative risk (URR) of industrial PAH exposures with BaP as a guide substance. The URR gives the relative risk at 100 μg BaP/m^3 years. From the data from all 39 studies the average estimated URR was 1.20. A calculation limited to the 22 studies on coke oven, gas works and aluminium production workers led to an average estimated URR of 1.17. As it is conceivable that extrapolation from high-exposure studies to much lower exposures leads to an underestimation of risk (see below), it is of interest to make a risk assessment restricted to studies in which exposures were not above 20 μg BaP/m³ years. Based on seven studies on coke oven, gas works and aluminium production workers in which exposures were not above 20 μg BaP/m³ years, Armstrong et al. (2004) found an average estimated URR of 1.87. Using the equation adapted from Armstrong et al. (2004), we converted the ULR factor into the unit lifetime risk:

$$URR = 1 + \left[ULR \left(\frac{1}{70}\right)\left(\frac{10}{23}\right)\left(\frac{230}{365}\right)\left(\frac{1}{0.09}\right)\right],$$

where ULR is the unit lifetime risk per nanogram per cubic metre continuous (23 m³/day vs. 10 m³/work day) exposure (365 days vs. 230 work days) over 70 years, assuming, as did Armstrong et al. (2004), a 9% baseline lifetime risk. Using a linear model, we calculated that 0.418 BaP adducts per 10^6 nucleotides as a marker of exposure to a mixture of PAHs such as coke oven emissions could be associated with a lung cancer risk of 39×10^{-3}–200×10^{-3} (risks derived from URRs equal to 1.17 and 1.87, respectively). It should be noted that Armstrong et al. (2004) found higher but less precisely estimated URRs when they included all 39 studies on cancer risk associated with PAH exposures in industry and did not restrict their calculations to coke oven, gas works and aluminium production workers. They found an estimated URR of 1.20 based on all 39 studies and an estimated URR of 4.54 based on 21 studies concerning exposures not exceeding 20 μg/m³ years.

Besides lung cancer, exposure to coke oven emissions was also associated with kidney cancer, prostate cancer and genitourinary cancer in male workers. Mandell et al. (1995) found a relative risk of 1.7 for kidney cancer in male workers in the blast furnace or coke oven industry (95% confidence interval, CI, 1.1–2.7). Costantino et al. (1995) reported significantly elevated relative risks for lung cancer and for prostate cancer and genitourinary cancer in general (1.95, CI 1.59–2.33; 1.57, CI 1.09–2.30; 1.49, CI 1.11–2.01, respectively). From the number of cancer deaths and the relative risks reported by Costantino et al. (1995) it can be calculated that in a population of 5,321 coke oven workers the induction of 124 excess cases of lung cancer by exposure to coke oven emissions was accompanied by the induction of 28 cases of genitourinary cancers. So 0.418 BaP adducts per 10^6 nucleotides as a marker of exposure to a mixture of PAHs such as coke oven emissions could be associated with a genitourinary cancer risk of 9×10^{-3}–45×10^{-3} and a cancer risk of 48×10^{-3}–245×10^{-3} for lung and genitourinary cancer combined.

Based on a linear model, DNA adduct levels in peripheral blood cells of the order of 0.05 BaP adducts per 10^6 nucleotides (such as observed in non-smokers, Table 8.4) could, if they resulted from exposure to air polluted by PAHs, be associated with cancer risks on the order of 5×10^{-3}–25×10^{-3}. To the extent that BaP-DNA adduct levels are due to exposures through ingestion, lung cancer risk from PAHs will be lower. This will mainly be relevant for non-smokers.

8.4.2 Supralinear Dose–Effect Relationship

Observations by Danheiser et al. (1989) and most (but not all) data collected for this review suggest a supralinear dose–effect relationship (low-dose hypersensitivity) in vitro between BaP exposure, on the one hand, and damage to DNA and mutations, on the other. Such a supralinear dose–effect relationship is also found in epidemiologic observations on humans (Ostro 2004) and the review by Armstrong et al. (2004) shows that the calculation of the risk associated with exposure to PAHs not exceeding 20 μg BaP/m^3 years (corresponding to an exposure to 1 μg BaP/m^3 over 20 years) leads to a higher value than calculations also including more intensive exposures. It is interesting to consider carcinogenic effects of environmental or other low-intensity exposures to PAHs in terms of existing knowledge of the induction of DNA repair by genotoxic agents such as ionizing radiation. There is substantial evidence that exposure to a first dose (adaptive dose) of a few milligrays up to a few hundred milligrays (up to 500 mGy in some circumstances; Broome et al. 2002) leads to a reduction in the mutagenic effects of a second, higher dose (Wolf 1998; UNSCEAR 2000). This phenomenon called adaptation, which has also been described for genotoxic substances, is probably due to the induction of DNA repair mechanisms (Wolf 1998). The critical dose for induction of the adaptive response is probably on the order of 1 mGy, with no induction of DNA double strand break repair or adaptive response occurring below this dose (Broome et al. 2002; Rothkamm and Löbrich 2003). The adaptive response takes one or more hours to be established and disappears after about 40 h (Wolf 1998). Exposures to ionizing radiation in doses below the threshold level for an adaptive response could be, per unit dose, more efficient in the induction of mutations than higher doses (low dose hypersensitivity) (Joiner et al. 1996).

Lifelong exposure of a human population to ionizing radiation at a level corresponding to the threshold dose per time period during which the adaptive response persists, i.e. 1 mGy per 40 h or 15.3 Gy per 70 years, can be estimated to lead to a risk of death from cancer of 100% before age 70. Indeed, according to UNSCEAR (2000) the risk of death from all solid cancers combined amounts to 11% following an acute dose of 1 Sv, with total solid cancer incident risks roughly twice those for mortality, while chronic exposure might lead to only 50% of the aforementioned risk. That exposures to very low doses of ionizing radiation do lead to the induction of cancer in humans is demonstrated by observations concerning radon in houses (Doll 1998), by observations made after the Chernobyl disaster (Okeanov et al. 2004) and by the fact that low-dose γ-radiation induced cytogenetic effects in residents of buildings (Chang et al. 1997). Risks associated with environmental and most occupational exposures to BaP and complex PAH mixtures are far lower than the risk associated with exposure to ionizing radiation at the threshold dose per time period during which the adaptive response persists.

It is likely that these exposures to PAHs do not reach a threshold level for induction of DNA repair and that consequently a higher efficiency in the induction of biological and health effects (compared with high experimental or occupational doses) might be expected. This might explain supralinear effects suggested by some epidemiological data.

PAHs such as BaP play an important role in the induction of lung cancer by tobacco smoke and by particulate matter in ambient air (Denissenko et al. 1996; Petruzzelli et al. 1998). A recent study from Norway (Bjartveit and Tverdal 2005) found that smoking one to four cigarettes per day is associated with an almost threefold increase in relative risk of dying from cardiovascular disease or from lung cancer, whereas Vineis et al. (2000) found that the increase of lung cancer risk in heavy smokers levelled of above a relative risk of 20. According to the WHO, the response per dose unit in the induction of lung cancer and of cardiopulmonary disease by particulate matter in ambient air is higher for low exposures than for more intensive exposures (Ostro 2004).

Acknowledgement: This work was realized in the framework of the activities of the Focal Point Environmental Health (Steunpunt Milieu en Gezondheid) of the Flemish Community. This work does not, though, reflect the position of the Focal Point. It only reflects the opinion of the authors.

References

Albert RE, Miller ML, Cody TE, Talaska G, Underwood P, Andringa A (1996) Epidermal cytokinetics, DNA adducts, and dermal inflammation in the mouse skin in response to repeated benzo(a)pyrene exposures. Toxicol Appl Pharmacol 136:67–74

Alexandrov K, Rojas M, Geneste O, Castegnaro M, Camus AM, Pertuzzelli S, Giuntini C, Bartsch H (1992) An improved fluorometric assay for dosimetry of benzo(a)pyrene diol-epoxide-DNA adducts in smokers' lung: comparisons with total bulky adducts and aryl hydrocarbon hydroxylase activity. Cancer Res 52:6248–6253

Alexandrov K, Sala M, Rojas M (1988) Differences in the DNA adducts formed in cultured rabbit and rat dermal fibroblasts by benzo(a)pyrene and (–)-benzo(a)pyrene-7,8-diol. Cancer Res 48:7132–7139

Allen-Hoffmann BL, Rheinwald JG (1984) Polycyclic aromatic hydrocarbon mutagenesis of human epidermal keratinocytes in culture. Proc Natl Acad Sci USA 81:7802–7806

Armstrong B, Hutchinson E, Unwin J, Fletcher T (2004) Lung cancer risk after exposure to polycyclic aromatic hydrocarbons: a review and meta-analysis. Environ Health Perspect 112:970–978

Arnould JP, Verhoest P, Bach V, Libert JP, Belegaud J (1997) Detection of benzo(a)pyrene-DNA adducts in human placenta and umbilical cord blood. Hum Exp Toxicol 16:716–721

Arnould JP, Pfohl-Leszkowicz A, Bach V, Libert JP, Belegau, J (1999) Biological monitoring exposure of workers from plant producing carbon electrodes: quantification of benzo(a)pyrene DNA-adducts in leukocytes, by a 32P-postlabelling method and an immunoassay. Hum Exp Toxicol 18:314–321

Ashurst SW, Cohen GM (1980) A benzo(a)pyrene-7,8-dihydrodiol-9,10-epoxide is the major metabolite involved in the binding of benzo(a)pyrene to DNA in isolated viable rat hepatocytes. Chem Biol Interact: 29:117–127

Ashurst SW, Cohen GM (1981) In vivo formation of benzo(a)pyrene diol epoxide-deoxyadenosine adducts in the skin of mice susceptible to benzo(a)pyrene-induced carcinogenesis. Int J Cancer 27:357–364

Aust AE, Falahee KJ, Maher VM, McCormick J (1980) Human cell-mediated benzo(a)pyrene cytotoxicity and mutagenicity in human diploid fibroblasts. Cancer Res, 40:4070–4075
Banasiewicz M, Nelson G, Swank A, Grubor N, Ross J, Nesnow S, Köfeler H, Small J, Jankowiak R (2004) Identification and quantitation of benzo(a)pyrene-derived DNA adducts formed at low adduction level in mice lung tissue. Anal Biochem 334:390–400
Bartley J, Baratholomew JC, Stampfer MR (1982) Metabolism of benzo(a)pyrene by human epithelial and fibroblastic cells: metabolite patterns and DNA adduct formation. J Cell Biochem 18:135–148
Bianchini F, Montesano R, Shuker DEG, Cuzick J, Wild CP (1993) Quantification of 7-methyldeoxyguanosine using immunoaffinity purification and HPLC with electrochemical detection. Carcinogenesis 14:1677–1682
Binková B, Giguère Y, Rössner P Jr, Dostál M, Šrám RJ (2000) The effect of dibenzo(a,l)pyrene and benzo(a)pyrene on human diploid lung fibroblasts: the induction of DNA adducts, expression of p53 and p21^{WAF1} proteins and cell cycle distribution. Mutat Res 471:57–70
Bjartveit K, Tverdal A (2005) Health consequences of smoking 1-4 cigarettes per day. Tob Control 14:315–320
Bodell WJ, Devanesan PD, Rogan EG, Cavalieri EL (1989) 32P-postlabeling analysis of benzo(a)pyrene-DNA adducts formed in vitro and in vivo. Chem Res Toxicol 2:312–315
Brooks RA, Gooderham NJ, Edwards RJ, Boobis A,. Winton DJ (1999) The mutagenicity of benzo(a)pyrene in mouse small intestine. Carcinogenesis 20:109–114
Broome EJ, Brown DL, Mitchel RE (2002) Dose responses for adaption to low doses of (60)Co gamma rays and (3)H beta particles in normal human fibroblasts. Radiat Res 158:181–186
Buss P, Caviezel M, Lutz WK (1990) Linear dose-response relationship for DNA adducts in rat liver from chronic exposure to aflatoxin B1. Carcinogenesis 11:2133–2135
Canova S, Viezzer C, Venier P, Navazio L, Revoltella R, Celotti L (1996) Metabolic activation of benzo(a)pyrene in two fetal mouse hepatocyte lines: induction of DNA adducts and micronuclei. Mutat Res 367:135–141
Chang WP, Hwang BF, Wang D, Wang JD (1997) Cytogenetic effect of chronic low-dose, low-dose-rate γ-radiation in residents of irradiated buildings. Lancet 350:330–333
Chen L, Devanesan PD, Higginbotham S, Ariese F, Jankowiak R, Small G., Rogan EG, Cavalieri EL (1996) Expanded analysis of benzo(a)pyrene-DNA adducts formed in vitro and in mouse skin: their significance in tumor initiation. Chem Res Toxicol 9:897–903
Chung FL, Zhang L, Ocando JE, Nath RG (1999) Role of 1,N^2-propanodeoxyguanonsine adducts as endogenous DNA lesions in rodents and humans. In: Singer B, Bartsch H (eds) IARC scientific publication no 150: Exocyclic DNA adducts in mutagenesis and carcinogenesis. International Agency on Cancer, Lyons
Chung FL, Nath RG, Ocando J, Nishikawa A, Zhang L (2000) Deoxyguanosine adducts of t-4-hydroxy-2-nonenal are endogenous DNA lesions in rodents and humans: detection and potential sources. Cancer Res 60:1507–1511
Costantino JP, Redmond CK, Bearden A (1995) Occupationally related cancer risk among coke oven workers: 30 years of follow-up. J Occup Environ Med 37:597–604
Culp SJ, Warbritton AR, Smith B., Li EE, Beland FA (2000) DNA adduct measurements, cell proliferation, and tumor mutation induction in relation to tumor formation in B6C3F1 mice fed coal tar or benzo(a)pyrene. Carcinogenesis 21:1433–1440
Cunningham ML, Ringrose PS (1983) Benzo(a)pyrene and aniline increase sister chromatid exchanges in cultured rat liver fibroblasts without addition of activating enzymes. Toxicol Lett 16:235–239
Danheiser SL, Liber SL, Thilly WG (1989) Long-term, low-dose benzo(a)pyrene-induced mutation in human lymphoblasts competent in xenobiotic metabolism. Mutat Res 201:143–147
Dean SW, Coates A, Brooks TM, Burlinson B (1998) Benzo(a)pyrene site of contact mutagenicity in skin of Muta™ mouse. Mutagenesis 13:515–518
De Bont R, van Larebeke N (2004) Endogenous DNA damage in humans: a review of quantitative data. Mutagenesis 19:169–185

Deckers-Schmelzle B, Klaus E, Kahl R, Kahl GF (1978) Binding of benzo(a)pyrene metabolites to cellular DNA in perfused rat lungs. Naunyn-Schmiedebert's Arch Pharmacol 303:303–307
Dedon PC, Plastaras JP, Rouzer CA, Marnett LJ (1998) Indirect mutagenesis by oxidative DNA damage: formation of the pyrimidopurinone adduct of deoxyguanosine by base propenal. Proc Natl Acad Sci USA 95:11113–11116
Denissenko MF, Pao A, Tang M, Pfeifer GP (1996) Preferential formation of benzo(a)pyrene adducts at lung cancer mutational hotspots in p53. Science 274:430–432
Devanesan PD, RamaKrishna NVS, Todorovic R, Rogan EG, Cavalieri EL, Jeong H, Jankowiak R, Small GJ (1992) Identification and quantitation of benzo(a)pyrene-DNA adducts formed by rat liver microsomes in vitro. Chem Res Toxicol 5:302–309
DiGiovanni J, Decina PC, Prichett WP, Fisher EP, Aalfs KK (1985) Formation and disappearance of benzo(a)pyrene DNA-adducts in mouse epidermis. Carcinogenesis 6:741–747
Divi RL, Beland FA, Fu PP, Von Tungeln LS, Schoket B, Camara JE, Ghei M, Rothman N, Sinha R, Poirier MC (2002) Highly sensitive chemiluminescence immunoassay for benzo(a)pyrene-DNA adducts: validation by comparison with other methods, and use in human biomonitoring. Carcinogenesis 23:2043–2049
Doll R (1998) Effects of small doses of ionising radiation. J Radiol Protect 18:163–174
Dosaka H, Abe S, Sasaka M, Miyamoto H, Kawakami Y (1987) Sister chromatid exchange induction by benzo(a)pyrene in cultured peripheral blood lymphocytes of lung cancer patients and healthy individuals with or without familial history of neoplasms. Int J Cancer 39:329–332
Föst U, Marczynski B, Kasemann R, Peter H (1989) Determination of 7-(2-hydroxyethyl)guanine with gas chromatography/mass spectrometry as a parameter for genotoxicity of ethylene oxide. Arch Toxicol Suppl 13:250–253
Fraga CG, Shigenaga MK, Park JW, Degan P, Ames BN (1990) Oxidative damage to DNA during aging: 8-hydroxy-2′-deoxyguanosine in rat organ DNA and urine. Proc Natl Acad Sci USA 87:4533–4537
Godschalk RWL, Vermeer TM, Kriek E, Floot B, Schilderman PAEL, Moonen EJC, Kleinjans JCS, van Schooten FJ (1997) Comparison of 32P-postlabeling and HPLC-FD analysis of DNA adducts in rats acutely exposed to benzo(a)pyrene. Chem Biol Interact 104:41–54
Hakura A, Tsutsui Y, Sonoda J, Kai J, Imade T, Shimada M, Sugihara Y, Mikami T (1998) Comparison between in vivo mutagenicity and carcinogenicity in multiple organs by benzo(a)pyrene in the lacZ transgenic mouse (Muta™ mouse). Mutat Res 398:123–130
Hanelt S, Helbig R, Hartmann A, Lang M, Seidel A, Speit G (1997) A comparative investigation of DNA adducts, DNA strand breaks and gene mutations induced by benzo(a)pyrene and (±)-anti-benzo(a)pyrene-7,8-diol 9,10-oxide in cultured human cells. Mutat Res 390:179–188
Hartmann A, Speit G (1995) Genotoxic effects of chemicals in the single cell gel (SCG) test with human blood cells in relation to the induction of sister-chromatid exchanges (SCE). Mutat Res 346:49–56
Haugen A, Becher G, Benestad C, Vahakangas K, Trivers GE, Newman MJ, Harris CC (1986) Determination of polycyclic aromatic hydrocarbons in the urine, benzo(a)pyrene diol epoxide-DNA adducts in lymphocyte DNA, and antibodies to the adducts in sera from coke oven workers exposed to measured amounts of polycyclic aromatic hydrocarbons in the work atmosphere. Cancer Res 46:4178–4183
Huberman E, Sachs L, Yang SK, Gelboin HV (1976) Identification of mutagenic metabolites of benzo(a)pyrene in mammalian cells. Proc Natl Acad Sci USA 73:607–611
Inoue K, Shibata T, Abe T (1983) Induction of sister-chromatid exchanges in human lymphocytes by indirect carcinogens with and without metabolic activation. Mutat Res 117:301–309
Jacono FL, Stecca C, Duverger M (1992) Mutagenic activation of benzo(a)pyrene by human red blood cells. Mutat Res 268:21–26
Joiner MC, Lambin P, Malaise EP, Robson T, Arrand JE, Skov KA, Marples B (1996) Hypersensitivity to very-low single radiation doses: its relationship to the adaptive response and induced radioresistance. Mutat Res 358:171–183
Li D, Wang M., Cheng L, Spitz MR, Hittelman WN, Wei Q (1996) In vitro induction of benzo(a)pyrene diol epoxide-DNA adducts in peripheral lymphocytes as a susceptibility marker for human lung cancer. Cancer Res 56:3638–3641

Liehr JG (2000) Is estradiol a genotoxic mutagenic carcinogen? Endocrine Rev 21:40–54
Loureiro APM, Marques SA, Garcia CCM, Di Mascio P, Medeiros HG (2002) Development of an on-line liquid chromatography-electrospray tandem mass spectrometry assay to quantitatively determine 1,N^2-etheno-2′-deoxyguanosine in DNA. Chem Res Toxicol 15:1302–1308
Machanoff R, O'Neill J.P, Hsie AW (1981) Quantitative analysis of cytotoxicity and mutagenicity of benzo(a)pyrene in mammalian cells (CHO/HGPRT system). Chem Biol Interact 34:1–10
Manchester DK, Weston A, Choi JS, Trivers GE, Fennessey PV, Quitana E, Farmer PB, Mann DL, Harris CC (1988) Detection of benzo(a)pyrene diol epoxide-DNA adducts in human placenta. Proc Natl Acad Sci USA 85:9243–9247
Mandell JS, McLaughlin JK, Schlehofer B, Mellemgaard A, Helmert U, Lindblad P, McCredie M, Adami HO (1995) International renal-cell cancer study. IV. Occup Int J Cancer 61:601–605
Miller RL, Garfinkel R, Horton M, Camann D, Perera FP, Whyatt RM, Kinney PL (2004) Polycyclic aromatic hydrocarbons, environmental tobacco smoke, and respiratory symptoms in an inner-city birth cohort. Chest 126:1071–1078
Monroe JJ, Kort KL, Miller JE, Marino DR, Skopek TR (1998) A comparative study of in vivo mutation assays: analysis of hprt, lacI, and cII/cI as mutational targets for N-nitroso-N-methylurea and benzo(a)pyrene in big blue mice. Mutat Res 421:121–136
Moore CJ, Pruess-Schwartz D, Mauthe RJ, Gould MN, Baird WM (1987) Interspecies differences in the major DNA adducts formed from benzo(a)pyrene but not 7,12-dimethylbenz(a)anthracene in rat and human mammary cell cultures. Cancer Res 47:4402–4406
Muir C, Waterhouse J, Mack T, Powell J, Whelan S (1987) Cancer incidence in five continents, vol V. IARC scientific publication 88. International Agency for Research on Cancer, Lyons
Nair J, Barbin A, Guichard Y, Bartsch H (1995) 1,N^6-Ethenodeoxyadenosine and 3,N^4-ethenodeoxycytidine in liver DNA from humans and untreated rodents detected by immunoaffinity/32P-postlabelling. Carcinogenesis 16:613–617
Nath RG, Chung FL (1994) Detection of exocyclic 1,N^2-propanodeoxyguanosine adducts as common DNA lesions in rodents and humans. Proc Natl Acad Sci USA 91:7491–7495
Nowak D, Meyer A, Schmidt-Preuss U, Gatzemeier U, Magnussen H, Rüdiger HW (1992) Formation of benzo(a)pyrene-DNA adducts in blood monocytes from lung cancer patients with a familial history of lung cancer. J Cancer Res Clin Oncol 118:67–71
Oberly TJ, Huffman DM, Scheuring JC, Garriott ML (1993) An evaluation of 6 chromosomal mutagens in the AS52/XPRT mutation assay utilizing suspension culture and soft agar cloning. Mutat Res 319:179–187
OEHHA (1994) Benzo(a)pyrene as a toxic air contaminant. Prepared by the Staffs of the California Air Resources Board and the Office of Environmental Health Hazard Assessment
Okeanov AE, Sosnovskaya EY, Priatkina OP (2004) A national cancer registry to assess trends after the Chernobyl accident. Swiss Med Wkly 134:645–649
Ostro B (2004) Outdoor air pollution: assessing the environmental burden of disease at national and local levels. WHO environmental burden of disease series no 5. World Health Organization, Geneva
Pavanello S, Favretto D, Brugnone F, Mastrangelo G, Dal Pra G, Clonfero E (1999) HPLC/fluorescence determination of anti-BPDE-DNA adducts in mononuclear white blood cells from PAH-exposed humans. Carcinogenesis 20:431–435
Pavanello S, Pulliero A, Siwinska E, Mielzynska D, Clonfero E (2005) Reduced nucleotide excision repair and GSTM1-null genotypes influence anti-BPDE-DNA adduct levels in mononuclear white blood cells of highly PAH-exposed coke oven workers. Carcinogenesis 26:169–175
Petruzzelli S, Celi A, Pulera N, Baliva F, Vlegi G, Carrozzi L, Ciacchini G, Bottai M, Di Pede F, Paoletti P, Giuntini C (1998) Serum antibodies to benzo(a)pyrene diol epoxide-DNA adducts in the general population: effects of air pollution, tobacco smoking, and family history of lung diseases. Cancer Res 58:4122–4126
Pot F, Heinrich U (1990) Relative significance of different hydrocarbons for the carcinogenic potency of emissions from various incomplete combustion processes. IARC Sci Publ 104:288–297
Ramos KS, Moorthy B (2005) Bioactivation of polycyclic aromatic hydrocarbon carcinogens within the vascular wall: implications for human atherogenesis. Drug Metab Rev 37:595–610

Ray-Chaudhuri R, Kelley S, Iype PT (1980) Induction of sister chromatid exchanges by carcinogens mediated through cultured rat liver epithelial cells. Carcinogenesis 1:779–786
Roberts MJ, Wondrak GT, Laurean DC, Jacobson MK, Jacobson EL (2003) DNA damage by carbonyl stress in human skin cells. Mutat Res 522:45–56
Rogan EG, Devanesan PD, RamaKrishna NVS, Higginbotham S, Padmavathi NS, Chapman K, Cavalieri EL, Jeong H, Jankowiak R, Small GJ (1993) Identification and quantitation of benzo(a)pyrene-DNA adducts formed in mouse skin. Chem Res Toxicol 6:356–363
Roggeband R, Wolterbeek APM, Rutten AAJJL, Baan RA (1993) Comparative 32P-postlabeling analysis of benzo(a)pyrene-DNA adducts formed in vivo upon activation of benzo(a)pyrene by human, rabbit and rodent liver microsomes. Carcinogenesis 14:1945–1950
Roggeband R, Van Den Berg PTM, Van Der Wulp CJM, Baan RA (1994a) Detection of DNA adducts in basal and non-basal cells of the hamster trachea exposed to benzo(a)pyrene in organ culture. J Histochem Cytochem 42:1427–1434
Roggeband R, Wolterbeek APM, van den Berg PTM, Baan RA (1994b) DNA adducts in hamster and rat tracheas exposed to benzo(a)pyrene in vitro. Toxicol Lett 72:105–111
Rojas M, Alexandrov K, van Schooten FJ, Hillebrand M, Kriek E, Barsch H (1994) Validation of a new fluorometric assay for benzo(a)pyrene diolepoxide-DNA adducts in human white blood cells: comparisons with 32P-postlabeling and ELISA. Carcinogenesis 15:557–560
Ross J, Nelson G, Kligerman A, Erexson G, Bryant M, Earley K, Gupta R, Nesnow S (1990) Formation and persistence of novel benzo(a)pyrene adducts in rat lung, liver, and peripheral blood lymphocyte DNA. Cancer Res 50:5088–5094
Rothkamm K, Löbrich M (2003) Evidence for a lack of DNA double-strand break repair in human cells exposed to very low x-ray doses. Proc Natl Acad Sci USA 100:5057–5062
Rüdiger HW, Marxen J, Kohl FV, Melderis H, van Wichert P (1979) Metabolism and formation of DNA adducts of benzo(a)pyrene in human diploid fibroblasts. Cancer Res 39:1083–1088
Rüdiger HW, Nowak D, Hartmann K, Cerutti P (1985) Enhanced formation of benzo(a) pyrene:DNA adducts in monocytes of patients with a presumed predisposition to lung cancer. Cancer Res 45:5890–5894
Salama SA, Sierra-Torres CH, Oh HY, Hamada FA, Au WW (2001) Variant metabolizing gene alleles determine the genotoxicity of benzo(a)pyrene. Environ Mol Mutagen 37:17–26
Schuler D, Otteneder M, Saglsdorff P, Eder E, Gupta RC, Lutz WK (1997) Comparative analysis of 8-oxo-2′-deoxyguanosine in DNA by 32P- and 33P-postlabeling and electrochemical detection. Carcinogenesis 18:2367–2371
Shamsuddin AKM, Sinopoli NT, Hemminki K, Boesch RR, Harris CC (1985) Detection of benzo(a)pyrene:DNA adducts in human white blood cells. Cancer Res 45:66–68
Shane S, de Boer J, Watson DE, Haseman JK, Glickman BW, Tindall KR (2000) Lac I mutation spectra following benzo(a)pyrene treatment of Big Blue mice. Carcinogenesis 21:715–725
Shugart L (1985) Covalent binding of benzo(a)pyrene diol epoxide to DNA of mouse skin: in vivo persistence of adducts formation. J Toxicol Environ Health 15:255–263
Stoner GD, Daniel FB, Schenck KM, Schut HAJ, Goldblatt PJ, Sandwisch DW (1982) Metabolism and DNA binding of benzo(a)pyrene in cultured human bladder and bronchus. Carcinogenesis 3:195–201
Talaska G, Jaeger M, Reilman R, Collins T and Warshawsky D (1996) Chronic, topical exposure to benzo(a)pyrene induces relatively high steady-state levels of DNA adducts in target tissues and alters kinetics of adduct loss. Proc Natl Acad Sci USA 93:7789–7793
Theall G, Eisinger M, Grunberger D (1981) Metabolism of benzo(a)pyrene and DNA adduct formation in cultured human epidermal keratinocytes. Carcinogenesis 2:581–587
Tong C, Brat V, Williams GM (1981) Sister-chromatid exchange induction by polycyclic aromatic hydrocarbons in an intact cell system of adult rat-liver epithelial cells. Mutat Res 91:467–473
UNSCEAR (2000) Sources and effects of ionising radiation. United Nations Scientific Committee on the Effects of Atomic Radiation UNSCEAR 2000 report to the General Assembly, with scientific annexes. Volume II: Effects. http://www.unscear.org/reports/2000_2.html

Van Schooten FJ, Hillebrand MJX, Scherer E, den Engelse L, Kriek E (1991) Immunocytochemical visualization of DNA adducts in mouse tissues and human white blood cells following treatment with benzo(a)pyrene or its diol epoxide. A quantitative approach. Carcinogenesis 12:427–433

Van Schooten FJ, van Leeuwen FE, Hillebrand MJX, Rijke ME, Hart AAM, van Veen HG, Oosterink S, Kriek E (1990) Determination of benzo(a)pyrene-diol-epoxide-DNA adducts in white blood cell DNA from coke oven workers: the impact of smoking, J Natl Cancer Inst 82:927–933

Vineis P, Kogevinas M, Simonato L, Brennan P, Boffetta P (2000) Levelling-off of the risk of lung and bladder cancer in heavy smokers: an analysis based on multicentric case-control studies and a metabolic interpretation. Mutat Res 463:103–110

Walker VE, Fennel TR, Upton, B, MacNeela JP, Swenberg JA (1993) Molecular dosimetry of DNA and hemoglobin adducts in mice and rats exposed to ethylene oxide. Environ. Health Perspect 99:11-17

Weston A, Bowman ED (1991) Fluorescence detection of benzo(a)pyrene-DNA adducts in human lung. Carcinogenesis 12:1445–1449

Weston A, Rowe ML, Manchester DK, Farmer PB, Mann DL, and Harris CC (1989) Fluorescence and mass spectral evidence for the formation of benzo(a)pyrene anti-diol-epoxide-DNA and -hemoglobin adducts in humans. Carcinogenesis 10:251–257

Wiencke J, McDowell ML, Bodell WJ (1990) Molecular dosimetry of DNA adducts and sister chromatid exchanges in human lymphocytes treated with benzo(a)pyrene. Carcinogenesis 11:1497–1502

Wolff S (1998) The adaptive response in radiobiology: evolving insights and implications. Environ Health Perspect 106(Suppl 1):277–283

Wu KY, Ranasinghe A, Upton PB, Walker VE, Swenberg JA (1999) Molecular dosimetry of endogenous and ethylene oxide-induced N7-(2-hydroxyethyl)guanine formation in tissues of rodents. Carcinogenesis 20:1787–1792

9 Analysis of DNA Double-Strand Breaks by Means of γ-H2AX Foci

Irena Szumiel

Abstract

DNA double strand break (DSB) quantification with the use of physicochemical methods is now often replaced by analysis of γ-H2AX foci in the nuclei of cells damaged with DSB-inducing agents. The H2A histone variant, H2AX, becomes phosphorylated by the phosphatidylinositol 3-kinase-like protein kinases, ATM and ATR, within 1–3 min after γ-irradiation. The phosphorylated form, called γ-H2AX, can be visualised as discrete foci with the use of specific antibodies with fluorescent tags and directly counted in the fluorescence microscope. More advanced techniques include flow cytometry or the use of confocal microscopy, image collection and computational analysis of distributions of foci size, fluorescence intensity and number per cell. Modification of the H2AX molecules is necessary to alter the local chromatin organisation in the surrounding of a DSB and thus to increase the accessibility of DNA to various repair factors which are colocalised in the foci. The foci are dispensable for the initial recruitment of proteins and for the early repair and signalling events. Afterwards, however, foci are needed to concentrate and retain the repair proteins in the vicinity of DSB. This seems to be important for repair of persistent lesions. The kinetics of DSB repair and that of formation and disappearance of foci do not correspond closely. Foci also are occasionally formed in S-phase cells as a result of broken replication forks. Nevertheless, γ-H2AX foci are now generally accepted as a DSB marker and are used to estimate radiation sensitivity.

9.1 Introduction

DNA double-strand breaks (DSB) are the most deleterious lesions directly or indirectly induced by DNA-damaging agents, and are occasionally formed during DNA replication owing to broken (collapsed) replication forks. Until recently, DSB quantification was time consuming and based on relations between physicochemical properties of isolated DNA or DNA in chromatin and the extent of DNA fragmentation under non-denaturing conditions. The DSB number was rather difficult to accurately estimate. The discovery that

Günter Obe and Vijayalaxmi (Eds.)
Chromosomal Alterations: Methods, Results and Importance in Human Health

discrete foci containing phosphorylated histone H2AX and some other DNA repair related proteins can be seen in the nuclei of irradiated cells and that there is an approximately 1:1 relation of foci to the calculated number of DSB (Sedelnikova et al. 2002; Rothkam and Loebrich 2003) created a basis for direct DSB estimation in single cells. Essentially, this is feasible; however, there are many confounding factors to be considered. Nevertheless, the method proved to be appropriate for detection of low damage levels and very useful in investigations of the DNA repair mechanisms; its application supplied much new information on the cellular response to DNA damage, as reviewed in this chapter.

9.2 Histone H2AX

Histones in vivo are subject to various modifications, such as phosphorylation, acetylation, methylation and ubiquitination. Phosphate moieties are transferred by nuclear enzymes on specific serine and threonine residues, usually located near the amino or carboxyl termini ("tails") of the histone molecules. Thus, their chemical properties are altered and this brings about changes in their interaction with DNA and with other proteins, and, in consequence, in the chromatin structure. Such changes are a necessary requirement for various functions of DNA performed in cells.

It has long been known that histones are heterogenous, but not all aspects of this heterogeneity are understood. An example of special interest for investigators of DNA damage is the family of H2A histones. It consists of three subfamilies present in varying proportions in various cell types: H2A1-H2A2, H2AZ and H2AX. In mammals, histone H2AX represents 2–25% of the H2A family (Redon et al. 2002). A specific function in DNA repair has been ascribed to the phosphorylated form of histone H2AX, called γ-H2AX (Rogakou et al. 1998). The current state of knowledge of the biological role of histone γ-H2AX is outlined in the following sections.

9.2.1 Histone H2AX Is Phosphorylated in Response to Induction of DNA Double-Strand Breaks

One effect of ionising radiation (or other agents that introduce DSB into DNA) in vivo and in vitro is induction of posttranslational modifications of proteins which can be detected by electrophoresis on two-dimensional gels on the basis of altered electrophoretic mobility and molecular mass. Labelling with ^{32}P can be used to indicate freshly phosphorylated molecules. In this way, phosphorylation of histone H2AX at serine 139 was identified (at the C-terminus of the molecule) and the modified form was named γ-H2AX (Rogakou et al. 1998).

The same modification of H2AX takes place in cells undergoing DNA fragmentation due to apoptosis (Rogakou et al. 2000). DSB are transient intermediates in genomic rearrangements within immunoglobulin and T-cell receptor genes (Chen et al. 2000; Petersen et al. 2001) and these processes also are accompanied by generation of γ-H2AX. Likewise, treatment with drugs that form DNA–protein cross-links or DNA adducts and present a steric hindrance for replication forks induces DSB and H2AX phosphorylation (Furuta et al. 2003; Meng et al. 2005). Altogether, at present, it is generally accepted that γ-H2AX serves as a marker of DSB that are both directly and indirectly generated. Detection of γ-H2AX in cells on microscope slides by immunofluorescence with the use of specific antibodies became a standard procedure in radiobiology and cytotoxicology.

When detected in nuclear extracts by western blotting, γ-H2AX molecules appear very soon after exposure of cells to ionising radiation; in some cells, half-maximal amounts are reached after 1 min and maximal amounts usually after 10–30 min. For example, γ-H2AX formation was studied in Chinese hamster ovary (CHO) cells (Rogakou et al. 1998), irradiated with 200 Gy of γ- rays on ice and then returned to the incubator for recovery at 37°C. As soon as 20 s after return to the incubator, γ-H2AX was detected, and the maximal level of γ-H2AX was reached after 10 min and then it gradually declined during the following 90 min. In the same paper, it was shown that γ-H2AX is evolutionarily conserved from yeast to human, indicating that its formation is part of a universal response to DSB.

9.2.2 γ-H2AX Is Present in the Nuclear Foci

When DSB are induced in DNA, a complex cellular response is triggered. One component of this response is accumulation of proteins involved in DNA repair or damage signalling. They can be seen as nuclear foci in the fluorescence microscope when specific antibodies with fluorescent tags are applied (Maser et al. 1997). With the use of an anti-γ-H2AX antibody with a fluorescent tag, a pattern of γ-H2AX foci can be revealed within 1–3 min after γ-irradiation (Paull et al. 2000) with doses 0.6–2 Gy, considerably lower than that used for the biochemical studies mentioned above (Rogakou et al. 1998). An example of γ-H2AX foci in non-irradiated and X-irradiated CHO cells is shown in Fig. 9.1.

9.2.3 The Phosphatidylinositol 3-Kinase-Like Protein Kinases Are the Main Enzymes that Phosphorylate Histone H2AX

The presence of DSB in nuclear DNA activates three related protein kinases, all belonging to the family of phosphatidylinositol-3-like kinases: DNA-dependent protein kinase (DNA-PK), ataxia telangiectasia mutated (ATM) – its function is impaired in the heritable hypersensitivity to ionising radiation) – and ATM and

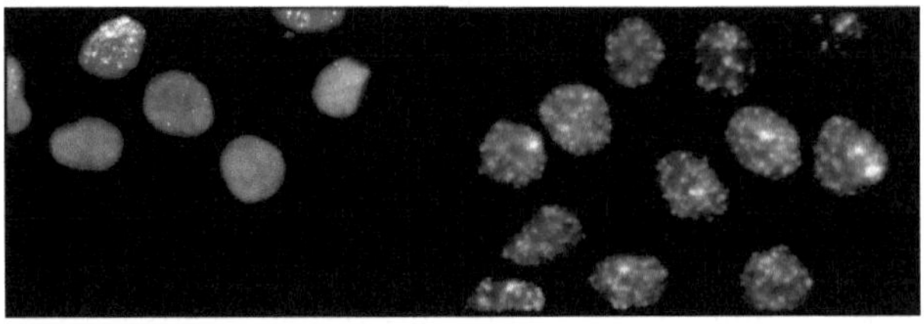

Fig. 9.1 γ-H2AX foci in CHO-K1 Chinese hamster ovary cells, **a** control and **b** X-irradiated (1 Gy, 20 min after exposure), as seen in the fluorescence microscope at × 400 magnification. (Courtesy of Maria Wojewódzka)

Rad3 related (ATR) – activated by arrest of DNA replication. These kinases form part of a monitoring system that detects DNA damage, generates alarm-signalling cascades and recruits repair enzymes to the damaged sites (reviewed in Szumiel 1998). The substrate specificities of these kinases in vitro are similar, but in vivo they have distinct substrates according to their functions.

All three kinases may carry out phosphorylation of H2AX in vivo in a redundant, overlapping manner (Stiff et al. 2004; Wang et al. 2005), but H2AX phosphorylation is practically absent in Atm–/– cells (devoid of both alleles of the Atm gene), whereas it is normal in DNA-PKcs–/– cells (lacking the catalytic subunit of the DNA-PK kinase) (Burma et al. 2001). Apparently, ATM is the main kinase that modifies H2AX, whereas DNA-PK is responsible for the minimal levels of H2AX phosphorylation in Atm–/– cells, since low doses of wortmannin, an inhibitor of phosphatidylinositol-3-like kinases, hence, also of DNA-PK, eliminate this trace amount of γ-H2AX. This observation is consistent with the finding that a fraction of nuclear ATM colocalises with γ-H2AX at sites of DSB in response to DNA damage (Andegeko et al. 2001). Foci with phosphorylated ATM (thus, with the kinase function activated) can be detected after irradiation; they appear and disappear with similar kinetics as γ-H2AX foci (Suzuki et al. 2006). Some studies implicate the contribution of additional kinases to H2AX phosphorylation (Wang et al. 2005).

When irradiated cells are treated with wortmannin, there is no effect on the formation of foci. In contrast, treatment of cells before irradiation completely prevents the accumulation of all components of the foci. These data show that H2AX phosphorylation takes place immediately after damage infliction and precedes sequential assembly of the studied repair proteins into repair foci (Paull et al. 2000). So, the likely scenario after irradiation is that a signal is produced to indicate that a DSB is present and must be repaired, ATM is activated at the site of the DSB and it immediately phosphorylates histone H2AX at the site of the break. This may then initiate the recruitment and assembly into foci of DNA repair, cell cycle checkpoint and damage-signalling factors (Burma et al. 2001; Fernandez-Capetillo et al. 2004).

9.2.4 Megabase Chromatin Domains Alter Conformation After H2AX Phosphorylation

As discussed by Rogakou et al. (1999) and Paull et al. (2000), modification of the H2AX molecules is necessary to alter the local chromatin organisation in the surrounding of a DSB and thus to increase the accessibility of DNA to various repair factors. Rogakou et al. (1998) calculated that approximately 1% of the H2AX becomes γ-phosphorylated per gray of ionising radiation. This radiation dose induces about 35 DSB per mammalian G_1 genome, which in 6×10^9 bp generates γ- H2AX distributed over 1% of the chromatin, i.e. about 0.03% of the H2AX histone appears to be modified per DSB, corresponding to about 2×10^6 bp of DNA per DSB. Several cell lines were examined which differed in the amount of the H2AX histone; nevertheless, the percentage of modified H2AX per DSB was approximately the same in all cells studied (0.03–0.06).

9.3 Repair and Checkpoint Factors Are Recruited into γ-H2AX Foci

After exposure to ionising radiation, γ-H2AX foci in the nuclei appear within minutes (Paull et al. 2000). Other components of the repair and checkpoint system (visualised via antibodies with different fluorescent tags) also can be detected by immunofluorescence as discrete foci that colocalise with γ-H2AX foci. Observation of the time course of localisation of various proteins in the foci indicates that they are assembled in a defined sequence. Mediator of DNA damage checkpoint protein 1 (MDC1) and p53 binding protein 1 (53BP1) colocalise very rapidly – 5–15 min after irradiation (Schultz et al. 2000; Stucki et al. 2005). MDC1 plays an important role in assembling other repair and checkpoint proteins (Xu and Stern 2003). Stucki and Jackson (2006) indicate that the interacting MDC1 and γ-H2AX molecules form an "affinity platform" on which other proteins are assembled and retained (Lukas et al. 2004; Stucki et al. 2005).

At a later time (about 1 h) γ-H2AX becomes colocalised with breast cancer 1 (BRCA1) followed by accumulation of RAD 50 or RAD 51 in different subsets of cells. RAD 51 is a component of the homologous recombination repair (HRR) system. RAD 50 is a part of the MRN complex (MRE11–RAD 50–NBS1, where NBS1 is Nijmengen breakage syndrome 1), which is a DSB sensor and hence is present at the site of the DSB immediately after damage generation (Lee and Paull 2005); however, it is also active in both HRR and non-homologous end joining (NHEJ). The components of the complex rapidly accumulate in the nuclear foci (Celeste et al. 2003a). The fork head associated (FHA) and the BRCA1 C-terminal (BRCT) domains of NBS1 are essential for physical interaction with γ-H2AX (Kobayashi et al. 2002) and thus for the localisation in the foci. Likewise, the BRCT repeat domain of MDC1 binds directly to γ-H2AX. Interestingly, RAD 50 and RAD 51 foci are not present in the same cell after damage induction (Maser et al. 1997), indicating that the two repair pathways, NHEJ and HRR, mutually exclude one another in individual cells.

9.4 Early DSB Rejoining Can Proceed Without Foci Formation

The recruitment of repair and checkpoint proteins to the sites of DSB is not dependent on the previous generation of the γ-H2AX foci, as concluded from the fact that X-irradiated H2AX-deficient cells are able to repair DSB and arrest progression through the cell cycle. Obviously, the foci are dispensable for the initial recruitment of proteins and for the early repair and signalling events. Afterwards, however, foci are needed to concentrate and retain the proteins in the vicinity of DSB (Celeste et al. 2003a). This seems to be important for repair of persistent lesions, since the H2AX-deficient cells are more radiosensitive than the wild-type cells, and mutant mice show defects in class switch recombination and spermatogenesis, decreased genomic stability and enhanced susceptibility to carcinogens and to ionising radiation (Celeste et al. 2003 a, b; Pilch et al. 2003). In agreement with these statements is the finding that the number of γ-H2AX foci gradually increases during the first 30–60 min after irradiation (Han et al. 2006; Wykes et al. 2006), whereas during that time a considerable fraction of DSB are repaired as judged from the results of the neutral comet assay or pulsed-field gel electrophoresis (PFGE).

9.5 Foci May Be Present Long After DSB Are Rejoined

Attempts were made to establish relations between the kinetics of DSB rejoining and foci numbers also in the late repair periods. In theory, repaired DSB should give a signal for dephoshorylation of γ-H2AX molecules and disintegration of the focus. Indeed, protein phosphatase 2A (PP2A) is involved in removing γ-H2AX foci. The PP2A catalytic subunit and γ-H2AX coimmunoprecipitate and colocalise in DNA damage foci and PP2A dephosphorylates the histone in vitro (Chowdhury et al. 2005). Hence, a decrease in foci frequency or in the amount of γ-H2AX detected by flow cytometry or western blotting should reflect the repair kinetics and correlate with survival; however, these relations are not so straightforward:

- DSB rejoining measured by PFGE or comet assay proceeds much faster than the decrease in foci frequency (Banath et al. 2004).
- Decrease in foci frequency is not associated with a decrease in the global γ-H2AX amount (estimated by flow cytometry or western blotting), suggesting that the modified histone is redistributed in the chromatin (Bouquet et al. 2006).
- DSB rejoining is not correlated with a decrease in the global γ-H2AX amount. For example, human melanoma cells with mutated ATM show a constant γ-H2AX level for 2 h after irradiation with 2 Gy X-rays, whereas the DSB rejoining proceeds normally (MacPhail et al. 2003a).
- When PP2A is inhibited or silenced by RNA interference, γ-H2AX foci persist, and the cells are hypersensitive to DNA damage.

Generally, it is assumed that persistent foci mark unrepaired or misrepaired DSB. Banath et al. (2004) pointed out that residual unrepaired DSB may be left undetected by the comet assay, which has a detection limit of 50 DSB per cell. Thus, the presence of only a few DSB per cell may keep the alarm system of the cell switched on and prevent the disassembly of the foci. Stucki and Jackson (2006) postulate that foci serve to amplify the signalling for cell cycle checkpoints in order to keep the cell safely arrested until all the residual damage has been repaired.

9.6 Practical Aspects of γ-H2AX Estimation

A description of the techniques for γ-H2AX histone detection can be found in Nakamura et al. (2006). Figure 9.2 shows an example of a procedure applied to obtain microscopic preparations suitable for foci scoring. The procedure must be modified according to the properties of the cell type studied, but the basic steps are the same: fixation, permeabilisation, blocking, incubation with the primary antibody followed by a secondary antibody. It should be added that since the early period of direct foci scoring with the use of an epi-fluorescence microscope, much more sophisticated approaches have become available: the use of confocal microscopy (especially important with less flat cells), image collection and computational analysis (Costes et al. 2006), including distributions of foci size, fluorescence intensity and number per cell. Figure 9.3 shows an example of foci analysis.

An alternative for foci scoring is flow cytometry. The detection of γ-H2AX is based on indirect immunofluorescence with the use of a secondary antibody tagged with fluorescein isothiocyanate (FITC), whereas DNA is counter-stained with propidium iodide (PI). Then, bivariate analysis of the intensity of cellular green (FITC) and red (PI) fluorescence data can be used to assess the relation between the presence of γ-H2AX as a marker of DSB and the cell cycle phase (Huang and Darzynkiewicz 2006). Figure 9.4 shows examples of the flow cytometry estimation of γ-H2AX.

- Seed 1x10^5 cells into 6-well culture plate containing a glass cover slip in each well
- Treat with a DNA damaging agent
- Fix in 4% paraformaldehyde for 15 min
- Wash with PBS
- Permeabilize in 0.2% Triton X-100
- Block with blocking serum for 1.5 h
- Incubate with a mouse monoclonal anti-γH2AX antibody (1:1000) for 2 h
- Incubate with FITC-conjugated goat-anti-mouse secondary antibody (1:500) for 1h
- Add DAPI (4', 6-diamidino-2-phenylindole) to the cells and incubate for 15 min
- Remove the cover slips from the plate and mount on to glass slides
- To prevent bias in selection of cells that display foci, count all cells in the field of vision (at least 50 cells)

Fig. 9.2 Example of the experimental procedure for preparation of microscope slides for examination of nuclear foci according to the description in Kaneko et al. (2005)

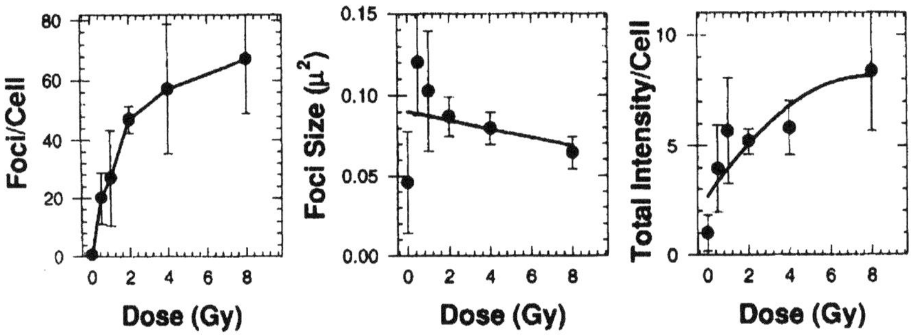

Fig. 9.3 Characterisation of γ-H2AX foci detected 30 min after various doses of X-rays in Chinese hamster V79 cells: average foci number per cell, average foci area per cell and average total cell fluorescence (mean and standard deviation, $n = 5–10$ cells). (Reproduced from MacPhail et al. 2003 with permission of Taylor and Francis Ltd.)

9.6.1 Foci Size and Phosphorylation Heterogeneity

As stated already, γ-H2AX serves as a generally acknowledged marker of DSB. Nevertheless, with the use of more advanced detection techniques, variation in size and differential distribution in cell cycle phases can be found even in

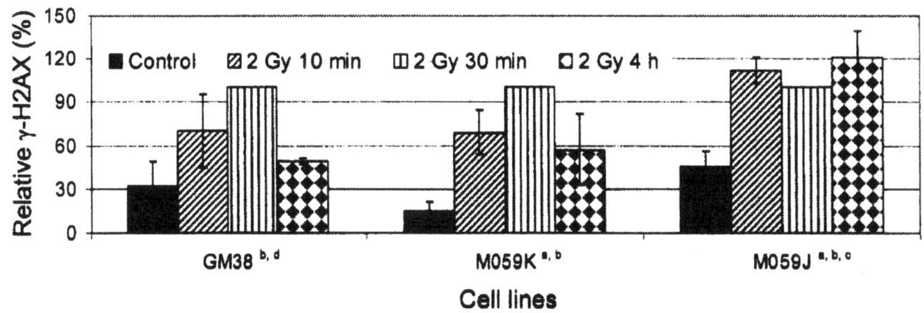

Fig. 9.4 Relative γ-H2AX phosphorylation levels determined after irradiation with 2-Gy γ-rays by flow cytometry and calculated from data obtained from three experiments on human fibroblasts GM38 and human glioma M059K and M059J cells, the latter with defective DNA dependent protein kinase dependent non-homologous end-joining repair. Values for the fluorescence signal intensity of cells stained with anti-γ-H2AX antibody were normalised to the value of the corresponding 30-min sample, arbitrarily set as 100%). Statistically significant differences: *a* $P<0.05$, control compared with 2 Gy, 10 min; *b* $P<0.05$, control compared with 2 Gy, 30 min; *c* $P<0.05$, control compared with 2 Gy, 4 h; *d* $P<0.05$, 2 Gy, 30 min compared with 2 Gy, 4 h; Student's *t* test. (Reproduced from Han et al. 2006 with permission of the Radiation Research Society)

cells not treated with DNA-damaging agents. Also the degree of phosphorylation considerably differs between individual foci.

Small foci are present in S-phase control cells and in all cell cycle phases early after low dose, low linear energy transfer (LET) irradiation (MacPhail et al. 2003b; Costes et al. 2006). It is assumed that replication forks occasionally collapse and the DSB thus generated evokes a normal cellular response aimed at repair. This accounts for the S-phase foci but not for those in other cell cycle phases. According to Han et al. (2006) in the analysed human cells, most of the DSB-unrelated foci (apart from those in early S-phase cells) have a volume smaller than 0.04 μm^3 and they persist after irradiation in unchanged numbers. Their origin and role are unknown. Figure 9.5 shows the numbers of DSB-related and DSB-unrelated foci with S-phase cells excluded from the analysis (although this exclusion does not change the general conclusions from these data; Han et al. 2006). It shows that the numbers of DSB-unrelated foci are very high and that without a careful analysis, the data on γ-H2AX foci can lead to erroneous conclusions.

Large foci can be found at a high incidence in non-irradiated senescent human cells and it is plausible to assume that they mark permanent DNA damage, since they also are found 2 h after high LET irradiation (Costes et al. 2006). In about 20–30% of foci, at higher magnification, with more sophisticated equipment it is possible to discern two to three spots within one large focus, seen as bright centres encircled with a dim halo. The number of spots better corresponds with the DSB number than that of the foci and may explain the often noted underestimation of foci number in relation to DSB (Han et al. 2006).

The extent of γ-H2AX phoshorylation and hence of fluorescence intensity of individual foci considerably differs, as shown in Fig. 9.6. It may be

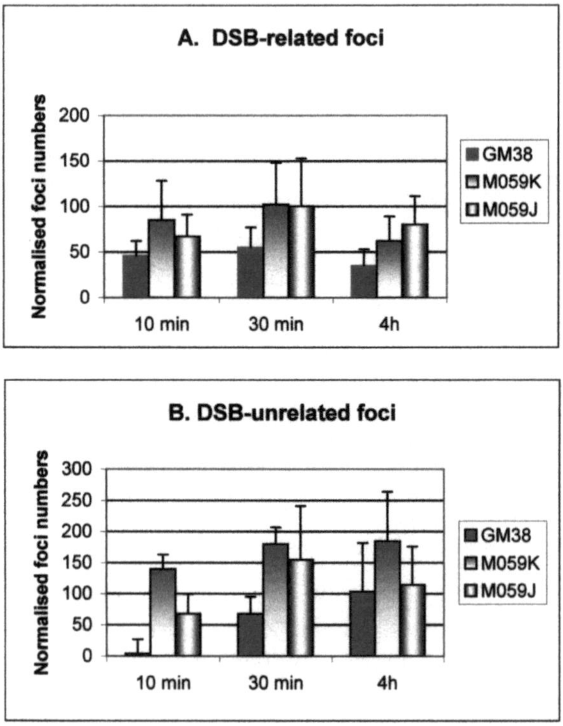

Fig. 9.5 Average numbers of double strand break (*DSB*) related **a** and DSB-unrelated **b** γ-H2AX foci after irradiation with 2 Gy γ-rays (mean ± standard deviation) with S-phase cells excluded from the analysis. Numerical data (numbers of foci normalised to the average nuclear volume at 30 min after irradiation) for human fibroblasts GM38 and human glioma M059K and M059J cells were taken from Han et al. (2006)

expected that as the foci are formed after damage infliction or persist for some time after DSB repair, the histone becomes gradually phosphorylated or dephosphorylated. The kinetics of these processes may depend on the complexity of the damage at individual sites.

Taking into account these confounding factors, it is advisable to set threshold levels of fluorescence intensity (e.g. one tenth of the maximal intensity) and of volume (e.g. 0.04 μm^3) (Han et al. 2006) in order to exclude DSB-unrelated foci from the analysis. The threshold values should be optimised for each cell type.

9.6.2 γ-H2AX Foci Frequency As a Measure of Radiosensitivity

The formation of foci is proportional to radiation dose and is similar in cells with differential radiosensitivity (Wykes et al. 2006). In contrast, the relation

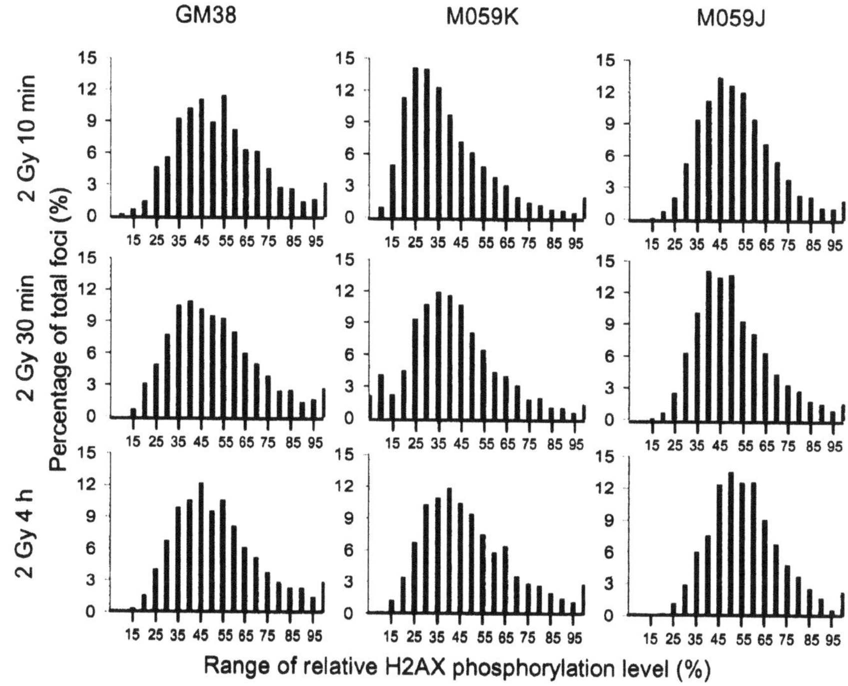

Fig. 9.6 Histogram of the relative H2AX phosphorylation level within DSB-related γ-H2AX foci in human fibroblasts GM38 and human glioma M059K and M059J cells. The *x*-axis represents the series of ranges of numerical value of the relative H2AX phosphorylation level (percentages of the phosphorylation level of the most highly phosphorylated foci in each individual cell) into which the data are sorted in a histogram analysis, and the *y*-axis represents the percentages of total DSB-related γ-H2AX foci within this range. (Reproduced from Han et al. 2006 with permission of the Radiation Research Society)

between the decrease in foci number and repair is by no means clear, as mentioned in Sect. 9.5. Then, what features of the cellular response to DSB are reflected by foci frequency at later times after DSB infliction?

Although the kinetics of focus formation and of the fast component of DSB repair (half-life 15–30 min) are not compatible, the slow component of DSB repair (1–4 h or more) would be consistent with the persistent accumulation of repair factors in foci. In this context, it is important that slowly repaired residual damage is better correlated with cell survival than the rate of fast repair (Badie et al. 1995). Also, correlations can be found between foci frequencies after 4 or 24 h, and radiation sensitivity expressed as surviving fraction after irradiation with 2 Gy (SF2). As discussed by Olive and Banath (2004), the latter parameter usually corresponds to the residual damage. Accordingly, such a correlation between the half-life of γ-H2AX loss and SF2 for a series of normal and tumour cell lines and murine tissues with a large variation in SF2 values was found by McPhail et al. (2003a),

Olive and Banath (2004) and Banath et al. (2004), whereas Wykes et al. (2006) presented a correlation between the mean focus number 4 h after irradiation and SF2 in three pairs of isogenic cell lines with wild type genotype or DSB repair defects.

It should be added that the aforementioned correlation differs between the series of p53-wild-type and p53-mutated human cell lines examined: in p53-wild-type cells the rate of loss of γ-H2AX is faster and this does not seem to depend on p53-controlled apoptosis. Another possibility is that p53 status affects fidelity of repair and the foci are retained at the misrejoined DSB (Banath et al. 2004). Hence, a conclusion, shared by some authors (Karlsson and Stenerlow 2004; Ichijima et al. 2005) that foci also mark chromatin sites not necessarily related to DNA discontinuity.

By comparing the foci numbers at later time intervals, 4 or 24 h, with those at 30 min or 1 h after damage infliction, it is possible to discern severe cellular deficiencies in DSB repair. Figures 9.4 and 9.5 include the data for a pair of glioma cell lines, one of which is DSB repair deficient. It can be seen that both the total γ-H2AX level determined by flow cytometry (Fig. 9.4) and the number of γ-H2AX foci (Fig. 9.5a) in the radiation-sensitive and repair-defective M059J cells do not decrease significantly between 30 min and 4 h after γ-irradiation with 2 Gy. This is taken as indication of a high residual damage level in M059J cells as compared with the repair competent M059K cells. A similar approach was taken to differentiate between wild-type murine cells (Atm+/+) and heterozygous cells (Atm+/–) with an intermediate radiosensitivity between normal and that of the homozygous recessive mutant (Atm–/–). No sharp distinction could be made after acute (2.5 Gy/min) irradiation, but 24-h exposure at 10 cGy/h allowed the radiosensitivity differences between the genotypes examined to be clearly distinguished, based on the foci frequencies (Kato et al. 2006).

Based on the same principle, an attempt was made to apply foci frequency as a predictor of radiosensitivity (Taneja et al. 2004). In a series of human tumour cell lines and xenograft tumours in athymic nude mice, γ-irradiation (3 Gy) induced γ-H2AX foci to a similar extent in all cell types, but the kinetics of focus removal were distinct and clearly dependent on radiosensitivity. Moreover, in a model for fractionated radiotherapy, accumulation of damage was estimated after a single dose or sequential doses of γ-radiation (a single 9-Gy dose or three fractions of 3 Gy at 24-h intervals. Divided doses enhanced the differences in the persistence of γ-H2AX foci between the head and neck cancer cell lines, radioresistant SQ-20B and radiosensitive SCC-61 cells: in SQ-20B cells, the foci disappeared with similar kinetics after each 3-Gy dose, whereas in SCC-61 cells, their number, size and fluorescence intensity increased. Western blot analysis of SQ-20B and SCC-61 cell extracts gave the same response pattern. Thus, γ-H2AX clearance may serve as a predictive test of tumour radiosensitivity to be applied in radiotherapy. Estimation of γ-H2AX accumulation by serial biopsy would help to establish "individually tailored therapy" by choosing the radiation dose and fractionation regimen according to radiosensitivity of the patient's tumour.

9.6.3 γ-H2AX Foci Size and Frequency Depending on Radiation Quality

In a sharp contrast to the uniform distribution of foci in cells exposed to X-rays, in the nuclei of cells irradiated with high LET radiation the foci are clustered (Goto et al. 2002). This provides direct support for other observations which indicate that high LET radiation produces highly clustered DNA damage, including complex lesions. In consequence, the repair proceeds at a slower rate than that of damage caused by low LET radiation (Pinto et al. 2005; Rydberg et al. 2005). Delayed disappearance of foci is, therefore, to be expected. Indeed, such observations were made by Karlsson and Stenerlow (2006) and Leatherbarrow et al. (2006): the relative numbers of residual foci correlated well with the residual damage estimated with the use of PFGE.

The formation of high and low LET radiation induced foci was studied by Costes et al. (2006). HCA2 human foreskin diploid fibroblasts in a confluent state were exposed to 30 cGy of 130 keV/μm nitrogen ions or X-rays (320 kV). The frequencies of the foci were initially similar for both kinds of radiation, but differences were seen both in the plateau level and in the time it was reached: 45 min after X-irradiation and 15 min after nitrogen ion irradiation. The frequency at the plateau was about 3.5 foci per 100 μm^2 for low LET radiation and about 2.0 foci per 100 μm^2 for high LET radiation. Another difference was in the focus size: in the 120 min after exposure, the average size increased from 0.20 to 0.45 μm^2 for high LET radiation and stayed in the same range from 0.22 to 0.30 μm^2 for low LET radiation. The size increase was accompanied by almost doubled fluorescence intensity, indicating a further phosphorylation of the histone during the 2-h period. Analysis of the size distribution allowed the authors to assign the role of high LET radiation to the very large (more than 0.7 μm^2) foci, the proportion of which rose from about 2% 5 min after exposure to about 20% 2 h after exposure to nitrogen ions.

9.7 Concluding Remarks

As outlined in this chapter, analysis of γ-H2AX foci in the nuclei of cells damaged with DSB-inducing agents allows us to identify the locality of the damage, its severity, repair and – when additional repair and checkpoint proteins are examined in overlapping foci – also to acquire new information on the molecular processes that take place at the damaged sites. Although γ-H2AX foci are now generally accepted as a DSB marker, there are background foci differing in size and number in various cell types and growth conditions (Banath et al. 2004) and this may complicate the analysis.

Recently, ATM foci have been proposed as an alternative method of DSB detection, more specific for ionising radiation induced damage (Suzuki et al. 2006). These foci rarely form in non-irradiated S-phase cells, are induced immediately after irradiation in a linear relation to dose in a dose range from 10 mGy to 1 Gy, and do not differ in the distribution of the mean focus size and the kinetics of disappearance in this dose range. With all these merits there is one disadvantage: for reasons unknown, in irradiated cells the size of S-phase foci is about half that of G1- or G2-phase foci, whereas the foci number is almost 2 times higher. So, for the use for exponentially growing cell cultures this difference must be taken into consideration. Clearly, the choice of marker for damage foci should be made depending on the aim of the experiment, not forgetting the confounding factors.

References

Andegeko Y, Moyal L, Mittelman L, Tsarfaty I, Shiloh Y, Rotman G (2001) Nuclear retention of ATM at sites of DNA double strand breaks. J Biol Chem 276:38224–3830

Badie C, Iliakis G, Foray N, Alsbeih G, Cedervall B, Chavaudra N, Pantelias G, Arlett C, Malaise EP (1995) Induction and rejoining of DNA double-strand breaks and interphase chromosome breaks after exposure to X rays in one normal and two hypersensitive human fibroblast cell lines. Radiat Res 144:26–35

Banath JP, MacPhail SH, Olive PL (2004) Radiation sensitivity, H2AX phosphorylation, and kinetics of repair of DNA strand breaks in irradiated cervical cancer cell lines. Cancer Res 64:7144–7149

Bouquet F, Muller C, Salles B (2006) The loss of gammaH2AX signal is a marker of DNA double strand breaks repair only at low levels of DNA damage. Cell Cycle 5:1116–1122

Burma S, Chen BP, Murphy M, Kurimasa A, Chen DJ (2001) ATM phosphorylates histone H2AX in response to DNA double-strand breaks. J Biol Chem 276:42462–42467

Celeste A, Fernandez-Capetillo O, Kruhlak MJ, Pilch DR, Staudt DW, Lee A, Bonner RF, Bonner WM, Nussenzweig A (2003a) Histone H2AX phosphorylation is dispensable for the initial recognition of DNA breaks. Nat Cell Biol 5:675–679

Celeste A, Difilippantonio S, Difilippantonio MJ, Fernandez-Capetillo O, Pilch DR, Sedelnikova OA, Eckhaus M, Ried T, Bonner WM, Nussenzweig A. (2003b) H2AX haploinsufficiency modifies genomic stability and tumor susceptibility. Cell 114:371–383

Chen HT, Bhandoola A, Difilippantonio MJ, Zhu J, Brown MJ, Tai X, Rogakou EP, Brotz TM, Bonner WM, Ried T, Nussenzweig A (2000) Response to RAG-mediated VDJ cleavage by NBS1 and gamma-H2AX. Science 29:1962–1965

Chowdhury D, Keogh MC, Ishii H, Peterson CL, Buratowski S, Lieberman J (2005) gamma-H2AX dephosphorylation by protein phosphatase 2A facilitates DNA double-strand break repair. Mol Cell 20:801–809

Costes SV, Boissiere A, Ravani S, Romano R, Parvin B, Barcellos-Hoff MH (2006) Imaging features that discriminate between foci induced by high- and low-LET radiation in human fibroblasts. Radiat Res 165:505–515

Fernandez-Capetillo O, Lee A, Nussenzweig M, Nussenzweig A (2004) H2AX: the histone guardian of the genome. DNA Repair 3:959–967

Furuta T, Takemura H, Liao ZY, Aune GJ, Redon C, Sedelnikova OA, Pilch DR, Rogakou EP, Celeste A, Chen HT, Nussenzweig A, Aladjem MI, Bonner WM, Pommier Y (2003) Phosphorylation of histone H2AX and activation of Mre11, Rad50, and Nbs1 in response to replication-dependent DNA double-strand breaks induced by mammalian DNA topoisomerase I cleavage complexes. J Biol Chem 278:20303–20312

Goto S, Watanabe M, Yatagai F (2002) Delayed cell cycle progression in human lymphoblastoid cells after exposure to high-LET radiation correlates with extremely localized DNA damage. Radiat Res 158:678–686

Han J, Hendzel MJ, Allalunis-Turner J (2006) Quantitative analysis reveals asynchronous and more than DSB-associated histone H2AX phosphorylation after exposure to ionizing radiation. Radiat Res 165:283–292

Huang X, Darzynkiewicz Z (2006) Cytometric assessment of histone H2AX phosphorylation: a reporter of DNA damage. Methods Mol Biol 314:73–80

Ichijima Y, Sakasai R, Okita N, Asahina K, Mizutani S, Teraoka H (2005) Phosphorylation of histone H2AX at M phase in human cells without DNA damage response. Biochem Biophys Res Commun 336:807–812

Kaneko H, Igarashi K, Kataoka K, Miura M (2005) Heat shock induces phosphorylation of histone H2AX in mammalian cells. Biochem Biophys Res Commun 328:1101–1106

Karlsson KH, Stenerlöw B (2004) Focus formation of DNA repair proteins in normal and repair-deficient cells irradiated with high-LET ions. Radiat Res 161:517–527

Kato TA, Nagasawa H, Weil MM, Genik PC, Little JB, Bedford JS (2006) γ-H2AX foci after low-dose-rate irradiation reveal *Atm* haploinsufficiency in mice. Radiat Res 166:47–54

Kobayashi J, Tauchi H, Sakamoto S, Nakamura A, Morishima K, Matsuura S, Kobayashi T, Tamai K, Tanimoto K, Komatsu K (2002) NBS1 localizes to gamma-H2AX foci through interaction with the FHA/BRCT domain. Curr Biol 12:1846–1851

Leatherbarrow EL, Harper JV, Cucinotta FA, O'Neill P (2006) Induction and quantification of gamma-H2AX foci following low and high LET-irradiation. Int J Radiat Biol 82:111–118

Lee JH, Paull TT (2005) ATM activation by DNA double-strand breaks through the Mre11-Rad50-Nbs1 complex. Science 308:551–554

Lukas C, Melander F, Stucki M, Falck J, Bekker-Jensen S, Goldberg M, Lerenthal Y, Jackson SP, Bartek J, Lukas J (2004) Mdc1 couples DNA double-strand break recognition by Nbs1 with its H2AX-dependent chromatin retention. EMBO J 23:2674–2683

MacPhail SH, Banath JP, Yu TY, Chu EH, Lambur H, Olive PL (2003a) Expression of phosphorylated histone H2AX in cultured cell lines following exposure to X-rays. Int J Radiat Biol 79:351–358

MacPhail SH, Banath JP, Yu Y, Chu E, Olive PL (2003b) Cell cycle-dependent expression of phosphorylated histone H2AX: reduced expression in unirradiated but not X-irradiated G1-phase cells. Radiat Res 159:759–767

Maser RS, Monsen KJ, Nelms BE, Petrini JH (1997) hMre11 and hRad50 nuclear foci are induced during the normal cellular response to DNA double-strand breaks. Mol Cell Biol 17:6087–6096

Meng LH, Kohlhagen G, Liao ZY, Antony S, Sausville E, Pommier Y (2005) DNA-protein cross-links and replication-dependent histone H2AX phosphorylation induced by aminoflavone (NSC 686288), a novel anticancer agent active against human breast cancer cells. Cancer Res 65:5337–5343

Nakamura A, Sedelnikova OA, Redon C, Pilch DR, Sinogeeva NI, Shroff R, Lichten M, Bonner WM (2006) Techniques for gamma-H2AX detection. Methods Enzymol 409:236–250

Olive PL, Banath JP (2004) Phosphorylation of histone H2AX as a measure of radiosensitivity. Int J Radiat Oncol Biol Phys 58:331–335

Paull TT, Rogakou EP, Yamazaki V, Kirchgessner CU, Gellert M, Bonner WM (2000) A critical role for histone H2AX in recruitment of repair factors to nuclear foci after DNA damage. Curr Biol 10:886–895

Petersen S, Casellas R, Reina-San-Martin B, Chen HT, Difilippantonio MJ, Wilson PC, Hanitsch L, Celeste A, Muramatsu M, Pilch DR, Redon C, Ried T, Bonner WM, Honjo T, Nussenzweig MC, Nussenzweig A (2001) AID is required to initiate Nbs1/gamma-H2AX focus formation and mutations at sites of class switching. Nature 414:660–665

Pilch DR, Sedelnikova OA, Redon C, Celeste A, Nussenzweig A, Bonner WM (2003) Characteristics of gamma-H2AX foci at DNA double-strand breaks sites. Biochem Cell Biol 81:123–129

Pinto M, Prise KM, Michael BD (2005) Evidence for complexity at the nanometer scale of radiation-induced DNA DSBs as a determinant of rejoining kinetics. Radiat Res 164:73–85

Redon C, Pilch D, Rogakou E, Sedelnikova O, Newrock K, Bonner W (2002) Histone H2A variants H2AX and H2AZ. Curr Opin Genet Dev 12:162–169

Rogakou EP, Pilch DR, Orr AH, Ivanova VS, Bonner WM (1998) DNA double-stranded breaks induce histone H2AX phosphorylation on serine 139. J Biol Chem 273:5858–5868

Rogakou EP, Boon C, Redon C, Bonner WM (1999) Megabase chromatin domains involved in DNA double-strand breaks in vivo. J Cell Biol 146:905–916

Rogakou EP, Nieves-Neira W, Boon C, Pommier Y, Bonner WM (2000) Initiation of DNA fragmentation during apoptosis induces phosphorylation of H2AX histone at serine 139. J Biol Chem 275:9390–9395

Rothkamm K, Lobrich M (2003) Evidence for a lack of DNA double-strand break repair in human cells exposed to very low x-ray doses. Proc Natl Acad Sci USA 100:5057–5062

Rydberg B, Cooper B, Cooper PK, Holley WR, Chatterjee A (2005) Dose-dependent misrejoining of radiation-induced DNA double-strand breaks in human fibroblasts: experimental and theoretical study for high- and low-LET radiation. Radiat Res 163:526–534

Schultz LB, Chehab NH, Malikzay A, Halazonetis TD (2000) p53 binding protein 1 (53BP1) is an early participant in the cellular response to DNA double-strand breaks. J Cell Biol 151:1381–1390

Sedelnikova OA, Rogakou EP, Panyutin IG, Bonner WM (2002) Quantitative detection of (125)IdU-induced DNA double-strand breaks with gamma-H2AX antibody. Radiat Res 158:486–492

Stiff T, O'Driscoll M, Rief N, Iwabuchi K, Lobrich M, Jeggo PA (2004) ATM and DNA-PK function redundantly to phosphorylate H2AX after exposure to ionizing radiation. Cancer Res 64:2390–2396

Stucki M, Jackson SP (2006) γH2AX and MDC1: Anchoring the DNA-damage-response machinery to broken chromosomes. DNA Repair 5:534–543

Stucki M, Clapperton JA, Mohammad D, Yaffe MB, Smerdon SJ, Jackson SP (2005) MDC1 directly binds phosphorylated histone H2AX to regulate cellular responses to DNA double-strand breaks. Cell 123:1213–1226

Suzuki K, Okada H, Yamauchi M, Oka Y, Kodama S, Watanabe M (2006) Qualitative and quantitative analysis of phosphorylated ATM foci induced by low-dose ionizing radiation. Radiat Res 165:499–504

Szumiel I (1998) Monitoring and signaling of radiation-induced damage in mammalian cells. Radiat Res 150:S92–101

Taneja N, Davis M, Choy JS, Beckett MA, Singh R, Kron SJ, Weichselbaum RR (2004) Histone H2AX phosphorylation as a predictor of radiosensitivity and target for radiotherapy. J Biol Chem 279:2273–2280

Wang H, Wang M, Wang H, Bocker W, Iliakis G (2005) Complex H2AX phosphorylation patterns by multiple kinases including ATM and DNA-PK in human cells exposed to ionizing radiation and treated with kinase inhibitors. J Cell Physiol 202:492–502

Wykes SM, Piasentin E, Joiner MC, Wilson GD, Marples B (2006) Low-dose hyper-radiosensitivity is not caused by a failure to recognize DNA double-strand breaks. Radiat Res 165:516–524

Xu X, Stern DF (2003) NFBD1/MDC1 regulates ionizing radiation-induced focus formation by DNA checkpoint signaling and repair factors. FASEB J 17:1842–1848

10 Comet Assay

WOLFGANG-ULRICH MÜLLER

Abstract

The "comet assay" was designed by Ostling and Johanson in 1984. Its name was derived from the shape of the nuclear DNA after some steps of manipulation. The comet assay can be applied to all cells, tissues and organs as long as it is possible to generate single cells. The concept of the assay is comparatively simple. In order to study DNA damage either directly after exposure to an agent or after repair has taken place, single cells are placed in a matrix on a microscope slide, all non-DNA material is removed, and an electrophoresis run in a weak electrical field is carried out. Depending on the experimental conditions, various types of DNA damage result in negatively charged loops and fragments which are drawn in the direction of the anode and either form the tail of the "comet" or, in the case of small fragments, migrate into the anode material. The comet assay in its various modifications definitely is a valuable tool for the assessment of DNA damage and is now used in many fields of research. A major problem is that many researchers underestimate the difficulties involved with the proper execution of this assay. The greatest danger for any seemingly "simple" assay is an uncritical use of the method resulting in a lot of contradictory results with the final result of abandoning the assay, because it is "unreliable". Hopefully, the comet assay will not meet that fate.

10.1 Introduction

The "comet assay" was designed by Ostling and Johanson (1984). Its name was derived from the shape of the nuclear DNA (Fig. 10.1) after some steps of manipulation, which will be outlined below. Meanwhile, the name "single cell gel electrophoresis" (SCGE) is used quite often, although this term is somewhat misleading as electrophoresis is not done with "single cells", but with the DNA of single cells.

The comet assay can be applied to all cells, tissues and organs as long as it is possible to generate single cells. The concept of the assay is comparatively simple. In order to study DNA damage either directly after exposure to an

Günter Obe and Vijayalaxmi (Eds.)
Chromosomal Alterations: Methods, Results and Importance in Human Health

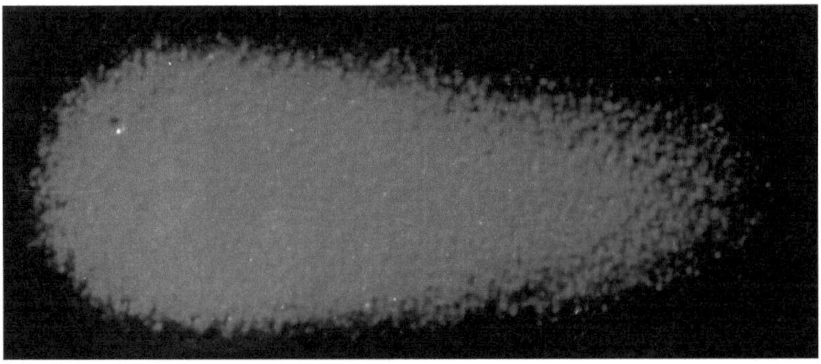

Fig. 10.1 Comet after exposure of lymphocytes to 10-Gy X-rays

agent or after repair has taken place, single cells are placed in a matrix on a microscope slide, all non-DNA material is removed (some nuclear-core proteins will persist) and an electrophoresis run in a weak electrical field is carried out. Depending on the experimental conditions, various types of DNA damage result in negatively charged loops and fragments which are drawn in the direction of the anode and either form the tail of the "comet" or, in the case of small fragments, migrate into the anode material. Thus, the release of DNA from a highly supercoiled DNA–protein complex is actually measured in the comet assay and there is a lot of evidence that the comet tail consists of DNA loops and not of DNA fragments (Collins et al. 1997b).

The comet assay is now used in many fields of research. In the following, the most important aspects of this assay will be outlined and some crucial issues that must be kept in mind, when running this assay, will be addressed.

10.2 Method

10.2.1 Basic Aspects

Many modifications of the method originally introduced by Ostling and Johanson (1984) have been suggested (Singh et al. 1988; Olive et al. 1990; Müller et al. 1994). Detailed guidelines with in-depth discussion of all the methodological aspects have been published (Rojas et al. 1999; Tice et al. 2000; Olive 2002; Collins 2004; Müller et al. 2004).

The principal steps of the procedure, however, have not changed:

- Use single cells
- Embed the cells into agarose on a microscope slide
- Remove all non-DNA material
- Unwind DNA
- Perform an electrophoresis run

- Stain the comets
- Evaluate

Basically, two approaches (with various modifications) are used: unwinding and electrophoresis are done either under more or less neutral or under alkaline conditions. Irrespective of the protocol chosen, it is of utmost importance to standardize the experimental conditions as far as possible. This is particularly important with regard to temperature, pH and voltage. An increase in voltage, which is usually in the range between 0.5 and 5 V/cm, can induce comet tails even in unexposed cells. In addition, one has to keep in mind that different cell types may need different assay conditions, which have to be identified before the experiments proper are started.

10.2.2 Scoring of Comets

Various methods are used for the quantitative assessment of the comets. Essentially, two different approaches have been introduced: visual scoring (Garcia et al. 2004) and image analysis (Böcker et al. 1997). In both cases, there are many different end points that are used to achieve a quantitative result. End points in visual scoring are, for example, the length of the comet tail or grouping into two to five categories according to the estimation by eye of the amount of DNA in the comet tail. In image analysis, fluorescence of the DNA in the comet is measured quantitatively and the results are expressed as, for example, tail-to-head ratio, percentage of DNA in the tail, or tail moment (i.e. the product of tail length and tail intensity). The last parameter is very popular, although, as Collins (2004) pointed out correctly, it is hard to understand why. In particular, the parameter is not linear with respect to dose, and the shape and, thus, the length of the comet tail very much depends on the experimental conditions chosen. A further disadvantage of the tail moment was addressed by Kumaraval (2006): tail moment is measured in arbitrary units and different image-analysis systems give different values, so comparisons are difficult to carry out.

Visual scoring has, of course, the big advantage that no expensive image-analysis hardware and software is required. On the other hand, the subjective factor is much higher in the case of visual scoring compared with image analysis. It is therefore particularly important to evaluate in a double-blind mode when visual scoring is applied. In addition, there are indications that the sensitivity of image-analysis systems is better (Ladeira et al. 2005).

10.2.3 Methods Aimed at the Detection of Specific Types of DNA Damage

A problem, and one which is consistently ignored by most newcomers to the field, is the fact that the comet assay does not detect a specific type of DNA damage, but reflects several types of damage. Among these are single-strand and double-strand breaks, base damage, incomplete excision-repair sites (in particular, after exposure to UV radiation), alkali-labile sites, cross-links and conformational changes. A lot of the modifications of the original method mentioned before aim at the detection of specific DNA lesions, like single-strand or double-strand breaks. One should keep in mind, however, that such simple allocations like "neutral version means detection of double-strand breaks" and "alkaline version means detection of single-strand breaks" are always wrong, because of the unavoidable presence of other modifications of the DNA.

What one can achieve, however, is a shift in the direction of a specific type, but one can never exclude all the other types completely. For example, the alkaline version will always include in addition to single-strand breaks (and some other types of damage) double-strand breaks. Owing to the much higher number of single-strand breaks compared with double-strand breaks, there is some justification to associate the alkaline version with the detection of single-strand breaks, but one should never forget that other types of damage contribute to the comet tail.

A useful categorization in terms of pH of unwinding and electrophoresis has been mentioned by Kumaravel (2006):

- Neutral (pH 7–8): double-strand breaks and cross-links
- Alkaline (pH 12.1–12.4): single-strand and double-strand breaks, incomplete excision-repair sites, cross-links
- Strongly alkaline (pH>12.6): alkali-labile sites in addition to all types listed above

One must be somewhat cautious with too simple classifications. The original method of Östling and Johanson (1984) was based on a pH of 9.5 and clearly picked up also single-strand breaks.

Descriptions of the methods aimed at the predominant detection of a specific type of damage can be obtained in the following publications:

- Single-strand breaks (Olive et al. 1990).
- Single strand breaks plus alkali-labile sites (Singh et al. 1988; McKelvey et al. 1990).
- Double-strand breaks (Olive et al. 1991).
- Base damage (Collins 2004). Enzymes are used that recognize specific damaged bases and introduce breaks at these sites. For example, endonuclease III (Endo III) recognizes oxidized pyrimidines and formamidopyrimidine DNA glycosylase (FPG) identifies 8-oxo-guanine and other damaged purines.
- Cross-links (Pfuhler and Wolf 1996; McKenna et al. 2003). Whereas the other types of DNA damage increase the amount of DNA in the comet tail, cross-links diminish it.
- Apoptosis (Olive et al. 1993b). DNA of apoptotic cells is assumed to show a very characteristic picture in the comet assay: extremely small or non-existent

heads and large, diffuse tails. This interpretation of the comet shape, however, has been doubted, because it can be shown that such comets disappear as a result of repair (which is not possible in apoptotic cells) and the DNA fragments generated during apoptosis are so small that they would get lost during lysis or electrophoresis (Collins 2004).
- Defects in specific DNA sequences (Santos et al. 1997; McKelvey-Martin et al. 1998; McKenna et al. 2003). A particularly interesting amplification of the comet assay is the application of the fluorescence in situ hybridization technique. This approach allows the detection of damage within defined DNA sequences and, thus, also in specific genes.

10.2.4 Problem of Standardization of Methods

Various suggestions have been made to standardize the method of the comet assay in order to be able to compare results more easily (Tice et al. 2000; Hartmann et al. 2003). On the one hand, however, the suggestions made for standardization are rather vague in most instances and, on the other hand, experience in many fields shows that it is almost impossible to convince scientists to change their methods. Thus, it will take quite a while before some sort of agreement might be reached, if it will be possible at all.

In any case, it is important that all methodological details are outlined precisely in the publications and that, at least, the methods are standardized in one and the same laboratory. This is particularly true for all temperatures, pH values and voltages.

10.2.5 Detection Limits

In particular in the field of radiation research, some information is available with regard to the lowest detection limit of the assay. One has to be very careful in this context to distinguish between results obtained with or without repair. Exclusion of repair is only possible in vitro, because you can prevent repair just by cooling the cells. Owing to the rapid repair of DNA damage, you will determine in vivo only the unrepaired fraction of DNA a couple of hours after the exposure. Even within minutes, the amount of damaged DNA will be markedly lower than immediately after exposure.

Modifications specifically intended to increase the sensitivity of the alkaline version of the assay made possible the detection of DNA damage in lymphocytes after X-ray doses in the range of 50 mGy (Singh et al. 1994). Effects of even lower doses down to 6 mGy after γ-ray exposure have been demonstrated (Malyapa et al. 1998). This corresponds to fewer than ten lesions per diploid genome (under the assumption that 1 Gy induces 1,160 strand breaks per diploid cell) (Moller 2006b). And finally, Singh et al. (1995)

reported 1.02 mGy as lowest detection level after external exposure of lymphocytes to technetium-99m.

The effects of such low doses cannot be detected under neutral conditions, simply because the number of damaged DNA sites that can be identified under neutral conditions is much lower (see before). Thus, at least, about 5 Gy is necessary to induce a detectable response in the neutral version of the assay that is related to double-strand breaks. Sometimes, an initial increase in the dose–response curve between 0 and 2 Gy is attributed to double-strand breaks; this increase, however, is due to a loss of DNA supercoiling by single-strand breaks (Olive 1999).

One should be careful with very high doses, because an extensive number of strand breaks will finally induce a lot of small fragments; and fragments smaller than about 50 kb may be lost during lysis and electrophoresis (Olive 1999).

There is another factor that has to be taken into consideration when proliferating cells are analysed: the number of cells in S phase of the cell cycle. The sensitivity of the neutral assay is much lower when cells in S phase are present (Olive et al. 1992), whereas there seems to be no effect of S-phase cells in the alkaline version (Olive et al. 1992; Malyapa et al. 1998).

10.3 Factors That Might Affect the Outcome of Comet Analyses

Several factors have been studied for possible effects on the comet assay. For an extensive review see Moller et al. (2000). It is very striking that many conflicting results have been published with regard to most of the factors mentioned in the following. This raises suspicion as to the quality of some of the published data.

10.3.1 Age

Moller (2006a) summarizing published data concluded that increased DNA migration is observable with increasing age.

10.3.2 Gender

Moller (2006a) concluded that no effect of gender on the comet assay is detectable, when one looks at all the data published. If at all, there is some indication that men show slightly higher amounts of DNA damage in the conventional alkaline version of the comet assay, whereas such a trend is not detectable for Endo III or FPG sensitive sites.

10.3.3 Smoking

In a meta-analysis, Hoffmann et al. (2005) summarized the data on the effect of smoking on the comet assay. Overall, they found an indication that the levels of DNA damage were higher in smokers compared with those in non-smokers. This result, however, is mainly due to a minority of studies not using image analysis, but arbitrary damage scores, because when the analysis was restricted to studies using image analysis, there was only a tendency for effects on the comet assay. Moller (2006a) concluded from a literature survey that no effect of smoking is detectable in the comet assay.

10.3.4 Alcohol

The comparatively few studies aimed at the role of alcohol in the comet assay are extremely contradictory. Definitely, additional studies are required.

10.3.5 Season

When running the comet assay one should be aware of the possibility that seasonal variation can occur (Moller 2006b). At least part of this variation is due to different sunlight exposure.

10.3.6 Physical Exercise

Extensive physical exercise has been reported to induce DNA damage that is detectable in the comet assay. This was shown for rickshaw pullers (Pandey et al. 2006), treadmill runners (Hartmann et al. 1994), short-distance triathletes (Hartmann et al. 1998) and marathon runners (Tsai et al. 2001). The results are somewhat conflicting, because Tsai et al. (2001) reported that oxidized pyrimidines (detected by Endo III) were responsible for most of the damage, whereas Hartmann et al. (1998) found more breaks and explicitly no increase in oxidized DNA bases (detected by FPG). (By the way, the author of this chapter did not find an effect in the comet assay after a marathon; but perhaps he did not run hard enough.)

10.3.7 Environmental Exposures

Air pollution and diet might have an impact on the results of the comet assay (Moller 2006b).

10.4 Advantages/Disadvantages

A number of advantages make the comet assay very attractive:

- It is possible to determine initial DNA damage and DNA repair. Initial damage can be monitored at low temperatures (e.g. on ice) in order to prevent repair processes; repair kinetics can be obtained after various times after exposure at 37°C.
- The assay is rapid. One can obtain results within a couple of hours after sampling of the cells.
- Few cells are necessary to obtain meaningful results. Many assays testing genotoxic effects need some tens or some hundreds of thousands cells; the comet assay can be done with ten to some hundred cells.
- Cell culturing is not required (although, of course, one can use cell cultures). When cells have to be cultured in order to run a specific assay, there always looms the danger that some cells will not survive under the culture conditions chosen. This can result in a serious selection bias.
- Cell proliferation is not necessary; the only prerequisite is the viability of the cells. In particular, several cytogenetic assays (dicentrics, reciprocal translocations, micronuclei) require mitosis. Thus, the result of these assays can be seriously affected by conditions that interfere with cell proliferation, for example by selecting against severely damaged cells, which are no longer able to reach mitosis.
- One can determine distributions not just means. Quite a number of assay systems are only able to give averages of populations of cells. In many cases, however, it is much more important to obtain an idea of the distribution of the characteristics of single cells. If one looks, for example, at two different tumours whose cells show on average the same repair capabilities, it makes a huge difference if the cells in one tumour all show more or less the same repair capacity and in the other tumour some cells with a very efficient repair are counterbalanced by a similar number of cells with poor repair. Those cells with the efficient repair may survive and result in a relapse of the tumour.
- The assay can be made very sensitive, so that the effects of comparatively low doses or concentrations can be assessed: This is a clear advantage over pulsed-field gel electrophoresis. For example, the effects of repair and the repair kinetics after non-lethal radiation doses can easily be detected in the comet assay; this is particularly important with respect to studies in the context of radiation therapy.
- With the exception of the image-analysis system and the software, the costs are low. When one uses visual scoring, one can avoid even the high costs of the image-analysis system. Visual scoring, however, has some drawbacks (see above).

One should not conceal that there are also some disadvantages.

- Various types of DNA damage are involved. This problem was discussed in detail in Sect. 10.2.3.
- The fidelity of the repair processes cannot be checked. In the comet assay the amount of DNA in the comet tail is taken as a measure of DNA damage. If the comet tail is reduced in the course of time after exposure to a damaging agent, this is seen as the result of repair processes. It is not possible, however, to draw conclusions as to the correct execution of repair.
- The assay is not possible with fixed material. This means that archived material cannot be used. It is, however, of course, possible to freeze and thaw cells as long as it is done without affecting the viability of the cells.

10.5 Applications

The comet assay has been introduced in many fields. One of the crucial questions still is which cell type is most suitable for the assay. Frequently, lymphocytes are used, because they offer some advantages: they can easily be obtained; they are single cells by nature; they are all in G_0 phase of the cell cycle. Some problems, however, should not be forgotten: there are several lymphocyte subpopulations with different sensitivities; it is not clear whether lymphocyte results are representative for other cell types; they are not target cells for cancer.

When proliferating cells are used in the assay, one has to keep in mind that cells in various cell cycle stages may behave differently in the comet assay, so the distribution of cells among the different stages may heavily affect the outcome of the assay.

There are several review articles on applications of the various types of the comet assay:

- Radiobiology (Olive 1999)
- Biomonitoring (Collins et al. 1997a; Kassie et al. 2000; Moller et al. 2000; Faust et al. 2004)
- Aquatic biomonitoring (Mitchelmore and Chipman 1998)
- Genotoxicity testing (Brendler-Schwaab et al. 2005; Moller 2005)
- Carcinogen monitoring (Anderson et al. 1998)
- Genetic ecotoxicology (Cotelle and Ferard 1999) (with the application of the comet assay in plants, worms, molluscs, fish, amphibians and mammalians)
- Various applications (Rojas et al. 1999; Collins 2004)

Some of these applications will be highlighted in the following sections.

10.5.1 Individual Radiosensitivity

People do not show identical or, at least, similar radiosensitivities. This is particularly true for some genetic diseases, but it also applies to the general population (ICRP 1999; NRPB 1999). Such differences play a crucial role after radiation accidents and in radiotherapy. DNA repair capacity of the individual is one important element that determines individual radiosensitivity, but differences in the initial DNA damage may also play a role. It is, therefore, no surprise that the comet assay has been used to predict individual radiosensitivity (Müller et al. 1999, 2001; Alapetite et al. 1999; McKeown et al. 2003).

10.5.2 Hypoxic Cells

In tumour radiotherapy the fraction of hypoxic cells plays an important role, because these cells are markedly less radiosensitive compared with euoxic cells. It is possible to identify hypoxic cells by the comet assay (Olive and Durand 1992; Olive et al. 1993a).

10.5.3 Genotoxicity of Chemicals and Identification of Carcinogenic Substances

A tremendous number of genotoxicity studies have been carried out (e.g. see Table 2 in Rojas et al. 1999, which lists almost 200 chemicals that had been tested in the comet assay until 1999). About 85% of the chemicals tested in the studies summarized by Rojas et al. showed a genotoxic effect; one must, however, keep in mind that most of these chemicals were chosen because there was already some information on genotoxic efficiency.

Numerous chemicals were studied in an in vivo comet assay by Sasaki et al. (2000). The major conclusion of the authors was that the comet assay used in various organs of mice and rats is a suitable technique to evaluate the in vivo genotoxicity of in vitro genotoxic agents. The assay had a high response ratio for rodent genotoxic carcinogens and a high negative response ratio for rodent genotoxic non-carcinogens.

10.5.4 Predictor of Cancer Risk

Baseline, benzopyrene diol epoxide and γ-radiation-induced DNA damage in peripheral blood lymphocytes was significantly higher in (untreated) patients with oesophageal cancer (Shao et al. 2005), bladder cancer (Schabath et al. 2003) or lung cancer (Zhang et al. 2000) compared with that in healthy individuals. The authors attributed these results to an increased genomic instability of the cancer cells. Similar results have been obtained for breast cancer cases

(Smith et al. 2003) and various cancer types (Sigurdson et al. 2005). In a very large study (397 women) Rajeswari et al. (2000) showed that lymphocytes of women with breast tumour showed the highest amount of DNA in the comet tails, followed by first-degree female relatives of breast cancer patients and the lowest values were shown by women without tumours without an enhanced family risk.

It remains to be elucidated, however, whether the comet assay has predictive power before cancer starts to develop or whether the observed effects are just a characteristic of already existing cancer cells. Thus, an extensive prospective study is required.

No effect on the comet assay was observed when lymphoblastoid cell lines with a heterozygous BRCA1 mutation were analysed (Trenz et al. 2005).

10.5.5 Biomonitoring

The possibilities to apply the comet assay in the environment for biomonitoring are virtually unlimited. Various possible genotoxic surroundings have been checked in the past (e.g. waste grounds, polluted air and water, areas contaminated with heavy metals) and publications with new fields of analysis appear all the time.

10.5.6 Assessment of Exposure to Ionizing Radiation in Vivo

There have been some attempts to estimate exposure to ionizing radiation of human beings. The dominant lesions after radiation exposure that are measured in the conventional comet assay are strand breaks. Owing to the comparatively rapid repair processes, already a couple of hours after radiation exposure only unrepaired DNA breaks (in addition to unrepaired DNA non-break lesions, which are identified by the conventional assay) can be detected by the comet assay. As one cannot prevent repair in vivo by, for example, cooling, it is not surprising that no effects could be demonstrated in people living in areas contaminated by the Chernobyl accident (Frenzilli et al. 1998; Garcia and Mandina 2005). Similarly, there was no indication that radiotherapy affects the outcome of the comet assay when lymphocytes were studied using the only slightly modified Ostling and Johanson technique (Müller et al. 2002).

There is, however, an indication that lymphocytes of people chronically exposed to ionizing radiation show a reduced repair capacity after an additional radiation exposure in vitro; this could be shown for patients subjected to radioiodine therapy and for children and adults living in the Chernobyl region (Plappert et al. 1997). In addition, Wojewodzka et al. (1998) reported that strand breakage was enhanced in workers exposed to chronic low-dose irradiation (unfortunately, without mentioning the doses received), but no increase in base damage was observed (Kruszewski et al. 1998). The authors mentioned, however, marked interindividual variability in the exposed as well as in the control group. Very high variability was also observed in a study on radon exposure (Hellman et al. 1999): there was a borderline significance ($P = 0.05$) for the relation between indoor radon exposure and tail moment in the alkaline comet assay for the whole group, but, for example, the white blood cells of the individual with the highest radon exposure had almost the lowest tail moment. This result was commented on by the authors in the following way: "Ionising radiation is obviously only one of many factors (environmental as well as constitutional) determining the individual background level of DNA damage".

10.5.7 UV Damage

UV radiation does not induce strand breaks. Nevertheless, the comet assay can be used to estimate damage induced by UV radiation, because base repair is carried out by excision of the affected DNA sites, thus introducing strand breaks. Therefore, in contrast to most other damaging agents, no comets can be detected directly after UV exposure, but an increase in the DNA in the comet tail is observed during the minutes and hours after the exposure owing to base repair. More detailed information can be obtained from the following publications: Gedik et al. (1992), Green et al. (1992) and Arlett et al. (1993).

10.6 Final Remark

The comet assay in its various modifications definitely is a valuable tool for the assessment of DNA damage. A major problem is that many researchers underestimate the difficulties involved with the proper execution of this assay. A lot of basic knowledge and a lot of experience is required to avoid mistakes ending up in seriously wrong conclusions. The greatest danger for any seemingly "simple" assay is an uncritical use of the method resulting in a lot of contradictory results with the final outcome of abandoning the assay, because it is "unreliable". Hopefully, the comet assay will not meet that fate.

References

Alapetite C, Thirion P, de la Rochefordière A, Cosset J-M, Moustacchi E (1999) Analysis by alkaline comet assay of cancer patients with severe reactions to radiotherapy: Defective rejoining of radioinduced DNA strand breaks in lymphocytes of breast cancer patients. Int J Cancer 83:83–90

Anderson D, Yu TW, McGregor DB (1998) Comet assay responses as indicators of carcinogen exposure. Mutagenesis 13:539–555

Arlett CF, Lowe JE, Harcourt SA, Waugh APW, Cole J, Roza L, Diffey BL, Mori T, Nikaido O, Green MHL (1993) Hypersensitivity of human lymphocytes to UV-B and solar irradiation. Cancer Res 53:609–614

Böcker W, Bauch T, Müller W-U, Streffer C (1997) Image analysis of comet assay measurements. Int J Radiat Biol 72:449–460

Brendler-Schwaab S, Hartmann A, Pfuhler S, Speit G (2005) The in vivo comet assay: use and status in genotoxicity testing. Mutagenesis 20:245–254

Collins A, Dusinska M, Franklin M, Somorovska M, Petrovska H, Duthie S, Fillion L, Panayiotidis M, Raslova K, Vaughan N (1997a) Comet assay in human biomonitoring studies: reliability, validation, and applications. Environ Mol Mutagen 30:139–146

Collins AR (2004) The comet assay for DNA damage and repair: principles, applications, and limitations. Mol Biotechnol 26:249–261

Collins AR, Dobson VL, Dusinska M, Kennedy G, Stetina R (1997b) The comet assay: what can it really tell us? Mutat Res 375:183–193

Cotelle S, Ferard JF (1999) Comet assay in genetic ecotoxicology: a review. Environ Mol Mutagen 34:246–255

Faust F, Kassie F, Knasmuller S, Boedecker RH, Mann M, Mersch-Sundermann V (2004) The use of the alkaline comet assay with lymphocytes in human biomonitoring studies. Mutat Res 566:209–229

Frenzilli G, Lori A, Panasiuk G, Ferdeghini M, Barale R (1998) Comet assay on children's leukocytes 8 years after the Chernobyl disaster. Mutat Res 415:151–158

Garcia O, Mandina T (2005) DNA damage evaluated by the comet assay in lymphocytes of children with 137Cs internal contamination caused by the Chernobyl accident. Mutat Res 565:191–197

Garcia O, Mandina T, Lamadrid AI, Diaz A, Remigio A, Gonzalez Y, Piloto J, Gonzalez JE, Alvarez A (2004) Sensitivity and variability of visual scoring in the comet assay. Results of an inter-laboratory scoring exercise with the use of silver staining. Mutat Res 556:25–34

Gedik CM, Ewen SWB, Collins AR (1992) Single-cell gel electrophoresis applied to the analysis of UV-C damage and its repair in human cells. Int J Radiat Biol 62:313–320

Green MHL, Lowe JE, Harcourt SA, Akinluyi P, Rowe T, Cole J, Anstey AV, Arlett CF (1992) UV-C sensitivity of unstimulated and stimulated human lymphocytes from normal and xeroderma pigmentosum donors in the comet assay: A potential diagnostic technique. Mutat Res 273:137–144

Hartmann A, Plappert U, Raddatz K, Grünert-Fuchs M, Speit G (1994) Does physical activity induce DNA damage? Mutagenesis 9:269–272

Hartmann A, Pfuhler S, Dennog C, Germadnik D, Pilger A, Speit G (1998) Exercise-induced DNA effects in human leukocytes are not accompanied by increased formation of 8-hydroxy-2′-deoxyguanosine or induction of micronuclei. Free Radic Biol Med 24:245–251

Hartmann A, Agurell E, Beevers C, Brendler-Schwaab S, Burlinson B, Clay P, Collins A, Smith A, Speit G, Thybaud V, Tice RR (2003) Recommendations for conducting the in vivo alkaline comet assay. 4th international comet assay workshop. Mutagenesis 18:45–51

Hellman B, Friis L, Vaghef H, Edling C (1999) Alkaline single cell gel electrophoresis and human biomonitoring for genotoxicity: a study on subjects with residential exposure to radon. Mutat Res 442:121–132

Hoffmann H, Hogel J, Speit G (2005) The effect of smoking on DNA effects in the comet assay: a meta-analysis. Mutagenesis 20:455–466

ICRP (1999) Genetic susceptibility to cancer. ICRP publication 79. Pergamon, Oxford

Kassie F, Parzefall W, Knasmuller S (2000) Single cell gel electrophoresis assay: a new technique for human biomonitoring studies. Mutat Res 463:13–31

Kruszewski M, Wojewodzka M, Iwanenko T, Collins AR, Szumiel I (1998) Application of the comet assay for monitoring DNA damage in workers exposed to chronic low-dose irradiation. II. Base damage. Mutat Res 416:37–57

Kumaravel TS, Jha AN (2006) Reliable Comet assay measurements for detecting DNA damage induced by ionising radiation and chemicals. Mutat Res 605:7–16

Ladeira MS, Rodrigues MA, Freire-Maia DV, Salvadori DM (2005) Use of Comet assay to assess DNA damage in patients infected by Helicobacter pylori: comparisons between visual and image analyses. Mutat Res 586:76–86

Malyapa RS, Bi C, Ahern EW, Roti Roti JL (1998) Detection of DNA damage by the alkaline comet assay after exposure to low-dose gamma radiation. Radiat Res 149:396–400

McKelvey VJ, Tice RR, Singh NP (1990) A technique for measuring DNA single strand breaks and alkali labile sites in individual cells. Br J Cancer 62:529–530

McKelvey-Martin VJ, Ho ET, McKeown SR, Johnston SR, McCarthy PJ, Rajab NF, Downes CS (1998) Emerging applications of the single cell gel electrophoresis (comet) assay. I. Management of invasive transitional cell human bladder carcinoma. II. Fluorescent in situ hybridization comets for the identification of damaged and repaired DNA sequences in individual cells. Mutagenesis 13:1–8

McKenna DJ, Gallus M, McKeown SR, Downes CS, McKelvey-Martin VJ (2003) Modification of the alkaline Comet assay to allow simultaneous evaluation of mitomycin C-induced DNA cross-link damage and repair of specific DNA sequences in RT4 cells. DNA Repair (Amst) 2:879–890

McKeown SR, Robson T, Price ME, Ho ET, Hirst DG, McKelvey-Martin VJ (2003) Potential use of the alkaline comet assay as a predictor of bladder tumour response to radiation. Br J Cancer 89:2264–2270

Mitchelmore CL, Chipman JK (1998) DNA strand breakage in aquatic organisms and the potential value of the comet assay in environmental monitoring. Mutat Res 399:135–147

Moller P (2005) Genotoxicity of environmental agents assessed by the alkaline comet assay. Basic Clin Pharmacol Toxicol 96(Suppl 1):1–42

Moller P (2006a) Assessment of reference values for DNA damage detected by the comet assay in human blood cell DNA. Mutat Res 612:84–104

Moller P (2006b) The alkaline comet assay: towards validation in biomonitoring of DNA damaging exposures. Basic Clin Pharmacol Toxicol 98:336–345

Moller P, Knudsen LE, Loft S, Wallin H (2000) The comet assay as a rapid test in biomonitoring occupational exposure to DNA-damaging agents and effect of confounding factors. Cancer Epidemiol Biomarkers Prev 9:1005–1015

Müller W-U, Bauch T, Streffer C, Niedereichholz F, Böcker W (1994) Comet assay studies of radiation-induced DNA damage and repair in various tumour cell lines. Int J Radiat Biol 65:315–319

Müller W-U, Bauch T, Streffer C (1999) The application of comet and micronucleus assay for the assessment of individual radiation risk. In: Baumstark-Khan C, Kozubek S, Horneck G (eds): Fundamentals for the assessment of risks from environmental radiation. Kluwer, Dordrecht, pp 127–134

Müller W-U, Bauch T, Stüben G, Sack H, Streffer C (2001) Radiation sensitivity of lymphocytes of healthy individuals and cancer patients as measured by the comet assay. Radiat Environ Biophys 40:83–89

Müller W-U, Bauch T, Streffer C, von Mallek D (2002) Does radiotherapy affect the outcome of the comet assay? Br J Radiol 75:608–614

Müller W-U, Ciborovius J, Bauch T, Johannes C, Schunck C, Mallek U, Böcker W, Obe G, Streffer C (2004) Analysis of the action of the restriction endonuclease AluI using three different comet assay protocols. Strahlenther Onkol 180:655–664

NRPB (1999) Genetic heterogeneity in the population and its implications for radiation risk. National Radiological Protection Board, Chilton

Olive PL (1999) DNA damage and repair in individual cells: applications of the comet assay in radiobiology. Int J Radiat Biol 75:395–405

Olive PL (2002) The comet assay. An overview of techniques. Methods Mol Biol 203:179–194

Olive PL, Durand RE (1992) Detection of hypoxic cells in a murine tumor with the use of the comet assay. J Nat Canc Inst 84:707–711

Olive PL, Banáth J, Durand RE (1990) Heterogeneity in radiation-induced DNA damage and repair in tumor and normal cells measured using the "comet" assay. Radiat Res 122:86–94

Olive PL, Wlodek D, Banáth JP (1991) DNA double-strand breaks measured in individual cells subjected to gel electrophoresis. Cancer Res 51:4671–4676
Olive PL, Wlodek D, Durand RE, Banáth JP (1992) Factors influencing DNA migration from individual cells subjected to gel electrophoresis. Exp Cell Res 198:259–267
Olive PL, Durand RE, Le Riche J, Olivotto IA, Jackson SM (1993a) Gel electrophoresis of individual cells to quantify hypoxic fraction in human breast cancers. Cancer Res 53:733–736
Olive PL, Frazer G, Banáth JP (1993b) Radiation-induced apoptosis measured in TK6 human B lymphoblast cells using the comet assay. Radiat Res 136:130–136
Ostling O, Johanson KJ (1984) Microelectrophoretic study of radiation-induced DNA damages in individual mammalian cells. Biochem Biophys Res Commun 123:291–298
Pandey AK, Bajpayee M, Parmar D, Rastogi SK, Mathur N, Seth PK, Dhawan A (2006) DNA damage in lymphocytes of Indian rickshaw pullers as measured by the alkaline comet assay. Environ Mol Mutagen 47:25–30
Pfuhler S, Wolf HU (1996) Detection of DNA-crosslinking agents with the alkaline comet assay. Environ Mol Mutagen 27:196–201
Plappert UG, Stocker B, Fender H, Fliedner TM (1997) Changes in the repair capacity of blood cells as a biomarker for chronic low-dose exposure to ionizing radiation. Environ Mol Mutagen 30:153–160
Rajeswari N, Ahuja YR, Malini U, Chandrashekar S, Balakrishna N, Rao KV, Khar A (2000) Risk assessment in first degree female relatives of breast cancer patients using the alkaline comet assay. Carcinogenesis 21:557–561
Rojas E, Lopez MC, Valverde M (1999) Single cell gel electrophoresis assay: methodology and applications. J Chromatogr B Biomed Sci Appl 722:225–254
Santos SJ, Singh NP, Natarajan AT (1997) Fluorescence in situ hybridization with comets. Exp Cell Res 232:407–411
Sasaki YF, Sekihashi K, Izumiyama F, Nishidate E, Saga A, Ishida K, Tsuda S (2000) The comet assay with multiple mouse organs: comparison of comet assay results and carcinogenicity with 208 chemicals selected from the IARC monographs and US NTP Carcinogenicity Database. Crit Rev Toxicol 30:629–799
Schabath MB, Spitz MR, Grossman HB, Zhang K, Dinney CP, Zheng PJ, Wu X (2003) Genetic instability in bladder cancer assessed by the comet assay. J Natl Cancer Inst 95:540–547
Shao L, Lin J, Huang M, Ajani JA, Wu X (2005) Predictors of esophageal cancer risk: assessment of susceptibility to DNA damage using comet assay. Genes Chromosomes Cancer 44:415–422
Sigurdson AJ, Hauptmann M, Alexander BH, Doody MM, Thomas CB, Struewing JP, Jones IM (2005) DNA damage among thyroid cancer and multiple cancer cases, controls, and long-lived individuals. Mutat Res 586:173–188
Singh NP, Graham MM, Singh V, Khan A (1995) Induction of DNA single-strand breaks in human lymphocytes by low doses of gamma-rays. Int J Radiat Biol 68:563–569
Singh NP, McCoy MT, Tice RR, Schneider EL (1988) A simple technique for quantitation of low levels of DNA damage in individual cells. Exp Cell Res 175:184–191
Singh NP, Stephens RE, Schneider EL (1994) Modifications of alkaline microgel electrophoresis for sensitive detection of DNA damage. Int J Radiat Biol 66:23–28
Smith TR, Miller MS, Lohman KK, Case LD, Hu JJ (2003) DNA damage and breast cancer risk. Carcinogenesis 24:883–889
Tice RR, Agurell E, Anderson D, Burlinson B, Hartmann A, Kobayashi H, Miyamae Y, Rojas E, Ryu JC, Sasaki YF (2000) Single cell gel/comet assay: guidelines for in vitro and in vivo genetic toxicology testing. Environ Mol Mutagen 35:206–221
Trenz K, Schutz P, Speit G (2005) Radiosensitivity of lymphoblastoid cell lines with a heterozygous BRCA1 mutation is not detected by the comet assay and pulsed field gel electrophoresis. Mutagenesis 20:131–137
Tsai K, Hsu TG, Hsu KM, Cheng H, Liu TY, Hsu CF, Kong CW (2001) Oxidative DNA damage in human peripheral leukocytes induced by massive aerobic exercise. Free Radic Biol Med 31:1465–1472

Wojewodzka M, Kruszewski M, Iwanenko T, Collins AR, Szumiel I (1998) Application of the comet assay for monitoring DNA damage in workers exposed to chronic low-dose irradiation. I. Strand breakage. Mutat Res 416:21–35
Zhang H, Buchholz TA, Hancock D, Spitz MR, Wu X (2000) Gamma-radiation-induced single cell DNA damage as a measure of susceptibility to lung cancer: a preliminary report. Int J Oncol 17:399–404

11 Origin of Chromosome Aberrations: Mechanisms

PETER E. BRYANT

Abstract

After more than 60 years of observation and research, the molecular mechanisms underlying radiation-induced chromosome aberration formation are still controversial. Here I review the classical and recent models for the origin of aberrations and some of the current controversies. The experimental evidence to date strongly suggests that the mechanisms underlying the two aberration types, chromosome and chromatid aberrations, differ. It is also likely that two separate mechanisms exist for the formation of chromosome exchanges; both contributing to the overall yield. Recent evidence supports the previously postulated existence of an alternative pathway to the classical 'two-hit' mechanism for chromosome exchanges, involving interaction of two DNA double-strand breaks (DSB). This alternative pathway involves a cell's response to a single DSB, perhaps of a relatively complex nature, that interacts with DNA of an undamaged chromosome. Like chromosome aberrations, chromatid aberrations comprise both breaks and exchanges that could have different underlying mechanisms. Recent studies have focussed on the formation and origin of chromatid breaks. It is known that these breaks are formed from single DSB, but are unlikely to be formed (as has been postulated previously) simply by an expansion of the DSB. Here I argue that chromatid breaks are most likely to result from chromatin rearrangements, initiated by a DSB, but are formed by a mechanism involving misjoining of chromatin ends by topoisomerase II.

11.1 Introduction

Chromosome aberrations originate in somatic cells as a result of exposure to physical agents such as ionizing radiation or to genotoxic chemicals. In this review I will focus on the action of ionizing radiation, although the mechanisms discussed may apply also to the formation of spontaneous and some types of chemically induced aberrations, especially those produced by radiomimetic agents such as bleomycin and zeocin. Ionizing radiation and these chemical clastogens produce both chromosome-type and chromatid-type aberrations,

Günter Obe and Vijayalaxmi (Eds.)
Chromosomal Alterations: Methods, Results and Importance in Human Health

although the mechanisms underlying the two aberration types appear to be different.

Let us begin by defining what we mean by chromosome and chromatid aberrations. Chromosome aberrations are formed when cells are exposed to radiation whilst in either the G1 or the G0 (quiescent) phase when the chromosomes are unduplicated. As a consequence, aberrations are observed in the metaphase chromosomes after the cell has progressed from the G1 phase of the cell-cycle, through the S phase and eventually arrived at mitosis. In the case of G0 cells (e.g. human peripheral blood T lymphocytes) it is necessary to stimulate cells back into cell division with phytohaemagglutinin so that the cells enter the cell cycle and proceed to mitosis, where the metaphase chromosomes can be observed. Chromatid aberrations are formed in duplicated chromosomes as a result of exposure of cells to genotoxic agents in either the S phase or the G2 phase of the cell cycle. One or both chromatids may be involved in an aberration.

Chromosome aberrations observed at metaphase fall into two major groups: breaks, resulting in chromosome fragments, and 'exchanges' that have originated via the misjoining of two chromosomes or the two broken ends of the same chromosome. Since they lack a centromere, chromosome fragments often become excluded from the main nucleus as the nuclear envelope reforms, and then appear as 'micronuclei' in the cell cytoplasm (Fig. 11.1).

Chromosome exchanges fall into two categories; firstly the unstable type. These are formed by the misjoining of two broken chromosomes and result

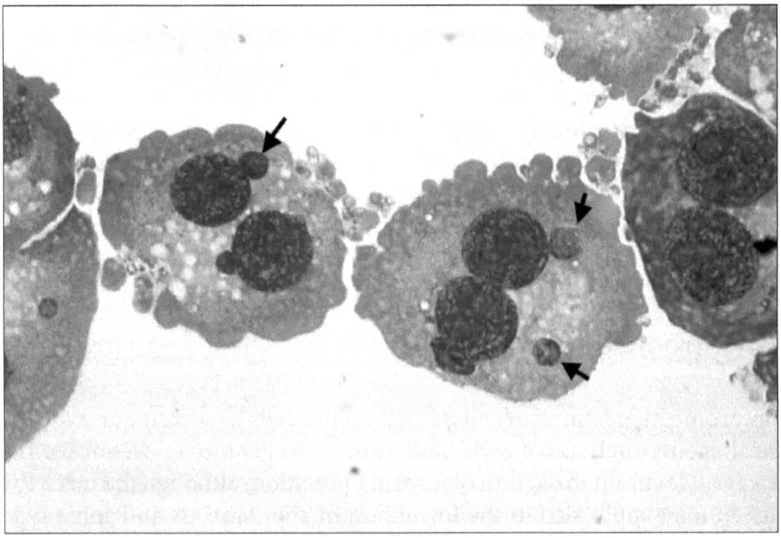

Fig. 11.1 Binucleate (Chinese hamster) cells showing multiple micronuclei (*arrows*) following irradiation and treatment with cytochalasin B

in linkage of two centromeres (dicentrics) or formation of a covalently linked ring within a single chromosome containing one centromere. In either of these cases the remaining acentric chromatin fragments also become linked, and are duplicated during the S phase. Unstable chromosome aberrations are usually lethal to a cell.

Stable chromosome rearrangments occur when two broken chromosomes misjoin and exchange material in a reciprocal fashion, allowing the cell to survive. The consequences of stable rearrangements may be the truncation and/or activation of DNA sequencies, e.g. an oncogene as in the case of the Philadelphia chromosome associated with chronic myeloid leukaemia. The mechanisms of chromosome aberrations are discussed in detail later.

Similarly, chromatid aberrations comprise both break and exchange types. However, in 'normal' human or rodent cells, following low doses of ionizing radiation, the single chromatid break (CB) predominates. At higher doses, and in certain mutant cell strains deficient in DNA repair, exchanges occur at higher frequencies. There are a variety of such exchanges, some intrachromosomal (e.g. sister-chromatid union) and some interchromosomal, i.e. formed between chromatids of two different chromosomes.

Chromosome and chromatid aberrations can be observed without the requirement of the cell to enter mitosis. The technique of premature chromosome condensation (Johnson and Rao 1970; Pandita and Hittleman 1992) uses the fusion of a mitotic cell with the test G1 or G0 cell, or more recently treatment of cells with calyculin A (Gotoh et al. 1999; Kawata et al. 2004) leading to the premature condensation of the interphase cell chromosomes. This technique, as we shall see, has resulted in the acquisition of valuable clues as to the mechanism of chromosome aberrations.

11.2 Primary DNA Damage

In reviewing the mechanisms of chromosome aberrations we have to first consider the initiating cellular damage that leads to their formation. Although it is known and accepted that ionizing radiation produces many types of chemical lesions in the DNA molecules of cells, such as strand breaks, base damage and cross-links, it is generally accepted that the DNA double-strand break (DSB) is the critical initiating lesion. The evidence for this comes primarily from the experiments of Natarajan et al. (1980), who used a poration technique employing Sendai virus to introduce *Neurospora* single-strand endonuclease into living cells to convert some of the induced single-strand breaks into DSB. It was found that this treatment led to a significant increase in the frequency of DSB in cells exposed to a prior dose of X-rays, and a corresponding increase in the number of chromosome aberrations was observed. These experiments were followed up by techniques in which various restriction endonucleases were introduced into dividing cells by various porating agents (Bryant 1984; Natarajan and Obe 1984; Winegar and Preston 1988). It was found that restriction endonuclease treatments (producing DSB with 3′-hydroxyl and 5′-phosphoryl end structures of various types) led to the formation of chromosome aberrations of all types. It was noted moreover that the 'blunt'-ended variety of DSB caused by enzymes such as *Pvu*II and *Eco*RV were much more efficient than the cohesive type produced by enzymes such as *Bam*H1, with a four-base 5′-overhang (Bryant 1984; Bryant and Christie 1989). It was deduced from these experiments that radiation induction of chromosome aberrations stemmed principally from the formation of DSB and that it was possible that DSB with overlapping ends might be rejoined by a different mechanism (e.g. single strand break rejoining) than that of blunt-ended DSB, which are known to be mainly rejoined by the non-homologous end-joining (NHEJ) pathway. Hence, it is likely that the reason why cohesive-ended DSB are much less effective at causing chromosome aberrations is because the rejoining rate for single-strand breaks is extremely fast (i.e. a lower probability of interactions leading to aberrations occurring). It was recognized that the end groups of DSB induced by restriction endonucleases were most probably in the majority of cases different from those induced by ionizing radiation. The reason for this is that the energy imparted by ionizing events in biological structures is many times greater than the energy of a covalent bond. Hence, following irradiation of a cell any bond within its DNA structure is susceptible to breakage, leaving untidy or 'dirty' ends of a DSB.

DSB are subject to rejoining by three known mechanisms: homologous recombinational repair (HRR), homologous single-strand annealing (SSA) and NHEJ (Thacker and Zdzienicka 2004). The formation of chromosome exchanges involves one or more of these mechanisms (Griffin and Thacker 2004; Griffin et al. 2005). Rejoining of DSB can be monitored by techniques such as low-voltage agarose-gel electrophoresis, and it is found that DSB rejoin with biphasic kinetics (Iliakis et al. 2004) (half-times of repair 10 min and 2h) which could possibly reflect the rejoining of different types of DSB. DSB can also be monitored using the disappearance of the γ-H2AX foci (Rotkamm and Lobrich 2003; MacPhail et al. 2003); these are large (megabase-pair) domains of histone H2AX phosphorylation around the sites of DSB (Rogakou et al. 1999). Chromosome exchanges are reported to be formed rapidly following irradiation (Holmberg and Gumauskas 1986; Holmberg 1990); however, the process is reported to be dose-dependent – rapid formation at low doses and slower formation at higher doses (Darroudi et al. 1998)

The frequency of chromosome aberrations depends on the rate and extent of rejoining of DSB since DSB-rejoining deficient mutant strains (e.g. xrs) show elevated levels of chromosome aberrations when treated with either radiation or restriction endonucleases (Kemp and Jeggo 1986; Bryant et al. 1987). However, other cell lines, notably those derived from ataxia telangiectasia cases show elevated chromosome aberration frequencies against a background of normal (or faster) rejoining of DSB compared with normals (Mozdarani and

Bryant 1989b; Foray et al. 1995). Thus, the conversion of DSB into chromosome aberrations is a complex process (particularly for chromatid aberrations; see later) not always dependent on the rate of rejoining of DSB.

11.3 Models of Chromosome Aberration Formation

Essentially, three models have been proposed for the formation of chromosome aberrations; these are the classical 'breakage-first' or 'breakage-and-reunion' model (Sax 1940; Lea and Catcheside 1942; Bender et al. 1974), the molecular recombination model (Chadwick and Leenhouts 1978) and the more recent transcription-based model (Radford 2002). It should be noted that some authors have included Revell's 'exchange' model (Revell 1959) as an alternative to the breakage-first model. However, this model was originally proposed to account only for chromatid aberrations although some authors have taken it as applying also to chromosome aberrations (Neary 1965; Savage 1998). I have chosen to discuss the model only as it applies to chromatid aberrations.

11.3.1 The Breakage-First Model

The breakage-first model is generally applied to the formation of *both* chromosome and chromatid aberrations; however, as mentioned already, evidence indicates that the mechanism involved in the formation of chromosome aberrations (i.e. in G0/G1 cells) is different from that for chromatid aberrations formed in the G2 phase of the cell cycle. Therefore, I will discuss the mechanisms for chromosome and chromatid aberrations separately.

The breakage-first model, in its molecular form (Bender et al. 1974), maintains that the single DNA molecule has to be broken (i.e. by a DSB) before interaction can take place with another DSB, leading to a misjoined structure that can give rise to an 'exchange' aberration, which may be of an unstable type (i.e. likely to result in a missegregation of chromatin between daughter cells at mitosis) such as a dicentric or ring with an associated acentric fragment, or a stable type such as a reciprocal translocation, which could allow the cell to continue proliferating, perhaps with the overexpression of a novel truncated oncogene, leading to neoplastic transformation. An example of this is the Philadelphia chromosome, a chromosome translocation [t(9;22)(q34;q11)] found in the bone-marrow disease chronic myeloid leukaemia, where the translocation of the *ABL* protooncogene from its normal site on chromosome 9q34 to chromosome 22q11 where it forms a fusion hybrid gene with the breakage cluster region. The hybrid gene expresses a novel tyrosine kinase, leading to loss of cell-cycle control, promotion of cell growth and inhibition of DNA repair, leading to genomic instability. A similar case is Bukitt lymphoma, a B-cell malignancy characterized by a chromosome 8;14 translocation. In this case the translocation of the *MYC* protooncogene (normally located at 8q24), most often to 14q32, results in *MYC* activation by enhancers of the heavy chain immunoglobulin gene located at 14q32. Such specific translocations are a feature of many malignancies (Rabbitts 1994) and the progression of cells towards the neoplastic state and subsequent metastasis may involve more than one such translocation resulting from genomic instability in cells as they move towards the fully malignant state.

One of the problems facing proponents of the breakage-first model has been the finding that chromosomes in interphase nuclei occupy seemingly discrete 'domains', adherent to the inner surface of the nucleus (Cremer et al. 1996) and therefore it is difficult to visualize how chromosomes can interact over the relatively large distances of several microns, around or across the nucleus of a cell. Early work in *Drosophila* (Serebrovsky 1929) proposed that the chromosomes needed to be actually physically touching before interaction (misjoining) can take place, sometimes referred to as the 'contact-first' hypothesis, and it has been pointed out by Savage (2000) that preexisting proximity of chromosome segments, if not direct contact, is likely to be the reason for the occurrence of such typical translocations as the Philadelphia chromosome in chronic myeloid leukaemia (see above) and the RET/H4 intrachange in many radiation-induced thyroid tumours (Nikiforova et al. 2000).

Under the breakage-first model, the failure of a DSB to rejoin would, as the cell enters metaphase, give rise to two chromosome fragments (one centric and one acentric) which will have been, like exchange aberrations, duplicated during semiconservative replication. Owing to the absence of a centromere, acentric fragments, although comprising two chromatid fragments, cannot be segregated into the two daughter cells at mitosis, and are often excluded from the cell's nucleus during the process of envelope reformation. Such excluded fragments become enclosed in a smaller nuclear envelope and are termed 'micronuclei'. Occasionally, the centric fragment, or even a whole lagging chromosome, may become excluded from the nucleus.

In some of the earliest work on induction of chromosome and chromatid aberrations in *Tradescantia* microspores, Sax (1939, 1940) made the distinction between 'one-hit' and 'two-hit' aberrations; one-hit aberrations being classified as terminal deletions (unrejoined breaks) and two-hit aberrations as fusions (or exchanges) between two chromosomes. Sax reported that if the dose rate was kept constant, the yield of two-hit aberrations is proportional to the square of the radiation dose, whereas one-hit aberrations were linear with dose. The quadratic induction of (two-hit) exchanges is easily understood in molecular terms, since the probability of inducing two DSB in two chromosomes increases as the radiation dose increases. It is now generally recognized that, at least for mammalian cells, there is also a linear (single-hit) component of the dose response for exchanges, resulting from the fact that, even with X-rays, some clusters of ionizing events are large enough to induce two DSB in two chromosomes. For low linear energy transfer (LET) radiation such as X-rays this component is relatively small. However, for high LET

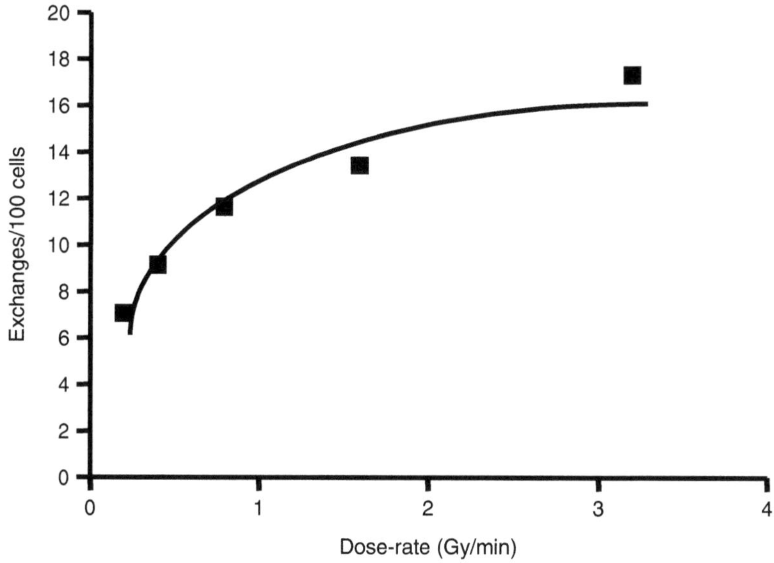

Fig. 11.2 The dose-rate effect for chromosome exchanges in *Tradescantia* microspores following a dose of 3.2 Gy. (Drawn from data of Sax 1939)

radiation (e.g. α-particles) the linear component of exchange induction dominates. Sax also recognized that by increasing the time of exposure to a given dose of radiation, the yield of two-hit aberrations would decrease (Fig. 11.2). This dose-rate effect is very relevant today in clarifying the mechanism of chromosome exchanges and I will return to this point later.

The breakage-first model was seriously challenged by the finding that very low energy X-rays (carbon-K X-rays that produce electrons with tracks of less 7 nm tracks) efficiently induce simple exchange-type aberrations and the dose–effect response for these aberrations is linear with dose (Virsik et al. 1980; Thacker et al. 1986; Griffin et al. 1998). The implication of this work is that exchanges can be induced by single ionizing events, and thus can arise from a single DSB. In support of the single DSB pathway to chromosome exchanges, Ludwikóv et al. (2002) used the property of the late replication of a region (Xq) of the Chinese hamster chromosome to label the bulk of chromosomes with ^{125}I while leaving Xq free of incorporated label. The frequency of exchanges between the unlabelled region and labelled chromosomes was found to be half that between different labelled chromosomes. The results support the existence of an alternative (single DSB) mechanism for the formation of exchanges.

Simple chromosome exchanges, i.e. those defined as being formed by interactions between two chromosomes carrying one DSB each (Savage 2002), should, according to the breakage-first model, be induced as a quadratic function of radiation dose. The single-hit–no-hit concept of chromosome exchanges would predict a completely linear dose–effect curve. However, the data for simple exchange low LET radiation is controversial, and dose–effect curves are linear–quadratic with exponents of dose for total simple exchanges around 1.35 (Edwards et al. 1999; Loucas and Cornforth 2001). A further important finding is that the frequency of formation of chromosome exchanges by low LET radiation is dose rate dependent and is reduced also by fractionating the dose with a time interval for repair. These properties were already noted by Sax (1939) in *Tradescantia* microspores (see above and Fig. 11.2). Sax used this as an argument for the requirement for the interaction of two lesions in one or two chromosomes. Sax realized that as the dose rate was decreased there was more time for repair of one of the lesions before a second was induced. Relevant to the above discussion of the breakage-first versus one-hit–no-hit formation of exchanges is that a striking dose-rate effect is seen for the formation of simple exchanges in human primary fibroblasts (Loucas et al. 2004). At very low dose rate, induction of simple exchanges is linear, possibly indicating that a limited fraction of the simple exchanges are formed via a one-DSB mechanism.

Other lines of evidence point to the requirement for two interacting DSB in the formation of chromosome exchanges. For example Cornforth (1990) elegantly demonstrated that by fusing cells together and thereby mixing irradiated chromosomes from one cell with unirradiated chromosomes from another, only a few interactions between the two sets of chromosomes occurred. Essentially only interactions leading to exchange aberrations occurred between the irradiated cell's chromosomes. Also, Wang et al. (1997) showed that exchanges (translocations) induced by the radiomimetic bleomycin in the *aprt* gene of Chinese hamster cells were almost all (six out of seven) reciprocal exchanges between the aprt sequence and an unrelated sequence on a different chromosome, consistent with a homology-independent (NHEJ) pathway and the breakage-first model. Support for the breakage-first model also came from the experiments of Richardson and Jasin (2000), who examined several recombinant embryonic stem (ES) cell clones in which recombination events had occurred between DSB in inserted sequences engineered to contain one or two I-*Sce*I sites, and cut by endogenous expression of the I-*Sce*I endonuclease. In support of the breakage-first model, only those recombinant clones resulting from interaction of DSB in cells with two I-*Sce*I sites (when incised by transiently expressed omeganuclease this yields a DSB with 3′-cohesive ends with four-base overlap) showed translocations, whereas those with only a single site did not. Analysis of the recombinational joins showed only a small degree of homology between the two sites, suggestive of SSA rather than homologous recombinational rejoining of the two sites. Whether such misjoining is representative of global misjoining of DSB in irradiated cells is not known, and it is possible that radiation-induced DSB, which are of a more complex nature (e.g. in the case of DSB induced by carbon-K X-rays, the electrons of which cause dense clusters of ionizations) may be misjoined in a different way from those produced by omeganuclease at I-*Sce*I sites.

Further recent evidence in support of the breakage-first model comes from experiments demonstrating movement together and apparent fusion of γ-H2AX foci within minutes of the passage of α-particle tracks or microbeams of ultrasoft X-rays (Aten et al. 2004; Hamada et al. 2006). This clustering of phosphorylated histone domains following DSB formation was shown to also implicate the MRE11 complex, which colocalizes with γ-H2AX clusters (Aten et al. 2004). These findings argue for a breakage-first mechanism of chromosome exchange formation and could possibly be an explanation for the formation of complex chromosome translocations involving interactions between breaks in several chromosomes as revealed by multicolour fluorescence in situ hybridization (FISH) (Griffin et al. 1995, 1996, 1998; Loucas et al. 2004; Anderson et al. 2002). However, this view of chromatin movement following induction of DSB has been challenged recently by Kruhlak et al. (2006), who demonstrated limited mobility and a simple expansion or diffusion of chromatin neighbouring a DSB following its induction by UV laser light. However, the question arises as to whether the difference in the types of damage induced by UV radiation (e.g. pyrimidine dimers which distort chromatin structure) as compared with α-particle or ultrasoft X-ray tracks (Aten et al. 2004; Hamada et al. 2006) could play a role in the modification of chromatin following exposure.

11.3.2 The Molecular Misrecombination Model

The misrecombination model for the formation of chromosome aberrations (Chadwick and Leenhouts 1978) invokes the Resnick (1976) model of homologous recombinational rejoining of DSB, which is prevalent in yeast but is thought to be of much lesser importance in mammalian cells (Essers et al. 2000; Wang et al. 2001; Iliakis et al. 2004). Essentially, the misrecombination model proposes that a DSB in one chromosome undergoing HRR with its homologue leads to misjoining with the homologous chromosome. At first glance this scheme would seem a likely explanation for single-hit exchanges; however, data do not appear to support the model, at least for normal cells. First of all, in normal (wild-type) cells, exchanges do not appear to be preferentially formed between homologous chromosome pairs (Savage 2002; Cornforth et al. 2002). For example, the frequency of exchanges between two human homologues in a hamster–human hybrid was the same as that in a hybrid containing only one human chromosome, i.e. there was no preferential interaction between the two homologues (Marcon et al. 2000). The frequency of chromosome exchanges between chromosomes 21 and 22 in Down syndrome (trisomy 21) as compared with normals (Grigorova and Natarajan 1998) was found to be similar, suggesting that homologous recombination does *not* play a role in the formation of exchanges in these cells. In contrast to these studies, Plan et al. (2005) reported an excess of homologous exchanges in human lymphocytes over that expected on the basis of random (non-homologous) interactions. Although the vast majority of dicentrics were heterologous, suggesting that homologous misrecombination is not the dominant mechanism, a role for this type of misjoining could not be ruled out entirely. Plan et al. suggest the excess could feasibly result from the relative proximity of homologues within the cell nucleus.

In *Xrcc2*$^{-/-}$ (a gene involved in the homologous recombination pathway) knockout mouse embryonic fibroblasts, the frequency of exchanges between homologues was increased dramatically, with cells showing a sevenfold increase in spontaneous homologous rearrangements (Deans et al. 2003). Also, in *Xrcc2*$^{-/-}$ ES cells the frequency of both spontaneous and radiation-induced exchanges was increased dramatically (Griffin et al. 2005). Thus, HRR is clearly important in the maintenance of genomic stability and in *preventing* formation of interchromosome rearrangements, so its proposed role in misrecombination leading to the formation of radiation-induced chromosome exchanges (Chadwick and Leenhouts 1978) is therefore doubtful. More plausible seems the proposal that misrecombination between chromosomes, leading to rearrangments, could arise via homologous SSA (Griffin and Thacker 2004) whereby minihomologies or microhomologies between heterologous chromosomes (e.g. satellite or Alu sequences which are common between all chromosomes) lead to misjoining. Such a process might conceivably occur, for example, during replication, or in the G1 phase during active gene expression where homologous sequences might be susceptible to strand invasion by a damaged strand and followed by strand-nicking to yield an interchromosomal exchange. Otherwise two DSB would be required to interact pairwise as in conventional SSA-mediated joining of DSB (Lin et al. 1984). Data of Griffin et al. (2005) show that deletion of the *Ercc1* gene (involved in SSA repair) results in elevated fragmentation of chromosomes after X-irradiation, whereas the frequency of simple and complex exchanges is only marginally increased, possibly indicating a role for SSA in the formation of exchanges.

11.3.3 The Transcription-Based Model

The idea that chromosomes, normally visualized by FISH during interphase as discrete concave domains attached to the nuclear matrix (Cremer and Cremer 2001), can interact over relatively large distances within the cell nucleus (Cremer et al. 1996; Savage 1998, 2000) raised the possibility that large loop domains may come together via channels within the nucleus to link transcription 'factories' (Iborra et al. 1996; Cook 1999). The idea of such interaction centres, corresponding to transcription factories, forms one component of the transcription-based model for the formation of chromosome aberrations (Radford 2002). The other component is the observation that some 80% of chromosome exchange aberrations occur in the light G-bands (R-bands) of X-irradiated normal human chromosomes (Holmberg and Jonasson 1973; Mattei et al. 1979; Nakagome et al. 1983; Radford et al. 1988) and in Chinese hamster chromosomes (Slijepcevic and Natarajan 1994;

Folle et al. 1998). These light bands are gene-rich and transcriptionally active regions of the chromosomes (Musio et al. 2002). The model proposes that a DSB induced in a loop from one chromosome becomes associated with a topoisomerase I (topo I) at a stalled RNA polymerase site. This trapped DNA–topo I then interacts with another topo I bound to a transcription unit on another chromosome. Radford proposes that the two DNA–topo I cleavages are then reversed, and in doing so misjoin the two chromosomes. A further cleavage would be required in order to resolve the structure to give a dicentric chromosome and a fragment or a reciprocal translocation.

As mentioned above, the idea of interactions between chromosomes at transcription factories and involving extruded loop domains has been proposed previously (Griffin et al. 1996; Savage 2000) and it seems feasible that regions of DNA undergoing active transcription at such factories, and hence having a more open structure, would be more vulnerable to cleavage and/or interaction via topoisomerases or indeed by homologous SSA (see before). Griffin et al. (2005) showed that deletion of the *Csb* gene ($Csb^{-/-}$ knockout in ES cells), a gene involved in transcription-coupled repair allowing more rapid nucleotide excision repair in the transcribed damaged DNA strand in active genes (van Hoffen et al. 2003), led to a significant reduction in complex exchanges relative to that in wild-type cells. The reduction in complex exchanges in $Csb^{-/-}$ cells was especially marked following α-particle irradiation. Griffin et al. (2005) suggest that the absence of Csb protein might hinder access of DSB to interactive transcription sites, or might simply block the multiple interactions, leading to complex exchanges.

Thus, while there is no direct evidence for the transcription-based model as such, DSB in gene-rich (transcribed) regions of chromosomes are more likely to undergo exchange. Furthermore, transcription sites in the cell are implicated in the formation of exchanges between DSB, possibly with unbroken strands, or between two or more DSB to form complex aberrations.

In summary, these studies indicate that although the major mechanism for formation of exchanges requires breakage first (i.e. two interacting DSB) it seems probable that at an alternative pathway may exist whereby some DSB, possibly those of a more complex nature (Griffin et al. 1998), can interact with a homologous sequence on an undamaged heterologous chromosome, leading to an exchange aberration.

11.4 Models of Chromatid Aberration Formation

11.4.1 Formation and Disappearance of Chromatid Breaks

Chromosome aberrations formed in the G2 phase of the cell cycle, i.e. where DNA has been duplicated, can occur in one or other or both of the sister chromatids. In human and other mammalian cell lines, the G2 phase is a highly radiosensitive phase and relatively small doses of radiation lead to significant blockade of the cell's progression into mitosis. Therefore, relatively small radiation doses are generally used (less than 1 Gy) and the frequency of exchanges in normal cells is very low. For example, in normal human lymphocytes the comparative frequencies were found to be 2.5 exchanges per 100 cells per gray compared with the maximum frequency of breaks 451 breaks per 100 cells per gray, giving a break-to-exchange ratio of 180 (MacLeod and Bryant 1992). Most recent studies of chromatid aberrations in mammalian cells have focussed on formation of chromatid breaks (CB). CB are induced linearly with dose, but the frequency of CB is not constant with time; the maximum frequency occurs at around 0.5 h following irradiation, with an exponential decline in the frequency thereafter (MacLeod and Bryant 1992; Parshad 1996) In contrast, the frequency of exchanges is more or less constant between 0.5 and 4h following irradiation.

The exponential decline in CB with time after irradiation has been variously interpreted in terms of repair or processing of DNA lesions such as base damage (Preston 1980) or DSB (Mozdarani and Bryant 1989a; MacLeod and Bryant 1992; Bryant and Slijepcevic 1993; Parshad 1996).

The induction of CB is linear with dose. This was demonstrated by Sax (1940) using *Tradescantia* microspores (Fig. 11.3), and when the maximum frequency of CB in human or rodent cells was taken as an end point, the induction of CB (Fig. 11.3) was also found to be linear (Bryant 1998; Bryant et al. 1998; Mozdarani et al. 2001). This view of induction and rejoining of CB has been challenged (Savage 2004) on the grounds that variability in the transit times of individual cells through the G2 phase will render measurements of breaks at various times after irradiation too imprecise to interpret them in this way. However, using the rapid premature chromosome condensation

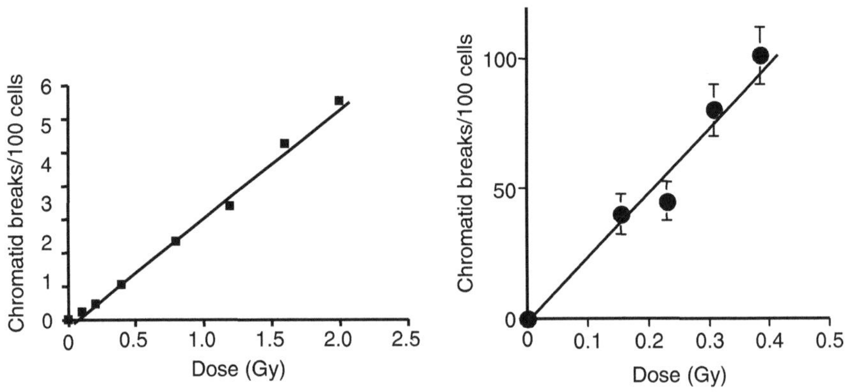

Fig. 11.3 *Left*: Frequency of chromatid breaks in *Tradescantia* microspores as a function of radiation dose. *Right*: Frequency of (G2) chromatid breaks in human lymphocytes. (*Left* drawn from data of Sax 1940; *right* from Bryant 1998)

induced by calyculin A (with which the induction of G2 CB can be studied without the putative complication of cell-cycle kinetics) the dose–effect relationship for CB was similarly shown to be linear (Gotoh et al. 1999; Kawata et al. 2004), confirming the single-event (DSB) origin of CB. Taken together, the decline in CB frequency with time during the G2 phase, both with a standard G2 assay (MacLeod and Bryant 1990, 1992; Parshad 1996) and with the premature chromosome condensation technique, which is independent of cell-cycle parameters (Pandita and Hittelman 1992; Gotoh et al. 1999; Kawata et al. 2000, 2004), suggests that for human and other mammalian cell lines, the kinetics of disappearance of CB does indeed represent a rejoining process. Moreover, the fact that treatment of cells with nucleoside analogues (arabinoside cytosine and adenine) following irradiation abrogates the rejoining or in some cases actually increases the frequencies of CB (Preston 1980; Mozdarani and Bryant 1989a, b; Antoccia et al. 1994; Parshad 1996) suggests a metabolically active rejoining process underlies the decline in CB with time.

11.4.2 Models of Chromatid Breakage

Four different models have been applied to account for CB. These are the breakage-first model (Sax 1940; Bender et al. 1974) the 'exchange' model (Revell 1959; Savage 1986; Harvey and Savage 1997) the misrecombinational model (Chadwick and Leenhouts 1978, 1998) and the signal model (Bryant 1998; Bryant et al. 2004).

The breakage-first model appears at first to fit much of the data for CB on the assumption that a CB can be equated with DSB (Bender et al. 1974; Parshad et al. 1990; Parshad 1996): linear induction of CB, and disappearance ('rejoining') of CB with time, as outlined already. However, several examples show that rejoining of CB does not accurately reflect the rejoining of DSB. For example, the rate of decline in frequency of CB with time in a DSB repair mutant (xrs5) was the same as in its wild-type (CHOK1) parental line, although the frequency of CB was more than 3 times that in CHOK1 (MacLeod and Bryant 1990; Bryant and Slijepcevic 1993). Also, in cell lines mutated in *Atm*, *Xrcc2* and *Xrcc8* genes, radiation-induced CB are significantly elevated (Taylor 1978; Mozdarani and Bryant 1989; Mozdarani et al. 2001; Bryant et al. 1993), while these cell lines all show normal or even faster rejoining of bulk DSB (Jones et al. 1990; Thacker and Ganesh 1990; Foray et al. 1995). A further point made previously (Bryant 1998; Bryant et al. 2004) is that of the huge difference in size between a DSB (where only a few base pairs are thought to be deleted;Pfeiffer et al. 2005) and a CB, which can be 5–20 Mbp (up to about 10% of the length of an average hamster chromosome). Two DSB might interact to produce such deletions, but this would yield a quadratic dose–effect relationship for CB instead of the observed linear induction (see before). Finally, it was shown that treatment of Chinese hamster ovary (CHO) cells with the nucleoside analogue arabinoside adenine at 100 μM abrogates the rejoining of CB without affecting the rejoining of DSB (Bryant et al. 2004).

The fact that *Xrcc2* (HRR pathway), *Xrcc5* (NHEJ pathway) and *Atm* gene mutations all lead to enhanced CB induction but do not affect rejoining indicates that these genes normally protect the cell from CB induction and are not actively involved in the production or processing of CB following irradiation with X-rays or γ-rays.

Thus, the breakage-first model does not satisfactorily account for the observed CB data. This conclusion is supported by evidence that indicates that the mechanism for CB is different from that for chromosome aberrations, where, as we have seen, breakage-first is the major mechanism. For example, Scott et al. (1999) showed a complete lack of correspondence between the frequencies of CB and micronuclei (a measure of chromosome fragment frequency – these arising by breaks and from unstable exchanges: rings and dicentrics) in peripheral blood lymphocytes of a group of breast cancer patients with variable individual chromosome radiosensitivity. Further evidence for this difference between the mechanisms of chromosome fragments and CB was provided by experiments of Vral et al. (2002) that showed the relative biological effectiveness (RBE) for neutrons for induction of micronuclei to be 5, whereas the RBE for CB was 1.0. These experiments suggest that the quality of DSB (i.e. the complexity) is an important determinant of chromosome aberrations, but does not appear to substantially affect the formation of CB.

11.4.3 The Revell 'Exchange' Model

The breakage-first model was challenged by Revell (1959), who proposed a model for chromatid aberrations (Fig. 11.4) under which the initiating lesions (unspecified) are those formed close together at the cross-over points of chromatin loops (Savage 1986). These lesions would interact, i.e. undergo subsequent breakage and misjoining, to give exchange-type aberrations, or if the rearrangements were incomplete they would lead to the various types of CB observed. Thus, in terms of Revell's model, CB (or terminal deletions) are a secondary product of an incomplete rearrangement. The model predicts that CB (i.e. a break in one chromatid, and not including isochromatid breaks) will be of five different types (Fig. 11.4).

Two of these types involve a rearrangement between sister chromatids and can be detected by a colour switch at the breakpoint in harlequin (fluorescence plus Giemsa) stained chromosomes (Fig. 11.5).

The frequency of such interarm rearrangements in irradiated and restriction endonuclease treated mammalian cells, in unirradiated cells and in omeganuclease-treated cells containing I-*Sce*I break sites (Rogers-Bald et al. 2000) is reported to be around 16%, and a constant ratio (Harvey and Savage 1997). This percentage is, of course, much lower than that originally predicted by Revell, who argued that all CB types would be produced at equal frequencies, thus yielding two fifths (or 40%) of the single CB appearing as

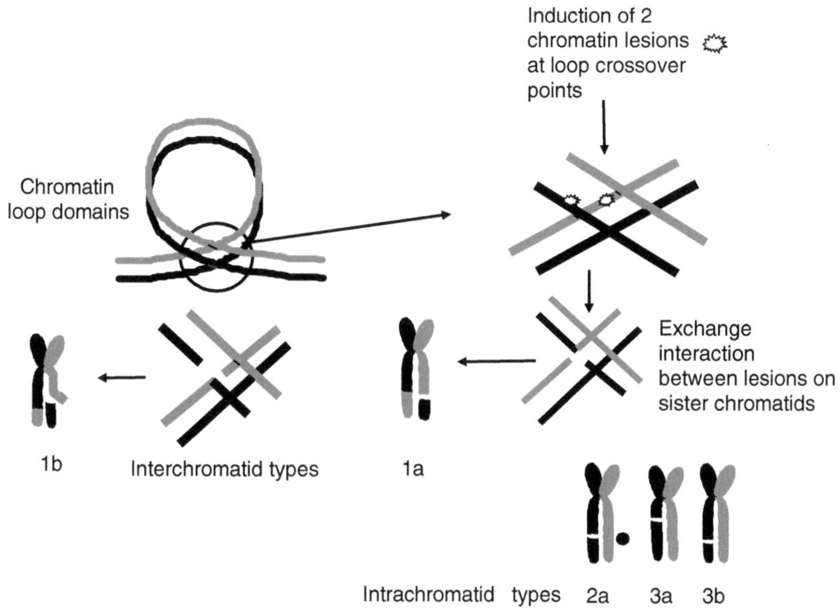

Fig. 11.4 Revell's exchange model. Note that the lesions at loop crossover points were not specified by Revell, but presumably would at some point have to be converted into double-strand breaks

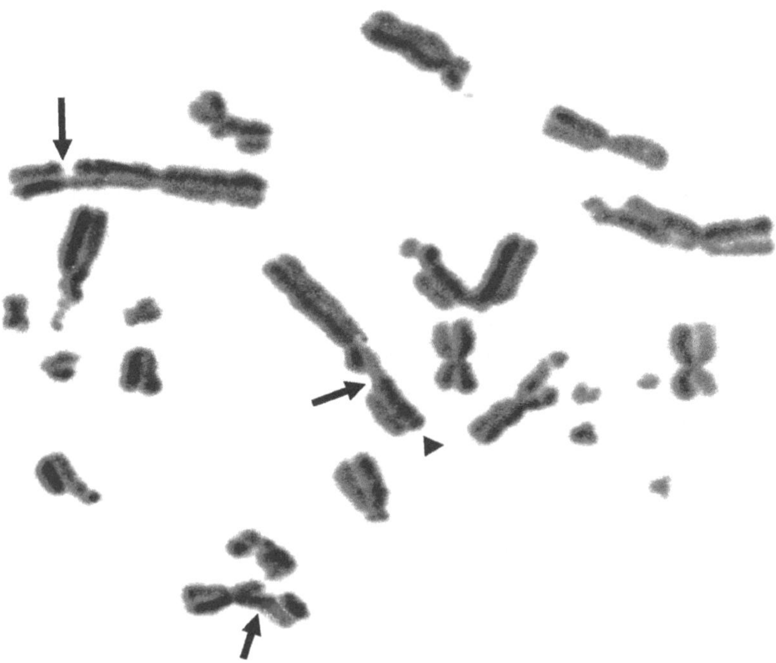

Fig. 11.5 Harlequin (fluorescence plus Giemsa) stained Chinese hamster chromosomes with interchromatid 'colour-switch' breaks (arrows). Other non-colour-switch breaks are considered to be formed by intrachromatid rearrangements. (From Bryant 1998)

'colour-switch breaks'. Although some measurements in earlier experiments with plant materials agreed with this ratio (Heddle et al. 1969), later studies found lower ratios (Heddle and Bodycote 1970; Wolff and Bodycote 1975). It was also claimed that the breakage-first model could accommodate such colour-switch events since the theory allowed incomplete exchanges between sister chromatids (Heddle 1994; Savage 1998). However, a colour-switch ratio of around 16% has also been reported for spontaneously occurring CB (Harvey and Savage 1997), where the probability of two interacting DSB occurring in sister chromatids at the same locus would be vanishingly small. It is therefore most likely that colour-switch breaks do not arise by a breakage-first mechanism.

The main problem with the Revell model as applied to human or rodent chromosomes is that it its basic premise is the pairwise interaction of two lesions at the cross-over points of looped domains, so predicting a quadratic or a linear–quadratic (assuming a proportion of the two-hit events will be caused by clusters of ionizations within one electron track) dose–effect response for the induction of CB. As mentioned above, the dose effect for CB for human and rodent cells was found to be linear.

11.4.4 The Misrecombination Model

Chadwick and Leenhouts (1978, 1998) included chromatid aberrations in their misrecombinational model. Under their model a single unrejoined DSB would lead to a non-colour-switch CB, and a stalled or incomplete inter-chromatid recombinational event, resulting from an attempt by the cell to use the undamaged sister chromatid to perform recombinational repair of the DSB, would lead to a colour-switch CB. This model seems unlikely to be correct, firstly because if it were the only or main mechanism of CB formation, the colour-switch ratio would be much higher than actually observed; possibly near 100%. Secondly, recombination-deficient cells are reported to show increased frequencies of colour-switch breaks as compared with those of wild-type controls (Mozdarani et al. 2001), indicating that the *Xrcc2* gene product normally has a role in preventing the formation of this type of CB; although the method of break scoring in this study has been challenged, in that chromatid gaps smaller than the width of a chromatid had been included in the colour-switch-break scores (Thacker and Griffin 2002). The arguments for including all clear gaps in CB scores have been rehearsed previously (see reply to Thacker and Griffin by Bryant et al. 2002a and in Bryant et al. 2002b). A further obvious point is that because colour switches (associated with colour-switch breaks) occur precisely at those gap sites that are equal to or *smaller* than the width of a chromatid in the same way as at gaps that are *larger* than the chromatid width, this shows that the narrower gaps are similarly the result of breakage and rearrangement of chromatin between the sister chromatids.

11.4.5 The Signal Model

The signal model (Bryant 1998; Bryant et al. 2004) was based on the loop domain idea of the Revell model (Revell 1959). It was proposed that radiation damage (a single DSB) occurring within a large mega-base-pair looped chromatin domain, portrayed as a simple single loop (Fig. 11.6) but probably much more complex and possibly equivalent to a transcription factory, generates a damage signal. Such a damage signal could be either *trans*- or *cis*-acting. One possible candidate for such a *trans*-acting damage signal is the rapid and prolonged Ser139 phosphorylation of histone H2AX by Atm protein over a megabase domain surrounding a DSB following irradiation (Rogakou et al. 1999; Burma et al. 2001). However, we have observed that the homozygous mutation of the *Atm* gene results in a twofold *elevation* of CB frequency (Mozdarani and Bryant 1989). H2AX-phosphorylation by ATM seems not be required for initial recruitment of DNA repair factors to DSB sites (Celeste et al. 2003) and more recently it has been proposed that DSB are initially recognized by 53BP1 protein owing to its property of interaction with constitutively methylated (at Lys79) histone H3, which may be exposed at DSB sites by local opening of the higher-order chromatin structure (Huyen et al. 2004; Zgheib et al. 2005). Local chromatin modification at DSB sites could feasibly be the signal reducing the fidelity of chromatid decatenation by topoisomerase II as cells progress towards mitosis.

Topoisomerase II is known to be subject to error-prone joining of cut DNA strands following treatment with agents inducing reactive oxygen species

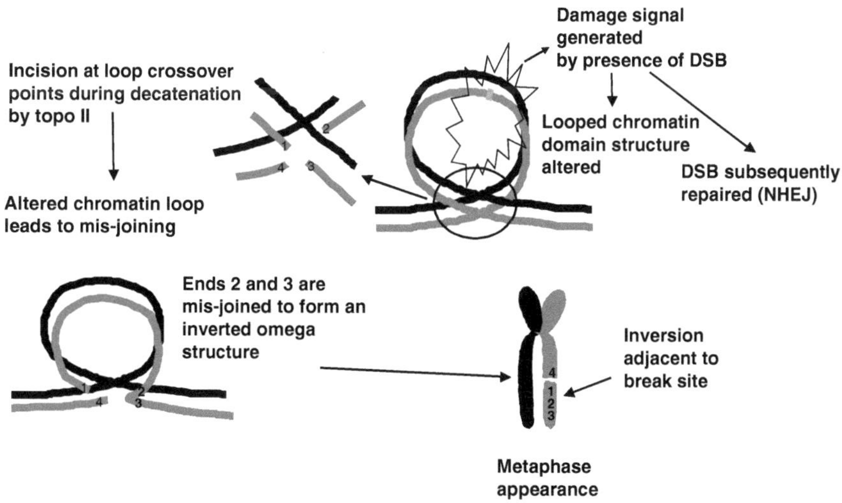

Fig. 11.6 A suggested scheme for the steps involved in the signal mechanism as applied to incomplete intrachromatid rearrangements, leading to chromatid breaks at metaphase. *DSB* double-strand breaks, *NHEJ* non-homologous end-joining, *topo II* topoisomerase II

(Li et al 1999). If incomplete, such misjoining of chromatin ends would lead to a CB; one of the several types proposed under the Revell model.

It is important to note that the signal model differs fundamentally from the Revell model in that it proposes only a *single* DSB is required to produce a CB, a proposal that is supported by data showing that a single (I-*Sce*I site) DSB in a wild-type Chinese hamster background is sufficient to induce a CB (Rogers-Bald et al. 2000). Also, CB are induced linearly with dose by carbon-K X-rays, where the track lengths of electrons produced are too short to 'hit' more than one DNA strand (Bryant et al. 2003).

The disappearance of CB with time (MacLeod and Bryant 1990, 1992) under the signal model would be explained by the sequential completion of the topoisomerase II misjoining and resultant incorporation of the chromatin rearrangement. Subsequently, the initiating DSB would be rejoined (independently of the rearrangement process) by one of the three known DSB rejoining pathways (HRR, SSA or NHEJ).

Thus, the signal model predicts the observed linear induction of CB, provides a possible framework with which to explain the lack of direct correspondence between the DSB rejoining and disappearance of CB with time after irradiation.

11.5 Conclusions

The mechanisms of formation of chromosome aberrations remain very controversial; however, some conclusions can be drawn from studies to date:

1. It is likely that chromosome aberrations are formed by a different mechanism from that for chromatid aberrations.
2. There may be two different mechanisms for the induction of chromosome exchanges; one requiring only a single DSB interacting with a microhomology on another undamaged chromosome, and the other involving two or more DSB interacting via a breakage-first mechanism.
3. CB are formed from single DSB.
4. CB are not formed by the breakage-first mechanism, but by a more complex mechanism possibly involving DSB signalling.

References

Anderson RM, Stevens DL, Goodhead DT (2002) M-FISH analysis shows that complex chromosome aberrations induced by alpha -particle tracks are cumulative products of localized rearrangements. Proc Natl Acad Sci USA 99:12167–12172

Antoccia A, Palitti F, Raggi T, Catena C, Tanzarella C (1994) Lack of effect of inhibitors of DNA synthesis/repair on the ionizing radiation-induced chromosomal damage in G2 stage of ataxia telangiectasia cells. Int J Radiat Biol 66:309–17

Aten JA, Stap J, Krawczyk PM, van Oven CH, Hoebe RA, Essers J, Kanaar R (2004) Science 303:92–95
Bender AM, Griggs HG, Bedford JS (1974) Mechanisms of chromosomal aberrations production III: Chemicals and ionizing radiation. Mutat Res 23:197–212
Bryant PE (1998) The signal model: a possible explanation for the conversion of DNA double-strand breaks into chromatid breaks Int J Radiat Biol 73:243–251
Bryant PE (1984) Enzymatic restriction of mammalian cell DNA using *Pvu II* and *Bam H1*: Evidence for double strand break origin of chromosome aberrations. Int J Radiat Biol 46:57–65
Bryant PE, Christie AF (1989) Induction of chromosomal aberration in CHO cells by restriction endonucleases: effects of blunt- and cohesive-ended double strand breaks in cells treated by "pellet" methods. Mutat Res 213:233–241
Bryant PE, Slijepcevic P (1993) G2 chromatid aberrations: kinetics and possible mechanisms. Environ Mol Mutagen 22:250–256
Bryant PE, Jeggo P, Birch D (1987) High chromosomal sensitivity of Chinese hamster xrs 5 cells to restriction endonuclease induced DNA double-strand breaks. Int J Radiat Biol 52:537–554
Bryant PE, Jones NJ, Liu N (1993) Radiosensitive Chinese hamster irs 2 cells show enhanced chromosomal sensitivity to ionising radiation and restriction endonuclease induced blunt-ended double-strand breaks. Mutagenesis 8:141–147
Bryant PE, Finnegan C, Swaffield L, Mozdarani H (1998) G2 chromatid breaks in murine scid cells. Mutagenesis 13:481–485
Bryant PE, Mozdarani H, Liu N, Jones NJ (2002a) Reply to the letter from Thacker and Griffin: Problems in scoring radiation-induced chromatid breaks. Int J Radiat Biol 78:947–948
Bryant PE, Gray L, Riches AC, Steel CM, Finnon P, Howe O, Kesterton I, Vral A, Curwen GB, Smart V, Tawn EJ, Whitehouse CA (2002b) The G2 assay: a technical report. Int J Radiat Biol 78:863–866
Bryant PE, Jones C, Armstrong G, Frankenberg-Schwager M, Frankenberg D (2003) Induction of chromatid breaks by carbon-K ultra-soft X-rays. Radiat Res 159:247–250
Bryant PE, Gray LJ, Peresse N (2004) Progress towards understanding the nature of chromatid breakage. Cytogenet Genome Res 104:65–71
Burma S, Chen BP, Murphy M, Kurimasa A, Chen D (2001) ATM phosphorylates histone H2AX in response to DNA double-strand breaks. J Biol Chem 276:42462–42467
Celeste A, Fernandez-Capetillo O, Kruhlak MJ, Pilch DR, Staudt DW, Lee A, Bonner RF, Bonner WM, Nussenzweig A (2003) Histone H2AX phosphorylation is dispensible for the initial recognition of DNA breaks. Nat Cell Biol 5:675–679
Chadwick KH, Leenhouts HP (1978) The rejoining of DNA double-strand breaks and a model for the formation of chromosomal rearrangements. Int J Radiat Biol 33:517–529
Chadwick KH, Leenhouts HP (1998) Radiation induced chromosome aberrations: some biophysical considerations. Mutat Res 404:113–117
Cook PR (1999) The organization of replication and transcription. Science 284:1790–1795
Cornforth MN (1990) Testing the notion of the one-hit exchange. Radiat Res 121:21–27
Cornforth MN, Greulich-Bode KM, Loucas BD, Arsuaga J, Vazquez M, Sachs RK, Bruckner M, Molls M, Hahnfelt P, Hlatky L, Brenner DJ (2002) Chromosomes are predominantly located randomly with respect to each other in interphase human cells. J Cell Biol 159:237–244
Cremer T, Cremer C (2001) Chromosome territories, nuclear architecture and gene regulation in mammalian cells. Nat Rev Genet 2:292–301
Cremer C, Munkel C, Granzow M, Jauch A, Dietzel S, Eils R, Guan XY, Meltzer RS, Trent JM, Langowski J, Cremer T (1996) Nuclear architecture and the induction of chromosomal aberrations. Mutat Res 366:97–116
Darroudi F, Fomina J, Meijers M, Natarajan AT (1998) Kinetics of the formation of chromosome aberrations in X-irradiated human lymphocytes, using PCC and FISH. Mutat Res 404:55–66
Deans B, Griffin CS, O'Regan P, Jasin M, Thacker J (2003) Homologous recombination deficiency leads to profound genetic instability in cells derived from *Xrcc2*-knockout mice. Cancer Res 63:8181–8187

Edwards AA, Hone PA, Moquet JE, Lloyd DC (1999) Simple chromosome exchanges are not linear with dose. Int J Radiat Biol 75:1113–1117

Essers J, van Steeg H, de Wit J, Swagemakers SM, Vermeij M, Hoeijmakers JH, Kanaar R (2000) Homologous and non-homologous recombination differentially affect DNA damage repair in mice. EMBO J 19:1703–1710

Folle GA, Marinez-Lopez, Boccardo E, Obe G (1998) Localization of chromosome breakpoints: implication of the chromatin structure and nuclear architecture. Mutat Res 404:17–26

Foray N, Arlett CF, Malaise EP (1995) Dose-rate effect on induction and repair of radiation-induced DNA double-strand breaks in a normal and an ataxia-telangiectasia human fibroblast cell line. Biochemie 77:900–905

Gotoh E, Kawata T, Durante M (1999) Chromatid break rejoining and exchange aberration formation following γ-ray exposure: analysis in G2 human fibroblasts by chemical-induced premature chromosome condensation. Int J Radiat Biol 75:1129–1135

Griffin C, Thacker J (2004) The role of homologous recombination repair in the formation of chromosome aberrations. Cytogenet Genome Res 104:21–27

Griffin C, Marsden SJ, Stevens DL, Simpson P, Savage JRK (1995) Frequencies of complex chromosome exchange aberrations induced by 238Pu α-particles and detected by fluorescence in situ hybridization using single chromosome-specific probes. Int J Radiat Biol 67:431–439

Griffin C, Stevens DL, Savage JRK (1996) Ultrasoft 1.5 keV Aluminium K X-rays are efficient producers of complex chromosome exchange aberrations as revealed by fluorescent in situ hybridization. Radiat Res 146:144–150

Griffin C, Hill MA, Papworth DG, Townsend KMS, Savage JRK, Goodhead DT (1998) Effectiveness of 0.28keV carbon K ultrasoft X-rays at producing simple and complex chromosome exchanges in human firbroblasts in vitro detected using FISH. Int J Radiat Biol 73:591–598

Griffin C, de Waard H, Deans B, Thacker J (2005) The involvement of key DNA repair pathways in the formation of chromosome rearrangements in embryonic stem cells. DNA Repair 4:1019–1027

Grigorova M, Natarajan AT (1998) Relative involvement of chromosome #21 in radiation induced exchange aberrations in lymphocytes of Down syndrome patients. Mutat Res 404:67–75

Hamada N, Schettino G, Kashino G, Vaid M, Suzuki K, Kodama S, Vojnovic B, Folkard M, Watanabe M, Michael BD, Prise KM (2006) Histone H2AX phosphorylation in normal human cells irradiated with focused ultrasoft X-rays: evidence for chromatin movement during repair. Radiat Res 166:31–38

Harvey AN, Savage JRK (1997) Investigating the nature of chromatid breaks produced by restriction endonucleases. Int J Radiat Biol 71:21–28

Heddle J (1994) Revelling in cytogenetics. Environ Mol Mutagen 23:(Suppl 24):35–38

Heddle J, Bodycote DJ (1970) On the formation of chromosomal aberrations. Mutat Res 9:117–126

Heddle J, Whissell, D, Bodycote DJ (1969) Changes in chromosome structure induced by radiations: a test of the two chief hypotheses. Nature 221:1158–1160

Holmberg M (1990) On the time course of the interactions between DNA breaks in the production of a radiation-induced chromosome exchange aberration. Mutat Res 232:267–272

Holmberg M, Gumauskas E (1986) The role of short-lived DNA lesions in the production of chromosome exchange aberrations. Mutat Res 160:221–229

Holmberg M, Jonasson J (1973) Preferential location of X-ray induced chromosome breakage in the R-bands of human chromosomes. Hereditas 73:57–67

Huyen Y, Zgheib O, DiTullio RA Jr, Gorgoulis VG, Zacaratos P, Petty TJ, Sheston EA, Mellert HS, Stavridi ES, Halazonetis TD (2004) Methylated lysine 79 of histone H3 targets 53BP1 to DNA double-strand breaks. Nature 432:406–411

Iborra FJ, Pombo A, Jackson DA, Cook PR (1996) Active RNA polymerases are localised within discrete transcription 'factories' in human nuclei. J Cell Sci 109:1–14.

Iliakis G, Wang H, Perrault AR, Boecker W, Rosidi B, Windhofer F, Wu W, Guan J, Terzoudi G, Pantelias G (2004) Mechanisms of DNA double-strand break repair and chromosome aberration formation. Cytogenet Genome Res 104:14–20

Johnson RT, Rao PN (1970) Mammalian cell fusion: induction of premature chromosome condensation in interphase nuclei. Nature 226:717–722
Jones NK, Stewart SA, Thompson LH (1990) Biochemical and genetic analysis of the Chinese hamster mutants irs 1 and irs 2 and their comparison to cultured ataxia telangiectasia cells. Mutagenesis 5:15–23
Kawata T, Gotoh E, Durante M, Wu H, George K, Furusawa Y, Cucinotta FA (2000) High-LET radiation-induced aberrations in prematurely condensed G2 chromosomes of human fibroblasts. Int J Radiat Biol 76:929–937
Kawata T, Ito H, Uno T, Saito M, Yamamoto S, Furusawa Y, Durante M, George K, Wu H, Cucinotta FA (2004) G2 chromatid damage and repair kinetics in normal human fibroblast cells exposed to low- or high-LET radiation. Cytogenet Genom Res 104:211–215
Kemp LM, Jeggo PA (1986) Radiation-induced chromosome damage in X-ray sensitive mutants (xrs) of the Chinese hamster ovary cell line. Mutat Res 166:255–263
Kruhlak MJ, Celeste A, Dellaire G, Fernandez-Capetillo O, Müller WG, McNally JG, Bazett-Jones DP, Nussenzweig A (2006) Changes in chromatin structure and mobility in living cells at sites of DNA double-strand breaks. J Cell Biol 172:823–834
Lea DE, Catcheside DG (1942) The mechanism of the induction by radiation of chromosome aberrations in *Tradescantia*. J Genet 44:216–222
Li T, Chen AY, Mao Y, Wang H, Liu LF (1999) Activation of topoisomerase II-mediated excision of chromosomal DNA loops during oxidatie stress. Genes Dev 13:1553–1560
Lin F-L, Sperle K, Sternberg N (1984) Model for homologous recombination during transfer of DNA into mouse L cells: role for DNA ends in the recombination process. Mol Cell Biol 4:1020–1034
Loucas BD, Cornforth MN (2001) Complex chromosome exchanges induced by gamma rays in human lymphocytes: an mFISH study. Radiat Res 155:660–671
Loucas BD, Eberle R, Bailey SM, Cornforth MN (2004) Influence of dose-rate on the induction of simple and complex chromosome exchanges by gamma-rays. Radiat Res 162:339–349
Ludwików G, Xiao Y, Hoebe RA, Franken NAP, Darroudi F, Stap J, van Oven CH, van Noorden CJF, Aten JA (2002) Induction of chromosome aberrations in unirradiated chromatin after partial irradidiation of a cell nucleus. Int J Radiat Biol 78:239–247
Macleod RAF, Bryant PE (1990) Similar kinetics of chromatid aberrations in X-irradiated xrs 5 and wild-type Chinese hamster cells. Mutagenesis 5:407–410
Macleod RAF, Bryant PE (1992) Effects of adenine arabinoside and coformycin of the kinetics of G2 chromatid aberrations in X-irradiated human lymphocytes. Mutagenesis 7:285–290
MacPhail SH, Banath JP, Yu TY, Chu EHM, Lambur H, Olive PL (2003) Expression of phosphorylated histone H2AX in cultured cell lines following exposure to X-rays. Int J Radiat Biol 79:1–8
Marcon F, Boei JJWA, Natarajan AT (2000) Recombination between two homologous chromosomes does not play a dominant role in the formation of chromosome aberrations. Int J Radiat Biol 76:1343–1348
Mattei MG, Ayme S, Mattei JF, Aurran Y, Giraud AF (1979) Distribution of spontaneous chromosome breaks in man. Cytogenet Cell Genet 23:95–102
Mozdarani H, Bryant PE (1989) Kinetics of chromatid aberrations in G2 ataxia telangiectasia cells exposed to X-rays and ara A. Int J Radiat Biol 55:71–84
Mozdarani H, Liu N, Jones NJ, Bryant PE (2001) The *XRCC2* human repair gene influences recombinational rearrangments leading to chromatid breaks. Int J Radiat Biol 77:859–865
Musio A, Mariani T, Vezzoni P, Frattini A (2002) Heterogenous gene distribution reflects human genome complexity as detected at the cytogenetic level. Cancer Genet Cytogenet 134:168–171
Nakagome Y, Matsubara T, Fujita H (1983) Distribution of break-points in human structural rearrangements. Am J Hum Genet 35:288–300
Natarajan AT, Obe G (1984) Molecular mechanisms involved in the production of chromosomal aberrations III. Restriction endonucleases. Chromosoma 90:120–127
Natarajan AT, Obe G, van Zeeland AA, Palitti F, Meijers M, Verdegaal-Immerzell EAM (1980) Molecular mechanisms involved in the production of chromosomal aberrations; II Utilization

of Neurospora endonuclease for the study of aberration production by X-rays in G1 and G2 stages of the cell cycle. Mutation Res 69:293–305
Neary GJ (1965) Chromosome aberrations and the theory of RBE: 1. General considerations. Int J Radiat Biol 9:477–502
Nikiforova MN, Stringer JR, Blough MRM, Fagin JA, Nikiforova YE (2000) Proximity of chromosomal loci that participate in radiation induced rearrangements in human cells. Science 290:138–141
Pandita TK, Hittelman WN (1992) Initial chromosome damage but not DNA damage is greater in ataxia telangiectasia cells. Radiat Res 130:94–103
Parshad R (1996) Deficient DNA repair capacity, a predisposing factor in breast cancer. Br J Cancer 74:1–5
Parshad R, Sanford KK, Jones GM (1990) Chromatid damage after G2 phase X-irradiation of cells from cancer prone individuals implicates deficiency in DNA repair. Proc Natl Acad Sci USA 80:5612–5616
Pfeiffer P, Feldmann E, Odersky A, Kuhfittig-Kulle S, Goedecke W (2005) Analysis of DNA double strand break repair by nonhomologous end joining in cell-free extracts from mammalian cells. Methods Mol Biol 291:351–71
Plan Y, Hlatky L, Hahnfeldt P, Sachs R, Loucas B, Cornforth M (2005) Full-color painting reveals an excess of radiation-induced dicentrics involving homologous chromosomes. Int J Radiat Biol 81:613–620
Preston RJ (1980) The effect of cytosine arabinofuranoside on the frequency of X-ray induced chromosome aberrations in normal human leukocytes, Mutat Res 69:71–79
Rabbitts TH (1994) Chromosomal translocations in human cancer. Nature 372:143–149
Radford IR (2002) Transcription-based model for the induction of interchromosomal exchange events by ionizing irradiation in mammalian cell lines that undergo necrosis. Int J Radiat Biol 78:1081–1093
Radford IR, Hodgson GS, Matthews JP (1988) Critical DNA target size model of ionizing radiation-induced mammalian cell death. Int J Radiat Biol 54:63–79
Resnick MA (1976) The repair of double-strand breaks in DNA: A model involving recombination. J Theor Biol 59:97–106
Revell SH (1959) The accurate estimation of chromatid breakage, and its relevance to a new interpretation of chromatid aberrations by ionizing radiations. Proc R Soc Lond Ser B 150:563–589
Richardson C, Jasin M (2000) Frequent chromosome translocations induced by DNA double-strand breaks. Nature 405:697–700
Rogakou EP, Boon C, Redon C, Bonner WM (1999) Megabase chromatin domains involved in DNA double-strand breaks in vivo. J Cell Biol 146:905–916
Rogers-Bald M, Sargent RG, Bryant PE (2000) Production of chromatid breaks by single dsb: evidence supporting the signal model. Int J Radiat Biol 76:23–29
Rotkamm K, Lobrich M (2003) Evidence for a lack of DNA double-strand break repair in human cells exposed to very low X-ray doses. Proc Natl Acad Sci USA 100:5057–5062
Savage JRK (1986) The 'Revell loop'. Clin Cytogenet Bull 1:192–196
Savage JRK (1998) A brief survey of aberration origin theories. Mutat Res 404:139–147
Savage JRK (2000) Proximity matters. Science 290:62–63
Savage JRK (2002) Reflections and meditations upon complex chromosomal exchanges. Mutat Res 512:93–109
Savage JRK (2004) On the nature of visible chromosomal gaps and breaks. Cytogenet Genome Res 104:46–55
Sax K (1939) The time factor in X-ray production of chromosome aberrations. Proc Nat Acad Sci USA 25:225–233
Sax K (1940) An analysis of X-ray induced chromosomal aberrations in *Tradescantia*. Genetics 25:41–68
Scott D, Barber JBP, Spreadborough AR, Burrill W, Roberts SA (1999) Increased radiosensitivity in breast cancer patients: a comparison of two assays. Int J Radiat Biol 75:1–10

Serebrovsky AS (1929) A general scheme for the origin of mutations. Am Nat 63:374–378
Slijepcevic P, Natarajan AT (1994) Distribution of radiation-induced G1 exchange and terminal deletion breakpoints in Chinese hamster chromosomes as detected by G banding. Int J Radiat Biol 66:747–755
Taylor AMR (1978) Unrepaired DNA strand breaks in irradiated ataxia telangiectasia lymphocytes suggested from cytogenetic observations. Mutat Res 50:407–418
Thacker J, Ganesh AN (1990) DNA break repair, radioresistance of DNA synthesis. and camptoghecin sensitivity in the radiation-sensitive irs mutants: Comparisons to ataxia-telangiectasia cells. Mutat Res 235:49–58
Thacker J, Griffin CS (2002) Problems in scoring radiation-induced chromatid breaks. Comments on the paper: The XRCC2 human repair gene influences recombinational rearrangements leading to chromatid breaks. Int J Radiat Biol 78:945–952
Thacker J, Wilkinson RE, Goodhead DT (1986) The induction of chromosome exchange aberrations by carbon ultrasoft X-rays in V79 hamster cells. Int J Radiat Biol 49:645–656
Thacker J, Zdzienicka MZ (2004) The *XRCC* genes: expanding roles in DNA double-strand break repair. DNA Repair 3:1081–1090
van Hoffen A, Balajee AS, van Zeeland AA, Mullenders LH (2003) Nucleotide excision repair and its interplay with transcription. Toxicology 193:79–90
Virsik RP, Schafer C, Harder D, Goodhead DT, Cox R, Thacker J (1980) Chromosome aberrations induced in human lymphocytes by ultrasoft Al (K) and C (K) X-rays. Int J Radiat Biol 38:545–557
Vral A, Thierens H, Bryant PE, De Ridder L (2001) A higher micronucleus yield in B- versus T-cells after low-dose gamma-irradiation is not linked with defective Ku86 protein. Int J Radiat Biol 77:329–339
Wang H, Zeng Z-C, Bui T-A, Sonada E, Takata M, Takeda S, Iliakis G (2001) Efficient rejoining of radiation-induced DNA double-strand breaks in vertebrate cells deficient in genes of the RAD52 epistasis group. Oncogene 20:2212–2224
Wang P, Zhou R-H, Zou Y, Jackson-Cook CK, Povirk L (1997) Highly conservative reciprocal translocations formed by apparent joining of exchanged DNA double-strand break ends. Proc Natl Acad Sci USA 94:12018–12023
Winegar RA, Preston JR (1988) The induction of chromosome aberrations by restriction endonucleases that produce blunt-end or cohesive-end double-strand breaks. Mutat Res 197:141–149
Wolff S, Bodycote J (1975) The induction of chromatid deletions in accord with the breakage-and-reunion hypothesis. Mutat Res 29:85–91
Zgheib O, Huyen Y, DiTullo RA Jr, Snyder A, Venere M, Stavridi ES, Halazonetis TD (2005) ATM signaling and 53BP1. Radiother Oncol 76:119–122

12 Chromatin Structure and the Formation of Chromosomal Alterations

GREGOR KRETH AND CHRISTOPH CREMER

Abstract

Current experimental studies on the induction of chromosomal aberrations suggest the influence of different structural features of chromosome organization which are responsible for deviations from the predicted aberration frequencies when only the DNA content of the chromosomes involved is considered. The nonrandom neighbourhood relationship resulting from the experimentally observed radial positioning of chromosome territories in lymphocyte nuclei was reported as one important factor. As an additional, second contribution the compaction level of a chromosome territory also has a major impact on aberration frequencies. Different compaction levels were observed, e.g. for the active and inactive X-chromosome territories in amniotic fluid or fibroblast nuclei. A recent study of Branco and Pombo (*PLOS Biol.* 4:e138, 2006) revealed, however, also for other chromosome territories of similar DNA content, remarkable differences in the volumes between gene-rich and gene-poor chromosome territories. In the present study we summarize computer simulations of nuclear structure where we quantify the effects of these nuclear features on aberration frequencies for the example of the active and the inactive X chromosome.

12.1 Introduction

Current experimental studies on the induction of chromosomal aberrations suggest the influence of different structural features of chromosome organization which are responsible for deviations from the predicted aberration frequencies when only the DNA content of the chromosomes involved is considered. Two geometrical constraints of interphase chromosome organization might be responsible and are discussed in this chapter: the nonrandom neighbourhood relationship of chromosome territories (CTs) derived from the experimentally observed radial arrangement of gene-dense and gene-poor CTs (Cremer et al. 2001; Boyle et al. 2001; Tanabe et al. 2002), and the presence of different compaction levels of CTs according to their gene densities/transcriptional activities (Branco and Pombo 2006). Arsuaga et al.

Günter Obe and Vijayalaxmi (Eds.)
Chromosomal Alterations: Methods, Results and Importance in Human Health

(2004) reported that the aberration frequencies of at least two clusters of chromosomes show a statistically significant deviation from a random chromosome association. These clusters comprise the nucleolus-organizing CTs {13, 14, 15, 21, 22} and a cluster of gene-rich chromosomes {1, 16, 17, 19, 22} which are located in the interior of lymphocyte nuclei. We quantified such effects by computer simulations where we regarded explicitly the gene density correlated radial arrangement of CTs in simulated nuclei (Kreth et al. 2004, 2007). For these calculations, a direct correlation between DNA content and volume was assumed and no specific assumptions about a more decondensed compaction of gene-rich chromosomes were made.

This is, however, in contrast to the recent study of Branco and Pombo (2006). They presented evidence for a gene density/transcriptional activity dependent compaction level of CTs. The experimentally observed remarkable difference in the aberration frequencies between (1;21) and (1;22) [the frequency of (1;22) is nearly 3 times higher than that of (1;21)] could be explained by an increased volume of CT 22. Both chromosomes possess nearly the same DNA content but CT 22 revealed nearly twice the volume of CT 21. To quantify the effect of such different compaction levels, in the present study we performed quantitative computer simulations of the active and the inactive X-chromosome territories, taking into account the different experimentally observed compaction levels and also different localizations for the inactive X-chromosome territory in the nuclear volume.

12.2 Modelling of Virtual Radiation Events

To estimate the effects of the higher-order spatial chromosome organization on aberration frequencies, we used an earlier version of the already published spherical 1-Mbp chromatin domain computer model of nuclear structure (Kreth et al. 2004). Again, each chromosome was modelled by a linear chain of spherical domains of 1 Mbp (with a diameter of 500 nm each). To account for the observed dynamic variability in the distances between the domains, zigzag random walk linkers (e.g. four linker segments of 300-nm length) were used. The territory compaction was modelled with the assumption of an elastic "tensegrity force" F_a acting on the 1-Mbp domains and directed towards the bary centre ("centre of gravity") of the entire territory. Between spatially interacting domains, a hard-sphere potential was assumed, allowing a minimal distance of a half domain radius (i.e. 125 nm). To achieve a higher compaction level for the inactive X chromosome (Fig. 12.1) in comparison with that of the active X chromosome, F_a was adjusted in such a way that after the virtual imaging and the segmentation process (Kreth et al. 2002) the experimentally obtained volume and enveloping surface ratios were obtained (see also Sect. 12.3).

Under the assumption of a random distribution of double-strand breaks (DSBs) within the DNA for low linear energy transfer (LET) radiation, the

Fig. 12.1 Surface rendering of a simulated nucleus according to the earlier version of the spherical 1-Mbp chromatin domain model (Kreth et al. 2002). For the visualization, only the active (the *light grey* chromosome) and the inactive (at the top of the nucleus, the *dark grey* chromosome) were chosen. The compaction level of the inactive X-chromosome territory is increased in comparison with that of the active X-chromosome territory to correspond to the experimental volume and enveloping surface ratios

probability of a break occurring within a certain 1-Mbp domain was modelled using Poisson distribution mathematics. The yield of DSBs was adapted to ensure a mean number of 50 DSBs per gray per nucleus according to experimental observations (referred to in Cornforth et al. 2002). The DSBs within the 1-Mbp domains were placed into randomly chosen locations. To determine an exchange (interchange/intrachange) between two DSBs in two 1-Mbp domains, only those 1-Mbp domains containing DSBs were considered which revealed a distance (the distance between the midpoints of both 1-Mbp domains) smaller than or equal to 500 nm. An exchange event depending on the distance between the two DSBs in the two domains was counted according to a normalized probability function (Kreth et al. 2004, 2007).

12.3 Modelling Results

The influence of the geometry of the spatial organization of interphase chromosomes on the yield of aberration frequencies can be estimated in a first step assuming a proportional relationship yield (contact surface) between two adjacent CTs. This implies also that the yield is proportional to the cross-section, which is proportional to r^2, of a single CT. With the assumption that the DNA content of a chromosome is direct proportional to its volume, and therefore the DNA content is proportional to r^3, we obtained the previously reported relation that the yield is proportional the two-thirds power of the DNA content (Kreth et al. 1998, 2004; Cornforth et al. 2002). When a so-called statistical arrangement of the CTs is assumed to exist, this relation holds true for all CTs. (Kreth et al. 1998, 2004, 2007). In mammalian cells, however, preferential radial arrangements were reported which suggest a gene density correlated arrangement of CTs in lymphocyte nuclei (Boyle et al. 2001) and a size-dependent arrangement in fibroblast nuclei (Bolzer et al. 2005). Such a nonrandom positioning has the consequence that CTs which are located preferentially near the nuclear periphery have a reduced contact surface to other CTs, because the part of the surface of such a CT which is located near the periphery cannot participate in a misrejoining event. The second important contribution is the so-called compaction level (ratio of DNA content to CT volume). CTs revealed obviously different compactions depending on their gene density or transcriptional activity. In such cases the linear relationship between DNA content and volume no longer holds true (Branco and Pombo 2006). One important example where different volumes were observed for the same DNA content is the experimentally observed difference between the active and inactive X-chromosome territories in female fibroblast and amniotic fluid cells (Fig. 12.2). Eils et al. (1996) determined the mean ratios between active and inactive X-chromosome territories in these cell nuclei and found for the enveloping surface (A_{env})

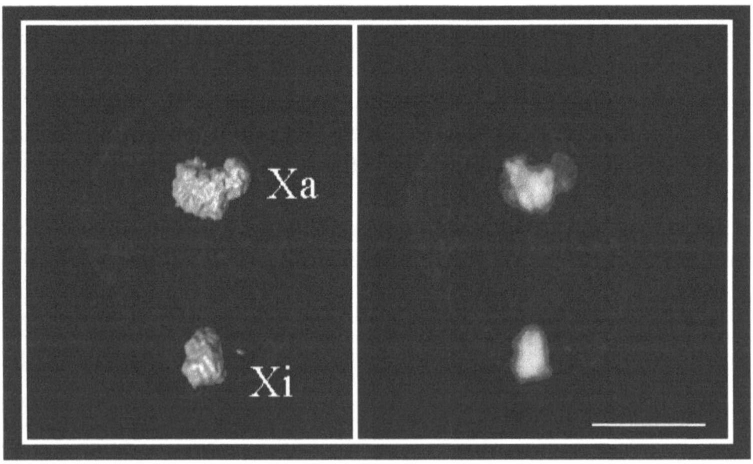

Fig. 12.2 Reconstruction of the active and inactive X-chromosome territories of an amniotic fluid cell of a confocal data stack. *Left*: Volume rendering. *Right*: Optical section. Cells were kindly provided by I. Solovei, University of Munich, Germany. *Xa* active X-chromosome territory, *Xi* inactive X-chromosome territory. Bar denotes 10μm

the ratio $A_{env}(Xa)/A_{env}(Xi)\approx 1.4$ and for the volume enclosed by the enveloping surface the ratio $V(Xa)/V(Xi)\approx 1.2$, where Xa and Xi refer to the active and inactive CTs, respectively. For the segmentation process, however, they chose the same threshold for both X-chromosome territories; this resulted in a slight underestimation of the surface of the active X-chromosome territory in contrast to that of the inactive X-chromosome territory. The reason for this underestimation of the surface of the active X-chromosome territory and the better conservation of the surface of the inactive X-chromosome territory surface is the sharper increase of the intensity profile of the inactive X-chromosome territory. In this way, lower intensities of the brighter-intensity profile of the active X-chromosome territory are lost, while the higher intensities of the inactive X-chromosome territory are conserved, when the same threshold is applied for both X-chromosome territories. This underestimation was quantified by model calculations. In this model, each CT was described by a linear chain of spherical 500-nm domains with a DNA content of 1 Mbp. The number of domains corresponds directly to the DNA content of each chromosome. The morphological difference of both X-chromosome territories was adjusted in this model to correspond to the experimental volume and enveloping surface ratios when the same threshold and segmentation process for the virtually convolved simulated data stacks were used (Kreth et al. 2002). Because of the underestimation, the real volume and the enveloping surfaces revealed slightly higher differences: $A_{env\ true}(Xa)/A_{env\ true}(Xi)=1.77$ and $V_{true}(Xa)/V_{true}(Xi)=1.67$. With use of these "true" values for the creation of virtual confocal microscopy images evaluated by image-analysis algorithms used for the experimental data, good agreement with the latter was obtained.

Taking into account the different morphologies for the active and the inactive X-chromosome in the computer model (Kreth et al. 2002), we created a random DSB pattern in a simulated nucleus according to low-LET γ-radiation at 3 Gy. In all, 400 statistically independent cell configurations were calculated and 32 different DSB distributions per nucleus were determined to end up with a sample of 12,800 configurations. For this study, the autosomes and the active X-chromosome were arranged in a statistical manner in the simulated nucleus. No preferential radial arrangements like in lymphocyte or fibroblast nuclei (Kreth et al. 2004) were assumed. For the inactive X-chromosome, however, different assumptions were tested: (1) a localization in the centre of the nuclear volume, (2) a statistical arrangement and (3) a localization at the nuclear envelope. The one-chromosome yields (an aberration event was assigned to each chromosome involved in an interchange aberration) were normalized in such a way that the sum of the one-chromosome yields for all chromosomes together amounted to 1. The resulting one-chromosome yields were plotted against the DNA content of each chromosome (Fig. 12.3). As mentioned before, the aberration yields for the statistically distributed autosomes and the active X-chromosome are close to the exponential fitting curve with exponent 2/3. The reduced enveloping surface and the different assumed

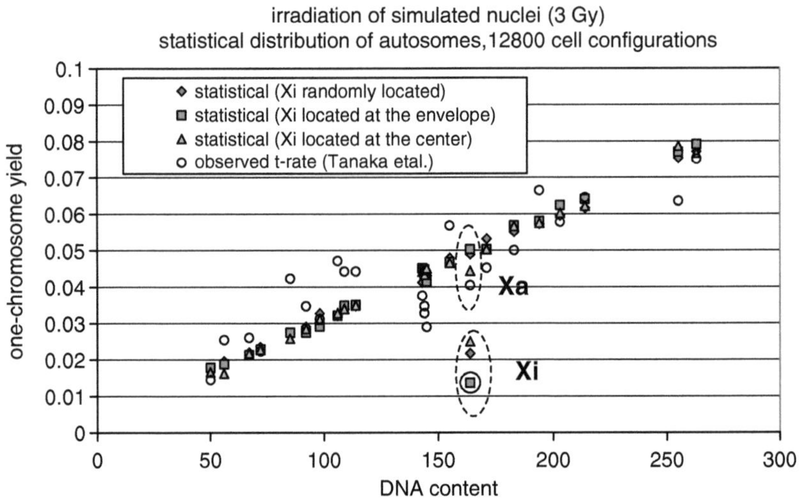

Fig. 12.3 Comparison between the calculated one-chromosome yields and the observed translocation rate (*t-rate*) of Hiroshima A-bomb survivors (Tanaka et al. 1983) of female cells. For the simulated data, the experimentally observed difference in the morphology was explicitly considered. Additionally different positions in the nuclear volume were tested for the inactive X-chromosome. All other simulated chromosome territories were placed randomly. For the comparison, the calculated one-chromosome yields and the observed translocation rate were normalized to 1

positionings for the inactive X-chromosome territory, however, show a clear effect on the aberration yields.

For the three different positionings we observed the following:

1. For the central positioning of the inactive X-chromosome territory (the active X-chromosome territory was located statistically), the aberration yield of the inactive X-chromosome territory shows the highest value because of the permanent accessibility of the complete surface of the chromosome; the aberration frequency ratio between the active and the inactive X-chromosome territories is $f(\mathrm{Xa})/f(\mathrm{Xi}) = 1.78$.
2. For the statistical arrangement (active X-chromosome territory and inactive X-chromosome territory located statistically) the yield of the inactive X-chromosome territory was reduced because of the probability of it being localized also at the nuclear periphery. Consequently the aberration frequency ratio was increased to $f(\mathrm{Xa})/f(\mathrm{Xi}) = 2.26$.
3. For the localization of the inactive X-chromosome territory at the nuclear envelope, however, the yield was drastically reduced; the reduced contact surface had a significant effect on the aberration yields. Here, the aberration frequency ratio between the statistically distributed active X-chromosome territory and the inactive X-chromosome territory was increased and $f(\mathrm{Xa})/f(\mathrm{Xi}) = 3.71$ was obtained.

The comparison with an observed translocation rate (again the sum of the one-chromosome yields for all chromosomes together was normalized to 1) from Hiroshima A-bomb survivors (Tanaka et al. 1983) revealed complete agreement with case 3 for the inactive X-chromosome territory. The scattering of the data points of the Hiroshima A-bomb survivors around the exponential fitting curve might mirror insufficient statistics but could also be the result of a nonrandom positioning of the other CTs. This possibility is suggested by simulations where a nonstatistical arrangement was assumed, e.g. for lymphocyte cells (Kreth et al. 2004, 2007).

In another approach, also for other CTs a close relationship between the contact surfaces, or an equivalent measure the intermingling volume, was reported. Branco and Pombo (2006) developed a novel fluorescence in situ hybridization (FISH) procedure for ultrathin cryosections which allows the use of wide-field microscopy with no reduction in the axial resolution. This allows a more precise evaluation of the volume of each CT and also of the intermingling volumes between CTs. Remarkably, they found that the volumes of most CTs were not only correlated with the DNA content but also with its gene density/transcriptional activity, which obviously corresponds to the example of the active and inactive X-chromosome territories. One prominent example is the observed volumes for CTs 21 and 22. Despite their similar DNA contents of 49 Mbp for CT 22 and 46 Mbp for CT 21, CT 22, which is a very gene rich chromosome, showed nearly twice of the volume of CT 21. They compared these results with aberration frequencies obtained by multicolour FISH (Arsuaga et al. 2004) and found a close relationship between the intermingling volume of two CTs and the evaluated aberration frequencies of Arsuaga et al. This means, for example, that the aberration yield for (1;22) is 3 times higher than that for (1;21).

12.4 Conclusions

To understand the influence of geometrical constraints of the nuclear architecture on aberration frequencies, we used a computer model which explicitly describes the polymer structure of the diploid human female genome, taking into account additionally the different morphologies of the active and inactive X-chromosome territories. As an additional constraint, we considered also different positionings in the nuclear volume for the inactive X-chromosome. Based on calculations of low-LET γ-radiation at 3 Gy, one chromosome yields were calculated for the different positionings. The expected results were confirmed: a central positioning of the inactive X-chromosome territory revealed the highest value for the aberration frequency because of the permanent accessibility of the complete contact surface. In the case of a random localization of the inactive X-chromosome territory, the yield was reduced slightly because of the possibility of it being located also at the periphery with a certain probability. The positioning of the inactive X-chromosome territory at the periphery revealed a drastic reduction of the inactive X-chromosome aberration frequency, which reflects the presence of only a part of the permanent reduction of a part of the surface of the territory. Remarkably this yield is in complete agreement with the observed yield for the inactive X-chromosome from survivors of the Hiroshima A-bomb. The experimentally observed peripheral positioning of the inactive chromosome in fibroblast and amniotic fluid cells would confirm this result. In addition to our previous studies where the effect of the non-random positioning of all CTs on aberration frequencies was investigated (Kreth et al. 2004), in the present study we tested for a single CT, the inactive X-chromosome territory, and the resulting aberration yields, when different locations for this chromosome are assumed, and additionally the different compaction level of the inactive X-chromosome territory in comparison with that of the active X-chromosome territory was considered.

Acknowledgements: This work was supported by the EC (contract no. F16R-CT-2003-508842, RISC-RAD). We also acknowledge Jan Boei and Thomas Cremer for stimulating discussions.

References

Arsuaga J, Greulich-Bode KM, Vazquez M, Bruckner M, Hahnfeldt P, Brenner DJ, Sachs R, Hlatky L (2004) Chromosome spatial clustering inferred from radiogenic aberrations. Int J Radiat Biol 80:507–515

Bolzer A, Kreth G, Solovei I, Koehler D, Saracoglu K, Fauth C, Müller S, Eils R, Cremer C, Speicher M, Cremer T (2005) Three-dimensional maps of all chromosomes in human male fibroblast nuclei and prometaphase rosettes. PLoS Biol 3:e157

Boyle S, Gilchrist S, Bridger JM, Mahy NL, Ellis JA, Bickmore WA (2001) The spatial organization of human chromosomes within the nuclei of normal and emerin-mutant cells. Hum Mol Genet 10:211–219

Branco MR, Pombo A (2006) Intermingling of chromosome territories in interphase suggests role in translocations and transcription-dependent associations. PLOS Biol 4:e138

Cornforth MN, Greulich-Bode KM, Loucas BD, Arsuaga J, Vazquez M, Sachs RK, Bruckner M, Molls M, Hahnfeldt P, Hlatky L, Brenner DJ (2002) Chromosomes are predominantly located randomly with respect to each other in interphase human cells. J Cell Biol 159: 237–244

Cremer M, von Hase J, Volm T, Brero A, Kreth G, Walter J, Fischer C, Solovei I, Cremer C, Cremer T (2001) Non-random radial higher-order chromatin arrangements in nuclei of diploid human cells. Chromosome Res 9:541–567

Eils R, Dietzel S, Bertin E, Granzow M, Schröck E, Speicher MR, Volm T, Ried T, Robert-Nicoud M, Cremer C, Cremer T (1996) Three-dimensional reconstruction of painted human interphase chromosomes: active and inactive X-chromosome territories have similar volumes but differ in shape and surface structure. J Cell Biol 135:1427–1440

Kreth G, Münkel C, Langowski J, Cremer T, Cremer C (1998) Chromatin structure and chromosome aberrations: modeling of damage induced by isotropic and localized irradiation. Mutat Res 404:77–88

Kreth G, Edelmann P, Münkel C, Langowski J, Cremer C (2002) Translocation frequencies for X and Y chromosomes predicted by computer simulations of nuclear structure. In: Sobti RC,

Obe G, Athwal RS (eds) Some aspects of chromosome structure and functions. Narosa, New Delhi, pp 57–71
Kreth G, Finsterle J, Cremer C (2004) Virtual radiation biophysics: implications of nuclear structure. Cytogenet Genome Res 104:157–161
Kreth G, Pazhanisamy SK, Hausmann M, Cremer C (2007) Cell type specific quantitative predictions of radiation-induced chromosome aberrations: a computer mode approach. Radiat Res (in press)
Tanabe H, Müller S, Neusser M, von Hase J, Calcagno E, Cremer M, Solovei I, Cremer C, Cremer T (2002) Evolutionary conservation of chromosome territory arrangements in cell nuclei from higher primates. Proc Natl Acad Sci USA 99:4424–4429
Tanaka K, Kamada N, Ohkita T, Kuramoto A (1983) Nonrandom distribution of chromosome breaks in lymphocytes of atomic bomb survivors. J Radiat Res 24:291–304

13 Role of Chromatin Structure and Activity in the Induction of Chromosomal Aberrations

Wilner Martínez-López, María Vittoria Di Tomaso, Leticia Méndez-Acuña, and María Mühlmann

Abstract

It is assumed that unrepaired or misrepaired DNA double-strand breaks lead to chromosomal aberrations. However, little information is available about the role of chromatin structure and remodelling in the formation of double-strand breaks and their processing into chromosome damage. Open chromatin conformation, recognized as euchromatin, possesses transcriptional activity, while tighter chromatin conformation, named heterochromatin, remains transcriptionally inactive. A higher order of differential chromatin structure can be observed at the chromosome level as banding patterns. In order to study the role of chromatin organization and activity in the induction of chromosomal aberrations, different approaches were employed: (1) analysis of the distribution of chromosome breakpoints induced by ionizing radiation (neutrons or γ-rays) in relation to the CHO G-banding pattern; (2) correlation of the CHO chromosome H4 acetylation pattern with radiation-induced breakpoint localization; (3) study of the influence of chromatin remodelling processes related to DNA replication and cell cycle progression analysing etoposide-induced and methyl methanesulfonate induced breakpoints distributed along Xp (euchromatic) and Xq (heterochromatic) CHO X chromosome arms; (4) localization of γ-ray-induced breakpoints to euchromatin and heterochromatin by exposing human fibroblast lines with different numbers of X chromosomes. Several factors, such as chromatin activity, chromatin remodelling processes associated with DNA synthesis or cell cycle progression as well as chromatin compartmentalization in euchromatin, heterochromatin and G-bands, have been found associated with the non-random distribution of chromosome damage.

13.1 Introduction

Almost every process occurring within the nuclei of eukaryotic cells is affected by chromatin structure (Jenuwein and Allis 2001). During the last few years, several studies have been carried out to understand how DNA replication, DNA repair and gene transcription can be regulated by chromatin

Günter Obe and Vijayalaxmi (Eds.)
Chromosomal Alterations: Methods, Results and Importance in Human Health

structure (Felsenfeld and Groudine 2003; Morrison et al. 2004; Polo and Almouzni 2005). Histone modifications (acetylation, phosphorylation and methylation) establish the so-called epigenetic histone code (Strahl and Allis 2000). Histone acetylation/deacetylation (controlled by acetyltransferases and deacetylases) is the most frequent epigenetic process, determining either gene activation or inactivation, respectively. Histone acetylation allows chromatin accessibility by reducing positive charges of lysines located in histone amino tails, diminishing the attractive force between the nucleosome core and negatively charged DNA (Jeppesen 1997; Turner 2000). The dynamic chromatin organization controls the access to DNA of many proteins, such as transcription factors, and influences DNA damage induction and processing (Green and Almouzni 2002; Rubbi and Milner 2003; Akhtar and Cavalli 2005; Chai et al. 2005).

DNA double-strand breaks (DSB) are the ultimate lesions for the formation of chromosomal aberrations (Obe et al. 2002), but the role of chromatin structure in the formation of DSB and their processing into chromosome damage is not very well defined.

Two major types of chromatin occur in all eukaryotic cells, namely, euchromatin and heterochromatin. Euchromatin is characterized by an open chromatin conformation, transcriptional activity and sensitivity to DNase I. Heterochromatin is held in a tighter chromatin conformation and is transcriptionally inactive as well as insensitive to DNase I (Mühlmann-Díaz and Bedford 1994). A further order of chromatin organization can be observed at microscope level as different banding patterns. A Giemsa banding pattern can be obtained, for example, after exposing fixed metaphase chromosomes to trypsin followed by Giemsa staining (G-banding, G-bands). G-bands are darkly stained bands (G-dark bands) at fixed position in the chromosomes with unstained areas between them (G-light bands). On the other hand, treatments of chromosome spreads with hot salt reveal the presence of a reverse staining pattern (R-banding, R-bands). G-light bands or R-bands correspond to early replicating GC-rich regions that contain housekeeping genes and short interspersed repeated elements. G-dark bands contain tissue-specific genes, late replicating AT-rich DNA and long interspersed repeated elements (Korenberg and Rykowsky 1988; Goldman et al. 1984). Constitutively heterochromatic C-bands, which mainly occur in the centromere regions, contain tandem non-coding highly repeated satellite DNA sequences. AT-rich DNA stretches of scaffold-associated regions mainly bind to the chromosome scaffold in G-dark band regions (Saitoh and Laemmli 1994). During S phase, G- and C-bands transiently decondense and recondense after DNA synthesis (Manuelidis and Chen 1990). Eukaryotic type II DNA topoisomerase is very active during replication and cell cycle related chromatin remodelling by transitorily cutting DNA strands to alleviate torsional stress (Champoux 2001). The clastogenic anti-topoisomerase II agents, such as etoposide, trap the cleavable complex of topoisomerase II and hinder the resealing of DNA strand breaks introduced by the enzyme (Palitti 1993; Palitti et al. 1994; Degrassi et al. 2004).

We analysed the role of chromatin structure and activity in the induction of chromosomal aberrations using different approaches:

1. Treating Chinese hamster ovary (CHO) cells with ionizing radiation (neutrons or γ-rays) to assign induced chromosome breakpoints to G-light and G-dark bands.
2. Immunollabelling of CHO chromosomes with antibodies against acetylated H4 histones to correlate radiation-induced breakpoints with the histone H4 acetylation pattern.
3. Studying the influence of chromatin remodelling processes related to DNA replication and cell cycle progression on chromosome damage by the analysis of the distribution of breakpoints induced by etoposide and the monofunctional alkylating agent methyl methanesulfonate (MMS) on the euchromatic Xp and heterochromatic Xq arms of the X chromosome of CHO cells during early and late S or G2 phases of the cell cycle.
4. Exposing lines of human fibroblasts with different numbers of X chromosomes to γ-rays and localizing induced breakpoints to euchromatin and heterochromatin.

13.2 Methodological Aspects

13.2.1 Localization of Radiation-Induced Breakpoints on CHO G-banded Chromosomes

CHO cells were irradiated either with 2 Gy γ-rays or 2 Gy fission neutrons. Chromosome slides were exposed to a 0.025% trypsin solution to obtain G-bands. Three hundred breakpoints for each type of irradiation were scored. Breakpoints sites were assigned to G-light or G-dark bands on 262-band resolution G-band idiograms drawn according to relative chromosome and band length.

13.2.2 Immunolabelling of CHO Acetylated H4 Histones and Densitometric Analysis

CHO metaphases were prepared for immunolabelling by cytocentrifugation and exposed to rabbit antiserum R5/12, which recognizes acetylated H4 histone ($H4^{+a}$) at Lys12 and binds predominantly to diacetylated, triacetylated and tetraacetylated H4 (Turner and Fellows 1989). Indirect immunofluorescent labelling with rabbit antisera was carried out as described by Jeppesen and Turner (1993). Slides were counterstained with Hoechst 33258.

Images of chromosomes were captured with a Zeiss Axioskop fluorescence microscope fitted with a Chroma 83000 triple band pass filter set and a Photometrics CH250 CCD camera, using Digital Scientific capture and analysis extensions to IPLab Spectrum v3.1. The chromosome images were used to construct densitometric curves.

13.2.3 Etoposide and MMS Treatments on CHO Cells: Breakpoint Mapping on the CHO X Chromosome

Exponentially growing cells were exposed for 40 min with either etoposide (10 or 20 μM) or MMS (10 or 20 mM) (Sigma) and were simultaneously pulse-labelled with 30 μM 5-bromo-2′-deoxyuridine (BrdUrd) (Sigma). BrdUrd incorporation was detected by immunolabelling.

The distribution of 1,200 breakpoints in Xp and Xq was analysed in damaged BrdUrd immunostained CHO metaphases, treated with etoposide in early and late S and G2 phases or with MMS in early and late S phases.

The numbers of breakpoints per aberration were determined by Obe and Winkel (1985) as follows: chromatid breaks 1; isochromatid breaks 2; quadriradials and triradials (chromatid interchanges) 2; interstitial deletions 2; duplication deletions 3; intrachanges 2; dicentrics 2; translocations 2; and rings 2. Xp and Xq expected breakpoint frequencies were estimated according to arm length.

13.2.4 Radiation Exposure of Human Cell Lines with Different Numbers of X Chromosomes

The following human fibroblast cell lines were used: AG1521 (46,XY) and AG1522 (46,XY); GM2149 (46,XX); GM4626 (47,XXX); GM00157 (49, XXXXY). Confluent fibroblasts were exposed to γ-rays (^{137}Cs) and first division cells were analysed.

13.2.4.1 Fluorescence in Situ Hybridization

Metaphase spreads were "painted" by the method of fluorescence in situ hybridization (FISH) using chromosome-specific DNA probe libraries for X chromosomes following the method of Pinkel et al. (1988) and Lucas et al. (1992).

Aberrations in the X chromosomes were scored in bicoloured chromosomes with one (translocations) or two (dicentric chromosomes) centromeres. In addition, acentric fragments were scored.

13.2.4.2 BrdUrd Incorporation and Immunolabelling

To ascertain late replicating patterns of X chromosomes, BrdUrd was added to cultures 3 h before addition of colcemid, and mitotic cells were collected and fixed 2 h later. The X chromosomes containing BrdUrd were detected by immunolabelling.

FISH was used to recognize X chromosomes with a biotinylated X-specific centromere probe which was postlabelled with fluorescein isothiocyanate (FITC)–avidin.

13.2.4.3 In Situ Nick Translation

DNase I sensitivity of fixed metaphase chromosomes from cell lines with multiple X chromosomes was determined by in situ nick translation according to the method of Kerem et al. (1983). Chromosome spreads were incubated in a nick translation buffer containing a mixture of DNase I/polymerase I, deoxyribonucleoside triphosphates and biotin-2′-deoxyuridine-5′-triphosphates. Slides were treated with a biotinylated X-specific centromere probe and detected by avidin conjugated with FITC. Propidium iodide was used for counterstaining.

13.3 Distribution of Chromosome Breakpoints in CHO Cells

13.3.1 Mapping of Radiation-Induced Breakpoints

Considerable differences have been reported with respect to the pattern of non-random chromosomal distribution of radiation-induced breakpoints in human cells analysed with G-banding techniques (Cooke et al. 1975; Buckton 1976; Dubos et al. 1978; Bauchinger and Götz 1979; Lee and Kamra 1981; Tanaka et al. 1983; Barrios et al. 1989; Johnson et al. 1999). Although non-random localizations of radiation-induced breakpoints were confirmed with chromosome painting techniques, contradictory results were reported with respect to the involvement of specific chromosomes and chromosome regions in aberrations (Tucker and Senft 1994; Dominguez et al. 1996; Grigorova et al. 1998; Griffin et al. 1995; Luomahaara et al. 1999; Cigarran et al. 1998; Kiuru et al. 2000; Knehr et al. 1994, 1996; Xiao and Natarajan 1999; Schleiermacher et al. 2003).

In CHO cells we found that breakpoints induced by γ-rays or neutrons are non-randomly distributed with respect to chromosome length and G-bands in a similar manner (Martínez-López et al. 1998). Breakpoints occurred preferentially in G-light bands and 22% clustered in 7% of the total chromosome length (Folle et al. 1998).

Similar results were found in hamster cells following exposure to X-rays (Slijepcevic and Natarajan 1994a, b) and to endonucleases (Folle et al. 1997, 1998), suggesting that chromatin conformation and nuclear structure play a role in the distribution of breakpoints.

G-light band regions containing active genes can be expected to have a more open conformation, making DNA more accessible to endonucleases and oxygen radicals induced by radiation (Löbrich et al. 1996; Forrester and Radford 2004). The width of the minor groove of DNA could also have an influence on the induction of lesions by influencing the accessibility of hydrogen atoms in sugar moieties to OH radicals (Sy et al. 1997).

The open conformation of transcribed chromatin might allow greater mobility of broken DNA ends, which may increase the chance of rejoining wrong partners, leading to chromosomal aberrations. In contrast, inactive chromatin is so tightly packaged that the movement of broken DNA ends is very constrained, so exchanges are unlikely to occur and restitution is the rule (Savage 1998).

Differential repair of randomly induced primary lesions could also explain the non-random distribution of induced breakpoints (Pellicia et al. 1985; Pantelias 1994; Sanders et al. 2004). More experimental evidence is needed to understand the possible role of repair mechanisms in chromosome breakpoint distribution.

13.3.2 Histone Acetylation Pattern and Clustering of Breakpoints

Transcriptionally active chromatin has a more accessible structure than transcriptionally inactive chromatin, as indicated by its increased sensitivity to nuclease digestion (Hebbes et al. 1994). The loosening of nucleosome structure caused by histone acetylation may facilitate the binding of initiating factors and RNA polymerase transcription complex to DNA. Immunolabelling of human metaphase chromosomes with fluorescent antibodies against acetylated histone H4 results in hyperacetylated (R-bands), relatively underacetylated (G-bands) and hypoacetylated (heterochromatin) regions (Jeppesen 1997; Brinkman et al. 2006). Underacetylation of inactive X chromosomes and constitutive heterochromatic chromosome regions was shown by Jeppesen and Turner (1993). Thus, the pattern of H4 hyperacetylation ($H4^{+a}$) can be considered a chromosomal marker for gene expression.

In CHO chromosomes we confirmed that the distribution of $H4^{+a}$ corresponds to an R-band pattern. The distribution of $H4^{+a}$ in CHO chromosomes and the underacetylated heterochromatic long arm of the X chromosome are shown in Fig. 13.1 (Martínez-López et al. 2001).

Radiation-induced breakpoints are clustered in highly acetylated chromosome regions (Fig. 13.1), confirming that chromosomal aberrations occur more frequently in active than in inactive chromatin (Martínez-López et al. 2001). Similar to human chromosomes (Jeppesen and Turner 1993), active genes and induced breakpoints seem to concentrate in the subtelomeric and telomeric regions in most CHO chromosomes (Folle et al. 1997, 1998; Martínez-López et al. 1998).

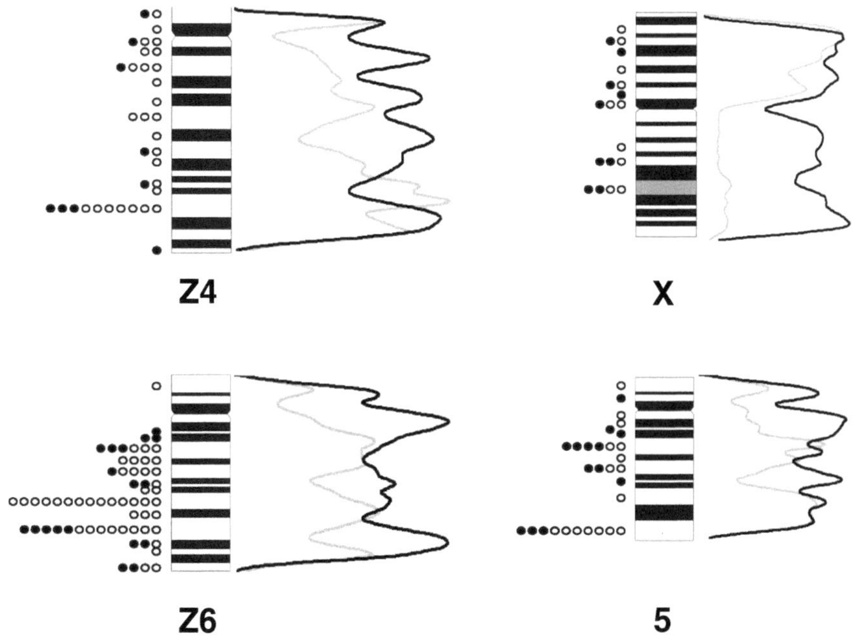

Fig. 13.1 Ideograms of Chinese hamster ovary chromosomes 5, Z4, Z6 and X employed in the analysis of the intrachromosomal localization of ionizing radiation induced breakpoints. On the *right side of the ideograms* representative densitometric tracings of histone H4 acetylation pattern (*grey*) and Hoechst 33258 (*black*) fluorescence of these chromosomes are presented. *Empty circles* correspond to breakpoints induced by neutrons and *filled circles* correspond to γ-rays. In chromosomes 5 and Z6, both breakage-prone bands map to the same G-light segments since the two terminal G-dark bands from Z6 originate in a reciprocal translocation involving the distal segments of 1p and 5q. The secondary constriction of the X chromosome is indicated by a *shaded area*

13.3.3 Distribution of Breakpoints Induced by Etoposide and MMS in the X Chromosome of CHO Cells

The advantage of inducing DSB by trapping the cleavable complex formed by topoisomerase II, make the anti-topoisomerase II compound etoposide a useful tool to study the influence of chromatin remodelling processes on the distribution of chromosome damage (Martínez-López et al. 2004).

The CHO X chromosome has a short euchromatic arm (Xp) and a long heterochromatic arm (Xq), which are early- and late-replicating, respectively; as revealed by BrdUrd pulse-labelling. The differential euchromatin/heterochromatin replication timing of Xp and Xq cells provides a valuable model to analyse the influence of DNA synthesis and chromatin remodelling on the distribution of induced breakpoints.

Xq is more sensitive to etoposide during late S phase (Martínez-López et al. 2004). To assess the influence of DNA replication and chromatin remodelling on breakpoint distribution, we studied breakpoints produced by etoposide and the alkylating agent MMS in Xp and Xq arms of the CHO X chromosome during early and late S phase. Additionally cells were exposed to etoposide in early and late G2 phase.

Breakpoints induced by etoposide and MMS in Xp or Xq during S phase are shown in Fig. 13.2. During early S phase (replication of Xp) etoposide- and MMS-induced breakpoints clustered in Xp. Similarly, during late S phase (when Xq replicates) breakpoints produced by both agents concentrated in Xq (Di Tomaso et al. 2006).

This shows that DNA replication plays an important role in breakpoint distribution. In the case of MMS, increased DNA accessibility due to chromatin decondensation associated with replication may explain the breakpoint distribution. The results with etoposide are explainable by the high activity of topoisomerase II and its inhibition during DNA synthesis.

Breakpoints induced by etoposide in early G2 phase were distributed randomly but mapped preferentially to Xp in late G2 phase (Fig. 13.2).

Studies performed in different cell cycle phases of CHO cells (G1 or G2) with ionizing radiation or restriction endonucleases revealed conflicting results with respect to localization of breakpoints in Xp or Xq (Slijepcevic and Natarajan 1994a, b; Folle et al. 1997; Grigorova et al. 1998; Xiao and Natarajan 1999). Contradictory findings could be due to subtle variations in cell cycle stages at the time of treatment.

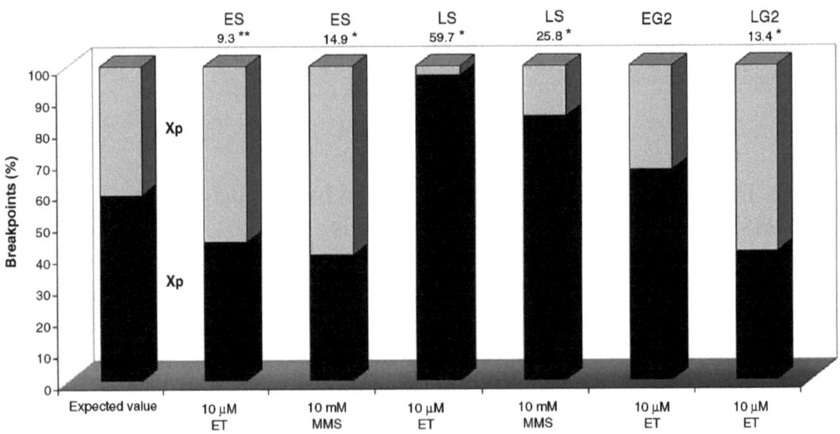

Fig. 13.2 The distribution (as percentages) of chromosome breakpoints induced by etoposide (*ET*; 10 μM) and methyl methanesulfonate (*MMS*; 10 mM) along Xp (*grey*) and Xq (*black*) in early S phase (*ES*) and late S phase (*LS*) as well as induced only by etoposide (10 μM) during early G2 phase (*EG2*) or late G2 phase (*LG2*). The *bar on the left* indicates expected frequencies of breakpoints according to Xp and Xq relative length. χ^2 values are indicated on the *top of the bars*. *p0.0005; **$p<0.005$

13.4 Chromosome Damage in Human Cell Lines with Multiple X Chromosomes

13.4.1 Replication Timing of Active and Inactive Human X Chromosomes

Studies on replication timing during S phase showed that each pair of human chromosomes has the same replication pattern, except for the X chromosomes in female cells. The inactive X chromosome is late-replicating as are extra X chromosomes in cells with multiple inactive X chromosomes; therefore, replication patterns seen after BrdUrd incorporation (replication bands) allow us to decide in which stage of the S phase the cell was when BrdUrd was added (Drouin et al. 1990).

To determine late labelling patterns of X chromosomes in cells with multiple X chromosomes, an X-centromere specific probe in combination with BrdUrd immunolabelling was applied. Male cells had only a small late-replicating pericentromeric region in their X chromosome. Normal female cells have one entirely late replicating X chromosome. The 47,XXX and 49,XXXXY cells exhibited two and three late-replicating X chromosomes, respectively.

13.4.2 Nuclease Sensitivity of Human Metaphase Chromosomes

It is well known that active or potentially active genes have a preferential DNase I sensitivity probably due to a less condensed chromatin conformation at the transcribed site (Weintraub and Groudine 1976). By in situ nick translation it is possible to distinguish between active and inactive X chromosomes on fixed metaphases (Kerem et al. 1983).

In order to determine whether the chromatin structure on both active and inactive X chromosomes presents uneven sensitivities to DNase I, in situ nick translation was performed on metaphase spreads from human cell lines containing different numbers of X chromosomes. The late-replicating supernumerary X chromosomes in 47,XXX and 49,XXXXY cells appeared to be very resistant to DNase I digestion.

It is worth mentioning that digestion of metaphase chromosomes with *Alu*I, *Bam*HI or DNase I followed by Giemsa staining produced G-bands in CHO cells (Folle and Obe 1995). A differential sensitivity of heterochromatin and euchromatin to endonuclease digestion was also observed by Chan et al. (2000). In CHO chromosomes the heterochromatic long arm of the X chromosome remained almost unaffected by enzyme treatment, showing that heterochromatin is less accessible to nucleases (Folle et al. 1997; Folle and Obe 1995).

Table 13.1 Total frequency of aberrations induced by γ-rays in cell lines containing different numbers of X chromosomes as well as the frequency of aberrations involving only X chromosomes

Cell line	Karyotype	X status	Total aberrations scored per cell (average 1.754)	Observed X aberrations per cell	Normalized X aberrations per cell[a]	Expected X aberrations per cell[b]
AG 1521	46,XY	1 (Xa)	1.89	0.18	0.17	0.17
AG 1522	46,XY	1 (Xa)	1.77	0.16	0.16	0.17
GM2149	46,XX	2 (1 Xa+1 Xi)	1.86	0.28	0.26	0.34
GM04626	47,XXX	3 (1 Xa+2 Xi)	1.66	0.34	0.36	0.51
GM00157	49,XXXXY	4 (1 Xa+3 Xi)	1.59	0.22	0.24	0.68

Xa active X chromosome, *Xi* inactive X chromosome
[a]Normalization was done to correct for small differences in cellular radiosensitivity among cell lines. The corrected or normalized values were estimated as follows:

$$\frac{\text{average of the total aberrations}}{\text{total aberrations per cell}} \times \text{aberrations in X chromosomes per cell.}$$

[b]Expected X aberrations per cell were estimated based on average aberrations involving active X chromosomes.

13.4.3 Chromosomal Aberrations Induced by γ-rays in Human X Chromosomes

Seabright (1973), employing G-banding, reported fewer X-ray induced aberrations in female human X chromosomes than expected from the length.

The frequencies of γ-ray induced aberrations involving the X chromosome in various cell lines with an active X chromosome and different numbers of inactive X chromosomes were measured by employing a specific chromosome painting probe. No increase in the yield of aberrations in the X chromosomes was found in 47,XXX or 49,XXXXY cells compared with 46,XX cells (Table 13.1), indicating that the chromatin structure of extra X chromosomes or their activity status affects the induction of chromosomal aberrations (Mühlmann-Díaz and Bedford 1994).

13.5 Conclusion

Chromosomal aberrations are the result of different cellular mechanisms interacting with DNA lesions. Most of these lesions are repaired in such a way that they do not lead to chromosome damage. Only a few DNA lesions (unrepaired or misrepaired) are transformed into chromosomal aberrations as observed by light microscopy.

Complex interplay of factors such as DNA conformation, transcription and replication can lead to a non-random interchromosomal and intrachromosomal distribution of chromosome breakpoints. Chromatin activity and remodelling processes associated with DNA replication and cell cycle progression are also involved in the topology of breakpoints. Differential distribution of euchromatin and heterochromatin along chromosomes as well as the compartmentalization of chromatin structure in G-light and G-dark bands can also affect the localization of breakpoints.

Acknowledgements: This work was supported in part by PDT/BID-DINACYT (project 91/29) and PEDECIBA from Uruguay and by CONICET and SECYT from Argentina (project PICTR 2002-00121).

References

Akhtar A, Cavalli G (2005) The epigenome network of excellence. PLoS Biol 3:762–763

Barrios L, Miró R, Caballín MR, Fuster C, Guedea F, Subias A, Egozcue J (1989) Cytogenetic effects of radiotherapy breakpoint distribution in induced chromosome aberrations. Cancer Genet Cytogenet 41:61–70

Bauchinger M, Götz G (1979) Distribution of radiation induced lesions in human chromosomes and dose-effect relation analysed with G-banding. Radiat Environ Biophys 16:335–366

Brinkman AB, Roelofsen T, Pennings SW, Martens JH, Jenuwein T, Stunnenberg HG (2006) Histone modification patterns associated with the human X chromosome. EMBO Rep 7:628–634

Buckton KE (1976) Identification with G and R banding of the position of breakage points induced in human chromosomes by in vitro X-irradiation. Int J Radiat Biol 29:475–488
Chai B, Huang J, Cairns BR and Laurent BC (2005) Distinct roles for the RSC and Swi/Snf ATP-dependent chromatin remodellers in DNA double-strand break repair. Genes Dev 19: 1656–1661
Champoux JJ (2001) DNA topoisomerases: structure, function, and mechanism. Ann Rev Biochem 70:369–413
Chan JK, Park PC, De Boni U (2000). Association of DNAse sensitive chromatin domains with the nuclear periphery in 3T3 cells in vitro. Biochem Cell Biol 78:67–78
Cigarran S, Barrios L, Barquinero JF, Caballin MR, Ribas M, Egozcue J (1998) Relationship between the DNA content of human chromosomes and involvement in radiation-induced structural aberrations, analysed by painting. Int J Radiat Biol 74:449–455
Cooke P, Seabright M, Wheeler M (1975) The differential distribution of X-rays induced chromosome lesions in trypsin-banded preparations from human subjects. Humangenetik 28:221–231
Degrassi F, Fiore M, Palitti F (2004) Chromosomal aberrations and genomic instability induced by topoisomerase-targeted antitumor drugs. Curr Med Chem Anti-Cancer Agents 4:317–325
Di Tomaso MV, Martínez-López W, Folle GA, Palitti F (2006) Modulation of chromosome damage localisation by DNA replication timing. Int J Radiat Biol 82:877–886
Dominguez I, Boei JJWA, Balajee AS, Natarajan AT (1996) Analysis of radiation-induced chromosome aberrations in Chinese hamster cells by FISH using chromosome-specific DNA libraries. Int J Radiat Biol 70:199–208
Drouin R, Lemieux N, Richer CL (1990) Analysis of DNA replication during S-phase by means of dynamic chromosome banding at high resolution. Chromosoma 99:273–280
Dubos C, Pequignot EV, Dutrillaux B (1978) Localisation of gamma-rays induced chromatid breaks using a three consecutive staining techniques. Mutat Res 49:127–131
Felsenfeld G, Groudine M (2003) Controlling the double helix. Nature 421:448–453
Folle GA, Obe G (1995) Localization of chromosome breakpoints induced by AluI and BamHI in Chinese hamster ovary (CHO) cells treated in the G1 phase of the cell cycle. Int J Radiat Biol 68:437–445
Folle GA, Boccardo E, Obe G (1997) Localization of chromosome breakpoints induced by DNase I in Chinese hamster ovary (CHO) cells. Chromosoma 106:391–399
Folle GA, Martínez-López W, Boccardo E, Obe G (1998) Localization of chromosome breakpoints: implication of the chromatin structure and nuclear architecture. Mutat Rese 404:17–26
Forrester HB, Radford IR (2004) Ionizing radiation-induced chromosomal rearrangements occur in transcriptionally active regions of the genome. Int J Radiat Biol 80:757–767
Goldman MA, Holmquist GP, Gray MC, Caston LA, Nag A (1984) Replication timing of mammalian genes and middle repetitive sequences. Science 224:686–692
Green CM, Almouzni G (2002) When repair meets chromatin. EMBO Rep 31:28–33
Griffin CS, Marsden SJ, Stevens DL, Simpson P, Savage JR (1995) Frequencies of complex chromosome exchange aberrations induced by 238Pu alpha-particles and detected by fluorescence in situ hybridization using single chromosome-specific probes, Int J Radiat Biol 67:431–439
Grigorova M, Brand R, Xiao Y, Natarajan AT (1998) Frequencies and types of exchange aberrations induced by X-rays and neutrons in Chinese hamster splenocytes detected by FISH using chromosome-specific DNA libraries. Int J Radiat Biol 74:297–314
Hebbes TR, Clayton AL, Thorne AW, Crane-Robinson C (1994) Core histone hyperacetylation co-maps with generalized DNase I sensitivity in the chicken beta-globin chromosomal domain. EMBO J 13:1823–1830
Jenuwein T, Allis CD (2001) Translating the histone code. Science 293:1074–1080
Jeppesen P (1997) Histone acetylation: a possible mechanism for the inheritance of cell memory at mitosis. BioEssays 19:67–74
Jeppesen J, Turner BM (1993) The inactive X chromosome in female mammals is distinguished by a lack of histone H4 acetylation, a cytogenetic marker for gene expression. Cell 74:281–289

Johnson KL, Brenner DJ, Nath J, Tucker JD, Geard CR (1999) Radiation-induced breakpoint misrejoining in human chromosomes: random or non-random?. Int J Radiat Biol 75:131–141
Kerem BS, Goitein R, Richler C, Marcus M, Cedar H (1983) In situ nick translation distinguishes between active and inactive X chromosomes. Nature 304:88–90
Kiuru A, Lindholm C, Auvinen A, Salomaa S (2000) Localization of radiation-induced chromosomal breakpoints along human chromosome 1 using a combination of G-banding and FISH. Int J Radiat Biol 76:667–672
Knehr S, Zitzelsberger H, Braselmann H, Bauchinger M (1994) Analysis for DNA-proportional distribution of radiation-induced chromosome aberrations in various triple combinations of human chromosomes using fluorescence in situ hybridization. Int J Radiat Biol 65:683–690
Knehr S, Zitzelsberger H, Braselmann U, Nahrstedt U, Bauchinger M (1996) Chromosome analysis by fluorescence in situ hybridization: further indications for a non-DNA-proportional involvement of single chromosomes in radiation-induced structural aberrations. Int J Radiat Biol 70:385–392
Korenberg JR, Rykowski MC (1988) Human genome organization: Alu, Lines, and the molecular structure of metaphase chromosome bands. Cell 53:391–400
Lee C, Kamra O (1981) The pattern of radiation induced transmissible aberrations in a human cell culture. Hum Genet 57:380–384
Löbrich M, Cooper PD, Rydberg B (1996) Non-random distribution of DNA double-strand breaks induced by particle irradiation. Int J Radiat Biol 70:493–503
Lucas JN, Awa A, Straume T, Poggensee M, Kodama Y Nakano M, Ohtaki K, Weier HU Pinkel D, Gray J (1992) Rapid translocation frequency analysis in humans decades after exposure to ionizing radiation. Int J Radiat Biol 62:53–63
Luomahaara S, Lindholm C, Mustonen R, Salomaa S (1999) Distribution of radiation-induced exchange aberrations in human chromosomes 1, 2 and 4. Int J Radiat Biol 75:1551–1556
Manuelidis L, Chen TL (1990) A unified model of eukaryotic chromosome. Cytometry 11:8–25
Martínez-López W, Boccardo E, Folle GA, Porro V, Obe G (1998) Intrachromosomal localization of aberration breakpoints induced by neutrons and γ-rays in Chinese hamster ovary (CHO) cells. Radiat Res 150:585–592
Martínez-López W, Folle GA, Obe G, Jeppesen P (2001) Chromosome regions enriched in hyperacetylated histone H4 are preferred sites for endonuclease- and radiation-induced breakpoints. Chromosome Res 9:69–75
Martínez-López W, Folle GA, Cassina G, Méndez-Acuña L, Di Tomaso MV, Obe G, Palitti F (2004) Distribution of breakpoints induced by etoposide and X-rays along the CHO X chromosome. Cytogenet Genome Res 104:182–187
Morrison A, Highland J, Krogan NJ, Arbel-Eden A, Greenblatt JF, Haber JE, Shen X (2004) INO80 and gamma-H2AX interaction links ATP-dependent chromatin remodelling to DNA damage repair. Cell 119:767–775
Mühlmann-Díaz MC, Bedford JS (1994) A comparison of radiation-induced aberrations in human cells involving early and late replication X-chromosomes. In: Obe G, Natarajan AT (eds) Chromosomal alterations. Origin and significance. Springer, Berlin, pp 125–131
Obe G, Winkel EU (1985) The chromosome breaking activity of the restriction endonuclease AluI in CHO cells is independent of the S-phase of the cell cycle. Mutat Res 152:25–29
Obe G, Savage JRK, Johannes C, Goedecke W, Jeppesen P, Natarajan AT, Martínez-Lopez W, Folle GA, Drets ME (2002) Chromosomal aberrations: formation, identification and distribution. Mutat Res 504:17–36
Palitti F (1993) Mechanism of induction of chromosomal aberrations by inhibitors of DNA topoisomerases. Environ Mol Mutagen 22:275–277
Palitti F, Mosesso P, Di Chiara D, Schinoppi A, Fiore M, Bassi L (1994) Use of antitopoisomerase drugs to study the mechanisms of induction of chromosomal damage. In: Obe G, Natarajan AT (eds) Chromosomal alterations. Origin and significance. Springer, Berlin, pp 103–115
Pantelias GE (1994) Factors determining the yields of radiation-induced chromosomal aberrations as visualized by means of premature chromosome condensation in interphase

cells. In: Obe G, Natarajan AT (eds) Chromosomal alterations. Origin and significance. Springer, Berlin, pp 140–149

Pellicia F, Micheli A, Olivieri G (1985) Inter- and intra-chromosomal distribution of chromatid breaks induced by X-rays during G2 in human lymphocytes. Mutat Res 150:293–298

Pinkel D, Landegent J, Collins C, Fuscoe J, Segraves R, Lucas J, Gray J (1988) Fluorescence in situ hybridization with human chromosome specific libraries: Detection of trisomy 21 and translocations of chromosome 4. Proc Natl Aacd Sci USA 85:9138–9142

Polo SE, Almouzni G (2005) Histone metabolic pathways and chromatin assembly factors as proliferation markers. Cancer Lett 220:1–9

Rubbi CP, Milner J (2003) p53 is a chromatin accessibility factor for nucleotide excision repair of DNA damage. EMBO J 22:975–986

Saitoh Y, Laemmli UK (1994) Metaphase chromosome structure: bands arise from a differential folding path of the highly AT-rich scaffold. Cell 76:609–622

Sanders MH, Bates SE, Wilbur BS, Holmquist GP (2004) Repair rates of R-band, G-band and C-band DNA in murine and human cultured cells. Cytogenet Genome Res 104:1–4

Savage JRK (1998) A brief survey of aberration origin theories. Mutat Rese 404:139–147

Schleiermacher G, Janoueix-Lerosey I, Combaret V, Derre J, Couturier J, Aurias A, Delattre O (2003) Combined 24-color karyotyping and comparative genomic hybridization analysis indicates predominant rearrangements of early replicating chromosome regions in neuroblastoma. Cancer Genet Cytogenet 141:32–42

Seabright M (1973) Noninvolvement of the human X chromosome in X-ray-induced exchanges. Cytogenet Cell Genet 12:342–356

Slijepcevic P, Natarajan AT (1994a) Distribution of X-ray-induced G2 chromatid damage among Chinese hamster chromosomes: influence of chromatin conformation. Mutat Res 323:113–119

Slijepcevic P, Natarajan AT (1994b) Distribution of radiation-induced G1 exchange and terminal deletion breakpoints in Chinese hamster chromosomes as detected by G banding. Int J Radiat Biol 66:747–755

Strahl BD, Allis CD (2000) The language of covalent histone modifications. Nature 403:41–45

Sy D, Savoye C, Begusova M, Michalik V, Charlier M, Spotheim-Maurizot M (1997) Sequence-dependent variations of DNA structure modulate radiation-induced strand breakage. Int J Radiat Biol 72:147–155

Tanaka K, Kamada N, Ohkita T, Kuramoto, A (1983) Non-random distribution of chromosome breaks in lymphocytes of atomic bomb survivors. J Radiat Rese 24:291–304

Tucker JD, Senft JR (1994) Analysis of naturally occurring and radiation-induced breakpoint locations in human chromosomes 1, 2 and 4. Radiat Rese 140:31–36

Turner BM (2000) Histone acetylation and an epigenetic code. BioEssays 22:836–845

Turner BM, Fellows G (1989) Specific antibodies reveal ordered and cell-cycle-related use of histone-H4 acetylation sites in mammalian cells. Eur J Biochem 179:131–139

Weintraub H, Groudine M (1976) Chromosomal subunits in active genes have an altered conformation. Science 193:848–856

Xiao Y, Natarajan AT (1999) Non-proportional involvement of Chinese hamster chromosomes 3, 4, 8 and 9 in X-rays-induced chromosomal aberrations. Int J Radiat Biol 75:943–951

14 A Biomedical Perspective of Telomere Structure and Function

MÁXIMO E. DRETS AND FEDERICO F. SANTIÑAQUE

Abstract

Telomeres are specialized nucleoproteic terminal structures of eukaryotic chromosomes that play an essential role in the stability of the genetic material within the cell, preventing its exonucleolytic degradation, end-to-end fusions and other chromosome-lethal events. Interest in this multifunctional chromosome segment has increased considerably since it intervenes in cell senescence, gene transcriptional regulation, premature aging syndromes and cancer, which are areas of great biomedical importance. Numerous "cryptic" microdeletions and rearrangements occurring in the subtelomeric region have been detected by fluorescent in situ hybridization and are related to mental retardation, congenital malformations and other severe clinical human conditions. Microphotometric scanning of T-banded subtelomeric segments of normal human and Chinese hamster ovary (CHO) chromosomes produces highly detailed graphic images allowing the detection of a differential distribution of high-density chromatin patterns between sister chromatids, namely, of similar size in both sister chromatids, predominating in one of the chromatids or concentrated in only one chromatid; these observations have been confirmed by scanning endoreduplicated CHO chromosomes. Furthermore, minute sister-chromatid exchanges were detected in normal human and CHO chromosomes. This precise analytical method additionally revealed that the highest chromatin densities of each chromosome arm were located in opposite chromatids in the majority of the chromosomes analyzed. The significance of the major aspects of function and molecular composition of the telomeric/subtelomeric segment and their relationship to biomedical problems recently reported, as well as some microphotometrical observations on the human and CHO chromatin structure are briefly reviewed.

14.1 Introduction

Telomeres are specialized DNA/protein structures located at the end of eukaryotic chromosomes that play an essential role in the stable maintenance of the genetic material within the cell. By capping chromosome ends, telomeres

Günter Obe and Vijayalaxmi (Eds.)
Chromosomal Alterations: Methods, Results and Importance in Human Health

prevent its exonucleolytic degradation, end-to-end fusions and other potential chromosome-lethal events (Blackburn 2001; Zakian 1995).

The study of telomeric area is increasing its biomedical importance since this multifunctional chromosome segment intervenes not only in keeping constant the number and structure of the chromosomes but it is also involved in cell senescence, gene transcriptional regulation, premature aging syndromes and cancer (Blackburn and Greider 1995; Blasco 2005; Baur et al. 2001; Greider 1998). It has also been shown that microdeletions and subtle rearrangements occurring in the subtelomeric region can cause mental retardation, congenital abnormalities and other severe clinical conditions.

In this chapter, we describe major aspects of the molecular composition and function of the telomeric/subtelomeric segment considering its possible relationship to some microscopic observations made on the structure of these areas in human and Chinese hamster ovary (CHO) chromosomes. Finally, we briefly discuss some aspects not yet well clarified on this critical chromosome segment.

14.2 The Complex Molecular Structure of the Terminal Chromosome Region

14.2.1 Telomeric DNA and Telomerase: Cell Senescence and Cancer

Mammalian telomeres are composed of tandem arrays of TTAGGG sequences (Blackburn and Gall 1978) and of several associated telomeric proteins, including some related to DNA repair/recombination (Hande 2004). The physical end of telomeric DNA presents a long single-stranded overhang of the G-rich strand that is found in all chromosome ends, irrespective of whether they were replicated by the lagging or the leading DNA synthesis (Makarov et al. 1997). This particular configuration of the telomeric DNA is conserved in eukaryotic cells and is involved in the stabilization of a specialized telomeric structure called a t-loop that is observable by electron microscopy in telomeric chromatin fibres (Griffith et al. 1999; Nikitina and Woodcock 2004). The t-loop model postulates that the terminal DNA forms a large loop capping the 3′ single-stranded overhang in the telomeric double-stranded DNA.

The length of telomeric DNA is established during zygotic development (Graakjaer et al. 2004) but it degrades during each cycle of DNA synthesis owing to incomplete telomere replication by conventional DNA polymerases and owing to the action of nucleases (Hug and Lingner 2006). The ribo nucleoprotein telomerase, a specific reverse transcriptase that extends the terminal repeated DNA, employs its specific RNA as a template for the addition of TTAGGG sequences to chromosome ends (Lingner et al. 1997); therefore, telomerase plays a key role in telomere length maintenance, compensating the progressive shortening of chromosome ends. However, it has been proposed that the progressive shortening of telomeric DNA produced in the absence of telomerase functions as a “biological clock” regulating the time of cell senescence or apoptosis (Harley 1995). Telomere shortening is observed in normal somatic cells of adult organisms but not in germinal cells or during the first stages of the embryonic development where it is compensated by the action of telomerase (Cong et al. 2002). The telomerase activity is detected in 90% of cancer cells, enabling an unlimited proliferation of neoplastic cells (Hiyama and Hiyama 2002). Since short telomeres lead to cell senescence or chromosome alterations, the length of telomeric DNA is a critical factor for the stability of telomeres, cell aging and cancer (Rodier et al. 2005). Recent findings indicate that chronic stress, cigarette smoking and obesity are significantly associated with oxidative damage and accelerated shortening of telomeres in women (Epel et al. 2004; Valdes et al. 2005).

14.2.2 Telomeric Chromatin

The telomeric chromatin contains structural proteins that bind sequence-specifically to the telomeric double-stranded DNA (TRF1 and TRF2) or to the G-rich single-stranded DNA (POT1) (Chong et al. 1995; van Steensel and de Lange 1997; Baumann and Cech 2001). These proteins interact with other telomeric proteins, including TIN2, Rap1 and Tpp1, to form a specific high-order chromatin structure that is associated with telomeres throughout the cell cycle (Smith and de Lange 1997; Liu et al. 2004). This protein complex that preserves telomeres formed by the subunits TRF1, TRF2, TIN2, Rap1, Tpp1 and POT1 has been collectively named “shelterin” (de Lange 2005).

Telomeric proteins form specialized structures with an elevated high-order complexity in the telomeric DNA. The protein TRF2 is capable of forming large duplex loops in the telomeric DNA (t-loops) (Stansel et al. 2001; Yoshimura et al. 2004). Furthermore, the main component of the telomeric chromatin, TRF1, binds to the telomeric DNA with a high spatial flexibility which could bend the telomeric DNA, to form a loop or, with an elevated tendency, pairing of telomeric duplexes (Bianchi et al. 1997, 1999; Griffith et al. 1998). The intratelomeric pairing of (TTAGGG)*n* repeats induced by TRF1 is stimulated when the TIN2 protein associates with the homodimeric complex (Kim et al. 2003). Griffith et al. (1998) suggested that the intratelomeric pairing produced by TRF1 is responsible for the telomeric DNA coiled configuration that is observed by electron microscopy in mitotic chromosomes (Steinmüller et al. 1993).

Telomerase accessibility to telomeric DNA is highly regulated by the telomeric proteins and by the level of chromatin condensation (Fang and Cech 1995; Blasco 2004). However, the most important function of the telomeric proteins appears to be telomere protection. Mutations in the protein TRF2 produce end-to-end fusions, nucleolytic degradation and other aberrant telomere phenotypes, suggesting that this protein plays an important role in the protective activity of telomeres (van Steensel et al. 1998). Griffith et al. (1999) proposed that TRF2 switches telomeres into a non-recombinogenic state through the formation of the t-loops. Besides, mice with a targeted deletion of TRF1 or deficiency for TIN2 show embryonic lethality (Karlseder et al. 2003; Chiang et al. 2004). Several recent findings suggested that an altered expression or function of the telomeric proteins is present in some human premature aging syndromes and tumours (Munoz et al. 2006; Garcia-Aranda et al. 2006; Oh et al. 2005).

14.3 Cryptic Subtelomeric Aberrations: Their Importance in Human Health

The subtelomeric region is composed of a mixture of unique DNA sequences together with different families of polymorphic middle repetitive DNA (Henderson 1995; Riethman 2005). This area possesses the highest gene concentration of the human genome (Saccone et al. 1992). The region close to the chromosome termini also contains highly polymorphic repetitive DNA sequences which are extraordinarily variable, dynamic and of rapid evolution (Der-Sarkissian et al. 2002). Besides, the subtelomeric region is enriched 25-fold in short internal (TTAGGG)*n* sequences in relation to the rest of the genome, which could play a potential role in subtelomeric recombination and telomere healing (Riethman et al. 2004).

It has been reported that the subtelomeric area is the site of a high number of interchanges that are able to produce phenotypic variability and embryonic developmental alterations (Badge et al. 2000). Using fluorescent in situ hybridization (FISH), Knight and Flint (2000) reported a considerable number of minute aberrations in the subtelomeric region in patients with mental retardation and congenital malformation syndromes.

The chapter of this syndromatic association started being clarified when Lejeune et al. (1959) discovered trisomy 21 and Sutherland (1977) detected the X chromosome fragility. However, a considerable number of congenital conditions, presenting mental retardation and malformations were obscure and not diagnosed at all by that time, being considered "idiopathic". However, the use of FISH made it possible to detect in the subtelomeric region minute aberrations called "cryptic" and believed to be responsible for 7.5% of congenital malformation and mental retardation (Sutton et al. 2002). Moreover, cryptic aberrations have also been found in other severe medical conditions such as in haemato-oncology, cancerology, gynaecology and psychiatry, showing the practical importance of these studies for clinical medicine and prognosis. The chromosome picture of these cryptic chromosome syndromes observed up to now in different areas of pathology is so complex

Table 14.1 Examples of complex cryptic aberrations detected in the subtelomeric region of human chromosomes associated with syndromes of mental retardation and congenital malformations

der(16)t(16;19)(p13.3;p13.3); Brown et al. (2000)	del(3)(p25.3p26.2); Cargile et al. (2002)
4p, dup(4)(p16.3p15.3); Cotter et al. (2001)	t(8;20)(p23;p13); de Vries et al. (2001)
t(2;6)(q37.3;q26); Giardino et al. (2003)	t(3;16)(q29;p13.3); Holisnki-Feder et al. (2000)
t(18;21)(q2.11q21.3); Horn et al. (2003)	46,XX,der(18)t(2;18)(p25;p11.2); Horsley et al. (1998)
dup(17p11/15q11); Keller et al. (2003)	(46,XX,der(13)t(8;13)(q24;q34); Kleeftra et al. (2000)
der(14)t(9;14)(q34.3;q32.33); Leube et al. (2003)	46,XX,inv(6)(q22.1q27); Lorda-Sanchez et al. (2000)
del(9)(q34.3)dup19(13.3); Quigley et al. (2004)	del(5q35.3); Rauch et al. (2003)
der(18)t(5;18)pat; Reddy and Fugate (1999)	t(11;18)(q25;q23); Schultz et al. (2001)
17p [46,XX,der(6)t(6;17)(q25.3;p13)]; Souter et al. (2003)	t(2,7);(q37;q35)/7q35→qter; Speleman et al. (2000)
t(4q35;10q26); van Deutekom et al. (1996)	46,XX,der(18)t(17;18)(q25;p11.32); Velagaleti et al. (2003)
t(17p13.3)(20q13.33); Walter et al. (2004)	46,XX,der(16)t(16;1); Warburton et al. (2000)

that their description escapes the possibility to describe them in detail in this chapter (Table 14.1). Therefore, the reader is referred to general reports published on this topic, particularly the extensive study of Raynan et al. (2006) that covers 11,688 cases.

14.4 Light Microscope Methods for the Study of Telomere Structure

14.4.1 T-banding

The telomere/subtelomeric chromatin structure and its DNA GC richness confer to the terminal chromosome region a high resistance to heat denaturation, being selectively stained with the T-banding method (Dutrillaux 1973). This technique is based upon an incubation of chromosomes in a hot buffer followed by Giemsa staining, leading to the appearance of T-bands as darkly stained segments in a lightly stained chromosome. T-bands are considered to be a subset of GC-rich R-bands highly resistant to digestion with the restriction endonuclease *Mse*I (Ludeña et al. 1991). Scanning electron microscopy shows that this region corresponds to areas of highly aggregated fibres (Jack et al. 1986).

14.4.2 Telomere Microdensitometric Scanning

A microphotometer system composed of a Zeiss (Oberkochen, Germany) microscope system processor MSP65 and a fast-scanning stage (0.25-μm step) coupled to a MicroVAX 3300 computer (Digital Equipment Corporation, Maynard, MA, USA) and a Netstation 4211 graphics terminal (Tektronix, Beaverton, OR, USA) was used for the study of the telomere structure.

Each scanning of T-banded telomeres covered an object area of 100 × 100 stage steps. One hundred densitometric measurements were made per scanning step, totalling one million measurements per microscope object. The data obtained per step were averaged in order to reduce random noise produced by the electronic components. The instrument setting resulted from a compromise between scanning speed and noise disturbance (object measurement to noise ratio greater than 10), which increased the reliability and precision of the system in all image-analysis procedures. For scan conversion, absorbances were represented as two-dimensional image units (pixels) in the discrete pixel image plane of the graphics terminal. The picture element size corresponded approximately to the resolving power of the light microscope (1 pixel being approximately 0.2 μm). Screen coordinates and pixel size assignment resulted from the magnification of the photomicrograph, the number of scanning steps and the final size of the telomere image displayed on the colour graphics terminal screen.

14.4.3 Computer Graphics Image Processing

Changes of contrast and brightness allowed the pixeled images to be displayed on a 0–100 grey scale corresponding to the absorbance measurement limits of the Zeiss module. For image discrimination and threshold densities, T-banded telomeres were sorted into 12 classes ranging from 0 to 100 absorbance units. Absorbance lower than 5 units corresponded to the background by default. Arbitrary colours were used to differentiate sorted classes, allowing a visual enhancement of the chromosome regional densities displayed. Binary filtering (shrink and expand operation) reduced unwanted pixels displayed in the background surface resulting from cytoplasm debris or stain precipitates. Data interpolation was used to obtain an image resolution of 600 × 600 compared with the original scanning of the 100 × 100 measurement matrix, enabling a more detailed image of chromosomal segments, since the staircase effect was eliminated. The number of pixels of the subtelomeric T-banded chromosome region corresponding to each sorted rank per chromatid

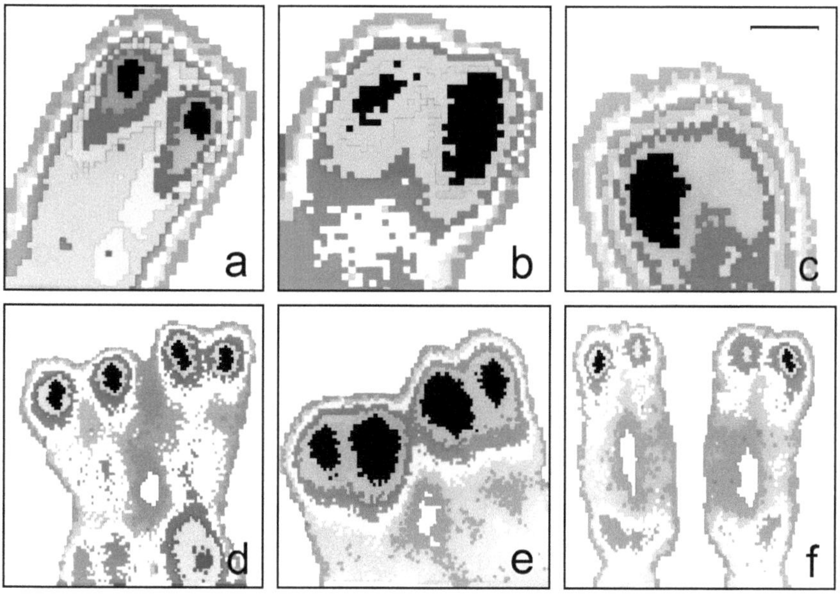

Fig. 14.1 Pixeled graphic images of high-density patterns detected in normal Chinese hamster ovary (CHO) telomeres (**a**–**c**) and in endoreduplicated CHO chromosomes (**d**–**f**). Note the similarity of the three basic high-density patterns found in normal and endoreduplicated chromosomes, indicating that only preexisting structures could replicate in the same maner. High-density areas are denoted in *black*. *Bar* 1 μm

was counted by drawing a line on the displayed image to define the chromatid borders using the graphic input rubberbanding feature of the Netstation 4211 graphics terminal. This feature allowed very precise quantitative measurements of the size of the high-density subtelomeric areas analysed to be obtained. Reprocessing of pixeled or interpolated colour graphic images allowed black-and-white illustrations to be obtained (Figs. 14.1–14.3) (see Drets et al. 1995a for a detailed description of the procedures).

14.5 Microdensitometric Analysis of T-banded Human and CHO Chromosomes

14.5.1 Differential Distribution of the Highest Chromatin Densities Between Sister Telomeres

The microphotometric analytical method developed in our laboratory was used for the study of the telomere structure of T-banded human and CHO chromosomes. This method allowed the detection of a differential distribution of high-density chromatin between sister telomeres (Drets et al. 1992)

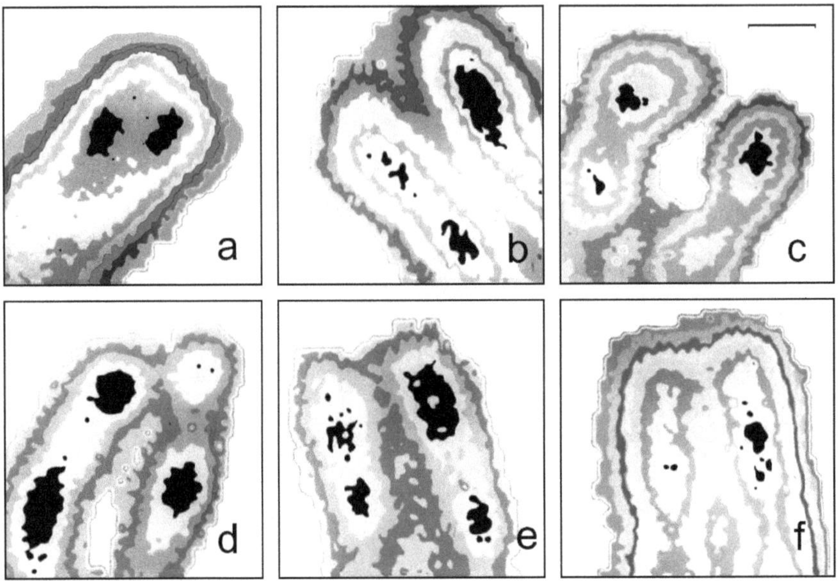

Fig. 14.2 Different examples of minute sister-chromatid exchanges of high-density segments detected in CHO (**a–c**) and human (**d–f**) chromosomes by means of microphotometrical scanning. Graphic images were obtained by interpolating microphotometric data. Exchanged blocks of high-density chromatin appear in **a** and **d.** Very tiny exchanges are seen in **b, e** and **f.** *Bar* 1 μm

that was arbitrarily classified into three types of density patterns, namely type I, high-density segments of similar size in both sister chromatids (Fig. 14.1a); type II, predominating in one of the chromatids (Fig. 14.1b); and type III, concentrated in only one chromatid (Fig. 14.1c). This unexpected distribution of high-density chromatin was confirmed by scanning T-banded endoreduplicated CHO chromosomes, since similar types of density patterns appeared in both sister chromosomes (Drets and Mendizábal 1998; Fig. 14.1d–f).

Interestingly, a prolonged incubation of chromosomes in the T-banding buffer induces the appearance of tiny holes in subtelomeric regions of sister chromatids. These holes were observed in normal and aberrant CHO chromosomes and were located in areas comparable to those of high-density telomeric areas, suggesting that a specific removal of high-density chromatin was produced (Drets et al. 1995b).

Recently, we studied the distribution of high-density chromatin at both ends of T-banded CHO chromosomes using our microdensitometric method and found that in the majority of the chromosomes that exhibited, in both ends, an asymmetrical distribution of high-density chromatin between sister telomeres (clearly noticeable in cases of the high-density pattern type III; Fig. 14.1c),

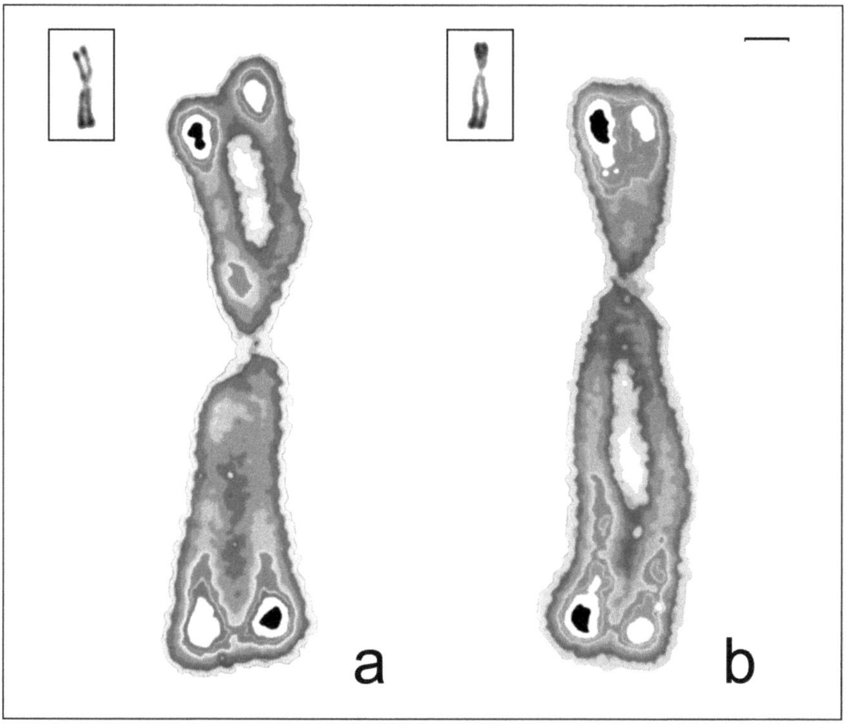

Fig. 14.3 Distribution of the high-density segments at opposite ends of T-banded CHO chromosomes. **a** High-density blocks are seen located in opposite chromatids (*trans* configuration). **b** High-density segments appearing in the same chromatids (*cis* configuration). The *insets* illustrate scanned chromosomes. *Bar* 1 μm

the highest-density segments of each arm were located in opposite chromatids. Preliminary observations showed that this configuration (*trans*) was detected in approximately 73% of the chromosomes analysed, and in a lower number of chromosomes (27%), high-density segments were found in the same chromatid (*cis* configuration). Figure 14.3 illustrates an example of both types of high-density segmentary configurations as detected by microdensitometric scanning of entire T-banded CHO chromosomes.

14.5.2 Minute Telomeric Sister-Chromatid Exchanges

Microphotometrical scanning of T-banded chromosomes also allowed the detection of a differential distribution of high-density areas between sister chromatids which looked like minute sister-chromatid exchanges that we named t-SCEs (Drets et al. 1992; Fig 14.2).

Several authors (Bailey et al. 2004; Bechter et al. 2004; Laud et al. 2005; Londoño-Vallejo et al. 2004; Wang et al. 2005), employing chromosome-orientation FISH, also detected minute sister-chromatid exchanges in telomere segments which are quite similar to the t-SCEs that we described earlier, but these authors used the term T-SCE to describing their observations. Since the use of different terms to describe similar, if not identical, structures detected with different cytological methods is undesirable, we proposed the use of the term ter-SCEs in order to standardize the nomenclature of these events (Drets et al. 2006). This new nomenclature would avoid any confusion with other terms generally used in human cytogenetics.

14.6 Final Remarks

The different types of graphic images obtained with our system closely corresponded to the T-banded segments observed with the light microscope. However, the computer graphic method used produced highly detailed images that allowed a more enhanced visualization of the chromatin structure of the terminal chromosome region than with the usual light microscope. This procedure revealed a complex high-density chromatin organization and distribution between sister chromatids in the terminal segments of human and CHO chromosomes. This finding was confirmed by scanning endoreduplicated CHO chromosomes, since sister chromosomes exhibited similar patterns of high-density chromatin distribution. Besides, the detection of a "diagonal" distribution of high-density areas at opposite ends of T-banded CHO chromosomes by microphotometry suggests that these observations represented a telomeric non-random phenomenon.

It is of interest to mention that Bekaert et al. (2002) using three-dimensional confocal microscopy and quantitative FISH found significant differences in telomere signal intensity between sister chromatids of human lymphocyte chromosomes. These observations are comparable to the density patterns found by us. This view may be additionally supported by the observation that high-density areas colocalize with telomere probe signals in human metaphase chromosomes (Drets 2004), which are reportedly embedded in the subterminal region of the chromatids (Moyzis et al. 1988; Steinmüller et al. 1993).

Pardue and DeBaryshe (1999) stressed the point that several findings suggest that eukaryotic telomeres may have functions other than chromosome end protection and recognition as intact chromosomes. Evidence supporting the high functional activity of the terminal region is the detection of minute sister chromatid exchanges (t-SCEs) in T-banded human and CHO chromosomes. Owing to the fact that with our method, 5-bromo-2′-deoxyuridine is not used for detecting t-SCEs, and hence the DNA composition is intact in these minute SCEs, they may represent in some way an expression of the diverse functionality prevailing in the region (Obe et al. 2002; Drets 2004).

In this connection, Jin and Ikushima (2004) considered that the frequent occurrence of the tiny t-SCEs observed in the T-banded terminal segments of CHO and human chromosomes is a visualization of telomere recombination, a view that supports the fact that many rearrangements occur more frequently in subtelomeric domains than in other regions of the human genome and undergo high rates of spontaneous recombination as reported by Cornforth and Eberle (2001).

The differential distribution of high-density chromatin at telomeres is revealed by the T-banding procedure and it is not clear up to now how the underlying chromosomal organization responds to the banding phenomena that allow the detection of the complex high-density segments. It is a well-established fact that the main banding procedures (C-, G-, R- and T-banding) reveal specific chromosome structures related to local variations of the DNA composition, but all of them depended on a controlled chromosome denaturation and on the complex chemical composition of the Giemsa stain. In this connection, Sumner (1990) pointed out that in chromosome banding dye accessibility may be modified either by proteins or by variation of the DNA base pair composition (or both) along the genome. An interesting approach was reported by Saitoh and Laemmli (1994), who described that the bands arise from a differential folding path of the chromosome scaffold.

Even though considerable information on the molecular analysis of the telomeric/subtelomeric DNA and associated protein complexes is available nowadays, a clear picture at the high level of organization of this metaphase chromosome segment is still missing. In this respect, we believe that a combined molecular and detailed microscope study on this complex chromosome segment may lead to a better understanding of its structural role and function within the nucleus. Besides, it is expected that a deeper insight into aspects of the telomeric/subtelomeric region obtained through new molecular procedures or improved microscope analytical methods will also produce a considerable impact in medicine since it will contribute to clarifying many still obscure pathological conditions, permitting a more precise diagnosis, treatment and prevention of illnesses severely affecting human populations.

Acknowledgement: This work was supported in part by PEDECIBA (Uruguay).

References

Badge RM, Yardley J, Jeffreys AJ, Armour JA (2000) Crossover breakpoint mapping identifies a subtelomeric hotspot for male meiotic recombination. Hum Mol Genet 9:1239–1244

Bailey SM, Brenneman MA, Goodwin EH (2004) Frequent recombination in telomeric DNA may extend the proliferative life of telomerase-negative cells. Nucleic Acids Res 32:3743–3751

Baumann P, Cech TR (2001) Pot1, the putative telomere end-binding protein in fission yeast and humans. Science 292:1171–1175

Baur JA, Zou Y, Shay JW, Wright WE (2001) Telomere position effect in human cells. Science 292:2075–2077

Bechter OE, Zou Y, Walker W, Wright WE, Shay JW (2004) Telomeric recombination in mismatch repair deficient human colon cancer cells after telomerase inhibition. Cancer Res 64:3444–3451

Bekaert S, Koll S, Thas O, Van Oostveldt P (2002) Comparing telomere length of sister chromatids in human lymphocytes using three-dimensional confocal microscopy. Cytometry 48:34–44

Bianchi A, Smith S, Chong L, Elias P, de Lange T (1997) TRF1 is a dimer and bends telomeric DNA. EMBO J 16:1785–1794

Bianchi A, Stansel RM, Fairall L, Griffith JD, Rhodes D, de Lange T (1999) TRF1 binds a bipartite telomeric site with extreme spatial flexibility. EMBO J 18:5735–5744

Blackburn EH (2001) Switching and signaling at the telomere. Cell 106:661–673

Blackburn EH, Gall JG (1978) A tandemly repeated sequence at the termini of the extrachromosomal ribosomal RNA genes in Tetrahymena. J Mol Biol 120:33–53

Blackburn EH, Greider CW (1995) Telomeres. Cold Spring Harbor Laboratory Press, Cold Spring Harbor

Blasco MA (2004) Telomere epigenetics: a high-order control of telomere length in mammalian cells. Carcinogenesis 25:1083–1087

Blasco MA (2005) Telomeres and human disease: ageing, cancer and beyond. Nat Rev Genet 6:611–622

Brown J, Horsley SW, Jung C, Saracoglu K, Janssen B, Brough M, Daschner M, Beedgen B, Kerkhoffs G, Eils R, Harris PC, Jauch A, Kearney L (2000) Identification of a subtle t(16;19)(p13.3;p13.3) in an infant with multiple congenital abnormalities using a 12-colour multiplex FISH telomere assay, M-TEL. Eur J Hum Genet 8:903–910

Cargile CB, Goh DL, Goodman BK, Chen XN, Korenberg JR, Semenza GL, Thomas GH (2002) Molecular cytogenetic characterization of a subtle interstitial del(3)(p25.3p26.2) in a patient with deletion 3p syndrome. Am J Med Genet 109:133–138

Chiang YJ, Kim SH, Tessarolo L, Campisi J, Hodes RJ (2004) Telomere-associated protein TIN2 is essential for early embryonic development through a telomerase-independent pathway. Mol Cell Biol 24:6631–6634

Chong L, van Steensel B, Broccoli D, Erdjument-Bromage H, Hanish, J, Tempst P, de Lange T (1995) A human telomeric protein. Science 270:1663–1667

Cong Y, Wright WE, Shay JW (2002) Human telomerase and its regulation. Microbiol Mol Biol Rev 66:407–425

Cornforth MN, Eberle RL (2001) Termini of human chromosomes display elevated rates of mitotic recombination. Mutagenesis 16:85–89

Cotter PD, Kaffe S, Li L, Gershin IF, Hirschhorn K (2001) Loss of subtelomeric sequence associated with a terminal inversion duplication of the short arm of chromosome 4. Am J Med Genet 102:76–80

de Lange T (2005) Shelterin: the protein complex that shapes and safeguards human telomeres Genes Dev 19:2100–2110

de Vries BB, Lees M, Knight SJ, Regan R, Corney D, Flint J, Barnicoat A, Winter RM (2001) Submicroscopic 8pter deletion, mild mental retardation, and behavioral problems caused by a familial t(8;20)(p23;p13) Am J Med Genet 99:314–319

Der-Sarkissian H, Vergnaud G, Borde YM, Thomas G, Londoño-Vallejo JA (2002) Segmental polymorphism in the proterminal regions of a subset of human chromosomes. Genome Res 12:1673–1678

Drets ME (2004) Cytological indications on the complex structure of the subtelomeric region. Cytogenet Genome Res 104:137–141

Drets ME, Mendizábal M (1998) The underlying structure of the subtelomeric segments detected by microphotometrical scanning and graphic image analysis. Mutat Res 404:13–16

Drets ME, Obe G, Monteverde FJ, Folle GA, Medina II, De Galvez MG, Duarte JE, Mechoso BH (1992) Computerized graphic and light microscope analyses of T-banded chromosome segments of Chinese hamster ovary cells and human lymphocytes. Biol Zentralbl 111:204–214

Drets ME, Drets GA, Queirolo PJ, Monteverde FJ (1995a) Computer graphics as a tool in cytogenetic research and education. Comput Appl Biosci 11:463–468
Drets ME, Mendizábal M, Boccardo EM, Bonomi R (1995b) Further analyses of subtelomeric and paracentric holes induced in human and Chinese hamster ovary cell chromosomes. Biol Zentralbl 114:329–338
Drets ME, Santiñaque FF, Obe G (2006) A proposal of a standardised nomenclature for terminal minute sister chromatid exchanges. Genet Mol Biol 29:446–447
Dutrillaux B (1973) Nouveau système de marquage chromosomique: Les bandes T. Chromosoma 41:395–402
Epel ES, Blackburn EH, Lin J, Dhabhar FS, Adler NE, Morrow JD, Cawthon RM (2004) Accelerated telomeres shortening in response to life stress. Proc Natl Acad Sci USA 101:17312–17315
Fang G, Cech TR (1995) Telomere proteins. In: Blackburn EH, Greider CW (eds) Telomeres. Cold Spring Harbor Laboratory Press, Cold Spring Harbor, pp 69–105
Garcia-Aranda C, de Juan C, Diaz-Lopez A, Sanchez-Pernaute A, Torres AJ, Diaz-Rubio E, Balibrea JL, Benito M, Iniesta P (2006) Correlations of telomeres length, telomerase activity, and telomeric-repeat binding factor 1 expression in colorectal carcinoma. Cancer 106:541–551
Giardino D, Finelli P, Gottardi G, De Canal G, Della Monica M, Lonardo F, Scarano G, Larizza L (2003) Narrowing the candidate region of Albright hereditary osteodystrophy syndrome by deletion mapping in a patient with unbalanced cryptic translocation t(2;6)(q37.3;q26). Am J Med Genet 122:261–265
Graakjaer J, Pascoe L, Der-Sarkisian H, Thomas G, Kolvraa S, Christensen K, Londoño-Vallejo JA (2004) The relative lengths of individual telomeres are defined in the zygote and strictly maintained during life. Aging Cell 3:97–102
Greider CW (1998) Telomerase activity, cell proliferation, and cancer. Proc Natl Acad Sci USA 95:90–92
Griffith J, Bianchi A, de Lange T (1998) TRF1 promotes parallel pairing of telomeric tracts in vitro. J Mol Biol 278:79–88
Griffith JD, Comeau L, Rosenfield S, Stansel RM, Bianchi A, Moss H, de Lange T (1999) Mammalian telomeres end in a large duplex loop. Cell 97:503–514
Hande MP (2004) DNA repair factors and telomere-chromosome integrity in mammalian cells. Cytogenet Genome Res 104:116–122
Harley CB (1995) Telomeres and aging. In: Blackburn EH, Greider CW (eds) Telomeres. Cold Spring Harbor Laboratory Press, Cold Spring Harbor, pp 247–263
Hiyama E, Hiyama K (2002) Clinical utility of telomerase in cancer. Oncogene 21:643–649
Henderson E (1995) Telomere DNA structure. In: Blackburn EH, Greider CW (eds) Telomeres. Cold Spring Harbor Laboratory Press, Cold Spring Harbor, pp 11–34
Holinski-Feder E, Reyniers E, Uhrig S, Golla A, Wauters J, Kroisel P, Bossuyt P, Rost I, Jedele K, Zierler H, Schwab S, Wildenauer D, Speicher MR, Willems PJ, Meitinger T, Kooy RF (2000) Familial mental retardation syndrome ATR-16 due to an inherited cryptic subtelomeric translocation, t(3;16)(q29;p13.3). Am J Hum Genet 66:16–25
Horn D, Neitzel H, Tonnies H, Kalscheuer V, Kunze J, Hinkel GK, Bartsch O (2003) Familial MCA/MR syndrome due to inherited submicroscopic translocation t(18;21)(q2.11q21.3) with breakpoint at the Down syndrome critical region. Am J Med Genet 117:236–244
Horsley SW, Knight SJ, Nixon J, Huson S, Fitchett M, Boone RA, Hilton-Jones D, Flint J, Kearney L (1998) Del(18p) shown to be a cryptic translocation using a multiprobe FISH assay for subtelomeric chromosome rearrangements. J Med Genet 35:722–726
Hug N, Lingner J (2006) Telomere length homeostasis. Chromosoma 115:413–425
Jack EM, Harrison CJ, Allen TD, Harris R (1986) A structural basis for R- and T-banding: a scanning electron microscopy study. Chromosoma 94:395–402
Jin G, Ikushima T (2004) Frequent occurrence of UVB-induced sister chromatid exchanges in telomere regions and its implication to telomere maintenance. Cytogenet Genome Res 104:310–314
Karlseder J, Kachatrian L, Takai H, Mercer K, Hingorani S, Jacks T, de Lange T (2003) Targeted deletion reveals an essential function for the telomere length regulator Trf1. Mol Cell Biol 23:6533–6541

Keller K, Williams C, Wharton P, Paulk M, Bent-Williams A, Gray B, Ward A, Stalker H, Wallace M, Carter R, Zori R (2003) Routine cytogenetic and FISH studies for 17p11/15q11 duplications and subtelomeric rearrangements studies in children with autism spectrum disorders. Am J Med Genet 117:105–111

Kleefstra T, van de Zande G, Merkx G, Mieloo H, Hoovers JM, Smeets D (2000) Identification of an unbalanced cryptic translocation between the chromosomes 8 and 13 in two sisters with mild mental retardation accompanied by mild dysmorphic features. Eur J Hum Genet 8:637–640

Kim S, Han S, You Y, Chen DJ, Campisi J (2003) The human telomere-associated protein TIN2 stimulates interactions between telomeric DNA tracts *in vitro*. EMBO Rep 4:685–691

Knight SJL, Flint J (2000) Perfect endings: a review of subtelomeric probes and their use in clinical diagnosis. J Med Genet 37:401–409

Laud PR, Multani AS, Bailey SM, Wu L, Ma J, Kingsley C, Lebel M, Pathak S, DePinho RA, Chang S (2005) Elevated telomere-telomere recombination in WRN-deficient, telomere dysfunctional cells promotes escape from senescence and engagement of the ALT pathway. Genes Dev 19:2560–2570

Lejeune J, Gauthier M, Turpin R (1959) Étude des chromosomes somatiques de neuf enfants mongoliens. C R Acad Sci Paris 248:1721–1722

Leube B, Majewski F, Drechsler M, Royer-Pokora B (2003) Unbalanced cryptic translocation der(14)t(9;14)(q34.3;q32.33) identified by subtelomeric FISH. Clin Dysmorphol 12: 261–265

Lingner J, Hughes TR, Shevchenko A, Mann M, Lundblad V, Cech TR (1997) Reverse transcriptase motifs in the catalytic subunit of telomerase. Science 276:561–567

Liu D, O'Connor MS, Qin J, Songyang Z (2004) Telosome, a mammalian telomere-associated complex formed by multiple telomeric proteins. J Biol Chem 279:51338–51342

Londoño-Vallejo JA, Der-Sarkissian H, Cazes L, Bacchetti S, Reddel RR (2004) Alternative lengthening of telomeres is characterized by high rates of telomeric exchange. Cancer Res 64:2324–2327

Lorda-Sanchez I, Lopez-Pajares I, Roche MC, Sanz R, Rodriguez De Alba M, Gonzalez-Gonzalez MC, Ibanez A, Ramos C, Ayuso C (2000) Cryptic 6q subtelomeric deletion associated with a paracentric inversion in a mildly retarded child. Am J Med Genet 95:336–338

Ludeña P, Sentis C, De Cabo F, Velazquez M, Fernandez-Piqueras J (1991) Visualization of R-bands in human metaphase chromosomes by the restriction endonuclease MseI. Cytogenet Cell Genet 57:82–86

Makarov VL, Hirose Y, Langmore JP (1997) Long G tails at both ends of human chromosomes suggest a C strand degradation mechanism for telomere shortening. Cell 88:657–666

Moyzis RK, Buckinham JM, Cram LS, Dani M, Deaven LL, Jones MD, Meyne J, Ratliff RL, Wu JR (1988) A highly conserved repetitive DNA sequence (TTAGGG)n present at the telomeres of human chromosomes. Proc Natl Acad Sci USA 85:6622–6626

Munoz P, Blanco R, Blasco MA (2006) Role of the TRF2 telomeric protein in cancer and ageing. Cell Cycle 5:718–721

Nikitina T, Woodcock CL (2004) Closed chromatin loops at the ends of chromosomes. J Cell Biol 166:161–165

Obe G, Pfeiffer P, Savage JRK, Johannes C, Goedecke W, Jeppesen P, Natarajan AT, Martínez-López W, Folle GA, Drets ME (2002) Chromosomal aberrations: formation, identification and distribution. Mutat Res 504:17–36

Oh BK, Kim YJ, Park C, Park YN (2005) Up-regulation of telomere-binding proteins, TRF1, TRF2, and TIN2 is related to telomere shortening during human multistep hepatocarcinogenesis. Am J Pathol 166:73–80

Pardue ML, DeBaryshe PG (1999) Telomeres and telomerase: more than the end of the line. Chromosoma 108:73–82

Quigley DI, Kaiser-Rogers K, Aylsworth AS, Rao KW (2004) Submicroscopic deletion 9(q34.3) and duplication 19(p13.3): identified by subtelomere FISH probes. Am J Med Genet 125:67–72

Rauch A, Beese M, Mayatepek E, Dorr HG, Wenzel D, Reis A, Trautmann U (2003) A novel 5q35.3 subtelomeric deletion syndrome. Am J Med Genet 121:1–8

Raynan JB, Tepperberg JH, Papenhausen P, Lamb AN, Hedrick J, Eash D, Ledbetter DH, Martin CL (2006) Subtelomere FISH analysis of 11,688 cases: an evaluation of the frequency and pattern of subtelomere rearrangements in individuals with developmental disabilities. J Med Genet 43:478–489

Reddy KS, Fugate JK (1999) A half cryptic derivative der(18)t(5;18)pat identified by M-FISH and subtelomere probes: clinical findings and review of subtelomeric rearrangements. Clin Genet 56:328–332

Riethman H (2005) Human subtelomere structure and variation. Chromosome Res 13:505–515

Riethman H, Ambrosini A, Castaneda C, Finklestein J, Hu XL, Mudunuri U, Paul S, Wei J (2004) Mapping and initial analysis of human subtelomeric sequence assemblies. Genome Res 14:18–28

Rodier F, Kim SH, Nijjar T, Yaswen P, Campisi J (2005) Cancer and aging: the importance of telomeres in genome maintenance. Int J Biochem Cell Biol 37:977–990

Saccone S, De Sario A, Della Valle G, Bernardi G (1992) The highest gene concentrations in the human genome are in T bands of metaphase chromosomes. Proc Natl Acad Sci USA 89:4913–4917

Saitoh Y, Laemmli UK (1994) Metaphase chromosome structure: bands arise from a differential folding path of the highly AT-rich scaffold. Cell 76:609–622

Schultz LN, Schmidt P, Tabor A, Bryndorf T, Christesen B, Ludsteen C (2001) Cryptic familial t(11;18)(q25;q23) incidentally detected by interphase FISH. Clin Genet 59:279–283

Smith S, de Lange T (1997) TRF1, a mammalian telomeric protein. Trends Genet 13:21–26

Souter VL, Glass IA, Chapman DB, Raff ML, Parisi MA, Opheim KE, Disteche CM (2003) Multiple fetal anomalies associated with subtle subtelomeric chromosomal rearrangements. Ultrasound Obstet Gynecol 21:609–615

Speleman F, Callens B, Logghe K, Van Roy N, Horsley SW, Jauch A, Verschraegen-Spac MR, Leroy JG (2000) Subtelomeric familial translocation t(2;7)(q37;q35) leading to partial trisomy 7q35→qter: molecular cytogenetic analysis and clinical phenotype in two generations. Am J Med Genet 93:349–354

Stansel R, de Lange T, Griffith JD (2001) T-loop assembly *in vitro* involves binding of TRF2 near the 3′ telomeric overhang. EMBO J 20: 5532–5540

Steinmüller J, Schleiermacher E, Scherthan H (1993) Direct detection of repetitive whole chromosome paint and telomere DNA probes by immunogold electron microscopy. Chromosome Res 1:45–51

Sumner AT (1990) Chromosome banding. Unwin Hyman, London

Sutherland GR (1977) Fragile sites on human chromosomes: demonstration of their dependence on the type of tissue culture medium. Science 197:265–266

Sutton VR, Coveler KJ, Lalani SR, Kashork CD, Shaffer LG (2002) Subtelomeric FISH uncovers trisomy 14p32: lessons for imprinted regions, cryptic rearrangements and variant acrocentric short arms. Am J Med Genet 112:23–27

Valdes AM, Andrew T, Gardner JP, Kimura M, Oelsner E, Cherkas LF, Aviv A, Spector TD (2005) Obesity, cigarette smoking, and telomeres length in women. Lancet 366:662–664

van Deutekom JC, Bakker E, Lemmers RJ, van der Wielen MJR, Bik E, Hofker MH, Padberg GW, Frants RR (1996) Evidence for subtelomeric exchange of 3.3 kb tandemly repeated units between chromosomes 4q35 and 10q26: implications for genetic counselling and etiology of FSHD1. Hum Mol Genet 5:1997–2003

van Steensel B, de Lange T (1997) Control of telomere length by the human telomeric protein TRF1. Nature 385:740–743

van Steensel B, Smogorzewska A, de Lange T (1998) TRF2 protects human telomeres from end-to-end fusions. Cell 92:401–413

Velagaleti GV, Jalal SM, Michaelis RC, Rowe TF, Nichols JR, Lockhart LH (2003) Molecular cytogenetic characterization of a *de novo* unbalanced translocation leading to trisomy

17q25→qter and monosomy 18p11.3→pter in a girl with dysmorphic features. Clin Dysmorphol 12:29–33

Walter S, Sandig K, Hinkel GK, Mitulla B, Ounap K, Sims G, Sitska M, Utermann B, Viertel P, Kalscheuer V, Bartsch O (2004) Subtelomere FISH in 50 children with mental retardation and minor anomalies, identified by a checklist, detects 10 rearrangements including a *de novo* balanced translocation of chromosomes 17p13.3 and 20q13.33. Am J Med Genet 128:364–373

Wang RC, Smogorzewska A, de Lange T (2004) Homologous recombination generates T-loop-sized deletions at human telomeres. Cell 119:355–368

Warburton P, Mohammed S, Ogilvie CM (2000) Detection of submicroscopic subtelomeric chromosome translocations: a new case study. Am J Med Genet 91:51–55

Yoshimura SH, Maruyama H, Ishikawa F, Ohki R, Takeyasu K (2004) Molecular mechanisms of DNA end-loop formation by TRF2. Genes Cells 9:205–218

Zakian VA (1995) Telomeres: beginning to understand the end. Science 270:1601–1607

15 Cytokinesis-Block Micronucleus Assay: a Comprehensive "Cytome" Approach for Measuring Chromosomal Instability, Mitotic Dysfunction and Cell Death Simultaneously in One Assay

Michael Fenech

Abstract

The cytokinesis-block micronucleus (CBMN) assay was originally developed as an ideal system for measuring micronuclei; however, it can also be used to measure nucleoplasmic bridges (NPBs), nuclear buds, cell death (necrosis or apoptosis) and nuclear division rate. Current evidence suggests that (1) NPBs originate from dicentric chromosomes in which the centromeres have been pulled to the opposite poles of the cell at anaphase and are therefore indicative of DNA misrepair, chromosome rearrangement or telomere end-fusions, (2) NPBs may break to form micronuclei, (3) the nuclear budding process is the mechanism by which cells remove amplified and/or excess DNA and is therefore a marker of gene amplification and/or altered gene dose, (4) cell cycle checkpoint defects result in micronucleus formation, and (5) hypomethylation of DNA, induced nutritionally or by inhibition of DNA methyltransferase can lead to micronucleus formation either via chromosome loss or via chromosome breakage. The strong correlation between micronucleus formation, nuclear budding and NPBs ($r = 0.75–0.77$, $P<0.001$) induced by either folic acid deficiency or exposure to ionising radiation is supportive of the hypothesis that folic acid deficiency and/or ionising radiation cause genomic instability and gene amplification by the initiation of breakage–fusion–bridge cycles. In its comprehensive mode the CBMN assay measures all cells, including necrotic and apoptotic cells, as well as number of nuclei per cell to provide a measure of cytotoxicity and mitotic activity. The CBMN assay has in fact evolved into a "cytome" method for measuring comprehensively chromosomal instability phenotype and cellular viability caused by genetic defects and/or nutritional deficiencies and/or exogenous genotoxins, opening up an exciting future for the use of this methodology in the emerging fields of nutrigenomics and toxicogenomics and their combinations.

15.1 Introduction

Micronuclei originate from chromosome fragments or whole chromosomes that lag behind at anaphase during nuclear division (Fenech 2000a). The micronucleus index in human cells has become one of the standard cytogenetic tests for genetic toxicology testing. The cytokinesis-block micronucleus

Günter Obe and Vijayalaxmi (Eds.)
Chromosomal Alterations: Methods, Results and Importance in Human Health

(CBMN) assay is the preferred method for measuring micronuclei in cultured human and/or mammalian cells because scoring is specifically restricted to once-divided cells. These cells are recognised by their binucleated appearance after inhibition of cytokinesis by cytochalasin B (Fenech 2000a). Restricting scoring of micronuclei in binucleated cells prevents confounding effects caused by suboptimal cell division kinetics, which is a major variable in this assay (Fenech 2000a). Over the past 17 years the CBMN assay has evolved into a comprehensive method for measuring chromosome breakage, chromosome loss, non-disjunction, necrosis, apoptosis and cytostasis (Fenech 2000a; Fig. 15.1). In combination with the use of centromeric probes or kinetochore antibodies it is also possible to distinguish between chromosome breakage and loss as the mechanism of micronucleus formation and to measure malsegregation of chromosomes amongst the two nuclei in binucleated cells. However, new developments now indicate that this method can

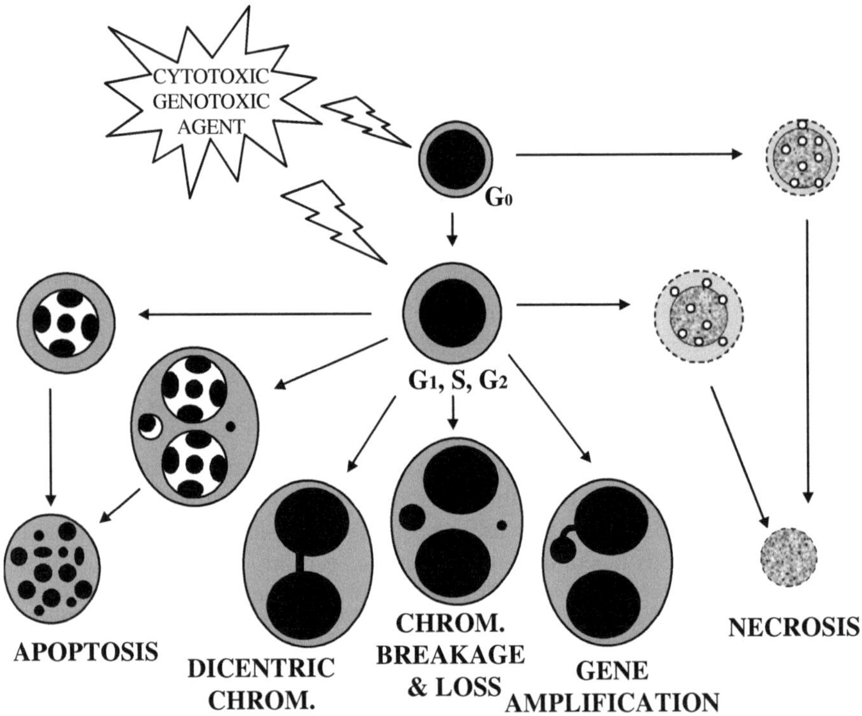

Fig. 15.1 The various possible fates of cultured cytokinesis-blocked cells following exposure to cytotoxic/genotoxic agents. Using these biomarkers within the cytokinesis-block micronucleus (CBMN) assay it is possible to measure the frequency of chromosome breakage (micronucleus), chromosome loss (micronucleus), chromosome rearrangement, e.g. dicentric chromosomes (nucleoplasmic bridges), gene amplification (nuclear buds), necrosis and apoptosis. In addition, cytostatic effects are readily estimated from the ratio of mononucleated, binucleated and multinucleated cells. Chromosome loss can be distinguished from chromosome breakage using pancentromeric probes or anti-kinetochore antibodies. In addition, non-disjunction (malsegregation of chromosomes) can also be measured in binucleated cells using chromosome-specific centromeric probes

also be used to measure nucleoplasmic bridges (NPBs), a biomarker of dicentric chromosomes resulting from telomere end-fusions or DNA misrepair, and to measure nuclear buds, a biomarker to detect gene amplification and altered gene dose events (Fenech 2000a; Umegaki and Fenech 2000; Haaf et al. 1999). The significance of these developments and the concept of the CBMN assay as a "cytome" assay of chromosomal instability are elaborated in the following sections. The "cytome" concept implies that every cell in the system studied is scored cytologically for its viability status (necrosis, apoptosis), its mitotic status (mononucleated, metaphase, anaphase, binucleated, multinucleated) and its chromosomal instability or damage status (presence of micronuclei, NPBs, nuclear buds and the number of centromere probe signals amongst nuclei of binucleated cells if such molecular tools are used in combination with the assay) (Fig. 15.2).

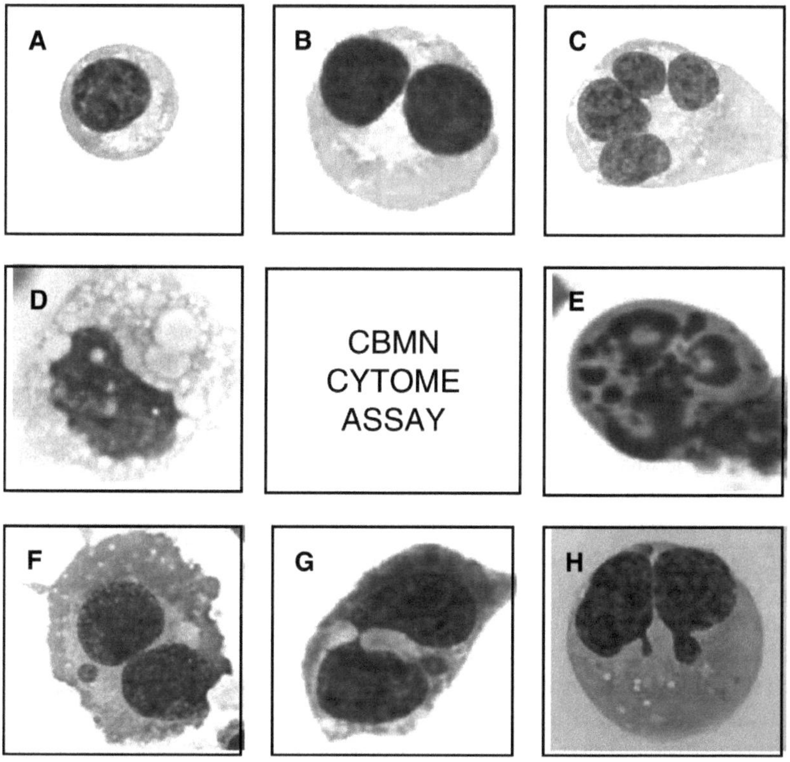

Fig. 15.2 Photomicrographs of the cells scored in the CBMN "cytome" assay. **a** Mononucleated cell, **b** binucleated cell, **c** multinucleated cell, **d** necrotic cell, **e** apoptotic cell, **f** binucleated cell containing one or more micronuclei, **g** binucleated containing a nucleoplasmic bridge and a micronucleus, **h** binucleated cell containing nuclear buds. The ratios of mononucleated, binucleated, multinucleated, necrotic and apoptotic cells are used to determine the mitotic division rate or nuclear division index (a measure of cytostasis) and cell death (cytotoxicity). The ratios of binucleated cells with micronuclei, nucleoplasmic bridges or nuclear buds relative to the total number of binucleated cells provides a measure of genome damage and/or chromosomal instability. For more details of the scoring criteria refer to Fenech et al. (2003)

15.2 Nucleoplasmic Bridges

We have proposed that NPBs between nuclei in binucleated cells should also be scored in the CBMN assay because they provide a measure of chromosome rearrangement which is otherwise not measured in this assay if only micronuclei are scored (Umegaki and Fenech 2000). The NPBs are assumed to occur when the centromeres of dicentric chromosomes are pulled to opposite poles of the cell at anaphase. It is rarely possible to observe dicentric anaphase bridges before the nuclear membrane is formed, because cells proceed through anaphase and telophase rapidly, completing cytokinesis and ultimately breakage of the NPB when the daughter cells separate. However, in the CBMN assay binucleated cells with NPBs are allowed to accumulate because cytokinesis is inhibited and the nuclear membrane is eventually formed around the chromosomes, allowing a NPB to be observed.

We have recently validated the use of NPBs as a biomarker of DNA damage in human WIL2-NS cells treated with hydrogen peroxide, superoxide or after coincubation with activated human neutrophils (Umegaki and Fenech 2000). NPB frequency in binucleated cells increased up to 20-fold relative to control cells in a dose-related manner with these agents that are known to induce DNA strand breaks. NPBs were positively correlated with micronucleus frequency in the same binucleated cells ($r > 0.82$, $P < 0.0001$). More than 60% of the binucleated cells with one or more NPBs were also observed to contain one or more micronuclei, which fits with the fact that acentric chromosome fragments (expressed as micronuclei) and dicentric chromosomes (expressed as NPBs) are known to be expressed simultaneously within a cell following exposure to agents such as ionising radiation that induce double-stranded DNA breaks (IAEA 1986).

The various mechanisms that could lead to NPB formation following DNA misrepair of strand breaks in DNA are illustrated and explained in detail in Fig. 15.3. An alternative mechanism for NPB formation is due to telomere end-fusions caused by telomere shortening, but in this case the NPB is not necessarily accompanied by the presence of an acentric chromosome fragment. The formation of anaphase and NPBs has been observed in models of rodent and human intestinal cancer in vivo and was shown to correlate positively with telomere length, indicating that NPB formation can also be used as a surrogate measure of critically short telomeres (Rudolph et al. 2001). More recent real-time in vitro studies have focussed on the fate of anaphase bridges and showed that during cytokinesis the bridges may be broken in more than one region, leading to the formation of a micronucleus (Hoffelder et al. 2004; Shimizu et al. 2005). Therefore, some micronuclei may also originate from anaphase bridges, although whether this actually happens in cytokinesis-blocked cells remains unclear.

The importance of scoring NPBs should not be underestimated because it provides direct evidence of genome damage resulting from misrepaired DNA

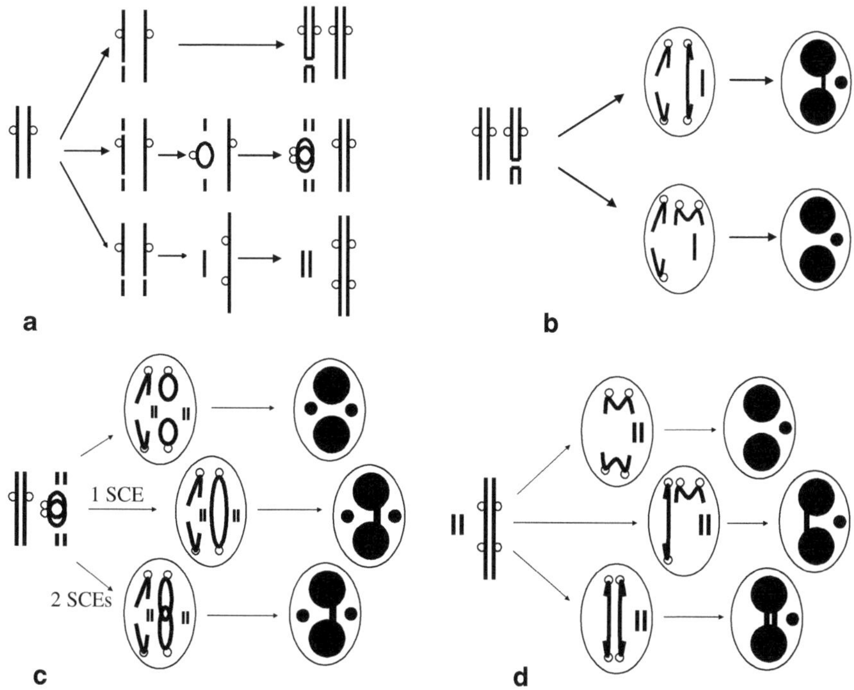

Fig. 15.3 **a** Initiating events in the formation of a nucleoplasmic bridge. *Top*: Double-strand DNA break is induced in one chromosome but remains unrepaired until after replication. Misjoining of the broken ends after replication leads to the formation of a dicentric chromatid and an acentric fragment. *Middle*: Two double-strand breaks on either side of the centromere of one chromosome are induced. The broken ends are misrepaired, producing a ring chromatid and two acentric fragments which are subsequently replicated. *Bottom*: Double-strand breaks are induced in two homologous or non-homologous chromosomes. Misjoining of the broken ends leads to the formation of a dicentric chromatid and acentric fragment which are subsequently replicated. **b** *Top*: Centromeres of a dicentric chromatid are pulled to opposite poles of the cell at anaphase to form a nucleoplasmic bridge and the acentric chromatid fragment lags behind to form a micronucleus. *Bottom*: Centromeres of a dicentric chromatid are both pulled to the same pole of the cell and no nucleoplasmic bridge is formed; however, the lagging acentric chromatid fragment results in the formation of a micronucleus. **c** *Top*: The ring chromatids are segregated normally but the acentric fragments lag at anaphase to form two micronuclei. *Middle*: The two ring chromatids, after completing one sister-chromatid exchange (*SCE*), are transformed into one large dicentric ring chromatid, which leads to the formation of a nucleoplasmic bridge when the centromeres are pulled to the opposite poles of the cell at anaphase. The accompanying acentric fragments lag at anaphase to form two micronuclei. *Bottom*: The two ring chromatids, after completing two SCEs, become concatenated, which leads to the formation of a nucleoplasmic bridge when the centromeres are pulled to the opposite poles of the cell at anaphase. The accompanying acentric fragments lag at anaphase to form two micronuclei. **d** *Top*: Centromeres of dicentric chromatids travel to the same pole of the cell and no nucleoplasmic bridges are formed. *Middle*: Centromeres of one of the dicentric chromatids are pulled to opposite poles of the cell, leading to the formation of one nucleoplasmic bridge. *Bottom*: Centromeres of both dicentric chromatids are pulled to opposite poles of the cell, leading to the formation of two nucleoplasmic bridges. In each of the above cases one micronucleus is produced from the lagging acentric chromosome fragment accompanying the dicentric chromosome

breaks which is otherwise not possible to deduce by scoring micronuclei only which could originate from acentric fragments of chromosome loss. Furthermore, following radiation-challenge or oxidant-challenge tests, the measurement of NPBs could be used to identify defects in recombinational DNA-repair pathways (Fig. 15.4) or to characterise and differentiate between the genotoxic mechanism of different agents. For example, we demonstrated that the NPB-to-micronucleus ratio in WIL2-NS cells induced by activated neutrophils is much higher than the ratios for hydrogen peroxide and superoxide, which were not different, indicating that the oxidants produced by activated neutrophils have a greater propensity to induce lesions leading to misrepair of DNA strand breaks and or possibly telomere end-shortening (Fig. 15.5).

15.3 Nuclear Buds

Over the past decade another unique mechanism of micronucleus, known as nuclear budding, has emerged. This process has been observed in cultures grown under strong selective conditions (Toledo et al. 1992; Shimura et al.

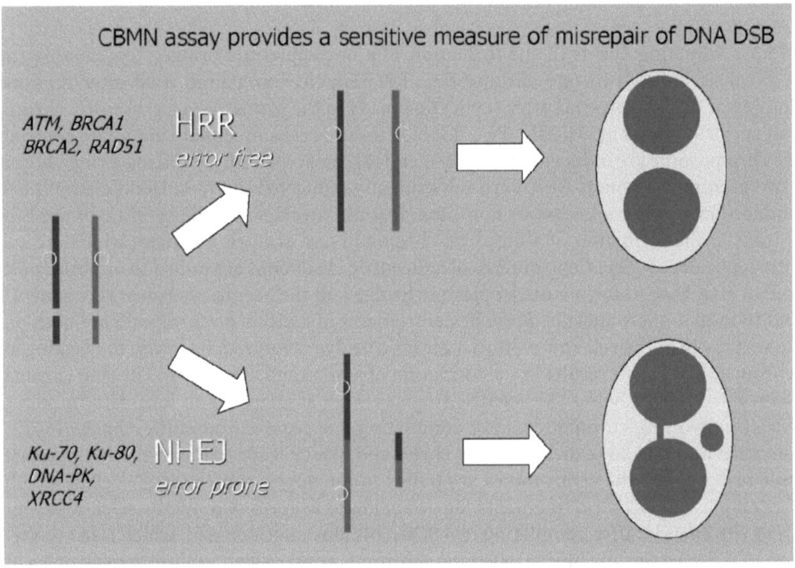

Fig. 15.4 Defects in genes involved in homologous recombinational repair (*HRR*) genes (e.g. *ATM*, *BRCA1*, *BRCA2*, *RAD51*) could be detected using the CBMN assay because under these conditions it may be anticipated that the cell would resort to the error-prone non-homologous end-joining (*NHEJ*) repair mechanism that is likely to lead to misrepair of double-strand breaks in DNA, formation of dicentric chromosomes and ultimately nucleoplasmic bridges which are detectable in the CBMN assay

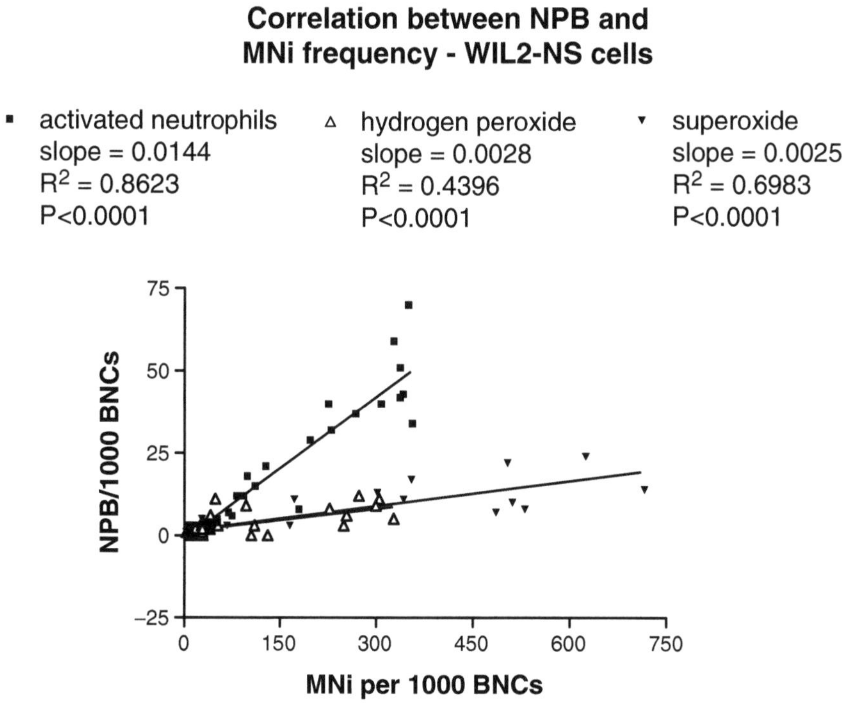

Fig. 15.5 Correlation between nucleoplasmic bridges and micronuclei in WIL2-NS cultures exposed to increasing doses of hydrogen peroxide or superoxide or activated neutrophils. (Data derived from experiments described in Umegaki and Fenech 2000). *MNi* micronuclei, *NPB* nucleoplasmic bridge, *BNC* binucleated cell

1999; Ma et al. 1993) which induce gene amplification. Shimizu et al. (1998, 2000) showed, using in vitro experiments with mammalian cells, that amplified DNA is selectively localised to specific sites at the periphery of the nucleus and eliminated via nuclear budding to form micronuclei during S phase of mitosis. Furthermore, DNA synthesis inhibitors such as hydroxyurea were shown to increase the rate of elimination of amplified DNA via this process. Amplified DNA may be eliminated through recombination between homologous regions within amplified sequences forming minicircles of acentric and atelomeric DNA (double minutes) which localise to distinct regions within the nucleus or through the excision of amplified sequences after segregation to distinct regions of the nucleus. The above suggests the hypothesis that the nucleus may have a capacity to sense excess DNA that does not fit well within the nuclear matrix, indicating a higher-order DNA repair or nuclear housekeeping process. Shimizu et al. (1998, 2000) have suggested that the nucleus eliminates excess amplified DNA by an active process that concentrates the amplified DNA to a peripheral point in the nucleus, following which this surplus DNA is budded out to form a micronucleus and eventually excluded from the cell altogether by extrusion of the micronucleus from the cytoplasm, leading to the formation of a "mini-cell". The process of nuclear budding occurs during S phase and the nuclear buds are characterised by having the same morphology as a micronucleus, with the exception that they are linked to the nucleus by a narrow or wide stalk of nucleoplasmic material depending on the stage of the budding process (Figs. 15.1, 15.2). The duration of the nuclear budding process and the extrusion of the resulting micronucleus from the cell remain largely unknown. It is also reported that micronuclei may also be formed by a budding process following exposure to γ-radiation (Haaf et al. 1999). In this process, Rad51-recombination protein complexes are detectable throughout the entire nucleus 3 h after irradiation and then concentrate into distinct foci before being extruded from the nucleus into micronuclei. Using end-labelling, Haaf et al. (1999) showed these micronuclei exhibit DNA fragmentation analogous to that observed in apoptotic nuclei (Haaf et al. 1999). Serrano-Garcia and Montero-Montaya (2001) have shown, in vitro and in vivo in human lymphocytes, that nuclear buds and micronuclei are expressed simultaneously following genotoxic exposure in human cells with positive correlation factors between these two biomarkers ranging from 0.73 to 0.84.

15.4 Micronuclei, NPBs and Nuclear Buds Induced by Folic Acid Deficiency

In a series of studies on folic acid deficiency in long-term primary human lymphocyte cultures we carefully quantified the interrelationship between micronuclei, NPBs and nuclear buds in an attempt to validate the use of these biomarkers and to determine more comprehensively the impact of folic acid deficiency on various aspects of genomic stability (Crott et al. 2001a, b; Kimura et al. 2004; Beetstra et al. 2006). Briefly, lymphocytes were cultured in duplicate in RPMI-1640 medium containing 5% dialysed foetal calf serum, 10 U/ml interleukin-2 and 12, 24, 60 or 120 nM folic acid for 9 days. The medium was refreshed at days 3 and 6, cytokinesis was inhibited on day 8 and cells were transferred to microscope slides 24 h later (for the detailed protocol see Crott et al. 2001a, b). We analysed separately the pooled data from this study and verified that folic acid concentration correlated significantly ($P<0.0001$) and negatively ($r = -0.63$ to -0.74) with all these markers of chromosome damage, which were minimised at 60–120 nM folic acid, the latter being greater than the concentration of folate normally observed in plasma (10–30n M). However, even more interestingly, we observed that the frequency of NPBs and nuclear buds correlated significantly and negatively with folic acid dose, suggesting that chromosome rearrangement and gene amplification are induced by folic acid deficiency. The strong cross-correlation between micronucleus, NPB and nuclear bud frequency ($r = 0.65–0.77$, $P<0.001$) suggests a common mechanism initiated by folic acid deficiency induced DNA breaks, although coincidence of effects with other DNA damage events (e.g. folic acid deficiency induced CpG hypomethylation) cannot be excluded.

15.5 Breakage–Fusion–Bridge Cycles Explain Micronuclei, NPBs and Nuclear Buds Induced by Folic Acid Deficiency

Because folic acid deficiency is known to cause gene amplification and chromosome damage such as double-stranded breaks (Blount et al. 1997; Fenech 2001; Melnyk et al. 1999), it is likely that these events contribute to the formation of micronuclei, NPBs and nuclear buds in this system. Gene amplification is thought to be a key event in cellular resistance to drugs like methotrexate (Biedler and Spengler 1976) and in tumour progression (Brison 1993). Several plausible models of gene amplification have been proposed (reviewed in Windle and Wahl 1992; Stark 1993); however, the presence of NPBs in the cytokinesis-blocked cells provides support for the "breakage–fusion–bridge (BFB) cycle" model described in the seminal work of McClintock (1942) in maize. According to this theory, sister chromatids which have both undergone double-stranded breakage fuse at a distal position (possibly telomeric) forming a dicentric chromosome which has two copies of homologous genes positioned between the two centromeres. During anaphase these dicentric chromosomes are drawn towards both poles and form (nucleoplasmic) bridges. During cytokinesis these dicentric chromosomes, which span both daughter nuclei, are thought to break unevenly and may form a chromosome with two copies of one or more genes and a chromosome (fragment) with no copies of these genes. The chromatids with multiple copy number of these genes may fuse again during interphase forming a dicentric chromosome (doubling again the gene copy number within the chromosome), which is then replicated during the next nuclear division leading to the next bridge–breakage–fusion cycle and further gene amplification. The BFB cycle mechanism is illustrated in Fig. 15.6.

The BFB cycle model has been validated as a mechanism for gene amplification in various systems, including (1) coformycin-induced amplification of the adenylate deaminase gene in Chinese hamster cells (Toledo et al. 1992), (2) *N*-(phosphonyacetyl)-L-aspartate induced amplification of the CAD gene in human fibrosarcoma cells which is enhanced by the expression of *Vpr*, a HIV accessory gene (Shimura et al. 1999), (3) methotrexate-induced

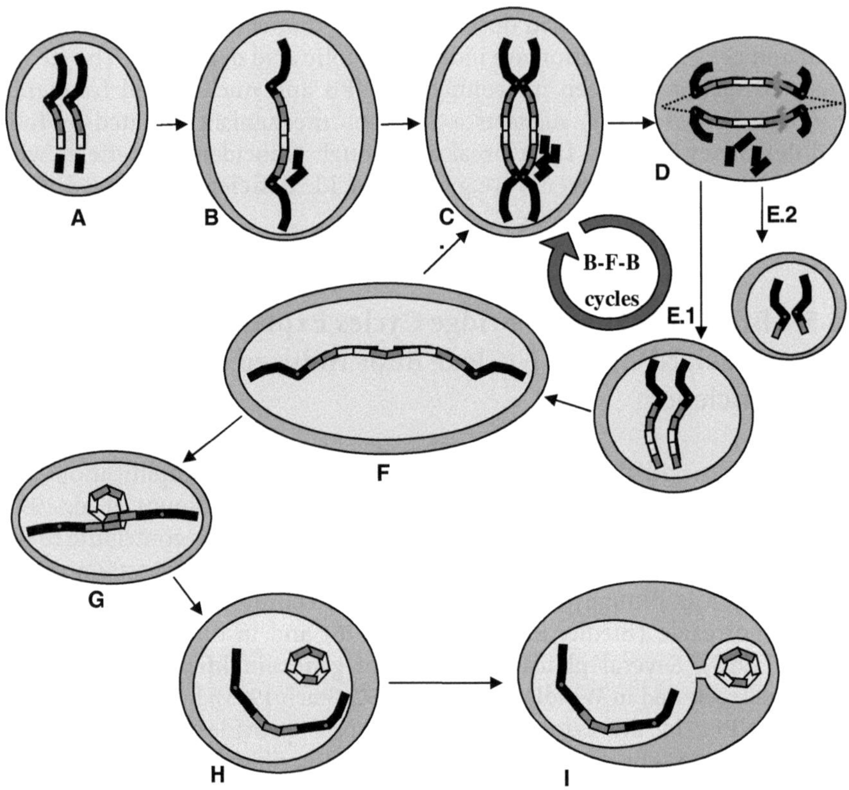

Fig. 15.6 Gene amplification via breakage-fusion-bridge cycles (*B-F-B*). Broken sister chromosomes **(A)** fuse at the "sticky" ends to form a dicentric chromosome **(B)**. The dicentric chromosome is replicated in S phase **(C)** and drawn to the cell's poles during anaphase (bridge) **(D)**. In this example, the dicentric chromosomes break in a non-central region and one daughter cell will receive chromosomes that contain two copies of the *yellow* and *red* genes **(E.1)**, while the other cell's chromosomes have a deletion in these genes **(E.2)**. The chromosomes with multiple copy number may fuse again **(F)** to propagate the amplification cycle (back to step C). Alternatively, recombination may occur between homologous sequences **(G)** and result in the "looping-out" of circular acentric DNA fragments or double minutes **(H)** which are subsequently extruded from the nucleus via budding **(I)**

amplification of the dihydrofolate reductase gene in Chinese hamster cells (Ma et al. 1993) and (4) actinomycin D induced amplification of the multidrug resistance 1 gene in Chinese hamster cells through the induction of fragile sites which determined the initiation and size of amplicons (Coquelle et al. 1997). Folic acid deficiency induced fragile site expression and DNA hypomethylation may have also contributed to the promotion of gene amplification and resulted in elimination of this DNA by nuclear budding in our system. For example, the induction of hypomethylation by 5-azacytidine has been reported to enhance *N*-(phosphonylacetyl)-L-aspartate induced amplification of the CAD gene in Syrian hamster kidney cells (Perry et al. 1992).

15.6 Micronucleus Formation Caused by Hypomethylation of Heterochromatin and Silencing of Cell Cycle Checkpoint Genes

The DNA methylation inibitor 5-azacytidine also induces distinct undercondensation of the heterochromatin regions of chromosomes 1, 9, 15, 16 and Y, and the specific loss of these chromosomes as micronuclei in human lymphocytes in vitro (Guttenbach and Schmid 1994). Similarly it has been shown that 5-azacytidine causes decondensation in the homogenously staining region in the highly methylated heteromorphic chromosome 15 and loss of this genetic material as micronuclei in human diploid lung fibroblast cell line TIG-7 (Satoh et al. 2004). The immunodeficiency–centromeric instability–facial anomalies (ICF) syndrome, which is caused by mutation in the DNA methyltransferase gene, is characterised by despiralisation of heterochromatin of chromosomes 1, 9 and 16 and loss of this chromatin into micronuclei and nuclear blebs (Xu et al. 1999). These events are likely to be relevant to the ageing process because Suzuki et al. (2002) demonstrated that in vitro ageing of normal human fibroblasts results in concomitant demethylation of satellite 2 and satellite 3 DNA which is abundant in the juxtacentromeric DNA of chromosomes 1, 9 and 16 and the increasing frequency of micronuclei that specifically contained these sequences. We have also recently demonstrated that folate deficiency caused a significant 26–35% increment in frequency of aneuploidy of chromosome 17 ($P = 0.0017$) and aneupoidy of chromosome 21 ($P = 0.0008$) relative to 120 nM 5-methyltetrahydrofolate or folic acid. The pattern of aneuploidy in binucleated cells was significantly correlated with that observed in mononucleated cells ($R = 0.51–0.75$, $P<0.0004$) and was consistent with a model based on chromosome loss or partial aneusomy rescue as the cause rather than non-disjunction, although the latter mechanism could not be excluded (Wang et al. 2004). Figure 15.7 illustrates possible mechanisms for the relationship between genome methylation and micronucleus formation.

It is evident that micronucleus expression could be increased as a result of hypomethylation of satellite DNA; however, it is also possible that increased genome damage may be caused by hypermethylation of CpG islands within or adjacent to the promoter regions of housekeeping genes involved in cell cycle checkpoints and DNA repair. For example CpG island hypermethylation of the mitotic spindle checkpoint genes such as *APC*, *BUB1* and *HCDC4* could reduce their expression and therefore increase the possibility of chromosome malsegregation leading to micronucleus formation (Rajagopalan et al. 2004; Rao et al. 2005).

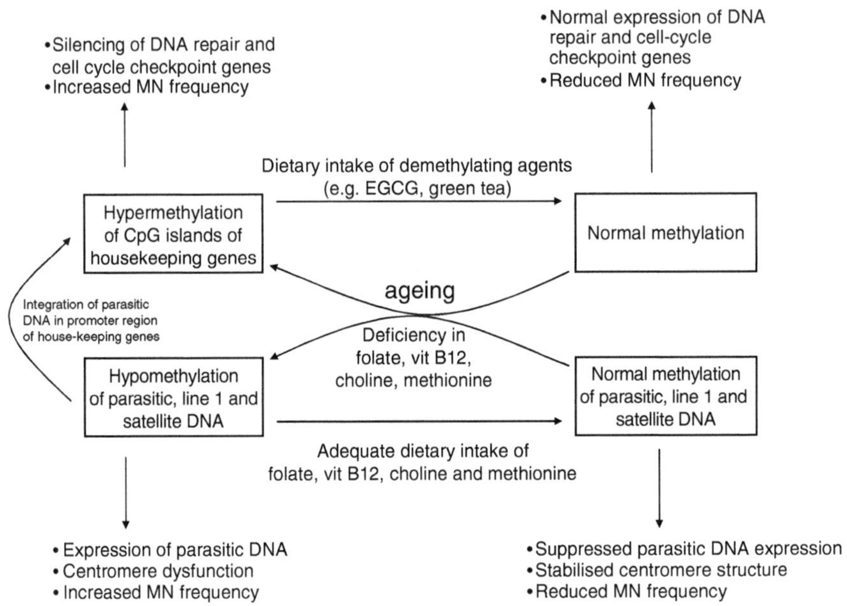

Fig. 15.7 The possible effects of dietary factors on methylation of CpG islands, parasitic DNA, gene expression, centromere function and micronucleus (*MN*) formation. *EGCG* epigallocatechin gallate. For further explanation refer to Fenech (2001, 2003) and Kimura et al. (2004)

15.7 Conclusion

In conclusion, the CBMN method has evolved into an efficient "cytome" assay of chromosomal instability, cell death and cytostasis enabling direct and/or indirect measurement of various aspects of cellular and nuclear dysfunction such as:

- Chromosomal instability phenotype
- BFB cycles
- DNA misrepair, chromosome breaks and asymmetrical rearrangement
- Telomere end-fusions
- Cell cycle checkpoint malfunction
- Malsegregation of chromosomes (when used in combination with pancentromeric or chromosome-specific probes)
- Gene amplification and nuclear elimination of excess DNA
- DNA hypomethylation (if combined with centromeric probes for chromosomes 1, 9 and 16)
- Altered mitotic activity and/or cytostasis
- Cell death by necrosis or apoptosis

The Micronucleus Assay Working Group of the International Workshop on Genotoxicity Test Procedures concluded that demonstration of cell proliferation in both control and treated cells is compulsory for the acceptance of the test because micronucleus expression is compromised when the proportion of dividing cells is reduced by cytostatic or poor tissue culture conditions (Kirsch-Volders et al. 2000, 2003; Fenech 2000b). It is evident that the best way to eliminate any possible confounding effects of altered cell division kinetics is to use the cytokinesis-block method regardless of the cell type used by scoring micronuclei in binucleated cells, which in addition allows measurement of NPBs as well as measurement of nuclear buds as well as cytostatic and cytotoxic effects. The choice not to use cytochalasin-B is only justifiable in those rare situations when cytokinesis defects are being specifically investigated.

The CBMN assay has already been proven to be an effective tool for the study of cellular and nuclear dysfunction caused by in vitro or in vivo ageing, micronutrient deficiency or excess, genotoxin exposure and genetic defects in genome maintenance. More recent studies indicate that this method is likely to also prove fruitful in the emerging fields of nutrigenomics and toxicogenomics and their combinations as it becomes increasingly clear that nutrient status also impacts on sensitivity to exogenous genotoxins (Beetstra et al. 2005). The ultimate test of the CBMN cytome assay is its ability to predict disease outcomes prospectively. These anticipated results, which were presented at the EEMS 2005 Kos Meeting and the HUMN Workshop at the 9th ICEM in San Francisco in September 2005, are indicative that the micronucleus index in lymphocytes is indeed predictive of cancer risk. Whether the use of the additional biomarkers in the CBMN assay (e.g. NPBs, nuclear buds) improves the predictive capacity of the assay remains to be determined, but a study by El-Zein et al. (2006) shows that the micronucleus, NPB and nuclear bud indices are sensitive indicators of lung cancer risk in smokers. As continuous variables, these measures were associated with 2-, 29- and 6-fold increases in the risk for spontaneous levels of the micronucleus, NPB and nuclear bud biomarkers and 2.3-, 45.5- and 10-fold increases in cancer risk for NNK-induced micronuclei, NPBs and nuclear buds, respectively [NNK is nicotine-derived nitrosamine 4-(methylnitrosamino)-1-(3-pyridyl)-1-butanone, a potent inducer of lung adenocarcinoma] (El-Zein et al. 2006).

References

Beetstra S, Thomas P, Salisbury C, Turner J, Fenech M (2005) Folic acid deficiency increases chromosomal instability, chromosome 21 aneuploidy and sensitivity to radiation-induced micronuclei. Mutat Res 578:317–326

Beetstra S, Salisbury C, Turner J, Altree M, McKinnon R, Suthers G, Fenech M (2006) Lymphocytes of BRCA1 and BRCA2 germ-line mutation carriers, with or without breast cancer, are not abnormally sensitive to the chromosome damaging effect of moderate folate deficiency. Carcinogenesis 27:517–524

Biedler JL, Spengler BA (1976) Metaphase chromosome anomaly: association with drug resistance and cell-specific products. Science 191:185-187

Blount BC, Mack MM, Wehr CM, MacGregor JT, Hiatt RA, Wang G, Wickramasinghe SN, Everson RB, Ames BN (1997) Folic acid deficiency causes uracil misincorporation into human DNA and chromosome breakage: implications for cancer and neuronal damage. Proc Natl Acad Sci USA 94:3290–3295

Brison O (1993) Gene amplification and tumor progression. Biochim Biophys Acta 1155:25–41

Coquelle A, Pipiras E, Toledo F, Buttin G, Debattisse M (1997) Expression of fragile sites triggers intrachromosomal mammalian gene amplification and sets boundaries to early amplicons. Cell 89:215–225

Crott JW, Mashiyama ST, Ames BN, Fenech M (2001a) Folic acid deficiency increases chromosome breakage and rearrangement, gene amplification and DNA-uracil content in human lymphocytes in vitro: effect of MTHFR C677T polymorphism. Cancer Epidemiol Biomarkers Prev 10:1089–1096

Crott JW, Mashiyama ST, Ames BN, Fenech M (2001b) Methylenetetrahydrofolate reductase C677T polymorphism does not alter folic acid deficiency-induced uracil incorporation into primary human lymphocyte DNA in vitro. Carcinogenesis 22:1019–1025

El-Zein RA, Schabath MB, Etzel CJ, Lopez MS, Franklin JD, Spitz MR (2006) Cytokinesis-blocked micronucleus assay as a novel biomarker for lung cancer risk. Cancer Res 66:6449–6456

Fenech M (2000a) The in vitro micronucleus technique. Mutat Res 455:81–95

Fenech M (2000b) A mathematical model of the in vitro micronucleus assay predicts false negative results if micronuclei are not specifically scored in binucleated cells or in cells that have completed one nuclear division. Mutagenesis 15:329–336

Fenech M (2001) The role of folic acid and vitamin B12 in genomic stability of human cells. Mutat Res 475:57–68

Fenech M (2003) Nutritional treatment of genome instability: a paradigm shift in disease prevention and in the setting of recommended dietary allowances. Nutr Res Rev 16:109–122

Fenech M, Chang WP, Kirsch-Volders M, Holland N, Bonassi S, Zeiger E (2003) HUMN project: detailed description of the scoring criteria for the cytokinesis-block micronucleus assay using isolated human lymphocyte cultures. Mutat Res 534:65–75

Guttenbach M, Schmid M (1994) Exclusion of specific human chromosomes into micronuclei by 5-azacytidine treatment of lymphocyte ceultures. Exp Cell Res 211:127–132

Haaf T, Raderschall E, Reddy G, Ward DC, Radding CM, Golub EI (1999) Sequestration of mammalian Rad51-recombination protein into micronuclei. J Cell Biol 144:11–20

Hoffelder DR, Luo L, Burke NA, Watkins SC, Gollin SM, Saunders WS (2004) Resolution of anaphase bridges in cancer cells. Chromosoma 112:389–397

IAEA (1986) Biological dosimetry: chromosomal aberrations analysis for dose assessment. Technical reports series no 260. International Atomic Energy Agency, Vienna, pp 13–27

Kirsch-Volders M, Sofuni T, Aardema M, Albertini S, Eastmond D, Fenech M, Ishidate M Jr, Kirchner S, Lorge E, Morita T, Norppa H, Surralles, Vanhauwaert A, Wakata A (2003) Report from the in vitro micronucleus assay working group. Mutat Res 540:153–163

Kimura M, Umegaki K, Higuchi H, Thomas P, Fenech M (2004) MTHFR C677T polymorphism, folic acid and riboflavin are important determinants of genome stability in cultured human lymphocytes. J Nutr 134:48–56

Kirsch-Volders M, Sofuni T, Aardema M, Albertini S, Eastmond D, Fenech M, Ishidate M, Lorge E, Norppa H, Surrales J, von der Hude W, Wakata A (2000) Report from the In Vitro Micronucleus Assay Working Group. Environ Mol Mutagen 35:167–172

Ma C, Martin S, Trask B, Hamlin JL (1993) Sister chromatid fusion initiates amplification of the dihydrofolate reductase gene in Chinese hamster cells. Genes Dev 7:605–620

McClintock B (1942) The fusion of broken ends of chromosomes following nuclear fusion. Proc Natl Acad Sci USA 28:458–463

Melnyk S, Pogribna M, Miller BJ, Basnakian AG, Pogribny IP, James SJ (1999) Uracil misincorporation, DNA strand breaks, and gene amplification are associated with tumorigenic cell

transformation in folic acid deficient/repleted Chinese hamster ovary cells. Cancer Lett 146:35–44
Perry ME, Rolfe M, McIntyre P, Commane M, Stark GR (1992) Induction of gene amplification by 5-aza-2′-deoxycytidine. Mutat Res 276:189–197
Rajagopalan H, Jallepalli PV, Rago C, Velculescu VE, Kinzler KW, Vogelstein B, Lengauer C (2004) Inactivation of hCDC4 can cause chromosomal instability. Nature 428:77–81
Rao CV, Yang YM, Swamy MV, Liu T, Fang Y, Mahmood R, Jhanwar-Uniyal M, Dai W. Colonic tumorigenesis in BubR1+/–ApcMin/+ compound mutant mice is linked to premature separation of sister chromatids and enhanced genomic instability. Proc Natl Acad Sci USA 102:4365–4370
Rudolph KL, Millard M, Bosenberg MW, DePinho RA (2001) Telomere dysfunction and evolution of intestinal carcinoma in mice and humans. Nat Genet 28:155–159
Satoh T, Yamamoto K, Miura KF, Sofuni T (2004) Region-specific chromatin decondensation and micronucleus formation induced by 5-azacytidine in human TIG-7 cells. Cytogenet Genome Res 104:289–294
Serrano-Garcia L, Montero-Montaya R (2001) Micronuclei and chromatid buds are the result of related genotoxic events. Env Mol Mutagenesis 38:38–45
Shimizu N, Itoh N, Utiyama H, Wahl GM (1998) Selective entrapment of extrachromosomally amplified DNA by nuclear budding and micronucleation during S phase. J Cell Biol 140: 1307–1320
Shimizu N, Shimuara T, Tanaka T (2000) Selective elimination of acentric double minutes from cancer cells through the extrusion of MNi. Mutat Res 448:81–90
Shimizu N, Shingaki K, Kaneko-Sasaguri Y, Hashizume T, Kanda T (2005) When, where and how the bridge breaks: anaphase bridge breakage plays a crucial role in gene amplification and HSR generation. Exp Cell Res 302:233–243
Shimura M, Onozuka Y, Yamaguchi T, Hatake K, Takaku F, Ishizaka Y (1999) MNi formation with chromosome breaks and gene amplification caused by Vpr, an accessory gene of human immunodeficiency virus. Cancer Res 59:2259–2264
Stark GR (1993) Regulation and mechanisms of mammalian gene amplification. Adv Cancer Res 61:87–113
Suzuki T, Fujii M, Ayusawa D (2002) Demethylation of classical satellite 2 and 3 DNA with chromosomal instability in senescent human fibroblasts. Exp Gerontol 37:1005–1014
Toledo F, Le Roscouet D, Buttin G, Debatisse M (1992) Co-amplified markers alternate in megabase long inverted repeats and cluster independently in interphase nuclei at early steps of mammalian gene amplification. EMBO J 11:2665–2673
Umegaki K, Fenech M (2000) Cytokinesis-block micronucleus assay in WIL2-NS cells: a sensitive system to detect chromosomal damage induced by reactive oxygen species and activated human neutrophils. Mutagenesis 15:261–269
Wang X, Thomas P, Xue J, Fenech M (2004) Folate deficiency induces aneuploidy in human lymphocytes in vitro-evidence using cytokinesis-blocked cells and probes specific for chromosomes 17 and 21. Mutat Res 551:167–180
Windle BE, Wahl GM (1992) Molecular dissection of mammalian gene amplification: new mechanistic insights revealed by analyses of very early events. Mutat Res 276:199–224
Xu GL, Bestor TH, Bourc'his D, Hsieh CL, Tommerup N, Bugge M, Hulten M, Qu X, Russo JJ, Veigas-Peguinot E (1999) Chromosome instability and immunodeficiency syndrome caused by mutations in a DNA methyltransferase gene. Nature 402:187–191

16 In Vivo Rodent Micronucleus Assay

MAKOTO HAYASHI

Abstract

Genotoxicity plays an important role for the safety evaluation of chemicals. Chromosomal aberration is one of two major end points of genotoxicity. The rodent haematopoietic cell micronucleus assay is most widely used as an in vivo test to evaluate structural and numerical chromosomal aberrations. The historical aspects of the development of the in vivo micronucleus test, the mechanism of micronucleus formation, and the characteristics of the test are reviewed. One of the limitations of the micronucleus assay is the use haematopoietic cells and it is desired to develop the genotoxic method targeted at cells other than haematopoietic cells. The micronucleus assay using cells other than bone marrow cells, for example using liver, skin, testis, and colon cells, has been developed and validated successfully. The validation studies have been organized as collaborative studies of the Mammalian Mutagenicity Study Group belonging to the Japanese Environmental Mutagen Society and have been published. The other recent development of the micronucleus test is automated evaluation systems. Image analysis and flow cytometry were introduced to the test system with great success. These automated methods are being validated for regulatory purposes. The in vivo multiple end point assay system, which is more efficient, is also discussed here. When we use transgenic mutagenicity model animals in the micronucleus assay, we can obtain information on both gene mutation and chromosomal aberrations concomitantly, and possibly on DNA damage when the micronucleus assay is combined with the comet assay. Such a multiple end point genotoxicity assay system should be highly appreciated from the viewpoint of animal welfare.

16.1 Introduction

Genotoxicity plays an important role for the safety evaluation of chemicals. It is well known that there are in vitro and in vivo assay systems for the evaluation of chemical genotoxicity on different end points. The bacterial gene mutation test and the chromosomal aberrations test using mammalian cultured cells

Günter Obe and Vijayalaxmi (Eds.)
Chromosomal Alterations: Methods, Results and Importance in Human Health

are representative examples. It can be said that the genotoxicology field has been developed on the basis of in vitro assay systems. It is, however, apparent that there are limitations of in vitro assay systems for chemical safety evaluation and risk assessment for human health, and in vivo assay systems are becoming more important from the viewpoint of weight of evidence and risk assessment. Several in vivo assay systems have been developed and used for different end points. Among these, the rodent micronucleus test using haematopoietic cells has been most widely and frequently used to detect induction of chromosomal aberration. The historical aspects of the development of the in vivo micronucleus test and the characteristics of the test will be reviewed.

In addition, the current development of the micronucleus test and future perspectives of the assay will be summarized and discussed. As previously mentioned, the in vivo rodent micronucleus test is widely used to assess the clastogenic and possibly the aneugenic potential of chemicals for regulatory purposes. Haematopoietic cells, however, have been mainly targeted to evaluate cytogenetic effects and immature bone marrow erythrocytes or reticulocytes in peripheral blood cells have been observed. Bone marrow cells are effectively exposed by the test chemical after treatment by intraperitoneal injection, gavage, and intravenous injection because the blood–bone marrow barrier is not effective. Some chemicals, however, are metabolized in liver to active metabolites; some of those metabolites have a very short lifespan and cannot reach bone marrow cells. This is one of the limitations of the micronucleus assay using haematopoietic cells. It is desired to develop the genotoxic method to target organs other than haematopoietic organs. Several test systems, e.g. transgenic animal model, comet assay, and also the micronucleus test, have been developed, but only a few systems have been validated to be used for assessment of chemical safety.

Micronucleus assay using cells other than bone marrow cells, e.g. using liver, skin, testis, and colon cells, have been developed and validated successfully. The validation studies have been organized as collaborative studies of the Mammalian Mutagenicity Study Group (MMS) belonging to the Japanese Environmental Mutagen Society (JEMS) and some of them have been published. The technical issues of the micronucleus assay using colon and skin were established and the methods were evaluated by some model colon and skin carcinogens, respectively. The methods could detect the model chemicals as micronucleus inducers. The assay systems using testis and liver cells have almost been established and validated. For the testis assay, we introduced double staining with 4′,6-diamidino-2-phenylindole and acridine orange. The staining made identification of spermatocyte from other cells at the different spermiogenesis stages easy. For the liver micronucleus assay, the partial hepatectomy method has been used, but we do not know the precise influence of hepatectomy. We developed the assay using young rat hepatocytes without partial hepatectomy and other pretreatment. We evaluated the system using some model liver carcinogens and the method could detect these chemicals as micronucleus inducers.

The other recent development of the micronucleus test is automated evaluation systems. Image analysis and flow cytometry were introduced to the test system with great success. Now these automated methods are being validated for regulatory purposes. The spleen function of many mammalian species, including rat and human, eliminates the majority of micronucleated erythrocytes from the peripheral blood and the peripheral blood of such species was thought not to be suitable for the micronucleus test. Because flow cytometry can analyse a large number of cells and detect small increases of micronucleated erythrocytes, the peripheral blood of rats and humans became usable in the micronucleus test.

Finally, I will propose the in vivo multiple end point assay system. When we use transgenic mutagenicity models in the micronucleus assay, we can obtain information on both gene mutation and chromosomal aberrations concomitantly, and possibly DNA damage by the comet assay. Such a multiple end point genotoxicity assay system should be highly appreciated.

16.2 Historical Overview

Since the nineteenth century, the micronucleus has been known as the Howell–Jolly body in peripheral blood of mammals. In the 1970s, the micronucleus assay was established as an alternative method to chromosomal aberration assessment by metaphase analysis. Now, the assay is most widely used as an in vivo assay and is a member of a standard core battery of genotoxicity assays. The major historical events in connection with the micronucleus are as follows:

Nineteenth century Howell, Jolly; small body in erythrocytes of malignant anaemia cat and rat.

1959 *Vicia faba* roots (fast neutrons and γ-rays); chromosomal aberration evaluation by the induction of micronuclei (Evans et al. 1959).

1970–1975 Chinese hamster, mouse (Trenimon and other strong alkylating agents); chromosomal aberration evaluation test system: nomenclature "Microkern test"; establishment of rodent micronucleus test (Boller und Schmid 1970; Heddle 1973; Schmid 1975).

1976 Primary culture of human peripheral lymphocytes; human material/in vitro (Countryman and Heddle 1976).

1979 Transplacental application; embryonic micronucleus assay (Cole et al. 1979; King and Wild 1979).

1980 Mouse peripheral blood micronucleus assay; spleen function of rat and other species (MacGregor et al. 1980).

1980 Aneugens can be detected by the micronucleus test by size difference/centromere staining (Yamamoto and Kikuchi 1980; Högstedt and Karlsson 1985; Degrassi and Tanzarella 1988; Thomson and Perry 1988; Miller et al. 1991).

1980 Guidelines included the micronucleus test as a member of the standard battery.
1981–1983 Male germ cell micronucleus test (Lähdetie and Parvinen 1981; Tates et al. 1983).
1982- Automation of micronucleus analysis by flow cytometry/image analysis (Hutter and Stöhr 1982; Hayashi et al. 1992a, b; Romagna and Staniforth 1989; Tates et al. 1990; Asano et al. 1998; Dertinger et al. 1996).
1983 First report of the Gene Tox Program; IPCS project on evaluation of short-term test (Heddle et al. 1983).
1983 Introduction of fluorescent staining; acridine orange/pyronin Y and Hoechst 33258 (Hayashi et al. 1983; MacGregor et al. 1983).
1984 JEMS·MMS started the Collaborative Study Group for the Micronucleus Test (CSGMT) (reviewed in Sutou 1996).
1985 Human lymphocytes and other nucleated cells; introduction of cytochalasin B (Fenech and Morley 1985).
1990 Acridine orange supravital staining method for mouse peripheral blood micronucleus assay (Hayashi et al. 1990).
1993 First IWGT in Melbourne, Australia (Hayashi et al. 1994).
1997 Revision of OECD test guideline 474 (OECD 1997).
1999 Second IWGT in Washington, DC, USA (Hayashi et al. 2000).
2005 Fourth IWGT in San Francisco, CA, USA (Hayashi et al. 2007).

16.3 The Rodent Micronucleus Assay for Evaluation of Chromosomal Aberration Induction of Agents

Evaluation of micronucleus frequency in rodents, the micronucleus assay, is the assay in a battery of genotoxicity tests which is primarily selected as an in vivo assay. It is, however, applicable to any experimental cell systems that undergo the cell cycle in vitro and in vivo, e.g. rodent haematopoietic cells, in vitro cultured mammalian cells, plants. The micronucleus assay is not only an alternative method to the chromosomal aberration assay analysing metaphase chromosomes but also has its own characteristics as an assay to evaluate genotoxicity of agents. The characteristics of the assay are summarized as follows:

- Any dividing cell population with any karyotype can be used.
- Accurate data can be obtained because of the simple end point.
- Wider window for detection can be expected.
- Spindle poisons can be also detected.
- Background levels of micronucleated erythrocytes are low and stable.
- No additional treatment, e.g. colcemid or 5-bromo-2′-deoxyuridine, is required.

But:

- Types of chromosomal aberration cannot be classified.
- Pseudomicronuclei are observed under some circumstances.

In this chapter, only the in vivo rodent erythrocyte micronucleus is focused upon. For the assay using other materials, the reader is referred to the references.

The mechanism of formation of micronuclei from structural chromosomal aberrations and the correlation between frequencies of micronucleated erythrocytes and cells with chromosomal aberrations have been reported (Hayashi et al. 1984). The acentric fragments of chromatid/chromosome resulting from chromosomal aberration form micronuclei efficiently. The flow of micronucleus formation and the relationship between chromosomal aberration and the micronucleus is shown schematically in Fig. 16.1. The *k* value in the figure shows the relationship between frequencies of micronucleated cells and cells with chromosomal aberrations. The micronuclei are also the result of the interference of the mitotic spindle and whole chromosomes remaining in the cytoplasm when new daughter nuclei were formed form the micronucleus beside the main (daughter) nuclei (Fig. 16.2).

16.3.1 Rodent Bone Marrow Micronucleus Assay

Femoral marrow cells are usually suspended in foetal bovine serum and smeared on clean glass slides. Then these cells are fixed, stained and the frequency of micronucleated polychromatic erythrocytes is evaluated. To improve the identification of young erythrocytes, acridine orange and Hoechst 33258 plus pyronin Y fluorescent staining methods have been introduced to identify RNA containing erythrocytes as young erythrocytes (Hayashi et al. 1983, MacGregor et al. 1983). The acridine orange method has been reevaluated for its usefulness in the micronucleus assay (Tinwell and Ashby 1989). The precise protocol of this method was defined by the guidelines (e.g. OECD 1997) after extensive discussion among experts at the IWGT meetings (Hayashi et al. 1994, 2000). These guidelines and references should be referred to for more detail methodological aspects.

16.3.2 Rodent Peripheral Blood Micronucleus Assay

The usefulness of mouse peripheral blood erythrocytes as experimental material for the micronucleus assay has been reported (MacGregor et al. 1980). Bone marrow cells, however, are still the major target cells for studying micronucleus induction by clastogens. If peripheral blood erythrocytes could be used instead of bone marrow cells without any reduction in the sensitivity

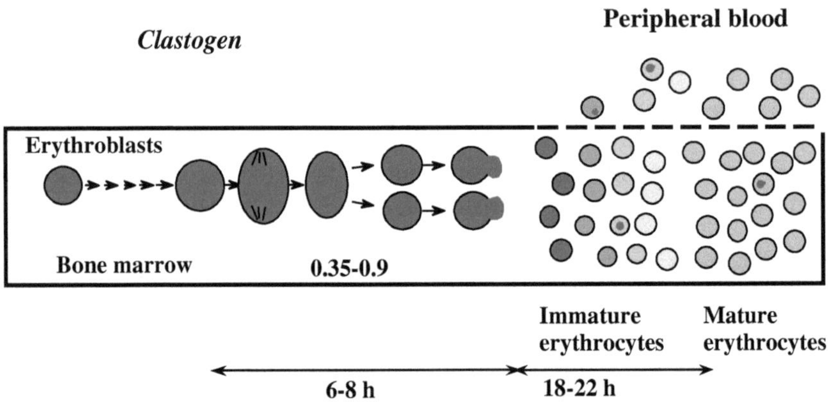

Fig. 16.1 Model of micronucleated erythrocyte formation in bone marrow by clastogens and the relationship between chromosomal aberration and the micronucleus. The *k* value shows the relationship between frequencies of micronucleated cells and cells with chromosomal aberrations

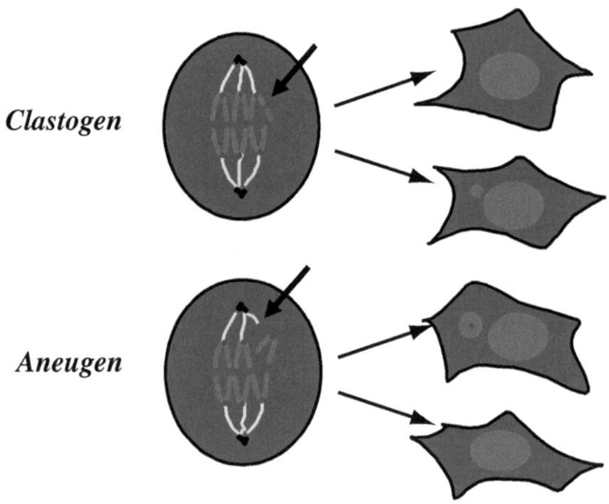

Fig. 16.2 Model of micronucleus induction by clastogens and aneugens. The micronuclei induced by aneugens are larger in size than those induced by clastogens and also have a centromere

of the test to identify the clastogenic effects of test chemicals, there should be several advantages. Firstly, the cell population is more uniform and, therefore, easier to score; therefore, there is a big possibility to automate the scoring process by image analysis and flow cytometry. Secondly, a small number of blood cells could repeatedly be taken from the tail of mice without killing the exposed animals; therefore, it would be possible to make preparations at the optimal sampling time after treatment(s) by periodic sampling from the same individual. Thus, no pilot experiments to establish optimal sampling times will be required. It is also possible to take a very small amount of blood from other toxicity experiments without interfering with the study.

16.4 The Micronucleus Assay Using Tissues Other Than Bone Marrow

As previously described, the micronucleus assay can apply any cell populations when they are dividing. Haematopoietic cells, i.e. bone marrow cells and peripheral blood erythrocytes, have been routinely used as target tissue/cells of this assay. Because the haematopoietic cell population in bone marrow is always dividing, researchers in the cytogenetic field preferred to use this cell population for their study historically. Recently, however, for assessment of human health the tendency is more for risk assessment than hazard identification. If this is the case, not only bone marrow and peripheral blood cells but also the site of contact of the tissues/cells and accumulation of the test chemical and also the target site of carcinogenicity should be evaluated to understand the mechanism of action.

16.4.1 Liver

It has been pointed out that several compounds which are metabolized to a biologically active form give a negative response in the bone marrow micronucleus assay (Morita et al. 1997). Some active metabolites have a very short lifespan and do not reach bone marrow at sufficient concentrations to induce micronuclei. In fact, some rodent liver carcinogens, including dialkyl nitrosamines, nitro aromatic compounds, and azo derivatives, gave negative results in a bone marrow assay (Angelosanto 1995). It is worthwhile therefore to consider the use of liver for detecting liver carcinogens.

Owing to low mitotic activity in adult animals, procedures to mitogenize liver cells were introduced. Tates et al. (1980) reported a method incorporating partial hepatectomy before or after chemical treatment in the liver micronucleus assay. The division of hepatocytes is stimulated by partial hepatectomy, and positive responses were observed in induction of micronuclei in liver cells after treatment with liver carcinogens that gave negative results in bone marrow. There are two shortcomings in the partial hepatectomy method: the first one is that it is not easy to perform hepatectomy on all the animals used in the assay, and the second is the activity level of enzymes such as cytochrome P450, styrene monooxygenase, epoxide hydrolase, and glutathione *S*-epoxide transferase 12 h after partial hepatectomy (Rossi et al.

1987). Another method used a treatment with mitogens for liver cells, e.g. 4-acetyl aminofluorene (Braithwaite and Ashby 1988).

Use of proliferating tissue is a prerequisite for the micronucleus assay and the hepatocytes of young rat (i.e. 4-week-old rats) are still proliferating, and the percentages of S-phase cells are more than 40 times greater than in adult rat liver (Parton and Garriott 1997). Sipes and Gandolfi (1993) reported that P450 levels reach a maximum for rats at 4 weeks of age. For example, diethylnitrosamine (DEN), which is a well-known rodent liver carcinogen, was used as a model compound. A negative response for DEN was obtained in the bone marrow micronucleus assay (Morita et al. 1997), but there was a positive response in the liver micronucleus assay using the partial hepatectomy method (Tates et al. 1980). The method using young rat liver showed an unambiguous positive response in induction of micronuclei.

16.4.2 Colon

For the colon micronucleus test, classic paraffin sections have been used to evaluate induction of micronuclei or nuclear anomalies in epithelial cells of intact colon (Wargovich et al. 1983; Migliore et al. 1989). To overcome the disadvantages of the section method, e.g. the improper position of sections which might miss micronuclei, steps were taken to develop a method using single cell suspensions of cells from mouse colon (Blakey et al. 1985; Ohyama and Tokumitsu 1996; Zhurkov et al. 1996). One of the methods developed by Tokumitsu et al. modified the method for the chromosomal aberration test (Tokumitsu 1976; Ohyama et al. 1991, 1992, 1993). This method was easier to perform and gave more accurate and reproducible data. The most probable reason was the analysis of intact cells using fluorescent staining (Hayashi et al. 1983). The method was evaluated by the collaborative study of JEMS/MMS CSGMT and showed good performance and interlaboratory reproducibility (Ohyama et al. 2002).

16.4.3 Skin

The skin is a major interface between the body and the environment; therefore, many mutagens and carcinogens are exposed directly. Accordingly, an in vivo skin micronucleus assay should be a useful tool for assessing genotoxic risk in certain regulatory situations, e.g. for the assessment of cosmetic ingredients. The assay using rodent skin has been evaluated and a validation study organized by CSGMT is being undertaken. The assay was characterized by four improvements: (1) cold-treating the epidermis with trypsin, (2) peeling it off with fine forceps, (3) treating it in hypotonic solution, and (4) staining it with acridine orange (Nishikawa et al. 1999, 2002, 2005). The data showed that skin epidermal basal cells (which are termed "basal + suprabasal" cells) could be used to evaluate genotoxicity following epidermal application of chemicals.

16.4.4 Testes

The assessment of the cytogenetic effect of agents is one of the major points for human health assessment in terms of genotoxicity. The use of a micronucleus assay in mammalian germ cells was proposed in the early 1980s by Lähdetie and Parvinen (1981) and by Tates et al. (1983). Lähdetie and Parvinen's method is referred to as the "dissection method". It takes advantage of the organization of seminiferous tubules, where different spermatogenic stages are specifically associated. On the other hand, the method proposed by Tates et al. (1983) was based on the preparation of a cell suspension from testicular tubules. The number of papers reporting data obtained by the spermatid micronucleus assay is still limited, compared with the large amount of information available from assays on somatic cells. A summary of the most recent data (from 1994) obtained with the spermatid micronucleus assay in rats and mice is presented in Table 2 of the IWGT report (Hayashi et al. 2000).

16.5 Automation of the Micronucleus Assay

The end point of the micronucleus assay is clearly defined and easily recognized; therefore, there is a possibility to easily automate it by image analysis and also by flow cytometry. Manual microscopic evaluation of micronuclei is still tedious and time-consuming and requires well-trained laboratory personnel, although it is faster and easier than metaphase analysis. To obtain more objective data quickly, automation techniques have been developed both with computerized image analysis (Tates et al. 1990; Romagna and Staniforth 1989; Parton et al. 1996; Asano et al. 1998) and with flow cytometry (Hutter and Stöehr 1982; Hayashi et al. 1992a, b; Krishna et al. 1998; Tometsko et al. 1993; Dertinger et al. 1996, 2003). Romagna and Staniforth (1989) emphasized the importance of careful pretreatment of samples when using bone marrow and May–Grünwalt Giemsa staining for the micronucleus assay. The use of peripheral blood and acridine orange supravital staining eliminates the need for pretreatment, but high-quality preparation is important for automated analysis. Adequate colour intensity and well-spread blood cells are necessary to distinguish the micronucleated reticulocytes accurately. The image-analysis system was applied to the peripheral blood micronucleus assay using the acridine orange coated glass slide method (Asano et al. 1998).

Unlike microscopy-based image analysis, flow cytometry analyses cells in aqueous suspension with high speed. The early systems were not capable of differentiating micronucleated immature erythrocytes accurately from micronucleated mature erythrocytes. A significant advance in this area came when Grawé et al. (1992, 1997a, b, 1998) described a flow cytometry method which was capable of scoring micronuclei in the immature erythrocyte population. Dertinger et al. (1996, 2003) developed a procedure which is based on RNase treatment in conjunction with propidium iodide and anti-CD71–fluorescein isothiocyanate staining. This protocol is capable of resolving all four erythrocyte populations of interest: mature and immature erythrocytes, with and without micronuclei. An advantage of flow cytometry scoring over manual scoring and image analysis is its extremely high analysis rates. Modern flow cytometers are able to evaluate tens of thousands of immature and hundreds of thousands of mature erythrocytes for micronuclei in several minutes. The increment of the number of cells analysed increases the sensitivity of the micronucleus assay from a statistical viewpoint (Abramsson-Zetterber et al. 1995; Abramsson-Zetterber 2003; Torous et al. 2006; Asano et al. 2006). The methods are almost ready for the assay to be used for regulatory purposes.

16.6 Multiple End Point Assays

Genotoxicity plays an important role for the safety evaluation of chemicals. When the carcinogenicity of a chemical evident, the threshold can be estimated only when the genotoxic mechanism does not operate for carcinogenesis, otherwise a threshold cannot be set. Without a genotoxic mechanism – non-genotoxic carcinogen – the threshold can be estimated but with the genotoxic mechanism – genotoxic carcinogen – it cannot be estimated. It is well known that there are in vitro and in vivo assay systems for the evaluation of chemical genotoxicity on different end points. The bacterial gene mutation test and the chromosomal aberrations test using mammalian cultured cells are representative examples. It can be said that the genotoxicology field has been developed on the basis of in vitro assay systems. It is, however, apparent that there are limitations of in vitro assay systems for chemical safety evaluation and risk assessment for human health and in vivo assay systems are becoming more important from the viewpoint of weight of evidence. Several in vivo assay systems have been developed and used for different end points. Among these, the rodent micronucleus test using haematopoietic cells has been most widely and frequently used to detect induction of chromosomal aberration. However, global assessment of genotoxic risk is desired for human health risk evaluation. For this purpose, in vivo assays came into the spotlight because of concern over weight of evidence and also mode of action. On the other hand, however, the animal welfare movement and the 3R (replace, reduce, refine) concern are affecting the increase of animal experimentation. The peripheral blood micronucleus assay does not require killing animals and can be integrated into other toxicological studies (Hayashi et al. 2000). If the transgenic animals are used for the general toxicological study, then we can incorporate the comet assay to assess DNA damage, gene mutation assessment, the micronucleus assay to evaluate cytogenetic effect together with the general toxicity assessment (Suzuki et al. 1997; Kohara et al. 2001,

2002a, b). Such a multiple end point assay system will be important and it can provide information of more global aspects, including genotoxicity. Moreover, the technique can be incorporated into the general toxicological study using rats and provide much information about the toxicity of the test chemical for global assessment (Hayashi et al. 2000; Wakata 1998; Hamada et al. 2001). The combination of the toxicological studies will reduce the need for experimental animals.

References

Abramsson-Zetterberg L (2003) The dose-response relationship at very low doses of acrylamide is linear in the flow cytometer-based mouse micronucleus assay. Mutat Res 535:215–222

Abramsson-Zetterberg L, Grawé J, Zetterberg G (1995) Flow cytometric analysis of micronucleus induction in mice by internal exposure to 137 Cs at very low dose rates. Int J Radiat Biol 67:29–36

Angelosanto FA (1995) Tissues other than bone marrow that can be used for cytogenetic analyses. Environ Mol Mutagen 25:338–343

Asano N, Katsuma Y, Tamura H, Higashikuni N, Hayashi M (1998) An automated new technique for scoring the rodent micronucleus assay: computerized image analysis of acridine orange supravitally stained peripheral blood cells. Mutat Res 404:149–154

Asano N, Torous DK, Tometsko CR, Dertinger SD, Morita T, Hayashi M (2006) Practical threshold for micronucleated reticulocyte induction observed for low doses of mitomycin C, Ara-C and colchicine. Mutagenesis 21:15–20

Blakey DH, Duncan AMV, Wargovich MJ, Goldverg MT, Bruce WR, Heddle JA (1985) Detection of nuclear anomalies in colonic epithelium of the mouse. Cancer Res 45:242–249

Boller K, Schmid W (1970) Chemische Mutagenese beim Sauger. Das Knochemark des Chinesischen Hasmters als in vivo-Testsystem. Haematologische Befunde nach Behandlung mit Trenimon. Humangenetik 11:35–54

Braithwaite I, Ashby J (1988) A non-invasive micronucleus assay in the rat liver. Mutat Res 203:23–32

Cole RJ, Taylor NA, Cole J, Arlette CF (1979) Transplacental effects of chemical mutagens detected by the micronucleus test. Nature 227:317–318

Countryman PI, Heddle JA (1976) The production of micronuclei from chromosome aberrations in irradiated cultures of human lymphocytes. Mutat Res 41:321–332

Degrassi F, Tanzarella C (1988) Immunofluorescent staining of kinetochores in micronuclei: a new assay for the detection of aneuploidy. Mutat Res 203:339–345

Dertinger SD, Chen Y, Miller RK, Brewer KJ, Smudzin T, Torous DK, Hall NE, Olvany KA, Murante FG, Tometsko CR (2003) Micronucleated CD71-positive reticulocytes: a blood-based endpoint of cytogenetic damage in humans. Mutat Res 542:77–87

Dertinger SD, Torous DK, Tometsko KR (1996) Simple and reliable enumeration of micronucleated reticulocytes with a single-laser flow cytometer. Mutat Res 371:283–292

Evans HJ, Neary GJ, Williamson FS (1959) The relative biological efficiency of single doses of fast neutrons and gamma-rays in *Vicia faba* roots and the effect of oxygen: Part II. Chromosome damage: the production of micronuclei. Int J Radiat Biol 3:216–229

Fenech M, Morley AA (1985) Measurement of micronuclei in lymphocytes. Mutat Res 147:35–54

Grawé J, Zetterberg G, Amneus H (1992) Flow-cytometric enumeration of micronucleated polychromatic erythrocytes in mouse peripheral blood. Cytometry 13:750–758

Grawé J, Adler I-D, Nüsse M (1997a) Quantitative and qualitative studies of micronucleus induction in mouse erythrocytes using flow cytometry. II. Analysis of micronuclei of aneugenic and clastogenic origin by dual color FISH on populations of bone marrow PCEs flow sorted on the basis of their relative DNA content. Mutagenesis 12:9–15

Grawé J, Nüsse M, Adler I-D (1997b) Quantitative and qualitative studies of micronucleus induction in mouse erythrocytes using flow cytometry: I. Measurement of micronucleus induction in peripheral blood polychromatic erythrocytes by chemicals with known and suspected genotoxicity. Mutagenesis 12:1–8

Grawé J, Abramsson-Zetterberg L, Zetterberg G (1998) Low dose effects of chemicals as assessed by the flow cytometric in vivo micronucleus assay. Mutat Res 405:199–208

Hamada S, Sutou S, Morita T, Wakata A, Asanami S, Hosoya S, Ozawa S, Kondo K, Nakajima M, Shimada H, Osawa K, Kondo Y, Asano N, Sato S, Tamura H, Yajima N, Marshall R, Moore C, Blakey DH, Schechtman LM, Weaver JL, Torous DK, Proudlock R, Ito S, Namiki C, Hayashi M (2001) Evaluation of the rodent micronucleus assay by a 28-day-treatment protocol: summary of the 13th collaborative study by CSGMT/JEMS·MMS. Environ Mol Mutagen 37:93–110

Hayashi M, Sofuni T, Ishidate M Jr (1983) An application of acridine orange fluorescent staining to the micronucleus test. Mutat Res 120:241–247

Hayashi M, Sofuni T, Ishidate M Jr (1984) Kinetics of micronucleus formation in relation to chromosomal aberrations in mouse bone marrow. Mutat Res 127:129–137

Hayashi M, Morita T, Kodama Y, Sofuni T, Ishidate M Jr (1990) The micronucleus assay with mouse peripheral blood reticulocytes using acridine orange coated slides. Mutat Res 245:245–249

Hayashi M, Norppa H, Sofuni T, Ishidate M Jr (1992a) Mouse bone marrow micronucleus test using flow cytometry. Mutagenesis 7:251–256

Hayashi M, Norppa H, Sofuni T, Ishidate M Jr (1992b) Flow cytometric micronucleus test with mouse peripheral erythrocytes. Mutagenesis 7:257–264

Hayashi M, Tice RR, MacGregor JT, Anderson D, Blakey DH, Kirsch-Volders M, Oleson FB Jr, Pacchierotti F, Romagna F, Shimada H, Sutou S, Vannier B (1994) *In vivo* rodent erythrocyte micronucleus assay. Mutat Res 312:293–304

Hayashi M, MacGregor JT, Gatehouse DG, Adler I-D, Blakey DH, Dertinger SD, Krishna G, Morita T, Russo A, Sutou S (2000) *In vivo* rodent erythrocyte micronucleus assay. II. Some aspects of protocol design including repeated treatments, integration with toxicity testing, and automated scoring. Environ Mol Mutagen 35:234–252

Hayashi M, MacGregor JT, Gatehouse DG, Blakey DH, Dertinger SD, Abramsson-Zetterberg L, Krishna G, Morita T, Russo A, Asano N, Suzuki H, Ohyama W, Gibson D (2007) In vivo erythrocyte micronucleus assay: III. Validation and regulatory acceptance of automated scoring and the use of rat peripheral blood reticulocytes, with discussion of non-hematopoietic target cells and a single dose-level limit test. Mutat Res 627:10–30

Heddle JA (1973) A rapid in vivo test for chromosomal damage. Mutat Res 18:187–190

Heddle JA, Hite M, Kirkhart B, Mavourmin K, MacGregor JT, Newell GW, Salamone MF (1983) The induction of micronuclei as a measure of genotoxicity: A report of the U.S. Environmental Protection Agency Gene-Tox program. Mutat Res 123:527–549

Hutter K-J, Stöehr M (1982) Rapid detection of mutagen induced micronucleated erythrocytes by flow cytometry. Histochemistry 75:353–362

Högstedt B, Karlssson (1985) A The size of micronuclei in human lymphocytes varies according to inducing agent used. Mutat Res 156:229–232

King MT, Wild D (1979) Transplacental mutagenesis: The micronucleus test on fetal mouse blood. Hum Genet 51:183–194

Kohara A, Suzuki T, Honma M, Hirano N, Ohsawa K, Ohwada T, Hayashi M (2001) Mutation spectrum of o-aminoazotoluene in the cII gene of lambda/*lacZ* transgenic mice (Muta™ mouse). Mutat Res 491:211–220

Kohara A, Suzuki T, Honma M, Oomori T, Ohwada T, Hayashi M (2002a) Dinitropyrenes induce gene mutations in multiple organs of the lambda/*lacZ* transgenic mouse (Muta™ mouse). Mutat Res 515:73–83

Kohara A, Suzuki T, Honma M, Ohwada T, Hayashi M (2002b) Mutagenicity of aristolochic acid in the lambda/*lacZ* transgenic mouse (Muta™ mouse). Mutat Res 515:63–72

Krishna G, Urda G, Theiss J (1998) Principles and practices of integrating genotoxicity evaluation into routine toxicology studies: A pharmaceutical industry perspective. Environ Mol Mutagen 32:115–120
Lähdetie J, Parvinen M (1981) Meiotic micronuclei induced by X-ray in early spermatids of the rat. Mutat Res 81:103–115
MacGregor JT, Wehr CM, Gould DH (1980) Clastogen-induced micronuclei in peripheral blood erythrocytes: the basis of a improve micronucleus test. Environ Mutagen 2:509–514
MacGregor JT, Wehr CM, Langlois RG (1983) A simple fluorescent staining procedure for micronuclei and RNA in erythrocytes using Hoechst 33258 and pyronin Y. Mutat Res 120:269–275
Migliore L, Ventura L, Barale R, Loprieno N, Castellino S, Pulci R (1989) Micronuclei and nuclear anomalies induced in the gastro-intestinal epithelium of rat treated with formaldehyde. Mutagenesis 4:327–334
Miller BM, Zitzelsberger HF, Weier HU, Adler I-D (1991) Classification of micronuclei in mouse erythrocytes: immunofluorescent staining using CREST antibodies compared to in situ hybridization with biotinylated gamma satellite DNA. Mutagenesis 6:297–302
Morita T, Asano N, Awogi T, Sasaki YF, Sato S, Shimada H, Sutou S, Suzuki T, Wakata A, Sofuni T, Hayashi M (1997) Evaluation of the rodent micronucleus assay in the screening of IARC carcinogens (groups 1, 2A, and 2B) The summary report of the 6th collaborative study by CSGMT/JEMS·MMS. Mutat Res 389:3–122
Nishikawa T, Haresaku M, Adachi K, Masuda M, Hayashi M (1999) Study of a rat skin in vivo micronucleus test: data generated by mitomycin C and methyl methanesulfonate. Mutat Res 444:159–166
Nishikawa T, Hasekura M, Fukushima A, Nakamura T, Adachi K, Masuda M, Hayashi M (2002) Further evaluation of an in vivo micronucleus test on rat and mouse skin: results with five skin carcinogens. Mutat Res 513:93–102
Nishikawa T, Nakamura T, Fukushima A, Takagi Y (2005) Further evaluation of the skin micronucleus test: Results obtained using 10 polycyclic aromatic hydrocarbons. Mutat Res 588:58–63
OECD (1997) Guideline for the testing of chemicals. mammalian erythrocyte micronucleus test. Guideline 474, July 1997
Ohyama W, Tokumitsu T (1996) An in vivo micronucleus test using colonic epithelial cells of mice. Environ Mutat Res Commun 17:265–269
Ohyama W, Urakawa M, Tokumistu T (1991) Chromosome observations of mouse intestinal epithelial cells. Mutat Res 252:258
Ohyama W, Urakawa M, Tokumistu T (1992) Chromosome aberrations induced by MMC in the intestinal epithelial cells of mice. Mutat Res 272:278
Ohyama W, Urakawa M, Tokumistu T (1993) Chromosome aberrations in the intestinal epithelial cells of mice treated with 1,2-dimethylhydrazine. Mutat Res 292: 287
Ohyama W, Gonda M, Miyajima H, Kondo K, Noguchi T, Yoshida J, Hatkeyama S, Watabe E, Ueno Y, Hayashi M, Tokumitsu T (2002) Collaborative validation study of the in vivo micronucleus test using mouse colonic epithelial cells. Mutat Res 518:39–45
Parton JW, Garriott ML (1997) An evaluation of micronucleus induction in bone marrow and in hepatocytes isolated from collagenase perfused liver or from formalin-fixed liver using four-week-old rats treated with known clastogens. Environ Mol Mutagen 29:379–385
Parton JW, Hoffman WP, Garriott ML (1996) Validation of an automated image analysis micronucleus scoring system. Mutat Res 370:65–73
Romagna F, Staniforth CD (1989) The automated bone marrow micronucleus test. Mutat Res 213:91–104
Rossi AM, Romano M, Zaccaro L, Pulci R, Salmona M (1987) DNA synthesis, mitotic index, drug-metabolising systems and cytogenetic analysis in regenerating rat liver. Mutat Res 182:75–82
Schmid W (1975) The micronucleus test. Mutat Res 31:9–15
Sipes GI, Gandolfi AJ (1993) Biotransformation of toxicants. In: Amdur MO, Doull J, Klassen CD (eds) Casarett and Doull's toxicology, the basic science of poisons. McGraw-Hill, New York, pp 88–126

Sutou S (1996) Achievements by CSGMT/JEMS·MMS: the Collaborative Study Group for the Micronucleus Test in the Mammalian Mutagenesis Study Group of the Environmental Mutagen Society of Japan. Mutat Res 340:151–174
Suzuki T, Hayashi M, Wang X, Yamamoto K, Ono T, Myhr BC, Sofuni T (1997) A comparison of the genotoxicity of ethylnitrosourea and ethyl methanesulfonate in *lacZ* transgenic mice (Muta™ mouse). Mutat Res 395:75–82
Tates AD, Neuteboom I, Hofker M, den Englse L (1980) A micronucleus technique for detecting clastogenic effects of mutagens/carcinogens (DEN, DMN) in hepatocytes of rat liver in vivo. Mutat Res 74:1–20
Tates AD, Dietrich JJ, de Vogel N, Newteboom I, Bos A (1983) A micronucleus method for detection of meiotic micronuclei in male germ cells of mammals. Mutat Res 121:131–138
Tates AD, van Welie MT, Ploem JS (1990) The present state of the automated micronucleus test for lymphocytes. Int J Radiat Biol 58:813–825
Thomson EJ, Perry PE (1988) The identification of micronucleated chromosomes: a possible assay for aneuploidy. Mutagenesis 3:415–418
Tinwell H, Ashby J (1989) Comparison of acridine orange and Giemsa stains in several mouse bone marrow micronucleus assay – including a triple dose study. Mutagenesis 4:476–481
Tokumitsu T (1976) Chromosomal study on mouse intestinal mucosal cells, with a note on aberrations induced by a carcinogen, MNNG. Proc Jpn Acad 52:513–516
Tometsko AM, Torous DK, Dertinger SD (1993) Analysis of micronucleated cells by flow cytometry. 3. Advanced technology for detecting clastogenic activity. Mutat Res 292:145–153
Torous D, Asano N, Tometsuko C, Sugunan S, Dertinger S, Morita T, Hayashi M (2006) Performance of flow cytometric analysis for the micronucleus assay – a reconstruction model using serial dilutions of malaria-infected cells with normal mouse peripheral blood. Mutagenesis 21:11–13
Wakata A, Miyamae Y, Sato S, Suzuki T, Morita T, Asano N, Awogi T, Kondo K, Hayashi M (1998) Evaluation of the rat micronucleus test with bone marrow and peripheral blood: The summary of the 9th collaborative study by CSGMT/MMS·JEMS. Environ Mol Mutagen 32:84–100
Wargovich MJ, Goldberg MT, Newmark HL, Bruce WR (1983) Nuclear aberrations as a short-term test for genotoxicity to the colon: Evaluation of nineteen agents in mice. J Natl Cancer Inst 71:133–137
Yamamoto KI, Kikuchi Y (1980) A comparison of diameters of micronuclei induced by clastogens and spindle poisons. Mutat Res 71:127–131
Zhurkov VS, Sycheva LP, Salamatova O, Vyskubenko IF, Feldt EG, Sherenesheva NI (1996) Selective induction of micronuclei in the rat/mouse colon and liver by 1,2-dimethylhydrazine: a seven-tissue comparative study. Mutat Res 368:115–120

17 Sister-Chromatid Exchanges

ANDRZEJ WOJCIK AND GÜNTER OBE

Abstract

The sister chromatid exchange (SCE) test is a relatively simple assay, sensitive to most mutagenic compounds. Its major disadvantage is that in order to detect SCEs cells must be either treated with the label for one round of replication prior to treatment with the mutagen studied or treated simultaneously, but fixed and harvested following two rounds of replication. Thus, one can never be sure if the label (usually 5-bromo-2′-deoxyuridine, BrdU) used does not interact with the mutagen studied or if the scoring of the second posttreatment mitoses is not a source of bias in the estimation of the SCE frequency. A further problem is the necessity to allow the cells to divide twice before mitotic cells are harvested. Given the fact that cells in culture do not grow in a synchronized manner, perturbations of the cell cycle kinetics provide a potential source of error in estimating the true SCE frequency. The aim of this chapter is to discuss some problems of the SCE assay, especially those related to the use of BrdU.

17.1 Introduction

Sister-chromatid exchanges (SCEs) reflect an interchange mechanism between sister chromatids of mitotic chromosomes (Latt 1981). The visualization of SCEs in linear chromosomes requires that the sister chromatids are stained differentially (Fig. 17.1). The most widely applied method with which this can be accomplished involves the use of the halogenated thymidine analogue 5-bromo-2′-deoxyuridine (BrdU) (Latt 1973; Perry and Wolff 1974).

In ring chromosomes, SCEs can be detected without differential staining of the chromatids, as first described in 1938 by Barbara McClintock (1938) in *Zea mais*. She noticed that sister chromatids of ring-shaped chromosomes may recombine to form a single, double-sized ring. The only explanation for this was to assume that an interchange occurred between the chromatids of the original ring chromosome. In 1957, Taylor et al. (1957) used autoradiography to visualize the incorporation of tritiated thymidine into the DNA of plant cells. Cells were labelled for one cell cycle with tritiated thymidine, the

Günter Obe and Vijayalaxmi (Eds.)
Chromosomal Alterations: Methods, Results and Importance in Human Health

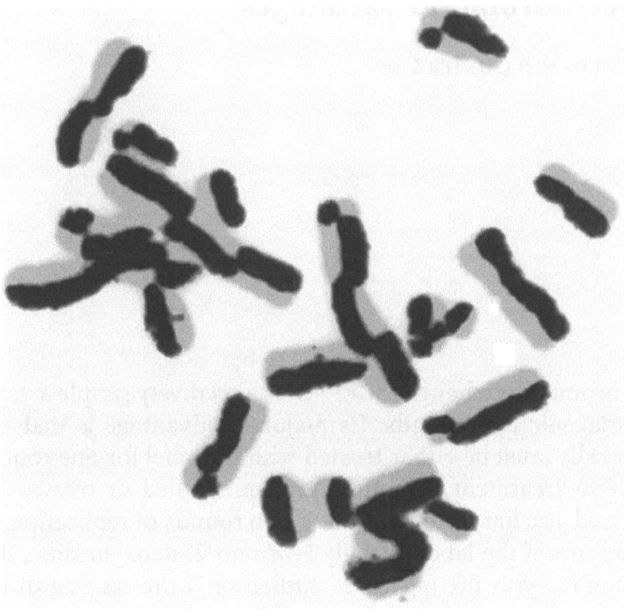

Fig. 17.1 A Chinese hamster ovary cell differentially labelled with 5-bromo-2′-deoxyuridine (BrdU) and stained by the fluorescence plus Giemsa technique with a fluorochrome-labelled monoclonal antibody

ensuing mitoses were suppressed with colchizine and during the next cell cycle cells were grown in the presence of thymidine. Following autoradiography, the authors analysed tetraploid metaphases in which the first cell cycle was exposed to tritiated thymidine and the second to thymidine. The chromatids of the chromosomes in tetraploid metaphases were tritiated thymidine–thymidine and thymidine–thymidine. In autoradiographs tritiated thymidine–thymidine but not thymidine–thymidine chromatids were covered with silver grains. Taylor et al. found SCEs between labelled and unlabelled chromatids. Some SCEs occurred at the same sites in two sister chromosomes of the tetraploid metaphases (twin SCEs) and in single chromosomes (single SCEs). One twin SCE corresponds to two single SCEs, indicating that twin SCEs result from replication of a SCE which occurred in S phase of the first cell cycle, and single SCEs occurred in S phase of the second cell cycle. Similar results were obtained with mammalian cells labelled with BrdU (Wolff and Perry 1975; Bruckmann et al. 1999b). The analysis of SCEs in autoradiographs is difficult because the differentiation of small SCEs is often impossible; therefore, the SCE test was not widely used. This situation changed after the development of the fluorescence plus Giemsa (FPG) method by Perry and Wolff (1974).

Since the introduction of FPG staining, numerous studies have provided evidence that SCEs are induced by a great majority of mutagens in cells that can incorporate labels such as BrdU for one replication cycle and then undergo a second replication, in which the presence of BrdU is optional (Latt 1981; Natarajan and Mullenders 1987). SCEs are quite easy to score and, thanks to the fact that their spontaneous number is around five per metaphase (Sutou 1981; Natarajan et al. 1986; Bender et al. 1992), only a low number of cells (usually between 30 and 50) need to be scored in order to detect a statistically significant enhancement of the SCE frequency. This makes the SCE system a favourite end point in mutagenesis studies.

However, in order to detect SCEs, cells must be either treated with BrdU for one round of replication prior to treatment with the mutagen studied or treated simultaneously, but fixed and harvested following two rounds of replication (Fig. 17.2). Thus, when compounds are tested for their mutagenic potential, one can never exclude the possibility that an enhanced SCE frequency is due to the interaction of the compound with BrdU. A further problem is the necessity to allow the cells to divide twice before mitotic cells are harvested. Given the fact that cells in culture do not grow in a synchronized manner, perturbations of the cell cycle kinetics provide a potential source of error in estimating the true SCE frequency.

The aim of this chapter is to discuss some problems of the SCE assay, especially those related to the use of BrdU.

17.2 The Influence of BrdU on the Spontaneous SCE Level

The mechanisms of differential staining by the BrdU–FPG method have been described in detail (Goto et al. 1978; Holmquist 1987). The impact of BrdU on the spontaneous level of SCE was originally studied by employing very low concentrations of BrdU. With the FPG technique, optimal quality of differential staining is achieved when the concentration of BrdU in the medium is about 20 μM (Goto et al. 1978). Much lower concentrations of BrdU can be applied when a monoclonal antibody against BrdU is used. Pinkel et al. (1985) showed for the first time that SCEs can be detected in Chinese hamster ovary (CHO) cells when the DNA is substituted with 5-bromouracil (BrU) at a level of 0.6%. These authors found that the baseline level of SCEs was about six per cell and that this frequency did not change up to a substitution level of 20%, but increased with higher substitution levels. Somewhat different results were reported by Natarajan et al. (1986), who incubated CHO cells in medium containing a mixture of 10 μM BrdU, 10 μM thymidine, 1 μM 5-fluorodeoxyuridine and 100 μM deoxycytidine. The frequency of SCEs obtained at the lowest level of BrdU substitution (1%) was about four per cell and increased steadily with increasing level of BrdU, reaching about seven

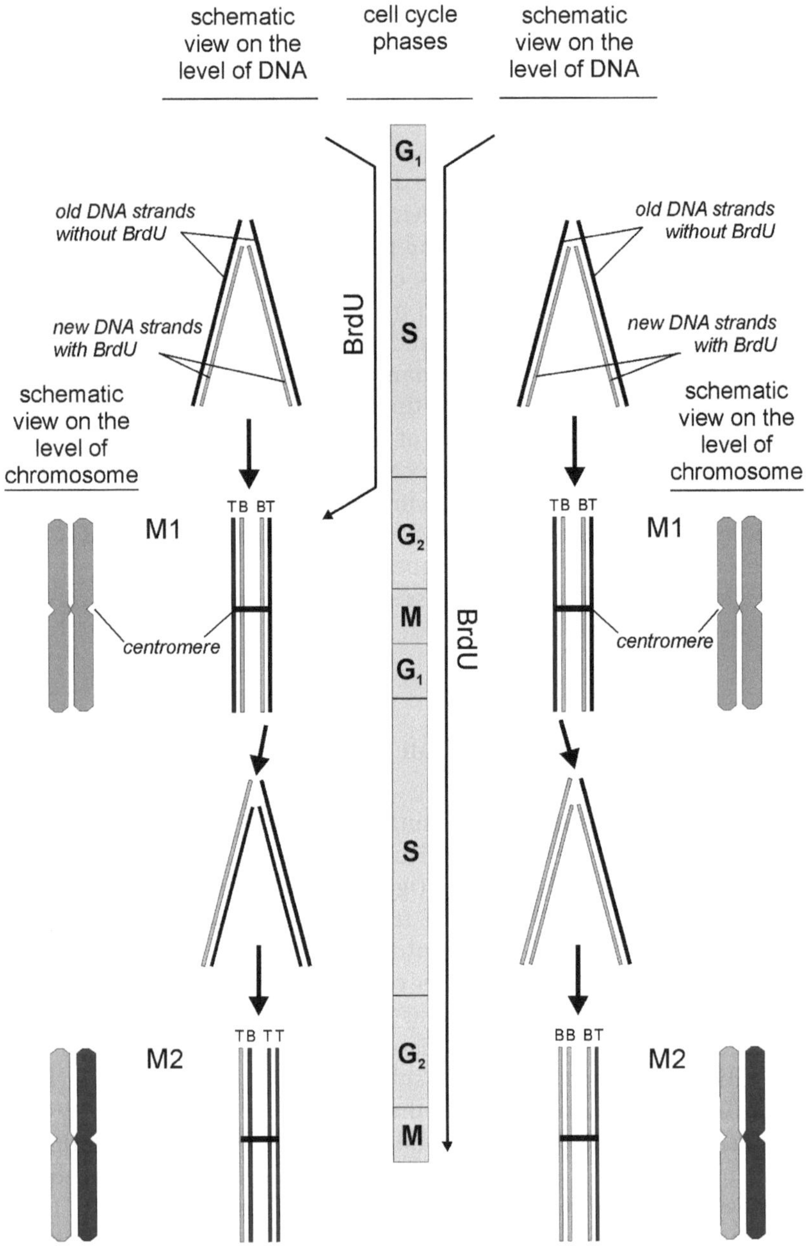

Fig. 17.2 Labelling scheme with BrdU of a cell which leads to differential staining of sister chromatids. *Left*: Labelling for one round of replication. *Right*: Labelling for two rounds of replication

per cell at a substitution level of 20%. A value of four SCEs was also calculated by Sutou (1981), who analysed ring chromosomes in CHO cells not substituted with BrdU. It has been suggested that the spontaneous value of about five SCEs per cell is an indirect manifestation of endogenous double-strand breaks (DSBs) in cellular DNA which are repaired by homologous recombination during replication (Vilenchik and Knudson 2003).

An alternative method of differential staining was developed by Bruckmann et al. (1999a, b). These authors electroporated CHO cells in the presence of biotin-2′-deoxyuridine-5′-triphosphate (dUTP), which was detected on microscope slides by immunological methods. The chemical structure of biotin-dUTP suggests that, unlike BrdU, it does not give rise to radicals. The SCE frequency of about six per cell was determined in all chromosomes, excluding X, Z10 and the small chromosomes Z12 and Z13, which were generally not labelled over their whole length. Since the omitted chromosomes comprise approximately 12% of the total chromosome length, the corrected, total genomic frequency of SCE was about seven. The higher level of SCE found with the biotin-dUTP method compared with the SCE level obtained with the lowest concentration of BrdU may indicate that biotin-dUTP induces a few additional SCEs, possibly by steric hindrance during DNA replication (Bruckmann et al. 1999a).

17.3 BrdU As a Source of DNA Damage Leading to SCE Formation

As discussed in the previous section a disadvantage of using BrdU for differential staining is that it induces SCEs by itself. The source of these SCEs is DNA damage, which arises owing to spontaneous or induced formation of uracilyl radicals and bromine atoms (Morris 1991; Fig. 17.3). One factor that causes the disintegration of BrdU is UV light (von Sonntag 2005). This is the reason why cell cultures should be kept in the dark following the addition of BrdU. An interesting question concerns the type of DNA damage which is responsible for SCEs in cells prelabelled with BrdU. The major photochemical

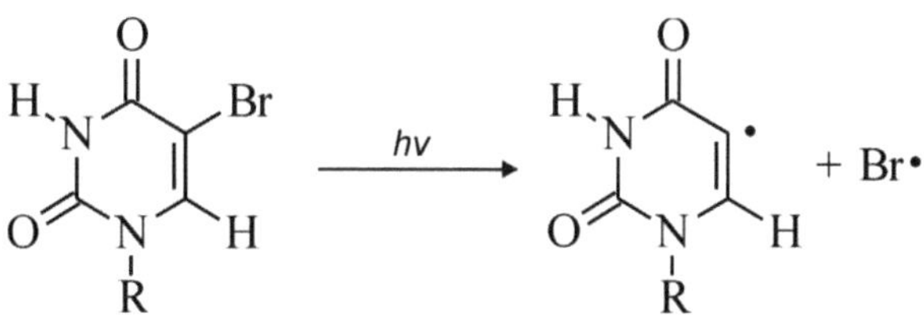

Fig. 17.3 Disintegration of 5-bromouracil leading to the formation of a uracilyl radical and a bromine atom

lesion in BrdU-substituted DNA is a single-strand break (SSB) (Hutchinson 1973); however, studies with the alkaline comet assay indicate that SSBs are repaired very quickly (Olive et al. 1990; Wojcik et al. 1996) and it was shown that UV radiation induces SCEs when cells prelabelled with BrdU are exposed in G1 phase of the cell cycle (Wojcik et al. 2003). Since SCEs are formed when DNA replication is stalled at a damaged template (Wolff et al. 1974; Painter 1980; Cortes et al. 1993; Wojcik et al. 2004), it appears unlikely that SSBs induced in G1 phase prevail in a non-repaired form until S phase, where they would give rise to SCEs. Furthermore, it has been shown that ionizing radiation, which gives rise to many SSBs, is a very poor inducer of SCEs (Littlefield et al. 1979). In fact it was demonstrated that following irradiation of CHO cells with X-rays, no SCEs are formed in cells not prelabelled with BrdU (Bruckmann et al. 1999b). This shows that it is the radiation damage to BrdU which is responsible for SCE formation. Considering the above, it appears unlikely that SSBs are the major lesions responsible for SCE formation.

Chemical agents that form DNA interstrand cross-links (ICLs) are among the most potent inducers of SCEs (Latt 1981). Wojcik et al. (2003) have shown that BrdU-induced damage is responsible for more than 80% of the SCEs formed in UV-irradiated cells unifilarily labelled with BrdU. Moreover, the authors pointed out that in DNA unifilarily labelled with BrdU and exposed to UV the bromine atom can, with a given likelihood, oxidize a purine base of the opposing DNA strand. This will yield two radicals at opposing strands (uracilyl radical and the oxidized purine) in the very neighbourhood, which is the precondition for an ICL (Wojcik et al. 2003; von Sonntag 2005). Indeed, the results of a recent study performed with the modified comet assay of Merk and Speit (1999) demonstrate that ICLs are formed in CHO cells unifilarily labelled with BrdU and exposed to UV radiation (Wojcik et al. 2006). This result was substantiated by the observation that CL-V4B cells exposed to BrdU and UV-C, but not to UV-C alone showed a very high level of chromosomal damage (Wojcik et al. 2006). CL-V4B cells have a deficient Rad51C paralogue that renders them extremely sensitive towards ICLs (Godthelp et al. 2002). Thus, it appears that ICLs induced by the disintegration of BrU are the source of SCEs.

17.4 The Influence of BrdU on the SCE Level Induced by Mutagenic Compounds

Owing to the lack of a straightforward alternative label to BrdU, only limited data exist on the impact of BrdU on the induction of SCEs by chemical and physical mutagens. Several authors reported that BrdU influences the SCE frequency induced by different mutagens (Popescu et al. 1980; Ockey 1981; Schvartzman and Tice 1982; Morgan and Wolff 1984; Zhao et al. 1992; Gonzalez-Beltran and Morales-Ramirez 2003). Given the fact that BrdU also inhibits the repair of DNA damage (Lawrence et al. 1995; Wang et al. 1994a, b) it may be difficult to judge whether its presence modifies the level of induced damage or the fidelity of repair.

From the considerations outlined in the previous section it can be assumed that BrdU will strongly enhance the action of agents which induce free-radical formation and promote the disintegration of BrU. As already mentioned, BrdU strongly enhances the SCE frequency induced by UV radiation (Wojcik et al. 2003). With respect to ionizing radiation, SCEs are only induced when cells prelabelled with BrdU are exposed (Bruckmann et al. 1999b). Stoilov et al. (2002) have applied the biotin-dUTP labelling technique to study the role of BrdU in the induction of SCEs by DNase I, Alu I and mitomycin C (MMC). CHO cells were treated in G1 phase of the cell cycle. Following treatment with Alu I, similar frequencies of SCEs were observed in cells prelabelled with BrdU and biotin-dUTP. DNase I induced significantly more SCEs in cells prelabelled with BrdU than with biotin-dUTP. MMC induced slightly more SCEs in cells labelled with biotin-dUTP, but the difference was not significant. The authors give two possible explanations for the observation that DNase I induced significantly more SCEs in cells prelabelled with BrdU than with biotin-dUTP. Firstly, it can be assumed that biotin-dUTP hinders the DNA incision activity of DNase I. The second explanation relies on the observations that DNase I induces SSBs which, when close together, lead to the formation of DSBs (Suck and Oefner 1986) and that the disintegration of BrU also may lead to the formation of SSBs (Morris 1991). If a SSB induced by DNase I lies in close proximity to a SSB induced by BrU, a DSB will be formed. In this sense, BrdU and DNase I would act synergistically. This assumes that DSB may be involved in SCE formation when formed in G1 phase of the cell cycle. The fact that restriction enzymes, neocarcinostatin and bleomycin effectively induce SCEs appears to support this (Natarajan et al. 1985; Obe et al. 1994; Schunck and Obe 1995). The mechanisms by which this occurs are not known, although newer data suggest that these SCEs are "false", in that they result from exchange-type chromosomal aberrations (Sayed Aly et al. 2002). The problem of "false" SCEs following treatment with S-phase-independent agents is discussed in the next section. In any case, no synergistic effect of BrdU would be expected with the DSB-inducing agent Alu I and the intercalating agent MMC, and this was confirmed experimentally (Stoilov et al. 2002).

17.5 The Problem of "False" and "True" SCE

Chemical mutagens, especially those capable of alkylating or cross-linking DNA, induce SCEs very effectively (Latt 1981; Perry and Thomson 1984). Such mutagens do not directly induce DSBs in the DNA and are described as "S-phase-dependent". The reason for this is that before DSBs are formed (which eventually leads to chromosomal aberrations), a cell must pass through an S phase. The chromosomal aberrations induced by S-phase-dependent agents are of chromatid-type (Savage 1975). In contrast, S-phase-independent agents are capable of directly inducing DSBs and they are a source of chromosome-type aberrations when cells are treated in G1 phase of the cell cycle. S-phase-independent agents are ionizing radiation, endonucleases such as DNase I, benzene nuclease, restriction endonucleases (RE) and a few antibiotics such as bleomycin or neocarcinostatin..

The observation that endonucleases induce SCEs was first reported by Natarajan et al. (1985) and Stoilov et al. (1986) for RE and Obe et al. (1994) for DNase I and benzene nuclease. Endonucleases are only effective in inducing SCEs when cells unifilarily labelled with BrdU are treated in G1 phase of the cell cycle (Morgan et al. 1989; Folle et al. 1992). The nature of DNA lesions responsible for SCE induction by endonucleases and other S-phase-independent agents has been a matter of debate because it is difficult to envisage how DSBs induced in G1 phase can lead to SCEs in S phase. Folle et al. (1992) proposed that apart from DSBs, RE could induce SSBs in canonical structures of chromosomal DNA or by cleavage of DNA–messenger RNA hybrids in G1 phase and that some of these SSBs could persist until S phase, where they would give rise to SCEs. If this were true, endonucleases should induce chromatid-type aberrations in cells treated in G1 phase; however, this is very rarely the case (Obe et al. 1987; Sayed Aly et al. 2002). In addition, Morgan et al. (1989) and Obe et al. (1994) have shown that when cells are treated with endonucleases in S phase, chromatid-type aberrations are induced but not SCEs.

The above considerations indicate that no other DNA lesions but the DSBs are the initial events leading to production of SCEs by endonucleases. Because an unrepaired DSB is a lethal event (Dikomey et al. 1998; Obe et al. 1992), cells possess efficient signalling and repair mechanisms preventing them from entering S phase before DSBs are repaired (Szumiel and Sochanowicz 1998; Pfeiffer et al. 2000; Obe et al. 2002). This clearly points towards G1 as the phase of the cell cycle in which SCE are produced by DSBs. However, how can DSBs lead to SCEs in G1 phase? A simple explanation is the assumption that the SCEs result from chromosomal exchange-type aberrations, and are therefore "false" SCEs (Fig. 17.4), as opposed to "true" SCEs, which are formed as a result of DNA recombination events during S phase. The existence of "false" SCEs was first postulated by Wolff et al. (1974) and Mühlmann-Diaz and Bedford (1995) for SCEs induced by X-rays. Bruckmann et al. (1999b) confirmed that only "false" SCEs are induced by X-rays when cells prelabelled with biotin-dUTP (but not BrdU: here true SCEs are formed owing to the disintegration of BrU – see Sect. 4) are exposed in G1 phase of the cell cycle. Sayed Aly et al. (2002) analysed the relationship between chromosomal aberrations and SCEs in CHO cells treated in G1 phase with DNase I and observed a very good, linear correlation between aberrations and SCEs. At the same time, no clear dose–response curve for SCEs was seen in cells without any chromosomal aberrations, suggesting that the SCEs induced by the endonuclease are all "false".

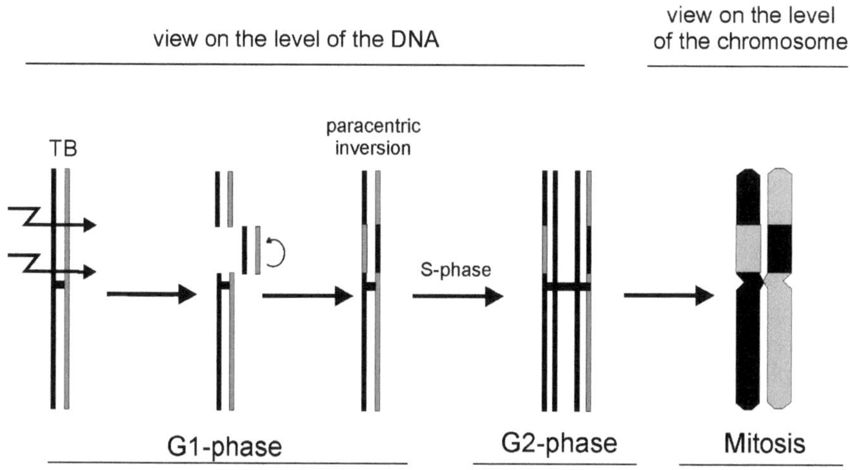

Fig. 17.4 How a paracentric inversion induced in G1 phase of the cell cycle can lead to the formation of a "false" sister-chromatid exchange

17.6 The Problem of Scoring SCEs in the Second Posttreatment Mitoses

SCEs can be visualized by growing cells for either two rounds of replication in the presence of BrdU or for one round with BrdU and the next without (Fig. 17.2). Both experimental schemes have advantages and disadvantages. If one wishes to analyse the first posttreatment mitoses, cells unifilarily labelled with BrdU must be treated. The disadvantage of this scheme is that, as discussed already, BrdU can influence the SCE-inducing capacity of the mutagen tested.

If BrdU is added simultaneously, or after treating cells with a DNA-damaging agent, the effect on SCEs can only be analysed in the second post-treatment mitoses. Two main problems are associated with this scheme. One problem is that most mutagens induce both SCEs and chromosomal aberrations and this may lead to mitotic death of cells when the mutagen is added together with BrdU (Carrano and Heddle 1973; Kirkland 1992). Thus, by scoring second posttreatment metaphases, one cannot be sure as to the extent of the elimination of damaged cells before they can be analysed for the presence of SCEs. This can be a source of bias in the estimation of the frequency of induced SCEs.

Another problem is related to the observation that DNA lesions induced by mutagens can persist for a number of cell cycles, giving rise to an enhanced SCE frequency during several posttreatment mitoses. This effect was studied extensively by Kaina and Aurich (1985), who pulse-treated V79 cells with different alkylating agents and scored SCEs in the first (M1), second (M2) and third (M3) posttreatment mitoses. The authors observed that following treatment with the methylating agents *N*-methyl-*N*-nitrosourea, *N*-methyl-*N'*-nitro-*N*-nitrosoguanidine and 1,3-dimethyl-3-phenyl-1-nitrosourea high SCE frequencies were detected in M2 and M3 at doses which induced no SCEs in M1. Ethyl methanesulphonate (EMS), *N*-ethyl-*N*-nitrosourea and methyl methanesulphonate (MMS) induced the highest SCE frequency in M2, followed by M3 and M1. MMC induced similar SCE frequencies in M1 and M2, a result very similar to that observed by Wojcik et al. (2005). An enhanced frequency of SCEs over two cell cycles following treatment with MMS and EMS was also reported by Gonzalez-Beltran and Morales-Ramirez (2003). Although nothing is known about the nature of the persistent lesions, their occurrence points out the necessity of using a uniform experimental protocol when planning interlaboratory comparison studies.

17.7 Conclusions

The SCE test is a relatively simple assay, sensitive to most mutagenic compounds. Its major disadvantage is that in order to detect SCEs, cells must be either treated with the label for one round of replication prior to treatment with the mutagen studied or treated simultaneously, but fixed and harvested following two rounds of replication. Thus, one can never be sure if the label used (usually BrdU) does not interact with the mutagen studied, or if the scoring of the second posttreatment mitoses is not a source of bias in the estimation of the SCE frequency. These factors must always be considered in the interpretation of the results obtained.

References

Bender MA, Preston RJ, Leonard RC, Pyatt BE, Gooch PC (1992) On the distribution of spontaneous SCE in human peripheral blood lymphocytes. Mutat Res 281:227–232

Bruckmann E, Wojcik A, Obe G (1999a) Sister chromatid differentiation with biotin-dUTP. Chromosome Res 7:185–189

Bruckmann E, Wojcik A, Obe G (1999b) X-irradiation of G1 CHO cells induces SCE which are both true and false in BrdU-substituted cells but only false in biotin-dUTP-substituted cells. Chromosome Res 7:277–288

Carrano AV, Heddle JA (1973) The fate of chromosome aberrations. J Theor Biol 38:289–304

Cortes F, Pinero J, Ortiz T (1993) Importance of replication fork progression for the induction of chromosome damage and SCE by inhibitors of topoisomerases. Chromosome Res 4:540–544

Dikomey E, Dahm-Daphi J, Brammer I, Martensen R, Kaina B (1998) Correlation between cellular radiosensitivity and non-repaired double-strand breaks studied in nine mammalian cell lines. Int J Radiat Biol 73:269–278

Folle GA, Johannes C, Mechoso BH, Medina II, Obe G (1992) The restriction endonuclease AluI induces sister chromatid exchange in Chinese hamster ovary cells. Mutagenesis 7:291–294

Godthelp BC, Wiegant WW, van Duijn-Goedhart A, Schaerer OD, van Buul PPW, Kanaar R, Zdzienicka MZ (2002) Mammalian Rad51C contributes to DNA cross-link resistance, sister chromatid cohesion and genomic stability. Nucleic Acids Res 30:2172–2182

Gonzalez-Beltran F, Morales-Ramirez P (2003) Repairability during G1 of lesions eliciting sister chromatid exchanges induced by methylmethanesulfonate or ethylmethanesulfonate in bromodeoxyuridine-substituted and unsubstituted DNA strands. Mutagenesis 18:13–17

Goto K, Maeda S, Kano Y, Sugiyama T (1978) Factors involved in differential giemsa-staining of sister chromatids. Chromosoma 66:351–359

Holmquist GE (1987) The magic of cytogenetic staining. In: Obe G, Basler A (eds) Cytogenetics. Springer, Berlin, pp 30–47

Hutchinson F (1973) The lesions produced by ultraviolet light in DNA containing 5-bromouracil. Q Rev Biophys 6:201–246

Kaina B, Aurich O (1985) Dependency of the yield of sister-chromaid exchanges induced by alylating agents on fixation time. Possible involvement of secondary lesions in sister-chromatid exchange induction. Mutat Res 149:451–461

Kirkland D. (1992) Chromosomal aberration tests in vitro: problems with protocol design and interpretation of results. Mutagenesis 7:95–106

Latt S (1973) Microfluorometric detection of deoxyribonucleic acid replication in human metaphase chromosomes. Proc Natl Acad Sci USA 70:3395–3399

Latt S (1981) Sister chromatid exchange formation. Annu Rev Genet 15:11–55

Lawrence TS, Davis MA, Normolle DP (1995) Effect of bromodeoxyuridine on radiation-induced DNA damage and repair based on DNA fragment size using pulsed-field gel electrophoresis. Radiat Res 144:282–287

Littlefield LG, Colyer SP, Joiner EJ, DuFrain RJ (1979) Sister chromatid exchanges in human lymphocytes exposed to ionizing radiation during G_0. Radiat Res 78:514–521

McClintock B (1938) The production of homozygous deficient tissues with mutant characteristics by means of aberrant mitotic behaviour of ring-shaped chromosomes. Genetics 23:315–376

Merk O, Speit G (1999) Detection of crosslinks with the comet assay in relationship to genotoxicity and cytotoxicity. Environ Mol Mutagen 33:167–172

Morgan WF, Wolff S (1984) Effect of 5-bromodeoxyuridine substitution on sister chromatid exchange induction by chemicals. Chromosoma 89:285–289

Morgan WF, Chung HW, Phillips JW, Winegar RA (1989) Restriction endonucleases do not induce sister-chromatid exchanges in Chinese hamster ovary cells. Mutat Res 226:203–209

Morris SM (1991) The genetic toxicology of 5-bromodeoxyuridine in mammalian cells. Mutat Res 258:161–188

Mühlmann-Diaz MC, Bedford JS (1995) Comparison of gamma-induced chromosome ring and inversion frequencies. Radiat Res 143:175–180

Natarajan AT, Mullenders LHF (1987) Sister chromatid exchanges. In: Obe G, Basler A (eds) Cytogenetics. Springer, Berlin, pp 339–344

Natarajan AT, Mullenders LHF, Meijers M, Mukherjee M (1985) Induction of sister-chromatid exchanges by restriction endonucleases. Mutat Res 144:33–39

Natarajan AT, Rotteveel AHM, van Pieterson J, Schliermann MG (1986) Influence of incorporated 5-bromodeoxyuridine on the frequencies of spontaneous and induced sister-chromatid exchanges, detected by immunological methods. Mutat Res 163:51–55

Obe G, Vasudev V, Johannes C (1987) Chromosome aberrations induced by restriction endonucleases. In: Obe G, Basler A (eds) Cytogenetics. Springer, Berlin, pp 300–314

Obe G, Johannes C, Schulte-Frohlinde D (1992) DNA double-strand breaks induced by sparsely ionizing radiation and endonucleases as critical lesions for cell death, chromosomal aberrations, mutations and oncogenic transformation. Mutagenesis 7:3–12

Obe G, Schunck C, Johannes C (1994) Induction of sister-chromatid exchanges by ALU I, DNase I, benzon nuclease and bleomycin in Chinese hamster ovary (CHO) cells. Mutat Res 307:315–321

Obe G, Pfeiffer P, Savage JRK, Johannes C, Goedecke W, Jeppesen P, Natarajan AT, Martinez-Lopez W, Folle GA, Drets M (2002) Chromosomal aberrations: formation, identification and distribution. Mutat Res 504:17–36
Ockey CH (1981) Methyl methane-sulfonate (MMS) induced SCEs are reduced by the BrdU used to visualize them. Chromosoma 84:243–256
Olive PL, Banath JP, Durand RE (1990) Heterogeneity in radiation-induced DNA damage and repair in tumor and normal cells measured using the "comet" assay. Radiat Res 122:86–94
Painter RB (1980) A replication model for sister-chromatid exchange. Mutat Res 70:337–341
Perry PE, Thomson EJ (1984) The methodology of sister chromatid exchanges. In: Kilbey BJ, Legator M, Nichols W, Ramel C (eds), Handbook of mutagenicity test procedures. Elsevier, Amsterdam, pp 495–529
Perry P, Wolff S (1974) New Giemsa method for differential staining of sister chromatids. Nature 251:156–157
Pfeiffer P, Goedecke W, Obe G (2000) Mechanisms of DNA double-strand break repair and their potential to induce chromosomal aberrations. Mutagenesis 15:289–302
Pinkel D, Thompson LH, Gray JW, Vanderlaan M (1985) Measurement of sister chromatid exchanges at very low bromodeoxyuridine substitution levels using a monoclonal antibody in Chinese hamster ovary cells. Cancer Res 45:5795–5798
Popescu NC, Amnnbaugh SA, DiPaolo JA (1980) Reduced N-methyl-N-nitro-N-nitrosoguanidine sister chromatid exchange induction in Chinese hamster V79 cells pre-exposed to 5-bromodeoxyuridine. Chromosoma 76:329–338
Savage JRK (1975) Classification and relationship of induced chromosomal structural changes. J Med Genet 12:103–122
Sayed Aly M, Wojcik A, Schunck C, Obe G (2002) Correlation of chromosomal aberrations and sister chromatid exchanges in individual CHO cells pre-labelled with BrdU and treated with DNaseI or X-rays. Int J Radiat Biol 78:1037–1044
Schunck C, Obe G (1995) Neocarzinostatin induces chromosomal aberrations and sister chromatid exchanges in Chinese hamster ovary cells. Mutagenesis 10:37–42
Schvartzman JB, Tice RR (1982) 5-Bromodeoxyuridine and its role in the production of sister chromatid exchange. In: Sandberg AA (ed) Sister chromatid exchange. Liss, New York, pp 123–134
Stoilov L, Mullenders LHF, Natarajan AT (1986) Influence of bromodeoxyuridine substitution of thymidine on sister-chromatid exchanges and chromosomal aberrations induced by restriction endonucleases. Mutat Res 174:295–301
Stoilov L, Wojcik A, Giri A, Obe G (2002) SCE formation after exposure of CHO cells pre-labelled with BrdU or biotin-dUTP to various DNA-damaging agents. Mutagenesis 17:399–403
Suck D, Oefner C (1986) Structure of DNase I at 2.0 A resolution suggets a mechanism for binding to and cutting DNA. Nature 321:620–625
Sutou S (1981) Spontaneous sister-chromatid exchanges in Chinese hamster cells in vivo and in vitro. Mutat Res 82:331–341
Szumiel I, Sochanowicz B (1998) Signal transfer in the cellular response to ionizing radiation. Nukleonika 43:133–146
Taylor JH, Woods PS, Hughes ML (1957) The organisation and duplication of chromosomes as revealed by autoradiographic studies using tritium-labeled thymidine. Proc Natl Acad Sci USA 43:122–128
Vilenchik MM, Knudson AG (2003) Endogenous DNA double-strand breaks: production, fidelity of repair, and induction of cancer. Proc Natl Acad Sci USA 100:12871–12876
von Sonntag C (2005) Free-radical-induced DNA damage and its repair. A chemical perspective. Springer, Heidelberg
Wang Y, Pantelias GE, Iliakis G (1994a) Mechanism of radiosensitization by halogenated pyrimidines: the contribution of excess DNA and chromosome damage in BrdU radiosensitization may be minimal in plateau-phase cells. Int J Radiat Biol 66:133–142

Wang Y, Pantelias GE, Iliakis G (1994b) The effect of scavenging of eaq- by acetone on the survival of CHO cells prelabelled with BrdU and exposed to X-rays. Radiat Res 66:133–142
Wojcik A, Sauer K, Zölzer F, Bauch T, Müller W-U (1996) Analysis of DNA damage recovery processes in the adaptive response to ionizing radiation in human lymphocytes. Mutagenesis 11:291–297
Wojcik A, von Sonntag C, Obe G (2003) Application of the biotin-dUTP chromosome labelling technique to study the role of 5-bromo-2′-deoxyuridine in the formation of UV-induced sister chromatid exchanges in CHO cells. J Photochem Photobiol B 69:139–144
Wojcik A, Bruckmann E, Obe G (2004) Insights into the mechanisms of sister chromatid exchange formation. Cytogenet Genome Res 104:304–309
Wojcik A, Stoilov L, Szumiel I, Legerski RJ, Obe G (2005) Rad51C-deficient CL-V4B cells exhibit normal levels of mitomycin C-induced SCEs but reduced levels of UVC-induced SCEs. Biochem Biophys Res Commun 326:805–810
Wojcik A, Bochenek A, Lankoff A, Lisowska H, Padjas A, Szumiel I, von Sonntag C, Obe G (2006) DNA interstrand crosslinks are induced in cells prelabelled with 5-bromo-2-deoxyuridine and exposed to UVC radiation. J Photochem Photobiol B 84:15–20
Wolff S, Perry P (1975) Insights on chromosome structure from sister chromatid exchange ratios and the lack of both isolabelling and heterolabelling as determined by the FPG technique. Exp Cell Res 93:23–30
Wolff S, Bodycote J, Painter RB (1974) Sister chromatid exchanges induced in Chinese hamster cells by UV irradiation of different stages of the cell cycle: the necessity for cells to pass through S. Mutat Res 25:73–81
Zhao JH, Tohda H, Oikawa A (1992) Camptothecin-induced sister chromatid exchange is dependent of the presence of bromodeoxyuridine and the phase of the cell-cycle. Mutat Res 282:49–54

18 Fluorescence in Situ Hybridisation

ILSE CHUDOBA

Abstract

The method of fluorescence in situ hybridisation (FISH) is based on the fundamental property of DNA strands to anneal to one another in a complementary fashion. However, there are different classes of DNA probes available for FISH investigations, but not all of them are suitable for all targets. Locus-specific probes are derived from single-copy DNA and produce a specific signal. In contrast, painting probes are heterogeneous DNA libraries, composed of different DNA classes, namely single copy sequences as well as repetitive sequences. Furthermore, DNA probes constituted of repetitive sequences only target specific structures of the chromosomes, i.e. the centromeres and the telomeres. All these probes have in common that they are directly or indirectly labelled and after hybridisation are visualised under an epifluorescence microscope. Comparative genomic hybridisation (CGH), a method to determine genetic imbalances with a resolution between 2 and 10 Mb, uses normal metaphase spreads as the target and the fluorescently labelled test DNA is hybridised together with differently labelled normal reference DNA. The resolution of CGH has been improved tremendously by the introduction of array-CGH, where the target chromosomes are replaced by DNA from, for example, bacterial artificial chromosomes with a spacing of 1 Mb and beyond. FISH has developed into a valuable tool for genetic investigations in clinical as well as tumour cytogenetics for detection of microdeletions, amplifications, specific fusion genes and others.

18.1 Introduction

Fluorescence in situ hybridisation (FISH) was introduced in 1986 (Pinkel et al. 1986). In general, FISH is applied to slides prepared from standard cytogenetic cell suspensions, but it can be carried out on tissue sections, blood or bone marrow smears as well. The method of FISH is based on the fundamental property of DNA strands to anneal to one another in a complementary fashion (complementary base pairing). Owing to this basic feature of single-stranded DNA, a specific DNA probe will hybridise to its complementary target DNA

Günter Obe and Vijayalaxmi (Eds.)
Chromosomal Alterations: Methods, Results and Importance in Human Health

sequence, no matter whether it is present in a metaphase chromosome, an interphase nucleus or as naked filamentary DNA (fibre-FISH).

There are different classes of DNA probes available for FISH investigations, but not all of them are suitable for all targets. For example, a whole chromosome painting probe is constituted of a library of different DNA sequences from the same chromosome, but specific for a variety of different loci on the respective chromosome. Such a probe is informative on metaphase chromosomes, but will contribute only little to no information on a DNA fibre.

FISH is carried out for different purposes. By using locus-specific probes, the presence or absence of a specific DNA sequence can be determined, or specific DNA probes can be assigned to particular chromosomal sites. In clinical cytogenetics, FISH investigations provide clarification of aberrations that were observed using standard cytogenetic techniques (GTG-banding). Furthermore, FISH is able to detect cryptic or submicroscopic chromosomal aberrations (e.g. microdeletions).

18.2 The Different Classes of DNA Probes

The human DNA is composed of different DNA classes (Table 18.1), which are of variable abundance. The basic nature of these DNA classes is of importance for the technical aspect of FISH experiments. A single copy sequence is present only once in the human haploid genome and, thus, will produce a specific signal.

In contrast, painting probes are heterogeneous DNA libraries and include all DNA classes. In order to avoid cross-hybridisation with repetitive DNA

Table 18.1 The different DNA classes of the human genome

Class	Copy number	Distribution	Percentage of the human genome
Single copy sequence	1	Unique copies throughout the whole genome	50
Low copy number repeat	2–20	Mainly throughout the euchromatin	10
Moderately repeated DNA	50–5,000	Scattered throughout the entire genome	25
Highly repeated DNA	10,000–500,000	Clustered in heterochromatin, like satellite DNA but also interspersed throughout the genome	15

and to improve the specificity of hybridisation, a so-called preannealing step using Cot-1 DNA has to be performed (Cremer et al. 1990).

Cot-1 DNA represents highly repetitive sequences among the nuclear DNA. It is obtained from total genomic DNA by selecting for rapidly reassociating DNA sequences during renaturation of DNA. During the preannealing step, Cot-1 DNA blocks repetitive sequences in the probe and, thus, prevents hybridisation labelled probe sequences to the target DNA.

18.3 Chromosome Painting Probes

Chromosome painting probes are DNA libraries which recognise specifically the sequences spanning the length of a particular chromosome (Cremer et al. 1988). They can be generated either by flow sorting or by microdissection.

In the case of flow sorting, a dispersed suspension of chromosomes is stained with fluorochromes and passed through a flow cytometer. The sorting is carried out according to the fluorescence intensity of each chromosome, which depends on its DNA content and the AT-to-GC ratio.

Whole chromosomes can be isolated from metaphase spreads by using a technique which is similar to microdissection of chromosomes. This can be either laser-mediated (Monajembashi et al. 1986) or carried out using extended glass needles (Senger et al. 1990; Chudoba and Senger 2002). In both cases, the chromosomal DNA has to be amplified subsequently. A powerful method for representative amplification of small amounts of DNA of which the specific sequences are unknown is the degenerate oligonucleotide-primed polymerase chain reaction (PCR) method (Telenius et al. 1992). Here, the complete DNA is amplified in a PCR with a set of degenerate primers. Consequently, not only single copy sequences are present in the final product, but all kinds of repetitive DNA as well. Therefore, the preannealing with Cot-1 DNA is particularly important when hybridising chromosome painting probes.

Chromosome painting probes can be specific for either a complete chromosome or only parts of it, like arm-specific partial painting probes or even smaller parts such as region-specific or band-specific painting probes. Whole chromosome painting probes hybridise to both chromosome homologues in metaphases. If a translocation is present, the FISH signal of one chromosome is split and distributed among the two translocation partners (Fig. 18.1a). In clinical cytogenetics and tumour cytogenetics, whole chromosome painting probes are mainly used in order to confirm results obtained by other cytogenetic methods.

A more sophisticated use of whole chromosome painting probes was implemented by the methods of multicolour karyotyping: multiplex FISH (mFISH) (Speicher et al. 1996) and spectral karyotyping (Schröck et al. 1996). Both approaches use different image capturing and analysis strategies,

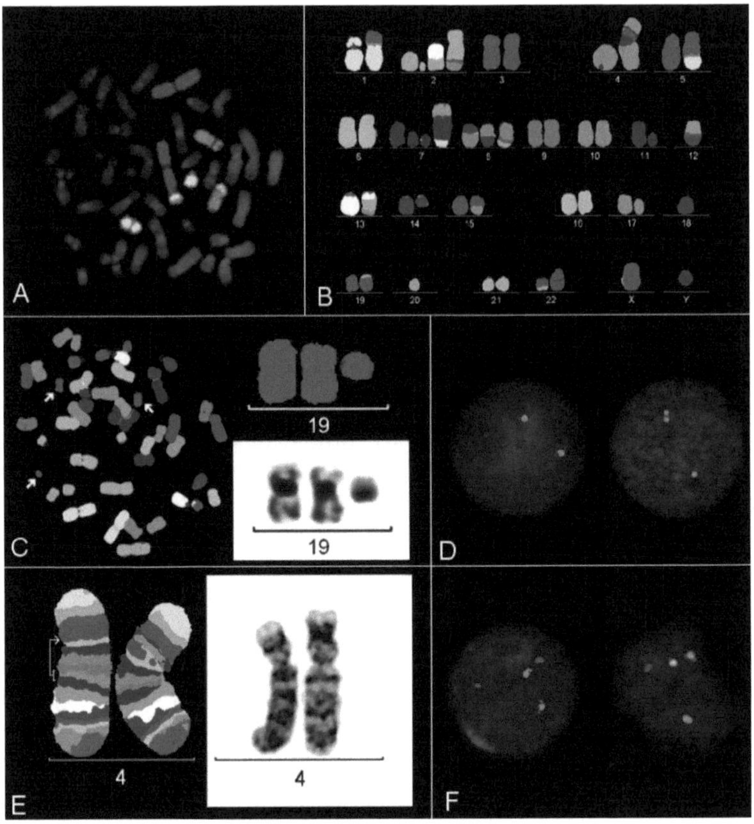

Fig. 18.1 **a** A metaphase after hybridisation with whole chromosome painting probes. Chromosome 1 is labelled in *red*, chromosome 12 in *green* and chromosome 16 in *red* and *green*, resulting in a *yellow* signal. A complex translocation occurred in which fragments of chromosomes 12 and 16 were transferred to chromosome 1 and a fragment from 16 was transferred to chromosome 12. **b** A karyogram in false colour display after multiplex fluorescence in situ hybridisation (FISH). Multiple chromosomal aberrations are observed. **c** Another example for multicolor FISH, where a supernumerary marker chromosome was identified as a derivative of chromosome 19. The metaphase spread and the respective chromosomes 19 in false colour (*upper row*) and inverted 4′,6-diamidino-2-phenylindole display (*lower row*) are shown. **d** Two normal interphase nuclei showing hybridisation signals of a locus-specific probe labelled with fluorescein isothiocyanate (*green spots*). Both nuclei show two spots; however, one of the signals in the nucleus on the *right* is split into two separate spots. **e** A chromosomal aberration of chromosome 4. The original interpretation after GTG-banding (*right*) explained the derivative chromosome 4 as a pericentric inversion. Multicolour banding analysis clearly revealed that a region from the long arm of chromosome 4, comprising the *grey band* and parts of the *purple band* (indicated by the *square brackets* on the chromosome normal 4, *left*), was excised and reinserted into the short arm of the same chromosome, which results in a splitting of the red colour band. **f** Two interphase nuclei after hybridisation with a *BCR/ABL* dual-colour dual-fusion probe. *Left*: A normal signal constellation. *Right*: A fusion between *BCR* and *ABL* is indicated by two fusion signals

but very similar probe kits. They are mixtures of whole chromosome painting probes for all chromosomes and each whole chromosome painting probe is labelled with a combination of different fluorochromes. For a human mFISH kit, five different fluorochromes are necessary to provide a minimum of 24 different combinations (Table 18.2). Multicolour karyotyping has been shown to be a valuable tool in the precise characterisation of complex chromosomal aberrations (Schoch et al. 2002, Fig. 18.1b, c).

Intrachromosomal aberrations cannot be detected using whole chromosome painting probes. For a more specific analysis of, for example, pericentric inversions, hybridisation with arm-specific probes is more suitable. The original

Table 18.2 Example of a labelling scheme for a probe kit for multicolour karyotyping. Here, the scheme for the probe kit 24 XCyte (MetaSystems, Germany) is shown

Chromosome number	DEAC	FITC	Spectrum Orange™	Texas Red®	Cy™5
1					X
2	X				
3				X	
4		X			
5			X		
6		X			X
7	X				X
8				X	X
9			X		X
10	X	X			
11		X		X	
12		X	X		
13	X			X	
14	X		X		
15			X	X	
16	X	X			X
17		X		X	X
18		X	X		X
19	X			X	X
20	X		X		X
21			X	X	X
22	X	X		X	
X	X	X	X		
Y	X		X	X	

DEAC (diethylamino)coumarin, *FITC* fluorescein isothiocyanate pattern of a normal chromosome with its short arm labelled in red and long arm in green will results in an alternating pattern of red–green–red–green.

Microdissection of region-specific partial chromosome painting probes is the basis for a chromosome-specific multicolour banding, details of which were first published in 1999 (Chudoba et al. 1999). A specific feature of these probe sets is that each partial chromosome painting probe overlaps with its adjacent partial chromosome painting probe. They are labelled with different fluorochromes and are hybridised together. Owing to the overlapping of neighbouring probes, a multitude of different fluorescence intensity ratios are obtained. Fluorescence intensity ratios within defined regions are measured and a false colour is assigned to a certain ratio, which leads to a characteristic and reproducible colour banding pattern (Fig. 18.1e). Multicolour banding analyses are carried out in order to precisely define chromosomal aberrations in clinical (Lemke et al. 2001) and tumour (Lestou et al. 2003) cytogenetics. Furthermore, it was demonstrated that this method provides important information in experimental cytogenetics (Johannes et al. 2004).

18.4 FISH Probes Composed of Repetitive DNA Sequences

There are two different kinds of probes that hybridise to specific chromosomal structures featuring repetitive sequences, namely the centromeres and the telomeres of chromosomes. A preannealing step with Cot-1 DNA should be avoided, because this would suppress the sequences of interest.

18.4.1 Centromere-Specific Probes

Normal human centromeres consist of megabases of alpha or beta satellite DNA; a repeat family based on monomers that are approximately 171-bp long (Willard and Waye 1987). Centromere-specific FISH probes hybridise to these highly repetitive sequences.

Centromere probes are available for most human chromosomes; however, owing to the minimal differences in the centromere DNA sequences of chromosomes 5 and 19, 13 and 21 as well as 14 and 22, there cannot be distinguished by FISH. Usually, alpha/beta satellite centromere probes yield a large, bright signal which makes them useful for chromosome identification in metaphase preparations as well as chromosome enumeration in interphase nuclei.

Examples for hybridisations with centromere-specific probes are the detection of aneuploidies in interphase nuclei, the definition of the mosaic status of sex chromosome aneuploidies, the monitoring of bone marrow transplant in sex mismatch bone marrow transplant and the identification of the origin of marker chromosomes.

18.4.2 Telomere-Specific Probes

Telomeres are the tips of eukaryotic chromosomes. They have important functions in the protection, replication and stabilisation of chromosome ends. In most organisms studied, telomeres are composed of short repeated DNA sequences. These repeats are highly conserved; in fact all vertebrates appear to have the same simple sequence repeat in telomeres: $(TTAGGG)_n$ (Blackburn 1991).

The progressive shortening of telomeres is an important component of senescence and is involved in the control of the cell cycle (Dahse et al. 1997). In cancer cells, the telomere shortening is overcome, resulting in more or less unlimited cell divisions.

For quantification of telomeres, specific probes are available which comprise the aforementioned telomere-specific repeat. Advanced imaging systems are used in order to quantify the fluorescence signals of telomeres.

18.5 Locus-Specific Probes

Unique sequence probes hybridise to single copy DNA sequences in a specific chromosomal region or gene. This kind of probe is derived from cloned sequences. The different cloning vectors differ with respect to the size of the DNA fragment they can harbour, the cloning efficiency and further aspects such as stability and extractability of the DNA (Landegent et al. 1987; Sambrook et al. 2001).

The size and the intensity of fluorescence signals depend on the size of the DNA fragment that was cloned (Table 18.3). FISH probes based on yeast artificial chromosome result in signals that comparable to those of centromere signals in interphase nuclei regarding size and fluorescence intensity. On metaphase chromosomes the signals appear as separated spots on each chromatid. In the interphase state, one spot is usually observed for each locus. If the same locus is present on both chromosome homologues, two spots are observed in interphases. Deletions are indicated by a loss of a spot, and accordingly a gain of the same locus is indicated by an extra spot. Depending

Table 18.3 Types of vectors suitable for cloning

Vector		Maximum size of insert (kb)
YAC	Yeast artificial chromosome	1,000
BAC	Bacterial artificial chromosome	300
PAC	P1 (bacteriophage) derived artificial chromosome	100
Cosmid	A plasmid vector that carries COS sites	45

on the state of the cell cycle, even normal cells occasionally exhibit two separated spots that are derived from a normal locus. This indicates that the cell has undergone the S phase of the cell cycle and the respective region has been duplicated already (Fig. 18.1d).

18.6 Comparative Genomic Hybridisation and Array Comparative Genomic Hybridisation

The DNA probes described already are generally hybridised onto metaphase spreads or interphase nuclei prepared from a patient's blood or other tissues. However, in tumour cytogenetics, especially of solid tumours, the number of metaphases is often limited and the chromosomal morphology quite poor. Here comparative genomic hybridisation (CGH) can provide information about gains and losses of chromosomes and chromosomal regions (Kallioniemi et al. 1992). In contrast to standard cytogenetics techniques, dividing cells for metaphase preparation are not required. The DNA necessary for the experiment can be isolated from fresh or frozen tumour samples or even from archival formalin-embedded tissue samples (Speicher et al. 1993).

The basic prerequisites are the fluorescently labelled DNA from the tumour (test DNA) and a differently labelled normal DNA (control DNA). The two DNA fractions are mixed together in equal proportions and hybridised to metaphase spreads from normal tissue. Test and control DNA compete with each other for hybridisation sites on the target chromosomes. Adapted imaging systems compare the fluorescence intensities of the two different fluorochromes along the individual chromosomes. The resulting ratios are usually displayed as profiles and compared to the ISCN ideogram. Profiles exceeding defined limits in one or the other direction indicate gains and losses for the respective chromosomal regions of the tumour sample.

Even though the original idea of CGH was to gain more insight into tumours that are inaccessible with standard cytogenetic techniques, CGH proved to be a valuable tool as well in the determination of imbalanced constitutional chromosome aberrations (Tönnies et al. 2001). Not only the chromosomal origin of supernumerary marker chromosomes or de novo unbalanced translocations can be defined, but even the band-specific origin can be determined. The major limitation of CGH is the fact that only chromosomal imbalances can be detected and balanced aberrations remain undetected. Furthermore, the presence of normal cells mixed with the tumour tissue or the existence of a supernumerary marker as a mosaic reduce the effect and might even hide the actual presence of a chromosomal imbalance.

In experienced hands the resolution of CGH is rated to be between 3 and 10 Mb (Kirchhoff et al. 1999; Tönnies et al. 2001). Factors influencing the resolution are the quality of the hybridisation (signal background ratio), the quality of the DNA (homogeneity of the signals) and the condensation grade of the target chromosomes.

In order to improve the resolution of CGH recent developments resulted in the establishment of the so-called array-CGH or matrix-CGH (Solinas-Toldo et al. 1997; Pinkel et al. 1998). Metaphase spreads are replaced by specific DNA sequences, spotted in arrays on glass slides. The procedure is quite similar to the already described method of CGH: test DNA and control DNA are hybridised to the array. For genomic analysis bacterial artificial chromosome (BAC) arrays are used in which DNA from human BACs are spotted. The resolution depends on the number of BACs used and the spacing between neighbouring ones. A clone set has been established which comprises more than 2,500 BACs and which results in a resolution of approximately 1 Mb across the whole genome (Fiegler et al. 2003).

18.7 Probe Labelling

In order to make a probe suitable for FISH, the DNA has to be labelled. Labelling can be carried out by use of so-called reporter molecules (indirect labelling) or fluorescent dyes (direct labelling).

18.7.1 Indirect Labelling

In indirect labelling, a probe DNA is marked with reporter molecules. This is achieved by synthesising the complementary DNA strand in the presence of nucleotides conjugated with a reporter molecule. Common reporter molecules are biotin, digoxigenin, oestrogen and dinitrophenyl. The labelled nucleotides are introduced by nick-translation (Langer et al. 1981) or PCR. After hybridisation an indirectly labelled probe has to be detected either with specific antibodies carrying a fluorescent dye or, in the case of biotin, with fluorescently labelled avidin or streptavidin. These are proteins isolated from chicken egg white and have a particular specific affinity to biotin. Both primary and secondary antibodies have been used for the detection of indirectly labelled FISH probes. A primary antibody directly detects the incorporated reporter molecule. If a suitable antibody conjugate is not available, secondary antibodies can be used. In this case, the secondary antibody carries the fluorophore and recognises the unlabelled primary antibody. A major disadvantage of indirect labelling is the time-consuming detection after the hybridisation. However, a clear advantage is the signal amplification, which is especially favourable for relatively small probes (only several kilobases).

18.7.2 Direct Labelling

In the case of direct labelling, fluorescent molecules are directly attached to the DNA. This is the method of choice for most applications, because it is a straightforward and very convenient procedure. After the hybridisation only a stringent wash has to be carried out in order to eliminate the surplus of probe DNA.

18.8 The Hybridisation Procedure

A probe is hybridised to its target DNA under conditions that allow reannealing of specific complementary DNA strands. It is essential that the probe and target DNAs are denatured prior to hybridisation.

The hybridisation temperature is determined by the melting point of DNA. The melting point (T_m) of a given DNA is defined as the temperature where half of the DNA is present in a single-strand form (Sambrook et al. 2001). T_m depends on the GC content of the DNA: the higher the molar ratio of GC, the higher T_m. Sodium ions (monovalent cations) interact electrostatically with nucleic acids, i.e. salt concentrations below 0.4 M decrease the stability of the double-strand form. Divalent cations, on the other hand, strongly stabilise double-strand DNA, and they can be complexed by adding citrate to the hybridisation mixture.

DNA denatures at 90-100°C in presence of 0.1–0.2 M Na+. However, the morphology of target cells or chromosomes is better retained at lower temperatures; therefore, denaturation is usually carried out in the presence of the chaotropic compound formamide. T_m of double-strand DNA is reduced by 0.72°C for each percent of formamide; however, the rate of renaturation decreases in the presence of formamide (Casey and Davidson 1977).

Other factors influencing hybridisation are pH, the probe DNA fragment length and the probe concentration. The rate of hybridisation is relatively independent of the pH in the range between pH 5 and pH 9. However, some of the fluorochromes (e.g. fluorescein isothiocyanate) used for direct labelling are quite sensitive to pH changes and therefore a pH around 7 is recommended. The probe fragment length should be between 150 and 400 bp. The longer the fragments, the more difficult it is for them to penetrate cell plasma structures and the three-dimensional nuclear scaffold to reach the target DNA. Furthermore, long fragments (more than 800 bp) are difficult to dissolve and result in spotlike precipitations of labelled probe on the slide. These precipitates impair the interpretation of results and very often they exhibit high fluorescence intensities which hamper correct image acquisition. For the probe concentration the following rule applies: the higher the probe concentration, the higher the reannealing rate; however, the optimal probe concentration depends largely on the nature of the probe and varies between 2 and 100 ng of DNA per microlitre of hybridisation mix.

According to the factors influencing denaturation and hybridisation, some standard conditions have evolved. Denaturation of the target is usually carried out using 70% formamide in the presence of 2 × saline–sodium citrate (SCC) at pH 7–7.5. SSC is a widely used buffer for hybridisations. It is prepared as a 20 × concentrate of 3 M NaCl and 0.3 M sodium citrate. The pH is adjusted to 7.0. The denaturation temperature varies between 68 and 72°C. It is recommended to prepare the denaturation mixture fresh.

The probe is usually dissolved in the hybridisation mixture. The mixture contains formamide as well in a concentration of 50–70%. SSC is present in a 2 × concentration and a phosphate buffer is added to stabilise the pH. Dextran sulphate is added and acts as a hybridisation enhancer. In aqueous solutions, dextran sulphate is highly hydrated and thus reduces the availability of water for the DNA. This effect results in an apparent increase in probe concentration and consequently higher hybridisation rates.

18.9 Applications

There is a vast diversity of applications for FISH. In clinical cytogenetics, FISH with whole chromosome painting probes as well as locus-specific probes is used. Locus-specific probes are especially of interest in the diagnostic field of microdeletion syndromes (e.g. Smith–Magenis syndrome, DiGeorge syndrome). Here, deleted regions which are below the microscopic resolution of metaphase chromosomes are characterised by locus-specific probes. In general, a reference probe is included which hybridises to another locus or region of the same chromosome and which is labelled in a different colour. The examination is carried out on chromosomes and the reference probe helps to identify the relevant chromosome. Absence of the relevant hybridisation signal on one of the two homologue chromosomes indicates the occurrence of the respective microdeletion syndrome.

FISH on interphase nuclei plays an important role in prenatal diagnosis. Uncultured amniocytes are fixed and spread on microscope slides. FISH is applied using a combination of centromere-specific and locus-specific probes. The probes identify the chromosomes which are most frequently involved in numerical aberrations (13, 18, 21, X and Y).

In tumour cytogenetics, FISH has become an important tool in identifying specific tumour-related aberrations. The benefits of FISH are obvious: on the one hand, the quality of conventional banding preparations in tumour cytogenetics is often suboptimal and, on the other hand, complex chromosomal aberrations obscure important tumour-related alterations. The number of specific aberrations that have been identified has increased tremendously over the last few years and so has the number of specific FISH assays. Furthermore, there is a diversity of mechanisms of aberrations in tumour cytogenetics, such as deletions of tumour-suppressor genes which can be identified in the interphase by loss of a hybridisation signal. Amplification of tumour genes like *Her2/NEU* can be identified in interphases through quantification of the fluorescence signal.

Specific fusion genes like *BCR/ABL* in the Philadelphia chromosome, which is a specific aberrations in chronic myeloid leukaemia (Nowell and Hungerford 1960; Rowley 1973), can be detected by analysing the pattern of differently labelled probes that are specific for the relevant chromosomal loci (Tkachuk et al. 1990). For a translocation two chromosomal breaks must occur. In our example, one occurs in the *BCR* locus and the other in the *ABL* locus. A part of the *ABL* gene is translocated to a part of the *BCR* gene, creating a fusion gene and vice versa. A translocation is identified by the presence of a fusion signal (colocalisation of one of each of the differently labelled spots). However, with increasing knowledge about the underlying complex mechanism accompanying a *BCR/ABL* fusion more sophisticated FISH assays have evolved. The so-called dual colour dual fusion not only identifies the fusion gene by detecting the fusion signal, but also detects deletions eventually occurring at the translocation sites. The selected probes are designed in such a way that they span the breakpoint regions, leading to split signals when a translocation occurs. The two probes are labelled with different colours and a double fusion occurs if a translocation is present. At the same time the total number of spots for each colour increases from two in a normal cell to three in a cell with a simple translocation (Fig. 18.1f). Deletions may occur at either of the derivative chromosomes and can be identified by loss of one of the spots in cells with a translocation. Identification of additional deletions is of utmost importance because deletions on the derivative chromosome 9 characterise a group of patients with significantly higher risk of disease progression (Reid et al. 2003).

In radiation biology, FISH is widely applied for the assessment of chromosomal breaks. Hybridisations are carried out after in vitro exposure to different kinds of damaging radiation (Knehr et al. 1999; Durante et al. 2003; Romm and Stephan 2004). In general, whole chromosome painting probes are used to quantify the DNA damage by counting the occurrence of chromosome breaks. For these applications a representative mix of three whole chromosome painting probes is used and high numbers (several hundreds) of metaphase spreads have to be evaluated in order to establish dose–response curves.

These correlations provide the basis for the interpretation of results obtained from FISH analysis performed on individuals who were exposed to radiation (Edwards et al. 2005). Long-lived, sensitive and specific biomarkers of particular mutagenic agents are much sought after and potentially have broad applications in the fields of cancer biology, epidemiology and prevention (Hande et al. 2005). Therefore, numerous retrospective investigations have been performed on individuals who have been exposed to radiation owing to accidents like the nuclear blast at Chernobyl (Snigiryova et al. 1997; Hande et al. 2003).

18.10 Conclusion

Since its introduction in the 1980s, FISH has undergone a still continuing refinement of its methodology. Thanks to the specificity of the probes and its sensitivity, FISH has developed into a powerful tool for diagnostics and research with capabilities far beyond those of classical cytogenetics. In the early 1980s, cytogenetics could not arouse much excitement anymore and was regarded as an old-fashioned method. But since FISH expanded into standard cytogenetics, the gap between chromosomes and molecular genetics has been bridged.

References

Blackburn EH (1991) Structure and function of telomeres. Nature 350:569–573

Casey J, Davidson N (1977) Rates of formation and thermal stabilities of RNA:DNA and DNA:DNA duplexes at high concentrations of formamide. Nucleic Acids Res 4:1539–1552

Chudoba I, Senger G (2002) Microdissection of chromosomes and reverse FISH. In: Rautenstrauß BW, Liehr T (eds) FISH technology. Springer, Berlin, pp 388–407

Chudoba I, Plesch A, Lörch T, Lemke J, Claussen U, Senger G (1999) High resolution multicolor-banding: a new technique for refined FISH analysis of human chromosomes. Cytogenet Cell Genet 84:156–160

Cremer T, Lichter P, Borden J, Ward DC, Manuelidis L (1988) Detection of chromosome aberrations in metaphase and interphase tumor cells by in situ hybridization using chromosome-specific library probes. Hum Genet 80:235

Cremer T, Popp S, Emmerich P, Lichter P, Cremer C (1990) Rapid metaphase and interphase detection of radiation-induced chromosome aberrations in human lymphocytes by chromosomal suppression in situ hybridization. Cytometry 11:110–118

Dahse R, Fiedler W, Ernst G (1997) Telomeres and telomerase: biological and clinical importance. Clin Chem 43:708–714

Durante M, Snigiryova G, Akaeva E, Bogomazova A, Druzhinin S, Fedorenko B, Greco O, Novitskaya N, Rubanovich A, Shevchenko V, Von Recklinghausen U, Obe G (2003) Chromosome aberration dosimetry in cosmonauts after single or multiple space flights. Cytogenet Genome Res 103:40–46

Edwards AA, Lindholm C, Darroudi F, Stephan G, Romm H, Barquinero J, Barrios L, Caballin MR, Roy L, Whitehouse CA, Tawn EJ, Moquet J, Lloyd DC, Voisin P (2005) Review of translocations detected by FISH for retrospective biological dosimetry applications. Radiat Prot Dosimetry 113:396–402

Fiegler H, Carr P, Douglas EJ, Burford DC, Hunt S, Scott CE, Smith J, Vetrie D, Gorman P, Tomlinson IP, Carter NP (2003) DNA microarrays for comparative genomic hybridization based on DOP-PCR amplification of BAC and PAC clones. Genes Chromosomes Cancer 36:361–374

Hande MP, Azizova TV, Geard CR, Burak LE, Mitchell CR, Khokhryakov VF, Vasilenko EK, Brenner DJ (2003) Past exposure to densely ionizing radiation leaves a unique permanent signature in the genome. Am J Hum Genet 72:1162–1170

Hande MP, Azizova TV, Burak LE, Khokhryakov VF, Geard CR, Brenner DJ (2005) Complex chromosome aberrations persist in individuals many years after occupational exposure to densely ionizing radiation: an mFISH study. Genes Chromosomes Cancer 44:1–9

Johannes C, Horstmann M, Durante M, Chudoba I, Obe G (2004) Chromosome intrachanges and interchanges detected by multicolor banding in lymphocytes: searching for clastogen signatures in the human genome. Radiat Res 161:540–548

Kallioniemi A, Kallioniemi OP, Sudar D, Rutovitz D, Gray JW, Waldman F, Pinkel D (1992) Comparative genomic hybridization for molecular cytogenetic analysis of solid tumors. Science 258:818–821

Kirchhoff M, Gerdes T, Maahr J, Rose H, Bentz M, Dohner H, Lundsteen C (1999) Deletions below 10 megabasepairs are detected in comparative genomic hybridization by standard reference intervals. Genes Chromosomes Cancer 4:410–413

Knehr S, Huber R, Braselmann H, Schraube H, Bauchinger M (1999) Multicolour FISH painting for the analysis of chromosomal aberrations induced by 220 kV X-rays and fission neutrons. Int J Radiat Biol 75:407–418

Landegent JE, Jansen in de Wal N, Dirks RW, Baao F, van der Ploeg M (1987) Use of whole cosmid cloned genomic sequences for chromosomal localization by non-radioactive in situ hybridization. Hum Genet 77:366–370

Langer PR, Waldrop AA, Ward DC (1981) Enzymatic synthesis of biotin-labeled polynucleotides: novel nucleic acid affinity probes. Proc Natl Acad Sci USA 78:6633–6637

Lemke J, Chudoba I, Senger G, Stumm M, Loncarevic IF, Henry C, Zabel B, Claussen U (2001) Improved definition of chromosomal breakpoints using high-resolution multicolour banding. Hum Genet 108:478–483

Lestou VS, Ludkovski O, Connors JM, Gascoyne RD, Lam WL, Horsman DE (2003) Characterization of the recurrent translocation t(1;1)(p36.3;q21.1-2) in non-Hodgkin lymphoma by multicolor banding and fluorescence in situ hybridization analysis. Genes Chromosomes Cancer 36:375–381

Monajembashi S, Cremer C, Cremer T, Wolfrum J, Greulich KO (1986) Microdissection of human chromosomes by a laser microbeam. Exp Cell Res 167:262

Nowell PC, Hungerford DA (1960) A minute chromosome in human granulocyte leukemia. Science 132:1497

Pinkel D, Straume T, Gray JW (1986) Cytogenetic analysis using quantitative, high-sensitivity, fluorescence hybridization. Proc Natl Acad Sci USA 83:2934–2938

Pinkel D, Segraves R, Sudar D, Clark S, Poole I, Kowbel D, Collins C, Kuo WL, Chen C, Zhai Y, Dairkee SH, Ljung BM, Gray JW, Albertson DG (1998) High resolution analysis of DNA copy number variation using comparative genomic hybridization to microarrays. Nat Genet 20:207–211

Reid AG, Huntly BJ, Grace C, Green AR, Nacheva EP (2003) Survival implications of molecular heterogeneity in variant Philadelphia-positive chronic myeloid leukaemia.Br J Haematol 121:419–427

Romm H, Stephan G (2004) Dose dependency of FISH-detected translocations in stable and unstable cells after Cs gamma irradiation of human lymphocytes in vitro. Cytogenet Genome Res 104:162–167

Rowley JD (1973) A new consistent chromosomal abnormality in chronic myelogenous leukaemia identified by quinacrin fluorescence and Giemsa staining. Nature 243:290–293

Sambrook J, Fritsch EF, Maniatis T (2001) Molecular cloning: a laboratory manual. Cold Spring Harbor Laboratories, Cold Spring Harbor

Schoch C, Haferlach T, Bursch S, Gerstner D, Schnittger S, Dugas M, Kern W, Löffler H, Hiddemann W (2002) Loss of genetic material is more common than gain in acute myeloid leukemia with complex aberrant karyotype: a detailed analysis of 125 cases using conventional chromosome analysis and fluorescence in situ hybridization including 24-color FISH. Genes Chromosomes Cancer 35:20–29

Schröck E, du Manoir S, Veldman T, Schoell B, Wienberg J, Ferguson-Smith MA, Ning Y, Ledbetter DH, Bar-Am I, Soenksen D, Garini Y, Ried T (1996) Multicolor spectral karyotyping of human chromosomes. Science 273:494–497

Senger G, Lüdecke H-J, Horsthemke B, Claussen U (1990) Microdissection of banded human chromosomes. Hum Genet 84:507–511

Snigiryova G, Braselmann H, Salassidis K, Shevchenko V, Bauchinger M (1997) Retrospective biodosimetry of Chernobyl clean-up workers using chromosome painting and conventional chromosome analysis. Int J Radiat Biol 71:119–127

Solinas-Toldo S, Lampel S, Stilgenbauer S, Nickolenko J, Benner A, Dohner H, Cremer T, Lichter P (1997) Matrix-based comparative genomic hybridization: biochips to screen for genomic imbalances. Genes Chromosomes Cancer 20:399–407

Speicher MR, du Manoir S, Schrock E, Holtgreve-Grez H, Schoell B, Lengauer C, Cremer T, Ried T (1993) Molecular cytogenetic analysis of formalin-fixed, paraffin-embedded solid tumors by comparative genomic hybridization after universal DNA-amplification. Hum Mol Genet 11:1907–1914

Speicher MR, Gwyn Ballard S, Ward DC (1996) Karyotyping human chromosomes by combinatorial multi-fluor FISH. Nat Genet 12:368–375

Tkachuk DC, Westbrook CA, Andreeff M, Donlon TA, Cleary ML, Suryanarayan K, Homge M, Redner A, Gray J, Pinkel D (1990) Detection of bcr-abl fusion in chronic myelogeneous leukemia by in situ hybridization. Science 250:559–562

Telenius H, Carter NP, Bebb CE, Nordenskjold M, Ponder BA, Tunnacliffe A (1992) Degenerate oligonucleotide-primed PCR: General amplification of target DNA by a single degenerate primer. Genomics 13:718–725

Tönnies H, Stumm M, Wegner RD, Chudoba I, Kalscheuer V, Neitzel H (2001) Comparative genomic hybridization based strategy for the analysis of different chromosome imbalances detected in conventional cytogenetic diagnostics. Cytogenet Cell Genet 93:188–194

Willard HF, Waye JS (1987) Chromosome-specific subsets of human alpha satellite DNA: analysis of sequence divergence within and between chromosomal subsets and evidence for an ancestral pentameric repeat. J Mol Evol 25:207–214

19 The CAB and S&S Systems, and an Approach Towards the Classification of Complex Chromosome Exchanges

John R.K. Savage

Abstract

From the results of fluorescence in situ hybridization (FISH) "painting" of induced *chromosome-type* exchanges, it is now clear that very many of them arise from the interaction between three or more breaks in two or more chromosomes, i.e. they are actually "complex exchanges". This is not obvious from the usual solid-staining methods, so nomenclature, scoring systems and theory have developed on the assumption that two-break exchanges are the norm. Such systems are therefore proving inadequate for recording "painted" exchanges. For general purposes, complex exchanges can be classified using the CAB system. However, every CAB can have many break-end interactions and so each generates a family of different exchange patterns, all of which fall within the same designation. This renders such a method of classification rather crude. If, for a given multicolour FISH painted exchange, the single-paint pattern for each participating chromosome is classified according to the S&S system, the resulting string of pattern codes forms a useful recording/scoring tool (the S&S descriptor) which turns out to be much more specific than the CAB. Moreover, such descriptors usually contain enough information to make a diagrammatic reconstruction of the original overall exchange pattern. This chapter presents a table of all the possible CAB families up to CAB 6/6/6, and drawings of the complete S&S single-paint codes up to CAB 5/5/5. The formation of an S&S descriptor is illustrated, and some of its limitations are discussed.

19.1 Introduction

With the arrival of fluorescence in situ hybridization (FISH) "painting" of individual homologue pairs came the realization that many of the *chromosome-type* exchanges, normally assumed to be simple two-break events, were, in fact, the result of interactions between three or more breaks in two or more chromosomes – i.e. they were "complex exchanges" (Savage 2002). Such complex exchanges are well known for *chromatid-type* aberrations, but were generally believed to be rare for *chromosome-type* ones.

Günter Obe and Vijayalaxmi (Eds.)
Chromosomal Alterations: Methods, Results and Importance in Human Health

Initially, only a few chromosome-specific fluorochromes were available, so most of the observed anomalous patterns came from exchanges with just one participating chromosome painted, and this situation conditioned the early classification attempts.

Now that it is possible to "paint" all the homologue pairs distinctively (multicolour FISH, mFISH, and equivalent techniques; Speicher et al. 1996), the situation is exacerbated because it is clear than many chromosomes and breaks may be involved in a single exchange (Cornforth 2001).

We are now faced with the problem of the classification and recording of these new anomalous patterns, since the established scoring schemes, which are based upon solid-staining and a two-break origin inference (Savage 1976), are no longer valid. Many different combinations of *interchanges* and *intrachanges* can occur within a single complex exchange.

This chapter suggests one possible step towards the solution of this problem.

19.2 Configuration Versus Pattern

A very important principle has emerged from painting that has been largely overlooked in all our previous aberration studies.

Let us define the actual complete structural exchange, with all its related components as the *configuration*.

What we actually see down the microscope is a *pattern*, derived from this configuration. Now, how much of the configuration is revealed in the pattern depends upon the staining protocol we use. With the majority of protocols in regular use, we almost always see only part of the configuration. Since our qualitative and quantitative scoring can only be done from the pattern, our inferences must frequently be wrong. Obviously we have been underestimating the level of chromosome damage produced by aberration-inducing agents, and considerable revision of our past ideas is in order.

Only when every participating chromosome is distinctively coloured are we able to see (in most, but not all cases) the actual configuration and assess the nature of the exchange and the relationships of the participants.

Unfortunately, the commonest protocol, solid-staining, is the least efficient for *chromosome-type* exchanges. Much information is lost, and false identification is frequent. This principle is illustrated in Fig. 19.1.

Nevertheless, within a cell system, quantitative scores of solid-stained *chromosome-type* aberrations have proved to be consistent in relation to radiation dose, and have been found to be very reliable for techniques such as biological dosimetry. It is only when morphology and scores are applied to fundamental studies on the formation and origin of aberrations that the inefficiency assumes importance (Savage 1998). It must always be remembered that any treatment which influences the formation of complex exchanges will inevitably alter the recovered frequencies of solid-stained aberrations.

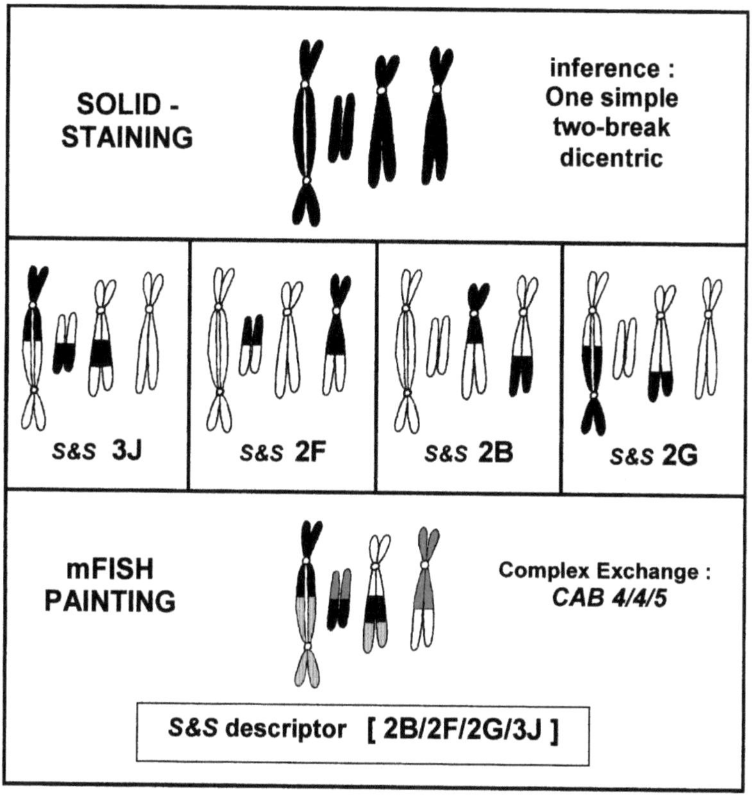

Fig. 19.1 *Top*: A collection of four chromosomes, solid-stained. One asymmetrical exchange (a "dicentric") is present and one infers that the accompanying acentric fragment is related to it, being formed by fusion of the two terminal deletions, i.e. a simple two-break interchange. Other symmetrical changes may be present, but these are not normally seen with solid-staining. *Middle*: Each of the four chromosomes can yield a single-paint S&S pattern, (painted segments shown *solid*), but in an experimental situation, only one will be seen, for each individual cell is unique. *Inference has to be drawn solely from the pattern you happen to get.* Pattern 1 (*3J*) has two parts expected for a simple dicentric, but has an additional insertion in another chromosome; thus, the dicentric cannot be simple but is a complex exchange involving a minimum of four breaks. Patterns 2 and 4 (*2F*, *2G*) are anomalous, each requiring at least three breaks, and again confirm that the dicentric is not simple. Pattern 3 (*2B*) is interesting. It suggests a simple reciprocal translocation, but reference to the multicolour fluorescence in situ hybridization (mFISH) pattern shows the unpainted segments derive from thee other chromosomes! This is a classic example of a *pseudosimple* pattern. *Bottom*: mFISH staining shows the configuration generating the patterns of the other staining protocols. It indicates that all four chromosomes are involved in one complex exchange, something impossible to deduce from the other protocols. The $CAB_{observed}$ is 4/4/5. For practical purposes we have to assume that this is also the CAB_{actual} but there may be non-registering hidden changes. Taken together, the panels demonstrate the principle that most staining protocols only reveal part of the actual exchange configuration, and result in an underestimate of the actual damage present. The S&S descriptor is the string of the total S&S codes present in the observed exchange, and is discussed in the text

19.3 The CAB System

Since complexes involve several breaks, it is possible to make a theoretical classification based on the arrangement of these breaks in relation to the chromosome arms (Savage and Simpson 1994). For any given number of breaks, there is always a limited number of ways in which they can be allocated to arms: three ways for three breaks, eight ways for four breaks, 15 ways for five breaks, 31 ways for six breaks and so on.

These possibilities can be expressed by giving, in order, the number of chromosomes in the exchange, the number of arms involved and the total number of breaks (CAB). Thus, CAB 2/4/5 is an exchange derived from five breaks distributed between four arms of two chromosomes. In some cases, more than one arrangement is possible, and then families are distinguished by the addition of a suffix, e.g. 2/3/5(A), 2/3/5(B).

Table 19.1 gives the arrangement of breaks for all CABs from 2/2/3 up to 6/6/6.

Table 19.1 The CAB families classification

$CAB_{initial}$	Arrangement of breaks	$CAB_{initial}$	Arrangement of breaks
2/2/3	0◆2; 0◆1	2/2/6(A)	0◆3; 0◆3
2/3/3	1◆1; 0◆1	2/2/6(B)	0◆4; 0◆2
3/3/3	0◆1; 0◆1; 0◆1	2/2/6(C)	0◆5; 0◆1
		2/3/6(A)	2◆2; 0◆2
2/2/4(A)	0◆2; 0◆2	2/3/6(B)	1◆3; 0◆2
2/2/4(B)	0◆3; 0◆1	2/3/6(C)	2◆3; 0◆1
2/3/4(A)	1◆1; 0◆2	2/3/6(D)	1◆4; 0◆1
2/3/4(B)	1◆2; 0◆1	2/3/6(E)	1◆2; 0◆3
2/4/4	1◆1; 1◆1	2/3/6(F)	1◆1; 0◆4
3/3/4	0◆2; 0◆1; 0◆1	2/4/6(A)	1◆2; 1◆2
3/4/4	1◆1; 0◆1; 0◆1	2/4/6(B)	2◆2; 1◆1
4/4/4	0◆1; 0◆1; 0◆1; 0◆1	2/4/6(C)	1◆3; 1◆1
2/2/5(A)	0◆3; 0◆2	3/3/6(A)	0◆2; 0◆2; 0◆2
2/2/5(B)	0◆4; 0◆1	3/3/6(B)	0◆3; 0◆2; 0◆1
2/3/5(A)	1◆1; 0◆3	3/3/6(C)	0◆4; 0◆1; 0◆1
2/3/5(B)	1◆2; 0◆2	3/4/6(A)	1◆1; 0◆2; 0◆2
2/3/5(C)	2◆2; 0◆1	3/4/6(B)	2◆2; 0◆1; 0◆1
2/3/5(D)	1◆3; 0◆1	3/4/6(C)	1◆2; 0◆2; 0◆1
2/4/5	1◆2; 1◆1	3/4/6(D)	1◆3; 0◆1; 0◆1
		3/4/6(E)	1◆1; 0◆3; 0◆1
		3/5/6(A)	1◆1; 1◆1; 0◆2

(Continued)

Table 19.1 The CAB families classification—Cont'd

$CAB_{initial}$	Arrangement of breaks	$CAB_{initial}$	Arrangement of breaks
3/3/5(A)	0◆2; 0◆2; 0◆1	3/5/6(B)	1◆2; 1◆1; 0◆1
3/3/5(B)	0◆3; 0◆1; 0◆1	3/6/6	1◆1; 1◆1; 1◆1
3/4/5(A)	1◆1; 0◆2; 0◆1		
3/4/5(B)	1◆2; 0◆1; 0◆1	4/4/6(A)	0◆2; 0◆2; 0◆1; 0◆1
3/5/5	1◆1; 1◆1; 0◆1	4/4/6(B)	0◆3; 0◆1; 0◆1; 0◆1
		4/5/6(A)	1◆1; 0◆2; 0◆1; 0◆1
4/4/5	0◆2; 0◆1; 0◆1; 0◆1	4/5/6(B)	1◆2; 0◆1; 0◆1; 0◆1
4/5/5	1◆1; 0◆1; 0◆1; 0◆1	4/6/6	1◆1; 1◆1; 0◆1; 0◆1
5/5/5	0◆1; 0◆1; 0◆1; 0◆1; 0◆1	5/5/6	0◆2; 0◆1; 0◆1; 0◆1; 0◆1
		5/6/6	1◆1; 0◆1; 0◆1; 0◆1; 0◆1
		6/6/6	0◆1; 0◆1; 0◆1; 0◆1; 0◆1; 0◆1

◆=centromere

In practice, because of visual limitations, we have to recognize the existence of three distinct CABs:

1. $CAB_{initial}$ is the arrangement of breaks prior to the start of the exchange process. This is used for theoretical calculations, but will always remain hypothetical since some break-end junctions may not register visually.
2. CAB_{actual} is the arrangement of the breaks that do not restitute, but actually take part in illegitimate rejoining.
3. $CAB_{observed}$ is that deduced from the rejoin junctions (usually, but not necessarily bicoloured) observed in the microscopical exchange pattern.

For practical purposes, one has to assume that $CAB_{observed} = CAB_{actual}$, but same-colour hidden junctions (SCHJ) frequently mean that the observed is less than the actual.

The CAB system has proved to be a useful general classification for complex exchange patterns, but remains rather broad.

19.4 The S&S System

Each CAB defines a family of configurations, the number in the family being determined by the number of break-ends available for interaction.

Each interaction removes two break-ends. Thus, for C/A/3 (six break-ends) there are $5 \times 3 \times 1 = 15$ possible configurations; for C/A/4 (eight break-ends) $7 \times 5 \times 3 \times 1 = 105$ possible configurations; for C/A/5 945 possible configurations; for C/A/6 10,395 possible configurations, and so on (in general, $(2b)!/b!2^b$, where b is the number of interactive breaks). These numbers include configurations with break-end restitutions, so not all configurations will be complex exchanges, and some will not reflect the $CAB_{initial}$ (Sect. 19.6.).

It was quickly realized that using a single-chromosome paint, a given configuration can produce a variety of different patterns, some of which may lead to false inferences (*pseudosimple* patterns). This is illustrated in Fig. 19.1. Also it was realized that the same single-paint pattern can arise from totally different configurations within the same or from different CABs (Savage 2002).

Therefore, it became of interest to know just how many unique single-paint patterns existed, and if any were diagnostic for a particular CAB.

From CAB 2/2/3 to CAB 5/5/5 there are 26 families which generate between them 15,060 configurations and 41,895 single-chromosome paintings. All of these configurations and paintings were analysed and only 203 *unique* single-paint patterns emerged. These represent 0.0048 (about 0.5%) of the total, so it is obvious that a large number of patters recur, and none can be regarded as diagnostic for a particular CAB. However, in many cases a minimum number of interacting breaks is required before a particular pattern will be encountered.

Each of these unique single-paint patterns was given an alphanumeric code. Those containing a centric-ring (*interarm intrachange*) have the prefix CR followed by a number. The rest have a numerical prefix corresponding to the actual number of painted elements present, and are distinguished from each other by an alphabetical suffix. These codes comprise the S&S system (Savage and Tucker 1996).

Part of the system has been published (Savage and Simpson 1994), but the complete code series (up to CAB 5/5/5) has hitherto only been available from an unpublished monograph. It is now presented in its entirety in Tables 19.2–19.7.

The patterns illustrated are for structurally complete forms where all painted segments (shown solid in the tables) are potentially visible. For each pattern, the *minimum* number of breaks in the *CAB* required to produce it is shown in the top right-hand corner of each square.

There are additional patterns if "one-way" exchanges are included, which, although known, are not considered here. There may also be some new patterns which will be discovered in higher-order CABs (i.e. above 5/5/5), but the enormous number of configurations which these generate preclude detailed analysis.

19.5 mFISH Painting

It has now become possible to paint all the homologue pairs in the human karyotype distinctively, using techniques like mFISH (Speicher et al. 1996) and its equivalents. So now, for the first time, we can see the complete configuration (or a large part of it) which is underlying and generating the observed exchange painting pattern, and so get a much fuller picture of the relationship of its parts (Fig. 19.1).

Table 19.2 S&S single-paint codes. Painted segments shown *solid*

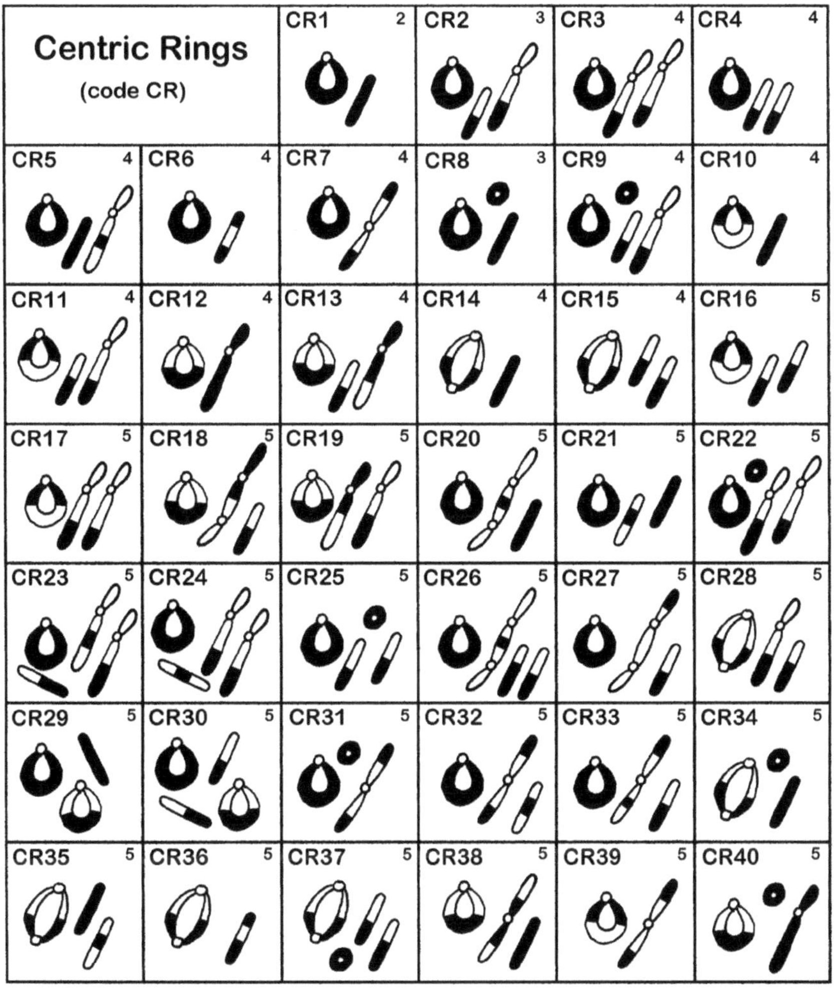

Unfortunately, this increased precision has introduced a whole raft of problems which we have not had to consider before, not least that complex exchanges are more common, and more complicated than we have hitherto realized (Savage 2002).

One pressing problem is how to classify and record the exchange patterns we see. At the moment, for quantitative purposes, counting the total number of bicolour junctions (each of which must represent at least one rejoining event) will give a crude estimate of overall damage, although it must be recognized that valid rejoining events within a given chromosome, or when homologues are involved in the same exchange, will be missed. This approach, however, does not help in recording the form of individual exchanges.

Table 19.3 S&S single-paint codes. Painted segments shown *solid*

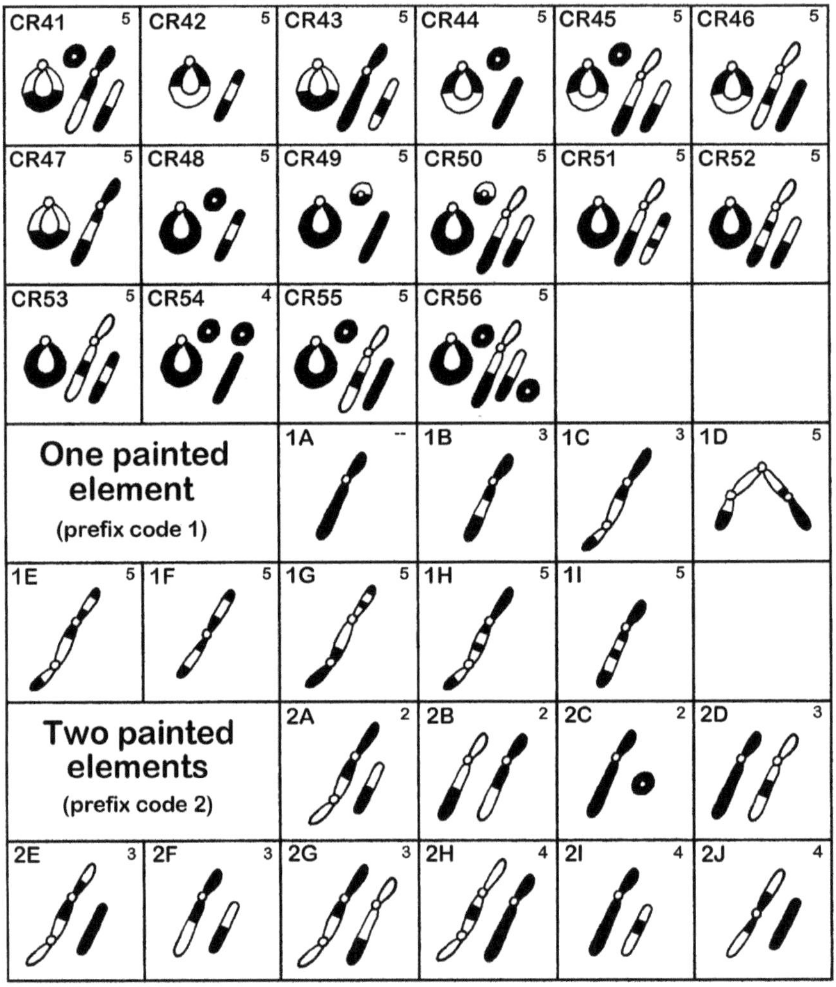

The old scoring schemes, based on solid staining, allowed convenient names for the various types seen, so the scoring criteria like "dicentrics" or "centric-rings" were universally understood, and could be used for valid interlaboratory comparisons (Savage 1976). But now, with mFISH, there is such a plethora of different patterns that simple names for each are out of the question.

We can, of course, classify any exchange pattern using $CAB_{observed}$, but as there are relatively few CAB families, many quite different exchange patterns will be covered by such a designation.

By ensuring that each participating chromosome is identified by a code, a complete description for each painted element in an exchange pattern could

Table 19.4 S&S single-paint codes. Painted segments shown *solid*

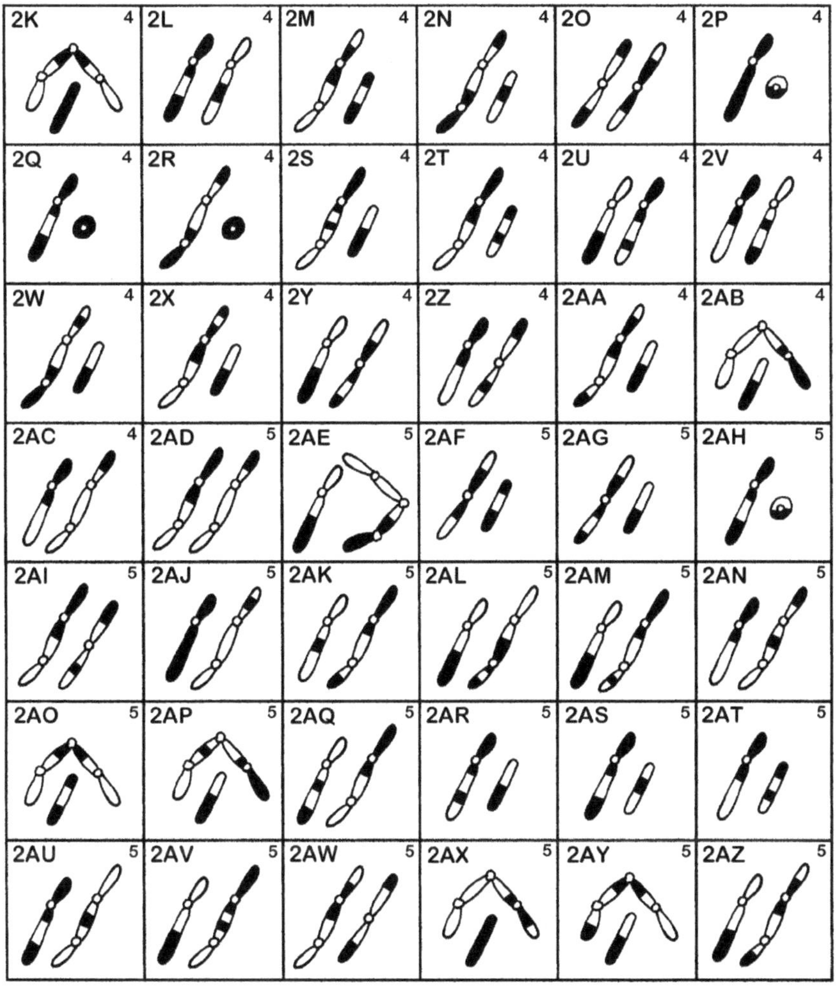

be provided by the PAINT nomenclature (Tucker et al. 1995). The linked elements would then form a descriptor, but this soon becomes very cumbersome when large exchanges are present in a mFISH situation.

19.6 The S&S Descriptor

A more refined approach is to make use of the S&S single-paint codes to construct a descriptor for each observed exchange pattern in its entirety.

Provided that all the chromosomes participating in an exchange are distinctively painted, it is possible to consider the pattern of each one individually, and

Table 19.5 S&S single-paint codes. Painted segments shown *solid*

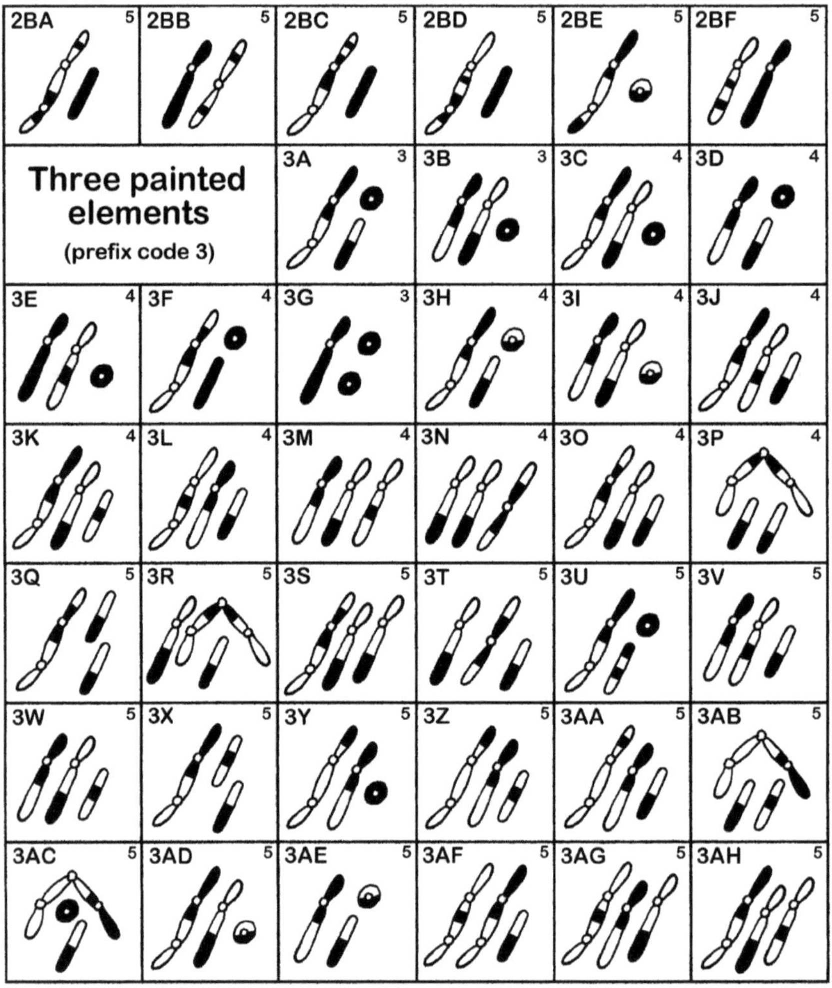

assign to it an S&S single-paint code. The string of codes thus obtained (expressed in the order found in Tables 19.2–19.7) forms a simple and useful descriptor for the whole exchange pattern. An example is given in Fig. 19.1, where the complete exchange from a CAB 4/4/5 becomes a [2B/2F/2G/3J]. The relevant S&S patterns and codes can be found in Tables 19.3 and 19.5.

This descriptor not only facilitates the scoring and classification of complex exchanges, but provides sufficient data for their diagrammatic reconstruction.

It will be noted that for any exchange pattern, the S&S single-paint codes present are not wholly independent. Certain collections are permissible, others impossible. This will be obvious if we consider any two-chromosome exchange

Table 19.6 S&S single-paint codes. Painted segments shown *solid*

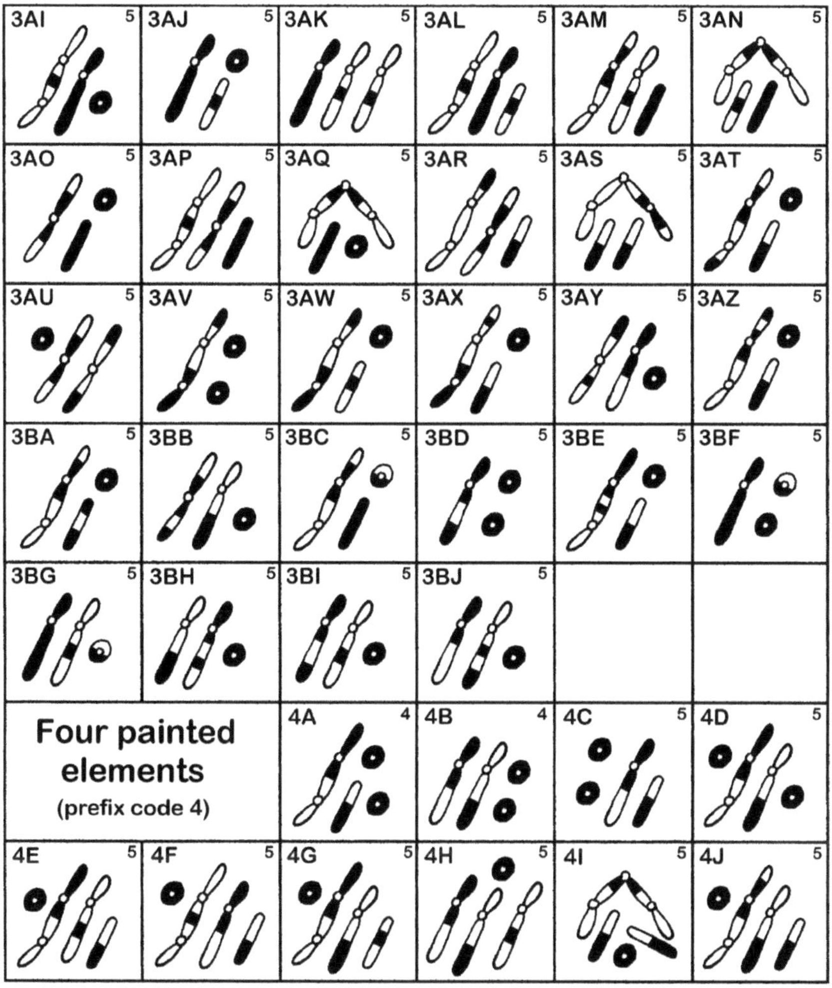

pattern, for here only two S&S codes form a descriptor, and once one is fixed, so also is the other. Consequently, a limited number of recurrent pairs like 2A/2A, 2M/2N and CR2/2F always occur, whilst other combinations like 2A/2F are precluded.

For CABs with several chromosomes, many more code combinations become possible, and the same code may now occur within different descriptors, but the limitation to permissible combinations still holds.

As a consequence of these restrictions, all the theoretical configurations for any CAB fall into a limited number of recurrent descriptors. Some code

Table 19.7 S&S single-paint codes. Painted segments shown *solid*

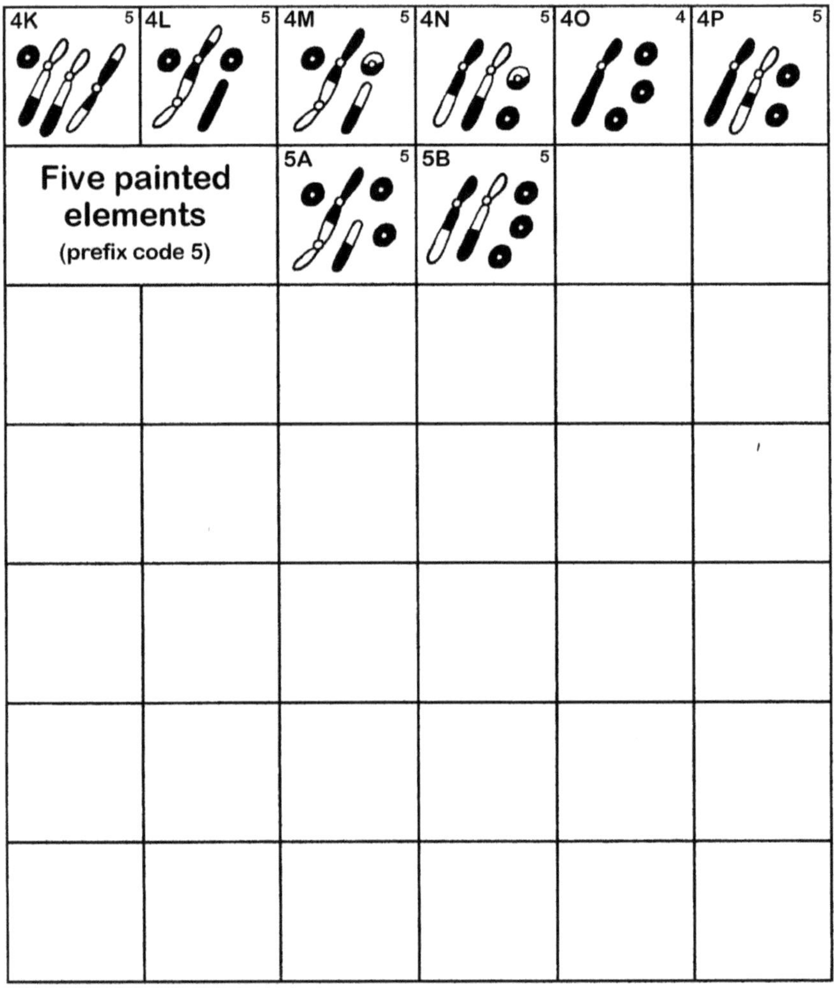

combinations are very common (e.g. 2A/2F/2G), and are found in several families, whilst a few are confined to one family. Within a family, the actual number of configurations falling within each descriptor varies considerably, depending largely on the number of different break-end interactions possible to satisfy a particular exchange pattern.

The permissible descriptors which classify each configuration can be derived theoretically for each CABinitial family, and it has been shown that these fall into three different categories:

1. *Restitution-reduced configurations.* Any restitution occurring as part of the exchange process effectively lowers the $CAB_{initial}$ by introducing apparently normal chromosomes (code 1A) and/or by producing SCHJ.

As the observer will be unaware of these restitution interactions, the net effect will be to record a lower-order $CAB_{observed}$.

The theoretical number of configurations which involve one or more restitution events can be readily derived form a full expansion of a CAB. This number is solely dependent upon the available number of interacting breaks. Thus, of the 15 configurations possible for CA3 families, seven of them have restitutions. Similarly, CA4 families have 45 in 105, and CA5 families have 401 in 945.

2. *Integrated configurations.* No restitutions; all rejoins are illegitimate, with the majority observable as bicolour junctions. However, even here, internal chromosome interactions, or interactions between any homologues present, can lead to SCHJ, with concomitant underestimation of CAB_{actual}.
3. *Disintegrated configurations.* When the number of breaks equals or exceeds 4, then some complex exchanges can arise from two or more apparently independent lower-order exchanges. These are known as sequential exchange complexes (Savage 2002). Sometimes this can cause physical breakup of a complex exchange, and the lower-order exchanges will be observed and recorded as separate events. For example, in CAB 3/5/5, the theoretical descriptor CR1/1C/2E will actually appear as two separate exchanges, CR1 and 1C/2E. In theory, disintegrated descriptors should be relatively rare, forming only a few percent of the total configurations within a CAB family.

19.7 Conclusions

The S&S descriptor provides an easy method for recording the complete observed pattern of any complex exchange where all the participating chromosomes are distinctively painted. It can be made even more precise by adding the actual chromosome number to each S&S code in the descriptor. The data thus provided enable one to make a fairly accurate diagrammatic reconstruction of the entire pattern.

Whilst the number of legitimate descriptors is limited, it is still very large, but this is inevitable, bearing in mind the enormous random combinatorial possibilities that exist for sets of proximate break-ends.

How useful this approach will prove for numerical studies must await the results of its practical application in experimental situations.

References

Cornforth, MN (2001) Analyzing radiation-induced complex chromosome rearrangements by combinatorial painting. Radiat Rese 155:643–659

Savage JRK (1976) Annotation: classification and relationships of induced chromosomal structural changes. J Med Genet 13:103–122

Savage JRK (1998) A brief survey of aberration origin theories. Mutat Res 404:139–147
Savage JRK (2002) Reflections and meditations upon complex chromosomal exchanges. Mutat Rese 512:93–109
Savage JRK, Simpson P (1994) FISH "painting" patterns resulting from complex exchanges. Mutat Res 312:51–60
Savage JRK, Tucker JD (1996) Nomenclature systems for FISH-painted chromosome aberrations. Mutat Res 366:153–161
Speicher MR, Ballard SG, Ward DC (1996) Karyotyping human chromosomes by combinatorial multi-fluor FISH. Nat Genet 12:368–375
Tucker JD, Morgan WF, Awa AA, Bauchinger M, Blakey D, Cornforth MN, Littlefield GL, Natarajan AT, Shassere C (1995) A proposed system for scoring structural aberrations detected by chromosome painting. Cytogenet Cell Genet 68:211–221

20 Aberration Patterns and Cell Cycle Progression Following Exposure of Lymphocytes to the Alkylating Agent Trenimon

Ursula von Recklinghausen, Christian Johannes, Lothar Riedel, and Günter Obe

Abstract

Chromosomal aberrations are generally analyzed in first metaphases (M1) following exposure to chromosome-breaking agents. During subsequent cell cycles there is strong selection against aberrant cells. Aberrant cells which survive selection exhibit changed aberration patterns. Data on chromosomal aberrations induced with the chemical mutagen Trenimon in the G1 phase of human lymphocytes in culture are discussed. Trenimon leads to various types of DNA damage, but not to DNA double-strand breaks; it is an S phase-dependent chromosome-breaking agent. Aberrations are analyzed in first (M1), second (M2) and third and further (M3+) posttreatment metaphases using preparations block-stained with Giemsa stain or with multicolour fluorescence in situ hybridization. In M1 cells, chromatid-type aberrations, such as chromatid breaks and chromatid translocations occur. In M2 and M3+ cells, the aberration frequencies diminish considerably and in aberrant cells which survived selection, chromosome-type aberrations such as dicentric chromosomes and chromosome translocations appear. Asymmetrical chromatid translocations, where the two centromeres of the participating chromosomes are connected, give rise to dicentric chromosomes and to normal chromosomes. Symmetrical chromatid translocations, in which the centromeres of the participating chromosomes are not connected, give rise to reciprocal translocations and to normal chromosomes or to two types of nonreciprocal translocations. Cells with reciprocal, but not with nonreciprocal translocations are expected to survive because no genetic information is lost. Cells with micronuclei and polyploid cells preferably occur at later divisions and will then be selected.

20.1 Introduction

Chromosomal aberrations should be analyzed in first in vitro metaphases (M1) after exposure to a chromosome-breaking agent. Two major reasons for this are (1) selection of cells with chromosomal aberrations leads to reduced frequencies of chromosomal aberrations in second (M2) or third and later

Günter Obe and Vijayalaxmi (Eds.)
Chromosomal Alterations: Methods, Results and Importance in Human Health

(M3+) metaphases and (2) when cells with chromosomal aberrations divide the aberration pattern changes. Both effects have been repeatedly analyzed in cells exposed to radiation or chemical mutagens (Beek et al. 1980; Manti et al. 2006; Marshall and Obe 1998; Obe and Beek 1982; Riedel 1989; Riedel and Obe 1980). Mutagens which do not directly induce DNA double-strand breaks lead to chromatid-type chromosomal aberrations in M1 when applied before or during DNA synthesis. During subsequent cell cycles, chromatid-type chromosomal aberrations are transformed to chromosome-type chromosomal aberrations which are then visible in M2 or M3+. Transformation of aberration types in connection with selection of most cells with chromosomal aberrations would lead to wrong assumptions on the types of lesions and on the frequencies of chromosomal aberrations induced by the test agent in question.

In this chapter we discuss results of experiments in which lymphocytes in G1 phase of the cell cycle were exposed to the trifunctional alkylating agent Trenimon [2, 3, 5-tris(1-aziridinyl)-1,4-benzoquinone, TRN]. TRN is an S-dependently chromosome breaking agent (Obe and Beek 1979; Rademacher and Obe 1995). Exposure of cells before or during S phase to TRN leads to chromatid-type aberrations such as chromatid interchanges in M1; in M2 and M3+ chromatid-type aberrations are replaced by chromosome-type aberrations, such as dicentric chromosomes (DIC) and translocations (Beek et al. 1980; Marshall and Obe 1998; Obe and Beek 1982; Riedel and Obe 1980). Asymmetrical chromatid interchanges (RB′a) give rise to DIC and acentric fragments; symmetrical chromatid interchanges (RB′s) give rise to reciprocal or nonreciprocal translocations (Marshall and Obe 1998). Many types of human cancers are characterized by specific translocations (Mitelman et al. 1997) and it is interesting that they may occur by transformation of chromatid-type exchanges.

20.2 Methods

The results of two sets of experiments using human lymphocytes are presented. In one study cytogenetic effects were analyzed in Giemsa-stained preparations. In a second study chromosomal aberrations were analyzed in preparations painted with the method of multicolor fluorescence in situ hybridization (mFISH).

20.2.1 Giemsa Staining

Standard lymphocyte cultures of 5 ml were set up with heparinized whole blood (0.4 ml) from a 32-year-old male donor supplemented with Ham's F-10 medium (4 ml), fetal calf serum (0.5 ml), phytohemagglutinin (0.12 ml) and antibiotics (penicillin, streptomycin, 0.02 ml). For each fixation time four cultures were set up, and two of them received 5-bromdeoxyuridine (BrdUrd; final concentration 20 μmol/l) at a culture time of 22 h. Exposure to TRN (0.5 μmol/l) was for 5 h beginning 15 h after the start of culture. After exposure, TRN was washed out by discarding the medium and substitution with medium from untreated cultures set up for this purpose (conditioned medium). All cultures received colcemid (final concentration 0.08 μg/ml) 4 h before fixation. Culture times were 55–119 h in steps of 4 h. Cultures were set up in such a way that fixations and preparations could be performed during the day. Metaphases were prepared using routine procedures.

Preparations obtained from cultures without BrdUrd were block-stained with Giemsa solution; those from cultures with BrdUrd were differentially stained for sister-chromatid differentiation to analyze the frequencies of M1, M2 and M3+ in 200 metaphases (100 per parallel culture per fixation time). Chromosomal aberrations were analyzed in preparations from cultures without BrdUrd. At each fixation time, 500 metaphases (250 per parallel culture) were scored for chromosomal aberrations, 2,000 mononucleated interphases for micronuclei (MN) (1,000 per parallel culture), 1,000 metaphases for polyploid and endoreduplicated cells (500 per parallel culture) and 1,000 metaphases (500 per parallel culture) for premature chromosome condensation (PCC) in MN (MN-PCC).

20.2.2 Multicolor Fluorescence in Situ Hybridization

Lymphocyte cultures were set up using heparinized whole blood of a female donor principally as described in Sect. 20.2.1, but Mc Coy's 5A medium and 20% fetal calf serum were used. At a culture time of 18 h, TRN (final concentration 0.5 μmol/l) was added to the cultures for 2 h. After exposure, cells were washed and incubated in conditioned medium for the remaining culture times of 55, 60, 74, 122, 194 and 242-h. The 192- and 242-h cultures received conditioned medium for a second time at 122 h, which resulted in a better mitotic index. To determine the proportion of M1, M2 and M3+ metaphases after different incubation times, one culture each received BrdUrd (final concentration 20 μmol/l). After adding colcemid (final concentration 0.08 μg/ml) for the last 4 h of incubation, metaphases were prepared using routine procedures. Slides from cultures with BrdUrd were differentially stained in order to analyze the frequencies of M1, M2 and M3+ cells. The other slides were painted with mFISH probes and counterstained with 4′,6-diamidino-2-phenylindole, which allows all chromosomes to be recognized by their specific color spectrum.

Pooled data from two independent experiments are given.

20.2.3 Description of Aberration Types Analyzed

Gaps: Gaps are unstained areas in chromatids which are not longer than the diameter of the chromatids in which they occur. This structural change is usually not considered as a real chromosomal aberration.

Chromatid breaks (B′): Breaks in chromatids with the broken part more or less dislocated from the chromatids from which they descend.

Isochromatid breaks (B″): Both chromatids of a chromosome are broken at the same site and by this produce fragments without centromeres. B″ can also arise from chromosome breaks which occur in unreplicated chromosome regions.

Chromatid translocations (RB′): In most cases RB′ are exchanges between broken chromatids of two chromosomes. In rare cases more than two chromosomes can participate in the formation of an RB′. In RB′ with two chromosomes, two types can be differentiated, namely, symmetrical (RB′s) and asymmetrical (RB′a) ones.

RB′s: Reciprocal exchanges between the broken chromatids with all chromatids linked to a centromere (X-type fusion). The chromatids of an RB′s separate in anaphase without difficulties, but may give rise to different types of translocations in subsequent cell cycles.

RB′a: Exchange between broken chromatids in such a way that both centromeres are connected by the fused chromatids of both chromosomes (U-type fusion). The chromatid fragments without centromeres generally also fuse. In anaphase RB′a may lead to DIC or MN. A rare variety of chromatid exchanges are triradials where a B′ in one chromosome fuses with a B″ in a second chromosome.

DIC: DIC are chromosomes with two centromeres associated with a B″. DIC are generally formed by fusion of two centromere-bearing fragments of two unreplicated chromosomes. Both centromere-less fragments may fuse and by this form a B″. After treatment with TRN, DIC can also be formed when cells with chromatid interchanges divide.

Ring chromosomes (rings): Rings may arise from two breaks within one unreplicated chromosome. They may contain centromeres (centric ring) or not (acentric rings). Centric rings are associated with fragments (B″). Rings are quite rare and may also represent derived aberration types.

Polyploids and endoreduplications: Polyploid metaphases contain more than two sets of chromosomes. One reason for this may be the formation of endoreduplications, in which DNA synthesis is not followed by a segregation of the chromosome set, or by other disturbances of the mitotic mechanism.

MN: When cells divide, chromosome fragments without centromeres are generally not distributed to the cell poles, but enclose themselves with their own nuclear membrane. Whole chromosomes when not properly distributed can also give rise to MN.

MN-PCC: Compared with that of the main nucleus, the cell cycle of MN is slowed. When main nuclei undergo mitosis most MN are in S phase and very few are in G2 or G1 phases. When the main nucleus undergoes mitosis, mitotic factors stimulate MN to enter a prophaselike stage in which the chromosomes are visible. MN-PCC is generally located at the periphery of the metaphases. G1-PCC are long single-stranded structures; G2-PCC are double-stranded structures which are generally Giemsa-banded. S-PCC, by far the most frequent MN-PCC type, are characterized by a pulverized appearance (Obe and Beek 1982).

Translocations: Translocations can only rarely be detected in block-stained preparations. They are easily detectable in FISH-painted preparations. Translocations can be reciprocal (chromosomal material is mutually exchanged between two chromosomes) or nonreciprocal (chromosomal material from one chromosome is transferred to another chromosome, but not vice versa).

20.3 Results

20.3.1 Giemsa Staining

At 55 h, 95% of the metaphases were M1 and only 5% were M2. The maximal frequencies of M2 occurred at fixation times of 75 and 79 h with 48 and 50%, respectively. At these fixation times 32 and 24% of the cells were still in M1, and 28 and 38.5% were already in M3+. High frequencies of M3+ cells occurred at fixation times of 103 h (85%) up to 119 h (94%). B′ including B″ and RB′ (including few triradials) had their maximum at 55 h with 51.0 and 16.0 per 100 metaphases, respectively. The frequencies of B′ as well as RB′ (including triradials) were rare or absent at fixation times of 79–119 h. Gaps are not considered as chromosomal aberrations, but they show a similar pattern to chromatid-type chromosomal aberrations in that their frequencies are high at early fixation times (55 and 59 h: 15.4 and 12.2 per 100 metaphases, respectively) and much lower at later fixation times (63 h: 5.3 per 100 metaphases; 111 h: 0.3 per 100 metaphases). B″ occurred at all fixation times. Their frequencies dropped from 35.2 per 100 metaphases at 59 h to 10.0 or less per 100 metaphases at fixation times of 79 h and later. The decrease of the frequencies of B″ was not as pronounced as that found for B′ and RB′. This indicates that they are not uniform aberration types. B″ can be of chromosome or chromatid origin and in addition they can be derived from RB′a, DIC and centric ring chromosomes. DIC have a maximum frequency of 6.4 per 100 cells at 83 h; at 55 h only 2.0 DIC per 100 cells were found. Surviving cells with DIC can lead to daughter cells with DIC and this may explain why low frequencies of DIC are found at later fixation times. Rings were very rare and had a maximum frequency of 0.8 per 100 metaphases at 111 and 115 h and of 1.0 at 119 h. MN occurred at all fixation times and had their maximum frequency of 70 per 1,000 interphases at 79 h. The maximum frequencies of MN-PCC were 61 per 1,000 metaphases at 83 h and occurred at all fixation times at low frequencies (8–56 per 1000 metaphases). Polyploid metaphases including endoreduplications are very rare at fixation times of 55–63 h (zero to three per 1,000 metaphases) and reach a maximal value of 19 per 1,000 metaphases at 91 h. Endoreduplications do not occur before 83 h of cultivation, but they are very rare at all fixation times (two to five per 1,000 metaphases). The frequency of metaphases with chromosomal aberrations excluding gaps is

39.2% at 55 h and continuously drops to 6.6% at 119 h. The maximum frequencies of chromosomal alterations were found as follows: gaps, B′ including B″, RB′ (including triradials) at 55 h, DIC including rings at 83 h, MN at 79 h, MN-PCC at 83 h, polyploid metaphases and endoreduplications at 91 h. The data clearly show that TRN is an S-dependently acting chromosome breaking agent. It induces mainly chromatid-type aberrations in M1. In M2 and M3+, chromatid aberrations continuously diminish and derived aberrations appear. All data obtained from Giemsa-stained preparations are from Riedel (1989).

20.3.2 Multicolor Fluorescence in Situ Hybridization

The frequencies of M1, M2 and M3+ cells at different fixation times as determined in differentially stained metaphases from cultures with BrdUrd are shown in Fig. 20.1.

At 55 h, 95% M1 and 5% M2 cells were found. The number of M1 cells continuously decreased and they were not detected at 194 h. The first M3+ cells were found at 74 h and reached a value of 96% at 194 h. Preparations

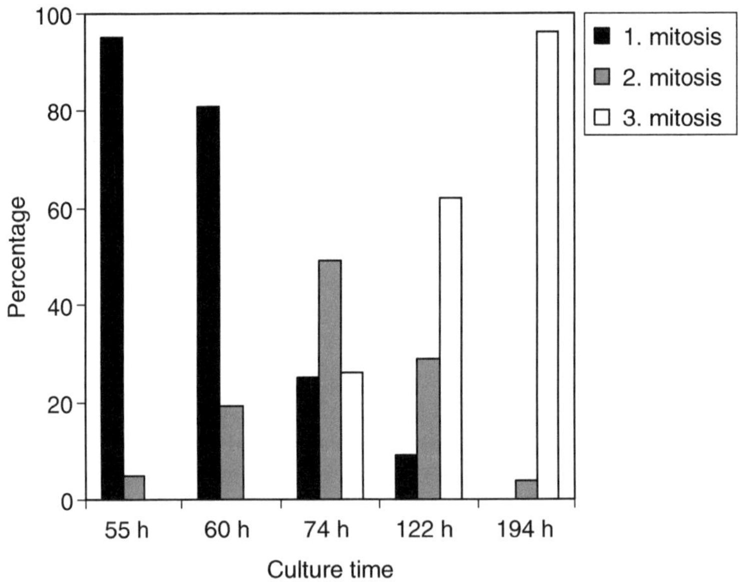

Fig. 20.1 First (M1, uniformly darkly stained chromosomes), second (M2, all chromosomes differentially stained) and third (M3+, different proportion of differentially and uniformly lightly stained chromosomes) metaphases after Trenimon treatment following different culture times. Data are from the multicolor fluorescence in situ hybridization (mFISH) study

Table 20.1 Exchange aberrations per 100 metaphases after Trenimon treatment at different cultivation times following painting with multicolor fluorescence in situ hybridization

Culture time (h)	Number of cells scored	Chromatid interchanges	Dicentrics	Translocations
74	258	10	6	30
122	307	1	17	35
194	309	0	12	33
242	65	0	9	17

painted with mFISH are especially suited for the analysis of exchange-type aberrations. Table 20.1 show the frequencies of exchange-type aberrations at fixation times of 74–242 h. RB′ predominantly occurred at 74 h and only one was found at 122 h. Both DIC and translocations had their maximum at 122 h, but occurred at all fixation times. Translocations were always more frequent than DIC.

Examples of different types of exchange-type aberrations are shown in Fig. 20.2.

In accordance with our data obtained with Giemsa staining (Sect. 20.3.2) the mFISH data show that exposure of human lymphocytes in G1 phase of the cell cycle to TRN leads to chromatid-type aberrations in M1 which give rise to chromosome-type aberrations in M2 and M3+. DIC are derived from RB′a; reciprocal and nonreciprocal translocations are derived from RB′s (Marshall and Obe 1998). Upon mitotic division a cell with an RB′a will give rise to a normal cell and a cell with a DIC plus associated fragment. Assuming a random distribution of chromatids, a dividing cell with an RB′s will lead to undamaged cells, cells with reciprocal translocations or cells with nonreciprocal translocations (Marshall and Obe 1998).

In contrast to nonreciprocal translocations derived from RB′s, translocations directly induced by X-rays are probably all reciprocal, although some of them appear to be nonreciprocal (Obe et al. 2002). Since cells with nonreciprocal translocations are genetically imbalanced, they are prone to be selected during cell divisions. Cells with reciprocal translocations in which no deleterious mutations were induced during their origin and which do not contain other types of chromosomal aberrations are genetically balanced and will survive during ongoing divisions. This seems to be the case in radiation-exposed people in whose lymphocytes the frequencies of reciprocal translocations are used for retrospective dosimetry (Chap. 24 by Edwards et al.). Experimental studies in which DIC and translocations were analyzed in splenocytes of mice up to 120 days after X-irradiation show that DIC disappear much faster than translocations (Natarajan and Obe 1999).

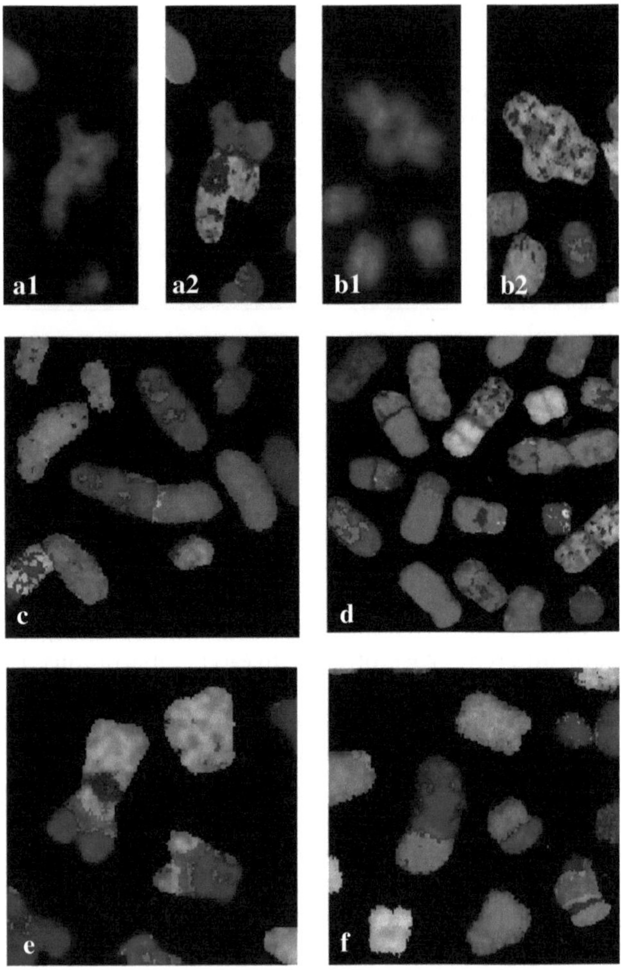

Fig. 20.2 Metaphases of human chromosomes showing Trenimon-induced aberrations stained with mFISH. **a1** Asymmetric chromatid translocation (RB′a) as visible with 4′, 6-diamidino-2-phenylindole (DAPI) counterstain. **a2** Same aberration painted with mFISH shows the participation of chromosomes 1 and 5. **b1** Symmetric chromatid translocation (RB′s), DAPI-stained. **b2** Same aberration painted with mFISH shows the participation of both chromosomes 1. **c** Dicentric chromosome formed by chromosomes 5 and X. **d** Five chromosomes involved in translocations. **e** Reciprocal translocation between chromosomes 1 and 5. **f** Reciprocal translocation between chromosomes 4 and 15

20.3.3 Conclusions

We analyzed chromosomal aberrations induced in G1 lymphocytes by the polyfunctional alkylating agent TRN and found typical changes of the aberrations pattern at different recovery times. A strong selection against

21.3 Standard Method

All the genetic toxicology laboratories of the Hygiene Service of the Czech Republic have been working according to the standard operating procedures published by the Reference Laboratory of Genetic Toxicology (Rössner et al. 1998, 2000). Since 2001, the Reference Laboratory has met the requirements of the CSN EN ISO/IEC 17025 standard and holds the Certificate of Accreditation of the Czech Accreditation Institute.

The modified conventional Hungerford method was used on short-term 48-h cultures with all cells being in the first division. Cultures of PBLs from whole venous blood were established within 24 h after the collection. The cultures were prepared in RPMI 1640 Sevac medium supplemented with 20% calf serum Opavac and 1% phytohemagglutinin (Murex, UK). Two identical cultures were set up from each sample, cultivated at 37°C, and harvested after 48 h of incubation. Two hours before the end of cultivation 0.1 μg/ml colchicine (Fluka) was added. Cells were collected by centrifugation, resuspended in a prewarmed hypotonic solution (0.075 M KCl) for 20 min and fixed in acetic acid/methanol (1:3, v/v) on slides. These were air-dried and stained with 5% Giemsa solution (pH 6.8). Slides from each culture were numbered randomly and scored as “blind” in numerical order. At least 100 well-spread metaphases with 46 ± 1 centromeres were examined per donor.

The following four categories of chromosomal aberrations were evaluated: chromatid and chromosome breaks, and chromatid and chromosome exchanges. Gaps were recorded, but they were not scored as aberrations. Cells bearing breaks or exchanges were marked as aberrant cells.

21.4 Chromosomal Aberrations and Maximum Allowable Concentrations

Cytogenetic analysis of chromosomal aberrations has been suggested as a tool for checking whether a safe maximum allowable concentration (MAC) has really been established (Sram and Kuleshov 1980; Sram 1981). The MAC of a chemical is defined as the maximum concentration of a gas, vapor or aerosol in the working atmosphere that, according to contemporary scientific knowledge, does not adversely affect the health of exposed humans.

The curve in Fig. 21.1 represents a simplified model situation in which the exposure (concentration of the chemical in the working atmosphere integrated over time) is related to a measurable effect such as the frequency of aberrant cells. It is assumed that a clastogenic exposure that does not increase the frequency of adverse effects over the level found in a suitable control

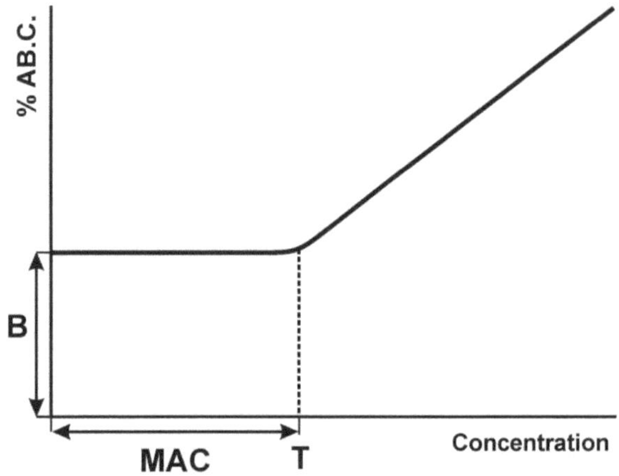

Fig. 21.1 Principle of establishing the maximum allowable concentration (*MAC*). *B* background frequency of aberrant cells determined in controls, *T* threshold limit of clastogenic action, *AB.C* aberrant cells

(unexposed) population is expected to be safe. According to this reasoning, the MAC should be set below the threshold level for clastogenic action.

It is, however, a controversial issue if any threshold can be established for carcinogenic and/or mutagenic effects. We argued in favor of a threshold level in the following way:

> Use of this method for checking MAC safety is based on the assumption that some mutagenic/carcinogenic molecules must be metabolically activated before reaching the target-cell DNA. Moreover, the DNA damage caused by these chemicals may be still eliminated by repair mechanisms. So we have good reasons to believe that also for chemical mutagens/carcinogens a threshold value can be found. This threshold, however, may vary from person to person . . . (Sram 1981).

In spite of the fact that the determination of chromosomal aberrations in PBLs has been conceptualized as a biomarker of early effects, in common practice we have interpreted it solely as a biomarker of exposure objectively indicating a need for reducing occupational exposure.

21.5 Cytogenetic Analysis and Public Health Practice

In the late 1970s several papers demonstrated that occupational exposure to vinyl chloride monomer (VCM) (Ducatman et al. 1975; Funes-Cravioto et al. 1975; Hansteen et al. 1978), chloroprene (Katosova 1973) and styrene (Meretoja et al. 1977) significantly increased the frequency of chromosomal aberrations in PBLs. A decrease in exposure to VCM (Hansteen et al. 1978) and chloroprene (Katosova 1973) was paralleled by a decrease in the frequency of aberrant cells to the level found in unexposed control groups. These findings resulted in the setting of new and safer exposure limits at a level below the observed threshold for clastogenic effects and set a precedent followed by public health authorities in the Czech Republic.

To provide public health authorities with a firm base for the decision-making process, the Reference Laboratory estimated the frequency of chromosomal aberrations in the Czech general population. This estimate, based on data from unexposed controls, was 1.2–2.0% aberrant cells. Svandova and Rössner (1989) analyzed data from 1,201 unexposed controls, collected in the frame of the Czech Hygiene Service in the period 1980–1988. The mean frequency of aberrant cells was 1.75%. More recently, Rössner et al. (1998) summarized the data on the spontaneous frequency of chromosomal aberrations in the general population of subjects in the period 1977–1997 ($N = 5{,}430$). The mean frequency of aberrant cells found in the subset of individuals of age from 20 to 63 years ($N = 2{,}801$) was 1.59 ± 1.34%. Both estimates were based on the analyses of at least 100 well-spread metaphases per subject.

The occupationally exposed groups with a mean frequency of aberrant cells above this reference value were then categorized by the frequencies of chromosomal aberrations (Sram et al. 2004a):

1. A group at increased genotoxic risk bearing on average 2–4% aberrant cells: exposure to acrylonitrile, asbestos, benzene, chromium-welding, cytostatics, dimethylformamide, ECHH, ethylbenzene, formaldehyde, halothane, methyl mercury, polycyclic aromatic hydrocarbons (PAHs): coke oven, glassware, rubber production, soft coal open-cast mining, soft coal pressure gasification, soot production, soft coal underground mining, PCB, propane–butane, radon: ore mining, uranium mining, uranium dressing plant, styrene, synthetic resins, toluene (Fig. 21.2).
2. A group at high genotoxic risk bearing more than 4% aberrant cells: exposure to BCME, cytostatics, mixture of styrene and methyl metacrylate, nickel, nitrous gases, olachindox, PAHs: coal tar production, glassware, VCM: PVC production (Fig. 21.3).

The discovery of occupational groups with elevated levels of chromosomal aberrations (usually initiated by a thorough hygiene inspection of workplaces), coupled with industrial hygiene measurements of the quality of the working environment, allowed estimation of the level of occupational exposure to carcinogens or mutagens. Simultaneously, a database and literature search for the toxicological properties of all agents identified in the working environment was begun. Even if technical and/or organizational measures reduced occupational exposure below the MACs in effect at the time, if repeated cytogenetic analyses still indicated an increase in the frequency of aberrant cells in the same group, then the public health authorities reassessed the safety of the MACs in use, and these were subsequently set below the

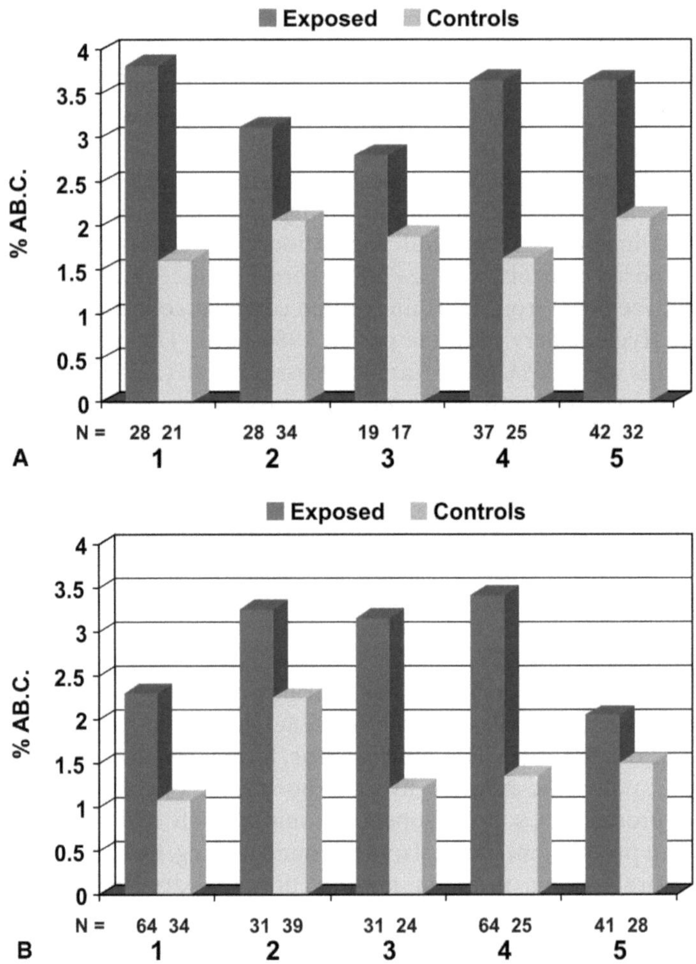

Fig. 21.2 Results of cytogenetic analyses in groups with an increased genotoxic risk. **a** *1* dimethylformamide, *2* epichlorohydrin, *3* styrene, *4* synthetic resins, *5* toluene (printing). **b** Exposure to polycyclic aromatic hydrocarbons: *1* coke oven, *2* rubber production, *3* soft coal open-cast mining, *4* soft coal pressure gasification, *5* soot production

threshold for clastogenic effects. Strict measures were enforced particularly in the case of groups at high genotoxic risk.

The target for public health intervention was not only technological improvements at critical workplaces but also behavioral changes in the exposed workers. This usually promoted positive changes such as more systematic care about the working environment, the voluntary use of personal protective equipment and so on.

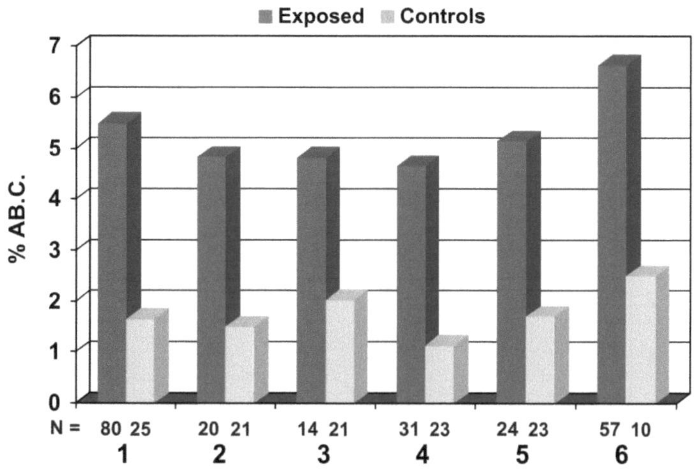

Fig. 21.3 Results of cytogenetic analysis in groups with high genotoxic risk. *1* bis(chloromethyl) ether, *2* coal tar production, *3* cytostatics, *4* glassware production, *5* styrene + methyl methacrylate (button production), *6* PVC production

21.6 Examples of Occupational Exposures to Carcinogens

21.6.1 Chemical Industry

21.6.1.1 Bis(chloromethyl) Ether, Chloromethyl Methyl Ether

BCME and chloromethyl methyl ether (CMME) were used as chlormethylating agents in the production of anionic and cationic ion exchangers. BCME causes lung cancer and is listed among human carcinogens (IARC 1987).

The ion-exchange unit in which workers were exposed to both halogenated ethers was in operation from 1973 to 1987. In 1974 it became the very first workplace in which chromosomal aberrations were used as a biomarker of exposure to evaluate the biological significance of occupational exposure. The results obtained were presented as evidence that ion-exchanger production poses a high genotoxic risk for workers. Therefore, since 1976 cytogenetic analysis has become part of their periodic checkups, and 80–110 workers have been examined each year (Zudova and Landa 1977; Roznickova et al. 1987).

The workers involved in ion-exchanger production were considered to be at high carcinogenic risk. When all available technological and organizational corrective measures were implemented at the workplaces but exposure to BCME still remained, this group became the first target for ascorbic acid supplementation (Sram et al. 1983a; Fig. 21.4). Gebhart et al. (1985) showed a clear-cut ability of ascorbic acid to block the clastogenic action of alkylating

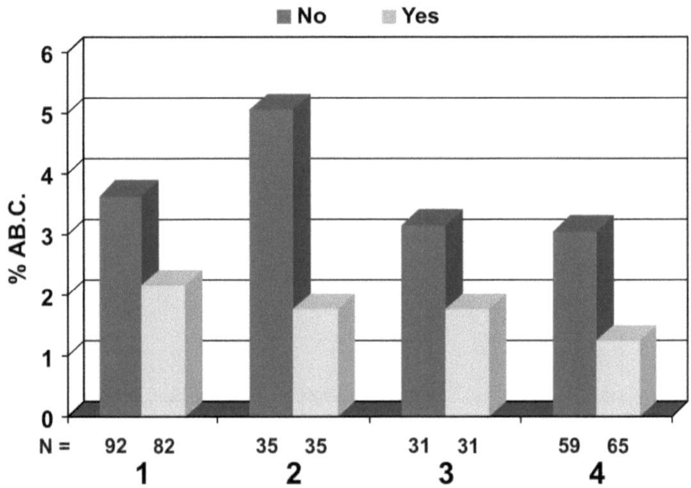

Fig. 21.4 Effect of the ascorbic acid supplementation on the frequency of aberrant cells

agents: therefore, in the early 1980s we believed it appropriate to administer ascorbic acid as a natural protector against the clastogenic (and mutagenic/carcinogenic) effects of environmental chemicals. When supplementation stopped in 1986, the frequency of aberrant cells increased to 4.35 ± 0.42%. The unit was closed in February 1987. After another 6 months, the frequency of aberrant cells decreased in the originally exposed workers (who meanwhile had moved to other production units in the same company) to 2.90 ± 0.15%, and in 1988, after another 12 months, the frequency decreased to 2.07 ± 0.15%.

21.6.1.2 Epichlorohydrin

Sram et al. (1980) carried out cytogenetic analysis in three groups of subjects: 28 workers occupationally exposed to ECHH for 4 years in a new production unit, 34 matched controls and 21 external controls (members of the general population). The mean concentration of ECHH of 0.936 mg/m^3 (0.20–1.85) found at workplaces in 1978 was below the MAC of 1 mg/m^3: We found 3.12% aberrant cells in the ECHH-exposed group, 2.06% aberrant cells in matched controls and 1.33% aberrant cells in the external controls. It was concluded that a MAC of 1.0 mg ECHH/m^3 was still too high to avoid the genotoxic risk associated with exposure to ECHH.

This occupational group was followed up for an additional 4 years. During this time ECHH exposure decreased to a mean concentration of 0.384 mg/m^3 (0.05–1.11). At this exposure level, ECHH did not exhibit any significant clastogenic effects. In a group of 33 workers we observed 2.00% aberrant cells vs. 1.68% aberrant cells in a group of 25 controls (Sram et al. 1983b).

21.6.1.3 Vinyl Chloride Monomer

Rössner et al. (1980) studied a group of workers ($N = 31$) in a VCM production unit. At the time of sampling, they had been exposed for 2–3 years to VCM levels below the Czechoslovakian MAC of 10 mg/m^3. In exposed workers they observed 1.60% aberrant cells in 1977 and 1.05% in 1978. In controls ($N = 35$) they found 1.12% aberrant cells.

The authors concluded that if VCM air concentrations in the working environment did not exceed 10 mg/m^3, then no increase in chromosomal aberrations in exposed workers seemed to be detectable.

21.6.1.4 Petrochemical Industry

Acrylonitrile The process of polymerization of India rubber is associated with exposure to acrylonitrile, butadiene and styrene. The most significant chemical regarding human health is considered to be acrylonitrile, because its concentrations in occupational settings often exceed the MAC (0.5 mg/m^3). A study carried out in a chemical plant in 1988 indicated a significant increase in the frequency of aberrant cells in the exposed group ($N = 54$, 2.96% aberrant cells) vs. controls ($N = 25$, 1.63% aberrant cells $P<0.01$). It was concluded that exposure to acrylonitrile posed a genotoxic risk, and this occupational group was examined for chromosomal aberrations every second year (Rössner et al. 1989).

1,3-Butadiene The association between occupational exposure to 1,3-butadiene (BD) and chromosomal damage in PBLs was studied in monomer production and polymerization units (Sorsa et al. 1994).

The clastogenic effects of occupational exposure to BD were studied in 1994 in 19 male workers from the monomer production unit and 19 control subjects from the heat production unit (Sram et al. 1998). The individuals' exposures to BD were measured by passive personal monitors. It was demonstrated that exposure to a median BD concentration of 0.53 mg/m^3 significantly increased the percentage of aberrant cells. The mean frequency of aberrant cells reported in exposed workers was 3.11%, and in controls it was 2.03% ($P<0.01$). The results presented by Sram et al. (1998) showed that although exposure levels in modern plants were substantially reduced, a significant induction of cytogenetic damage could still be demonstrated.

Ethylbenzene Ethylbenzene is produced by benzene alkylation by ethylene. During the production process, workers are exposed to a mixture of several chemicals. The cytogenetic effects in the staff members involved in an ethylbenzene production unit exposed to benzene, ethylbenzene, diethylbenzene, xylene and toluene were studied in 2002 (Sram et al. 2004b). The concentrations of ethylbenzene ranged from 0.2 to 13.1 mg/m^3 and the concentrations of benzene ranged from 0.4 to 15.1 mg/m^3. The mean frequency of aberrant cells reported in exposed workers was 3.28% ($N=39$), and that in controls was 2.05% ($N=55$, $P<0.001$). Using later a new technology (closed system), no significant differences in the frequency of aberrant cells were observed between the exposed and control groups.

21.6.2 Mining

21.6.2.1 Ore Mining

The increased incidence of lung cancer attracted the attention of the Hygiene Service to a tin mine in the early 1970s (Roznickova et al. 1987). In addition to other measures, cytogenetic analysis has been employed as one of the control measures since 1976. The estimated individual exposure did not exceed 15 mSv/year and 8 mSv/3 months.

The differences in the magnitude of chromosomal damage were dependent on radon dose. If a miner's exposure exceeded the yearly exposure limit of 15 mSv/year ($N = 36$) then the mean frequency of aberrant cells was 5.7%. Even higher values were seen in miners who had been excessively exposed for 3 months prior to sampling ($N = 20$, 8.2% aberrant cells). The reported number of aberrant cells in the group of miners below the limit ($N = 199$) was 4.1%.

Thirty-six cases of cancer were reported in this cohort of miners. Cox regression models, which took into account the age at the time of the first cytogenetic assay, radon exposure and smoking, showed significant associations between cancer incidence and the frequency of chromatid breaks and the frequency of aberrant cells. An 1% increase in the frequency of aberrant cells in miners was paralleled by a 62% increase in the relative risk of cancer ($P<0.001$) (Smerhovsky et al. 2002).

21.6.2.2 Soft Coal Open-Cast Mining

In the 1980s, soft coal open-cast mining operations were characterized by ever-present fumes and fire hazards, resulting in a significant exposure to PAHs. In 1984, two miner groups were examined: group 1 – engaged in stripping operations approximately 20–50 m under the mine surface – and group 2 – engaged in digging operations approximately 50–80 m under the mine surface. The miners were the operators of belt haulage conveyors, giant excavators and overburden dumping machines. Group 1 miners ($N = 27$) showed 2.41% aberrant cells, and group 2 miners showed ($N = 31$) 3.16% aberrant cells. In comparison, the workers in a coal cleaning plant ($N = 23$) had 1.30% aberrant cells and matched controls ($N = 24$) had 1.21% (Sram et al. 1985).

21.6.2.3 *Soft Coal Underground Mining*

In 1985, miners from deep mines using two different blast technologies were examined. In miners using a coyote-hole blast, resulting in a short exposure to nitrous gases, the frequency of aberrant cells in the group studied ($N = 31$) increased to a mean of 2.97%. The miners ($N = 22$) using a wall blast were not exposed to explosive combustion products, and the frequency of aberrant cells found in this group was 1.82%. The clastogenic effect observed in miners using a coyote-hole blast was significantly increased in comparison with a control group ($N = 20$, 1.55% aberrant cells, $P<0.05$) (Roznickova et al. 1987).

21.6.2.4 *Uranium Mining*

A study focused on chromosomal damage in uranium miners was carried out in 1990 in three mines (Hamr, $N = 20$, Tachov, $N = 23$, and Pribram, $N = 23$) located in three different regions of the Czech Republic. The control group was recruited in the town of Pribram ($N = 23$) and the capital, Prague ($N = 33$). The observed frequencies of aberrant cells were as follows: Hamr 3.44%, Tachov 2.69%, Pribram 2.65%. The frequency of aberrant cells in controls selected in Pribram was 2.16%, and that of controls selected in Prague was 1.52%. In miners from Hamr, the frequency of aberrant cells was significantly increased when compared with both control groups ($P<0.05$). Contrary to expectations, no increase in the frequency of dicentric chromosomes was observed (Sram et al. 1993a).

The uranium mines studied were very deep (1,200–1,600 m) with high temperature and humidity, providing suitable conditions for the growth of different molds. In throat swab samples taken from miners ($N = 116$), molds producing mycotoxins were identified. In contrast, mycological examination in controls ($N = 78$) was negative (Sram et al. 1993a). The same molds found in miners were also detected in samples from mine walls and timbers (Sram et al. 1993b). It was hypothesized that mutagenic mycotoxins might be another risk factor responsible for the high cancer risk in uranium miners. All three mines were shut down in 1991; therefore, no further investigations were possible.

21.6.3 Health Care

21.6.3.1 *Cytostatics*

Since 1981, the original cytostatic drug Platidiam (*cis*-diaminodichlor platinum complex, *cis*-DDP) has been produced in the former Czechoslovakia. The health care providers involved in the clinical testing of this drug began complaining about side effects very similar to those seen in treated patients. Because of a lack of other suitable methods to objectify occupational exposure to *cis*-DDP, a decision was taken to use cytogenetic analysis. The first results showed a mean frequency of aberrant cells exceeding 6.0% in the exposed staff. This finding initiated a systematic follow-up of this group and thorough inspection of all wards where antineoplastic drugs were administered. As corrective measures, the Hygiene Service required that infusions be prepared in biohazard hoods and that personal protective equipment such as gloves and masks be used during drug preparation and administration (Rössner et al. 1985; Zudova et al. 1985). Since that time, health care providers administering or preparing antineoplastic drugs have been treated as a group at high carcinogenic risk. Cytogenetic analysis of PBLs has become a component of their periodic checkups. At present, the health care providers involved in the preparation or administration of antineoplastic drugs represent one of the largest groups occupationally exposed to carcinogens (Smerhovsky et al. 2001).

21.6.3.2 Halothane

Srb et al. (1987) used cytogenetic analysis to study the effects of occupational exposure in anesthesiologists in 1981 and 1982. The halothane concentrations in the breathing zone of anesthesiologists measured during pediatric surgery ranged from 76 to 907 mg/m^3, during orthopedic surgery from 31 to 106 mg/m^3, and during neurosurgery from 30 to 49 mg/m^3. The levels of chromosomal damage in this group (N = 31) were 1.95% aberrant cells and 2.45% aberrant cells in 1981 and 1982, respectively. In controls, Srb et al. (1987) found 0.93% aberrant cells and 1.61% aberrant cells in 1981 and 1982, respectively (P<0.01 and P<0.05). These results initiated improvements in the construction of inhalation anesthetic equipment and the installation of air conditioning in operating theaters. Nevertheless, similar frequencies of aberrant cells were observed in anesthesiologists in a hospital in the town of Jihlava many years later (Hanzl 1991, unpublished results given in Sram et al. 2004a).

21.6.3.3 Microbiology

Srb and Kubzova (1985) investigated a group of subjects working in microbiological laboratories, predominantly exposed to various infectious agents. The exposed group (N = 21) showed 4.07% aberrant cells, and the control subjects (N = 65) showed 1.55% aberrant cells (P<0.001). The authors concluded that this type of exposure was associated with a high genotoxic risk.

21.7 Conclusions

We reviewed all scientific papers and reports dealing with cytogenetic analyses published by Czech authors in the period from 1974 to 2005, emphasizing the importance of the cytogenetic method not only for the Czech public health

(Sram et al. 2004a). During this period of 33 years, the method has proved its ability to detect an occupational hazard. The application of cytogenetic analysis was particularly fruitful for checking the clastogenic effects of mixtures of different chemicals and under conditions in which, besides inhalation exposure, other pathways contributed to human exposure. We observed a significant association between the increased frequency of aberrant cells and occupational exposure to carcinogens and mutagens at a group level.

As a biomarker of exposure, cytogenetic analysis wrote a great chapter in the history of Czech public health and in many cases played a crucial role in improving working conditions in the Czech Republic. An excellent example is the case of soft coal pressure gasification (Sram et al. 2004a). The gasification of soft coal was a chemical process used extensively in the Czech Republic because of its large coal reserves. The repeated findings of elevated levels of chromosomal aberrations in workers engaged in this operation became the main impetus to include this technology in the list of proven human carcinogens by the Czech Ministry of Health (1985) (group I according to IARC). Furthermore, Czech genetic toxicologists first reported the genotoxicity of soft coal open-cast mining (Sram et al. 1985), platinum cytostatics in health care providers (Zudova et al. 1985) and asbestos in workers (Srb 1984; Srb et al. 1985).

At the present time we are witnessing the final step in validating conventional cytogenetic analysis as a biomarker of early effects. During the 1990s several European countries accumulated sets of data large enough to test the cancer predictivity of the frequency of chromosomal aberrations (Hagmar et al. 1998, 2004; Bonassi et al. 2000, 2004). The Czech Republic has also contributed to this effort. From the archives of Czech genetic toxicology laboratories, we retrieved data on approximately 20,000 cytogenetic analyses performed on an occupational group of almost 12,000 individuals (Rössner et al. 2005). In this context, we would like to emphasize the fact that contrary to other studies, the incidence of cancer in exposed groups is relatively low, but still significant. We believe this is also due to the activities of the Czech Hygiene Service, which very efficiently used cytogenetic data to put into place effective preventive measures. Perhaps we may speculate that the low cancer incidence observed in our cohort is the best evidence for the importance of cytogenetic analysis for public health.

References

Albertini RJ, Anderson D, Douglas GR, Hagmar L, Hemminki K, Merlo F, Natarajan AT, Norppa H, Shuker DE, Tice R, Waters MD, Aition A (2000) IPCS guidelines for the monitoring of genotoxic effects of carcinogens in humans. Mutat Res 463:111–172

Bochkov NP, Zhurkov VS, Jakovenko KN, Kuleshov NP (1974) Culture of lymphocytes as a material to studying genetic effects in persons exposed to mutagens. Dokl Akad Nauk SSSR 218:463–465 (in Russian)

Bonassi S, Hagmar L, Strömberg U, Montagud AH, Tinnerberg H, Forni A, Heikkilä P, Wanders S, Wilhardt P, Hansteen IL, Knudsen L, Norppa H (2000) Chromosomal aberrations in lymphocytes predict human cancer independently of exposure to carcinogens. Cancer Res 60:1619–1625

Bonassi S, Znaor A, Norppa H, Hagmar J (2004) Chromosomal aberrations and risk of cancer in humans: an epidemiologic perspective. Cytogenet Genome Res 104:376–382
Buckton KE, Evans HJ (1973) Methods for the analysis of human chromosome aberrations. WHO, Geneva
Carrano A, Natarajan AT (1988) Considerations for population monitoring using cytogenetic techniques. Mutat. Res 204:379–406
Czech Ministry of Health (1985) Hygienic guidelines for the work with chemical carcinogens. Decree of the Ministry of Health of the Czech Republic no 64/1985
Ducatman A, Hirschorn K, Selikoff IJ (1975) Vinyl chloride exposure and human chromosome aberrations. Mutat Res 31:163–168
Funes-Cravioto F, Lambert B, Linsten J, Ehrenberg L, Natarajan AT, Osterman-Golkar S (1975) Chromosome aberrations in workers exposed to vinyl chloride. Lancet I 459
Gebhart E, Wagner H, Crziwok K, Behnsen H (1985) The action of anticlastogens in human lymphocyte cultures and their modification by rat-liver S 9 mix. II. Studies with vitamins C and E. Mutat Res 149:83–94
Hagmar L, Bonassi S, Stromberg U, Brogger A, Knudsen L, Norppa H, Reuterwall C (1998) Chromosomal aberrations in lymphocytes predict human cancer: a report from the European Study Group on Cytogenetic Biomarkers and Health (ESCH). Cancer Res 58:4117–4121
Hagmar L, Stromberg U, Bonassi S, Hansteen IL, Knudsen L, Lindholm C, Norppa H (2004) Impact of types of lymphocyte chromosomal aberrations on human cancer risk: results from Nordic and Italian cohorts. Cancer Res 64:2258–2263
Hansteen IL, Hillestad L, Thils-Evensen E, Heldaas SS (1978) Effects of vinyl chloride in man. A cytogenetic follow-up study. Mutat Res 51:271–278
IARC (1987) IARC monographs on the evaluation of carcinogenic risk to humans, overall evaluations of carcinogenicity: an updating of IARC monographs, vols 1–42, supplement 7. IARC, Lyons
Katosova LD (1973) Cytogenetic analysis of peripheral blood of workers engaged in the production of chloroprene. Gig Tr Prof Zabol 17:30–32 (in Russian)
Kilian DJ, Picciano D (1976) Cytogenetic surveillance of industrial populations. In: Hollander A (ed) Chemical mutagens. Principles and methods for their detection, vol 4. Plenum, New York, pp 321–339
Meretoja T, Vainio H, Sorsa M, Härkkonen H (1977) Occupational exposure and chromosomal aberrations. Mutat Res 56:193–198
Natarajan AT, Obe G (1980) Screening of human populations for mutations induced by environmental pollutants: Use of human lymphocyte system. Toxicol Environ Saf 4:468–481
Purchase IFH (1978) Chromosomal analysis of exposed populations: a review of industrial problems. In: Evans HJ, Lloyd DC (eds) Mutagen-induced chromosome damage in man. Edinburgh University Press, Edinburgh, pp 258–267
Rössner P (1985) Genetic impact of chemical compounds present in the environment. Final research report P17-335-457-05. Institute of Hygiene and Epidemiology, Prague (in Czech)
Rössner P (1989) Genetic impact of chemical compounds present in the environment. Research report P12-335-87-31.5. Institute of Hygiene and Epidemiology, Prague (in Czech)
Rössner P, Sram RJ, Novakova J, Lambl V (1980) Cytogenetic analysis in workers occupationally exposed to vinyl chloride. Mutat Res 43:425–427
Rössner P, Cerna M, Bavorova H, Pastorkova A, Ocadlikova D (1995) Monitoring of human exposure to occupational genotoxicants. Cent Eur J Publ Health 3:219–223
Rössner P, Sram RJ, Bavorova H, Ocadlikova D, Cerna M, Svandova E (1998) Spontaneous level of chromosomal aberrations in peripheral blood lymphocytes of control individuals of the Czech Republic population. Toxicol Lett 96,97:137–142
Rössner P, Bavorova H, Ocadlikova D (2000) Methods on biologic monitoring of genotoxic effects of environment – standard methods: cytogenetic analysis of peripheral lymphocytes. AHEM 5:333 (in Czech)
Rössner P, Boffetta P, Ceppi M, Bonassi S, Smerhovsky Z, Landa K, Juzova D, Sram RJ (2005) Chromosomal aberrations in lymphocytes of healthy subjects and risk of cancer. Environ Health Perspect 113:517–520

Roznickova I (1987) Possibilities of the study on genotoxic risk in occupational exposure. Research report. District Institute of Hygiene, Teplice (in Czech)

Smerhovsky Z, Landa K, Rössner P, Brabec M, Zudova, Hola N, Pokorna Z, Mareckova J, Hurychova D (2001) Risk of cancer in an occupationally exposed cohort with increased level of chromosomal aberrations. Environ Health Perspect 109:41–45

Smerhovsky Z, Landa K, Rössner P, Juzova D, Brabec M, Zudova Z, Hola N, Zarska N, Navsimalova E (2002) Increased risk of cancer in radon-exposed miners with elevated frequency of chromosomal aberrations. Mutat Res 514:165–176

Sorsa M, Wilbourn J, Vainio H (1992) Human cytogenetic damage as a predictor of cancer risk. IARC Sci Publ 543–554

Sorsa M, Autio K, Demopolous NA, Järventaus H, Rössner P, Sram RJ, Stephanou G, Vladimiropoulos D (1994) Human cytogenetic biomonitoring of occupational exposure to 1,3-butadiene. Mutat Res 309:321–326

Sram RJ (1981) Cytogenetic analysis of peripheral lymphocytes as a method for monitoring environmental levels of mutagens. In: Gut I, Cikrt M, Plaa GL (eds) Industrial and environmental xenobiotics: metabolism and pharmacokinetics of organic chemicals and methods. Springer, Berlin, pp 187–193

Sram RJ, Kuleshov NP (1980) Monitoring of the occupational exposure to mutagens by the cytogenetic analysis of human peripheral lymphocytes in vivo. Arch Toxicol Suppl 4:11–1

Sram RJ, Zudova Z, Kuleshov NP (1980) Cytogenetic analysis of peripheral lymphocytes in workers occupationally exposed to epichlorohydrin. Mutat Res 70:115–120

Sram RJ, Samkova I, Hola N (1983a) High-dose ascorbic acid prophylaxis in workers occupationally exposed to halogenated ethers. J Hyg Epidemiol Microbiol Immunol 27:305–318

Sram R, Landa K, Samkova I (1983b) Effect of occupational exposure to epichlorohydrin on the frequency of chromosome aberrations in peripheral lymphocytes. Mutat Res 122:59–64

Sram RJ, Hola N, Kotesovec F, Vavra R (1985) Chromosomal abnormalities in soft coal opencast mining workers. Mutat Res 144:271–275

Sram RJ, Binkova B, Dobias L, Rössner P, Topinka J, Vesela D, Vesely D, Stejskalova J, Bavorova H, Rericha V (1993a) Monitoring genotoxic exposure in uranium miners. Environ Health Perspect 99:303–305

Sram RJ, Dobias L, Rössner P, Vesela D, Vesely D, Rakusova R, Rericha V (1993b) Monitoring genotoxic exposure in uranium mines. Environ Health Perspect 101 (Suppl 3): 155–158

Sram RJ, Rössner P, Peltonen K, Podrazilová K, Mračková G, Demopoulos NA, Stephanou G, Vladimiropoulos D, Darroudi F, Tates AD (1998) Chromosomal aberrations, sister-chromatid exchanges, cells with high frequency of SCE, micronuclei and comet assay parameters in 1,3-butadiene exposed workers. Mutat Res 419:145–154

Sram RJ, Rössner P, Smerhovsky Z (2004a) Cytogenetic analysis and occupational health in the Czech Republic. Mutat Res 566:21–48

Sram RJ, Beskid O, Binkova B, Rössner P, Smerhovsky Z (2004b) Cytogenetic analysis using fluorescence in situ hybridization (FISH) to evaluate occupational exposure to carcinogens. Toxicol Lett 149:335–344

Srb V (1984) Genotoxic acitivity testing in production units with asbestos risk. I. Cytogenetic examination of peripheral lymphocytes. Prac Lek 36:175–178 (in Czech)

Srb V, Kubzova E (1985) Genetic risk at work in microbiological laboratory (In Czech). Cas Lek Ces 124:686–689

Srb V, Kubzova E, Musil M, Havel V, Kubikova K, Hartman M (1985) Testing the genotoxic activity in plant with asbestos risk. II. Cytogenetic monitoring of workers in 1982–1983. Prac Lek 37:76–80 (in Czech)

Srb V, Prochazkova J, Chylkova V, Turkova M et al (1987) Biologic-medical monitoring of members of surgical teams working in total anaesthetic hazards. Collec Sci Commun Charles Univ Hradec Kralove 30:337–412

Svandova E, Rössner P (1989) The contribution to the assessment of spontaneous level of chromosomal aberrations in Czechoslovak population. Prac Lek 41:216–219 (in Czech)

Zudova Z, Koudela K, Cerna M, Kapickova M, Mihula J, Suskavicova L (1985) Occupational exposure of medical personnel in selected health establishments. AHEM Suppl 3:230–235 (in Czech)
Zudova Z, Landa K (1977) Genetic risk of occupational exposure to haloethers. Mutat Res 46:242–243

22 Biological Dosimetry

GÜNTER STEPHAN, URSULA OESTREICHER, AND HORST ROMM

Abstract

Chromosome analysis is the method of choice in case of biological dosimetry, used for the quantification of exposures to ionising radiation. The advantages and disadvantages of dicentric chromosomes and symmetrical translocations are described. In general, confounding factors on the lower detectable dose limit and scoring criteria for symmetrical translocations are discussed. In the case of acute exposures, scoring of dicentric chromosomes is preferred when the blood sample can be taken shortly after the exposure. The dicentric chromosome is characteristic for ionising radiation and the spontaneous frequency is very low in the healthy general population (about one dicentric per 1,000 cells) and may be influenced by smoking. After whole-body exposure with low linear energy transfer radiation, doses down to about 100 mGy are detectable. In the case of chronic exposure or if the exposure happened years before blood sampling, the analysis of symmetrical translocations in stable cells is favoured. In contrast to dicentric chromosomes, which fade with time, symmetrical translocations can persist, being transmitted from blood-forming tissue to the peripheral blood. The background frequency of symmetrical translocations shows a clear age-dependency, resulting in an about tenfold higher spontaneous frequency and in higher individual variation. Using appropriate calibration curves, one can detect whole-body doses of about 300 mGy for individuals under 40 years of age and about 500 mGy for individuals over 40 years of age.

22.1 Introduction

Biological dosimetry is the quantification of exposures to ionising radiation by means of a biological indicator. Such indicators can provide a means of estimating an individual's exposure to ionising radiation and of predicting possible consequences. Such biological estimates of exposure would enable suitable medical treatment to be planned when information from physical dosimetry cannot be clearly related to radiation exposure or is not available.

Günter Obe and Vijayalaxmi (Eds.)
Chromosomal Alterations: Methods, Results and Importance in Human Health

Over a number of years several biological responses to radiation have been investigated, such as the induction of chromosome aberrations in peripheral blood lymphocytes and the increased level of micronuclei. In addition, other radiation-induced changes have been described using haematological, cytochemical, immunological and biochemical techniques.

Cytogenetic changes in the indicator system ‘peripheral blood’ have achieved significance for radiological protection purposes. The main breakthrough occurred in 1960 when Nowell (1960) reported a method for stimulating peripheral blood lymphocytes to divide in culture. With the availability of lymphocytes and the ease with which they can be cultured and dropped onto slides, these cells became the material of choice for biological dosimetry purposes. In 1962 it was suggested by Bender and Gooch (1962) that dicentric chromosomes in peripheral blood lymphocytes could well be used for the detection and dose assessment of human radiation exposures and these authors (Bender and Gooch 1966) used this method for the first time in the sense of biological dosimetry on the occasion of the so-called Recuplex criticality accident at Hanford, USA. A further significant development of the method was harlequin staining (Perry and Wolf 1974). With this technique it became possible to distinguish for the first time subsequent mitotic divisions after culture initiation. This discrimination of first and following cell divisions is insofar important as dicentric chromosomes are lost at cell division. Following the development of the fluorescence in situ hybridisation (FISH) technique (Pinkel et al. 1986), symmetrical translocations can easily be detected. This aberration type has the advantage that it can pass cell divisions without being selected. In consequence, dicentric chromosomes are the indicator of choice for recent and acute radiation exposures and symmetrical translocations for past and chronic radiation conditions.

This chapter describes the advantages and disadvantages of dicentric chromosomes and symmetrical translocations for the purpose of biological dosimetry considering confounding factors on the lower detectable dose limit. Furthermore, scoring criteria for symmetrical translocations are discussed.

22.2 Blood Sampling and Culture Conditions

A blood sample should be taken as soon as possible after radiation exposure. The best anticoagulant for lymphocyte cultures is preservative-free lithium heparin. In the case of partial-body or non-uniform exposure, the sample should be taken earliest about 24 h after the exposure in order to ensure a homogeneous mixture of irradiated and unirradiated lymphocytes. For the analysis of dicentric chromosomes, blood should be sampled within 4 weeks after exposure since after this time the frequency begins to decrease (IAEA 2001), but after exposure with lower doses the sampling time can be extended to about 6 months.

T lymphocytes are generally stimulated in vitro to proliferate by the mitogen phytohaemagglutinin during a culture period of about 48 h at 37°C. In order to discriminate between first and following cell divisions, bromodeoxyuridine is added to the culture medium. After the induction of a mitotic block by colcemid for the last 3 h of culture and subsequent fixation, the suspension can be dropped onto slides. The metaphase preparations can be further processed for conventional Giemsa staining (fluorescence plus Giemsa) or for FISH. A detailed protocol for culture setup is given by IAEA (2001).

22.3 Background Frequency of Asymmetrical and Symmetrical Chromosome Exchanges

An important prerequisite for dose reconstruction of exposed persons is the knowledge of the background levels of the corresponding chromosome aberration types determined in blood samples of healthy unexposed persons. Therefore a careful analysis of the baseline frequency of chromosome aberrations is important since this information has a direct bearing on biological dosimetry, especially in cases where individuals are suspected to have been overexposed, but no preexposed background aberration frequency is known.

Numerous studies have been performed to evaluate spontaneous background levels; however, the frequencies vary extensively among different studies for asymmetrical exchanges (dicentrics) (reviewed in Lloyd et al. 1980; Bauchinger 1995). Values for dicentric yields between 0.09 per 1,000 cells (Sevan'kaev et al. 1974) and 2.99 per 1,000 cells (Ganguly 1993) have been reported. Also symmetrical exchanges (translocations) vary among laboratories. It was expected that the use of harlequin staining together with standardisation of culture conditions would eliminate most interlaboratory variations, especially for dicentric chromosomes. It is evident, however, that most of the interlaboratory differences remain unexplained and should be attributed, at least partly, to the scoring criteria of the observer. In consequence, for biological dosimetry purposes each laboratory should have its own control data and its own dose–response curves. If so, in the case of a chromosome analysis of an exposed person, all conditions (culture, preparation, scoring) are identical for the three samples (control, dose–response curve, exposed person). Since the quantification of radiation exposure in the low-dose range is mainly influenced by the background frequency of chromosome aberrations, complete misinterpretation of data obtained for subjects who have received low-level exposure may occur depending to the background levels selected from the literature.

From Fig. 22.1 it becomes clear that most of the 53 individuals investigated show zero and one dicentric per 1,000 cells. Individuals with three and four dicentrics per 1,000 cells are smokers with a daily consumption of more than 30 cigarettes. The background level of this group (2.5 ± 0.6 dicentrics per 1,000 cells)

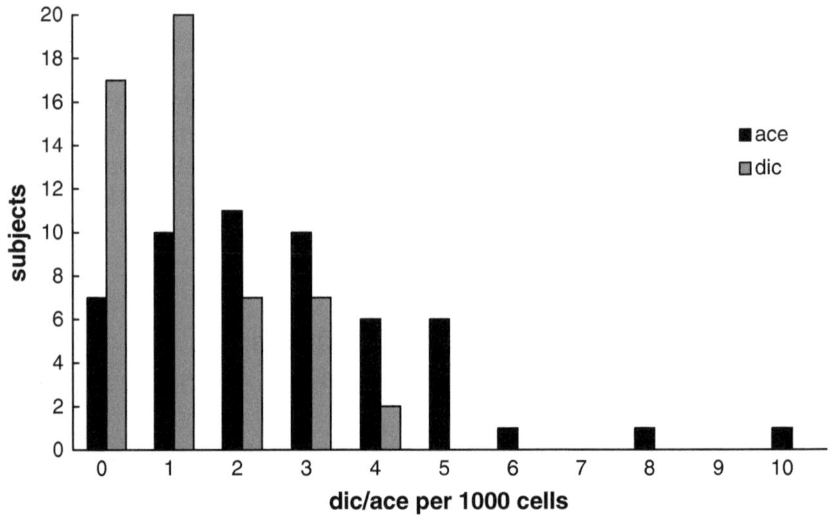

Fig. 22.1 Variation of the background frequency of dicentric chromosomes (*dic*) and acentric fragments (*ace*) between individuals (Stephan and Pressl 1999)

is significantly higher than that for non-smokers (0.9 ± 0.2 dicentrics per 1,000 cells). In the control group, with an age range from 20 to 73 years, no relation to age was observed; furthermore, there was no difference between men and women in the frequency of dicentric chromosomes (Stephan and Pressl 1999). In children, however, the background frequency of dicentrics was found to be lower (0.4 ± 0.2 dicentrics per 1,000 cells) than in adults (unpublished data).

The background frequency of symmetrical translocations is about 10 times higher than the dicentric frequency (Pressl et al. 1999). Whitehouse et al. (2005) have summarised background levels of eight working groups. It became evident that gender has no influence, smoking has only a marginal influence on the background frequency of symmetrical translocations, but age shows a very obvious effect. There is a significant difference in the background level between people younger than 40 years and those older than 40 years. As can be seen from Fig. 22.2, with increasing age the variability between individuals also increases, which, of course, influences the lowest detectable dose limit.

Sometimes the question is posed whether an increased level of chromosome aberrations has some health consequences. Since symmetrical translocations can be transmitted to descendent cells they may bear a potential risk and it is well documented that cancer is generally recognised to involve chromosomal changes. It is, however, not possible to predict some risk associated with an increased level of radiation-induced chromosome aberrations on the individual level. It was, however, shown that in populations individuals with an

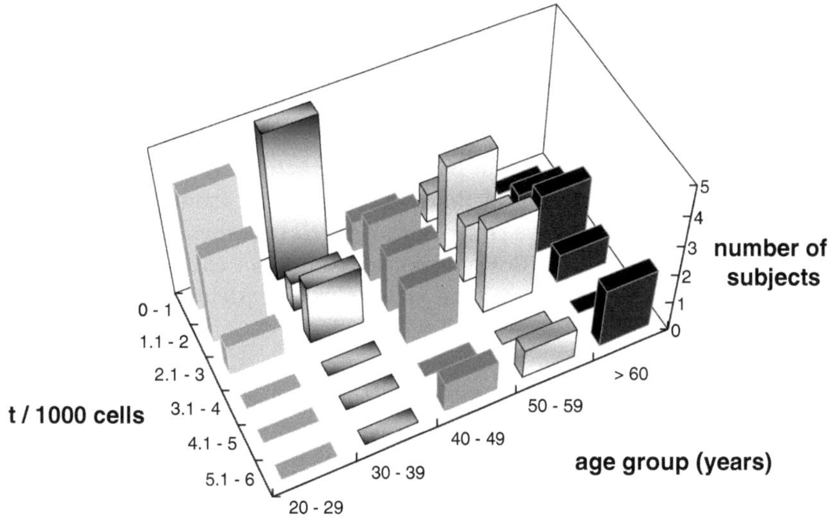

Fig. 22.2 Variation of the background frequency of symmetrical translocations (t) between individuals and dependency on age (Pressl et al. 1999)

elevated background frequency of aberrations have a higher risk of developing a carcinoma than individuals with a lower frequency (Hagmar et al. 1998).

22.4 Dose–Response Relationships

Dose–response curves for the yield of exchange aberrations are generated by irradiation of unstimulated lymphocytes in vitro. The relation between acute doses of low linear energy transfer (LET) radiation and yield of exchange aberrations can be described by the following equation:

$$y = c + \alpha D + \beta D^2,$$

where y is the yield of aberrations per cell, c is the background frequency, α is the coefficient for exchange aberrations produced by single electron tracks, β is the corresponding coefficient for aberrations produced as a consequence of two electron tracks and D is the absorbed dose in the cell. The quotient α/β is equal to the dose at which the linear and quadratic components contribute equally to the formation of exchange aberrations. Below this dose, the majority of aberrations will be produced by single tracks. As the interacting lesions must be produced almost simultaneously when the ionising particle traverses the nucleus, aberrations associated with the linear term are independent of the effects of dose rate or fractionation. The effect of fractionating the exposure or modifying the dose rate of low-LET radiation will affect the yield of aberrations mainly at higher doses where the damage is associated with the dose-squared term (βD^2) of the above equation. The reason for this is that the two breaks in two different chromosomes which are the supposition for forming exchange aberrations must be induced temporally and in space close enough for a random pairwise interaction. With increasing LET there is an increased likelihood that two breaks are produced by a single ionising track and the relative contribution of αD becomes more prominent. For high-LET radiations, the dose–response relation for exchange aberrations tends therefore to be linear and the yield is independent of exposure time and fractionation:

$$y = c + \alpha D.$$

It is of course of practical importance to know what are the relative efficiencies of the types of radiation of various qualities that are in use, or are likely to be in use, in medicine, or are in everyday use in industry. The relative biological effectiveness (RBE)–dose relations for the production of dicentric chromosomes are shown in Fig. 22.3 for various radiation qualities. It is obvious that the indicated X-ray quality possesses a higher effectivity for the induction of dicentric chromosomes in comparison with Cs γ-rays. The biological efficiency of X-rays is influenced by filtration since different filters/filter combinations produce various photon energy spectra and the biological effectivity of photon radiation is dependent on photon energy (Krumrey et al. 2004). It is well established that neutrons have a higher RBE than photon radiation, but also the RBE of neutrons is influenced by their energy (Nolte et al. 2005). In consequence, for biological dosimetry purposes an appropriate calibration curve must be used for dose estimation.

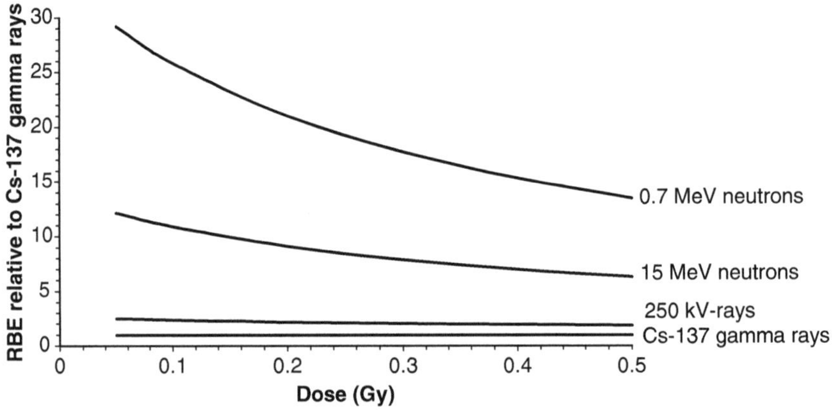

Fig. 22.3 Dose—relative biological effectiveness (*RBE*) relation for the induction of dicentric chromosomes by different radiation qualities

22.5 In Vitro and in Vivo Radiation-Induced Aberration Frequencies

The advantages of measuring radiation-induced chromosome aberrations produced in human circulating lymphocytes are obvious. The cells are in the same phase of the cell cycle at the time the blood sample is taken; therefore, only chromosome-type aberrations are produced. This fact makes quantification of the data reasonably simple. The circulating lymphocytes offer the opportunity to perform direct quantitative studies of the effect of ionising radiation on chromosomes. For the conversion of an observed frequency of aberrations into a dose, calibration curves are used which are established by in vitro exposures of whole blood. In order to extrapolate to humans from this system with confidence, results from controlled comparisons of in vivo and in vitro exposures must be known. There are some data which show that there is no significant difference in the aberration frequencies after whole-body irradiation of patients and their blood in vitro with X-rays (Buckton et al. 1969; Leonard et al. 1995). In comparative in vivo and in vitro investigations after ^{60}Co γ-irradiation, again, comparable results were obtained for the two different exposure conditions (Schmid et al. 1974). In a more recent paper (Stephan et al. 2005), results obtained in ankylosing spondylitis patients who had been treated with ^{224}Ra were in good agreement with published dose–response curves established by in vitro exposure to α-particles (Edwards et al. 1980; Schmid et al. 1996). These data demonstrate that calibration curves established by in vitro exposure can be used for the conversion of an observed frequency of exchange aberrations in an exposed individual into a dose.

22.6 Dicentric Chromosomes or Symmetrical Translocations for Dose Estimation

Biological dosimetry using the measurement of the yield of dicentric chromosomes in human lymphocytes is an established technique. It works well for acute radiation exposures when the blood sample is taken within a few weeks after exposure. This technique is capable of estimating whole-body doses of low-LET radiation down to about 100 mGy. It is less sensitive, however, for chronic exposures or if the exposure occurred years prior to the blood sampling. This is because the yield of dicentric chromosomes decreases with time after exposure depending on the cell turnover characteristics of the exposed person. A half-time of about 3 years can be applied to estimate the initial dicentric yield for dose estimation. But those estimates may be affected by uncertainties influenced by individual factors like those stimulating the cell turnover. Cells containing symmetrical translocations are not substantially eliminated from circulating blood with increasing postexposure time, since translocations do not impose a direct mechanistic hindrance to cell division and therefore they can be transmitted from blood-forming tissue to peripheral lymphocytes. For this reason, it would appear that the most reliable retrospective biological dosimetry can be expected from determining frequencies of this aberration type. Translocations can be recognised by FISH with composite whole-chromosome-specific DNA probes (Pinkel et al. 1986). There are a number of studies which have shown results suggesting that symmetrical translocations do not persist with time (Tucker et al 1997; Matsumoto et al. 1998; Natarajan et al. 1998), whereas others confirm persistence (Salassidis et al. 1995; Lucas et al. 1996; Stephan and Pressl 1997; Lloyd et al. 1998). An explanation for the conflicting results may be that in all the studies quoted only the painted chromosomes were scored. This means that translocations occurring in cells which carry also dicentric chromosomes, acentric fragments and centric rings (unstable cells) in the counterstained material are lost at cell division. Lindholm and Edwards (2004) showed, however, that when translocations in only stable cells were scored, the frequency remained at the same level for years. These results imply that retrospective dosimetry and calibration should be performed using stable cells only. In contrast to dicentric chromosomes, symmetrical translocations show an about tenfold higher background frequency and also a high interindividual variation (Fig. 22.2) and, furthermore, they show a distinct dependency on age. In consequence, the lower detectable dose limit is about 300 mGy in individuals under 40 years and about 500 mGy for individuals older than 40 years.

22.7 Summary

Scoring of chromosome aberrations in metaphase preparations of human T lymphocytes is an established method for estimating individual whole-body doses of radiation exposures. For acute exposures where the blood sample can be taken shortly after the exposure, scoring of dicentric chromosomes is the method of choice. For retrospective biological dosimetry, scoring of dicentric chromosomes is less precise since this aberration type is eliminated from peripheral blood with increasing postexposure time. Symmetrical translocations can pass cell division; therefore, they can be transmitted from blood-forming tissue to the peripheral blood and the frequency remains at the same level for years. In contrast to dicentric chromosomes, which show a low background frequency (about one dicentric per 1,000 cells), symmetrical translocations have a about tenfold higher spontaneous frequency and a higher individual variation. Furthermore, symmetrical translocations show a strong dependency on age and have an influence on the lowest detectable dose limit. Using appropriate calibration curves, we can quantify whole-body doses of about 100 mGy of low-LET radiation by dicentric chromosomes. Symmetrical translocations can detect doses to about 300 mGy for individuals under 40 years of age and about 500 mGy for individuals over 40 years of age.

References

Bauchinger M (1995) Quantification of low-level radiation exposure by conventional chromosome aberration analysis. Mutat Res 339:177–189

Bender MA, Gooch PC (1962) Types and rates of x-ray-induced chromosome aberrations in human blood irradiated in vitro. Proc Natl Acad Sci USA 48:522–532

Bender MA, Gooch PC (1966) Somatic chromosome aberrations induced by human whole-body irradiation: the "Recuplex" criticality accident. Radiat Res 29:568–582

Buckton KE, Langlands AO, Smith PG, Looby PC, Woodcock GE, McLelland J (1969) Chromosome aberrations induced in human peripheral blood by 2 MeV x irradiation to the whole body and in vitro. In: IAEA symposium on radiation induced cancer. Proceedings of the symposium in Athens. International Atomic Energy Agency, pp 135–146

Edwards AA, Purrott RJ, Prosser JS, Lloyd DC (1980) The induction of chromosome aberrations in human lymphocytes by alpha-radiation. Int J Radiat Biol 38:83–91

Ganguly BB (1993) Cell division, chromosomal damage and micronucleus formation in peripheral lymphocytes of healthy donors: related to donor's age. Mutat Res 295:135–148

Hagmar L, Bonassi S, Stromberg U, Brogger A, Knudsen LE, Norppa H, Reuterwall C (1998) Chromosomal aberrations in lymphocytes predict human cancer: a report from the European Study Group on Cytogenetic Biomarkers and Health. Cancer Res 58:4117–4121

IAEA (2001) Cytogenetic analysis for radiation dose assessment. Technical report series no 405. International Atomic Energy Agency, Vienna

Krumrey M, Ulm G, Schmid E (2004) Dicentric chromosomes in monolayers of human lymphocytes produced by monochromatized synchroton radiation with photon energies from 1.83 keV to 17.4 keV. Radiat Environ Biophys 43:1–6

Leonard A, Baltus I, Leonard ED, Gerber GB, Richard F, Wambersie A (1995) Dose-effect relationship for in vivo and in vitro induction of dicentric aberrations in blood lymphocytes of children. Radiat Res 141:95–98

Lindholm C, Edwards A (2004) Long-term persistence of translocations in stable lymphocytes from victims of a radiological accident. Int J Radiat Biol 80:559–566

Lloyd DC, Purrot RJ, Reeder EJ (1980) The incidence of unstable chromosome aberrations in peripheral blood lymphocytes from unirradiated and occupationally exposed people. Mutat Res 72:523–532

Lloyd DC, Moquet JE, Oram S, Edwards AA, Lucas JN (1998) Accidental intake of tritiated water: a cytogenetic follow-up case on translocation stability and dose reconstruction. Int J Radiat Biol 73:543–547

Lucas JN, Hill FS, Burk CE, Cox AB, Straume T (1996) Stability of the translocation frequency following whole-body irradiation measured in rhesus monkeys. Int J Radiat Biol 70:309–318

Matsumoto K, Ramsey MJ, Nelson DO, Tucker JD (1998) Persistence of radiation-induced translocations in human peripheral blood determined by chromosome painting Radiat Res 149:602–613

Natarajan AT, Santos SJ, Darroudi F, Hadjidikova V, Vermeulen S, Chatterjee S, Berg M, Grigorova M, Sakamoto-Hojo ET, Granath F, Ramalho AT, Curado MP (1998) 137 Cesium-induced chromosome aberrations analysed by fluorescence in situ hybridization: eight years follow up of the Goiania radiation accident victims. Mutat Res 400:299–312

Nolte R, Mühlbradt KH, Meulders JP, Stephan G, Haney M, Schmid E (2005) RBE of quasi-monoenergetic 60 MeV neutron radiation for induction of dicentric chromosomes in human lymphocytes. Radiat Environ Biophys 44:210–209

Nowell PC (1960) Phytohemagglutinin: an initiator of mitosis in cultures of normal human leukocytes. Cancer Res 20:462–466
Perry P, Wolff S (1974) New Giemsa method for differential staining of sister chromatids. Nature 251:156–158
Pinkel D, Straume T, Gray JW (1986) Cytogenetic analysis using quantitative high-sensitivity fluorescence hybridisation. Proc Nat Acad Sci USA 83:2934–2938
Pressl S, Edwards A, Stephan G (1999) The influence of age, sex and smoking habits on the background level of fish-detected translocations. Mutat Res 442:89–95
Salassidis K, Georgiadou-Schumacher V, Braselmann H, Müller P, Peter RU, Bauchinger M (1995) Chromosome painting in highly irradiated Chernobyl victims: a follow-up study to evaluate the stability of symmetrical translocations and the influence of clonal aberrations for retrospective dose estimation. Int J Radiat Biol 68:257–262
Schmid E, Bauchinger M, Bunde E, Ferbert HF, v Lieven H (1974) Comparison of the chromosome damage and its dose response after medical whole-body exposure to 60Co-gamma-rays and irradiation of blood in vitro. Int J Radiat Biol Relat Stud Phys Chem Med 26:31–37
Schmid E, Hieber L, Heinzmann U, Roos H, Kellerer AM (1996) Analysis of chromosome aberrations in human peripheral lymphocytes induced by in vitro alpha-particle irradiation. Radiat Environ Biophys 35:179-184
Sevan'kaev AV, Kozlov VM, Guseev GG, Izmailova NN (1974) Frequency of spontaneous chromosome aberrations in the culture of human lymphocytes. Genetics (USSR) 10:114–120
Stephan G, Pressl S (1997) Chromosome aberrations in human lymphocytes analysed by fluorescence in situ hybridization after in vitro irradiation, and in radiation workers, 11 years after an accidental radiation exposure. Int J Radiat Biol 71:293–299
Stephan G, Pressl S (1999) Chromosomal aberrations in peripheral lymphocytes from healthy subjects as detected in first cell division. Mutat Res 446:231–237
Stephan G, Kampen WU, Nosske D, Roos H (2005) Chromosomal aberrations in peripheral lymphocytes of patients treated with radium-224 for ankylosing spondylitis. Radiat Environ Biophys 44:23–28
Tucker JD, Breneman JW, Briner JF, Eveleth GG, Langlois RG, Moore DH 2nd (1997) Persistence of radiation-induced translocations in rat peripheral blood determined by chromosome painting. Environ Mol Mutagen 30:264–272
Whitehouse CA, Edwards AA, Tawn EJ, Stephan G, Oestreicher U, Moquet JE, Lloyd DC, Roy L, Voisin P, Lindholm C, Barquinero J, Barrios L, Caballin MR, Darroudi F, Fomina J (2005) Translocation yields in peripheral blood lymphocytes from control populations. Int J Radiat Biol 81:139–145

23 Statistical Methods for Biological Dosimetry

MARTA SZŁUIŃSKA, ALAN A. EDWARDS, AND DAVID C. LLOYD

Abstract

Statistical methods most commonly employed in biological dosimetry by analysis of chromosomal aberrations in blood lymphocytes are reviewed and discussed. They are presented, together with worked examples, to illustrate the principles underlying the approaches to assessing data from differing types of radiation overexposures. These techniques have been developed from experience gained over many years from dealing with many radiological incidents and with in vitro experiments that explore possible irradiation scenarios. Together, they serve to illustrate practical aspects of the techniques, ranging from fitting curves to dose–response calibration data through to specifying uncertainties on individual dose estimates. A novel method of computation of dose and the corresponding 95% limits, taking into account both the Poisson-derived uncertainties in induced aberration yield as well as uncertainties associated with the calibration curve, is presented. A very good agreement of this stringent Bayesian-like calculation approach with a more simplified one, which merely takes into consideration the Poisson uncertainties in yield, validates the approximations used over many years for dose assessment.

23.1 Introduction

Human T lymphocytes are long-lived cells that can be easily obtained from a blood sample. They comprise a population of G_0-phase cells, which however can be stimulated in vitro by the mitogen phytohaemagglutinin (Nowell 1960) and can be stopped at their first metaphase with colcemid after about 45 h of culture at 37°C. Slides are then prepared and stained with Giemsa or fluorescence in situ hybridisation (FISH) probes and scored for aberrations. A check for contamination with second-division metaphases is recommended and is carried out by incorporating bromodeoxyuridine in conjunction with fluorescence plus Giemsa staining. The full details of these techniques can be found in an IAEA (2001) manual.

With conventional Giemsa staining, the scored aberrations comprise mainly dicentrics, centric rings and acentric fragments. For biological dosimetry,

Günter Obe and Vijayalaxmi (Eds.)
Chromosomal Alterations: Methods, Results and Importance in Human Health

dicentrics are the preferred aberration owing to their low background incidence, approximately 0.5–1 dicentric per 1,000 cells, and the reliability of detection. Centric rings are approximately 5% as frequent as dicentrics (Lloyd and Purrott 1981). The main disadvantage of the dicentric assay is that the damage is unstable as damaged cells are removed from the pool of peripheral blood lymphocytes at the rate with which cells are replaced. It is presently believed that for persons with normal haematological parameters, the disappearance rate is exponential with a half-life of approximately 3 years (Lloyd et al. 1980), although significant individual variation is observed. In this respect, the introduction of the FISH technique (Pinkel et al. 1986) permits the more persistent aberrations, such as stable translocations, to be scored, providing a reliable retrospective and long-term exposure dosimetry system (Lindholm and Edwards 2004; Darroudi 2000). The progress in biological dosimetry using FISH cytogenetics has recently been reviewed by Edwards et al. (2005) and Whitehouse et al. (2005).

An additional technique occasionally employed in biodosimetry is the cytokinesis blocked micronucleus assay (IAEA 2001). This identifies fragments, or sometimes whole chromosomes, that do not segregate properly at mitosis to daughter nuclei. They are excluded into the cytoplasm as discrete chromatin positive bodies seen alongside the two daughter nuclei. The most important advantage of this technique stems from the speed with which micronuclei can be scored, which, it has been suggested, is particularly convenient in the case of initial triage of large numbers of persons involved in a major radiation incident (Kóteles 1996). However, the value of the technique is limited mainly by the larger and more variable background frequency of micronuclei as compared with dicentrics.

The aim of this chapter is to review and discuss the statistical methods employed at the Health Protection Agency (HPA) (formerly the National Radiological Protection Board) for the analysis of chromosome aberrations in blood lymphocytes. Worked examples are included, which serve to illustrate practical aspects of the techniques. A novel method of computation of dose and the corresponding 95% limits, taking into account both the Poisson-derived uncertainties in induced aberration yield as well as uncertainties associated with the calibration curve, is also presented. A very good agreement of the latter approach with a more simplified one, which merely takes into consideration the Poisson uncertainties in yield, further validates the approximations used over many years for dose assessment.

23.2 Statistical Approach

The treatment of cytogenetics data invariably relies on tests obtained from standard textbooks of statistics. Thus, methods such as χ^2 and t tests are used when testing for homogeneity or establishing differences or trends between data points or patients. Uncertainties may be expressed as standard deviations, standard errors of the mean or confidence limits based on Poisson, binomial or normal distributions. These may involve conventional or Bayesian techniques. As always, the key to data analysis is to know which is the appropriate test for the purpose and how to interpret the result. This most certainly applies to the use of cytogenetics in radiation biodosimetry and this chapter attempts to summarise experience gained over several decades.

23.2.1 Dose–Response Calibration Curves

In the majority of cases, accidental overexposures are to X-radiation or γ-radiation sources; therefore, the determination of in vitro dose–response calibration curves for these types of radiation are the priority for laboratories involved in biological dosimetry and ought to be supported by reliable and accurate physical dosimetry. Important considerations, discussed in detail in IAEA (2001), to bear in mind are (1) the uniform irradiation of the blood sample, which can be ensured by adjusting the distance between the source and the sample to be sufficiently long; (2) the production of charged particle equilibrium, achieved by surrounding the sample with materials of similar composition to blood; (3) the dose rate for the in vitro simulation of an acute exposure, which should be chosen such that the required dose is delivered in less than 15 min; and (4) the geometry and materials of and surrounding the ionisation chamber should be selected such that the corrections for distance, absorption, scattering and the mismatch at the blood–holder interface are minimised.

From a statistical point of view, in order to carry out a satisfactory calibration curve fitting procedure, a sufficient number of degrees of freedom (DF) is necessary. This translates to around ten doses in the range 0.1–5.0 Gy for low linear energy transfer (LET) radiation such as X-rays or γ-rays. Doses higher than 5 Gy show progressive deviation from a curve of simple form, eventually producing saturation of the aberration yield (see the discussion at the end of this section; Lloyd and Edwards 1983). It is still a matter of some debate as to how to treat the background level of aberrations in the fitting procedure. The problem arises owing to the very low incidence of background dicentrics that may give a zero yield in a scored data point. The inclusion of the point of zero in the curve fitting may lead to a fitted negative control level, which is not biologically meaningful. The data point in question could be omitted from the fitting procedure and the curve forced to pass through the origin; however, this too is unrealistic because there is a large body of control data showing that, while low, the background is positive. Therefore, at HPA, the solution that has been chosen is to fit the dose–response curve with a small but positive yield with a large uncertainty assigned to the zero-dose data point.

The yield of chromosome aberrations (Y) is related to dose (D) by the following equation:

$$Y = c + \alpha D + \beta D^2, \quad (23.1)$$

Table 23.1 Summary of values of α and β and their standard errors (*SE*) for different types of radiation

Reference	Radiation source	$\alpha \pm$ SE (Gy^{-1})	$\beta \pm$ SE (Gy^{-2})
Lloyd et al. (1986)	250-kVp X-rays	0.046 ± 0.005	0.065 ± 0.003
Lloyd et al. (1986)	^{60}Co γ-rays	0.014 ± 0.004	0.076 ± 0.003
Prosser et al. (1983)	^{3}H β-particles	0.054 ± 0.005	0.067 ± 0.004
Lloyd et al. (1984)	14.9-MeV neutrons	0.195 ± 0.018	0.119 ± 0.020
Lloyd et al. (1976)	Cyclotron-generated neutrons ($\bar{E}$ = 7.6 MeV)	0.482 ± 0.040	0.064 ± 0.022
Lloyd et al. (1978)	^{252}Cf fission neutrons ($\bar{E}$ = 2.13 MeV)	0.600 ± 0.019	–
Lloyd et al. (1976)	Fission neutrons ($\bar{E}$ = 0.9 MeV)	0.733 ± 0.024	–
Lloyd et al. (1976)	Fission neutrons ($\bar{E}$ = 0.7 MeV)	0.832 ± 0.032	–
Lloyd et al. (1988)	24-keV neutrons	0.821 ± 0.031	–
Edwards et al. (1980)	^{242}Cm α-particles	0.286 ± 0.015	–
Purrott et al. (1980)	^{239}Pu α-particles	0.375 ± 0.024	–
Edwards et al. (1986)	8.7-MeV protons (L = 5.1 keV μm−1)	0.044 ± 0.075	0.058 ± 0.057
Edwards et al. (1986)	23.5-MeV ^{3}He ions (L = 22 keV μm^{-1})	0.394 ± 0.018	–

$\bar{E}$ mean energy, L linear energy transfer

where c, α and β are the fitted coefficients. The control value c for dicentrics is approximately 0.0005–0.001. Some values of α and β measured at HPA are summarised in Table 23.1; however in practice, a synthesis of various experimental sets of data, judged to be the best to date, is used to produce the coefficients routinely used for the purposes of biological dosimetry.

Chromosome aberrations are a good example of a biological effect caused by ionising radiation where it is possible to discriminate differences between radiation qualities that are generally termed low LET. These are qualities for which in radiological protection a deliberate decision has been taken to ascribe a weighting factor (w_R) of 1; thus, no attempt is made to distinguish between their biological effectiveness (ICRP 1990). Nevertheless, with dicentrics, differences can be seen and are reflected in the linear yield coefficient α and are thus most evident at low doses. In practice, laboratories tend to calibrate with orthovoltage X-rays (200–250 kVp) and with a convenient γ-ray source, usually cobalt-60. However, when confronted with the need to estimate a dose in a case that involves exposure to other low-LET sources, the practice at HPA is to assume an α coefficient by extrapolating or interpolating between established calibration data. For example, when dealing with lower-energy diagnostic X-rays, such as 30 kVp, an α coefficient of about 0.05 is assumed. For cobalt-60 γ-ray sources ($\bar{E}$ = 1.2 MeV) an α value of 0.02 is used, but for lower-energy sources, also commonly used in industrial radiography, such as caesium-137 ($\bar{E} = 660$ keV) or iridium-192 ($\bar{E} = 317$ keV), α values of 0.025 and 0.03 are used, respectively.

Values of α and β from Table 23.1 have been used to produce an illustrative comparison of calibration curves for some chosen types of radiation shown in Fig. 23.1. It is worth noting that while the dose–response curves depend very strongly on the energy in the case of neutrons, they are rather similar in the case of photon radiation, but, as discussed earlier, with γ-rays being less effective at producing aberrations at low doses than X-rays. The reason for this stems from differences in the actual physical interaction of ionising radiation of different qualities with tissue. In the case of high doses of low-LET, or sparsely ionisation radiation, the ionisations are randomly distributed among cells and a large number of tracks are present; hence, assuming an equal probability that any DNA damage can interact with a nearby similar break and be converted into an aberration, aberrations should be randomly distributed among cells. Indeed, it has been shown that aberrations for low-LET radiation conform closely to the Poisson distribution; however, this is not the case for high-LET radiation (e.g. neutrons, α-particles), where more cells with multiple and no aberrations are observed than expected from the Poisson distribution. This is because ionisations are non-randomly distributed across the cell, as high-LET energy is deposited in discrete tracks. From this it follows that while aberrations produced by a single track have a frequency that is proportional to a linear function of dose (β in Eq. 23.1 equals zero; e.g. for lower-energy neutrons), those induced by two independent tracks have a frequency proportional to the square of dose. The ratio of α/β of

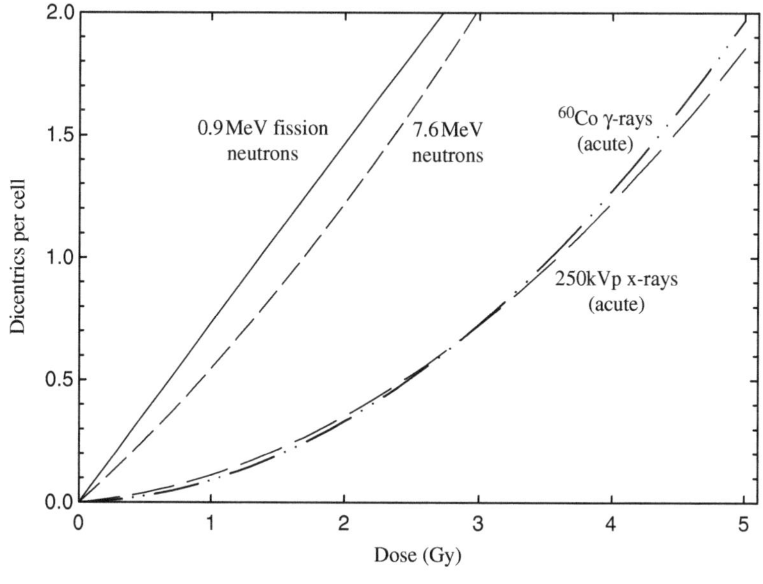

Fig. 23.1 Dose–response curves for four selected radiation qualities

the coefficients in Eq. 23.1 is equivalent to the dose at which linear and quadratic components contribute equally to the formation of aberrations.

In practice, prior to fitting a polynomial defined by Eq. 23.1 to the data points given at each dose, a test on the distribution of dicentrics is carried out in order to determine whether the yields at each data point conform to the Poisson distribution. Following the method of Savage (1970), the so-called *u* test of the goodness of fit is employed (CtamPoiss computer program available from HPA). This relies on the value of the dispersion index, which provides a measure of how well the sample's variance corresponds to that predicted by the Poisson distribution. Therefore, any deviation of the variance-to-mean ratio from unity (i.e. underdispersion or overdispersion compared with the Poisson distribution) can subsequently be taken into account in the fitting procedure by adjusting the weights of the data points appropriately (by multiplying the Poisson-derived variance of the data points by the ratio of the variance and the mean). Although some laboratories tend to use the individual ratios of the variance and the mean in their curve fitting, at HPA it is common practice to check for any trends in the values of variance-to-mean ratios with increasing dose. This is preferable because the ratios themselves have large statistical uncertainties. If no trends are observed, then a mean ratio is calculated and applied to all the data points.

The importance of weighting in curve fitting cannot be overstressed, as more accurate points have to be considered more seriously than the inaccurate ones and, indeed, many commercially available curve-fitting programs simply ascribe equal weights to each point. This has led to erroneous dicentric response curves being published. The most common approach is to minimise the sum of squared differences (SSD) between the observed and fitted values, with weights of the data points given by the inverse of their variance. It was suggested by Papworth (1975) to base the estimates of variance on the value of the fitted yield rather than the observed one. This can be done iteratively until a self-consistent solution is found (MlPol computer program available from HPA). This method of iteratively reweighted least squares is equivalent to that of the maximum likelihood (Papworth 1975). The goodness of the fit can be tested assuming that SSD follows a χ^2 distribution with DF equal to the number of data points less the number of fitted parameters (normally three: c, α, β). A close correspondence between SSD and DF implies a good fit. Whilst values of SSD significantly greater than DF suggest an uncertainty due to the lack of fit, SSD considerably less than DF hint to the data points being too close, within their errors, to the fitted curve. Generally, the uncertainties in the parameters should be based on the Poisson estimate of the variance; however, in the case of lack of fit, where SSD>DF, the uncertainties based on Poisson statistics should be increased by $(\mathrm{SSD/DF})^{1/2}$.

The consequence of the inclusion of data points obtained at 5 Gy and above in the fitting procedure can be demonstrated using data from an early classic experiment of Norman and Sasaki (1966), who irradiated human

Table 23.2 The effect of sequentially removing the higher dose levels on the results of fitting the data of Norman and Sasaki (1966)

Highest dose (Gy)	$\alpha \pm \text{SE} \times 10^{-3}\ \text{Gy}^{-1}$	$\beta \pm \text{SE} \times 10^{-2}\ \text{Gy}^{-2}$	χ^2	Degrees of freedom
30	180.0 ± 37.0	0.95 ± 0.24	205	10
25	149.0 ± 34.0	1.35 ± 0.28	142	9
20	120.0 ± 31.0	1.79 ± 0.32	99	8
12	49.0 ± 21.0	3.20 ± 0.35	30	7
8	16.0 ± 15.0	4.18 ± 0.36	11	6
6	3.4 ± 13.1	4.67 ± 0.37	5.9	5
5	−5.9 ± 4.4	5.16 ± 0.44	2.5	4

lymphocytes with X-rays to doses up to 30 Gy. As mentioned before, approximately from 5 Gy onwards, a saturation of dose–response curves is observed and, therefore, special care must be taken in the selection of the data points used in the fit. As shown in Table 23.2, the inclusion of higher-dose data points in the fit to the quadratic model (Eq. 23.1) using the maximum-likelihood method leads to a decrease in the values of the β coefficient and a simultaneous increase in the α coefficient. This is due to the fact that α and β are correlated and, hence, lower values of one must be compensated by enlarged values of the other. From Table 23.2, it can be deduced that the best fit in this case is obtained when doses up to 6 Gy are included in the fit.

23.2.2 Dose Assessment in Whole-Body Exposure

Deriving the dose from a given yield of aberrations does not present any difficulty as it follows directly from Eq. 23.1 that

$$D=\left(-\alpha+\sqrt{\left[\alpha^2+4\beta\left(Y-c\right)\right]}\right)\Big/2\beta. \tag{23.2}$$

However, there are a number of approaches that permit calculation of uncertainties. These are customarily expressed as confidence limits, with 95% being the limit chosen most often. The 95% confidence limits define an interval that embraces the true dose on 95% of occasions. The difficulty in determining the confidence limits arises from the fact that there are two components to the uncertainty: one originates from the Poisson nature of aberrations seen in blood samples from irradiated persons, and the other is associated with uncertainties in the calibration curve. This problem was addressed by Savage and Papworth (2000) and resulted in two fairly complex methods of estimating the confidence limits. In the first approach, the confidence limits can be obtained from differentiation of Eq. 23.2 and expressing the variance in dose (var$D$) in terms of variances and covariances of the fitted parameters α, β and c. Then, 95% confidence limits are given by $D \pm 1.96\text{SE}(D)$, where $\text{SE}(D) = \sqrt{\text{var}\, D}$. The second approach is similar, but relies on uncertainties in yield rather than dose. Full details of the two methods can be found in IAEA (2001).

In this chapter, it is intended to emphasise the two methods of the confidence limit estimation that are routinely employed at HPA. It has recently been shown that the simplified approach, described in detail below, produces results that are in very good agreement with a more sophisticated explicit calculation. The simplified approach, illustrated in Fig. 23.2, is a modification of the method proposed by Merkle (1983) and involves the following steps:

- Determining the yields corresponding to the lower and upper 95% confidence limit on the observed yield (Y_L and Y_U), assuming the Poisson distribution.
- Determining the dose at which Y_L crosses the dose–response curve. This corresponds to the lower confidence limit (D_L).
- Determining the dose at which Y_U crosses the dose–response curve. This corresponds to the upper confidence limit (D_U).

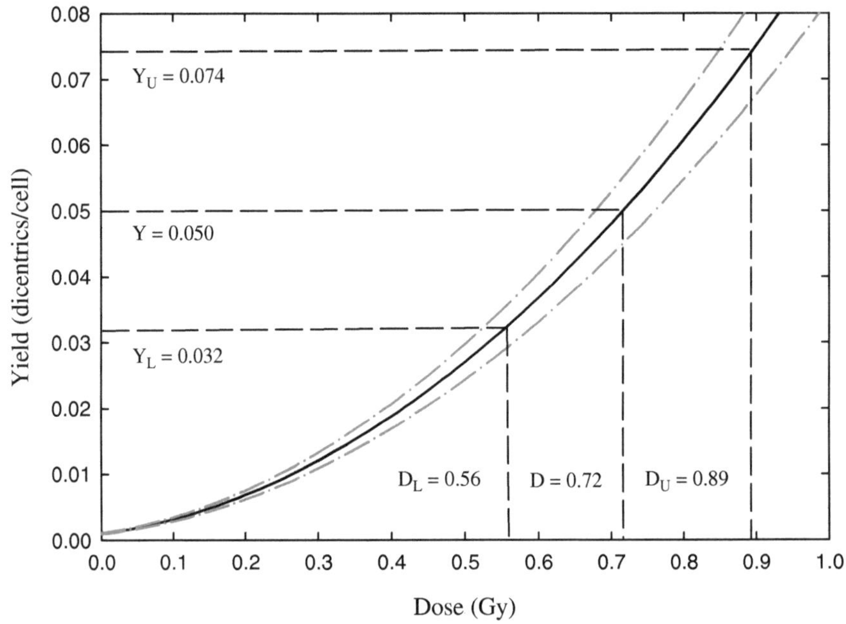

Fig. 23.2 A dose-response calibration curve used to estimate uncertainties for the simplified approach. The *dot-dashed lines* represent 95% confidence limits of the calibration curve. *Y* yield, *D* dose, *U* upper 95% confidence limits, *L* lower 95% confidence limit

Example: Twenty-five dicentrics are observed in 500 cells from a person overexposed to γ-rays. The parameters of the γ-ray calibration curve, shown in Fig. 23.2, are $c = 0.001$, $\alpha = 0.014$ and $\beta = 0.076$.

- From standard statistical tables or using a computer spreadsheet, the 95% lower and upper confidence limits on 25 observations are 16.18 and 36.91, respectively; hence, $Y_L = 16.18/500 = 0.032$ and $Y_U = 36.91/500 = 0.074$ dicentrics per cell.
- Y_L intersects the calibration curve at $D_L = 0.56$ Gy.
- Y_U intersects the calibration curve at $D_U = 0.89$ Gy.

It is important to recognise that the simplified approach does not take account of uncertainties associated with the dose–response curve. An initial approach to take into consideration both the uncertainties in the calibration curve as well as the Poisson nature of the yield was made by Merkle (1983) by reading off the value of dose for which $Y_L(Y_U)$ intersects the upper (lower) 95% confidence limit of the dose–response calibration curve. This approach overestimated the effect of uncertainties in the calibration curve.

The second approach discussed here, developed specifically to test the departure of the results of the simplified method from an explicit calculation accounting for all uncertainties, relies on a systematic convolution of the possible values of yield with the possible values of the calibration curve. This is a Bayesian type of approach as uncertainties in the observed yield and in the calibration curve are expressed as continuous probability distributions. More specifically, as shown schematically in Fig. 23.3, the values of possible yields are described by the Poisson likelihood function for the particular number of observed dicentrics (dark grey curve in the Fig. 23.3, in this case 25 dicentrics seen in 500 cells). This Poisson likelihood function is then convoluted with a distribution of possible values of dose given, for each value of yield (Y), by dose likelihood functions (light grey curves in Fig. 23.3). These are assumed to be normal and are determined using the knowledge of the standard errors on each data point of the calibration curve (obtained in turn from the curve-fitting procedure, described earlier in Sect. 23.2.1). The standard deviation of the normal distribution can be approximated by $\sigma = (D_{UL} - D_{LL})/2$, where D_{LL} and D_{UL} are the lower and upper limits of dose computed from the standard error in Y (Fig. 23.3). In the case of γ-rays, the standard errors on the yields at individual data points incorporated in the calculation varied from very large at very low doses to approximately 2–3% on doses from 0.8 Gy onwards. The summation of all the dose likelihood functions, weighted according to the likelihood given in dark grey, produces the final distribution of dose (thick black curve), from which a mean, a most likely or a median value, as well as 95% limits, can be calculated. The development of the computer program required checks of convergence of the dose results depending on the increments in yield and dose used for the calculation. Moreover, in order to simulate a biologically

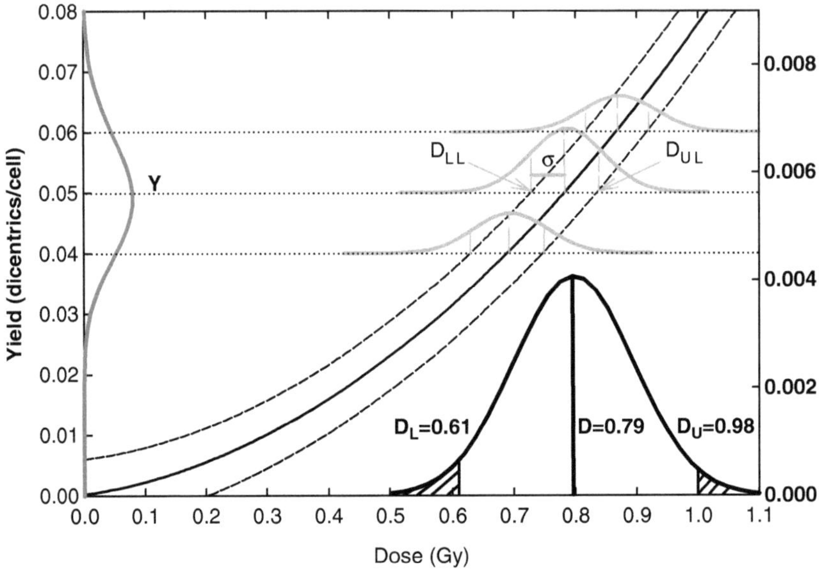

Fig. 23.3 Illustration of the explicit dose calculation which takes into account both the Poisson nature of the yield distribution and the uncertainties in the calibration curves: *solid black curve* γ-ray calibration curve $Y = 0.0005(\pm 0.0004) + 0.0128(\pm 0.0031)\,D + 0.064(\pm 0.0021)\,D^2$; *dashed black curves* -standard error on the calibration curve (exaggerated for clarity); *dark grey curve* the yield likelihood function for 25 dicentrics scored in 500 cells; *light grey curves* dose likelihood functions; *thick black curve* dose distribution. For more details see the text

realistic situation, it has been assumed that all negative dose values correspond to zero dose. In Table 23.3, the estimated doses obtained employing the simplified method are compared with those from the explicit calculation. Strictly, the 95% confidence limits in the simplified method and the 95% limits in the calculation are not equivalent; nevertheless, a very good agreement between the two set of data for the commonly encountered numbers of aberrations entirely justifies the use of the simplified method for the purposes of biological dosimetry. However, for very high and very low yields of aberrations, notable departures from the results of the simplified approach may be expected. Whilst at low doses the difference is due to negative doses not being permitted in the Bayesian calculation, at high doses it is the errors in the curve that become increasingly significant. It was concluded by Lloyd et al. (2007) that by the traditional method of scoring 1,000 metaphases, the lowest measurable γ-ray dose is around 100 mGy. The present investigation has also confirmed the general feeling that the uncertainties on a *good* calibration curve contribute little to the total convoluted uncertainty of the estimated dose, in contrast to the uncertainties associated with the Poisson nature of the yield distribution. Indeed in the case of γ-rays for the yields considered in Table 23.3, it has been verified that σ

Table 23.3 A comparison between the dose and the 95% lower (*LCL*; *LL*) and upper (*UCL*; *UL*) limits obtained using the simplified approach and the explicit calculation for various numbers of dicentrics scored in 500 cells in the case of γ-ray and fission neutron irradiation

	Simplified method			Explicit calculation		
Dicentrics	95% LCL	Mean dose (Gy)	95% UCL	95% LL	Mean dose (Gy)	95% UL
γ-rays: $Y = 0.0005(\pm 0.0004) + 0.0128(\pm 0.0031)D + 0.064\ (\pm 0.0021)D^2$						
0	0	0	0.242	0	0.070	0.249
5	0.129	0.297	0.505	0.164	0.326	0.506
20	0.519	0.692	0.883	0.538	0.704	0.882
50	0.978	1.151	1.336	0.990	1.154	1.334
Neutrons: $Y = 0.0005(\pm 0.0005) + 0.832(\pm 0.009)D$						
0	0	0	0.008	0	0.002	0.008
5	0.003	0.011	0.028	0.005	0.013	0.025
20	0.029	0.048	0.074	0.028	0.046	0.068
50	0.089	0.120	0.158	0.083	0.112	0.148

would have to be increased by a factor larger than 3 in order to observe considerable differences in the 95% limits of dose.

The procedures discussed above for the dicentric assay, such as principles of curve fitting, distribution testing and derivation of uncertainties, apply equally to the FISH assay as well as the micronucleus assay. In the latter case, it is worth noting that micronuclei are inherently overdispersed. For FISH, statistical analysis using the frequency of aberrations detected with chromosome painting probes depends on, amongst others, the number of cells analysed, the cocktail of chromosomes used and the exposure time. The full genome translocation frequency is given by $F_G = F_p/2.05f_p(1 - f_p)$, where F_p is the aberration frequency detected by FISH and f_p is the fraction of genome hybridised (Lucas 1992) taking into account the gender of the subject. Values of f_p can be obtained from Morton (1991). It is important to note that the control level of aberrations in the case of FISH analysis is much higher than in the case of the dicentric assay and is variable, with age being the main confounding factor. This is primarily because FISH is a retrospective technique and thus reflects a cumulative lifetime dose to the bone marrow. In Fig. 23.4, the results of the most comprehensive study to date of the dependence of the background levels of stable translocations on age is shown, where the age group of 80 years or more produced approximately 15 translocations per 1,000 genome-equivalent cells. To summarise, the calibration curves used in the case of FISH for the dose assessment differ from those for the dicentric assay in that the *c* coefficient in Eq. 23.2 is age-dependent and the quadratic term is equal to zero for all qualities of radiation as the exposure is protracted, i.e. the probability of a translocation being produced by two or more different tracks is negligible.

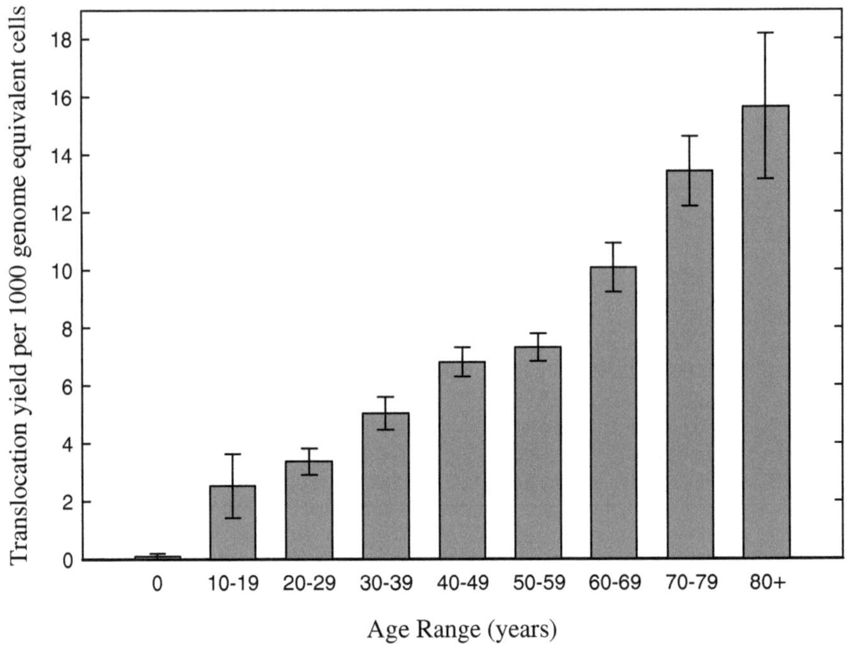

Fig. 23.4 Dependence of the background level of stable translocations on age measured with fluorescence in situ hybridisation (Whitehouse et al. 2005)

23.3 Examples of Dose Estimation

23.3.1 Criticality

In a criticality accident the body is irradiated by both neutrons and γ-rays. Since the ratio of doses due to neutrons and γ-rays is often known from physical measurement, the separate doses due to the two types of radiation can be obtained iteratively as follows:

1. Initially, all aberrations are assumed to be due to neutrons, and from the measured yield of dicentrics, a dose is estimated.
2. The ratio of neutrons to γ-rays is used to estimate the dose due to γ-rays.
3. From the estimate of γ-ray dose, a yield of dicentrics due to γ-rays is obtained.
4. The neutron yield is obtained by subtracting the yield of dicentrics in stage 3 from the measured yield.
5. A new estimate of neutron dose is made, and stages 2–5 are repeated until a self-consistent solution is found.

Example: Following a criticality accident 100 cells are scored and 100 dicentrics are observed. The neutron to γ-ray ratio available from physical measurements is 2:3. Cytogenetic dose estimates are to be made with 0.7-MeV fission neutron ($Y = 0.0005 + 0.83D$) and ^{60}Co γ-ray ($Y = 0.0005 + 0.014D + 0.076D^2$) calibration curves:

1. 100 dicentrics per 100 cells = 1.0 dicentric per cell, which is equivalent to 1.20 Gy neutron dose.
2. 1.20 × 3/2 = 1.81 Gy γ-rays.
3. 1.81 Gy γ-rays is equivalent to 0.275 dicentrics per cell.
4. 1.0–0.275 = 0.725, which is the dicentric yield attributable to neutrons.
5. 0.725 dicentrics per cell is equivalent to 0.87 Gy neutron dose.

In Table 23.4, the sequence of steps 1–5 is laid out. After a few iterations a solution is found with doses of approximately 1.0 and 1.5 Gy due to neutrons and γ-rays, respectively.

An alternative approach developed for criticality incidents involves an elaborate Bayesian type of calculation (Brame and Groer 2003), where uncertainties in all parameters can be rigorously taken into account by expressing them in terms of probability distributions. This technique, although time-consuming, is particularly useful in situations where the γ-ray to neutron ratio is uncertain (Voisin et al. 2004). However, the two methods give results in good agreement.

23.3.2 Low Dose Overexposure

Example: A case of a worker who returned a thermoluminescence dosimeter badge which recorded a dose of 66 mSv is considered. This was reported to be due to γ-rays of energy around 100 keV and it was not possible to discount it as a false reading owing to various circumstances. Subsequently, the dose was allocated to the worker's dose record with a consequence of the worker being removed from radiological work. This case was referred for a cytogenetic test, which resulted in an observation of one dicentric in 1,000 metaphases. From the calibration curve $Y = 0.001 + 0.04D + 0.06D^2$ a dose estimate of zero with

Table 23.4 Sequence of steps for estimation of doses for mixed neutron and γ-ray irradiation

Steps 1 and 5: neutron dose (Gy)	Step 2: γ-ray dose (Gy)	Step 3: γ-ray yield (dicentrics/cell)	Step 4: neutron yield (dicentrics/cell)
1.20	1.81	0.275	0.725
0.87	1.31	0.149	0.851
1.02	1.54	0.202	0.798
0.96	1.44	0.178	0.822
0.99	1.48	0.188	0.812

95% confidence limits ranging from 0 to 100 mGy was obtained and, thus, a dose of 66 mSv could not be discounted. However, there is a different way in which results can be presented. If only two possibilities are taken into consideration, i.e. zero dose or 66 mGy, then the odds in favour of zero dose come out as 4.5:1. This is obtained from the relative probabilities of observing one dicentric from the two doses. If the dose was zero, then the expected number of dicentrics to be observed is 1 in 1,000 (the presumed background). For a dose of 66 mGy, 3.9 dicentrics are expected in 1,000 cells by reference to the calibration curve. Therefore, the chances of observing one dicentric when one and 3.9 are expected, derived from a Poisson distribution, are $e^{-1} = 0.368$ and $3.9e^{-3.9} = 0.08$, respectively. The ratio of the two values is approximately 4.5.

In this particular case, the latter approach of presenting the results might be more appropriate because on the balance of probabilities this was a dose received by the badge and not the person. Whilst long-term health effects at such low doses are unlikely, there are legal and social consequences. In Table 23.5, 95% confidence limits for various low yields of dicentrics in 1,000 cells are shown together with odds ratios representing the probability of 0 or 100 mGy dose.

Experience has shown that when results of biodosimetry investigations are reported to physicians or even directly to patients, it is often difficult for them to comprehend the concept of confidence limits. The odds method, when appropriate, is much more easily understood. However, for situations where the odds method cannot be used, a new approach expressing a normalised distribution in dose (Fig. 23.5) has been developed in an effort to visually aid the presentation of the results (Lloyd et al. 2006). The method for deriving this distribution is to set up the probability distribution in yield which could have produced the observed number of dicentrics, which is

Table 23.5 Ninety-five percent upper and lower confidence limits (*UCL*; *LCL*) on various low yields of dicentrics in 1,000 cells and the odds ratios showing the likelihood (p) of the observation given that the dose is 0 or 100 mGy. The assumed calibration is $Y = 0.001 + 0.04D + 0.06D^2$

Observations (dicentrics)	Dose (Gy) LCL	Mean	UCL	Odds ratio p(0 Gy)/p(0.1 Gy)
0	–	0	0.06	180:1
1	0	0	0.10	25:1
2	0	0.02	0.13	4:1
3	0	0.05	0.16	1:1
4	0	0.07	0.18	1:9
5	0.02	0.10	0.21	1:50
6	0.03	0.11	0.23	1:310
7	0.04	0.13	0.25	1:1,900

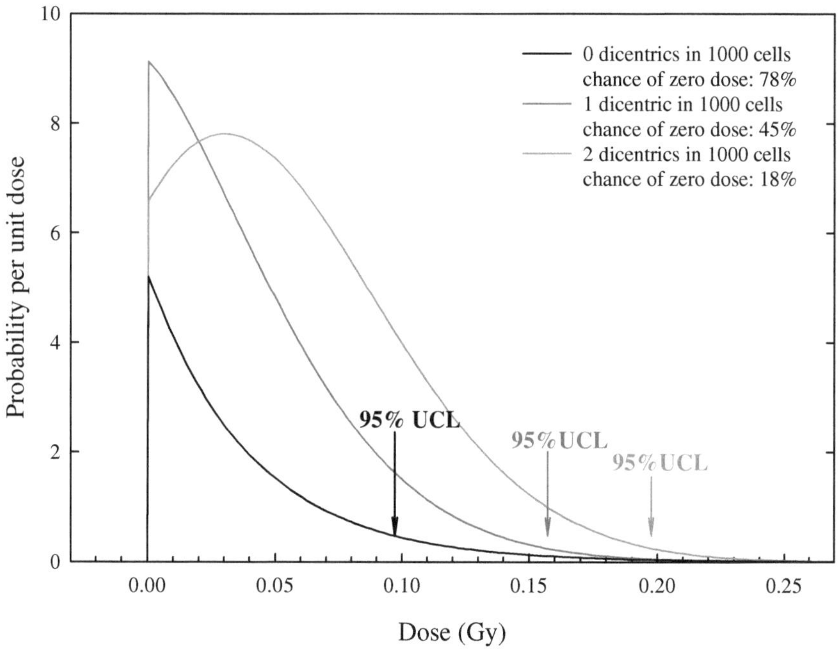

Fig. 23.5 A probability distribution in dose derived from the observed number of dicentrics in 1,000 cells, calculated for high-energy γ-rays (e.g. cobalt-60). *UCL* upper confidence limit

numerically identical to the likelihood function. Then, for each value of yield, a value of dose can be assigned, leading to the distribution in dose. Arbitrarily, all situations where the yield is less than the control level have been assigned a dose of zero and, thus, the proportion of the full bell-shaped curve that represents doses below zero has not been drawn.

23.3.3 Partial Body Exposure

In the case of a partial body exposure a more realistic estimate of dose may be given than simply stating the mean dose to the whole body. For partial body exposures, where a high dose is received inhomogeneously over the body, the distribution of dicentrics amongst all scored cells is overdispersed and this is used to estimate the extent of partial body irradiation. The observed distribution of cells is assumed to comprise (1) a Poisson distribution, which represents the irradiated fraction of the body, and (2) the remaining unexposed part. It should be stressed that it is a simplified way of approximating a real situation where a gradient of doses is expected. Undamaged cells include two subpopulations, those from the unexposed fraction and irradiated cells representing the first term of the Poisson series (e^{-Y}, i.e. irradiated but no dicentrics observed). Maximum-likelihood estimates of the fraction, f, of cells scored which were irradiated and the corresponding mean yield, Y, may be made using the following equations (Lloyd et al. 1991):

$$\frac{Y}{1-e^{-Y}} = \frac{X}{N-n_0} \tag{23.3}$$

and

$$f = \frac{X}{NY}, \tag{23.4}$$

where N is the number of cells scored, X is the number of dicentrics observed and n_0 is the number of cells free of dicentrics. Equation 23.3 can only be solved iteratively. Then, the mean dose to the irradiated fraction of the body can be estimated using the standard dose–response curves. In order to derive an actual fraction of body exposed to radiation, f needs to be corrected for the effects of mitotic delay and interphase death (apoptosis), as these are the factors reducing the ability of cells to reach metaphase in a 48-h culture. If the fraction of irradiated cells which reach metaphase was p, the fraction of the body irradiated, F, is given by

$$F = \frac{f/p}{1-f+f/p}. \tag{23.5}$$

At HPA, the computer program CtamPoiss (available on request) is used to solve the contaminated Poisson problems. For more details on the limitations of the aforementioned approach, see IAEA (2001).

An alternative approach to estimate a dose delivered in a partial body exposure was proposed by Sasaki and Miyata (1968) and is known as the Qdr method. In this case, the yield of dicentrics and rings in just the damaged cells is considered, assuming that these cells were present at the time of the irradiation. Then, Qdr is the expected yield of dicentrics and rings (X) amongst the damaged cells (N_d) and is given by

$$\mathrm{Qdr} = X/N_d = Y_1/\left[1-\exp\left(-Y_1-Y_2\right)\right], \tag{23.6}$$

where Y_1 and Y_2 are yields of dicentrics plus rings and excess acentrics, respectively. Despite the limitations of the original Qdr method, which include the lack of any information on the size of the fraction irradiated as well as incorrect assumptions that excess acentric aberrations conform to a Poisson distribution and that the background frequency of dicentrics plus rings is zero, the agreement between the two methods described is good.

Additionally, in an attempt to reduce the oversimplification of regarding the exposure geometry as two fractions, an unirradiated and a uniformly irradiated component, Sasaki (2003) suggested an approach whereby the chromosome aberration distribution is unfolded into a mixed Poisson distribution and, from that, into a dose-distribution profile.

Example: An in vitro experiment was performed to simulate partial body exposure, whereby 50% of irradiated cells (5 Gy, 250 kVp X-ray) were mixed with unirradiated ones from the same donor (Lloyd et al. 1991). The number of dicentrics scored was 45 in 247 cells, with the following distribution: 225 cells with zero, four with one, 13 with two and five with three dicentrics. This produced the ratio of variance to mean of 2; therefore, an overdispersion expected from partial body exposure was revealed. The application of the contaminated Poisson method using the dose–effect curve of $Y = 0.0005 + 0.04D + 0.06D^2$ and the value of $D_0 = 3$ Gy ($p = e^{-D/D_0}$), which represents the relative selection against irradiated cells reaching metaphase, resulted in an estimate of the mean dose of (4.9 ± 0.5) Gy delivered to an estimated (40 ± 6)% of the body. Only (0.11 ± 0.02)% of the scored cells appeared to have been irradiated, which demonstrates the need to compensate for the reduction due to interphase death and mitotic delay. In order to carry out the Qdr analysis, the number of dicentrics plus rings, 48 in this case, as well as the number of damaged cells, 31, and the following dose–response calibration curves are required: $Y_{\text{dicentrics + rings}} = 0.0005 + 0.043D + 0.063D^2$ and $Y_{\text{acentrics}} = 0.0035 + 0.051D + 0.11D^2$. By substitution into Eq. 23.6,

$$\text{Qdr} = 48/31 = 1.55 = \frac{0.0005 + 0.043D + 0.063D^2}{1 - \exp\left(-0.0035 - 0.051D - 0.11D^2\right)}. \tag{23.7}$$

Equation 23.7 can be solved for D iteratively and gives an estimate of 4.4 Gy. Sasaki and Miyata (1968) did not state how uncertainties on dose could be calculated and it is not at all obvious how this should be done. The problem lies in the fact that both the number of dicentrics and rings as well as the number of damaged cells carry uncertainties, which are correlated. It might be possible to use Monte Carlo simulation techniques to solve the problem but this has not, to our knowledge, been done. This problem becomes even more complex when the errors in the calibration curves are considered. Nevertheless, the contaminated Poisson approach and the Qdr method yield consistent results.

23.3.4 Protracted and Fractionated Exposure

Protraction or fractionation of exposure can result in reduced numbers of observed chromosome aberrations. Whilst it is not the case for high-LET radiation, because the dose–response curve is linear, in the case of low-LET radiation the β coefficient in Eq. 23.1 is reduced. This is due to repair mechanisms that have time to operate during the course of the protracted exposure. It has been shown that the decrease in the frequencies of aberrations appears to follow an exponential function with a mean time of approximately 2 h.

Lea and Catcheside (1942) considered a time-dependant factor, known as the G function, to facilitate the modification of the dose-squared coefficient as follows:

$$Y = \alpha D + \beta G(x)D^2, \tag{23.8}$$

where $G(x) = 2[x - 1 + \exp(-x)]/x^2$ and $x = t/t_0$, where t is the time over which the radiation occurred and t_0 is the mean lifetime of the breaks (about 2 h). Applying these ideas to protracted and fractionated cases requires a certain amount of information about the incident, such as the time over which exposure happened and the constancy of the dose rate. The analysis is only worthwhile in instances where large doses and long exposure times (up to few days) are experienced. For a dose being delivered over a very long period of time, $G(x)$ reaches zero since Eq. 23.1 reduces to $Y = c + \alpha D$. For interfraction times shorter than about 6 h, the G function can be written as $\exp(-t_1/t_0)$, where t_1 is the time between fractions.

Example: In 1998 a serious accident occurred in Turkey when a ^{60}Co teletherapy source was sold as scrap metal. Ten persons were exposed, mainly during 1 day, with exposure times ranging from 2 to 7 h. Here, a case of one of the subjects exposed during 6 h is considered. The pooled results from three laboratories yielded 474 dicentrics scored in 1,363 cells. From the acute dose–response curve $Y = 0.001 + 0.02D + 0.06D^2$, the acute dose (± standard error) of 2.2 ± 0.1 Gy is obtained. The application of the G function with $x = t/t_0 = 6/2 = 3$ gives $G(x) = 0.46$; hence, the dose–response curve becomes $Y = 0.001 + 0.02D + 0.028D^2$ and the dicentric frequency corresponds to a 6 h exposure of 3.2 ± 0.1 Gy.

23.4 Summary

Statistical methods in biological dosimetry are well developed as over the years a wealth of information has been gleaned through the experience of dealing with a number of different radiological incidents as well as in vitro experiments designed to explore possible irradiation scenarios. In this chapter, appropriate statistical techniques have been reviewed and discussed in order to highlight the scope of their application. These were accompanied by worked examples, often extracted from actual incidents, which further elucidate the principles underlying each of the approaches. It is worth emphasising that different radiological scenarios require different sets of information in order to carry out meaningful analyses. It is also pleasing to confirm that approximations used routinely in the calculation of 95% confidence limits on dose are in very good agreement with a stringent Bayesian-like calculation which takes into account both the Poisson-derived uncertainties in yield as well as uncertainties in the dose–response calibration curve.

References

Brame RS, Groer PG (2003) Bayesian methods for chromosome dosimetry following a criticality accident. Radiat Prot Dosim 104:61–63

Darroudi F (2000) The use of FISH translocations analyses for retrospective biological dosimetry: How stable are stable chromosome aberrations? Radiat Prot Dosim 88:101-109

Edwards AA, Purrott RJ, Prosser JS, Lloyd DC (1980) The induction of chromosome aberrations in human lymphocytes by alpha-radiation. Int J Radiat Biol 38:83–91

Edwards AA, Lloyd DC, Prosser JS, Finnon P, Moquet JE (1986) Chromosome aberrations induced in human lymphocytes by 8.7 MeV protons and 23.5 MeV helium-3 ions. Int J Radiat Biol 50:137–145

Edwards AA, Lindholm C, Darroudi F, Stephan G, Romm H, Barquinero J, Barrios L, Caballin MR, Roy L, Whitehouse CA, Tawn EJ, Moquet J, Lloyd DC, Voisin P (2005) Review of translocations detected by FISH for retrospective biological dosimetry applications. Radiat Prot Dosim 113:396–402

IAEA (2001) Cytogenetic analysis for radiation dose assessment. Technical report series no. 405. International Atomic Energy Agency, Vienna

ICRP (1990) Recommendations of the international commission on radiological protection. (Publication 60). Ann ICRP 21

Kóteles GJ (1996) The human lymphocytes micronucleus assay. A review on its applications in occupational and environmental medicine. Cent Eur J Occup Environ Med 2:12–30

Lea DE, Catcheside DG (1942) The mechanism of the induction by radiation of chromosome aberrations in Tradescantia. J Genet 44:216–245

Lindholm C, Edwards AA (2004) Long-term persistence of translocations in stable lymphocytes from victims of a radiological accident. Int J Radiat Biol 80:559–566

Lloyd DC, Edwards AA (1983) Chromosome aberrations in human lymphocytes: effect of radiation quality, dose and dose rate. In: Ishihara T, Sasaki MS (eds) Radiation-induced chromosome damage in man. Liss, New York, pp 23–49

Lloyd DC, Purrott RJ (1981) Chromosome aberration analysis in radiological protection dosimetry. Radiat Prot Dosim 1:19–28

Lloyd DC, Purrott RJ, Dolphin GW, Edwards AA (1976) Chromosome aberrations induced in human lymphocytes by neutron irradiation. Int J Radiat Biol 29:169–182

Lloyd DC, Purrott RJ, Reeder EJ, Edwards AA, Dolphin GW (1978) Chromosome aberrations induced in human lymphocytes by radiation from 252Cf. Int J Radiat Biol 34:177–186

Lloyd DC, Purrott RJ, Reeder EJ (1980) The incidence of unstable chromosome aberrations in peripheral blood lymphocytes from unirradiated and occupationally exposed people. Mutat Res 72:523–532

Lloyd DC, Edwards AA, Prosser JS, Bolton D, Sherwin AG (1984) Chromosome aberrations induced in human lymphocytes by D-T neutrons. Radiat Res 98:561–573

Lloyd DC, Edwards AA, Prosser JS (1986) Chromosome aberrations induced in human lymphocytes by in vitro acute X and gamma radiation. Radiat Prot Dosim 15:83–88

Lloyd DC, Edwards AA, Prosser JS, Finnon P, Moquet JE (1988) In vitro induction of chromosomal aberrations in human lymphocytes, with and without boron 10, by radiations concerned in boron neutron capture therapy. Br J Radiol 61:1136–1141

Lloyd DC, Edwards AA, Natarajan AT, Darroudi F (1991) Biological dosimetry applied to in vitro simulated partial body irradiation. In: CEC report. Treatment and biological dosimetry of exposed persons. European Commission, Luxembourg

Lloyd DC, Edwards AA, Moquet JE, Hone PA, Szłuińska M (2006) Doses in radiation accidents investigated by chromosomal aberration analysis XXIV: review of cases investigated, 2003–2005. HPA-RPD-012. Health Protection Agency, Chilton

Lloyd DC, Edwards AA, Szłuińska M (2007) The minimum detectable dose by biodosimetry in a radiation accident. Mutat Res (in press)

Lucas JN (1992) Rapid translocation frequency analysis in humans decades after exposure to ionising radiation. Int J Radiat Biol 62:53–63

Merkle W (1983) Statistical methods in regression and calibration analysis of chromosome aberration data. Radiat Environ Biophys 21:217–233
Morton NE (1991) Parameters of the human genome. Proc Natl Acad Sci USA 88:7474–7476
Norman A, Sasaki MS (1966) Chromosome-exchange aberrations in human lymphocytes. Int J Radiat Biol 11:321–328
Nowell PC (1960). Photohemagglutinin – an initiator of mitosis in cultures of normal human leukocytes. Cancer Res 20:462–466
Papworth D (1975) Curve fitting by maximum likelihood. Radiat Bot 15:127–131
Pinkel D, Straume T, Gray JW (1986) Cytogenetic analysis using quantitative high-sensitivity fluorescent hybridisation. Proc Nat Acad Sci USA 83:2934–2938
Prosser JS, Lloyd DC, Edwards AA, Stather JW (1983) The induction of chromosome aberrations in human lymphocytes by exposure to tritiated water in vitro. Radiat Prot Dosim 4:21–26
Purrott RJ, Edwards AA, Lloyd DC, Stather JW (1980) The induction of chromosome aberrations in human lymphocytes by in vitro irradiation with α-particles from plutonium-239. Int J Radiat Biol 38:277–284
Sasaki MS (2003) Chromosomal biodosimetry by unfolding a mixed Poisson distribution: a generalized model. Int J Radiat Biol 79:83–97
Sasaki MS, Miyata H (1968) Biological dosimetry in atom bomb survivors. Nature 220:1189–1193
Savage JRK (1970) Sites of radiation induced chromosome exchanges. In: Ebert M, Howard A (eds) Current topics in radiation research vol 6. North-Holland, Amsterdam, pp 129–194
Savage JRK, Papworth DG (2000) Constructing a 2B calibration curve for retrospective dose reconstruction. Radiat Prot Dosim 88:69–76
Voisin P, Roy L, Hone PA, Edwards AA, Lloyd DC, Stephan G, Romm H, Groer PG, Brame R (2004) Criticality accident dosimetry by chromosomal analysis. Radiat Prot Dosim 110:443–447
Whitehouse CA, Edwards AA, Tawn EJ, Stephan G, Oestreicher U, Moquet JE, Lloyd DC, Roy L, Voisin P, Lindholm C, Barquinero J, Barrios L, Caballin MR, Darroudi F, Fomina J (2005) Translocations yields in peripheral blood lymphocytes from control populations. Int J Radiat Biol 81:139–145

24 Retrospective Biological Dosimetry by FISH

Alan A. Edwards, David C. Lloyd, and Marta Szłuińska

Abstract

This chapter reviews the present status of applying fluorescence in situ hybridisation (FISH) for retrospective biological dosimetry by chromosomal analysis. The technique helps to overcome a major drawback, the limited persistence of signal, of the more traditional method of dicentric analysis. FISH enables the more persistent so-called stable translocations to be visualised. Criteria developed by a consensus of European laboratories are described. They attempt to define which cells and which types of translocations to score for both in vitro calibration and case investigation. Also reviewed are the control levels of translocations which are highly age dependent and constitute the major determinant of the low-dose sensitivity of the method. Evidence for the stability of the technique with time is discussed. This needs further verification by following up new accident cases for many years.

24.1 Introduction

For many years, ionising radiation has been known to cause chromosome aberrations in cells. Their application for biological dosimetry has centred on the dicentric aberration in human lymphocytes and this has been used successfully for estimating doses following radiation accidents. The one disadvantage of the dicentric is that it is unstable and does not pass successfully though cell division. This means that the dicentric yield will decrease with time following the accident, making it unsuitable for exposures received either many years before blood sampling or continuously over decades.

The translocation does not suffer the same fate and the advent of fluorescence in situ hybridisation (FISH) has made the scoring of translocations much easier than with the earlier banding techniques. It is now easier to investigate the ability of cells containing translocations to pass through cell division. However, more complex rearrangements have also been revealed, which has created additional problems in deciding how to use FISH techniques for retrospective biological dosimetry. This decision requires more information on persistence of translocations and international agreement

Günter Obe and Vijayalaxmi (Eds.)
Chromosomal Alterations: Methods, Results and Importance in Human Health

concerning the various judgements involved. A special issue of a journal edited by Edwards (2000) gave preliminary views of a number of laboratories attempting to solve these problems. Updated information has led to consensus views on how to use FISH for biological dosimetry (Edwards et al. 2005; Whitehouse et al. 2005). A follow-up of an accident which occurred in Estonia has proved particularly helpful (Lindholm et al. 1998; Lindholm and Salomaa 2000; Lindholm and Edwards 2004). This chapter recalls the consensus views and the reasons for those views.

24.2 Method

A blood specimen is taken from a person for whom a dose estimate is required and the cells are cultured using techniques that enable the chromosomes to be observed. These have been published repeatedly and described in detail (IAEA 2001). In brief, phytohaemagglutinin is added to stimulate the small lymphocytes to divide and after 45 h of culture at 37°C, colcemid is added to stop the cells at their first metaphase. At about 48 h, metaphase spreads are harvested and slides prepared according to standard techniques. Chromosomes are then stained by FISH techniques using whole-chromosome and pancentromeric probes (IAEA 2001).

24.3 Which Chromosomes To Paint?

To date, most laboratories agree that the centromeres of each chromosome should be clearly visible to distinguish dicentrics from translocations. Early work (Nakano et al. 1993) showed that some dicentrics could be misclassified as translocations when viewed with chromosome paints alone. Some laboratories can identify the centromere because the two chromatids of each chromosome and the single indentation or "waist" are clearly visible. Some consider the extra brightness of the centromere when using 4′,6-diamidino-2-phenylindole counterstain is sufficient for its location and yet others prefer to specifically FISH-paint the centromeres. Apart from that, most laboratories usually paint three of the larger chromosome pairs in the groups A-C i.e., numbers 1–12, either in a single or in multiple colours, and counterstain the remainder. When about 20% of the genome is painted in a single colour, this permits 30–35% of the total translocations present to be detected. One laboratory painted chromosome pairs 1–6, three in one colour and three in another, leading to an improved translocation detection efficiency of about 56% (Tucker 2000). There were early indications that certain chromosomes did not take part in exchanges in proportion to that expected from their DNA content (Knehr et al. 1996; Boei et al. 1997). Later, more comprehensive experiments indicated better agreement with the expected values (Barquinero et al. 1998). Thus, the formulae of Lucas and Deng (2000), which rely upon the probability that a chromosome will take part in an exchange being proportional to its DNA content, have been shown to give acceptable conversions to full genome yields. The full genome translocation frequency is given by $F_G = F_p/2.05f_p(1-f_p)$, where F_p is the aberration frequency detected by FISH and f_p is the fraction of genome hybridised in a single colour taking into account the gender of the subject. Values of f_p can be obtained from Morton (1991). This enables direct comparisons to be made between laboratories that routinely paint different chromosome combinations and the equation can be expanded to account for painting in two or more colours.

24.4 What To Score

Differences in the scoring criteria used strongly influence interlaboratory variability in reported results. The scoring of translocations requires two judgements. The first involves whether or not to score the cell and the second concerns which particular aberrations relate to a retrospective dose.

Experience has shown that in a suspected case of past exposure, at least 1,000 genome-equivalent cells would need to be scored, corresponding to about 3,000 cells when three of the larger chromosomes are painted. It would be quite inefficient to select cells and score them using the same criteria as for dicentrics, i.e. only cells with exactly 46 chromosomes (IAEA 2001). A more rapid and therefore less rigorous approach has evolved whereby, when searching for suitable metaphases at low microscope magnification, most scorers ensure at a glance that there are about 46 chromosomes in each metaphase. Then at high magnification most insist that the chromosomes are well spread, that the painted chromosomes (usually six) are clear and that any that are shorter than they should be are accompanied by the deleted fragment, either as a separate object or attached to another chromosome. In other words the total length of each painted chromosome is present somewhere in the cell. Some laboratories have allowed the scoring of cells containing shortened painted chromosomes provided the corresponding centromere is present. It has now been demonstrated that including these cells that are deficient in some of the painted material can lead to abnormally high translocation yields, particularly when considering low induced frequencies near control levels (Lloyd et al. 2001; Edwards et al. 2002).

Most laboratories record two-way and one-way translocations separately, because of an early report that two-way translocation yields are more stable with time than the one-way variety (Lindholm et al. 1998). The terminology of “one-way” and “two-way” was invented as a purely descriptive way of naming aberrations without implying any underlying mechanism. Later research with improved resolution showed that many of the apparently one-way translocations were in fact also reciprocal (Fomina et al. 2000). Many laboratories record only those aberrations that are seen to involve the painted chromosomes. Some (Finnon et al. 1995; Rodriguez et al. 2004), however, have adopted a routine of recording all the aberrations seen, including those involved in the counterstained material. This enables them to distinguish stable from unstable cells using full genome information. Interestingly this concept has reinvented the designations of C_u and C_s introduced many years ago (Buckton et al. 1967) and used, for example, in the classic follow-up studies of the irradiated ankylosing spondylitis patients. With the realisation that lymphocytes sampled many years after irradiation were derived from stem cells when irradiated, it is strictly only the stable daughter cells from the irradiated population that have survived through the subsequent divisions. So strictly it is translocations in stable cells which should be measured, and when this is done translocation frequencies for an exposed individual show better constancy with time (see later.) To save scoring time when yields in the region of control levels are measured, only cells that contain aberrations involving the painted chromosomes are scanned for other unstable aberrations among the remaining chromosomes. The inclusion of a small number of unstable cells with cells that have no visible translocations has only a marginal effect on the overall yield and so pragmatically any bias can be ignored.

24.5 Control Levels

Control levels of translocations have been measured by several laboratories and the data have been reviewed and analysed (Sorokine-Durm et al. 2000). This database has been recently updated, with stricter inclusion criteria (Whitehouse et al. 2005), and subjected to a more detailed analysis. It might be expected that dicentric yields in unirradiated persons should not vary substantially with time because an approximate equilibrium would be established. The rate of production of dicentrics is balanced by the rate of loss. By contrast, translocation yields should show a continuous increase with time, and therefore age, because translocations are not removed. This has indeed been found and applies to both one-way and two-way translocations (Lucas and Deng 2000; Tucker 2000; Sorokine-Durm et al. 2000). The same applies to translocations in cells containing only simple aberrations (Whitehouse et al. 2005). The increase is from about three translocations per 1,000 genome-equivalent cells at age 20 to around 15 at age 80 (Whitehouse et al. 2005); thus, if translocation frequencies are to be used for retrospective biological dosimetry, then an age-dependent control level must be taken into account. Even after correcting for age, there still exists extra variation in translocation yield between unexposed individuals, and the causes of this require further investigation (Sorokine-Durm et al. 2000; Whitehouse et al. 2005). The effect that smoking habits have on control translocation yields is equivocal; One survey involving seven laboratories reported no observable effect (Whitehouse et al. 2005), whereas another laboratory reported a small effect in the region of 30% (Jones et al. 2002). Overall, smoking is not a serious confounding factor that significantly elevates the background frequency of translocations. Lymphocytes from control persons generally do not contain many unstable aberrations, so it makes little difference whether translocations are measured in stable cells or in all cells and this greatly simplifies the microscope analysis. However, the inclusion of cells with incomplete painted material can cause a serious overestimate of control translocation yields (Lloyd et al. 2001; Edwards et al. 2002).

24.6 Persistence of Translocations

For retrospective biological dosimetry by FISH, failure for translocation yield to persist in vivo has two possible causes. The first is that on passage through a cell division aberrations that are unstable are likely to disappear because the daughter cells are not viable; therefore, if translocations are more likely to occur in unstable than in stable cells, this will cause a decrease in yield. The second is that the sensitivity for the production of aberrations in stem cells in bone marrow is not the same as in mature lymphocytes. Initially one-way and two-way translocations were distinguished because the two-way type appeared to be more stable. Later, both types of translocations appeared more constant with time provided scoring was limited to stable cells (Lindholm and Edwards 2004). A review of data on persistence (Darroudi 2000) indicated that experiments which followed cells through several cycles showed that translocations are very stable, particularly the two-way variety. There is some evidence that the one-way types decrease either because they are associated with complex rearrangements, most of which are unstable, or because they are incomplete exchanges with associated fragments and so the cell as a whole is unstable. Experiments in vivo with laboratory animals or the small amount of human experience have led to conflicting impressions. A follow-up of the victims of the Goiania accident shows a large reduction (about a factor of 3) in translocation yield compared with dicentric yield measured at the time of the accident. This applied to those persons with doses of 1 Gy and above, but below 0.8 Gy, dicentric and translocation yields were much closer. However, no attempt was made to distinguish one-way and two-way translocations or stable and unstable cells. A follow-up of reactor crew from the Chernobyl accident showed reasonable consistency between translocation yield 10–13 years after the accident and dicentric measurements made at the time for doses up to 3 Gy. Above that, the dose estimate from translocations was low, implying some reduction in translocation yield. Again no attempt to distinguish one-way and two-way translocations or stable and unstable cells was made (Sevan'kaev et al. 2005).

Three persons from the Estonian accident, which occurred in 1994, could be followed using FISH techniques from the beginning. The follow-up shows that both one-way and two-way translocations decrease with time, but that one-way translocations decrease faster (Lindholm et al. 1998; Lindholm and Salomaa 2000). The Estonian cases have now been followed for 7 years and the translocation yield is seen to level off after about 2 years following the initial reduction (Lindholm and Edwards 2004). However, a further observation was that if translocations in stable cells are scored, then the yield does not fall with time after exposure. The yield in stable cells up to 2 years following exposure is the same as the observations in all cells in samples taken several years after exposure (Lindholm and Edwards 2004). This is understandable because at later times cell division will have removed most unstable cells. Follow-up of a tritiated water intake by a worker has shown no decrease of translocation yield with respect to the original dicentric yield for either one-way or two-way translocations over a period of 11 years (Lloyd et al. 1998). An overall message is emerging that translocation yields are stable with time provided that the translocations are measured in stable cells. There is in vitro evidence that, following simulated part-body exposures, both types of translocation yield reduce because translocations are confined to the exposed group of cells, which are more likely to be unstable by virtue of also containing dicentrics and fragments (Guerrero-Carbajal et al. 1998). This might explain, at least in part, why some translocation yields in the Goiania subjects reduced so much with time. The message that comes from this limited set of human data, accumulated with differing scoring criteria, is that future accidents need to be followed up using FISH techniques, noting on the score sheets whether the damaged cells are stable or not. It is essential to obtain further evidence for this stable cell hypothesis.

A few experiments with mice and monkeys have shown a decrease in translocation yield with time, but some of these experiments involved partial body exposure and most doses were high. Acute whole-body doses up to 2 Gy show good persistence of two-way translocations (Darroudi 2000). Clearly further work on persistence of translocations is necessary, concentrating on translocations in stable cells.

24.7 Calibration

As has been pointed out previously, there is no guarantee that the induced yield of translocations per unit dose in stem cells is the same as in mature lymphocytes. However, as experiments are now suggesting that translocations do persist well in vivo following whole-body exposure, particularly when scored in stable cells, this implies near equality of sensitivity between stem cells and mature lymphocytes. This is because at early times the lymphocytes scored are those that were irradiated, while at later times they are derived from irradiated stem cells which have passed through many cell divisions. With this assumption it is therefore possible to calibrate a system of retrospective biological dosimetry using translocations induced in vitro in lymphocytes. The literature now contains quite a lot of dose–response curves, albeit not all confined to stable cells. A frequent problem, however, is to generate enough data at sufficiently low doses to define the linear term with a reasonable statistical uncertainty. It is of course the retrospective discrimination of relatively low doses or high doses received at low dose rates which will be the greatest application for FISH biodosimetry. High acute doses tend to be appreciated early because of tissue injury signs well within the time frame for successful use of the simpler dicentric assay. Calibrating for low doses requires a lot of cells to be scored and a lowest dose of 0.5 Gy is not sufficient.

Despite early reports that immediately after irradiation translocations considerably outnumber dicentrics (Schmid et al. 1992), there is now a consensus view (e.g. Nakano et al. 1993; Bauchinger et al. 1993) that reciprocal translocations and dicentrics are induced with a more-or-less equal frequency. Moreover, one laboratory has scored translocations and all other aberrations involving painted and unpainted chromosomes so that full genome correction formulae and ratios of translocations to dicentrics may be checked (Finnon et al. 1995). These results are encouraging because the correction for the full genome has been shown to apply also to dicentric yields, at least for the chosen chromosome combination. It has also been shown that the ratio of induced two-way and one-way translocations to all dicentrics is constant, independent of dose. This being the case, the linear term for translocations may be derived from the relatively better defined linear term for dicentrics without the need for a large amount of scoring at low doses. In the longer term one would wish to see FISH calibrations taken down to lower doses, but certainly as an interim measure a laboratory could use the analogy with dicentrics with reasonable confidence.

24.8 Sensitivity

Based on knowledge to date and the assumptions about calibration, an estimate of the lowest dose that can be measured by translocations may be made. Converted to genome equivalents, the total (two-way plus one-way) translocation yield produced by 1 Gy chronic irradiation from cobalt-60 is about 15 per 1,000 cells. For a person aged 60 years, the control level is about nine translocations per 1,000 cells; thus, a dose of about 0.5 Gy would nearly double the control level and it could be argued that it is not reasonable to claim to have measured a dose unless the control level is doubled. The control level itself is probably known to an accuracy of about 20% at any age, although some extra variation from person to person still exists. Assuming that Poisson statistics can be used, it is necessary to score about 1,000 genome-equivalent cells (about 3,000 actual cells with six painted chromosomes) from a 60-year-old person exposed to 0.5 Gy to have a translocation yield which approaches being significantly higher than that in a control. A similar calculation for a young person, under 20 years, where the control yield is around two translocations per 1,000 genome-equivalent cells, shows that a dose of about 0.3 Gy can be measured by scoring around 1,000 genome-equivalent cells.

24.9 Conclusion

There are decades of experience to show that the measurement of dicentrics in human lymphocytes produces a reasonable estimate of recent absorbed dose to a person. For exposures which occur some years prior to blood sampling and which may additionally be delivered over a long period of time, the dicentric is not so useful. The persistence of translocation yields in stable cells is indicated, but more data from humans, particularly from the follow-up of accidents, are required. More work on the causes of extra variation in control subjects, apart from age, is also required. The removal of cells which do not have the full complement of the painted chromosomes has helped improve consistency in control groups. The protocol for the scoring of translocations for retrospective biological dosimetry has reached the consensus stage, at least within laboratories in Europe.

References

Barquinero JF, Knehr S, Braselmann H, Figel M, Bauchinger M (1998) DNA-proportional distribution of radiation-induced chromosome aberrations analysed by fluorescence *in situ* hybridization painting of all chromosomes of a human female karyotype. Int J Radiat Biol 74:315–323

Bauchinger M, Schmid E, Zitzelberger H, Braselmann H, Nahrstedt U (1993) Radiation-induced chromosome aberrations analysed by two colour fluorescence in situ hybridisation with composite whole chromosome-specific DNA probes and a pancentromeric DNA probe. Int J Radiat Biol 64:179–184

Boei JWA, Vermeulen S, Natarajan AT (1997) Different involvement of chromosomes 1 and 4 in the formation of chromosomal aberrations in human lymphocytes after X-irradiation. Int J Radiat Biol 72:139–145

Buckton K, Smith PG, Court Brown WM (1967) The estimation of lymphocyte lifespan from studies on males treated with X-rays for ankylosing spondylitis. In Evans HJ, Court Brown WM, McLean AS (eds) Human radiation cytogenetics. North Holland, Amsterdam, pp 106–114

Darroudi F (2000) The use of FISH translocations analyses for retrospective biological dosimetry: How stable are stable chromosome aberrations? Radiat Prot Dosim 88:101–109

Edwards AA (ed) (2000) Fluorescence *in situ* hybridisation (FISH) biological dosimetry. Radiat Prot Dosim 88:1–112

Edwards AA, Lindholm C, Darroudi F, Stephan G, Romm H, Barquinero J, Barrios L, Caballin M-R, Roy L, Whitehouse CA, Tawn EJ, Moquet JE, Lloyd DC, Voisin P (2005) Review of translocations detected by FISH for retrospective biological dosimetry applications. Radiat Prot Dosim 113:396–402

Edwards A, Maznik N, Moquet J, Hone P, Vinnikov V, Lloyd D, Cox R (2002) Choosing metaphases for biological dosimetry by fluorescence in situ hybridization (FISH). Radiat Res 157:467–471

Finnon P, Moquet JE, Edwards AA, Lloyd DC (1995) The ^{60}Co gamma ray dose-response for chromosomal aberrations in human lymphocytes analysed by FISH; applicability to biological dosimetry. Int J Radiat Biol 75:1215–1222

Fomina J, Darroudi F, Boie JWA, Natarajan AT (2000) Discrimination between complete and incomplete chromosome exchanges in X-irradiated human lymphocytes using FISH with pan-centromeric and chromosome specific DNA probes in combination with telomeric DNA probe. Int J Radiat Biol 76:807–813

Guerrero-Carbajal YC, Moquet JE, Edwards AA, Lloyd DC (1998) The persistence of FISH translocations for retrospective biological dosimetry after simulated whole or partial body irradiation. Radiat Prot Dosim 76:159–168

IAEA (2001) Cytogenetic analysis for radiation dose assessment. A manual. Technical reports series no 405. IAEA, Vienna

Jones IM, Galick H, Kato P, Langlois RG, Mendelsohn MI, Murphy G, Pleshnov P, Ramsey MJ, Thomas CB, Tucker JD, Tureva L, Vorobsova I, Nelson DG (2002) Three somatic genetic biomarkers and covariates in radiation-exposed Russian cleanup workers of the Chernobyl nuclear reactor 6-13 years after exposure. Radiat Res 158:424–442

Knehr S, Zitzelberger H, Braselmann H, Nahrstedt U, Bauchinger M (1996) Chromosome analysis by fluorescence *in situ* hybidization: further indications for a non-DNA proportional involvement of single chromosomes in radiation-induced structural aberrations. Int. J Radiat Biol 70:385–392

Lindholm C, Edwards AA (2004) Long-term persistence of translocations in stable lymphocytes from victims of a radiological accident. Int J Radiat Biol 80:559–566

Lindholm C, Salomaa S (2000) Dose assessment of past accidental or chronic exposure using FISH chromosome painting. Radiat Prot Dosim 88:21–25

Lindholm C, Tekkel M, Veidebaum T, Ilus T, Salomaa S (1998) Persistence of translocations after accidental exposure to ionizing radiation. Int J Radiat Biol 74:565–571

Lloyd DC, Moquet JE, Oram S, Edwards AA, Lucas JN (1998) Accidental intake of tritiated water: a cytogenetic follow-up case on translocation stability and dose reconstruction. Int J Radiat Biol 73:543–547

Lloyd DC, Lucas JN, Edwards AA, Deng W, Valente E, Hone PA, Moquet JE (2001) A study to verify a reported excess of chromosomal aberrations in blood lymphocytes of Namibian uranium miners. Radiat Res 155:809–817

Lucas JN, Deng W (2000) Views on issues in radiation biodosimetry based on chromosome translocations measured by FISH. Radiat Prot Dosim 88:77–86

Morton NE (1991) Parameters of the human genome. Proc Natl Acad Sci 88:7474–7476

Nakano M, Nakashima E, Pawel DJ, Kodama Y, Awa A (1993) Frequency of reciprocal translocations and dicentrics induced in human blood lymphocytes by X-irradiation as determined by fluorescence in situ hybridization. Int J Radiat Biol 64:565–569

Rodriguez P, Montoro A, Barquinero JF, Caballín MR, Villaescusa I, Barrios L (2004) Analysis of translocations in stable cells and their implication in retrospective biological dosimetry. Radiat Res 162:31–38

Schmid E, Zitzelberger H, Braselmann H, Gray JW, Bauchinger M (1992) Radiation-induced chromosome aberrations analysed by fluorescence in situ hybridisation with a triple combination of composite whole chromosome-specific DNA probes. Int J Radiat Biol 62:673–678

Sevan'kaev AV, Lloyd DC, Edwards AA, Khvostunov IK, Mikhailova GF, Golub EV, Shepel NN, Nadejina NM, Galstian IA, Nugis VY, Barrios L, Caballin MR, Barquinero JF (2005) A cytogenetic

follow-up of some highly irradiated victims of the Chernobyl accident. Radiat Prot Dosim 113:152–161
Sorokine-Durm I, Whitehouse C, Edwards A (2000) The variability of translocation yields amongst control populations. Radiat Prot Dosim 88:93–99
Tucker JD (2000) Evaluation of chromosome translocations by FISH for radiation biodosimetry: a view from one laboratory. Radiat Prot Dosim 88:87–92
Whitehouse CA, Edwards AA, Tawn EJ, Stephan G, Oestreicher U, Moquet JE, Lloyd, DC, Roy L, Voisin P, Lindholm C, Barquinero J, Barrios L, Caballin MR Darroudi F, Fomina J (2005) Translocation yields in peripheral blood lymphocytes from control populations. Int J Radiat Biol 81:139–145

25 Chromosomal Aberrations in Astronauts

Kerry George and Marco Durante

Abstract

During spaceflight, astronauts are exposed to very penetrating radiation consisting primarily of high-energy charged particles, mostly protons, and secondary radiation created by interaction of nuclei with shielding material. Significant uncertainty still exists regarding the biological effects of exposure to space radiation. Cytogenetic biodosimetry analyses provide direct measurements of space radiation induced biological damage that take into account individual radiosensitivity and the influence of stress conditions, including microgravity. A review of current data on in vivo induced chromosome damage in the blood lymphocytes of astronauts strongly indicates that space missions of a few months or more induce measurable increases in the yield of chromosome damage that can be used to estimate equivalent dose and radiation risk. Recent studies indicate that biodosimetry estimates from single spaceflights lie within the range expected from physical dosimetry and biophysical models, but single individual measurements have large uncertainties and the total sample population remains low. Retrospective doses are more difficult to estimate because of the fairly rapid time-dependent loss of "stable" aberrations in blood lymphocytes. Also, biodosimetry estimates from individuals who participate in repeated missions, or very long missions, may be complicated by an adaptive response to space radiation and/or changes in lymphocyte survival and repopulation. No cytogenetic biomarkers of space radiation exposure have been uncovered so far, owing in part to the statistical uncertainties inherent in these techniques. A discussion of published data is presented and specific issues related to space radiation biodosimetry protocols are discussed.

25.1 Introduction

The duration of manned spaceflight missions has steadily increased over the years. From brief forays during the Gemini and Apollo era that lasted up to only a few days, missions have extended to weeks on board the Space Shuttle, and months aboard Mir and the International Space Station (ISS). In addition,

Günter Obe and Vijayalaxmi (Eds.)
Chromosomal Alterations: Methods, Results and Importance in Human Health

some proposed future Mars exploration missions could extend to as long as 3 years. This increased time in space means that crewmembers are exposed to higher radiation doses, and concerns have been raised about the potential health effects from these exposures. Radiation exposure limits are based on the lifetime excess cancer risk, because cancer is the main health concern related to space radiation exposure (Cucinotta and Durante 2006). Currently accepted radiation exposure limits restrict excess lifetime cancer mortality risk to no more than 3% (NCRP 2000).

The space radiation environment is a complex mixture of different types of radiation that is very different from the terrestrial radiation environment. The primary sources of radiation in outer space derive from galactic cosmic rays, protons and electrons trapped in the Earth's magnetic field, and infrequent solar particle events. In addition, secondary radiation, such as neutrons and light charged particles, are created by interactions of high-charge (Z) and high-energy (E) nuclei (HZE particles) with spacecraft shielding material and with the human body itself (NCRP 2000). Although HZE particles contribute only a few percent to the particle flux in space, the significantly higher ionizing energy and greater penetration power of this type of radiation could contribute significantly to biological effects.

Health risks to crewmembers are currently assessed from physical measurements of radiation dose measured using passive thermoluminescent dosimeters and plastic nuclear track detectors. These measurements provide accumulated absorbed skin doses that are combined with computerized anatomical models and radiation transport codes that estimate the effects of shielding. These data are used to calculate radiation doses to specific organs and to assess the corresponding health risk. However, substantial limitations arise with this process that lead to significant uncertainty in estimating the biologically important dose (Cucinotta and Durante 2006). Quality factors for space radiation risk coefficients (mostly derived from the A-bomb survivors database), and the dose and dose-rate effectiveness factor are all quantities affected by large uncertainty. In addition, the location of crew dosimeters cannot always be assured, individual radiosensitivity is totally neglected in the risk models, and synergism between microgravity and radiation is still poorly understood. Recently, biodosimetry programs have been introduced that derive equivalent dose or risk from direct measurements of cytogenetic damage in the peripheral blood lymphocytes of crewmembers (Durante 2005). These data have the added advantage of taking into account individual radiosensitivity in addition to the influence of microgravity effects and other stress conditions, and can be very useful to benchmark the current models used to estimate the risk from physical dosimetry measurements (NCRP 2002).

Cytogenetic biodosimetry methods have been used extensively for assessing radiation exposures, and remain the most sensitive in vivo indicator of dose available to date (IAEA 2001). Dicentric chromosomes have been used for many years to evaluate accidental radiation exposures because they can be easily identified in Giemsa-stained chromosome preparations and preexposure background levels are very low. However, the astronaut population is unique in that astronauts consent to considerably higher occupational radiation exposures than terrestrial radiation workers and, unlike the situation with accident victims, preexposure levels of chromosome damage can be accurately assessed in astronauts, allowing for greater sensitivity in biological dose estimations.

25.2 Measured Yields of Exchanges in the Peripheral Blood Lymphocytes of Astronauts

The first in vivo studies of chromosome damage in the peripheral blood lymphocytes of crewmembers were conducted in the 1960s during the Gemini program. Bender et al. (1967, 1968) used solid Giemsa staining techniques to assess dicentrics, rings, fragments, and chromatid aberrations in blood samples obtained from astronauts who participated in the Gemini-III and Gemini-XI missions, which lasted about 5 h and 3 days, respectively. No significant increase in damage was detected when samples collected before and after flight were compared. More recently these techniques were tested on crewmembers participating in longer-duration flights on board the Mir space station (Testard et al. 1996; Obe et al. 1997; Federenko et al. 2001). Testard et al. (1996) studied seven crewmembers, two of whom participated in flights lasting 3 weeks or less and showed no significant elevation of chromosome damage after flight, and five cosmonauts of whom participated in a 6-month mission and had elevated chromosome damage after flight. Biological dose estimates for the longer-duration flights were calculated using established in vitro calibration curves for X-rays, and varied among individuals, with values ranging from 130 to 455 mGy. Obe et al. (1997) studied crewmembers who participated in spaceflights lasting from 120 to 198 days and reported significant elevation in chromosome damage after flight for six of the seven individuals studied. Pooled data for the six individuals who had increased chromosome damage showed a post-flight dicentric yield that was roughly 3.5 times preflight values. Individual biological doses were estimated using published in vitro X-ray-induced dose–response curves, and values ranged from 50 to 290 mGy depending on the individual. Limited physical dose measurements were reported for the two studies described above, but data indicated that the frequencies of aberrations roughly correlated with absorbed radiation dose. A study by Federenko et al. (2001) of 22 individuals who participated in space missions lasting an average of 4–6 months also reported that the mean yields of dicentrics were about threefold greater after flight than before flight; however, large interindividual variations were reported for this group, with some individuals having no appreciable increase in dicentric yield after flight. The biological dose estimates for the individual crewmembers with increased dicentric yields varied from 20 to 280 mGy. No comparisons with physical dose estimates were given for this study.

Evidence from these cytogenetic studies strongly indicates that the radiation dose accumulated during a long-duration flight can induce measurable increases in the yield of chromosome damage, whereas the doses encountered during shorter missions of a few weeks or less are below the threshold for detection using this technique; however, the type of chromosome aberration evaluated is an important factor when considering these data. All the studies described above focus on the detection of dicentric chromosomes, and it has been well documented that radiation-induced dicentrics undergo time-dependent decay in blood lymphocytes. Dicentrics have been estimated to have an average half-life of about 3 years in human peripheral blood lymphocytes, but recent studies indicate large variability in decay rate among individuals (Durante 2005). It would follow that the use of dicentric aberrations for biodosimetry analysis after missions lasting several months or more may not be reliable owing to the loss of aberrant cells during the mission. In addition, any significant individual variability in the rate of loss of cells with dicentrics would indicate that yields could not be accurately corrected for elimination of this type of exchange from the blood.

The introduction of the fluorescence in situ hybridization (FISH) chromosome painting technique radically improved the ease and accuracy of assessing yields of monocentric aberrations such as translocations (Lucas 1997). These types of so-called stable chromosome aberrations have a higher probability of survival after cell division than unstable aberrations, and are therefore commonly believed to persist in peripheral blood cells for many years. This indicates that translocations would be a more appropriate tool than dicentric analysis for assessing the effect of chronic radiation exposures such as those encountered during long-duration missions, as well as for assessing retrospective radiation doses. Background levels of translocations are higher than those of dicentrics; they are known to increase with age and show considerable variation between individuals (Tucker 2001). Although this can be a challenge for situations where preexposure levels of translocations are not available, such as after accidental radiation exposure, it is not a problem for spaceflight biodosimetry because samples can be obtained before flight.

George et al. (2001a) used chromosome painting to assess the yield of chromosome exchanges in the peripheral lymphocytes of six crewmembers before and after they participated in space missions lasting 3 months or more, and in two crewmembers who participated in a 10-day Space Shuttle mission. Postflight increases in chromosome damage were detected in all the crewmembers of the longer-duration missions, whereas no significant change was detected in crewmembers of the shorter Space Shuttle mission. Doses were determined using preflight in vitro dose–response curves for γ-rays that were generated using the individual crewmembers' blood lymphocytes.

Preflight levels of translocations were shown to vary considerably between individuals. Doses estimated from total chromosome damage ranged from 31 to 166 mGy, depending on the individual, and these values roughly correlated with the physical measurements of dose. Average relative biological effectiveness (RBE) was estimated by comparing biological dose estimates with measured physical doses. Average RBE values were in excellent agreement with the average quality factor of 2.5 that was calculated from physical readings corrected for the HZE component of the exposure and adjusted to representative doses at organ locations. These data confirm the results of an earlier study by Yang et al. (1997) in which FISH chromosome painting was used to assess aberrations in three crewmembers who participated in a space mission lasting 115 days.

25.3 Biologically Based Risk Estimates

As we noted in Sect. 25.1, astronauts are exposed to radiation in space under conditions substantially different from those experienced on Earth (Fig. 25.1). The radiation field is totally different, and the crewmembers are simultaneously exposed to several other stressors, such as microgravity. Given these differences, energy deposition (physical dose) may not be representative of the true space radiation risk, and this is why the use of biomarkers of risk is so important for astronauts (Durante 2005). A direct comparison of risk estimates from physical dose measurements and application of models (NCRP 2000) with biological measurements is possible because chromosome aberrations in peripheral blood lymphocytes have been validated as biomarkers of cancer risk (see Chap. 31 by Bonassi).

Risk projections for long-term missions in low Earth orbit (LEO) can be obtained from biodosimetry data using the quantitative approach derived from the European Study Group in Cytogenetic Biomarkers and Health (ESCH) database (Bonassi et al. 2005). The postflight relative increase in aberration frequency can then be used directly to estimate the cancer risk, and this biologically based estimate can be compared with conventional risk estimates based on physical dosimetry and models (NCRP 2000). Through analysis of the cytogenetic data described above, it has been determined that the average excess relative risk (ERR) of cancer induction after long-term missions in LEO is 0.09, with large interindividual variations (ranging from −0.08 to 0.42) (Durante et al. 2001; Durante 2005). Individual risk estimates in this same cohort using conventional assessment leads to an average ERR of 0.05, with a range from 0.02 to 0.12. Despite the large spread in the "biological" values, it is remarkable that the two estimates lie in the same range. In comparison with equivalent doses calculations in the NASA study (George et al. 2001a, c), these results suggest that current average risk estimates for astronauts in LEO are accurate to within a factor of 3 or less (Durante 2005).

Fig. 25.1 An astronaut involved in extravehicular activities during STS-121 mission. The astronauts are exposed to radiation plus many other stressors (including microgravity) during spaceflights. (Photo courtesy of NASA)

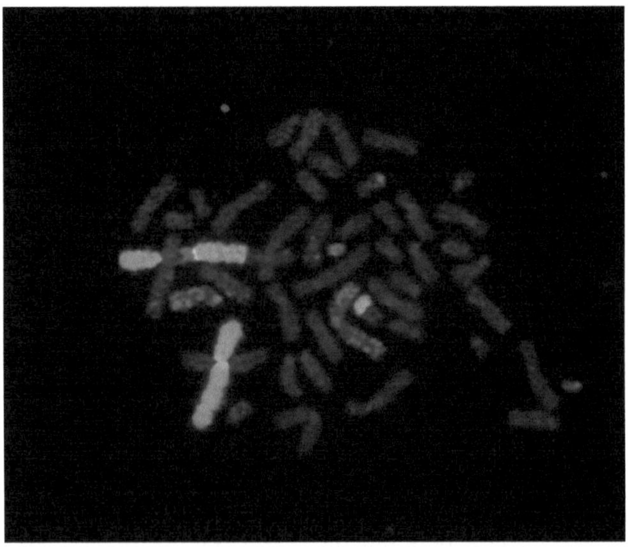

Fig. 25.2 Complex-type exchange involving chromosomes 1 and 2, detected by fluorescence in situ hybridization (FISH) painting in a blood lymphocyte of an astronaut after a long-term space mission

ERRs derived from chromosome aberrations present a wider scatter than those calculated from dose. Certainly, statistical error involving aberration measurements accounts for most of the widened distribution, but interindividual variability in risk may also contribute to this variance.

25.4 Cytogenetic Signatures of Space Radiation Exposure

In addition to the detection of simple reciprocal-type chromosome exchanges, chromosome painting can be used to detect complex-type exchanges (Fig. 25.2), which are defined as exchanges involving three or more breaks in two or more chromosomes. Many studies have demonstrated that no significant increase in complex exchanges is detected after low linear energy transfer (LET) radiation such as γ-rays and X-rays at exposures below 2 Gy, whereas low doses of high-LET radiation, such as the HZE ions in space, are capable of producing extremely complex patterns of chromosome damage (George et al. 2003a; Durante et al. 2002). This evidence suggests that complex-type damage can be induced by a single particle's traversal of a cell and indicates that this type of damage can be a biomarker of exposure to low doses of high-LET radiation (Anderson et al. 2002). It follows, then, that any increase in complex-type chromosome damage detected in blood lymphocytes of crewmembers after spaceflight would be due entirely to the high-LET component of the exposure.

However, one confounding factor in the assessment of high-LET radiation-induced chromosome damage is the influence of delays in progression of the cell cycle from G2 to metaphase. Recent ground-based studies have indicated that high-LET-induced chromosome damage undergoes a complicated pattern of mitotic delay (George et al. 2001b), and it is assumed that this is due to the less homogenous nature of this type of exposure compared with X-ray or γ-ray irradiation. When chromosomes in the G2 phase of the cell cycle are used for the analysis of high-LET-induced cytogenetic damage, much higher yields of damage are obtained than with standard metaphase analysis. This is due to extensive G2–metaphase cell-cycle delays. It is therefore possible that postflight expression of complex chromosome damage in crewmembers' lymphocytes could be influenced by a complicated pattern of mitotic delay after exposure to the mixed radiation field encountered in space and consequently it could be underestimated in metaphase samples collected at one time point. To ensure that chromosome damage is not underestimated in astronauts' blood lymphocytes, it may be more appropriate to use the chemically induced premature chromosome condensation (PCC) technique for analysis, in which chromosomes in the G2 phase as well as metaphase are assessed.

Cytogenetic analysis using FISH painting of lymphocytes from ten astronauts after long-term space missions of 3 months or more revealed that the frequency of complex damage was very low and in most cases only one or two complex exchanges were detected in several thousand PCC samples analyzed for each individual (George et al. 2003b). A low background level of complex damage was also detected in preflight samples, hampering statistical significance. Similar results were obtained in a cohort of Russian cosmonauts (Durante et al. 2004). In a different study, Horstmann et al. (2005) used a multiplex FISH technique (MFISH) to assess all chromosomes simultaneously in each cell and detected no complex-type exchanges in a total of 3,590 lymphocyte cells pooled from astronauts involved in short- and long-duration space missions. However, this study was conducted using standard metaphase analysis and therefore does not discount the possibility that complex damage was not expressed owing to extensive cell-cycle delay. Testard et al. (1996) and Obe et al. (1997) also reported a very low frequency of highly damaged metaphase cells after long-duration missions; however, these studies were limited to so-called rogue cells containing multiple dicentric and acentric fragments because analysis was conducted using solid staining techniques, and again cells were analyzed using standard metaphase techniques. The nature of the "rogue" cells is still unclear, but they are generally considered to be unrelated to radiation exposure (Mustonen et al. 1998).

Intrachromosomal exchanges (inversions) have also been suggested as a biomarker of exposure to densely ionizing radiation (Hande et al. 2003). Only one group so far (Horstmann et al. 2005) has attempted to assess the frequency of intrachromosomal exchanges in blood lymphocytes after spaceflight. For this study, a high-resolution multicolor fluorescence banding (mBAND) probe for chromosome 5 was used, and no inversions were detected in 2,800 cells pooled from 11 astronauts involved in long- and short-duration missions.

25.5 Persistence of Space Radiation Induced Cytogenetic Damage

It was commonly believed that chromosome translocations would persist in the blood lymphocytes for many years; however, as more data become available, evidence suggests that this may not always be the case (reviewed in Tucker 2001; IAEA 2001; Durante 2005). Some limited data are available on the persistence of chromosome aberrations in astronaut peripheral blood lymphocytes. Durante et al. (2003) clearly demonstrated that dicentric yields in astronauts decay with time after flight. The authors could not measure the complete time dependence of translocations using FISH painting, but their results show a lack of correlation between time in space and translocation yields (Fig. 25.3). No significant correlation was observed between yields of dicentrics or translocations and duration of flight or absorbed dose. The yield of aberrations declined rapidly during the time interval between flights, and after repeated flights translocation yields were similar to background levels before the first mission (Fig. 25.4).

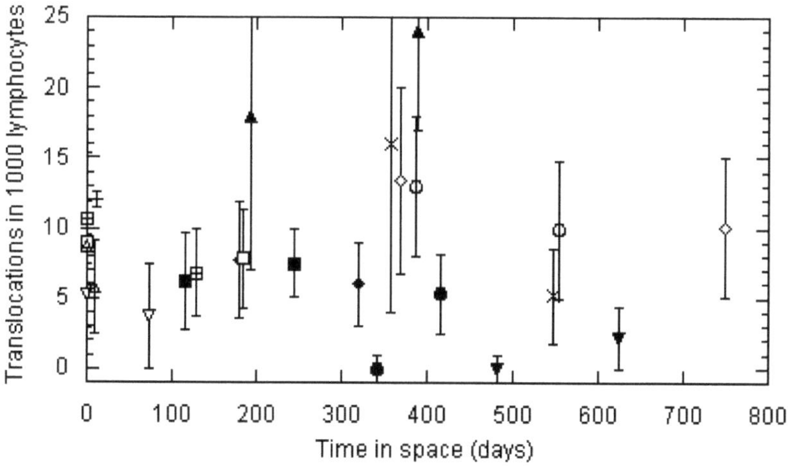

Fig. 25.3 Relationship between translocation frequency (measured by FISH and converted to whole-genome equivalents) and total duration of space sojourns for cosmonauts involved in multiple spaceflights. Each *symbol* represents a different cosmonaut. (Plots from Durante et al. 2003, reproduced with permission of S. Karger AG, Basel)

George et al. (2005) analyzed blood samples collected at various times, beginning directly after return from space to more than 5 years after flight. For five of the six astronauts studied, the analysis of individual time courses for translocations revealed a temporal decline of yields with various half-lives ranging from 10 to 58 months. Since the level of stable aberrations depends on the interplay between natural loss of circulating T lymphocytes and replenishment from the stem or progenitor cells, the differences in the rates of decay could be explained by interindividual variation in lymphocyte turnover. The results indicate possible complications with the use of stable aberrations for retrospective dose reconstruction or risk estimates from biological data. It should be stressed that space is a unique environment and these results may not be indicative of terrestrial radiation exposures. It is possible that confounding factors from exposure to the space environment, such as stress and microgravity, could induce cell instability. It has been shown that spaceflights induce drastic changes in the immune system, and the ability to respond to challenges such as infection is reduced in the space environment.

The complex time dependence of the chromosome aberration yield is apparently a severe limitation on the use of cytogenetic end points to estimate equivalent dose or risk. In fact, the result of the biological test will be strongly dependent on the time after sampling. However, it can be argued that the "time factor" does not necessarily reflect a technical limitation of the cytogenetic biomarker, but rather provides an *individual* risk assessment rather than the *average* risk estimate based solely on absorbed dose (Durante 2005). As in the case of individuals who quit smoking at a certain age after

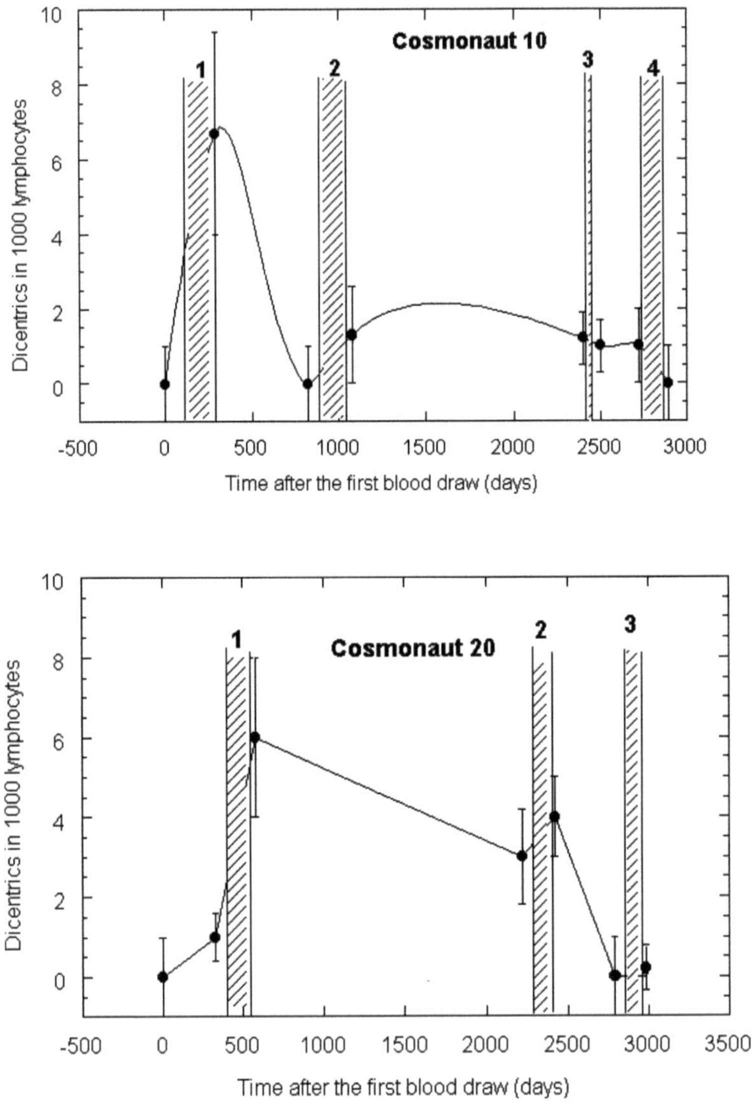

Fig. 25.4 Time course of dicentrics in two Russian cosmonauts involved in multiple spaceflights. The dicentric frequency, measured in metaphases of lymphocytes after Giemsa staining, is plotted vs. the time from the blood draw used for the first cytogenetic test. *Shaded areas* indicate the space mission time, and the *numbers between the bars* the flight number for the cosmonaut. In a period of about 10 years, cosmonaut 10 made four spaceflights and cosmonaut 20 three flights. The *line connecting the datapoints* is a guide for the eye. (Plots from Durante et al. 2003, reproduced with permission of S. Karger AG, Basel)

years of chronic exposure to carcinogens in the cigarette smoke, radiation relative risk decreases with time after exposure (as shown in the recent analysis of A-bomb survivors; Preston et al. 2003), and the decrease is likely to show a large interindividual variability, caused by physiological, environmental, and genetic factors. Since these factors are not taken into account in biophysical models of radiation action, a reliable picture of the relative risk for an individual at a given time after exposure may be derived only by biomarkers of risk.

25.6 Methodology Considerations and Suggestions for Standardized Protocols

After all available data have been considered, it appears that cytogenetic biodosimetry can be used reliably to reconstruct radiation dose after protracted exposure to space radiation, and these data can potentially be used to assess individual health risk. However, it is becoming clear that space radiation biodosimetry has some unique challenges, compared with terrestrial-based methods, which must be addressed before a standard procedure can be established. It is apparent that several major areas should be considered, as addressed in the following sections.

25.6.1 Collection of Chromosome Samples

Lymphocytes must be examined in blood samples that were collected before flight to establish accurate preflight yields of damage. This not only provides an individual baseline for translocations, which are highly variable between individuals, but also increases the dose sensitivity of the assay. Considering that chromosome damage may undergo a fairly rapid time-dependent decay, postflight samples must be collected within 1 or 2 weeks after the mission. Some studies have used samples collected the day after return from space; however, it has been noted that response of lymphocytes to phytohemagglutinin (PHA), a compound that is used to induce lymphocytes to undergo cell division, is severely suppressed on the day of return from space. The mitotic index was drastically lower in samples collected from three crewmembers on the day of return from a 3-month space mission than it was in preflight samples collected from the same individuals (Yang et al. 1997). The number of cycling cells was much lower, and progression through the cell cycle was much slower. This phenomenon, which may be related to changes observed in the immune response and/or other stress and microgravity effects, seems to last only a few days. The growth kinetics of PHA-stimulated lymphocytes collected 9 days after flight were almost the same as those in preflight samples. Given the challenges of collecting chromosome samples within a few days of return from flight, a later sampling time (7–14 days after flight) is preferable.

The analysis of chemically induced PCC samples, which contain G2-phase as well as metaphase cells, is preferable to standard metaphase analysis. PCC analysis will ensure that any cells undergoing extensive G2–metaphase cell-cycle delay will still be included in the analysis. As mentioned above, this method has been shown to help avoid population bias after high-LET radiation in vitro and may ensure correct yields of complex damage induced in vivo, although this has yet to be proven. In addition, PCC samples generally have a higher index of analyzable cells, so using them makes it easier to obtain sufficient cells from samples that are consistently difficult to culture.

25.6.2 Assessment of Chromosome Damage

Since it has been well documented that radiation-induced dicentrics undergo fairly fast time-dependent decays and will certainly be lost from the blood during a longer spaceflight mission, stable monocentric aberrations will provide more accurate assessments of dose. A combination of three FISH painting probes for the larger chromosome pairs that comprise about 20% of the genome is recommended for analysis. This technique allows damage to be scored quickly, and if each chromosome is painted with a different color, the detection of complex damage will be maximized, while exchanges within the painted chromosomes can also be detected. Data for analysis of individual combinations of chromosomes can be extrapolated to whole-genome-equivalent values using the appropriate calculations (George et al. 2001a). Use of MFISH techniques is not recommended for routine biodosimetry analysis because analysis of individual cells takes a considerable amount of time.

The analysis of complex-type exchanges could provide valuable information about the high-LET component of radiation exposure, as any significant increase will likely be due to heavy ions. However, it appears this type of exchange is quite rare and it is unlikely that valuable data can be obtained using the average number of cells analyzed for these tests.

One other important consideration when scoring chromosome damage for dose estimations is that clonal exchanges must be identified and excluded. Several recent studies have indicated that some individuals' blood lymphocytes possess cells with balanced chromosome abnormalities that have clonally expanded in vivo, resulting in a population of cells with identical rearrangements. Including these clonal exchanges can result in an overestimation of radiation-induced damage and, therefore, influence the accuracy of dose calculations. Clonal exchanges may be more prevalent than previously thought. In fact, in one study clonal aberrations were identified in three of 12 astronauts (George et al. 2004). These clonal cells were present in samples collected both before and after spaceflight, and yields were higher than those previously reported for healthy individuals in this age range (40–52 years of age). The individuals with clonal aberrations were followed up over several months and the yields of all clones decreased during this period. It is possible that the inclusion of clonal exchanges could explain some of the interindividual biological differences in dose estimates for astronauts. Since clonal aberrations may be associated with increased risk of tumorigenesis, it is important to accurately identify cells containing clonal rearrangements for risk assessment as well as biodosimetry.

25.6.3 Dose Limitations

The radiation dose incurred during shorter spaceflight missions is below the threshold for currently available cytogenetic biodosimetry techniques. The lowest dose that can be assessed seems to be around 10 cGy. This generally restricts the use of biodosimetry to missions lasting at least 3 months. Using the FISH technique, several thousand cells should be analyzed in both preflight and postflight samples to obtain statistically significant data, although exact numbers can vary depending on the individual's background frequencies of damage.

25.6.4 Individual Sensitivity

Recent data indicate that in vitro dose response for chromosome damage may vary for some individuals (Fig. 25.5). If this is the case, then one standard calibration curve cannot be used for all individuals and it is necessary to generate a preflight in vitro calibration curve for blood lymphocytes from each crewmember before their participation in their respective space mission. This will indicate if the particular individual has a greater predisposed sensitivity to radiation and should be monitored more closely. In addition, if the same combination of chromosomes is assessed before and after flight and individual dose–response curves are used, then the results can be compared directly without the need for subtracting background yield of chromosome damage. This results in a substantially decreased error in dose estimations.

Recent estimates have put the overall percentage of individuals heterozygous for the ataxia telangiectasia (A-T) gene at around 3% of the human population. These individuals have defective DNA repair capabilities that result in a higher sensitivity to radiation and they are clearly at greater risk of health effects due to radiation exposure. If these percentages are correct, it is very likely that the astronaut population includes some A-T heterozygous individuals. Clearly the issue of individual sensitivity to space radiation is very important and data are limited. More study is required in this area.

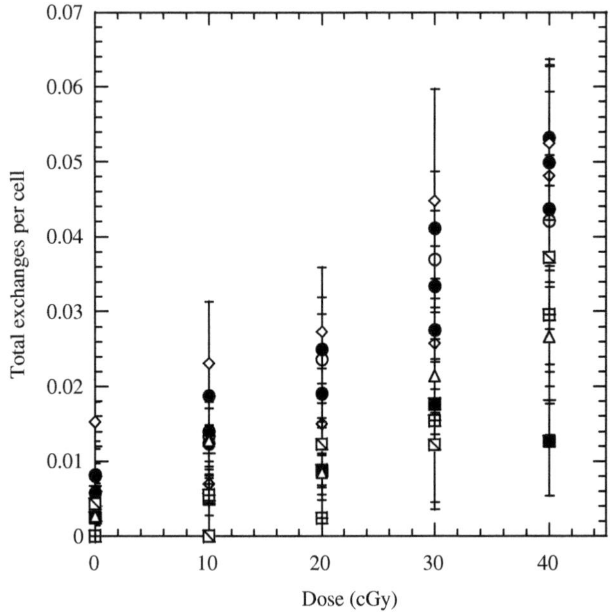

Fig. 25.5 Dose–response curves for chromosome damage in astronauts' peripheral blood lymphocytes induced after in vitro irradiation with γ-rays. Each *symbol* represents a different crewmember. *Bars* present standard errors of the mean values

25.6.6 Dose Estimations

As mentioned already, if the same combination of chromosomes is assessed for each sample, then postflight increases can be obtained by subtracting preflight background yields from the postflight yield and biologically determined dose can be calculated directly using the curve fit for the preflight calibration curve. At least two doses below 10 cGy should be used for the in vitro calibration to increase the accuracy of the curve fitting for doses within the range encountered during spaceflight. At least five different doses are required for the calibration curve, and using the probes that comprise 20% of the whole genome, several thousand cells must be analyzed to generate sufficient data for curve fitting.

25.6.7 Influence of Repeat Flights

The recent study of crewmembers after repeated flights (Durante et al. 2003) indicates that aberrations in individuals who have been involved in several missions may not be additive (Fig. 25.4), perhaps because of adaptive response to space radiation exposure or loss of radiosensitive cell subpopulations.

Given that these individuals may have less of an increase in chromosome damage than expected after repeated flights, it may not be possible to accurately assess their dose using these methods. It is important, therefore, to research an individual's previous radiation exposure from earlier spaceflights and possibly medical procedures, and to take this into consideration when evaluating biodosimetry data. This study of adaptive response is very interesting and it is important that the phenomenon be investigated further.

25.7 Conclusions

It now appears that if blood is collected within 1 or 2 weeks of return from space, cytogenetic biodosimetry can be used reliably to estimate equivalent dose and radiation risk after protracted exposure to space radiation for a few months or more. Results indicate that biodosimetry estimates from single spaceflights lie within the range expected from physical dosimetry and biophysical models. However, very large uncertainties are associated with single individual measurements and the total sample population remains low.

Retrospective doses may be more difficult to estimate because of the fairly rapid time-dependent loss of "stable" aberrations in blood lymphocytes. Also, biodosimetry estimates from individuals who participate in repeated missions, or very long (interplanetary) missions, may be complicated by an adaptive response to space radiation and/or changes in lymphocyte survival and repopulation. Suggested protocols for space radiation biodosimetry and issues that need to be addressed were discussed.

References

Anderson RM, Stevens DL, Goodhead DT (2002) M-FISH analysis shows that complex chromosome aberrations induced by alpha -particle tracks are cumulative products of localized rearrangements. Proc Natl Acad Sci USA 99:12167–12172

Bender MA, Gooch PC, Kondo S (1967) The Gemini-3 S-4 spaceflight-radiation interaction experiment. Radiat Res 31:91–111

Bender MA, Gooch PC, Kondo S (1968) The Gemini XI S-4 spaceflight-radiation interaction experiment: the human blood experiment. Radiat Res 34:228–238

Bonassi S, Ugolini D, Kirsh-Volders M, Stromberg U, Vermeulen R, Tucker JD (2005) Human population studies with cytogenetic biomarkers: review of the literature and future prospectives. Environ Mol Mutagen 45:258–270

Cucinotta, FA, Durante M (2006) Cancer risk from exposure to galactic cosmic rays: implications for space exploration by human beings. Lancet Oncol 7:431–435

Durante M, Bonassi S, George K, Cucinotta FA (2001) Risk estimation based on chromosomal aberrations induced by radiation. Radiat Res 156:662–667

Durante M, George K, Wu H, Cucinotta FA (2002) Karyotypes of human lymphocytes exposed to high-energy iron ions. Radiat Res 158:581–590

Durante M, Snigiryova G, Akaeva E, Bogomazova S, Druzhinin S, Ferenfo B. Greco O, Novitskaya N, Rubanovich A, Shevchenko V, von Recklinghausen U, Obe G (2003) Chromsoome aberration dosimetry in cosmonauts after single or multiple flights. Cytogenet Genome Res 103:40–46
Durante M, Ando K, Furusawa Y, Obe G, George K, Cucinotta FA (2004) Complex chromosomal rearrangements induced in vivo by heavy ions. Cytogenet Genome Res 104:240–244
Durante M (2005) Biomarkers of space radiation risk. Radiat Res 164:467–473
Federenko B, Druzhinin S, Yudaeva L, Petrov V, Akatov Y, Snigiryova G, Novitskaya N, Shevchenko V, Rubanovich (2001) Cytogenetic studies of Blood lymphocytes from cosmonauts after long-term space flights on Mir station. Adv Space Res 27:355-359
George K, Durante M, Wu H, Willingham V, Badhwar G, Cucinotta, FA (2001a) Chromosome aberrations in astronauts' blood lymphocytes after space flight. Radiat Res 156:731–738
George K, Wu H, Willingham V, Furusawa Y, Kawata T, Cucinotta FA (2001b) High- and low-LET induced chromsome damage in human lymphocytes: a time course of aberrations in metaphase and interphase. Int J Radiat Biol 77:175–183
George K, Wu H, Willingham V, Cucinotta FA (2001c) The effect of space radiation on the induction of chromosome damage. Phys Med 17:224–228
George K, Durante M, Willingham V, Wu, Yang TC, Cucinotta FA (2003a) Biological effectiveness of accelerted particles for the induction of chomosome damage measured in metphase and interphase human lymphoyctes. Radiat Res 160:425–435
George K, Durante M, Wu H, Willingham V, Cucinotta FA (2003b) In vivo and in vitro measurements of complex-type chromosomal exchanges induced by heavy ions. Adv Space Res 31:1525–1535
George K, Durante M, Willingham V, Cucinotta FA (2004) Chromosome aberrations of clonal origin are present in astronauts' blood lymphocytes. Cytogenet Genome Res 104:245–251
George K, Willingham V, Cucinotta FA (2005) Stability of chromosome aberrations in the blood lymphocytes of astronauts after space flight by FISH chromosome painting. Radiat Res 164:474–480
Hande, MP, Azizova TV, Geard CR, Burak LE, Mitchell CR, Khokhryakov VF, Vasilenko EK, Brenner DJ (2003) Past exposure to densely ionizing radiation leaves a unique permanent signature in the genome. Am J Hum Genet 72:1162–1170
Hortsmann M, Duante M, Johannes C, Pieper R, Obe G (2005) Space radiation does not induce a significant increase of intrachromosomal exchanges in astronauts' lymphocytes. Radiat Environ Biophysics 44:219–224
IAEA (2001) Cytogenetic analysis for radiation dose assessment. Technical report no 405. International Atomic Energy Agency, Vienna
Lucas JN (1997) Dose reconstruction for individuals exposed to ionizing radiation using chromosome painting. Radiat Res 148:33–38
Mustonen R, Lindholm C, Tawn EJ, Sabatier L, Salomaa S (1998) The incidence of cytogenetically abnormal rogue cells in peripheral blood. Int J Radiat Biol 74:781–785
NCRP (2000) Radiation protection guidance for activities in low-Earth orbit. NCRP report no. 132. NCRP, Bethesda
NCRP (2002) Operational radiation safety program for astronauts in low-Earth orbit: a basic framework. NCRP report no 142. NCRP, Bethesda
Obe G, Johannes I, Johannes C, Hallman K, Reitz G, Facius R (1997) Chromosomal aberrations in blood lymphocytes of astronauts after long-term space flights. Int J Radiat Biol 72:727–734
Preston DL, Pierce DA, Shimizu Y, Ron E, Mabuchi K (2003) Dose response and temporal patterns of radiation-associated solid cancer risks. Health Phys 85:43–46
Testard L, Ricoul M, Hoffschir F, Flury-Herard A, Dutrillaux B, Federenko B, Gerasimenko V, Sabatier L (1996) Radiation-induced chromosome damage in astronauts' lymphocytes. Int J Radiat Biol 70:403–411
Tucker JD (2001) FISH cytogenetics and the future of radiation biodosimetry. Radiat Prot Dosim 9755–60
Yang TC, George K, Johnson AS, Durante M, Fedorenko BS (1997) Biodosimetry results from space flight Mir-18. Radiat Res 48:17–23

26 Internet Information System and Literature Database on Biomedical Effects of Electromagnetic Fields: Cancer and Cytogenetic

JIRI SILNY, ROMAN WIENERT, FRANK KLUBERTZ, AND SARAH DRIESSEN

Abstract

In order to assess potential health effects due to electromagnetic field exposure the outcome of in vitro studies must be compared with the results from existing epidemiological investigations, animal and in vivo experiments. For this purpose, sometimes several hundred publications related to specific exposure conditions must be researched and evaluated in a time-consuming process. The knowledge-based scientific database "*EMF-Portal*" presented here facilitates this process by providing standardized descriptors of exposure characteristics, material, method and results which have been extracted from individual biological/medical as well as epidemiological publications. In this chapter, an evaluation process concerning the endpoint cancer/genotoxicity, particularly in connection with mobile phone exposure, is presented as an example. Some predominantly epidemiological and a few in vitro investigations have reported to have found effects. According to our assessment, however, there is no clear consistency within the categories (in vitro, in vivo and epidemiology). The *EMF-Portal* currently contains more than 10,000 publications mainly from peer-reviewed scientific journals; on average, more than 600 new articles are added per year. A comprehensive glossary and a number of additional tools provide the user with valuable medical and technical assistance. A query structure, alternatively general or more detailed, allows the scope of the literature search to be narrowed down to individual publications regarding one particular subject. The *EMF-Portal* is already available in German and English; other languages can be added. All information collected is available via Internet access and free of charge at http://www.emf-portal.org.

26.1 Introduction

In the general populations of industrialized countries there are growing concerns and even fears of adverse health effects due to technical electromagnetic fields (EMF) occurring in everyday life. With the introduction of new techniques and technologies, the exposure of humans to technical EMF is becoming more and more diversified and the total intensity is increasing gradually.

Günter Obe and Vijayalaxmi (Eds.)
Chromosomal Alterations: Methods, Results and Importance in Human Health

Everyday fields are sometimes associated with subjective feelings of discomfort such as headache, sleeplessness and dizziness but also with serious illnesses like multiple sclerosis, Parkinson's disease or even cancer. This association is maintained and increasingly circulated in public appearances by influential social groups and commercial companies and institutes, without a critical discussion based on actual current scientific knowledge taking place. Business with public fears of EMF is a booming market; already enormous amounts of money are spent on unnecessary measurements conducted by so-called building biologists and on pointless countermeasures such as "protective" foils and chips.

There is a critical debate of these alleged health hazards by scientists and the general public but also by other groups, such as physicians, lawyers and politicians, who are increasingly confronted with these problems and must deal with them.

The most reliable basis for the assessment of current knowledge is the worldwide scientific literature which so far already amounts to more than 10,000 publications, even when counting only those published in peer-reviewed journals.

Investigations on chromosomal changes due to electromagnetic exposure are an important cornerstone in the demonstration of health effects. Their results must be compared with the results from epidemiological studies and animal experiments, particularly in case of the frequently discussed endpoint "cancer". Taken together, there are several hundred publications which must be traced and obtained prior to scrutiny and evaluation.

Among other information sources, the Internet-based *EMF-Portal* (accessible free of charge at http://www.emf-portal.org) provides valuable assistance in the assessment of the current status of knowledge in the research area of bioelectromagnetic interaction with about 10,000 included publications so far. In a first step, these publications must be divided according to physical criteria and medical/biological endpoints end as well as according to the investigation methods applied into categories and classes before a comprehensive evaluation of the results in individual publications and categories can be carried out. However, a majority of these are of a more technical and theoretical nature and deal predominantly with questions of dosimetry as well as with hypotheses on mechanisms of action.

Following the explanation of the necessary medical and physical differentiation of the publications, it is subsequently illustrated how the *EMF-Portal* can be advantageously used for literature research by laymen as well as by experts.

26.2 Classification of Publications

26.2.1 Physical Properties

Electromagnetic waves represent a major group of physical phenomena which follow comparable physical rules but must be differentiated with regard to their potential effects on the organism in several ways. Following the established effects of electromagnetic waves in the organism, one must distinguish primarily between non-ionizing and ionizing electromagnetic waves. The elementary energy of electromagnetic waves increases with frequency; in case of X-rays or γ-rays it reaches a point where the binding energy of biological molecules and atoms is exceeded. These ionizing waves damage the bindings of biomolecules and initiate irreversible biochemical reactions. Specific medical problems are associated with ionizing electromagnetic waves; they are not treated here further.

The subject matters of this chapter are non-ionizing electromagnetic waves with frequencies lower than 300 GHz. They are, according to their frequencies and their established effects in the organism, divided into low-frequency electric and magnetic fields (0–10 MHz) as well as radiofrequency fields and microwaves (10 MHz–300 GHz), as shown in Fig. 26.1.

Strong low-frequency electromagnetic waves have been shown to stimulate nerves, muscles and sensory receptors, whereas strong radiofrequency fields lead to heating of the tissue. Indeed, the effects of low-frequency and radiofrequency EMF are very different; both, however, can initiate a substantial health disturbance or even a life-threatening situation when the applied intensity is very high. The thresholds for effects depend on a number of features of the applied EMF, for example frequency, modulation, strength, spatial extension and duration of the exposure; their precise definition is a prerequisite for an equivalent comparison of publications.

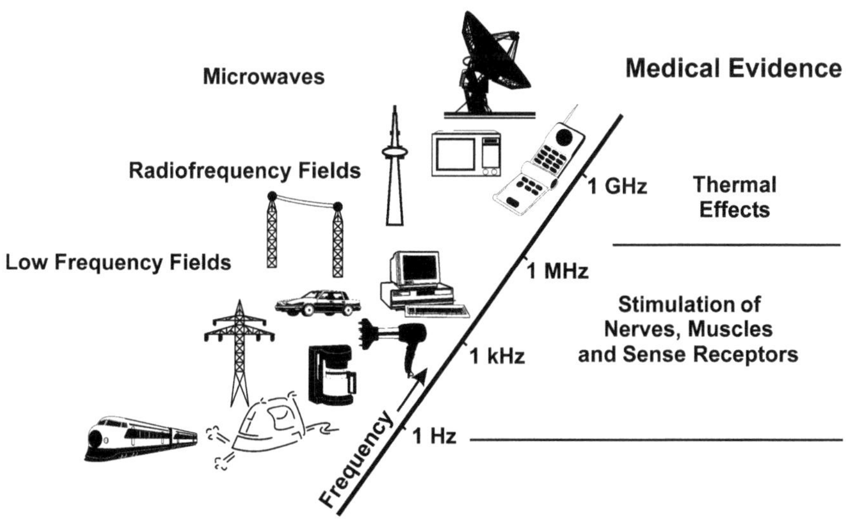

Fig. 26.1 Differentiation of non-ionizing electromagnetic fields according to their possible effects in humans

26.2.2 Medical/Biological Discrimination

For the medical/biological studies a working hypothesis is phrased on the basis of postulated working mechanisms or of scientific results but also on the basis of preliminary reports. The working hypothesis must contain an assumption that defined EMF can promote or initiate an effect or illness (Fig. 26.2).

A confirmation or non-confirmation of the working hypothesis in a single study alone is not sufficient for its acceptance or rejection. Biological variations or even systematic or random failures could be a reason for false results; therefore, only concurrent findings regarding predefined fields, a medical endpoint

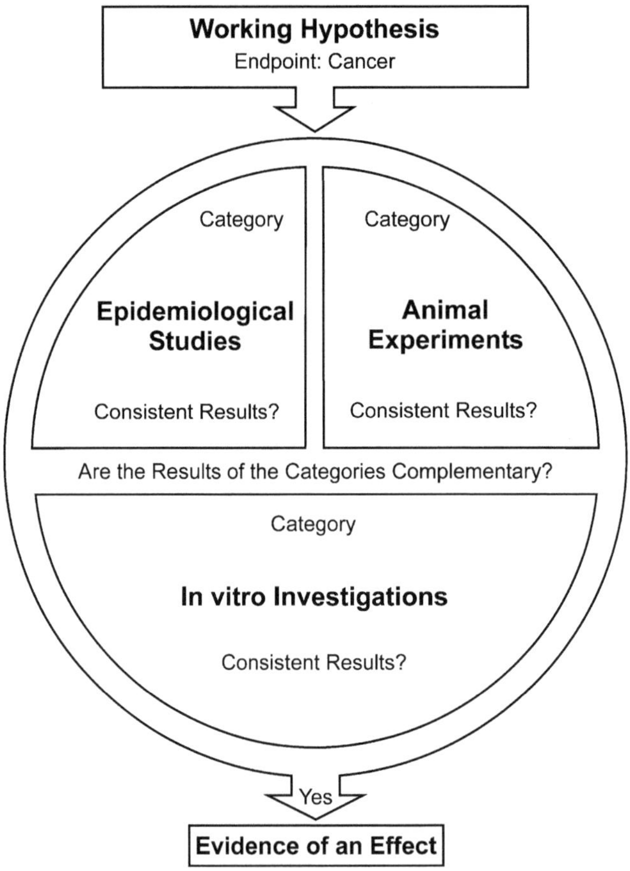

Fig. 26.2 The contribution of publications, originating from the use of different investigation methods (categories) and with the same medical endpoint, to current knowledge. The literature shows clear evidence of an effect only if the results are consistent in each category and complementary between the categories

and literature categories elaborated by a number of independent research groups can be accepted as sustainable results. For this reason the first step of the review is to collect all publications pertaining to one subject, consulting medical databases, for example PubMed, or problem-oriented databases such as IVI's (Information Ventures' *EMF Database*) *EMF Database*, the WHO EMF Research Database or our femu's (Research Center of Bioelectromagnetic Interaction) *EMF-Portal*, which are available at different conditions.

According to the existing literature categories, a broad division is carried out between epidemiological investigations, provocation studies, animal experiments and in vitro investigations. All these investigation classes have certain advantages and disadvantages which must be taken into account in the final evaluation. For example, the advantage of epidemiological and provocation studies is their focus on humans. Their weakness generally lies in their shortcomings regarding the retrospective data collection, the statistical evaluation in small groups, as well as in the missing causality of an observed effect. Therefore, animal experiments or in vitro investigations must always be carried out additionally in order to corroborate the causality of an effect that was found in an epidemiological study.

Animal experiments allow studies with big populations of homogeneous groups in well-standardized attempts, thus yielding more reliable results. A disadvantage of animal studies is the problem of transferability of the results from animal to human.

The cause of an effect and the medical/biological mechanisms can be investigated advantageously in in vitro investigations. On the other hand, results from in vitro investigations must be reviewed with regard to their transferability to animals and, ultimately, humans.

Each of the previously stated classes of biological investigations utilizes different procedures to which in each case a greater or lesser explanatory power individually or in combination with one another can be attributed.

In the following sections the evaluation of the category "in vitro investigations" with the subject "chromosomal changes" (or "genotoxicity" in general, respectively) within the scope of the endpoint "cancer" is demonstrated exemplarily utilizing our *EMF-Portal*.

26.3 The *EMF-Portal*: Internet Information System on the Effects of Electromagnetic Fields

Within the framework of the Internet Information Portal on the effects of EMF on humans and the environment (which is accessible free of charge at http://www.emf-portal.org), scientific research results are summarized comprehensibly (*EMF-Portal* 2006). An "introduction to the fundamentals of the effects of EMF", a "database of exposure sources occurring in everyday life" and a "glossary" are available as supporting tools.

Scientific investigations are an important basis for the assessment of current knowledge on the consequences of electromagnetic exposure for the health of persons or the environment. But with scientific studies one often meets the following problems:

- They are not readily available to everyone
- They are not easy to understand
- Individual studies frequently contradict each other
- Some publications contain an abundance of expert terminology which tells the scientist from other research areas only little and the layman nothing at all
- They occur often in a greater context of high complexity which requires extensive background knowledge.

Moreover, it is difficult and time-consuming to find suitable publications in a collection of more than 10,000 studies and the constantly growing body of literature is too specific regarding the subjects currently discussed and the variety of questions posed.

Currently, only the *EMF-Portal* introduced here offers concise, comparable and comprehensible summaries of scientific research results (in English and German) covering studies published in the last 25 years.

26.3.1 Literature Database – the Principal Item

The principal item of the *EMF-Portal* is the literature database. Within the scope of the *EMF-Portal*, articles in the area of bioelectromagnetic interaction of low-frequency and radiofrequency fields with the organism are researched in more than 1,300 scientific journals on a regular basis, and more than 600 studies are added per year. Moreover, included in the database are mainly published articles from peer-reviewed journals or detailed working reports and research reports if they describe the methodology applied and the experiments conducted comprehensively.

In the literature database these publications are sorted according to study types, categorized according to thematic profiles and represented in different levels of detail.

Individual articles can be found via the menu item "publication query" (Sect. 26.3.2).

26.3.2 How to Find a Publication?

There are two different ways to find publications. In the "standard query", for example, one can enter terms like "mobile phone" and "chromosome aberration". The search engine looks for related concepts and adds them to the query (e.g. "cellular phone", "handheld phone", "GSM", "TETRA", or "aberration", "chromosome break", "inversion", "translocation" and so on, respectively). In addition, the frequency utilized by the device (mobile phone from 890 MHz to 2.1 GHz) is included in the query automatically by the search engine.

The "detailed query" makes it possible to narrow down the contents of the query precisely. In a first step, a restriction of the query can be made according to bibliographical information (author, title, abstract, journal, year and keywords). By selecting the items "exposure parameters" and/or "general data" additional choices become available and the query can be limited further.

For bibliographical data, articles are searched in all collected publications (currently about 10,000 articles). In contrast, by inclusion of the "exposure parameters" and/or "general data", the search is restricted to the publications which have already been summarized. The query within the "exposure parameters" is possible within about 3,000 publications, while the query within the "general data" is possible within about 1,400 publications.

Found publications are returned in the form of a "hit list". Figure 26.3 shows the first three publications found by the query on the subject of "mobile phone" (including the search terms "mobile phone", "cellular phone", "GSM" and "UMTS") and "genotoxicity". For a better overview, the list contains not only the bibliographical data but also some important summarized details of the respective study (aim, endpoint, frequency range).

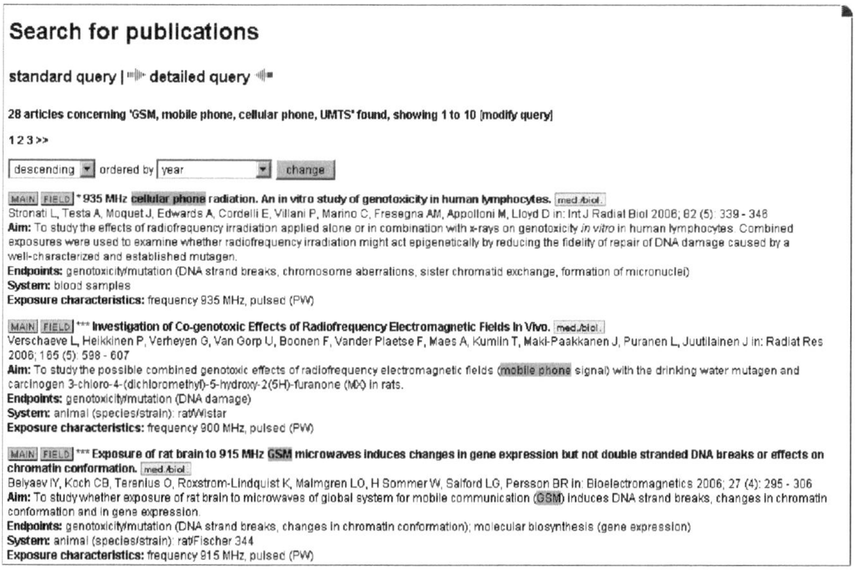

Fig. 26.3 Generated result list (of the first three results) following a detailed query for "mobile phone" (with the search terms "mobile phone", "cellular phone", "GSM" and "UMTS") and "genotoxicity"

The associated summaries can be viewed by selecting (clicking) an article from the "hit list". Figure 26.4 shows the representation of the first article from the result list exemplarily.

Using the German or English summaries of individual publications, the user gets a first impression concerning the essential contents of the processed publications without the need for access to the original full text.

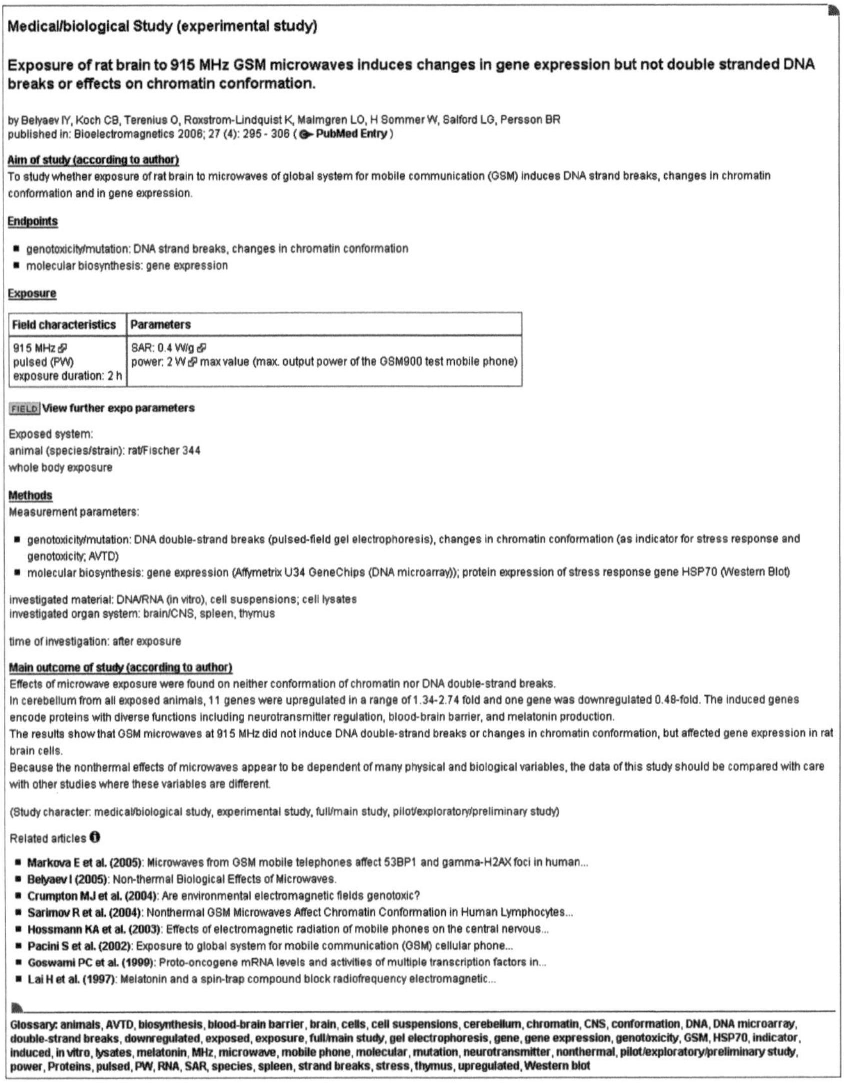

Medical/biological Study (experimental study)

Exposure of rat brain to 915 MHz GSM microwaves induces changes in gene expression but not double stranded DNA breaks or effects on chromatin conformation.

by Belyaev IY, Koch CB, Terenius O, Roxstrom-Lindquist K, Malmgren LO, H Sommer W, Salford LG, Persson BR
published in: Bioelectromagnetics 2006; 27 (4): 295 - 306 (**PubMed Entry**)

Aim of study (according to author)
To study whether exposure of rat brain to microwaves of global system for mobile communication (GSM) induces DNA strand breaks, changes in chromatin conformation and in gene expression.

Endpoints

- genotoxicity/mutation: DNA strand breaks, changes in chromatin conformation
- molecular biosynthesis: gene expression

Exposure

Field characteristics	Parameters
915 MHz pulsed (PW) exposure duration: 2 h	SAR: 0.4 W/g power: 2 W max value (max. output power of the GSM900 test mobile phone)

FIELD **View further expo parameters**

Exposed system:
animal (species/strain): rat/Fischer 344
whole body exposure

Methods
Measurement parameters:

- genotoxicity/mutation: DNA double-strand breaks (pulsed-field gel electrophoresis), changes in chromatin conformation (as indicator for stress response and genotoxicity; AVTD)
- molecular biosynthesis: gene expression (Affymetrix U34 GeneChips (DNA microarray)); protein expression of stress response gene HSP70 (Western Blot)

investigated material: DNA/RNA (in vitro), cell suspensions; cell lysates
investigated organ system: brain/CNS, spleen, thymus

time of investigation: after exposure

Main outcome of study (according to author)
Effects of microwave exposure were found on neither conformation of chromatin nor DNA double-strand breaks.
In cerebellum from all exposed animals, 11 genes were upregulated in a range of 1.34-2.74 fold and one gene was downregulated 0.48-fold. The induced genes encode proteins with diverse functions including neurotransmitter regulation, blood-brain barrier, and melatonin production.
The results show that GSM microwaves at 915 MHz did not induce DNA double-strand breaks or changes in chromatin conformation, but affected gene expression in rat brain cells.
Because the nonthermal effects of microwaves appear to be dependent of many physical and biological variables, the data of this study should be compared with care with other studies where these variables are different.

(Study character: medical/biological study, experimental study, full/main study, pilot/exploratory/preliminary study)

Related articles

- **Markova E et al. (2005)**: Microwaves from GSM mobile telephones affect 53BP1 and gamma-H2AX foci in human...
- **Belyaev I (2005)**: Non-thermal Biological Effects of Microwaves.
- **Crumpton MJ et al. (2004)**: Are environmental electromagnetic fields genotoxic?
- **Sarimov R et al. (2004)**: Nonthermal GSM Microwaves Affect Chromatin Conformation in Human Lymphocytes...
- **Hossmann KA et al. (2003)**: Effects of electromagnetic radiation of mobile phones on the central nervous...
- **Pacini S et al. (2002)**: Exposure to global system for mobile communication (GSM) cellular phone...
- **Goswami PC et al. (1999)**: Proto-oncogene mRNA levels and activities of multiple transcription factors in...
- **Lai H et al. (1997)**: Melatonin and a spin-trap compound block radiofrequency electromagnetic...

Glossary: animals, AVTD, biosynthesis, blood-brain barrier, brain, cells, cell suspensions, cerebellum, chromatin, CNS, conformation, DNA, DNA microarray, double-strand breaks, downregulated, exposed, exposure, full/main study, gel electrophoresis, gene, gene expression, genotoxicity, GSM, HSP70, indicator, induced, in vitro, lysates, melatonin, MHz, microwave, mobile phone, molecular, mutation, neurotransmitter, nonthermal, pilot/exploratory/preliminary study, power, Proteins, pulsed, PW, RNA, SAR, species, spleen, strand breaks, stress, thymus, upregulated, Western blot

Fig. 26.4 Example of the summary of an experimental study

So far, epidemiological and experimental studies from the medical/biological area have been evaluated. The structure of the representation of the epidemiological studies or the experimental investigations, respectively, is always the same. This serves the purpose of a quick orientation as well as an optimum comparability of the studies with one another.

In the case of reviews (survey studies), only the main subject focus and frequency range are extracted so that they can be found more easily and allow the user to decide quickly whether the respective review is of interest.

26.3.3 An Example on the Subject of "Genotoxicity/Cancer"

Currently, there are 204 experimental studies in the database from different frequency ranges concerning the topic "genotoxicity" (Fig. 26.5).

If one wants to put the subject "genotoxicity" in a broader context in order to be able to evaluate the possible effects described in the studies more globally, regarding, for example, "cancer", it is advisable to differentiate between in vitro investigations, animal experiments and epidemiological studies (Sect. 26.2).

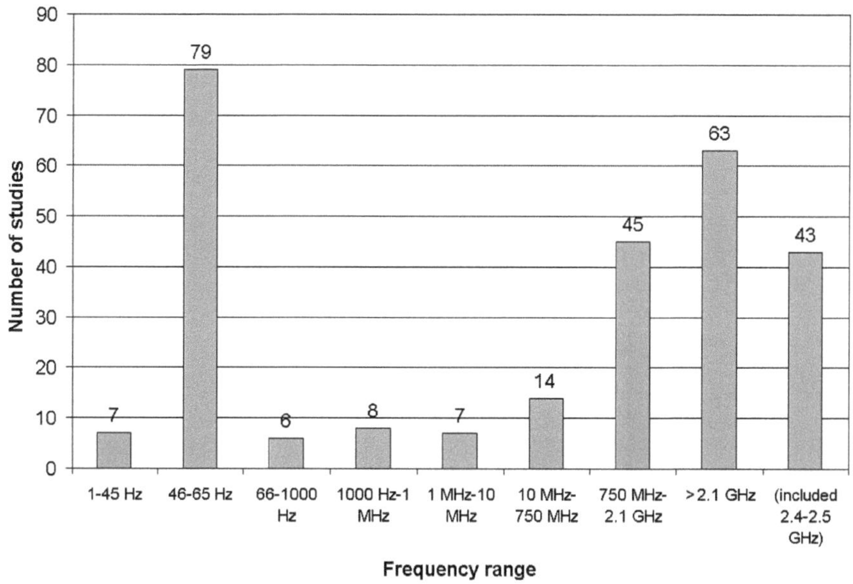

Fig. 26.5 Studies with the main topic "genotoxicity", distributed according to the frequency applied. If several frequencies or broader frequency bands were used in individual studies, the studies of each frequency range are added separately; hence, the sum of all studies listed in the chart is more than 204 (actual sum of existing studies)

With the help of the detailed query (Sect. 26.3.2) one can search for different exposure types and frequency ranges, for different exposed materials (in vitro systems, animals, plants or humans) and for different endpoints.

Figure 26.6 shows, for example, the results of different detailed queries:

- For mobile phone (search words "GSM", "cellular phone", "mobile phone" or "UMTS")
- In each case divided into categories according to the exposed in vitro system or animal
- For the biological endpoints "cancer" or "genotoxicity".

In total there are currently 52 studies on the effects of EMF in the mobile phone frequency range regarding the endpoints "genotoxicity" or "cancer": 18 in vitro investigations and seven animal studies deal with the endpoint "genotoxicity", 17 experimental animal studies and 35 epidemiological studies address the endpoint "cancer".

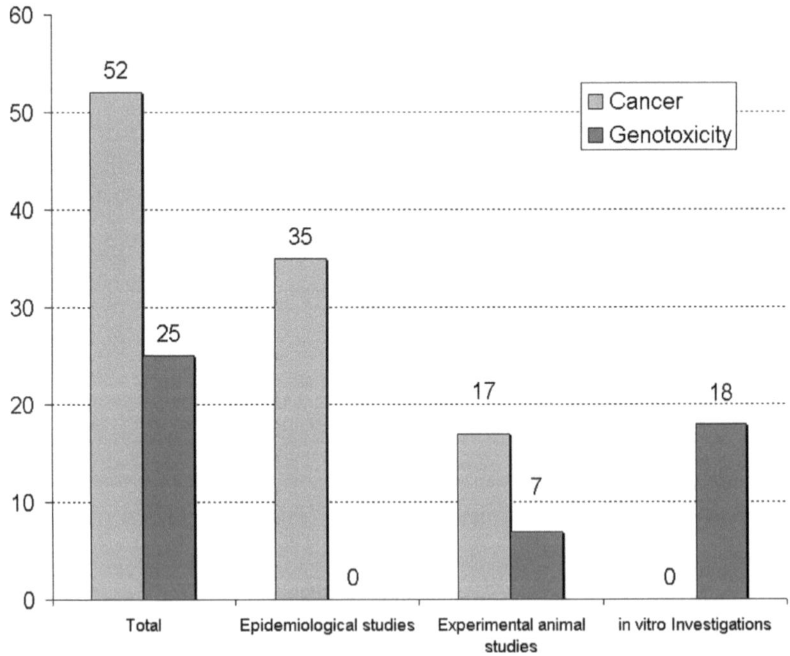

Fig. 26.6 Results of different detailed queries for "cancer/genotoxicity" and "mobile phone", divided into in vitro investigations, experimental animal studies and epidemiological studies

- There are no consistent data from the in vitro investigations regarding the endpoint "genotoxicity". There are about the same number of publications that report effects of exposure to EMF in the frequency range related to mobile phones and that do not. The studies differ greatly with respect to their frequency ranges (835–1,900 MHz), the different types of modulation (TDMA, CDMA, FDMA, etc.) and specific absorption rates (SAR; from 0.005 to 10 W/kg) as well as the different cell systems applied and measurement parameters investigated (chromosome alteration, sister-chromatid exchange, micronuclei formation, DNA stability, DNA strand breaks). An evaluation and a comparison of these inconsistent data are very difficult and nearly impossible for non-experts. Standardized investigation methods should be required for future studies.
- Additionally, there are seven experimental animal studies regarding the endpoint "genotoxicity". These studies also differ greatly with respect to their frequency ranges (900–1,747 MHz), the applied SAR values (0.07–33 W/kg, whole body), the mobile phone systems and types of modulation (GSM, PDC, IRIDIUM, FDMA, TDMA) that are taken into account, the different animals investigated (a variety of mouse and rat strains) as well as their utilized measurement parameters (DNA strand breaks, chromatin structure, formation of micronuclei, DNA sequence analysis, chromosomal alterations). Moreover, the exposure duration varies between 2 h/day for 1 day and for 2 years. The results of the animal studies are equally inconsistent. Three out of seven studies report observed effects; four studies do not. Only experts in this field are competent enough to compare and interpret these data adequately (Sect. 26.4).
- In contrast, the 17 experimental animal studies addressing the endpoint "cancer" provide more consistent data. Although the frequency ranges (835–1,439 MHz) investigated as well as the modulation and mobile phone systems (GSM, PDC, CDMA, FDMA, TDMA, NMT, UMTS) differ in the various studies, the applied SAR values (whole body) were in the range 0.4–1 W/kg in the majority of investigations. Even though different mouse and rat strains as well as different types of cancer (lymphoma, skin cancer, mammary carcinoma, liver cancer) were investigated, all publications (with the exception of one) conclude that EMF exposure in the frequency range related to mobile phones neither initiates nor promotes or copromotes cancer.
- Of the 35 epidemiological studies, 15 have found a weak but significantly increased risk of malignant brain tumours, acoustic neuroma, leukaemia and uveal melanoma in the users of mobile phones. However, 14 of these studies were published by the same Swedish group, and they predominantly address a since-outdated analogue mobile phone system. The remaining 20 studies from different research groups have not shown any association between the short- or medium-term exposure by microwaves emitted by mobile phones and brain tumours, acoustic neuroma, cancer of the salivary gland, leukaemia or malignant melanomas of the eye.

26.4 Discussion

The body of scientific literature regarding chromosomal alterations due to EMF provides important and detailed insights into possible effects and influence mechanisms. But for the evaluation of potential health effects a prudent comparison of the findings of epidemiological studies, animal experiments and in vitro investigations resulting from comparable EMF exposure is necessary.

The review of health effects due to EMF is a demanding and time-consuming process even if the physical and medical expert knowledge is available. In the course of this process all publications dealing with a defined EMF range and a particular medical endpoint are collected and evaluated individually. Today, literature research is generally carried out utilizing Internet databases which – beyond the mere bibliographical data – provide a variety of additional information and are available under specific terms and conditions. For example, through the journal literature search system *PubMed* (2006), the by far largest medical database, virtually anyone can access more than 15,000,000 biomedical journal citations of publications since 1966. Currently, it contains about 9,000 articles related to EMF.

As an alternative, specific databases pertaining to EMF can be used. IVI's *EMF Database* (2006) provides about the same scope of information as *PubMed*, but additionally collects records of poster sessions and meeting abstracts. However, the use of its Internet interface is limited to a number of basic operations and the viewing of a handful of sample records; the use of the complete CD version containing third-party abstracts is subject to a fee.

In contrast, the Internet-based WHO *EMF Project Database* (2006) of the WHO International EMF Project and the IEEE *EMF Citation List* (2006), are free of charge and provide additional descriptors such as the type of the study and the endpoints pursued; unfortunately, however, as yet they have only a small stock of records. The Swiss database *ELMAR* (2006), also contains additional descriptors regarding endpoints, materials, methods and results, but is small and is available only in German. Currently, only femu's *EMF-Portal* contains comprehensive information beyond bibliographical standard data as it includes detailed descriptions of the biological or medical studies conducted as well as their results and exposure parameters. Moreover, the Internet access to the information covering practically the whole body of literature pertaining to bioelectromagnetic interaction since 1980 is completely free of charge and access restrictions.

Individual studies can be evaluated objectively under the condition that only explicitly cited data from published works are taken into account. But the next step, the validation of the consistency of a postulated effect in the profiles of literature resulting from comparable investigations is - to a certain extent – a subjective process in which a high level of expert knowledge and the ability to interpret scientific results are essential.

The same problems and requirements arise when trying to assess the complementarities of the results reported between the categories; therefore a comprehensive evaluation like that should preferably be conducted independently by more than one expert, if possible.

Owing to the rapid growth of the body of literature in this area of research with an annual increment of more than 600 publications, this evaluation process should be repeated at least every 4 or 5 years.

The question whether – or if so to what extent – EMF can promote or even initiate cancer is of crucial interest for highly industrialized societies. Here, scientific publications answering the questions concerning chromosomal alterations in EMF are of paramount importance. Several hundred epidemiological studies, animal studies as well as in vitro investigations have been carried out. Some predominantly epidemiological and a few in vitro investigations report to have actually found effects. But there is no clear consistency within the categories and there is no complementary support between the groups. This conclusion can be derived for the low-frequency as well as the radiofrequency range of EMF (Gezondheidsraad Jaarbericht 2005; Health Protection Agency 2006; NRPB 2004). So far, the scientific literature does not provide any sustainable evidence for an association between EMF occurring in everyday life and the various forms of cancer.

References

ELMAR (2006) Dokumentationsstelle *ELMAR* (ELectroMAgnetic Radiation), Basel, Switzerland. http://www.elmar.unibas.ch/elmar_suche.html. Cited 19 July 2006

IVI's *EMF Database* (2006) Information Ventures, Philadelphia, USA. http://infoventures.com/emf/database/. Cited 19 July 2006

EMF-Portal (2006) Research Center for Bioelectromagnetic Interaction – femu, Aachen. http://www.emf-portal.org/_index.php?l=e. Cited 19 July 2006

WHO *EMF Project Database* (2006) World Health Organization, Geneva, Switzerland. http://www.who.int/peh-emf/research/database/emfstudies. Cited 19 July 2006

Gezondheidsraad Health Council of the Netherlands (2005) Electromagnetic fields: annual update 2005. Jaarbericht van de Gezondheidsraad no 2005/14. Gezondheidsraad Health Council of the Netherlands, The Hague

Health Protection Agency (2006) Power frequency electromagnetic fields, melatonin and the risk of breast cancer. Report of an independent advisory group on non-ionising radiation. Documents of the Health Protection Agency. Series B: radiation, chemical and environmental hazards. HPA series B:RCE(1). Health Protection Agency, Chilton, pp 1–180

IEEE EMF Citation List (2006) http://www.who.int/peh-emf/research/database/IEEEdatabase. Cited 19 July 2006

NRPB (2004) Mobile phones and health 2004. Report by the board of NRPB. Doc NRPB 15(5):1–114

PubMed (2006) US National Library of Medicine, Bethesda. http://www.ncbi.nlm.nih.gov/entrez/query.fcgi?DB=pubmed. Cited 19 July 2006

27 Interpreting Studies on Magnetic Fields and the Risk of Childhood Leukaemia: an Ongoing Challenge

Joachim Schüz and Tracy Lightfoot

Abstract

Childhood cancer under the age of 15 years is a rare event accounting for less than 1% of all malignancies diagnosed each year in developed countries. Overall the risk of developing acute leukaemia before the age of 15 years is around 1:2,000. Acute leukaemia is a heterogeneous disease and the advent of improved molecular diagnostic techniques has provided greater insight into its underlying molecular biology. The presence of chromosomal abnormalities in neonatal blood spots from children with leukaemia supports the hypothesis for a prenatal origin of the disease. Furthermore, identification of these abnormalities at a rate higher than the proportion of children who will develop the associated leukaemia subtype has provided evidence that childhood leukaemia is at least a two-step process with a need for additional genetic events and/or exposures. Improved knowledge of chromosomal alterations and methods for their analysis offers new opportunities for understanding the relevance of environmental exposures and the importance of gene–environment interactions in the development of the disease. A prominent example is the ongoing challenge of interpreting findings from epidemiological studies on exposure to magnetic fields and childhood leukaemia risk, where a doubling of leukaemia risk was consistently seen across most studies at average magnetic field exposures of 0.3–0.4 μT. The interpretation of the magnetic-field-related findings is particularly delicate because possible explanations range from methodological artefacts to a causal link. To date, neither of these two contradictory explanations can be completely ruled out.

27.1 Introduction

In 2002, within the framework of their Monography Program on the evaluation of carcinogenic risks to humans, the International Agency for Research on Cancer (IARC) classified extremely low-frequency magnetic fields as a possible carcinogen to humans (IARC 2002). This classification was based on epidemiological studies that investigated the relationship between exposure to residential magnetic fields and the risk of childhood leukaemia. In these

Günter Obe and Vijayalaxmi (Eds.)
Chromosomal Alterations: Methods, Results and Importance in Human Health

studies, about a doubling of the leukaemia risk was found with average exposure above 0.3–0.4 μT (Greenland et al. 2000; Ahlbom et al. 2000). These findings were considered as being credible, but chance, bias and confounding could not be ruled out with reasonable confidence. No support came from either experimental animal or cell studies and no plausible explanation for the epidemiological findings has been put forward (IARC 2002). Thus, the overall evidence was considered to be too weak to provide sufficient evidence of carcinogenicity; however, the epidemiological findings were not dismissed and merit clarification. The disagreement between results from the observational and experimental studies could be due to the failure of experimental studies to provide an appropriate model for the complex origin of childhood leukaemia; this will be discussed in Sect. 27.2. Alternatively, it could be due to methodological flaws of the observational studies or other mechanisms unrelated to magnetic fields that nevertheless create a correlation between exposure and disease risk; this will be discussed in Sect. 27.3 following an account of the epidemiological evidence. Finally, overall conclusions will be drawn and the role of the identification of chromosomal alterations in clarifying this issue will be presented (Sect. 27.4).

27.2 Leukaemia in Children

27.2.1 Description of Epidemiology and Biological Heterogeneity

Cancer in children under the age of 15 years is a rare event accounting for less than 1% of all malignancies diagnosed each year in developed countries. Of those cancers seen in children, leukaemias, in particular acute leukaemias, are the most common, accounting for a third of all diagnoses. The incidence rate of acute leukaemias ranges from 30–50 per million children per year (Steliarova-Foucher et al. 2004). Overall the risk of developing acute leukaemia before the age of 15 years is around 1:2,000 (Greaves 2002), with the disease being more prevalent in boys than in girls (UKCCS 2000). The distinctive age distribution characterised by a rapid increase in incidence after birth, peaking between 2–5 years of age before declining to a steady rate seen for childhood leukaemia, has resulted in many aetiological hypotheses (UKCCS 2000), most notably those involving exposure to infections (Greaves 2002; Kinlen 1995). The most recent data on incidence trends indicate a small annual increase of 0.7% in European countries in the last three decades (Steliarova-Foucher et al. 2004); however, this increase is likely to be attributable to improvements in the classification and registration of childhood cancers.

The biological heterogeneity of childhood leukaemia is well documented (Biondi et al. 2000) and it is therefore unlikely that there is a single causal exposure or mechanism. The major subtypes diagnosed in children are acute lymphoblastic leukaemia (ALL) and acute myeloid leukaemia (AML). They account for 80% and around 15–17% of cases in white populations, respectively, with a slightly higher proportion of AML cases reported in Asia and black populations of North America (Parkin et al. 1988). By contrast, chronic forms of leukaemia in children are extremely rare in all populations and seldom exceed 4% of all leukaemia diagnoses.

Historically, one of the problems faced when trying to unravel the potential determinants of childhood leukaemia was the inability to classify different subtypes of leukaemia such that all cases were pooled together, which may have masked any potential associations. However, over the last 25 years our knowledge of immunology has increased, leading to an improvement in diagnostic techniques. Furthermore, these advances in classifying the different biological subtypes combined with the recognition of prognostic variables and improved chemotherapeutic agents has resulted in more than 80% of children diagnosed with acute leukaemia surviving into adulthood (Pui et al. 2004); however, some children remain unresponsive to treatment (Biondi et al. 2000), and the long-term health consequences of survivors is the subject of much ongoing research.

Subtypes of leukaemia are characterised by genetic alterations including point mutations, deletions and inversions resulting in gross chromosome changes such as translocations or hyperdiploidy or hypodiploidy (Look 1997). As with other cancers, chromosomal translocations are initiated by DNA double-strand breaks. In contrast to epithelial carcinomas, however, the translocations are generally balanced or reciprocal and often result in chimeric or fusion genes which generate hybrid proteins with altered properties such as activated kinase activity or novel transcriptional regulation (Look 1997). Although over 200 genes have been associated with the occurrence of chromosomal abnormalities, some are more frequently observed than others and are related to particular morphologic or phenotypic subtypes (Greaves and Wiemels 2003; Table 27.1). Those most commonly seen include *MLL*, *TEL* and *AML1* – all of which have been shown to be essential for normal haemopoiesis in mice knockout studies (Okudo et al. 1996) [the acronyms display gene symbols that represent the positional mapping of a gene, like *MLL* being located on 11q23, which is band 2 subband 3 on the long arm (q) of chromosome 11].

In infants under 12 months old, a distinct type of ALL with a pro-B phenotype has been identified such that over 80% of cases have 11q23 chromosomal translocations. These translocations involve the breakage and recombination of the *MLL* gene, which is involved in regulation of gene expression and maintenance of chromatin stability (Yu et al. 1998), with one of several potential partner genes (Heerema et al. 1994). The marked peak in incidence seen between 2 and 5 years of age is associated with common B-cell precursor ALL where the most common abnormalities reported are hyperdiploidy and the *TEL-AML1* gene fusion which account for 35% and 25% of cases, respectively (Greaves 2005). The question arises as to how and when these translocations occur. However, there is now substantive evidence that some, if not all, of the translocations seen in childhood leukaemia originate in utero.

Table 27.1 The major biological subtypes of childhood leukaemia and associated chromosomal changes. (Adapted from Greaves and Wiemels 2003)

Subtype	Cell type involved	Chromosome abnormality	Molecular lesion	Frequency (%)	Functional product
ALL	B-cell progenitor-monocytic[a] (infants)	11q23 translocations	*MLL-AF4*	~85 (of infant ALL)	Modified transcription factor[b]
			MLL-ENL and other fusions	~5 (of total ALL)	
	B-cell precursor	Hyperdiploidy	Increased gene dose	~35 (of B-cell precursor ALL)	Unknown
		t(12;21)(p13;q22)	*TEL-AML1* fusion	~20 (of B-cell precursor ALL)	Chimeric transcription factor[c]
		t(1;19)(q23;p13)	*E2A-PBX1* fusion	~5 (of B-cell precursor ALL)	Chimeric transcription factor
		t(9;22)(q34;q11)	*BCR-ABL* fusion	~5 (of B-cell precursor ALL)	Activated kinase
	T-cell precursor	1q deletion; t(1;14)(p32;q11)	*SIL-SCL* fusion	~25 (of T-cell precursor ALL)	Dysregulated transcription factor *(SCL/TAL1)*
AML	In infants	11q23 translocations	*MLL-AF6*, *MLL-AF9*, *MLL-AF10* or other fusions	~50 (of infant AML)	Modified transcription factor[b]
		t(8;21)(q22;q22)	*AML-ETO* fusion	~15 (of total AML)	Chimeric transcription factor[c]

ALL acute lymphoblastic leukaemia, *AML* acute myeloid leukaemia
[a]This subtype has phenotypic features of both early B-lineage progenitors and monocytes.
[b]*MLL* may modify chromatin structure and control the expression of key developmental genes (such as *HOX*).
[c]*AML-1*, which is an important positive transcriptional factor, may be switched to a transcriptional repressor in these fusions.

27.2.2 Origins of Childhood Leukaemia

Leukaemia is a clonal disease arising from the neoplastic transformation of a single cell which is likely to be either a pluripotent stem cell or a more mature progenitor cell (Greaves 2005). The single cell, via repeated cell division, generates an expanding clone of neoplastic cells which is highly malignant. As well as the nature of the leukaemic clone, the stage of disease progression at the time the pathological symptoms are recognised, the diagnosis made and treatment regimen implemented are also important for the clinical outcome of the disease. Any delay in diagnosis increases the probability that the clone will have progressed to the point where additional mutations will have accrued, including those involved with drug resistance, which can make treatment more difficult.

Identification of the origin of chromosomal translocations has been facilitated by the design of analytical tests for specific fusion genes which can detect as few as one to 20 leukaemic cells in a blood spot. The presence of the same fusion gene sequence in neonatal blood spots or Guthrie cards and the patient's leukaemic cells at diagnosis provides compelling evidence that chromosomal translocations can originate in utero and that leukaemia can be initiated prenatally. Indeed at least six different types of paediatric leukaemia have been identified where the chromosomal rearrangements, including hyperdiploidy, have been shown to originate in utero (Greaves 2005; McHale and Smith 2004). Data from studies of identical twins and children with concordant ALL support this hypothesis (Greaves 2006).

Screening of newborn cord blood samples suggests that around 1% have a functional *TEL-AML1* fusion gene, which is 100 times the proportion of children that will develop ALL with a *TEL-AML1* gene fusion before the age of 15 years (Mori et al. 2002). This implies that the frequency of conversion of the preleukaemic clone to overt disease is low, and that development of ALL is a multistep process requiring at least one prenatal initiating event in combination with other prenatal and/or postnatal events (Fig. 27.1). These events may involve additional genetic changes arising either before or after the formation of the chromosomal translocation, as well as exposure to biological, chemical

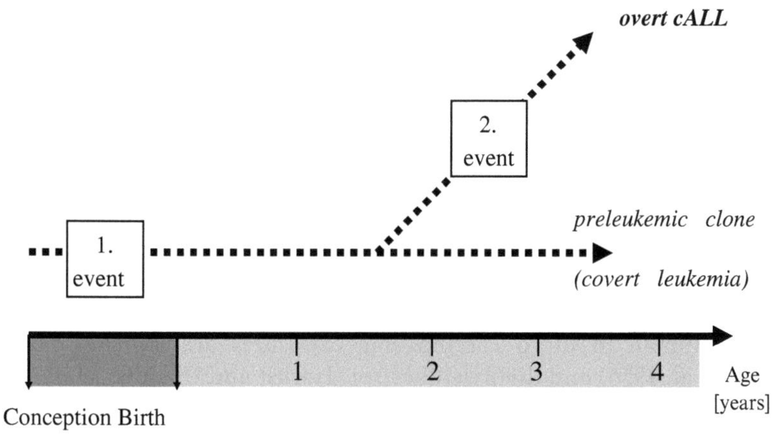

Fig. 27.1 A two-step model for the natural history of the most common subtype of childhood leukaemia, common acute lymphoblastic leukaemia (*cALL*). (Slightly modified from Greaves 1999)

or physical agents (Greaves 2006; Roman et al. 2006, 2007). Whilst the initiating in utero event is believed to be common, it is thought that the subsequent events act as the rate-determining steps in disease development. Assuming that the "two-hit" model for childhood leukaemia is correct, Greaves (2002) suggests that for every child diagnosed with ALL, there are at least 20 "healthy" children who also have a chromosomal translocation generated in utero. There is clearly a need to identify the possible exposures or molecular events that cause these chromosome breaks which are inefficiently or incorrectly repaired. Owing to the biological diversity of childhood leukaemia it is, however, unlikely that there is a single cause for the disease.

27.2.3 Aetiology of Childhood Leukaemia

In adults, carcinogens can accumulate over a long period of time, whereas in children the potential exposure period is much shorter. Accordingly, exposures acting before birth and early in life have long been thought to be important determinants of leukaemia, and the list of suspected chemical, physical and biological agents continues to increase. There are several mechanisms by which exposure to both exogenous and endogenous agents may impact on the risk of childhood leukaemia.

With respect to parental exposures prior to conception, exposure to possible carcinogenic substances early in the mother's life may be important since oocyte maturation begins during gestation. Accordingly, germ-cell mutations arising in the offspring could be attributable to exposures acting on the maternal grandmother while she was pregnant with the index child's mother. Conversely, paternal exposures may play a more direct role since spermatogenesis continues from puberty to old age. Thus, there is a greater period of time during which the accumulation of mutations may occur.

During pregnancy, exposures to agents such as ionising radiation may act directly, whilst others may act indirectly via placental transfer. However, it is not only the timing of exposure that is important but also the mother's and the embryo's ability to metabolise and detoxify the agents to which they are exposed. Furthermore, maternal preconceptional exposure to agents that have a slow metabolic clearance could result in the mother retaining embryotoxic doses into the early stages of pregnancy.

Whilst animal models have shown that preconceptional exposure of germ cells to carcinogens and mutagens results in excess tumours in offspring, no such links to cancer in humans have been confirmed. Indeed the only accepted human in utero carcinogenic exposures are ionising radiation (Stewart et al. 1956) and diethylstilbestrol (Herbst and Ulfelder 1971).

After birth, the infant can still be affected by parental exposures, which in the case of breast-fed infants may have occurred many years in the past, as well as being influenced by direct environmental exposures such as magnetic fields, which will be dealt with separately in Sect. 27.3.

To date the major focus has been with respect to infection. Indeed, the role of infection in the development of childhood ALL is the focus of much current research. The "Kinlen" hypothesis speculates that acute leukaemia may arise as a consequence of exposure to a specific infection, particularly evident at times of unusual population movement and mixing (Kinlen 1995). However, whilst data continue to accrue in this area, as yet no virus-dependent mechanism has been positively identified. By contrast, the "Greaves" hypothesis proposes that a deficit of exposure to infectious agents in infancy and subsequent "delayed" infectious challenge may be causal in the development of B-cell precursor common ALL (c-ALL) (Greaves 2006). However, recent findings suggest that increased exposure to infections early in life may, in some children, also play a critical role (Roman et al. 2006). Much has also been documented regarding the effects of breast-feeding on childhood leukemia risk since immunological benefits are provided to the offspring that may contribute to their response to infection. Whilst many studies have reported weak protective associations between breast-feeding and childhood ALL, data from the UK Childhood Cancer Study and the Oxford Survey of Childhood Cancers failed to support the hypothesis (Lightfoot and Roman 2004). The interpretation of these findings is not straightforward since breast-feeding may also allow transfer of genotoxins along with viruses from the mother to the child.

Many parental exposures have also been suggested as determinants of childhood leukaemia, including general lifestyle factors such as smoking and alcohol intake and exposure to DNA topoisomerase inhibitors and folic acid (Lightfoot and Roman 2004). The adverse health effects of cigarette smoking and excessive alcohol consumption are well documented – both with respect to the health of the individual and with respect to the health of the offspring; however, whilst many studies have reported on this topic there is no consistent evidence to confirm links with childhood leukaemia (Lightfoot and Roman 2004). *MLL* rearrangements are frequently seen in adult cases of leukaemia following chemotherapy treatment with topoisomerase II inhibitors. It is therefore possible that in utero exposure to topoisomerase inhibitors (including pesticides, benzene metabolites, bioflavanoids, some laxatives and antibiotics) could also be important in infant leukaemia; however, the data are sparse and rely entirely on self-report (Lightfoot and Roman 2004).

Folic acid levels and polymorphisms in folate-dependent enzymes have been linked to various types of cancer, including ALL in both adults and children. Folic acid deficiency and polymorphisms which mimic the conditions of folic acid deficiency alter patterns of DNA methylation and impair DNA synthesis and repair and may therefore influence leukaemogenesis. Research has focused on the polymorphic enzyme methylenetetrahydrofolate reductase which is key in regulating the balance between DNA methylation and DNA synthesis. The number of studies to date that have investigated methylenetetrahydrofolate reductase polymorphisms in relation to childhood ALL are limited and based on small numbers (Lightfoot and Roman 2004).

Unfortunately, the evidence regarding the majority of suggested exposures is limited and often contradictory and the search for the candidate exposure – assuming it exists – continues.

27.3 Exposure to Magnetic Fields and the Risk of Childhood Leukaemia

27.3.1 Summary of the Epidemiological Findings

Magnetic fields have been studied as a risk factor for childhood leukaemia since the late 1970s (Wertheimer and Leeper 1979). Since then more than 20 epidemiological studies have investigated this topic, with significant improvements in methods of exposure assessment over time. These include (1) simple distance estimations between power lines and the residence, (2) classifications of different types of power lines and their nearest distance to the residence (wire code), (3) magnetic field calculations based on power load data of power lines and (4) various measurement protocols, ranging from outdoor spot measurements to 24–48-h measurements in the child's bedroom and the use of personal dosimeters. Recently, these studies have been pooled in two separate meta-analyses. Greenland et al. (2000) pooled all available studies in this area and used the original data to estimate a pooled relative risk estimate; however, they had to combine various exposure indices into one single metric; hence, the microtesla estimate for the child's exposure in this pooling study is based on the somewhat different exposure assessment methods used across studies. Greenland et al. (2000) reported a combined relative risk estimate for leukaemia of 1.7 (95% confidence interval, CI, 1.2–2.3) in children exposed to average magnetic fields above 0.3 μT compared with those exposed below 0.1 μT. Ahlbom et al. (2000) only pooled studies that fulfilled certain inclusion criteria such as a defined population base for case ascertainment and control recruitment and usage of long-term measurements or historical magnetic field calculations for exposure assessment. Ahlbom et al. (2000) reported a relative risk estimate of 2.0 (95% CI 1.3–3.1) for exposures of 0.4 μT and above compared with exposures below 0.1 μT. The studies demonstrate a high degree of consistency; however, a major contribution to the overall relative risk estimate comes from a US study (Linet et al. 1997) and no association was seen in a large study in the UK (UKCCS 1999) (Table 27.2). A shortcoming was that the possibilities for dose–response analysis were limited, as only very few children were exposed to average magnetic fields of 0.4–1 μT or higher; hence, the data were too sparse to reliably predict the course of a dose–response curve for magnetic fields higher than 0.4 μT and, based on available data, this course is more or less compatible with trends ranging from a further increase in risk, to a constant risk or a downward gradient. Findings from these two pooled analyses

Table 27.2 Results from nine studies included in a pooled analysis of studies on residential magnetic fields and the risk of childhood leukaemia. (Adapted from Ahlbom et al. 2000)

	Odds ratio (95% CI) ≥ 0.4 μT vs. <0.1 μT	Leukaemia cases	
		Observed ≥ 0.4 μT	Expected[a] ≥ 0.4 μT
Canada	1.55 (0.65–3.68)	13	10
USA	3.44 (1.24–9.54)	17	5
UK	1.00 (0.30–3.37)	4	4
Norway	0 cases, 10 controls	0	3
Germany[b]	3.53 (1.01–12.3)	7	2
Sweden	3.74 (1.23–11.4)	5	2
Finland	6.21 (0.68–56.9)	1	0
Denmark	2 cases, 0 controls	2	0
New Zealand	0 cases, 0 controls	0	0
Total[b]	2.04 (1.33–3.15)	49	26

CI confidence interval
[a]Rounded to whole numbers
[b]Update of Ahlbom et al. (2000) with a second German study (Schüz et al. 2001)

were those referred to in recent reviews of claims of consistency in the epidemiological evidence (IARC 2002). In particular, the approach by Ahlbom et al. (2000) seemed to strengthen the evidence, since by introducing more uniformity into the analytical strategies, there was more consistency in the meta-analysis than in the originally published results. However, if studies suffer from the same systematic bias, this bias will be amplified in a pooled analysis rather than eliminated.

27.3.2 Interpreting the Empirical Association

27.3.2.1 Methodological Artefacts

In an IARC (2002) review it was stated that chance, bias and confounding cannot be ruled out with reasonable confidence as an explanation for the epidemiological association between exposure to magnetic fields and the risk of childhood leukaemia. While chance alone seems unlikely, given the results of the two meta-analyses (Ahlbom et al. 2000; Greenland et al. 2000), further investigations are needed as to whether confounding, exposure misclassification and selection bias have a strong impact.

With respect to confounding, which is where the effect of the exposure of interest itself is affected by another factor, many candidates have been proposed, including traffic density, use of pesticides in the vicinity of power lines, environmental tobacco smoke, to name only a few. However, none have been identified as true confounders to date (Schüz 2007). In statistical simulations it has been shown that a plausible confounder must itself be quite a strong risk factor for childhood leukaemia and, at the same time, must be highly correlated with magnetic field strengths from all sources of elevated magnetic fields (Langholz 2001). No such a factor has currently been identified (Sect. 27.2.3). While confounding by an unknown factor is always a possibility taking the currently available knowledge into consideration, this seems to be an unlikely explanation.

With regard to exposure assessment methods there is huge potential for exposure misclassification, for example classifying children as unexposed when they were in reality exposed or vice versa. Even long-term measurements in case-control studies only extend over at most 2 days and averages of such measurements are used to categorise past exposure of children. However, it is unlikely that the magnitude of misclassification from most of the exposure assessment methods was associated with disease status, i.e. the probability of misclassification being different for cases and controls. In such situations of nondifferential exposure misclassification, the expected bias in the relative risk estimate is towards an underestimation of an association; hence, it is very unlikely that the exposure misclassification apparent in the epidemiological studies was producing a spurious positive association. Conversely, should the association between magnetic fields and childhood leukaemia be causal, it will be a concern that exposure misclassification led to an underestimation of the strength of the association.

With regard to selection bias, the low participation rates in many studies were of concern, especially in studies using measurements for exposure assessment (Mezei and Kheifets 2006; Schüz 2007). Data suggested that families with a lower social status were particularly underrepresented among controls. As the likelihood of measuring higher fields increased with the number of apartments per building, and buildings with a higher number of apartments were more often inhabited by families of lower social status, the association may have been overestimated. In a simulation analysis using a German case-control data set (Schüz et al. 2001), selection bias explained about 66% of the risk increase observed in the study. A considerable impact of selection bias has also been demonstrated for a large US study (Hatch et al. 2000). These findings leave no doubt that selection bias plays a role in studies using measurement devices. The studies using calculated magnetic fields were not affected by selection bias, as no contact with the families was necessary and participation was virtually complete. These studies also confirm an association between magnetic fields and the risk of childhood leukaemia, however based on much smaller numbers of exposed children (see Table 3 in Ahlbom et al. 2000).

Taking all these considerations into account, the probability is that selection bias alone is not sufficient to explain the entire association. However, since the small increased risk is based on relatively small numbers of exposed children, a combination of selection bias, confounding and chance cannot be ruled out as a plausible explanation for the observed association.

27.3.2.2 Alternative Explanations

The absence of a reproducible mechanism for the interaction of magnetic fields below 100 μT with biological systems has led researchers to suggest an indirect involvement of magnetic fields in leukaemogenesis; however, hypotheses proposed during the last few decades have only been speculative. One recent hypothesis is based on the assumption that exposure to magnetic fields suppresses the nocturnal production and release of the hormone melatonin, which is assumed to have oncostatic capabilities (Henshaw and Reiter 2005); however, studies regarding a magnetic-field-related suppression of melatonin are controversial (IARC 2002) and a recent pooled analysis on nighttime exposure to magnetic fields and the risk of childhood leukaemia found no support for the melatonin hypothesis (Schüz et al. 2007). A second recent hypothesis claims that contact currents occur with greater probability in residences with higher magnetic fields and that contact currents via contact with metallic water fixtures during bathing of the child lead to higher bone marrow doses of induced currents rather than magnetic field exposures alone (Kavet 2005); however, so far there have been no studies to support the role of contact currents in leukaemogenesis. Whilst this hypothesis fits very well with the characteristics of the US power distribution system, it is unclear how well it corresponds to the situation in Europe, where the same association between magnetic fields and leukaemia risk has been observed. This indirect mechanism must also be extremely strong to be detected with only a weakly related level of residential magnetic field exposure.

27.3.2.3 A Causal Link

As mentioned in Sect. 27.2.2, the development of childhood leukaemia is a unique process and it is not certain whether the animal models used in genotoxicity or cell proliferation studies are appropriate. The long-term animal carcinogenicity studies provide little evidence that exposure to magnetic fields alone could induce any type of cancer, including haemopoietic cancers (IARC 2002). Results from studies examining the growth of transplanted tumour cells and coexposure studies with known carcinogens have also tended to be negative. However, recent findings on the stimulation of cell proliferation and inhibition of apoptosis have been of interest since environmental agents can also promote the development of cancer through non-genotoxic mechanisms. Unfortunately, these results are not fully understood and as yet have not been reproduced (Fedrowitz et al. 2002; Kumlin et al. 2002). Furthermore, a recent review of in vitro studies provided no further clarification (Vijayalaxmi and Obe 2005). Data from a series of in vitro studies carried out by a European consortium (Reflex project), showed the occurrence of chromosomal damage in some, but not all cell types (including lymphocytes), and with intermittent rather than continuous exposures. A recent meta-analysis of in vitro studies and short-term animal studies that combined magnetic fields with known carcinogens or other toxic physical or chemical agents found some positive results. However, since most of the studies used magnetic fields of 100 μT or higher, the findings do not provide an adequate explanation for the epidemiological findings as the associations seen were based on much lower exposures (Juutilainen et al. 2006). Overall, evidence from animal studies or in vitro studies neither supports nor contradicts the epidemiological findings.

27.4 Conclusions

Childhood leukaemia is a biologically heterogeneous disease. The advent of more powerful molecular diagnostic techniques has provided greater insight into the underlying molecular biology of the disease, which has been critical for improving the survival of children with acute leukaemia. Furthermore, the knowledge that different subtypes of ALL tend to segregate by age has perhaps provided some rationale as to why there are marked prognostic differences between infants, children and adults.

Whilst over 200 genes have been found to be involved in chromosomal abnormalities, many are rare and only certain ones have been associated with childhood leukaemia. However, the identification of specific chromosomal translocations has been of great importance since the chimeric fusion genes derived act as unique markers for each patient's leukaemic clone, and do not only function as prognostic markers but also enable the origins of the disease to be tracked.

The retrospective analyses of Guthrie cards from children diagnosed with leukaemia for the presence of chromosomal abnormalities has provided evidence for a prenatal origin of the disease. Furthermore, identification of chromosomal translocations in Guthrie cards at a rate much higher than the proportion of children who will develop the associated subtype of leukaemia demonstrated that childhood leukaemia is at least a two-step process with a need for additional genetic events and/or exposures – the search for which has continued over many years.

Although much has been learnt from the identification of specific chromosomal translocations associated with subtypes of acute leukaemia for tailoring therapies and improving outcome, progress in understanding the determinants of childhood leukaemia has not been as profitable. Since many of the translocations are rare, some studies have not been sufficiently large enough to enable analyses of these subgroups. Furthermore, there is widespread geographical variation in routine analysis of biological samples; thus, collection of these parameters has not been a feature of all epidemiological studies. This has made it difficult, in fact in many instances impossible, to pool data meaningfully in order to look at exposures such as magnetic fields. Hence, many of the findings with respect to the cause of childhood leukaemia are based only on ALL and AML and do not take into account specific features associated with chromosomal translocations. Nevertheless improved knowledge of chromosomal alterations offers new opportunities for understanding the relevance of environmental exposures and the importance of gene–environment interactions in determining the development of the disease.

A prominent example is the ongoing challenge of interpreting findings from epidemiological studies on exposure to magnetic fields and the risk of childhood leukaemia. Whilst the finding of a doubling of the leukaemia risk at average magnetic field exposures of 0.3–0.4 μT is relatively consistent across studies, it is not clear whether there is a dose–response relationship and if one were to exist what would it be. Sparse data prevent any firm conclusions about the magnitude of risk at field levels higher than 0.4 μT. Furthermore, for exposures in the range from 0.1 to 0.4 μT, the data are ambiguous, which is compatible with the suggestion of a dose–response trend or a threshold effect at 0.3–0.4 μT. The interpretation of the magnetic-field-related findings is particularly delicate because possible explanations range from methodological artefacts to a causal link. To date neither of these two contradictory explanations can be completely ruled out; hence, it becomes imperative to combine knowledge of the biology of the disease with evaluations of the accuracy of the observational studies findings.

In Fig. 27.2 an attempt is made to summarise the current state of knowledge by clarifying the likelihood of the various possible explanations. With regard to methodological flaws, it is clear that the association between magnetic fields

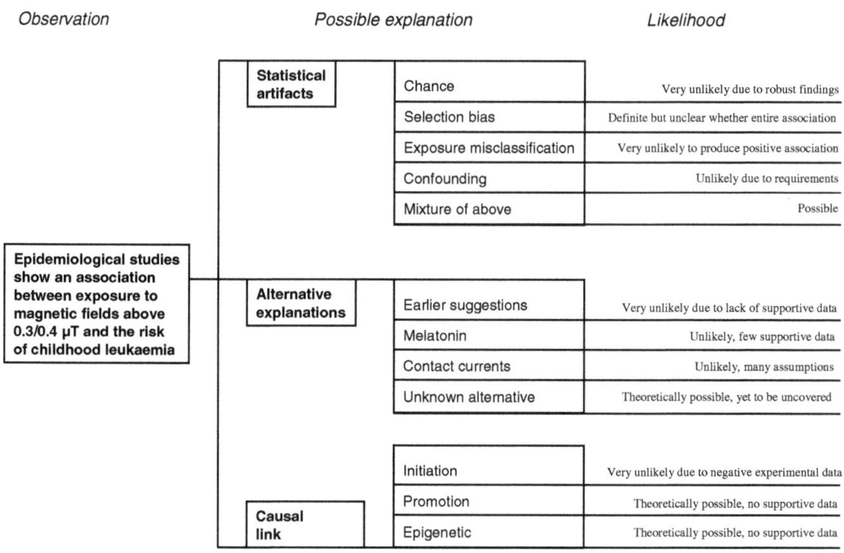

Fig. 27.2 Summary of possible mechanisms to explain the epidemiological association between exposure to magnetic fields and the risk of childhood leukaemia

and leukaemia risk is overestimated owing to selection bias, but it is uncertain whether this bias accounts for the association in its entirety. Furthermore, chance, bias from exposure misclassification and confounding are unlikely on their own to be responsible for producing a spurious positive association; however, a combination of all factors is possible. With regard to indirect mechanisms, many are speculative and the most recent hypotheses on contact currents and the disruption of the melatonin production process are mainly theoretical considerations and are not well supported by data. Although experimental studies to date have failed to provide supportive evidence, they do not contradict the epidemiological results, and it is somewhat disconcerting that no additional evidence for or against a causal link has emerged from the various studies. Whilst there is some evidence against an initiation of leukaemia, either promotion of the disease or an epigenetic mechanism is theoretically possible; however, there are few supportive data for this assumption. Conversely, it must be noted that it is more or less accepted that low-level magnetic fields do not transfer enough energy to cells to directly cause DNA damage. Furthermore, at field levels of 0.3–0.4 μT no other interaction with biological systems apart from the epidemiological childhood leukaemia findings has been identified. It is also important to note that the average exposure levels to the Earth's natural magnetic field are much higher than average exposures to man-made magnetic fields (but only the latter one is alternating, of course).

Be that as it may, there are important lessons to learn, with respect to both the cause of childhood leukaemia and the need for improving epidemiological methods to identify presumably weak associations. Thus, the identification of a severe bias producing such a robust statistical association would have a huge impact on the interpretation of other epidemiological findings, not only in the area of childhood leukaemia (Sect. 27.2.3) but also in other fields. In all, the most promising approach is intensified research into the cause of childhood leukaemia, with the goals to increase our understanding of the biological diversity of the disease, as reflected in the various chromosomal alterations, and to identify interactions of multiple exposures and their timing in the development of this complex disease.

References

Ahlbom A, Day N, Feychting M, Roman E, Skinner J, Dockerty J, Linet M, McBride M, Michaelis J, Olsen JH, Tynes T, Verkasalo PK (2000) A pooled analysis of magnetic fields and childhood leukaemia. Br J Cancer 83:692–698

Biondi A, Cimino G, Pieters R, Pui CH (2000) Biological and therapeutic aspects of infant leukemia. Blood 96:24–33

Fedrowitz M, Westermann J, Löscher W (2002) Magnetic field exposure increases cell proliferation but does not affect melatonin levels in the mammary gland of female Sprague Dawley rats. Cancer Res 62:1356–1363

Greaves M (1999) Molecular genetics, natural history and the demise of childhood leukaemia. Eur J Cancer 35:1941–1953

Greaves M (2002) Childhood leukaemia. BMJ 324:283–287
Greaves M (2005) In utero origins of childhood leukaemia. Early Hum Dev 81:123–129
Greaves M (2006) Infection, immune responses and the aetiology of childhood leukaemia. Nat Rev Cancer 6:193–203
Greaves MF, Wiemels J (2003) Origins of chromosome translocations in childhood leukaemia. Nat Rev Cancer 3:1–11
Greenland S, Sheppard AR, Kaune WT, Poole C, Kelsh MA (2000) A pooled analysis of magnetic fields, wire codes, and childhood leukemia. Childhood Leukemia-EMF Study Group. Epidemiology 11:624-634
Hatch EE, Kleinerman RA, Linet MS, Tarone RE, Kaune WT, Auvinen A, Baris D, Robison LL, Wacholder S (2000) Do confounding or selection factors of residential wiring codes and magnetic fields distort findings of electromagnetic fields studies? Epidemiology 11:189–198
Heerema NA, Arthur DC, Sather H, Albo V, Feusner J, Lange BJ, Steinherz PG, Zeltzer P, Hammond D, Reaman GH (1994) Cytogenetic features of infants less than 12 months of age at diagnosis of acute lymphoblastic leukemia: impact of the 11q23 breakpoint on outcome: a report of the Childrens Cancer Group. Blood 83:2274–2284
Henshaw DL, Reiter RJ (2005) Do magnetic fields cause increased risk of childhood leukemia via melatonin disruption? Bioelectromagnetics 7:S86–97
Herbst AL, Ulfelder H (1971) Adenocarcinoma of the vagina: association of maternal stilbestrol therapy with tumour appearance in young women. N Engl J Med 284:446–450
IARC (2002) Monographs on the evaluation of carcinogenic risks to humans, vol 80: non-ionizing radiation, part 1: static and extremely low-frequency (ELF) electric and magnetic fields. IARC Press, Lyon
Juutilainen J, Kumlin T, Naarala J (2006) Do extremely low frequency magnetic fields enhance the effects of environmental carcinogens? A meta-analysis of experimental studies. Int J Radiat Biol 82:1–12
Kavet R (2005) Contact current hypothesis: summary of results to date. Bioelectromagnetics 7:S75–85
Kinlen LJ (1995) Epidemiological evidence for an infective basis in childhood leukaemia. Br J Cancer 71:1–5
Kumlin T, Heikkinen P, Kosma VM, Alhonen L, Janne J, Juutilainen J (2002) p53-independent apoptosis in UV-irradiated mouse skin: possible inhibition by 50 Hz magnetic fields. Radiat Environ Biophys 41:155–158
Langholz B (2001) Factors that explain the power line configuration wiring code-childhood leukemia association: what would they look like? Bioelectromagnetics 5:S19–31
Lightfoot TJ, Roman E (2004) Causes of childhood leukaemia and lymphoma. Toxicol Appl Pharmacol 199:104–17
Linet MS, Hatch EE, Kleinerman RA, Robison LL, Kaune WT, Friedman DR, Severson RK, Haines CM, Hartsock CT, Niwa S, Wacholder S, Tarone RE (1997) Residential exposure to magnetic fields and acute lymphoblastic leukemia in children. N Engl J Med 337:1–7
Look AT (1997) Oncogenic transcription factors in the human acute leukaemias. Science 278:1059–1064
McHale CM, Smith MT (2004) Prenatal origin of chromosomal translocations in acute childhood leukaemia: implications and future directions. Am J Hematol 75:254–257
Mezei G, Kheifets L (2006) Selection bias and its implications for case-control studies: a case study of magnetic field exposure and childhood leukaemia. Int J Epidemiol 35:397–406
Mori H, Colman SM, Xiao Z, Ford AM, Healy LE, Donaldson C, Hows JM, Navarrete C, Greaves M (2002) Chromosome translocations and covert leukemic clones are generated during normal fetal development. Proc Natl Acad Sci USA 99:8242–8247
Okuda T, van Deursen J, Hiebert SW, Grosveld G, Downing JR (1996) AML1, the target of multiple chromosomal translocations in human leukaemia, is essential for normal fetal liver hematopoiesis. Cell 84:321–330
Parkin DM, Stiller CA, Draper GJ, Bieber CA, Terracini B, Young JL (1988) International incidence of childhood cancer. IARC scientific publications 87. IARC Press, Lyon

Pui CH, Relling MV, Downing JR (2004) Acute lymphoblastic leukaemia. N Engl J Med 350:1535–1548

Roman E, Doyle P, Lightfoot T, Ansell P, Simpson J, Allan JM, Eden TOB (2006) Molar pregnancy, childhood cancer and genomic imprinting - is there a link? Hum Fertil 9:171–174

Roman E, Simpson J, Ansell P, Kinsey S, Mitchell CD, McKinney PA, Birch JM, Greaves M, Eden TOB (2007) Childhood acute lymphoblastic leukaemia and infections in the first year of life: a report from the UKCCS. Am J Epidemiol 165:496–504

Schüz J (2007) Implications on protection guidelines from epidemiologic studies on magnetic fields and the risk of childhood leukemia. Health Phys (in press)

Schüz J, Grigat JP, Brinkmann K, Michaelis J (2001) Residential magnetic fields as a risk factor for childhood acute leukaemia: results from a German population-based case-control study. Int J Cancer 91:728–735

Schüz J, Svendsen AL, Linet MS, McBride ML, Roman E, Feychting M, Kheifets L, Lightfoot T, Mezei G, Simpson J, Ahlbom A (2007) Night-time exposure to electromagnetic fields and childhood leukemia: an extended pooled analysis. Am J Epidemiol (in press)

Steliarova-Foucher E, Stiller C, Kaatsch P, Berrino F, Coebergh JW, Lacour B, Parkin M (2004) Geographical patterns and time trends of cancer incidence and survival among children and adolescents in Europe since the 1970s (the ACCIS project): an epidemiological study. Lancet 364:2097–105

Stewart A, Webb J, Giles D, Hewitt D (1956) Malignant disease in childhood and diagnostic radiation in utero. Lancet 2:447

UKCCS (1999) Exposure to power-frequency magnetic fields and the risk of childhood cancer. Lancet 354:1925–1931

UKCCS (2000) The United Kingdom Childhood Cancer Study: objectives, materials and methods. Br J Cancer 82:1073–1102

Vijayalaxmi, Obe G (2005) Controversial cytogenetic observations in mammalian somatic cells exposed to extremely low frequency electromagnetic radiation: a review and future research recommendations. Bioelectromagnetics 26:412–430

Wertheimer N, Leeper E (1979) Electrical wiring configurations and childhood cancer. Am J Epidemiol 109:273–284

Yu BD, Hanson RD, Hess JL, Horning SE, Korsmeyer SJ (1998) MLL, a mammalian trithorax-group gene, functions as a transcriptional maintenance factor in morphogenesis Proc Natl Acad Sci USA 95:10632–10636

28 Cytogenetic and Carcinogenic Effects of Exposure to Radiofrequency Radiation

JAMES P. MCNAMEE AND PASCALE V. BELLIER

Abstract

Radiofrequency radiation (RFR) is a portion of the electromagnetic spectrum with frequencies of 3 kHz–300 GHz. RFR is produced by many man-made sources, including mobile phones and base stations, television and radio broadcasting facilities, radar, medical equipment, microwave ovens, radiofrequency heaters as well as a diverse variety of other electronic devices within our living and working environments. Owing to ongoing public concern and the increasing prevalence of RFR-emitting devices, a great deal of research has been conducted over the past 50 years to evaluate the biological and/or health effects of thermalizing and non-thermalizing RFR exposures. In the absence of decisive epidemiological evidence to support or refute an association between RFR exposure and cancer risk, laboratory studies of possible mechanisms of carcinogenesis by RFR are important. The scientific literature on this subject is full of conflicting results and the question of whether RFR exposure can contribute to cancer risk remains unresolved. This chapter contains a literature review of the evidence for RFR-induced cytogenetic effects, but also a critique of the literature, highlighting deficiencies in the design of some studies that should be taken into account when assessing the health risk of RFR.

28.1 Introduction

Electromagnetic radiation is emitted by many natural and man-made sources and is a fundamental aspect of our lives. We are warmed by electromagnetic radiation emitted from the sun and our eyes can detect the visible light portion of the electromagnetic spectrum. The source of all electromagnetic radiation is accelerating charges and electromagnetic radiation manifests itself as oscillating electric and magnetic fields. The frequency of this oscillation (measured in hertz; 1 Hz is one oscillation per second) determines the physical properties of these fields. Radiofrequency radiation (RFR) is a portion of the electromagnetic spectrum with frequencies of 3 kHz–300 GHz, which is below that of visible light and above that of extremely low frequency fields (Fig. 28.1). RFR is

Günter Obe and Vijayalaxmi (Eds.)
Chromosomal Alterations: Methods, Results and Importance in Human Health

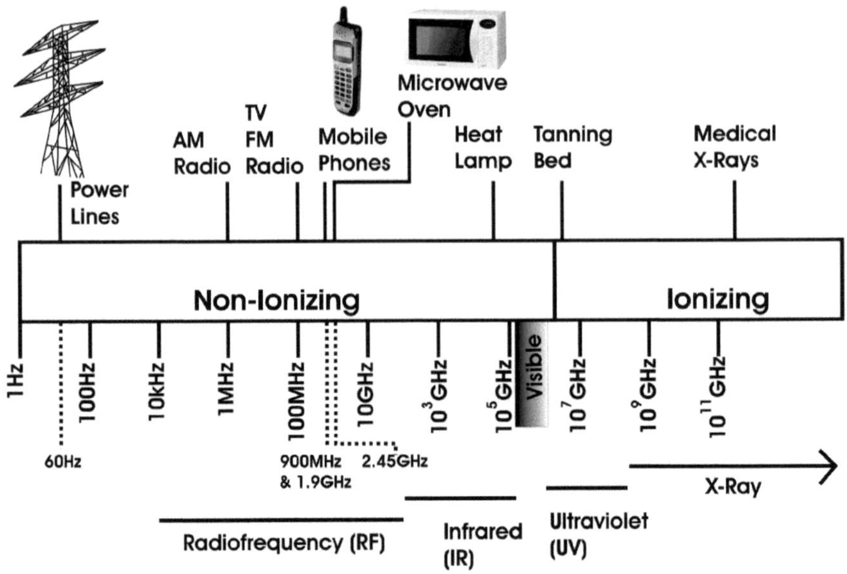

Fig. 28.1 In the electromagnetic spectrum, radiofrequency radiation (RFR) encompasses frequencies ranging from 3 kHz to 300 GHz. Typical man-made sources of RFR include broadcast AM/FM/TV, mobile phones and base stations, microwave ovens and radar

produced by many man-made sources, including mobile phones and base stations, television and radio broadcasting facilities, radar, medical equipment, microwave ovens, radiofrequency heaters as well as a diverse variety of other electronic devices within our living and working environments.

Public anxiety about exposure to RFR has existed for over 80 years. Initially, concern focused on RFR emissions from broadcast AM (530 kHz–1.6 MHz) and FM (88–108 MHz) RFR. This was followed by concerns over RFR emissions from radar installations (1–12.5 GHz), broadcast TV (54–890 MHz), microwave ovens (2.45 GHz) and video-display terminals (3–30 kHz). During the past decade, the increased use of mobile phones (850 MHz–1.9 GHz) and the emergence of new wireless technologies have reinvigorated public apprehension about the safety of human exposure to RFR. Owing to ongoing public concern and the prevalence of RFR-emitting devices, a great deal of research has been conducted over the past 50 years to evaluate the biological and/or health effects of thermalizing and non-thermalizing RFR exposures.

Unlike ionizing radiation, RFR does not have sufficient energy to directly break chemical bonds (Meltz 2003). However, sufficiently intense RFR can cause heating of materials with finite conductivity, including biological tissue. A number of well-established biological effects and adverse health effects from acute exposure to intense RFR have been documented. For the most part, these effects relate to localized heating of sensitive organs/tissues from intense RFR exposure; however, non-thermal effects such as nerve and muscle stimulation are also known to occur. The specific biological responses to RFR are generally related to the rate of RFR energy absorbed. The rate and distribution of RFR energy absorbed depends strongly on the frequency, intensity and orientation of the incident fields as well as the body size and its constitutive properties (dielectric constant and conductivity). At frequencies above 1 MHz, RFR absorption is commonly described in terms of the specific absorption rate (SAR), which is a measure of the rate of energy deposition per unit mass of body tissue and is usually expressed in units of watts per kilogram. Based on a large amount of historical knowledge, national and international exposure limits have been established to protect the general public against adverse effects associated with acute RFR exposures. However, the issue of safety of exposure to long-term, low-level RFR remains controversial and the risk of development of cancer remains the primary public health concern.

At present, there is a relative paucity of epidemiological evidence related to cancer risks associated with long-term, low-level RFR exposure. In the absence of decisive epidemiological evidence to support or refute an association between RFR exposure and cancer risk, laboratory studies of possible mechanisms of carcinogenesis by RFR are important. The scientific literature on this subject is full of conflicting results and the question of whether RFR exposure can contribute to cancer risk remains unresolved. This chapter will review the current state of the literature with respect to cytogenetic and carcinogenic effects of RFR. Since the thermal effects of RFR are well understood and cytogenetic alterations are known to be induced by heat, this chapter will focus on the evidence for cytogenetic effects from low-level, non-thermalizing RFR.

28.2 Challenges When Conducting RFR Research

There are many inherent challenges that must be overcome when conducting RFR bioeffects research which relate to the physical properties of this radiation. One of the most important interactions between RFR and matter is the rotational friction caused when small polar molecules (such as those in water and biological tissues) reorient themselves in the presence of the oscillating field, thereby producing heat. The rate at which heat energy is applied to the sample must not exceed the rate at which it is removed, otherwise the temperature within the sample will rise and thermal confounding may occur. If RFR energy is applied to biological samples at a low rate (e.g., below 1 W/kg) or for a short duration, then passive cooling and/or sweating may permit the tissue/body/sample temperature to remain within a normal physiological range. However, if higher SARs (above 1 W/kg) are employed and/or prolonged, then some form of active cooling mechanism will likely be required to ensure that excessive sample heating does not occur. When performing RFR experiments, it is important to include appropriate temperature controls within the experimental setup to alleviate thermal confounding in the study results. To this end, the use of non-perturbing temperature probes and active cooling with either forced air or water is particularly beneficial.

Another significant challenge that must be addressed when studying the biological effects of RFR relates to the homogeneity of energy absorption within the sample (SAR distribution) and the maintenance of a relatively homogeneous, non-perturbing temperature throughout the biological sample. These are two separate, yet related, phenomena. While controlling volume-averaged temperature within the biological sample is important, it is also important to ensure that RFR energy is deposited in a homogeneous manner or "hot spots" within the biological sample can occur. This phenomenon is akin to focusing light energy on an object, whereby the localized temperature at the focal point is considerably elevated, while that of the surrounding area is not. In such a situation with RFR, the local temperature for some part of the sample may be greatly elevated, while the volume-averaged temperature remains relatively unaffected owing to heat convection and diffusion. With some in vitro exposure systems, heterogeneous RFR absorption can also cause convective flow within aqueous biological samples owing to the presence of temperature gradients within the sample. Such convection currents in a RFR-exposed sample, relative to a control, may be a source of confounding owing to a resultant stirring of the RFR-exposed samples that does not occur within the matched sham control.

Since RFR behaves in a similar fashion to light, with respect to its ability to reverberate/reflect off metallic surfaces, care must be taken to ensure that samples are exposed in a reverberant-free environment or at least that all sources of reflection are taken into account in the dosimetry analysis. This is particularly problematic for in vitro studies, where cells must be maintained under appropriate conditions (37°C, 50–60% humidity, 5% CO_2/95% air) in a tissue culture incubator. Most incubators are built with a stainless-steel-coated interior which acts as an ideal reflector for RFR. As a result, when conducting in vitro RFR research, great care must be taken to either contain RFR within a closed exposure system inside a standard tissue culture incubator or design alternative non-RFR perturbing tissue culture environments. In either case, maintenance of sample temperature, humidity, osmolarity and pH is a challenge and must be carefully controlled.

Owing to the above considerations, it is critical that RFR bioeffects studies encompass certain experimental controls to ensure that confounding variables do not result in artefactual biological observations. For in vitro studies, monitoring the temperature, pH and cell viability within the sample is paramount. For both in vivo and in vitro studies, empirical and/or numerical analysis of SAR distribution patterns within the sample is fundamentally important. The inclusion of unexposed (negative) controls, sham (non-RFR exposed) controls and positive controls is also imperative to ensure that the assay methodology is responding appropriately and experimental conditions are identical. Inclusion of these controls is particularly important when evaluating studies where RFR-induced effects were observed. On the other hand, analysis of statistical power in studies where no evidence of RFR-induced effects were observed is important to determine whether a sufficient number of samples (independent experiments) were employed to allow the detection of significant differences between groups. For all studies, the investigators should be blinded with respect to the exposure status of the samples until all laboratory experiments, data acquisition and statistical analysis are complete.

In summary, when evaluating the literature related to RFR-induced cytogenetic and carcinogenic effects, all of the above considerations should be taken into account. This chapter contains not only a literature review but also a critique of the literature, highlighting certain deficiencies in the design of some of the studies that should be taken into account when evaluating the significance of certain RFR research findings.

28.3 Cytogenetic Effects of RFR

28.3.1 In Vitro Studies

Changes involving DNA and/or chromosome structure in somatic cells are of great relevance owing to their association with cancer and other diseases. Despite the general consensus that RFR is not sufficiently energetic to directly damage DNA, a large number of in vitro studies have been conducted to assess the genotoxic potential of RFR at a variety of frequencies. It is generally agreed that when the in vitro RFR cytogenetic studies published over the past 30 years are taken together and subjected to a weight-of-evidence evaluation, there is no compelling information to suggest that RFR exposure has any direct mutagenic, genotoxic or clastogenic effects (Brusick et al. 1998; Vijayalaxmi and Obe 2004; Moulder et al. 2005; Verschaeve 2005). In this section, the scientific literature relating to the evidence of in vitro cytogenetic effects of RFR will be reviewed and evaluated with particular focus on recent papers and/or pivotal studies (Table 28.1).

The majority of studies examining the cytogenetic effects of RFR during the last 5 years have been conducted in human blood cultures. Of these studies, only a few have observed a "positive" effect resulting from RFR exposures using the cytokinesis block micronucleus (CBMN) assay. Studies by d'Ambrosio et al. (2002), Tice et al. (2002) and Zotti-Martelli et al. (2000, 2005) have all reported increased micronuclei (MN) in RFR-exposed human blood cultures using the CBMN assay. In the study by Tice et al. (2002), significant increases in MN frequency were observed following exposure of unstimulated blood cultures exposed to 837 MHz and 1.9 GHz RFR (with various paradigms) at SARs of 5 and 10 W/kg at the longest RFR exposure time of 24 h. However, the authors identified thermal confounding as a possible explanation for these results as higher localized temperatures may have occurred within the sample region containing the blood cells. It is very

Table 28.1 Summary of in vitro studies that have investigated the possible cytogenetic effects of radiofrequency radiation

Authors	RFR conditions	SAR; duration	Tissue /cell line	End point	Results	Comments
Zotti-Martelli et al. (2000, 2005)	1.8, 2.45 or 7.7 GHz; CW	No SAR given; 15-, 30-, 60-, 120- or 180-min exposures	Human whole blood	MN	Increased MN	Insufficient details on exposure design and dosimetry; no temperature monitoring in actual samples; insufficient details on control conditions
Maes et al. (2001)	900 MHz; various modulations	SARs ranging from 0.4 to 10 W/kg; 2-h exposure	Human whole blood	CA and SCE	No effect	Low sample number for most exposures ($N = 2$); no sham or positive controls
Vijayalaxmi et al. (2001a, b)	835.62 or 874.74 MHz; CW	SARs ranging from 4.4 to 5.5 W/kg; 24-h exposure	Diluted human blood	MN and CA	No effect	Sample number ($N = 2$ for each SAR tested)
d'Ambrosio et al. (2002)	1.748 GHz; CW and phase-modulated	Average SAR of 2.25 W/kg; 15-min exposure	Mitogen-stimulated diluted human blood	MN	Increased MN following exposure to phase-modulated RFR	No sham or positive controls
Tice et al. (2002)	837 MHz, 1.9 GHz; analogue, PCS, CDMA and TDMA	SARs ranging from 1 to 10 W/kg; 3- and 24-h exposures	Diluted human blood	MN	Increased MN at highest SARs after 24-h exposure period	Low sample number ($N = 1$ for each condition); possible thermal confounding due to localized hot spots

Bisht et al. (2002)	847 MHz CDMA-modulated and 835 MHz FDMA-modulated	SARs ranging from 3.2 to 5.1 W/kg; 3-, 8-, 16- or 24-h exposures	C3H 10T1/2 cells	MN	No effect	
McNamee et al. (2002a, b, 2003)	1.9 GHz; CW and pulse-modulated	SARs ranging from 0.1 to 10 W/kg; 2- and 24-h exposures	Diluted human blood	MN	No effect	No comment
Zeni et al. (2003, 2005)	900 or 925 MHz; CWand GSM-modulated	SARs of 0.2, 1 or 1.6 W/kg; various exposure periods	Whole or diluted human blood and mitogen-stimulated lymphocytes	MN, CA and SCE	No effect	No sham or positive controls in the 2003 study
Mashevich et al. (2003)	830 MHz; CW	SARs ranging from 1.6 to 8.8 W/kg; 72-h exposure	Mitogen-stimulated diluted human blood	Aneuploidy	Increase in chromosome 17 aneuploidy	Non-homogenous exposure; sample flasks were thermally insulated from the incubator and no active cooling was employed, resulting in a possible thermal effect
Koyama et al. (2003)	2.45 GHz	SARs ranging from 13 to 100 W/kg; 18-h exposure	CHO-K1 cells, following incubation with cytochalasin B for 24 h	MN	Increased MN at SARs of 78 and 100 W/kg; no effect below 50 W/kg	High SAR levels with little relevance towards risk assessment; did not report the number of repeat experiments

(Continued)

Table 28.1 Summary of in vitro studies that have investigated the possible cytogenetic effects of radiofrequency radiation—Cont'd

Authors	RFR conditions	SAR; duration	Tissue /cell line	End point	Results	Comments
Komatsubara et al. (2005)	2.45 GHz CW and pulse-modulated	SARs ranging from 5 to 100 W/kg; 2-h exposure	Mouse m5S cells	CA	No effect	High SAR levels with little relevance towards risk assessment; apparently a single sample for each RFR exposure condition
Stronati et al. (2006)	935 MHz; GSM-modulated	SARs of 1 and 2 W/kg; 24-h exposure	Mitogen-stimulated human whole blood	MN, CA and SCE	No effect	

CA chromosome aberrations, *CDMA* code-division multiple access, *CHO* Chinese hamster ovary, *CW* continuous wave, *FDMA* frequency-division multiple access, *GSM* global system for mobile communication, *MN* micronuclei, *PCS* personal communication services, *RFR* radiofrequency radiation, *SAR* specific absorption rate, *SCE* sister-chromatid exchange, *TDMA* time-division multiple access important to note that blood samples from only on to two donors were used for each experimental condition; therefore, only limited conclusions can be drawn from this study. Zotti-Martelli et al. (2000, 2005) also observed increased MN in the CBMN assay following exposure of unstimulated whole-blood cultures to 1.8, 2.45 and 7.7 GHz continuous-wave (CW) RFR for up to 3 h at power densities ranging from 5 to 30 mW/cm^2. However, very limited information regarding the RFR exposure system and dosimetry was provided for these studies. Furthermore, while the authors measured the temperature variation within "pseudosamples" during RFR exposure, water and not whole blood was used in these experiments. This is problematic since water has very different dielectric and conductivity properties from whole blood, and much more energy would be absorbed within whole-blood cultures than in water. As such, at the RFR exposure levels described by the authors, it is possible that the observed increases in MN frequency may have been due to hyperthermia. In a well-controlled study by d'Ambrosio et al. (2002), increased MN frequency was reported in phytohemagglutinin (PHA) stimulated human lymphocytes exposed to phase-modulated 1.748 GHz RFR, but not to 1.748 GHz CW RFR. This result suggested that the cellular response to RFR may depend upon the modulation characteristics of the applied RFR field.

On the other hand, many other research groups have failed to find any evidence of RFR-induced cytogenetic effects in human blood cell cultures regardless of the modulation paradigm. Vijayalaxmi et al. (2001a, b) exposed unstimulated whole-blood cultures to 835.62 MHz frequency division multiple access (FDMA) modulated and 847.74 MHz code division multiple access (CDMA) modulated RFR for 24 h at mean SARs of either 4.4 or 5.0 W/kg and 4.9 or 5.5 W/kg, respectively. The authors found no significant changes in either cytogenetic end point between the sham and RFR-exposed samples for either of these RFR exposure conditions. Similarly, McNamee et al. (2002a, b, 2003) exposed unstimulated human whole-blood cultures to 1.9 GHz CW and pulse-modulated RFR for 2–24 h at SARs ranging from 0.1 to 10 W/kg. The formation of MN was assessed using the CBMN assay and no significant changes were found in the binucleated cell frequency, the frequency of binucleated cells with MN or the total MN incidence at any SAR level tested when compared with the sham cells. Unlike many previous studies, the studies by McNamee et al. incorporated exposure of human blood samples from five separate donors to sham and five different RFR exposure levels, allowing the possibility of a dose–response relationship to be assessed. Furthermore, real-time temperature measurements were performed every 60 s within each of the RFR-exposed and sham cultures, thereby excluding thermal effects as a possible confounding variable.

Maes et al. (2001) and Zeni et al. (2003, 2005) examined the effect of 900 MHz global system for mobile communication (GSM) modulated RFR exposure at SARs ranging from 0.2 to 10 W/kg in either PHA-stimulated or unstimulated blood cultures and found similar results. However, while the studies by Maes et al. (2001) and Zeni et al. (2003) found no evidence that

RFR exposure caused any genotoxic effects; these studies did not include the necessary controls. Since both studies lacked positive controls, the relative sensitivity of the cytogenetic assays in these laboratories to detect an effect is difficult to interpret and therefore the significance of these negative results is diminished. Furthermore, these studies failed to include sham controls and relied only upon unexposed controls. Although the importance of a sham control is obvious when a positive effect is detected, it is important to point out that the appropriate comparison for all RFR-exposed samples should be the sham control, rather than the unexposed control, since the sham and RFR-exposed samples should have been treated identically except for the RFR exposure. This comparison ensures that the exposure system and exposure environment are not the cause of possible observed effects, but rather that the RFR exposure itself is responsible. In a follow-up study, Zeni et al. (2005) included all the necessary controls when exposing unstimulated whole-blood cultures to 900 MHz GSM-modulated RFR for 2 h at 1 W/kg. No evidence of chromosomal damage, as measured by analysis of chromosome aberrations and sister-chromatid exchange (SCE), was observed. Very recently, Stronati et al. (2006) exposed PHA-stimulated whole blood to 935 MHz GSM-modulated RFR at SARs of 1 and 2 W/kg for 24 h and observed no increase in chromosome aberrations, MN frequency or SCE in RFR-exposed samples when compared with sham controls.

Several in vitro studies have also assessed the effect of RFR exposure on cytogenetic end points using a variety of animal-derived cell lines. Bisht et al. (2002) exposed a mouse embryo fibroblast derived cell line (C3H 10T1/2) to 835.62 MHz FDMA-modulated or 847.74 MHz CDMA-modulated RFR for up to 24 h at SARs of 3.2–5.1 W/kg, in both the plateau phase and the exponential growth phase of these cells. The authors found no evidence that RFR exposure had any effect on MN frequency at any SAR, modulation paradigm or exposure duration. Koyama et al. (2003) exposed Chinese hamster ovary (CHO) K1 cells to 2.45 GHz CW RFR for 18 h at SARs ranging from 13 to 100 W/kg. While the authors reported no evidence of increased MN frequency in RFR samples exposed at SARs up to 50 W/kg, an increased incidence in MN frequency was reported in cells exposed at SARs of 78 and 100 W/kg. The authors measured the average temperature of the highest RFR exposed group (100 W/kg) to be approximately 39°C. When positive control samples (exposed to a temperature of 39°C for 18 h) were assessed, a similar increased incidence of MN was reported. Thus, the increased MN frequency at the higher SARs assessed in this study appears to be due to thermal effects. Another recent study by this group (Komatsubara et al. 2005) assessed the incidence of chromosome aberrations in murine (embryonal skin derived) m5S cells following a 2-h exposure to 2.45 GHz CW RFR at SARs of 5–100 W/kg or 2.45 GHz pulse-modulated RFR at SARs of 50 and 100 W/kg. The authors reported no evidence of increased chromatid-type or chromosome-type aberrations in RFR-exposed samples at any SAR tested when compared with sham cells. It is interesting to note that the shorter exposure time (2 h) to an elevated temperature of

39–41°C had no apparent clastogenic effect in this study, while the prolonged (18-h) exposure to 39°C in the study by Koyama et al. (2003) appeared to stimulate heat-induced MN formation. In any event, the high SAR levels employed in these studies have little relevance when assessing the relative risk of RFR towards the general population since such intense exposures are not typically encountered in our working or living environment.

In an interesting study by Mashevich et al. (2003), diluted human whole blood was stimulated with 3% PHA and then exposed to 830 MHz CW RFR for 72 h at SARs ranging from 1.6 to 8.8 W/kg. Following the culture period, the cells were harvested for cytogenetic analysis and the authors labeled the α-satellite DNA of chromosome 17 using fluorescence in situ hybridization. A linear increase in chromosome 17 aneuploidy as a function of increasing SAR was reported. The authors suggested that this effect was non-thermal in nature on the basis of their assumption that the temperature within the RFR-exposed samples did not exceed 38°C, while separate control experiments demonstrated that prolonged culturing at temperatures below 38.5°C did not affect chromosome 17 aneuploidy. However, Chou and Swicord (2003) argued the assumption that RFR sample temperatures remained below 38°C since the RFR-exposed flasks were enclosed in Plexiglas and were thereby thermally insulated from the incubator temperature. Chou and Swicord (2003) noted that such a situation would create non-uniform heating within the samples owing to inefficient heat exchange and the localized temperatures within the samples may have exceeded 40°C. As a result, the observed changes in chromosome 17 aneuploidy could have been related to thermal effects. Since no active cooling was employed in this study to ensure the sample temperature was maintained within an acceptable range and no temperature measurements were recorded during the actual exposure period, thermal confounding is a likely explanation for these observations.

28.3.2 In Vivo Studies

Despite an exhaustive number of in vitro RFR studies, there have been relatively few studies examining the cytogenetic effects in animals following in vivo RFR exposure (Table 28.2). Many early studies on RFR cytogenetic effects focused on the ability of 2.45 GHz microwave radiation to cause alterations in meiotic chromosomes in spermatogonia, as microwave hyperthermia was considered as a possible means of male contraception (Fahim et al. 1975). However, inadequate control of thermal effects has been identified as a possible confounding variable in many of the early studies that demonstrated RFR-induced genotoxic effects (Brusick et al. 1998; Royal Society of Canada 1999; Vijayalaxmi and Obe 2004; Moulder et al. 2005; Verschaeve 2005).

Among the early studies where thermal effects were unlikely, Manikowski-Czerska et al. (1985) evaluated the effect of 0.5–20 W/kg 2.45 GHz RFR exposure on mouse spermatogonia. Adult CBA/CAY mice were exposed for

Table 28.2 Summary of in vivo studies that have investigated the possible cytogenetic effects of radiofrequency radiation

Authors	RFR conditions	SAR; duration	Tissue/cell line	End point	Results	Comments
Manikowski-Czerska et al. (1985)	2.45 GHz; CW	Whole body average SAR ranging from 0.05 to 20 W/kg; 30 min/day, 6 days/week, 2 weeks	CBA/CAY mice	CA in sperm	Dose-response increase in translocations and univalent chromosomes	Restrained animals; no positive control
Beechey et al. (1986)	2.45 GHz	Estimated SARs of 0.05–20 W/kg; 30 min/day, 6 days/week, 2 weeks	Hybrid F1(C3H/HeH × 101/H) male mice	CA in sperm	No effect	Restrained animals; no positive control
Saunders et al. (1988)	2.45 GHz	SAR of 5 W/kg; 6 h/day, over 8 weeks, for total exposure of 120 h	Male C3H mice	CA in sperm	No effect	Animals partially restrained, no vertical movement; no positive control
Vijayalaxmi et al. (1997, 1998)	2.45 GHz; CW	Whole body average SAR of 1 W/kg; 20 h/day, 7 days/week, 18 months	C3H/HeJ mammary cancer prone female mice	MN-PCE in blood and bone marrow	Increase in MN-PCE in bone marrow and blood	Increase observed was not deemed to be biologically significant since it was small and within the normal spontaneous range of MN-PCE in mice
Vijayalaxmi et al. (2001c)	2.45 GHz; CW	Whole body average SAR of 12 W/kg; 24-h exposure	Male Sprague-Dawley rats	MN-PCE in blood and bone marrow	No effect	
Sykes et al. (2001)	900 MHz; pulse-modulated	Whole body average SAR	Spleen cells of pKZ1 transgenic mice	Somatic intra-chromosomal	Decrease in rate of spontaneous	Restrained animals; no positive control;

		of 4 W/kg; 30 min/day for 1, 5 or 25 days		recombination	recombination after 25 days of exposure	decrease observed at 25 days after exposure probably due to pooling of all non-exposed control data, thus comparisons between sham and exposed mice were not properly matched
Trosic et al. (2002, 2004)	2.45 GHz; CW	Whole body average SAR of 1–2 W/kg; 2 h/day, 7 days/week for up to 30 days	Male Wistar rats	MN-PCE in blood or bone marrow	Increase in MN-PCE after 8- and 15-day exposures	No positive control; MN-PCE levels in RFR-exposed and sham groups were within the normal range for rat peripheral blood
Vijayalaxmi et al. (2003)	1.6 GHz; CW	Brain-average SAR of 0.16 or 1.6 W/kg; 2 h/day, 2 years	Fisher 344 rats in utero and until 2 years after birth	MN-PCE in bone marrow	No effect	Restrained animals
Görlitz et al. (2005)	902 MHz GSM-modulated, and 1.747 GHz DCS-modulated	Whole body average SAR of 2.8–24.9 W/kg; 2 h/day, 5 days/week for 1 or 6 weeks	Male and female B6C3F1 mice	MN-PCE in blood (6 weeks only) and bone marrow; MN in keratinocytes and spleen lymphocytes	No effect	Restrained animals

DCS digital cordless standard, *MN-PCE* micronuclei in polychromatic erythrocytes

30 min/day, 6 days/week for two consecutive weeks and then chromosome number and structure were evaluated in isolated spermatogonia. At the end of the exposure period, a dose-responsive increase in the number of translocations and univalent chromosomes at meiotic metaphase were reported in the RFR-exposed mice, while no translocations were observed in sham mice. In an attempt to replicate these findings, Beechey et al. (1986) employed a similar irradiation protocol using hybrid male mice but observed no evidence of increased chromosome aberrations in spermatogonia from RFR-exposed mice, where increases in rectal temperatures did not exceed 1°C. Saunders et al. (1988) examined this effect further by exposing C3H mice to 2.45 GHz RFR for 6 h/day over an 8-week period at a SAR of approximately 5 W/kg. No evidence of increased reciprocal translocations or increased numbers of cells with autosome or sex chromosome univalents was observed.

The majority of in vivo cytogenetics studies on RFR have focused on evaluating the incidence of MN in rodent peripheral blood or bone marrow. Only a limited number of chronic and/or lifetime RFR exposure studies have been conducted in experimental animals. Of these, Vijayalaxmi et al. (1997) exposed mammary cancer-prone C3H/HeJ female mice to 2.45 GHz CW RFR at a whole-body SAR of 1 W/kg for 20 h/day, 7 days/week for 18 months. Peripheral blood and bone marrow smears were assessed for the incidence of MN in polychromatic erythrocytes (PCE) and normochromatic erythrocytes (NCE). At this time-point, a chronic level of MN-PCE and/or MN-NCE would be anticipated in both the peripheral blood and the bone marrow of RFR-exposed mice, if RFR exposure induced a clastogenic effect. The authors reported a small, yet statistically significant increase in the number of MN-PCE in both bone marrow and peripheral blood of RFR-exposed mice; however, the authors cautioned that this difference (an increase of 0.05%) was not biologically significant since the MN-PCE frequencies for both the RFR and the control groups were within the normal range of the spontaneous incidence of MN-PCE in mice (Vijayalaxmi et al. 1998). Vijayalaxmi et al. (2003) also evaluated the incidence of MN-PCE in peripheral blood and bone marrow smears of Fischer 344 rats exposed (since birth) for 2 h/day for 2 years to a 1.6 GHz Iridium wireless communication signal at brain-averaged SARs of 0.16 and 1.6 W/kg. In this study, no significant differences were observed between RFR-exposed male and female rats relative to the sham controls. More recently, Görlitz et al. (2005) exposed B6C3F1 mice to 902 MHz GSM-modulated or 1,747 MHz digital cordless standard (DCS) modulated RFR for either 1 week at 2 h/day at whole-body SARs of 3.7–33.2 W/kg or for 6 weeks at 2 h/day for 5 days/week at SARs of 2.8–24.9 W/kg. After either a 1-week or a 6-week exposure protocol, MN in bone marrow erythrocytes, keratinocytes and spleen lymphocytes were evaluated in both male and female mice. No significant differences in the incidence of MN were observed in any of the RFR-treated groups relative to the matched sham controls.

A series of studies have also evaluated the effect of acute and/or subacute in vivo RFR exposure on cytogenetic end points in experimental animals.

Vijayalaxmi et al. (2001c) exposed male Sprague-Dawley rats continuously for 24 h to 2.45 GHz CW RFR at a whole body average SAR of 12 W/kg. At 24 h after the end of RFR exposure, the frequency of MN-PCE in peripheral blood and bone marrow smears was evaluated. RFR-exposed rats displayed elevated, yet statistically insignificant, MN-PCE in both peripheral blood and bone marrow compared with sham animals, but once again the MN-PCE frequencies in both groups were within the normal expected range for rats. In a similar study, Trosic et al. (2002, 2004) exposed adult male Wistar rats for 2 h/day, 7 days/week for up to 30 days to 2.45 MHz CW RFR at an approximate SAR of 1–2 W/kg. The authors examined MN-PCE in peripheral blood and bone marrow at various times of exposure and observed a significant increase in MN-PCE between RFR-exposed animals relative to sham controls at 8 and 15 days of exposure, but no statistically significant differences were observed after 2 or 30 days' exposure. While the authors indicated that RFR exposure may have induced a genotoxic response at days 8–15 which was subsequently masked by an adaptive response, there are no experimental data to support this hypothesis. Since the relative MN-PCE levels in both the RFR-exposed and the sham groups were within the normal range for rat peripheral blood and there was no supporting evidence of elevated MN-PCE at other time-points, these data must be considered equivocal.

In an interesting study, Sykes et al. (2001) used pKZ1 transgenic mice, which possessed a β-galactosidase reporter gene in reverse orientation, to assess somatic intrachromosomal recombination inversion events. The authors exposed the mice to 900 MHz pulse-modulated RFR for 30 min/day for 1, 5 or 25 days at a whole body average SAR of 4 W/kg. While no effect of RFR exposure was observed on intrachromosomal recombination in spleen cells of pKZ1 mice after 1 or 5 days' exposure, a significant decrease in the rate of spontaneous recombination was observed after a 25-day RFR exposure period. However, this result appears to have arisen from the statistical approach employed in this study. The authors found no significant difference in the number of recombination events between the non-exposed control groups at days 1, 5 and 25 and therefore chose to combine control data from all time-points into a single pooled control group for comparison with the RFR-exposed groups at each time-point. As a result, the RFR-exposed samples for each time-point were not directly compared with their matched non-exposed time-point controls. Comparison of the RFR-exposed samples with the time-point matched controls would not likely have yielded significant differences. In any event, it is unclear what the biological significance of a reduced rate of spontaneous intrachromosomal recombination may be.

28.3.3 Human Studies

Only a limited number of studies have evaluated cytogenetic aberrations in human blood lymphocytes following occupational exposure to RFR (Table 28.3).

Table 28.3 Summary of human studies that have investigated the possible cytogenetic effects of radiofrequency radiation

Authors	RFR conditions	SAR; duration	Tissue/cell line	End point	Results	Comments
Garson et al. (1991)	0.4 MHz–20 GHz	SAR unknown, occupational exposure	Radio linemen and age-matched controls	CA	No effect	
Fucic et al. (1992)	1.25–1.35 GHz	SAR unknown, occupational exposure	Non-smoking men	MN	Increased MN frequency	
Garaj-Vrhovac and Fucic (1993)	1.25–1.35 GHz; pulsed	SAR unknown, occupational exposure	Male air traffic control workers	CA	Increased CA in all subjects	
Maes et al. (1995)	450–900 MHz	SAR unknown, occupational exposure	Male transmission antennae maintenance workers	CA	No effect	Pilot study, only 6 exposed and 6 control individuals
Lalic et al. (2001)	Frequencies reaching 8 GHz	SAR unknown, occupational exposure	Subjects working in radio-relay stations	CA	Increased CA relative to background, based on literature references	Authors did not assess the background aberration frequencies in their laboratory to ensure it was similar to literature values; insufficient information on exposure conditions

Gadhia et al. (2003)	890–960 MHz	SAR unknown, minimum 2-year exposure	Mobile-phone users (non-smoker, non-alcoholic or smoker-alcoholic) and matched controls	CA and SCE	Increase in CA and SCE for mobile phone users, compared with controls	Insufficient details on stratification of subjects
Maes et al. (2006)	150 MHz–40 GHz; various modulations	SAR unknown, occupational exposure	Radio-field engineers and administrative staff of mobile phone companies	SCE and CA	No effect	

Two early studies by Fucic et al. (1992) and Garaj-Vrhovac and Fucic (1993) reported an increased incidence of chromosome aberrations and MN in the blood of male subjects who were occupationally exposed to RFR (1.25–1.35 GHz). Gadhia et al. (2003) recently reported an increased incidence of chromosome aberrations and SCE frequencies in normal and smoker–alcoholic mobile-phone users (890–960 MHz). In contrast, Maes et al. (1995) found no evidence of excess aberrations in peripheral blood from antenna maintenance workers exposed to 450–950 MHz RFR, while Garson et al. (1991) found no evidence of increased aberrations in peripheral blood from radio linemen exposed to 0.4-MHz–20 GHz RFR. However, these studies are generally considered inconclusive owing to a variety of methodological shortcomings such as inadequate RFR exposure dosimetry, lack of detailed information regarding enrolled subjects and inadequate sample size based on statistical power analysis (Verschaeve 2005).

More recently, Lalic et al. (2001) reported increased chromosome aberrations (chromatid breaks, acentric fragments and dicentric chromosomes) in peripheral blood from subjects working in radio-relay stations. The authors reported increased levels of acentric fragments and dicentric chromosomes in the RFR-exposed subjects relative to literature reference values for the background incidence of these forms of aberrations in control populations. However, since the authors did not assess aberration frequency in any control samples within their laboratory, it cannot be determined whether the authors detected an RFR-induced cytogenetic effect or whether the background frequency for detecting these forms of aberrations was considerably higher in their laboratory relative to reference values from other laboratories. For these reasons, these data must be considered methodologically flawed.

Maes et al. (2006) evaluated a variety of cytogenetic end points in peripheral blood samples obtained from 38 radio field engineers and 11 administrative staff of a mobile-phone company and 25 unrelated controls. The radio field engineers were exposed to a variety of antennas where they were frequently exposed to a variety of frequencies ranging from 450 MHz to several gigahertz; administrative staff were exposed to RFR in the 150–170 MHz and 6–40 GHz range; while the controls were not occupationally exposed to RFR. No statistically significant increase in chromosome aberration (which included analysis of gaps, breaks, interchromatid exchanges, translocations, dicentric chromosomes and acentric fragments) or SCE frequencies was observed in RFR-exposed subjects, relative to control subjects. However, it is important to point out that no dosimetric analysis was carried out, although measurements indicated that RFR exposures were typically below the limits of the ICNIRP guidelines (International Commission on Non-Ionizing Radiation Protection 1998).

In conclusion, the vast majority of in vitro studies using human blood cultures and mammalian cell lines have found no consistent evidence of cytogenetic effects from non-thermal RFR exposures. Of the limited number of in vitro studies reporting RFR-induced cytogenetic effects, thermal confounding appears to be a likely explanation. Of the few studies which have assessed the effect of in vivo RFR exposure on cytogenetic end points in animals and humans, many suffer from obvious methodological and/or statistical shortcomings. Future well-characterized prospective human and animal RFR exposure studies would be beneficial to permit a more complete RFR health risk assessment.

28.4 DNA Strand Breaks

28.4.1 In Vitro Studies

Since the publication of in vivo studies by Lai and Singh (1995, 1996, 1997), a large number of studies have assessed the ability of in vitro RFR exposure at a variety of frequencies to induce DNA strand breaks in a variety of mammalian cell lines and in human peripheral blood (Table 28.4). Unlike the in vivo data, the results of in vitro investigations of RFR-induced DNA damage and/or repair have yielded nearly uniform negative results (Brusick 1998; Vijayalaxmi and Obe 2004; Moulder et al. 2005).

Phillips et al. (1998) reported alterations in primary DNA damage at clearly non-thermal RFR exposure levels in Molt-4 cells, following in vitro exposure to a variety of conditions (813–836 MHz; time division multiple access, TDMA, modulated or Integrated Digital Enhanced Network (iDEN) modulated; SAR 0.002–0.02 W/kg). However, this study has been criticized owing to a lack of internal consistency as some RFR frequency/modulation paradigms were reported to statistically increase the level of DNA damage, while others were found to significantly reduce the level of DNA damage. Furthermore, there was no dose dependency for these observations and the fold-differences in control values between experiments often exceeded the fold-differences between sham and RFR-exposed samples. Thus, the results from this study must be considered equivocal. Hook et al. (2004) re-evaluated the ability of RFR (813–847 MHz; CDMA-, FDMA-, iDEN- or TDMA-modulated; SAR 0.0024–3.2 W/kg) to elicit DNA damage in Molt-4 T-lymphoblastoid cells, but failed to find any evidence of a RFR-induced genotoxic effect at any frequency/modulation paradigm employed. Most other studies assessing the effect of in vitro RFR exposure on mammalian cell lines have also found no evidence of genotoxic effects. Malyapa et al. (1997a, b), Li et al. (2001) and LaGroye et al. (2004b) found no evidence of increased DNA damage, as determined by the alkaline comet assay, in either mouse C3H 10T1/2 fibroblasts or human glioblastoma U87MG cells following in vitro exposure for 2–24 h to 835 MHz–2.45 GHz RFR at a variety of modulations in the SAR range of 0.6–5.1 W/kg. More recently, Sakuma et al. (2006) exposed human glioblastoma A172 cells and normal human IMR-90 fibroblasts to 2.1425 GHz RFR for 2–24 h at SARs of 0.08–0.8 W/kg, but also found no evidence of any RFR-mediated effect on DNA damage.

Table 28.4 Summary of in vitro studies that have investigated the possible DNA-strand breaks effects of radiofrequency radiation

Authors	RFR conditions	SAR; duration	Tissue/cell line	End point	Results	Comments
Malyapa et al. (1997a, b)	2.45 GHz, 835.6 and 847.7 MHz; various modulations	SARs ranging from 0.6 to 1.9 W/kg; 2-, 4- and 24-h exposures	Human glioblastoma U87MG cells and mouse C3H 10T1/2 fibroblasts	DNA strand breaks	No effect	Exponentially growing and plateau-phase cells
Maes et al. (1997)	935.2 MHz; CW	SARs of 0.3–0.4 W/kg; 2-h exposure	Human whole blood	DNA strand breaks	No effect	
Phillips et al. (1998)	813.6 and 836.5 MHz; TDMA- and iDEN-modulated	SARs ranging from 0.002 to 0.026 W/kg; intermittent exposures of 2, 3 and 21 h	Molt-4 T-lymphoblastoid cells	DNA strand breaks	Increased/decreased DNA damage, data are equivocal	No positive controls; low sample number for some conditions (1 or 2); results internally inconsistent; inconsistency in control values
Vijayalaxmi et al. (2000)	2.45 GHz; pulse-modulated	SAR of 2.1 W/kg; 2-h exposure	Whole human blood	DNA strand breaks	No effect	
Li et al. (2001)	847.7 and 835.6 MHz; CDMA- and FDMA-modulated	SARs of 3.8 and 4.9 W/kg for CDMA and 5.1 W/kg for FDMA; 2-, 4- and 24-h exposures	Mouse C3H 10T1/2 fibroblasts	DNA strand breaks	No effect	Exponentially growing and plateau-phase cells
Tice et al. (2002)	837 MHz, 1.9 GHz; analogue, PCS, CDMA and TDMA	SARs ranging from 1 to 100 W/kg; 3- and 24-h exposures	Diluted human blood	DNA strand breaks	No effect	Low sample number ($N = 1$ or 2 for all conditions)
McNamee et al. (2002a, b, 2003)	1.9 GHz; CW and pulse-modulated	SARs ranging from 0.1 to 10 W/kg; 2- and 24-h exposures	Diluted human blood	DNA strand breaks	No effect	

LaGroye et al. (2004a)	2.45 GHz; CW	Average SAR of 1.9 W/kg; 2-h exposure	Mouse C3H 10T1/2 fibroblasts	DNA strand breaks	No effect	
Hook et al. (2004)	813.6–847.7 MHz; various modulations	SARs ranging from 0.002 to 3.2 W/kg; various exposure times up to 24 h	Molt-4 T-lymphoblastoid cells	DNA strand breaks	No effect	Exponentially growing cells
Diem et al. (2005)	1.8 GHz; intermittent or continuous CW and GSM	SARs of 1.2 and 2 W/kg; 4-, 16- or 24-h exposure	Human diploid fibroblasts and rat granulosa cells	DNA strand breaks	Increased single-strand and double-strand breaks in both cell types	Classification of comets into 5 categories followed by transformation by arbitrary factors; no control for differences in S/G2-phase or apoptotic cells which may have resulted in their classification in an elevated "damage" category
Zeni et al. (2005)	900 MHz; GSM-modulated	SARs of 0.3 and 1 W/kg; 2-h exposure	Human peripheral blood lymphocytes	DNA strand breaks	No effect	
Sakuma et al. (2006)	2.14 GHz; CW and W-CDMA-modulated	SARs of 0.08, 0.25 and 0.8 W/kg; 2- and 24-h exposures	Human glioblastoma A172 cells and human IMR-90 fibroblasts	DNA strand breaks	No effect	
Stronati et al. (2006)	935 MHZ; GSM-modulated	SARs of 1 and 2 W/kg; 24-h exposure	Human whole blood	DNA strand breaks	No effect	

iDEN Integrated Digital Enhanced Network, *W-CDMA* wideband CDMA

Studies by Maes et al. (1997), Vijayalaxmi et al. (2000), Tice et al. (2002), McNamee et al. (2002a, b, 2003), Zeni et al. (2005) and Stronati et al. (2006) have also assessed the ability of non-thermal RFR, at a variety of frequencies and modulations, to induce DNA damage in human peripheral lymphocytes using the alkaline comet assay. The study by Tice et al. (2002) possessed an inadequate number of independent experiments ($N = 1–2$ for each exposure paradigm) upon which to accurately assess a possible RFR-induced genotoxic effect. However, the remaining studies have provided strong evidence that non-thermal RFR exposure has no effect on inducing DNA damage in cultured human lymphocytes. In contrast, Diem et al. (2005) reported increased DNA damage (DNA single-strand and double-strand breaks) in human fibroblasts and rat granulosa cells following in vitro exposure of these cells to 1.8 GHz RFR at SARs of 1.2–2 W/kg for 4–24 h, under a variety of exposure conditions. In this study, the authors classified DNA "comets" into one of five categories based upon visible inspection, then applied a transformation factor to achieve a "normalized tail factor" for each sample. Vijayalaxmi et al. (2006) raised several concerns for this approach since such transformation of each classification group by arbitrary transformation factors did not control for differences in S/G2-phase or apoptotic cells. These cells would be classified into the "higher" damage categories owing to either the presence of DNA replication forks or orderly DNA fragmentation, respectively. As such, small differences in the rate of cell growth/death between control and RFR-exposed groups may have been responsible for the observed differences between the sham and RFR-exposed samples. In any event, additional studies are required to validate these observations.

In summary, there is no consistent evidence in the literature for non-thermal, RFR-induced DNA damage in either mammalian-derived cell lines or primary human lymphocytes. This information is supported by a large database of studies using traditional bacterial/mammalian cell lines and recent transgenic in vivo mutagenesis assays, where there is virtually no evidence of any mutagenic effects from RFR exposure.

28.4.2 In Vivo Studies

Analysis of RFR-induced DNA strand breaks (using the alkaline comet assay) in animal tissue has been much more controversial (Table 28.5). An early study by Sarker et al. (1994) reported RFR-induced alterations in *Hinf*I digested genomic DNA, relative to control DNA, in both brain and testis from mice exposed to 2.45 GHz RFR at a SAR of 1.18 W/kg for 2 h/day for 120–200 days. This was followed by a series of highly publicized studies by Lai and Singh (1995, 1996, 1997), who reported increased levels of primary DNA damage (which may have included DNA single-strand and double-strand breaks, alkali-labile sites and DNA cross-links) in rat brain cells at 0–4 h after a 2-h in vivo exposure to 2.45 GHz CW or pulse-modulated RFR at SARs

Table 28.5 Summary of in vivo studies that have investigated the possible DNA-strand breaks effects of radiofrequency radiation

Authors	RFR conditions	SAR; duration	Tissue/cell line	End point	Results	Comments
Sarker et al. (1994)	2.45 GHz; CW	Average SAR of 1.18 W/kg; 2 h/day, up to 200 days	Inbred male Swiss albino mice; isolated cells from brain and testis	DNA transformations	Distinct DNA alterations in exposed samples	Restrained animals; sample number $N = 2$ per time-point
Lai and Singh (1995, 1996, 1997)	2.45 GHz; PW or CW	Whole body average SAR 0.6 and 1.2 W/kg; 2-h exposure	Male Sprague-Dawley rats; isolated brain cells	DNA strand breaks immediately and 4 h after exposure	Increased DNA damage 4 h after CW and PW exposure and 0 h after CW exposure	Non-restrained animals; no positive control; injection of free-radical scavengers blocked RFR DNA damaging effect; used proteinase K in lysis buffer
Malyapa et al. (1998)	2.45 GHz; CW	Whole body average SAR of 1.2 W/kg; 2-h exposure	Male Sprague-Dawley rats; isolated brain cells	DNA strand breaks immediately and 4 h after exposure	No effect	Non-restrained animals; observed an increased background level of DNA damage in rats euthanized with CO_2 as done in the Lai and Singh studies; observed that the time from euthanasia to dissection needs to be minimized, where increased time results in increased damage; did not use proteinase K in lysis buffer

(*Continued*)

Table 28.5 Summary of in vivo studies that have investigated the possible DNA-strand breaks effects of radiofrequency radiation—Cont'd

Authors	RFR conditions	SAR; duration	Tissue/cell line	End point	Results	Comments
LaGroye et al. (2004b)	2.45 GHz; PW	Whole body average SAR of 1.2 W/kg; 2-h exposure	Male Sprague-Dawley rats; isolated brain cells	DNA strand breaks 4 h after exposure	No effect	Non-restrained animals; lysis buffer with and without proteinase K
Aitken et al. (2005)	900 MHz	Whole body average SAR of 0.09 W/kg; 12 h/day, 7 days	Male CD1 Swiss mice; isolated spermatozoa	DNA strand Breaks and DNA integrity	No effect on strand breaks, but observed DNA lesions in a segment of the β-globin gene and in the mitochondrial genome	Non-restrained animals; no positive control
Paulraj and Behari (2006)	2.45 and 16.5 GHz; amplitude modulated	Whole body average SAR of 1 or 2 W/kg; 2 h/day, 5 days/week, 7 weeks	Male Wistar rats; isolated brain cells	DNA strand breaks	Increased DNA damage in exposed rats	Restrained animals; no positive control
Belyaev et al. (2006)	915 MHz; GSM	Whole body average SAR of 0.4 W/kg; 2-h exposure	Male Fisher 344 rats; isolated cells from spleen, thymus and brain	DNA double-strand breaks and chromatin conformation	No effect	Non-restrained animals; no positive control

PW pulsed wave

of 0.6–1.2 W/kg. Of particular interest was the observation that injection of melatonin or the spin-trap compound *N-tert*-butyl-α-phenylnitrone (PBN), either immediately before or after RFR exposure, could prevent the occurrence of RFR-induced DNA damage, indicating a role for free radicals in this effect. These studies generated a flurry of controversy, alarming media reports and research to confirm these findings and to identify a mechanism by which RFR exposure may have influenced DNA damage formation and/or repair.

In an attempt to confirm these observations, Malyapa et al. (1998) exposed rats to 2.45 GHz CW RFR for 2 h at a SAR of 1.2 W/kg, but found no evidence of immediate or delayed RFR-induced DNA damage in the rodent brain. However, the authors did note that the background level of DNA damage increased in brain cells from CO_2-asphyxiated animals relative to that in guillotined animals and that the time from CO_2 asphyxiation until dissection was an important parameter in the extent of background DNA damage. This is considered an important observation since the studies by Lai and Singh (1995, 1996, 1997) employed CO_2 asphyxiation and included numerous wash steps to remove red blood cells, resulting in an apparent large time-lag between CO_2 asphyxiation and processing of samples via the alkaline and neutral comet assays. As a result, there is considerable potential that the observations by Lai and Singh were not related to RFR exposure, but rather were the result of post-euthanasia confounding by the time-dependent generation of free radicals in the tissue extracts, despite the inclusion of PBN in the mincing buffer. On the other hand, unlike Lai and Singh, Malyapa et al. (1998) did not use proteinase K in their comet assay protocol. Since proteinase K digestion removes nuclear proteins, DNA from nuclear-protein associations and DNA–protein cross-links would be free to migrate and the studies by Lai and Singh may have detected DNA damage that Malyapa's group could not observe. To address this possibility, LaGroye et al. (2004a) used both versions of the alkaline comet assay to evaluate the ability of in vivo 2.45 GHz pulsed-wave RFR exposure at a SAR of 1.2 W/kg to induce DNA damage in the rat brain at 4 h after an acute 2-h exposure. In this study, animals were killed one at a time by guillotine and no DNA damage was detected following RFR exposure, indicating that the comet assay methodology did not appear to be a significant factor in explaining the conflicting results between these laboratories.

More recently, Paulraj and Behari (2006) reported increased DNA damage in brain cells of Wistar rats following exposure to 2.45 or 16.5 GHz RFR for 35 days at 2 h/day at SARs of approximately 1.0 and 2.0 W/kg, respectively. In contrast, Belyaev et al. (2006) found no evidence of DNA double-strand breaks or changes in chromatin conformation in rat brain following a 2-h acute exposure to 915 MHz GSM-modulated RFR at 0.4 W/kg. Similarly, Aitken et al. (2005) observed no evidence of single-strand or double-strand DNA breaks in spermatozoa of mice exposed to 900 MHz RFR for 12 h/day for 7 days; however, Aitken et al. (2005) noted altered PCR amplification of a

9-kb fragment of the β-globin gene and a 10-kb fragment of the mitochondrial genome in RFR-exposed mouse spermatozoa, relative to control animals. The authors argued that these changes were indirect evidence of RFR-induced DNA lesions, which may have been undetectable using the conventional genotoxicity assays. Obviously, additional in vivo studies are required to delineate the inconsistencies in the literature.

28.5 Epigenetic Effects of RFR

Despite the apparent inability of non-thermal RFR to induce direct (genotoxic) DNA and/or cytogenetic damage, it is possible that RFR may interact with biological systems in an indirect (non-genotoxic or epigenetic) fashion to modulate the response of cells or tissues to known genotoxins. Proposed mechanisms for RFR-induced epigenetic effects include changes in the fidelity and/or efficiency of DNA repair processes, changes in cell-cycle progression and/or cell proliferation, changes in the expression of certain genes (such as heat-shock proteins and proto-oncogenes) and changes in cell survival. The evidence for these RFR-induced effects is contradictory and outside the focus on this review. This section will focus solely on the cytogenetic evidence of RFR-induced epigenetic effects.

Relatively few studies have examined the consequence of combined effects of exposure to RFR and known physical and/or chemical mutagens on cytogenetic alterations. Early studies by Balcer-Kubiczek and Harrison (1985, 1989, 1991) reported that 2.45 GHz RFR exposure at SARs of 0.1–4.4 W/kg for 24 h, could increase the rate of neoplastic transformation of C3H 10T1/2 cells when coexposed to either benzo[*a*]pyrene or X-rays in the presence of the tumor promoter 12-*O*-tetradecanoylphorbol-13-acetate (TPA). Subsequent studies by Cain et al. (1997) and Roti Roti et al. (2001) failed to observe similar effects, although these studies employed lower SARs and different RFR frequencies. More recently, Wang et al. (2005a) exposed C3H 10T1/2 cells to 2.45 GHz RFR at SARs of 5–200 W/kg for 2 h, either with RFR alone or in the presence of RFR and 3-methylcholanthrene. No evidence of RFR-induced neoplastic transformation was observed at any SAR level when cells were exposed to RFR in the presence or absence of TPA; however, increased transformation was observed in cells co-exposed to 3-methylcholanthrene and RFR at SAR levels greater than 100 W/kg. While the temperature in the 100–200 W/kg groups reached 40 and 44°C respectively, the authors argued that the RFR-induced increase in neoplastic transformation was not due to thermal effects since control samples were exposed to similar temperatures but no increase in neoplastic transformation was observed in these samples. It is important to point out that while the volume-averaged temperature was reported not to have exceeded 40–44°C, the presence of hot spots within the sample cannot be ruled out since insufficient information on SAR uniformity was reported.

Initial studies by Maes et al. (1996, 1997) reported a strong synergistic increase in the frequency of SCE in human whole-blood cultures following exposure to 900- or 954 MHz RFR at SARs of 0.3–1.5 W/kg and mitomycin C. Similar epigenetic effects were reported by Zhang et al. (2002) and Wang et al. (2005b), where increased levels of MN and primary DNA damage were observed in human whole-blood cultures following co-exposure to RFR and a variety of mutagenic compounds. Unfortunately, these studies suffer from a number serious deficiencies, including inadequate sample size, inappropriate statistics and insufficient detail relating to the exposure system design, dosimetry and SAR distribution to exclude the possibility of thermal effects. On the other hand, an increased frequency of MN was also reported in CHO-K1 cells following coexposure to bleomycin and 2.45 GHz RFR at high SARs (78–100 W/kg), in a well-controlled study by Koyama et al. (2003). It is interesting to note from this study that high RFR SAR exposures seem to induce non-thermal epigenetic cytogenetic effects; however, the relevance of these findings to health risk assessment is questionable owing to the magnitude of these exposures.

In contrast, several recent studies where dosimetry was well defined and possible thermal influences were eliminated failed to detect evidence of RFR-induced epigenetic effects on cytogenetic end points. In a follow-up study to their original work, Maes et al. (2001) exposed human blood cultures to mitomycin C or X-rays and 900 MHz RFR with a variety of modulation paradigms for 2–24 h at SARs of 0.4–10 W/kg, but failed to find any evidence of excess chromosome aberrations or SCE in RFR-exposed samples. Similarly, LaGroye et al. (2004b) reported no evidence of increased primary DNA damage (as measured by the alkaline comet assay) in C3H 10T1/2 cells co-exposed for 2 h to 2.45 GHz RFR at a SAR of 1.9 W/kg and either ^{137}Cs γ-rays or cisplatin. More recently, Stronati et al. (2006) assessed the ability of 935 MHz GSM-modulated RFR exposure at a SAR of 1–2 W/kg for 24 h to alter the magnitude of the cytogenetic and primary DNA damage induced by 1 Gy X-irradiation in human blood cultures. No evidence of increased unstable chromosome aberrations, SCE, MN or primary DNA damage was observed in the RFR plus X-ray co-exposed groups, relative to the X-ray only controls.

Recently, Verschaeve et al. (2006) were the first group to evaluate the ability of in vivo RFR exposure to modulate the cytogenetic effects induced by a known mutagen, namely, 3-chloro-4-(dichloromethyl)-5-hydroxy-2(5*H*)-furanone (MX). In this study, female Wistar rats were supplied with drinking water containing 19 μg/ml MX and were either sham or exposed to 900 MHz RFR for 2 years at 2 h/day, 5 days/week at SARs of 0.3 and 0.9 W/kg. Blood samples were taken at 3, 6 and 24 months of exposure and brain and liver samples were acquired at the end of the study. No evidence of either increased RFR-induced primary DNA damage or MN formation was observed when compared with MX-only controls. However, it is important to point out that MX treatment also had no significant effect on primary DNA damage levels in the blood or liver of exposed animals relative to controls and was only found to increase DNA damage in the rat brain samples at 24 months. Similarly, there was no evidence of an increased frequency of MN in PCEs in the MX-treated rats at any time-point relative to control animals, suggesting that the dose of MX applied to the drinking water may not have been optimal for evaluation of the epigenetic effects of RFR in vivo. Thus, while this study is novel in design and scope, the importance of this study is limited for the evaluation of the human health risks of RFR exposure.

28.6 Cancer Studies in Animals

Although the analysis of human populations has been a powerful tool for detecting previously unidentified human carcinogens, current epidemiological studies of RFR have been hindered by the relatively recent introduction (within the last 10 years) of a wide variety of RFR-emitting devices and the long latency phase (more than 20 years) for the development of most humans cancers. Furthermore, since new wireless technologies and applications are emerging at a tremendous rate, it is difficult for any epidemiology study to capture the effect of long-term human exposure to specific RFR conditions. As such, cancer studies in experimental animals are particularly important in evaluating whether an association between RFR exposure and cancer exists as laboratory animal studies provide an integrated system in which RFR experimental variables can be controlled, specific hypotheses can be investigated and exposures can be precisely measured. In this section, the evidence for a causal association between long-term, low-level RFR exposure and cancer in rodents will be reviewed (Tables 28.6–28.8).

Prior to 1997, only a limited number of studies assessed the effect of long-term, low-level RFR exposure in animals and most of these studies suffered from inadequate dosimetry, confinement or thermal stress and high rates of inherent disease in the animals (Moulder et al. 2005); however, two studies merit consideration. Chou et al. (1992) reported that following 25 months' exposure of Sprague-Dawley rats for 21.5 h/day to 2.45 GHz pulse-modulated RFR at a whole body averaged SAR of 0.15–0.4 W/kg, a significant increase in overall malignant tumor incidence was observed in the RFR-exposed rats when compared with the sham group. However, the authors cautioned that the biological significance of these results was limited as there were no accompanying differences in lifespan, total number of tumors or type of tumor. Wu et al. (1994) exposed Balb/c mice to 2.45 GHz RFR for 3 h/day, 6 days/week for 5 months at SARs of 10–12 W/kg in the presence of the known colon cancer inducing compound dimethylhydrazine. The authors did not find any evidence that RFR exposure had an effect on the incidence or size of chemically induced tumors, or on the incidence of protuberant or infiltrative tumors in tumor-bearing mice when compared with control groups.

Table 28.6 Summary of cancer studies in animals that have investigated the effects of radiofrequency radiation – 1992–2000

Authors	RFR conditions	SAR; duration	Animal model	End point	Results	Comments
Chou et al. (1992)	2.45 GHz; pulse-modulated	Whole body average SAR between 0.15 and 0.4 W/kg; 21.5 h/day for 25 months	Male Sprague-Dawley rats; 100 rats per group	Tumor incidence in all major organs	Increased incidence of primary malignant tumors in RFR-exposed rats	No differences in lifespan, total number of tumors or tumor type; unrestrained animals; all animals were histopathologically assessed; no positive control
Wu et al. (1994)	2.45 GHz; CW	SARs of 10–12 W/kg; 3 h/day, 6 days/week for 5 months	Dimethylhydrazine-induced colon cancer in Balb/c mice; 26–32 mice per group	Incidence of colon cancer	No effect	Animal movement partially restricted during exposure; all animals were histopathologically assessed; low number of animals per group; no positive control
Repacholi et al. (1997)	900 MHz; pulse-modulated	Average SARs ranged from 0.13 to 1.4 W/kg; 2×30 min/day for 18 months	Eμ-*Pim*1 (lymphoma-prone) mice; 100 mice per group	Incidence of lymphoma; tumor incidence in all major organs	Increased incidence of lymphoma in RFR-exposed mice	Unrestrained animals; not all animals were histopathologically assessed; non-homogeneous SAR between and within animals; no positive control

(*Continued*)

Table 28.6 Summary of cancer studies in animals that have investigated the effects of radiofrequency radiation – 1992–2000—Cont'd

Authors	RFR conditions	SAR; duration	Animal model	End point	Results	Comments
Toler et al. (1997)	435 MHz; pulse-modulated	Whole body average SAR of 0.32 W/kg; 22 h/day, 7 days/week for 21 months	Female C3H/HeJ (mammary tumor prone) mice; 200 mice per group	Tumor incidence in mammary glands and all major organs	No effect	Unrestrained animals; all animals were histopathologically assessed; no positive control
Frei et al. (1998a, b)	2.45 GHz; CW	Whole body average SAR of 0.3 and 1.0 W/kg; 20 h/day, 7 days/week for 18 months	Female C3H/HeJ (mammary tumor prone) mice; 100 mice per group	Tumor incidence in mammary glands and all major organs	No effect	Unrestrained animals; all animals were histopathologically assessed; no positive control
Adey et al. (1999, 2000)	836 MHz; CW or pulse-modulated	Brain-average SAR of 1.0–1.6 W/kg; in utero and postweaning exposure for 2 years	ENU-treated and untreated Fischer 344 rats; 60–90 rats per group	Incidence of primary CNS tumors	No effect	Restrained animals; histopathological assessment performed on CNS tissue from all animals

CNS central nervous system, *ENU* ethylnitrosourea

Table 28.7 Summary of cancer studies in animals that have investigated the effects of radiofrequency radiation – 2001–2002

Authors	RFR conditions	SAR; duration	Animal model	End point	Results	Comments
Zook and Simmens (2001, 2006)	860 MHz; CW and pulse-modulated	Brain average SAR of 1.0 W/kg; 6 h/day, 5 days/week for 2 years	ENU-treated and untreated Sprague-Dawley rats; 60 rats per group	Incidence of tumors in CNS and in all major organs	No effect	Restrained animals; all animals were histopathologically assessed
Heikkinen et al. (2001)	902 MHz; CW or GSM-modulated	Whole body average SARs of 0.35 and 1.5 W/kg; 1.5 h/day, 5 days/week for 78 weeks	Female CBA/S mice initiated with ionizing radiation; 50 mice per group	Incidence of neoplastic lesions in all major organs	No effect	Restrained animals; all animals were histopathologically assessed
Imaida et al. (2001)	1.5 GHz; pulse-modulated	Whole body average SAR of 0.08 W/kg; 90 min/day, 5 days/week for 19 weeks	DMBA-induced skin tumors in female ICR mice; 30–48 mice per group	Incidence of tumors on skin and in several major organs	No effect	Restrained animals; all animals were histopathologically assessed
Utteridge et al. (2002)	898.4 MHz; GSM-modulated	SARs of 0.25, 1.0, 2.0 and 4.0 W/kg; 1 h/day, 5 days/week for 104 weeks	Eμ-*Pim*1 (lymphoma-prone) mice; 120 mice per group	Incidence of lymphoma; tumor incidence in all major organs	No effect	Restrained animals; all animals were histopathologically assessed; ENU-positive control included
Bartsch et al. (2002)	900 MHz; GSM-modulated	Whole body average SAR ranging from 0.017 to 0.07 W/kg; continuous exposure for 1 year	DMBA-induced mammary tumors in female Sprague -Dawley rats; 60 rats per group	Incidence of mammary tumors	No effect	Unrestrained animals; histopathological assessment only performed when mammary tumors became 1–2 cm in diameter

DMBA 7,12-dimethybenz[*a*]anthracene

Table 28.8 Summary of cancer studies in animals that have investigated the effects of radiofrequency radiation – 2003–2006

Authors	RFR conditions	SAR; duration	Animal model	End point	Results	Comments
Anane et al. (2003)	900 MHz; GSM-modulated	Whole body average SARs of 0.1–3.5 W/kg; 2 h/day, 5 days/week for 9 weeks	DMBA-induced mammary tumors in female Sprague-Dawley rats; 16 rats per group	Incidence of mammary tumors	Equivocal results	Animal movement partially restricted during exposure; histopathological assessment only performed on mammary glands with visible (macroscopic) tumors; insufficient number of animals per group; inconsistency across replicate experiments
La Regina et al. (2003)	835 MHz FDMA- and 847 MHz CDMA-modulated	Brain-average SAR of 1.3 W/kg; 4 h/day, 5 days/week for 2 years	Fischer 344 rats; 80 rats per group	Incidence of tumors in all major organs	No effect	Restrained animals; no positive control; all animals were histopathologically assessed
Heikkinen et al. (2003)	849 or 902 MHz; pulse-modulated	Whole body average SAR of 0.5 W/kg; 1.5 h/day, 5 days/week for 52 weeks	Female transgenic and normal K2 mice coexposed to UV radiation; 8–27 mice per group	Incidence of skin tumors	No effect	Restrained animals; only macroscopic skin lesions were histopathologically assessed; low animal numbers (8–27 per group)
Sommer et al. (2004)	900 MHz; GSM-modulated	Whole body average SAR of 0.4 W/kg; continuous exposure for 46 weeks	Female AKR/J (lymphoma-prone) mice; 160 mice per group	Incidence of lymphoma	No effect	Unrestrained animals; all animals were histopathologically assessed; no positive control

Anderson et al. (2004)	1.6 GHz; Iridium signal	Brain-average SAR ranging from 0.16 to 1.6 W/kg; in utero and postweaning exposure (2 h/day) for 2 years	Fischer 344 rats; 90 rats per group	Incidence of tumors in CNS and all major organs	No effect	Restrained animals; all animals were histopathologically assessed; no positive control
Shirai et al. (2005)	1.6 GHz; pulse-modulated	Brain-average SAR of 0.67 and 2.0 W/kg; 90 min/day, 5 days/week for 104 weeks	ENU-treated and untreated Fischer 344 rats; 100 rats per group	Incidence of tumors in CNS and all major organs	No effect	Restrained animals; all animals were histopathologically assessed
Huang et al. (2005)	849 MHz or 1.763 GHz; CDMA-modulated	Whole body average SAR of 0.4 W/kg; 2×45 min/day, 5 days/week for 19 weeks	DMBA-induced skin tumors in male ICR mice; 20 mice per group	Incidence of skin tumors	No effect	Insufficient information on animal exposure conditions; low animal numbers (20 per group)
Yu et al. (2006)	900 MHz; GSM-modulated	Whole body average SARs of 0.44, 1.33 and 4.0 W/kg; 4 h/day, 5 days/week for 26 weeks	DMBA-induced mammary tumors in female Sprague-Dawley rats; 100 rats per group	Incidence of mammary tumors	No effect	Restrained animals; histopathological assessment performed on mammary glands from all animals

UV ultraviolet

Since 1997, a relatively large number of studies have assessed the effect of long-term, low-level RFR exposure on the induction or promotion of tumors in both normal and tumor-prone rodents. Repacholi et al. (1997) exposed unrestrained mice genetically predisposed to develop lymphomas (Eμ-*Pim1*) to 900 MHz pulse-modulated RFR for 2 × 30 min/day for up to 18 months at SARs ranging from 0.008–4.2 W/kg. In this highly publicized study, the authors reported that while lymphoblastic lymphoma was not significantly affected, an increased incidence of non-lymphoblastic lymphoma was observed in RFR-exposed mice when compared with sham controls. This study was criticized for its poor RFR exposure homogeneity (SAR range 0.008–4.2 W/kg) among the experimental animals and inadequate histopathological assessment (e.g., only ill or dead animals were examined pathologically), which may have inadvertently introduced bias into the analysis. In an effort to address these shortcomings, Utteridge et al. (2002) exposed restrained Eμ-*Pim1* transgenic mice and wild-type mice to 898.4 MHz GSM-modulated RFR for 1 h/day, 5 days/week for 104 weeks at SARs of 0.25, 1.0, 2.0 and 4.0 W/kg. Under tightly controlled exposure conditions and following pathological examination of all animals, the authors reported that long-term RFR exposure had no significant effect on lymphoma induction in normal or lymphoma-prone mice when compared with sham controls. While this study was also criticized since it was not a direct replication of the study by Repacholi et al. (1997) and therefore could not discount the previous findings by this group, the findings from the current study deserve merit in their own regard. Recently, Sommer et al. (2004) continuously exposed unrestrained lymphoma-prone AKR/J mice to 900 MHz GSM-modulated RFR for 46 weeks at a whole body average SAR of 0.4 W/kg. The AKR/J mouse strain was chosen since it is genetically predisposed to develop thymic lymphoblastic lymphoma as its genome carries the AK virus. The authors found no evidence of decreased survival rates or increased incidence of lymphoma in RFR-exposed mice, relative to sham mice, at any SAR level tested. While these results are consistent with those of Utteridge et al. (2002), it must be noted that Repacholi et al. (1997) reported an increased incidence of non-lymphoblastic lymphoma which was not evaluated in this study.

Six recent studies have assessed the ability of long-term RFR exposure to influence the development of mammary tumors in rodent models. Frei et al. (1998a, b) exposed C3H/HeJ mammary tumor-prone mice to 2.45 GHz CW RFR for 18 months at 20 h/day, 7 days/week at SARs of 0.3 and 1 W/kg. In both studies, there were no significant differences between RFR-exposed and sham mice with respect to tumor incidence, latency to tumor onset, growth rate of mammary tumors or animal longevity. In a related study by the same group, Toler et al. (1997) found no evidence that latency to tumor onset, growth rate of mammary tumors or animal longevity was affected by a 21-month exposure (22 h/day, 7 days/week) of C3H/HeJ mammary tumor-prone mice to 435 MHz pulse-modulated RFR at a SAR of 0.32 W/kg. In an attempt to investigate the possible tumor-promoting capability of RFR, several groups have used the well-characterized 7,12-dimethylbenz[*a*]anthracene (DMBA) induced mammary tumor model to study this issue in female Sprague-Dawley rats. Bartsch et al. (2002) reported the results of three independent experiments, initiated in three consecutive years, where unrestrained rats were continuously exposed to 900 MHz GSM-modulated RFR at whole body averaged SARs ranging from 0.017 to 0.07 W/kg. Histopathological evaluation of all tumors was performed as the tumors reached 1–2 cm in diameter during the course of the study and the milk line was verified for the presence of any additional tumors in all animals. The authors found a significantly delayed malignant tumor median latency in RFR-exposed rats in their first experiment that was absent in their other two studies. However, as a whole, the authors concluded that RFR exposure had no effect on either DMBA-induced mammary tumor latency or the cumulative tumor incidence since this prolonged latency to tumor onset was not replicated in their subsequent studies. In a similar study, Anane et al. (2003) performed two independent studies where female Sprague-Dawley rats with DMBA-induced mammary tumors were exposed to 900 MHz GSM-modulated RFR for 2 h/day, 5 days/week for 9 weeks at SARs of 0.1–3.5 W/kg. A total of 16 animals were included in each exposure group. All macroscopic mammary tumors were histopathologically evaluated; however, mammary glands without visible tumors were not evaluated and microscopic tumors may have been missed. The authors reported no differences in the latency to tumor onset between the sham and RFR-exposed groups in each study, nor between either of the two studies. A significant increase in tumor incidence was observed in rats exposed to RFR at SARs of 1.4 and 2.2 W/kg in the first experiment, but the opposite effect was observed in their second experiment, where a decreased tumor incidence was reported in rats exposed to RFR at a SAR of 1.4 W/kg. Furthermore, tumor multiplicity on rats bearing tumors was inconsistent in the 1.4-W/kg RFR-exposed groups between the two studies. In the second experiment, the tumor-bearing rats in the 1.4-W/kg RFR-exposed group had fewer malignant tumors than tumor-bearing rats in the sham group, whereas no such difference was observed in the first study. The inconsistency in the results, which may be partially attributable to low animal numbers per group, led the authors to conclude that this study added no new evidence as to the co-promoting effect of RFR. More recently, Yu et al. (2006) also investigated the effects of 900 MHz GSM-modulated RFR exposure on mammary tumor promotion in female rats with DMBA-induced mammary tumors. The authors exposed rats (100 per group) to cage conditions or to RFR for 4 h/day, 5 days/week for 26 weeks at SARs of 0, 0.44, 1.33 and 4.0 W/kg. All mammary glands were palpated weekly and analyzed histopathologically. The authors reported that no statistically significant differences were observed between RFR-exposed and sham rats for latency to tumor onset, mammary tumor incidence, tumor multiplicity or tumor size on tumor-bearing rats. However, the investigators did observe a statistically significant increase in body weight and tumor incidence in the cage control group and the latency to tumor onset was also significantly shorter in these animals compared with that in the sham and RFR-exposed groups. The authors mentioned that these differences may have been due to daily food deprivation of the sham and RFR-exposed groups, which may have led to reduced tumorigenesis in these groups relative to the cage controls. In summary, when these studies using mammary tumor prone mice and chemically induced mammary tumors are assessed collectively, there is no compelling evidence of altered mammary tumor incidence or development by RFR exposure.

The possible association between RFR exposure and brain tumor risk has been of great public and scientific concern for the past decade, owing to the large and ever-increasing number of mobile-phone users worldwide who are exposed to RFR in close proximity to the head. Since 2000, several animal cancer studies have investigated whether RFR exposure may influence brain tumor initiation and/or promotion. Adey et al. (2000) exposed ethylnitrosourea (ENU) treated and normal Fischer 344 rats continuously to 836 MHz CW RF fields in utero on gestational day 18 for 3–4 days until parturition, then at 10 days after weaning the pups were exposed intermittently to RFR for the remainder of their lifespan (approximately 2 years). The brain-averaged SARs were estimated to range from 1 to 1.2 W/kg. The authors found no evidence that RFR exposure had any effect on either spontaneous or ENU-induced brain tumor incidence. In an earlier study by the same group (Adey et al. 1999), the authors employed a similar experimental design but the rats were exposed to 836 MHz pulse-modulated RFR at brain-averaged SARs of 1.0–1.6 W/kg. In their earlier study, the authors also reported no evidence of any tumorigenic effects from RFR exposure, but did note that RFR exposure caused a trend towards a reduced incidence in ENU-induced brain tumors. More recently, Shirai et al. (2005) exposed normal and ENU-treated Fischer 344 rats to 1.439 GHz pulse-modulated RFR at brain-averaged SARs of 0.67 and 2 W/kg. RFR exposures commenced at 5 weeks of age for 90 min/day, 5 days/week for 104 weeks. The authors reported that there were no differences in survival rates, incidences or numbers of brain and/or spinal cord tumors or tumor types in ENU-treated rats among the RFR-exposed groups relative to sham animals. Anderson et al. (2004) conducted a similar study, where pregnant Fischer 344 rats were exposed to a 1.6 GHz Iridium signal. Similar to the previous studies, rats were exposed in utero on gestational day 19 until weaning (about 23 days) for 2 h/day, 7 days/week, then the offspring were exposed for 2 h/day for 5 days/week under near-field conditions to 1.6 GHz RFR at brain-averaged SARs ranging from 0.16–1.6 W/kg for 2 years. No statistically significant differences in the incidence of cancer in RFR-treated normal rats relative to sham rats was observed in this 2-year study.

La Regina et al. (2003) exposed normal adult Fischer 344 rats to either 835 or 847 MHz RFR for 4 h/day, 5 days/week for 2 years at a time-averaged brain SAR of 1.3 W/kg. At the end of the study, all major organs were examined grossly and histopathologically for the incidence of spontaneous tumor formation. The authors found no evidence of altered tumor formation or tissue hyperplasia in the brain or other organs of RFR-exposed rats, relative to sham controls. Zook and Simmens (2001) also examined the effect of long-term exposure to 860 MHz pulse-modulated or CW RFR on tumor initiation and promotion in the brains of rats during a 2-year bioassay. ENU-treated and untreated rats were exposed to RFR for 6 h/day, 5 days/week from 2 up to 24 months of age. The brain-averaged SAR was calculated to be approximately 1 W/kg. At the end of the exposure period, all but two rats (898 in total) were necropsied and all major tissues were studied histopathologically. The authors reported no statistically significant differences in tumor incidence, volume, multiplicity, histological type, malignancy or fatality due to brain tumors from RFR exposure. Furthermore, there was no evidence of RFR-induced neoplasia in any other tissue examined. However, animals exposed to 860 MHz pulse-modulated RFR in the highest-dosed ENU group tended to develop fatal brain tumors at a shortened latency compared with the sham group. Although the trend was not statistically significant, the authors further examined this possible RFR-promoting effect in a more recent study (Zook and Simmens 2006). In this study, a similar experimental design was employed, however an equal number of RFR-exposed and control animals were sacrificed at 30-day intervals between 171 and 325 days of exposure. As with their previous study, histopathological analysis was performed on all rats (1,080 in total). The authors again found no statistically significant differences in tumor incidence, volume, multiplicity, histological type, malignancy or fatality associated with any type of neurogenic tumor. Furthermore, Zook and Simmens (2006) found no evidence of RFR-induced shortening of brain tumor latency in this study. The authors concluded that their studies found no evidence that RFR could induce a tumor-promoting effect.

Heikkinen et al. (2001) examined the tumor-promoting ability of RFR on tumors initiated with ionizing radiation. The authors exposed 200 female CBA/S mice to 4-Gy ionizing radiation (given in weekly 1.33-Gy fractions) followed by 902 MHz GSM-modulated or CW RFR at SARs ranging from 0.35–1.5 W/kg for 1.5 h/day, 5 days per week for 78 weeks. No significant differences were reported in the incidence of any neoplastic lesions in mice exposed to RFR, relative to the sham control group. In a later study, Heikkinen et al. (2003) examined the ability of RFR to influence the promotion of ultraviolet (UV) light induced skin tumorigenesis in transgenic ornithine decarboxylase overexpressing mice and in normal mice. In this study, mice were exposed to UV radiation in the presence/absence of 849 or 902 MHz pulse-modulated RFR for 1.5 h/day, 5 days/week for 52 weeks. The authors observed no statistically significant changes due to RFR exposure in any of the parameters examined, although a trend of accelerated tumor development was seen in both RFR groups. However, the authors had low animal numbers in all of their groups, ranging from eight in their non-transgenic cage control group to only 27 in their non-transgenic GSM-modulated RFR group, thereby reducing the statistical power of the study to detect differences between groups. Imaida et al. (2001) investigated the co-promoting ability of 1.5 GHz pulse-modulated RFR (whole body average SAR 0.08 W/kg; peak skin SAR 2 W/kg) in ICR mice initiated with DMBA. The animals were exposed to RFR for 90 min/day, 5 days/week for 19 weeks. At 20 weeks, all animals were killed and skin tumors were analyzed histopathologically. No evidence of RFR-induced skin tumor co-promotion was observed. Similarly, Huang et al. (2005) assessed skin tumor co-promotion in ICR mice exposed to either 849 or 1,763 MHz CDMA RFR (whole body average SAR 0.4 W/kg) for 2 × 45 min/day, 5 days/week for 19 weeks following DMBA initiation and found no evidence of RFR-induced skin tumor promotion.

Overall, the results of animal cancer studies relating to RFR exposure are overwhelmingly negative, but it is important to note that these types of animal studies have inherent limitations. Rodent strains react differently to restraint stress, food deprivation, acclimatization as well as to environmental conditions in various testing laboratories. This phenomenon is readily observable by comparing the low reproducibility in the rate of tumor incidence by known carcinogens in these animal studies, even when conducted within the same laboratory. Therefore, the limited number of studies with positive effects should not be dismissed entirely without additional replication studies.

28.7 Summary

Over the past 30 years, hundreds of studies have been conducted in laboratories around the world to evaluate the ability of non-thermalizing RFR to induce DNA strand breaks, chromosome aberrations, MN, SCE and cancer in a variety of animal and human systems. The vast majority of in vitro, in vivo and human RFR investigations conducted to date have found no consistent evidence of RFR-induced cytogenetic, genotoxic or carcinogenic effects. Of the limited number of in vitro studies reporting RFR-induced cytogenetic effects, thermal effects appear to be a possible confounding variable. Relatively few studies have assessed cytogenetic end points in animal and human tissue following in vivo RFR exposure and obvious methodological and/or statistical shortcomings are evident in many of these studies. The results of animal cancer studies relating to RFR exposure have reported a nearly uniform lack of association between RFR exposure and cancer initiation and/or promotion. However, current epidemiological studies of RFR have been hindered by the relatively recent introduction (within the last 10 years) of a wide variety of RFR-emitting devices and the long latency phase (more than 20 years) for the occurrence of most humans cancers. Therefore, despite the extensive body of experimental studies conducted to date, the issue surrounding the safety of RFR will likely remain a significant public concern for years to come. In the absence of conclusive epidemiology, future well-characterized prospective human and animal RFR cytogenetics studies would be beneficial to permit a more complete RFR health risk assessment.

References

Adey WR, Byus CV, Cain CD, Higgins RJ, Jones RA, Kean CJ, Kuster N, MacMurray A, Stagg RB, Zimmerman G, Phillips JL, Haggren W (1999) Spontaneous and nitrosourea-induced primary tumors of the central nervous system in Fischer 344 rats chronically exposed to 836 MHz modulated microwaves. Radiat Res 152:293–302

Adey WR, Byus CV, Cain CD, Higgins RJ, Jones RA, Kean CJ, Kuster N, MacMurray A, Stagg RB, Zimmerman G (2000) Spontaneous and nitrosourea-induced primary tumors of the central nervous system in Fischer 344 rats chronically exposed to frequency-modulated microwave fields. Cancer Res 60:1857–1863

Aitken RJ, Bennetts LE, Sawyer D, Wiklendt AM, King BV (2005) Impact of radio frequency electromagnetic radiation on DNA integrity in the male germline. Int J Androl 28:171–179

Anane R, Dulou P-E, Taxile M, Geffard M, Crespeau F-L, Veyret B (2003) Effects of GSM-900 microwaves on DMBA-induced mammary gland tumors in female Sprague-Dawley rats. Radiat Res 160:492–497

Anderson LE, Sheen DM, Wilson BW, Grumbein SL, Creim JA, Sasser LB (2004) Two-year chronic bioassay study of rats exposed to a 1.6 GHz radiofrequency signal. Radiat Res 162:201–210

Balcer-Kubiczek EK, Harrison GH (1985) Evidence for microwave carcinogenesis in vitro. Carcinogenesis 6:859–864

Balcer-Kubiczek EK, Harrison GH (1989) Induction of neoplastic transformation in C3H/10T1/2 cells by 2.45 GHz microwaves and phorbol ester. Radiat Res 117:531–537

Balcer-Kubiczek EK, Harrison GH (1991) Neoplastic transformation of C3H/10T1/2 cells following exposure to 120-Hz modulated 2.45 GHz microwaves and phorbol ester tumor promoter. Radiat Res 126:65–72

Bartsch H, Bartsch C, Seebald E, Deerberg F, Dietz K, Vollrath L, Mecke D (2002) Chronic exposure to a GSM-like signal (mobile phone) does not stimulate the development of DMBA-induced mammary tumors in rats: Results of three consecutive studies. Radiat Res 157:183–190

Beechey CV, Brooker D, Kowalczuk CI, Saunders RD, Searle AG (1986) Cytogenetic effects of microwave irradiation on male germ cells of the mouse. Int J Radiat Biol Phys Chem Med 50:909–918

Belyaev IY, Koch CB, Terenius O, Roxstrom-Lindquist K, Malmgren LOG, Sommer WH, Salford LG, Persson BRR (2006) Exposure of rat brain to 915 MHz GSM microwaves induces changes in gene expression but not double-stranded DNA breaks or effects on chromatin conformation. Bioelectromagnetics 27:295–306

Bisht KS, Moros EG, Straube WL, Baty JD, Roti Roti JL (2002) The effect of 835.62 MHz FDMA or 847.74 MHz CDMA modulated radiofrequency radiation on the induction of micronuclei in C3H 10T1/2 cells. Radiat Res 157:506–515

Brusick D, Albertini R, McRee D, Peterson D, Williams G, Hanawalt P, Preston J (1998) Genotoxicity of radiofrequency radiation. Environ Mol Mutagen 32:1–16

Cain CD, Thomas DL, Adey WR (1997) Focus formation of C3H10T1/2 cells and exposure to a 836.55 MHz modulated radiofrequency field. Bioelectromagnetics 18:237–243

Chou C-K, Swicord M (2003) Comment on "Exposure of human peripheral blood lymphocytes to electromagnetic fields associated with cellular phones leads to chromosomal instability, by Mashevich et al." Bioelectromagnetics 2003:24(2):82–90. Bioelectromagnetics 24:582

Chou C-K, Guy AW, Kunz LL, Johnson RB, Crowley JJ, Krupp JH (1992) Long-term, low-level microwave irradiation of rats. Bioelectromagnetics 13:469–496

d'Ambrosio G, Massa R, Scarfì MR, Zeni O (2002) Cytogenetic damage in human lymphocytes following GMSK phase modulated microwave exposure. Bioelectromagnetics 23:7–13

Diem E, Schwartz C, Adlkofer F, Jahn O, Rudiger H (2005) Non-thermal DNA breakage by mobile-phone radiation (1800 MHz) in human fibroblasts and in transformed GFSH-R17 rat granulosa cells in vitro. Mutat Res 583:178–183

Fahim MS, Fahim Z, Der R, Hall DG, Harman J (1975) Heat in male contraception (hot water 60 degrees C, infrared, microwave, and ultrasound). Contraception 11:549–562

Frei MR, Berger RE, Dusch SJ, Guel V, Jauchem JR, Merritt JH, Stedham MA (1998a) Chronic exposure of cancer-prone mice to low-level 2450 MHz radiofrequency radiation. Bioelectromagnetics 19:20–31

Frei MR, Jauchem JR, Dusch SJ, Merritt JH, Berger RE, Stedham MA (1998b) Chronic, low-level (1.0 W/kg) exposure of mice prone to mammary cancer to 2450 MHz microwaves. Radiat Res 150:568–576

Fucic A, Garaj-Vrhovac V, Skara M, Dimitrovic B (1992) X-rays, microwaves and vinyl chloride monomer: their clastogenic and aneugenic activity, using the micronucleus assay on human lymphocytes. Mutat Res 282:265–271

Gadhia PK, Shah T, Mistry A, Pithawala M, Tamakuwala D (2003) A preliminary study to assess possible chromosomal damage among users of digital mobile phones. Electromagn Biol Med 22:149–159

Garaj-Vrhovac V, Fucic A (1993) The rate of elimination of chromosomal aberrations after accidental exposure to microwave radiation. Bioelectrochem Bioenerg 30:319–325

Garson OM, McRobert TL, Campbell LJ, Hocking BA, Gordon I (1991) A chromosomal study of workers with long-term exposure to radio-frequency radiation. Med J Aust 155: 289–292

Görlitz B-D, Müller M, Ebert S, Hecker H, Kuster N, Dasenbrock C (2005) Effects of 1-week and 6-week exposure to GSM/DCS radiofrequency radiation on micronucleus formation in B6C3F1 mice. Radiat Res 164:431–439

Heikkinen P, Kosma V-M, Hongisto T, Huuskonen H, Hyysalo P, Komulainen H, Kumlin T, Lahtinen T, Lang S, Puranen L, Juutilanen J (2001) Effects of mobile phone radiation on x-ray-induced tumorigenesis in mice. Radiat Res 156:775–785

Heikkinen P, Kosma V-M, Alhonen L, Huuskonen H, Komulainen H, Kumlin T, Laitinen JT, Lang S, Puranen L, Juutilainen J (2003) Effects of mobile phone radiation on UV-induced skin tumorigenesis in ornithine decarboxylase transgenic and non-transgenic mice. Int J Radiat Biol 79:221–233

Hook GJ, Zhang P, LaGroye I, Li L, Higashikubo R, Moros EG, Straube WL, Pickard WF, Baty JD, Roti Roti JL (2004) Measurement of DNA damage and apoptosis in Molt-4 cells after in vitro exposure to radiofrequency radiation. Radiat Res 161:193–200

Huang T-Q, Lee J-S, Kim T-H, Pack J-K, Jang J-J, Seo J-S (2005) Effect of radiofrequency radiation exposure on mouse skin tumorigenesis initiated by 7, 12-dimethybenz[α]anthracene. Int J Radiat Biol 81:861–867

Imaida K, Kuzutani K, Wang J, Fujiwara O, Ogiso T, Kato K, Shirai T (2001) Lack of promotion of 7,12-dimethylbenz[a]anthracene-initiated mouse skin carcinogenesis by 1.5 GHz electromagnetic near fields. Carcinogenesis 22:1837–1841

International Commission on Non-Ionizing Radiation Protection (1998) Guidelines for limiting exposure to time-varying electric, magnetic, and electromagnetic fields (up to 300 GHz). Health Phys 74:494–522

Komatsubara Y, Hirose H, Sakurai T, Koyama S, Suzuki Y, Taki M, Miyakoshi J (2005) Effect of high-frequency electromagnetic fields with a wide range of SARs on chromosomal aberrations in murine m5S cells. Mutat Res 587:114–119

Koyama S, Nakahara T, Wake K, Taki M, Isozumi Y, Miyakoshi J (2003) Effects of high frequency electromagnetic fields on micronucleus formation in CHO-K1 cells. Mutat Res 541:81–89

LaGroye I, Anane R, Wettring BA, Moros EG, Straube WL, LaRegina M, Niehoff M, Pickard WF, Baty J, Roti Roti JL (2004a) Measurement of DNA damage after acute exposure to pulsed-wave 2450 MHz microwaves in rat brain cells by two alkaline comet assay methods. Int J Radiat Biol 80:11–20

LaGroye I, Hook GJ, Wettring BA, Baty JD, Moros EG, Straube WL, Roti Roti JL (2004b) Measurements of alkali-labile DNA damage and protein-DNA crosslinks after 2450 MHz microwave and low-dose gamma irradiation in vitro. Radiat Res 161:201–214

Lai H, Singh NP (1995) Acute low-intensity microwave exposure increases DNA single-strand breaks in rat brain cells. Bioelectromagnetics 16:207–210
Lai H, Singh NP (1996) Single- and double-strand breaks in rat brain cells after acute exposure to radiofrequency electromagnetic radiation. Int J Radiat Biol 69:513–521
Lai H, Singh NP (1997) Melatonin and a spin-trap compound block radiofrequency electromagnetic radiation-induced DNA strand breaks in rat brain cells. Bioelectromagnetics 18:446–454
Lalic H, Lekic A, Radosevic-Stasic (2001) Comparison of chromosome aberrations in peripheral blood lymphocytes from people occupationally exposed to ionizing and radiofrequency radiation. Acta Med Okayama 55:117–127
La Regina M, Moros EG, Pickard WF, Straube WL, Baty J, Roti Roti JL (2003) Effect of chronic exposure to 835.62 MHz FDMA or 847.74 MHz CDMA radiofrequency radiation on the incidence of spontaneous tumors in rats. Radiat Res 160:143–151
Li L, Bisht KS, LaGroye I, Zhang P, Straube WL, Moros EG, Roti Roti JL (2001) Measurement of DNA damage in mammalian cells exposed in vitro to radiofrequency fields at SARs of 3-5 W/kg. Radiat Res 156:328–332
Maes A, Collier M, Slaets D, Verschaeve L (1995) Cytogenetic effects of microwaves from mobile communication frequencies (954 MHz) Electro-Magnetobiol 14:91–98
Maes A, Collier M, Slaets D, Verschaeve L (1996) 954 MHz microwaves enhance the mutagenic properties of mitomycin C. Environ Mol Mutagen 28:496–501
Maes A, Collier M, Van Gorp U, Vandoninck S, Verschaeve L (1997) Cytogenetic effects of 935.2-MHz (GSM) microwaves alone and in combination with mitomycin C. Mutat Res 393:151–156
Maes A, Collier M, Verschaeve L (2001) Cytogenetic effects of 900 MHz (GSM) microwaves on human lymphocytes. Bioelectromagnetics 22:91–96
Maes A, Van Gorp U, Verschaeve L (2006) Cytogenetic investigation of subjects professionally exposed to radiofrequency radiation. Mutagenesis 21:139–142
Malyapa RS, Ahern EW, Straube WL, Moros EG, Pickard WF, Roti Roti JL (1997a) Measurement of DNA damage after exposure to 2450 MHz electromagnetic radiation. Radiat Res 148:608–617
Malyapa RS, Ahern EW, Straube WL, Moros EG, Pickard WF, Roti Roti JL (1997b) Measurement of DNA damage after exposure to electromagnetic radiation in the cellular phone communication frequency band (835.62 and 847.74 MHz). Radiat Res 148:618–627
Malyapa RS, Ahern EW, Bi C, Straube WL, LaRegina M, Pickard WF, Roti Roti JL (1998) DNA damage in rat brain cells after in vivo exposure to 2450 MHz electromagnetic radiation and various methods of euthanasia. Radiat Res 149:637–645
Manikowski-Czerska E, Czerski P, Leach WM (1985) Effects of 2.45 GHz microwaves on meiotic chromosomes of male CBA/CAY mice. J Hered 76:71–73
Mashevich M, Folkman D, Kesar A, Barbul A, Korenstein R, Jerby E, Avivi L (2003) Exposure of human peripheral blood lymphocytes to electromagnetic fields associated with cellular phones leads to chromosomal instability. Bioelectromagnetics 24:82–90
McNamee JP, Bellier PV, Gajda GB, Miller SM, Lemay EP, Lavallee BF, Marro L, Thansandote A (2002a) DNA damage and micronucleus induction in human leukocytes after acute in vitro exposure to a 1.9 GHz continuous-wave radiofrequency field. Radiat Res., 158:523-533
McNamee JP, Bellier PV, Gajda GB, Lavallee BF, Lemay EP, Marro L, Thansandote A (2002b) DNA damage in human leukocytes after acute in vitro exposure to a 1.9 GHz pulse-modulated radiofrequency field. Radiat Res 158:534–537
McNamee JP, Bellier PV, Gajda GB, Lavallee BF, Marro L, Lemay E, Thansandote A (2003) No evidence for genotoxic effects from 24 h exposure of human leukocytes to 1.9 GHz radiofrequency fields. Radiat Res 159:693–697
Meltz ML (2003) Radiofrequency exposure and mammalian cell toxicity, genotoxicity, and transformation. Bioelectromagnetics Suppl 6:S196–213
Moulder JE, Foster KR, Erdreich LS, McNamee JP (2005) Mobile phones, mobile phone base stations and cancer: a review. Int J Radiat Biol 81:189–203
Paulraj R, Behari J (2006) Single stranded DNA breaks in rat brain cells exposed to microwave radiation. Mutat Res 596:76–80

Phillips JL, Ivaschuk O, Isida-Jones T, Jones RA, Campbell-Beachler M, Haggren W (1998) DNA damage in Molt-4 T-lymphoblastoid cells exposed to cellular telephone radiofrequency fields. Bioelectrochem. Bioenergetics 45:103–110

Repacholi MH, Basten A, Gebski V, Noonan D, Finnie J, Harris AW (1997) Lymphomas in Eμ-*Pim1* transgenic mice exposed to pulsed 900 MHz electromagnetic fields. Radiat Res 147:631–640

Roti Roti JL, Malyapa RS, Bisht KS, Ahern EW, Moros EG, Pickard WF, Straube WL (2001) Neoplastic transformation in C3H10T1/2 cells after exposure to 835.62 MHz FDMA and 847.74 MHz CDMA radiations. Radiat Res 155:239–247

Royal Society of Canada (1999) A review of the potential health risks of radiofrequency fields from wireless telecommunication devices. An expert panel report prepared at the request of the Royal Society of Canada for Health Canada. http://www.rsc.ca//files/publications/expert_panels/RF//RFreport-en.pdf

Sakuma N, Komatsubara Y, Takeda H, Hirose H, Sekijima M, Nojima T, Miyakoshi J (2006) DNA strand breaks are not induced in human cells exposed to 2.1425 GHz band CW and W-CDMA modulated radiofrequency fields allocated to mobile radio base stations. Bioelectromagnetics 27:51–57

Sarker S, Ali S, Behari J (1994) Effect of low power microwave on the mouse genome: A direct DNA analysis. Mutat Res 320:141–147

Saunders RD, Kowalczuk CI, Beechey CV, Dunford R (1988) Studies of the induction of dominant lethals and translocations in male mice after chronic exposure to microwave radiation. Int J Radiat Biol Phys Chem Med 53:983–992

Shirai T, Kawabe M, Ichihara T, Fujiwara O, Taki M, Watanabe S, Wake K, Yamanaka Y, Imaida K, Asamoto M, Tamano S (2005) Chronic exposure to a 1.439 GHz electromagnetic field used for cellular phones does not promote N-ethylnitrosourea induced central nervous system tumors in F344 rats. Bioelectromagnetics 26:59–68

Sommer AM, Streckert J, Bitz AK, Hansen VW, Lerchl A (2004) No effects of GSM-modulated 900 MHz electromagnetic fields on survival rate and spontaneous development of lymphoma in female AKR/J mice. BMC Cancer 4:77

Stronati L, Testa A, Moquet J, Edwards A, Corelli E, Villani P, Marino C, Fresegna AM, Appolloni M, Lloyd D (2006) 935 MHz cellular phone radiation. An in vitro study of genotoxicity in human lymphocytes. Int J Radiat Biol 82:339–346

Sykes PJ, McCallum BD, Bangay MJ, Hooker AM, Morley AA (2001) Effect of exposure to 900 MHz radiofrequency radiation on intrachromosomal recombination in pKZ1 mice. Radiat Res 156:495–502

Tice RR, Hook GG, Donner M, McRee DI, Guy AW (2002) Genotoxicity of radiofrequency signals. I. Investigation of DNA damage and micronuclei induction in cultured human blood cells. Bioelectromagnetics 23:113–126

Toler JC, Shelton WW, Frei MR, Merritt JH, Stedham MA (1997) Long-term, low-level exposure of mice prone to mammary tumors to 435 MHz radiofrequency radiation. Radiat Res 148:227–234

Trosic I, Busljeta I, Kasuba V, Rozgaj R (2002) Micronucleus induction after whole-body microwave irradiation of rats. Mutat Res 521:73–79

Trosic I, Busljeta I, Modlic B (2004) Investigation of the genotoxic effect of microwave irradiation in rat bone marrow cells: in vivo exposure. Mutagenesis 19:361–364

Utteridge TD, Gebski V, Finnie JW, Vernon-Roberts B, Kuchel TR (2002) Long-term exposure of Eμ-*Pim1* transgenic mice to 898.4 MHz microwaves does not increase lymphoma incidence. Radiat Res 158:357–364

Verschaeve L (2005) Genetic effects of radiofrequency radiation (RFR). Toxicol Appl Pharmacol 207:S336–S341

Verschaeve L, Heikkinen P, Verheyen G, Van Gorp U, Boonen F, Vander Plaetse F, Maes A, Kumlin T, Maki-Paakkanen J, Puranen, Juutilainen J (2006) Investigation of co-genotoxic effects of radiofrequency electromagnetic fields in vivo. Radiat Res 165:598–607

Vijayalaxmi, Obe G (2004) Controversial cytogenetic observations in mammalian somatic cells exposed to radiofrequency radiation. Radiat Res 162:481–496

Vijayalaxmi, Frei MR, Dusch SJ, Guel V, Meltz ML, Jauchem JR (1997) Frequency of micronuclei in the peripheral blood and bone marrow of cancer-prone mice chronically exposed to 2450 MHz radiofrequency radiation. Radiat Res 147:495–500

Vijayalaxmi, Frei MR, Dusch SJ, Guel V, Meltz ML, Jauchem JR (1998) Correction of an error in calculation in the article "Frequency of micronuclei in the peripheral blood and bone marrow of cancer-prone mice chronically exposed to 2450 MHz radiofrequency radiation" [Radiat. Res. 147:495–500]. Radiat Res 149:308–312

Vijayalaxmi, Leal BZ, Szilagyi M, Prihoda TJ, Meltz ML (2000) Primary DNA damage in human blood lymphocytes exposed in vitro to 2450 MHz radiofrequency radiation. Radiat Res 153:479–486

Vijayalaxmi, Pickard WF, Bisht KS, Leal BZ, Meltz ML, Roti Roti JL (2001a) Cytogenetic studies in human blood lymphocytes exposed in vitro to radiofrequency radiation at a cellular telephone frequency (835.62 MHz, FDMA). Radiat Res 155:113–121

Vijayalaxmi, Bisht KS, Pickard WF, Meltz ML, Roti Roti JL, Moros EG (2001b) Chromosome damage and micronucleus formation in human blood lymphocytes exposed in vitro to radiofrequency radiation at a cellular telephone frequency (847.74 MHz, CDMA). Radiat Res 156:430–432

Vijayalaxmi, Pickard WF, Bisht HS, Prihoda TJ, Meltz ML, La Regina MC, Roti Roti JL, Straube WL, Moros EG (2001c) Micronuclei in the peripheral blood and bone marrow cells of rats exposed to 2450 MHz radiofrequency radiation. Int J Radiat Biol 77:1109–1115

Vijayalaxmi, Sasser LB, Morris JE, Wilson BW, Anderson LE (2003) Genotoxic potential of 1.6 GHz wireless communication signal: In vivo two-year bioassay. Radiat Res 159:558–564

Vijayalaxmi, McNamee JP, Scarfi MR (2006) Comments on: "DNA strand breaks" by Diem et al. [Mutat. Res. 583 (2005) 178-183] and Ivancsits et al. [Mutat. Res. 583 (2005) 184-188]. Mutat Res 603:104–106

Wang J, Sakurai T, Koyama S, Komatubara Y, Suzuki Y, Taki M, Miyakoshi J (2005a) Effects of 2450 MHz electromagnetic fields with a wide range of SARs on methylcholanthrene-induced transformation in C3H10T1/2 cells. J Radiat Res 46:351–361

Wang B, He J, Jin L, Lu D, Zheng W, Lou J, Deng H (2005b) Studying the synergistic damage effects induced by 1.8 GHz radiofrequency field radiation (RFR) with four chemical mutagens on human lymphocyte DNA using comet assay in vitro. Mutat Res 578:149–157

Wu RY, Chiang H, Shao BJ, Li NG, Fu YD (1994) Effects of 2.45 GHz microwave radiation and phorbol ester 12-O-tetradecanoylphorbol-13-acetate on dimethylhydrazine-induced colon cancer in mice. Bioelectromagnetics 15:531–538

Yu D, Shen Y, Kuster N, Fu Y, Chiang H (2006) Effects of 900 MHz GSM wireless communication signals on DMBA-induced mammary tumors in rats. Radiat Res 165:174–180

Zhang M-B, He J-L, Jin L-F, Lu D-Q (2002) Study of low-intensity 2450 MHz microwave exposure enhancing the genotoxic effects of mitomycin C using micronucleus test and comet assay in vitro. Biomed Environ Sci 15:283–290

Zeni O, Chiavoni AS, Sannino A, Antolini A, Forigo D, Bersani F, Scarfi MR (2003) Lack of genotoxic effects (micronucleus induction) in human lymphocytes exposed in vitro to 900 MHz electromagnetic fields. Radiat Res 160:152–158

Zeni O, Romano M, Perrotta A, Lioi MB, Barbieri R, d'Ambrosio G, Massa R, Scarfi MR (2005) Evaluation of genotoxic effects in human peripheral blood leukocytes following an acute in vitro exposure to 900 MHz radiofrequency fields. Bioelectromagnetics 26:258–265

Zook BC, Simmens SJ (2001) The effects of 860 MHz radiofrequency radiation on the induction or promotion of brain tumors and other neoplasms in rats. Radiat Res 155:572–583

Zook BC, Simmens SJ (2006) The effects of pulsed 860 MHz radiofrequency radiation on the promotion of neurogenic tumors in rats. Radiat Res 165:608–615

Zotti-Martelli L, Peccatori M, Scarpato R, Migliore L (2000) Induction of micronuclei in human lymphocytes exposed in vitro to microwave radiation. Mutat Res 472:51–58

Zotti-Martelli L, Peccatori M, Maggini V, Ballardin M, Barale R (2005) Individual responsiveness to induction of micronuclei in human lymphocytes after exposure in vitro to 1800-MHz microwave radiation. Mutat Res 582:42–52

29 Radiofrequency Radiation and Replication Studies

Maria Rosaria Scarfì and Ferdinando Bersani

Abstract

The induction of genotoxic effects after exposure to radiofrequency radiation represents one of the important aspects in the risk assessment of human exposure because of a close correlation between DNA damage and carcinogenesis. Most of the papers on this topic, published in peer-reviewed scientific journals, refer to experiments conducted by independent investigations who employed different experimental conditions, such as exposure characteristics, biological material examined as well as laboratory techniques. In particular, genotoxicity has been evaluated using several cytogenetic techniques, such as chromosomal aberrations, sister-chromatid exchanges, micronuclei and single-cell gel electrophoresis/comet assay. Just a few studies were "confirmatory," carried out adopting exactly the same experimental conditions. In this chapter the studies reported in the literature on the genotoxic potential of *in vivo* and *in vitro* exposure to radiofrequency radiation where several authors employed similar experimental conditions are compared and discussed.

29.1 Introduction

The aim of this chapter is to compare and discuss the studies reported in the literature where several authors employed similar experimental conditions to investigate the genotoxic potential of *in vivo* and *in vitro* exposure to radiofrequency radiation (RFR) using the classical (chromosomal aberrations, CA, sister-chromatid exchanges, SCE, and micronuclei) and more recent (single-cell gel electrophoresis/comet assay) cytogenetic techniques. Genotoxicity represents one of the important aspects in the risk assessment of human exposure to RFR because of a close correlation between DNA damage and carcinogenesis. Most of the *in vivo* and *in vitro* studies which have been published in peer-reviewed scientific journals were conducted by independent investigators. A small number of these were 'replicate' studies conducted under similar experimental conditions, while just a few of them were 'confirmatory' studies which adopted exactly the same experimental conditions. In all of these studies, several variables were included in the experimentation, viz.

Günter Obe and Vijayalaxmi (Eds.)
Chromosomal Alterations: Methods, Results and Importance in Human Health

RFR exposure characteristics, biological material examined as well as laboratory techniques. Moreover, not all researchers have provided detailed information on the experimental procedures used to enable independent investigators to replicate and confirm their observations. In some publications the experimental protocols were not adequately described. This makes it extremely difficult to evaluate whether or not the replicate and confirmatory studies were 'real' replications and confirmations.

29.2 DNA Strand Breaks

Lai and Singh applied the comet assay to evaluate DNA damage in brain cells of Sprague-Dawley rats which were exposed for 2 h to 2,450 MHz RFR in a cylindrical waveguide. The authors found no significant increase in DNA single-strand breaks when the alkaline comet assay was applied immediately after exposure to pulsed-wave (PW; 2 μs width, 500 pulses/s) RFR, while a specific absorption rate (SAR) dependent (0.6 and 1.2 W/kg) increase in single-strand breaks was observed both immediately and at 4 h after exposure to 2,450 MHz continuous wave (CW) RFR (Lai and Singh 1995). In a second paper the authors confirmed their previous results on single-strand breaks and also reported the increased DNA double-strand breaks at 4 h after exposure to both CW and PW RFR (Lai and Singh 1996). Malyapa et al. (1998) repeated the experiments of Lai and Singh by exposing male Sprague-Dawley rats to 2,450 MHz RFR for 2 h in a cylindrical waveguide system. The animals were killed immediately and at 4 h after RFR or sham exposure. Cells from cerebral cortex and hippocampus were used to perform the alkaline comet assay. The results did not indicate any significant differences in single-strand breaks between RFR-exposed and sham-exposed rats. As an extension, the latter authors also applied two methods of euthanasia, viz. CO_2 asphyxia or decapitation by guillotine. No significant increase in DNA strand breaks were observed between RFR-exposed and sham-exposed controls. Malyapa et al. (1997a) also performed in vitro experiments where cultured mammalian cells were exposed to RFR. Exponentially growing human glioblastoma U87MG cells and rodent C3H 10T1/2 cells were exposed for 2 h to 2,450 MHz RFR (CW) in a radial transmission line at a calculated SAR of 0.7 and 1.9 W/kg. The cells were examined immediately after RFR exposure as well as at 4 h after exposure. No significant differences in DNA single-strand breaks were observed between RFR-exposed and sham-exposed cells. Malyapa et al. (1997b) also investigated DNA strand breaks in the same cell types (in plateau phase or in exponential growth stage) which were exposed for 2, 4 or 24 h to 835.62 MHz frequency modulated CW (FMCW) RFR and 847.74 MHz code division multiple access (CDMA) RFR, at a SAR of 0.6 W/kg. No significant RFR exposure effect on DNA single-strand breaks was observed in both plateau phase cells and in exponentially growing cells.

It is important to consider here the differences in the comet assay protocols used by Lai and Sing (1995, 1996) and Malyapa et al. (1997a, 1998) in the above mentioned *in vivo* and *in vitro* studies. Malyapa et al. (1997a, 1998) used a protocol which did not include the treatment of the cells with proteinase K, while Lai and Singh (1995, 1996) included the proteinase K digestion of the cells which removes possible DNA–protein cross-links that hinder the movement of DNA strand breaks into the comet tail. To address this issue, Lagroye et al (2004) used both methods (with and without proteinase K treatment of the cells) and evaluated DNA strand breaks, alkali-labile lesions and DNA–protein or DNA–DNA cross-links in one cell type used by Malyapa et al. (1997b). C3H 10T1/2 cells were exposed for 2 h to 2,450 MHz CW RFR at a SAR of 1.9 W/kg. Cells treated with 1 mM cisplatin, a well-known DNA cross-link inducer, were included as positive controls. The results obtained indicated no significant differences in DNA damage between RFR-exposed and sham-exposed cells and confirmed the observations of Malyapa et al. (1997b). These findings also indicate that the slight differences in the protocols adopted to perform the comet assay were not responsible for the earlier contradictory results (Lagroye et al. 2004).

In other investigations, Phillips et al. (1998) evaluated the induction of DNA single-strand breaks in Molt-4 T-lymphoblastoid cells exposed for 2, 3 and 21 h to 813.5 MHz (Integrated Digital Enhanced Network, iDEN) and 836.55 MHz (time-division multiple access, TDMA) RFR at SAR values of 2.4 and 26 mW/kg. The exposures/sham exposures were conducted in transverse electromagnetic (TEM) cells under controlled conditions of temperature. The authors reported an increase, decrease or no significant effects in DNA damage, depending on the experimental parameters investigated (signal, SAR, exposure duration). The authors suggested that the observed effects could be due to a balance between RFR-induced damage and DNA repair. Recently, Hook et al. (2004) repeated the experiments of Phillips et al. (1998) using the same Molt-4 T-lymphoblastoid cells and the same RFR-exposure conditions in terms of frequency, signals, SAR values, exposure duration and biological assay. The induction of DNA damage was evaluated by means of the alkaline comet assay, performed immediately after RFR exposure as well as at 4 h after exposure. For all the experimental conditions tested, Hook et al. (2004) failed to confirm the observations of Phillips et al. (1998).

29.3 Micronucleus Test

Tice et al. (2002) used the micronucleus (MN) technique to determine the genotoxic potential in human blood lymphocytes (G_0, unstimulated) exposed in vitro to several RFR (837 and 1,909.8 MHz) signals, such as CW, global system for mobile communications (GSM), TDMA and CDMA. RFR exposures for 3 and 24 h were conducted in a TEM cell exposure system. No significant induction of MN was observed when the cells were exposed to RFR at a SAR below 1 W/kg; however, a consistent and reproducible increase in the incidence of MN was detected when the cells were exposed for 24 h a SAR of 5 or 10 W/kg. The increased incidence of MN was independent of the RFR signal employed. Tice et al. (2002) stated that the increased MN could be due to the localized heat production with consequent temperature elevation during RFR exposure.

Several other research groups examined the incidence of MN in human blood lymphocytes under conditions similar to those used by Tice et al (2002), and with appropriate temperature controls during RFR exposure. No significant increase in the incidence of MN (and also of CA) was observed when human blood lymphocytes were exposed for 24 h to 835.5 MHz frequency division multiple access (FDMA) RFR or 847.7 MHz (CDMA) RFR (Vijayalaxmi et al. 2001a, b). McNamee et al. (2002a, b, 2003) also reported similar 'negative' results in human blood lymphocytes exposed for 24 h to 1,900 MHz (CW, PW) at SAR values ranging from 0 to 10 W/kg. In a more recent study, Tice et al. (2004) could not confirm their original observations when some of their experimental conditions were modified, such as the exposure system (radial transmission line instead of a TEM cell), the containers for blood exposures (Petri dishes instead of tubes) and the number of donors employed in the study (three instead of one). In the latter study, Tice et al. (2004) also evaluated the effect of maintaining unstimulated whole-blood cultures at different temperatures (35, 39, 42°C) for 24 h before stimulation with phytohemagglutinin, with and without RFR exposure. No increase in the incidence of MN was observed under these experimental conditions. The latter data indicated that the increase in temperature might not be responsible for the genotoxic effects observed in their previous investigation. Bisht et al. (2002) confirmed the lack of MN induction in C3H 10T1/2 cells (exponentially growing and plateau-phase cells) exposed for 3, 8, 16 or 24 h to 835.5 MHz (FDMA) and 847.7 MHz (CDMA) RFR at SARs ranging from 3.2 to 5.1 W/kg. More recently, Scarfì et al. (2006) repeated the experiment of Tice et al (2002) using a wire patch cell with four channels and a double-blind procedure. The data indicated an absence of MN induction in human peripheral blood lymphocytes from ten healthy donors exposed for 24 h to 900 MHz (TDMA in GSM modulation) RFR at lower SAR levels (0, 1, 5 or 10 W/kg peak values).

29.4 Combination Studies – RFR and Other Mutagens

The possibility that RFR could enhance the genotoxic effect of well-known mutagens was investigated by several researchers. Kerbacher et al. (1990) evaluated the induction of CA in Chinese hamster ovary cells after 2-h exposure to 2,450 MHz RFR at a SAR of 33.8 W/kg. They also evaluated the effects of simultaneous exposures to RFR and two genotoxic chemicals, viz. mitomycin C (MMC) and adriamycin (ADR). The results indicated no significant induction of CA in cells exposed to RFR alone or in cells exposed to RFR plus MMC/ADR. Under the same experimental conditions, Ciaravino et al (1991) confirmed the absence of effects of RFR alone and in combination with ADR, by evaluating the incidence of SCE.

Maes et al. (1996) conducted combination experiments using RFR and MMC. In the first investigation, human whole-blood lymphocytes (G_0, unstimulated) were exposed for 2 h to 954 MHz RFR emitted by an antenna from a GSM base station. During the exposure samples were positioned 5 cm from the antenna and the power input was 15 W, with a calculated SAR of 1.5 W/kg. The authors reported no genotoxic effects in terms of SCE induction in RFR-exposed samples, but a clear and reproducible synergistic effect was detected when RFR-exposed samples were co-cultured with MMC. The authors hypothesized that there may exist a microwave-induced sensitization of RFR-exposed cells when they are treated with MMC. Maes et al (1997) repeated the experiments where human whole blood (G_0, unstimulated) was exposed for 2 h to 935.2 MHz RFR in a TEM transmission line that allowed a homogeneous radiation field. The power input was lower that in the previous experiments (4.5 W), with a SAR of 0.3–0.4 W/kg. The genotoxic effect of RFR was investigated by evaluating three cytogenetic end points: CA, SCEs and single-strand breaks. The combination effect of RFR was determined using SCE frequency. The authors confirmed the lack of effects due to RFR alone, while a slight increase in SCE frequency was detected in cells exposed to RFR plus MMC; the increase in SCE frequency was lower than that reported in the previous experiments (Maes et al. 1996) and the authors hypothesized that the difference in the two sets of experiments may be due to the mode of exposure (continuous vs. pseudorandom). In further experiments, Maes et al. (2001) evaluated the induction of genotoxicity in human lymphocytes exposed to 900 MHz RFR (GSM) with and without MMC treatment (SCE) and X-irradiation to determine the incidence of SCE and CA, respectively. Three modes of exposure were applied: continuous carrier wave, pseudorandom signal (corresponding to a GSM signal as obtained during conversation) and dummy burst signal (corresponding to the signal in standby mode). In the experiments devoted to evaluation of the effects of RFR alone, for each exposure condition the calculated SARs were 0.4, 2.0, 3.5, 5.5 and 10 W/kg. The possible synergistic effect with MMC or X-rays was examined at SAR values of 2.0 and 3.5 W/kg. In all the experimental conditions investigated neither RFR exposure nor co-exposures induced genotoxic effects.

More recently, Stronati et al. (2006) confirmed the absence of synergistic genotoxic effects in human peripheral blood lymphocytes (G_0, unstimulated) exposed for 24 h to 935 MHz RFR alone or coexposed with X-rays. The tests employed were alkaline comet assay, CA, MN and SCE. The experimental conditions adopted were similar but not identical to those tested by Maes et al. (2001). In particular, exposures to a 935 MHz GSM basic signal were carried out at SAR values of 1 or 2 W/kg, while 1 Gy/min was applied immediately before or after RFR.

29.5 Discussion

The effects of RFR on biological systems are generally divided into two categories, viz. thermal and nonthermal. While there is sufficient knowledge of thermal effects of RFR, the nature of alleged nonthermal effects and even their existence is still largely controversial. On the other hand, if RFR exposure has genotoxic potential, it could be considered more due to a nonthermal effect or at the edge of a thermal (athermal) effect. In this context, owing to the absence of convincing and comprehensive hypotheses about the biophysical mechanisms underlying the nonthermal effects of RFR exposure, it is difficult to conduct hypothesis-driven experiments. The choice of the physical parameters of the RFR exposures (frequency, amplitude, modulation, etc.) are dictated either by an arbitrary choice among the huge variety of combinations (we can name this approach as a 'fishing' approach) or by the physical characteristics of the RFR field signals more often used in daily-life applications, such as in wireless technology.

The possibility that RFR can affect DNA integrity has often been challenged on the physical consideration that RFR-associated photon energy is too low to break chemical bonds. Indeed, if real genotoxic effects could be demonstrated, the mechanisms underlying such effects could be due a more subtle though as yet unknown mechanism. For example, it could be due to an indirect effect of RFR on the enzymes involved in the repair of DNA damage or on other processes potentially amplified by the cell machinery. In the context of this low plausibility, the experimental evidence for a significant genotoxic effect of RFR should be sound and unquestionable. Besides, high-quality experiments together with a high level of reproducibility is required. Furthermore, independent replications of positive findings are more a cogent requirement than an academic exercise.

In general, scientists do not like to simply 'repeat' experiments conducted by other investigators, since this appears like a pure technical and uncreative activity; they prefer to extend the previous observations by the examination of other genotoxic end-points or, at least, to significantly improve some technical details. As described above, true 'replication' and 'confirmation' studies on the genotoxic effects of RFR are rare.

It is important to evaluate the replication experiments of Lai and Singh (1995, 1996) with respect to DNA strand breaks. The investigation of Malyapa et al. (1998) was a replication with slight differences in the comet assay protocol (inclusion of proteinase K) but included a methodological extension of the use of two different kinds of euthanasia. Perhaps a true replication study was carried out by Lagroye et al (2004) in which the laboratory protocols used by the earlier two research groups were compared. The data confirmed those of Malyapa et al (1998) and strongly suggest an absence of genotoxic effects of RFR exposure. The 'positive' observations of Lai and Singh (1995, 1996) could thus be due to unknown reasons and not RFR exposure. The *in vitro* studies of Malyapa (1997a, b) represent a sort of extension or a parallel confirmation of their *in vivo* experiments. Bisht et al. (2002) extended these *in vitro* observations of Malyapa et al (1998) which included RFR exposure at higher SAR levels and another genotoxic end point, viz. MN. The data from these studies again were 'negative'.

There were several 'partial replication' investigations (Vijayalaxmi et al. 2001a, b; McNamee et al. 2002a, b, 2003) which did not confirm the original 'positive' MN data in human blood lymphocytes exposed to RFR at 5 or 10 W/kg (Tice et al. 2002). This again points to 'potential differences' in RFR exposure conditions used in the study of Tice et al. (2002) or to other unknown reasons but not directly related to RFR exposure.

It is difficult to draw definitive conclusions from replication/confirmation experiments. If two series of independent experiments conducted using similar experimental conditions report contradictory results, the final conclusion remains uncertain. One of the ways to resolve this uncertainty would be to find a reasonable rationale for different outcomes, which is often difficult to achieve. At the moment, only a restricted number of experiments have been replicated. It is suggested that a meta-analysis should be performed with all of the available cytogenetic data to resolve the open question of the genotoxic potential of RFR exposure. It is also desirable to initiate an international collaborative investigation with the participation of experienced cytogeneticists.

References

Bisht KS, Moros EG, Straube WL, Baty JD, Roti Roti JL (2002) The effect of 835.62 MHz FDMA or 847.74 MHz CDMA modulated radiofrequency radiation on the induction of micronuclei in C3H 10T(1/2) cells. Radiat Res 157:506–515

Ciaravino V, Meltz ML, Erwin DN (1991) Absence of a synergistic effect between moderate-power radio-frequency electromagnetic radiation and Adriamycin on cell-cycle progression and sister-chromatid exchange. Bioelectromagnetics 12:289–298

Hook GJ, Zhang P, Lagroye I, Li L, Higashikubo R, Moros EG, Straube WL, Pickard WF, Baty JD, Roti Roti JL (2004) Measurement of DNA damage and apoptosis in Molt-4 cells after in vitro exposure to radiofrequency radiation. Radiat Res 161:193–200

Kerbacher JJ, Meltz ML, Erwin DN (1990) Influence of radiofrequency radiation on chromosome aberrations in CHO cells and its interaction with DNA-damaging agents. Radiat Res 123:311–319

Lagroye I, Hook GJ, Wettring BA, Baty JD, Moros EG, Straube WL, Roti Roti JL (2004) Measurements of alkali-labile DNA damage and protein-DNA crosslinks after 2450 MHz microwave and low-dose gamma irradiation in vitro. Radiat Res 161:201–214

Lai H, Singh NP (1995) Acute low-intensity microwave exposure increases DNA single strand breaks in rat brain cells. Bioelectromagnetics 16:207–210

Lai H, Singh NP (1996) Single- and double-strand DNA breaks in rat brain cells after acute exposure to radiofrequency electromagnetic radiation. Int J Radiat Biol 69:513–521

Maes A, Collier M, Slaets D, Verschaeve L (1996) 954 MHz microwaves enhance the mutagenic properties of mitomycin C. Environ Mol Mutagen 28:26–30

Maes A, Collier M, Van Gorp U, Vandoninck S, Verschaeve L (1997) Cytogenetic effects of 935.2-MHz (GSM) microwaves alone and in combination with mitomycin C. Mutat Res 393:151–156

Maes A, Collier M, Verschaeve L (2001) Cytogenetic effects of 900 MHz (GSM) microwaves on human lymphocytes. Bioelectromagnetics 22:91–96

Malyapa RS, Ahern EW, Straube WL, Moros EG, Pickard WF, Roti Roti JL (1997a) Measurement of DNA damage after exposure to 2450 MHz electromagnetic radiation. Radiat Res 148:608–617

Malyapa RS, Ahern EW, Straube WL, Moros EG, Pickard WF, Roti Roti JL (1997b) Measurement of DNA damage after exposure to electromagnetic radiation in the cellular phone communication frequency band (835.62 and 847.74 MHz). Radiat Res 148:618–627

Malyapa RS, Ahern EW, Bi C, Straube WL, La Regina M, Pickard WF, Roti Roti JL (1998) DNA damage in rat brain cells after in vivo exposure to 2450 MHz electromagnetic radiation and various methods of euthanasia. Radiat Res 149:637–645

McNamee JP, Bellier PV, Gajda GB, Miller SM, Lemay EP, Lavallee BF, Marro L, Thansandote A (2002a) DNA damage and micronucleus induction in human leukocytes after acute in vitro exposure to a 1.9 GHz continuous-wave radiofrequency field. Radiat Res 158:523–533

McNamee JP, Bellier PV, Gajda GB, Lavallee BF, Lemay EP, Marro L, Thansandote A (2002b) DNA damage in human leukocytes after in vitro exposure to a 1.9 GHz pulse-modulated radiofrequency field. Radiat Res 158:534–537

McNamee JP, Bellier PV, Gajda GB, Lavallee BF, Marro L, Lemay EP, Thansandote A (2003) No evidence for genotoxic effects from 24 h exposure of human leukocytes to 1.9 GHz radiofrequency fields. Radiat Res 159:693–697

Phillips JL, Ivaschuk O, Ishida-Jones T, Jones RA, Campbell-Beachler M (1998) DNA damage in Molt-4 T-lymphoblastoid cells exposed to cellular telephone radiofrequency fields in vitro. Bioelectrochem Bioenerg 45:103–110

Scarfì MR, Fresegna AM, Villani P, Pinto R, Marino C, Sarti M, Sannino A, Altavista P, Lovisolo GA (2006) Exposure to radiofrequency radiation (900 MHz, GSM signal) does not affect micronucleus frequency and cell proliferation in human peripheral blood lymphocytes. Radiat Res 165:655–663

Stronati L, Testa A, Moquet J, Edwards A, Cordelli E, Villani P, Marino C, Fresegna AM, Appolloni M, Lloyd D (2006) 935 MHz cellular phone radiation. An in vitro study of genotoxicity in human lymphocytes. Int J Radiat Biol 82:339–346

Tice RR, Hook GG, Donner M, McRee DI, Guy AW (2002) Genotoxicity of radiofrequency signals. I. Investigation of DNA damage and micronuclei induction in cultured human blood cells. Bioelectromagnetics 23:113–126

Tice RR, Baldetti C, Kehl M, Levesque B, Winters J, Roti Roti J, Moros E, Mueller J, Cha B (2004) Effect of exposure to radiofrequency radiation (837 MHz RF) emitted by a TDMA cellular telephone on the frequency of micronuclei in cultured human blood lymphocytes. Tutorial session at the 26th annual meeting of the Bioelectromagnetic Society on research related to the effect of radiofrequency energy on micronucleus formation, Washington, 25 June 2004

Vijayalaxmi, Pickard WF, Bisht KS, Leal BZ, Meltz ML, Roti Roti JL, Straube WL, Moros EG (2001a) Cytogenetic studies in human blood lymphocytes exposed in vitro to radiofrequency radiation at a cellular telephone frequency (835.62 MHz, FDMA). Radiat Res 155:113–121

Vijayalaxmi, Bisht KS, Pickard WF, Meltz ML, Roti Roti JL, Moros EG (2001b) Chromosome damage and micronucleus formation in human blood lymphocytes exposed in vitro to radiofrequency radiation at a cellular telephone frequency (847.74 MHz, CDMA). Radiat Res 156:430–432

30 Chromosome Analysis in Cancer Patients: Applications and Limitations

Siddharth G. Adhvaryu

Abstract

Since altered genetic mechanisms lead to the development of cancer, chromosome analysis plays a very important role in diagnosis and treatment-monitoring of patients with various types of cancer. The precise methodology involves several steps. Short-term cell culture, arrest of cell division in metaphase, hypotonic treatment, and cell harvest help in getting a cell suspension rich in metaphase cells. Slide preparation and staining protocols (G-banding) allow microscopic analysis and precise identification of the chromosome complement. Malignant growth involving the hematopoietic system is categorized as myeloid or lymphoid, depending upon the origin of the cells. The myeloid neoplasms are further subgrouped into myeloproliferative disorders, myelodysplastic syndromes, and acute myeloid leukemia (AML). In the leukemic cells of a chronic myeloid leukemia (CML) patient, the presence of t(9;22)(q34;q11.2) determines the diagnosis. Just like CML, very specific chromosome abnormalities help in diagnosis of different types of myeloid leukemias. The lymphoid neoplasms include acute lymphoblastic leukemia (ALL), chronic lymphoproliferative disorders, and lymphoma. The detection of t(12;21)(p13;q22) or a chromosome count of more than 50 chromosomes is considered a good prognosis in children with ALL. Similarly, the presence of a specific chromosome abnormality helps in the classification of and treatment for lymphoma. The information on chromosome abnormalities in solid tumors has also been gaining importance in recent years. To address the limitation of resolution and to bridge the gap between classical cytogenetic analysis and molecular diagnosis, techniques such as fluorescence in situ hybridization (FISH), comparative genomic hybridization, multiplex FISH, and spectral karyotyping are now available and are discussed briefly in this chapter.

30.1 Introduction

Cancer develops by the clonal expansion of a cell having altered genetic mechanisms. These genetic defects are restricted to the tumor tissue and are not present in other cells of the body. The defects may be very subtle at gene

Günter Obe and Vijayalaxmi (Eds.)
Chromosomal Alterations: Methods, Results and Importance in Human Health

level or they may be easily detectable as chromosomal changes. Both benign and malignant tumors have been shown to exhibit chromosome abnormalities. The chromosomal changes can be numerical or structural (Mitelman et al. 2006) and have been classified into two main categories: primary and secondary abnormalities. Primary abnormalities are specific or causative for a specific tumor type, may be the only cytogenetic abnormality, and may play an important role in the earliest stages of tumor initiation. Secondary abnormalities are also nonrandom, but less disease specific, and are postulated to be later events contributing to the process of tumor progression. In the clinical practice, cytogenetic analysis is an invaluable tool for diagnosis, staging, prognosis, and treatment monitoring of hematological malignancies and many solid tumors.

A deleted G group chromosome described as Philadelphia (Ph) chromosome by Nowell and Hungerford (1960) in chronic myeloid leukemia (CML) was the first chromosome abnormality in any cancer. The rapid advances, however, started following the discovery of banding techniques by Casperson et al. (1969) and Seabright (1971) and with the availability of high-resolution banding protocols described by Yunis (1974). Since then a new era in cancer cytogenetics and a parallel progress in the application of molecular genetic techniques have not only increased our understanding of the pathogenetic mechanisms underlying the neoplastic process, but have also led to development of targeted therapies for CML (Druker 2001).

30.2 Methods

The basic requirement for chromosome analysis in any cancer tissue is the need to have actively dividing cells. If sufficient mitotic activity is present in the specimen collected (e.g., bone marrow, peripheral blood with circulating blasts, or fresh tumor tissue) cells can be harvested after a 1-h colcemid treatment or overnight incubation in the culture medium containing colcemid followed by cell harvest can be performed. If sufficient mitotic activity is not present, a short-term or long-term cell culture procedure is employed for propagating the cells. In clinical laboratories various protocols are used to obtain the needed cytogenetic information. These protocols mainly depend upon the tissue to be processed. When the cells are actively dividing, the cell division is blocked at metaphase stage by adding spindle poisons (e.g., colchicine, colcemid, vinblastine). The cells are subjected to hypotonic treatment so as to swell the cell and make the cell membrane easy to rupture when the cell suspension is dropped on a glass slide and to the fixation with acetic acid–alcohol (1:3) mixture to stop the subsequent enzyme degradation. The cell pellet obtained is optimally diluted and the suspension is dropped on cleaned slides. The optimal quality depends upon the humidity and temperature and many laboratories use a mechanical dropping chamber for this purpose. Then the slides are subjected to a combination of time and temperature treatment for aging before subjecting them to one of the banding protocols. The most commonly used is G-banding, which includes trypsin treatment followed by one of the nuclear stains (Giemsa, Wright, or Leishman) to reveal the characteristic banding patterns (Seabright 1971; Sumner 1971). Usually 15–20 metaphase cells are completely analyzed preferably from more than one culture source. This analysis is the most important step and the technologist performing the microscope analysis must have sufficient technical experience in identifying various abnormalities and in selection of cells for analysis. In many cancers the abnormal cells are not well banded and care must be exercised in not excluding them from analysis. It is desirable to have the involvement of more than one technologist in finishing analysis on any case. The technologist finishing the microscope analysis decides after close review if there are any clonal abnormalities. In recent years, multiple metaphase cells have been imaged using computerized imaging systems. Use of a darkroom has become almost obsolete since the late 1990s. Two karyotypes of the main clone and one of each side clone are prepared. If the analysis yields a lack of any clonal abnormalities, a careful review of clinical information is done by the qualified cytogeneticist to determine if further tests are needed. A final report is submitted to the referring physician explaining the details of findings and the clinical significance of these findings. It is impossible to encompass the methodology and applications of chromosome alterations in cancer patients in a few pages. Those interested in learning the details will find useful protocols in Barch et al. (1997), Keagle and Gersen (1999), and Rooney (2001).

Owing to technical reasons of easy tissue sampling and to the availability of repeat specimens, most cytogenetic knowledge initially came from hematological disorders. In terms of human morbidity and mortality, solid tumors are very important and our knowledge of cytogenetic changes in solid tumors has increased in recent years. In the following sections the cytogenetic aberrations have been grouped under headings of hematological disorders and solid tumors.

30.3 Hematologic Cancers

The French–American–British (FAB) classification had been used most extensively till the recently adopted WHO system (Jaffe et al. 2001) became available. According to the origin of the cells, hematologic neoplasms are described as myeloid or lymphoid neoplasms. The myeloid neoplasms have four major subgroups: (1) chronic myeloproliferative disorders (CMPDs), (2) myeloproliferative/myelodysplastic disorders, (3) myelodysplastic syndromes (MDS), and (4) acute myeloid leukemia (AML). The data on the lymphoid neoplasms (incidence of acute leukemia in different age groups) which are summarized

Table 30.1 Age distribution of acute myeloid leukemia (*AML*) and acute lymphocytic leukemia (*ALL*)

	Age groups	AML	ALL
Acute leukemia	Children	15%	85%
	Adults	80%	20%
Annual incidence	Children	1.2:100,000	9.8:100,000
	Age 50	3.5:100,000	12.8:100,000
	Age 70	15:100,000	
	Age 90	35:100,000	

in Table 30.1 indicate that acute lymphocytic leukemia (ALL) is predominantly a pediatric disease, whereas AML is more common in adults.

30.3.1 Myeloid Neoplasms

CMPDs are clonal stem cell disorders resulting in elevated white cell counts in the blood. CMPDs occur primarily in adults during their 50s and 70s. The classic CMPD is CML. This was the first malignancy reported to have a chromosomal marker in the form of t(9;22)(q34;q11.2). This rearrangement results in the formation of BCR/ABL chimeric gene and is present in 100% of CML cases. In about 2–5% of these CML cases, the translocation may not be visible at the chromosome level. However, the BCR/ABL rearrangement is detectable at the molecular level by polymerase chain reaction (PCR) or by fluorescence in situ hybridization (FISH). Juxtaposition of ABL to BCR results in enhanced tyrosine kinase activity and this knowledge led to the effective treatment for CML using the drug imanitib, which is a very effective inhibitor of the hybrid BCR/ABL protein product (Druker 2001). During the accelerated phase or blast crisis of CML, additional abnormalities such as +8, +Ph, iso(17q), or +19 are detected either alone or in combination (Heim and Mitelman 1995). The detection of chromosome abnormality in the case of CML establishes a marker for monitoring the treatment effectiveness, presence of minimal residual disease, and early prediction of relapse or progression to the accelerated phase. For treatment-monitoring, molecular techniques such as reverse transcription PCR have proven to be much more sensitive. Nonetheless, chromosome analysis is still the preferred technique for detecting the acute phase of CML or blast crisis of CLM where additional cytogenetic markers [+8, +19, iso(17q), or +Ph] are detected well before the clinical manifestation. Unlike CML, there are no specific chromosomal changes associated with the other myeloproliferative disorders. A variety of nonspecific changes are indeed observed with some frequency. These include deletions of the long arms of chromosomes 13 and 20, monosomy 7, and trisomies 8 and 9. Information about chromosome abnormalities in CMPDs is presented in Table 30.2.

Table 30.2 Chromosomal rearrangements in myeloproliferative disorders

	CML	CNL	CEL	PV	CIMF	ET
Specific chromosome abnormality	t(9;22)	None	None	None	None	None
Other common nonspecific abnormalities	+8[a], +Ph, iso(17q), +19	+8, +9, del(20q), del(11q)	+8, t(5;12)	+8, +9, del(20q), del(13q)	+8, del(20q),-7,del(7q), del(11q), del(13q)	+8, del(13q)

CEL chronic eosinophilic leukemia, *CIMF* chronic idiopathic myelofibrosis, *CNL* chronic neutrophilic leukemia, *ET* essential thrombocythemia, *PV* polycythemia vera
[a]Only during the acute phase/blast crisis of CML

The MDS are a heterogeneous group of disorders showing clonal stem cell abnormalities leading to dysplasia and disturbed hematopoesis in one or more myeloid cell lineages. This may result in the presence of some myeloblast (below the 20% threshold limit). Primary MDS usually occurs in older individuals; however, secondary MDS can occur at an earlier age as a result of effective chemotherapies for various other types of cancers. Among these are refractory anemia (RA), RA with ringed sideroblasts, RA with multilineage dysplasia, RA with excess blasts, unclassified MDS, and MDS due to 5q deletion. The latter, commonly identified as "5q– syndrome" is more common in women. As the name suggests the deletion of 5q is the sole abnormality and is associated with a longer survival rate. Additional cytogenetic abnormalities usually indicate progression to AML. According to the cytogenetic abnormalities detected, MDS is grouped as (1) normal cytogenetics, isolated del(5q), del(20q), and –Y as "good risk," (2) chromosome 7 abnormalities or more than three recurring abnormalities as "poor risk," and (3) all other chromosome abnormalities as "intermediate risk."

When the number of myeloblasts exceeds 20%, the condition is labeled as AML. Under WHO classification, AMLs have been grouped as (1) AML with recurrent genetic abnormality, (2) AML with multilineage dysplasia, (3) therapy-related or secondary AML/MDS, and (4) AML not otherwise categorized. This WHO system is based on morphology, immunophenotypic, genetic, and clinical features; however, the reader will find the FAB classification (AML M0–M7) being commonly used in laboratory settings. There is no scientific conflict between the two systems and those working solely in the laboratory setting will be better off knowing both systems. Different classes of myeloid neoplasms indicating the characteristic chromosome abnormalities and related gene rearrangements are summarized in Table 30.3.

AML with translocation involving 11q23 (MLL gene rearrangement) have been considered indicative of an adverse outcome, whereas t(8;21), t(15;17), and inv(16) in AML is considered good prognosis. During the disease

Table 30.3 Chromosomal rearrangements in myeloproliferative disorders

Chromosomal abnormality	Gene/s involved	Neoplasm
t(8;21)(q22;q22)	*AML1-ETO*	AML-M2, other AML
inv(16),t(16;16)	*CBFB-MYH11*	AML-M4Eo, other AML
t(15;17)(q24;q21)	*PML-RARa*	AML-M3 (APL)
11q23 translocation	*MLL*	MLL rearranged AML

progression in about 75–80% of the patients, secondary chromosome abnormalities develop. A loss of chromosome 7 or 7q deletion, loss of chromosome 5 or 5q deletion, 11q23 abnormalities, or 12p deletions are often detected as additional abnormalities during progression of myeloid neoplasms. Some of these abnormalities have been related to the exposure to the chemotherapeutic agents. In general, the presence of multiple clonal abnormalities is indicative of a more aggressive disease.

Some specific chromosome abnormalities in myeloid malignancies are presented in Fig. 30.1.

30.3.2 Lymphoid Neoplasms

These include ALLs (when 30% or more lymphoblasts are detected in the bone marrow), chronic lymphoproliferative disorders (CLPD) and lymphoma. Depending on the involvement of the abnormal cell type, lymphoid neoplasms are of B-cell or T-cell origin. The most common recurrent genetic abnormalities in ALL are indicated in Table 30.4.

ALL is a malignant proliferation of immature lymphoid cells in the bone marrow and is usually present in the peripheral circulation. Peripheral blood studies reveal anemia, thrombocytopenia, and usually increased numbers of lymphoblasts. Chromosomes from ALL cases often exhibit poor quality and

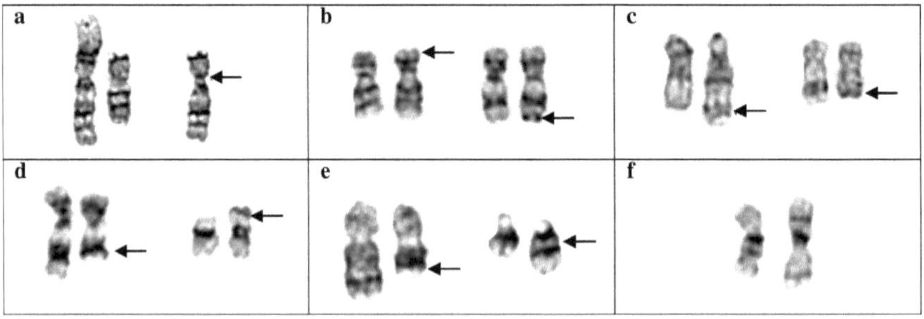

Fig. 30.1 Common abnormalities associated with acute myeloid leukemia. **a** Translocation (1;7)(q10;q10). **b** Translocation (9;11)(p21;q23). **c** Translocation (15;17)(q22;q11). **d** Translocation (11;19)(q23;p13). **e** Translocation (8;21)(q22;q22). **f** Inversion (16)(p13q22)

Table 30.4 Chromosomal rearrangements and genes involved in ALL

t(9;22)	*BCR-ABL1*	ALL
t(1;19)	*E2A-PBX1*	Pre-B ALL
t(12;21)	*TEL(ETV6)-AML1*	Common B ALL
t(4;11), 11(q23)	*MLL*	Infant ALL
t(8;14)	*c-MYC*	Burkitt/ALL-L3

poor staining, making them diagnostically challenging. Nevertheless, about two thirds of cases of ALL demonstrate recognizable cytogenetic changes, and the cytogenetic findings often have prognostic significance. Massive hyperdiploidy (50+ chromosomes), and especially with extra copies of chromosomes 4, 10, and 17 without structural anomalies, carries a good prognosis. Near-haploidy (a chromosome number of less than 30 and usually 23–28) is generally associated with a poor outcome. It is important not to overlook apparently broken metaphase spreads when performing the cytogenetic evaluation, as they could represent near-haploid cells. The translocation t(12;21)(p13;q22) is considered a good prognosis. This abnormality cannot be visualized cytogenetically; however, it can be easily confirmed with FISH probes. Another abnormality t(1;19)(q23;p13.3) is considered good prognosis with certain therapy regimens. Translocations involving a 14q11 breakpoint are seen with some frequency in T-cell ALL. Some examples are t(10;14)(q24;q11), t(11;14)(p13-15;q11), and t(8;14)(q24;q11). The T-cell receptor alpha (TCRA) locus resides at 14q11 and it appears that juxtaposition of it with a variety of oncogenes results in their activation. In general, T-cell ALL patients have a poorer prognosis than patients with B-cell ALL. The translocation t(9;22) (q34.1;q11) detected in ALL looks cytogenetically similar to that observed in CML; however, the breakpoint on chromosome 22 is different at the molecular level. This translocation is associated with an extremely poor prognosis and is seen more frequently as patient age increases. Any rearrangement involving 11q23 (MLL locus) seen in infants with ALL also carries a poor prognosis (Harrison 2001; Pui et al. 2004). When seen in adults, it is usually secondary to the treatment of their cancer with topoisomerase II inhibitors.

Among the CLPD, chronic lymphocytic leukemia (CLL) and multiple myeloma account for almost one third of all leukemias in the USA and Europe. CLL is a disease of middle-aged and elderly adults. The peripheral blood shows increased numbers of white blood cells, with mature lymphocytes predominating. Over 95% of CLLs are of B-cell origin; hence, the use of a B-cell mitogen is generally recommended for cell cultures. B-cell CLL has a good prognosis, whereas T-cell CLL carries a poor prognosis. CLL can transform, becoming an acute leukemia, most often ALL. Trisomy 12 and the translocation t(11;14)(q13;q32) are the most common abnormalities detected in B-cell CLL. Deletions involving the long arm of chromosome 13 and the short arm of chromosome 17 are fairly common and may require FISH analysis for detection.

The inversion, inv(14)(q11q32) is commonly seen in T-cell CLL as are other rearrangements involving q11 and q32 breakpoints on chromosome 14.

Multiple myeloma is a B-cell tumor in which there is a malignant proliferation of plasma cells that impairs bone marrow function. Multiple myeloma is also characterized by lytic bone lesions and impaired renal function. Plasma cells divide slowly and obviously care must be taken to improve the likelihood that the appropriate cells are being detected. FISH can be helpful by allowing large number of cells to be screened. Thirty to fifty percent of cases show cytogenetic changes including many numerical abnormalities, t(11;14)(q13;q32), 13q deletions, 6q deletions, or abnormalities of chromosome 1 (Harousseau et al. 2004).

Lymphomas can be broadly classifies as Hodgkin lymphoma and non-Hodgkin lymphoma. Hodgkin lymphoma is the most common lymphoma. Non-Hodgkin lymphoma is a complex group of neoplasms that over the years have been classified in numerous ways. Cytogenetic abnormalities are seen in 90% of cases, far greater than in any leukemia. The cytogenetics show highly specific and consistent cytogenetic changes; however, Burkitt lymphoma is an immunoglobulin-producing B-cell tumor. The Epstein–Barr virus might play a role in its development. Eighty percent of cases of Burkitt lymphoma show a specific translocation, t(8;14)(q24;q32), whereas the remaining 20% show involvement of 2p12 or 22q11.2, with 8q24 as variant translocations. These translocations juxtapose c-*myc* oncogene with the immunoglobulin gene promoter region, resulting in MYC activation. Secondary changes include abnormalities of the short arm of chromosome 1, seen in 30% of cases, and abnormalities involving a 13q34 breakpoint, seen in 15% of cases.

Some specific chromosome abnormalities in lymphoid malignancies are presented in Fig. 30.2.

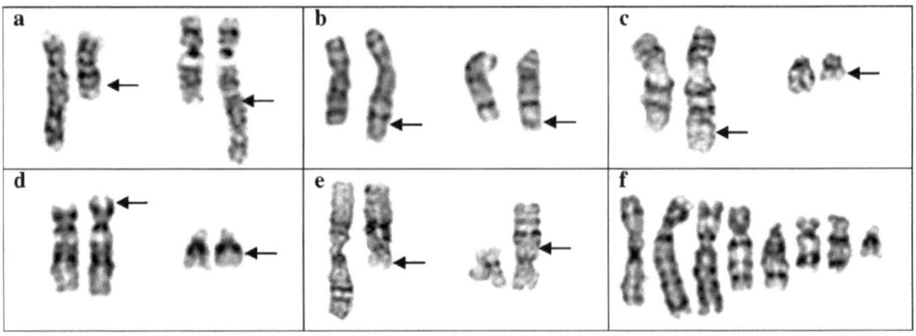

Fig. 30.2 Common abnormalities associated with lymphoma or lymphocytic leukemia. **a** Translocation (4;11)(q21;q23). **b** Translocation (8;14)(q24;q11). **c** Translocation (9;22)(q34;q11.2). **d** *Cryptic* translocation (12;21). The chromosomes pictured are from a case reported as normal in routine chromosome analysis but that was 95% positive for TEL/AML1 rearrangement by fluorescence in situ hybridization. This picture illustrates the difficulty in identifying this translocation in routine chromosome analysis. **e** Translocation (1;19)(q23;p13). **f** Common trisomies in hyperdiploidy (chromosomes X, 4, 6, 10, 14, 17, 18, and 21)

30.4 Solid Tumors

Compared with the details reported on chromosome analysis in hematological neoplasms, similar information on solid tumors is very limited. This is mainly due to reasons such as specimen collection requiring an operative procedure, symptoms only being noticeable when malignancy has already progressed, and repeated specimens not being available in most cases. Secondly, each solid tumor needs to be processed differently and may need different cell culture media, etc. In general, simple chromosome alterations are found in solid tumors of benign or initial malignant stages, whereas highly malignant tumors have more complex changes sometimes involving multiple abnormal cell lines. Important chromosomal aberrations, possible genetic alterations, and estimated prevalence in different solid tumor types are presented in Table 30.5.

Some specific chromosome abnormalities in solid tumors are presented in Fig. 30.3.

Cytogenetic analysis has become an integral part of the tumor diagnosis and can provide vital evidence for differential diagnosis. The (X;18)(p11;q11) is pathognomonic for synovial sarcoma and with the availability of FISH probes can help in making a diagnosis in the absence of dividing cells. The t(12;22)(q13;q12) is found as a characteristic abnormality in clear-cell sarcoma as well as in a minor subset of myxoid liposarcoma. In spite of recent advances in cytogenetic analysis of solid tumors, the impact on deciding prognosis is significantly less than for hematological neoplasms.

Table 30.5 Important chromosomal rearrangements in solid tumors

Tumor type	Translocation	Fusion transcript	Estimated prevalence (%)
Ewing sarcoma/peripheral neuroectodermal tumor	t(11;22)(q24;q12)	EWS-FL11	95
	t(21;22)(q22;q12)	EWS-ERG	5
Desmoplastic small round cell tumor	t(11;22)(p13;q12)	EWS-WT1	99
Myxoid/round-cell liposarcoma	t(12;16)(q13;p11)	TLS-CHOP	>95
	t(12;22)(q13;p12)	EWS-CHOP	<5
Extraskeletal myxoid chondrosarcoma	t(9;22)(q22;q12)	EWS-CHN	75
	t(9;17)(q22;q11)	TAF2N-CHN	25
Malignant melanoma of soft parts/ clear cell sarcoma	t(12;22)(q13;p12)	EWS-ATF1	>90
Synovial sarcoma	t(X;18)(p11.23;q11)	SYT-SSX1	65
	t(X;18)(p11.21;q11)	SYT-SSX2	35
Alveolar rhabdomyosarcoma	t(2;13)(q35;q14)	PAX3-FKHR	72
	t(1;13)(p36;q14)	PAX7-FKHR	9
Dermatofibrosarcoma protuberans	t(17;22)(q22;q13)	COL1A-PDGFB	>99
Congenital fibrosarcoma	t(12;15)(p13;q25)	ETV6-NTRK3	>99
Alveolar soft part sarcoma	t(X;17)(p11;q25)	ASPL-TFE3	>99

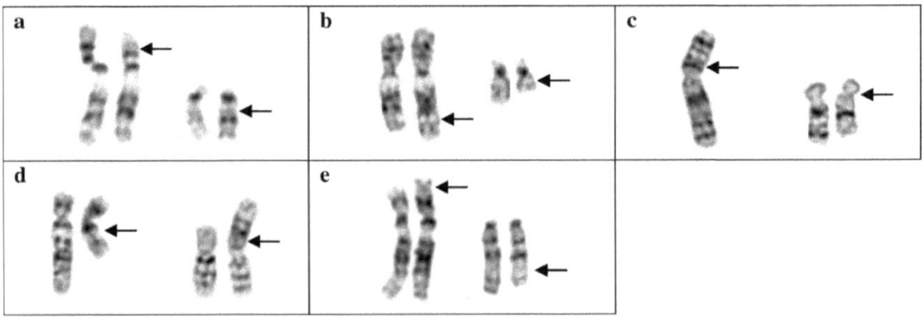

Fig. 30.3 Important chromosome abnormalities in solid tumors. **a** Translocation (11;22)(p13;q12) in desmoplastic round cell tumor. **b** Translocation (11;22)(q24;q12) in Ewing sarcoma. **c** Translocation (X;18)(p11.23;q11) in synovial sarcoma. **d** Translocation (12;16)(q13;p11) in liposarcoma. **e** Translocation (12;15)(p13;q25) in congenital fibrosarcoma and congenital mesoblastic nephroma

There are two major limitations of classical cytogenetic analysis: (1) it requires the presence of mitotic cells; (2) even in very good quality banding preparations, the resolution is very limited and it may not be possible to detect translocations such as t(12;21) in case of pre-B-cell ALL. At the 550-band level each band represents approximately 5×10^6–10×10^6 base pairs. The FISH technique provides a very useful advancement in overcoming resolution limitations of G-banded chromosome analysis. In recent years, techniques such as comparative genomic hybridization (CGH), multiplex FISH, and spectral karyotyping have become available and are briefly discussed next.

30.5 Molecular Cytogenetics

When conventional analysis fails to detect chromosomal defects with the use of banded chromosome spreads or fails to provide any mitotic activity, alternative techniques may reveal the presence of cytogenetic abnormalities or gene alterations commonly associated with them. These include molecular cytogenetics in the form of FISH (Pinkel et al. 1986) and the PCR, which, although plagued by their own limitations, circumvented some technical problems associated with routine cytogenetic analysis. These recent techniques have tremendously increased the sensitivity level for detection of genetic defects in cancer specimens (Fan 2002).

The FISH technique involves the hybridization of single-stranded DNA probe(s) to hybridize with homologous single-stranded sequence(s) in chromosomes of metaphase or interphase nuclei. Various types of FISH probes are available, such as whole-chromosome-specific or paint probes, centromere-specific or locus-specific. These probes are now available in different forms, such as dual color break apart probes, dual color single fusion, dual color dual fusion, and so on. Each type of probe has a specific ability and provides very specific answers about deletions or translocations or both (Fig. 30.4).

Probe Type	*Application*	*Signals*
LSI: Locus-Specific Identifier Dual-Color Break-Apart Rearrangement Probes	Useful for detecting multiple translocations involving a specific genetic breakpoint. Normal cells display a yellow fusion signal. A rearrangement will split the signal into a separate red and green signal	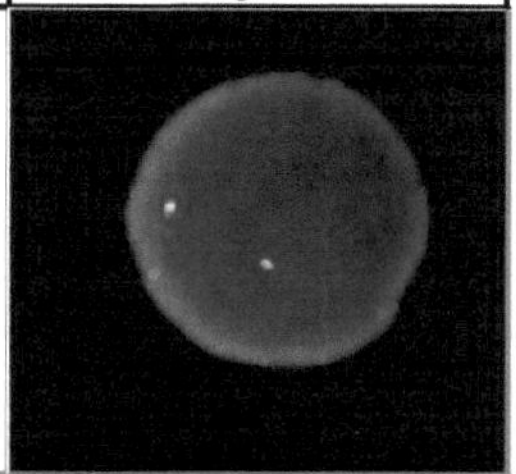
LSI: Locus-Specific Identifier Dual-Color, Single Fusion Probes.	Detects high percentages of cells possessing specific chromosomal translocations. Normal cells will display 2 red and 2 green signals. Translocations show as a single fusion signal.	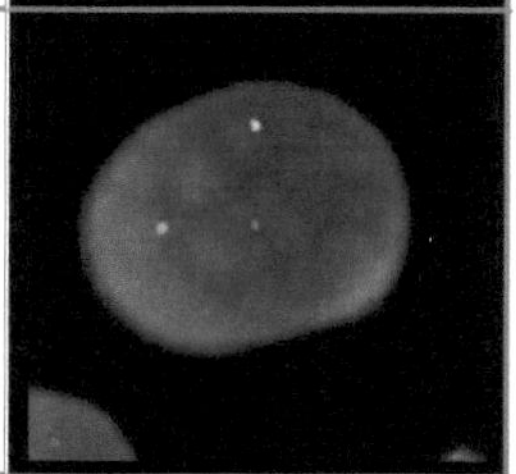
LSI: Locus-Specific Identifier Dual-Color, Dual Fusion Probes.	Detects low percentages of cells possessing specific chromosomal translocations. The dual fusion design greatly reduces false positive signals due to random co-localization of chromosomes. Translocations appear as 2 fusion signals. One on each derivative chromosome.	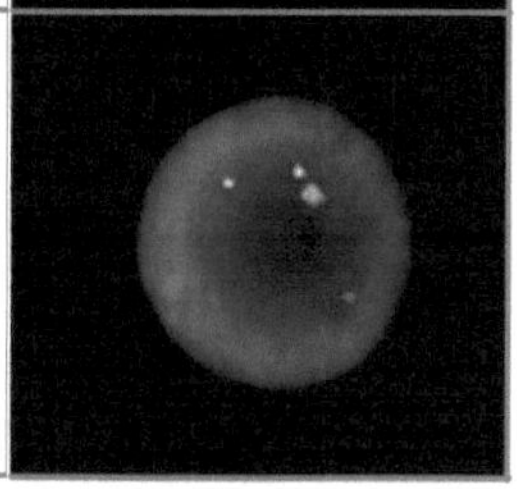

Probe Type	*Application*	*Signals*
CEP: Chromosome Enumeration Probe	Chromosome-specific probes which hybridize to satellite DNA sequences. Useful for enumeration of specific chromosomes in both interphase and metaphase cells.	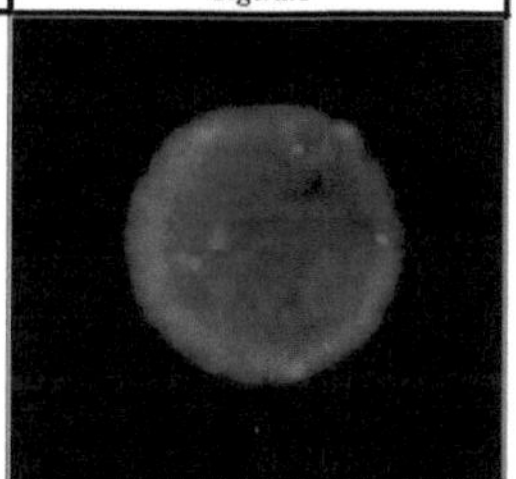
WCP: Whole Chromosome Paint Probes	Labels entire chromosome. The cell shows presence of t(9;22), t(9;11) and deletion of 17p in same cell with help of WCPs for 9, 11, 17 and 22. Useful for detection of complex or unknown translocations as well as identification of marker chromosomes.	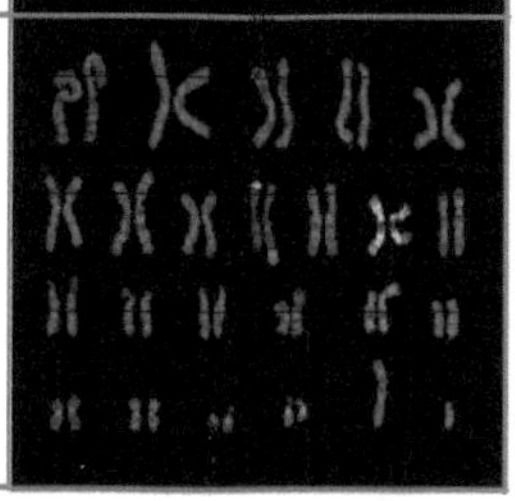

Fig. 30.4 Types of probes available and commonly used in molecular cytogenetics. Dual-color probes are labeled *red* (one locus) and *green* (second locus). When a translocation occurs the red and green signals come very close to each other on the derivative chromosome. This proximity (fusion) of red and green appears as a "*yellow*" signal under the fluorescence microscope. Normal is two red + two green, abnormal is one red + one green + one or two fusion signals on an interphase nucleus. In the break-apart probe the 3′ end of a locus is labeled green and the 5′ end of same locus is labeled red. So normal condition is a fusion (yellow) signal. When the locus is involved in translocation the 3′ and 5′ regions (green and red) separate and the signals on an abnormal interphase nucleus are one fusion (yellow), one green (3′), and one red (5′), separated owing to translocation. These probes are useful when the same locus is involved in translocation with several different chromosomes in causing clinically identical cancer

The molecular probes are fluorescently tagged either directly or by various "sandwich" methods and are detected with a fluorescence microscope. In this way, genetic alterations can be detected at the single-cell level, in a dividing or a nondividing cell, and correlated with morphology and immunophenotypic features of that cell. FISH is particularly advantageous in patients with acute leukemia in clinical remission. In this situation, cells carrying a patient's diagnostic cytogenetic aberration can be readily enumerated by counting fluorescence signals or hybrid (merged) signals when using translocation-specific probes. The major limitation of the FISH technique is that only those aberrations that are targeted by the probes used will be detected. In other words one can get an answer only specific to the question, whereas routine chromosome analysis provides visualization of all the abnormalities in the cell under the microscope. FISH is therefore not recommended for establishing an aberrant karyotype at the time of diagnosis, unless the clinical presentation is suggestive of a specific cytogenetic abnormality [e.g., abnormal eosinophilia suggesting inv(16)]. Although excellent for monitoring the t(9;22) in CML, for instance, when monitoring disappearance of the translocation under imatinib mesylate therapy, FISH will not detect the various cytogenetic abnormalities (+8, iso17q, +19, or +Ph) associated with blast crisis of CML.

The use of FISH technology involves tedious validation, quality assurance, and quality control issues as well as the availability of a fluorescence microscope and the imaging system. However, in addition to aiding in detecting a known abnormality on an interphase nucleus, the FISH technique helps very

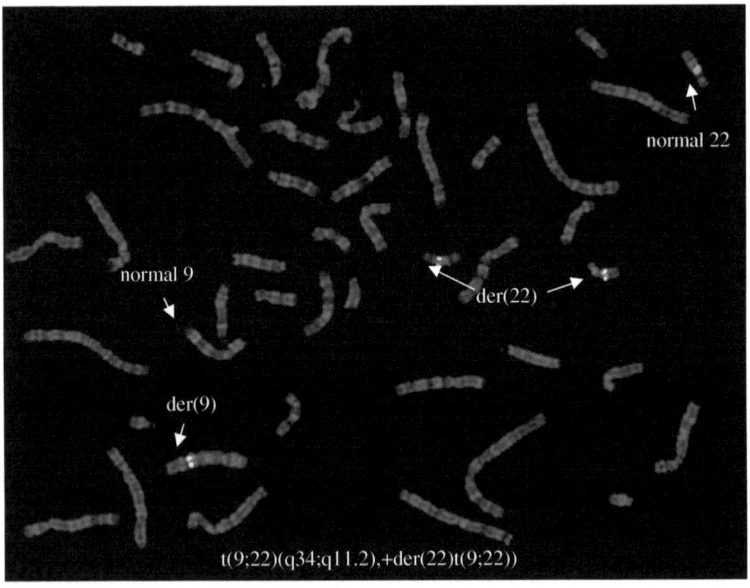

Fig. 30.5 Metaphase fluorescence in situ hybridization. A translocation (9;22) with an extra copy of the Philadelphia chromosome [*der(22)*]

effectively in characterization of complex abnormalities and in identification of the marker chromosomes. It has also proven very useful in the detection of submicroscopic deletions (RB1, p53), amplifications (e.g., MLL, MYC, HER2), or masked rearrangements (Fig. 30.5).

In addition to the FISH protocols briefly described here, other FISH-based protocols are being used in various situations. These include metaphase CGH, which can identify chromosomal losses or gains in tumor cells (Kallioniemi et al. 1992), multiplex FISH (Speicher et al. 1996), spectra karyotyping (Schrock et al. 1996), or array CGH (Schoch et al. 2004). Similarly

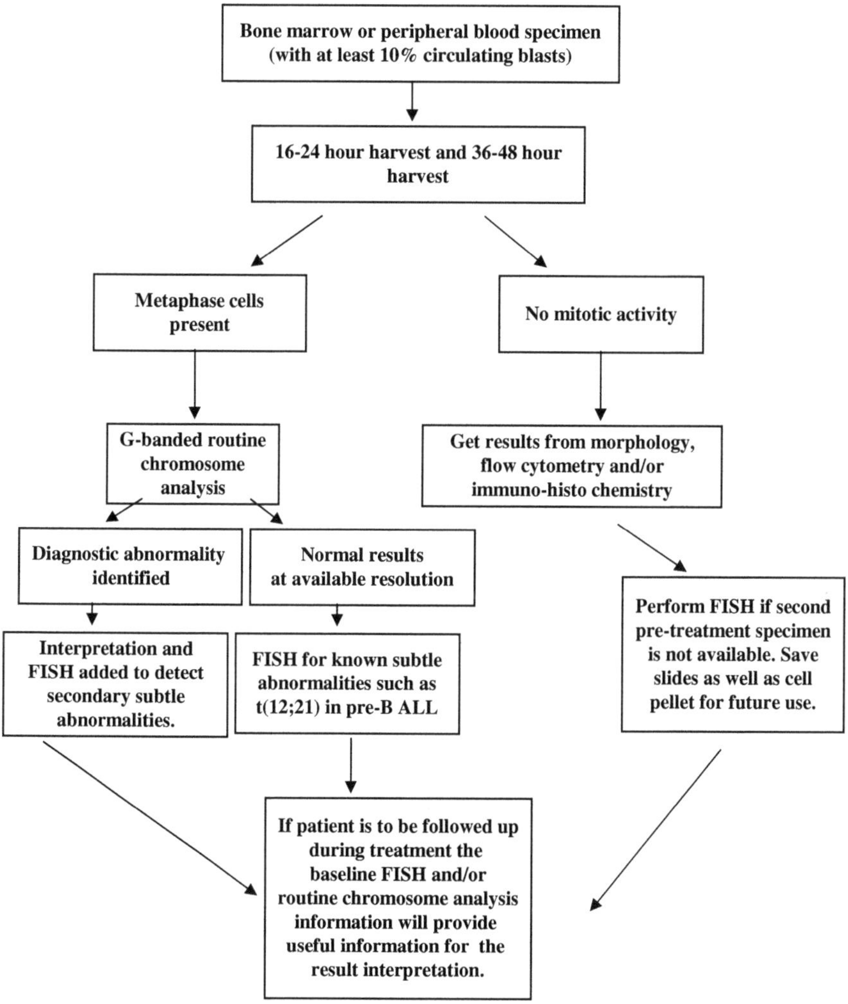

Fig. 30.6 Summary of useful strategies during initial diagnosis and subsequent treatment-monitoring of a patient. *FISH* fluorescence in situ hybridization, *ALL* acute lymphocytic leukemia

evolving newer techniques are coming into the clinical arena very rapidly. An emerging technology that addresses the limitations of metaphase CGH is microarray CGH. This technology offers multiple platforms and has revolutionized the field of conventional cytogenetics. Those readers interested in learning more of these newer technologies should refer to Fan (2002) and other recent review articles on these topics.

30.6 Future for Cytogenetics in Cancer Investigations

With increasing knowledge about the specific gene rearrangements and identification of abnormal proteins responsible for the neoplastic disease, more successful therapies will be available in the near future. This shall change the role of the cytogenetic laboratory from the primary diagnostic laboratory to one which helps in effective disease management. This will be possible by deciding on remission, presence of minimal residual disease, and early prediction of disease progression using routine and molecular cytogenetic techniques. The flow chart in Fig. 30.6 summarizes the strategy the laboratory may adopt during initial diagnosis and future monitoring of repeat specimens from the same patients.

Acknowledgement: Excellent technical help of Kenneth A. Hughes in the preparation of the figures is sincerely appreciated.

References

Barch MJ, Knutsen T, Spurbeck JL (1997) The AGT cytogenetics laboratory manual, 3rd edn Lippincott-Raven, New York

Casperson T, Zech L, Johansson C (1970) Differential binding of alkylating fluorochromes in human chromosomes. Exp Cell Res 60:315–319

Druker BJ, Talpaz M, Resta DJ et al (2001) Efficacy and safety of a specific inhibitor of the BCR-ABL tyrosine kinase in chronic myeloid leukemia. N Eng J Med 344:1031–1037

Fan Y-S (ed) (2002) Molecular cytogenetics, protocols and applications. Humana, Totowa

Harousseau J-L, Shaughnessy J Jr, Richardson P (2004) Multiple myeloma. Hematol Am Soc Hematol Educ Program 237–256

Harrison CJ (2001) Acute lymphoblastic leukemia. Best Practice Research Clin Haematol 14:593–607

Heim S, Mitelman F (1995) Cancer cytogenetics, 2nd edn. Wiley–Liss, New York

Jaffe ES, Harris NL, Stein H, Vardiman JW (eds) (2001) World Health Organization classification of tumours. Pathology and genetics of tumours of haematopoietic and lymphoid tissue. IARC, Lyons

Kallioniemi A, Kallioniemi O-P, Sudar D et al (1992) Comparative genomic hybridization for molecular cytogenetic analysis of solid tumors. Science 258:818–821

Keagle MB, Gersen SL (1999) Basic laboratory procedures. In: Gersen SL, Keagle MB (eds) The principles of clinical cytogenetics. Humana, Totowa, pp 71–89

Mitelman F, Johansson B, Mertens F (eds) (2006) Mitelman database of chromosome aberrations in cancer. http://cgap.nci.nih.gov/Chromosomes/Mitelman
Nowell PC, Hungerford DA (1960) A minute chromosome in human chronic granulocytic leukemia. Science 132:1497
Pinkel D, Gray J, Trask B, van den Engh G, Fusco J, van Dekken H (1986) Cytogenetic analysis by in situ hybridization with fluorescently labeled nucleic acid probes. Cold Spring Harbor Symp Quant Biol 51:151–157
Pui C-H, Schrappe M, Ribeiro RC, Niemeyer CM (2004) Childhood and adolescent lymphoid and myeloid leukemia. Hematol Am Soc Hematol Educ Program 118–145
Rooney DE (2001) Human cytogenetics: constitutional analysis, 3rd edn. Oxford University Press, Oxford
Schoch C, Dugas M, Kern W, Kohlmann A, Schnittger S, Haferlach T (2004) "Deep insight" into microarray technology. Atlas Genet Cytogenet Oncol Haematol. http://AtlasGeneticsOncology.org/Deep/MicroarraysID20045.html
Seabright MA (1971) Rapid banding technique for human chromosomes. Lancet 2:971–972
Schrock E, du Manoir S, Veldman T, Schoell B, Wienberg J, Ferguson-Smith MA, Ning Y, Ledbetter DH, Bar-Am I, Soenksen D, Garini Y, Ried T (1996) Multicolor spectral karyotyping of human chromosomes. Science 273:494–497
Speicher MR, Ballard SG, Ward DC (1996) Karyotyping human chromosomes by combinatorial multiflour FISH. Nat Genet 12:368–375
Sumner AT, Evans HJ, Buckland RA (1971) New technique for distinguishing between human chromosomes. Nat New Biol 232:31–32
Yunis J (1976) High resolution of human chromosomes. Science 191:1268–1270

31 Chromosomal Aberration in Peripheral Blood Lymphocytes of Healthy Subjects and Risk of Cancer

STEFANO BONASSI

Abstract

The pioneering papers of Theodor Boveri opened the way to extensive research on the mechanism linking chromosomal abnormalities to the pathogenesis of cancer. As a result of this effort, robust theoretical and empirical evidence accumulated, and an increased cancer risk was postulated in healthy subjects with high levels of chromosomal aberrations (CA). The first cohort study which directly evaluated the presence of this association was carried out within the framework of a collaborative study in northern Europe in 1994. Subjects classified as having high frequency of CA showed incidence rates that were more twice those of the general population. The results of this first study were replicated a few years later in a parallel research initiative carried out in Italy, and the subsequent pooled analysis of these two cohorts published in 1998 refined the quantitative estimate of the CA–cancer association. More recent results referring to cohorts in the Czech Republic (2005) and in five European countries (2007) contributed further evidence, suggesting a stronger association with chromosome-type aberrations. A pooled analysis of all cohort studies published involving more that 22,000 subjects from 11 countries has been carried out within the framework of the European project Cytogenetic Biomarkers and Human Cancer Risk, and preliminary results are discussed. The leading priorities to be addressed in the field are the application of existing evidence to the control of occupational and environmental exposure to genotoxic agents, and the evaluation of individual cancer risk assessment procedures based on markers of chromosome damage.

31.1 Biological Rationale

The first consistent theory postulating the importance of chromosomes for cancer was developed at the turn of the twentieth century by Theodore Boveri (1914), who elaborated the idea that loss or gain of chromosomes by aberrant mitoses may lead to cancer. This pioneering view was formalized nearly 50 years later, in the early 1960s, with the description of the first specific chromosome abnormality in a human primary tumor, i.e., the Philadelphia

Günter Obe and Vijayalaxmi (Eds.)
Chromosomal Alterations: Methods, Results and Importance in Human Health

chromosome (Novell and Hungerford 1960), and this was substantiated in the 1970s with the introduction of banding techniques.

During the 1980s, a number of new cancer-specific chromosomal rearrangements (many of them translocations) mostly occurring in hematological tumors were found. These alterations allowed a deeper insight into mechanistic aspects; in particular it was clear that many genes affected by chromosomal rearrangements were involved in some critical stages in cell growth, development, or survival. These results led to a better understanding of the mechanisms of formation of chromosomal alterations and their role in cancer development (Rowley 1998).

The model linking chromosome breakage to the activation/deactivation of genes relevant to early stages of carcinogenesis was further refined in the 1990s by the use of the fluorescence in situ hybridization technique. These studies contributed to the understanding of the role of DNA double-strand breaks (DSBs) in the formation of chromosomal aberrations (CA). A further refinement in the understanding of the origin of CA was provided by a number of papers published at the turn of the millennium, which revealed the major role of homology-dependent and homology-independent pathways of DSB repair (Richardson and Jasin 2000; Pfeiffer et al. 2000).

The clastogenic action of many environmental agents has well been documented and their role in turning on the process leading from DSB to altered function of cancer genes is generally recognized. It is also interesting to note that many DSBs are formed endogenously as a consequence of physiologic processes, such as replication, meiosis, V(D)J recombination, and DNA repair (Pfeiffer et al. 2000; Vilenchik and Knudson 2003). Although endogenous DSBs are repaired with high fidelity, errors in DSB repair are considered to contribute significantly to human cancer rate (Vilenchik and Knudson 2003).

The role of chromosomal alterations in the early stages of carcinogenesis provides support to observations, from both laboratory and human studies, linking the clastogenic effect of many compounds to their carcinogenic properties (Bonassi 1999). Furthermore, the availability of a reliable model of causality supports the biological plausibility of studies linking chromosomal damage to the risk of cancer, and is the driving force that justifies the use of the CA assay for cancer-prevention policies.

31.2 The Epidemiologic Approach

The identification of an accurate mechanistic model corroborates the presence of association between the frequency of CA in healthy subjects and the risk of cancer; however, the role of chromosome damage in carcinogenesis is substantiated by more evidence. The high frequencies of CA in patients affected by cancer, the parallel increase of CA and risk of cancer in subjects affected by genetic diseases interfering with DNA repair, and the already mentioned association between clastogenic and carcinogenic properties are the most convincing.

All these observations, despite their importance, provide only circumstantial evidence for a causal relationship and cannot provide any real data about the strength of the association between CA and cancer in the general population, which would be required from a public health perspective. The availability of risk estimates based on the frequency of a biomarker in healthy individuals is a prerequisite to validate the use of CA and other potential surrogate markers for cancer-prevention purposes. This approach is particularly appropriate when the biomarker studied is possibly affected by the presence of the disease, i.e., the so-called *reverse causality*; in this case prospective cohort studies are preferable for the biomarker validation (Rothman 1986).

Major problems in planning a cohort study based on biomarker frequency are the small size of most biomonitoring studies, the heterogeneity of technical protocols among laboratories and the differences in methods over the years, the high number of subjects without demographic data, and the easy misclassification due to the use of only a single (or a few) measure(s) in their life. A cohort approach is particularly problematic for those biomarkers, such as cytogenetic biomarkers, which need fresh tissues and therefore cannot be easily measured in frozen samples.

Most published cohort studies linking CA frequency to the risk of cancer adopted the methodology of the historical cohort (Rothman 1986).

31.3 Literature Review of Cohort Studies Linking CA Frequency to the Risk of Cancer

The first cohort study on healthy subjects screened for CA (but also for sister-chromatid exchanges, SCE, and micronuclei frequency) which had enough statistical power to evaluate the presence of a causal link was published in 1994 as a result of a collaborative initiative carried out in northern Europe (Hagmar et al. 1994). A group of 3,182 subjects from Finland, Sweden, Norway, and Denmark, free of cancer at the time of cytogenetic analysis, were followed up from 1970 to 1991. Eighty-five cancer cases were found. In order to standardize for the interlaboratory variation, the results were trichotomized for each laboratory into three strata: low (1–33 percentile), medium (34–66 percentile), or high (67–100 percentile). Subjects classified as having a high frequency of CA showed incidence rates that were more than twice those of the general population (standardized incidence ratio, SIR 2.1, 95% confidence interval, CI, 1.5–2.8), and a highly significant trend of SIRs by the frequency of CA was observed ($p = 0.0009$).

These findings were confirmed 1 year later in an independent cohort of 1,455 subjects gathered from 17 Italian laboratories and followed up from 1969 to 1994 for a total of 16,190 person-years. A significant increase in mortality rate for all cancers was found in subjects with high and medium frequency of CA (standardized mortality ratio, SMR, 1.79, 95% CI 1.02–2.90; and SMR 1.82, 95% CI 1.11–2.81, respectively) (Bonassi et al. 1995). In 1998 a first collaborative European study, pooling the cohorts from Nordic countries and Italy, produced a joint data analysis and a further extension of the follow-up. The results strengthened the confidence in the previous findings, while the survival analysis improved the understanding of the CA–cancer relationship, showing highly different patterns of survival in the three CA frequency categories (Hagmar et al. 1998). To investigate the possible effect of smoking habit and occupational exposures to carcinogens as confounders or effect modifiers of the relationship evaluated, a case-control study nested within the cohort was performed. The results showed that CA frequencies could predict cancer independently of exposure to major carcinogens, making the point that the predictivity of high rates of CA does not require such exposures (Bonassi et al. 2000). The first non-European report showing an association between CA and cancer risk was published in 1999 by a Taiwanese group (Liou et al. 1999). A case-control study nested in a small cohort of 686 residents in a blackfoot endemic area showed that a frequency of chromosome-type CA over the median is highly predictive of cancer risk (odds ratio 12.0, 95% CI 1.56–92.29). No association was found with chromatid-type CA and SCE.

A new independent cohort study was carried out in the Czech Republic, based on data from a large national program of occupational safety surveillance ongoing since 1975. Various groups of workers with heterogeneous exposures had been investigated. Overall, 11,834 subjects were followed up for cancer incidence and mortality from 1975 to 2000, accounting for a total of 113,967 person-years (Rossner et al. 2005). A significant association was found between the overall cancer incidence and the presence of chromosome-type aberrations (chromosome-type breaks, ring chromosomes, marker chromosomes, and dicentrics) (relative risk, RR, for high vs. low CA frequency 1.24, 95% CI 1.03–1.50), but not chromatid-type aberrations (chromatid-type breaks and chromatid exchanges). Stomach cancer showed a strong association with the frequency of total CA (RR 7.79, 95% CI 1.01–60.0). The predictivity of CA observed in subjects exposed to various classes of carcinogens did not significantly differ from that for the group of nonexposed subjects.

The issue whether subclasses of CA are more effective in predicting cancer incidence was specifically addressed in a paper of the Nordic–Italian cohorts, showing no evidence of a different predictivity (Hagmar et al. 2004). This result was contradicted by the latest European cohort study, which evaluated a cohort of 6,430 individuals from nine laboratories in Croatia, Hungary, Lithuania, Poland, and Slovakia, and included CA surveys performed during 1978–2002. In this study 224 cancer cases were observed. The RR of cancer was 1.57 (95% CI 1.08, 2.28) for a medium CA level and 1.75 (95% CI

1.20, 2.55) for a high level. The RR for the group with a high frequency of chromosome-type CA was 1.47 (95% CI 1.11, 1.93), while that for chromatid-type CA was 1.09 (95% CI 0.82, 1.45). The analyses of risk of specific cancers were limited by small numbers, but the association was stronger for stomach cancer (Boffetta et al. 2007).

The contrasting findings concerning different predictivity of subclasses of CA are possibly due to intrinsic limitations of data. From a mechanistic point of view, chromosome-type aberrations occur directly from DSB in G0 lymphocytes, or during repair of lesions other than DSB (e.g., adducts), and therefore seem to be more causally related to early stages of carcinogenesis than chromatid-type aberrations, mostly arising during replication of damaged DNA in culture. The interlaboratory difference in the criteria used for CA classification, and the measurement of specific chromosome damage in a surrogate tissue, made the historical cohort a poor study design for addressing this issue.

A summary of the major findings from cohort studies on CA frequency and risk of cancer is reported in Table 31.1.

31.4 Pooled Analysis of Published Data

The experience from clinical trials indicates that a summary of the published studies (meta-analysis) could be useful even in observational epidemiology and molecular genetic research to obtain organized information from the existing results and to address future studies. Another way to summarize data in observational studies is to pool individual records and reanalyze the data. This approach, despite the highest costs, has several advantages over the classic meta-analysis of published data, mainly the possibility to perform interaction tests and subgroup analysis including dose–response curves (Taioli and Bonassi 2003). The pooled approach has been widely used in those cohort studies described above to validate predictive biomarkers. In particular, a European effort to collect all available information about population studies using cytogenetic biomarkers has been conducted within the framework of the research project Cytogenetic Biomarkers and Human Cancer Risk (CRB). This European collaborative project was established in 2001 to further study cancer risk prediction by cytogenetic biomarkers. The project had a number of scientific objectives, which included the quantitative assessment of risk prediction, the role of CA subclasses, the association with specific cancer sites, and the role of genetic susceptibility. The most challenging aim of the project was to create a pooled cohort of all cohort studies published so far. This achievement required 5 years of work, and in Table 31.2 a general description of the pooled database is reported that will be analyzed to address all items formulated in the rationale of the CRB project.

Table 31.1 Overview of published cohort studies evaluating chromosomal aberrations as cancer-risk biomarkers (all studies categorized subjects in three groups of chromosomal aberration frequency, i.e., low, medium, and high, and relative risks in the table are reported accordingly)

Study	Country	Subjects	Person-years	Follow-up period	Cancer cases (deaths)	SIR (95% CI)	SMR (95% CI)	RR (95% CI) incidence	RR (95% CI) mortality
The Nordic Study Group on the Health Risk of Chromosome Damage (1990)	Sweden, Finland, Norway	1,787	7,547	1970–1985	26	0.9 (0.3–1.9) 0.9 (0.3–2.0) 1.8 (0.9–3.0)			
Hagmar et al. (1994)	Sweden, Finland, Norway, Denmark	1,979	17,666	1970–1988	66	0.9 (0.5–1.4) 0.7 (0.3–1.2) 2.1 (1.5–2.8)			
Bonassi et al. (1995)	Italy	1,455	16,190	1969–1994	(44)		0.8 (0.7–1.6) 1.8 (1.0–2.9) 1.8 (1.1–2.8)		
Hagmar et al. (1998)	Sweden, Finland, Norway, Denmark, Italy	3,541	43,827	1970–1995	91 (64)	0.8 (0.5–1.2) 0.8 (0.5–1.3) 1.5 (1.1–2.0)	0.8 (0.5–1.8) 1.2 (0.7–1.8) 2.0 (1.4–2.9)	1.0 (–) 1.3 (0.8–2.2) **2.1 (1.3–3.5)**	1.0 (–) 1.3 (0.7–2.6) **2.5 (1.3–4.7)**
Rossner et al. (2005)	Czech Republic	11,834	113,967	1975–2000	485			1.0 (–) 1.2 (0.9–1.5) 1.1 (0.9–1.4) 1.0 (–)[a] **1.2 (1.03–1.50)**	
Boffetta et al. (2007)	Croatia, Hungary, Poland, Slovakia, Lithuania	6,430	54,655	1978–2002	224			1.0 (–) **1.8 (1.29–2.7)** **1.8 (1.2–2.7)**	

Relative risk (*RR*) values in *bold* correspond to $p<0.05$.
SIR standardized incidence ratio, *CI* confidence interval, *SMR* standardized mortality ratio
[a]Chromosome-type chromosomal aberration (high vs. low)

Table 31.2 Selected characteristics of the study population included in the pooled European study on chromosomal aberration (*CA*) frequency and risk of cancer

Country	Subjects	Cancer cases	Period of test	Mean CA frequency (SD)	Mean age (SD)	Males (%)	Exposed (%)	Current smokers (%)
Croatia	1,320	24	1982–2000	3.5 (2.3)	44.4 (9.7)	55.8	100.0	44.8
Hungary	840	46	1978–2001	1.9 (2.1)	45.3 (12.5)	64.8	25.5	40.1
Lithuania	812	24	1981–2002	3.3 (2.0)	48.7 (9.4)	78.4	91.6	48.4
Slovakia	2,994	55	1985–2001	2.1 (1.7)	47.8 (10.5)	57.8	76.4	42.5
Poland	456	22	1981–2001	1.5 (1.8)	55.1 (14.5)	84.2	57.5	55.3
Czech Republic	11,991	363	1975–1999	2.4 (1.8)	47.5 (11.7)	55.4	84.5	39.2
Sweden	756	48	1970–1987	1.8 (1.8)	54.6 (11.7)	72.9	–	–
Norway	471	48	1970–1988	1.5 (1.3)	57.6 (14.4)	78.0	65.8	38.2
Denmark	197	6	1987	1.8 (1.5)	50.7 (10.8)	100.0	66.0	43.1
Finland	557	39	1974–1988	2.3 (2.1)	54.6 (9.6)	69.5	–	–
Total incidence	20,394	675	1970–2002	2.4 (1.9)	48.2 (11.7)	60.0	75.5	40.9
Italy (mortality)	1,964	99	1965–1995	3.6 (4.2)	56.3 (12.9)	72.4	62.1	29.8
Total	22,358	774	1965–2002	2.5 (2.2)	48.2 (11.7)	60.8	74.2	39.8

SD standard deviation

31.5 Preliminary Results and Open Issues

As evident from Table 31.2, the pooled cohort was largely unbalanced, mostly owing to the large size of the Czech cohort that represented by itself about 50% of the whole database. Despite this limitation and the obvious heterogeneity due to discrepancy in methods and different life-styles in various countries, reliable estimates of the association between CA frequency and the risk of cancer could be provided. This was done by comparing cancer incidence rates for the medium and high levels versus the low level, after adjusting for the confounding effects of age, gender, smoking habit, and occupational exposure to mutagens or carcinogens. A *random effect* term was then included in the models to adjust for the differences in cancer rates occurring among different countries. The presence of effect modification was tested by computing the log-likelihood ratio test for two hierarchical models, the first with and the second without interaction terms involving CA (Lindsey 1995).

Preliminary results showed a significant increase of cancer risk in subjects with medium and high frequencies of CA when compared with those in the lowest tertile of the CA distribution. The increase of RRs ranged between 30 and 46%, with the highest risks found in the Hungarian and Norwegian cohorts, 4.24 and 2.21, respectively. The higher risk described in the Czech (Rossner et al. 2005) and in the new central and eastern European countries (Boffetta et al. 2007) cohorts for stomach cancer was confirmed by the pooled reanalysis which in subjects in the highest tertile found a RR for this tumor 3 times higher than those in the lowest tertile ($p<0.01$). Finally, the evidence obtained from these two cohorts that chromosome-type CA are a better predictor for cancer risk than chromatid type CA was largely confirmed by the pooled analysis, which reported the maximum risk in subjects with medium and high frequency of this CA type (1.32 and 1.44, respectively, $p<0.01$).

The large amount of data available from the pooled analysis, however, cannot address all open issues. Among these, establishing the role of individual genetic variability and, as a consequence, the reliability of individual cancer risk assessment are the most urgent. This item was partially addressed by a work package of the CRB project which evaluated, in a subgroup of the Italian, Norwegian, and Danish cohorts, the potential modification effect induced by most common genetic polymorphisms. Preliminary results from this ongoing study revealed that the role of genetic polymorphisms – at least those of the glutathione S-transferase family – is marginal, and its influence on the association between CA frequency and risk of cancer is negligible. This study, nested within the main European cohort, revealed an unexpected finding concerning the role of technical variability. In the frame of the nested study it was shown that the strength of the association between CA and cancer in the subgroup evaluated in Pisa – where frozen slides were homogeneously restained and rescored – was much more evident than in any other previous study, with an odds ratio for subjects with high CA frequency of 13.9 ($p<0.001$).

31.6 Conclusion

After 25 years of research aimed at demonstrating that the frequency of CA in healthy subjects can be used to predict cancer some general conclusions have been reached. All cohort studies measuring the link between CA frequency and risk of cancer found positive results. In some cases these results were unequivocal, as in the Nordic/Italian experience, in other cases the association was not so strong, such as in the Czech and Slovak cohorts, which nevertheless showed a significant association of cancer risk with chromosome-type CA. These findings, supported by the strong theoretical background linking chromosomal rearrangements to early stages of carcinogenesis, provide a significant contribution to the understanding of early stages of carcinogenesis, as shown by the high number of citations (about 500 overall) of the cohort studies described above, and the dramatically increasing trends of publications on chromosome damage as biomarkers for cancer (Bonassi et al. 2005).

Despite many issues remaining to be defined, especially from the mechanistic point of view, the current priorities to be addressed in the field are the application of existing evidence to regulation in the management of occupational exposure to genotoxic carcinogens, and the evaluation of an individual cancer risk assessment based on the use of markers of chromosome damage.

Future studies in this direction, corroborated by similar evidence from studies performed on the micronucleus assay (Bonassi et al. 2007) (but not on SCE, Norppa et al. 2006) are necessary to validate the use of biomarkers of chromosomal alterations in comprehensive cancer-prevention programs.

References

Boffetta P, van der Hel O, Norppa H, Fabianova E, Fucic A, Gundy S, Lazutka J, Cebulska-Wasilewska A, Puskailerova D, Znaor A, Kelecsenyi Z, Kurtinaitis J, Rachtan J, Forni A, Vermeulen R, Bonassi S (2007) Chromosomal aberrations and cancer risk: results of a cohort study from Central Europe. Am J Epidemiol 165:36–43

Bonassi S (1999) Combining environmental exposure and genetic effect measurements in health outcome assessment. Mutat Res 428:177–185

Bonassi S, Abbondandolo A, Camurri L, Dal Pra' L, De Ferrari M, Degrassi F, Forni A, Lamberti L, Lando C, Padovani P, Sbrana I, Vecchio D, Puntoni R (1995) Are chromosome aberrations in circulating lymphocytes predictive of a future cancer onset in humans? Preliminary results of an Italian cohort study. Cancer Genet Cytogenet 79:133–135

Bonassi S, Hagmar L, Strömberg U, Huici Montagud A, Tinnerberg H, Forni A, Heikkilä P, Wanders S, Wilhardt P, Hansteen I-L, Knudsen LE, Norppa H, for the European Study Group on Cytogenetic Biomarkers and Health (ESCH) (2000) Chromosomal aberrations in lymphocytes predict human cancer independently of exposure to carcinogens. Cancer Res 60:1619–1625

Bonassi S, Ugolini D, Kirsch-Volders M, Strömberg U, Vermeulen R, Tucker JD (2005) Human population studies with cytogenetic biomarkers: review of the literature and future prospectives. Environ Mol Mutagen 54:258–270

Bonassi S, Znaor A, Ceppi C, Lando L, Chang WP, Holland N, Kirsch-Volders M, Zeiger E, Ban S, Barale R, Bigatti MP, Bolognesi C, Cebulska-Wasilewska A, Fabianova E, Fucic A, Hagmar

L, Joksic G, Martelli A, Migliore L, Mirkova E, Scarfi' MR, Zijno A, Norppa H, Fenech M (2007) An increased micronucleus frequency in peripheral blood lymphocytes predicts the risk of cancer in humans. Carcinogenesis (in press). doi:10.1093/carcin/bgl177

Boveri T (1914) Zur Frage der Entstehung maligner Tumoren. Fischer, Jena Hagmar L, Brøgger A, Hansteen I-L, Heim S, Högsted B, Knudsen L, Lambert B, Linnainmaa K, Mitelman F, Nordenson I, Reuterwall C, Salomaa S, Skerfving S, Sorsa M (1994) Cancer risk in humans predicted by increased levels of chromosomal aberrations in lymphocytes: Nordic study group on the health risk of chromosomal damage. Cancer Res 54:2919–2922

Hagmar L, Bonassi S, Strömberg U, Brøgger A, Knudsen L, Norppa H, Reuterwall C (1998) Chromosomal aberrations in lymphocytes predict human cancer – A report from the European Study Group on Cytogenetic Biomarkers and Health (ESCH). Cancer Res 58:4117–4121

Hagmar L, Brøgger A, Hansteen I-L, Heim S, Högsted B, Knudsen L, Lambert B, Linnainmaa K, Mitelman F, Nordenson I, Reuterwall C, Salomaa S, Skerfving S, Sorsa M (1994) Cancer risk in humans predicted by increased levels of chromosomal aberrations in lymphocytes: Nordic study group on the health risk of chromosomal damage. Cancer Res 54:2919–2922

Hagmar L, Strömberg U, Bonassi S, Hansteen I-L, Knudsen LE, Lindholm C, Norppa H (2004) Impact of types of lymphocyte chromosomal aberrations of human cancer risk: Results from Nordic and Italian cohorts. Cancer Res 64:2258–2263

Lindsey JK (1995) Modelling frequency and count data. Oxford University Press, Oxford

Liou S-H, Lung J-C, Chen Y-H, Yang T, Hsieh L-L, Chen C-J, Wu T-N (1999) Increased chromosome-type chromosome aberration frequencies as biomarkers of cancer risk in a blackfoot endemic area. Cancer Res 59:1481–1484

Nordic Study Group on the Health Risk of Chromosome Damage (1990) An inter-Nordic prospective study on cytogenetic endpoints and cancer risk. Cancer Genet Cytogenet 45:85–92

Norppa H, Bonassi S, Hansteen I-L, Hagmar L, Strömberg U, Rössner P, Boffetta P, Lindholm C, Gundy S, Lazutka J, Cebulska-Wasilewska A, Fabiánová E, Šrám RJ, Knudsen LE, Barale R, Fucic A (2006) Chromosomal aberrations and SCEs as biomarkers of cancer risk. Mutat Res 600:37–45

Nowell P, Hungerford D (1960) A minute chromosome in human granulocytic leukemia. Science 132:1497

Pfeiffer P, Goedecke W, Obe G (2000) Mechanisms of DNA double-strand break repair and their potential to induce chromosomal aberrations. Mutagenesis 15:289–302

Richardson C, Jasin M (2000) Frequent chromosomal translocations induced by DNA double-strand breaks. Nature 405:697

Rossner P, Boffetta P, Ceppi M, Bonassi S, Smerhovsky Z, Landa K, Juzova D, Sram RJ (2005) Chromosomal aberration in lymphocytes of healthy subjects and risk of cancer. Environ Health Perspect 113:517–520

Rothman KJ (1986) Modern epidemiology. Little, Brown, Boston

Rowley JD (1998) The critical role of chromosome translocations in human leukemias. Annu Rev Genet 32:495–519

Taioli E, Bonassi S (2002) Methodological issues in pooled analysis of biomarker studies. Mutat Res 512:85–92

Vilenchik MM, Knudson AG (2003) Endogenous DNA double-strand breaks: Production, fidelity of repair, and induction of cancer. Proc Natl Acad Sci USA 100:12871–12876

Index

MIX
Papier aus verantwortungsvollen Quellen
Paper from responsible sources
FSC® C105338

If you have any concerns about our products,
you can contact us on
ProductSafety@springernature.com

In case Publisher is established outside the EU,
the EU authorized representative is:
Springer Nature Customer Service Center GmbH
Europaplatz 3, 69115 Heidelberg, Germany

Printed by Libri Plureos GmbH
in Hamburg, Germany

Laminar-Turbulent Transition

International Union of Theoretical
and Applied Mechanics

R. Kobayashi (Ed.)

Laminar-Turbulent Transition

IUTAM Symposium, Sendai/Japan
September 5 – 9, 1994

Springer-Verlag
Berlin Heidelberg New York
London Paris Tokyo
Hong Kong Barcelona Budapest

Prof. Dr. Ryoji Kobayashi

Tohoku University
Dept. of Machine Intelligence
and Systems Engineering
Aramaki-aza-Aoba, Aoba-ku
Sendai 980-77
Japan

ISBN-13: 978-3-642-79767-5 e-ISBN-13: 978-3-642-79765-1
DOI: 10.1007/978-3-642-79765-1

Cip data applied for

Softcover reprint of the hardcover 1st edition 1995

SPIN: 10494332 61/3020 – 5 4 3 2 1 0 – Printed on acid-free paper.

Preface

The International Union of Theoretical and Applied Mechanics (IUTAM) decided in 1992 to sponsor the fourth Symposium on Laminar-Turbulent Transition, Sendai/Japan, 1994. The objectives of the present Symposium were to deepen the fundamental knowledge of stability and laminar-turbulent transition in three-dimensional and compressible flows and to contribute to recent developing technologies in the field. This Symposium followed the three previous IUTAM-Symposia (Stuttgart 1979, Novosibirsk 1984 and Toulouse 1989).

The Scientific Committee selected two keynote lectures and 62 technical papers. The Symposium was held on the 5th to 9th of September, 1994, at the Sendai International Center in Sendai. The participants were 82 scientists from 10 countries. The keynote lectures have critically reviewed recent development of researches concerning the laminar-to-turbulent transition phenomena from the fundamental and the application aspects. Many papers presented were concerned about the detailed mechanism of the boundary layer transition (receptivity, secondary instability, turbulent spot and bypass transition). Particular emphasis was further placed on the transition of three-dimensional boundary layers on rotation systems and on swept wings. Attention was also given to compressible hypersonic flows.

The Symposium was evaluated as successful, which results from the active cooperation of many persons concerned. We are grateful to the members of the Scientific Committee for the scientific operations before and during the symposium and to the Local Organizing Committee for the preparation of the Symposium. We would express our thanks to the session chairmen for the efficient way, to the speakers for the excellent presentations and all the delegates for having provided stimulating discussions. Thanks are also given to the societies concerned for the financial support. Finally we appreciate the Springer-Verlag for the publication of the Symposium Proceedings.

Sendai, December 1994

Ryoji Kobayashi
Chairman of Symposium

Scientific Committee

H. Alfredsson (Sweden)
D. Arnal (France)
H. Bippes (Germany)
M. Gaster (UK)
R. Kobayashi, Chairman (Japan)
V. Ya. Levchenko (Russia)
R. Narasimha (India)
M. Nishioka (Japan)
E. Reshotko (USA)
W. S. Saric (USA)
T. Tatsumi (Japan)
H. Zhou (China)

Local Organizing Committee

Y. Aihara (Tokai Univ.)
Y. Fukunishi, Secretary (Tohoku Univ.)
H. Hashimoto (Tohoku Univ.)
N. Itoh (National Aerospace Lab.)
R. Kobayashi , Chairman (Tohoku Univ.)
Y. Kohama, Secretary (Tohoku Univ.)
J. Kondo (Tohoku Univ.)
S. Masuda (Keio Univ.)
T. Matsui (Gifu Univ.)
K. Nakabayashi (Nagoya Inst. of Tech.)
M. Nishioka (Univ. of Osaka Pref.)
H. Sato (Inst. of Flow Research)
N. Shuto (Tohoku Univ.)
T. Tatsumi (Kyoto Inst. of Tech.)
Y. Toba (Tohoku Univ.)

Symposium Sponsor

International Union of Theoretical and
Apllied Mechanics (IUTAM)

Suppurted by
The Japan Ministry of Education, Science, and Culture
The Japan Society for the Promotion Science
The Commemorative Association for the Japan World Exposition (1970)
The International Science Foundation, Washington

Participants

Aihara Y.	Japan	Tokai Univ. Hiratsuka
Alfredsson P.H.	Sweden	Royal Inst. of Tech. Stockholm
Arakawa M.	Japan	Keio Univ. Yokohama
Araki Y.	Japan	Yamaha Motor Co. Iwata
Arnal D.	France	CERT/ONERA Toulouse
Asai M.	Japan	Tokyo Metropolitan Inst. of Tech. Hino, Tokyo
Azuma T.	Japan	Osaka City Univ. Osaka
Bertolotti F.P.	Germany	DLR Goettingen
Bippes H.	Germany	DLR Goettingen
Craik A.D.D.	U.K.	Univ. of St. Andrews St. Andrews, Scotland
Crighton D.G.	U.K.	Cambridge Univ. Cambridge
Crouch J.D.	U.S.A.	Boeing Com. Airplane Group Seattle
Daiguji H.	Japan	Tohoku Univ. Sendai
Draad A.A.	Netherlands	Delft Univ. of Tech. Delft
Elofsson P.A.	Sweden	Royal Inst. of Tech. Stockholm
Eppler R.	Germany	Univ. of Stuttgart Stuttgart
Fasel H.F.	U.S.A.	Univ. of Arizona Tucson
Fukunishi Y.	Japan	Tohoku Univ. Sendai
Hama F.R.	Germany	iPCFD Muenchen

Hammerton P.W.	U.K.	Cambridge Univ. Cambridge
Hanifi A.	Sweden	Royal Inst. of Tech. Stockholm
Healey J.J.	U.K.	Cambridge Univ. Cambridge
Henningson D.S.	Sweden	FFA Bromma
Hino M.	Japan	Chuo Univ. Hachiohji, Tokyo
Hori D.	Japan	Keio Univ. Yokohama
Hornung H.G.	U.S.A.	California Inst. of Tech. Pasadena
Inoue O.	Japan	Tohoku Univ. Sendai
Ito H.	Japan	Nihon Univ. Koriyama
Itoh N.	Japan	NAL Chofu, Tokyo
Jennings M.J.	U.K.	Cambridge Univ. Cambridge
Kachanov Y.S.	Russia	Russian Academy of Sci., SB Novosibirsk
Kamiyama S.	Japan	Tohoku Univ. Sendai
Kazakov V.A.	Russia	Zhukovsky Central Aero-Hydro. Inst. Moscow
Kobayashi R.	Japan	Tohoku Univ. Sendai
Kohama Y.	Japan	Tohoku Univ. Sendai
Kosinov A.D.	Russia	Russian Academy of Sci., SB Novosibirsk
Kosorygin V.S.	Russia	Russian Academy of Sci., SB Novosibirsk
Kozlov V.V.	Russia	Russian Academy of Sci., SB Novosibirsk
Kufner E.	Germany	DLR Goettingen

Landahl M.T.	U.S.A.	Massachusetts Inst. of Tech. Cambridge
Le Gal P.	France	Univ. d'Aix-Marseille II Marseille
Levchenko V.Ya.	Russia	Russian Academy of Sci., SB Novosibirsk
Lingwood R.J.	U.K.	Cambridge Univ. Cambridge
Lysenko V.I.	Russia	Russian Academy. of Sci., SB Novosibirsk
Maekawa H.	Japan	Univ. of Electro-Comm. Chofu, Tokyo
Malik M.R.	U.S.A.	High Tech. Corp. Hampton
Manuilovich S.V.	Russia	Zhukovsky Central Aero-Hydro. Inst. Moscow
Masuda S.	Japan	Keio Univ. Yokohama
Matsubara M.	Sweden	Royal Inst. of Tech. Stockholm
Matsui T.	Japan	Gifu Univ. Gifu
Matsunobu Y.	Japan	Keio Univ. Yokohama
Meitz H.L.	U.S.A.	Univ. of Arizona Tucson
Morinishi Y.	Japan	Nagoya Inst. of Tech. Nagoya
Mueller W.	Germany	Univ. of Stuttgart Stuttgart
Nakabayashi K.	Japan	Nagoya Inst. of Tech. Nagoya
Narasimha R.	India	Indian Inst. of Sci. Bangalore
Nishikawa T.	Japan	Kobe Steel Ltd. Takasago
Nishioka M.	Japan	Univ. of Osaka Pref. Sakai
Otaguro T.	Japan	Hitachi Co. Ltd. Tsuchiura

Perrin F.	France	Minis. of Defance STPA/EG Paris
Poll D.I.A.	U.K.	Univ. of Manchester Manchester
Prabhu A.	India	Indian Inst. of Sci. Bangalore
Reshotko E.	U.S.A.	Case Western Reserve Univ. Cleveland
Rist U.	Germany	Univ. of Stuttgart Stuttgart
Saric W.S.	U.S.A.	Arizona State Univ. Tempe
Sato H.	Japan	Inst. of Flow Research Tokyo
Shaikh F.N.	U.K.	Cambridge Univ. Cambridge
Shimizu H.	Japan	Hitachi Co. Ltd. Tsuchiura
Simen M.	Germany	DLR Goettingen
Sudani N.	Japan	NAL Chofu, Tokyo
Suttan J.	Germany	Tech. Univ. Berlin Berlin
Takagi S.	Japan	NAL Chofu, Tokyo
Tanaka K.	Japan	Osaka Univ. of Commerce Higashi-osaka, Osaka
Tatsumi T.	Japan	Kyoto Inst. of Tech. Kyoto
Toba Y.	Japan	Tohoku Univ. Sendai
Tsuchida Y.	Japan	Nagoya Inst. of Tech. Nagoya
van Hest B.F.A.	Netherlands	Delft Univ. of Tech. Delft
Westin K.J.A.	Sweden	Royal Inst. of Tech. Stockholm
Wu X.	U.K.	Imperial College London

Yamamoto K.	Japan	NAL Chofu, Tokyo
Yoshida K.	Japan	Kawasaki Heavy Ind. Ltd. Kakamigahara
Zhou H.	China	Tianjin Univ. Tianjin

Contributors

Airiau Ch.	France	CERT/ONERA Toulouse
Alfredsson P.H.	Sweden	Royal Inst. of Tech. Stockholm
Arakawa M.	Japan	Keio Univ. Yokohama
Arnal D.	France	CERT/ONERA Toulouse
Asai M.	Japan	Tokyo Metropolitan Inst. of Tech. Hino, Tokyo
Asmolov E.S.	Russia	Zhukovsky Central Aero-Hydro. Inst. Moscow
Azuma T.	Japan	Osaka City Univ. Osaka
Bakchinov A.A.	Russia	Russian Academy of Sci., SB Novosibirsk
Baumann M.	Germany	Tech. Univ. Berlin Berlin
Berlin S.	Sweden	Royal Inst. of Tech. Stockholm
Bertolotti F.P.	Germany	DLR Goettingen
Bestek H.	Germany	Univ. of Stuttgart Stuttgart
Bippes H.	Germany	DLR Goettingen
Boiko A.V.	Russia	Russian Academy of Sci., SB Novosibirsk
Casalis G.	France	CERT/ONERA Toulouse
Chauve M.P.	France	Univ. d'Aix-Marseille II Marseille
Craik A.D.D.	U.K.	Univ. of St. Andrews St. Andrews, Scotland
Crouch J.D.	U.S.A.	Boeing Com. Airplane Group Seattle
Dahlkild A.	Sweden	Royal Inst. of Tech. Stockholm

Dallmann U.	Germany	DLR Goettingen
Danks M.	U.K.	Univ. of Manchester Manchester
Draad A.A.	Netherlands	Delft Univ. of Tech. Delft
Elofsson P.A.	Sweden	Royal Inst. of Tech. Stockholm
Fasel H.F.	U.S.A.	Univ. of Arizona Tucson
Fukunishi Y.	Japan	Tohoku Univ. Sendai
Fujimura K.	Japan	Japan Atomic Energy Res. Inst. Tokai, Ibaraki
Gaponenko V.R.	Russia	Russian Academy of Sci., SB Novosibirsk
Gaponov S.A.	Russia	Russian Academy of Sci., SB Novosibirsk
Gaster M.	U.K.	Cambridge Univ. Cambridge
Germain P.	U.S.A.	California Inst. of Tech. Pasadena
Grek G.R.	Russia	Russian Academy of Sci., SB Novosibirsk
Hammerton P.W.	U.K.	Cambridge Univ. Cambridge
Hanifi A.	Sweden	Royal Inst. of Tech. Stockholm
Healey J.J.	U.K.	Cambridge Univ. Cambridge
Hein S.	Germany	DLR Goettingen
Henningson D.S.	Sweden	FFA Bromma
Holm E.S.	U.S.A.	Univ. of Wyoming Laramie
Hori D.	Japan	Keio Univ. Yokohama
Hornung H.G.	U.S.A.	California Inst. of Tech. Pasadena

Itoh N	Japan	NAL Chofu, Tokyo
Ivanov A.V.	Russia	Russian Academy of Sci., SB Novosibirsk
Jahanmiri M.	India	Indian Inst. of Sci. Bangalore
Jarre S.	France	Univ. d'Aix-Marseille II Marseille
Jennings M.J.	U.K.	Cambridge Univ. Cambridge
Juillen J.C.	France	CERT/ONERA Toulouse
Kachanov Y.S.	Russia	Russian Academy of Sci., SB Novosibirsk
Kato T.	Japan	Tohoku Univ. Sendai
Kanner H.	U.S.A.	George Washington Univ. Washington D.C.
Kazakov V.A.	Russia	Zhukovsky Central Aero-Hydro. Inst. Moscow
Klingmann B.G.B.	Switzland	DME-IMHEF Ecublens
Kloker M	Germany	Univ. of Stuttgart Stuttgart
Kobayashi R.	Japan	Tohoku Univ. Sendai
Kodashima Y.	Japan	Tohoku Univ. Sendai
Kohama Y.	Japan	Tohoku Univ. Sendai
Kosinov A.D.	Russia	Russian Academy of Sci., SB Novosibirsk
Kosorygin V.S.	Russia	Russian Academy of Sci., SB Novosibirsk
Kozlov V.V.	Russia	Russian Academy of Sci., SB Novosibirsk
Kufner E.	Germany	DLR Goettingen
Kuiken G.D.C.	Netherlands	Delft Univ. of Tech. Delft

Landahl M.T.	U.S.A.	Massachusetts Inst. of Tech. Cambridge
Le Gal P.	France	Univ. d'Aix-Marseille II Marseille
Levchenko V.Ya.	Russia	Russian Academy of Sci., SB Novosibirsk
Li F.	U.S.A	High Tech. Corp. Hampton
Lin R.S.	U.S.A.	High Tech. Corp. Hampton
Lingwood R.J.	U.K.	Cambridge Univ. Cambridge
Lundbladh A.	Sweden	FFA Bromma
Lysenko V.I.	Russia	Russian Academy of Sci., SB Novosibirsk
Maekawa H.	Japan	Univ. of Electro-Comm. Chofu, Tokyo
Malik M.R.	U.S.A.	High Tech. Corp. Hampton
Manuilovich S.V.	Russia	Zhukovsky Central Aero-Hydro. Inst. Moscow
Maslennikova I.I.	Rossia	Russian Academy of Sci., SB Novosibirsk
Masuda S.	Japan	Keio Univ. Yokohama
Matsubara M.	Sweden	Royal Inst. of Tech. Stockholm
Matunobu Y.	Japan	Keio Univ. Yokohama
Meitz H.L.	U.S.A.	Univ. of Arizona Tucson
Michalke A.	Germany	Tech. Univ. Berlin Berlin
Morinishi Y.	Japan	Nagoya Inst. of Tech. Nagoya
Mueller W.	Germany	Univ. of Stuttgart Stuttgart
Nakabayashi K.	Japan	Nagoya Inst. of Tech. Nagoya

Narasimha R.	India	Indian Inst. of Sci. Bangalore
Nieuwstadt F.T.M.	Netherlands	Delft Univ. of Tech. Delft
Nishikawa T.	Japan	Kobe Steel Ltd. Takasago
Nishioka M.	Japan	Univ. of Osaka Pref. Sakai
Nishioka T.	Japan	Hitachi Co. Ltd. Tuchiura
Nitsche W.	Germany	Tech. Univ. Berlin Berlin
Passchier D.M.	Netherlands	Delft Univ. of Tech. Delft
Poll D.I.A.	U.K.	Univ. of Manchester Manchester
Prabhu A.	India	Indian Inst. of Sci. Bangalore
Radeztsky R.H.,Jr.	U.S.A.	Arizona State Univ. Tempe
Rasmussen B.K.	U.S.A.	Arizona State Univ. Tempe
Reibert M.S.	U.S.A.	Arizona State Univ. Tempe
Rist U.	Germany	Univ. of Stuttgart Stuttgart
Saito H.	Japan	Inst. of Flow Research Tokyo
Saric W.S.	U.S.A.	Arizona State Univ. Tempe
Sato H.	Japan	Inst. of Flow Research Tokyo
Scherbakov V.A.	Russia	Russian Academy of Sci., SB Novosibirsk
Semionov N.V.	Russia	Russian Academy of Sci., SB Novosibirsk
Shaikh F.N.	U.K.	Cambridge Univ. Cambridge
Shevel'kov S.G.	Russia	Russian Academy of Sci., SB Novosibirsk

Simen M.	Germany	DLR Goettingen
Stewart P.A.	U.K.	Cambridge Univ. Cambridge
Suttan J.	Geramany	Tech. Univ. Berlin Berlin
Syzrantsev V.V.	Russia	Russian Academy of Sci., SB Novosibirsk
Takami H.	Japan	Univ. of Electro-Comm. Chofu, Tokyo
Takagi S.	Japan	NAL Chofu, Tokyo
Tanaka K.	Japan	Osaka Univ. of Commerce Higashi-osaka, Osaka
Theofilis V.	Netherlands	Univ. of Twente Enschede
Tsuchida Y.	Japan	Nagoya Inst. of Tech. Nagoya
van Hest B.F.A.	Netherlands	Delft Univ. of Tech. Delft
van Ingen J.L.	Netherlands	Delft Univ. of Tech. Delft
Watanabe H.	Japan	Tohoku Univ. Sendai
Wei W.	U.S.A.	Arizona State Univ. Tempe
Wendt V.	Germany	DLR Goettingen
Westin K.J.A.	Sweden	Royal Inst. of Tech. Stockholm
Wu X.	U.K.	Imperial College London
Yamamoto K.	Japan	NAL Chofu, Tokyo
Zhou H.	China	Tianjin Univ. Tianjin
Zheng Z.	Japan	Nagoya Inst. of Tech Nagoya

Contents

+) Paper not available

IUTAM Symposium on Laminar-Turbulent Transition, September 5-9, 1994, Sendai, Japan

Keynote Lectures

A Survey of the Transition Prediction Methods: from Analytical Criteria to PSE and DNS

D. Arnal - G. Casalis - J.C. Juillen
CERT/ONERA - Aerothermodynamics Department
2 avenue E. Belin - BP 4025 - 31055 Toulouse Cedex (FRANCE)

Summary

The objective of this paper is to give a survey of the tools which are currently available for transition prediction. The simplest methods (analytical criteria, data base methods) are first presented. Emphasis is then given on techniques based on the linear stability theory (local and PSE approaches). When non linear mechanisms need to be accounted for, it becomes necessary to use more complex methods such as non linear PSE or DNS. Typical examples of applications are presented for each of these methods.

1. Introduction

As laminar-turbulent transition plays a very important role in many practical applications, a large amount of theoretical, numerical and experimental work has been devoted to this subject for many years. There are in fact several paths to turbulence, which can be summarized as follows. If the amplitude of the forced disturbances (noise, free stream turbulence, roughness elements, ...) is large, non linear phenomena are immediately observed in the laminar boundary layer and transition occurs very rapidly. This mechanism is known as a "bypass". On the other side, forced disturbances of small amplitude can induce a so-called "natural transition". In this case, the first step of the transition process is the boundary layer receptivity. Let us recall that receptivity refers to the mechanisms that cause the environmental disturbances to generate unstable waves which are the eigenmodes of the viscous flow. The linear amplification of these waves constitutes the second step of the transition process. The third stage occurs when the unstable waves reach a finite amplitude, so that their behavior deviates from that predicted by the linear theory. This nonlinear evolution results eventually in the breakdown to turbulence.

The goal of the paper is not to analyze the theories which describe such a sequence of complicated phenomena, but to give an overview of the practical methods aimed at predicting the onset of transition. Of course, most of these methods do not claim to model the physics of all the possible routes to turbulence ; they only represent short term answers to engineering problems. Starting with the simplest techniques, paragraphs 2 and 3 provide some insight into recent development of analytical criteria and data base methods. An important part of the paper (paragraph 4) is devoted to the analysis of the advantages and of the shortcomings of the popular e^n method based on linear stability theory. As the n factor can be computed either from a local analysis or from a PSE (Parabolized Stability Equations) approach, similarities and differences between these different procedures will be highlighted.

The last two paragraphs deal with more complex methods which are able to take into account weakly non-linear mechanisms (non-linear PSE) or strongly non-linear mechanisms (DNS : Direct Numerical Simulations). Due to the space limitation, the discussion will be mainly restricted to results obtained at CERT/ONERA since the preceding IUTAM Symposium on transition (Toulouse, September 1989).

2. Analytical Criteria

In this section, the word criterion must be interpreted as a more or less empirical correlation between laminar boundary layer parameters at the onset of transition.

In the case of bypass-induced transition, many correlations have been proposed for engineering purposes. The so-called leading edge contamination constitutes a typical example of bypass. This phenomenon is likely to occur when a swept wing is in contact with a solid wall (fuselage). The relevant parameter is the attachment line Reynolds number which is denoted as $\overline{R}$. According to a very simple criterion (Pfenninger [1], Poll [2]), the leading edge is contaminated by the turbulent structures coming from the wall as soon as $\overline{R}$ exceeds a critical value close to 250.

As far as "natural" transition is concerned, the development of simple criteria presents an unquestionable practical interest. A review of the correlations available for two-dimensional flow can be found in [3]. Specific criteria were also proposed in the case of transition triggered by pure crossflow disturbances. As an example, figure 1 shows a criterion which was developed at CERT/ONERA [4]. This criterion is a correlation between the crossflow Reynolds number $R\delta_2$ (based on the integral of the crossflow profile) and the streamwise shape factor H.

An example of application of the preceding criterion is shown in figure 2. A Gaster bump is placed on the leading edge of a swept wing in order to prevent leading edge contamination. This bump is shaped in such a way that the turbulent boundary layer is brought to rest at a stagnation point whilst a "clean" laminar flow is generated on the rear part of the device. The figure presents a comparison between the measured and computed transition lines around the bump. Due to the strong deflection of the flow on both sides of the device, a powerful crossflow instability is generated, and transition takes place at a rather short distance from the attachment line. Further downstream (in the spanwise direction), the transition front moves at larger chordwise distances. The overall pattern is well predicted by the simple correlation of figure 1. Let us note, however, that discrepancies can arise when transition is not of the pure crossflow type.

3. Data Base Methods

It is well known that compressibility modifies the development of unstable waves leading to "natural" transition. As compressibility effects are difficult to take into account in simple analytical criteria, a possibility is to develop data base methods, the complexity of which is intermediate between empirical correlations and direct numerical solutions of the stability equations.

Up to now, the application of the published data base methods is restricted to two-dimensional flows. The principle is to compute the stability properties of self similar (or locally similar) basic flows and to interpolate between the tabulated results for any real situation in order to estimate the growth rates of the amplified eigenmodes. It is then possible to use the classical e^n method to estimate the transition location.

A data base method was proposed by van Ingen [5] for low speed flows and by Arnal [6] for Mach numbers between 0 and 1.3. The interest of the data base methods is that the computing time is reduced by several orders of magnitude by comparison with exact stability calculations. Figure 3 shows a comparison between exact stability computations and the results obtained with this simplified method. This example is related to experiments conducted on a two-dimensional airfoil which was tested in the transonic wind tunnel T2 at CERT/ONERA, in the framework of a cooperative work between ONERA and DLR. From 5 to 15 percent chord, the local Mach number decreases from 1.2 to 1.05, then it is nearly constant up to the trailing edge. The variations of the n factor computed with the data base method closely resemble those deduced from the exact stability calculations.

4. Methods based on the Linear Stability Theory : e^n Method(s)

The e^n method is widely used by aeronautical engineers for transition prediction. It was first developed by Smith and Gamberoni [7] and by van Ingen [8] for two-dimensional, low speed flows and then extended to more complex problems. The procedure for applying the e^n method requires three successive steps. The first step consists in the calculation of the laminar velocity and temperature profiles along the body. In the second step, the local growth rates of the unstable waves are computed for each of these profiles. This can be done by solving either the local stability equations or the Parabolized Stability Equations (PSE). Both approaches are described below. In the third step, the local growth rates are integrated in order to determine the n factor.

4.1. Local Theory

Any fluctuating quantity r' is expressed by :

$$r' = r(y) \exp [\, i(\alpha x + \beta z - \omega t) \,] \qquad (1)$$

The assumptions of the local theory are well known : α, β and ω are constant parameters, the disturbances are very small in order to linearize the Navier-Stokes equations, and the parallel flow approximation is used. As a consequence, stability at a particular station (x, z) is determined by local conditions at that station independent of all others. Numerically, one has to solve a system of ordinary differential equations for the amplitude functions. As these equations are homogeneous with homogeneous boundary conditions, the problem is an eigenvalue one, i.e. non trivial solutions only exist for certain combinations of α, β and ω (dispersion relation).

In this paper, the discussion will be restricted to spatially growing waves : ω is real, $\alpha = \alpha_r + i\alpha_i$, $\beta = \beta_r + i\beta_i$ are complex.

For two-dimensional, incompressible flows, β is set to zero, and the n factor is defined by :

$$n = \underset{f}{\mathrm{Max}} \left(\int_{x_0}^{x} -\alpha_i \, dx \right) \qquad (2)$$

with f denoting physical (dimensional) frequencies. x_0 is the streamwise distance where the waves of frequency f start to be amplified.

For two-dimensional, compressible flows, the problem becomes more complex, because the most unstable waves are often oblique waves. $\psi = \tan^{-1}(\beta_r/\alpha_r)$ represents the angle between the streamwise direction x and the wavenumber vector. It is still assumed that the amplification takes place in the x-direction only, i.e. $\beta_i = 0$, but β_r needs to be specified. To compute the total growth rate of a given frequency, the following procedure is used : at each streamwise location, one seeks for the direction ψ_M of the wave which gives the maximum value $-\alpha_{iM}$ of $-\alpha_i$, and the n factor is defined by :

$$n = \underset{f}{\mathrm{Max}} \left(\int_{x_0}^{x} -\alpha_{iM} \, dx \right) \equiv \underset{f}{\mathrm{Max}} \left(\int_{x_0}^{x} \underset{\psi}{\mathrm{Max}}(-\alpha_i) \, dx \right) \qquad (3)$$

This is the so-called envelope method.

The extension of the e^n method to three-dimensional flows is not straightforward. The first reason is that the assumption $\beta_i = 0$ is not necessarily correct. Hence β_i must be assigned or computed. Several solutions have been proposed to solve this problem (see review in [9]). For instance, it is possible to use the wave packet theory and to impose the ratio $\partial\alpha/\partial\beta$ to be real. A simpler solution is to assume that the growth direction is the group velocity direction or the potential flow direction. In the case of infinite swept wings, it is often assumed that there is no amplification in the spanwise direction z, i.e. $\beta_i = 0$.

After one of the previous assumptions for β_i has been adopted, one has to integrate the local growth rates in order to compute the n factor. Several strategies are available :

♦ envelope method : this strategy was previously described for two-dimensional, compressible flows.

♦ fixed frequency/fixed spanwise wavenumber method : the integrations are performed by following waves having a given dimensional frequency f and a constant value of the dimensional component β_r of the wavenumber vector in the z-direction. The n factor is computed by maximizing the total amplification with respect to both f and β_r.

♦ fixed frequency/fixed wavelength method and fixed frequency/fixed direction method : as a wave of fixed frequency moves downstream, the wavelength or the propagation direction of the disturbances is kept constant. These strategies resemble the previous one in this sense that the n factor represents the envelope of several envelope curves.

♦ streamwise n factor/crossflow n factor : the principle is to compute a n factor for Tollmien-Schlichting disturbances and another n factor for crossflow disturbances. Transition is assumed to occur for particular combinations of these parameters.

As it could be expected, each strategy gives a different value of the n factor at the onset of transition. To illustrate this point, figure 4 shows a comparison between the n factors calculated with the envelope method and with the fixed frequency/fixed spanwise wavenumber method. The computations were performed for a transonic swept wing in the framework of the ELFIN (European Laminar Flow INvestigation) project. It can be seen that the fixed frequency/fixed spanwise wavenumber strategy gives much lower n factors than the envelope method.

4.2. Linear PSE Approach

A new formulation of the stability analysis was proposed by Herbert, see overview in [10]. In this approach, the general expression of the disturbances is :

$$r' = r(x, y) \exp [i(\theta(x) + \beta z - \omega t)] \text{ with } \frac{d\theta}{dx} = \alpha(x) \qquad (4)$$

α is complex, β and ω are real and constant. In contrast to the local approach, the amplitude functions depend on y and x, and α depends on x. Substituting relation (4) into the stability equations, neglecting $\partial^2 r/\partial x^2$ and linearizing in r yields a partial differential equation of the form :

$$Lr + M\frac{\partial r}{\partial x} + \frac{d\alpha}{dx} Nr = 0 \qquad (5)$$

where L, M and N are operators in y with coefficients that depend on x and y. When α is computed from a so-called normalization condition, equation (5) can be solved using a marching procedure in x with prescribed initial conditions. This constitutes the PSE (Parabolized Stability Equation) approach. The interest of this procedure is that the nonparallel effects are taken into account.

4.3. Comparison between Local and PSE Approaches

A numerical code solving the PSE equations was developed by Airiau and Casalis [11]. PSE results were compared with the results of the local approach in the case of non similar flows involving varying streamwise pressure gradients. The objective of these comparisons was to analyze the non parallel effects.

A first series of comparisons was made for two-dimensional, low speed flows. Only two-dimensional waves ($\beta = 0$) were considered in the computations. As far as the n factor is concerned, the non parallel effects were found to be negligible, even for transitions occurring in strong positive pressure gradients. A complete account of these comparisons can be found in [12]. This means that for this type of flow, the e^n method provides similar results by using either the local approach or the PSE approach.

The conclusions are different for three-dimensional flows. In this case, meaningful comparisons need to be made with the same strategy of integration of the n factor. As β is real and constant in relation (4), it is clear that the PSE approach can be compared with the local approach only when the latter is using the fixed frequency/fixed spanwise wavenumber strategy (β constant) with no amplification in the spanwise direction z (β real).

Figures 5 and 6 show the total growth rates obtained by both procedures in the case of a swept wing which was studied at CERT/ONERA in a subsonic wind tunnel. The flow is accelerated all along the chord, so that crossflow disturbances play a major role in the transition process. 9 frequencies were considered, with 11 values of β for each frequency. The main result is that the PSE approach (figure 5) gives larger n factors than the local approach (figure 6). At the measured transition point ($X/C = 0.4$), the n factors are equal to 6 and 4, respectively. Thus non parallel effects are strongly destabilizing for crossflow disturbances. Let us note that smaller effects were found for the swept Hiemenz flow, see discussion in [13]. Systematic investigations are necessary in order to draw more precise conclusions.

4.4. Shortcomings of the e^n Method

As the e^n method is based on linear stability only, receptivity and non linear mechanisms are not taken into account. In addition, the non parallel effects are neglected in the local procedure, and it has been shown that they could be important for oblique waves.

Several particular problems arise for three-dimensional flows. The first one is to choose the "best" strategy of integration of the n factor when the local approach is used. The envelope method is widely used, but its physical meaning is not clear in many cases, especially for swept wing flows in the vicinity of the point of minimum pressure : rapid variations in ψ_M can be observed around this point when the dominant instability suddenly changes from the crossflow to the streamwise type. The other strategies are often used in order to avoid these discontinuous (and probably unphysical) evolutions. In particular this problem does not appear with PSE computations.

From a practical point of view, the most important issue is the value of the n factor at the onset of transition. Concerning two-dimensional flows, it is now admitted that the e^n method with $n \simeq 10$ can be applied to predict transition for a wide range of flows if the background disturbance level is small enough. In three-dimensional flows, the results depend, of course, on the strategy which is chosen to compute the n factor, but, even with the same strategy, the results are not very clear. In fact, all the strategies produce a large scatter in the values of n at the onset of transition. A possible reason is that crossflow and streamwise instabilities are not initiated by the same type of forced disturbances. For instance, the crossflow disturbances are very sensitive to micron-sized roughness elements which have no effect on streamwise disturbances. Another reason is that there exists a multiplicity of non-linear processes before the breakdown to turbulence, depending on the relative part of stationary and travelling unstable modes.

However, the e^n method remains a very practical and efficient tool, especially for parametric studies. For a given test model and for a given disturbance environment, it is often able to predict the variation of the transition location when changing a parameter which governs the stability properties of the mean flow.

5. Weakly Non Linear Approaches

Today, several approaches are available to analyse the behavior of the disturbances when they begin to deviate from the linear amplification regime. These numerical tools include, for instance, the method of multiple scales or the secondary instability theories. Direct Numerical Simulations are of course very helpful for improving the understanding of the non-linear mechanisms. For practical applications, non-linear PSE are particularly attractive ; although they only describe weakly non-linear phenomena, they are less time-consuming than DNS and they can be used for rather complex geometries.

The basic ideas of non-linear PSE have been explained in many papers, see for instance [10]. Let us recall that the disturbances are written as double Fourier expansions containing two- and three-dimensional discrete normal modes denoted as (n,m) modes :

$$r' = \sum_{n=-\infty}^{\infty} \sum_{m=-\infty}^{\infty} r_{n,m}(x,y) \exp\left(i \left[\int_{x_0}^{x} \alpha_{n,m}(\xi)\, d\xi + m\beta z - n\omega t/2 \right] \right) \qquad (6)$$

The (2,0) mode represents the primary wave. In order to simulate the onset of strong resonances, it is assumed that the phase speed is identical for all modes (tuned modes), i.e. :

$$\forall n, \forall m \qquad \alpha_{n,m}(\xi) = n\alpha(\xi)/2 + i a_{n,m}(\xi) \qquad (7)$$

where α is the real part of the wavenumber of the primary mode and $a_{n,m}$ describes the amplification or the damping of the waves. Introducing (6) into the Navier-Stokes equations leads to a system of coupled partial differential equations which are solved by a marching procedure.

Figure 7 illustrates the interest of non-linear PSE for transition prediction. We consider a flat plate, incompressible flow, and the initial conditions are such that a H-type resonance is likely to occur for a dimensionless frequency $F = 4\ 10^{-5}$. When the computations proceed downstream, the (0,0) mode starts to develop rapidly at the onset of resonance. As this mode corresponds to a steady distortion of the basic flow, the mean velocity profile is modified in such a way that the shape factor H begins to decrease and that the skin friction coefficient Cf begins to increase. The computation breaks down some distance downstream of the onset of this "numerical" transition. The concept of "critical n factor" does not exist for non-linear PSE, but the problem is now to choose adequate initial conditions for the primary and the secondary waves.

In another paper presented in the same meeting [14], it is shown that the non-linear PSE can be used for parametric studies. The initial conditions are optimized in order to fit the experimental data for a given pressure gradient ; then a good agreement with the measurements is achieved for five other pressure distributions by using the same initial data.

The previous examples deal with two-dimensional mean flows. A survey of the problems related to three-dimensional flows can be found in the excellent paper by Malik et al [15].

6. Direct Numerical Simulations (DNS)

The DNS are particularly useful for complex problems which cannot be treated by simple linear or weakly non-linear theories. These problems include strongly non-linear phenomena in "natural" transition, receptivity or bypass-induced transitions.

• *Strongly non-linear phenomena in "natural" transition.* It has been shown previously that calculations based on non-linear PSE stop when elliptic effects become important, i.e. when the wave amplitude exceeds some critical threshold. This limitation is not too severe with DNS, which are able to describe the flow development up to the breakdown to turbulence and even further downstream up to the fully turbulent regime. A good example of the current capabilities of DNS is given in the paper by Kloker and Fasel [16]. Let us note, however, that this kind of DNS needs to impose initial conditions for the disturbances, i.e. the problem is similar to that described for non linear PSE.

• *Receptivity.* The aim of the receptivity theory is to explain the origin of the boundary layer eigenmodes and to give their initial amplitude at branch I of the neutral curve. Many investigations have been devoted to this problem since about 15 years. Most of them are based on asymptotic theories, see for instance review in [17]. To validate the theoretical results, careful experiments as well as DNS computations are necessary. The work of Lin et al [18] gives a typical example of DNS calculations illustrating the concepts of leading edge and localized receptivity. Figure 8 shows DNS results obtained at CERT/ONERA [19]. A flat plate is placed in a periodic flow field expressed by :

$$U = U_\infty \, [1 + \varepsilon \cos\omega t], \quad \varepsilon \ll 1 \qquad (8)$$

The boundary layer flow is the sum of three components : a steady flow of the Blasius type, a Stokes layer and a periodic eigenmode of small amplitude. The amplitude distribution of this eigenmode is plotted in figure 8 for a streamwise location close to the leading edge. Its shape is similar to that of a Tollmien-Schlichting wave, as it can be seen by comparison with the solution of the Orr-Sommerfeld (OS) equation for the same frequency and for the same Reynolds number. The major progress is that the DNS provide us with the absolute value of the amplitude, whereas the OS solution is obtained with an arbitrary multiplicative constant (in figure 8, this constant was determined to fit the maximum amplitude of the DNS solution). This kind of DNS calculation makes it possible to compute receptivity coefficients and it could help to develop prediction methods based on disturbance amplitude instead of disturbance amplification.

• *Bypass.* As mentioned in paragraph 2, purely empirical criteria are often used to estimate the transition location when the forced disturbances are so large that the transition process is fully non-linear (bypass). It is clear that DNS are particularly helpful, not only to understand the physics of the phenomena but also to validate and to extend the practical prediction methods. An interesting example is given by the DNS performed by Spalart [20] for the problem of leading edge contamination. Spalart found that turbulence becomes self-sustaining along the attachment line of a swept wing for $\overline{R}$ between 200 and 250 - in agreement with the well known criterion proposed by Pfenninger and Poll, see paragraph 2. He also computed the evolution of the contamination Reynolds number when suction is applied around the leading edge. These numerical results were qualitatively confirmed some years later by two independent series of experiments (Poll [21], Juillen and Arnal [22]) which are presented in this Symposium. This example illustrates the fact that DNS and experiments can be complementary to each other for the improvement of engineering prediction methods.

7. Conclusion

A wide range of prediction methods is available to estimate the onset of transition for practical purposes. Of course, each of them presents interesting advantages and strong limitations.

It is very easy to implement <u>analytical criteria</u> in boundary layer or Navier-Stokes codes. However, as they are usually deduced from a limited number of experiments, they have to be applied in the "right" conditions.

To the authors knowledge, the available <u>data base</u> methods are valid for two-dimensional flows only. As compared with exact solutions of the stability equations, they can provide fairly good results with a very limited computing time.

The <u>e^n method</u> remains the most widely used method for transition prediction. Several difficulties have been pointed out : choice of the integration strategy, value of the n factor at transition, differences between local and PSE approaches, ... A pessimistic statement is that there are so many possible routes to turbulence that a "universal" value of n cannot exist. For aeronautical applications, however, it can be expected that the disturbance environment in flight conditions does not change very much, so that the scatter in the value of n could be acceptable.

As far as <u>non linear PSE</u> is concerned, the major improvement is that a numerical transition occurs "naturally", because the resonance mechanisms are now included in the model. The new problem is the choice of correct initial conditions (amplitude, frequency, ...) which are unknown for most of the real flow situations.

<u>DNS</u> play a more and more important role in transition research. In principle, they are able to describe all the details of the transition process, including receptivity and strongly non linear mechanisms. Although they cannot be used routinely for transition prediction, they are useful to check the validity and the range of application of the other techniques.

In all cases, careful experiments are needed to improve the transition prediction methods.

References

[1] Pfenninger, W. "Flow Phenomena at the Leading Edge of Swept Wings." AGARDograph 97 (May 1965)

[2] Poll, D.I.A., "Some Aspects of the Flow near a Swept Attachment Line with Particular Reference to Boundary Layer Transition", College of Aeronautics Report 7805 (August 1978)

[3] Arnal, D., "Description and Prediction of Transition in Two-Dimensional, Incompressible Flow", AGARD Report n° 709, (1984)

[4] Arnal, D., Habiballah, M., Coustols, E., "Laminar Instability Theory and Transition Criteria in Two- And Three-Dimensional Flows", La Recherche Aérospatiale n° 1984-2, (1984)

[5] Van Ingen, J.L., "Transition, Pressure Gradient, Suction, Separation and Stability Theory", AGARD CP 224, (1977)

[6] Arnal, D., "Transition Prediction in Transonic Flow", IUTAM Symp. "Transonicum III", Göttingen, Ed. Zierep and Oertel, Springer Verlag, (1988)

[7] Smith, A.M.O., Gamberoni, N., "Transition, Pressure Gradient and Stability Theory", Douglas Aircraft Co. Rept. ES 26388, El Segundo, Calif., (1956)

[8] Van Ingen, J.L., "A Suggested Semi-Empirical Method for the Calculation of the Boundary Layer Transition Region", Univ. of Techn., Dept. of Aero. Eng., Rept UTH-74, Delft, (1956)

[9] Arnal, D., "Boundary Layer Transition : Predictions based on Linear Theory", AGARD Report 793, (April 1994)

[10] Herbert, T., "Parabolized Stability Equations", AGARD Report 793, (April 1994)

[11] Airiau C., Casalis G., "Stabilité Linéaire de la Couche Limite par un Système d'Equations Paraboliques", La Recherche Aérospatiale, (Octobre 1993)

[12] Airiau, C., "Stabilité Linéaire et Faiblement Non Linéaire d'une Couche Limite Laminaire Incompressible par un Système d'Equations Parabolisé (PSE)", Thesis, ENSAE, Toulouse (June 1994)

[13] Malik, M.R., Li, F., "Three-Dimensional Boundary Layer Stability and Transition", SAE Paper 921991, (October 1992)

[14] Airiau, C., Casalis, G., "Non Linear PSE Compared with DNS and Experiment", IUTAM Symposium, Sendai (September 1994)

[15] Malik, M.R., Li, F., Chang, C.L., "Crossflow Disturbances in Three-Dimensional Boundary Layers : Non Linear Development, Wave Interaction and Secondary Instability", J. F.M. Vol. 286, (1994), 1-36

[16] Kloker, M., Fasel, H., "Direct Numerical Simulation of Boundary Layer Transition with Strong Adverse Pressure Gradient", IUTAM Symposium, Sendai, (September 1994)

[17] Kerschen, E.J., "Boundary Layer Receptivity", AIAA Paper 89-1109 (1989)

[18] Lin, N., Reed, H.L., Saric, W.S., "Leading Edge Receptivity : Navier-Stokes Computations", Boundary Layer Transition and Control Conference, Cambridge, (April 1991)

[19] Casalis G., Cantaloube B., "Receptivity by Direct Numerical Simulation", First ERCOFTAC Meeting on Direct and Large-Eddy Simulation, Kluwer Academic Publ. (1994)

[20] Spalart, P.R., "Numerical Study of Sink-Flow Boundary Layers", J.F.M. Vol. 172, (1986), 307-328

[21] Poll, D.I.A., Danks, M., "Relaminarization of the Swept Wing Attachment Line by Surface Suction", IUTAM Symposium, Sendai, (September 1994)

[22] Juillen, J.C., Arnal, D., "Experimental Study of Boundary Layer Suction Effects on Leading Edge Contamination along the Attachment Line of a Swept Wing", IUTAM Symposium, Sendai, (September 1994)

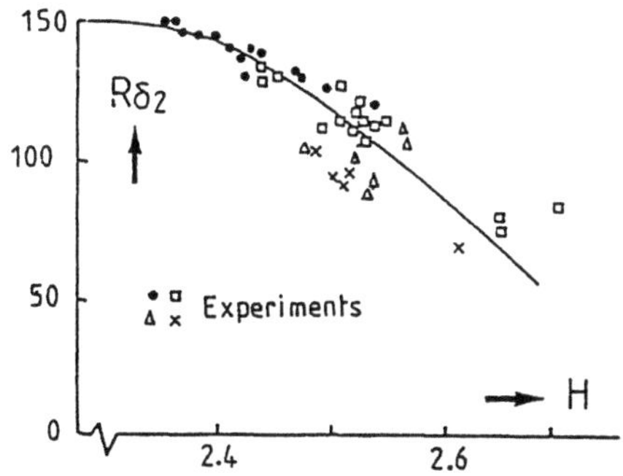

Fig. 1- Criterion for transition induced by crossflow instability

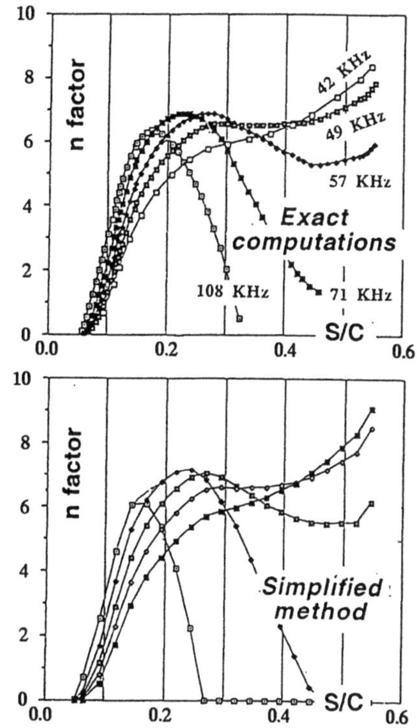

Fig. 3- Data base method
$M_\infty = 0.78$, $Rc = 3.3\ 10^6$

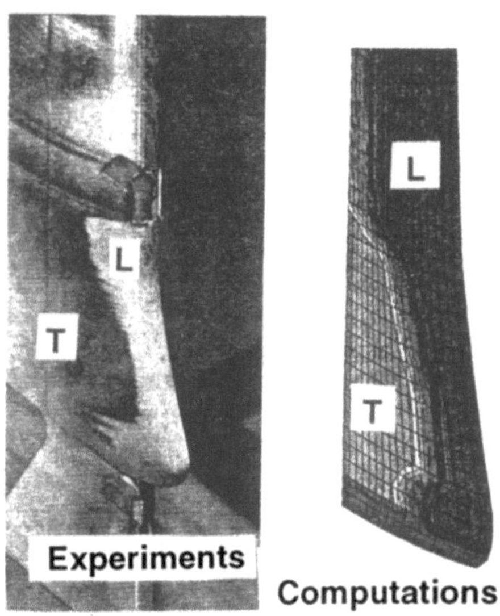

Fig. 2- Transition front around a Gaster bump (L: Laminar, T: Turbulent)

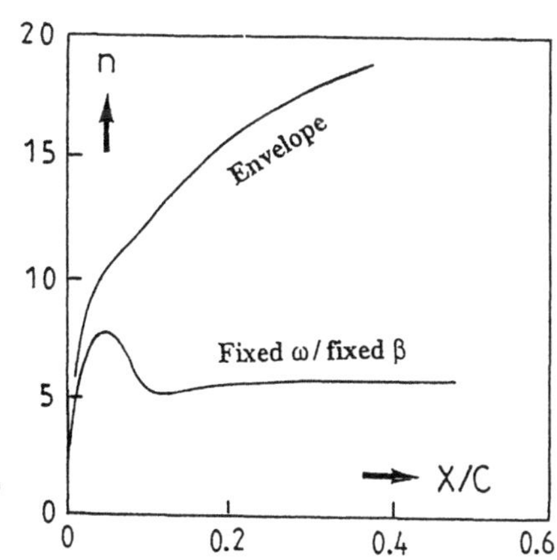

Fig. 4- Swept wing: n factors computed with two strategies ($M_\infty = 0.78$, $Rc = 20\ 10^6$, $\varphi = 27.5°$)

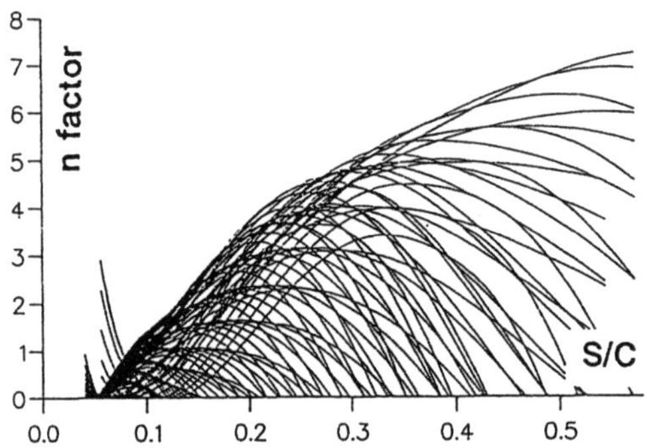

Fig. 5- n factor: PSE approach
(Q_∞ = 55 m/s, C = 0.35 m, φ= 60°)

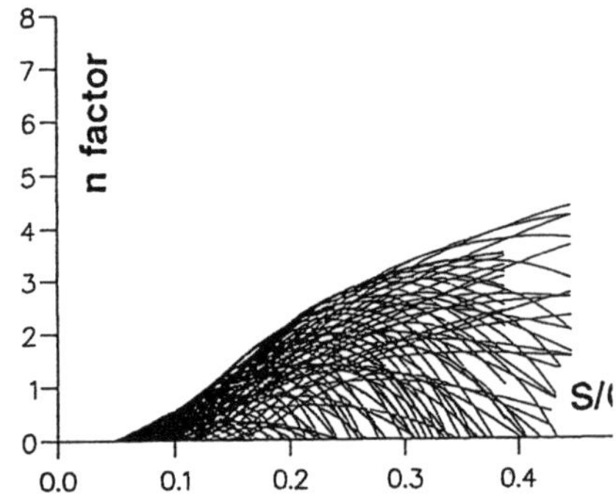

Fig. 6- n factor: local approach. Same conditions as in figure 5

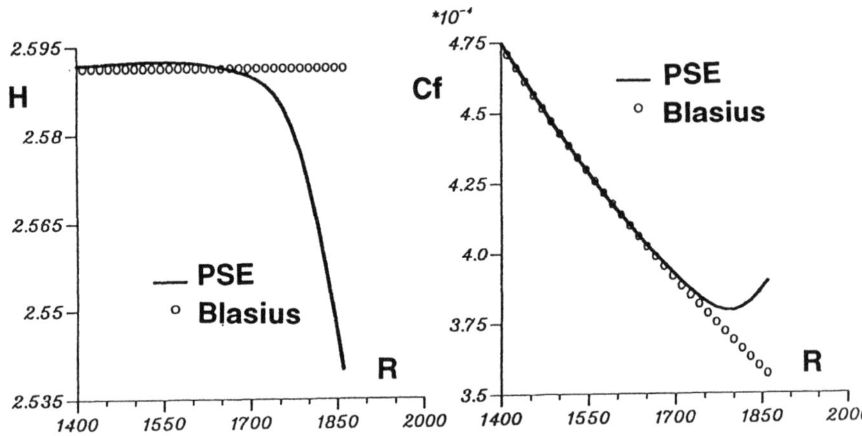

Fig. 7- Non linear PSE. R is the square root of the streamwise Reynolds number

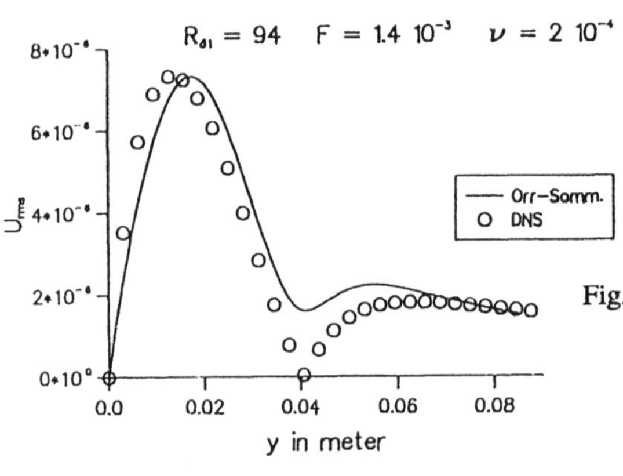

Fig. 8- DNS computations for receptivity problems on a flat plate

Some Fundamental Problems on Transition to Wall Turbulence

Michio Nishioka

Department of Aerospace Engineering, University of Osaka Prefecture
Sakai, Osaka 593, Japan

Summary
On the basis of experimental results for the ribbon-induced transition in plane Poiseuille flow, the present paper discusses (1) the residual turbulence in channel and its roles in the flow randomization process through the secondary instability and the high-frequency instability, and (2) the secondary instability caused by two phase-coupled T-S waves.

Key words: TS wave, Transition, Secondary Instability, High-Frequency Instability, Tertiary Instability, Background Turbulence, Flow Randomization.

1. Introduction

How does a laminar shear flow undergo transition to turbulence? A possible answer in general terms may be that turbulence is caused by a sequence of flow instabilities. For Blasius flow and plane Poiseuille flow, this approach has been very successful in studying the transition initially controlled by the spatial (convective) growth of two-dimensional (2-D) TS waves. Owing to a number of theoretical, experimental and numerical studies made since the experiments of Klebanoff, Kovasznay and Hama, our understanding of the transition has been much improved. On the basis of the secondary instability theory we are now able to predict the threshold (amplitude of TS wave) for the onset of the three-dimensional (3-D) wave growth which eventually leads to the breakdown into wall turbulence: see reviews [1-5].

Despite the great success of the secondary instability concept, however, the transition to wall turbulence is really so complicated that there still remain open questions of fundamental importance. Here, I would like to discuss one such problem related to the background residual turbulence. Although it is well recognized that in the open-flow systems under consideration the flow randomization should highly depend on the background turbulence, its actual roles at the stages of the secondary and higher instabilities, have not been clarified yet. In the present paper, on the basis of our experiments made in plane Poiseuille flow, it will be shown that the residual turbulence (consisting of continuous spectrum) really plays important roles even at the initial stage of the secondary instability, and beyond. Another problem to be discussed here is a process of the secondary instability which is caused by two phase-coupled TS waves of frequency $f_1(=4f_0)$ and $f_2(=5f_0)$. Our interest in this particular transition is to see how the two TS waves can set the stage for the secondary instability of the fundamental resonance.

2. Secondary Instability and High-Frequency Instability

Since, as will be shown later, the flow randomization proceeds through the secondary and high-frequency instabilities, a brief review is in order on the instabilities identified in plane Poiseuille flow. Under low background turbulence, the secondary instability in Blasius flow and plane Poiseuille flow may be of the sub-harmonic resonance or of the fundamental resonance, depending on the initial three-dimensionality in the basic mean flow. The subharmonic type may govern the transition when the basic flow is free of spanwise distortion, and the initial stage is characterized by the interaction between the primary 2-D TS wave and a pair of secondary oblique waves having the subharmonic frequency (of 2-D mode) with the nearly the same phase velocity as the primary, a necessary condition for the resonance. The flow randomization undergoes through the spectral filling-up process due to the nonlinear sum and difference interactions. This type of transition starts for the 2-D wave amplitude (in rms terms of the streamwise velocity component u) above a threshold between 0.2 to 0.6% of the freestream velocity. It should be noted that the threshold depends on whether the sub-harmonic resonance is due to Craik's mechanism or Herbert's: see [6]. When the basic flow is not purely two-dimensional but with appreciable distortions in the spanwise direction, the fundamental resonance governs the transition. This is the case in our channel flow.

In the case of the fundamental resonance, which highly depends on the spanwise distortion of the basic flow, a pair of oblique waves are first generated through the interaction of the primary 2-D TS wave with the initial mean flow distortion. Thus, without saying, the oblique waves have the same spanwise wavenumber as the initial mean flow distortion, and travel at the same phase velocity as the primary 2-D wave so that subsequently effective interactions may occur between the 2-D wave and the oblique waves to distort the mean flow with the same spanwise wavenumber. In our plane channel flow the initial spanwise distortion has almost no streamwise vorticity. However, for the amplitude of 2-D wave above a threshold about 1% of centerline velocity U_C, the streamwise vorticity is generated due to the above-mentioned interaction. Thus, for the 2-D amplitude above the threshold, the nonlinear interactions occur like resonance among the 2-D wave, the oblique waves, and the 3-D mean flow distortion to amplify themselves to form the so-called peak-valley wave structure:see [12]. The streamwise vorticity fields develop into a secondary-flow like structure with a pair of counter rotating streamwise vortices over a spanwise peak region, and the main spanwise vortex lines are convected away from the wall at spanwise peak positions and then stretched in the spanwise direction. This gives rise to a local high-shear layer riding over the secondary-flow structure at each peak region, and the instantaneous velocity distribution develops an inflection point at the position of the highest shear. The local high-shear layer appears once for each cycle of the primary wave, becoming appreciable at the 4% stage and pronounced as the wave system grows. It is of course three-dimensional, with a delta-wing like structure. As the wave system grows further, the shear layer is rapidly intensified and eventually breaks down into smaller-scale, high-frequency vortices called hairpin eddies. The process is not unlike a vortex roll-up process, and in the beginning a kink appears near the leading edge of the shear layer. So the process manifests itself by the appearance of the so-called spike, a large low-velocity pulse, which a hot-wire senses when the kink passes across it. At the stage where two spikes appear for each cycle, the first-formed kink portion have grown into a hairpin eddy: as for those details of the ribbon-induced transition in plane Poiseuille flow, see [7-12].

In order to examine the nature of the breakdown process, the high-shear layer at various stages were excited by artificial high-frequency disturbances in [8]. The wave structure and the growth rate of the excited small-amplitude, high-frequency disturbances were measured in detail. Without artificial disturbances, 'natural' small-amplitude, high-frequency disturbances were identified to grow into a distinct spike and then singled out to examine both the structure and the growth rate.

A simple stability calculation was also made on the instantaneous inflectional velocity distribution. All these results were compared in [8]. The comparison showed that the initial process of the breakdown can be interpreted as a process of inflectional instability similar to that of a conventional mixing layer, though it is by no means two-dimensional. In other words, it is similar to a typical instability event where a local sheet-like distribution of vorticity to develop into tube-like vortical structures, namely hairpin eddies in the present case. The high-frequency instability under consideration is also a kind of the secondary instability and may be termed the high-frequency secondary instability. Or it may be called the tertiary instability, considering the order of the occurrence.

3. Soliton Model and Wave Resonant Concept

Kachanov, Ryzhov and Smith [13] proposed a hypothesis that the spike may be interpreted as a kind of soliton given by the Benjamin-Ono equation. This approach is interesting for the possibility of describing the whole process from the formation of the high-shear layer to its breakdown into hairpin eddies, though this is not confirmed yet. They state that (1) the high-frequency instability actually takes place; however, (2) the generation of the flash-spikes (which in this context they mean hairpin eddies) is probably not connected immediately with the high-frequency instability; and (3) the spikes are almost certainly of soliton nature. According to Kachanov [5], the spikes represent strictly deterministic, periodic structures generated by the flow as a result of gradual amplification of definite 3-D harmonics of the fundamental wave and their phase synchronization. Relying on subharmonic-type four-wave resonant interactions (among higher harmonic components) as well as on the fundamental resonance, he expects the wave resonant concept to describe the mechanism for the formation of high-shear layer and its breakdown. If the phase synchronization in the strict sense is the necessary condition for the formation of the high-shear layer and its subsequent breakdown, the random background turbulence may not come into play in these processes. As will be shown later, this is not the case in our channel flow.

4. Role of Background Turbulence in Transition to Wall Turbulence

4.1 Experimental Procedure

The whole experiment is carried out in a long channel of rectangular cross section whose half depth h and total span are respectively 7.3 mm and 400 mm and thus the aspect ratio is 27.4. The test section is located beyond $600h$ downstream from the channel inlet.

A hot-wire anemometer measures the time-mean and fluctuation velocity U and u in the streamwise x-direction. The probe consisting of two hot-wires (24 mm apart in x-direction, for measuring the phase velocity) is fixed at $48h$ downstream from the vibrating ribbon, while it can be continuously traversed in the normal-to-wall y-direction and spanwise z-direction. Therefore, the wave growth in x-direction is simulated at the fixed-x observation position by varying the amplitude of the ribbon oscillation, as done in our previous experiments. The Reynolds number R is fixed at a subcritical value 5000, where R is defined on the center-plane velocity U_C (9.8 m/s) and the half depth h (7.3 mm). In the test section the laminar flow is fully developed. The residual turbulence is less than 0.05% of U_C in rms values and can be of order of 0.1% instantaneously. However, the basic flow is not purely two-dimensional but with a slight distortion. The mean velocity varies in the spanwise z-direction. The spanwise variation is, at most, of the order of 2% of U_C in terms of peak-to-peak values, and it is almost periodic with a dominant spanwise wavelength λ_z of about 25 mm. The slight mean flow distortion is found to be well represented by the three-dimensional, least damped, steady eigenmode without streamwise vorticity. This distortion is the origin of the peak-valley wave development leading to wall turbulence in our channel:see [12].

We exclusively rely on the FFT analysis in the present study. The hot-wire signal is stored in a micro-computer through 12-bit A/D conversion at such a rate that exactly 64 data are obtained for one period of the ribbon oscillation $1/f_{TS}$ and the number of total data is 8192 for a single conversion run, which corresponds to a time interval of $128/f_{TS}$. The ribbon frequency f_{TS} is set at 72.3 Hz near the least damped TS wave frequency at $R = 5000$. Thus the frequency resolution for FFT analysis is $\Delta f = 72.3/128 \approx 0.565$ Hz. It is noted that the AC components of hot-wire signal are amplified 10~1000 times for the A/D conversion to maintain the data accuracy for the FFT analysis.

4.2 Background Turbulence in Channel Flow

The background residual turbulence is first described. As shown in Fig.1, The residual turbulence in our channel is found to consist of continuous spectrum, except for the fan noise and hum noise which have no noticeable effects on the wave development under consideration. The continuous spectrum exists at frequencies below about 80 Hz and the amplitude is from 2×10^{-4} to 10^{-5} of U_C. From the measurement of the phase velocity and the y-distributions of the rms amplitude, it turns out that the lowest-frequency components below 25 Hz is of almost purely sound-wave like, and the amplitude distribution is well represented by the oscillating Poiseuille flow solution. This is confirmed for selected 0.5 ~ 3.96 Hz band, 4.52 ~ 7.91 Hz band and 12.43 ~ 23.73 Hz band components. On the other hand, a selected 62.70 ~ 68.90 Hz band component is identified as 2D-TS wave mode by comparing the amplitude distribution with the corresponding eigenfunction. Thus it appears that the residual turbulence in our channel consists of sound-wave modes and 2-D TS wave modes. In the flow randomization process, the lowest-frequency components of the sound-wave modes play an important role. Since the sound-wave modes with long wavelengths are like mean flow and may effectively interact with the ribbon-induced 2-D TS wave and the related 3-D waves to give rise to the relatives appearing as sidebands of these discrete wave modes.

4.3 Flow Randomization at Secondary Instability Stage

Even when the ribbon is excited to introduce the 2-D TS wave of $f_{TS} = 72.3$ Hz, the continuous spectrum does not change much for the wave amplitude less than the threshold about $0.01U_C$. Above the threshold, however, the continuous spectrum increases in the lowest-frequency band as well as in the sidebands of f_{TS} and its harmonics. To show this, Fig.2 illustrates the spectrum of typical u-fluctuations observed at $y/h = 0.34$ (the wall locates at $y = 0$) corresponding to the y-position of the high-shear layer, at the $u'_m/U_C = 6\%$ stage, where u'_m denotes the y-maximum of rms u-fluctuation at a typical spanwise peak position. The total u is not purely periodic. We see irregular time-variations in the amplitude of the V-shaped wave form. The total u can be decomposed into the periodic component (consisting of the line spectra at f_{TS} and its harmonics, with their total contribution of about 13% of U_C in peak-to-peak amplitude) and the irregular component (due to the continuous spectrum), and their wave forms are also compared in Fig.2.

Although all the frequency components of the continuous spectrum are less than $0.001U_C$ in amplitude, their total contribution can instantaneously exceed 6% of U_C in peak-to-peak terms, suggesting a great influence on the formation of the high-shear layer. It is quite interesting that the u-fluctuation due to the continuous spectrum does not look completely irregular. We see the positive and negative V-shaped wave forms only. Their appearance is almost periodic, and nearly synchronized to the periodic components, though the positive-negative sign and the amplitude change at random. These feature are very hard to imagine directly from the spectral distribution. One may say that all the frequency components (of the continuous spectrum) are nearly synchronized at least locally in time so as to generate the V-shaped wave forms only. Or one may also say that the V-shaped wave forms are independent to each other when being far apart in time, and generated locally in time and thus intermittently. The reason for the local (and approximate) synchronization is that some of the continuous-spectrum components are directly related to the periodic components, and the reason for the irregular and intermittent appearance (of the V-shaped wave form) is that surely involved is the lowest-frequency components which are originally due to the random residual turbulence.

Fig.3 displays an enlarged view of the spectrum given in Fig.2. We notice a remarkable similarity in shape between the spectral distributions in the lowest-frequency band and in the sidebands of the fundamental f_{TS} and its harmonics; see also Fig.2. To examine this similarity in detail, we first focus on the u-fluctuation due to the fundamental f_{TS} component together with its sideband components in the range of frequency from 55.3 to 89.3 Hz. As illustrated in Fig.3, the fluctuation is highly modulated. The modulated u-signal can be demodulated as written below:

$$u(t) = A_0\left(1 + A(t)\right)\cos\left(2\pi f_{TS}\, t + P(t)\right)$$

where $A(t)$ and $P(t)$ respectively denote the amplitude and phase modulations, which are easily calculated on the basis of the FFT data. In Fig.4, $A(t)$ and $P(t)$ are compared with the u-fluctuation, $L(t)$ due to the lowest-frequency components (in the range less than 17 Hz). Except for the phase shift of 180 °, they are almost exactly similar in wave form.

Due to the counter-rotating streamwise vortices generated by the fundamental resonance, the main spanwise vortex lines are convected away from the wall at a peak region. Thus it is reasonable to infer that $A(t)$ and $P(t)$ are of positive value for the time interval when the convection (and then the high-shear layer) is intensified. A remarkable similarity between $A(t)$, $P(t)$ and $L(t)$ supports this view and suggests that the lowest-frequency components can quite effectively modulate the intensity of the streamwise vorticies, and further that the the lowest-frequency components as well as the sideband components are mainly of oblique wave modes with the same spanwise wave number as that of the fundamental resonance; $2\pi/\lambda_Z$.

In this connection, Fig.5 compares the lowest-frequency components (in the range;0.5 to 11.3 Hz), the lower-sideband components (61.0 to 71.7 Hz) and the higher-sideband components (72.9 to 83.6 Hz), on their y-distributions of rms amplitude observed at the u'_m/U_C = 7% stage, at the same peak. Here, all the y-distributions are similar in shape, typical of oblique wave, taking maximum at the position of the high-shear layer. We may conclude that both the sideband-frequency components and the lowest-frequency components are of oblique modes. This is of course due to the nonlinear interactions among the wave components for the fundamental resonance and the lowest-frequency and sideband-frequency components; this is probably also the case for each higher harmonic components. As for the important phase velocity, the measurements are summarized here. The lower-sideband and higher-sideband components (in Fig.5) have the same phase velocity, which increases with y but remains nearly constant near and across the high-shear layer: $0.35U_C$ at $y/h = 0.22$, $0.42U_C$ at high-shear layer ($y/h = 0.38$) and $0.41U_C$ at $y/h = 0.50$. As for the lowest-frequency components, the phase velocity is $0.56 \sim 0.66U_C$ at $y/h = 0.34$, $0.43 \sim 0.53U_C$ at high-shear layer ($y/h = 0.38$), and $0.33 \sim$

$0.45U_C$ at $y/h = 0.50$. In this measurement we do not decompose these components into 2-D and 3-D waves. However, since it is reasonable to infer that the oblique wave modes are dominant across the high-shear layer, let us focus on the phase velocity there. We see that the three components travel at almost the same velocity as that of the high-shear layer, namely $0.40U_C$ at this 7% stage. These important features of the lowest-frequency and sideband-frequency components are ascertained by introducing a discrete sound wave of 14.7 Hz and measuring the rms amplitude distributions and phase velocities of these components.

As described thus far, the residual turbulence has important effects on the randomization of the flow at the stage of the secondary instability leading to the formation of the high-shear layer. The lowest-frequency and sideband-frequency components are strongly vortical as can be seen from the measurements of the phase velocity and the y-distribution of the rms amplitude. How to quantify their effects on the flow randomization? In this sense it is interesting to examine the rms values of $A(t)$ and $P(t)$, which are respectively called the amplitude index AI, and the phase index PI, at various y-positions, at various stages of the wave development. We have done this at the same peak. Simultaneously, the correlation coefficients between $L(t)$ and $A(t)$, and between $L(t)$ and $P(t)$ are calculated, and are respectively written as C_{LA} and C_{LP}. Because of the limited space here, only each y-maximum of AI, PI, C_{LA} and C_{LP} (examined for y/h = 0.1 to 0.7) are given below:

Flow Stage	:	0.5 %	2.0 %	6.0 %	8.0 %	1-spike	2-spike	3-spike
AI	:	0.066	0.051	0.105	0.131	0.21	0.31	0.19
PI (degree)	:	4.3	3.1	5.6	9.6	18.5	32.5	11.2
C_{LA}	:	0.33	0.80	0.96	0.97	0.96	0.89	0.96
C_{LP}	:	0.24	0.38	0.86	0.93	0.96	0.91	0.69

This table shows the effects of the lowest frequency components to be the most essential in the flow randomization, which proceeds through generating sideband frequency components of oblique wave modes by the nonlinear interactions mentioned earlier, and thus being pronounced at the high-shear layer (where, $C_{LA} \approx 1$) and the related high-shear region in valley, or the near wall high-shear regions at the later stages.

4.4 Growth of High-Frequency, Continuous-Spectrum Components

With the results presented thus far, we may say that the formation and development of the high-shear layer need not be strictly deterministic in the sense that continuous-spectrum components can contribute to the process significantly, although this does not exclude such a possibility that the process can be periodic and strictly deterministic under extremely low background turbulence. This view seems to be also true for the breakdown process of the high-shear layer into hairpin eddies as discussed below.

To show the growth of high-frequency, continuous-spectrum components in the breakdown, Fig.6 demonstrates the spectrum observed at the 2-spike stage and at y/h = 0.55, while Fig.7 illustrates the same results in terms of u-fluctuations, namely, the total u, the periodic u due to all the discrete spectra, the periodic u due to the discrete spectra higher than $4f_{TS}$, and the random u due to the continuous-spectrum components higher than $4f_{TS}$. The corresponding results are obtained at various y-positions and at 1-spike, 2-spike, 3-spike and multi-spike stages. As for the high-frequency components, the random u is dominant, compared with the periodic u, in generating hairpin eddies. The random u does not look totally irregular, and the continuous-spectrum components are organized or synchronized in phase to generate spike-wave form at each cycle of the fundamental f_{TS}.

Furthermore, in the the spectrum distribution given in Fig.6, we see a hillock of continuous spectrum not only in the frequency range of high-growth rate predicted by the high-frequency instability (see [8]), but also its harmonics frequency ranges. The discrete spectra also show the corresponding hillocks. We see all these features at 1-spike, 2-spike, 3-spike and multi-spike stages. The present writer interprets these results as supporting the concept of the high-frequency instability of the high-shear layer for the mechanism of generating hairpin eddies. This view does not exclude the possibility that the spike is of such soliton nature as advocated by Kachanov, because a hairpin eddy, once generated, moves towards the freestream, and if isolated in the freestream without shear, it only decays very slowly, and deforms gradually under the self-induction effect, so that it can maintain its structure while traveling downstream almost parallel to the wall for a considerably long distance.

In conclusion, may it be initially the oscillating Poiseuille flow or TS wave mode, the background, continuous-spectrum turbulence develops 3-D oblique wave modes in the lowest frequency range as well as in the sidebands of f_{TS} and its harmonics through the nonlinear interactions among the discrete wave modes, continuous-spectrum wave modes and the mean flow distortion modes. The continuous-spectrum components are thus transferred to high-frequency ranges at the stage of the secondary instability, and are further amplified at the stage of the high-frequency instability. The high-frequency continuous-spectrum components can make a dominant contribution to the formation of hairpin eddies. In this way the background turbulence plays significant roles in the flow randomization. Without any background turbulence, the chaos may set in, but we do not know exactly when and how it does in the transition under consideration.

5. Secondary Instability caused by Two Phase-Coupled TS Waves

We have done an experimental study where two phase-coupled TS waves of frequencies f_1 and f_2 near and around the frequency of the most unstable mode (actually, the least damped mode) are excited by means of the vibrating ribbon in plane Poiseuille flow at R = 5000. f_1 and f_2 are set at $4f_0$ and $5f_0$ respectively, where f_0 = 15.5 Hz. This is probably one of the simplest way of simulating natural transition where the wave interactions among different frequencies may take place to set the stage for the secondary instability. The ribbon is driven by a current proportional to ($a_1 \sin 2\pi f_1 t + a_2 \sin 2\pi f_2$). The ratio a_2/a_1 is fixed so that the two TS waves are of equal amplitude at the fixed-x observation position, 48h downstream from the ribbon. The vibrating ribbon is really a linear system. With the fixed a_2/a_1, the total ribbon current can be increased without changing the equal amplitude condition (at the observation position), unless the secondary instability sets in. When the y-maximum of total rms u is 0.3% of U_c, the corresponding maximum is 0.21% for u_1 (with f_1) and u_2 (with f_2), without significant nonlinear interactions: u_1 and u_2 are purely of TS mode. The total u is of course modulated at f_0, the beating frequency (f_2 - f_1). It appears that the secondary instability for the three-dimensional wave growth sets in locally in time when the total u (beating at f_0) becomes maximum. The threshold (of this maximum) is found to be almost the same as that for a TS wave of a single frequency of ($f_2 + f_1$)/2. At the threshold condition, the amplitudes of f_1 and f_2 TS waves (of 2-D mode) are about 0.5%. Among the possible 3-D wave modes which have the fundamental spanwise wavenumber due to the mean flow distortion, the f_0, f_1 and f_2 components are dominant and of almost equal amplitude. These clearly indicate that this particular secondary instability is caused by interactions in which the oblique waves of beating f_0 and the mean flow distortion mode play important roles to operate the fundamental-type and near fundamental-type resonances. The 3-D waves can grow for a time period (much shorter than $1/f_0$) in which the resonance (and/or near resonance) conditions are maintained, as illustrated in Fig.8. With increasing the ribbon current, the high-shear layer develops and eventually breaks down into hairpin eddies. These events occur at the beat frequency f_0, and seemingly not different from those observed for a single TS wave; see Fig.9.

Although we have not followed the spatial wave development, it may be conclude that the secondary instability, in the sense of the growth of 3-D waves, may occur under local flow conditions for resonant and nearly resonant wave interactions. This conclusion is in good harmony with the observations in section 4.3 on the important roles of the lowest-frequency and the sideband-frequency continuous-spectrum components.

The author wishes to express his cordial thanks to to Prof. M. Asai for his cooperation and to Profs. S. Iida and H. Sato for their continual encouragement.

References

[1] Herbert, Th.: Secondary instability of boundary layers, Ann. Rev. Fluid Mech. 20 (1988) 487-526.
[2] Bayly, B. J., Orszag, S. and Herbert, Th.: Instability mechanisms of boundary layer transition, Ann. Rev. Fluid Mech. 20 (1988) 359-391.
[3] Morkovin, M. V. and Reshotko, E.: Dialogue on Progress and issues in stability and transition research, in Laminar-Turbulent Transition (eds. D. Arnal and R. Michel), Springer (1990) 3-29.
[4] Kleiser, L. and Zang, T. A.: Numerical simulation of transition in wall bounded shear flows, Ann. Rev. Fluid Mech. 23 (1991) 495-537.
[5] Kachanov, Y. S.: Physical mechanisms of laminar-boundary-layer transition, Ann, Rev. Fluid Mech. 26 (1994) 411-482.
[6] Saric, W. S. and Thomas, A. S. W.: Experiments on subharmonic route to turbulence in boundary layers, in Turbulence and Chaotic Phenomena in Fluids (ed. T. Tatsumi), North-Holland (1984) 117-122.
[7] Nishioka,M., Iida, S and Ichikawa, Y.: An experimental investigation of the stability of plane Poiseuille flow, J. Fluid Mech. 72 (1975) 731-751.
[8] Nishioka, M., Asai, M. and Iida S.: An experimental investigation of the secondary instability, in Laminar-Turbulent Transition (ed. R. Eppler and H. Fasel), Springer (1980) 37-46.
[9] Nishioka, M., Asai, M. and Iida S.: Wall phenomena in the final stage of transition to turbulence, in Transition and Turbulence (ed. R. E. Meyer), Academic Press (1981) 113-126.
[10] Nishioka, M. and Asai, M.: Evolution of Tollmien-Schlichting waves into wall turbulence, in Turbulence and Chaotic Phenomena in Fluids (ed. T. Tatsumi), North-Holland (1984) 87-92.
[11] Nishioka, M. and Asai, M.: Three-dimensional wave disturbances in plane Poiseuille flow, in Laminar-Turbulent Transition (ed. V. V. Kozlov), Springer (1985) 173-182.
[12] Asai, M. and Nishioka, M.: Origin of the peak-valley wave structure leading to wall turbulence, J. Fluid Mech. 208 (1989) 1-23.
[13] Kachanov, Y. S., Ryzhov, O. S. and Smith, F. T.: Formation of solitons in transitional boundary layers : theory and experiment, J. Fluid Mech. 251 (1993) 273-297.

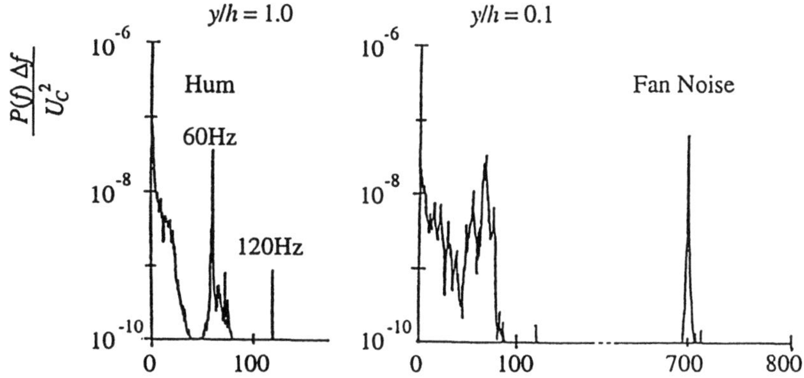

Fig. 1. Background turbulence in 2-D chanel. Spectra at y/h = 1.0 (channel center) and 0.1.

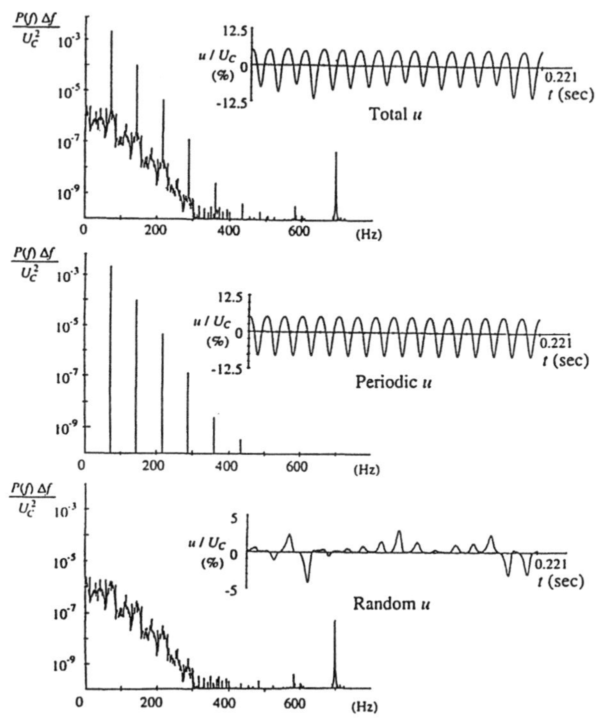

Fig. 2. Sepctral components and corresponding u-fluctuations at 6% stage.

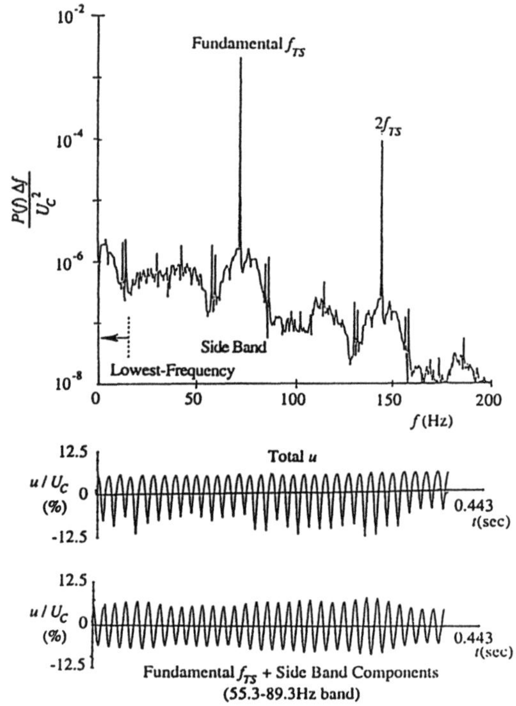

Fig. 3. Enlarged view of the spectrum in Fig. 2.
Effects of sideband components on fundamental waves are illustrated by comparing the amplitude modulated u with the total u.

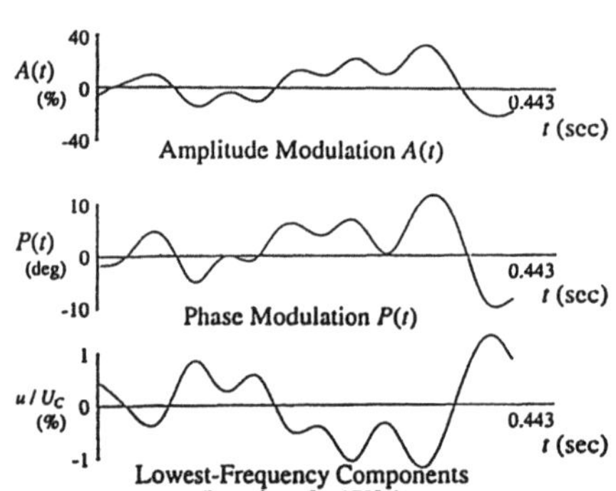

Fig. 4. Comparison among $A(t)$, $P(t)$ and $L(t)$.

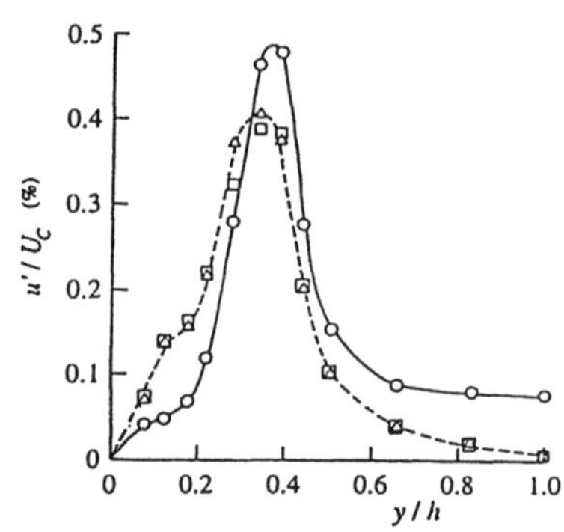

Fig.5. Amplitude (rms) distributions of the lowest-frequency(○:0.5-11.3Hz), the lower-sideband (△:61.0-71.7Hz) and the higher-sideband (□:72.9-83.6Hz) components.

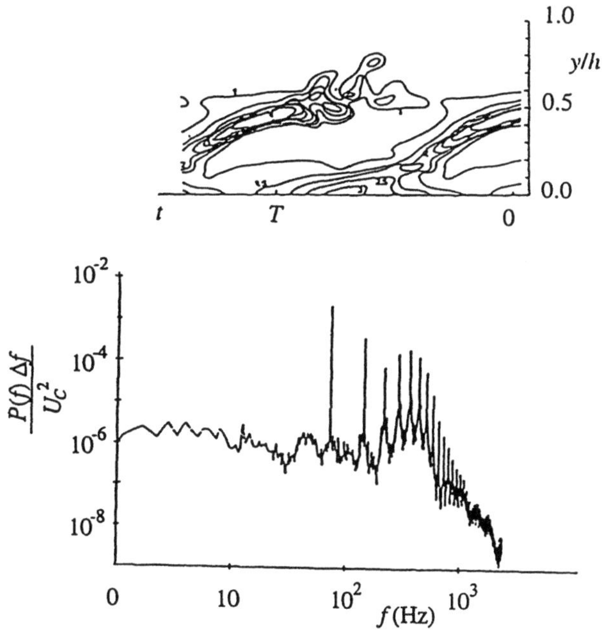

Fig. 6. High-shear layer (represented by equi-shear lines), and spectrum at $y/h = 0.55$, at 2-spike stage.

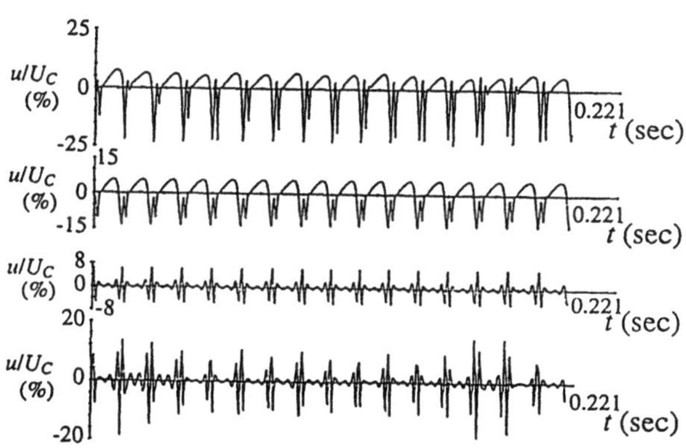

Fig. 7. Various u-fluctuations at $y/h = 0.55$, at 2-spike stage. From top, total u, periodic u due to all the discrete spectra, periodic u due to the discrete spectra higher than $4f_{TS}$, and random u due to the continuous-spectrum components higher than $4f_{TS}$.

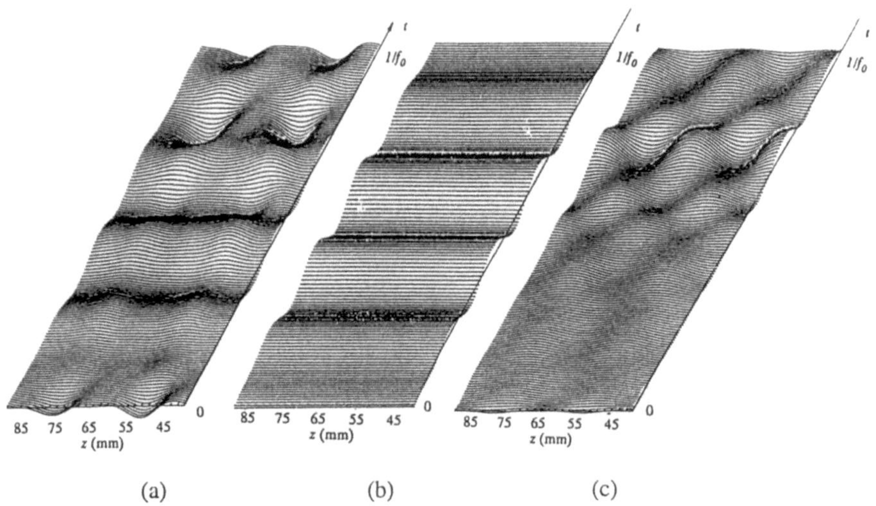

Fig. 8. The secondary instability caused by two phase-coupled TS waves, illustrated by means of u-fluctuations at $y/h = 0.185$, at $u'_m/U_C = 2\%$ stage. Total u in (a), 2-D waves in (b), and 3-D waves (with the spanwise wavenumber, $2\pi/\lambda_z$) in (c).

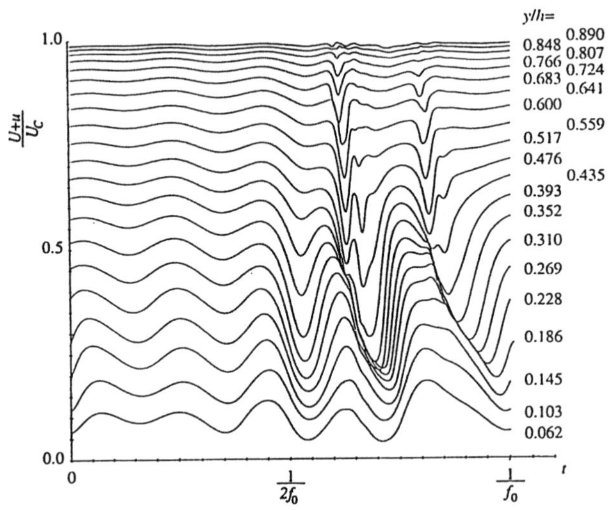

Fig. 9. Ensemble-averaged wave forms at 2-spike stage ($z = 77$mm). The transition is initially controlled by two phase-coupled TS waves. The formation of high-shear layer and its eventual breakdown into hairpin eddies occur in the same way as observed for the case of a single TS wave.

Session A: Fundamentals I

Is the Classic Weakly Nonlinear Theory of Hydrodynamic Stability Good Enough and How It Can Be Improved

H. Zhou

Department of Mechanics, Tianjin University, Tianjin 300072, China

K. Fujimura

Japan Atomic Energy Research Institute, Tokai Research Establishment, Tokai–mura, Naka–gun, Ibaraki–ken 319–11, Japan

Summary

In acknowledging that the classic, weakly nonlinear theory should be seen as a milestone in the developement of the theory of hydrodynamic stability, it has been re–examined. It was shown that its equillibrium solution must be divergent. Also, it was pointed out that the original formulation did not fit an arbitrary initial condition, and there was another time scale R other than the two time scales usually thought to exist. Ways for its improvement were suggested, and the results were in agreement with those obtained by numerical simulations, as long as the amplitude of the fundamental wave was not too large.

Key words: hydrodynamic stability, weakly nonlinear theory, improvements

1. Introduction

Weakly nonlinear theory of hydrodynamic stability has existed for more than 30 years and has been acknowledged as a classic theory now. However, very recently, Zhou (1991), Zhou and You (1992, 1993) have analysed this theory in detail again and found it did need some improvements. Later, the present authors pursued further in this direction and more new results have been obtained, which made the above mentioned analysis more convincing. In this paper, we shall take the plane Poiseuille flow as the prototype problem to show the main results.

The weakly nonlinear theory of hydrodynamic stability is essentially a perturbation theory, in which the solution is expanded as a series of a certain small parameter. Often, the amplitude of the fundamental disturbance wave, the Tollmien–Schlichting wave, is taken as the small parameter. The T–S wave can be written in the following form,

$$\{\mathbf{u}_1, p_1\}^T = \{\hat{\mathbf{u}}_1(y), \hat{p}_1(y)\}^T e^{i(\alpha x - \omega t)} + c.c.,$$

$$or = a(t)\{\hat{\mathbf{u}}_1(y), \hat{p}_1(y)\}^T e^\theta + c.c., \qquad \theta = \alpha x - \omega_r t, \tag{1}$$

where a is the amplitude, if $\hat{\mathbf{u}}_1$ *and* $\hat{p}_1$ have been suitably normalized, α the wavenumber, $\omega = \omega_r + i\omega_i$, the complex eigenvalue, to be solved together with $\hat{\mathbf{u}}_1$ and $\hat{p}_1$ from an eigenvalue problem, and c.c stands for the complex conjugate.

Since in equ.(1) $\{\mathbf{u}_1 p_1\}^T$ is written in the form depending on t only through a and θ, so in the original theory, all higher order terms are assumed also to be dependent on t implicitly through a and θ, and strictly in proportion to a proper power of a. Thus, the solution of the perturbation equation are sought in the form as

$$\{\mathbf{u}, p\}^T = \sum_{n=1}^{\infty} a^n \{\mathbf{u}_n(\theta, y), p_n(\theta, y)\}. \tag{2}$$

In the nonlinear case, a and θ are assumed to satisfy the following Landau equation

$$da / dt = \omega_i a + \sum_{s=2}^{\infty} A_s a^s, \qquad d\theta / dt = -\omega_r + \sum_{s=2}^{\infty} B_s a^{s-1}. \tag{3}$$

The Landau equation has been used mainly in two ways. First, by putting $da / dt = 0$, one can solve for a. Denote the so found a by a_e, it is the equillibrium amplitude of the fundamental. Second, Landau equation has been used as the nonlinear evolution equation for the disturbance.

However, we will show that one has to be careful in using the Landau equation in the above ways.

2. The new findings

Starting from the second order, there will be mean flow distortion. By picking up a special partial sum of the mean flow distortion in the solution, Zhou and You (1993) were able to prove that the radius of convergence of the solution must be smaller than $(\omega_i / A_3)^{1/2}$ in the case of constant pressure gradient, but somehow larger in the case of constant mass flux. Following the same idea, but instead of analysing the mean flow distortion, we analysed the terms having the factor $\exp(2i\theta)$. Also, by picking up a special partial sum of such terms, we were able to prove that the necessary condition for the convergence of the solution (2) is

$$\left| \sum_{s=1}^{N} A_s a^s / \omega_i a \right| < 1 \qquad \textit{for arbitrary but finite } N \tag{4}$$

which implies that if we calculate the equillibrium amplitude for the fundamental, by truncating the Landau equation at any finite order, then for this amplitude, the solution (2) must be divergent.

We can give an heuritic explanation to this phenomemum of divergence. The fundamental assumption for a perturbation analysis is that the higher order terms must be much smaller than the linear term. While talking about the equillibrium solution, then this assumption is violated, as we need the higher order terms to balance the linear term in the Landau equation. In the original formulation, people actually move higher order terms, which are not order of magnitude smaller than the linear term, into the higher order perturbation equations, thus break the principle that the higher order perturbation equations should only contain terms whose order of magnitude are compatible with that equation.

As pointed out first by Zhou (1991), the solution (2) has a very special structure, such that once the amplitude of the fundamental is fixed, all the higher order terms are fixed too, no matter what the real situation is.

In view of which, Zhou and You (1993) suggested that all the higher order terms should be kept as functions of t, y and x, and be solved in the way as initial value problems. Thus, the solution (2) should be replaced by the following expression

$$\{u,p\}^T = a\{\hat{u}_1(y),\hat{p}_1(y)\}^T e^{i\theta} + c.c. + \sum_{n=2}^{\infty}\{\mathbf{u}_n(t,x,y),p_n(t,x,y)\}^T \tag{5}$$

and the perturbation equations for u and p read:

$$\frac{\partial \mathbf{u}_n}{\partial t} + u_0\frac{\partial \mathbf{u}_n}{\partial x} + v_n\frac{d\mathbf{u}_0}{dy} + \nabla p_n - \frac{1}{R}\nabla^2\mathbf{u}_n = -\sum_{\substack{j,k\geq 1\\ j+k=n+1}}(\mathbf{u}_j\cdot\nabla)\mathbf{u}_k - A_n\frac{\partial \mathbf{u}_1}{\partial a} - B_n\frac{\partial \mathbf{u}_1}{\partial \theta}, \tag{6}$$

Then they proposed the way to simplify the solution and calculated the evolution of the disturbances for several super– as well as sub–critical cases, the results were in agreement with those obtained by numerical simulations, which were unable to achieve by the original theory.

However, there was a slip, though it had only minor effects on the results.

The correct way to express $\mathbf{u}_n$ and p_n is to write them as $\mathbf{u}_n(t,y,\theta)$ and $p_n(t,y,\theta)$, Thus, the proper perturbation equations for the higher order terms should be

$$\frac{\partial \mathbf{u}_n}{\partial t} - \omega_r\frac{\partial \mathbf{u}_n}{\partial \theta} + u_0\frac{\partial \mathbf{u}_A}{\partial x} v_n\frac{d\mathbf{u}_0}{dy} + \nabla p_n - \frac{1}{R}\nabla^2\mathbf{u}_n =$$

But for the case of $s=0$, we need more terms, depending on the accuracy we expect for the expansion (10).

We have calculated the equillibrium amplitude for two cases, the results are shown in Table 1. We can see that the present method does offer convergent solution, while the original method does not. To be sure that our convergent solution is correct, we carried out a numerical simulation for the evolution of the disturbance for the first case, starting from a pure T–S wave with an initial amplitude of 0.02 without higher order terms, using 8 Fourier mode in x direction and 64 Chybeshev in y direction. The time for approaching the equillibrium state was unexpectedly long. It took more than 33000 time units, and the final equillibrium amplitude for the T–S wave was 0.02653, yielding a descrepency of 0.7%. Considering that we have only 8 Fourier modes in x direction, the agreement must be considered as excellent.

We have also calculated the evolution of the disturbance for several cases, using fifth order approximation of the new method as well as the numerical simulation. The comparison of the results were quite good, as long as the amplitude of the T–S wave is not too large. To quote an example, which started with a pure T–S wav with an initial amplitude 0.05, the result is shown in Fig.1. The most striking feature is that it clearly shows that the mean flow distortion has its own time scale R for its evolution. Also, the sudden change of the gradient in the evolution curve at the point q is unexplanable by the original theory. In fact, it was due to the constant changing of the shape of the mean flow distortion, governed by a diffusion equation having a time scale R, as it appears in the equation. If we superimpose the two figures in Fig.1 together, they are virtually indistinguishable. But we should admit that in calculating the evolution rather than just the equillibrium amplitude, we would have to do much more amount of work than the original method, though it is still two order of magnitude faster than the numerical simulation if we only carry out up to the fifth order.

We have also compared the evolution of the disturbances with the same initial condition in their linear parts, i.e. the T–S waves, but with different nonlinear parts. For example, we have calculated the evolution of the disturbance with the same initial amplitude for the T–S wave as for the case shown in Fig.1, but with the second order terms not being zero, but calculated in accord with those given by the original weakly nonlinear theory. Then we found that the decay of the fundamental was significantly slower than those given in Fig.1. At $t=200$, it was 40% larger than those shown in Fig. 1. The main reason was in this case, the mean flow distortion was much larger than those just being generated from nothing, and their shape were quite different at all the time.

$$-\sum_{\substack{j,k\geq 1\\ j+k=n+1}}(\mathbf{u}_j\cdot\nabla)\mathbf{u}_k - A_n\frac{\partial \mathbf{u}_1}{\partial a} - \sum_{\substack{j,k\geq 1\\ j+k=n+1}} B_n\frac{\partial \mathbf{u}_1}{\partial \theta}. \tag{7}$$

In general, the perurbation equations we must solve are of the form

$$L\left(y,\frac{\partial}{\partial t},\frac{\partial}{\partial x},\frac{\partial}{\partial y}\right)\mathbf{v} = g(t)F(y)e^{is\theta} \tag{8}$$

where L is a linear operator. The solution procedure consists of three steps:
(1) Solve the eigenvalue problem posed by

$$L\mathbf{v}=0, \quad \text{with the same boundary condition,} \tag{9}$$

by assuming $\mathbf{v}=\hat{\mathbf{v}}(y)e^{(is\alpha x-\omega' t)}$. Denote its eigenvalues, arranged in the order of descending ω_i', and eigenfuctions by $\omega'_n, \hat{\mathbf{v}}_n$.
(2) Find the perticular solution of equ.(8) by putting g=1 and assuming $\mathbf{v}=\bar{\mathbf{v}}(y)e^{is\theta}$. By the expansion theorem, $\bar{\mathbf{v}}$ can be expanded into a series of $\hat{\mathbf{v}}_n$ as

$$\bar{\mathbf{v}} = \sum_{n=1}^{\infty} b_n\hat{\mathbf{v}}_n. \tag{10}$$

(3) Let the initial value of $\mathbf{v}$ be $\tilde{\mathbf{v}}(y)e^{is\alpha x}$, *and expand* $\tilde{\mathbf{v}}(y)$ as a series of $\hat{\mathbf{v}}_n$ as well

$$\tilde{\mathbf{v}} = \sum_{n=1}^{\infty} B_n\hat{\mathbf{v}}_n. \tag{11}$$

Then the solution of equ. (8) with the initial value $\tilde{\mathbf{v}}(y)e^{is\alpha x}$ can be expressed as

$$\mathbf{v} = g(t)\bar{\mathbf{v}}e^{is\theta} - \sum b'_n\hat{\mathbf{v}}_n e^{(is\alpha x - i\omega'_{nr}t)} \tag{12}$$

where we have $\omega'_n=\omega'_{nr}+i\omega'_{ni}$, and b'_n satisfies the following equation and initial condition, which is very easy to integrate

$$\frac{db'_n}{dt} = \omega'_{ni}b'_n + \frac{dg}{dt}b_n e^{i(\omega'_{nr}-n\omega_r)t}, \quad b'_n(0) = b_n g - B_n.$$

Since in general, ω'_{ni}has large negative value, except in the case of s=0. So under the condition $|dg/dt| << |\omega_{1i}|$,the series in (12) has a rather small value after a short period. Thus we can see it as having only the effect of delaying the $\mathbf{v}$ to approach $g\bar{\mathbf{v}}e^{is\theta}$, and the transient time depends only on ω'_{1i}. Thus the solution (12) can be greatly simplified as:

$$\mathbf{v} = B\bar{\mathbf{v}}e^{is\theta}, \quad B(0) = B_n, \quad \frac{dB}{dt} = \omega'_{1i}(B-g) \tag{13}$$

We would like also to mention that the weakly nonlinear theory has been extended to deal with the evolution of a wave packets, resulting in Ginzberg–Landau equation with constant coefficient at its nonlinear term. From our analysis, one must cast doubt on its legitimacy, as the coefficient can not be constant due to the constant changing of the shape of the higher order terms, starting from the second order, especially those for the mean flow distortion.

Table 1 The equillibrium amplitude

	case 1: α =0.8662, R=7800		case 2: α =0.7615, R=12000	
order of appro.	Present method	Ordinary method	Present method	Ordinary method
3	0.02057	0.02203	0.01815	0.01870
5	no solution	no solution	0.02195	0.02624
7	0.02525	0.02735	0.02041	0.02189
9	0.02764	0.02811	0.02063	0.02167
11	0.02654	0.02885	0.02059	0.02159
13	0.02864	0.02802	0.02059	0.02152
15	0.02669	0.02799	0.02059	0.02146
17	0.02671	0.02769		
19	0.02670	0.02790		

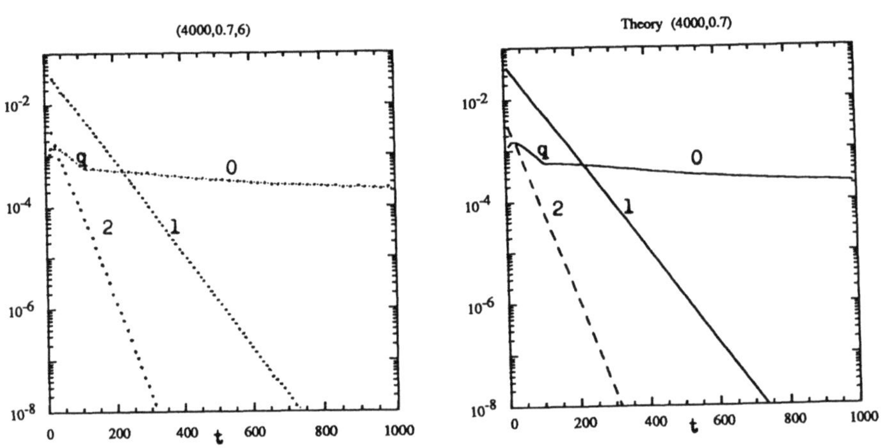

numerical simulation

present method, 5th order appro.

Fig.1 The evolution of the disturbances

1——the maximun amplitude of the wave with wavenumber α

2——the maximum amplitude of the wave with wavenumber 2α

0——the maximun amplitude of the mean flow distortion

3. Conclusions

As a summary, we can have the following conclusions:
(1) The original weakly nonlinear theory of hydrodynamic stability should still be seen as a milestone in the developement of the nonlinear theory of hydrodynamic stability.
(2) It can still be used to calculate the approximate equillibrium amplitude in the asymmptotic sense.
(3) To calculate the evolution of the disturbance, the new method should be used, because the original theory can not, in general, meet the initial condition, and the evolution of the disturbance is critically depending on its initial condition, especially for its short term evolution.
(4) Since there is another time scale R other than the two time scales usually thought to exist, it would be difficult, if not impossible, to apply the multiple scale analysis.

Acknowledgement The first author, H. Zhou, would like to thank the National Natural Science Foundation of China for their support. He would like also to thank Japan Atomic Energy Research Institute for the invitation of visiting their Institute, which made his cooperation with the second author, K. Fujimura, much more effective.

References

〔1〕 Zhou, H., A necessary modification for the amplitude equation of the weakly nonlinear theory of hydrodynamic stability, Chinese Science Bulletin, 1991, 36: 372–376
〔2〕 Zhou, H. & You, X., Weakly nonlinear theory of hydrodynamic stability revisited, Science in China (series A), 1992, 35: 52–61
〔3〕 Zhou, H. & You, X., On problems in the weakly nonlinear theory of hydrodynamic stability and its improvement, Acta Mechanica Sinica, 1993, 9: 1–12

Role of Non-Wave-Like Disturbances in Transition

M. T. Landahl
Department of Aeronautics and Astronautics,
Massachusetts Institute of Technology, Cambridge MA 02139

Summary

The space-time development of a three-dimensional initial localized disturbance is considered in three different stages, a linear shear-interaction stage, a linear viscous stage, and a nonlinear stage. The analytical solutions show the appearance of convected portions, with constant, and logarithmically increasing amplitudes, on which regions of high streamwise and spanwise shear may arise. Also, there is formed a streak of low or high-speed flow increasing in length in proportion to time.

Key words: Transition, by-pass transition, local disturbances, algebraic instability.

1. Introduction

How disturbances evolve in a shear flow of small viscosity has been a problem at the center of transition research from the early beginning of that field of inquiry since Rayleigh [1] showed that for an inviscid parallel shear flow a necessary condition for an infinite wave train to grow exponentially with time is the presence of an inflection point in the shear flow.. Most of the investigations to date have focused on how Tollmien-Schlichting wave-like disturbances evolve in time and space . However, as Gustavsson [2] and others have pointed out, discrete wave modes do not provide a complete description of the disturbance flow field in a boundary layer, but a continuous set must also be included which may be of increasing importance as the viscosity becomes smaller.

Ellingsen and Palm [3] were among the first to point out the possible role of different kinds of disturbances than the classical Tollmien-Schlichting waves in transition. They considered an inviscid initial disturbance which was independent of the streamwise direction and demonstrated that such a disturbance will grow linearly with time in its initial linear phase. Non-wave-like disturbances have in recent years attracted renewed interest for their possible importance for so-called by-pass transition, i.e., in cases in which the Tollmien-Schlichting wave disturbance growth before the flow breakdown may be weak or absent, as may be the case for strong initial disturbances, or for parallel flows supporting no linear growing waves.

As shown by Landahl [4], algebraic growth may arise for disturbances which are three-dimensional. He demonstrated that the disturbance energy of an initial inviscid three-dimensional disturbance having a nonzero net vertical momentum would grow at least linearly in time. Gustavsson's [2] calculations showed that such disturbances may attain a large total energy growth before decaying.

In order to show how non-wave-like disturbances behave in a parallel shear flow, show behavior that may lead to transition we consider here the evolution in space and time of a general disturbances in a parallel shear flow of small viscosity. The analysis will be carried out directly in the physical space following the idea set forth by Landahl [5], in which the flow development is considered in three different stages starting with a linear shear interaction stage, based on which viscous and nonlinear effects are added.

2. Space-time evolution of a localized three-dimensional inviscid disturbance in a parallel shear flow

We consider the evolution of the velocity field

$$U_i(x_j,t) = U(y)\,\delta_{i1} + u_i(x_j,t) = Sy + u_i(x_j,t) \tag{1}$$

due to a weak three-dimensional disturbance introduced at t=0 into a steady parallel shear flow, $U_1(x_j,t) = U(y)\delta_{i1}$. How such a disturbance could be created will be discussed later. We consider first the case of constant shear rate U'(y)=S, the possibility of including the effect of velocity profile curvature is to be considered later. The initial disturbance is assumed to be localized in space so that its initial velocity perturbations may be taken to be zero for $|x|>\Lambda_1$, $y<0, y>\Lambda_2$, $|z|>\Lambda_3$, where Λ_1, Λ_2, Λ_3 denote, respectively, the streamwise, vertical, and spanwise scales of the initial disturbance, all being of the same order. We will investigate the downstream evolution in time of the disturbance in a frame of reference that may be convected downstream by a velocity within the range $0 \le U_c \le U_{max}$, This flow is governed by

$$\frac{Du_i}{Dt} + U'(y)u_2\,\delta_{1i} = \frac{1}{\rho}\left[-\frac{\partial p}{\partial x_i} + \frac{\partial}{\partial x_j}\left(\mu\frac{\partial u_i}{\partial x_j} + \tau_{ij}\right)\right] \tag{2}$$

$$\frac{\partial u_i}{\partial x_i} = 0 \tag{3}$$

where

$$\frac{D}{Dt} = \frac{\partial}{\partial t} + U\frac{\partial}{\partial x_1},\quad \tau_{ij} = -\rho(u_i u_j - \langle u_i u_j \rangle) \tag{4}$$

Following Landahl [5] we assume the flow to evolve in three different stages, namely; I) a linear shear interaction stage, for $t = O(t_s), t_s = 1/U' = S^{-1}$; II) a viscous linear stage, $t = O(t_v)$, $t_v = (\Lambda_1{}^2/\nu S^2)^{1/3}$; III) an inviscid nonlinear stage $t = O(tn)$, $t_n = S\Lambda_1/v_o$. Of interest is also in addition a nonlinear viscous stage, characterized by $t = O(t_n) = O(t_v)$, but this will not be considered here.

On the assumption that $t_s \ll t_v \ll t_n$ the solution for the different stages may be obtained analytically. A rigid wall is assumed to be located at y=0. Its effect is accounted for in the initial inviscid stage through the addition of a mirror image of the initial v-velocity, and by setting S = -S for y<0. Velocity profile curvature may be included through an iterative procedure as described below.

3. Shear interaction stage, $t = O(t_s)$

It is convenient to formulate the problem in terms of then fluid element liftup **l**, and pressure impulse, **p**, defined as follows:

$$\mathbf{I} = \int_0^t \mathbf{v} Dt' \tag{5}$$

$$\mathbf{p} = \int_0^t p Dt' \tag{6}$$

Here $\int_0^t ..Dt'$ denotes time integration in a Lagrangian frame, i.e., with $\xi = x - U(y)t = x - Sy$ held constant. In terms of $\mathbf{I}$ and $\mathbf{p}$ one has

$$u = -\frac{1}{\rho}\frac{\partial \mathbf{p}}{\partial x} - S\,\mathbf{I} + u_0(\xi,y,z) \tag{7}$$

$$w = -\frac{1}{\rho}\frac{\partial \mathbf{p}}{\partial z} + w_0(\xi,y,z) \tag{8}$$

$$\frac{DL}{Dt} = U''\frac{\partial v}{\partial x} = 0 \tag{9}$$

where

$$L(x,y,z,t) = \nabla^2 v \tag{10}$$

Integration with respect to time gives

$$L(x,y,z,t) = L_0(\xi,y,z,t),\ \xi = x - yT\ ,\ T = St \tag{11}$$

Because of continuity the initial disturbance must satisfy

$$\int_{-\infty}^{\infty} d\xi' \int_{-\infty}^{\infty} dz'\, L_0(\xi',y',z') = 0 \tag{12}$$

The solution of (10) is given by

$$v = -\frac{1}{4\pi}\int_{-\infty}^{\infty} d\xi' \int_{-\infty}^{\infty} dy' \int_{-\infty}^{\infty} dz' L_0(\xi',y',z')\,\frac{1}{R'} \tag{13}$$

where $\xi' = x' - Ty'$. and where

$$R' = \sqrt{(x-x')^2+(y-y')^2+(z-z')^2} \equiv \sqrt{X^2+Y^2+Z^2} \tag{14}$$

For the pressure impulse one finds

$$\mathbf{p} \equiv \int_0^t p(\xi+Ut',y,z,t')dt' = \rho\nabla_h^{-2}[v_y - v_{0y}(\xi,y,z) - S\mathbf{I}_x] \tag{15}$$

where ∇_h^{-2} is the inverse horizontal Laplacian,

$$\nabla_h^{-2} f(x,z) = \frac{1}{2\pi}\int_{-\infty}^{\infty}\int_{-\infty}^{\infty} f(x',z')\ln\sqrt{(x-x')^2+(z-z')^2}\, dx'dz') \tag{16}$$

Integration of v over time gives, with $\mathbf{I} = 0$ for $t = 0$,

$$\mathbf{I} = \frac{1}{4\pi S}\int_{-\infty}^{\infty} dz' \int_{-\infty}^{\infty} d\xi' \int_{-\infty}^{\infty} \frac{L_0(\xi', y-Y, z')dY}{Y} \ln\left[\frac{\Xi+YT+\sqrt{(\Xi+YT)^2+Y^2+Z^2}}{\Xi+\sqrt{\Xi^2+Y^2+Z^2}}\right] \tag{17}$$

where $\Xi = \xi - \xi'$

3.1 Long-time behavior

To study the behavior of the solution for long times we consider its asymptotic behavior for $T \to \infty$. For this, we first carry out an integration of (17) by parts with respect to ξ'. With

$$f_0(x,y,z) = \int_{-\infty}^{x} L_0(\xi', y.z) d\xi' \tag{18}$$

and with $f_0(\Lambda_1, y, z) = f_\infty(y,z)$, this gives

$$\mathbf{I} = \frac{1}{4\pi S}\int_{-\infty}^{\infty} dz' \int_{-\infty}^{\infty} d\xi' \int_{-\infty}^{\infty} \frac{f_\infty(y-Y, z')dY}{Y} \ln\frac{x-\Lambda_1 - y'T + \sqrt{(x-\Lambda_1-y'T)^2+Y^2+Z^2}}{\xi - \Lambda_1 + \sqrt{(\xi-\Lambda_1)^2+Y^2+Z^2}}$$

$$- \frac{1}{4\pi S}\int_{-\infty}^{\infty} dz' \int_{-\infty}^{\infty} d\xi' \int_{-\infty}^{\infty} \frac{f_0(\xi', y-Y, z')dY}{Y}\left[\frac{1}{\sqrt{(\Xi+YT)^2+Y^2+Z^2}} - \frac{1}{\sqrt{\Xi^2+Y^2+Z^2}}\right] \tag{19}$$

An integration by parts of the last terms with respect to Y, gives, with $R_1 = \sqrt{\Xi^2 + Z^2}$,

$$\mathbf{I} = \frac{1}{4\pi S}\int_{-\infty}^{\infty} dz' \Bigg\{ \int_{-\infty}^{\infty} \frac{f_\infty(y-Y, z')dY}{Y} \ln\left[\frac{x - \Lambda_1 - y'T + \sqrt{(x-\Lambda_1-y'T)^2+Y^2+Z^2}}{\xi - \Lambda_1 + \sqrt{(\xi-\Lambda_1)^2+Y^2+Z^2}}\right.$$

$$\left.\int_{-\infty}^{\infty} d\xi' \int_{-\infty}^{\infty} dY f_{0y'}(\xi', y-Y, z') \frac{\operatorname{sgn} Y}{R_1} \ln\left[\frac{R_1 + \frac{\Xi YT}{R_1} + \sqrt{(1+T^2)Y^2 + 2Y\Xi T + R_1^2}}{R_1 + \sqrt{R_1^2+Y^2}}\right]\right\} \tag{20}$$

In the first double integral term in (20) we allow for x to be large by introducing, $\bar{x} = x/T$. Also, for $YT \to \infty$ and $\Xi \to T(\bar{x} - y) + \xi'$ the argument of the logarithmic term in the second triple integral tbecomes independent of y', so that the integral over y' may be carried out directly. Thereby, with $\Lambda_1/T \to 0$ the asymptotic expression for $\mathbf{I}$ simplifies to

$$I = \frac{1}{4\pi S}\int_{-\infty}^{\infty}dz' \int_{-\infty}^{\infty} \frac{f_\infty(y-Y,z')dY}{Y}\ln[\frac{|\bar{x}-y'|+\bar{x}-y' + (Y^2+Z^2)/(2T^2|\bar{x}-y'|)}{|\bar{x}-y|+\bar{x}-y+ (Y^2+Z^2)/(2T^2|\bar{x}-y|)}]$$

$$\int_{-\infty}^{\infty}d\xi' \int_{-\infty}^{\infty} dY f_{0y'}(\xi',y-Y,z')\frac{\text{sgn}Y}{R_1}\ln[\frac{R_1+\frac{\Xi YT}{R_1}+\sqrt{(1+T^2)Y^2+2Y\Xi T+R_1^2}}{R_1+\sqrt{R_1^2+Y^2}}]\} \tag{21}$$

Because of (12), the first term will give a nonzero contribution in the limit only if $\bar{x}$-y' and $\bar{x}-y= \xi/T \equiv \Xi$ have different signs. The limiting form of the liftup may therefore be written as

$$I \rightarrow f_1(x,y,z) + \ln T\, f_2(x,y,z) + f_3(\bar{x},y,z) \tag{22}$$

where

$$f_1(\xi,y,z) = -\frac{1}{4\pi S}\int_{-\infty}^{\infty}dz'\{ \int_{-\infty}^{\infty}d\xi' [\int_{-\infty}^{\infty}dy' f_{0y'}(\xi',y',z')\ln[1+\frac{1}{R_1}\sqrt{Y^2+R_1^2}\ ]-$$

$$-\frac{2f_0(\xi',y,z')}{\sqrt{(\xi-\xi')^2+(z-z')^2}}\ln[1+\frac{\xi-\xi'}{\sqrt{(\xi-\xi')^2+(z-z')^2}}]\} \tag{23}$$

$$f_2(\xi,y,z) = -\frac{1}{2\pi S}\int_{-\infty}^{\infty}\frac{f_\infty(y,z')}{\sqrt{(\xi-\xi')^2+(z-z')^2}})dz' \tag{24}$$

$$f_3(\Xi, y,z) = \frac{1}{4\pi S}\int_{-\infty}^{\bar{x}} dy' \frac{F(y,y',z')}{y-y'}, \text{for } \Xi<0;\quad f_3(\Xi, y,z) = \frac{1}{4\pi S}\int_{\bar{x}}^{\infty} dy' \frac{F(y,y',z')}{y-y'}, \text{for } \Xi>0 \tag{25}$$

where $F(y,y',z)= \int_{-\infty}^{\infty} f_\infty(y',z')\ln[(y-y')^2+(z-z')^2]dz'$

The terms with f_1 and f_2 represent the convected parts of the disturbance whereas f_3 describes the evolution of the continuously growing streak. Because of the singularities at y'=y, the y'-integrals in (25), (26) must be interpreted as Cauchy principal value ones. The y'-integral in (26) may be cast as one over positive values of y', only, by allowing for the effect of reflection in the rigid wall at y=0 by letting $v_0(x,-y,z)=-v_0(x,y,z)$ and replacing S by -S for y < 0.

The vertical velocity component is found by taking the substantial time derivative of the liftup. This gives

$$v = \frac{1}{t} f_2(\xi,y,z) + O(T^{-2}) \tag{27}$$

From v, and with $\partial v/\partial y = -TS\partial v/\partial\xi + O(T^{-1})$, the pressure impulse is obtained from (16) as

$$p = \rho\nabla_h^{-2}\{-S[(1+\ln T)f_{2\xi}+f_{1\xi}] - v_{0y}(\xi,y,z)\} \tag{28}$$

with the aid of which u and w may be obtained from (8) and (9). At the wall, I = 0, hence

$$p_{wall} = -\rho\nabla_h^{-2}\{v_{0y}(\xi,y,z)\} \tag{29}$$

Because the pressure gradients give rise to terms which are functions of ξ it follows that the long-time limits of u and w may be written as

$$u = u_\infty(\xi y,z) - Sf_3(\xi,y,z),\ w = w_\infty(\xi y,z) \tag{30}$$

For initial disturbances that are highly elongated in the streamwise direction, the streamwise velocity perturbation the pressure gradient contribution to the streamwise velocity may be neglected. Because the continuous streak elongation, the total perturbation energy of the disturbance can be obtained from the net square liftup and becomes proportional to T, in accordance with Landahl's analysis.

3.2 Vorticity

From the asymptotic long-time results for the velocity perturbation components one can determine the vorticity. One finds for the terms of leading order in T^{-1}

$$\omega_1 = -Tw_{\infty\xi}\ ,\ \omega_2 = -u_{\infty z},\ \omega_3 = Tw_{\infty\xi} \tag{31}$$

where subscript ξ denotes partial derivative. The intensification of ω_1 and ω_3 with time is due to stretching (in case of ω_3) and turning (in case of ω_1) of the basic spanwise vorticity, which continues forever in the inviscid convected disturbances. Viscosity will of course eventually make the convected disturbances decay, as will be seen below.

4. Viscous stage

In the subsequent viscous stage different parts of the disturbance will show different viscous behaviors. Near the wall, a Stokes' type viscous wall layer will form driven by the pressure disturbances associated with the inviscid disturbance evolution. The solution for the wall layer may be directly obtained by superposition of solutions for an impulsively moving wall, and will not be considered any further. Instead, we will consider the viscous decay of the convected velocity disturbance.

The streamwise velocity component is governed in the linearized case by

$$\frac{\partial u}{\partial t} + Sy\frac{\partial u}{\partial x} + Sv = -\frac{1}{\rho}\frac{\partial p}{\partial x} + \nu\frac{\partial^2 u}{\partial y} \tag{32}$$

The convected solution accounts for all the inviscid terms, including the pressure gradient term. To take viscosity into account we assume a solution of the form

$$u = f(\xi,y,z,\tau) \tag{33}$$

where $\tau=\tau(T)$ is a "slow" viscous time to be determined. In the absence of viscosity, u_∞satisfies (32). With viscosity, we may use u_∞as the initial condition for f. Substitution of f into (32) and assuming that the nonviscous terms balance we obtain, with $\partial f/\partial y = -ST\partial f/\partial\xi + O(\Lambda_2^{-1})$,

$$f_\tau\frac{\partial\tau}{\partial T} = \nu S^2T^2 f_{\xi\xi} \tag{34}$$

from which it follows that with $\tau = \frac{1}{3}\nu S^2 T^3$ (34) turns into the regular diffusion equation with the solution in τ and ξ with the solution

$$f = -\int_{-\infty}^{\infty} d\xi' \frac{u_\infty(\xi',y,z)}{\sqrt{2\pi\tau}} \exp\left[-\frac{(\xi-\xi')^2}{2\tau}\right] \tag{35}$$

5. Nonlinear stage

On neglecting viscosity one finds from (1)--(5) the following equation for v:

$$\frac{DL}{Dt} = U''\frac{\partial v}{\partial x} + \nabla^2 T_2 - \frac{\partial^2 T_i}{\partial x_i \partial x_2} \approx U''\frac{\partial v}{\partial x} + \frac{\partial^2}{\partial y^2}\left[\frac{\partial}{\partial x}(uv) + \frac{\partial}{\partial z}(vw)\right] + \ldots, T_i = \frac{1}{\rho}\frac{\partial t_{ij}}{\partial x_j} \tag{36}$$

Because of the shear intensification for large times due to $\partial/\partial y \approx -T\partial/\partial\xi$ the y-derivative terms will be the dominating ones. After integration over time in the Lagrangian frame one obtains

$$L = U''\frac{\partial l}{\partial x} + L_0(\xi,y,z) - \int_0^t [(uv)_x + (vw)_z]_{yy} Dt' \tag{37}$$

The term proportional to the velocity profile curvature may be taken into account through iteration as shown in Landahl [6]. The initial Laplacian may be corrected for the nonlinear terms in a similar manner. In a weak nonlinear treatment in which the nonlinearity is accounted for through an expansion in the perturbation velocity amplitudes the logarithmic time dependence, and the streak evolution, could be accounted for by the replacements of $f_\infty(y,z)$ in f_2 and f_3 by $f_\infty(y,z)+\Delta f_{\infty nl}$, where

$$\Delta f_{\infty nl} = \int_0^t (uv)_{xyy} Dt' \int_{-\infty}^{\infty} (wv)_{zyy}\, dx \tag{38}$$

In a spectral decomposition treatment with iteration in this manner one could be expected to find the appearance of increasingly higher harmonics in the spanwise Fourier variable at the zero streamwise wave number. It is shown in Landahl [6] that a patch of growing oblique instability waves could give rise to an initial disturbance of the same size as the patch showing algebraic instability. Hence, algebraic instability is likely to be an important part of the transition mechanism in the later stages of Tollmien-Schlichting-wave initiated transition when three-dimensionality sets in.

Strong nonlinearities may arise in two different ways, through a) secondary instability, and b) through singularities in the inviscid nonlinear evolution. The possibility of secondary instability of the convected disturbances is indicated by the inviscid long-time behavior, with $u = u_\infty(\xi,y,z)$, $w = w_\infty(\xi,y,z)$, from which

$$\frac{\partial u}{\partial y} = -Tu_{\infty\xi} + u_{\infty y}, \quad \frac{\partial w}{\partial y} = -Tw_{\infty\xi} + w_{\infty y} \tag{39}$$

For a two-dimensional flow, inflectional instability may arise when $\partial^2 u/\partial y^2 = 0$ and the shear is at a maximum. From (39) this means that instability may occur after a long time when $u_{\infty\xi}$ has a minimum that is, where there is a maximum streamwise deceleration of the flow. For the z-central portions of the disturbance, this would generally tend to occur on the back of the disturbance. For oblique waves one would also expect to find cross-flow instability due to inflection in the spanwise velocity. The latter may be instrumental in the spanwise spreading of the disturbance, which is absent in the linear stage.

The possibility of a singular behavior of the inviscid nonlinear solution has been investigated by Russell & Landahl [7], who found that for a "flat" eddy, for which the pressure gradient terms could be neglected in the inviscid evolution equations, this could occur for initial conditions leading to the convergence of the fluid elements along the wall. This instability manifests itself as a concentrated jet outflow from the wall.

6. Conclusions

The space-time evolution of a localized three-dimensional disturbance as obtained from the analysis in this work reveals many interesting features which are likely to be of importance in transition such as the formation of a streak, of regions of enhanced streamwise and spanwise vorticity, and their viscous decay. These could be expected to be of dominating influence in the mechanism of transition spot formation and other cases of by-pass transition. The results are believed to be of interest also in connection with fully developed shear-flow turbulence, however. Many of the dominant structures found in the near-wall region of a turbulent boundary layer, namely the streaks, the streamwise vortices, and the span-wise shear layers, are qualitatively the same as seen in the long-term evolution of the inviscid structures generated by a localized initial three-dimensional disturbance discussed here. A random superposition of inviscid streaky structures from localized initial disturbances will give rise to a momentum transport (Reynolds shear stress) that is proportional to the time rate of change of the mean-square liftup (c.f. mixing-length theory). This suggests that a meaningful turbulence model for the mean flow could possibly be constructed from the superposition of such randomly distributed localized initial disturbances.

References

[1] Rayleigh, Lord: On the stability, or instability, of certain fluid motions, Proc. London Math. Society 11 (1880) 57-70.

[2] Gustavsson, L.H.: Energy growth of three-dimensional disturbances in plane Poiseuille flow, J. Fluid Mech. 224 (1981) 243-51.

[3] Ellingsen, T. & Palm, E.: Stability of linear flows. Phys. Fluids. 18 (1975) 487-488.

[4] Landahl, M. T.: A note on algebraic instability of inviscid parallel shear flows, J. Fluid Mech. 98 (1980) 241-60.

[5] Landahl, M. T.: Model for the wall-layer structure of a turbulent shear flow. European J. Mechanics, B/Fluids, 12 (1993) 85-96.

[6] Landahl, M. T.: On sublayer streaks J. Fluid Mech. 212 (1990) 243-51.

[7] Russell, J. M. & Landahl, M.T.: The evolution of a flat eddy near a wall in an inviscid shear flow, Phys. Fluids, 27 (1984) 557-70.

Interaction of Phase-Locked Modes: A New Mechanism for the Rapid Growth of Three-Dimensional Disturbances

Xuesong Wu

Department of Mathematics, Imperical College,
180 Queens Gate, London SW7 2BZ, UK

Philip A. Stewart

DAMTP, University of Cambridge,
Silver Street, Cambridge CB3 9EW, UK

Abstract

In this paper, we identify a new mechanism which can promote rapid growth of three-dimensional disturbances in a plane shear flow. The mechanism involves the interaction between a planar mode and a single oblique mode which are *phase-locked* in the sense that they have the same phase speed. We show that such an interaction can cause the oblique mode to grow *super-exponentially*. The phase-locked interaction operates under much less restrictive conditions than the subharmonic resonant-triad mechanism, and hence can have wide application. It may be particularly relevant to transition in symmetric plane shear layers, e.g. a planar wake or jet, for which subharmonic resonance (of sinuous modes) is inactive.

Key words: Phase-locked modal interaction, Super-exponential growth, Transition, Turbulence, Wake.

1 Introduction

Laminar-turbulent transition is often initiated by instability. In incompressible (parallel or nearly parallel) shear flows, the initial disturbances may be predominantly two-dimensional, and follow linear instability theory. Further downstream, however, three-dimensional disturbances grow more rapidly and become dominant. This is certainly the result of nonlinear effects since linear theory predicts that the three-dimensional disturbances have smaller growth-rates than their two-dimensional counterparts.

Two mechanisms have been suggested to explain this rapid amplification of three-dimensional disturbances. The first is the subharmonic resonant-triad interaction, put forward in [1], [2]. The second mechanism is the secondary-instability theory (see e.g. [3] and references therein), which proposes that the mean flow plus a finite–amplitude two–dimensional disturbance can support three–dimensional instability waves. While this is an attractive, and physically valid notion, it still lacks a solid mathematical justification except in the case where an equilibrium two-dimensional wave exists (but see also [4]).

Recent studies of subharmonic resonant triads (e.g. [5]–[11]) have shed much light on the transition process. On the other hand, disturbances which do not form a resonant triad are also observed to develop quickly. Moreover, in symmetric shear flows such as a plane wake, the quadratic resonant–triad interaction is inactive, at least as far as sinuous modes are concerned, because the contributions from the two critical layers cancel ([9]). We must therefore seek an alternative explanation for the rapid amplification of three–dimensional disturbances which is observed in such flows ([12]–[14]).

2 A new mechanism: phase–locked modal interaction

The mechanism that we shall describe can operate in a general quasi–parallel shear flow which supports Rayleigh instability waves. We assume that the flow is described by Cartesian coordinates (x, y, z), where x, y and z are streamwise, transverse and spanwise coordinates respectively. We adopt the usual non-dimensionalization based on a typical thickness of the shear layer, say δ^*, and a suitable reference velocity U_0. The Reynolds number R is defined as

$$R = \frac{U_0 \delta^*}{\nu}$$

where ν is the kinematic viscosity. Throughout this paper, we shall assume that $R \gg 1$ so that a self–consistent analysis can be pursued. The basic flow has velocity

$$\left(\bar{U}(x_3, y), R^{-1}\bar{V}(x_3, y), 0\right) ,$$

where $x_3 = x/R$ is a 'slow' variable associated with the non-parallel nature of the flow. We denote the perturbed flow by

$$(U, V, W) = \left(\bar{U} + u, R^{-1}\bar{V} + v, w\right) , \qquad (2.1)$$

where u, v, w are functions of x, y, z and t.

Our proposed instability mechanism involves the interaction between a planar mode with streamwise wavenumber α_0 and phase speed c_0, and a single oblique mode with streamwise and spanwise wavenumbers $(\alpha, -\beta)$ and phase speed c. According to linear theory, an unstable mode eventually becomes neutral at some downstream location, because of the viscous spreading of the shear layer. It follows from Squire's transformation that if

$$\alpha^2 + \beta^2 = \alpha_0^2 , \tag{2.2}$$

the planar mode and the oblique mode become neutral at the same location. Moreover, when the condition (2.2) is satisfied, the two modes have the same phase velocity, i.e.

$$c_0 = c , \tag{2.3}$$

so that the modes are 'phase–locked' and have the same critical layer(s). This situation is important because it allows particularly efficient nonlinear interactions to take place in the common critical layer(s). (Note that the subharmonic resonant triad requires in addition that $\alpha = \alpha_0/2$, and so can only operate under much more restrictive conditions.)

We suppose that the initial amplitude of the disturbance is 'tuned' so that nonlinear effects first become important in the vicinity of the neutral location, where the linear growth–rate is small. It is then possible to employ a *weakly nonlinear* approach. Further, it is assumed that the disturbance amplitude expands in the form

$$\epsilon B(x_1)e^{\mathrm{i}\alpha_0\zeta} + \delta A(x_1)e^{\mathrm{i}(\alpha\zeta-\beta z)} + c.c. + \text{higher-order terms} , \tag{2.4}$$

where $\epsilon, \delta \ll 1$ represent the magnitudes of the planar and oblique modes respectively, and

$$\zeta = x - ct ,$$

is the coordinate travelling with the common phase velocity. $B(x_1)$ and $A(x_1)$ are the (order-one) amplitude functions of the planar and the oblique modes respectively. The slow variable x_1, which describes the streamwise development of the disturbance, is defined by

$$x_1 = \mu x ,$$

where $\mu \ll 1$ is a measure of the growth–rate.

We are interested in the evolution of the modes in the so–called *non-equilibrium critical–layer* régime[1] as this is believed to be of most relevance to experiments. For the present problem, it turns out that when

$$\epsilon = O(\mu^{\frac{7}{2}}) , \tag{2.5}$$

the planar mode affects the development of the oblique modes. Furthermore, provided that

$$\delta \ll O(\epsilon^{\frac{6}{7}}) , \tag{2.6}$$

[1]The reader is referred to [15], [16] for reviews of recent developments in this area.

the planar mode is itself unaffected by nonlinearity. Finally, if both viscous and non-equilibrium effects are to be significant in the critical layer, we require the 'Haberman parameter' ([17])

$$\lambda = R^{-1}\epsilon^{-\frac{6}{7}} . \tag{2.7}$$

to be of order one. After an appropriate rescaling, the equations governing the amplitudes $A(x_1)$ and $B(x_1)$ can be written as

$$\frac{\mathrm{d}\,A}{\mathrm{d}\,x_1} = \kappa A + \Gamma \int_0^{+\infty}\int_0^{+\infty} K(\xi,\eta)B(x_1-\sigma_f\xi)B^*(x_1-\xi-\sigma\eta)A(x_1-\xi-\eta)\,\mathrm{d}\,\xi\,\mathrm{d}\,\eta \,, \tag{2.8}$$

$$\frac{\mathrm{d}\,B}{\mathrm{d}\,x_1} = B \,, \tag{2.9}$$

where

$$K(\xi,\eta) = \xi^2\eta^2 e^{-s(\xi^3+\eta^3)} \,, \tag{2.10}$$

$$\sigma = \alpha_1/\alpha_0 \,, \quad \sigma_f = 1-\sigma \,, \quad s = \frac{1}{3}\lambda\alpha^2\bar{U}_c'^2\sigma_f^2 \,, \tag{2.11}$$

and the parameter Γ depends on the basic flow profile. The detailed derivation is rather complicated, and will be given in a forthcoming paper.

The amplitude equations (2.8), (2.9) predict that while the planar mode grows exponentially,

$$B = B_0 e^{x_1} \,, \tag{2.12}$$

the oblique mode experiences *super-exponential growth* at large distances downstream:

$$A(x_1) \to A_\infty \exp(\frac{1}{7}\kappa x_1 + a_0 e^{\frac{2}{7}x_1}) \quad \text{as } x_1 \to +\infty \,, \tag{2.13}$$

where B_0, A_∞ are complex constants, and

$$a_0 = (\frac{|B_0|^2|\Gamma|}{32})^{\frac{1}{7}}[7 + \mathrm{i}\arg\Gamma - \frac{1}{14}(\arg\Gamma)^2] \,. \tag{2.14}$$

Note that the real part of a_0 is always positive, which implies that super-exponential growth can always occur regardless of the value of Γ. Super-exponential growth can also arise through subharmonic resonance ([5], [10]) or through side-band instability ([18]). Here, we have identified another mechanism which can induce super-exponential growth, and so provide a possible explanation of how three-dimensional disturbances eventually overtake the planar wave and become dominant in the later stage of transition even though the disturbances are not in form of a resonant triad.

3 Application to a plane wake and numerical solutions

As mentioned earlier, the subharmonic resonance of sinuous modes is inactive in a plane wake, and hence cannot explain the rapid development of three-dimensional structures observed in this flow. Instead, the phase-locked modal interaction identified in this paper may provide a viable mechanism.

The basic-flow profile is taken to be

$$\bar{U} = U_0 - \mathrm{sech}^2 y \ . \qquad (3.15)$$

For such a symmetric profile, there exist both sinuous modes (with v even in y), and varicose modes (with v odd). We shall restrict attention to sinuous modes since they have a larger linear growth-rate, and indeed in experiments it is these modes which dominate in the early stages of transition. For the neutral planar mode,

$$\alpha_0 = 2 \ , \quad \bar{U}_c = c = U_0 - \frac{2}{3} \ , \quad \phi = \mathrm{sech}^2 y \ . \qquad (3.16)$$

The two critical levels coincide with the inflextion points at $y = \pm\frac{1}{2}\ln(2+\sqrt{3})$.

Solutions of (2.8) have been obtained in the case of $\tan^{-1}\frac{\beta}{\alpha} = 60^o$, i.e. $\alpha = 1$ and $\beta = \sqrt{3}$, which is appropriate to the experiments of [1]. In our calculations, we choose $\bar{U}_c = 3.0$, which corresponds to the velocity deficit 0.67, very close to the condition in these experiments. We also note that in most experiments, the velocity deficit in the far wake is about this size.

The development of the oblique wave is shown in Figure 1 for four different values of λ. The linear growth-rate of the oblique mode is only about one-tenth of that of the planar mode. However, once the planar mode has attained a certain magnitude, the growth of the oblique-mode amplitude is significantly enhanced through the phase-locked interaction, and ultimately tends to the super-exponential form (2.13). Viscosity has a stabilising effect in that it postpones the onset of the phase-locked interaction, but it cannot halt the ultimate super-exponential growth.

Figure 2 shows the instantaneous growth-rate of the oblique wave, $\Re(\bar{A}'/A)$. It is interesting to note that in the inviscid and moderately viscous cases, the growth-rate exhibits a step-like behaviour. That is, when the phase-locked interaction comes into play, the growth-rate 'jumps' to a higher value and remains more or less constant for some distance. Over this distance, the amplitude grows almost exponentially but with a growth-rate about twenty times the linear value. For the slightly viscous case, e.g. $\lambda = 5.0$, the initial effect of the phase-locked interaction is to cause a transient decay before enhancing the amplification. The above findings are in agreement with the observations of [1], where it was found that the oblique waves decayed slightly in the initial

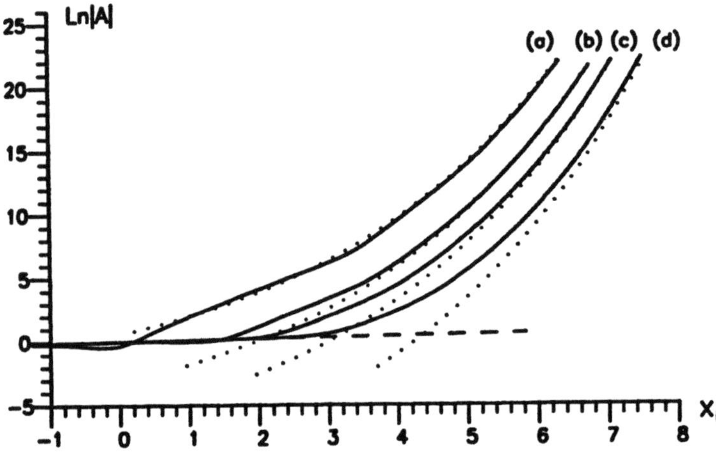

Fig.1 Ln|A| *vs.* x_1: (*a*) $\lambda = 0$, (*b*) $\lambda = 5$, (*c*) $\lambda = 15$, (*d*) $\lambda = 50$. The dotted and the dashed lines represent the asymptotic solution (2.13) and the linear solution respectively.

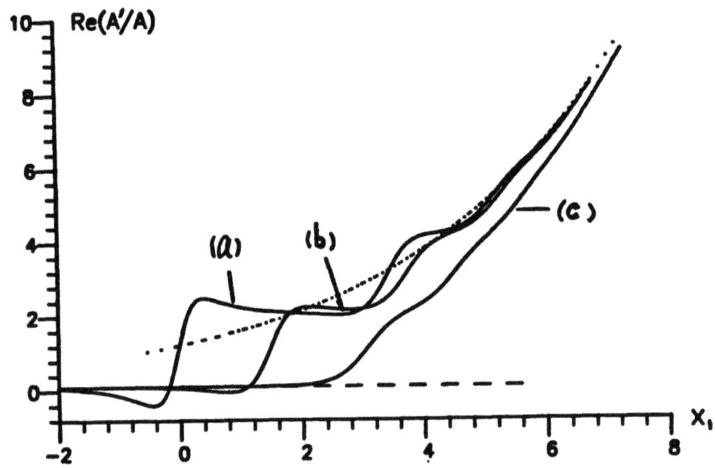

Fig.2 The instantaneous growth rate $\Re(A'/A)$ *vs.* x_1: (*a*) $\lambda = 0$, (*b*) $\lambda = 5$, (*c*) $\lambda = 50$. The dotted line represents the asymptotic result.

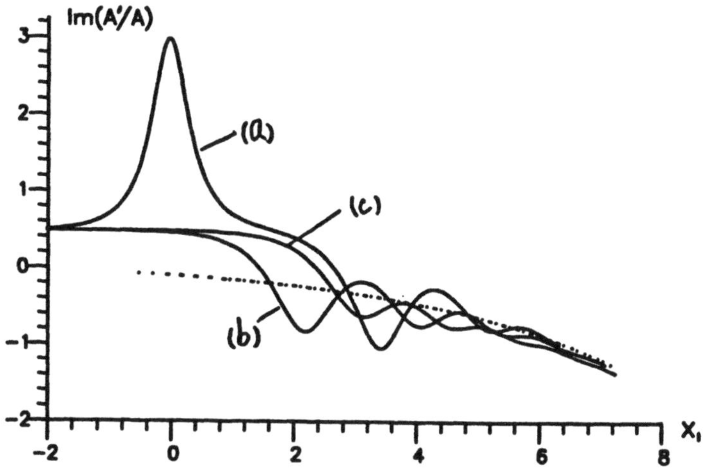

Fig.3 The wavenumber correction $\Im(A'/A)$ *vs.* x_1: (*a*) $\lambda = 0$, (*b*) $\lambda = 5$, (*c*) $\lambda = 50$. The dotted line represents the asymptotic result.

stage of the interaction, and then amplified exponentially with a growth–rate comparable to that of the planar mode.

Apart from causing a super–exponential growth in the amplitude, the phase–locked interaction also induces a wavenumber correction, $\Im(\bar{A}'/A)$, as shown in Figure 3. The simultaneous modulation of amplitude and wavenumber is characteristic of transition in a plane wake and in a mixing layer ([19], [20]).

In conclusion, we suggest that in cases where disturbances do not form a resonant triad, or the resonant triad is inactive, the phase–locked modal interaction identified in this paper provides an alternative mechanism for the observed rapid development of three-dimensional disturbances.

References

[1] Raetz, G.S. 1959 *Northrop Corp. NOR-59-383 BLC-121.*

[2] Craik, A.D.D. 1971 *J. Fluid Mech.* **50**, 393-413.

[3] Herbert, T. 1988 *Ann. Rev. Fluid Mech.* **20**, 487-526.

[4] Stewart, P.A. & Smith, F.T. 1987 *Proc. R. Soc. Lond.* A**409**, 229–248.

[5] Goldstein, M.E. & Lee, S.S. 1992 *J. Fluid Mech.* **245**, 523-551.

[6] Mankbadi, R.R., Wu, X. & Lee, S.S. 1993 *J. Fluid Mech.* **256**, 85-106.

[7] Wu, X. 1992 *J. Fluid Mech.*, **245**, 553-597.

[8] Wu, X. 1993 *Proc. R. Soc. Lond.* **A433**, 95-106.

[9] Wu, X. 1994 Viscous effects on the fully–coupled resonant triad interaction: an analytical approach. *submitted to J. Fluid Mech.*

[10] Wundrow, D.W., Hultgren, L.S. & Goldstein, M.E. 1994 *J. Fluid Mech.* **262**, 343.

[11] Lee, S.S. 1994 Critical–layer analysis of full coupled resonant–triad interaction in a boundary layer. *submitted to J. Fluid Mech.*

[12] Corke, T.C., Krull, J.D. & Ghassemi, M. 1992 *J. Fluid Mech.* **239**, 99-132.

[13] Williamson, C.H.K. & Prasad, A. 1994a *J. Fluid Mech.* **256**, 269-313.

[14] Williamson, C.H.K. & Prasad, A. 1994b *J. Fluid Mech.* **256**, 315-341.

[15] Goldstein, M.E. 1994 *Phys. Fluids*, **A6**, 42-65.

[16] Cowley, S.J.& Wu, X. 1994 Asymptotic approaches to transition modelling. In *Progress in Transition Modelling, AGARD report 793.*

[17] Haberman, R. 1972 *Studies in Appl. Maths.* **50**, 139.

[18] Wu, X. & Stewart, P.A. 1994 Side–band instability of growing waves. *to be submitted to Phys. Fluids.*

[19] Miksad, R.W., Jones, F.L., Powers, E.J. & Khadra 1982 *J. Fluid Mech.* **123**, 1-29.

[20] Miksad, R.W. 1972 *J. Fluid Mech.* **56**, 695.

Stability of Two- and Three-Dimensional Time-Dependent Flows with Locally Uniform Strain Rates

A.D.D. Craik
School of Mathematical & Computational Sciences
University of St.Andrews, Fife KY16 9SS,
Scotland, U.K.

Summary

Most steady flows with constant vorticity and elliptical streamlines are known to be unstable. These, and certain axisymmetric time-periodic flows, can be analysed by Floquet theory. However, Floquet theory is inapplicable to other time-periodic flows that yield disturbance equations containing a quasi-periodic, rather than periodic, function. A practical method for surmounting this difficulty was recently given by Bayly, Holm & Lifshitz [11]: we outline this and describe further flows that can be so treated.

Key words: flow instability, Floquet theory, quasi-periodic potentials

1. Introduction

The stability of steady, unbounded, incompressible flows with spatially-uniform strain rates has been studied by [1-7] and others. All such primary flows, whether steady or not, necessarily have the form (see [3,8])

$$\mathbf{U}(\mathbf{x},t) = \mathbf{S}(t)\mathbf{x} + \mathbf{U}^0(t) \qquad (1a)$$

where $\mathbf{S}(t)$ is a 3x3 matrix with zero trace and such that

$$d\mathbf{S}/dt + \mathbf{S}^2 = \mathbf{M}(t) \ \text{(symmetric)} \qquad (1b)$$

where $\mathbf{M}(t)$ is an arbitrary symmetic matrix. Here, $\mathbf{x}$ denotes Cartesian coordinates and t is time. Three-dimensional disturbances are taken in the form

$$\mathbf{u}'(\mathbf{x},t) = \mathrm{Re}\{\mathbf{v}(t)\exp[i\alpha(t).\mathbf{x}]\} \qquad (2)$$

for each given $\mathbf{v}(0)$, $\alpha(0)$. When a steady primary flow $\mathbf{U}(\mathbf{x})$ has closed elliptical streamlines, the wavenumber vector $\alpha(t)$ is found as an explicit periodic function, representing advection of the disturbance by the primary flow in such a way that material planes remain planes at all times t. The influence of kinematic viscosity ν is easily eliminated from the analysis by incorporating a multiplicative damping factor

$$\exp\{-\nu\int^{t} \alpha.\alpha \, dt\}.$$

The wavenumber advection equation is

$$d\alpha/dt = -S^{tr}\alpha \tag{3}$$

where the superscript 'tr' denotes transpose. Then, the amplitude components $\mathbf{v}(t)$ of the reduced inviscid problem are found from the solution of a Floquet problem

$$d\mathbf{B}/dt = \mathbf{T}(t)\mathbf{B}, \qquad \mathbf{B}(0) = \mathbf{I}_2, \tag{4a}$$

where $\mathbf{I}_2$ is the unit matrix and $\mathbf{T}(t)$ is a 2x2 time-periodic matrix known in terms of $\alpha(t)$. To calculate the Floquet exponents, and so determine whether the chosen initial disturbance is stable or unstable, it is necessary only to compute over one time period (say $0 < t < \tau$) to calculate the trace of $\mathbf{B}$ at $t = \tau$: see [4-6]. Alternatively, the problem may be cast as a single Schrödinger-type equation

$$d^2q/dt^2 + V(t)q = 0 \tag{4b}$$

where the potential $V(t)$ is τ-periodic. Typically, there are bands of instability corresponding to various internal resonances, much as for the Mathieu equation. Instability is suppressed only when these resonance criteria cannot be met for the flow in question.

A class of unbounded three-dimensional time-periodic primary flows was considered by [8] (see also [9]). This class comprises flows with spatially-uniform vorticity distorted by a pulsating stagnation-point flow (see below). Then, the wavenumber vector of the disturbance is doubly periodic (broadly speaking, with one period associated with advection by the mean rotation and the other with the imposed pulsations), and these separate periodicities are not necessarily rationally related. Then, the matrix $\mathbf{T}(t)$ and potential $V(t)$ are not necessarily t-periodic, but may instead be quasi-periodic containing separate periodic functions with incommensurate periods. Floquet theory can still be used when all the periods present are rationally related: but numerical integration must then be carried out over $0 < t < M$ where M is the lowest common multiple of the periods. If M is very large, this integration becomes impraticable; but when the periods are incommensurate (i.e. M is infinite) Floquet theory is unavailable. It was found in [8] that a sub-class of time-periodic flows that are axisymmetric is still amenable to Floquet analysis because the matrix $\mathbf{T}(t)$ and potential $V(t)$ are again τ-periodic even though $\alpha(t)$ is not. They also showed that, with weak but non-axisymmetric periodic pulsations, an approximate procedure yields the strongest of the instability bands. Other apparently time-periodic flows, within a rotating and precessing axisymmetric ellipsoid, are also reducible to Floquet analysis because a reference frame may be chosen within which these flows are steady [10].

2. Quasi-periodic potentials.

The question of how to cope with quasi-periodic forcing is an important one, for its resolution allows the stability of many more flows to be determined. This has recently been addressed by Bayly, Holm & Lifshitz [11]. Informed by theoretical studies of the Schrödinger equation with quasi-periodic potentials [12, 13], they have devised a convenient and convincing computational method. With (a, b) defined as (q, dq/dt), the Schrödinger equation is first recast as

$$\begin{pmatrix} da/dt \\ db/dt \end{pmatrix} = \begin{pmatrix} 0 & 1 \\ -V & 0 \end{pmatrix} \begin{pmatrix} a \\ b \end{pmatrix}$$

Then, polar variables c(t), d(t) are introduced such that

$$a(t) = \exp[d(t)] \sin[c(t)], \quad b(t) = \exp[d(t)] \cos[c(t)].$$

These satisfy

$$\begin{aligned} 2dc/dt &= (1 + V) + (1 - V) \cos(2c). \\ 2dd/dt &= (1 - V) \sin(2c). \end{aligned} \qquad (5)$$

Note that d(t) does not arise in the former. Clearly, the flow is unstable if, for some initial data, d(t) grows without bound as t approaches infinity.

Now, it is known from the theory of quasi-periodic potentials that, with rather weak restrictions on V(t), the limiting values

$$W = \lim \{c(t)/t\}, \quad I = \lim \{d(t)/t\}$$

exist and are independent of the initial values c(0), d(0). The limits W and I are respectively known as the 'winding number' and 'growth rate'. For instability, it is sufficient that $I > 0$. It is impossible, in general, to determine I by any calculation over a finite time interval; but [11] have convincingly shown that it can be approximated by calculations over finite intervals.

Two families of two-dimensional flows with time-periodic strain rates were examined in [11]. The first family, chosen for analytical convenience, contains steady elliptical flow as a limiting case but leads in general to a potential V(t) that contains functions with two periods. Illustrative examples are given where the periods are rationally and irrationally related. A single parameter that characterises the orientation of the initial three-dimensional wavenumber $\alpha(0)$ is the angle θ between it and the normal to the plane of the basic flow. Estimates of W and I were found, for various θ in $[0, \pi/2]$, by computing for times long enough that the limits are convincingly approached. Graphs show their estimates of W and I versus $\mu = \cos\theta$.

When the periods are rationally related, there are several distinct bands of instability, one of which is much stronger and wider than the rest. (The number of such bands, which can be predicted theoretically, is confirmed; but care is needed to discover the weakest of these, as they have very small growth rates.) Within each instability band, the winding number W takes a different constant value, and W increases monotonically with μ between the bands. Identical results may be found by Floquet theory.

When the periods are irrationally related (and Floquet theory is unavailable) estimates of W and I are plotted similarly, and several instability bands are again evident. However, it is now impossible to find all instability bands. Though these are infinite in number, and dense on the μ-axis, all but a few are too weak to be of physical significance, in view of the expected damping role of viscosity. Again, the winding number W is constant within each visible band of instability.

The second class of flows considered in [11] is the family of time-periodic uniform-vorticity cores associated with externally strained Kirchhoff-Kida vortices [14]. These elliptically-shaped vortices can change their shape as they rotate or librate. The 'geometrical optics' approximation is invoked, to justify treatment of these flows as spatially infinite when the scale of the disturbance wavelength is small compared with the minor axis of the vortex core. In similar manner, they demonstrate that almost all members of this family of inviscid flows are strongly unstable to three-dimensional plane-wave disturbances.

3. Other time-dependent flows

Other time-dependent primary flows are amenable to the method used in [11]. The first class is that described in [8]. These have time-periodic principal rates of strain a, b, -a-b constantly directed along the coordinate axes, so that the matrix **S** of (1a) is

$$\mathbf{S} = \begin{pmatrix} a(t) & -\omega_3(t) & \omega_2(t) \\ \omega_3(t) & b(t) & -\omega_1(t) \\ -\omega_2(t) & \omega_1(t) & -a(t) - b(t) \end{pmatrix} ,$$

$$a(t) = a_0 \cos\Omega t, \quad b(t) = b_0 \cos\Omega t, \quad \Omega \text{ constant.}$$

Because of (1b), the off-diagonal components $\omega = [\omega_1, \omega_2, \omega_3]$, which yield one half the vorticity, are

$$\omega_1 = \omega_{10}\exp(a_0\zeta), \quad \omega_2 = \omega_{20}\exp(b_0\zeta), \quad \omega_3 = \omega_{30}\exp[-(a_0 + b_0)\zeta],$$

where $\zeta(t) = \Omega^{-1}\sin\Omega t$ and the ω_{i0} are constants. Clearly, the periodic extension and contraction of vorticity is non-sinusoidal. Disturbances of the form (2) may now be introduced. The resulting evolution equations (3) for the three components of wavenumber $\alpha(t)$ have quasi-periodic solutions, with analytical expressions available only in the axisymmetric case $a_0 = b_0$, $\omega_{10} = \omega_{20} = 0$ studied in [8]. The corresponding amplitude equations (4a or b) in non-axisymmetric cases have a quasiperiodic matrix T(t) or potential V(t) expressible in terms of the α-components. Computation of the winding number W and growth rate I can then proceed as described above. This calculation is in hand, but details are not yet available. Almost all such flows are likely to be strongly unstable.

Another class that can be treated similarly are periodic flows within an ellipsoid, say $(x/a)^2 + (y/b)^2 + (z/c)^2 = \text{constant}$. Stability of steady flows within rotating and/or precessing ellipsoids has already been examined in [10,15]; but not periodic ones. If the 'geometrical optics' approximation is again made for small-scale disturbances, the above methods apply. Possible primary flows with spatially-uniform strain rates are $\mathbf{U} = \mathbf{S}(t)\mathbf{x}$ where

$$S(t) = \begin{pmatrix} 0 & a\omega_3/b & -a\omega_2/c \\ -b\omega_3/a & 0 & b\omega_1/c \\ c\omega_2/a & -c\omega_1/b & 0 \end{pmatrix}$$

where

$$d\omega_1/dt = \frac{c^2 - b^2}{c^2 + b^2}\,\omega_2\omega_3 \quad , \qquad d\omega_2/dt = \frac{a^2 - c^2}{a^2 + c^2}\,\omega_3\omega_1 \quad ,$$

$$d\omega_3/dt = \frac{b^2 - a^2}{b^2 + a^2}\,\omega_1\omega_2 \quad . \tag{6}$$

The wavenumber equation (3) becomes

$$d\beta/dt = -\omega \times \beta, \quad \beta = [a\alpha_1, b\alpha_2, c\alpha_3], \quad \omega = [\omega_1, \omega_2, \omega_3]. \tag{7}$$

The amplitude equation reduces to the form

$$\begin{pmatrix} dP/dt \\ dQ/dt \end{pmatrix} = \begin{pmatrix} r_{11} & r_{12} \\ r_{21} & r_{22} \end{pmatrix} \begin{pmatrix} P \\ Q \end{pmatrix} ,$$

where $P = \alpha_1 u_2 - \alpha_2 u_1$ and $Q = |\alpha(0)|u_2$. The quasi-periodic functions $r_{ij(t)}$ are lengthy expressions depending on the ω_i and α_i, which derive from (6) and (7). This amplitude equation is readily cast in the form (5) with suitably-defined modulus and phase variables. The winding number W and growth-rate I may then be computed for chosen initial data and the stability of the flow established.

4. Discussion.

The discovery [4-6] that uniform-vorticity elliptical vortex cores are strongly unstable -- unless stabilised by viscosity on small enough lengthscales -- has implications for the persistence, or otherwise, of coherent eddies in turbulent flows. The results of [8,11] on time-periodic flows reinforce the view that all but the smallest or most circular eddies are likely to be inherently unstable to a broad band of three-dimensional disturbances. A wide range of unstable wavelengths enables rapid transfer of energy from large-scale mean flows to small-scale disturbances. The only flows likely to be stable are those with net vorticity close to zero in an inertial frame. Exceptions should perhaps also be made of those geophysical flows where the Earth's atmosphere is regarded as a thin layer: for the constraint of two-dimensionality rules out potentially-unstable three-dimensional disturbances. But the instability is certainly relevant in other geophysical and astrophysical contexts. Indeed, Malkus' experiment [16] led him to suggest that precession might sustain the Earth's magnetic field, and this provided an impetus for his later experiment [17; see also 18,19] that clearly demonstrated elliptical instability. Also, a recent study [20] of self-gravitating Riemann ellipsoids confirms that this family of possible stellar structures are subject to elliptical instability. The universality of such parametric instability within a wide class of flows is now firmly established.

References

[1] Kelvin, Lord: Phil. Mag. 24 (5) (1887) 188-196.
[2] Lagnado, R.R., Phan-Thien, N. & Leal, L.G.: Phys. Fluids 27 (1984) 1094-1101.
[3] Craik, A.D.D. & Criminale, W.O.: Proc. R. Soc. Lond. A406 (1986) 13-26.
[4] Bayly, B.J.: Phys. Rev. Lett. 57 (1986) 2160-2171.
[5] Craik, A.D.D.: J. Fluid Mech. 198 (1989) 275-292.
[6] Waleffe, F.: Phys. Fluids A2 (1990) 76-80.
[7] Miyazaki, T. & Fukumoto, Y.: Phys. Fluids A4 (1992) 2515-2522.
[8] Craik, A.D.D. & Allen, H.R.: J. Fluid Mech. 234 (1992) 613-627.
[9] Mansour, N.N. & Lundgren, T.S.: Phys. Fluids A2 (1990) 2089-2091.
[10] Kerswell, R.R.: Geophys. Astrophys. Fluid Dyn. 72 (1993) 107-144.
[11] Bayly, B.J., Holm, D.D. & Lifshitz, A.: 'Stability of elliptical vortex columns.' Preprint (1994).
[12] Johnson, R. & Moser, J.: Commun. Math. Phys. 84 (1982) 403-438.
[13] Simon, B.: Advances Appl. Mech. 3 (1982) 463-490.
[14] Kida, S.: J. Phys. Soc. Japan 50 (1981) 3517-3520.
[15] Gledzer, E.B. & Ponomarev, V.M.: J. Fluid Mech. 240 (1992) 1-30.
[16] Malkus, W.V.R. : Science 160 (1968) 259-264.
[17] Malkus, W.V.R. : Geophys. Astrophys. Fluid Dyn. 48 (1989) 123-143.
[18] Vladimirov, V.A., Ribak, L.Ya. & Tarasov, V. : Prikl. Mech. Tekh. Fiz. 3 (1983) 61-69.
[19] Manasseh, R.: J. Fluid Mech. 243 (1992) 261-296.
[20] Lebowitz, N.R. & Lifshitz, A.: 'Short-wavelength instabilities of Riemann ellipsoids.' Preprint (1994).

Session B: Fundamentals II

The Incompressible Attachment-Line Boundary-Layer Stability

R.-S. Lin and M. R. Malik
High Technology Corporation
Hampton, VA 23666, USA

Summary

In this paper, the stability of the incompressible attachment-line boundary layer is investigated by solving a 2D *partial-differential* eigenvalue problem. This approach fully incorporates effects of streamline divergence and surface curvature. Also, it provides global information regarding general 2D and 3D perturbations. By applying the e^N method, the value N=10 is found to correlate well with most of the experimental transition data. The effects of leading-edge curvature and non-uniform suction are also examined. Results indicate that convex leading-edge curvature has a moderate stabilizing effect, while a practical engineering suction can have a significantly stabilizing effect.

Key words : attachment line, 2nd-order boundary layer, transition, laminar flow control.

1. Introduction

Evidence gathered from various wind tunnel and flight experiments [1–3] indicates that once the attachment-line boundary layer becomes turbulent the possibility of achieving a laminar swept wing is virtually eliminated. These results clearly demonstrate that the understanding of attachment-line boundary-layer stability represents a practical issue of primary importance in the design of advanced Laminar Flow Control swept wings.

Previous studies on this problem have been mostly experimental [4–7]. By examining all of the available experimental data, Poll [8] argued that the critical Reynolds number R_θ, based on the momentum thickness along the attachment line, was about 230. In the past, the theoretical studies on this problem have been scarce. Based on a parallel theory the critical Reynolds number (R_θ) was predicted to be close to 270 [6]. Later, Hall, Malik & Poll [9] (referred to herein as HMP) made a nonparallel stability analysis based on a family of 2D disturbances to the swept Hiemenz flow, in which the surface-curvature effect was ignored. The 2D disturbance was assumed to be a travelling wave of the form,

$$\left\{u', v', w'\right\} = \{x\hat{u}(y), \hat{v}(y), \hat{w}(y)\}e^{i[\beta z - \omega t]} \qquad (1)$$

where β is the spanwise wavenumber. They found the critical Reynolds number (R_θ) to be about 235, which agreed very well with the experimental data [5,7].

Recently, a generalized attachment-line stability approach has been developed by Lin & Malik [10]. This approach solves the 2D-eigenvalue problem resulting from the linearized partial-differential stability equations. The restriction imposed by Eq.(1) is lifted, therefore, disturbances of general structure are admitted. In this paper, firstly, this 2D-eigenvalue approach will be applied to the swept Hiemenz flow. Secondly, the method will be extended to study effects of leading-edge curvature and non-uniform suction. These issues are directly related to passive and active attachment-line boundary-layer control.

2. Problem Formulation

In Fig. 1, the coordinate system used is shown schematically; the x-axis is taken to be the chordwise direction along the surface, the y-axis is in the normal direction, and the z-axis is in the spanwise direction, where *r* is the local radius of wall curvature in the x-y plane, and ***a*** is the radius of curvature of the leading edge. We consider general infinitesimal disturbances propagating along the attachment line, so that the instantaneous velocities *u, v,* and *w* and pressure *p* can be expressed as $q(x,y,z,t) = \bar{Q}(x,y) + q'(x,y,z,t)$, where *q=(u,v,w,p)*, and barred and primed quantities represent basic-state and disturbance-state quantities, respectively.

For an infinite swept wing, the basic flow varies in both the wall normal (y) and the chordwise (x) directions, but not in z. Consequently, the disturbance solution is separable in the variables z and t, and can be expressed in the form $q'(x,y,z,t) = \hat{q}(x,y)e^{i[\beta z-\omega t]} + c.c.$. Under this normal-mode assumption, the set of nondimensional linear stability equations can be written as

$$\frac{r}{r+y}\frac{\partial \hat{u}}{\partial x} + \frac{\partial \hat{v}}{\partial y} + \frac{\hat{v}}{r+y} + i\beta\hat{w} = 0 \tag{2}$$

$$\begin{aligned} &-i\omega\hat{u} + \frac{r}{r+y}\bar{U}\frac{\partial \hat{u}}{\partial x} + \frac{r}{r+y}\hat{u}\frac{\partial \bar{U}}{\partial x} + \bar{V}\frac{\partial \hat{u}}{\partial y} + \hat{v}\frac{\partial \bar{U}}{\partial y} + \frac{\bar{V}}{r+y}\hat{u} + \frac{\bar{U}}{r+y}\hat{v} \\ &+ i\beta\bar{W}\hat{u} = -\frac{r}{r+y}\frac{\partial \hat{p}}{\partial x} + \frac{1}{\bar{R}}\left\{\left(\frac{r}{r+y}\right)^2\frac{\partial^2 \hat{u}}{\partial x^2} + \frac{ry}{(r+y)^3}\frac{dr}{dx}\frac{\partial \hat{u}}{\partial x} + \frac{\partial^2 \hat{u}}{\partial y^2}\right. \\ &\left. + \frac{1}{r+y}\frac{\partial \hat{u}}{\partial y} + \frac{2r}{(r+y)^2}\frac{\partial \hat{v}}{\partial x} - \frac{r}{(r+y)^3}\frac{dr}{dx}\hat{v}\left\{-\frac{\hat{u}}{(r+y)^2} - \beta^2\hat{u}\right\}\right. \end{aligned} \tag{3}$$

$$\begin{aligned} &-i\omega\hat{v} + \frac{r}{r+y}\bar{U}\frac{\partial \hat{v}}{\partial x} + \frac{r}{r+y}\hat{u}\frac{\partial \bar{V}}{\partial x} + \bar{V}\frac{\partial \hat{v}}{\partial y} + \hat{v}\frac{\partial \bar{V}}{\partial y} - \frac{2\bar{U}}{r+y}\hat{u} + i\beta\bar{W}\hat{v} \\ &= -\frac{\partial \hat{p}}{\partial y} + \frac{1}{\bar{R}}\left\{\left(\frac{r}{r+y}\right)^2\frac{\partial^2 \hat{v}}{\partial x^2} + \frac{ry}{(r+y)^3}\frac{dr}{dx}\frac{\partial \hat{v}}{\partial x} + \frac{\partial^2 \hat{v}}{\partial y^2} + \frac{1}{r+y}\frac{\partial \hat{v}}{\partial y}\right. \\ &- \frac{2r}{(r+y)^2}\frac{\partial \hat{u}}{\partial x}\left\{+\frac{r}{(r+y)^3}\frac{dr}{dx}\hat{u} - \frac{\hat{v}}{(r+y)^2} - \beta^2\hat{v}\right\} \end{aligned} \tag{4}$$

$$\begin{aligned} &-i\omega\hat{w} + \frac{r}{r+y}\bar{U}\frac{\partial \hat{w}}{\partial x} + \frac{r}{r+y}\hat{u}\frac{\partial \bar{W}}{\partial x} + \bar{V}\frac{\partial \hat{w}}{\partial y} + \hat{v}\frac{\partial \bar{W}}{\partial y} + i\beta\bar{W}\hat{w} = -i\beta\hat{p} \\ &+ \frac{1}{\bar{R}}\left\{\left(\frac{r}{r+y}\right)^2\frac{\partial^2 \hat{w}}{\partial x^2}\left\{+\frac{ry}{(r+y)^3}\frac{dr}{dx}\frac{\partial \hat{w}}{\partial x} + \frac{\partial^2 \hat{w}}{\partial y^2} + \frac{1}{r+y}\frac{\partial \hat{w}}{\partial y} - \beta^2\hat{w}\right\}\right. \end{aligned} \tag{5}$$

Here, the Reynolds number is defined as $\bar{R} = W_e\Delta/\nu$, where ν is the kinematic viscosity, W_e is the spanwsie component of the inviscid velocity at the edge of the boundary layer, while Δ is the boundary-layer length scale, $\Delta = \sqrt{\nu/(dU_e/dx)_{x=0}}$. As discussed by Lin & Malik [11], two types of disturbances are possible. They are the symmetric and antisymmetric solutions. Together with appropriate boundary conditions, Eqs.(2–5) form a *2D partial-differential* eigenvalue problem which is discretized by using a double Chebyshev collocation method. In the discrete sense, the generalized algebraic eigenvalue problem of the form $[\mathbf{A}]\vec{x} = \lambda[\mathbf{B}]\vec{x}$, where [A] is a complex non-Hermitian matrix and [B] is a real diagonal singular matrix, is then solved by the QR algorithm which yields all the eigenvalues.

3. Results

3.1 Swept Hiemenz Flow

If the leading-edge boundary-layer thickness is much smaller than the radius of curvature, $\delta << R$, then the surface near the attachment line can be considered locally flat and the use of the Cartesian coordinate system is justified. The N-S solution of such a configuration is known as the swept Hiemenz flow. Details regarding this flow can be found in HMP.

Fig. 2 shows the variation of the imaginary part of the complex phase speed (c_i) as a function of spanwise wavenumber (β) for Reynolds numbers ($\bar{R}$) equal to 650 and 800. It is noted that as the Reynolds number exceeds 650, the first unstable antisymmetric mode (A1) emerges from below. For $\bar{R}$=800, three extra branches of low-instability modes (A1, S2, A2) coexist with the most unstable S1 solution. It is conceivable that more branches of unstable travelling waves will arise as Reynolds number becomes even higher. However, current results show that the growth rates always follow the relation, *S1>A1>S2>A2>S3>...*, without exception.

Also plotted in Fig. 2 are results obtained by using the approach of HMP. The comparison indicates that the stability curve given by the HMP solution matches the stability data representing the S1 mode computed by the current approach. This clearly indicates that the disturbances of the form given in Eq.(1) are indeed the most amplified disturbances in the incompressible attachment-line boundary layer and, therefore, the most relevant one from the linear stability point of view. Spalart [12] reached the same conclusion but note that at the Reynolds number of his direct numerical simulation the present theory predicts only one unstable mode.

Fig. 3 shows constant phase lines of spanwise disturbances in the x-z plane at a fixed y. It can be seen that except for the S1 mode, all other modes display different degrees of phase variation in the x–direction. In other words, the most unstable S1 mode is a 2D disturbance, while other low-instability modes clearly are 3D perturbations.

By applying the e^N method, the transition locations of experimental data [7] are correlated with N=7, 10, and 13 as computed by the current method. To simplify the problem, the attachment-line boundary layer is assumed to be uniform, i.e. R_θ is constant. In this case, the computation of the total amplification factor, N, of the most unstable mode (S1) is very easy. Since the spatial growth rate will not depend on z, the N factor can be calculated as $N = -\int_0^{S_T} \beta_i \, dz = -\beta_i S_T$, where $-\beta_i$ is the spatial growth rate of the most unstable mode and is computed based on the spatial stability theory (in this case c_i=0 and β is complex) for a given R_θ, while

$z=S_T$ corresponds to the transition location measured from the upstream tip of the model. Fig. 4 presents the results of our computations together with experimental data. It is noted that N=10 correlates the transition location well with most of the experimental data. From the theoretical curves, the results suggest that the critical Reynolds number, with the surface-curvature effect neglected, is about $\bar{R} = 583.1$ (or $R_\theta = 236$) which has already been calculated by HMP.

3.2 Effects of Surface Curvature

For small leading-edge radius, one way to obtain mean flow is to solve the Navier-Stokes equations. Alternatively, higher-order boundary-layer theory can provide reasonable approximations. Here, we employ the latter approach. The technique of matched asymptotic expansions is used to find the 2nd-order corrections which consist of a *surface-curvature* term and a *displacement* term. The solution process is tedious, interested readers are referred to References [11, 13].

Here, a leading-edge Reynolds number R_a, is defined as $U_\infty a/\nu$, where U_∞ is the freestream velocity. In Fig. 5, comparisons of chordwise velocity profiles between the classical Prandtl (1st-order) solution and the 2nd-order solution are presented for 3 different leading-edge Reynolds numbers; $R_a=10^4$, 10^5, and 10^6. It is clear that the difference between 1st-order and 2nd-order solutions increases as the leading-edge Reynolds number decreases.

To demonstrate the effect of surface curvature, calculations using different order boundary-layer solutions and different stability formulations are performed at $\bar{R} = 700$, and the following comparisons are made:

(1) 1st-order boundary layer, Cartesian coordinates (xyz), $R_a=\infty$,
(2) 2nd-order boundary layer, body-intrinsic coordinates (bdy), $R_a=10^6$,
(3) same as (2), except $R_a=10^5$,
(4) same as (2), except $R_a=10^4$.

In Fig. 6, temporal growth rates are plotted vs. spanwise wavenumber β of the most unstable 2D travelling disturbances. The difference between curves (1) and (2,3,4) clearly shows that the effect of surface curvature is *stabilizing* and this contrasts sharply with the result reported by Wang et al. [14] recently. However, it may be argued that the approach used by Wang et al. does not correctly include surface-curvature effects either in the basic-state calculation (1st-order boundary-layer theory was used), nor in the stability analysis (stability was formulated in the Cartesian coordinate system). Again in Fig. 6, among curves (2), (3), and (4), as one would expect, when R_a increases, the stability curves approach to that of the swept Hiemenz flow. The difference between curves (1) and (4) is substantial; the maximum growth rate of curve (4) is 50% of that of curve (1).

To illustrate the stabilizing effect of convex surface curvature on the critical Reynolds number $\bar{R}$, the neutral curves associated with the 2D travelling mode are computed and are given in Fig. 7 for the leading-edge Reynolds numbers R_a of 10^4, 10^5, 10^6, and ∞ (swept Hiemenz flow). The increase in the critical Reynolds number is about 10% for R_a changing from infinite to 10^4. Further reducing the leading-edge Reynolds number might result in even higher critical Reynolds number; however, the accuracy of the 2nd-order boundary-layer theory itself may become questionable.

In practice, for a closed body, such as an airfoil, it is almost impossible to construct the 2nd-oder boundary-layer solution. Therefore, for design purposes, most of the time engineers will be forced to make the best use of the 1st-order boundary-layer solution. In Fig. 8, a comparison of computations using different principles is made. These calculations are performed based on the following strategies : (1) 1st-xyz: 1st-order boundary-layer solution is taken as the basic state and the stability equations are cast in the Cartesian coordinate system, i.e. the curvature effects are totally ignored; (2) 1st-bdy: the curvature effect is retained in the stability equations, see Eqs.(2–5), although the 1st-order basic state is used; (3) 2nd-bdy: the curvature effects are fully incorporated into both the basic state and the stability analysis. Of course, as one would expect, strategy (3) should give the most accurate results. The comparison shows that even without the 2nd-order boundary-layer solutions, the error associated with the stability analyses can be reduced by performing the analysis using the stability formulation of Eqs.(2–5).

3.3 Wall Suction

For a swept-wing leading edge, Poll's swept cylinder [15] represents a fair model with detailed information. At 60° sweep and a chord Reynolds number of about 1.3×10^6, the chordwise velocity distribution had been measured, and the data were smoothed by fitting a polynomial of the form:

$$\frac{u}{U_\infty} = 8.0148\left(\frac{x}{C_o}\right) - 21.746\left(\frac{x}{C_o}\right)^3 - 16.673\left(\frac{x}{C_o}\right)^5 + 112.8\left(\frac{x}{C_o}\right)^7 \tag{6}$$

where C_o(=0.4572m) is the chord of the model. The 2D Cp distribution is given in Fig. 9. It is clear that the surface pressure, p_s, has its maximum, $p_{a.l.}$, at the attachment line.

From a practical active control point of view, the vicinity of the leading edge is usually covered by a single suction flute with uniform chamber pressure p_f. A parameter called suction resistance (*SR*) is then defined as

$$SR = \frac{\Delta p}{\frac{1}{2}\rho v_s^2} \tag{7}$$

where $\Delta p = p_s - p_f$ is the local pressure drop across the surface, and v_s is the local suction velocity. Hence, with given chamber pressure p_f, the suction distribution can be calculated using Eq.(7). Here, for the sake of simplicity, we assume *SR* to have a constant value of 10^6, which in this case will allow a suction velocity on the order of $v_s/W_\infty \sim 10^{-4}$. The suction velocity profiles with various pressure ratios $p_f/p_{a.l.}$ are given in Fig. 10. Here, we assume that the flute only covers the surface up to where the suction velocity reaches zero. The 3D boundary-layer flow with prescribed suction distribution is obtained by solving the traditional non-similar boundary-layer equations using a mixed Chebyshev-spectral /5th-order finite-difference method.

In Fig. 11, we plot the neutral curves corresponding to $p_f/p_{a.l.}$=0.99, 0.97, and zero suction. It is noted that a strong stabilizing effect can be achieved with a moderate amount of non-uniform suction. For example, in the case with $p_f/p_{a.l.}$=0.97, where the maximum suction velocity is about $W_\infty/10^4$, the critical

Reynolds number is raised by as much as 50%. This suggests that active control such as surface suction is a much more effective way of controlling the attachment-line boundary-layer instability than a passive control which involves reshaping the leading-edge geometry. Of course, a smaller nose radius also implies smaller $\bar{R}$ and, hence, more stable flow.

4. Conclusions

In this paper, the 2D-eigenvalue method is first applied to the stability of the swept Hiemenz flow. New unstable 3D modes in addition to the 2D mode found by HMP are identified. It is verified that the 2D symmetric disturbance indeed represents the most unstable mode, hence is most relevant to the transition process from the linear stability point of view. By applying the e^N method, the value N=10 is found to correlate well with most of the experimental transition data. Secondly, two issues related to the control, passive and active, have been examined. Results indicate that the effect of leading-edge curvature is moderately stabilizing. On the other hand, the critical Reynolds number can be raised significantly by a small amount of non-uniform suction.

Acknowledgment

This work was supported by National Aeronautics and Space Administration under NASA Contract No. NAS1–19299.

References

[1] Pfenninger, W., AGARDograph 97, 1965.
[2] Gaster, M., The Aeronautical Quarterly, Vol. XVIII, Part 2 (1967) 165–184.
[3] Maddalon, D.V., Collier, F.S., Montoya, L.C., and Land, C.K., AIAA Paper 89–1893.
[4] Cumpsty, N., and Head, M.R., The Aeronautical Quarterly, Vol. XX (1969) 99–113.
[5] Pfenninger, W. and Bacon, J. W., Viscous Drag Reduction, ed. C. S. Wells, Plenum, 1969.
[6] Poll, D.I.A.,, College of Aeronautics Report 7805, 1978.
[7] Poll, D.I.A., The Aeronautical Quarterly, Vol. XXX, 1979, pp.607.
[8] Poll, D.I.A., IUTAM Symposium on Laminar-Turbulent Transition. Stuttgart, 1980, Springer-Verlag.
[9] Hall, P., Malik, M. R., and Poll, D.I.A., Proc. R. Soc. Lond., A395 (1984) 229–245.
[10] Lin, R.-S. and Malik, M.R., *Transition, Turbulence and Combustion*, Hussaini, et al. Ed., 1994, Kluwer Academic Pub.
[11] Lin, R.-S. and Malik, M.R., AIAA Paper 94–2372, 1994.
[12] Spalart, P. R., AGARD CP No. 438, 1988.
[13] Gersten, K. and Gross, J. F., Int. J. Heat Mass Transfer, Vol. 16 (1973) 65–79.
[14] Wang, M., Herbert, Th., and Stuckert, G.K., AIAA Paper 94–0180, 1994.
[15] Poll, D.I.A., J. Fluid Mech., Vol. 150. (1985) 329–356.

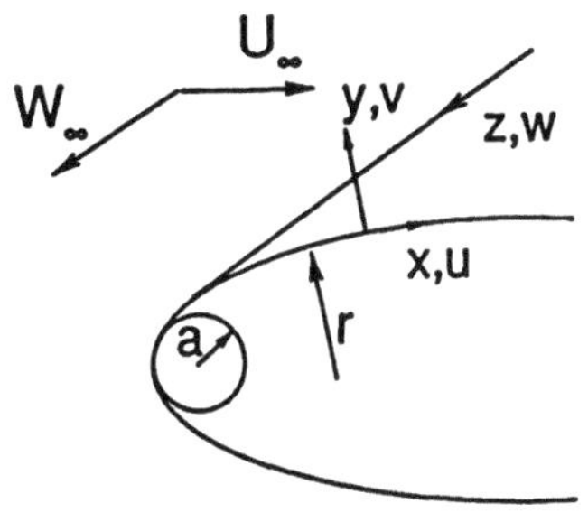

Fig. 1. Coordinate system for attachment-line boundary layer flow.

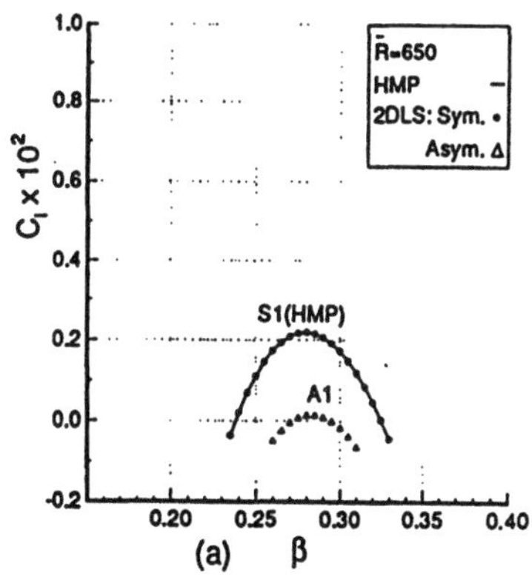

Fig. 2. Variation of c_i for symmetric and antisymmetric disturbances of swept Hiemenz flow at $\overline{R}$ = 650, 700, and 800. Solid line: HMP solution; Symbols: current results.

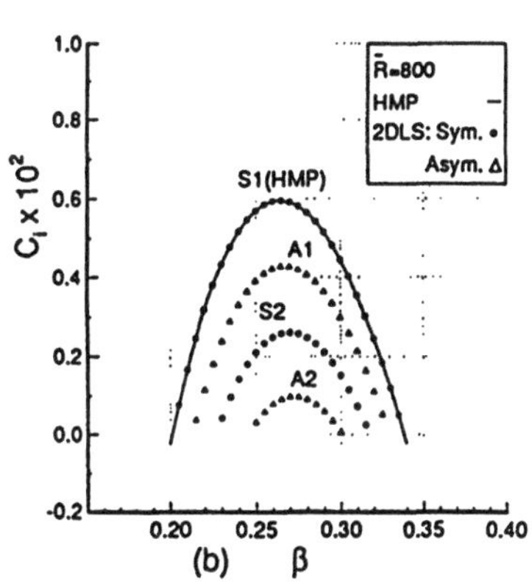

Fig. 2. Concluded.

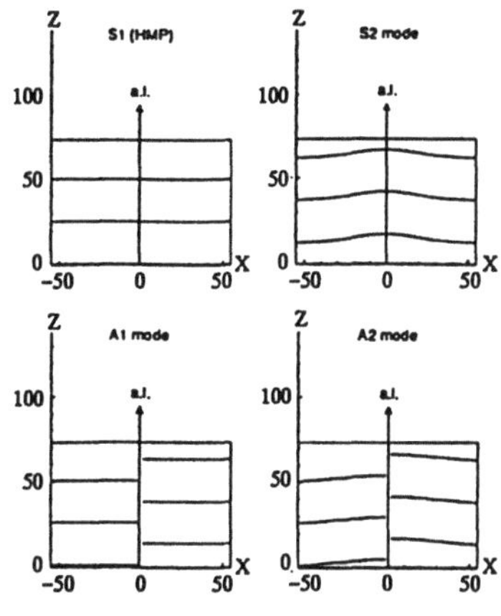

Fig. 3. Constant phase lines of w' disturbance in the x–z plane at a fixed y.

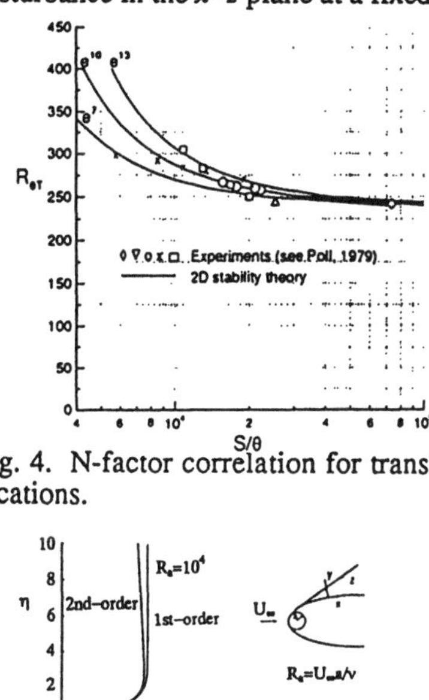

Fig. 4. N-factor correlation for transition locations.

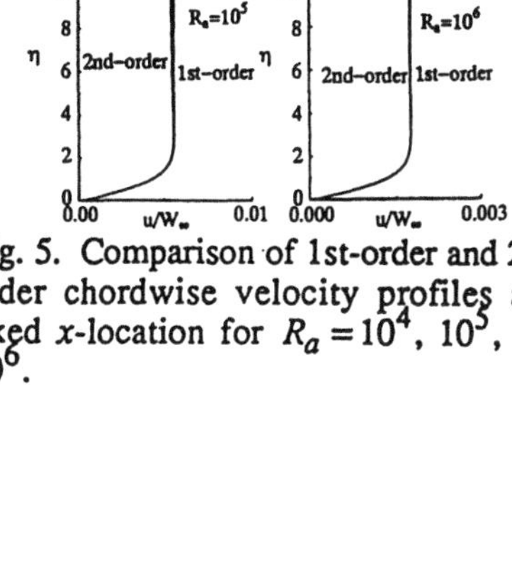

Fig. 5. Comparison of 1st-order and 2nd-order chordwise velocity profiles at a fixed x-location for $R_a = 10^4$, 10^5, and 10^6.

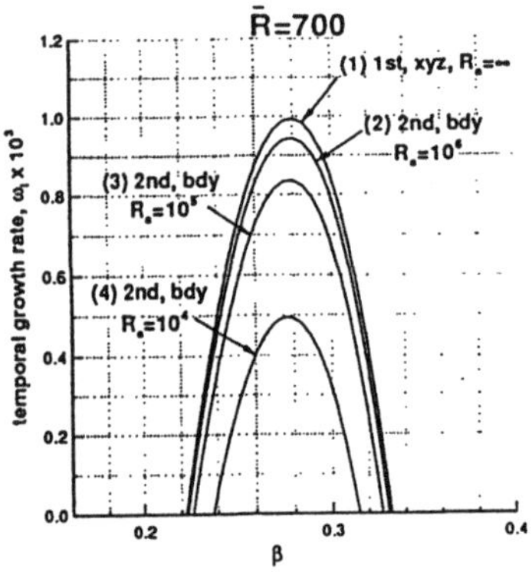

Fig. 6. Comparison of temporal growth rates for $\overline{R}$ = 700, and $R_a = \infty$, 10^6, 10^5, and 10^4.

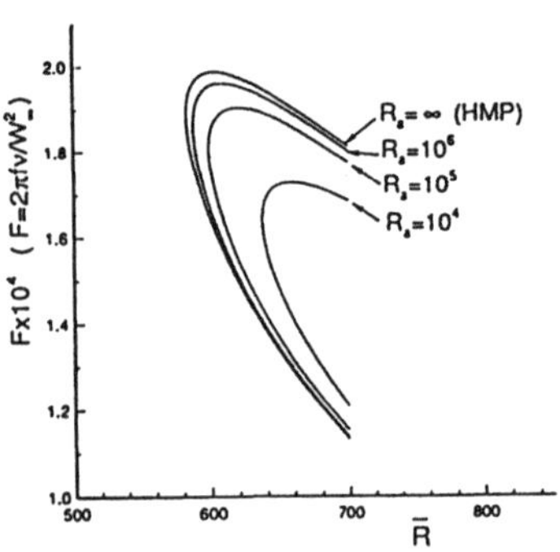

Fig. 7. Effect of surface curvature on neutral curves.

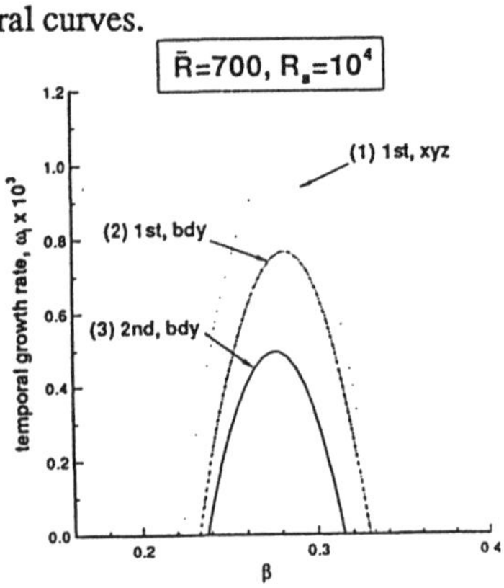

Fig. 8. Temporal growth rate computed by different strategies: (1) 1st-xyz; (2) 1st-bdy; (3) 2nd-bdy.

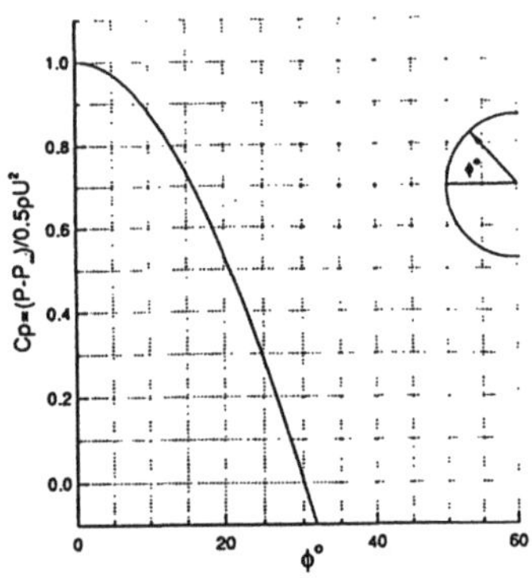

Fig. 9. *Cp* distribution of Poll's swept cylinder[15].

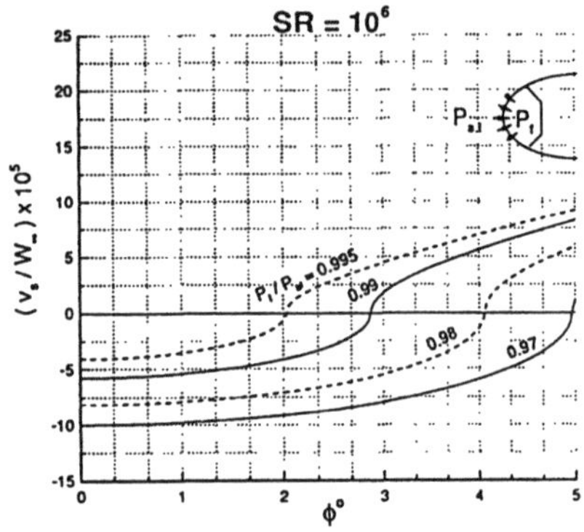

Fig. 10. Suction-velocity distribution with various pressure ratios, $P_f P_{a.l.}$; where P_f is the flute chamber pressure and $P_{a.l.}$ is the pressure at the attachment line.

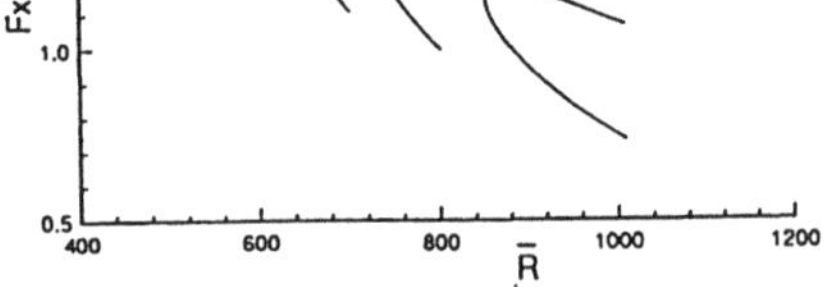

Fig. 11. Neutral curve of 2D wave with various levels of non-uniform suction.

Secondary Instability Associated with Streamwise Vortices in a Rotating Boundary Layer

S. Masuda, D. Hori
Department of Mechanical Engineering, Keio University, Yokohama, Japan
M. Matsubara
Department of Mechanics/Fluid Physics,Royal Institute of Technology, Stockholm, Sweden

Summary
The experiment on the secondary instability associated with Goertler vortices in rotating boundary layer is presented. The flow visualizations and hot-wire measurements show the instability with respect to spanwise disturbance resulting in the wavelength modulation. The instability with respect to streamwise disturbance appears downstream in the form of horseshoe and wavy modes, whose selection is related to the modulated spanwise wavelength of the nonlinear Goertler vortices.

Key words : Secondary instability, Streamwise vortex, Boundary layer, Transition, System rotation, Coriolis force

1. Introduction

Flow instability preceded by streamwise vortices, such as in curved and/or rotating boundary layers and channels, in a circular Couette flow or in a skewed boundary layer, form an important category of laminar–turbulent transition. The common feature of this type of transition is that the streamwise vortices make the base flow velocity profile inflectional, no matter how they have been originated, and the subsequent development of the disturbance is considered to be dominated by the inflectional instability.

Referring to the present understanding of the curved Goertler flow, the instability related to the rotating Goertler flow may be expected as illustrated in Fig. 1. Besides the interest for industrial applications, the present configuration has the experimental advantage of independent and continuous variations of the viscous and the body force parameters by varying the free stream velocity and the rotational speed, being especially suitable for parametric studies. The weak spanwise distur–

bance superimposed on the base Blasius boundary layer begins to grow if the Goertler number,

$$G_\delta \equiv \frac{U_w\delta}{\nu}\sqrt{\frac{\Omega\delta}{U_w}} = \frac{1}{(2\pi)^{3/2}}\left[\frac{U_w\lambda}{\nu}\left(\frac{\Omega\lambda}{U_w}\right)^{1/2}\right]\left(\frac{2\pi\delta}{\lambda}\right)^{3/2}$$

exceeds the critical value, forming the counter rotating Goertler vortices[1][2]. In this equation, U_w, δ, Ω, ν and λ denote free stream velocity, boundary layer thickness, angular velocity of system rotation, kinematic viscosity and spanwise wavelength of disturbance respectively. The parameter within the square bracket is the non-dimensional wavelength denoted by Λ hereafter, which is independent of streamwise position x, if λ does not vary downstream. When the disturbance amplitude reaches certain nonlinear level, the instability with respect to spanwise and streamwise disturbances may begin to appear. For curved Goertler flow, Guo and Finlay[3] studied the spanwise instability, called Eckhaus instability, using the parabolized spatial stability analysis and examined the mechanism of wavenumber modulation. The streamwise instability has been more extensively investigated and usually called as secondary instability in its narrow sense. The temporal theory[4][5], the spatial theory[6], DNS[7] and the experiment[8] all showed the existence of the horseshoe and the wavy modes, which stem from the inflectional profile caused by the primary Goertler vortices.

In this paper, we report the results of the experimental investigation of nonlinear development of the rotating Goertler vortices. The main interests are, 1) to examine the nonlinear instability with respect to spanwise disturbance and resulting wavenumber modulation, 2)to examine the nonlinear instability with respect to streamwise disturbance and resulting multi-mode behavior and 3)to determine the critical conditions for nonlinear instability and mode selection.

2. Experimental Apparatus and Procedure

The rotating wind tunnel and the coordinate systems employed in the present study is shown in Fig. 2. It is an open return type with the test section of 390mm×86mm ×760mm, which is mounted on a table of 2m diameter and rotates around the axis parallel to z with variable rotational speed up to 20rad/s. The free stream velocity is up to 10m/s and the free stream turbulence level is of the order of 1%. An alu–

minum test plate of 6mm thick is installed into the test section. The leading edge is shaped into 3.3deg wedge with rounded tip of 0.5mm diameter.

The artificial spanwise disturbance with the wavelength λ_i has been introduced by applying the spanwise array of rectangular protuberances of 0.3mm thick, 18mm long and 2mm wide at the streamwise position nearly corresponding to the neutral Goertler number. The velocity measurement is carried out by employing a single hot wire probe with CTA circuit mounted on the rotating table. The remote controlled traversing mechanism enables a continuous y– and z–positioning of the probe. For visualization, a sheet of paraffin mist was carefully introduced from the slit located at x=36mm. The photograph in (x–z)plane was taken by 35mm still camera mounted on the rotating table.

3. Experimental Results

3.1 Flow visualization

The instantaneous flow visualization photographs are given in Fig. 3 for three different wavelength of initial spanwise disturbances λ_i. The accumulated smoke streaks seen at the spanwise position of the disturbance pieces indicate the upwash by the action of streamwise vortices. For moderate wavelength of λ_i=12mm (Λ_i=396, Fig. 3a), the streamwise periodicity of the horseshoe pattern begins to appear near G_δ=160 followed by the breakdown to turbulence after a few streamwise periods. The spanwise wavelength is independent of x during this course of transition process for this initial wavelength. A new streak between the primary streaks plays only negligible role on transition. In case of longer wavelength of λ_i=16mm (Λ_i=787, Fig. 3b), a pair of new streaks appear between each pair of primary streaks, some of which being absorbed into the primary streaks while the other being transformed into the streamwise periodic pattern downstream. The remarkable feature is that a pair of streaks with large spanwise spacing tends to exhibit horseshoe pattern, while the other pair with small spacing tend to exhibit wavy pattern. The typical enlarged photographs are given in Fig. 4. For the initial disturbance with smaller wavelength of λ_i=4mm (Λ_i=67, Fig. 3c), the secondary streak does not appear and some primary streaks get closer and streak spacing changes irregularly downstream.

3.2 Velocity field

The iso–contours of mean streamwise velocity and its fluctuations in (y–z) plane at the streamwise position corresponding to G_δ=140 are given in Fig. 5. For λ_i=8mm (Λ_i=278, Fig. 5a), the low speed and high fluctuation region of the regular mushroom shape, precisely centered at the spanwise positions of the primary smoke streaks, can be observed. For long initial wavelength of λ_i=16mm (Λ_i=787, Fig. 5b), two additional regions of low speed and high fluctuation appear between the original pair. This result indicates the generation of two pairs of counter rotating vortices between primary pairs. This feature is similar to the vortex splitting found in curved Goertler flow with long wavelength[3]. Their calculation for Λ_i=600 based on the parabolized Navier–Stokes equations shows the generation of one small vortex pair between two pairs of primary vortices, the result being slightly different from the present result. For short wavelength λ_i=4mm (Λ_i=98.3, Fig. 5c), the velocity and its fluctuation at G_δ=140 indicate rather irregular patterns, not restricted to the spanwise position of the initial disturbance. The results is qualitatively coicident with the PSE result for curved Goertler flow for Λ_i=125[3].

In passing, the maximum value of spanwise variation of mean velocity $\Delta u / U_w$ is plotted agaist streamwise distance x in Fig. 6, as a measure of the growth of the streamwise vortices. Near x~0.24m(G_δ~96), the velocity difference begins to deviate from the initial linear growth and reaches its maximum near x~0.32m(G_δ ~117), indicating that the base Goertler vortices have reached significant nonlinear level.

3.3 Spanwise wavelength of developing Goertler vortices

The measured non–dimensional spanwise wavelength Λ of the Goertler vortices is plotted against the given initial wavelength Λ_i in Fig. 7. The data were compiled at the streamwise position, where the streamwise instability is ready to set in. For $100<\Lambda_i<500$, Λ is nearly equal to Λ_i, which means the Goertler vortices to develop preserving its wavelength of initial disturbance. In curved Goertler flow, the wavelength near Λ_i=210 has the largest growth rate[9][10], although much longer wavelength is often observed in the experiment. Guo and Finlay[3] attributed this discrepancy to the weakness of wavelength selection of primary instability due to result firmly supports this idea. For Λ_i<100, the measured wavelength is always greater than the initial one and tends to be doubled. For Λ_i>500, the measured wavelength is always smaller than the initial one and scatters around Λ=200. These results suggest that Eckhaus instability and resulting wavenumber modulation also exists in the same manner in the rotating Goertler flow.

3.4 Critical conditions for secondary instability and mode selection

In Fig. 8, the Goertler number at the streamwise position, where the streamwise periodicity is first observed in the photographs, is plotted against the non–dimensional spanwise wavenumber $2\pi\delta / \lambda$ at the same streamwise positions. The open and filled symbols indicate the horseshoe and wavy modes respectively. The lowest value of G_δ at which the spanwise and streamwise periodicity were observed in the present experiment were 95 and 113 respectively. These values may be considered as the lowest critical value for spanwise and streamwise instabilities, because of the high freestream turbulence level of the order of 1%. The selection of the secondary instability modes is well correlated with the non–dimensional spanwise wavelength Λ just upstream of the beginning of the streamwise instability. For Λ>200, the horseshoe mode mostly appears, while Λ<200, the wavy mode is more likely to be observed. For $\Lambda \approx 200$, the mixture of both modes is often observed. In the DNS for curved boundary layers[7], the wavy and horseshoe modes were found for Λ=450 and 650 respectively, showing qualitative similarity with the present result. Since the growth rate of the wavy mode in their simulation was larger than that of horseshoe mode, they concluded that the wavy mode has more dominant role on transition, which is in accordance with the experiment[8]. On the contrary to them, the present result indicates that wavy mode is not always dominant but depends on the wavelength of nonlinear Goertler vortices modified by the spanwise instability.

4. Concluding Remarks

The flow visualizations and hot–wire measurements of the flow field with Goertler vortices in a rotating boundary layer show the instability with respect to the span–wise disturbance resulting in the wavelength modulation. There is a broad range of spanwise wavelength of initial disturbance, within which the wavelength does not change downstream until breakdown occurs. Out of this range, the wavelength is modulated by merging and splitting of vortices due to instability with respect to the spanwise disturbance. Further downstream, the instability with respect to the streamwise disturbance manifests itself as horseshoe and wavy modes, whose selection is closely connected with the non-dimensional spanwise wave-length of the nonlinear Goertler vortices modulated by the spanwise instability.

Acknowledgment

The authors indebted to Dr. Obi for his valuable discussions and to Mr. Hamada for his effort for compiling the experimental data. The part of this work has been supported by the Ministry of Education, Science and Culture through a Grant-in-Aid for General Scientific Research, No.06650217.

References

[1]Masuda,S and Matsubara,M. : Visual Study of Boundary Layer Transition on Rotating Flat Plate, Laminar-Turbulent Transition, Proc. IUTAM Symp. Toulouse(1989)465-474.
[2]Matsubara,M. and Masuda,S. : Three-Dimensional Instability in Rotating Boundary Layer , Boundary Layer Stability and Transition to Turbulence, ASME FED-114(1991)103-108.
[3]Guo,Y. and Finlay,W.H. : Wavenumber Selection and Irregularity of Spatially Developing Nonlinear Dean and Goertler Vortices, J. Fluid Mech. 264(1994)1-40.
[4]Yu,X. and Liu, J.T.C. : The Secondary Instability in Goertler Flow, Phys. Fluids A3(1991) 1845-1847.
[5]Sarby, A.S. and Liu, J.T.C. : Longitudinal Vorticity Elements in Boundary Layers : Nonlinear Development from Initial Goertler Vortices as a Prototype problem, J. Fluid Mech. 231(1991) 615-663.
[6]Hall, P. and Horstman, N.J. : The Linear Inviscid Secondary Instability of Longitudinal Vortex Structures in Boundary Layers, J. Fluid Mech. 232 (1991) 357-375.
[7]Liu, W. and Domaradzki,J.A. : Direct Numerical Simulation of Transition to Turbulence in Goertler Flow, J. Fluid Mech. 246(1993)267-299.
[8]Swearingen, J.D. and Blackwelder, R.F. : The Growth and Breakdown of Streamwise Vortices in the Presence of a Wall, J. Fluid Mech. 182(1987)255-290.
[9]Floryan, J.M. and Saric, W.S. : Wavelength selection and growth of Goertler vortices, AIAA J. 22(1984)1529-1538.
[10]Hall, P. : The linear development of Goertler vortices in growing boundary layers, J. Fluid Mech. 130(1983)41-58.

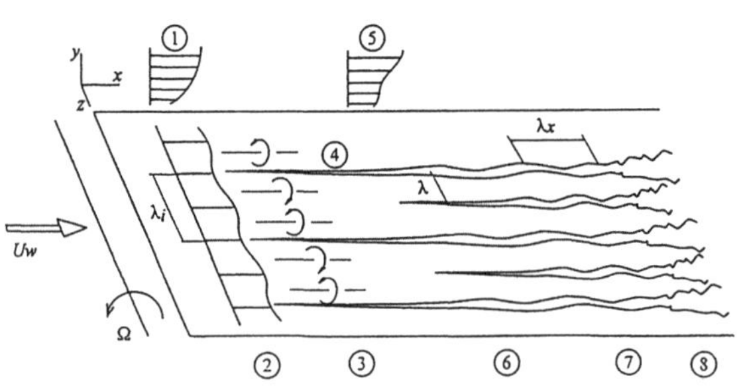

① base Blasius profile
② weak spanwise disturbance
③ primary instability (Goertler vortices)
④ low speed streak
⑤ inflectional profile
⑥ spanwise instability (wave number modulation)
⑦ streamwise instability (horseshoe/wavy modes)
⑧ breakdown to turbulence

Fig. 1 Sketch view of rotating Goertler flow

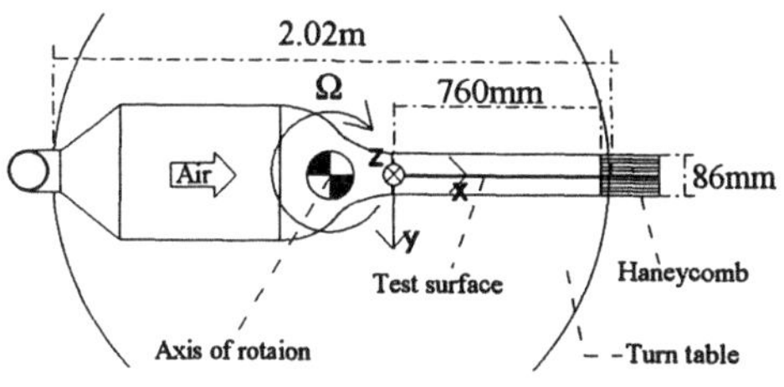

Fig. 2 Rotaing wind tunnel facility

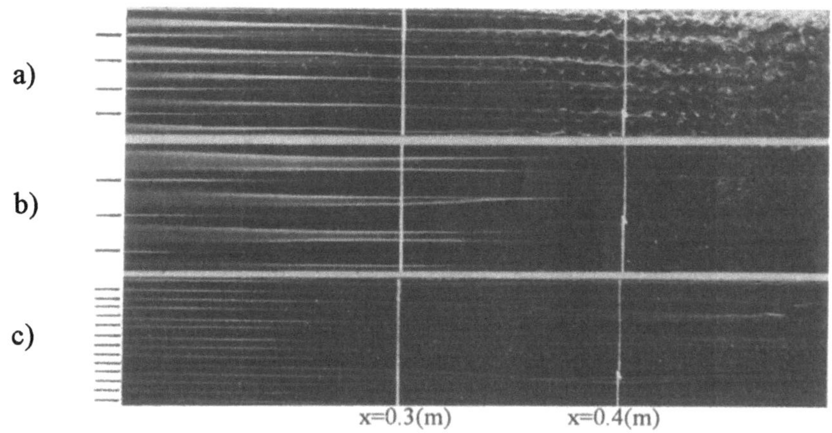

Fig. 3 Visualisation photographs in (x-z)plane,a)U_w=3m/s,Ω=6.8rad/s, λ_i=12mm(Λ_i=396),U_w=5m/s,Ω=6.8rad/s,λ_i=16mm(Λ_i=787), c)U_w=4m/s,Ω=6.8rad/s,λ_i=16mm(Λ_i=67), -: spanwise position of disturbance pieces.

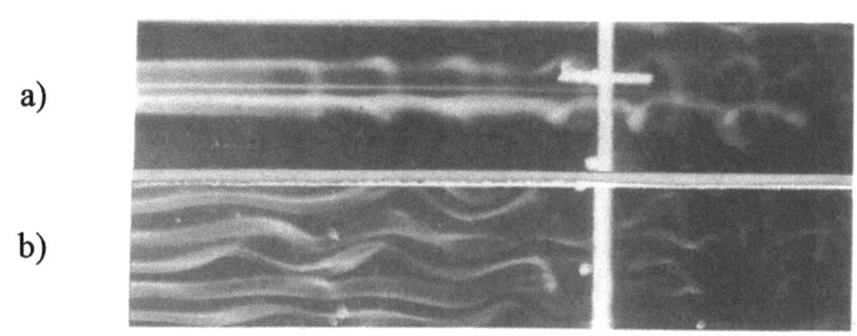

Fig. 4 Typical enlarged view of nonlinear instability with respect to streamwise disturbance, a) horseshoe mode, b) wavy mode.

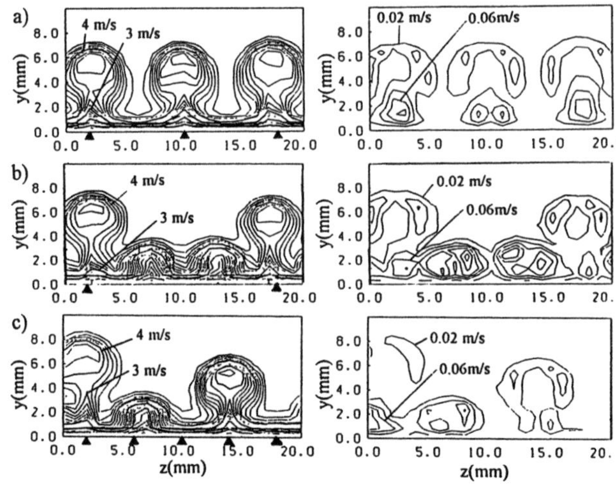

Fig. 5 Iso-contours of mean streamwise velocity (left) and its fluctuations (right) at G_δ=140, Uw = 5m/s, Ω=6.8rad/s, ▲:spanwise position of disturbance pieces, a)λ_i=8mm, Λ_i=278, b)λ_i=16mm, Λ_i= 787, c)λ_i=4mm, Λ_i=98.

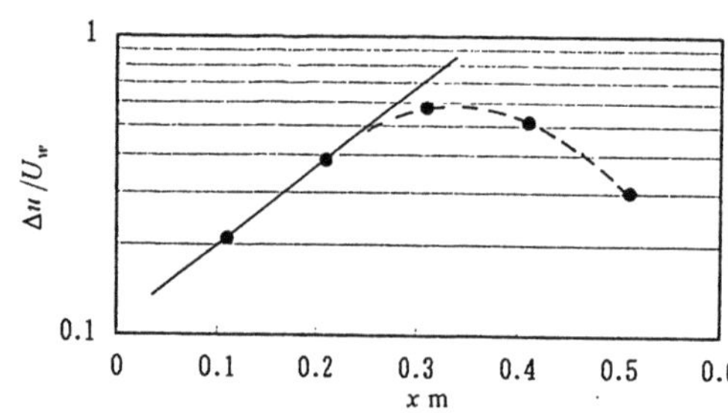

Fig. 6 Maximum value of spanwise variation of mean velocity, Uw=5m/s, Ω=5.2rad/s.

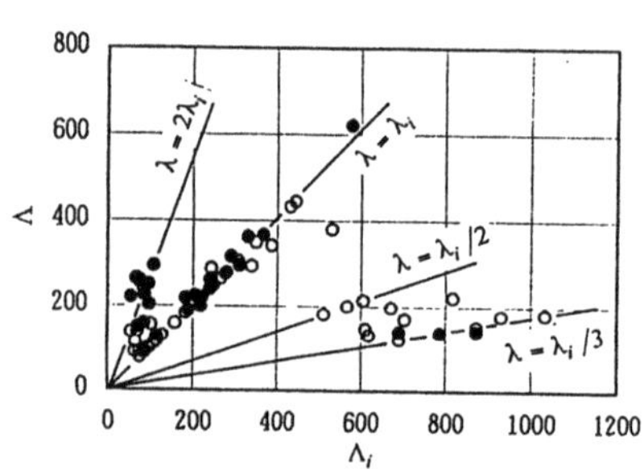

Fig. 7 Spanwise wavelength of non-linear Goertler vortices at x=0.3m, ●:Hot wire data, ○: visualization.

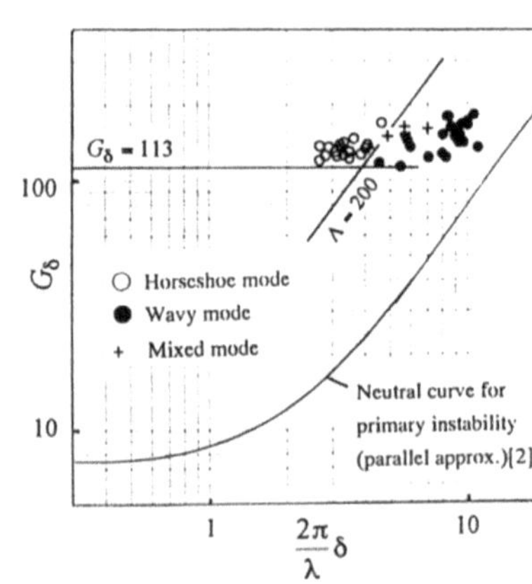

Fig. 8 Conditions for secondary instability and mode selection.

Laminar-Turbulent Transition in Spherical Couette Flow

K. Nakabayashi, Z. Zheng, Y. Tsuchida & Y. Morinishi
Department of Mechanical Engineering,
Nagoya Institute of Technology, Nagoya 466, Japan

Summary

The laminar–turbulent transition of spherical Couette flow with only the inner sphere rotating has been investigated by velocity measurements and simultaneous spectral and flow–visualization measurements for clearance ratios $\beta \leq 0.20$, where Taylor–Görtler (TG) vortices are detectable near the equator. The present study has focused on β dependence of the transition phenomena and on relaminarization for β=0.14. We have investigated transition Reynolds numbers, rotation frequencies of travelling waves and spiral TG vortices, the formation mechanism of spiral TG vortices, the law of the wall of mean velocity at the equator, and both the rms value and correlation dimension of fluctuating velocity leading up to the relaminarization.

Key words: Spherical Couette flow, Laminar–turbulent transition, Taylor–Görtler vortex, Clearance ratio, Relaminarization, Three–dimensional effect, Law of the wall, Scenario to relaminarization.

1. Introduction

We studied the laminar–turbulent transition of spherical Couette flow (SCF) for the case of a small clearance ratio. In the transition for very small clearance ratios, β<0.1, vortex structures observed near the equator are similar to those in circular Couette flow (CCF) with only the inner cylinder rotating. These are regarded as a kind of cylinder–type disturbance. However, the critical Reynolds number Rec of the Taylor instability for SCF does not agree with that for CCF [1–2]. The above influence on Rec is subject to the effect of three dimensionality which is caused by Ekman boundary flow. This 3D effect increases with β . For not very small clearance ratios, i.e., β=0.14 ~ 0.20, on the other hand, disk–type disturbances such as Stuart vortices [3] and shear waves [4], which are similar to class A or B wave in the Ekman boundary layer on a rotating disk [5], occur within the Ekman secondary flow, in addition to cylinder–type disturbances observed at the equator. The cylinder–type disturbances at the equator interact with the above disk–type disturbances.

We earlier reported the occurrence of relaminarization in the transition process for β=0.14 [6]. As Reynolds number increases, the rms value of velocity fluctuation first increases, then decreases, and becomes almost zero. This phenomenon is conjectured to be linked with the 3D effect as described above. However, the reason why the relaminarization occurs for β=0.14 has not been investigated.

In the present study, to find clues to the solution of the relaminarization mechanism, we first investigate the difference of transition phenomena between very small clearance ratios and not very small ones. Secondly, wave speed of travelling waves, mechanism of TG vortex formation and vortex structure near

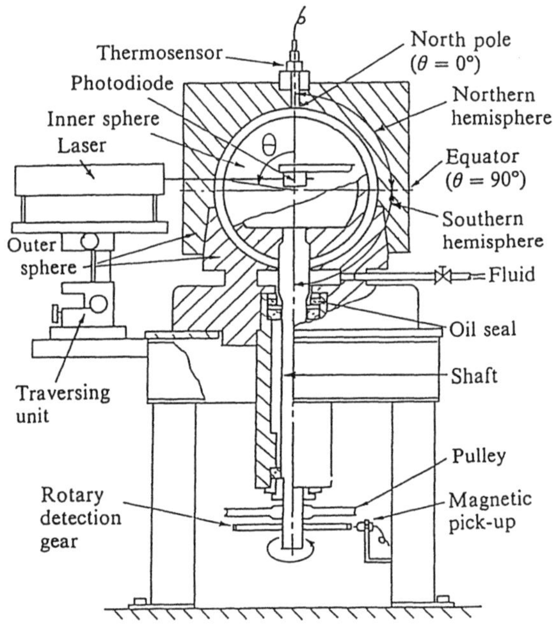

Fig. 1 Experimental apparatus.

the equator are considered. Next, we investigate the law of the wall for the mean velocity distribution at the equator in the wide range of Reynolds number from laminar to fully developed turbulent flow for β=0.14 where relaminarization occurs. Further, we investigate the evolution of rms value of velocity fluctuation, and the scenario leading up to the relaminarization.

2. Experimental Apparatus and Techniques

Fig. 1 shows the experimental apparatus [1]; the rotating inner sphere made of aluminum alloy was installed with an accuracy of within ±0.02 mm whip and ±0.02 mm concentricity in the stationary outer sphere composed of two transparent acrylic plastic blocks. We investigated the transitions for $\beta=(R_2-R_1)/R_1$=0.024, 0.06, 0.14 and 0.206 on condition that the Reynolds number Re ($=2\pi foR_1^2/\nu$) was quasi-statically increased from zero. R_1 and R_2 are the inner- and outer-sphere radii, respectively; fo is the frequency of the inner sphere rotation; and ν is the kinematic viscosity. Vortical flow structures, both wave numbers and fundamental frequencies of disturbances were identified by simultaneous spectral and flow-visualization measurements [6]. The rotation frequency of disturbance was obtained from the fundamental frequency divided by wave number. Azimuthal and meridian components of fluctuating velocity, u and w, respectively, were detected at a fixed point in the laboratory by LDV.

3. Experimental Results and Consideration

3.1 Effect of Three-dimensional Flow on Transition Process

Fig. 2 shows the relationship between onset Reynolds numbers Ret of disturbances and β [1-2,4,6]. As seen in Fig. 2, there is a clear difference of transition phenomenon between the range for β<0.1 (Range A) and that for β>0.17 (Range B). In Range A, cylinder-type disturbances such as spiral TG vortices and travelling waves occur, and their Ret values show a sharp decrease with an increase of β, corresponding to the curve for the Taylor instability [1]. In Range B, on the other hand, the cylinder-type disturbances do not occur. Instead, shear waves [4], spiral vortices (Stuart vortices) [3] and ring vortices occur. These disturbances are disk-type ones, as described previously. This is also confirmed by the result that the Ret values exist almost on the extrapolation curve of those in

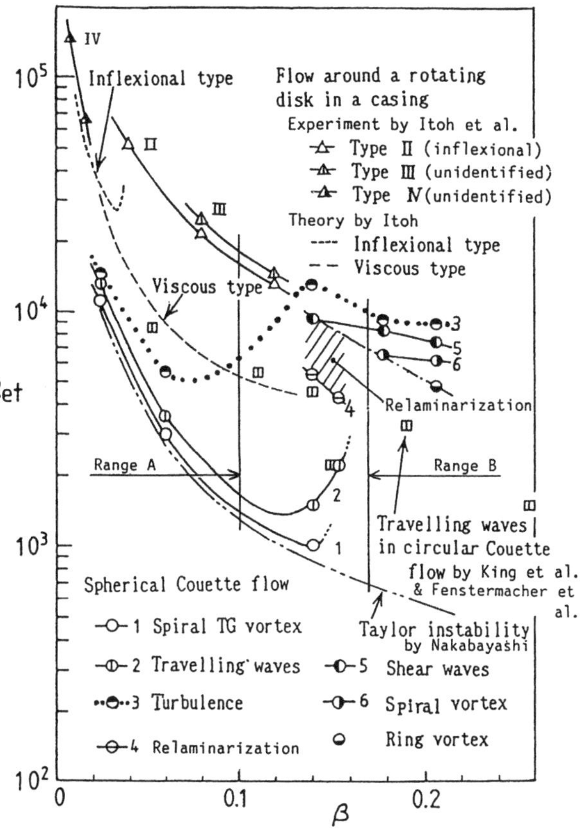

Fig. 2 Clearance–ratio dependence of the onset Reynolds numbers of disturbances.

the flow around a disk rotating in a casing (FDC) [7–8]. These disk–type disturbances do not occur in Range A, because turbulence develops prior to them, as seen in Fig. 2.

In the intermediate range between Ranges A and B, where both cylinder–type and disk–type disturbances occur, the relaminarizing flow is observed in the hatched region. Thus, with an increase of Re, cylinder–type disturbance (spiral TG vortices) disappears at the lower limit of the hatched region, and disk–type disturbance occurs at the upper limit. On the other hand, the onset Reynolds number of cylinder–type disturbance (line 2) in the intermediate range increases rapidly with β, i.e., development of the cylinder–type disturbance is suppressed with an increase of β. In the transition phenomenon in CCF [9–10], travelling waves (cylinder–type disturbance) occur in the whole range of β. The increase of β is linked with the 3D effect appearing at the equator. Therefore, the relaminarization (disappearance of cylinder–type disturbance) is related to the 3D effect.

Fig. 3 shows the β dependence of wave speed, f_w/m, of travelling waves, comparing with that in CCF [9], and the β dependence of rotation frequency, f_S/S_P, of spiral TG vortices. f_W and f_S, respectively, stand for dimensionless fundamental frequencies of the travelling

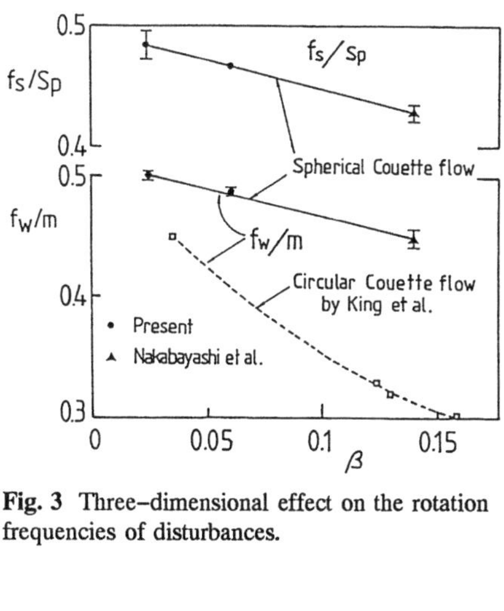

Fig. 3 Three–dimensional effect on the rotation frequencies of disturbances.

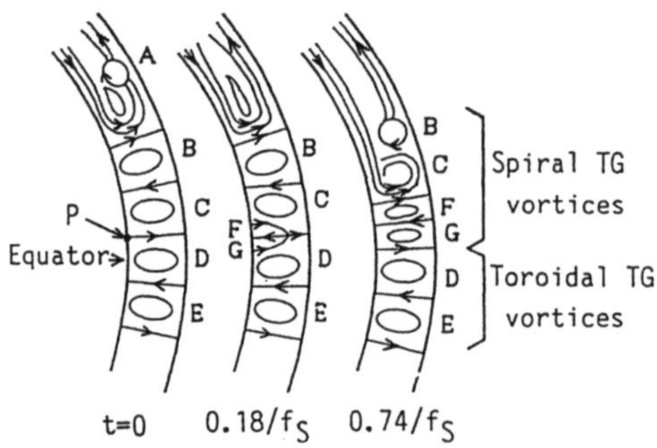

Fig. 4 Schematic of the occurrence of spiral TG vortices (β=0.14, R*=1.2).

waves and the spiral TG vortices, where the values are non-dimensionalized by $\hat{f}o$, and m and S_P are their wave numbers. Uncertainty bars show the range of the Re dependence of wave speed or rotation frequency. In SCF, travelling waves occur not only on toroidal but also on spiral TG vortices, and the azimuthal component of wave speed of the travelling waves on spiral TG vortices is equal to the wave speed (f_W/m) of travelling azimuthal waves on toroidal TG vortices. The f_W/m value for SCF is larger than that for CCF, and the difference between SCF and CCF becomes greater with β. Since the 3D effect appearing at the equator increases with β, the difference is thought to be due to the 3D effect. The f_S/S_P value, which is the rotation frequency of the starting point of the spiral TG vortices produced by the 3D effect on TG vortices, is almost equal to the f_W/m value.

Next, vortical structures of TG vortices are discussed. Fig. 4 shows schematically the formation mechanism of spiral TG vortex cells in a fixed meridian plane at R*=Re/Rec=1.2. The time t is non-dimensionalized by $1/\hat{f}o$. E and D show a pair of toroidal TG vortex cells, and C and B show a pair of spiral TG vortex cells. New spiral TG vortex cells G and F (at t=0.18/f_S and 0.74/f_S) occur in a pair at the detachment point P of the outflow boundary between toroidal and spiral TG vortices and then grow toward the outer sphere. Simultaneously, C and B cells are incorporated into the Ekman secondary flow. Consequently, C cell co-rotating with the Ekman secondary flow increases circulation of the secondary flow, and B cell, when counter-rotating, decays by viscosity. The above successive process is repeated with the period 1/f_S.

3.2 Mean Velocity Profile near Stationary Outer Sphere at β=0.14

In this section, we consider the 3D effect on the mean velocity at the equator for β=0.14 where the relaminarization occurs with increasing R*. The linear and logarithmic laws of the wall of the streamwise mean velocity are generically given in the following form.

$$U^+ = y^+ \tag{1a}$$

$$U^+ = \frac{1}{\kappa} \ln y^+ + C \tag{1b}$$

where κ and C are Karman and additive constants, respectively.

In the plane boundary layer flow, κ=0.41 and C=4.9 are given [11]. Smith & Townsend [12] obtained κ=0.41 and C=4.4 for the turbulent flow of CCF with only the inner cylinder rotating, where the effect of curvature was ignored on the law of the wall. For the above cases, $U^+=U/U_*$ and $y^+= U_* y/\nu$ are defined (U_* ; the friction velocity at the wall).

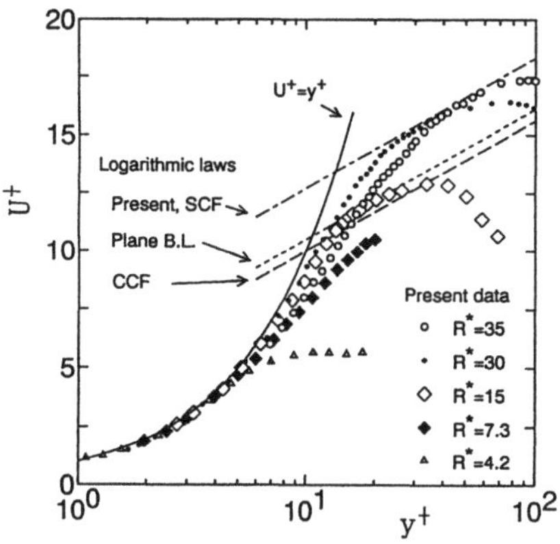

Fig. 5 Law of the wall of mean velocity components near the outer sphere wall.

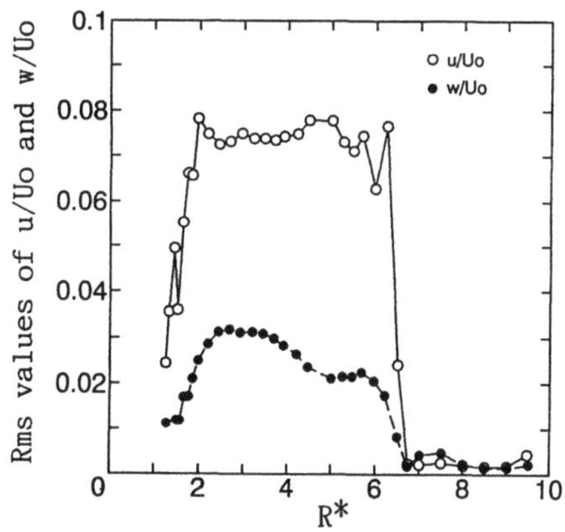

Fig. 6 Evolution of rms values of u/Uo and w/Uo.

In the present study, we obtained the law of the wall for the azimuthal velocity component near the stationary outer sphere wall at the equator, following Nakamura et al. [13]. Here, we define U^+ and y^+ as

$$U^+ = \left(\frac{U}{U_*}\right)\left(\frac{R_2}{r}\right) \tag{2a}$$

$$y^+ = \left(\frac{R_2 U_*}{3\nu}\right)\left(\left(\frac{R_2}{r}\right)^3-1\right) \tag{2b}$$

Fig. 5 shows the law of the wall of the present measurements in comparison with the logarithmic laws for the above-described plane boundary layer and CCF. Present data are measured by hot wire for several R* values. All the data agree with Eq.(1a) near the wall ($y^+<5$). As described below, the flow situations are respectively characterized as a chaotic regime at R*=4.2, a relaminarization regime at R*=7.3, and a transition regime at R*=15. Present data have the logarithmic mean velocity profile written in Eq.(1b) with κ=0.41 and C=7.1 for fully developed turbulent flow R*=30 and 35. The Karman constant 0.41 is the same as that for the plane boundary layer flow. The additive constant C differs from that for CCF with a rotating inner cylinder. The above remarkable difference between SCF and CCF is caused by the wall curvature in the meridian plane. This fact indicates the 3D effect in SCF.

3.3 Evolution of Rms Values and Scenario to Relaminarization

Fig. 6 shows the evolution of rms values of velocity fluctuation measured by LDV at the center of the clearance at the equator. Open and closed circles indicate the results of u/Uo and w/Uo, respectively ($Uo=2\pi foR_1$). Toroidal TG vortex is stationary, and the rms values are zero for 1.0<R*<1.3. The spiral TG vortex occurs at R*=1.3, and then velocity fluctuation of u increases. The sudden decrease of the rms values at R*=1.6 corresponds to a change in the number of spiral TG vortex from 3 to 2. The fluctuation begins to increase again at R*=1.8.

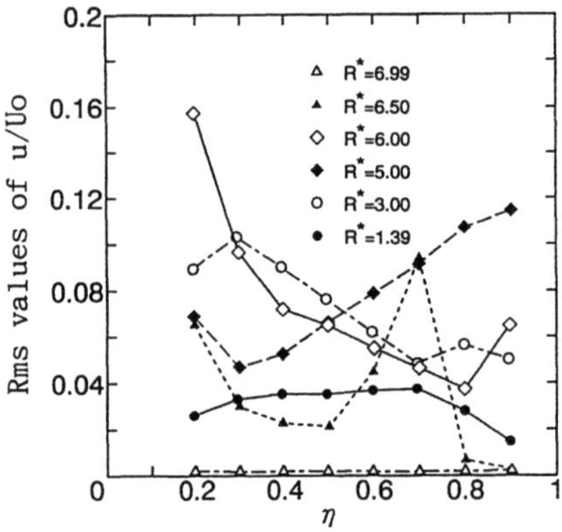

Fig. 7 Evolution of rms value distribution for u/Uo.

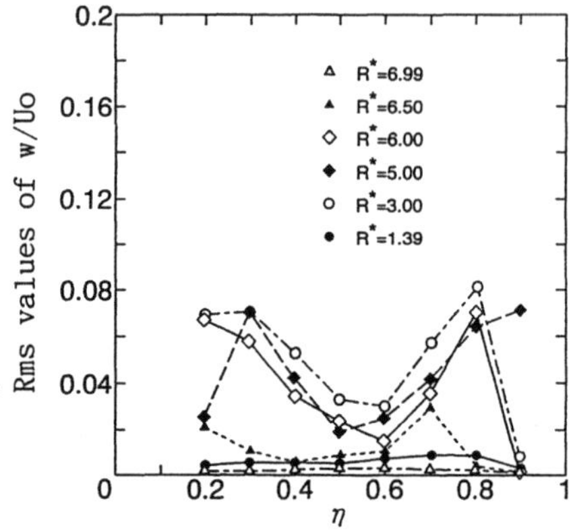

Fig. 8 Evolution of rms value distribution for w/Uo.

For 2<R*<6, the rms value of u remains constant until relaminarization occurs at R*=7. The rms values of w are about 1/3 times that of u, but the qualitative behavior of w is similar to that of u.

Fig. 7 shows the evolution of the rms value profiles of u in the clearance at the equator. At R*=1.39, the rms values are small and the profile is flat. At R*=3.00, there exist the maximum rms value at η=0.3 and a weak peak at η=0.8. For 3.0≤R*≤6.50, the behavior of the profile is more complex than at w. Due to the disappearance of f_{S3} at R*=6.99, the flow becomes steady state and the rms values becomes almost zero.

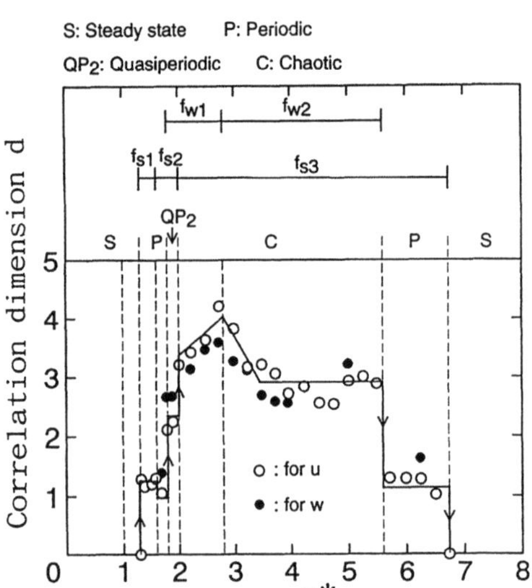

Fig. 9 Evolution of correlation dimension and scenario to relaminarization.

Fig. 8 shows also the evolution of the rms value profiles of w at the equator. The rms values are small in the central part of the clearance, but have peaks near the inner and outer sphere walls for 3.00≤R*≤6.50. This tendency agrees with the observations of vortex cell fluctuation.

f_{S1} = 1.32,	S_P = 3
f_{S2} = 0.86,	S_P = 2
f_{S3} = 0.43,	S_P = 1
f_{W1}= 2.65,	m = 6
f_{W2}= 2.24,	m = 5

Table 1. Fundamental frequencies for β=0.14.

Fig. 9 shows the evolution of correlation dimension d, the change in the fundamental frequencies and the scenario leading up to relaminarization [14] with increasing R*. Open and closed circles indicate the correlation dimensions for u and w, respectively. The correlation dimensions of u and w are in agreement, although the intensities of u and w are very different as described above.

The scenario is explained on the attractor of the flow as follows. In the region of 1.0<R*<1.3, the attractor is a fixed point (S, steady state), which corresponds to the toroidal TG vortex. After the first Hopf bifurcation at R*=1.3 (occurrence of f_{S1}), the attractor is a limit cycle (P, periodic) for 1.3<R*<1.8, while the transition from f_{S1} to f_{S2} occurs at R*=1.6. Values of fundamental frequencies for β=0.14 are shown in Table 1. At R*=1.8, the second Hopf bifurcation (occurrence of f_{w1}) takes place, leading to the appearance of the quasiperiodic (QP_2) regime with two fundamental frequencies which appear in the Fourier spectrum of velocity fluctuation shown in Fig. 10. The spectrum E(f) is transformed from the time series of u at the center of the clearance at the equator. E(f) is composed of f_{S2}, f_{W1}, and the linear combinations of these fundamental frequencies (i,j,k are integers in Fig. 10). The corresponding attractor is a torus T^2, which is obtained on the trajectory of u in the plane of u/Uo–(du/d$\hat{t}$)/(Uo$\hat{f}$o), as shown in Fig. 11. At the bifurcation near R*=2.0, the frequency f_{S2} changes to f_{S3}. The attractor is characteristic of a strange attractor corresponding to low-dimensional chaos (C). Although regarded as the third Hopf bifurcation, we can not observe any quasiperiodic (QP_3) regime just after this bifurcation. At R*=2.00~2.74, d increases with R* where the direct cascade takes place. After the transition from f_{w1} to f_{w2} at R*=2.74~3.00, d decreases with increasing R*, where a fourth bifurcation could occur. Next, bifurcation changes the chaos into the periodic motion of f_{S3} at R*=5.50~5.70 where the attractor is a limit cycle. At R*=6.74, d becomes almost zero, and the successive attractor is a fixed point (S).

4. Concluding remarks

The present study has focused on the 3D effect on the laminar–turbulent transition in SCF and the scenario leading up to the relaminarization at β=0.14.
(1) Vortical structures in the laminar–turbulent transition in the case β<0.1 (Range

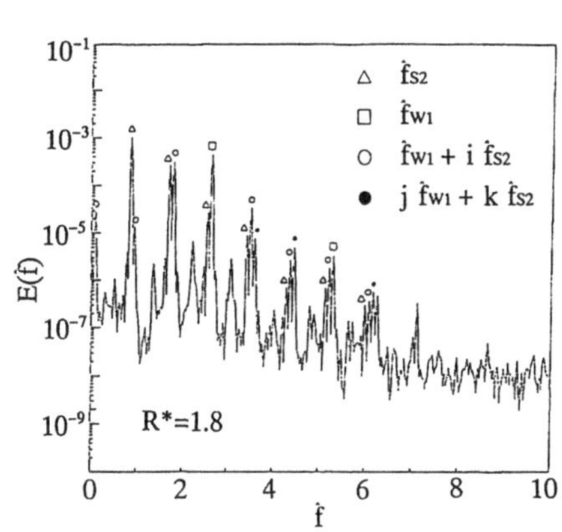

Fig. 10 Energy spectrum of u at R*=1.8.

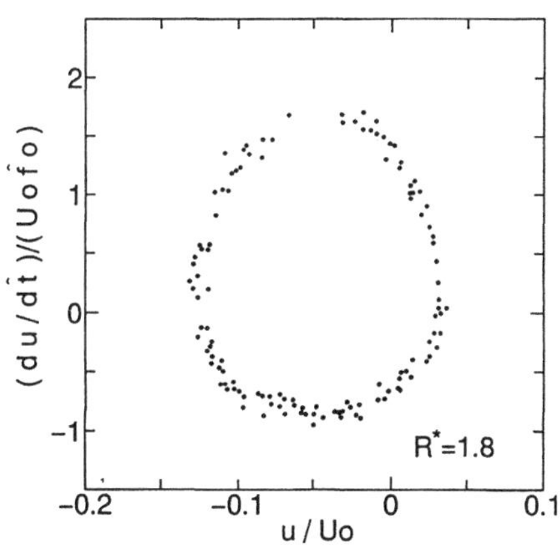

Fig. 11 Poincaré section of u at R*=1.8.

A) differ very much from the case β=0.18~0.20 (Range B).
(2) Relaminarization occurs only in a limited region between Ranges A and B.
(3) The effect of three-dimensionality near the equator affects the wave speed of traveling waves and the mechanism of vortex formation.
(4) Law of the wall was obtained for the azimuthal velocity component near the stationary outer sphere at the equator. In fully developed turbulent flow, the Karman constant of logarithmic law equals that for CCF, but the additive constant is greater than for the latter.
(5) Rms values of u and w increase around R*=2, and remain constant till they decrease to zero around R*=7. The value of w is about 1/3 times that of u.
(6) Rms value profiles of u in the clearance show complex change with increasing R*, but those of w have a peak near both the outer and inner sphere walls, respectively.
(7) The scenario leading up to relaminarization is as follows: Steady state -> Periodic state -> Quasiperiodic regime -> Chaotic regime -> Periodic state -> Steady state.

References

[1] Nakabayashi, K.: Transition of Taylor-Görtler Vortex Flow in Spherical Couette Flow, J. Fluid Mech. 132 (1983) 209-230.
[2] Yavorskaya, I. M., Belyaev, Yu. N., Monakhov, A. A., Astaf'eva, N. M., Scherbakov, S. A. and Vvedenskaya, N. D.: Stability, Nonuniqueness and Transition to Turbulence in the Flow between Two Rotating Spheres, Rep. 595, Space Research Institute of the Academy of Science, USSR (1980) 1-63.
[3] Wimmer, M.: Experiments on a Viscous Fluid Flow between Concentric Rotating Spheres, J. Fluid Mech. 78 (1976) 317-335.
[4] Bühler, K. and Zierep, J.: Dynamical Instabilities and Transition to Turbulence in Spherical Gap Flows, In Advances in Turbulence, Proc. 1st European Turbulence Conference (ed. G. Comte-Bellot and J. Mathieu), Springer (1987) 16-26.
[5] Greenspan, H. P.: The Theory of Rotating Fluids, Cambridge University Press (1968) 275-281.
[6] Nakabayashi, K. and Tsuchida, Y.: Spectral Study of the Laminar-turbulent Transition in Spherical Couette Flow, J. Fluid Mech. 194 (1988) 101-132.
[7] Itoh, M., Yamada, Y. and Kondo, N.: Visualization of the Flow Around a Disk Rotating in a Casing, J. Flow Visual. Soc. Japan (in Japanese) 2 (1982) 125-130.
[8] Itoh, M.: On the Stability of Flow between Parallel Rotating Conical Surfaces (1st Report, the Case of One Stationary Conical Surface), Trans. Japan Soc. Mech. Engrs. (in Japanese) 55 (1989) 3305-3311.
[9] King, G. P., Li, Y., Lee, W., Swinney, H. L. and Marcus, P. S.: Wave Speeds in Wavy Taylor-vortex Flow, J. Fluid Mech. 141 (1984) 365-390.
[10] Fenstermacher, P. R., Swinney, H. L. and Gollub, J. P.: Dynamical Instabilities and the Transition to Chaotic Taylor Vortex Flow, J. Fluid Mech. 94 (1979) 103-128.
[11] Hinze, J. O.: Turbulence, McGraw-Hill (1975) 626-629.
[12] Smith, G.P. and Townsend, A. A.: Turbulent Couette Flow between Concentric Cylinders at Large Taylor Numbers, J. Fluid Mech. 123 (1982) 187-217.
[13] Nakamura, I., Ueki, Y. and Yamashita, S.: The Turbulent Shear Flow on a Rotating Cylinder in a Quiescent Fluid, Trans. Japan Soc. Mech. Engrs. (in Japanese) 49-446 (1983) 2230-2235.
[14] Berge, P., Pomeau, Y. & Vidal, C.: Order Within Chaos, Hermann (1984) 144-190.

Nonlinear PSE Compared with DNS and Experiment

Ch. Airiau and G. Casalis
CERT/ONERA - Département d'Aérothermodynamique, 2, avenue Edouard Belin, B.P. 4025, 31055 Toulouse Cedex (France)

Summary

This paper is devoted to some examples of application of the non linear PSE approach. After a recall of the theory, examples are given for self-similar flows up to flows on more realistic airfoils. The results are limited to two-dimensional cases.

Key words : non linear, instability, pressure gradient, PSE.

1 Introduction

In a two-dimensional boundary layer, the linear development of the Tollmien-Schlichting (TS) waves is well known. However, the occurrence of transition in the non linear regime is the objective of many numerical studies, including the present one. Two kinds of three-dimensional peak-valley structure can appear depending on receptivity conditions such as initial disturbances amplitudes. The first one is the K-type : characterized by fundamental resonance, aligned structure, rather high disturbed environment, the second one is the C or the H-type : with subharmonic resonance, staggered structure, rather low disturbed environment. Many experiments have been carried out on that subject, essentially on a flat plate without incidence and two concepts may explain the non linear regimes : the secondary instability theory (SIT) and the concept of the resonant interacting waves.

Today, two numerical tools are available and seem to be very helpful for improving the understanding of these non linear regimes : the Direct Numerical Simulations (DNS) and the Parabolized Stability Equations approach (PSE). The latter was firstly proposed by Herbert in 1987 [1].

After a brief review of the theory, this paper gives three kinds of applications concerning the weakly non linear stability of a two-dimensional incompressible boundary layer. The first one contains some comparisons with controlled experiments and SIT. The second one is devoted to the effects of a strong adverse pressure gradient on the resonance process in comparison with Kloker's DNS. The last one gives calculations on non self-similar boundary layers by investigating some uncontrolled experiments performed at CERT-ONERA.

2 Theory

2.1 General assumptions

Let x be the streamwise coordinate, y the coordinate normal to the wall and z the spanwise coordinate. Length and velocity are written in non dimensional form using a constant length δ and the constant upstream velocity U_∞. The incompressible two dimensional main flow can be represented by :

$$Q_M(x,y) = (U(x,y), V(x,y), 0, P(x,y)) \tag{1}$$

with classical meaning for the velocity components U, V, W and the pressure P. A three-dimensional unsteady disturbance $\tilde{q}(x,y,z,t)$, where $\tilde{q}$ stands for $(\tilde{u},\tilde{v},\tilde{w},\tilde{p})$, is superimposed to Q_M. To simulate the wave interaction concept, $\tilde{q}$ is written as a double Fourier expansion containing two and three-dimensional discrete normal modes noted by (n,m)-mode. Each of them is characterized by the circular frequency $n\omega/2$ and the spanwise wave number $m\beta$:

$$\tilde{q}(x,y,z,t) = \sum_{n=-\infty}^{\infty} \sum_{m=-\infty}^{\infty} q_{n,m}(x,y) \; \exp i \left[\int_{x_0}^{x} \alpha_{n,m}(\xi) d\xi + m\beta z - n\omega t/2 \right] \tag{2}$$

In equation (2), $q_{n,m}$ is an amplitude function and $\alpha_{n,m}$ a complex streamwise wave number. The $(2,0)$-mode represents the primary TS wave, which corresponds to the linear theory. From experimental results, we know that, in controlled processes and before the onset of strong resonance, the phase speed has been found nearly identical for each mode and that tuned modes appear to be the most amplified ones. Therefore, we assume that :

$$\forall n, \forall m \quad \alpha_{n,m}(\xi) = n\alpha(\xi)/2 + ia_{n,m}(\xi) \tag{3}$$

where $\alpha = Re(\alpha_{2,0})$ is the real part of the streamwise wave number of the primary TS mode. In an uncontrolled experiment, the disturbance spectra are not as discrete as assumed in (2) but it seems that (2) and (3) remain realistic models even in these cases. Let us insist that by (2) and (3) the approach is restricted to the weakly non linear regime.

2.2 PSE assumptions

Non parallelism is represented in (1) and (2) by the x-dependance of the mean flow, of the amplitude function and of the streamwise wavenumber. We know that these dependances are weak and they are neglected in the classical parallel theory, leading thereby to the Orr-Sommerfeld equation. With the PSE, these weak non parallel effects are taken into account by neglecting the second derivative of these terms with respect to x. Let us note (A) this assumption. As done in [2], it is possible to express (A) more clearly in an asymptotic theory framework (the multiple scale analysis). Thus let us consider two different length scales : L, based on the chord position, and δ, based on a boundary layer thickness ; x and z are scaled with L and y with δ. A complete asymptotic analysis is possible with $\varepsilon = \delta/L$ as small parameter and it is easy to show that the PSE assumption (A) is strictly equivalent to consider simultaneously the two first orders (ε^0 and ε^1) equations. But, instead of the corresponding hierarchy, the assumption (A) leads to a unique partial differential equation (for each mode) which contains exactly the terms of the two first order equations.

2.3 PSE equations

The steady flow (1) and the superimposed fluctuation (2) are introduced into the Navier-Stokes equations, this leads to a system of coupled partial differential equations of the form :

$$L_{n,m}q_{n,m} + M_{n,m}\frac{\partial}{\partial x}q_{n,m} = \sum_r \sum_s \frac{A_{r,s}A_{n-r,m-s}}{A_{n,m}} NLT_{n,m}(q_{r,s}, q_{n-r,m-s}) \quad (4)$$

The left hand side corresponds to the linear PSE approach when the non linear terms (NLT) are neglected. Thus, $L_{n,m}$ contains the Orr-Sommerfeld operator and some non parallel terms. As shown in [2], the linear equation is parabolic (in the meaning of the characteristics method) only under some conditions which depend on the choosen formulation (primitive variables, vorticity ...). Finally in (4), $A_{r,s} = \exp - \int a_{r,s}(x)\, dx$ represents the amplitude of the (r,s)-mode.

With the system of equations (4) only, the PSE are not closed. Let us recall that the streamwise wave number is determined by a dispersion relation in the eigenvalue problem of the parallel theory ; with the PSE, supplementary relations called normalizations are necessary. They come from the ambiguity in the x dependance in equation (2) and they impose the assumption (A) numerically. Different normalizations can be choosen, however calculations have proved that this choice has an influence on $q_{n,m}$ and $\alpha_{n,m}$ but not on the physical quantity $\tilde{q}_{n,m}$ (see equation (2)). Furthermore, boundary conditions must be included in the PSE : for each x, the no-slip condition at the wall and fluctuation attenuation far away from the wall are required for each mode. Due to the derivative in x in (4), initial conditions (at $x = x_0$) must be given. They consist on the type of resonance, the amplitude functions, the complex wave numbers and the amplitudes of : the primary mode (2,0) and the secondary mode ((1,1) in the case of subharmonic resonance, (2,1) for the fundamental resonance).

Finally, the complete PSE, including equation (4), normalizations, boundary and initial conditions are numerically integrated with a marching procedure in x. But let us remark that although equation (4) is fully parabolized, numerical

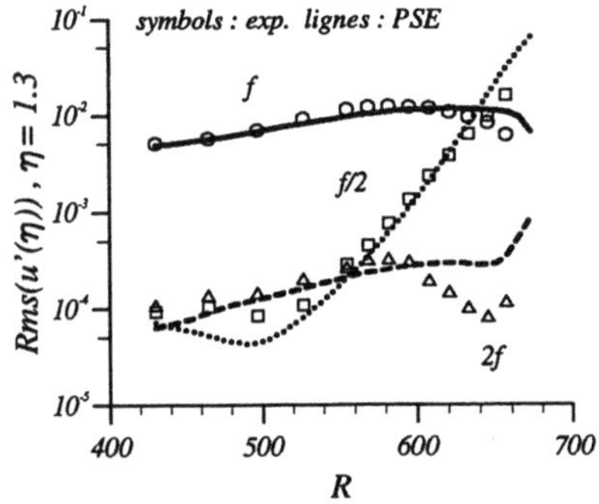

Figure 1: Kachanov and Levchenko's experiment, H-type resonance.

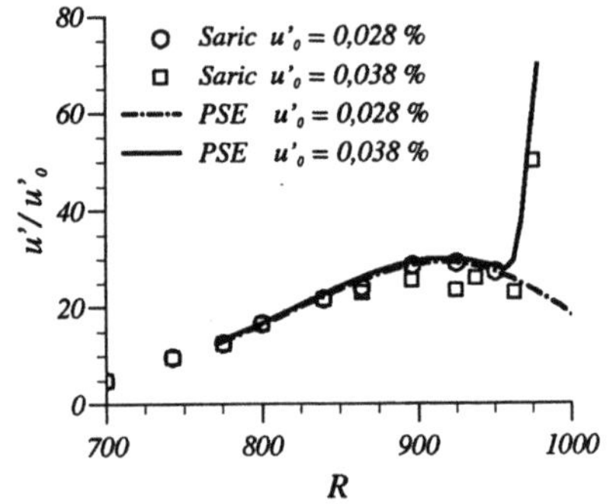

Figure 2: Saric's experiment, C-type of resonance.

instabilities can appear when the step size in x becomes sufficiently small. This leads us to consider that the complete PSE are probably not really parabolic.

3 Applications

3.1 Blasius boundary layer : PSE and experiment

Very good agreement is found for controlled experiments in a Blasius boundary layer. Figure (1) corresponds to Kachanov and Levchenko's experiment [3], for the H-type of resonance. The amplitudes are given at a given value of η, where : $\eta = y^*/l$ and $l = \sqrt{\nu x^*/U_\infty}$, as function of the Reynolds number R with $R = lU_\infty/\nu$. The reduced frequency is $F = 2\pi f \nu/U_\infty^2 = 124\ 10^{-6}$ and the non dimensionalized spanwise wave number $b = \beta^* 10^3 \nu/U_\infty$ is $b = 0.333$ which corresponds to a wave propagation angle ψ of 62°. On the other hand, figure (2) is an example of C-type resonance, PSE results are given in comparison with Saric *et al.* experiments [4]. Only the amplitude of the primary mode is represented for two initial amplitudes. The parameters are $F = 76\ 10^{-6}$ and $b = 0.1$, which gives a value of ψ close to 35°. In the case of figure (1), we showed in [5] that it is possible to obtain amplitude saturation just after branch II of the primary mode but this state is extremely sensitive to the initial amplitudes of the primary and secondary modes. Some cases given below have been obtained by a fine adjustement of these initial conditions.

3.2 Pressure gradient effect : comparison with DNS

Natural transition occuring by the subharmonic route on a two-dimensional boundary layer with a low pressure gradient was observed by Knapp *et al.* [6]. Non linear phenomena seem similar to those happening in cases without pressure gradient. When the mean-flow distorsion, i.e. the (0,0)-mode of (2), is added

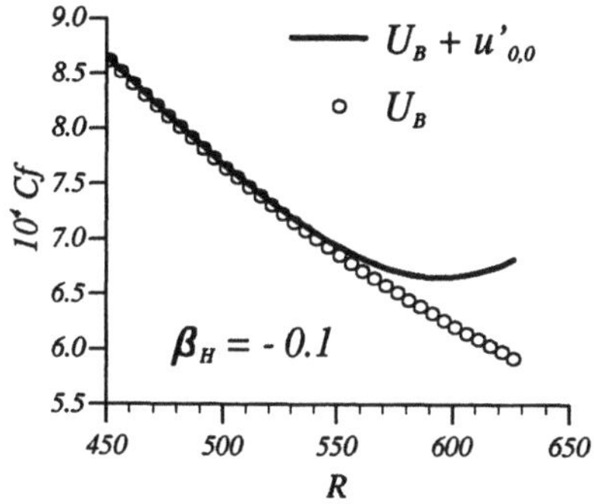

Figure 3: Skin Friction coefficient, K-type of resonance.

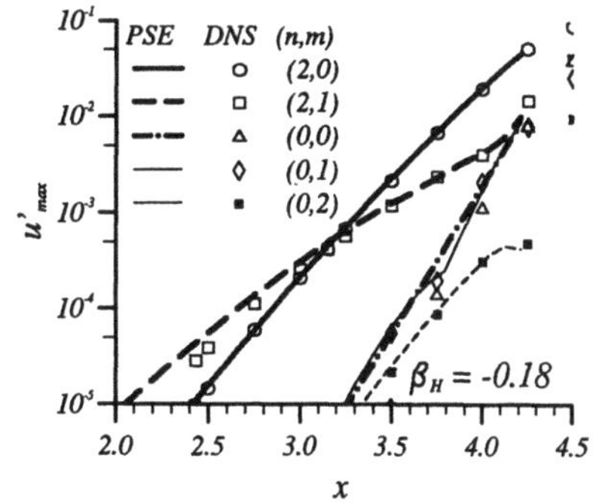

Figure 4: Comparison PSE - DNS (Kloker), K-type of resonance.

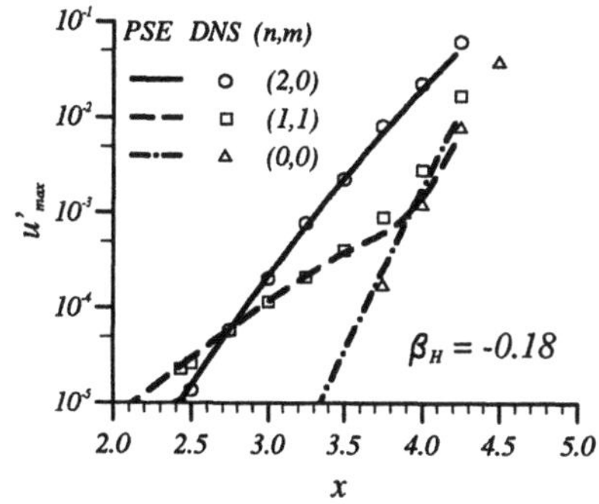

Figure 5: Comparison PSE - DNS (Kloker), H-type of resonance.

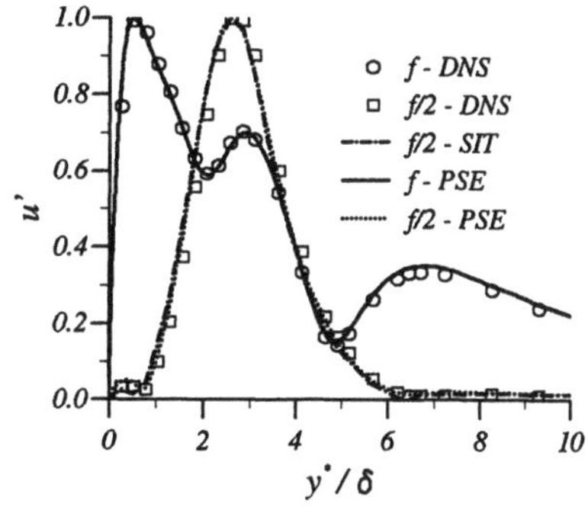

Figure 6: Amplitude functions for the primary and secondary modes.

to the main profile, it is possible to obtain changes in the mean flow ; figure (3) shows the skin-friction coefficient for a self-similar Falkner-Skan profile with a Hartree parameter equal to $\beta_H = -0.1$, an increase close to the onset of transition can be observed [5] as it is the case with the Blasius boundary layer.

But for strong positive pressure gradient, no experiment is available, we only know some calculations done by Herbert *et al.* in 1988 [7] with SIT. More recently, Kloker [8] has carried out some DNS with $\beta_H = -0.18$. As example, figure (4) shows a comparison between DNS and PSE in the fundamental resonance case. Both calculations agree very well and the onset of resonance is observed. But it seems that the convergence with PSE is more limited in amplitude : the maximum amplitude is 5%. However, despite the PSE approximations, the agreement concerning the mean-flow distorsions (the two-dimensional mode and the vortex modes) is noticeable. The same agreement is obtained with the H-type resonance as shown in figure (5). Figure (6) shows (for $R_{\delta_1} = 1634$) that the amplitude functions are in excellent agreement

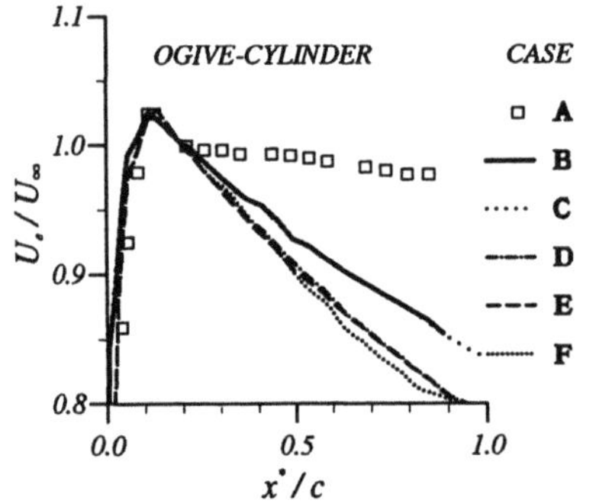

Figure 7: The six U_e distribution for the ogive-cylinder experiment.

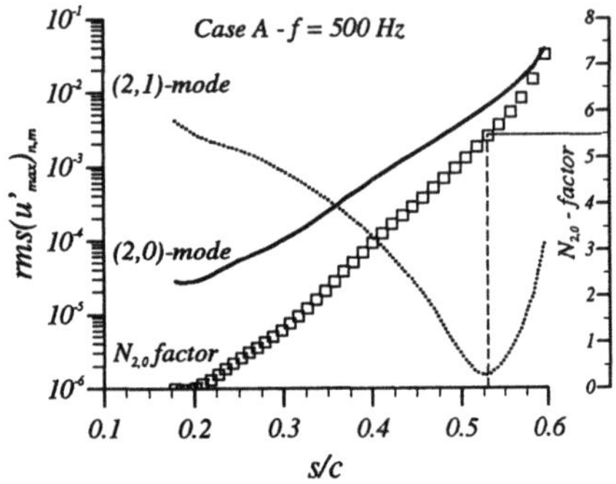

Figure 8: Ogive-cylinder case A, K-type of resonance.

too : DNS, SIT and PSE results are fully superposed. More generally, PSE and SIT are in perfect agreement as long as the amplitude of the primary mode is not modified in comparison with the linear evolution. In many cases, during the resonance, the secondary mode grows rapidly (as in figure (1)), modifying thereby the mean-flow distorsion and, if this latter is large enough, it modifies the primary mode.

3.3 Application with uncontrolled experiment

In 1979, Arnal *et al.* [9] performed experiments on an ogive-cylinder for decelerating flows, which can be interpreted as parametric studies for pressure gradient effects and unit Reynolds number effects. Figure (7) shows the external velocity distributions : six cases are considered. The pressure gradient for cases B and C is the same but the values of U_∞ are different, the same is true for cases D, E and F. Of course, the six cases have been studied in the same wind tunnel. The aim is to achieve non linear PSE calculations for these cases which are no longer self-similar boundary layers. To determine the initial conditions, the following procedure has been adopted. Firstly, taking into account the turbulence level of the wind tunnel (0.15%), we assume that transition occurs after a K-type of resonance. Then initial amplitudes of the primary and secondary modes are adjusted in order to reach 5% in fluctuation amplitude at the end of calculation, this corresponds to the convergence limit for the PSE. Let us note $N_{2,0}$ the linear N-factor of the $(2,0)$-mode. For these adjusted values, the $N_{2,0}$ factor is 7 at the last location which is close to the experimental onset of transition. For $\psi_{2,1}$, a value close to 60° was choosen for the secondary mode. Figure (8) gives the PSE results for this case A, only the primary and secondary modes are represented. The frequency $f = 500$ Hz is the most amplified one at the onset of transition, s is the curvilinear abscissa and c the chord. As indicated on the figure, resonance onset occurs for $N_{2,0} \simeq 5.5$. In a second step, we performed calculations for the five others cases with the assumption that the minimum

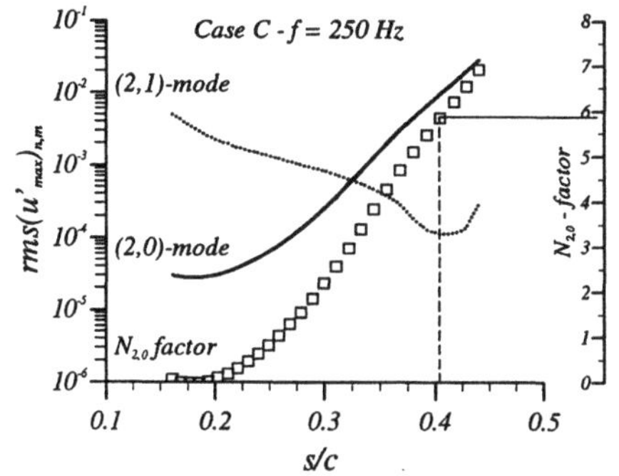

Figure 9: Ogive-cylinder case C, K-type of resonance.

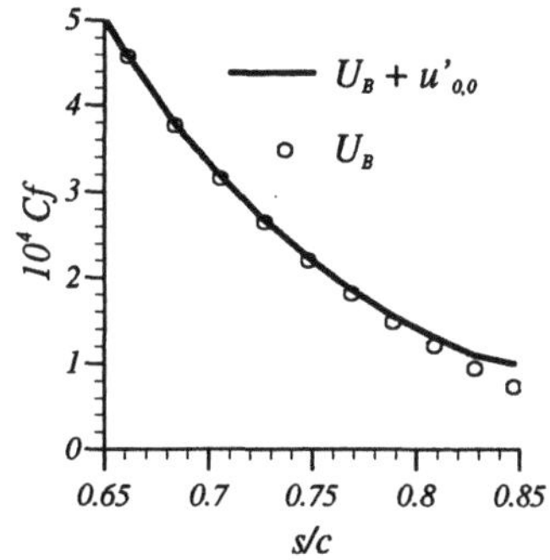

Figure 10: ONERA D airfoil, K-type of resonance.

values of the primary and secondary mode amplitudes are identical for these six cases (same receptivity conditions). Of course, in each case, we consider the most linearly dangerous frequency. Only a result is given here, figure (9), it corresponds to case C with the same representation as figure (8). Again, convergence stops for a 5% amplitude level at a location close to the experimental onset of transition (and where $N_{2,0}$ is close to 7). Moreover, resonance starts for $N_{2,0} \simeq 5.8$. A compilation of the six results leads to the following table where the shape factor H and the $N_{2,0}$ value at resonance onset are indicated :

	A	B	C	D	E	F
$N_{2,0}$	5.5	6.1	5.8	5.5	5.8	5.5
H	2.63	2.86	2.84	2.86	2.98	3.13

It appears that K-type of resonance always occurs for $N_{2,0}$ close to 5.5 : limit of the strictly linear behaviour (for 2D incompressible flows). Moreover, PSE convergence limit (5% for the fluctuation amplitude) corresponds to a location close to experimental onset of transition (and close to a value of 7 for the $N_{2,0}$ factor). To confirm these conclusions, a more difficult case has been considered : the flow on an ONERA D airfoil (without sweep) where experimental transition occurs at $s/c \simeq 0.85$, close to a separation bubble [10]. For this last case, we assume same initial conditions and we obtain the resonance onset for $N_{2,0} \simeq 5.6$ as before. Figure (10) represents the skin friction evolution corresponding to this calculation (as figure (3)) : a slow increase of the C_f value is obtained just before transition.

4 Conclusion

Non linear PSE seem to be an efficient tool for studying weak non linear effects. Very good agreement is obtained in comparison with experiment (Blasius flow) and DNS (strong adverse pressure gradient) for self-similar flows. Furthermore, with uncontrolled experiment, it has been possible to show that a linear N factor close to 5.5 seems to be the limit of the linear behaviour of the TS waves.

References

[1] Herbert Th. and Bertolotti F.P.: Stability analysis of nonparallel boundary layers, Bull. Am. Phys. Soc. vol 32 (1987).

[2] Airiau Ch : Stabilité linéaire et faiblement non linéaire d'une couche limite laminaire incompressible par un système d'équations parabolisé (PSE), thèse ENSAE (1994).

[3] Kachanov Y.S. and Levchenko V.Y.: The resonant interaction of disturbances at laminar-turbulent transition in a boundary layer, J.F.M. vol.138 (1984).

[4] Saric W.S., Levchenko V. Yu. and Kozlov V.V.: Forced and unforced subharmonic resonance in boundary layer transition, AIAA-84-0007 (1984).

[5] Airiau Ch. and Casalis G.: Weakly non linear instability of the laminar boundary layer calculated by a parabolic system of equations, Asymptotic modelling in fluid mechanics, J.P. Guiraud, Springer-Verlag (1994).

[6] Knapp C.F. and Roache P.J.: A combined visual and hot-wire anemometer investigation of boundary layer transition, AIAA Journal vol.6 (1968).

[7] Herbert Th. and Bertolotti F. P.: The effects of pressure gradients on the growth of subharmonic disturbances in boundary layers, Proc. Conf. Low Reynolds Number Airfoil Aerodyn., Ed. T. Muller (1985).

[8] Kloker M. and Bestek H.: Direct spatial simulation of laminar-turbulent transition in a boundary layer with strong adverse pressure gradient, DGLR-Jahrestagung (1992).

[9] Arnal D., Juillen J.C. and Michel R.: Analyse expérimentale de la transition de la couche limite avec gradient de pression nul ou positif, Israel Journal of Technology, vol. 17 (1979).

[10] Cousteix J. and Pailhas G.: Etude exploratoire d'un processus de transition laminaire turbulent au voisinage du décollement d'une couche limite laminaire, La Recherche Aérospatiale vol. 3 (1979).

Session C: Fundamentals III

Fine Structure of the 3-Dimensional Transition Region in the Wake behind a Thin-Shell Cylinder Placed Parallel to a Uniform Flow

Koji Tanaka+, Hironosuke Saito++ & Hiroshi Sato++

+ Dept.of Management,Osaka University of Commerce,Mikuriyasakae-cho, Higashi-Osaka 577,Japan.

++ Institute of Flow Research,6-10-39,Akasaka,Minato-ku,Tokyo 107, Japan.

Summary

Experimental investigations were made of the transition of the wake behind a thin-shell cylinder placed parallel to a uniform flow with sound excitation.The highly selective amplifications of the linearly unstable waves were observed. The 3-dimensionality becomes significant as the flow goes to downstream and this brought the gradual distortion of the anti-symmetrical profiles of the fundamental component. Measurements on the phase distributions in streamwise and azimuthal directions indicate a helical wave motion. The cycles of the present helical wave motion are different from that of an axismmetric slender body with full core.

Key words: Wake, Transition, Axisymmetric thin-shell cylinder.

1.Introduction.

Artificial control of the transition processes in the wake of a thin flat plate by an introduction of sound has succeeded remarkably as shown in Sato and Kuriki[1]. Fundamental transition mechanism of the two-dimensional wake behind a flat-plate was almost clarified by Sato and Saito[2].The transition process is categorized into the three subregions:linear ,nonlinear and randomization region.

Wake behind an axisymmetric 3-dimensional slender body was studied by Sato and Okada[3]. In their work,a helical vortex motion was indicated and their results have shown different manner from that of a 2-dimensional flat plate as shown in aforementioned papers[1][2].

Recent study of Monkewitz [4] emphasized the concept of absolute instability in the wakes and its relation to the large scale helical shedding. Oertel commented in his review paper[5] that the quantitative determination of absolute unstable wake region is the key to an efficient wake control. However their attention is focused on the structure of the near wake region and the relation to the stability theory.Their works are extremely important, but we are much interested in the full understanding of the streamwise development in a wake. It is important for us to answer the question why the randomization process is taken place from deterministic linear and nonlinear processes.

The main objectives of the present experimental investigation in the wake of a thin-shell cylinder placed parallel to a uniform flow are following two items.

(1) The differences from the wake of two-dimensional body.

(2) The differences from the axisymmetric 3-dimensional body with full core.

At the present time, we concentrate our study on the following three points.

(1) Initial stage of the transition processes, especially in linear and weakly interacted nonlinear region.

(2) The downstream evolution of the fundamental and harmonic components.

(3) Phase informations to conjecture a vortex motion.

2. Experimental method.

All experiments were conducted in a low-turbulence wind-tunnel. The wind-tunnnel is of straight, blow-out type. The test section is 40cm by 40cm cross-section and 3m length as shown in figure 1. A thin-shell cylinder is placed parallel to the uniform flow. The suspension of the cylinder in the center of cross section is made by fine wires of 0.1mm diameter from wind-tunnel walls. Since Reynolds number based on the diameter of the suspending wire was about 40 at the flow velocity of 6 m/sec, no shedding of vortex from the wire is detected. A loud-speaker is set on the right hand side from downstream.

Figure 2 shows the detailed dimensions of the specially designed thin-shell cylinder and co-ordinates systems. The cylinder is made of acrylic resin. The thickness is 2mm, 50mm outer diameter and 250mm length. The leading edge is smoothed out and the trailing edge is sharply tapered with an angle of about thirty degrees. The origin of the co-ordinates is at the top point of the circular trailing edge. X is a distance measured from the origin to downstream. Y is a vertical direction. For convenience, we termed positive Y region and negative Y region as outer-region and inner-region respectively.

Measurements were made by hot-wire anemometers. Two hot-wires are used to evaluate the phase angle from cross-correlations. The reference probe is fixed and the other probe is rotating in the azimuthal direction with same radius. Two signals from the sensors are recorded at each point simultaneously by two channnels of a digital audio tape recorder. Data aquisition system is connected to a personal computer and all signals are digitized to the process. The velocity of the uniform flow is about 6 m/s and Reynolds number based on the cylinder diamter is about 20,000. Experiments were made with sinusoidal sound excitation by the loud speaker with frequency of 313Hz . This frequency corresponds to the most unstable waves based on linear theory. The input level of the sound to the speaker is 1.5 V rms.

3. Experimental results

The normalized rate of amplification of the introduced sound against the various frequencies is shown in figure 3. The most amplified frequency of this experiment is 313Hz. If the frequency differs 3% around the

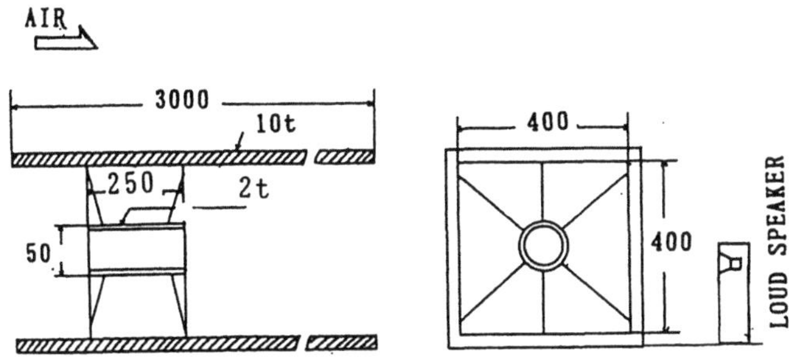

Fig. 1. Experimental setup and test section.

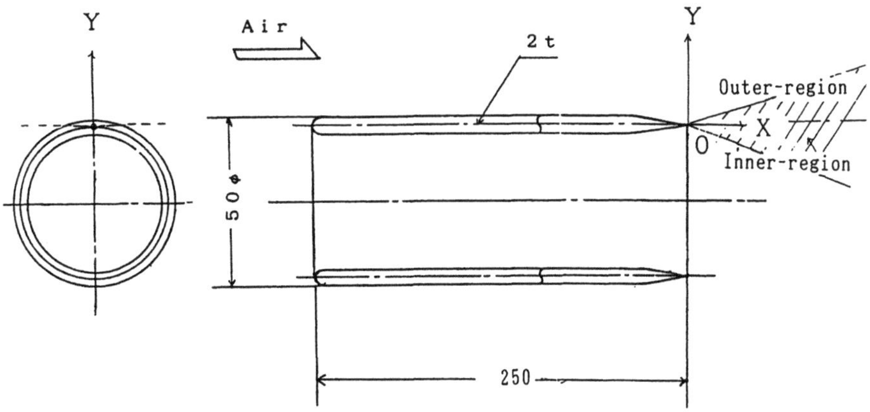

Fig. 2. Dimensions of the thin-shell cylinder
and co-ordinates system.

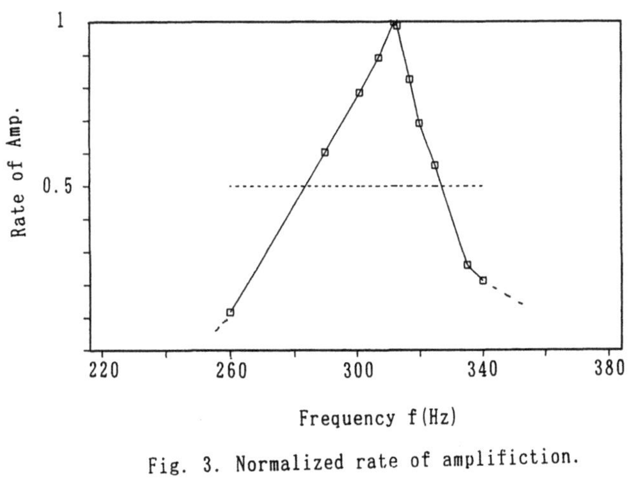

Fig. 3. Normalized rate of amplifiction.

dominant frequency, it is difficult to obtain the synchronized wave form. This highly selective amplification of the linearly unstable wave is an interesting finding of the present experiment. In the case of 2-dimennsional wake behind a flat plate, the frequency range of selective amplification is wider than that of the present case. It ranges about 45 % around the most amplified frequency.

Figure 4a shows the streamwise developments of the mean velocity profiles U/U_o, where U_o denotes the free-stream velocity. At X=40 to 50mm, the effects of the sound become significant and the defect of the mean velocity is decreasing. Correspondingly, the wake breadth widens rapidly.

In figure 4b, mean velocity distributions at two X-stations with magnified longitudinal scale are shown. At X=60mm, the wake breadth becomes wider than that of the upstream points. As the flow goes to downstream, the symmetrical profiles are distorted. This is a different manner from that of two-dimensional case. The Y position of the maximum velocity defect shifts from Y=0 at the downstream points.

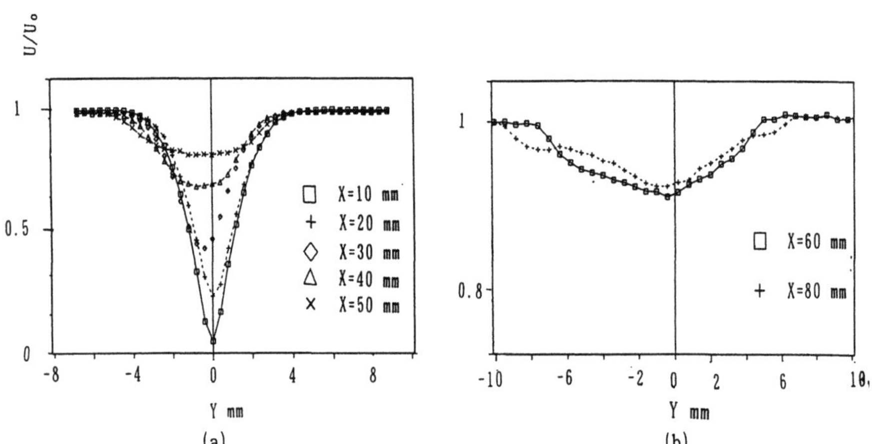

Fig. 4. Mean-velocity distribution.

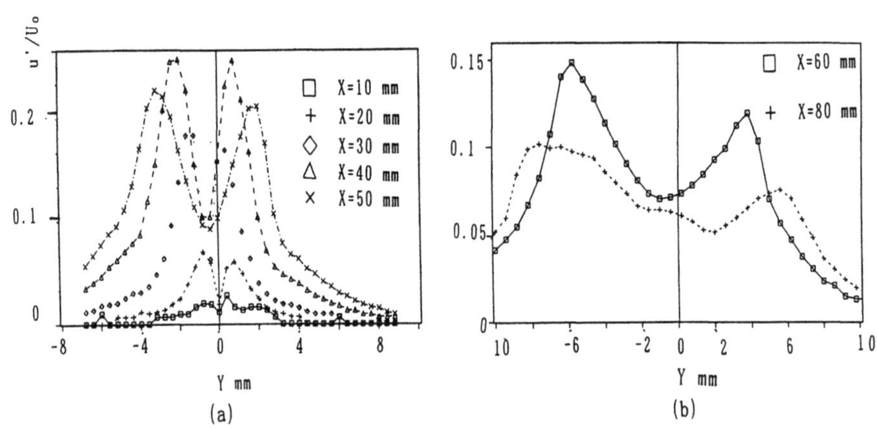

Fig. 5. Distribution of r.m.s value for u-fluctuation.

Figure 5a shows streamwise distribution of the non-dimensionalized r.m.s value of u-fluctuation u'/U_0. The symmetrical double peaks with respect to Y-axis are almost holding at X=10 to 50mm. In this X range, the highest magnitude of the r.m.s value reaches about 20 % that of U_0.
In figure 5b, the longitudinal scale of u'/U_0 is magnified and the distributions at two X stations are shown. At X=60mm, the symmetrical profiles with respect to Y-axis are distorted gradually. At X=80mm, the triple peakes appear and the maximum r.m.s value of u-fluctuation in the inner-region becomes larger than that of the outer-region.

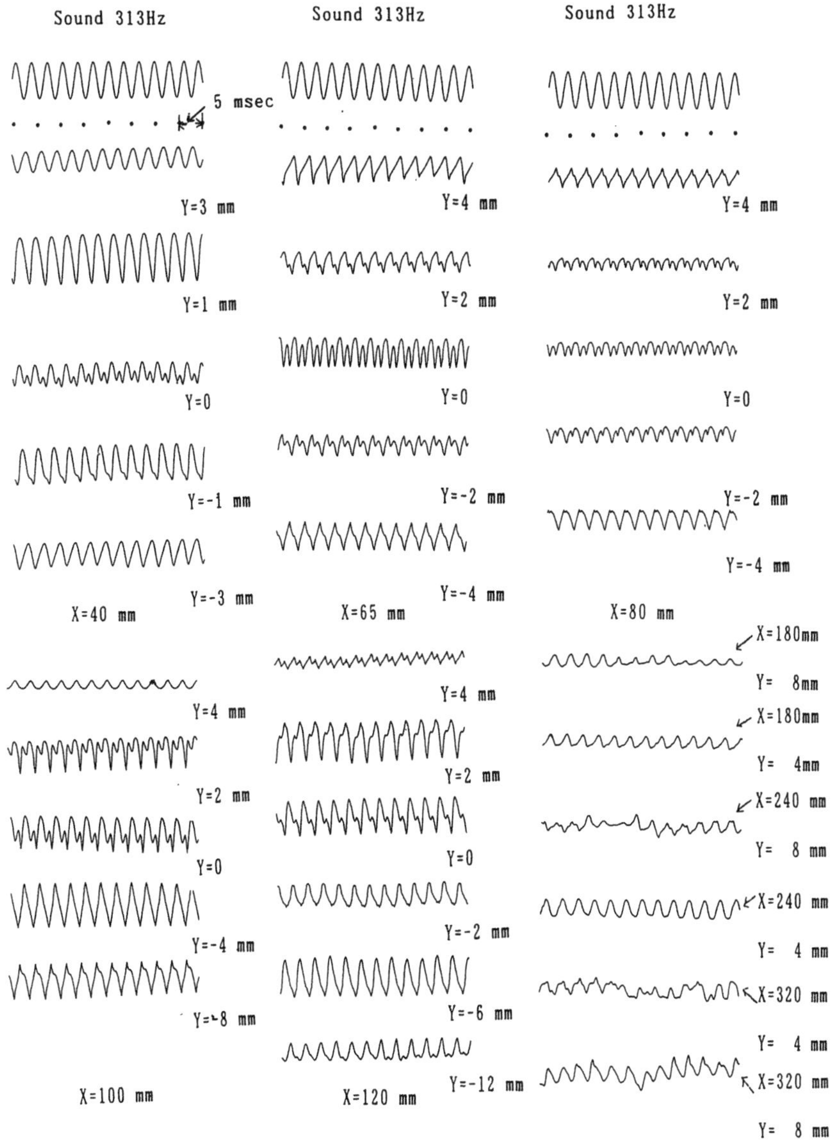

Fig. 6. Wave forms at various X-stations.

The typical wave forms at various X stations are presented in figure 6 to overview the whole flowfield. In this figure, the velocity increases upward. The time interval between dots is 5 msec. At the top of each column, the sinusoidal wave form corresponds to the introduced sound. At X=40mm, the weak nonlinearity influences around Y=0 and the wave form is significantly modified by the second harmonics. Almost similar wave form is observed at symmetrical points Y=2mm and Y=-2mm. At X=65mm, the triangular wave form is observed. Higher harmonics modulate the wave forms and the similarity of the wave forms are not observed at symmetrical Y positions. At X=80mm almost square wave form is observed. In the range between X=30mm and 80mm, various kinds of wave-forms are observed. However the flow is under control by the sound excitation and the modulation of wave forms by lower frequency components are not significant. Furher downstream, the regular wave-forms are gradually distorted. At X=240 to 330mm, the randomized wave forms are observed.
Figure 7 shows Y distributions of r.m.s. value of the fundamental component at various X stations. The ordinate is arbitrary but relative values are correctly expressed. At X=30 to 50mm, the anti-symmetrical profiles are observed. At X=65 to 80mm, the anti-symmetrical distributions are no longer holding. The magnitude of the fundamental component in inner-region is larger than that of outer-region.

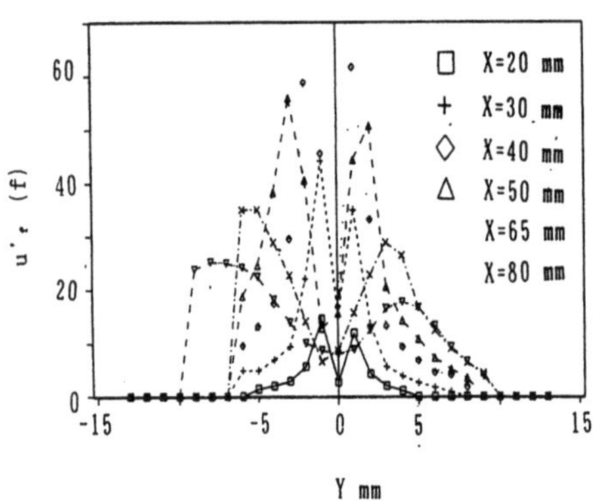

(a)

Fig. 7. Amplitude distribution of r.m.s value for fundamental component.

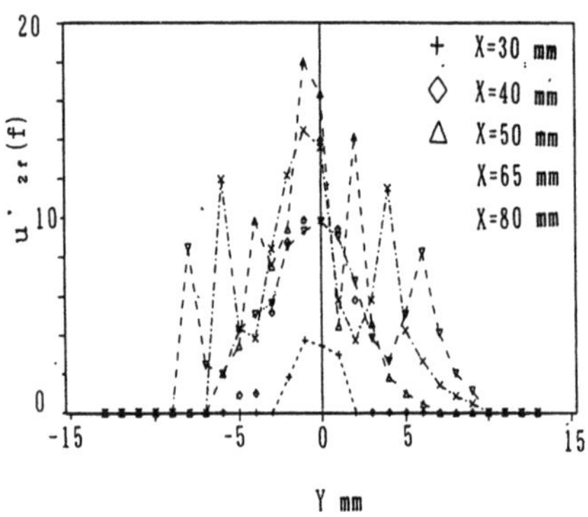

(b)

Fig. 8. Amplitude distribution of r.m.s value for second harmonics.

Figure 8 shows Y distributions of r.m.s value of the 2nd harmonics component at various X stations. The ordinate is arbitrary but relative magnitudes are correctly expressed. At X=30mm, the weak magnitude of the second higher harmonics component is observed in the vicinity of Y center.

The maximum magnitude of the 2nd harmonics component at that point is about 20 persent that of the fundamental component. As the flow goes to downstream, the symmetrical profiles are still remaining. At X=50 to 80mm, many peakes appear and they may bring the variation of the wave forms.

The magnitudes of the 2nd harmonics component are relatively smaller than those of fundamental component. Correspondingly, the contribution of the 2nd harmonics on the symmetrical distributions of the u-fluctuation are not influential. These results imply that the breaking of the symmetrial distribution is largely depending on the fundamental component. And it may be conjectured that in the inner-region 3-dimensionality becomes effective as the flow goes to downstream. It may be originated from the previous strong mixing of inner flow surrounded by the cylinder.

Figure 9 shows the azimuthal distribution of the phase angle of fundamental spectral component at X=30 mm, Y=3mm. The dashed line in figure shows expected value of the phase angle. As demonstrated in this figure, azimuthal angle change of 180 degrees results in the change of 360 degrees of the phase angle. In one turn two cycles are observble and it implies a helical wave motion in the linear region, but it is not stated with confidence.

As shown in figure 10, the same trend of the phase change as previous figure is observed at Y=-3mm. In both cases, observed wave length is about 19mm and the phase velocity of the fundamental component is about 6 m/s. This result implies that the fundamental component travels almost same velocity as that of the uniform flow. The observed values in the figures show quite large scattering. This results indicate that the flow is quite sensitive to the external disturbances and the carefull treatment of the correlation measurement is necessary.

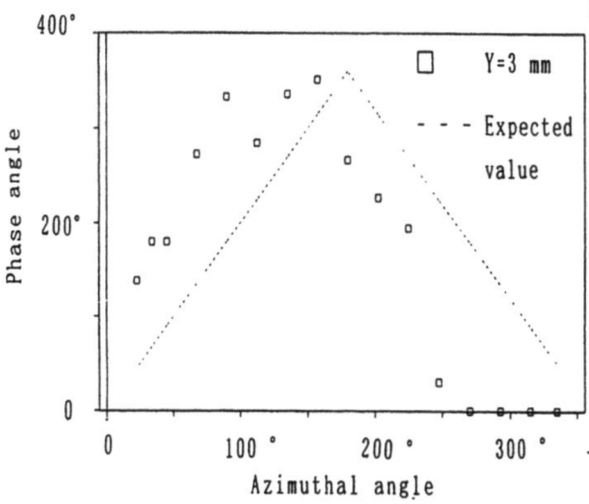

Fig. 9. Azimuthal change of phase angle. X=30 mm, Y=3mm.

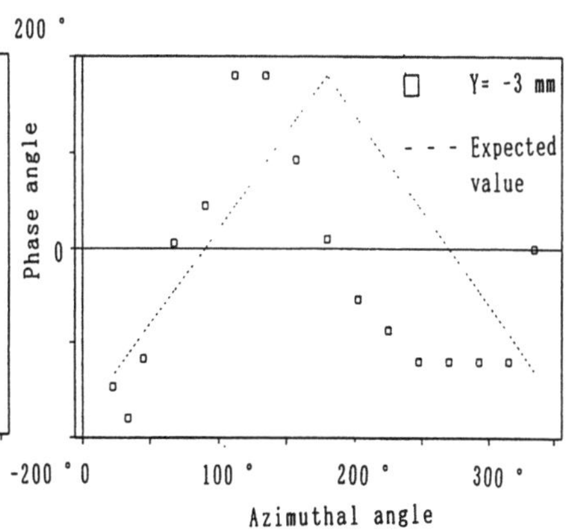

Fig. 10. Azimuthal change of phase angle. X=30 mm. Y=-3mm.

4.Conclusions

Following conclusions are obtained under the present experimental condition.
(1) In the linear region,a helical wave motion is implied..
(2) In the weakly nonlinear region,the symmetrical distribution of the fundamental component is gradually distorted and it results in the asymmetrical longitudinal distribution of u-fluctuation.
(3) Selective amplification of the linearly unstable waves in the narrow range of frequency is observed by the change of frequency of excitation.

References.

[1] Sato,H and Kuriki,K.:The mechanism of transition of a thin flat plate placed parallel to a uniform flow,J.Fluid Mec.11,part 3(1961) 321-352.
[2] Sato,H.and Saito,H.:Fine structure of energy spectra of velocity fluctuations in the transition region of a two-dimensional wake,J.Fluid.Mec. 67,part 3(1975) 539-559.
[3] Sato,H.and Okada,O.:The stability and transition of an axisymmetric wake, 26,part2(1966) 237-253.
[4] Monkevwitz,P.A.:A note on vortex shedding from axisymmetric bluff bodies.192(1988) 561-575.
[5] Oertel,H.Jr.:Wakes behind blunt bodies,Annu.Rev.Fluid.Mech.22(1990) 539-564.

Transition to Turbulence in Pipe Flow

Aswin A. Draad, Gerard D.C. Kuiken and Frans T.M. Nieuwstadt
Laboratory for Aero- & Hydrodynamics
Delft University of Technology
Rotterdamseweg 145 , 2628 AL Delft, The Netherlands

Summary

A horizontal pipe flow loop has been built for the experimental investigation of transition from laminar to turbulent flow. The experiments presented here have all been done with water. Laminar flow can be sustained up to Reynolds numbers of 40,000. The transition to turbulence was triggered by a small jet injected through the pipe wall, and was studied using Digital Particle Image Velocimetry (DPIV) at a Reynolds number of 5860. DPIV is a quantitative method for measuring instantaneous velocity fields whereas a technique like laser Doppler anemometry is only a one-point measurement technique. DPIV can be used to study the transition to turbulence in detail and it might well prove to be a useful tool in gaining more insight into the transition processes.

Key words: Pipe flow, Digital Particle Image Velocimetry, experiment, triggering.

1. Introduction

This investigation is part of a study on the transition from laminar to turbulent flow for non-Newtonian fluids. Before we start with experiments in these complex fluids, we have first performed experiments with water, i.e. with a Newtonian fluid. The results of these latter experiments will be presented here. For the geometry we have selected a pipe flow. Knowledge of the parameters governing the transition to turbulence in pipe flows is important since the type of flow, laminar or turbulent, is directly related to for instance pressure loss, efficiency of mixing and heat exchange. The transition to turbulence in pipe flows for Newtonian fluids is far from being understood and we anticipate that detailed measurements may shed more light on this problem.

To be able to investigate the parameters that govern the transition, laminar pipe flow at high Reynolds numbers must be available. Only in that case disturbances to the flow can be added in a controlled way, thus allowing the transition to be manipulated. In our case we used a small jet to trigger transition. Critical injection velocities that just trigger transition to turbulence were determined for a range of Reynolds numbers. The transition process was studied at one Reynolds number using Digital Particle Image Velocimetry (DPIV).

2. Experimental Setup

2.1 Flow Facility and Disturbance

We have constructed a new experimental pipe flow loop especially suited for transition studies. The horizontal pipe has a length of 34 m and an inner diameter (D=2R) of 40.0 mm. An overview of the facility is given in figure 1.

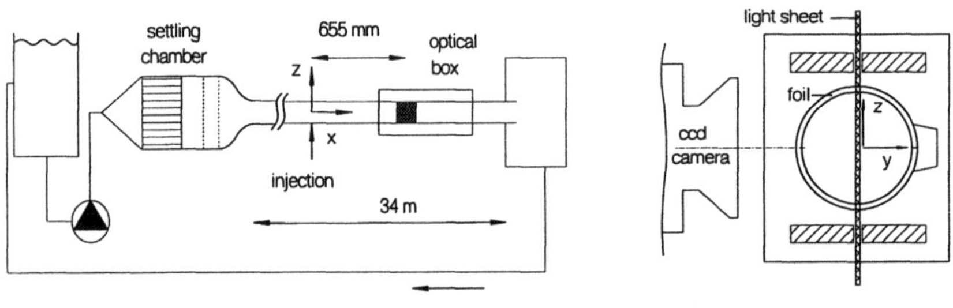

Figure 1: Horizontal pipe flow loop, with a cross section of the optical box, for laminar flow till Reynolds number 40,000. Measurements were done at 29.7 m from the pipe inlet. The flow was disturbed at Re=5860 using a small jet at 655 mm before the measurement position in either z-direction (parallel with light sheet) or y-direction (perpendicular to light sheet). The hatched parts inside the optical box are used to block reflections from the locations were the light sheet crosses the walls of the optical box.

Much care was devoted to avoid inlet disturbances. This was accomplished by fitting a smooth (5th order polynomial) contraction between the settling chamber (250 mm diameter) and the pipe inlet. The pipe is made up out of segments with a length of 2 m. The couplings of these segments were carefully designed (< 20 μm difference in wall position on both sides of the coupling) to avoid introduction of disturbances. To break up large scale flow disturbances in the settling chamber, we used a 20 cm long honeycomb constructed from 4 mm diameter drinking straws and 2 screens of 2 mm meshsize. The pipe setup sustains laminar flow up to Reynolds numbers of Re≈25,000, based on centerline velocity (U_{cl}) and pipe radius (R). This can be extended to Re≈40,000 if the last screen before the contraction is replaced with one having a meshsize of 0.5 mm. The length of the pipe is chosen such that fully developed Poiseuille flow (within 99%) can be reached up to Re≈15,000.

Disturbances are introduced into the flow in the form of a fluid jet injected through a 1 mm hole in the pipe wall. The jet is fed by a small container placed above the pipe and the injection time can be controlled by a programmable valve. The injection speed can be manipulated using a regulator.

LDA and DPIV measurements are performed with an 'optical box', i.e. a rectangular plexiglass box filled with water and placed around the pipe. Such a construction is necessary to reduce refraction effects due to the curvature of the pipe wall, which is a result of differences in the index of refraction for plexiglass (1.49), water (1.33) and air (1.00). Inside the optical box, the plexiglass pipe wall has been replaced at several places by a thin sheet to achieve a further reduction in refraction.

A special sheet, Teflon FEP 750A fluorocarbon film manufactured by Du Pont, was used which is 0.19 mm thick and has an index of refraction n=1.35. This is particularly useful for two-component LDA measurements in turbulent flows, since coincidence of the four laser beams is guaranteed till very close to the wall [1].

2.2 Measurement Technique

Particle Image Velocimetry yields the instantaneous two-dimensional velocity field in a planar cross section of a fluid. The motion of the particles is recorded in a multiple exposed image. The image is devided into small sub-images, and for each sub-image we compute the auto-correlation. All particle-image pairs that correspond to the motion of a tracer particle contribute to an off-center peak in the correlation. The (local average) displacement of the particle images is obtained from the centroid of this peak. Given the image magnification and time delay between de exposures we obtain the in-plane velocity.

The Digital Particle Image Velocimetry measurements were performed with an image acquisition system that is equipped with two CCD-cameras (Texas Instruments) having 1000 $\times$ 1016 square pixels. The cameras were operated at a frame rate of 10 Hz. The heart of the image acquisition system is a pair of two MaxVideo-20 (DataCube) pipeline processors which control the cameras and data transport. Each processor is equipped with four MegaStore-32 memory boards, which yields a total real-time storage capacity of 256 Mb. The entire system is controlled by a work station (Sun-Sparc). For the present experiments, only one camera was used.

For DPIV, the flow was seeded with small tracer particles (average diameter 30 μm) that were illuminated with a parallel scanning laser beam from an argon-ion laser operated at 2.2 Watt which was reduced to 500 mW at the output of the optical fiber. With the help of a rotating mirror and a cylindrical lens the laser beam of thickness 0.5 mm scanned parallel through the object plane. The time between two passages of the scanning laser beam was set to 1.725 ms. With an exposure time of 14.0 ms each particle is exposed 9 times. The camera was equipped with a lens having a focal length of 50 mm and we used an aperture of 2.8. The optical axis of the camera is perpendicular to the light sheet. The recording of a series of 131 frames was synchronized with the start of the injection. These images were then transferred to a workstation (HP-720) for data analysis. The images were interrogated in 32 $\times$ 32 pixel subimages with 50% overlap in each direction giving a grid of 59 $\times$ 60 vectors per frame. Zero-padding the signal before performing an FFT had to be applied as in a laminar flow both ends of the subimage correlate so well that the displacement is biased towards 32/3, 32/4, a.s.o. pixels. The size of the measurement volume then becomes 1.3 $\times$ 1.3 $\times$ 0.5 mm^3. For more details we refer to [2]. A detailed discussion on the image processing technique can be found in [3].

Laser Doppler measurements were performed with a two-component full backscatter probe (Dantec). Two-color laser light from an argon-ion laser (Spectra Physics 2020) was transmitted to the probe using optical fibers. The laser Doppler signal was analyzed using two Burst Spectrum Analyzers (Dantec). The size of the measurement volume was estimated to be 0.02 $\times$ 0.02 $\times$ 0.10 mm^3.

3. Velocity Profiles

The pipe flow facility is long enough to achieve fully developed Poiseuille flow at the chosen Reynolds number and measurement position. Although the flow was found to be laminar up to Reynolds numbers of 40,000, no Poiseuille profile could be obtained. Both the horizontal and the vertical profiles were found to be asymmetric as is shown in figure 2. We suspect that the reason for this deviation is caused by thermal convection. The flow loop is located in the basement, so temperature differences between day and night are already smaller than in other parts of the building. Nevertheless, during the night the temperature drops and as air warms up much faster during the day than water, the water temperature appears to lag the air temperature by approximately 1.5 °C. Given the low velocity in the settling chamber and the length of the pipe, this temperature difference is large enough to trigger convection rolls that may drastically deform the velocity profile. Insulation of the pipe and the settling chamber improved the profiles but they are still not completely symmetric. Particularly the asymmetry in the horizontal direction is puzzling because, if asymmetry is caused by convection only, the horizontal profile should be symmetric. Therefore, more study is required to find the cause of this deviation.

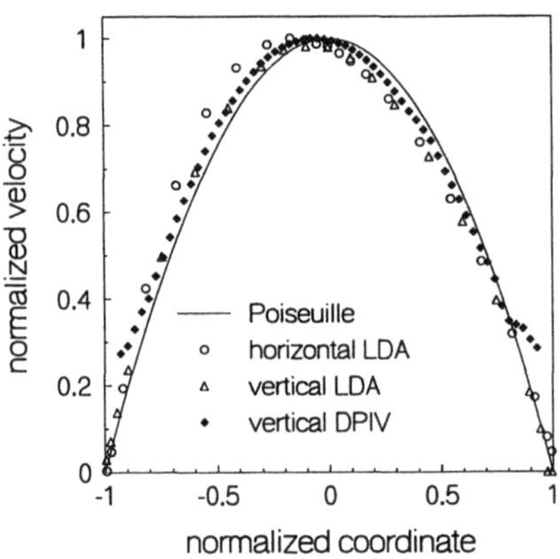

Figure 2: Measured asymmetric velocity profiles after insulation of the pipe; Re=5860

As can be seen in figure 2, the horizontal profile has an asymmetry that is stronger than the vertical profile. The vertical profiles for the LDA and DPIV measurements show the same asymmetry. Due to experimental limitations the DPIV and the LDA velocity profiles could not be measured simultaneously and the LDA measurement position is 0.5m downstream of the DPIV measurement location.

Close to the wall, DPIV measurements become impossible due to the small particle displacement. The velocity near the wall is so low that the particle images between two exposures overlap so that it is impossible to obtain the off-center correlation peak, on which the DPIV method is based. As a result one measures erroneous velocities near the wall. In order to measure closer to the wall, either higher resolution CCD-cameras are needed if the total cross section of the pipe is to be studied or alternatively one has to zoom in.

4. Disturbances

The flow was perturbed by injecting fluid periodically through a 1 mm hole in the pipe wall. Injection is possible at two positions in the circumference, i.e. perpendicular to the light sheet (negative y-direction) and parallel with the light sheet (positive z-direction) as shown in figure 1. This means that the measurement equipment does not have to be moved when the injection direction is changed. The injection velocity was determined by measuring the injected volume per injection. The flow was disturbed every 15 seconds with an injection period of 0.5 s. The critical injection velocity V_i, i.e. the magnitude of the jet velocity that just triggers transition to turbulence as observed using dye injection, has been measured for several Reynolds numbers. Figure 3 shows that V_i varies only slightly with the Reynolds number. Thus, the ratio of the injection velocity and centerline velocity, V_i/U_{cl}, is almost inversely proportional to Re and increases drastically from 1.2 at Re=23,500 to 8.9 at Re=3,600 (see figure 4). Pipe flow seems to become much more stable if the Reynolds number is reduced below approximately 7,000. Also, the results in figure 4 suggest that laminar pipe flow can indeed be stable till very high Reynolds numbers as suggested by [4].

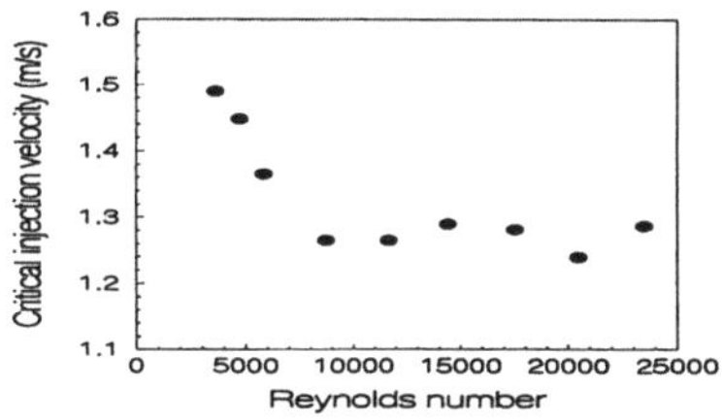

Figure 3: Dependence of critical injection velocity V_i on the Reynolds number in laminar pipe flow.

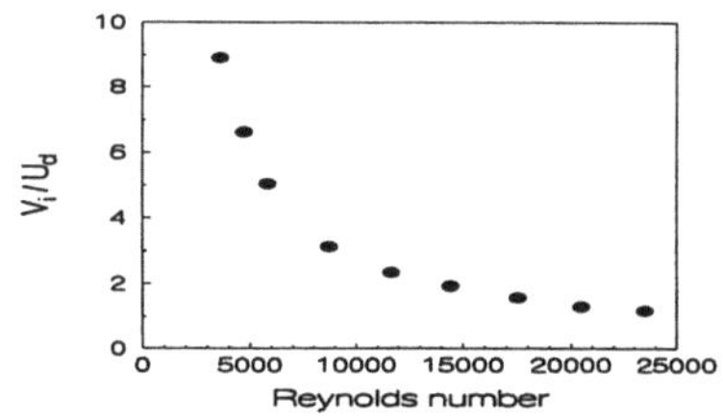

Figure 4: Variation of V_i/U_{cl} with the Reynolds number.

5. Results

During the DPIV experiments, the centerline velocity U_{cl} was adjusted to 0.272 m/s, giving a Reynolds number ($Re=U_{cl}R/\nu$) of 5860 at a water temperature of 23.3 °C. All quantities are made dimensionless using the pipe radius R and the centerline velocity U_{cl}. The injection velocity used in the experiments was slightly supercritical, $V_i/U_{cl}=5.2$. The injection was located at 29.0 m (1450 R) downstream of the pipe inlet. This point is referred to as x=0. Measurements were performed at 655 mm downstream of the injection, x=33. Two DPIV measurements will be presented, one in which the injection direction is parallel to the light sheet and one were the fluid is injected perpendicular to the light sheet. The entire cross section of the pipe is shown in the images and the center of the image field was located at x=33 (655 mm downstream of the injection point).

To follow the development of the transition, the velocity profiles are averaged over each frame and plotted versus time. The relative velocities were obtained by subtracting the laminar profile that was determined as an average of the profiles of the first 29 images.

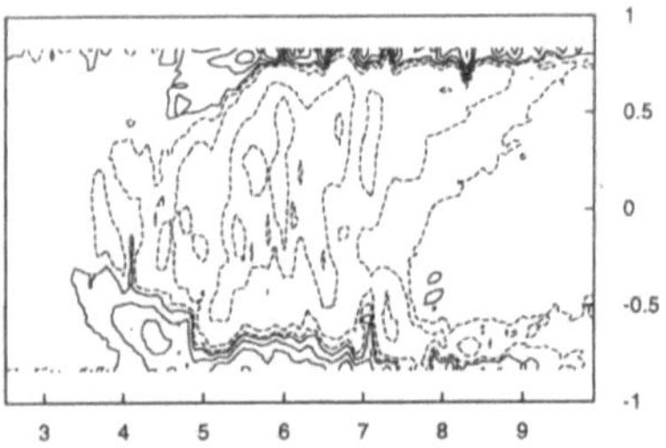

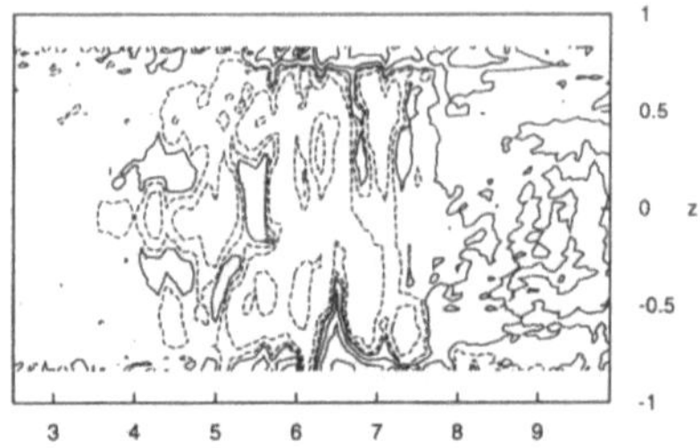

5a: levels: ±0.01, ±0.05, ±0.1, ±0.2, ±0.3,...
min: -0.344, max=0.389

6a: levels as in 6a; min=-0.311, max=0.318

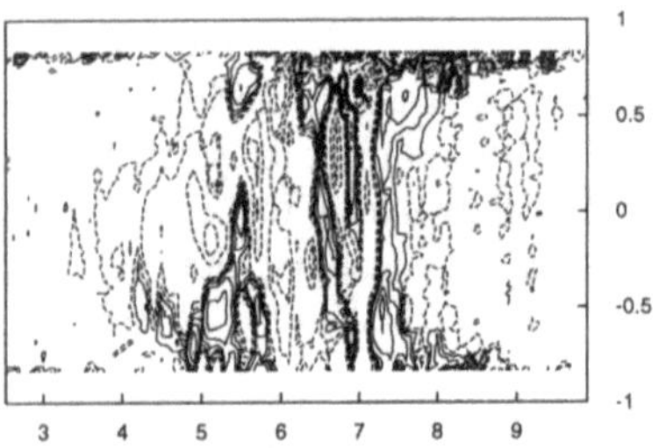

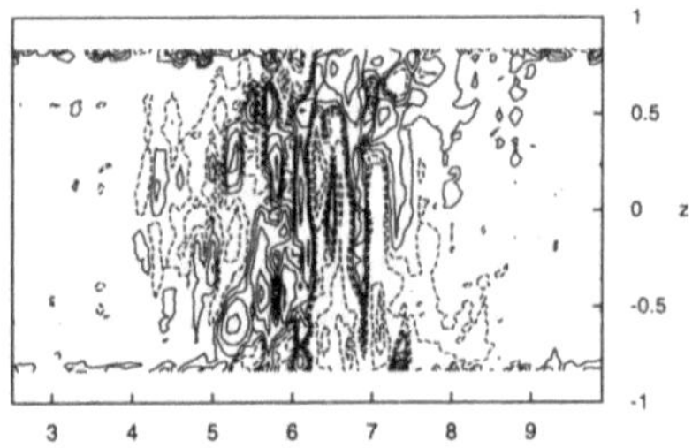

5b: levels: ±0.003, ±0.01, ±0.02, ±0.04, ±0.06, ...
min: -0.096, max=0.0536

6b: levels as in 6b; min=-0.058, max=0.081

Figure 5: Streamwise (a) and vertical disturbance (b) velocity profiles vs. time in seconds. Parallel injection; in z-direction. Solid lines are positive levels; dashed indicate negative.

Figure 6: As in figure 5 but now the injection direction is perpendicular to the light sheet; in negative y-direction.

In figures 5 and 6 the time sequence is shown for parallel and perpendicular injections respectively. The walls are located at the top and the bottom of the pictures. The disturbance velocities are plotted versus time so the transition to turbulence occurs at the left hand side whereas the relaminarization can be found at the right in the pictures. The data in the three rows closest to the walls has been eliminated as these are the areas in the flow were the velocity is too low for the DPIV technique to determine an off-center correlation peak for the displacement.

Injection results in a negative u-disturbance (streamwise) which for the parallel injection starts as early as 3.7 s. Almost simultaneously, a positive u disturbance starts to grow in the lower half of the pipe and grows towards the bottom wall. In the center the disturbances resembling small eddies (see figure 7a) also increase in strength having on average a vertical velocity directed downwards. When the region with eddy structures that result in positive u-disturbance reaches the bottom wall, a series of strongly counter-clockwise rotating eddies appear (t=4.9 s) that transport fluid away from the wall as is shown in figures 7b-7d. Shortly after this event, the top part of the flow, which was still relatively undisturbed, goes through transition rather quickly. Relaminarization is rather abrupt (t=7.3 s) in the center of the pipe and whereas a trail of disturbances extends close to the wall.

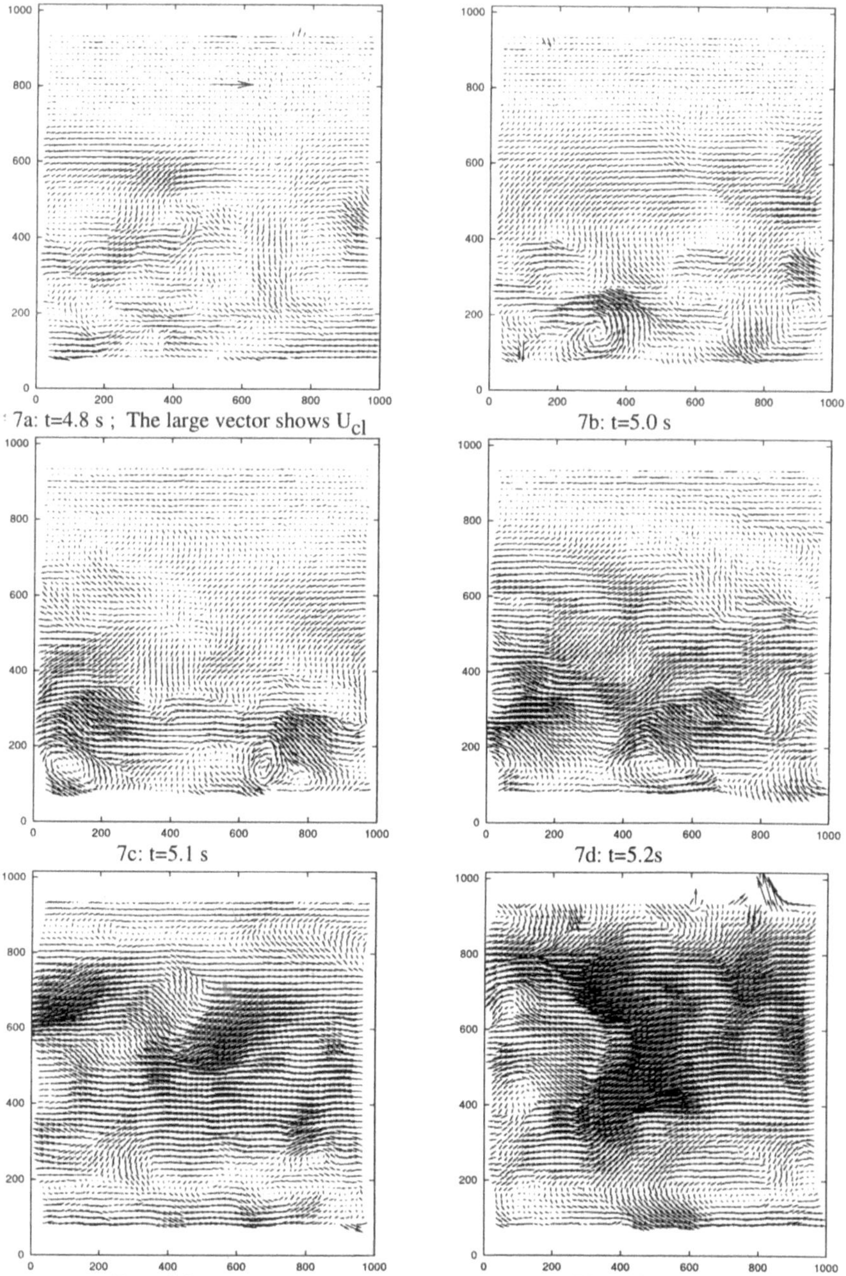

7a: t=4.8 s ; The large vector shows U_{cl}

7b: t=5.0 s

7c: t=5.1 s

7d: t=5.2s

7e: t=5.5 s

7f: t=6.0 s

Figure 7: Instantaneous disturbance velocity fields of a transition to turbulence triggered parallel to the image field at x=33 and y=0. Re=5860 and Vi/Ucl=5.2. Note the presence of the sequence of strong eddies shortly before the transition over the entire pipe sets in. The lines of small disturbances in figure 7e are the cross over lines between the laminar and turbulent velocity profiles.

When injected perpendicularly to the image plane in the direction of the camera, the flow is initially disturbed symmetrically (figure 6a). The jet again causes the flow to slow down in the center and forces fluid above and below to speed up and more side lobes are being formed. These patterns are similar to those found by Klingmann [5].

The break in symmetry occurs at approximately 5 s after injection and is likely to be related to the rapid transition after the appearance of the strong eddies shown in the parallel injection experiment.

In figure 7 several instantaneous velocity fields are shown for the in plane measurements. Clearly, much information on the small structures that are present in the flow is being erased when determining the average velocity profiles for each frame displayed the time evolution plots shown in figures 5 and 6.

6. Conclusions

The preliminary measurements presented in this paper show that Digital Particle Image Velocimetry is a powerful tool to study transition to turbulence. DPIV yields accurate instantaneous velocity fields showing detailed information on small structures that are present in the flow. The disturbances introduced by the injected jet first show up in the center part of the pipe and start growing towards the wall were the jet was injected from. When reaching this wall, a series of strong eddies form that transport fluid away from the wall. This seems to rapidly trigger transition in the entire cross section of the pipe and breaking symmetry in the plane perpendicular to the injection.

DPIV can be used to elucidate phenomena governing transition to turbulence that previously could not be studied quantitatively. Furthermore, the velocity fields that are obtained using DPIV give a much more detailed view of the flow and are more easily interpreted than contourplots of disturbance velocities.

This work is sponsored by Shell Research.

References

[1] Tahitu, G.J.R. 1994 *"The possibilities for laser-Doppler velocimetry in a turbulent pipe flow"*. Report MEAH-117 (Delft University of Technology). M.Sc.-thesis (in Dutch).

[2] Westerweel, J., Draad, A.A., Hoeven, J.G.Th. van der & Oord, J. van 1994 *"A Fast Data Acquisition System for Digital PIV; application to fully-developed and transitional turbulent pipe flow"*. Proc. 7th Int. Symp. Applications of Laser Techniques to Fluid Mechanics, Lisbon, Portugal, 11-14 July 1994.

[3] Westerweel, J. 1993 *"Digital Particle Image Velocimetry - Theory and application.* (PhD-thesis) University Press (Delft)

[4] Salwen, H. & Grosch, C.E. 1972 *"The stability of Poiseuille flow in a pipe of circular cross section"*. J. Fluid Mechanics, vol. 54, pp. 93-112.

[5] Klingmann, B.G.B. 1992 *"On transition due to three-dimensional disturbances in plane Poiseuille flow"*. J. Fluid Mech., vol. 240, pp. 167-195.

Subcritical Disturbance Growth Caused by Hairpin Eddies

Masahito Asai* and Michio Nishioka**

*Department of Aerospace Engineering, Tokyo Metropolitan Institute of Technology, Hino, Tokyo 191, Japan
**Department of Aerospace Engineering, University of Osaka Prefecture, Sakai, Osaka 593, Japan

Summary

In order to clarify the mechanism of wall turbulence generation as well as the related critical flow condition, we have been working on the subcritical boundary layer transition triggered by high-intensity hairpin eddies. In this paper, we present our recent results on the subcritical disturbance growth leading to wall turbulence, obtained by periodically exciting hairpin eddies through a small wall orifice. The minimum Reynolds number for the disturbance growth due to regeneration of hairpin eddies is found to be about 130 ~ 150 (based on the momentum thickness). Even when the energetic hairpin eddies are introduced, however, the regeneration of the hairpin eddies occurs only along a narrow strip (extending downstream from the orifice) at the subcritical Reynolds numbers.

Key words: Boundary Layer, Transition, Turbulence, Hairpin Eddies

1. Introduction

Our knowledge on the transition process in a boundary layer under weak disturbance conditions has been well improved through a number of the experimental, theoretical and numerical studies, in particular, on the ribbon-induced transition in Blasius flow and in plane Poiseuille flow (with almost the same instability characteristics as Blasius flow [1]); see the reviews [2,3]. In spite of the progress in understanding of the detailed process of transition in such typical wall-bounded flows, however, the minimum Reynolds number for generating wall turbulence and the related critical conditions in a boundary layer, are not sufficiently clarified. The present authors believe such critical conditions are closely related to the mechanism for sustaining the wall turbulence.

Through a series of experimental observations of the final stage of the ribbon-induced transition in plane Poiseuille flow, we first identified the disturbance directly triggering the near-wall bursts to be hairpin eddies coming from upstream [4,5]. According to our observation, after the passage of hairpin eddies resulting from the high-frequency secondary instability of the high-shear layer, the near-wall fluid is lifted up away from the wall to form a wall shear layer, which soon evolves (or breaks down) into hairpin eddies. The important role of hairpin eddies in generating wall turbulence is also recognized through recent studies on the wall turbulence [6,7].

In order to clarify the mechanism of wall turbulence generation as well as the related critical condition, we have been working on the boundary layer transition caused by high-intensity hairpin eddies at subcritical Reynolds numbers [8-10]. Through exciting hairpin eddies acoustically at the leading edge all over the extent in the spanwise direction [10], we showed that when the near wall flow is disturbed by the energetic hairpin eddies, the successive growth of hairpin eddies from the lifted-up near-wall shear layers can occur only beyond the Reynolds number (based on the momentum thickness of laminar boundary layer) $R_\theta \approx 130$, which is almost the same as the minimum transition R_θ found in plane Poiseuille flow with highly-disturbed inlet flow condition [11]. According to the linear stability theory, the critical R_θ is 200 for Blasius flow. In the present study, to clarify the critical flow in more detail, the disturbance growth is examined through introducing well-controlled hairpin eddies by using a wall orifice at the station below and around the cited critical Reynolds number.

2. Experimental procedure

The whole experiment is conducted in a wind tunnel of open jet, 200 mm×200mm in cross section. The schematic diagram of test section is illustrated in Fig.1. A boundary-layer plate (with sharp leading edge) set in the open test section is 540 mm long, 4 mm thick and 195 mm in span. A pair of large side walls maintain the two-dimensionality of the main stream. The orifice of 2 mm in diameter is drilled at the spanwise center position, 80 mm downstream of the leading edge, and is connected to a loudspeaker by a vinyl-hose. When the loudspeaker is driven, the forcing system can produce a local three-dimensional high-shear layer by periodically injecting and sucking air through the orifice. Such a high-shear layer quickly (almost at the same time as the formation of the high-shear layer) breaks down into high-frequency hairpin-shaped vortices.

A constant temperature hot-wire anemometer is used to measure the time-mean and fluctuation velocities in the streamwise (x) direction, U and u. The hot-wire probe is traversed in the x, y (normal to the upper surface) and z (spanwise) directions. The y-distributions of U and u' (r.m.s. value of u-fluctuation) are recorded on the X-Y recorder during each y-traverse. The hot-wire signal is also stored in a personal

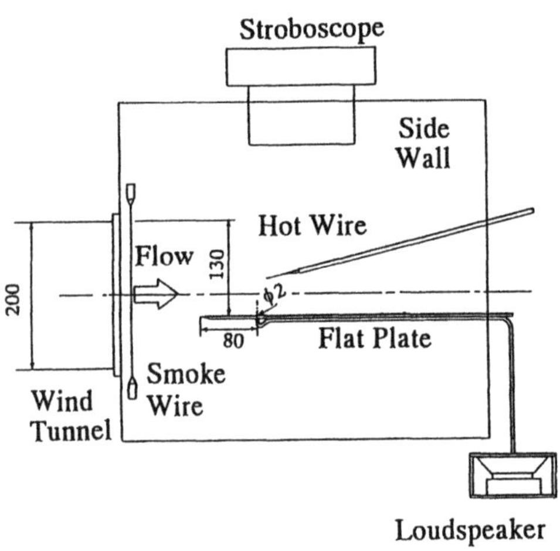

Fig. 1 Schematic diagram of test section (dimensions in mm).

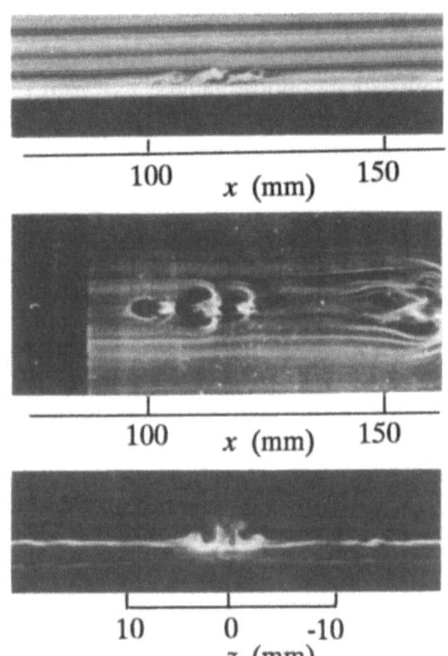

Fig. 2 Excitation of hairpin eddies. From top, side view, plan view, cross-section view (at x=110mm). U_∞ = 4.4 m/s, f = 52 Hz.

computer. The flow visualization is done by means of smoke-wire technique to identify the three-dimensional flow structures. The freestream velocity is in the range from 2.5 m/s to 6.0 m/s, for which the x-Reynolds number at the orifice station (x_d = 80 mm) R_{xd} ($=x_d U_\infty/\nu$; where ν is the kinematic viscosity) is from 1.4×10^4 to 3.4×10^4. Without the forcing, the boundary layer remains laminar over the whole observation region.

3. Results and discussion

The forcing system mentioned above may produce well-controlled hairpin eddies periodically in time. Fig. 2 demonstrates such orifice-generated hairpin eddies by the smoke-wire visualizations: The freestream velocity U_∞ is 4 m/s and the forcing frequency f is 52 Hz in this example. The three photographs (side-view, plan-view and cross-section view) are taken not simultaneously but the stroboscopic light is synchronized with the forcing signal so as to obtain the flow structure at the same phase of the forcing. These pictures no doubt show that three hairpin-shaped vortices are excited immediately downstream of the orifice. Here, the number of the excited hairpin eddies as well as their intensity can be controlled easily by adjusting the input current to the loudspeaker (therefore, the air injection through the orifice). Such well-controlled hairpin eddies can be excited for various freestream velocities (in the range from 2.5 m/s to 6.0 m/s), though the hairpin frequency increases with increasing the velocity U_∞ and vice versa. Fig. 3 compares the hairpin eddies excited at three different velocities U_∞ = 4.0, 4.4 and 5.0 m/s, in terms of waveforms of u-fluctuation near the orifice (at $x - x_d$ = 20 mm). Note that the forcing frequency f is 52 Hz in all the cases; the waveforms at U_∞ = 4.4 m/s correspond to the visualization picture given in Fig. 2. The high-frequency spikes (400 Hz ~ 500 Hz), observed in both the waveforms near and away from the wall, indicate the passage of hairpin eddies across the hot-wire sensor. Here, it should be noted that the forcing is well adjusted so as to generate hairpin eddies with almost the same intensity in the three cases as far as the associated u-fluctuation (i.e. high-frequency spikes) is concerned. In the following, we examine the boundary layer transition caused by these orifice-generated hairpin eddies.

Fig.4 compares the flow developments in the above three cases by means of smoke-wire visualization. Here, from the comparison of the visualized vortices near the orifice

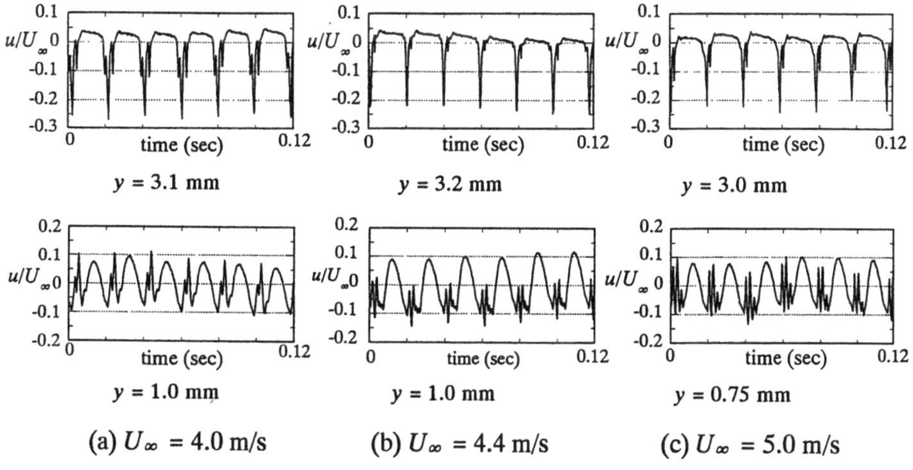

Fig. 3 Waveforms of u-fluctuation at $x - x_d$ = 20 mm.

it is also understood that the intensity of the initial hairpin eddies is almost the same for the three cases. At the lowest velocity $U_\infty = 4.0$ m/s, we see no peculiar phenomenon other than the periodic appearance and convection of the orifice-generated hairpin eddies, while at $U_\infty = 4.4$ m/s and 5.0 m/s, the successive generation of vortices is clearly observed beyond $x = 150$ mm. At the orifice station ($x_d = 80$ mm), the x-Reynolds number $R_x = 2.25$, 2.5, and 2.8×10^4 for $U_\infty = 4.0$, 4.4 and 5.0 m/s respectively. Thus, such slight increases in the Reynolds number can affect the successive generation of hairpin eddies.

To see the disturbance growth in such critical flows, Fig.5 shows the development of u'_m (the y-maximum of u') in the three cases corresponding to Fig.4; all the measurements are made at the spanwise center (just behind the orifice center). The figure clearly indicates the Reynolds number dependency of the disturbance growth, as suggested from Fig.4. For the two cases of $U_\infty = 4.4$ m/s and 5.0 m/s, the disturbance soon grows from the x-station immediately downstream of the orifice, i.e. from $x - x_d = 60$ mm ($R_x = 4.3\times10^4$) and 40 mm ($R_x = 4.2\times10^4$) at $U_\infty = 4.4$ m/s and 5.0 m/s

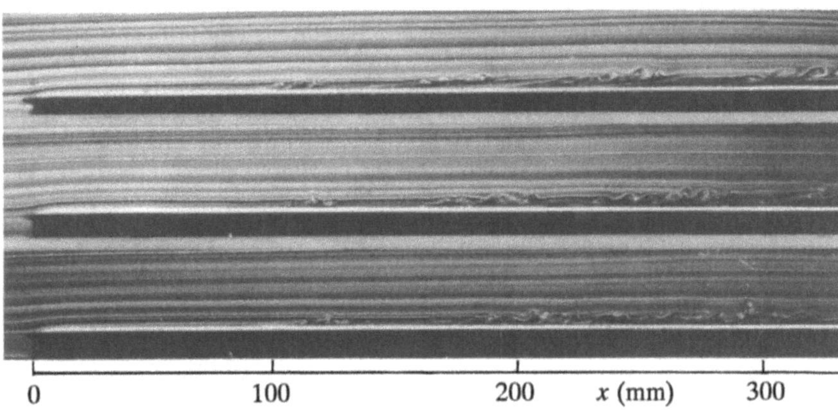

Fig. 4 Flow visualization of the subcritical boundary layer transition. From top photo. $U_\infty = 4.0$ m/s, 4.4 m/s and 5.0 m/s. Forcing frequency $f = 52$ Hz.

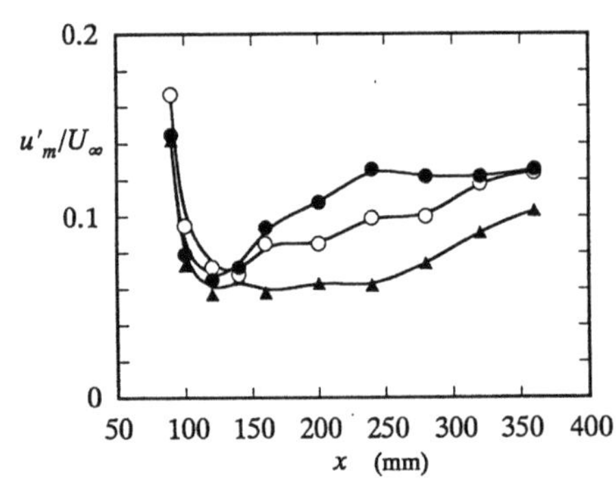

Fig. 5 u'_m. against x . ▲; $U_\infty = 4.0$ m/s, ●; 4.4 m/s, ○; 5.0 m/s.

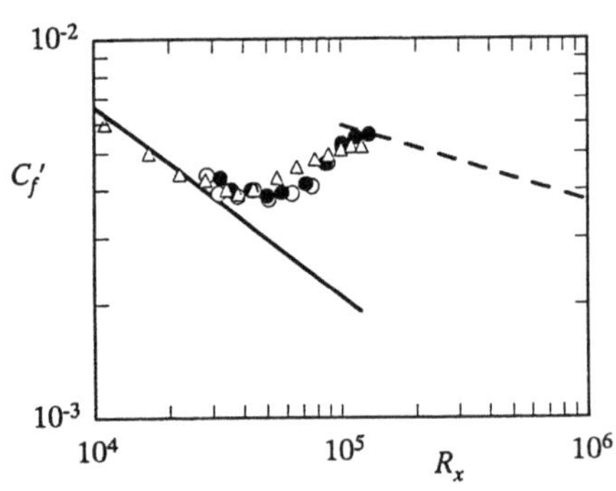

Fig. 6 C_f versus R_x . ○; $U_\infty = 4.4$ m/s, ●; 5.0 m/s, △; Ref. 10, ——; Laminar, - - - - -; Turbulent.

respectively, while at $U_\infty = 4.0$ m/s, the disturbance does not grow at the stations below $x = 250$ mm ($R_x = 7.0\times10^4$). The disturbance growth observed for the cases of $U_\infty = 4.4$ and 5.0 m/s is no doubt due to the successive generation (or regeneration) of hairpin eddies. In order to find the transition Reynolds number in the two cases $U_\infty = 4.4$ m/s and 5.0 m/s, we measure the coefficient of local skin friction C'_f at each x-station as plotted against R_x in Fig. 6. C'_f is obtained from careful measurements of the y-distributions of time-mean velocity U at each x-station. It is interesting that the transition Reynolds number is almost the same in both the cases, i.e. $R_x = 4\times10^4 \sim 5\times10^4$, where the momentum-thickness Reynolds number is found to be about 130 ~ 150. In the figure, our previous result [10] on the boundary layer transition due to the leading-edge-generated hairpin eddies is plotted for comparison, showing that the transition Reynolds number is almost the same as the present. It should be noted that for the case of $U_\infty = 4.0$ m/s, the development of disturbance is found to be highly intermittent even beyond $x = 240$ mm where the u'_m starts to increase. Even at $U_\infty = 4.0$ m/s, however, when the intensity of the initial disturbance is increased, the regeneration can occur continuously and start at the x-station much closer to the orifice. Indeed, in such a case (at $U_\infty = 4.0$ m/s), it is found that the disturbance growth starts at $x = 140$ mm ($R_x = 3.8\times10^4$) with wall bursts.

Through exciting such strong initial disturbances, we further examined the development of u-fluctuation at lower freestream velocities and found that only for $U_\infty > 3.0$ m/s, the disturbance growth can occur within the subcritical region. For instance, Fig. 7 shows the threshold behavior of the regeneration by means of flow visualization at $U_\infty = 2.55$, 2.75, 3.05 and 3.5 m/s. At 3.05 m/s and 3.5 m/s only, the successive generation of vortices behind the primary (orifice-generated) hairpins is observed, as in the cases $U_\infty = 4.4$ and 5.0 m/s (in Fig. 4). Indeed, Figs. 8(a) and (b) illustrate the disturbance developments at $U_\infty = 2.75$ m/s and 3.05 m/s respectively, in terms of the waveforms of u-fluctuation, showing that the near-wall flow exhibits

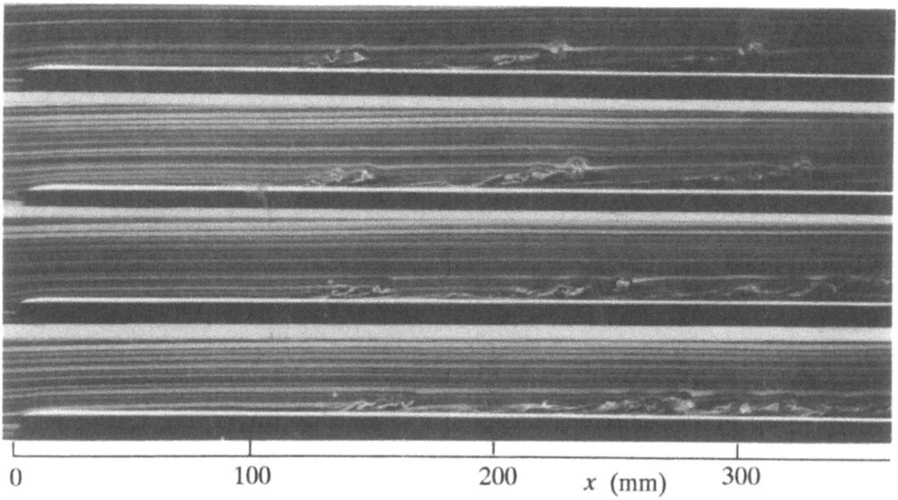

Fig. 7 Flow visualization of the disturbance development. From top photo, U_∞ = 2.55 m/s, 2.75 m/s, 3.05 and 3.5 m/s. Forcing frequency f = 30 Hz.

the occurrence of weak wall burst beyond $x - x_d = 120$ mm at $U_\infty = 3.05$ m/s (just above the critical flow speed), while at $U_\infty = 2.75$ m/s, the near-wall flow is rather calm in the whole observation region. Thus the flow at $U_\infty = 3.05$ m/s is really critical. Note that u'_m is found to increase beyond $x - x_d = 120$ mm. In this case, the R_θ is about 90 at the orifice station ($R_x = 1.7\times10^4$) and is about 140 at $x - x_d = 120$ mm ($R_x = 4.3\times10^4$). It is worth noting that the transition R_θ obtained in the present experiment is almost the same as those cited for the other wall-bounded flows; for instance, about 130 for plane Poiseuille flow [11] and about 100 for the attachment-line boundary layer [12,13].

Another important feature is observed for the lateral growth of the turbulent (active disturbance) region. Fig. 9 shows the plan view of smoke-wire visualization of the disturbance development at $U_\infty = 4.4$ m/s (corresponding to the side view in Fig.4). The lateral growth occurs through generating secondary vortices on both sides of convecting primary hairpins. It is also noted that on both sides of the turbulent patch, oblique waves are observed, but they do not break down into vortices in the observed subcritical region. The oblique waves are probably excited simultaneously with the hairpin eddies by the periodic air injection. Such a feature of the lateral contamination can also be understood from the cross-section view of smoke at $x = 170$ mm and 240 mm, Fig. 10. In these photographs, smoke is released from the station 50 mm upstream of the observation x to visualize the active vortical structures. Note that with stroboscopic light synchronized with the forcing signal, the photographs are taken at the same instance as those of Figs. 4 and 9 so that each cross section view (at x = 170 mm and 240 mm) visualizes the vortical structure just after the passage of the primary (orifice-generated) hairpins. In this regard, through detailed visualization studies on the regeneration process of single hairpin vortices, Smith et al [6] reveal that the secondary hairpins can be generated between the legs of the primary hairpins as well as just behind each leg by a vortex-surface interaction. At $x = 170$ mm, we see

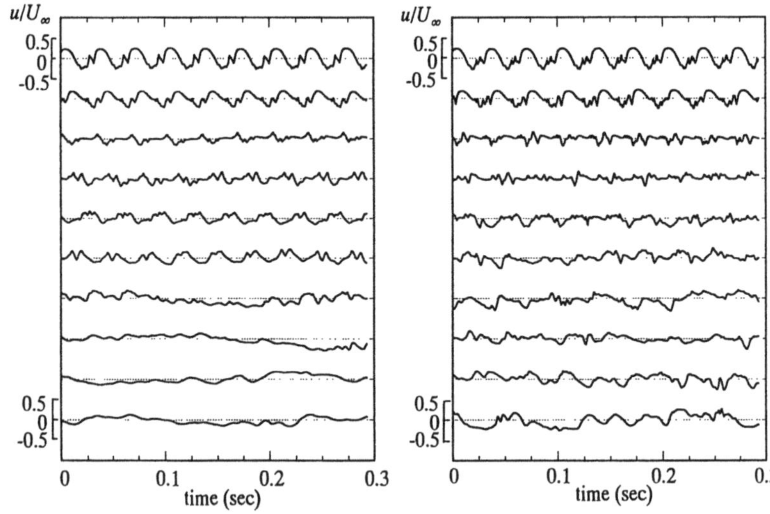

(a) $U_\infty = 2.75$ m/s (b) $U_\infty = 3.05$ m/s

Fig. 8 Development of u-fluctuation at $U_\infty = 2.75$ m/s and 3.05 m/s. Waveforms (at $y = 1.2$ mm); from top, $x - x_d$ = 10, 20, 40, 60, 80, 120, 160, 200, 240, 280 mm. Forcing frequency $f = 30$ Hz.

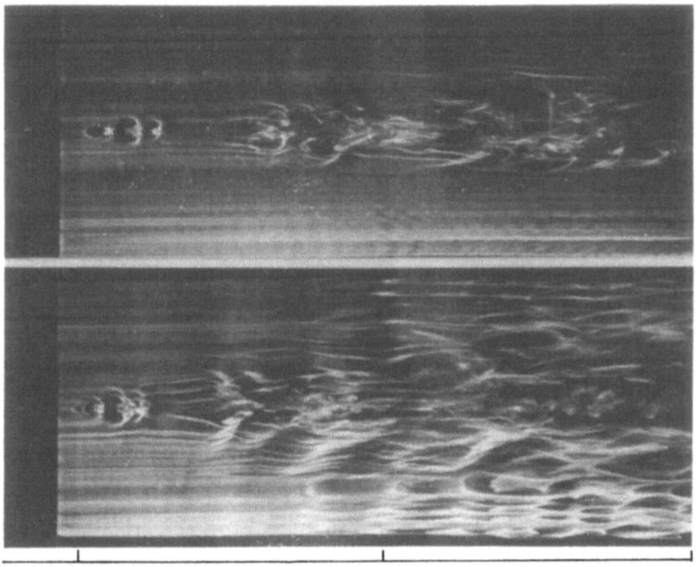

Fig. 9 Plan view of the disturbance development at $U_\infty = 4.4$m/s. (a) smoke is released from $y = 1.5$ mm, (b) smoke from $y = 0.7$ mm. Forcing frequency $f = 52$ Hz.

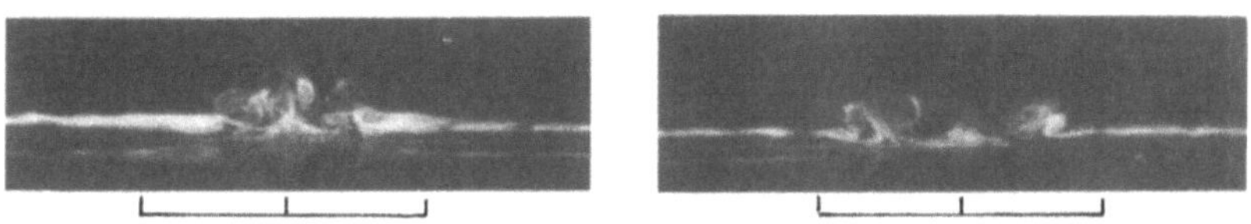

(a) $x = 170$ mm

(b) $x = 240$ mm

Fig. 10 Cross-section view of the disturbance development at $U_\infty = 4.4$m/s.

such a pair of secondary vortices of mushroom-type generated on the both sides of the primary (i.e. orifice-generated) hairpin. Importantly, however, the regeneration of hairpin eddies in the side regions is so weak in the present subcritical transition. Indeed, the half angle of lateral spreading of turbulent patch is about (or slightly less than) 2°, one order of magnitude smaller than that of the turbulent spot or wedge usually-observed at high Reynolds numbers. Due to strong viscous effects, it is more than 100 mm downstream from the orifice that the secondary vortices are created. At $x = 240$ mm where the secondary vortices distinctly appear on both sides of the primary hairpins (at $z = \pm 6$ mm), the flow at and around the spanwise center already approaches the turbulent stage (see Fig. 6); note that the spacing (6mm) is about 90 in local wall units, ν/u_τ, where u_τ denotes the friction velocity (at $x = 240$ mm). Thus, in the subcritical region, the lateral contamination due to the generation of secondary hairpin eddies is very slow and the disturbance growth seems to be essentially governed by the regeneration process immediately behind the primary hairpins.

4. Conclusions

As shown thus far, the energetic hairpin eddies, while decaying downstream, can trigger the transition at subcritical Reynolds numbers for R_{xd} $(= x_d U_\infty/\nu)$ beyond 1.7 $\times 10^4$, though the disturbance growth leading to wall turbulence highly depends on the intensity of the excited hairpin eddies at low subcritical Reynolds numbers, in particular, below $R_{xd} = 2.5\times10^4$. In all the cases, the successive generation or regeneration of hairpin eddies, which occurs after the passage of the hairpin eddies convecting from upstream, is responsible for the transition. The minimum Reynolds number for the subcritical disturbance growth due to hairpin eddies is examined and found to be $R_x = 4\times10^4 \sim 5\times10^4$, where R_θ is about 130~150 and almost the same as that found in our previous experiments on the subcritical transition due to the leading-edge-generated hairpins. Even when the energetic hairpin eddies are introduced to trigger the transition, the regeneration of the vortices occurs only along a narrow strip extending downstream from the orifice, with lateral spreading angle (half angle) of about 2°, that is, the lateral contamination is so weak at the subcritical Reynolds numbers.

The authors wish to express their cordial thanks to Profs. S. Iida and H. Sato for their continual encouragement. This work was in part supported by the Special Research Fund, Tokyo Metropolitan Government, as well as by Grant-in-Aid for Scientific Research (06651070), the Ministry of Education, Science and Culture Japan.

References

[1] Nishioka,M., Iida,S. and Ichikawa,Y.: An experimental investigation of the stability of plane Poiseuille flow, J. Fluid Mech. 72 (1975) 731-751.
[2] Morkovin, M.V. and Reshtko,E.: Dialogue on progress and issues in stability and transition research, in Laminar-Turbulent Transition (eds. D.Arnal and R.Michel), Springer (1990) 3-29.
[3] Kachanov, Yu.S. : Physical mechanisms of laminar-boundary-layer transition, Annu. Rev. Fluid Mech. **26** (1994) 411-482.
[4] Nishioka,M., Asai,M. and Iida,S.: Wall phenomena in the final stage of transition to turbulence, in Transition and Turbulence (ed. R.E.Meyer), Academic Press (1981) 113-126.
[5] Nishioka,M. and Asai,M.: Evolution of Tollmien-Schlichting waves into wall turbulence, in Turbulence and Chaotic Phenomena in Fluids (ed. T.Tatsumi), North-Holland (1984) 87-92.
[6] Smith,C.R., Walker,J.D.A., Haidari,A.H. and Sobrun,U.: On the dynamics of near-wall turbulence, Phil. Trans. R. Soc. Lond. A **336** (1991) 131-175.
[7] Robinson,S.K.: Coherent motions in the turbulent boundary layer, Annu. Rev. Fluid Mech. **23** (1991) 601-639.
[8] Asai, M. and Nishioka, M.: Development of wall turbulence in Blasius flow, in Laminar-Turbulent Transition (eds.D.Arnal and R.Michel), Springer (1990) 215-224.
[9] Nishioka,M., Asai,M. and Furumoto,S.: On the critical condition for wall turbulence generation, Acta Mechanica [suppl] **4** (1994) 87-94.
[10] Asai, M. and Nishioka, M.: Boundary layer transition caused bu hairpin eddies at subcritical Reynolds numbers, submitted to J. Fluid Mech.
[11] Nishioka, M. and Asai, M.: Some observations of the subcritical transition in plane Poiseuille flow, J. Fluid Mech. **150** (1985) 441-450.
[12] Gaster, M.: On the flow along swept leading edge, Aeronautical Quarterly **18** (1969) 165-184.
[13] Poll, D.I.A.: Some observations of the transition process on the windward face of a long yawed cylinder, J. Fluid Mech. **150** (1985) 329-356.

Session D:
Flow Measurements and Controls

In-Flight Measurements on a Laminar Wing Glove by Means of Surface Sensor Arrays

J. Suttan, M. Baumann, W. Nitsche
Institut für Luft- und Raumfahrt; Technische Universität Berlin
Marchstr. 14; D-10587 Berlin; Germany

SUMMARY

The paper reports on in-flight experiments with different multisensor piezofoil-arrays on a laminar wing glove. The piezofoil measures fluctuations of surface forces. By employing a compact high speed digital measurement system, it was possible to carry out flight tests concerning laminar-turbulent transition with a 48-sensor array on a motor-glider. The first array was primarily used for the detection of transition location and dynamic effects during transition, while the second array was aimed on measuring the expansion and development of an artificial excitation induced by a harmonic point source.

KEY WORDS: Piezoelectric Foil, Multisensor Arrays, Digital Measurement System, Flight Tests, Transition Detection, Transition Stimulation, Harmonic Point Source

INTRODUCTION

For the development and evaluation of modern transonic laminar airfoils with significantly large areas of laminar flow conditions, it is necessary to investigate in detail the mechanisms which lead to transition and, eventually, to turbulence. Future transport aircraft will make use of these airfoils to improve ecomomy as well as to reduce environmental pollution. In order to get a database for laminar-turbulent airfoil flows under full-scale Reynolds numbers and the low atmospheric turbulence levels, which can be compared to numerical solutions, these investigations have to be carried out in large windtunnels and, even more important, in free-flight.

In the investigations described here, a laminar wing glove was mounted on a Grob G109b-motorglider and equipped with two different piezofoil sensor arrays, which each consist of 48 discrete sensors. The piezofoil has proven its ability to measure unsteady surface forces for several years [1-3]. It can be fixed to arbitrary surfaces without influencing the flow. Sensor layouts can easily be customized to a given measuring case. Sensor numbers are only limited by the capabilities of the measurement and analysis system. For this reason, the piezofoil is especially suitable for measurements concerning dynamic surface forces in laminar-turbulent transition.

The tests with the first sensor array were carried out for different flight conditions to investigate the influences of Reynolds number, angle of attack and lift coefficient on the location of transition. For further investigations concerning the receptivity and development of 3D-instabilities, the wing glove was equipped with the second sensor array and a harmonic point source (miniature loudspeaker), which was able to induce an artificial excitation (sine waves with a fixed frequency) into the boundary layer. Different experiments [4,5] have shown the development of small disturbances into wedge-shaped turbulence areas. The design of the second array was aimed at a finer spatial resolution of the surface force fluctuations downstream of the point source.

EXPERIMENTAL APPARATUS

The layout of the two different piezofoil-sensorarrays is shown in **Figure 1**. On the first array, the sensors are arranged in four parallel lines to achieve a fine resolution in the x-direction, beginning at x/c=0.14 and ending at x/c=0.63 in steps of x/c=0.015 (20 mm). Additional sensors are positioned on the leading edge and at x/c=0.06. The second sensor array shows the arrangement of the sensors in a wedge-shape. Some sensors are positioned upstream of the point source to measure the undisturbed flow. The distance between each sensor (in x- as well as in y-direction) is 30 mm (x/c=0.023). Three miniature loudspeakers are used as harmonic point sources. They are working through fine holes with a diameter of 0.3 mm positioned on the centerline of the array, During the flight tests, the second speaker, located at x/c=0.235 and the third one, located at x/c=0.27, could be operated alternatively by means of a programmable function generator.

A digital measurement system for flight tests with a scanning frequency of 4 kHz for each sensor was employed. It is able to sample the data quasi-simultaneously from the 48 sensors [6]. After passing through individual amplifiers, a multiplexer and an A/D-converter, the data are transmitted to the intermediate storage unit through an electro-optical transmission line to avoid electronic noise. All elements of the system up to the electro-optical converter are positioned inside the wing glove to reduce the length of connecting wires and, by that, again reduce electromagnetic influences. The data were pre-analysed on a personal computer during the measurement and stored on a hard disk for further processing.

In addition to the surface sensors, a traversable shear stress probe (Computational Preston Tube Method, CPM3 [7]) was installed on the glove. The CPM3-method determines the local wall friction from dynamic pressures of three Preston tubes with different diameters. In flight, the CPM3-probe was operated fully automatically by means of a micro computer. Eight static pressure taps on the upper side of the glove allowed the measurement of pressure distributions.

In order to survey the outer flow conditions, a XP-probe [8], which is a combination of a X-wire-probe and a total-pressure sensor with a high dynamic resolution, was mounted on a retractable shaft in an underwing pod. A complete view of the measuring equipment is given in **Figure 2**.

DETECTION OF TRANSITION

The results of the tests with the first sensor array show a good correlation between the typical laminar-turbulent increase in wall shear stress, measured with the CPM3-probe, and the peak-values of the RMS-distribution computed from the time traces of the piezo sensors **(Fig. 3)**. A detailed analysis of the time traces clearly shows the appearance of high-frequency-disturbances as well as the formation and migration of turbulent spots **(Fig. 4a)**. Power spectra, which were derived from the time traces, show a general increase in signal strength with increasing chordwise position and significantly higher amplitudes between 450 and 750 Hz for sensors between 0,10<x/c<0,28. This indi-cates the onset of Tollmien-Schlichting instabilities in the laminar boundary layer. Relative amplification factors (Np*-factors, **Fig. 4b**) were calculated from these spectra by comparing the amplitudes of selected frequencies with the ones of the first sensor at x/c=0,06. The frequencies mentioned above show a linear increase in amplitude, as was to be expected from the linear stability theory.

TRANSITION STIMULATION

Pre-tests of the glove equipped with the second sensor array were carried out in a low-speed wind-tunnel. The results showed the ability of the harmonic point source operating at the instability frequency to stimulate effectively the prelocated onset of transition. The expected amplification of the induced frequency as well as the development and expansion of turbulent spots could clearly be observed.

After wind-tunnel testing, the complete test equipment was mounted on the wing of the motorglider. Different tests were flown with and as without artificial excitation through the point source. As an example, **Figure 5** shows the distribution of RMS-values for the same speed/angle of attack-combination (IAS=34,7m/s, alpha=6°). It is clearly visible that natural transition occurs on the last four sensor rows from x/c=0.44 to 0.51 with the point source switched off. With the point source operating at f=1000 Hz (equivalent to a dimensionless frequency F=0.5 at x/c=0.27), a shifting of higher RMS-values upstream can be observed. The highest RMS-values, which mark the laminar-turbulent breakdown, are situated on both sides of the transition wedge. The computation of power spectra and power density functions from the measured time traces show at first the expansion of the excitation frequency in the boundary layer downstream of the point source, followed by full-scale laminar-turbulent breakdown and finally turbulent flow **(Figure 6)**.

CONCLUSIONS

The presented results of flight tests employing two different piezoelectric sensor arrays clearly show the ability of the measurement system to measure the temporal as well as the spatial distribution of dynamic surface forces in transition. Furthermore, the harmonic point source was successfully tested in flight. Inducing an artificial excitation by operating the point source at the instability frequency stimulates the formation of a turbulence wedge downstream of the point source. Future activities will concentrate on sensor arrays with finer resolution (minimized sensor spacing) and improvements of the digital measurement system. More emphasis will be put on data analysis, e.g. on the calculation of phase speed and -angle from the measured time traces.

ACKNOWLEDGEMENTS

The authors wish to acknowledge the financial support for their work from the German Federal Ministry for Research and Technology (BMFT) and the German Research Foundation (DFG).

REFERENCES

[1] Nitsche, W.; Mirow, P. and Szodruch, J.: Piezoelectric Foils as a Means of Sensing Unsteady Surface Forces, Exp. in Fluids, Vol. 7, pp. 111 - 118, 1989

[2] Nitsche, W.; Mirow, P. and Dörfler, T.: Investigations on Flow Instabilities on Airfoils by Means of Piezofoil-Arrays, in: D. Arnal, R. Michel (Eds.), Laminar- Turbulent Transition, IUTAM Symposium Toulouse/France 1989, Springer-Verlag Berlin Heidelberg, 1990

[3] Ewald, B.; Durst, F; Krause, E. and Nitsche, W.: In-flight measurement techniques for laminar flow wing development, Z. Flugwiss. Weltraumforsch. 7 (1993), pp. 294 - 310

[4] Schubauer, G. B. and Klebanoff, P. S.: Contributions on the Mechanics of Boundary-Layer Transition, NACA-Rep. 1289, 1956

[5] Kachanov, Y. S.: Development of Spatial Wave Packets in Boundary Layer, in: V. V. Kozlov (Ed.), Laminar-Turbulent Transition, IUTAM Symposium Novosibirsk/USSR 1984, Springer-Verlag Berlin Heidelberg, 1985

[6] Devliotis, A.: Entwicklung und Erprobung eines flugtauglichen Digitalmeßsystems für die Untersuchung laminar-turbulenter Tragflügelumströmungen, Diss. TU Berlin, Fortschr.-Ber. VDI, Reihe 7, Nr. 202, VDI-Verlag, 1992

[7] Nitsche, W.; Thünker, R. and Haberland, C.: A Computational Preston Tube Method, in: L. J. S. Bradbury, F. Durst, B. E. Launder, F. W. Schmidt, J. H. Whitelaw (Eds.), Turbulent Shear Flows 4, Springer-Verlag Berlin Heidelberg, 1985

[8] Nasseri, M.: Entwicklung eines Sondenkonzepts zur simultanen Erfassung von Druck- und Geschwindigkeitsschwankungen in turbulenten Strömungen, Diss. TU Berlin, Fortschr.-Ber. VDI, Reihe 7, Nr. 220, VDI-Verlag, 1993

FIGURES

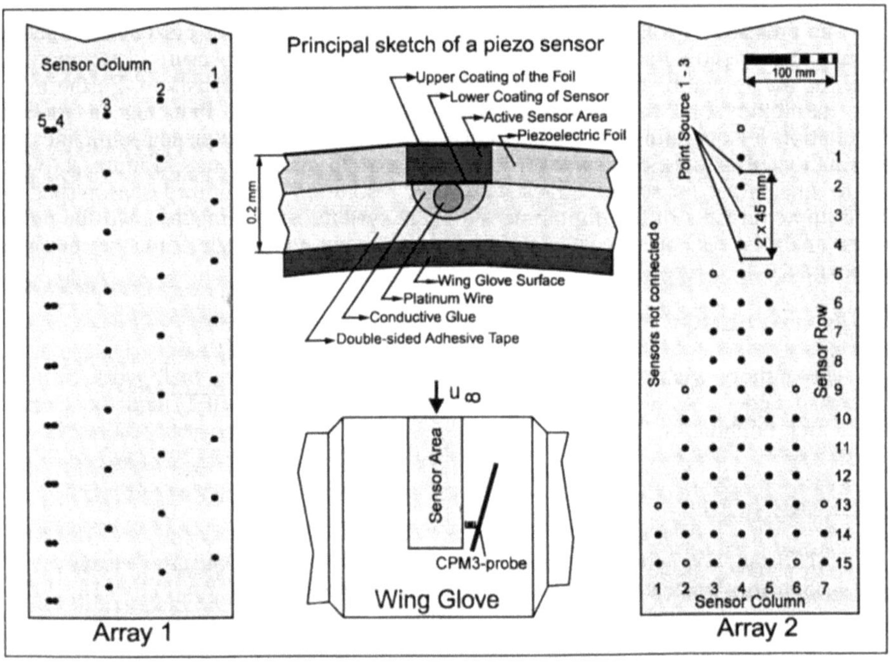

Fig. 1: Layout of Piezofoil-sensorarrays, principal sketch of single sensor,application of sensors on the wing glove

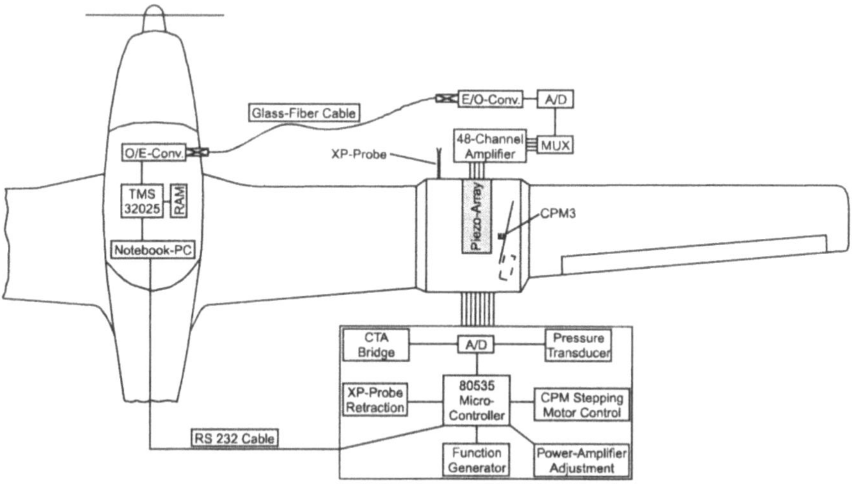

Figure 2: Measuring equipment for flight tests

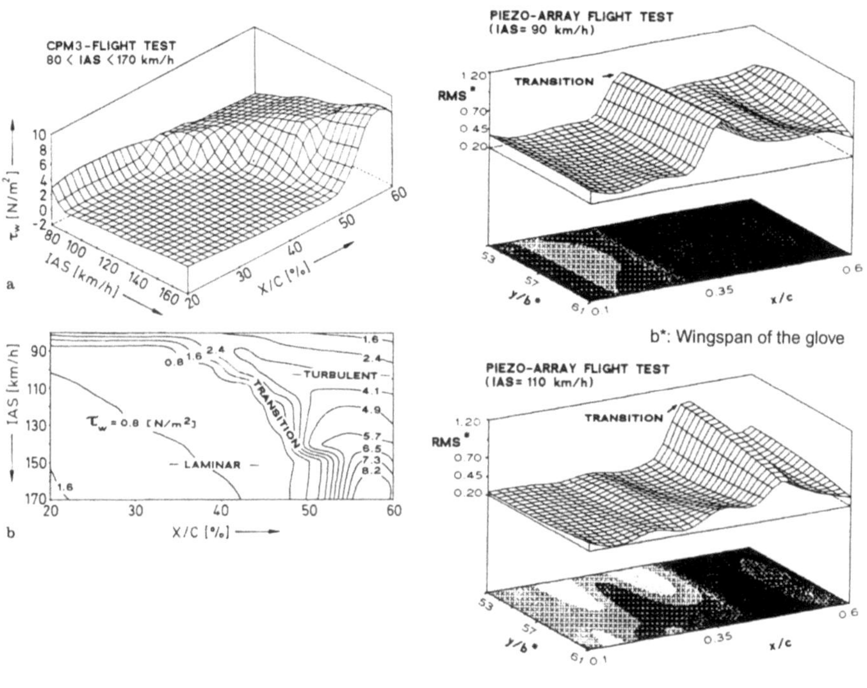

Figure 3: Comparison of CPM3- and piezofoil data

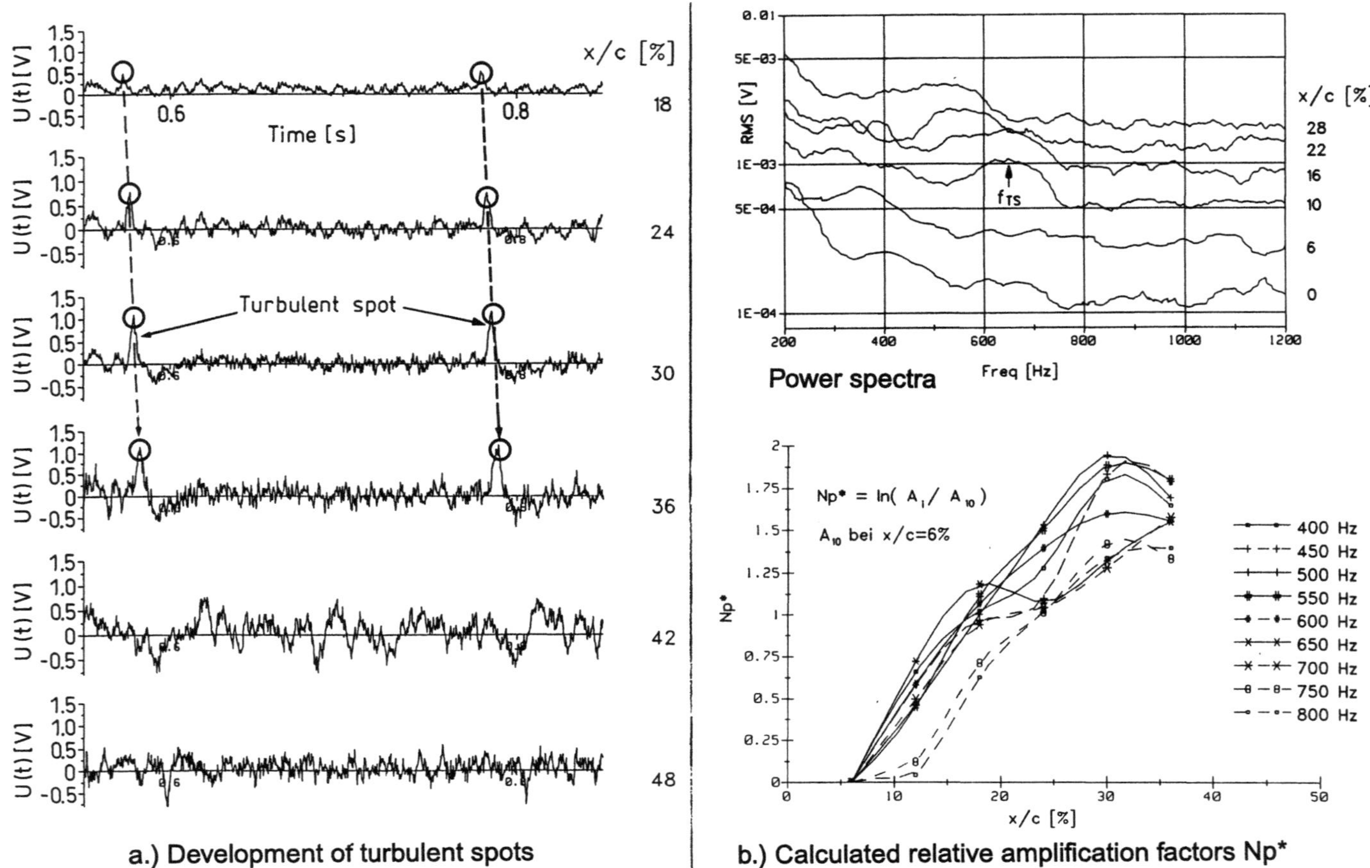

Figure 4.a: Development of turbulent spots
b: Power spectra and relative amplification factors

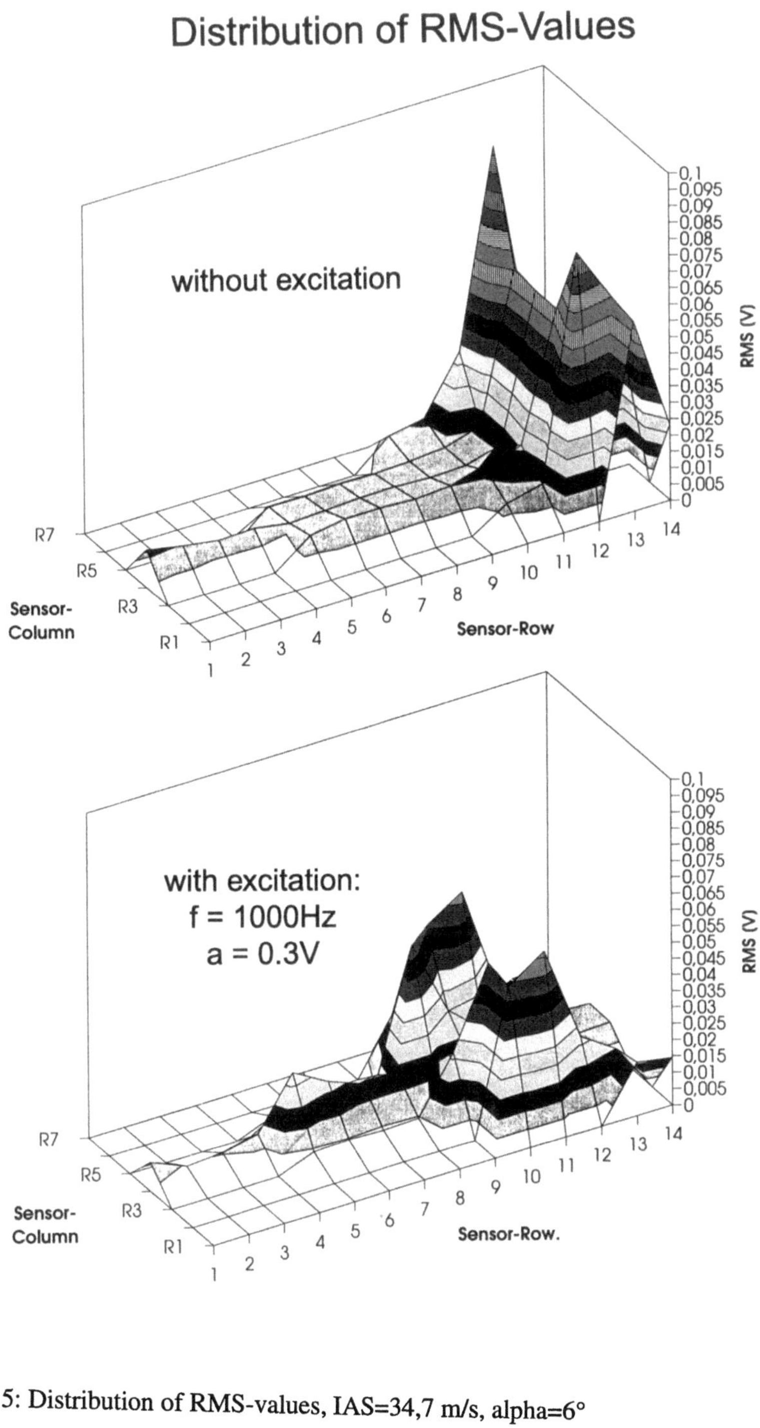

Figure 5: Distribution of RMS-values, IAS=34,7 m/s, alpha=6°

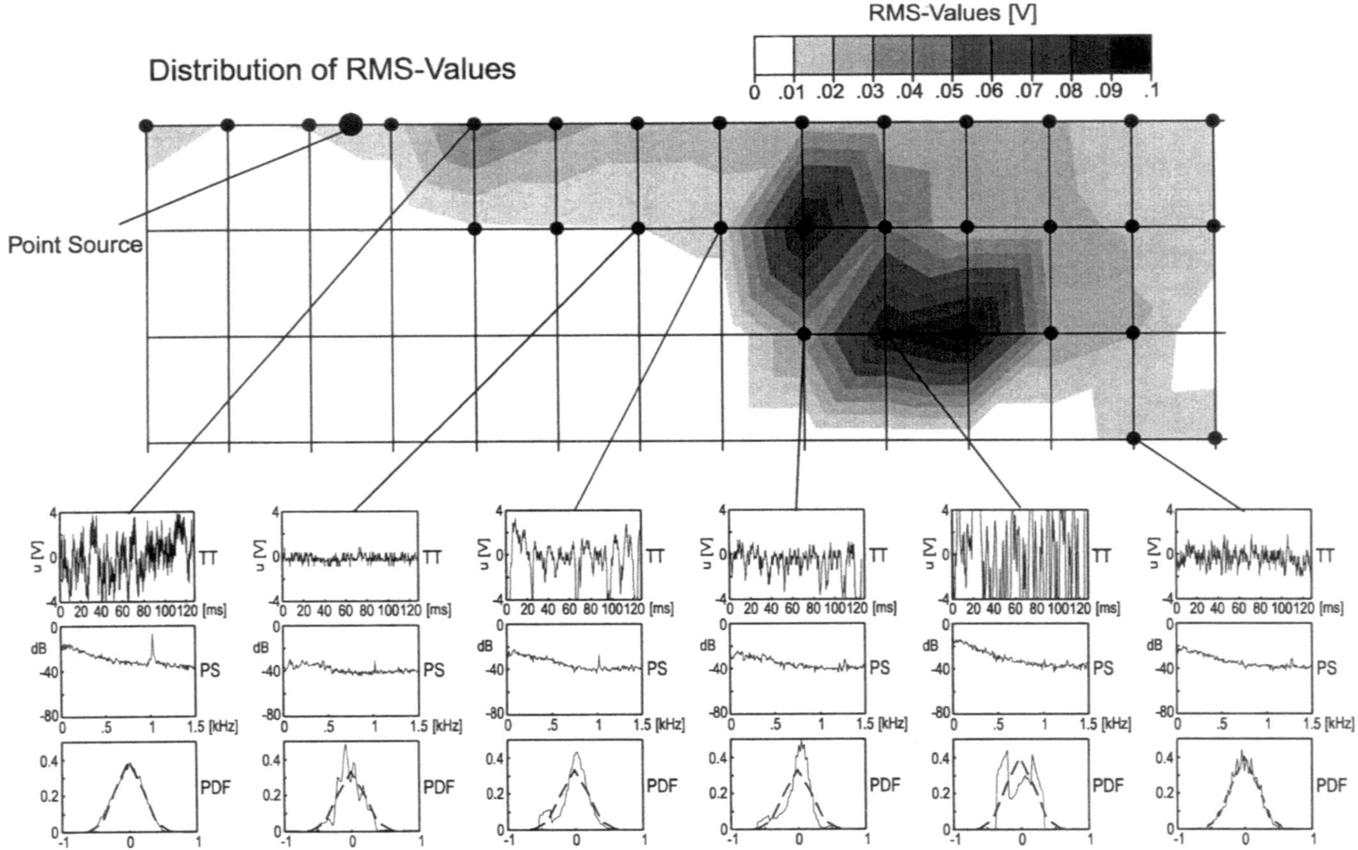

Figure 6: Time traces, power spectra and power density functions of selected sensors in the transition region, point source operating at f=1000Hz, a=0,3V

Laminarization of Three-Dimensional Compressible Boundary-Layer Flow Using Receptivity Control

S.V. Manuilovich
Central Aero-Hydrodynamics Institute
Zhukovsky, Moscow region, Russia

Summary

New methods for the laminarization of the boundary layer flow on swept wings are offered. The laminar-turbulent transition is assumed to be due to acoustic influence and roughness of wing surface. The basis of the methods is in mutual cancellation of unstable eigenmodes excited in the vicinity of leading edge of the wing and over the artificial unevenness of wing surface. This cancellation is ensured by special choice of the unevenness form. Calculations are carried out for demonstration the efficiency of the methods. A finite Reynolds number approach is used to obtain an adequate quantitative description of the phenomena in question.

Key Words: Laminarization, Receptivity control.

1. Introduction

It is well known that the first stage of laminar-turbulent transition consists in transformation of external disturbances into spatially growing unstable eigenmodes. This process is referred to as the boundary-layer receptivity. The most important external factors resulting in unstable mode generation are acoustic disturbances, free stream turbulence and vibration of wing surface. Instability waves are generated only via scattering of external disturbances by longitudinal inhomogeneities of the boundary layer flow, such as in the vicinity of leading edge and over roughness. Once generated, unstable modes undergo linear amplification and nonlinear interactions and then destroy laminar flow.

Two main types of methods for the attenuation of unstable oscillations exist. The first type contains methods based on stabilization of the boundary-layer mean flow. Classical examples of these methods are distributed suction and the cooling of a wing surface. The second type of methods, first proposed by Milling [1], applies the principle of T-S wave cancellation by means of the formation of artificial wave with the same frequency and amplitude but opposite phase. Essential disadvantage of the above active control is its complication. As was shown recently by Manuilovich [2] and Kosorygin & Polyakov [3], the laminarization problem for low-speed two-dimensional boundary-layer may be solved using receptivity control. The basis for the total receptivity relaxation is in mutual cancellation of T-S waves excited by sound in the vicinity of leading edge of wing and over the artificial two-dimensional unevenness of wing surface.

This paper is devoted to theoretical investigation of receptivity control in three-dimensional compressible boundary-layer flow. In order to attain an adequate quantitative description of the methods in question we use the finite Reynolds number approach [4],[5].

2. Cancellation of T-S Wave by Vibrator

Let us consider a three-dimensional flow over a swept wing with smooth rigid surface. We introduce a coordinate system, with the origin at some point on the wing surface, with a streamwise x-axis normal to attachment line, with y-axis normal to wing surface, and z-axis parallel to attachment line. In order to describe the process of unstable mode cancellation we shall study the local structure of disturbed flow in the vicinity of the origin. The longitudinal size of the local region involved is assumed to be small with respect to wing chord and large with respect to boundary layer thickness, so the main boundary-layer flow in this region is regarded to be homogeneous in x-direction. We suppose that boundary layer flow is laminar in the region involved.

We first suppose that the boundary-layer flow undergoes the influence of acoustic wave with frequency ω :

$$q \sim \varepsilon q_{\infty} \exp(i\alpha_s x + i\beta_s y + i\gamma_s z - i\omega t) + \text{c.c.}\ , \quad y \longrightarrow \infty\ , \quad |\varepsilon| \ll 1.$$

Here $q = u, v, w, \ldots$ denote parameters of flow disturbances. Complex amplitude of pressure oscillations at the origin is denoted by ε .

The local structure of the unsteady disturbances of the boundary-layer flow over a smooth rigid surface is given by the amplitude expansion

$$q = \varepsilon(q_s + \delta c_s q_0) \exp(i\gamma_s z - i\omega t) + \text{c.c.} + O(|\varepsilon|^2),\ 0<\delta\ll 1,\ |c_s| = O(1)$$

The first term in this expansion

$$q_s = q_s^*(y) \exp(i\alpha_s x)$$

describes the acoustic wave structure within boundary layer.

Furthermore, the local flow is disturbed by a superposition of discrete and continuous modes generated via scattering of sound by boundary-layer flow inhomogeneity in the vicinity of leading edge. From this set of free oscillations only the term

$$q_0 = q_0^*(y) \exp(i\alpha_{T\text{-}S} x)$$

due to T-S wave is retained in the expansion as being of interest for the following analysis.

In the quasi-parallel and quasi-homogeneous mean flow

approximation the local problem for reflecting acoustic wave and the eigenvalue problem are reduced to the boundary-value problems for Lees-Lin system.

Propagating farther downstream, the above 'natural' instability wave is amplified by unstable boundary-layer flow, and then causes laminar-turbulent transition.

In order to enlarge the length of the laminar boundary layer flow portion, Milling [1] used vibrating wire. The general idea of the method [1] consisted of the cancellation of T-S wave with the aid of the artificial one generated by two-dimensional vibrating hump. This cancellation became possible because of a special choice of the amplitude and phase of hump oscillations. Here we represent the analytical investigation of this method.

Simulating the action of artificial T-S wave generator, we assume that a localized portion of the wing surface oscillates in y-direction

$$y_w = \varepsilon\delta f(x)\, \exp(i\gamma_s z - i\omega t) + \text{c.c.} \tag{2.1}$$

Here the function f describes the longitudinal form of the surface oscillations. The flow disturbance generated by the simultaneous influence of the sound and the surface vibration takes the following form:

$$q = \varepsilon(q_s + \delta c_s q_0 + \delta q_v)\, \exp(i\gamma_s z - i\omega t) + \text{c.c.} + O(|\varepsilon|^2).$$

The third term in this expansion corresponds to the flow oscillations produced by the vibrator.

In order to calculate this term, it is convenient to use Fourier transformation with respect to the x-variable

$$q_v = (2\pi)^{-1}\int_{-\infty}^{+\infty} F\, q_v^*(y)\, \exp(ikx)\, dk \quad , \quad F(k) = \int_{-\infty}^{+\infty} f(x)\, \exp(-ikx)\, dx.$$

The normalized Fourier transforms satisfy the inhomogeneous boundary-value problem for the Lees-Lin system.

Using Cauchy's integral theorem, we expand Fourier integral into the superposition of discrete and continuous modes in the downstream region. The component corresponding to the excited artificial T-S wave is determined by the residue contribution due to the pole at $k = \alpha_{T\text{-}S}$.

Let us write out the full expressions for disturbances corresponding to T-S waves generated near the leading edge and over vibrator

$$q_{T\text{-}S}^{(n)} = \varepsilon\delta c_s q_0^*(y)\, \exp(i\alpha_{T\text{-}S}x + i\gamma_s z - i\omega t) + \text{c.c.}, \tag{2.2}$$

$$q_{T\text{-}S}^{(a)} = \varepsilon\delta F(\alpha_{T\text{-}S}) A_v q_0^*(y)\, \exp(i\alpha_{T\text{-}S}x + i\gamma_s z - i\omega t) + \text{c.c} \quad . \tag{2.3}$$

Here A_v denotes a specific coupling coefficient. Our purpose is to achieve this waves to cancel each other downstream from the vibrator. Such cancellation can be ensured by special choice of amplitude and phase of the vibration. Let

$$f = \sigma c_s g(x) , \quad \max|g| = 1 ,$$

where σ is a complex constant and g is a real function. The parameter $|\sigma|$ characterizes the ratio between the amplitudes of the vibration (2.1) and the amplitude of T-S wave (2.2) to cancel, the parameter $\arg\sigma$ is the phase difference between these two oscillations. The condition of mutual cancellation of the two waves (2.2) and (2.3)

$$q^{(n)}_{T-S} + q^{(a)}_{T-S} \equiv 0 \qquad (2.4)$$

will be satisfied if we assume that

$$\sigma = \frac{1}{G(\alpha_{T-S})\, A_v} ,$$

where G is the Fourier transform of the function f .

To illustrate the process of T-S wave cancellation, we use the complex function

$$\tau_c(x) = c_s \frac{\partial u_0}{\partial y}(x,0) + \frac{\partial u_v}{\partial y}(x,0)$$

and consider the case of a triangular vibrator

$$g(\xi) = \begin{cases} 1 - |\xi|/\ell , & |\xi| \le \ell \\ 0 & , |\xi| > \ell \end{cases} ,$$

where ℓ characterizes the longitudinal size of the vibrator.

We suppose that $c_s = 1$. This equality may by achieved by means of displacing the coordinate origin and redefining the dimensionless flow parameters.

The process of cancellation of two-dimensional T-S wave in Blasius boundary layer by vibrating hump is shown in Fig.1. for Reynolds number $R = 500$, $\omega = 0.05$, Mach number $M = 0.8$ (in the system of units adopted in [5]). The length of vibrator corresponds to $\ell = 20$. The calculations give the values $|\sigma| = 4.213$, $\arg\sigma = 2.578$.

The outlined results show that the disturbance really presents the incoming unstable wave in upstream region $\tau_c \sim \exp(i\alpha_{T-S}x)$, $\alpha_{T-S} = 0.1252 - i0.0029$. The location of the most intensive flow oscillations corresponds to the top of the vibrator. The amount of the disturbance under consideration is quickly damped downstream from the vibrator.

3. Cancellation of T-S Wave over Compliant Membrane

Essential disadvantage of the above active control is the complication arising due to necessity of using (a) a device for recording the parameters of canceled wave, (b) an unsteady generator of artificial disturbances, and (c) feedback control system. In order to overcome this drawback, Manuilovich [2] proposed to use an artificial flexibility of wing surface for excitation of 'artificial' T-S wave.

Suppose that a strip of wing surface is changed by compliant membrane. The membrane tension T/δ is supposed to be large with respect to the membrane mass and damping. This membrane performs oscillations under the exposure of acoustic pressure. The form of the oscillations satisfies the following boundary-value problem:

$$\frac{T}{\delta}\left(\frac{\partial^2 y_w}{\partial x^2} + \frac{\partial^2 y_w}{\partial y^2}\right) = \varepsilon \exp(i\alpha_s x + i\gamma_s z - i\omega t) + \text{c.c.}$$

$$y_w(t, X-\ell, z) = y_w(t, X+\ell, z) = 0$$

After determination of the oscillation form we reduce the problem of artificial T-S wave generation to that for the case of vibrator (2.1) with

$$f(x) = \frac{\exp(i\alpha_s X)}{T} g(x-X)$$

$$g(\xi) = (\alpha_s^2+\gamma_s^2)^{-1}\left[\frac{\cos \alpha_s \ell}{\cosh \gamma_s \ell} \cosh \gamma_s \xi + i \frac{\sin \alpha_s \ell}{\sinh \gamma_s \ell} \sinh \gamma_s \xi - \exp(i\alpha_s \xi)\right]$$

The amplitude of membrane oscillations may be controlled by varying the tension. The oscillation phase can't be controlled. Nevertheless, the phase of artificial T-S wave may be controlled by varying the membrane location X . Let

$$X = \frac{\pi + \arg G(\alpha_{T\text{-}S}) + \arg A_v - \arg c_s}{\text{Re}\, \alpha_{T\text{-}S} - \alpha_s},$$

$$T = \frac{|G(\alpha_{T\text{-}S})|\,|A_v|}{|c_s|} \exp(X \,\text{Im}\, \alpha_{T\text{-}S})$$

As a result the condition (2.4) of T-S wave cancellation is satisfied.

In Fig.2. the process of T-S wave cancellation over flexible membrane with $\ell = 20$ is illustrated. The flow parameters are chosen as in the above calculations. The acoustic wave propagates in x-direction; $\alpha_s = 0.0222$. In this case the cancellation membrane parameters are given by $X = 5.530$, $T = 58.1$.

4. Cancellation of T-S Wave over Unevenness

Consider another method of receptivity control. This method is based on using a two-dimensional roughness under acoustic field as the generator of artificial T-S wave

$$y_w = \delta f(x) \quad , \quad f = hg(x-X)$$

The phase and amplitude of such artificial wave may be controlled by varying the location X and the height h of unevenness, respectively.

In this case the cancellation parameters takes the form

$$X = \frac{\pi + \arg G(\alpha_{T-S} - \alpha_s) + \arg A_{sh} - \arg c_s}{\mathrm{Re}\,\alpha_{T-S} - \alpha_s},$$

$$h = \frac{|c_s|}{|G(\alpha_{T-S} - \alpha_s)|\,|A_{sh}|} \exp(-\,X\ \mathrm{Im}\,\alpha_{T-S}).$$

Here A_{sh} is a new specific coupling coefficient. Note that in the last two methods the cancellation parameters are independent of the complex amplitude ε so the cancellation condition is fulfilled regardless of the values of the amplitude and the phase of the acoustic wave. Being excited by the same acoustic wave, the natural T-S wave and artificial one are synchronized automatically.

5. Cancellation of Crossflow Instability Vortices over spanwise-periodic unevenness

Up to this point the methods of T-S wave cancellation have been considered. These methods are applicable for laminarization of boundary layer on a wing of small sweep angle. Except T-S instability, in three-dimensional boundary layer a cross-flow instability occurs. The importance of the last type of instability in laminar-turbulent transition increases with increase of sweep angle. One of the main cause of cross-flow instability vortex generation is roughness. The initial amplitude of these vortices may be large even in the case of random roughness with small height because the steady eigenmodes are excited via a direct scattering of main flow by surface roughness. For large sweep angle such stationary vortices may be the chief cause of laminar-turbulent transition.

Thus, random roughness in upstream region leads to formation of crossflow instability vortices, which in the local region involved takes the form:

$$q_{cf}^{(n)} = \delta c_r q_1^*(y) \exp(i\alpha_{cf} x + i\gamma z) + \text{c.c.}$$

This 'natural' system of vortices may be canceled by means of the artificial z-periodic unevenness

$$y_w = \delta\sigma c_r g(x)\ \exp(i\gamma z) + \text{c.c.},$$

where $|\sigma|$ characterizes the unevenness height, and $\arg\sigma$ characterizes the spanwise location. Scattering of the basic flow by such unevenness results in generation of 'artificial' system of vortices

$$q_{cf}^{(a)} = \delta c_r G(\alpha_{cf})\ A_{cf} q_1^*(y)\ \exp(i\alpha_{cf} x + i\gamma z) + \text{c.c.}$$

The cancellation condition

$$q_{cf}^{(n)} + q_{cf}^{(a)} \equiv 0$$

will be satisfied in downstream region if we choose the height and the location of the unevenness from the equality

$$\sigma = \frac{1}{G(\alpha_{cf})\ A_{cf}}.$$

References

[1] Milling, R.W.: Tollmien-Schlichting Wave Cancellation, Phys. Fluids **24**, No. 5 (1981) 979-981.

[2] Manuilovich, S.V.: On the Possibility of Cancellation of Tollmien-Schlichting Waves Generated by Sound, Dokl. Akad. Nauk SSSR **313**, No. 2 (1990) 280-282.

[3] Kosorygin, V.S. and Polyakov, N.F.: The Autodestruction of Unstable Waves in Laminar Boundary-Layer Flow, Preprint 11 (1990) Akad. Nauk SSSR, Siberian Div., Inst. Theor. Appl. Mech., Novosibirsk.

[4] Manuilovich, S.V.: On New Methods for Boundary-Layer Flow Laminarization, Journal of Theoretical and Computational Fluid Dynamics **6**, No. 1 (1994) 31-47.

[5] Manuilovich, S.V.: On the Receptivity of Subsonic Flow to Oscillating Effects Localized at the Base of the Boundary Layer, Izv. Akad. Nauk SSSR, Mekh. Zhidk. Gaza, No. 4 (1988) 63-69.

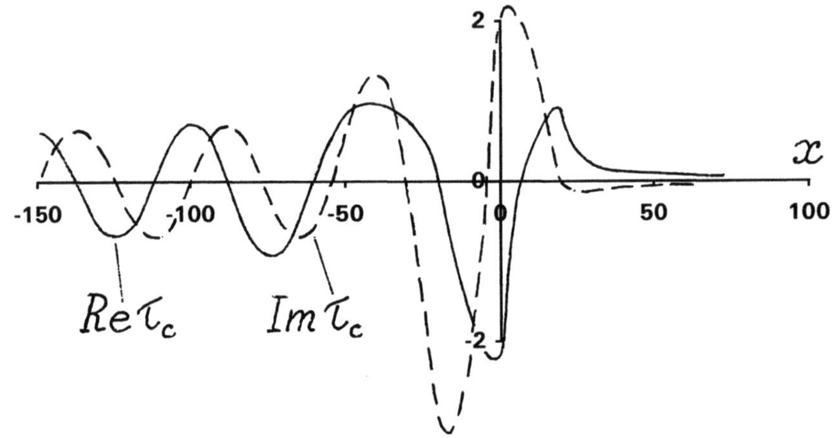

Figure 1. T-S wave cancellation by vibrator.

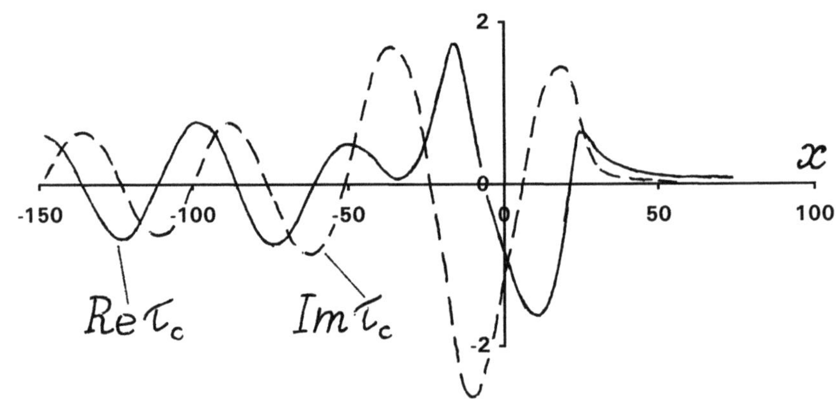

Figure 2. T-S wave cancellation over compliant membrane.

Relaminarisation of the Swept Wing Attachment-Line by Surface Suction

by D.I.A. Poll and M. Danks
Aerospace Division, Manchester School of Engineering,
University of Manchester, Manchester M13 9PL, England

Introduction

It is now well established that the leading edge is a critical region for boundary layer transition to turbulence on a swept-back wing. Under the appropriate conditions, transition can occur on the attachment line itself ($x=0$) and this is particularly important since, in the absence of subsequent relaminarisation, full chord, turbulent flow on both upper and lower surfaces will be produced. It has been demonstrated in previous papers [1,2] that, for an impervious surface, the attachment-line transition process can be instigated at Reynolds numbers which are very much lower than those associated with linear (and non-linear) stability theory. Moreover, there is a very clearly defined lower boundary for the transition Reynolds number in the presence of large forcing disturbances. Large disturbances can take the form of surface roughness or they can emanate from the wing fuselage junction. Under these circumstances the transition process is commonly referred to as "attachment-line contamination".

In the context of the laminar flow aircraft, attachment-line contamination is known to be a serious practical problem. Therefore, the determination of the effect of surface transpiration on attachment-line contamination in an important issue. However, at present, there are no accurate experimental data available - although Spalart [3] has published some results from direct numerical simulations of the turbulent attachment-line flow subjected to surface transpiration which give a hint of the behaviour.

The present work is aimed at clarifying the physics of the relaminarisation phenomenon and providing accurate, quantitative data.

Key Words : attachment line, relaminarisation, suction

1. The attachment-line flow

The external streamline pattern for the flow over a swept, leading edge is shown schematically in Figure 1. Feature A-A represents the demarkation between flow which passes over the upper surface of the wing and that which passes over the lower surface. If the wing has no sweep, A-A is the locus of the stagnation points. However, when the wing has sweep, A-A has a flow component along it and a velocity profile will be developed. In this case, the feature A-A is no longer a locus of stagnation points (stagnation line), and it is better described as an "attachment line". The state of the attachment-line boundary layer can be either laminar, transitional or turbulent depending upon condition.

When the wing is infinite swept, the boundary layer which develops along the attachment line exhibits no spanwise variation and, for incompressible flow, may be characterised by the following parameters

- a viscous length scale $\eta = \left[\dfrac{\nu}{dU_e/dx} \right]^{1/2}_{x=o}$

- a Reynolds number $\bar{R} = \dfrac{V_e \eta}{\nu}$

and, when there is transpiration at the wall, w(0),

- a transpiration parameter $\dfrac{w(0)}{V_e}$.

2. Experimental Arrangement

Tests have been conducted on a large, swept-cylinder model - see Figure 2. This has a "teardrop" cross section, with the leading edge being a circular cylinder and the rear half being a wooden fairing. The leading edge sweep is nominally 60^0, the normal-to-leading-edge chord is 0.813m and the leading edge radius is 0.403m. In order to provide a perforated surface for transpirtation work, the cylinder surface is made from 1.2mm thick titanium sheet. Prior to model construction, this sheet was laser perforated with holes of 50μm diameter with a hole-to-hole and row-to-row spacing of 400mm - see Figure 3. The perforated sheet was rolled and attached to aluminium formers before plenum chambers were attached to the inner surface. Transpiration at the attachment line is provided by a strip of perforation which begins 1.8m from the upstream tip and is 0.9m long in the spanwise direction. The strip extends 10^0 either side of the attachment line and has a single plenum chamber, which is connected, via values and rotameter flow meters, to a common, low-pressure tank. The whole system is mounted in the Goldstein Research Laboratory, low-speed, closed-return circuit, wind tunnel which has a 2.74m x 2.13m test section and a maximum velocity of 70m/sec. By varying the sweep angle by $\pm 5^o$ it is possible to obtain values of $\bar{R}$ in the range $200 < \bar{R} < 1100$, whilst the mean through surface velocity, w(0), can be set in the range 0 to - 0.01Ve.

3. Results

a) Effect of suction on a fully developed turbulent boundary layer.

For these experiments, a 4mm diameter circular trip-wire was wrapped around the model at a distance of 8 cms from the upstream tip. As expected on the basis of previous work (reference 1) when $\bar{R}$ was less than 245 the flow over the impervious portion of the model was laminar. However, beyond 245 transition flow occurs and, at values $\bar{R}$ in excess of approximately 280, observations of hot-wire signals indicated that the flow was fully turbulent. In order to investigate the relaminarisation process, a single hot-wire was placed at various spanwise distances, s, from the beginning of the suction strip. A particular tunnel free-stream speed was set and the suction rate was varied. The values of suction rate at which relaminarisation onset and end (fully laminar flow) occurred were noted.

Results for the onset and end of relaminarisation are presented in Figures 4 and 5. There is a tendency for the required suction levels to decrease as s increases and, for the end of relaminarisation, the data exhibit asymptotic behaviour is the limit of large s. The asymptotic limit for the end of relaminarisation is presented in Figure 6. In the figure, comparisons are made with measurements made on two separate occasions (repeatability) and indications of the end of relaminarisation as determined by surface, Pitot-tube measurements. In general, the Pitot-tube measurements indicate that for a given $\bar{R}$ slightly less suction is required. This is to be expected since, in order to obtain an indication with the Pitot tube, some turbulent bursting must be present. However, with the hot-wire the indicated value corresponds to a condition with no turbulent bursting at all. Also given are the Spalart relaminarisation data obtained from direct numerical simulation. The argeement between experiment and calculation is surprisingly good. All the data are well represented by the following, simple, algebraic relations

a) Hot-wire

$$\bar{R} = 245 \left\{1.07 \left[\bar{R} \cdot \frac{w(0)}{Ve}\right]^2 - 0.48 \left[\bar{R} \cdot \frac{w(0)}{Ve}\right] + 1\right\}^{1/2}$$

and b) Preston tube

$$\bar{R} = 255 \left\{1.15 \left[\bar{R} \cdot \frac{w(0)}{Ve}\right]^2 - 0.50 \left[\bar{R} \cdot \frac{w(0)}{Ve}\right] + 1\right\}^{1/2}$$

In Figure 7, the hot-wire data presented in Figure 5 are normalised with the data given in Figure 6 i.e. w' is the ratio of the suction required at a particular value of s/η to that needed when s/η tends to infinity at the same R. A reasonable collapse is obtained which can be approximated by

$$w' = 1 + EXP(2.0 - 0.0025\ s/\eta) \quad for\ s/\eta > 500.$$

The data show that, for asymptotic conditions to be established, the suction must be applied for 2000η (approximately 200 boundary layer thicknesses). A particularly interesting feature is revealed when the data of Figure 7 are replotted in the form shown in Figure 8 i.e. $\bar{R}$ versus w(0)/Ve. As the level of suction in increased, the R for the end of relaminarisation rises very rapidly, so that, when w(0)/Ve exceeds a critical level, the flow is laminar for all values of R i.e. turbulent flow is impossible.

From the expressions given above, these critical suction levels are found to be

a) Hot-wire $\frac{w(0)}{Ve} = -0.0039$ and b) Preston tube $\frac{w(o)}{Ve} = -0.0036$

b) Effect of suction when a large trip wire is located on the suction surface

In this test, the hot-wire was fixed at the furtherest downstream location on the suction strip. The trip-wire near the upstream tip was removed and, in this case, transition in the-attachment line boundary layer was found to begin when $\bar{R}$ was approximately 700. The 4mm diameter trip wire was then placed on the suction strip at a distance, s, upstream of the hot-wire and the measurements were made using the same technique as before.

Results are shown in Figure 8, together with corresponding results from the previous tests. It is apparent that there is some variation of w' with $\bar{R}$. However, as $\bar{R}$ increases the trip wire data approach those for the relaminarisation of the fully developed turbulent flow. Moreover, there is still an asymptotic limit as s/η tends to large values (greater than 2000) and this limit is the same as that shown in figure 7.

c) Attachment-line velocity profiles.

Velocity profile information has been deduced by using the results obtained by surface Pitot tubes. A set of tubes having external diameters varying from 0.4mm to 3mm was used. Each tube was fixed to the surface and the impact pressure recorded for a range of tunnel speeds and suction velocities. By converting the pressures to velocity ratios, results of the type shown in Figure 9 were obtained. The velocity profiles for fixed R and fixed wall transpiration rate were then obtained by interpolation. Typical results are presented in Figures 10, 11, 12 and 13.

4. References

[1] Poll, D.I.A. : Transition in the Infinite Swept Attachment-Line Boundary Layer, The Aeronautical Quarterly, Vol.XXX, November 1979.

[2] Hall, P., Malik, M.R. and Poll, D.I.A. : On the Stability of an Infinite Swept Attachment-Line Boundary Layer, Proceedings of the Royal Society, Series A, Volume 395, October 1984.

[3] Spalart, P. : Direct Numerical Study of Leading-Edge Contamination, Paper No.5, AGARD CP438, October 1988.

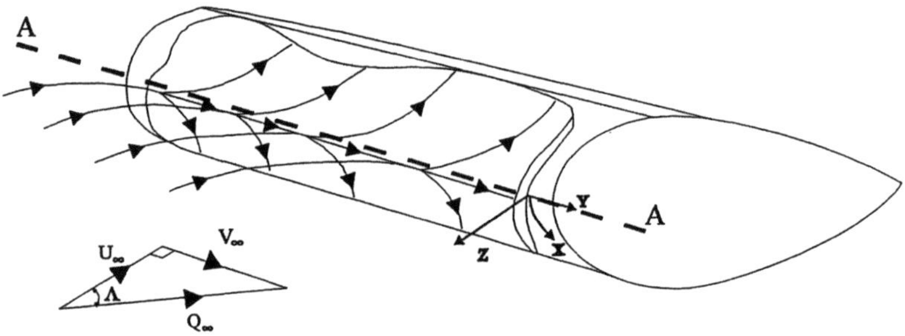

Figure 1 Flow in the vicinity of a swept attachment line.

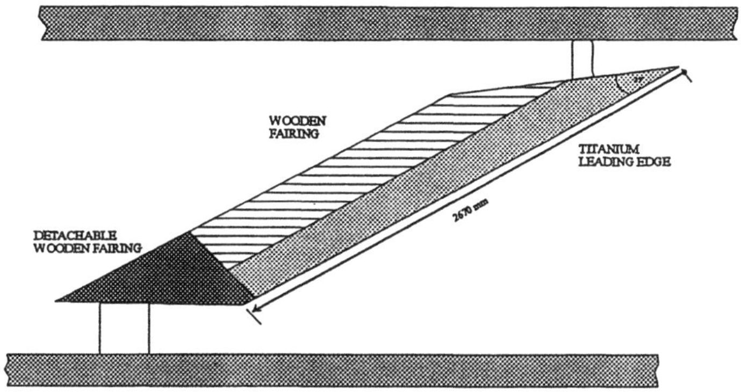

Figure 2 Sketch of model in wind-tunnel

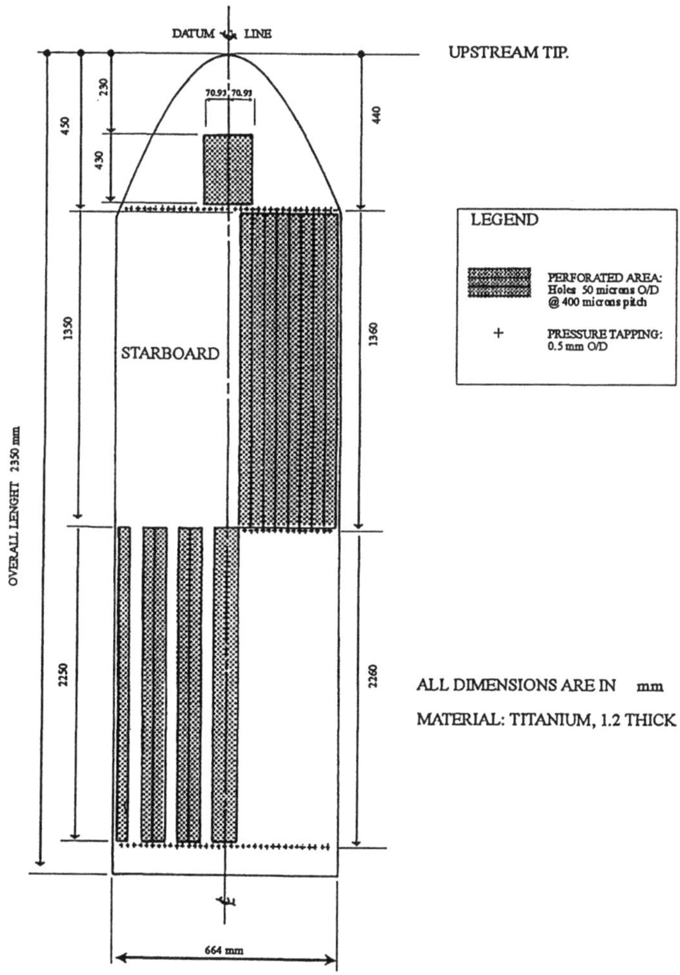

Figure 3 Layout of the suction surface

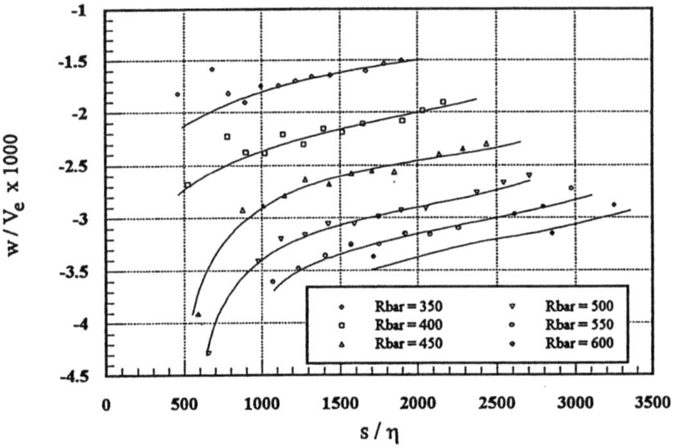

Figure 4
Critical suction rate for the onset of relaminarisation

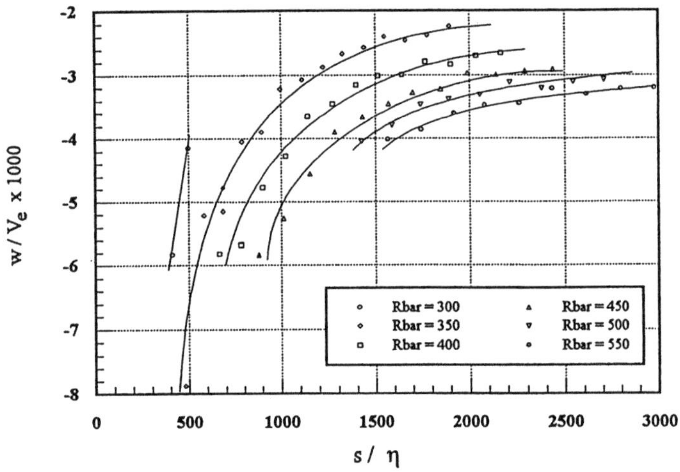

Figure 5
Critical suction rate for the end of relaminarisation

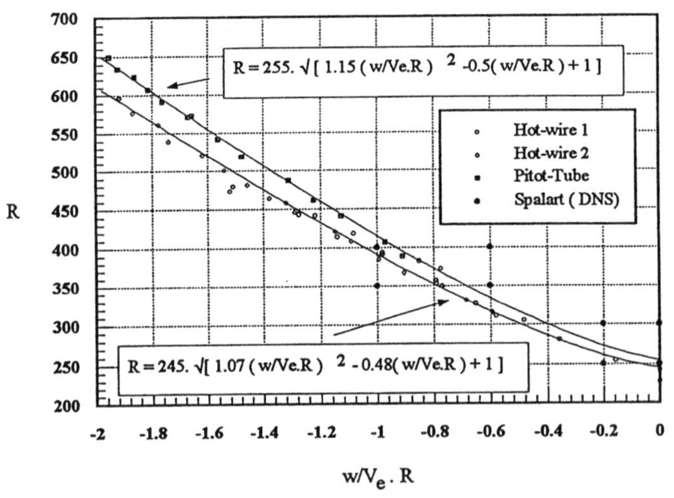

Figure 6
Asymptotic results for the end of relaminarisation

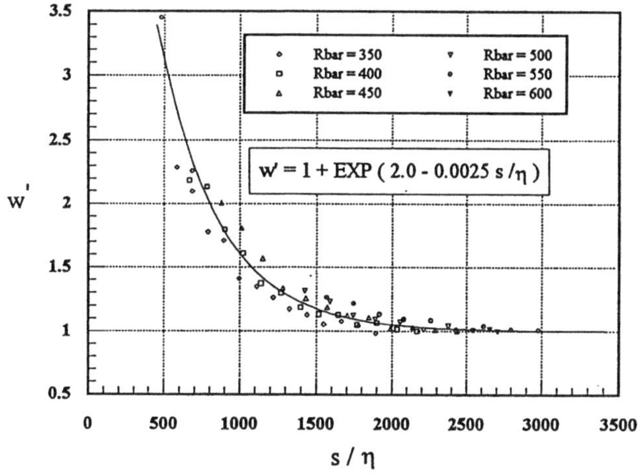

Figure 7
Normalised suction rate as a function of streamwise distance

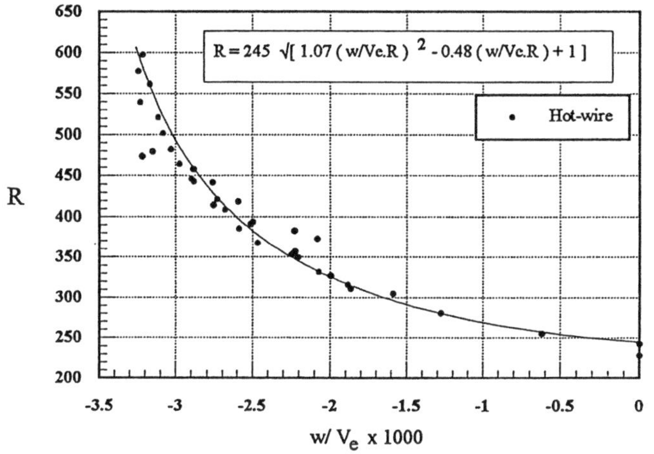

Figure 8
Critical conditions for the end of relaminarisation

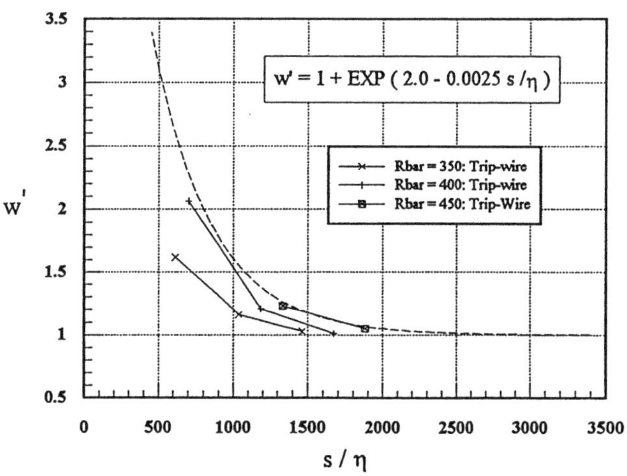

Figure 9
Normalised suction rate for the end of relaminarisation when a large trip wire is on the suction surface.

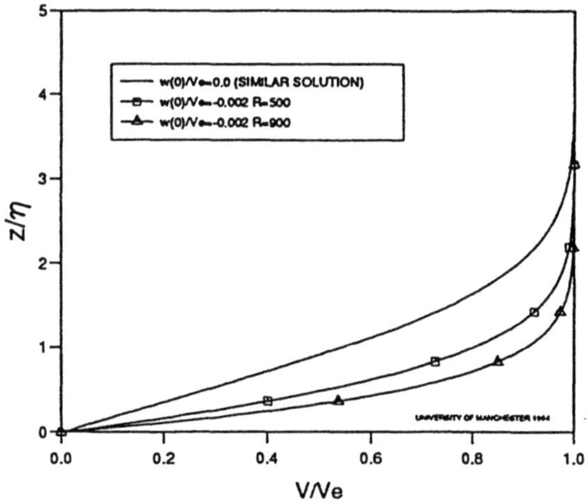

Figure 10
Laminar flow
profiles

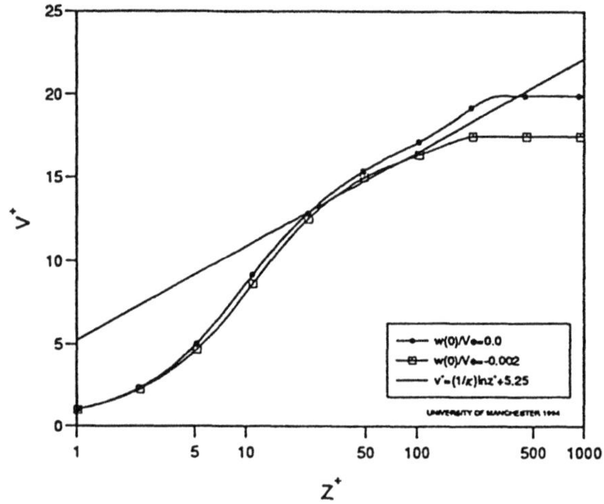

Figure 11
Turbulent flow
R=500

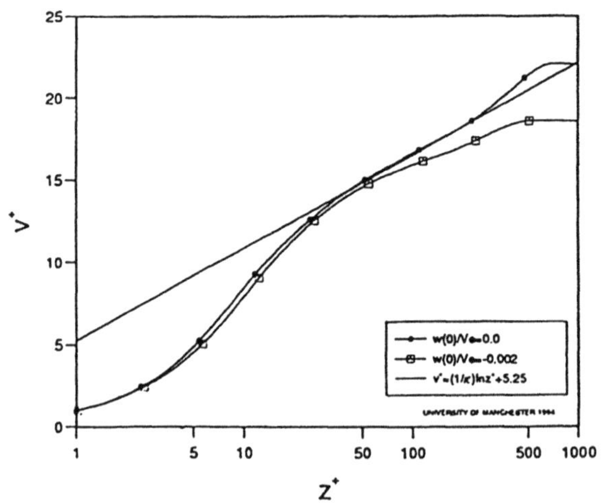

Figure 12
Turbulent flow
R=900

Session E: Compressible Flows

Nonlocal and Local Instability Analysis of Hypersonic Flows

M. Simen, S. Hein, F. Bertolotti, V. Wendt
Institut für Strömungsmechanik,
Deutsche Forschungsanstalt für Luft- und Raumfahrt,
Bunsenstr. 10, D-37073 Göttingen, Germany,
A. Hanifi, D. Henningson
Kungliga Tekniska Högskolan,
Department of Mechanics, S-100 44 Stockholm, Sweden

Summary

A nonlocal, linear instability theory is used to model convective amplification of waves with weakly divergent or curved wave-rays and wave-fronts, propagating in a weakly nonuniform flow. The compressible, nonlocal stability equations, which represent an eighth order system of parabolic differential equations in generalized coordinates, are derived from a consistent scaling method. Amplification rates from spatial DNS and nonlocal instability analysis of unstable first mode disturbances in hypersonic flow past a flat plate are shown to be in excellent agreement. Conical Divergence is shown to be stabilizing on first mode disturbances in nonuniform, nonparallel hypersonic flow past a pointed cone. Convex wall camber is found to have a strongly stabilizing effect, concave wall camber a strongly destabilizing effect on first and second mode disturbances. Due to concave curvature a third mode unstable region develops merged with the second mode.
Key words: Nonlocal, Instability, Divergence, Curvature, Hypersonics

Introduction

Nonlocal instability theory is a rather general, flexible and computationally efficient method to analyze convective disturbance wave growth. It consistently models wave propagation in realistic flows, which is characterized by wave-ray divergence and curvature as well as by nonuniform basic and disturbance flow properties. Nonlocal instability means that local disturbance growth is influenced by the local and upstream flow conditions as well as the change in amplitude due to divergent or convergent wave rays. Local instability is only affected by local flow conditions and the local formulation is limited to parallel flows and wave rays. Finally in globally unstable flows the entire flowfield influences local disturbance growth.

In nonlocal stability theory two additional assumptions are made different from classical, local formulations: As a first additional step, a more general, nonlocal wave ansatz is introduced. Opposed to a local ansatz it can also describe waves with divergent wave rays and curved wave fronts, as well as disturbance propagation and amplification in a nonuniform basic flow. As a second step, a nonlocal approximation is introduced, which only allows for divergence, curvature and nonuniformity within the order of magnitude of a small scaling parameter 1/Re. If only terms of O(1) and O(1/Re) are considered, for compressible flows an eighth order system of parabolic equations is obtained. This system can be solved efficiently by marching, once an auxiliary equation to obtain a unique solution is introduced. The scaling approach represents a consistent, 'rational' derivation procedure, similar to the multiple scales method, as e.g. described by El-Hady [2]. The parabolic type of the nonlocal equations is similar to the 'Parabolized Stability Equations (PSE)'. In this method based on ad hoc, physical assumptions certain 'elliptic' terms are dropped. The 'uniqueness condition' corresponds to the 'normalization' condition first introduced by Herbert & Bertolotti [1].

In this paper some additional and modified aspects of the nonlocal formulation originally introduced by Simen [3], [4] are discussed. Divergence, curvature and nonuniform flow effects on unstable hypersonic boundary layer flows are then analyzed. The instability analysis is based on the nonlocal, linear code NOLOT recently developed by Hein, Bertolotti, Simen at DLR in cooperation with Hanifi and Henningson at KTH/FFA in Stockholm, Sweden, as part of the DLR 'NOLOT'-project dealing with NOnLOcal Transition analysis and prediction of industry relevant flows.

Nonlocal Linear Stability Theory

The nonlocal stability equations are derived from the equations of conservation of mass, momentum and energy and the equation of state, governing the flow of viscous, compressible, ideal gas, formulated in primitive variables and written in generalized, orthogonal curvilinear coordinates. In these governing equations all flow and material quantities are decomposed into a steady basic flow $\bar{q}$ plus an unsteady disturbance flow component $\tilde{q}$ according to

$$q(x^i, t) = \bar{q}(x^i) + \tilde{q}(x^i, t). \tag{1}$$

The x^i are the orthogonal curvilinear coordinates with $i = 1, 2, 3$. x^3 is normal to the direction of spatial amplification x^1. Waves propagte in x^1, x^2 planes. t stands for time. The basic flow terms are subtracted to obtain the disturbance equations. For simplicity but not as a basic restriction, we have assumed that both the basic and disturbance flow is independent of the x^2 coordinate, i.e. we restrict ourselves to quasi three dimensional flow. Based on these disturbance equations, two fundamental steps are required to derive the nonlocal stability equations:

1. To describe wavy disturbances with divergent or curved wave rays and wave fronts propagating in a nonuniform flow, we introduce a nonlocal normal mode ansatz of the general form

$$\tilde{q} = \hat{q}(x^1, x^3)e^{\imath\Theta}, \tag{2}$$

where $\hat{q}$ stands for a complex, vectorial or scalar amplitude function, $\imath$ for the imaginary unit,

$$\Theta = \int k^j dx^j g_{jj} - \omega t \tag{3}$$

for the complex phase function and

$$k^1 = \frac{k(1)}{h_1} = \frac{1}{h_1}\frac{\alpha(x^1)}{h_1}, \quad k^2 = \frac{k(2)}{h_2} = \frac{1}{h_2}\frac{\beta}{h_2}, \quad k^3 = 0 \tag{4}$$

for the contravariant components of the complex wave vector. Frequency is given by ω, $k(i)$ denotes the physical components of the wave vector, α, β are wave numbers and $g_{jj} = h_j^2(x^1, x^3), j = 1, 2, 3$ is the metric tensor. The physical disturbance is obtained from the real part of $\tilde{q}$.

Divergent wave rays are described by the scale factors h_i in equation (4). As is well known from e.g. spherical acoustic waves, divergence causes a decay in amplitude in direction of wave propagation due to energy conservation. This is expressed in equation (2) by an amplitude function $\hat{q}$ that can vary along rays in x^1 direction. Disturbance propagation in a nonuniform basic flow is also described by the x^1, x^3 dependent amplitude functions.

Curved wave rays are described by the scale factors as well. They result in kinematic changes of the local physical wavelengths given by $\lambda(1) = 2\pi h_1/\alpha_r, \lambda(2) = 2\pi h_2/\beta_r$. Locally the wave fronts are orthogonal to the wave rays. Hence divergent wave rays yield curved wave fronts and vice versa. Also the wave fronts become distorted if the wave angle $\psi = arctan[(\beta h_1)/(\alpha h_2)]$ varies in x^3 direction as for any wave propagating oblique to the direction of principal wave ray curvature.

2. As a second step we only allow for weakly divergent or curved wave rays as well as a weakly nonuniform disturbance and basic flow. This is introduced into the stability equations by using a scaling method. The x^1 dependance of the basic and disturbance flow, the nonparallel basic flow velocity component $\bar{v}^3$, the curvature and divergence $\frac{\partial h_i}{\partial x^j}$ are assumed to scale with the small parameter $1/Re$ (Re: Reynolds number based on Blasius length). This can be written as $(i = 1, 2, 3; j = 1, 3)$:

$$x^1 = x_T^1 Re, \qquad \bar{v}^3 = \bar{v}_T^3/Re, \qquad \frac{\partial h_i}{\partial x^j} = \left(\frac{\partial h_i}{\partial x^j}\right)_T/Re, \tag{5}$$

whereas

$$x^3 = x_T^3, \qquad \bar{q} = \bar{q}_T, \qquad h_i = h_{iT},$$

$$\hat{q} = \hat{q}_T, \qquad \omega = \omega_T, \qquad \alpha = \alpha_T, \qquad \beta = \beta_T,$$

which means that these quantities are assumed to be of $O(1)$. With the above scaling introduced into the governing equations, considering only terms which scale with powers of Reynolds number up to $1/Re$, we obtain the nonlocal, linear stability equations. They describe the weakly nonuniform propagation and amplification of disturbances with divergent or curved wave rays in a nonuniform basic flow. There are two important results if the derivation is based on the above scaling method:

a) It provides a consistent, 'rational' formalism to control the order of the system of nonlocal differential equations. For compressible flow an eighth order system is obtained as in the classical, local parallel case, if terms up to the order of $1/Re$ are kept. If also terms of $O(1/Re^2)$ are considered, or only second derivatives in x^1 direction are neglected as in the PSE-method, a stiff ninth order system is obtained. This is because the ninth eigenvalue is an order of magnitude smaller than the others and becomes even zero at the wall, as it is linked to the wall normal basic flow velocity . The order of the system can be formally detected by transforming the nonlocal stability equations into a system of first order differential equations. One obtains eight equations in analogy to the combined first and second order equations of the multiple scales method. Thus eight boundary conditions are required for both the local and nonlocal problem.

b) It provides a consistent, 'rational' formalism to control the differential equation type. If terms up to $O(1)$ are kept, ordinary differential equations of the Rayleigh equation type are obtained. They describe the inviscid instability of parallel flows. If terms up to $O(1/Re)$ are kept, nonlocal stability equations are obtained. Divergence, curvature and nonuniform flow effects are consistently taken into account within the $O(1/Re)$ terms. The equations are primarily of parabolic type, as no second derivates in x^1 direction appear. Hence using the scaling method the parabolic type is an a posteriori result of the derivation, not an a priori assumption as in the PSE approximation. In this approximation based on physical arguments second derivatives and products of first derivatives in x^1 direction are not considered.

Within the generalized wave ansatz both the amplitude and phase function depend on the x^1 direction, which is the direction of spatial amplification. This ambiguity can be removed by specifying various forms of 'uniqueness' conditions such as $\frac{\partial \hat{q}}{\partial x^1}|_{x^3_m} = 0$, $\int_0^{x^3_b} \hat{q}^\dagger \frac{\partial \hat{q}}{\partial x^1} dx^3 = 0$ (†: complex conjugate, x^3_m: position where $\hat{q}$ has a maximum, x^3_b: outer boundary) etc., as first presented by Bertolotti, Herbert & Spalart [1] under the name 'normalization' condition.

The wavenumber β is obtained from the kinematic condition of wave propagation, that the gradient of the phase function Θ is in direction of the wave vector. This yields together with the quasi three dimensional flow assumption $\frac{\partial \beta}{\partial x^1} = 0$. The wavenumber α is obtained by satisfying iteratively the 'uniqueness' condition in the nonlocal case or an additional boundary condition in the local case.

In the nonlocal case the wavenumber in x^1 direction becomes

$$\alpha_N := -\imath\frac{1}{\tilde{q}}\frac{\partial\tilde{q}}{\partial x^1} = \alpha - \imath\frac{1}{\hat{q}}\frac{\partial\hat{q}}{\partial x^1}. \quad (6)$$

Then the physical waveangle ψ, which causes wave front distortion or wave ray curvature as ψ can vary both in x^1 and x^3 direction, is given by $\psi = arctan[\frac{\beta h_1}{\alpha_N h_2}]$.

The nonlocal phase function Θ_N can be written analogous to eqs. (3), however α is replaced by α_N, which is a function of both x^1 and x^3 due to the contribution of the amplitude function. Hence from $\vec{k}_N = grad\theta_N$ we obtain

$$k_N^1 = \frac{k_N(1)}{h_1} = \frac{1}{h_1}\frac{\alpha_N}{h_1}, \quad k_N^2 = \frac{1}{h_2}\frac{\beta}{h_2}, \quad k_N^3 = \frac{1}{h_3}\frac{\gamma}{h_3} = \frac{-\imath\frac{1}{\hat{q}}\frac{\partial\hat{q}}{\partial x^3}}{h_3^2}. \quad (7)$$

The normal component γ formally represents the change of phase in x^3-direction. From (7) the physical wavenumber and amplification rate in direction of spatial amplification x^1 is given by

$$\alpha_{r,P} = Re[k_N(1)] = \frac{1}{h_1}[\alpha_r + Im(\frac{1}{\hat{q}}\frac{\partial\hat{q}}{\partial x^1})], \quad (8)$$

$$-\alpha_{i,P} = Im[k_N(1)] = \frac{1}{h_1}[-\alpha_i + Re(\frac{1}{\hat{q}}\frac{\partial\hat{q}}{\partial x^1})]. \quad (9)$$

We note that both wavenumber and amplification rate depend on wave-ray curvature in x^1-direction as given by h_1. A decay in amplitude due to wave-ray divergence is represented by the term $Re(\frac{1}{\hat{q}}\frac{\partial\hat{q}}{\partial x^1})$.

Within the NOLOT-code the nonlocal stability equations are discretized by 4th order compact finite differences in the normal (x^3) direction and by first or second order backward finite differences in amplification (x^1) direction. Details of the numerical method and solution algorithm will be published elsewhere.

Wave-ray Divergence & Curvature Effects, Nonuniform Flow Effects

Among manifold activities to validate nonlocal instability analysis using the NOLOT code Figure 1 shows growth rates from spatial direct numerical simulation (DNS) by Eissler [5] in comparison with nonlocal growth rates.

We investigate unstable, hypersonic boundary layer flow past a flat plate at a boundary layer edge Mach number $Ma_e = 4.8$ and edge temperature $T_e = 55.41K$. The Prandtl number was set to $Pr = 0.71$ and Sutherland's law was used to compute viscosity. Here and in all following computations similarity solutions were used to generate basic flow profiles. The initialized first mode disturbance has a reduced frequency $F = \omega/Re = 5 \times 10^{-5}$ and a spanwise reduced wavenumber $b = \beta/Re = 0.08$. The disturbance evolution is difficult to compute as in the Reynolds number range investigated it represents a damped and highly oblique wave. Even more surprising is the excellent agreement of

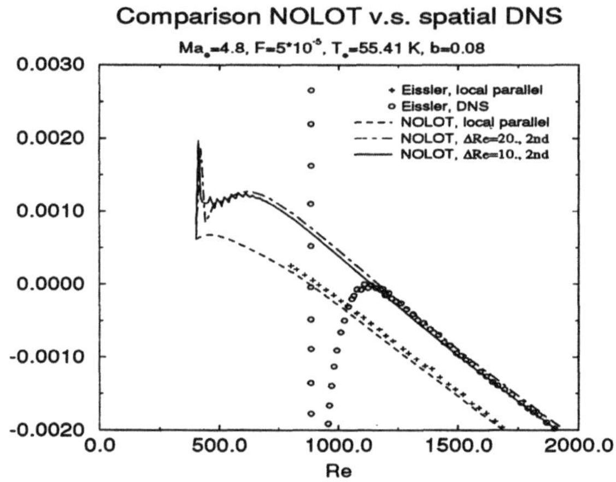

Figure 1: Nonlocal and local instability analysis in comparison with direct numerical simulation.

nonlocal instability analysis and spatial DNS after a transient phase of the simulation. Second order backward finite differences were used to compute the streamwise derivatives of the nonuniform disturbance propagation. The streamwise step size was varied between equal Reynolds number increments of $\Delta Re = 20$ and $\Delta Re = 10$ with a negligible effect on the growth rates.

A systematic study of the effects of wave-ray divergence, wave-ray curvature and flow nonuniformity on first, second and higher mode disturbance amplification in unstable, hypersonic boundary layer flow was performed. As model configurations pointed cones and cambered plates were selected.

From well known solutions to the wave equation for conical, acoustic waves it is known that the amplitude decreases with distance from the origin due to wave-ray divergence. It was interesting to see if a stabilizing effect of conical divergence could also be observed for an oblique wave propagating in nonuniform, i.e. nonparallel, hypersonic boundary layer flow past a pointed cone. The cone half angle $\Theta = 7^o$, the Mach number $M = 6.83$, the temperature $T = 70.45K$ at the boundary layer edge was chosen according to Stetson's experiment. A first mode wave with $F = 2.95 \times 10^{-5}$, $n = 4$ was selected. Here n is the number of modes in circumferential direction, which was held constant. As a result the waveangle decreases from $\psi \sim 56^o$ at $Re = 800$ to $\psi \sim 4^o$ at $Re = 4000$. Figure 2 a,b show that indeed at lower Reynolds numbers i.e. higher divergence rates, the growth rates are strongly reduced if divergence is taken into account or increased from a 3° to a 7° cone half angle. Compared to a classical locally parallel instability analysis with or without local conical divergence included, the nonlocal growth rates are significantly larger at higher Reynolds numbers due to nonparallel flow effects.

The effect of convex and concave curvature in combination with nonparallel flow effects was studied for hypersonic flow past a flat plate according to the DLR Ludwieg tube conditions, $M = 5, T = 61.2K, Re_{unit} = 7.5 \times 10^6 1/m$.

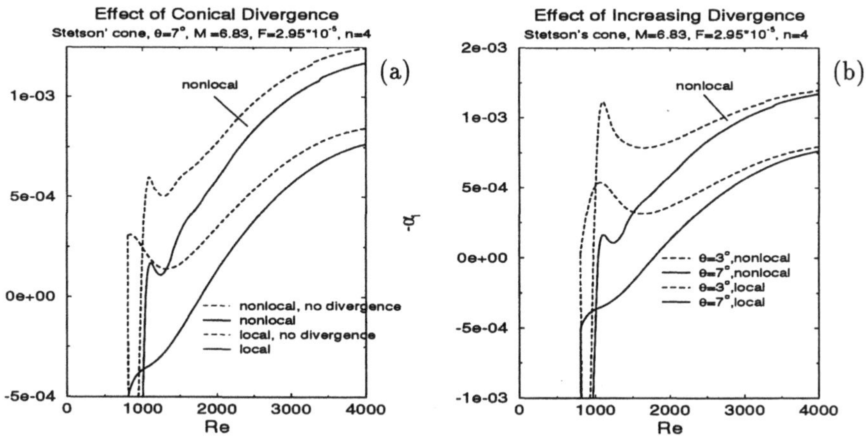

Figure 2: Effect of wave-ray divergence and nonparallel basic flow on nonlocal and local instability of hypersonic boundary layer flow past a cone.

An oblique first mode wave with $F = 3 \times 10^{-5}, b = 9 \times 10^{-5}$ and a second mode wave with $F = 1 \times 10^{-4}, b = 0$ was initialized. Figure 3 a,b shows a strong stabilizing effect of convex curvature for both first and second mode disturbances. This holds even for a radius of curvature $r = 1m$ larger than the plate length, which is $\sim 0.25m$ for a typical Reynolds number of $Re = 1300$ under the given experimental conditions. It is also interesting to note that the nonparallel effect can be stabilizing for the first mode in combination with concave wall camber. Quite surprising is the development of an additional unstable region after the second mode region for concave curvature.

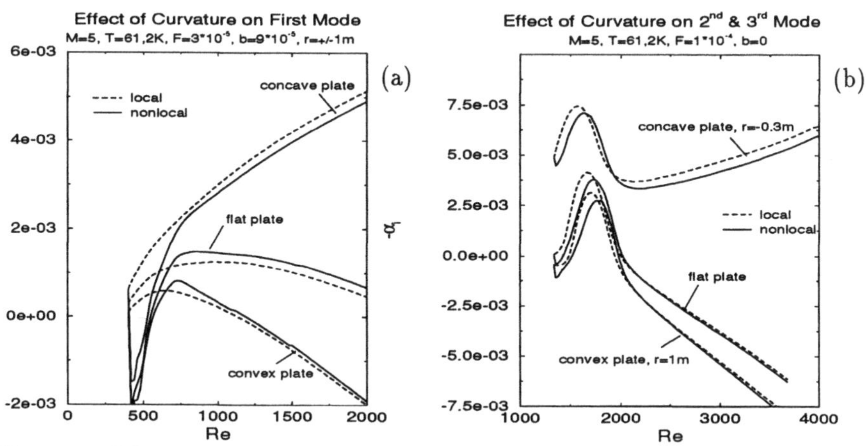

Figure 3: Effect of concave and convex wave-ray curvature, nonparallel basic flow on first and second mode waves in hypersonic flow past cambered plates.

To identify the nature of this instability pressure disturbance distributions are shown in Figure 4 for various Reynolds numbers. It can be observed that starting with one phase shift at $Re = 1600$ for a typical second mode amplitude distribution, an additional phase shift develops beyond $Re = 3600$ due to a second crossing of the $\tilde{p} = 0$ line. Hence, according to Mack's nomenclature this instability wave represents a third mode with its unstable region shifted to lower Mach numbers and frequencies due to concave wall curvature.

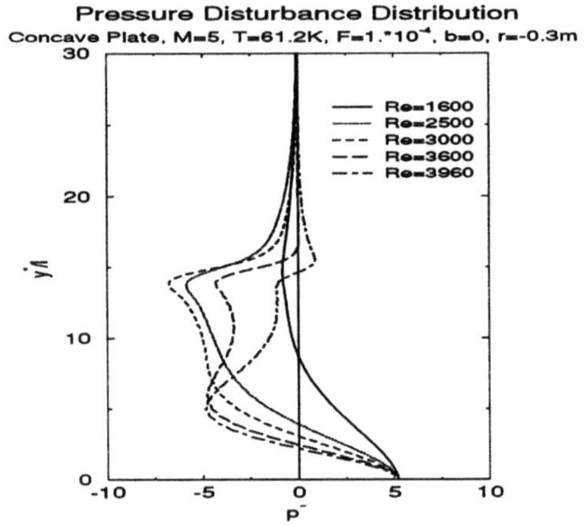

Figure 4: Evolution of pressure disturbances in hypersonic flow past a concave plate.

References

[1] Bertolotti, F. P., Herbert, Th., Spalart, P. R.
Linear and Nonlinear Stability of the Blasius Boundary Layer. J. Fluid Mech., Vol. 242, pp.441–474, 1992.

[2] El-Hady, N. M.
Nonparallel Instability of Supersonic and Hypersonic Boundary Layers. Phys. Fluids, A3(9), pp.2164–2178, 1991.

[3] Simen, M.
Local and Nonlocal Stability Theory of Spatially Varying Flows Instability, Transition and Turbulence, Springer Verlag, pp.181–201, 1992.

[4] Simen, M.
Lokale und Nichtlokale Instabilität hypersonischer Grenzschichtströmungen DLR FB 93-31, 1993.

[5] Eissler, W., Bestek, H., *Spatial Numerical Simulation of Nonlinear Transition Phenomena in Supersonic Boundary Layers* Proc. ASME Fluids Engineering Conference, Washington, D.C., 1993.

An Exploratory Study of Transition on a Slender Cone in Hypervelocity Flow

H. G. Hornung and P. Germain
Graduate Aeronautical Laboratories
California Institute of Technology
Pasadena, CA 91125, USA

Summary

Results are presented from an experimental exploratory study of laminar, transitional and turbulent boundary layers on a slender, sharp cone at zero incidence in hypervelocity flow. The study was conducted in the hypervelocity shock tunnel T5, whose high operating pressure permits achieving the necessary high Reynolds numbers at high total enthalpy. Surface heat transfer rate and resonantly enhanced flow visualization were used to determine transition location and flow structure. Results indicate that the transition mechanism at the conditions tested is via the Tollmien-Schlichting instability. No detailed information about the noise spectrum in the facility is available. A strong, gas-dependent increase of the transition Reynolds number evaluated at the reference temperature on specific total enthalpy was observed. This effect increases monotonically with the dissociation energy of the gas.

Key words: Hypervelocity, transition, shock tunnel, sharp cone.

1. Introduction

One of the most serious gaps in the knowledge of hypervelocity flows is that concerning transition to turbulence in the boundary layer. Previous experimental laboratory investigations on hypersonic transition were limited to cold hypersonics (see notably [1-4]), in which high Mach number is achieved, not so much by raising the speed, but mainly by lowering the speed of sound. Although this permits studies to be performed under conditions that have been characterized much more completely than the shock tunnel flow, it also misses important phenomena associated with the high temperatures encountered at conditions when the free-stream kinetic energy is comparable to one of the dissociation energies of the gas. Studies on hypervelocity transition have been almost entirely limited to numerical work (see *e.g.* [5]). These build on the extensive work of Mack [6]. A recent paper by Reshotko [7] provides an up-to-date overview.

Our work had the goal to make use of the fact that T5 provides high-Reynolds number capability, at specific reservoir enthalpy up to 15 MJ/kg, in order to explore the behavior of transition in this regime. The configuration chosen was the 5 deg. half-angle sharp cone, a model that has been widely used in previous experiments. In order to detect real-gas effects more effectively, the experiments were designed to be carried out with various gases.

2. Experimental Setup

2.1 The Free-Piston Shock Tunnel T5

T5 is a shock tunnel in which the driver gas is heated by an adiabatic compression with a heavy piston that is accelerated by compressed air. This method provides a pressure amplification of typically 1000 and a temperature amplification of typically 15. In this manner, the high pressure and speed of sound of the driver gas that are necessary to produce the high shock speeds into atmospheric air to generate hypervelocity flows, can be reached with minimal thermal load and energy requirements.

Fig. 1 shows a schematic diagram of the facility with enlargements of the junctions between the various parts. The piston is accelerated in the compression tube by compressed air initially contained in the secondary air reservoir, thus compressing the driver gas until the diaphragm burst pressure ($\simeq$ 90 MPa) is reached. The piston speed at rupture has to be sufficiently high ($\simeq$ 170 m/s) to maintain almost constant pressure after diaphragm rupture for a short time ($\simeq$2 ms). Thus, the free-piston driver is a constant-pressure driver, in contrast to the constant-volume driver of the conventional shock tunnel.

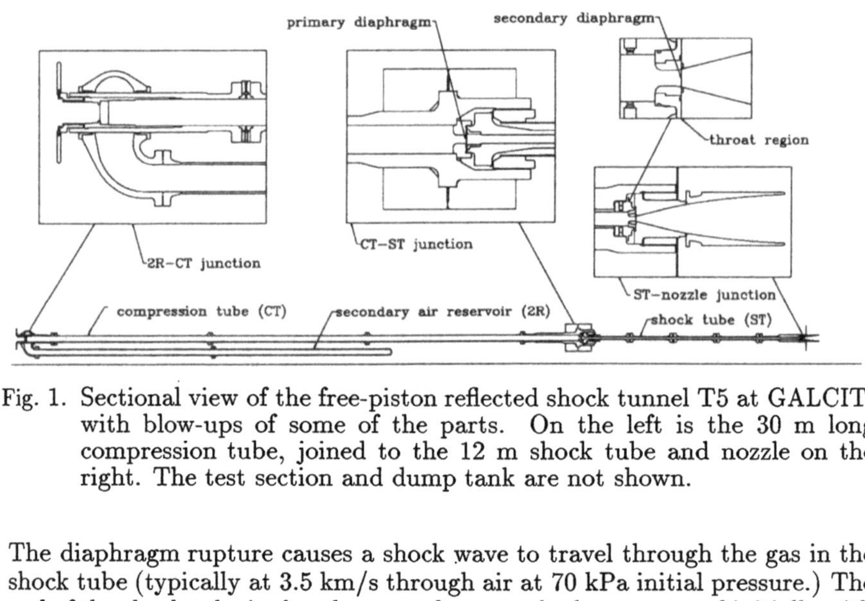

Fig. 1. Sectional view of the free-piston reflected shock tunnel T5 at GALCIT, with blow-ups of some of the parts. On the left is the 30 m long compression tube, joined to the 12 m shock tube and nozzle on the right. The test section and dump tank are not shown.

The diaphragm rupture causes a shock wave to travel through the gas in the shock tube (typically at 3.5 km/s through air at 70 kPa initial pressure.) The end of the shock tube is closed, except for a nozzle throat covered initially with a thin diaphragm. This thin diaphragm is ruptured by the arrival of the shock, and the shock is reflected from the end. This heats and compresses the gas to conditions suitable for steady nozzle expansion to specific total enthalpy h_0 of 12 MJ/kg with the above shock speed. The enthalpy and reservoir pressure ranges of T5 with air or nitrogen as test gas are:

$$3\,\mathrm{MJ/kg} < h_0 < 25\,\mathrm{MJ/kg} \quad \text{and} \quad 20\,\mathrm{MPa} < p_0 < 100\,\mathrm{MPa}.$$

In all the experiments of the present investigation, a contoured nozzle of area ratio 100 was used. At some of the conditions tested, this is operated significantly off design. This is particularly true for the helium tests, as will be pointed out later. In air and nitrogen, this gives Mach numbers in the range 4.7 to 6.5 at the edge of the cone boundary layer, depending on h_0.

2.2 The 5 deg Half-Angle Cone

The 5 deg. half-angle cone used throughout the experiments is 1 m long and consists of three sections, see Fig. 2. The tip section is made from solid Waspalloy, a stainless steel with good toughness at high temperature, and is 76 mm long. The tip was actually found to melt under the heating load of the flow, but reached a stable radius of about 0.15 mm from an initial value of 0.06 mm. No effect of this increase could be detected in the transition location. The next piece is made of aluminum and is 127 mm long. The remainder of the cone is made of two parts of aluminum, hollowed out to accommodate the cables for the instrumentation.

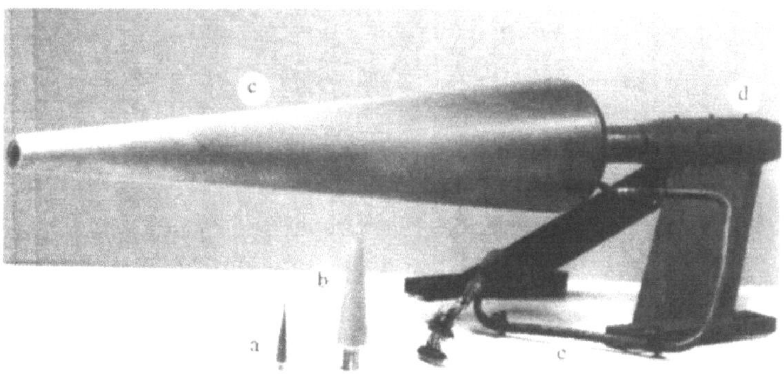

Fig. 2. Photograph of the model partly disassembled to show the tip (a), the second piece (b), and the main body (c). The sting mount is shown as (d), and (e) shows the copper tubing that shields the cables.

The cone is instrumented with 14 thermocouple heat transfer gauges distributed approximately logarithmically with distance from the tip, x, between $x = 241$ mm and 889 mm. The thermocouples are fast-response chromel-constantan devices made by Medtherm. The cone was mounted in the test section and accurately aligned parallel to the nozzle axis. The cone and nozzle axes were displaced relative to each other by 6.4 mm. The axial location of the cone was such that the tip was well inside the test rhombus, and the trailing characteristic of the test rhombus impinged on the cone around $x = 800$ mm.

2.3 Flow Visualization

The flow visualization system was set up for differential interferometry, but this was sufficiently sensitive to detect any structure only at the lowest enthalpies tested. We therefore resorted to resonantly enhanced visualization. This method uses the fact that the absorption of light within a spectral line of the gas is very high. If, therefore, the flow is illuminated with parallel light tuned accurately to a wavelength near the peak of a spectral line, regions of higher density will absorb more light. If the light is slightly off the peak, the refractive index becomes very high, and enhanced schlieren or interferometric effects may be used.

In our setup, we made use of one of the sodium D-lines. An accurately tunable dye laser was built for this purpose, and the flow was seeded with sodium as follows. A small quantity of aqueous solution of salt was deposited on the tip of the cone, or at a point of special interest. The flow causes the salt to heat up, be ablated, and dissociated. Since it is mainly present in the boundary layer, it enhances the visibility of the boundary layer structure preferentially. Only very small concentrations of sodium are necessary. Even 50 ppb are detectable, at T5 density. Control experiments showed that the presence of the salt did not change the heat flux distribution. Fig. 3 shows the arrangement of the T5 window relative to the cone and the frame of the view shown in the pictures of the next section.

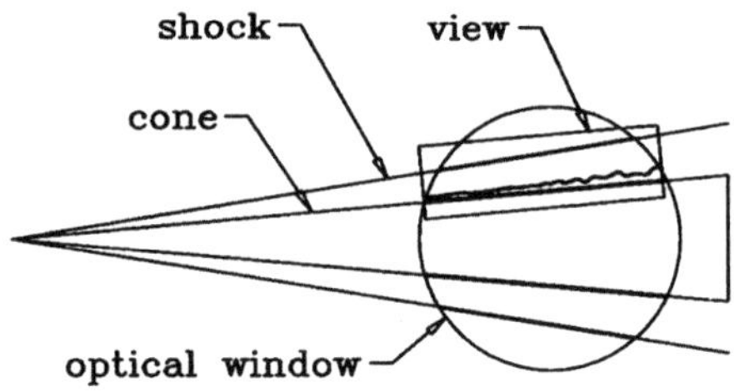

Fig. 3. Schematic diagram of cone with windows.

3. Results

3.1 Heat Flux Distribution

All the measured heat flux values were normalized to form the Stanton number, St according to

$$St = \frac{\dot{q}_w(x)}{\rho_e u_e (h_{aw} - h_w)},$$

where h_{aw} is the adiabatic wall enthalpy and h_w is the enthalpy at the wall. The latter is always very small compared to the former ion our experiments. $\dot{q}_w$ is the heat flux at the wall, and ρ_e and u_e are density and velocity at the boundary layer edge. The adiabatic wall temperature is taken to be

$$h_{aw} = h_0 - \frac{{u_e}^2}{2}(1 - r),$$

with the recovery factor approximately equal to the square root or one third power of the Prandtl number for laminar and turbulent flow respectively. The Reynolds number $Re(x)$ is evaluated using the distance from the tip, and either at the conditions at the edge of the boundary layer, $Re(x) = \rho_e u_e x/\mu_e$, or at the reference temperature, $Re^*(x) = \rho^* u_e x/\mu^*$. The reference temperature is defined by

$$\frac{T^*}{T_e} = 0.5 + 0.039 M_e^2 + 0.5 \frac{T_w}{T_e},$$

where the subscripts define the respective temperatures and M is the Mach number.

An important effect in high-enthalpy heat flux is the increase that may occur because of recombination at the model surface. This can be observed in our results by comparing laminar heat flux values taken at low pressure with values taken at high pressure. The higher pressure accelerates the recombination rate, so that a larger effect is expected. Fig. 4 shows the heat flux distributions on the cone at medium h_0 and low p_0 as well as at high h_0 and high p_0, in comparison with theoretically expected values. Clearly, the effects of wall recombination appear at the higher temperature and pressure. The large scatter of the data is associated with the uncertainties of deducing heat flux from surface temperature history. However, there is a systematic variation in these data also, which could be reduced by individual calibration of each of the gauges.

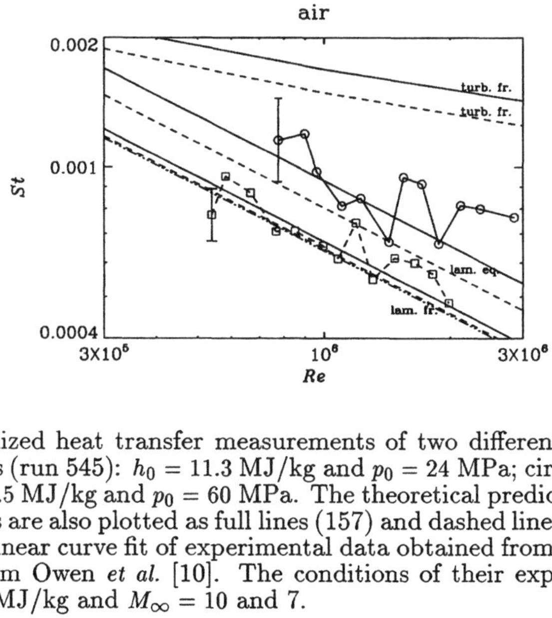

Fig. 4. Normalized heat transfer measurements of two different shots in T5. Squares (run 545): $h_0 = 11.3$ MJ/kg and $p_0 = 24$ MPa; circles (run 157): $h_0 = 18.5$ MJ/kg and $p_0 = 60$ MPa. The theoretical predictions for these air runs are also plotted as full lines (157) and dashed lines (545). ——··· ——: linear curve fit of experimental data obtained from DiCristina [9] and from Owen *et al.* [10]. The conditions of their experiments were $h_0 < 1$ MJ/kg and $M_\infty = 10$ and 7.

Fig. 5 illustrates how the heat flux data may be used to determine the transition location. The fairly sudden increase of the heat flux at transition defines the transition Reynolds number well. In later plots of the transition Reynolds number, the values shown were obtained by systematic fits of three straight lines to the experimental results.

3.2 Flow Visualization

In order to illustrate the effectiveness of the resonant enhancement, Fig. 6 and Fig. 7 show a transitional and a turbulent boundary layer respectively. To our knowledge, these pictures are the first such illustrations of the structure of high-enthalpy boundary layers. In Fig. 6 the flow begins with a smooth streakline of sodium at the left, which becomes wavy as it proceeds downstream, range of distances from the surface. In order to remedy this effect, we intend to improve the method by visualizing fluid in a sheet in the near future.

Fig. 7. Resonantly enhanced flow visualization of a turbulent boundary layer

3.3 Transition Reynolds Number

The transition Reynolds number that is observed in our experiments at enthalpies up to 12 MJ/kg in air and up to 15 MJ/kg in nitrogen is in the range 2.5 - 3.1 million. Above these enthalpies the value rises fairly rapidly, so that we are not able to reach transition even at 4 million in air at $h_0 = 14$ MJ/kg. Since the behavior of laminar boundary layers may be rendered almost independent of Mach number, specific heat ratio and wall temperature ratio if the material properties are evaluated at the reference temperature, we decided to plot the transition value of Re^* against enthalpy. The result is shown in Fig. 8. The previous data scatter over almost a decade (they come from different kinds of facilities), and range only up to h_0 = 3.5 MJ/kg, which is where our results start. There appears to be a steady increase of Re^*_{tr} with enthalpy, which is significantly stronger in air than in nitrogen. Clearly this exponential increase cannot continue, and the curves have to reach some kind of limit when h_0 exceeds the dissociation energy of the gas. However, this limit has not been reached, and the slope may still be steepening as indicated by the flagged symbols that represent cases where transition was not reached.

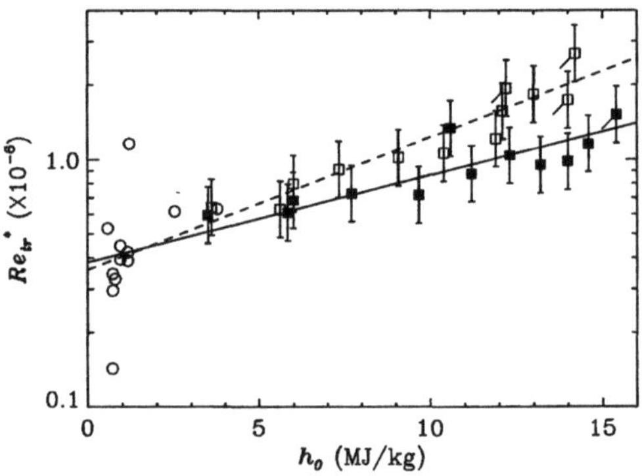

Fig. 8. Dependance on enthalpy of transition Reynolds number evaluated at the reference temperature. Open squares: air, filled symbols: nitrogen. The open circles indicate collections of previous data, Demetriades [8], DiCristina [9].

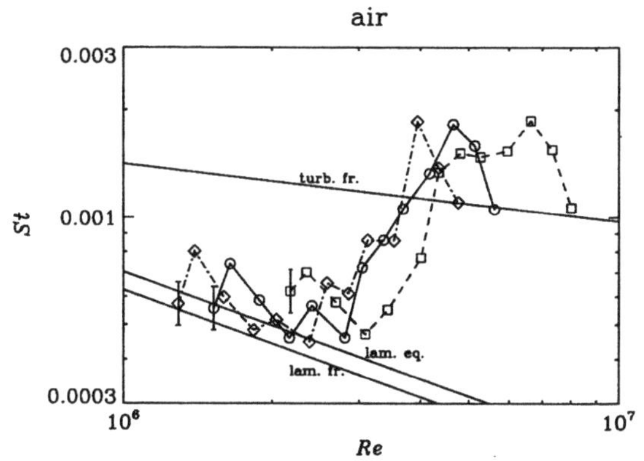

Fig. 5. Normalized heat transfer measurements of three different shots in T5. Squares (run 541): $h_0 = 7.3$ MJ/kg and $p_0 = 55$ MPa; circles (run 549): $h_0 = 9.0$ MJ/kg and $p_0 = 52$ MPa; diamonds (run 556, refer to run 537 in Table 2): $h_0 = 10.4$ MJ/kg and $p_0 = 55$ MPa. The theoretical predictions for run 549 are also plotted.

and finally becomes transitional as it leaves the picture on the right. The horizontal extent of the picture is approximately 150 mm. The wavelength of this wave is approximately 15 mm, which, with half of u_e gives a frequency of 150 kHz. This corresponds to the estimated Tollmien-Schlichting frequency at our conditions, and is very much lower than the estimated frequency of the second mode (1.5 MHz). In fact, the second-mode frequency is so high, that mechanical vibrations and flow oscillations can be practically ruled out as noise sources. Even the nozzle-wall boundary layer noise probably has a lower cut-off. On the other hand, the noise level at 150 kHz may be expected to be extremely high at our conditions. The fact that we observe the TS mode is therefore very possibly connected with the noise spectrum of the facility.

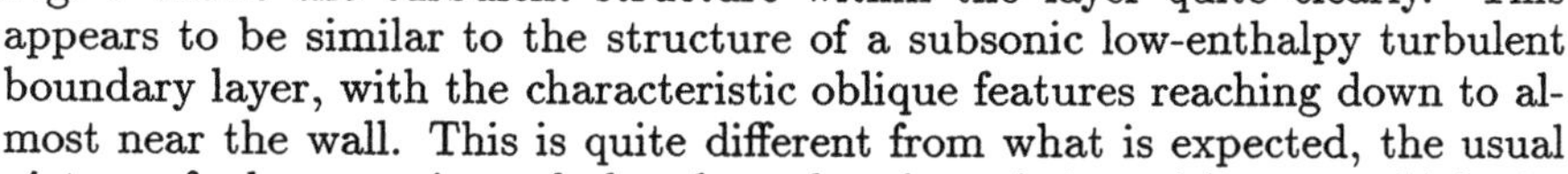

Fig. 6. Resonantly enhanced flow visualization of a transitional boundary layer.

Fig. 7 shows the turbulent structure within the layer quite clearly. This appears to be similar to the structure of a subsonic low-enthalpy turbulent boundary layer, with the characteristic oblique features reaching down to almost near the wall. This is quite different from what is expected, the usual picture of a hypersonic, turbulent boundary layer being with a very thick viscous sublayer. The interpretation of the picture is, however, not unambiguous. It may be, of course, that the hypervelocity boundary layer with an extremely cold wall is very different in structure from the (perhaps erroneous) picture we have all learned from cold hypersonics. On the other hand, we have to be careful with this interpretation, because the visualization method is line-of-sight integrating, and we are dealing with an axisymmetric model. Each point on the image has therefore been produced from information located at a

The difference in the slopes of the lines for air and nitrogen suggest that one should plot Re_{tr}^* against a dimensionless enthalpy, normalizing h_0 by the dissociation energy of the gas. Since the oxygen in air has a lower dissociation energy than nitrogen, this goes in the right direction to make the two curves collapse. However, it is not clear how one should assign a single representative dissociation energy to the gas mixture air.

To test the idea that the slope of the line depends on a characteristic energy of the gas, we did a few experiments with carbon dioxide and with helium. The carbon dioxide results give a much steeper increase of Re_{tr}^* with h_0 than air or nitrogen (giving $Re_{tr}^* = 6 \times 10^6$ at $h_0 = 6$ MJ/kg), and the helium does not show a change at all even at h_0=26 MJ/kg.This is consistent with the much lower dissociation energy of carbon dioxide, and the very much higher ionization energy of helium. These results are speculative at this point, more data will be needed to test the idea. Also, in the case of helium, transition was not reached, and the contoured nozzle was operated significantly off design.

Acknowledgements

This work was supported by AFOSR URI grant F49620-93-1-0338 (Dr. J. Tishkoff) and AFOSR grant F49610-92-J-0110 (Dr. L. Sakell).

References

[1] Demetriades, A.: Hypersonic viscous flow over a slender cone, Part III: Laminar instability and transition, *AIAA 7th Fluid and Plasma Dynamics Conference, AIAA Paper* 74-535, Palo Alto, California 1974.

[2] Kendall, J. M.: Wind tunnel experiments relating to supersonic and hypersonic boundary-layer transition, *AIAA Journal* **13** (3) 1975, 290-299.

[3] Potter, J. L.: Boundary-layer transition on supersonic cones in an aeroballistic range, *AIAA Journal* **13** (3) 1975, 270-277.

[4] Stetson, K. F.: On non-linear aspects of hypersonic boundary layer stability, *AIAA Journal* **26** (7) 1988, 883-885.

[5] Malik, M. R. and Anderson, E. C.: Real gas effects on hypersonic boundary layer stability, *Physics of Fluids A* **3** 1991, 803-821.

[6] Mack, L. M.: Boundary layer stability, *Rept.* 900-277 *Rev.* A, 1969, *Jet Propulsion Laboratory*, Pasadena, California.

[7] Reshotko, Eli: Boundary Layer Instability, Transition and Control. The 1994 AIAA Dryden Lecture in Research. *AIAA Paper* 94-0001, 1994.

[8] Demetriades, A.: Laminar boundary layer stability measurements at Mach 7 including wall temperature effects, Rept. AFOSR-TR-77-1311, Air Force Office of Scientific Research, Washington, D. C. 1977

[9] DiCristina, V.: Three-dimensional laminar boundary-layer transition on a sharp 8° cone at Mach 10, *AIAA Journal* **8** (5) 1970, 852-856.

[10] Owen, F. K., Horstman, C. C., Stainback, P. C.: Comparison of wind tunnel transition and freestream disturbance measurements, *AIAA Journal*

Session F: Poster Session I (Fundamentals, Flow Measurements, Controls, Compressible Flows and Turbulent Spot)

Turbulence in Pulsatile Flow through the Tube with Elastic Wall

M. Arakawa and Y. Matunobu
Department of Physics, Keio University
4-1-1, Hiyoshi, Kohoku-ku, Yokohama 223 Japan

Summary

Effects of the radial wall motion of a thin-walled straight rubber tube conveying pulsatile flow of water on the fluctuations of flow velocity were studied experimentally using hot-film anemometer. In order to detect the radial motion of outer wall of tube, a non-contacting gap sensor was used. It was shown that the compliance of wall of test tube depends on the transmural pressure of tube(i.e. inner minus outer pressures). The results show that the correlation between fluctuation of flow velocity and the wall motion of tube is classified roughly into two parts according to the mean transmural pressure. In the region that the mean transmural pressure is positive, the tube wall deforms elastically, and motion of tube wall suppresses the disturbance of flow. While, in the region where the mean transmural pressure is zero or negative, the motion of tube wall becomes more complicate. Non-elastic deformation of tube wall is accompanied by partially elastic deformation and the wall motion increases the disturbance of flow.

Key words : Turbulence, Elastic tube, Wall motion, Pulsatile flow, Compliance.

1. Introduction

The instability of flow through compliant flexible tube has been of interest to many investigators. To research this problem is not only of fundamental interest in fluid mechanics but also of practical importance in engineering fields. Further, in physiological fluid dynamics, the compliance of blood vessel is considered to be of vital importance to living body and it is significant to understand the dynamical behaviour of the blood flow through a vessel with compliant wall.

In our previous work [1], it was shown experimentally that in the steady flow through flexible rubber tubes, the critical Reynolds number at laminar-to-turbulent transition was higher than that in the flow through rigid tubes and the fluctuations of flow velocity were smaller than those of rigid tubes. Further, at higher Reynolds numbers, the outer wall of tube conveying water vibrated passively with small amplitude caused by the fluctuating flow and its frequency range was negatively correlated to that of fluctuation of flow velocity. Further, experimental studies were carried out in the state that the outer wall was activated by the external stimulus during passive vibration [2]. Then, it was shown that the fluctuation of flow velocity had small frequency range compared with that of no stimulation and that when the tube wall vibrated with the most susceptible frequency in response to the tube wall, the flow disturbance attained a minimum value. Therefore, it could be inferred that the wall vibration played an important role to control the flow disturbance.

Many experimental studies of blood flow have been investigated on a steady flow through rigid tube made of glass or synthetic regins as a model. However, the walls of artery have elastical character and it has been observed that the diameter of large arteries changes by $\pm 5\%$ over one heart rate in blood flow [3]. Then, it is considered to be impossible to disregard the effects of radial wall motion of blood vessel due to pulsatile flow in understanding the dynamical behaviour of blood flow. Main purpose of this paper, consists in further clarifying experimentally the effects of radial wall motion of compliant tube on fluctuation of flow velocity using thin-walled rubber tubes conveying pulsatile flow of water.

2. Experimental Procedure

Figure 1 shows schematically the arrangement of measurement system. Observations were made on the transitional laminar-to-turbulent flows as well as full developed turbulent flows using soft latex tube of 35 cm length, 10 mm inner diameter and 0.5 mm wall thickness. Polyurethane foam sheet on the adjustable platform was used to level and the test section was alined horizontally. Steady flow of water was supplied to the test section by constant head tank through entrance and the sinusoidal pressure pulses to be superimposed on this steady flow was generated by piston-cylinder system connected with variable-speed motor. The flow rate was measured with the flow meter located downstream to the pulsating pump. The pressure difference across the test section was measured with two electrical pressure transducers.

Fluctuation of flow velocity was recorded with hot-film probe positioned at the tube outlet on the center line of tube. To detect the displacement of tube wall, a non-contacting gap sensor near the tube surface was used at the middle of tube length and a thin aluminum foil (6×15mm) was pasted at the measured position as a electric conductor. Displacement of the outer wall of tube was measured by detecting the eddy currents corresponding to the gap width between sensor and the aluminum foil.

Reynolds number $Re = \rho d_e U_m/\mu$ was determined based on the entrance diameter d_e, the mean flow velocity U_m, estimated from flow rate, and the density ρ and the viscosity μ. The amplitude of the stroke volume of pulsating pump was 5 ml and frequency of pulsation was fixed at 1 Hz. Then, the range of relative amplitude of pulsating component of mean flow, $\Delta Q/Q_m$, was from 0.3 to 1. Frequency parameter $\alpha[= (d_e/2)(\sqrt{2\pi f/\nu})]$, characterized pulsatile flows, was 13 where f is the frequency of pulsation, and $\nu(= \mu/\rho)$ the kinematic viscosity of water.

3. Results and Discussion

The elastic modulus of tube wall depends on transmural pressure of tube(i.e. inner minus outer pressures). The static characteristics of rubber tube, u sed in this study, is shown in Fig. 2, where the diameter of tube d, estimated from volume change, is plotted against the transmural pressure P_t. Typical cross-sectional shapes are illustrated along the curve. When the transmural pressure is nearly zero, the tube begins to collapse whose circular cross-sectional shape changes into elliptical one. As the transmural pressure decreases down to a negative value, the deformation of tube becomes large, and the shape of cross section is like a cocoon or barbell. In this region which the transmural pressure is zero or negative value, tube wall may behave like non-elastic body. While when the transmural pressure is positive value, the tube has a circular cross section and expands gradually with increasing transmural pressure and the elasticity of tube wall becomes larger.

Typical records of pulsatile flow through rubber tube are shown in Fig. 3, at Reynolds number 5×10^3, where P_u represents entrance pressure, P_d the exit pressure, d the outer diameter of tube, Q the flow rate, and ΔU the velocity fluctuation of flow. At the same Reynolds number, the flow shows different behaviour according to downstream(distal) resistance of flow adjusting by the valve. In this experiment, the wall motion was varied over a considerably wide range by controlling a distal resistance of flow. In this figure, the distal resistance becomes larger from left to right and the changing process of deformation in cross section of tube is shown schematically at lower part of the figure.

The turbulent intensity was estimated based on the ensemble average of fluctuations of flow velocity. Figure 4 shows how to define the turbulent intensity for pulsatile flow. The top in this figure, No.1 is a typical trace of instant flow velocity U and No.2 its ensemble average $\langle U \rangle$ for 32 data segments. No.3 is the difference between No.1 and No.2, that is, the turbulent velocity. The square root of time average of this turbulent velocity is adopted as the mean turbulent intensity. Mean turbulent intensities at various Reynolds numbers are shown in Fig. 5, where they increase with increasing Reynolds numbers and at the same Reynolds number, they increase as the distal resistance becomes smaller. If these data are re-plotted against the mean internal pressure difference $\sqrt{\overline{(\Delta P)^2}}$, it is the square root of time average of the pressure difference ΔP,where $\Delta P = P_u - P_d$, it can be shown that the data for different distal resistance fall nearly on a common curve as shown Fig. 6. This fact can be interpreted that the pressure drop at high Reynolds numb ers is caused mainly in the energy dissipation due to turbulent motion irrespective of flow conditions near the wall.

Figure 7 shows the fluctuation of wall motion at various mean transmural pressures $\sqrt{\langle P_t \rangle^2}$, defined by the square root of time average. The fluctuations of wall motion are represented by the ratio of the amplitude Δd (peak to peak) of the varying diameters to the mean diameter. Acoording to this figure, the fluctuation of wall motion has a decreasing tendency with increasing transmural pressure of tube.

The turbulent intensity versus the fluctuation of wall motion is shown in Fig. 8. The turbulent intensity decreases with the increase of fluctuation of wall motion up to about 8%, and beyond this value, the turbulent intensity increases, on the contrary. This situation can be classified roughly into two parts according to the mean transmural pressure and can be explained as follows (Fig. 9).

When the fluctuation of wall motion is about 8%, the turbulent intensity reaches a minimum value. This point corresponds to point A shown in Fig. 9, which is the same as Fig. 2 at which the mean transmural pressure is nearly zero. The deformation character of tube wall is different in the right and left sides to the point A. If the volume pulse(i.e. diameter pulse) due to pulsatile flow passes the point A, the pressure pulse is severely modulated like (a). When the variation range of tube diameter happens to be in the right side to the point A, the tube wall will deform elastically. While, when the variation range of tube diameter happens to be in the left side to the point A, the tube wall is supposed to deform non-elastically.

The first range(1); in the range of fluctuation of wall motion up to 8%, the mean transmural pressure of tube is positive. The second range (2); in the range of the fluctuation of wall motion exceed 8%, the mean transmural pressure is zero or negative. In the range(1), the tube wall conveying pulsatile flow deforms elastically, and the turbulent intensity of flow decreases with the increase of the fluctuation of wall motion. In this region, elastic wall motion seems to suppress the flow disturbance, like in the case of steady flow. While, in the range(2), the wall motion due to the pulsatile flow is more complicated and the non-elastic deformation is accompanied by partially elastic deformation. In this region, the wall motion excites the flow disturbance and the turbulent intensity increases with increasing of the fluctuation of wall motion.

It may be a rough idea to interpret the dynamical states in terms of the static curve. But, the essential points of the problem can be obtained qualitatively at least by this idea.

In the present investigation, the effects of wall motion of thin-walled straight flexible rubber tube conveying pulsatile flow on fluctuation of flow velocity were carried out. The results are as follows.

(1) The correlation between the wall motion of tube due to pulsatile flow and the

fluctuation of flow velocity can be classified roughly into two parts according to the transmural pressure. At the point which the wall motion is near 8%, represented by the ratio of the diameter amplitude Δd to the mean diameter d_{m}, the turbulent intensity attains a minimum value and this point is considered to be critical in separating two parts.
(2) First part : in the range which the wall motion of tube up to 8%, the mean transmural pressure is positive and the turbulent intensity of flow decreases with increasing wall motion. In this region, the tube wall deforms elastically, and wall motion seems to suppress the flow disturbance.
(3) Second part : in the range which the wall motion exceeds 8%, the mean transmural pressure is zero or negative and the turbulent intensity of flow increases with increasing wall motion. In this region, the wall motion of tube is more complicated and the non-elastic deformation is accompanied by partially elastic deformation, and the wall motion excites the flow disturbance.

Then, the our experimental results show that in the pulsatile flow through the elastic tube, the tube wall compliant to the flow pulsation plays an important role in the control of the disturbance of flow.

References

[1] Arakawa, M. and Matunobu, Y. : Transitional Process of Turbulent Flow through Flexible Tubes, Fluid Dynamics Reserch, 6 (1990) 107-110.
[2] Arakawa, M. and Matunobu, Y. : Effects of Active Wall Motion of Flexible Tubes on Flow Disturbance, ASME, FED-114 (1991) 191-195.
[3] McDonald, D.A. : Blood Flow in Arteries. Edward Arnold, London, (1974)

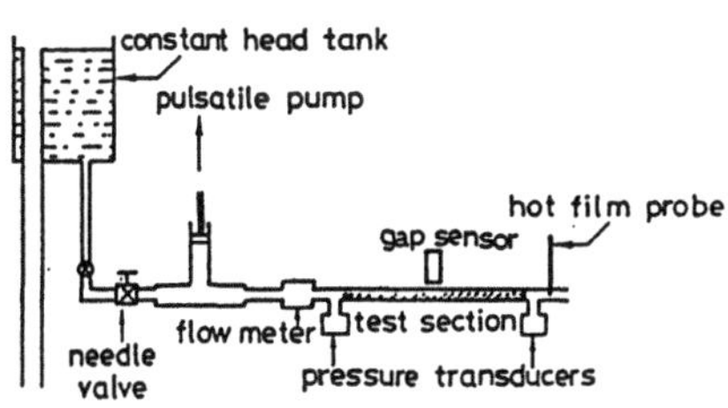

Fig.1. Schematic view of experimental apparatus.

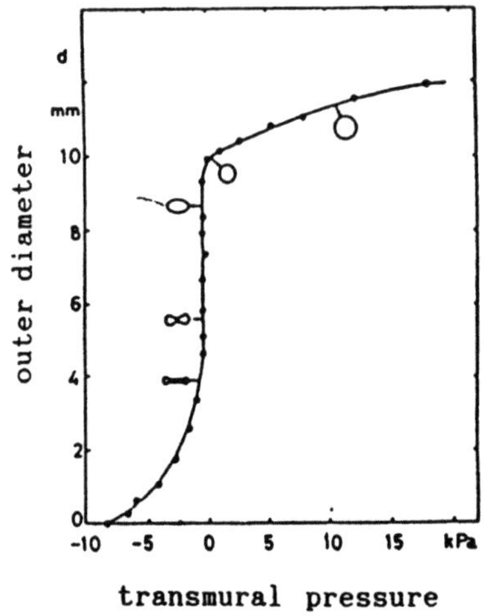

Fig.2. Static characteristics of rubber tube.
Typical cross-sectional shapes are illustrated.
d : tube diameter,
Pt : transmural pressure.

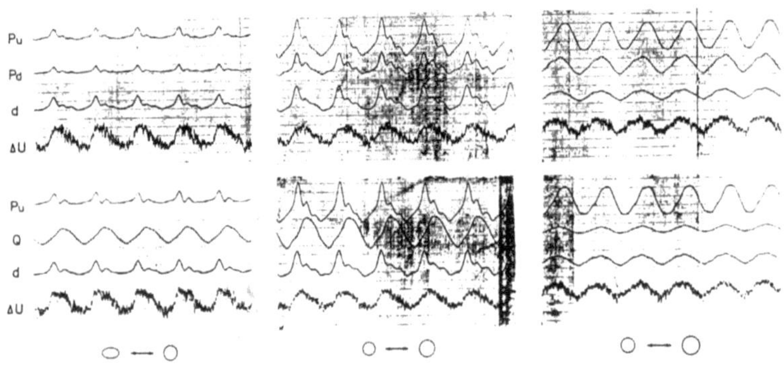

Fig.3. Typical records for pulsatile flow ($Re = 5\times10^3$).
Pu : upstream pressure, Pd : downstream pressure,
d : tube diameter, Q : flow rate, ΔU : fluctuation of flow velocity.

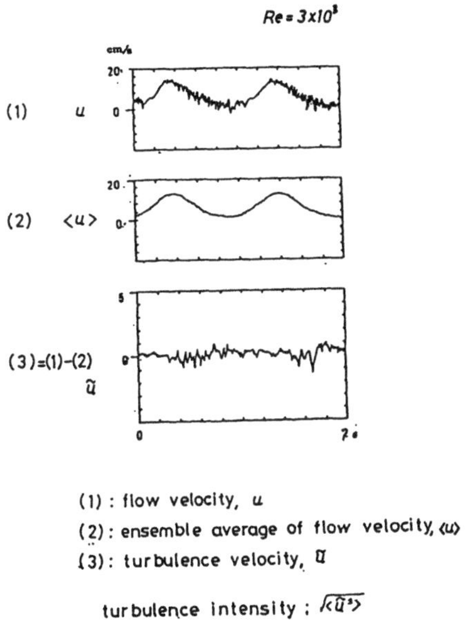

Fig.4. Turbulent intensity.

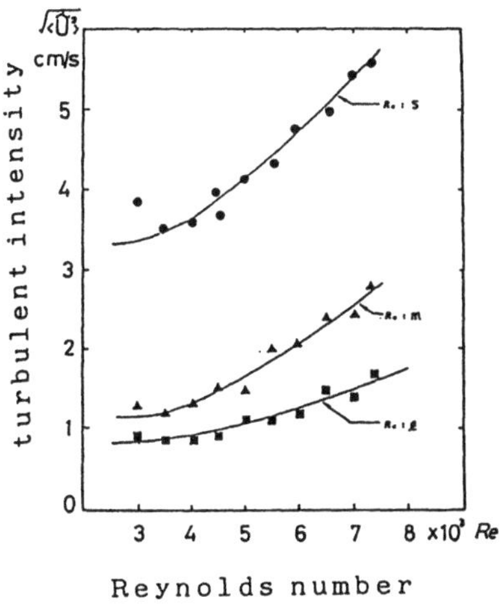

Fig.5. The turbulent intensity at various Reynolds numbers. Rd : downstream resistance.

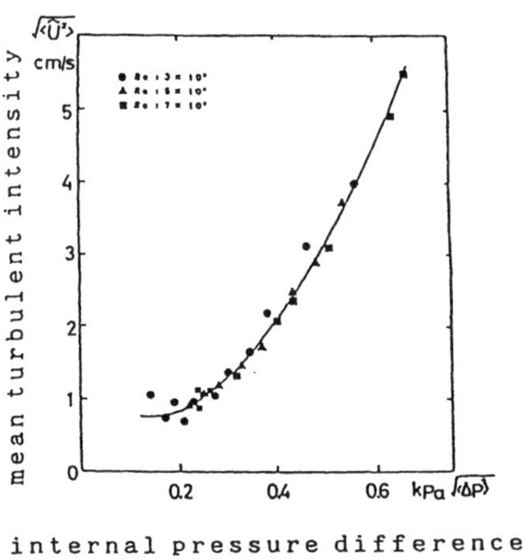

Fig.6. The turbulent intensity at various internal pressure differences.

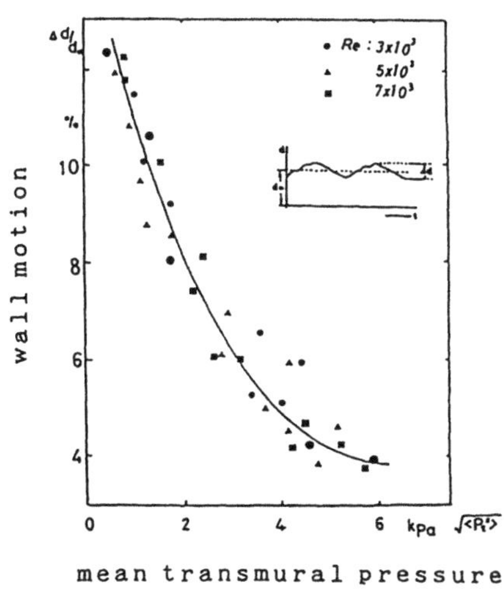

Fig.7. Wall motion at various mean transmural pressures.

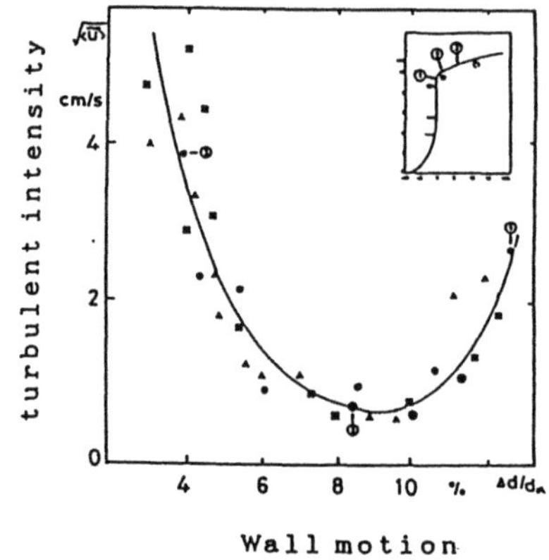

Fig.8. The turbulent intensity at various wall motions.

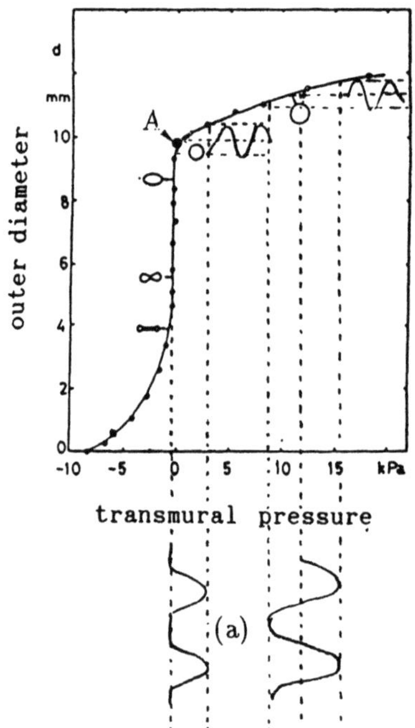

Fig.9. Interpretation of the relation between motion and compliance of tube wall.

Experimental Study of Boundary Layer Suction Effects on Leading Edge Contamination along the Attachment Line of a Swept Wing

J.C. Juillen - D. Arnal
CERT/ONERA - Aerothermodynamics Department
2 avenue E. Belin - BP 4025 - 31055 Toulouse Cedex (FRANCE)

Summary

An experimental study on a swept wing is performed to determine the effects of boundary layer suction on the leading edge contamination. The results are compared with the direct numerical simulations of Spalart.

Key words : attachment line, contamination, suction.

1. Introduction

The so-called leading edge contamination is likely to occur when a swept back wing is attached to a solid wall on which the flow is turbulent (fuselage, wind tunnel wall, ...). Many experimental data have shown that the boundary layer developing along the leading edge remains laminar provided the attachment line Reynolds number $\overline{R}$ is lower than a certain critical value. This Reynolds number is formed with the velocity along the attachment line $W_e = Q_\infty \sin\varphi$ and the length scale $\eta = \left[\frac{\nu}{dU_e/dX}\right]^{1/2}_{X=0}$, $\overline{R} = \frac{W_e \eta}{\nu}$. U_e is the external boundary layer velocity in the direction normal to the leading edge and X is the curvilinear abscissa in the same direction. If $\overline{R}$ exceeds a critical value, the attachment line flow is contaminated by the large turbulent structures coming from the wall. This phenomenon is fully non linear and constitutes a typical example of "bypass", in the sense that the linear mechanisms which control "natural" transition do not play any role. Pfenninger [1] observed that the critical value of $\overline{R}$ was close to 250. This very simple criterion was verified by many authors [2]. If $\overline{R}$ is greater than 250, turbulence spreads over the whole wing and the benefits of the laminar flow control systems (shaping, suction, ...) are lost.

As a consequence, it is necessary to develop specific tools in order to delay the onset of leading edge contamination. One possibility is to use passive devices such as Gaster bumps which are able to delay the onset of leading edge contamination up to values of $\overline{R}$ of the order of 400 [3]. Another solution is to apply a localized suction along the attachment line of a contaminated wing. At present, no available experimental data exist ; only some results from direct numerical simulations were published by Spalart [4]. These results are presented in figure 1. K is the suction parameter defined by : $K = \frac{V_W \overline{R}}{Q_\infty \sin\varphi}$. It seems that when K is lower than -1, an asymptotic value close to $\overline{R} = 370$ is reached.

The objective of the present investigation is to obtain experimental data with spanwise uniform suction velocity distribution, which could support the numerical results published by Spalart. This work was supported by Airbus Industrie.

2. Experimental Set-Up - Instrumentation

2.1. Wind Tunnel

The wind tunnel (devoted to transition studies in threedimensional incompressible flows) is of open return circuit type. The test section is 2.5 m long, 0.6 m wide and 0.35 m high. The maximum velocity in the empty test section is about 85 ms^{-1}, with a turbulence level close to 0.07 % and a noise level $\tilde{p}'/q_\infty$ close to 0.5 %, in the frequency bandwidth 3 Hz - 25 kHz.

2.2. Suction Model

It consists of a circular cylinder nose (D = 0.120 m) which constitutes the windward face of the model, whilst the leeside has a triangular section in order to prevent the formation of unsteady wakes (figure 2). The span is 1.5 m and the chord is equal to 0.36 m, so the aspect ratio (chord/diameter) is equal to 3. The model can be placed in the wind tunnel test section at zero angle of attack, with sweep angles of 50° and 60°. Six independent suction chambers (200 mm long and 20 mm wide) are located along the leading edge. Each of them is connected to a sonic nozzle flowmeter. The porous wall is a perforated nickel sheet , 0.1 mm thick. The holes diameters are 0.06 mm and their centre-to-centre spacing is 0.32 mm. To increase pressure loss and to prevent "blowing off" due to the pressure gradient in the direction normal to the leading edge, moss is introduced in the chambers on the internal side of the perforated sheet. Each chamber is equipped with a static pressure tap. Two rows of static pressure holes (diameter = 0.4 mm) are drilled at Z/D = 3.06 and Z/D = 6.83 respectively.

2.3. Measurement Techniques

Different kinds of measurements were performed during the tests :

- static pressure distributions,
- suction mass flow rates,
- hot films records (wall shear stress fluctuations only),
- sublimation visualisations.

The hot films are glued on the model at different Z/D spanwise locations off the suction area (X = 16 mm), X is the curvilinear abscissa measured from the attachment line. They are operating in the constant temperature (130°C) mode. It was verified that for a given Z/D position, the hot film gives the same results as another one glued on the attachment line.

3. Measured Velocity Distributions

A distribution of the velocity W_e measured on the attachment line in the spanwise direction Z/D (Z/D = 0 at the wing wall junction) is plotted in figure 3. The curve exhibits a rather strong positive gradient in the region between Z/D = 0 and Z/D = 2.5. This gradient becomes less intense for values of Z/D greater than 2.5. The infinite swept wing conditions ($dW_e/dZ = 0$) are never exactly reached on the model due to the fact that the span length is too short. The following table gives, for two values of Z/D, the values of the effective swept angle φ_{eff} deduced from the measured values of W_e, $\varphi_{eff} = \sin^{-1} \frac{W_e}{Q_\infty}$.

$\varphi = 50°$		$\varphi = 50°$	
φ_{eff} Z/D = 3.06	φ_{eff} Z/D = 6.83	φ_{eff} Z/D = 3.06	φ_{eff} Z/D = 6.83
57.3°	61.5°	71°	75°

The measured velocities normal to the leading edge (U_e) distribution are plotted as function of X in figure 4. It is possible to draw a straight line and to determine the velocity gradient $\left[\frac{dU_e}{dX}\right]_{X=0}$ on the attachment line. The determination of this gradient and of the values of φ_{eff} allows to calculate the values of $\overline{R}$. Due to the high blockage ratio value (0.29) of the model, the maximum mean velocity in the test section is limited to 65 ms^{-1}, so the maximum value of $\overline{R}$ is near 500.

4. Results

4.1. Contamination Determination

For given values of V_w, the instantaneous signals delivered by each hot film are recorded, then the rms values (rms = root mean square) are plotted as functions of Q_∞ (figures 5 and 6). The values of Q_∞ at which the first turbulent spots appear correspond to a strong change in the slope of the curves, determining the values of $Q_{\infty cont.}$ at which leading edge contamination occurs (see also instantaneous records).

The results are summarized in figures 7 and 8 for the two values of the sweep angle (50° and 60°). It can be observed that suction has a large effect : for the last station (Z/D = 6 or Z/D = 6.9) with the upper values of V_w, the values of the velocity at which contamination occurs are about three times larger than the value without suction. Due to the fact that the infinite swept wing conditions are not respected, the values of Q_∞ at which the contamination occurs are lower near the wing-wall junction than at the tip of the wing.

4.2. Visualisations

Sublimation visualisations have been performed in order to detect the transition location when the attachment line is not contaminated, i.e. when suction is applied. An example of results is shown in figure 9. An interpretation of this visualisation is given in figure 10. The white areas correspond to a laminar state of the boundary layer and the black ones to a turbulent state. The non-uniformity in spanwise direction is clearly apparent. The "natural" transition location by crossflow instability is in qualitative agreement with the position predicted using an empirical criterion. These experiments demonstrate that the "new" laminar boundary layer seems to be "clean". However more detailed measurements are necessary to clarify this point.

4.3. Comparison with Spalart Results

The experimental values of $\overline{R}$ obtained for first turbulent spots onset on the last hot film are plotted in figure 11 as function of the suction parameter K ; the horizontal dashed line indicates the critical value $\overline{R} = 250$. The bars represent the deviation of Spalart DNS results (maxSpalart and minSpalart). For values of K lower than -1, the agreement between present experiments and DNS results is good, but the asymptotic trend indicated by Spalart is not evident.

5. Conclusions

These experiments demonstrated that suction around the leading edge is efficient to prevent attachment line contamination. They also demonstrated the existence of a "natural" transition by crossflow instability downstream of the relaminarized region.

The agreement with Spalart numerical simulations is fairly good ; however the asymptotic trend ($\overline{R} = 370$) indicated by Spalart is not obvious in the experiments.

References

[1]Pfenninger, W., "Flow Phenomena at the Leading Edge of Swept Wings" - Recent Developments in Boundary Layer Research, Part IV - AGARDograph 97 (May 1965)

[2] Arnal, D., "Boundary Layer Transition : Prediction, Application to Drag Reduction" - AGARD/FDP VKI Special Course on Skin Friction Drag Reduction, Brussels, Belgium (2-6 March 1992)

[3] Reneaux, J., Preist, J., Arnal, D., Juillen, J.C., "Design and Testing of Two Gaster Bumps for the A320 Laminar Flow" - Internal Report (September 1992)

[4] Spalart, P.R., "Direct Numerical Study of Leading Edge Contamination", AGARD CP N° 438 (1988)

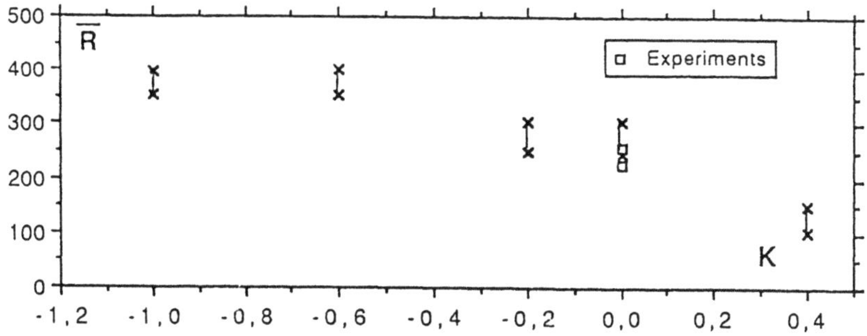

Figure1.SPALART direct numerical simulation results

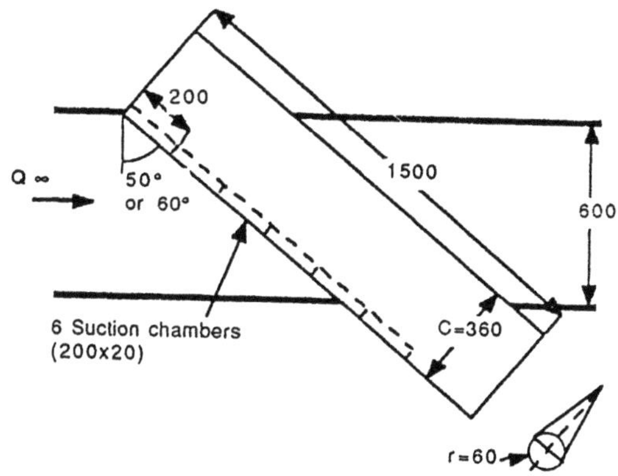

Figure2. Suction model dimensions(mm)

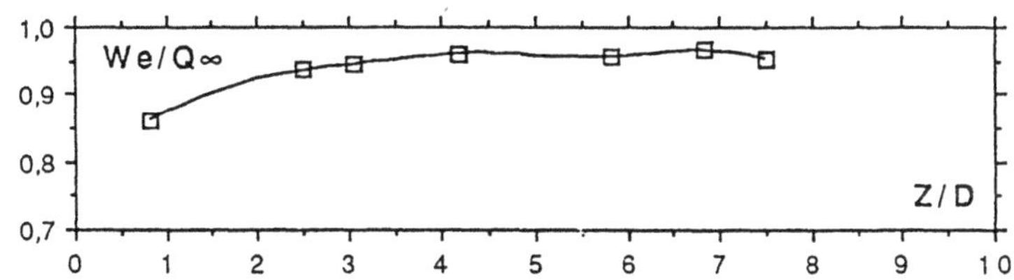

Figure3.Velocity distribution along the attachment line

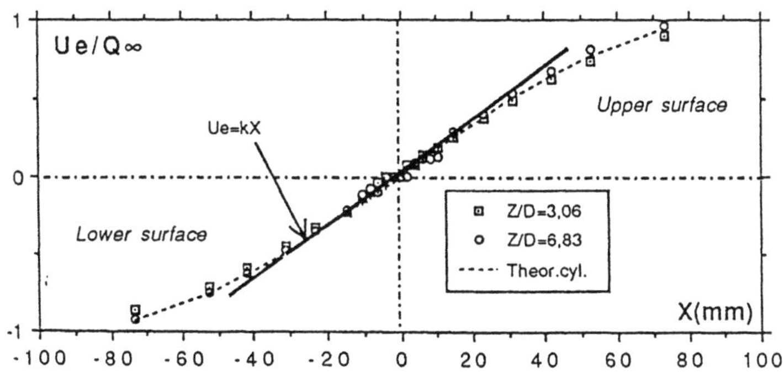

Figure4.Distribution of the velocity normal to the leading edge

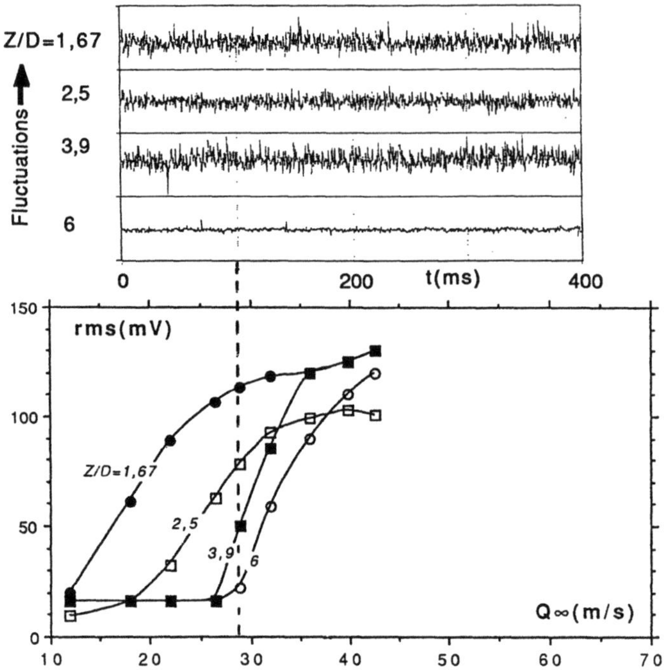

Figure5.rms fluctuations evolutions.(Vw=0)

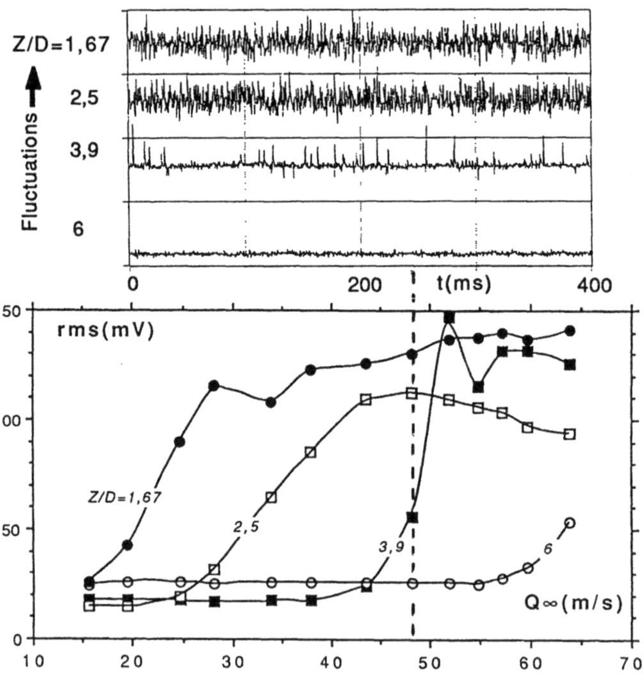

Figure6.rms evolutions of fluctuations.(-Vw=0,05m/s)

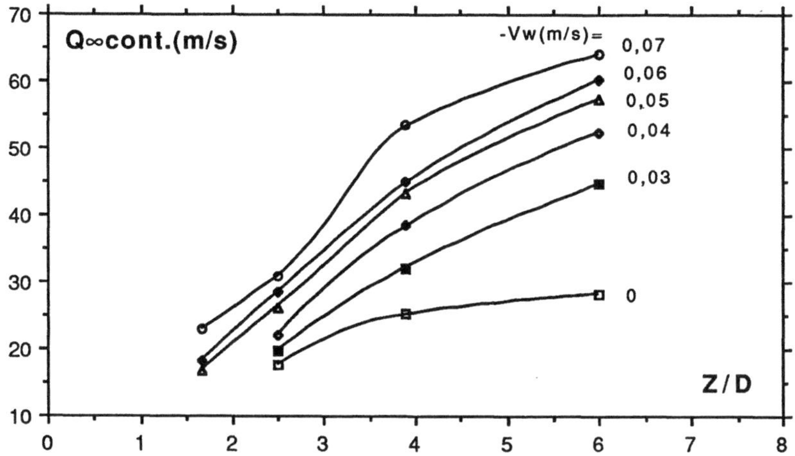

Figure7.Evolutions of Q∞cont.(phi=50°)

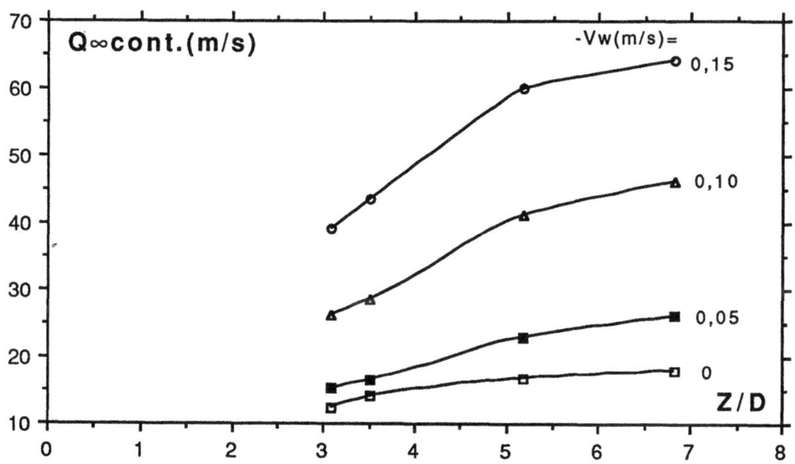

Figure8.Evolutions of Q∞cont.(phi=60°)

Figure9.Sublimation visualisation.
(phi=50°;Q∞=40m/s;-Vw=0,07m/s)

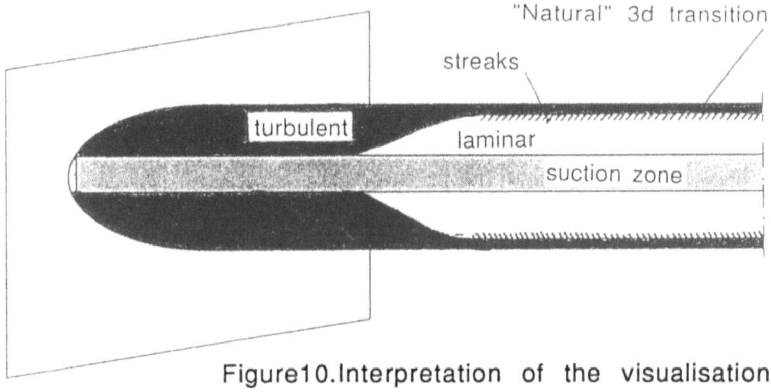

Figure10.Interpretation of the visualisation

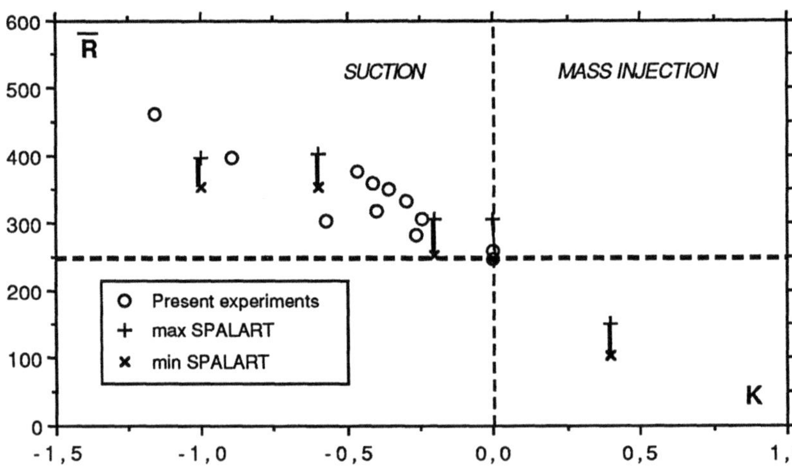

Figure11.Comparison between experiments and SPALART calculation

Nonlinear Development of Waves in the Supersonic Boundary Layer

S.A.Gaponov, A.D.Kosinov, I.I.Maslennikova, N.V.Semionov, S.G.Shevel'kov
Institute of Theoretical and Applied Mechanics,
Russian Academy of Sciences, 630090 Novosibirsk, Russia

Summary

In the paper theoretical and experimental results of investigation of the linear and nonlinear development of waves in the supersonic boundary layer are given.

Key words : Supersonic boundary layer, linear and nonlinear stability.

1. Introduction

Last years the great advances were achieved in current knowledge of transition mechanisms in low-speed boundary layer flows [1]. There are two distinct mechanisms that have been identified for low-disturbance environments: nonlinear evolution of sufficiently intensive 2-D Tollmien-Schlichting (T-S) wave (K-type transition), and resonant interaction of low-amplitude 2-D TS wave with 3-D background subharmonics (S-type transition). The latter case was investigated, in particular, on the bases of first-order weakly nonlinear theory [2], and good quantitative agreement with practically all available observations and direct Navier-Stokes simulations was achieved. In contrary, transitional mechanisms in supersonic boundary layers remain poorly understood, except linear theory [1,3]. Due to problems in experimental investigations, the validation of proposed theoretical models is very difficult.

Now a limited number of theoretical and numerical works on nonlinear instability of compressible flows is available [4-8]. Some of them [6,7] concerns secondary instability, the relevance of which to supersonic Mach numbers, where 3-D disturbances are more unstable than 2-D ones, is completely unclear. Also some direct numerical simulations [5,7,8] became available. It may be noted that results [8] are in qualitative correspondence with experiments [3,9]. In this work a nonlinear phenomena in the supersonic boundary layer are studied both experimentally and theoretically.

2. Theory Remarks

For weakly nonlinear interaction of disturbances the solution of the nonstationary Navier-Stokes equations can be taken as a sum of stationary part and a small but finite nonstationary quantity. We shall study the disturbance development in the boundary-layer for large Reynolds numbers. In this case the stationary solution is changed slowly in the main flow direction. In truth it depends only on the normal to a surface co-ordinates. The full solution has the form:

$$\mathbf{Q}(x,y,z,t)=\mathbf{Q}_0(\varepsilon_1,x,y)+\varepsilon\mathbf{Z}(x,y,z,t) \quad (1)$$

Considering only the weak quadratic nonlinearity one can obtain the system:

$$L\mathbf{Z}=\varepsilon M(\mathbf{Z},\mathbf{Z}) \quad (2)$$

Here L and M are linear and nonlinear operators, **Z** is the eight-dimensional vector. Its components are velocities in x and z directions and their derivative with respect to y, velocity in y-direction, pressure, temperature and there derivative with respect to y. For the boundary-layer on a flat plate we use non-dimensional spatial and time co-ordinates:

$$\eta = y\sqrt{\frac{U_e}{\nu_e x}}, \xi = 2\sqrt{\frac{U_e x}{\nu_e}}, \zeta = \frac{U_e}{\nu_e}z, \tau = \frac{U_e^2}{\nu_e}t \quad (3)$$

The solution of (2) we search in the form

$$\mathbf{Z}=\mathbf{Z}'+\Delta z$$

$$\mathbf{Z}'= \sum z_j \exp[i\vartheta_j], \vartheta_j = \beta_j\zeta - F_j\tau \quad (4)$$

For given ξ, z_j is approximated by the sum of the eigenfunctions of a discrete spectrum of the linear homogeneous problem in which we use transformations: $\partial/\partial\zeta = \beta_j$, $\partial/\partial\tau = F_j$, $\partial/\partial\xi = \alpha_j^k(\xi)$. If the main flow is parallel α_j^k is constant. For flows slowly changed in x-direction $\alpha_j^k(\xi)$ can be obtained using necessary corrections. Then

$$z_j = \sum_k A_j^k Z_j^k \exp[i\int\alpha_j^k d\xi] \quad (5)$$

Taking into consideration (5), (4) and condition $||\Delta z|| << ||z||$ one can obtain the linear system respect to Δz

$$L\Delta z=\varepsilon g(\xi,\eta,\zeta,\tau) \quad (6)$$

The structure of the right part is as followed

$$g = B(\xi,\eta)\sum_j \frac{\partial z_j}{\partial\xi}e^{i\vartheta_j} + \sum_{p,1} M(z_p, z_1)e^{i(\vartheta_p+\vartheta_1)}$$

For resonant case $\beta_1=\beta_2+\beta_3$, $\omega_1=\omega_2+\omega_3$ three subsystems can be separated from (6)

$$L(F_j,\beta_j)\Delta z_j = g_j,\ j=1,2,3 \quad (7)$$

where:

$$g_1 = B\frac{\partial z_1}{\partial\xi} + \varepsilon M(z_2,z_3);\ g_2 = B\frac{\partial z_2}{\partial\xi} + \varepsilon M(z_1,z_3^*);\ g_3 = B\frac{\partial z_3}{\partial\xi} + \varepsilon M(z_1,z_2^*);$$

Here complex conjugated values are marked by a star. If the solution of (6) are searched as the sum of the eigenfunctions of considerated homogenous problem the solvability condition determines the amplitude equations for A_j^k. The obtained system is complex yet for simulation of the disturbances development in the space. Therefore we consider a more simple problem, when one of the waves exceeds others. In this case $z_j = A_j\exp(i\int\alpha_j d\xi)$ and the amplitude equations have the form:

$$\frac{dA_1}{d\xi} = \varepsilon a_1 \exp[i\Delta_1(\xi)]A_2A_3$$
$$\frac{dA_2}{d\xi} = \varepsilon a_2 \exp[i\Delta_2(\xi)]A_1A_3^* \quad (8)$$
$$\frac{dA_3}{d\xi} = \varepsilon a_3 \exp[i\Delta_3(\xi)]A_1A_2^*$$

The system (8) is similar to the system obtained in [10]. There are some differences in the presence of detuning:

$$\Delta_1 = \int(\alpha_2 + \alpha_3 - \alpha_1)d\xi$$
$$\Delta_2 = \int(\alpha_1 - \alpha_2 - \alpha_3^*)d\xi \quad (9)$$
$$\Delta_3 = \int(\alpha_1 - \alpha_2^* - \alpha_3)d\xi$$

For resonant case [9] $\Delta_j=0$. The real $\Delta_j \neq 0$ were taken in a consideration in [2]. In our case Δ_j are complex. The system (8) is correct both for $\Delta_j=0$ and $\Delta_j \neq 0$. It is important for a case of the disturbances development in the boundary layer, were Δ_j change with ξ. Unfortunately we did not carry out calculations in the full volume. But we investigated the disturbances structure in the supersonic boundary layers concerned to the linear approach. We have primary results on the nonlinear interaction.

2.1. Results and Comparison with Experiment

First of all the linear stability of the supersonic boundary layer on the flat plate is studied both for parallel and nonparallel flows. There are Mach and Reynolds numbers $M=4$, $Re=\sqrt{U_e x / \nu_e}=600$, Prandl number $Pr=0.72$, the ratio of the specific heats $\gamma=1.4$ in the calculations. The Sutherland's viscosity law was used. Theoretical results are compared with experimental data [11] in Fig.1. It is shown that for $\chi=tg^{-1}(\beta/\alpha_r)<70°$ there is a good agreement between the theory and experiments. But for $\chi>80°$ there is the disagreement. We also studied the influence of wave numbers β on a phase velocity C and discovered that C depended on β very slowly. It means that resonant conditions have place for a wide region of

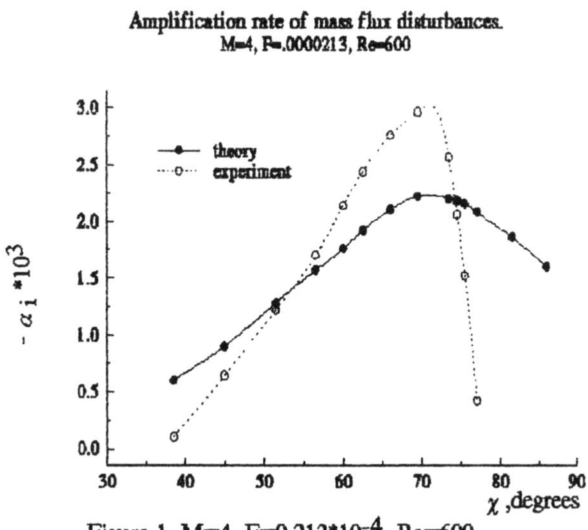

Figure 1. M=4. F=0.213*10^{-4}, Re=600

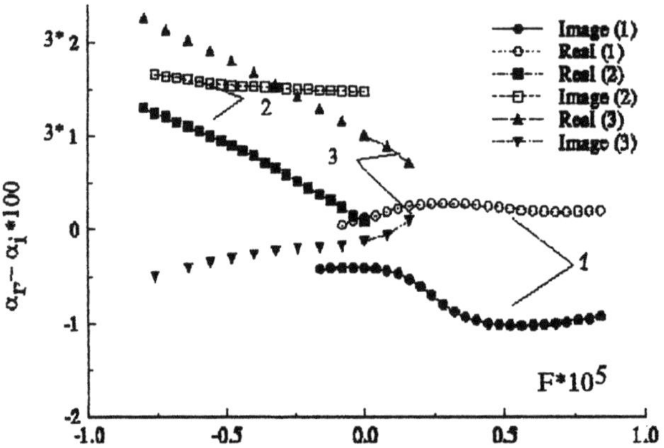

Figure 2. Real(α) and Im(α) versus F for β=0

β. Experiments confirm this results. Another kinds of disturbances are studied too. In Fig.2 the real and imaginary parts of the wave numbers as a function of the frequency are given for three additional types of disturbances. Our investigation indicate that for the cases 1 and 3 disturbances are incoming.The waves of the third type generate into the almost periodical in a space disturbances when F→0.The stationary periodical waves are Mach lines. The disturbances energy propagates along this lines to the boundary layer and are absorbed by it. This disturbances generate the periodical boundary layer nonuniformity of the high amplitude. So in the conventional wind tunnels the stationary flow in the boundary layer of models is periodical.

Interaction of this flow with the T-S waves leads to disturbances of the form $\Pi(\xi)\exp[i(\alpha\xi-F\tau)]$ where Π is periodical function with the period length $\lambda_0=2\pi/\lambda_0$, α, F are the wave number and frequency of the T-S wave. The similar structure of disturbances was observed in [3]. Disturbances in the case 2 are outgoing and they propagate upstream with a fade. This phenomena is discovered in experiments [11].

3. Experiment

The experiment on nonlinear instability of waves in the supersonic boundary-layer concerning to Mach 2.0 was described in [9]. Experimental investigations of the waves parameters confirm the existence of three-dimensional resonance between fundamental and subharmonic waves. This triad is not symmetrical because of the fundamental wave is oblique. The nonsymmetrical wave triad comprises 3-D waves propagating with angles from ±(55°÷65°) for fundamental (20 kHz) to ±(78°÷83°) and ±(60°÷70°) for subharmonic (10 kHz) waves.

Increasing of initial disturbances (particularly for Mach number M=3.0) leads to another type of the disturbances interaction. Oscillograms in this case are similar to K-type transition in subsonic boundary layer. Increasing disturbances amplitude of 2-D waves are found. Waves with multiple frequencies of the subharmonic mode are observed downstream. In the paper experimental results will be discussed in detail.

3.1. Experimental Equipment

The experiments were performed in the supersonic wind tunnel T-325 of the Institute of Theoretical and Applied Mechanics of the Siberian Division of the Russian Academy of Sciences. Test section dimensions is 200×200×600 mm.

Mach numbers were 2.0 and 3.0 and unit Reynolds numbers (Re_1) were 6.6×10^6 and 11.21×10^6 (1/m). Measurements were performed in the boundary layer of a flat plate mounted at the zero angle of attack. The model was 10 mm in thickness, 450 mm in length, 200 mm in width. The leading edge thickness was b=0.02 mm. By design, the source of disturbances was the same as in [3]. The disturbances from the spark discharge chamber penetrated to the flow through a small hole in the plate surface. The diameter of the hole was 0.42 mm. The coordinates of the disturbances source were X=38 mm, Z=0 mm.

The disturbances in the flow were registered using a constant-temperature hot-wire anemometer. The probes were constructed of tungsten wire, which had dimension of 6 μm in diameter and 1.2 mm in length. By mean of a traversing gear, the probe was moved in directions X,Y,Z. The coordinates of the hot-wire were measured in steps of 0.1 mm for X and Z and 0.01 mm for Y. Signal fluctuation from the hot-wire was recorded by a 10 bit 1 MHz A/D converter. Synchronous summation of the signal up to 200 points was carried out in the experiments for improvement of signal/noise. The amplitude and phase of frequency harmonics were defined and controlled in experiments. This allowed precisely enough to define the boundary of artificially disturbed flow. To define the spectra of the wave numbers, discrete Fourier transformation was used.

3.2. Results & Discussion

Development of periodic disturbances of large initial amplitude at Mach 3.0. Electric discharge was ignited in the experiments at a frequency of 20000 sparks per second. For example in Fig.3 initial data for modulated generation at Mach 3.0 are shown. These are constant amplitude lines and view of the oscillograms. As it was shown by us in [9] disturbances spectra consist of 3-D waves in nonlinear region of waves interaction, and the main energy part for subharmonic wave is at approximately 80 degrees wave inclination. However experiments show that main energy part may be at small angles of wave inclination for large initial disturbances amplitudes. This result is presented in Fig.4 as disturbances development from linear to nonlinear regions at Mach 3, Re_1=11,210,000 m^{-1} and x=60, 130 mm. Here β is wave number in z-direction.

The amplitude β-spectra obtained at x=100 mm show that the subharmonic resonance is mechanism of the fast transformation of disturbances. We observed also the disturbances with large inclination angles. Similar results at Mach 2.0 for large initial amplitude was found too. We propose that the parametrical amplification of the disturbances leads to intensive sound radiation from the supersonic boundary layer because β-spectrum at x=130 mm is similar to one obtained in free flow over flat plate for acoustic waves [12]. We plan to continue investigation to clarify nature of the acoustic radiation from the supersonic boundary layer. Another detailed data about features of nonlinear interaction in the supersonic boundary layer will be processed. Both theoretical and experimental research the nonlinear phenomena in the supersonic boundary layer will be continued.

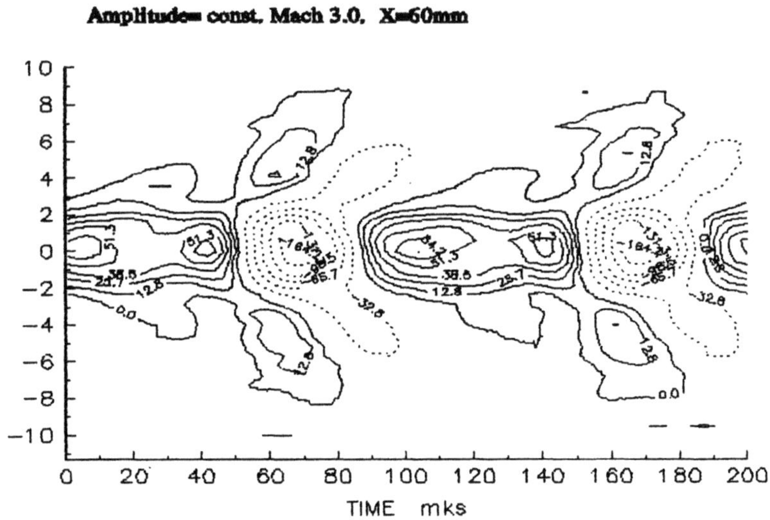

Figure 3a.

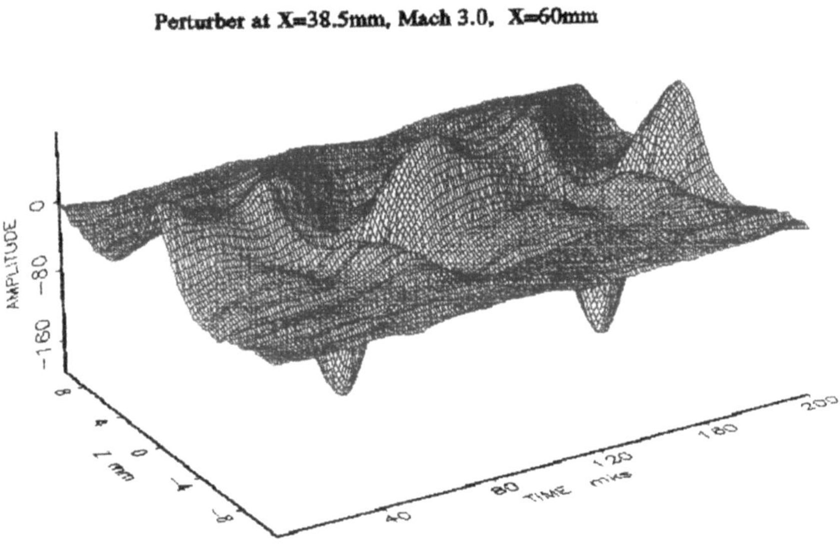

Figure 3b.

AMPLITUDE SPECTRA, X=60mm, M=3.0, U=610,1 m/s

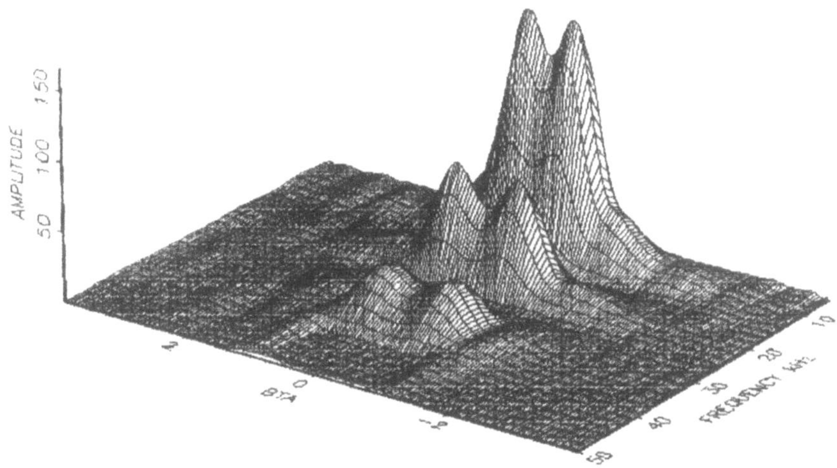

Figure 4a.

AMPLITUDE SPECTRA, X=130mm, M=3.0, U=610,1 m/s

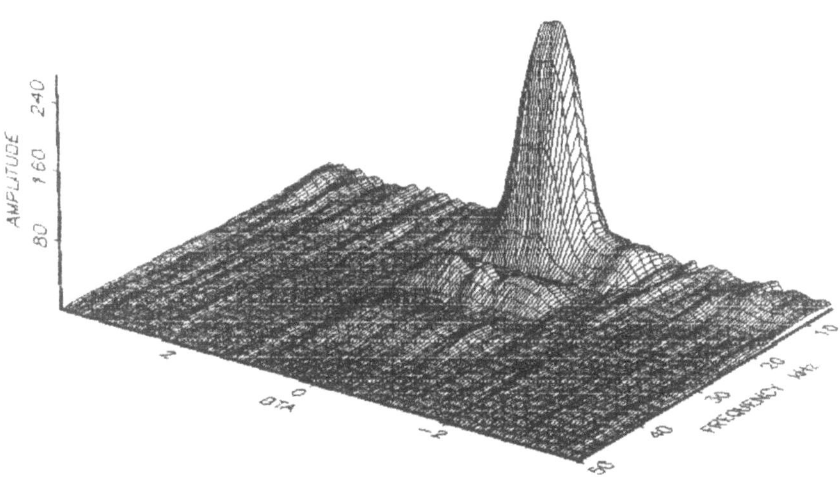

Figure 4b.

Acknowledgements.

This work is supported by Russian Fond of Fundamental Investigations under Grant N 94-01-00052-a. The paper was prepared using James Kendall's PC and software from Steven P. Schneider. The experiments is supported by Sibirian Division of Russian Academy of Sciences and by ISF.

References:

[1] Morkovin M.V., Reshotko E.: Dialog on progress and issues in stability and transition research. In: Laminar-Turbulent Transition, Springer-Verlag, (1990) 3-29.
[2] Zelman M.B., Maslennikova I.I.: Tollmien-Schlichting-wave resonant mechanism for subharmonic-tipe transition J.FLuid Mech. 252 (1993) 449-478.
[3] Kosinov A.D., Maslov A.A., Shevel'kov S.G.: Experiments on the Stability of Supersonic Laminar Boundary layers, J.Fluid Mech. 219 (1990) 621-633.
[4] Masad J.A., Nayfeh A.H.: On the subharmonic instability of compressible boundary layers. In:Laminar-Turbulent Transition Springer-Verlag, (1990) 271-278.
[5] Thumm A., Wolz W., Fasel H.: Numerical simulation of spatially growing three_dimentional disturbance waves in compressible boundary layers. Springer-Verlag, (1990) 303-310.
[6] El-Hady N.M.: Secondary instability of compressible boundary layer to subharmonic three- dimentional disturbances. (1989) AIAA Paper No.89-0035.
[7] Chang C.L., Malik M.R.: Nonlinear stability of a Mach 1.6 boundary layer. In: Nonlinear Instability of Nonparallel Flows, IUTAM Symp. Potsdam, USA (1990) p.50.
[8] Adams N.A., Sandham N.D.: Numerical Simulation of Boundary Layer Transition at Mach Two, Applied Scientific Research. 51. (1993) 371-375.
[9] Kosinov A.D., Semionov N.V., Shevel'kov S.G., Zinin O.I.: Experiments on the Nonlinear Instability of Supersonic Boundary Layer, In: Nonlinear Instability of Nonparallel Flows, IUTAM Symp. Potsdam, USA, (1993) 46.
[10] Craik, A.D.D.: Non-linear resonant instability in boundary layers. J.Fluid Mech., 50, (1971) 393-413.
[11] Kosinov A.D., Semionov N.V., Shevelkov S.G.: Special features of generation and development of a harmonic wave in a supersonic boundary layer. In: Recent Advances in Experimental Fluid Mechanics, IAP Beijing (1992) 53-58.
[12] Maslov A.A. & Semionov N.V.: Acoustic disturbances and supersonic boundary layer. In: Problems of Nonlinear Acoustics, Novosibirsk (1987) 132-134.

Analysis of the Formation of a Turbulent Spot in a Laminar Boundary Layer

V.A. Kazakov
Central Aero-Hydrodynamic Institute, Zhukovsky, Moscow reg., Russia

Summary

The process of the formation of a turbulent spot in a laminar boundary layer on a flat plate in incompressible fluid has been analyzed numerically. The initial disturbance is introduced by a local distortion of the plate. The analysis has been made for large Reynolds numbers within the framework of triple-deck theory. All stages of the flow evolution has been traced - through the formation of a linear wave packet from the unstable Tollmien-Schlichting waves and then its nonlinear deformation and disintegration. This process results in localized ejections of the near-wall fluid (bursts) and leads to the formation of a characteristic arrow shape of a turbulent spot observed in experiments.

Key words : Turbulent spot, Nonlinear wave packet, Burst.

1. Introduction

A turbulent spot in a laminar boundary layer [1-6 and references therein] is of great interest first of all due to its role in a natural turbulization of the flow. Turbulent patches (spots) appear in various places of the transition region and spread downstream, enlarging and merging to form a fully developed turbulent boundary layer. The other reason to study the spot is that the internal properties of the turbulent region in a developed spot are the same as in a turbulent boundary layer. The turbulent spot arises and develops in a laminar flow, from a very small local disturbance, so we have a much better opportunity to trace the process of turbulization, and this constitutes the third attractive feature in studying a turbulent spot. In this connection, numerical experiments on powerful computers to model the generation and evolution of a turbulent spot become of particular interest and importance. A sufficiently well posed initial-boundary problem may be solved, and this has been done in the present paper. It is not possible to formulate such a problem for a fully developed turbulent flow. The numerical analysis based on triple-deck theory [7-8] makes it possible to somewhat simplify the investigation of the phenomenon in comparison with the solution of the full Navier-Stokes equations, although obvious restrictions of this approach are conserved.

2. Formulation of the problem

A three-dimensional nonstationary problem is formulated in the following way. On a flat plate placed parallel with an incompressible fluid flow moving with the velocity U_∞, a slight local initial perturbation (a hump) rises at some place at a distance L from the leading edge of the plate. The slightness of the disturbance is defined by the characteristic scales of a triple-deck model of boundary-layer asymptotic theory. We define a Reynolds number $Re = U_\infty L/\nu >> 1$, ν is the kinematic viscosity, and introduce a small parameter $\varepsilon = Re^{-1/8} << 1$. The height of a surface distortion is assumed to be of the order $O(L\varepsilon^5)$, the streamwise and spanwise dimensions $O(L\varepsilon^3)$, the characteristic time $O(L/U_\infty\varepsilon^2)$.

The flow will be treated in a rectangular coordinate system with its origin at the center of the surface disturbance, the x^* axis directed downstream along the plate, y^* normal to the plate, z^* in the transverse direction. We denote t_* the time, u^*, v_*, w^* velocity components along the axes x^*, y^*, z^*, ρ^* the density, p the pressure. Nondimensional variables (quantities without asterisks) are introduced:

$$(x^*, y^*, z^*) = L(\varepsilon^3 x, \varepsilon^5 y, \varepsilon^3 z) , \qquad t^* = (L/U_\infty)\varepsilon^2 t ,$$
$$(u^*, v^*, w^*) = U_\infty(\varepsilon u, \varepsilon^3 v, \varepsilon w) , \qquad p^* - p_\infty = \rho_\infty U_\infty^2 \varepsilon^2 p .$$

In accordance with asymptotic theory of free interaction, in the wall region of a boundary layer with the streamwise and transverse dimensions $O(L\varepsilon^3)$ and normal to the surface $O(L\varepsilon^5)$, the flow is described by the system of equations of Prandtl, which can be written after Prandtl's transformation as

$$u_t + uu_x + vu_y + wu_z = - p_x + u_{yy} ,$$
$$w_t + uw_x + vw_y + ww_z = - p_z + w_{yy} ,$$
$$u_x + v_y + w_z = 0 , \qquad p_y = 0 .$$

At the outer boundary, as $y \longrightarrow \infty$, matching with the external inviscid flow yields an interaction condition, that links the unknown pressure with the boundary-layer thickness function $A(t,x,z)$ determined by the flow field in the wall layer,

$$p(t,x,z) = - \frac{1}{2\pi} \int_{-\infty}^{\infty} \int_{-\infty}^{\infty} \frac{\partial^2 A/\partial\xi^2}{[(x-\xi)^2 + (z-\zeta)^2]^{1/2}} \, d\xi d\zeta .$$

The solution must also satisfy the boundary conditions:

$$u = v = w = 0 \quad \text{at } y = 0 ,$$
$$u \longrightarrow y , \quad v, w, p \longrightarrow 0 \text{ as } x \longrightarrow -\infty ,$$
$$u - y \longrightarrow A(t,x,z) + f(t,x,z) , \quad w \longrightarrow 0 \text{ as } y \longrightarrow \infty ,$$

the function $f(t,x,z)$ describes the shape of distorted surface. In computations, the boundary conditions are implemented at finite, sufficiently remote distances. We also assume the symmetry of the flow about the plane $z = 0$.

The governing equations are approximated by finite differences on an nonuniform grid (x_i, y_j, z_k) with the coordinate stretching near the zero point. The interaction integral used as the upper boundary condition is determined as a sum of integrals over the cells with the apexes at the middle points of the grid cells. For the solution of the system of linear equations resulting from the discretization, the iterative plane Gauss-Seidel method is used for planes z_k with the strongly implicit procedure [9,10] in each plane z_k.

3. Numerical analysis

The initial perturbing deformation of the plate was specified by the function $f(t,x,z) = H(t)/ch(4r)$, $r=(x^2+z^2)^{1/2}$, noticeably different from zero only in the circle of unit radius. The height of the hump $H(t)$ reached its maximum value h linearly in time during the interval $0 \le t \le 0.1$ and then disappeared also linearly during the same time. The solution was based upon the experience of analogous computations in investigating wave packets in a plane boundary layer in earlier works [11,12]. Results presented below have been obtained on a nonuniform grid with 150x30x25 cells in the domain $-7 \le x \le 14$, $0 \le y \le 22$, $0 \le z \le 5$ with the point $x_i=0$ at $i=50$, minimum values of grid steps near the zero point $\Delta x = \Delta y = \Delta z = 0.1$ and the time step $\Delta t = 0.01$. Because the disturbances developing in the boundary layer were convected far downstream, the computational domain was also transferred after them in a certain time interval.

Some moments in the flow evolution from the surface distortion of a small maximum height, $h=0.1$, are represented in Fig. 1. In all spatial views, the x axis is directed downstream along the long side of the graphs, and z in the transverse direction. At the linear stage, from the wide spectrum of harmonics of starting-up disturbances, Tollmien-Schlichting (TS) waves from the vicinity of the most unstable wave number are amplifying and forming a wave packet with a set of crests and troughs. The packet resembles much the one constructed on the basis of experimental data in [13]. The largest amplitude of the waves is reached on the center-line. With increasing disturbances and the development of nonlinearity, the wave packet distorts (Fig. 1c), and in the amplitude spectrum, there appear a region of subharmonics and a pair of symmetric lateral sections corresponding to rapidly growing oblique waves. The maximum amplitudes of oscillations are concentrating in the following after each other lateral regions of positive and negative (somewhat downstream) pressure on the waves crests and troughs (Fig. 1c,d). The growing peaks start to noticeably overtake the waves which have given them rise. As the streamwise friction signature shows, a local reverse flow is forming in the region of the minimum pressure. The limiting streamlines, turning round both sides of the zone of higher pressure, converge into a singular point of a node type, a stable node (Fig. 2a). On the opposite side of the region of a reverse flow, the streamlines are separated by a saddle point. The stable node appears near the minimum pressure location. Somewhat later, when pressure peaks reaches their largest values (Fig. 1d), the local zone of a reverse flow is widening (Fig. 2b), and the limiting streamlines are forming two successive nodes, stable (a nodal point

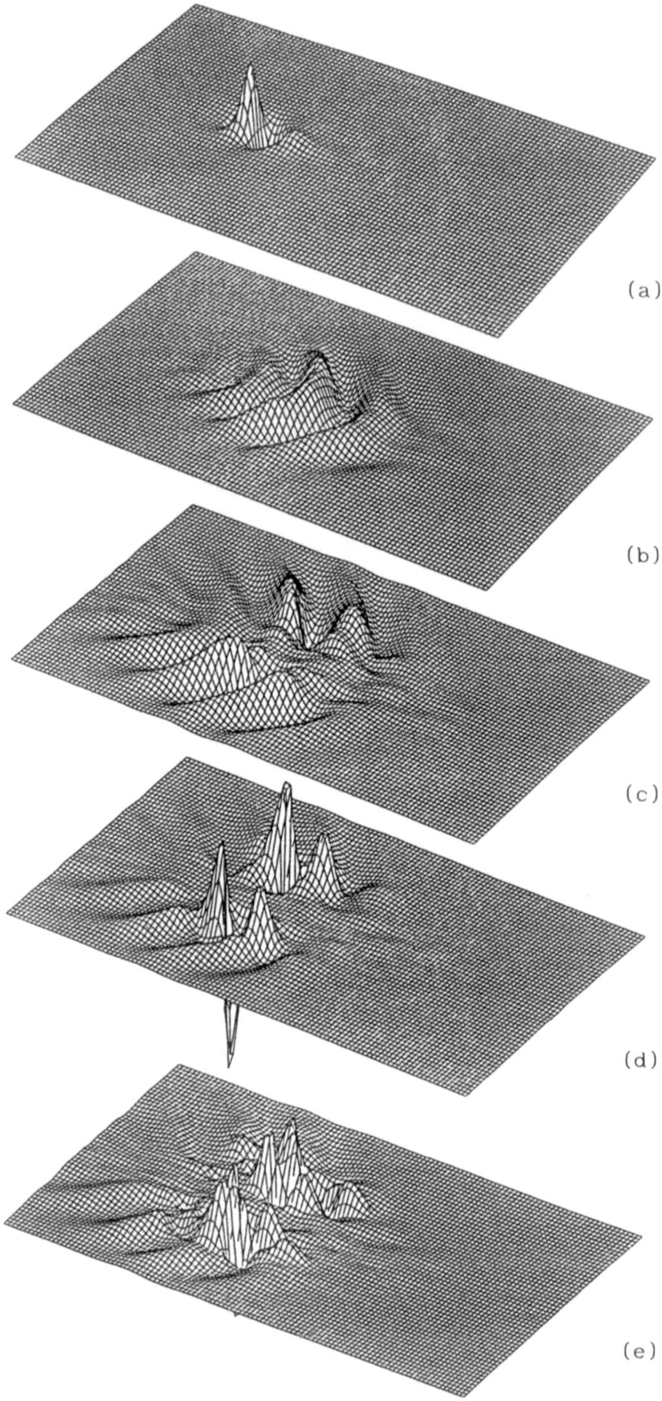

Fig. 1(a-e). For caption see the next page.

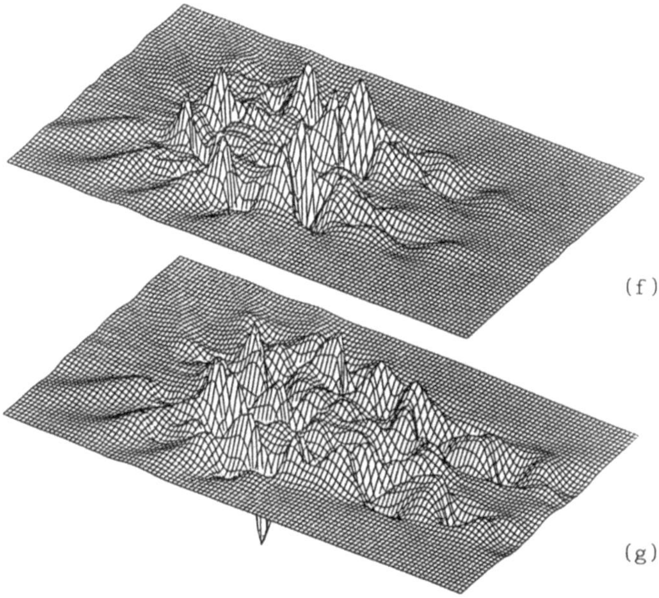

Fig. 1. Perspective views of the wave packet evolution, function $p(t,x,z)$; $-5 \leq z \leq 5$; (a) $t=0.2$, $-5 \leq x \leq 11$, $-0.083 \leq p \leq 0.117$; (b) $t=3$, $0 \leq x \leq 16$, $-0.071 \leq p \leq 0.071$; (c) $t=6$, $9 \leq x \leq 25$, $-0.666 \leq p \leq 0.549$; (d) $t=6.8$, $12 \leq x \leq 28$, $-9.12 \leq p \leq 4.93$; (e) $t=7.0$, $13 \leq x \leq 29$, $-5.53 \leq p \leq 3.29$; (f) $t=7.3$, $14 \leq x \leq 30$, $-5.15 \leq p \leq 3.54$; (g) $t=7.5$, $14 \leq x \leq 30$, $-5.33 \leq p \leq 2.67$.

of separation) and unstable (a nodal point of attachment), and two saddle points at the sides. In more detail the flow pattern in this place is shown in Fig. 2c. Such converging of streamlines into a stable node suggests their departure from the surface. In fact, the flow visualization (Fig. 3) reveals a concentrated ejection of fluid particles up from the plate into the main body of the flow. Individual ejections of the same kind were clearly recorded in a cross-section of a turbulent spot in the photograph in experiments in [4].

The phenomenon observed can be considered as the main "unit" event in the process of turbulization of a boundary layer, resulting from the development of wave disturbances. As will be seen further, bursts of such kind, although not so strong in most cases, constitute the basis of the formation, spreading and supporting of the turbulent spot. The ejection of fluid upward results in intensive mixing of the boundary layer - one of the main features of the turbulent flow. Entering the upper layers the fluid particles are accelerated, so that the disturbed region propagates fast downstream, overtaking the wave packet, from which it has originated and on which it is "leant on". The streamlines which form the burst come from the wall shear layer, so, as Fig. 3 demonstrates, having lifted into rapidly moving layers, they exhibit convective instability in a sort of a wave motion and, furthermore, within a very narrow band. Such an oscillatory motion, from one or two waves ahead of the turbulent spot, was recorded in cine films [5]. The narrowness of the

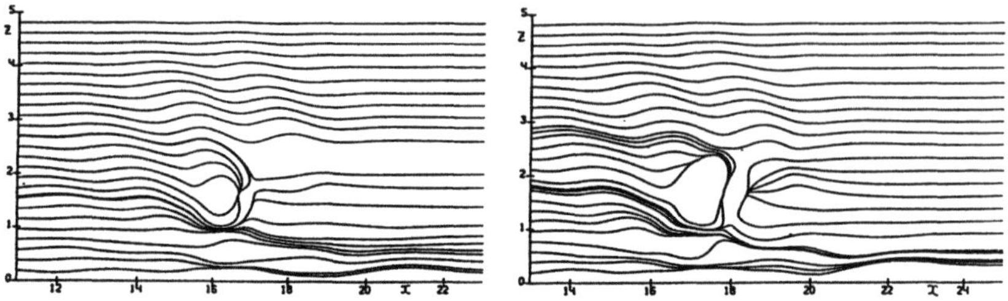

(a) (b)

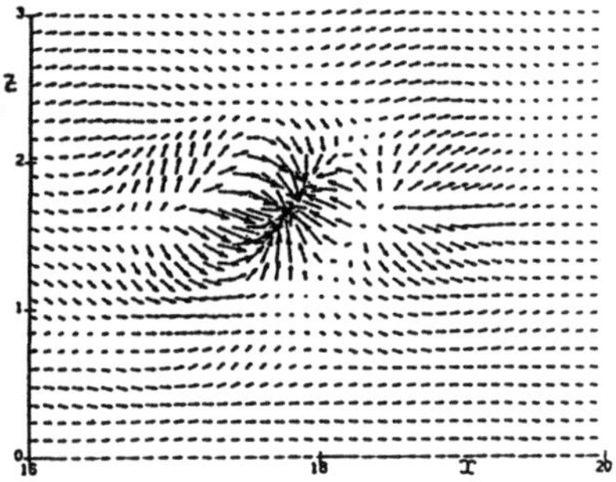

(c)

Fig. 2. Limiting streamlines, (a) t=6.5, (b) t=6.8; (c) skin-friction vector field (τ_x, τ_z), t=6.8.

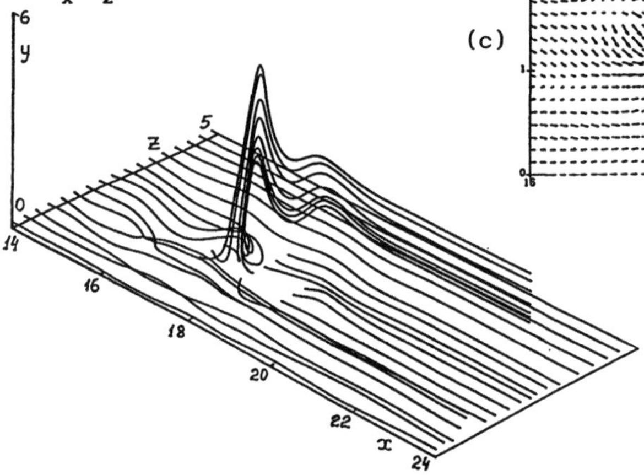

Fig. 3. Burst, perspective view of the streamlines, t=6.8.

band of oscillations, as now is clear, is explained by the concentration of streamlines in the vicinity of the node of separation, the location of the burst.

In a two-dimensional wave packet [11,12], when the amplitude and pressure gradients become sufficiently large, there appeared small-scale oscillations, "spikes", observed in the K-type of transition. In a three-dimensional wave packet, the forming positive and negative pressure peaks split in the transverse direction, giving rise to two pairs of new peaks of lower amplitude (Fig. 4). The region of a reverse flow is also divided into two small zones for each pair of pressure peaks. Closer to the center-line, a stable node and a saddle point form, i.e. the same configuration as earlier in Fig. 2a; somewhat farther, two nodes, stable and unstable, and two saddles appear, as in Fig. 2b. At these places, the fluid is ejected upward, though not so strongly as shown in Fig. 3 at large pressure amplitudes. The splitting of the growing pressure peaks is associated with the formation of new harmonics of nonlinearly interacting oblique waves.
Parallel with this process, new local regions of higher pressure value emerge, the precursors of new bursts. Their appearances can be seen at the internal extremities of the original TS waves behind and outside (in the spanwise direction) of the strongly disturbed

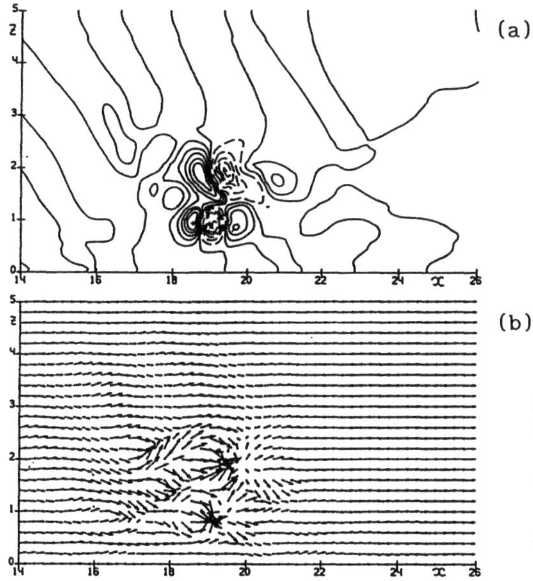

Fig. 4. $t=7$; (a) pressure contours; $-5.53 \le p \le 3.29$; ——, $p \ge 0$; – –, $p<0$; +, *max p*; ×, *min p*; (b) skin-friction vector field (τ_x, τ_z).

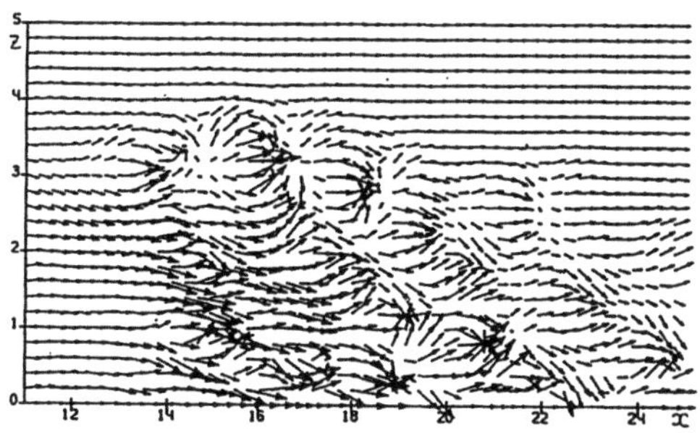

Fig. 5. Skin-friction vector field (τ_x, τ_z); the case of the initial disturbance with maximum amplitude $h=1$; $t=5.5$.

region, in its rear part, and also in front of the main pressure peaks at the "remnants" of the forward TS waves (Fig. 1f,g). The fluid ejected into the upper layers in the forward bursts overtakes the main disturbed region and, moving in an oscillatory manner, as shown in Fig. 3, destabilizes the appropriate lower parts of a laminar layer. These waves are evident in the forward part of the spatial views in Fig. 1f,g. The following bursts in the middle part, still farther from the center-line, form the lateral boundary of the arising turbulent spot. Its characteristic arrowhead shape results in (Fig. 5). The streamlines from the burst in the rear now experience not two but 4-5 oscillations and rise still higher, passing downstream over the location of the forward burst. Thus, a pronounced "overhang" from oscillating vortical fluid is forming over the wall laminar layer. This feature was observed in experiments, e.g. [2,5].

Thus, the results of computational experiments evidence that, in the process of the formation, spreading and supporting of the turbulent spot, the fundamental role belongs to TS waves generated in the unstable boundary layer. Their nonlinear disintegration leading to turbulent bursts, on the one hand, is the cause of the lateral, "discrete" spreading of the turbulent region and the spot (at its wing tips), and, on the other hand, supports the turbulent region in the rear on account of generation of new bursts. In turn, these bursts arise as a result of appearance and rapid rising of local zones of higher pressure amplitude on the crests and troughs of oblique TS waves adjacent to the spot, which has a character of a wave resonance. The main events of breakdown to turbulence take place in the viscous wall layer, but are connected with the outer flow by means of self-induced pressure. In the case of a large initial disturbance (h=3), the process of the formation of a turbulent spot is nonlinear from the very beginning. Again oblique waves form at the sides of the strongly disturbed central region, and their disintegration leads to generation of turbulent bursts.

References

[1] Emmons,H.W.:The laminar-turbulent transition in a boundary layer - Part I, J. Aero. Sci. 18 (1951) 490-498.

[2] Schubauer,G.B. and Klebanoff,P.S.:Contributions on the mechanics of boundary layer transition, NACA Rep. no. 1289 (1956).

[3] Elder,J.W.:An experimental investigation of turbulent spots and breakdown to turbulence, J. Fluid Mech. 9 (1960) 235-246.

[4] Perry,A.E., Lim,T.T. and Teh,E.W.:A visual study of turbulent spots, J. Fluid Mech. 104 (1981) 387-405.

[5] Gad-el-Hak,M., Blackwelder,R.F. and Riley,J.J.: On the growth of turbulent regions in laminar boundary layers, J. Fluid Mech. 110 (1981) 73-95.

[6] Glezer,A., Katz,Y. and Wygnanski,I.:On the breakdown of the wave packet trailing a turbulent spot in a laminar boundary layer, J. Fluid Mech. 198 (1989) 1-26.

[7] Stewartson,K.:Multistructured boundary layers on flat plates and related bodies, Advances in Appl. Mech. 14 (1974) 145-239.

[8] Ryzhov,O.S. and Terent'ev,E.D.:On the nonstationary boundary layer with self-induced pressure (in Russian), Prikl. matem. i mekh. 41 (1977) 1007-1023.

[9] Stone,H.L.:Iterative solution of implicit approximations of multidimensional partial equations, SIAM J. Numer. Analys. 5 (1968) 530-558.

[10] Kazakov,V.A.:Strongly implicit alternate-triangular procedure for the solution to the problems of asymptotic theory of boundary layer (in Russian), Zh. vychisl. matem. i matem. fiz. 25 (1985) 1382-1390.

[11] Kazakov,V.A.:On spreading disturbances and generation of vortices in a laminar boundary layer (in Russian), 6th All-Union Congress on Theor. and Appl. Mech. Tashkent (1986) 320.

[12] Kazakov,V.A.:Numerical investigation of the formation of waves and generation of vortices in a laminar boundary layer, Russian J. Comput. Mech. 2 (1994) (in press).

[13] Gaster,M. and Grant,I.:An experimental investigation of the formation and development of a wave packet in a laminar boundary layer, Proc. Roy. Soc. A347 (1975) 253-269.

Entropy- and Boundary Layer Instability of Hypersonic Cone Flows-Effects of Mean Flow Variations

Ewald Kufner, Uwe Dallmann
DLR - Institute of Fluid Mechanics
Bunsenstrasse 10, D-37073 Göttingen, Germany

Summary

The present paper demonstrates results of hypersonic mean flow calculations and instability analyses with a major strength on the effects of mean flow modifications. The investigated geomtries are sharp and blunt circular cones. For both of these geometries the calculated disturbance amplifications on the basis of primary local linear stability theory and an ideal gas assumption show clear deviations to experimental results. The performed mean flow modifications contain variations of the wall temperature conditions by prescribing adiabatic, heated, cooled and radiation adiabatic temperatures as well as the consideration of various bluntnesses.

The general effect of wall cooling is to amplify second mode waves and to damp first mode instabilities as known from flat plate investigations. The new observations are that the wall temperature effect is more pronounced on the first mode than on the second mode and that the wave angle of maximum amplification of the first mode is increasing with cooling.

The provocation of transition reversal by bluntness increasing has not been successfull. However, an unexpected vanishing of the first mode appeared. Derived from such cases and cases of a complete damping of the first mode due to wall cooling an expansion of the double generalized inflection point criterion is proposed for finite local Reynolds numbers.

More detailed investigations of the entropy layer in the hypersonic flow past blunt cones show two maxima of the first derivative of the local Mach number profile towards the wall normal coordinate, indicating two generalized inflection points and two possible entropy layer instabilities. Indeed for the first time two different entropy layer instability types can be introduced and detailed information about it is presented.

Keywords

Sharp and blunt cones, calculation of hypersonic mean flows, instability analysis, wall temperature and bluntness effects, new inflection point criterion, two different entropy layer instabilities.

Investigated Geometries and Flow Conditions

The option of the present investigations is to perform studies to an optimization of the bluntness of hypersonic space planes. Herein an accurate determination of the whole transition process from the laminar to the turbulent boundary-layer state is very important. The bluntness effect on stability and transition along aerospace plane bodies is here theoretically investigated by a generic simplification of such geometries to a sharp and various circular blunt cones at zero angle of attack. All the cones have a half angle of 7°, the freestream Mach number is $M_\infty = 8.0$. In the case of the sharp cone we choose a freestream unit Reynolds number of $Re_\infty/ft. = 1.0 \cdot 10^6$, a freestream temperature of $T_\infty = 52.753K$ (Stetson's et al. case [11]) and adiabatic wall boundary conditions. Its total length is $L = 1.9896m(78.312in.)$.
The blunt cone investigations refer to a freestream unit Reynolds number of $Re_\infty/ft. = 2.5 \cdot 10^6$ and a freestream temperature of $T_\infty = 54.347K$. The performed mean flow variations contain modifications of the bluntness and the wall temperature conditions as given in table 1

Table 1: Investigated Geometries and Wall Temperature Conditions of Blunt Cones

RN	3.81 mm (0.15 in.)	17.78 mm (0.70 in.)	42.672 mm (1.68 in.)	6.35 mm (0.25 in.)
Re_∞:	$3.1250 \cdot 10^4$	$1.4583 \cdot 10^5$	$3.5000 \cdot 10^5$	$5.2083 \cdot 10^4$
Wall Temp. T_W	adiabatic radiation adiabatic, $\epsilon = 0.85$	adiabatic	adiabatic	adiabatic
Isothermal Wall Temp. T_W/Tad at $\overline{S}/\overline{RN} = 175.6$	1.2 0.9 0.8 0.6 0.3			
Length [m]: $= 266 \cdot \overline{RN}$	1.0135 (40. in.)	4.7295	11.3507	1.6891

Instability investigations of the present geometries require a very accurate solution of the laminar mean flow which is calculated by Thin-Layer Navier-Stokes codes of B. Müller [7] and S. Riedelbauch [8]. The instability analysis is based on the primary local linear stability theory code of Simen [9].

Results

Comparisons of Theoretical and Experimental Data: Stability investigations for a mean flow solution with the ideal gas assumption on the basis of primary local linear theory show in general larger spatial amplification rates than provided by the experimental investigations of Stetson et al. [10], [11]. The frequency of maximum amplification is especially in the case of a calculated second mode shifted to slightly smaller values. As **figure 1** (sharp cone, position

$\overline{S}/\overline{L}$ = 0.34, second mode only) and **figure 2** (blunt cone, $\overline{RN}$ = 3.81mm, position $\overline{S}/\overline{RN}$ = 175.6) demonstrate this is valid for sharp cones as well as for blunt cones.
The single symbol in figure 1 refers to the maximum amplification of the second mode determined by Chang & Malik [1] on the basis of a PNS- (Parabolized Navier Stokes Equations) mean flow solution. The good agreement of Simen's [9] theoretical results and Stetson's experimental results is meanwhile proved to be caused by a mean flow inaccuracy in Simen's Thin-Layer Navier-Stokes solutions and was accidental.
In figure 2 we find besides Stetson's et al. [10] and the present [4] solutions also amplification rates which are based on a second order euler/boundary layer mean flow solution of Stilla [3], Esfahanian's results (Thin-Layer Navier-Stokes mean flow) and the data of Malik et al. [6]. Especially within the second mode amplification range the agreement of the theoretical data is very good however, with clear discrepances to the experimental data.

Modifications of the Wall Temperature: The effect of wall temperature modifications on spatial amplification rates for the example of the blunt cone with a nose radius of $\overline{RN}$ = 3.81mm (compare table 1) is given in **figure 3**. These results refer to a downstream position of $\overline{S}/\overline{RN} \approx 175$ and show the typical wall temperature influences on disturbance growth rates as already known from Mack's [5] flat plate investigations. In comparison to the adiabatic wall conditions wall cooling causes an additional amplification of second mode instabilities and shifts them to larger frequencies. First mode instabilities are damped in combination with a reduction of the range of amplified frequencies. It should be noticed that the relative modification of the ratio of the maximum amplification rate at a certain wall temperature to the value at an adiabatic temperature is larger for the first mode than for the second mode. For a wall temperature of 0.3 times the adiabatic temperature the first mode is completely damped. Nevertheless the double generalized inflection point criterion, which is valid for inviscid instabilities, indicates the possible existence of an inflectional mode (not to be seen in the figure). Comparable situations will also be mentioned later in this paper with an empirical postulation of a new criterion. Besides this our investigations show, that the critical Reynolds number of first mode instabilities increases with wall cooling while the critical Reynolds number of second modes is hardly affected. For the adiabatic case in contrast to flat plate investigations of Mack the critical Reynolds number of the first mode is smaller than for the second mode.
According to the knowledge of the author not yet mentioned in the literature is an increase of the wave angle of maximum amplification of the first mode with cooling as given in table 2.

Bluntness Modifications: Within the present stability analyses of the mean flows past blunt cones with increased nose radius of $\overline{RN}$ = 17.78mm and $\overline{RN}$ = 42.672mm no indication of any tendency to a transition reversal phenomenon

Table 2: Wave Angle of Maximum Amplification of the First Mode, Blunt Cone, $\overline{RN} = 3.81mm$, $\overline{S}/\overline{RN} \simeq 175$

$\frac{T_W}{T_{ad}}$	1.2	1	0.9	str. adiabat	0.6
$\psi_{-\alpha_{i,max}}$	61.678°	64.405°	65.262°	68.095°	70.068°

could be noticed as was especially expected from the latter case . Astonishingly these bluntnesses do not show first mode instabilities either but only a second mode. Investigations of the double generalized inflection point criterion, which is an expansion of the generalized inflection point criterion for axisymmetric flows would indicate the possible existence of inflectional instabilities. Such a situation was already mentioned in connection with the damping of first mode instabilities due to wall cooling in the previous chapter.
Let ψ be the wave angle, M_e the boundary layer edge Mach number and U the wall tangential mean flow velocity in downstream direction, then the wall normal coordinate $Y_{0,\psi}$ of the sonic neutral point under consideration of the wave angle of maximum amplification ψ is defined by $\frac{c_0}{\cos\psi} = 1 - \frac{1}{M_e \cos\psi} = U(Y_{0,\psi})$. c_0 is the phase velocity of the wave at the sonic neutral point. The double generalized inflection point criterion indicates the possible existence of a subsonic neutral wave if the wall normal coordinate of the inflection point Y_S is larger than $Y_{0,\psi}$, where Y_S is defined by $\left(\frac{d}{dY}(\varrho\frac{dU}{dY}) - m_{23}\varrho\frac{dU}{dY}\right)|_{Y_S} = 0$. m_{23} is the metric term which expresses the dependency of the local radius of the circumferential curvature on the wall normal coordinate. As mentioned this criterion is valid for inviscid instabilities. According to the present investigations a better applicable criterion for the possible existence of first mode instabilities ($\psi \approx 60^o$) at finite local Reynolds numbers is the condition $Y_K > Y_{0,\psi}$, where Y_K is the smallest wall normal coordinate with

$$(\frac{\partial}{\partial Y}(\frac{\partial}{\partial Y}(\varrho\frac{\partial U}{\partial Y}) - m_{23}\varrho\frac{\partial U}{\partial Y}))|_{Y=Y_K} = 0 \quad .$$

Entropy Layer Instabilities: In the experiments of Stetson et al. [10] entropy layer instabilities could be found only for blunt cones with a nose radius which exceeded a certain threshold value. In the present case detailed mean flow investigations were performed, which did not indicate any reason why there should be no entropy layer instability for smaller nose radius like for instance $\overline{RN} = 3.81mm$ (compare table 1). In the range of the flow fields where the entropy layer thickness is much larger than the boundary layer thickness the here calculated entropy layers show two generalized inflection points outside of the boundary layer, giving a hint that two different types of instabilities might be present.
Indeed all flows past the here investigated blunt cones show two different instabilities within the entropy layers with maxima in the amplitude functions outside of the boundary layer. It is here the first time that such detailed information about entropy layer instabilities can be provided.

Figure 4 shows spatial amplification rates of the two different entropy layer instability types for the example of the position $\overline{S}/\overline{RN} = 60.25$ of the cone with a nose radius of $\overline{RN} = 42.672mm$. The amplifications in the frequency range of $100Hz < \overline{f} < 450Hz$ are of clearly subsonic character, while those of $0Hz < \overline{f} < 52Hz$ are slightly supersonic. It may be posssible that the latter instabilities correspond with the mode called *IA* by Duck & Shaw. This mode appeares in combination with quite small wave angles of $0 < \psi < 41°$. The subsonic mode has large wave angles of $\psi > 80^O$. These large wave angles collidate with the parallel flow assumption within the here applied local stability theory so that the presented results may be somewhat inaccurate. Therefore, in future more detailed investigations by nonlocal methods are planned. Nevertheless, especially the amplifications of the subsonic mode show a clear decay in downstream direction towards the swallowing point as given in **figure 5** for three different amplified frequencies of the corresponding instability mode of the mean flow past the cone with a nose radius of $\overline{RN} = 17.78mm$.
More detailed information about the here demonstrated results can especially be found in the thesis of the author [4].

References

[1] Chang, C. L., Malik, M. R.
Effects of Shock on the Stability of Hypersonic Boundary Layers. AIAA Paper 90-1448, 1990.

[2] Kahn, M. M. S., Reshotko, E.
Stability of the Laminar Boundary Layer on a Blunted Plate in Supersonic Flow. Thesis, Case Western Reserve University, Cleveland, Ohio, FTAS/TR-79-142, 1980

[3] Kufner, E., Dallmann, U., Stilla, J.
Instability of Hypersonic Flow Past Blunt Cones -Effects of Mean Flow Variations-. AIAA Paper 93-2983, 1993.

[4] Kufner, E.
Numerische Untersuchungen der Strömungsinstabilitäten an spitzen und stumpfen Kegeln bei hypersonischen Machzahlen. Thesis, University of Stuttgart, 1994.

[5] Mack, L. M.
Boundary Layer Linear Stability Theory. AGARD Report 709, 1984, pp.3-1 bis 3-81.

[6] Malik, M. R., Spall, R. E.,Chang, C. L.
Effect of Nose Bluntness on Boundary Layer Stability and Transition. AIAA Paper 90-0112, 1990.

[7] Müller, B.
Upwind Relaxation Method for Hypersonic Flow Simulation. DLR-FB 91-36, 1991.

[8] Riedelbauch, S.
Aerothermodynamische Eigenschaften von Hyperschallströmungen über strahlungsadiabaten Oberflächen. DLR-FB 91-42 and Thesis, TU München, 1991.

[9] Simen, M.
Lokale und nichtlokale Instabilität hypersonischer Grenzschichtströmungen. DLR-FB 93-31 and Thesis, University of Stuttgart, 1993.

[10] Stetson, K. F., Thompson, E. R., Donaldson, J. C., Siler, L. G.
Laminar Boundary Layer Stability Experiments on a Cone at Mach 8, Part 2: Blunt Cone. AIAA Paper 84-0006, 1984.

[11] Stetson, K. F., Thompson, E. R., Donaldson, J. C., Siler, L. G.
Laminar Boundary Layer Stability Experiments on a Cone at Mach 8, Part 1: Sharp Cone. AIAA Paper 83-1761, 1983.

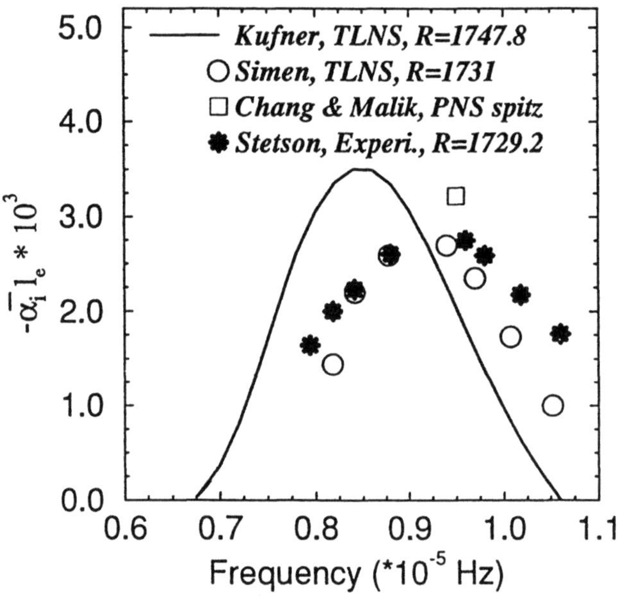

Figure 1: Comparison of experimental and theoretical spatial amplification rates of the second mode. Sharp cone, position $\overline{S}/\overline{S}_{max} = 0.34$, adiabatic wall.

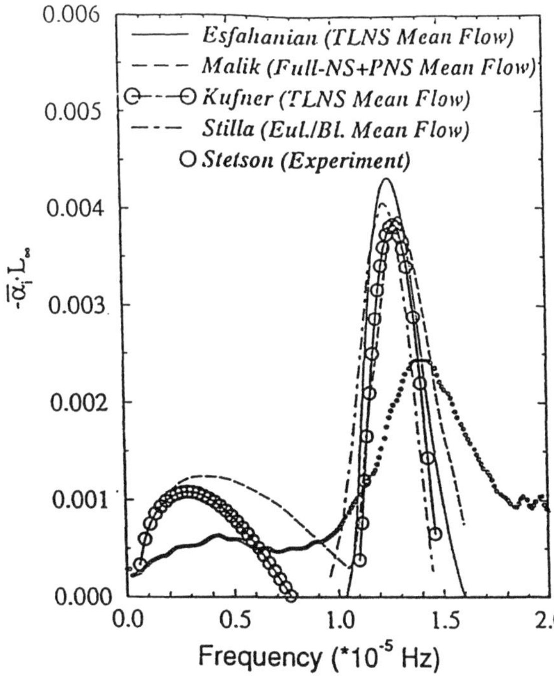

Figure 2: Comparison of spatial amplification rates according to different authors. Blunt cone, $\overline{RN} = 3.81mm$, position $\overline{S}/\overline{RN} \approx 175$, adiabatic wall.

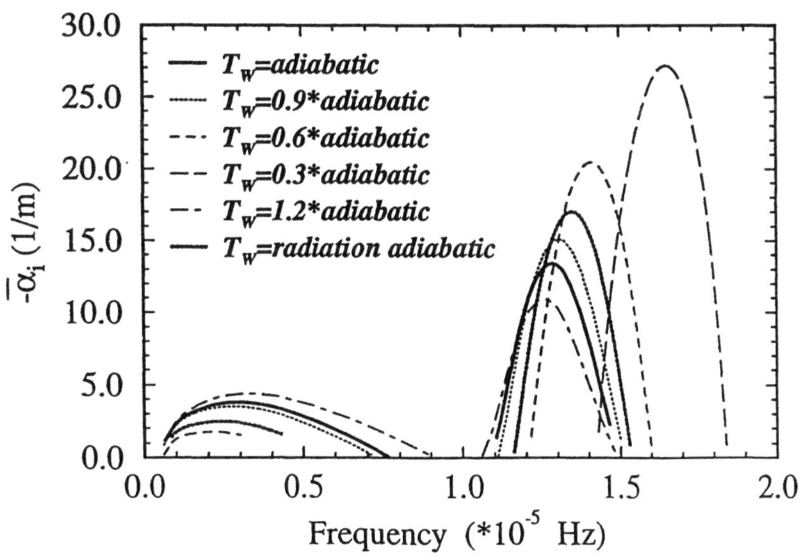

Figure 3: Effect of the wall temperature on spatial amplification rates. Blunt cone, $\overline{RN} = 3.81mm$, position $\overline{S}/\overline{RN} \approx 175$.

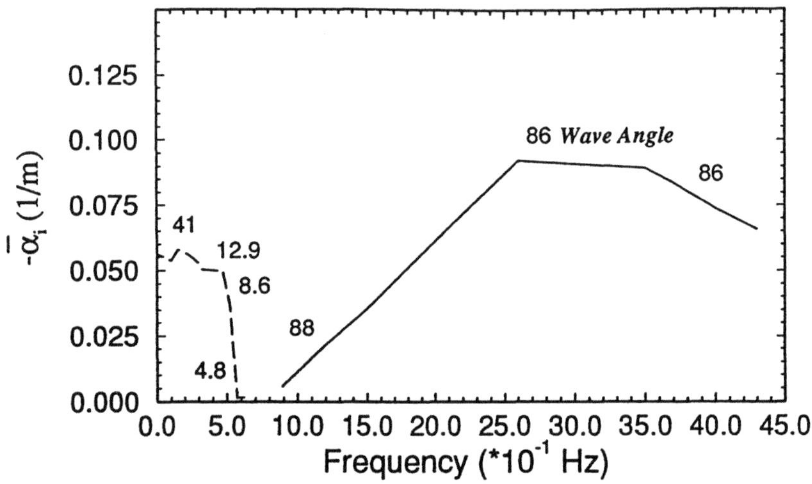

Figure 4: Spatial amplification rates of entropy layer instabilities. Blunt cone, $\overline{RN} = 42.672mm$, position $\overline{S}/\overline{RN} = 60.25$. Wave angles are given in degree.

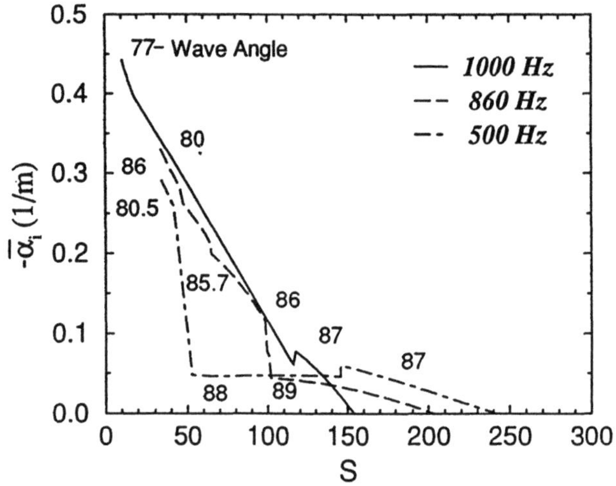

Figure 5: Spatial amplification rates of subsonic entropy layer instability, followed in downstream direction. Blunt cone, $\overline{RN} = 17.78mm$. Wave angles are given in degree.

Mathematical Nature of Parabolized Stability Equations

Fei Li and Mujeeb R. Malik
High Technology Corporation
Hampton, VA 23666 USA

Summary

The primitive-variable formulation of Parabolized Stability Equations is ill-posed due to the ellipticity introduced by $\partial\hat{p}/\partial x$ term and marching solution eventually blows up for sufficiently small step size. It is shown that this difficulty can be overcome if the minimum step size is greater than the inverse of the real part of the streamwise wavenumber, α_r. An alternative is to drop the $\partial\hat{p}/\partial x$ term, in which case the residual ellipticity is of no consequence for marching computations with much smaller but practical step sizes.

Key words: initial value problem, PSE, boundary-layer stability, marching step-size.

1. Introduction

The laminar-turbulent transition process in a boundary layer is of great fundamental and practical interest in fluid mechanics. In many cases, transition occurs via the destabilization and subsequent growth of wave structures in the boundary layer. Classical theories concerning the amplification of these waves use quasi-parallel assumption and ignore the growth of the boundary layer. Other theories (e.g., Gaster [1]) can deal with the boundary-layer growth locally for given problems. Direct Navier-Stokes solutions (Fasel & Konzelmann [2]) could give very satisfactory results but at the cost of much more CPU time and larger memory size. The recently developed method of parabolized stability equations (PSE) (Herbert [3], Bertolotti et al. [4], Chang et al. [5]) offers a fast and efficient way to analyze the spatial growth of instabilities in the boundary layer.

In the linear PSE method, a wave-like disturbance, ϕ, representing velocity, and pressure is assumed to be of the form

$$\phi(x,y,z,t) = \hat{\phi}(x,y)E \quad ; \quad E = \exp\left(i\left(\int_{x_0}^{x} \alpha(\xi)d\xi + \beta z - \omega t\right)\right). \tag{1a}$$

where α and β are the x and z wave numbers, ω is the disturbance frequency and $\hat{\phi}(x,y)$ is the shape function. The physical quantities (u,v, etc.) can be obtained by adding the complex conjugate component. In the above decomposition it has been assumed that the mean flow is independent of the spanwise coordinate z. The first and second derivatives of ϕ with respect to x can be written as

$$\phi_x = \left(i\alpha\hat{\phi} + \hat{\phi}_x\right)E \tag{1b}$$

$$\phi_{xx} = \left(-\alpha^2\hat{\phi} + i\alpha_x\hat{\phi} + 2i\alpha\hat{\phi}_x + \hat{\phi}_{xx}\right)E \tag{1c}$$

Substituting (1) and y-derivatives of ϕ into the linearized Navier-Stokes equations, we obtain a set of equations with the $\hat{\phi}(x,y)$'s and α as unknowns. Since there is now one more unknown (namely, $\alpha(x)$) than the equations, another condition is needed for the closure of the system. We take advantage of the slow variation of the mean flow in the streamwise direction and impose a condition on $\alpha(x)$ such that "most" of the waviness and growth of the disturbance are absorbed into the exponential function E, making the shape function $\hat{\phi}(x,y)$ slowly varying with x. Hence, the term containing $\hat{\phi}_{xx}$ (in (1c)) can be dropped and we arrive at a set of new equations in which the only second-order derivatives are those with respect to y. These parabolized stability equations may be written as

$$L_0\hat{\phi} + L_1\frac{\partial\hat{\phi}}{\partial x} + L_2\frac{\partial\hat{\phi}}{\partial y} + L_3\frac{\partial^2\hat{\phi}}{\partial y^2} = 0 \tag{2a}$$

where $\hat{\phi} = (\hat{u},\hat{v},\hat{w},\hat{p})^T$ and L_0–L_3 are square matrices given, for example, in Malik et al. [6]. Equation (2a) is to be solved subject to the constraint (imposing the condition that the shape function varies slowly with x)

$$F\left(\alpha,\hat{\phi}\right) = 0. \tag{2b}$$

Equation (2a) is solved numerically by marching from the initial station at $x = x_0$ with some initial condition, $\hat{\phi}(x_0,y)$. The solution at $x = x_0 + \delta x$ is computed with $\alpha\left(x_0 + \delta x\right) = \alpha\left(x_0\right)$ as a first approximation, then a new α is calculated using Eq. (2b). Equation (2a) is solved again with the new value of α. This process continues until the solution converges. The marching is then carried to the next x-station. In this paper, we show that this procedure is ill-posed and solution eventually blows up for sufficiently small marching step size. We also derive the minimum step size above which a stable numerical solution for the PSE formulation can be obtained.

2. Condition for Well-Posed Initial Value Problem

For well-posedness, we follow the procedure given by Kreiss and Lorenz [7] and consider an initial value problem

$$\frac{\partial u}{\partial x} = A\frac{\partial u}{\partial y} + Bu \tag{3}$$

$$u(0,y) = f(y) \tag{4}$$

where $f,u \in C^n$, $A,B \in C^{n\times n}$ and $y \in \Re$. The solution can be constructed by using Fourier transform in y and has the form

$$u(x,y)=\sum_{j=1}^{n}\frac{1}{2\pi}\int_{-\infty}^{\infty}A_j e^{i\eta y}g_j(\eta)e^{\lambda_j x}\,d\eta\ . \tag{5}$$

where λ_j are the eigenvalues of the system and η is the y wavenumber. If $\left|e^{\lambda_j x}\right|$ is unbounded for $|\eta|\to\infty$, the integral in Eq. (5) may not exist, and the initial value problem defined by Eqs. (3) and (4) is said to be ill-posed (Kreiss & Lorenz [7]).

In finding the eigenvalues $\lambda_j(\eta)$, both matrices A and B are used. In fact, subject to the condition that the eigenvalues of A are non-zero and distinct, the well-posedness of Eq. (3) is determined by the eigenvalues of A only. However, we must stress that, if some of the eigenvalues of A are zero, then all matrices in the governing equation must be used to obtain the correct result.

We now consider and analyze the well-posedness of the initial value problem associated with the PSE equations. The system consisting of Eqs. (2a) and (2b) is nonlinear since α appears in the coefficient matrices. In order to simplify the numerical stability analysis of the marching procedure, we will assume that α is known a priori. We choose a simple two-dimensional basic flow with constant velocity-components and apply Fourier transform in y-direction, i.e., let $\hat{u}(x,y)=\tilde{u}(x)\exp(i\eta y)$. After some algebra, we can write Eq. (2a) as follows,

$$\frac{d}{dx}\begin{pmatrix}\tilde{u}\\ \tilde{v}\\ \tilde{p}\end{pmatrix}=\begin{bmatrix}-i\alpha & -i\eta & 0\\ 0 & -\dfrac{C}{D} & -i\dfrac{\eta}{D}\\ -C+i\alpha D & i\eta D & -i\alpha\end{bmatrix}\begin{pmatrix}\tilde{u}\\ \tilde{v}\\ \tilde{p}\end{pmatrix} \tag{6}$$

where $C=i\alpha U-i\omega-i\dfrac{d\alpha}{dx}+\dfrac{\alpha^2}{R}+i\eta V+\dfrac{\eta^2}{R}$ and $D=U-2i\dfrac{\alpha}{R}$. The three eigenvalues of the matrix in Eq. (6) are given by

$$\lambda_1=\eta-i\alpha,\quad \lambda_2=-\eta-i\alpha,\quad \lambda_3=-\frac{C}{D}, \tag{7}$$

For a well-posed initial value problem, it is required that Real $\lambda_j<0$ for $|\eta|\to\infty$. Since η can take any value for general initial conditions, λ_1 and λ_2 have positive real parts for some η. Therefore, Eq. (6) is not well-posed for initial value problem and some approximation must be made for a stable marching solution. We will discuss it in the next section.

Due to the complicated nature of λ_3, it is not easy to find the sign of its real part. However, in order to determine the well-posedness, we only need to know the behavior of λ_3 as $|\eta|\to\infty$, provided that the coefficient of the highest order term in η is not purely imaginary. In this limit, we have

$$\mathrm{Real}(\lambda_3)\to-\frac{\eta^2}{R}\frac{U+2\alpha_i/R}{(U+2\alpha_i/R)^2+4\alpha_r^2/R^2} \tag{8}$$

Hence, for growing waves ($\alpha_i<0$), real $\lambda_3>0$ for sufficiently small U, i.e.,

$$U < \frac{-2\alpha_i}{R} \tag{9}$$

For large Reynolds number and small α_i (typically $O(10^2)$ and $O(10^{-2})$, respectively), condition (9) is satisfied only in a very small neighborhood of the wall. From a practical view point, difficulties in numerical solution never arise from λ_3 for wall-bounded flows.

3. Stability of the Marching Procedure

The marching procedure can be made stable by using a sufficiently large step size, δx, as in PNS for mean flow, and we will derive a condition for stability. If the eigenvalues are distinct, then Eq. (6) can be wrriten as

$$\frac{d\varphi}{dx} = \lambda\varphi \tag{10}$$

where φ represents a linear combination of $\tilde{u}$, $\tilde{v}$ and $\tilde{p}$. If this equation is solved with backward difference with step-size δx, von Neumann analysis leads to an amplification factor given by

$$\gamma = \frac{1}{1-\delta x\lambda} \tag{11}$$

Numerical stability requires $|\gamma| < 1$, which can be satisfied if $(1-\lambda_r\delta x)^2 + \lambda_i^2\delta x^2 > 1$ and hence, $\delta x > 2\lambda_r / (\lambda_r^2 + \lambda_i^2)$. Since λ depends on η, then minimum step size is given by

$$\delta x_{\min} = \sup_{\eta}\left(\frac{2\lambda_r}{\lambda_r^2 + \lambda_i^2}\right) \tag{12}$$

Equation (12) implies that there is no step-size restriction if $\lambda_r \le 0$, otherwise the step size has to be greater than some finite value if the marching is to be stable. For example, von Neumann analysis of, say, the equation associated with λ_2 leads to an amplification factor given by

$$\gamma_2 = \frac{1}{(1-(\eta-\alpha_i)\delta x) + i\alpha_r\delta x} \tag{13}$$

Numerical stability requires $|\gamma_2| < 1$ for all η. This can be achieved only if

$$\delta x > \frac{1}{|\alpha_r|}, \tag{14}$$

which is graphically depicted in Fig. 1 where $|\gamma_2(\eta)|$ is plotted. The minimum step size restriction implies that a maximum of 2π steps per disturbance wavelength are allowed for the marching. This is consistent with the behavior of the numerical solution first reported by Chang et al. [5]. We should note that, in PSE, we solve for the slow varying shape functions and so this step-size restriction does not cause problems in terms of accuracy; indeed, the numerical examples given in Joslin et al. [9] show that the PSE solution obtained with 3 steps per wavelength for Tollmien-Schlichting waves in a Blasius boundary layer is in excellent agreement with very accurate direct simulation of Navier-Stokes equations using 60 grid points per TS wavelength.

When step-size smaller than $\delta x_{\min} = 1/|\alpha_r|$ is required for either higher resolution or for the convergence of nonlinear terms in an implicit numerical scheme, a further approximation is sometimes made, i.e., dropping the pressure gradient term $\partial\hat{p}/\partial x$ (see Chang et al. [5], Haj-Hariri [10]). We point out, however, that most of the physical pressure gradient is absorbed into the term $i\alpha\hat{p}$, and that $\partial\hat{p}/\partial x$ is very small in comparison (refer to Eq. (2c)). After $\partial\hat{p}/\partial x$ is dropped, Eq. (6) can be reduced to a set of 2 equations by eliminating $\hat{p}$. Carrying out the well-posedness analysis, we obtain two eigenvalues

$$\lambda_1 = -i\left(\alpha + \frac{\eta^2}{\alpha}\right), \quad \lambda_2 = -\frac{C}{D}$$

where C and D are the same as those in Eq. (7). Here, λ_2 gives rise to ill-posedness due to viscosity, which is insignificant as mentioned before. We note that λ_1 represents ill-posedness if $\alpha_i < 0$ (i.e., for growing waves). Using Eq. (12), and assuming that $|\alpha_i| << |\alpha_r|$, we obtain the minimum step-size required for a stable marching scheme,

$$\delta x_{\min} \approx \frac{-\alpha_i}{4|\alpha|} \cdot \frac{1}{|\alpha_r|} + O\left(\frac{\alpha_i^3}{\alpha_r^3}\right) \tag{15}$$

For small $\alpha_i/|\alpha|$, we see that the restriction on step-size is relaxed by at least an order of magnitude in comparison with the condition given above.

4. Application to Blasius Boundary Layer

The above analysis was simplified in that the mean flow (U,V) was considered constant. We now apply the results of the above analysis to a realistic problem, i.e., the Blasius boundary-layer which is subject to Tollmien-Schlichting (TS) instability. We consider a disturbance of fixed frequency $F = 0.7 \times 10^{-4}$ where

$$F = \frac{2\pi\nu}{U_e^2}\bar{f}$$

with $\bar{f}$ being the dimensional frequency in Hertz.

Calculations were started at a streamwise location x_0, corresponding to $R = U_e L/\nu = 500$, where $L = \sqrt{\nu x_0/U_e}$. For this disturbance, the initial wavenumber at $R = 500$ is $\alpha_r = .106$. As the wave travels downstream, the wavenumber gradually changes to approximately $\alpha_r = .103$. According to Eq. (15), the minimum step-size is $\Delta x_{\min} \approx 9.5$ based upon the length scale at $R = 500$. We ran our PSE code with $\partial\hat{p}/\partial x$ term included and $\Delta x = 10$ (greater than $\Delta x_{\min}$). The growth rate plotted in Fig. 2 shows a smooth solution. The growth rate in this figure has been made nondimensional with local length scale which varies with x; however, a constant length scale was used in the marching computations. Next we use $\Delta x = 9$ (smaller than Δx_{min}), the solution eventually blows up. Now, we drop $\partial \hat{p} / \partial x$ and further reduce the step-size to 5, a smooth solution is again obtained. These calculations confirm the estimate for the minimum step-size and show that the dropping of $\partial \hat{p} / \partial x$ relaxed the step-size restriction and show that the stability condition derived in the previous section provides a remarkably good prediction of the behavior of the PSE solution. The comparison of the solution with $\Delta x = 5$ and 10 shows that the approximation $\partial \hat{p} / \partial x = 0$ does not introduce any error, but this is not necessarily the case in other flows.

Figure 3 shows the growth rate results for the same frequency disturbance and $\partial \hat{p} / \partial x = 0$. Three different step sizes were chosen: $\Delta x = .3$, .25 and .2. With $\Delta x =$ .3, the solution is generally smooth but appears to develop some very slight wiggles near the end. However, with $\Delta x = .25$ and .2 solution blows up. The numerical instability occurs earlier for the smaller step size.

The $\Delta x = .2$ calculation is repeated (Fig. 4) with $i\alpha/R$ dropped to eliminate elliptic effect due to viscosity. The behavior of the solution is unaffected. This indicates that these oscillations are not caused by ellipticity due to the viscous terms. Finally, we eliminate the iteration for α, and do the calculation with fixed real α (the value at $R = 500$). The growth rate shown in Fig. 4 is inaccurate, but the growth rate curve is smooth. For the present case of $(-\alpha_i)_{max} \approx .0055$ and $\alpha_r \approx .1$, it follows from (15) that $\Delta x_{min} \approx .13$ which is not far from the value of .25 or .2 used in Figs. 3(b) and (c), respectively. The fact that smooth solution was obtained in Fig. 3(c) with α fixed does not necessarily suggest that (19) is not operative since real (α) was used and with $\alpha_i = 0$ Eq. (15) gives $\Delta x_{min} \approx 0$.

In order to clarify whether (15) is operative, we perform additional computations and report the results in Fig. 4 which contains the correct solution from Fig. 2 as well as the smooth solution from Fig. 3(c) with $\alpha_i = 0$ and $\Delta x = .2$. Also included are solutions using fixed $\alpha = .104 - i\,.001$ and $\alpha = .104 - i\,.004$ ($\hat{p}_x$ and $i\alpha/R$ have been dropped). Equation (15) for $\alpha_i = 0$, $\alpha_i = -.001$ and $-.004$ yield minimum step size of 0, .023 and .092, respectively. In the first two cases, the step size of $\Delta x = .2$ used in the computation is much larger than the critical value given by (15) and therefore the solution is smooth. Since initially the value of α_i is close to the correct solution given by (2b), the computed growth rate is close to the correct growth rate. However, further downstream the computed growth rate is in error since (2b) is not satisfied. With $\alpha_i = -.004$, the computational step size is not very far from the critical value and, therefore, solution becomes oscillatory and eventually blows up. In this case, since initially $\alpha_i = -.004$ is not close to the solution of (2b) the growth rate is in error at small Reynolds numbers. However, as the correct growth rate approaches the value of $-.004$, the solution with $\Delta x = .2$ and without imposition of (2b) approaches the correct solution even though it is oscillatory due to the numerical instability. This example clearly suggests that numerical instability observed with $\Delta x = .2$ is associated with the condition given by (15).

We point out that the values of $\Delta x = .3$ and .2 correspond to about 200 to 300 marching steps per TS wavelength. Clearly, in no linear or nonlinear applications one needs to take such a small step and, therefore, Eq. (15) is of no significance for most practical computations.

7. Conclusions

The mathematical nature of parabolized stability equations (PSE) is studied. It is shown that the primitive variable formulation is mathematically ill-posed due to the pressure-gradient term. The condition for the stable marching solution is derived. Examples from the linear stability of Blasius boundary-layer are given to show that this condition gives reasonable estimates of the numerical behavior of the parabolized equations. The results also show that the ellipticity associated with the viscous term is insignificant. It is also found that, contrary to the findings of Haj-Hariri, dropping $\partial \hat{p} / \partial x$ term does not completely remove ellipticity.

Acknowledgments

This work was performed under AFOSR contract F49620-91-C-0014. The second author wishes to thank Professor David Gottlieb for useful discussions and for referring him to the book of Kreiss and Lorenz on initial boundary value problems.

References

[1] Gaster, M.: On the Effects of Boundary Layer Growth on Flow Stability, J. Fluid Mech. 66, (1974) 465–480.
[2] Fasel, H. F. and Konzelmann, U.: Non-parallel Stability of a Flat Plate Boundary Layer using the Complete Navier-Stokes Equations, J. Fluid Mech. 221, (1990) 311–347.
[3] Herbert , Th.: Boundary-Layer Transition — Analysis and Prediction Revisited, AIAA Paper 91-0737 (1991).
[4] Bertolotti, F. P., Herbert, Th. and Spalart, P. R.: Linear and Nonlinear Stability of the Blasius Boundary Layer, J. Fluid Mech. 242 (1992) 441-474.
[5] Chang, C.-L., Malik, M. R., Erlebacher, G. and Hussaini, M. Y.: Compressible Stability of Growing Boundary Layers Using Parabolized Stability Equations, AIAA Paper 91-1636 (1991).
[6] Malik, M. R., Li, F. and Chang, C.-L.: Crossflow Disturbances in Three-Dimensional Boundary Layers: Nonlinear Development, Wave Interaction and Secondary Instability, J. Fluid Mech. 268 (1994) 1–36.
[7] Kreiss, H.-O. and Lorenz, J.: Initial-Boundary Value Problems and the Navier-Stokes Equations, Academic Press (1989).
[8] Golub, G. H. and VanLoan, C. F.: Matrix Computations, The Johns Hopkins University Press (1983) 200.
[9] Joslin, R. D., Streett, C. L., and Chang, C.-L.: 3-D Incompressible Spatial Direct Numerical Simulation Code Validation Study — A Comparison with Linear Stability & Parabolic Stability Equation Theories for Boundary-Layer Transition on a Flat Plate, NASA TP-3205 (1992).
[10] Haj-Hariri, H.: Characteristics Analysis of the Parabolized Stability Equations, Studies in Applied Mathematics 92 (1994) 41–53.

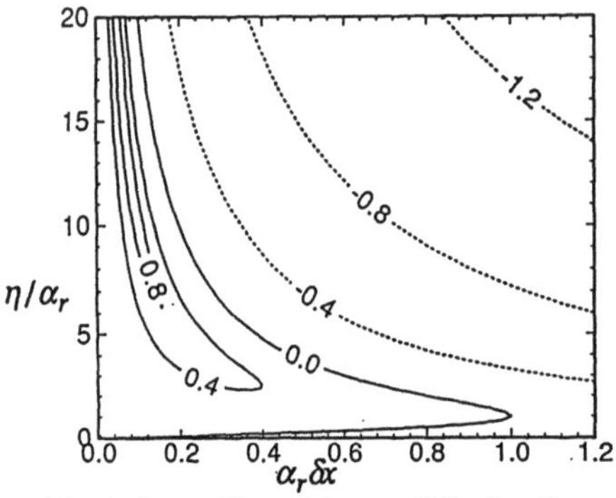

Fig. 1. Logarithm of the amplification factor γ for real α. Solid lines represent numerical instability.

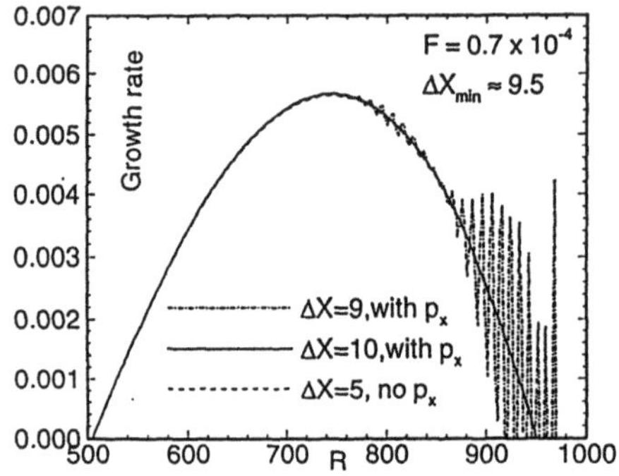

Fig. 2. Effect of marching step size on the growth rate of the TS disturbance of frequency $F = .7\times10^{-4}$. The critical step size for stable solution is $\Delta x \approx 9.5$.

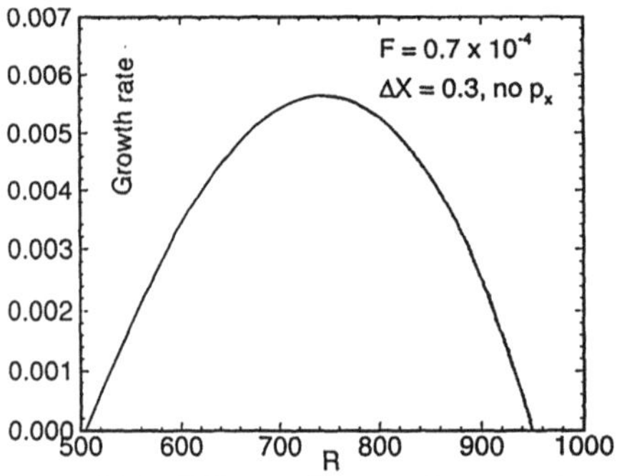

Fig. 3(a). Effect of step size on the growth rate of the TS disturbance ($F = .7\times10^{-4}$) with $\partial\hat{p}/\partial x = 0$.

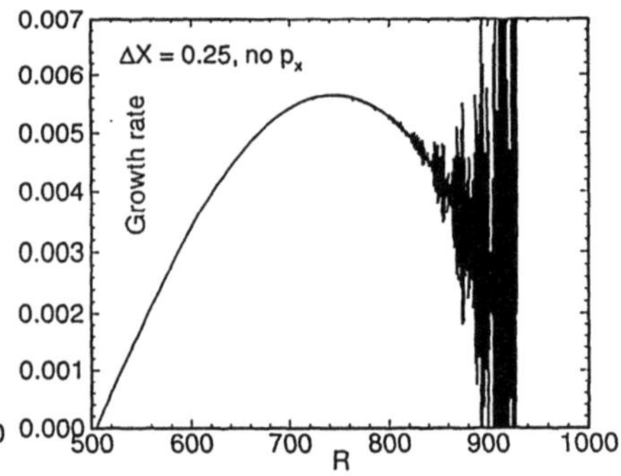

Fig. 3(b). Continued.

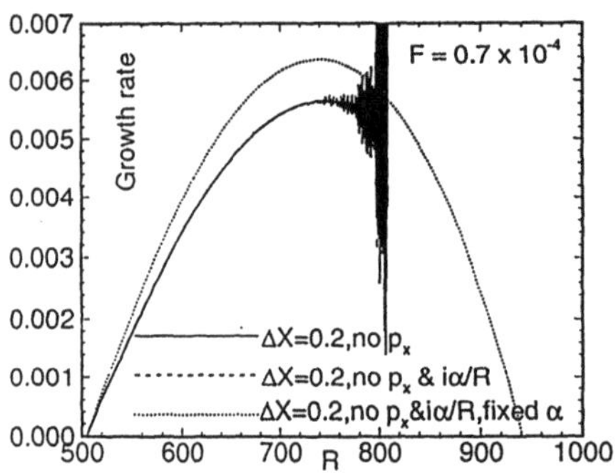

Fig. 3(c). Concluded.

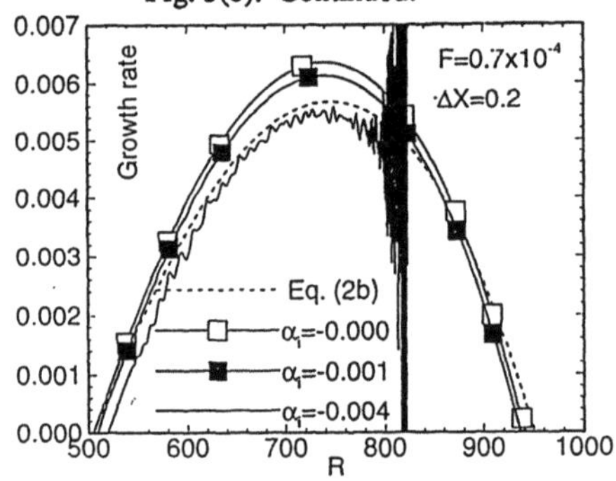

Fig. 4. Growth rate with $\partial\hat{p}/\partial x = 0$, $i\alpha/R = 0$, and fixed $\alpha_r = 0.104$.

High-Speed Boundary-Layer Stability and Transition

V.I. Lysenko
Institute of Theoretical and Applied Mechanics of Russian Academy of Sciences, Siberian Branch, Novosibirsk 630090, Russia

Summary
The experimental and numerical investigations of the high-speed boundary-layer stability and transition were carried out. The hotshot, nitrogen, wind tunnels and shock tube were used. The influence of Mach number, temperature factor, ratio of specific heats, entropy layer, dissociation, ionization and magnetic field were studied. The problem of turbulization of a stable hypersonic boundary layer by means of an obstacle was considered as well. The stability and laminar/turbulent transition in high-speed wakes behind a flat plate were studied experimentally too.

Key words: supersonic, hypersonic, boundary layer, stability, transition.

Results and Discussion

The calculations of the boundary-layer stability and its transition were performed using the e^n method (n=9) at different values of the Mach number M and the temperature factor T_w (the wall-to-adiabatic wall temperature ratio). The calculations were carried out for disturbances of the first (analogous to the well-known Tollmien-Schlichting waves of an incompressible fluid) and the second (a variety of acoustic resonance in a shear flow) unstable modes. For the numerical integration the system of stability equations in the Dunn-Lin approximation was used. The calculations were performed for the most unstable three-dimensional disturbances of the first mode, and for the two-dimensional (most unstable) disturbances of the second mode. The results for a cone at zero incidence are presented in Figs.1 and 2. R_t is the transition Reynolds number. It follows from Fig.1 that for M more than 7 and T_w=1, with increasing M the Reynolds number of transition, caused by both the first and the second (high-frequency) modes, increases. The surface cooling (Fig.2) stabilizes the first-mode disturbances and destabilizes the second-mode disturbances.

The influence of the stagnation temperature and velocity (pressure) gradient on the boundary-layer stability and the transition was also studied. This influence is shown in Figs.3 and 4. F is the dimensionless frequency parameter and α_i are the disturbances amplification rates. The law of viscosity as a function of temperature according to Sutherland was used in the calculations. A local similarity in the boundary-layer flow is assumed. Figs.3 and 4 illustrate the stabilizing effect of the stagnation temperature and the negative pressure gradient on both the first- and the second-mode disturbances. In particular, the effect on the second mode is more pronounced than on the first mode.

The boundary-layer stability on a flat plate within the range of Mach numbers (at the boundary-layer bound) M_e=11-16 was studied experimentally in the nitrogen tunnel T-327. The measurements of the boundary-layer stability were carried out with the help of the hot-wire anemometer of constant current. The results of the theory for the second (high-frequency) disturbances mode were confirmed qualitatively. It is shown in Fig.5 that the growth of Mach number exerts a stabilizing influence on the boundary-layer stability. The decrease in temperature factor (Fig.6) exerts a destabilizing effect on the boundary-layer stability.

In the same nitrogen tunnel T-327 at M_e=16 a study was carried out of the stable boundary-layer turbulization by means of an obstacle of a rectangular shape. The same plate served as a model but the obstacle was located at the distance of 60 mm apart from the leading edge. The obstacle height varied and had the following values: 0,1,2,3,4,5,8,10 and 15 mm. The thickness of the boundary layer on the plate without turbulator at the longitudinal coordinate x=60 mm was 11 mm, approximately. It was observed in the experiments (Figs.7 and 8) that even a very stable boundary layer at large Mach numbers on a flat plate can be turbulized by an obstacle.

The following experiments were carried out in the wind tunnel T-325 on a flat plate, insulated, with an obstacle for M_e=4. The hot-wire anemometer of constant current and the probes with tungsten wire of 6 micrometers in diameter and 1.5 mm in length were used. The profiles of mass flow disturbances (Fig.9) and the profiles of average value of velocity/viscosity relation were obtained. A rectangular obstacle was located 41 mm apart from the leading edge. The obstacle height h had the following values: 0,0.1,0.2,0.4,0.8 and 1.2 mm (the thickness of the boundary layer Y on the plate without turbulator at x=41 mm was about 0.8 mm). The measurements were carried out in the section at x=123 mm. It was observed that with an increasing h the thickness of the disturbed layer grew considerably and the profiles of velocity/viscosity relation were straightened. The maximum of the pulsations of the mass flow is situated at y/Y=0.7 for the laminar boundary layer. For the turbulent boundary layer (when h=0) it is at y/Y=0.5. When h grows, the position of maximum (regarding the boundary of disturbed layer) changes only a bit. By the way, for all h the normal coordinate of this maximum is more than h. The maximum intensity of the mass flow pulsations decreases with the transition from the laminar boundary layer to the turbulent one. then h/Y grows up to 1, remains on the same level (though an insignificant rising occurs), and decreases again with the further growth of h so that the profiles of the mass flow pulsations become steeper.

The following experiments were carried out at M=4 in the wind tunnel T-325 on an insulated plate using the hot-wire anemometer of constant current. The leading-edge bluntness of the flat plate had the following values: 0.1, 1, 1.5, 2, 3, 5 and 10 mm. The carried out experiments showed that the growth of an entropy layer with the increasing bluntness of the model destabilizes both the disturbances in the entropy layer (Figs.10 and 11) and (beginning at some bluntness) the first-mode disturbances in the boundary layer (Fig.12). Thus at M=4 the whole shock layer is destabilized. A clearly expressed reverse of the transition was observed in the dependence of the transition Reynolds number on the leading edge bluntness (Fig.13). The comparison with the data for M=8 allows to assume that with increasing Mach number the role of the entropy-layer instability increases in a general instability of the whole shock layer.

The experiments concerning the determination of the boundary-layer stability near the plate surface and near the critical layer (inside the boundary layer) were carried out in the wind tunnel T-325. The stability of the boundary layer and the transition position were determined by means of the constant-current hot-wire anemometer and five hot-wire probes located on the plate surface (on ebonite substrates) at the distance of 10 mm one from another. It turned out that at M=2 the experimental amplification rates of the disturbances near the model surface were less than those inside the boundary layer near the critical layer. At M=4 (Fig.14) they were approximately equivalent. The measurement results were also compared with the results of calculations (by Gaponov) which were executed on the basis of the linearized Navier-Stokes equations with due regard to the weak nonparallelism of the main flow.

The experiments and calculations regarding the influence of the ratio of specific heats on the stability and transition of the boundary layer were also carried out. The experiments were performed in the hotshot tunnel IT-302M at the Mach number $M_e=4$, temperature factor $T_w=0.18$ and stagnation temperature $T_0=1800$ K. The air, carbon dioxide and argon were the test gases (the values of ratio of specific heats were 1.41, 1.2 and 1.67, respectively). A polished steel flat plate was used as a test model. An optical method was employed for fixation of the transition location. . Besides experiments, the calculations of boundary layer stability and its transition were performed by using the e^n method at different values of the ratio of specific heats. The calculations were carried out for two-dimensional disturbances of the second unstable mode (for $M_e=4$ and $T_w=0.18$ precisely, the second mode determines the location of the boundary-layer transition, and the most unstable are two-dimensional disturbances). The calculations have shown that with the decreasing ratio of specific heats the disturbances amplification rates increase considerably. The frequency of disturbances leading to the transition increases as well. Both experiments and calculations have shown (Figs.15 and 16) that with the decreasing ratio of specific heats, the Reynolds number of transition, depending on the disturbances of the second mode, decreases considerably.

The following experiments were carried out aimed at the determination of the influence of dissociation and ionization of gases on the boundary-layer transition. The effect of dissociation was studied in the hotshot tunnel 302M. The model and the instrumentation for transition fixation remained the same as in the previous series of experiments, but the free stream parameters varied. Air and nitrogen were the test gases. The free stream Mach number varied from 5 to 7 depending on the stagnation temperature (in particular, for temperatures higher than that of the oxygen dissociation, M=5). The flow stagnation temperature T_0 varied within the range from 1100 up to 3600 K. The temperature factor $T_w=0.32$-0.10. At stagnation temperature more about 2600 K, the oxygen, being a part of the air, started dissociating, and the dissociation of nitrogen did not occur within the range of the used temperatures. For these data of M and T_w, the flow instability was caused by the second mode disturbances. Experiments have shown (Fig.17) that the gas dissociation at M=5 and Tw=0.1 leads to a decrease in the transition Reynolds number. It qualitatively corresponds to the computation results of Petrov for a case of external flow dissociation at moderate Mach number less than about 6 (when the dissociation stabilizes the first-mode oscillations and, on the contrary, destabilizes the second-mode disturbances).

In the shock tube UT-3 the effect of gas ionization on the transition location on tube walls was examined. The test gases were argon and xenon. They are monoatomic gases with close ratios of specific heats (at the normal conditions equal to 1.67), but different ionization potentials (argon has a higher value of 15.76ev as compared with 12.13ev for xenon). The ionization of xenon began at the shock-wave Mach number M_s=8.6, and that of argon at M_s=9.5. The performed experiments have shown (Fig.18) that the ionization of the flow at M_s=8-11 leads to a weak increase in the transition Reynolds number on the shock-tube walls. It qualitatively corresponds to the computation results of Petrov for the case of external flow dissociation at large Mach number M more than 8 approximately (when the dissociation already started exerting some stabilizing effect on the second mode disturbances).

The effect of the longitudinal magnetic field on the boundary-layer transition on a flat plate with an obstacle in the ionized gas was studied in the tunnel T-327 at M_e=16. It was observed (Fig.19) that the longitudinal magnetic field stabilizes the boundary-layer disturbances, and the transition is delayed.

The determination of position of laminar/turbulent transition of the high-speed wake after flat plate was studied in the hotshot tunnel "Transit" at M=4, 5 and 7. The performed experiments have shown that the wake transition position depends on the state (laminar, turbulent or transitional) of the plate boundary layer. The growth of Mach number exerts a stabilizing influence on the wake transition. The experiments concerning the determination of the wake (behind the flat plate with the sharp leading edge and blunt back edge) stability at Mach number 4 were carried out in the wind tunnel T-325 also. Characteristic persisting peaks in the power spectra in a free shear layer and wake after the plate are observed. On the whole (Fig.20), the wake is more unstable than the plate boundary layer (near back edge) and the free shear layer (after this plate).

Conclusions:

(a) With the increasing Mach number and stagnation temperature the hypersonic boundary-layer stability increases, but the decreasing temperature factor and ratio of specific heats destabilizes the second-mode disturbances.

(b) Such hypersonic effects as dissociation, ionization and the influence of a longitudinal magnetic field exert the stabilizing effect on the boundary-layer stability and the transition for M more than 8. In other words, if in the calculations without consideration of these effects the boundary layer is laminar, it will be laminar in the calculations where those effects are included.

(c) The very stable (at M approximately equal to 20) hypersonic boundary layer can be turbulized by means of an obstacle on the flowed model surface.

(d) The growth of an entropy layer with the increasing bluntness of the model at M=4 destabilizes both the disturbances in the entropy layer and (beginning at some bluntness) the first-mode disturbances in the boundary layer.

(e) The growth of Mach number exerts a stabilizing influence on the wake (behind the flat plate) transition.

(f) On the whole, the wake (behind the flat plate with the sharp leading edge and blunt back edge) is more unstable than the plate boundary layer (near back edge) and the free shear layer (after this plate).

Fig. 1

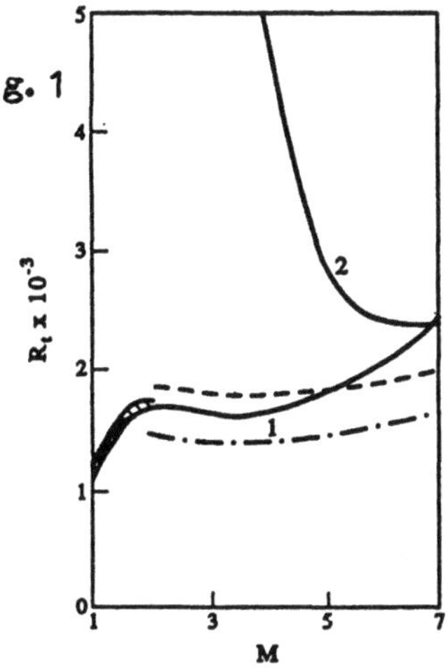

Fig. 2

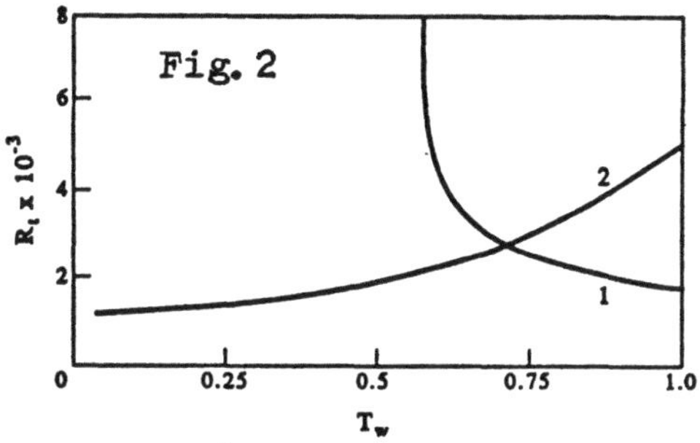

Fig. 3

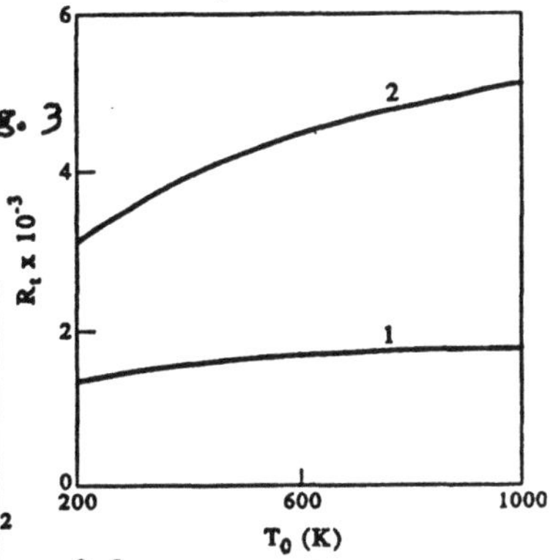

Fig. 4

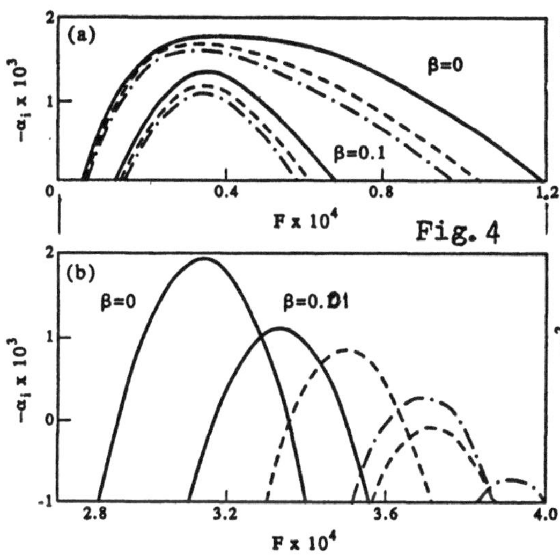

Fig. 5

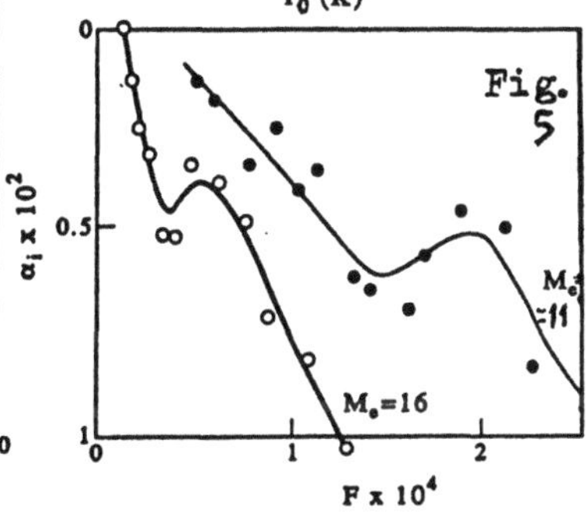

FIG. 1. Numerical Reynolds number of the transition on an insulated cone as a function of Mach number ($T_e = -50$°C): (1) for the first mode of disturbances; (2) for the second mode, ▨, the flight data of Fisher and Dougherty: (–·–), the ballistic data of Beckwith and Bertram :

D. F. FISHER and N. S. DOUGHERTY, JR, In-flight transition measurement on a 10° cone at Mach numbers from 0.5 to 2.0. NASA Technical Publication No. NASA-TP-1971 (1982).

I. E. BECKWITH and M. H. BERTRAM, A survey of NASA Langley studies on high-speed transition and the quiet tunnel. NASA Technical Report No. NASA-TM-X-2566 (1972).

FIG. 2. Numerical transition Reynolds number as a function of temperature factor ($M = 4$, $T_0 = 937$ K): (1) for the first unstable mode; (2) for the second mode.

FIG. 3. Numerical transition Reynolds number as a function of stagnation temperature ($M = 4$, $T_W = 1$): (1) for the first mode; (2) for the second mode of disturbances.

FIG. 4. Numerical disturbance amplification rates as functions of the frequency parameter at different negative pressure gradients (positive β) and different stagnation temperatures ($M = 4$, $T_W = 1$, $R = 780$): (a) first mode; (b) second mode; (——), $T_0 = 300$ K; (– – –), 600 K; (–·–), 900 K.

FIG. 5. Experimental disturbance amplification rates as functions of the frequency parameter for $T_W = 0.28$–0.30, $R = 481$–558, $M_e = 11$ and 16.

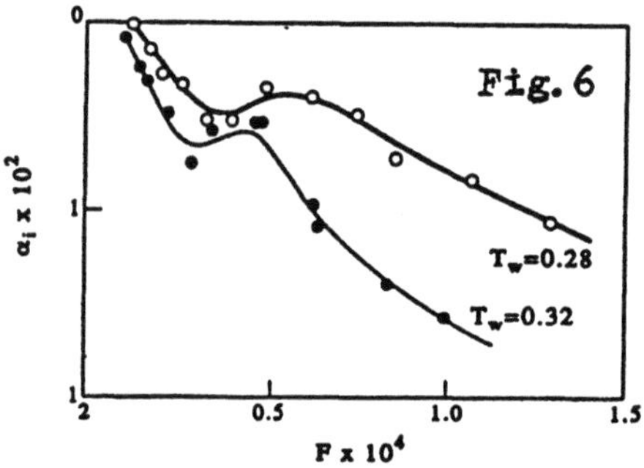

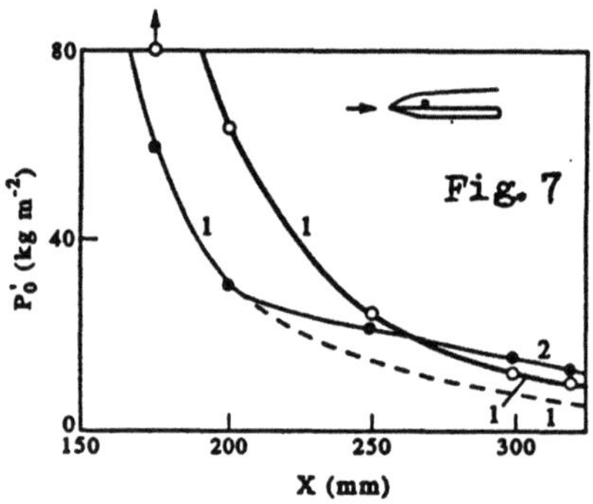

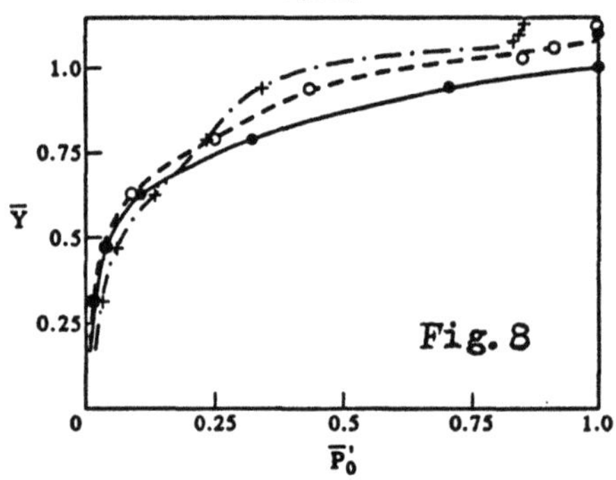

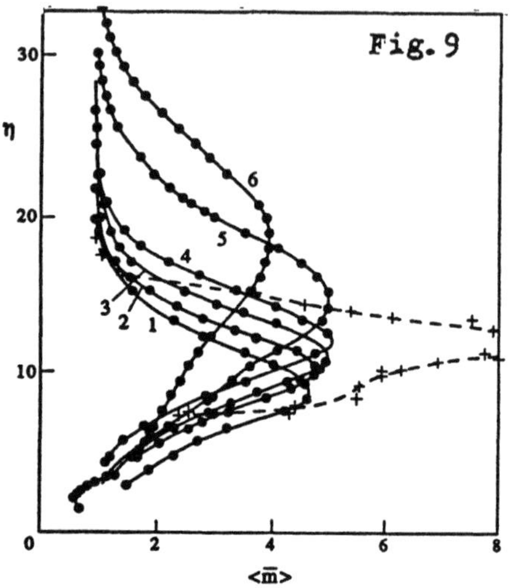

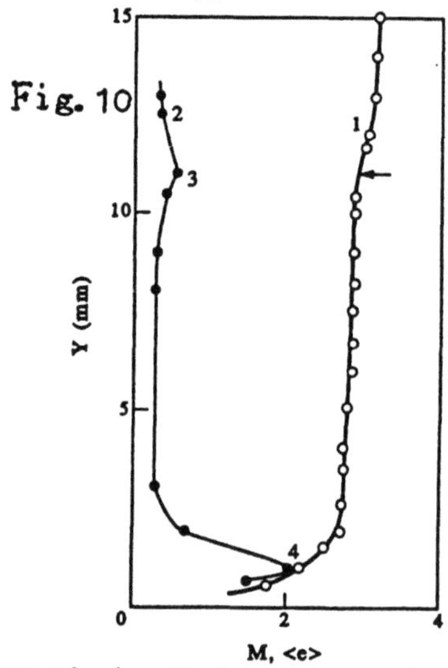

FIG. 6. Experimental amplification rates of disturbances as functions of the frequency parameter for $M_e = 16$, $R = 558$–600, $T_W = 0.28$ and 0.32.

FIG. 7. Distribution of the local Pitot total pressure along the streamwise coordinate for $M_e = 16$, the normal coordinate $y = 20$ mm and different obstacle height values: (○), $h = 0$; (●), $h = 15$ mm; (1) laminar boundary layer; (2) similar to turbulent boundary layer.

FIG. 8. Distribution of the local Pitot total pressure across the boundary layer at different values of obstacle height, $M_e = 16$, $(Re_1)_\infty = (1.03$–$1.13) \times 10^6\ \mathrm{m}^{-1}$, $x = 295$ mm: (●), $h = 0$; (○), 8 mm; (+), 15 mm.

FIG. 9. Distribution of the mass-flow disturbances across the boundary layer at different dimensionless obstacle heights; $M_e = 4$, $(Re_1)_\infty = 11 \times 10^6\ \mathrm{m}^{-1}$, $x = 123$ mm: (+, ---), laminar boundary layer; (●, ——), turbulent boundary layer; (1) $\bar{h} = 0$; (2) 0.12; (3) 0.25; (4) 0.5; (5) 1.0; (6) 1.5.

FIG. 10. Profiles of Mach number (1) and the hot-wire voltage fluctuations (2) in a shock layer for the plate leading-edge bluntness $b = 5$ mm, $M_\infty = 4$, $(Re_1)_\infty = 25.3 \times 10^6\ \mathrm{m}^{-1}$, $x = 80$ mm: (3) entropy layer; (4) boundary layer.

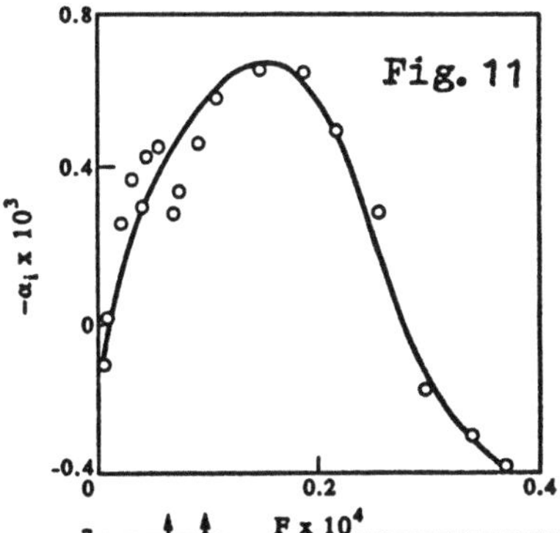

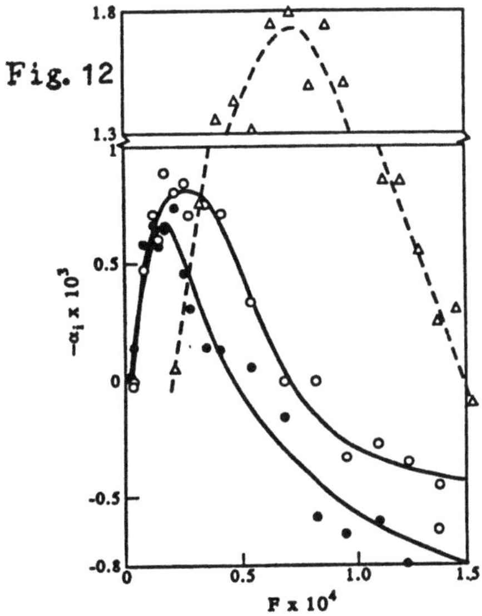

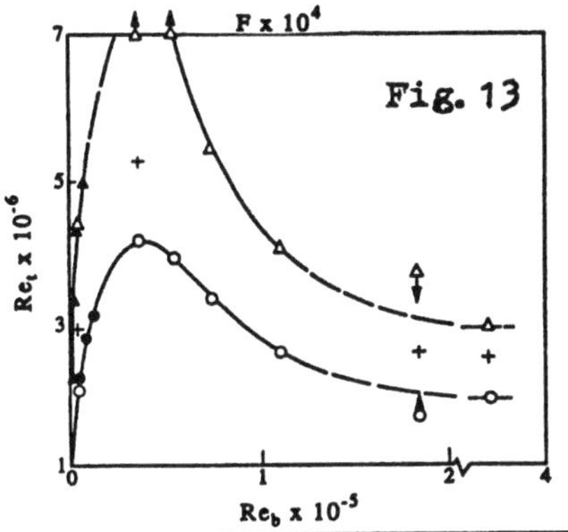

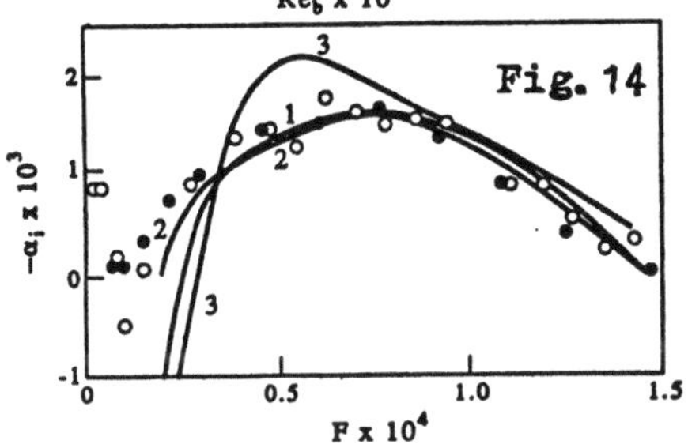

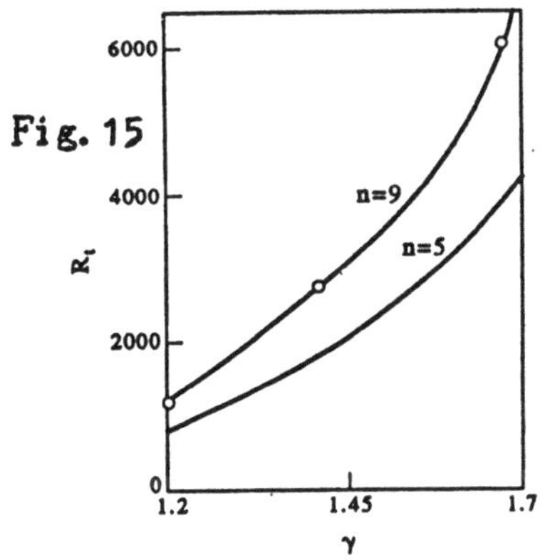

FIG. 11. Experimental amplification rates of the disturbances in an entropy layer as functions of the frequency parameter for $M_\infty = 4$, $(Re_1)_\infty = 25.3 \times 10^6\ \text{m}^{-1}$, $x = 80$ mm, $b = 5$ mm.

FIG. 12. Experimental boundary-layer disturbance amplification rates as functions of the frequency parameter for $M_\infty = 4$, $M_e = 2.7$, $(Re_1)_e \approx 7.8 \times 10^6\ \text{m}^{-1}$, $R_e = 765$–780, for different leading-edge bluntnesses: (○), $b = 5$ mm; (●), 3 mm; (△), 0.02 mm.

FIG. 13. Transition Reynolds number as function of the bluntness Reynolds number, $M_\infty = 4$, $(Re_1)_\infty = 37 \times 10^6\ \text{m}^{-1}$: (○, ●), full-pressure gauge transition-onset data; (△, ▲), full-pressure gauge transition-end data; (+), hot-wire transition data.

FIG. 14. Experimental and numerical disturbance amplification rates inside a boundary layer and near a surface as functions of the frequency parameter at $M = 4$, $R = 780$–795: (○, ●), hot-wire experiments; (———), calculations; (○), inside the boundary layer near a critical layer; (●), near the plate surface; (1) maximum (in the boundary layer) mass-flow disturbances; (2) mass-flow disturbances near the surface; (3) temperature and stagnation-temperature disturbances near the plate surface.

FIG. 15. Numerical Reynolds number of the flat-plate transition as a function of the ratio of specific heats at two values of disturbance-amplification coefficient ($n = 5$ and 9) at $M = 4$ for second-mode disturbances: (———), calculations at Prandtl number $\sigma = 0.72$ and Sutherland constant $T_s = 110$ K; (○), calculations for carbon dioxide ($\gamma = 1.2$, $\sigma = 0.64$, $T_s = 240$ K), air ($\gamma = 1.41$, $\sigma = 0.72$, $T_s = 110$ K) and argon ($\gamma = 1.67$, $\sigma = 0.75$, $T_s = 142$ K).

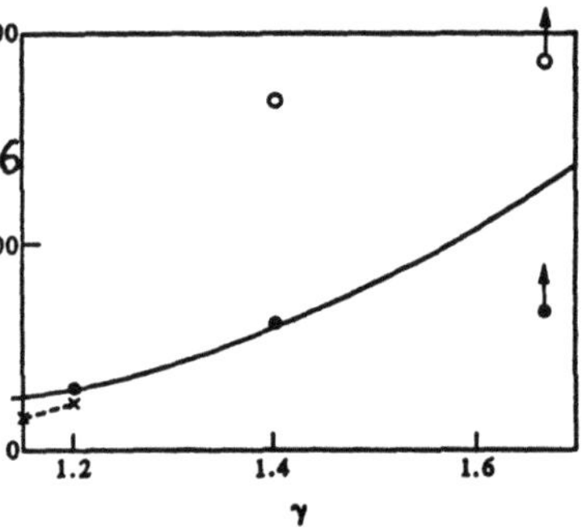

Fig. 16

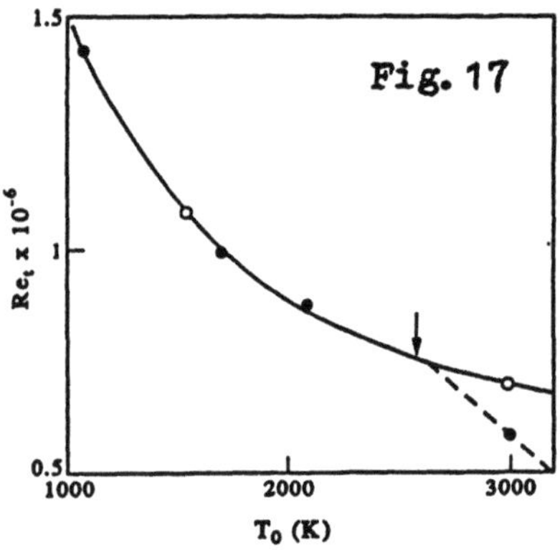

Fig. 17

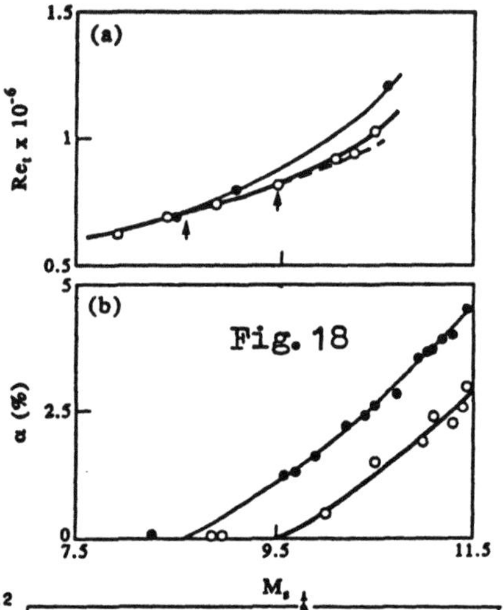

Fig. 18

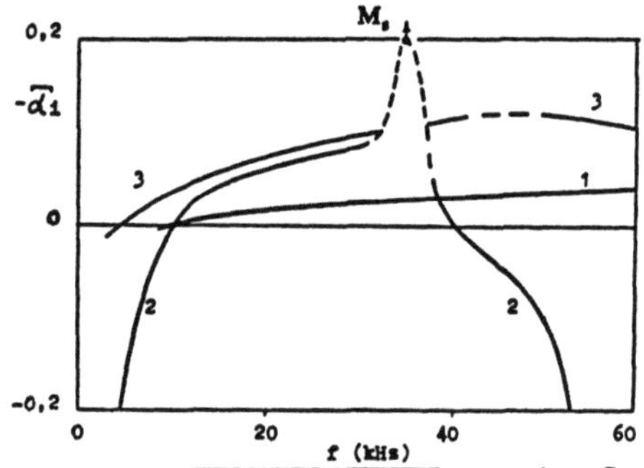

Fig. 20

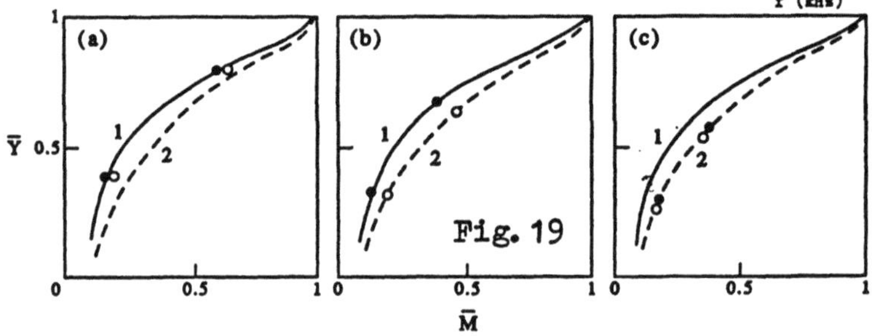

Fig. 19

FIG. 16. Experimental and numerical transition Reynolds number as a function of the ratio of specific heats, $M_e = 4$: (●), experiment, $Re_1 = 4.5 \times 10^6$ m^{-1}; (○), experiment, $Re_1 = 40 \times 10^6$ m^{-1}; (———), calculation at $n = 2.5$; (×), the numerical data of Petrov on absolute instability: G. V. PETROV, The effect of a shock wave limited by a hypersonic shock layer on boundary-layer stability, in *Instability of Sub- and Supersonic Flows*, pp. 25–34. ITAM SO AN SSSR, Novosibirsk (in Russian) (1982).

FIG. 17. Comparison of transition Reynolds numbers for nitrogen and air at different stagnation temperatures: (○), nitrogen; (●), air; (↓), temperature of oxygen dissociation.

FIG. 18. Comparison of transition Reynolds numbers (a) and ionization rates (b) for argon and xenon at different shock-wave Mach numbers: (○), argon; (●), xenon; (↑), start of ionization (for xenon and argon).

FIG. 19. Experimental profiles of Mach number in the boundary layer on a flat plate with an obstacle and data of experiments without and with a longitudinal magnetic field, $M_e = 16$: (1) laminar boundary layer; (2) turbulent boundary layer; (○), without magnetic field: (●), with magnetic field; (a) $x = 200$ mm; (b) 250 mm; (c) 300 mm.

FIG. 20. Experimental disturbance amplification rates as functions of the frequency ($M_\infty = 4$, $Re_{1\infty} = 9 \cdot 10^6$ m^{-1}) for: (1) plate boundary layer (near back edge), (2) free shear layer, (3) wake.

"Vortex Dynamics" of a Vortex Ring Introduced into a Laminar Boundary Layer

Hiroshi Maekawa, Hajime Takami
Department of Mechanical and Control Engineering,
The University of Electro-Communications, Chofu, Tokyo, 182, Japan
Takahiro Nishioka
Hitachi Co. Ltd., Tuchiura, Ibaragi, 315, Japan

Summary

In the present work, the motion of a single vortex ring introduced in a laminar boundary layer was observed by means of flow visualization. The turbulent flow evolves from the vortex ring depends on the initial configuration of the vortex ring in a laminar boundary layer. The initial evolution process is strongly dependent on the direction of circulation of the vortex ring in the layer. Furthermore, the turbulent region evolves from the primary vortex was investigated using pseudo-synchronized hot-wire measurements. Breakdown of the vortex ring in a laminar boundary layer leads to the high fluctuation velocity fields within the turbulent bloblike structure.

Key Words:Vortex Ring, Boundary Layer, Transition, Vortex Dynamics

1. Introduction

When a localized disturbance evolves downstream in a laminar boundary layer, the growth of the disturbance results in a localized turbulent region. Since Emmons[1] was concerned with the role of spots in the transition process, a majority of subsequent studies have concentrated on a characteristic feature of the spot. Recent works have revealed the internal structures of the spot(see Gad-el Hak[2], Wygnanski[3], Matusi[4], Cantwell[5] and Sokolov[6]). In another experiment, Gaster and Grant[7] studied the evolution of a wave packet in a laminar boundary layer produced by a localized perturbation at the wall boundary.

The initial development of a large amplitude disturbance is strongly dependent on the initial disturbance. From a practical point of view, therefore, the understanding of the initial development of a localized disturbance is very important. In order to control the transition process of laminar boundary layers, well manipulated coherent disturbances with the specified degree to which the initiated disturbance becomes turbulent have an advantage. Vortex rings of various magnitudes of circulation and diameters, well controlled 'coherent motion', can be generated. Vortex dynamics of a single vortex ring within the rollers of a plane mixing layer, Maekawa and Nishioka[8] investigated effects of the vortex ring introduced on pairing process of the mixing layer, has led to helical pairing at a shorter distance than that by subharmonic forcing.

Objective of the present work is to present 'vortex dynamics' of a vortex ring in a laminar boundary layer which has some advantages to control the boundary layer transition process. Nonlinear interaction of a vortex loop in a laminar boundary layer leads to breakdown of the primary vortices[9]. The localized turbulent region evolves from the vortex ring in a laminar boundary is studied.

2. Experimental method

The laminar boundary layer develops on one of the walls of a flat plate mounted parallel to the free stream in a wind tunnel or of a towing flat plate[8]. Figure 1 shows the experimental setup. Alternate measurement using a towing flat plate is employed for the case of small circulation of the vortex ring. Vortex rings are produced by an audio speaker driven by a square wave with a definite gradient and duration. Circulation of the vortex ring is controlled by the gradient and duration of the square wave. Two cases of the vortex ring configuration in the boundary layer shear are studied: The vortex ring flies upstream(Case 1) or downstream(Case 2), as shown in Fig.1.

A single hot-wire probe was mounted on a traversing unit. Measurements have been made on the centerline of the turbulent spotlike structure at six distances from the collision point(nominally, $x_s = -100, -50, 0, 50, 100, 150mm$), and also six distances in the spanwise direction from the centerline (nominally, $z = 0, 10, 20, 30, 40, 50mm$). To check the quality of the laminar boundary layer, velocity profiles were measured at several distances of x_s, without the vortex ring. The resulting distributions for zero pressure gradient indicated reasonable agreement with the theoretical Blasius distributions. Hot wire measurements were synchronized with the vortex ring. The data, thus, are stored during the event using the square wave as a reference.

Furthermore, the evolution of the flow was investigated by means of flow visualization using a smoke wire technique. Flow pictures associated with the event of a side/end view were taken by a CCD camera.

3. Results and Discussions

3.1 Vortex ring and boundary layer

Reynolds number of the vortex ring employed, based on the initial circulation, was 15000/2860, which is larger than the threshold value of 1600 obtained by Orlandi[10]. The diameter of the vortex ring was 40mm. Reynolds number of the laminar boundary layer based on the displacement thickness ($\delta_1 = 6mm$) at $x_s = 0mm$ (x=400mm from the leading edge of the flat plate) was 228.

3.2 Visualized structures

3.2.1 Side view

Figure 2 shows the consecutive flow pictures for Case 1. In Fig. 2, the vortex ring is coming from the left hand side and the free stream flows to left. At t=0.1 (sec), we observe that the vortex ring has reached on the wall boundary. The free stream visualized by smoke lines are led to the boundary layer because of the induced velocity by the vortex ring. A part of vortex ring entered into the boundary layer marches upstream and induce a secondary vortex. As we

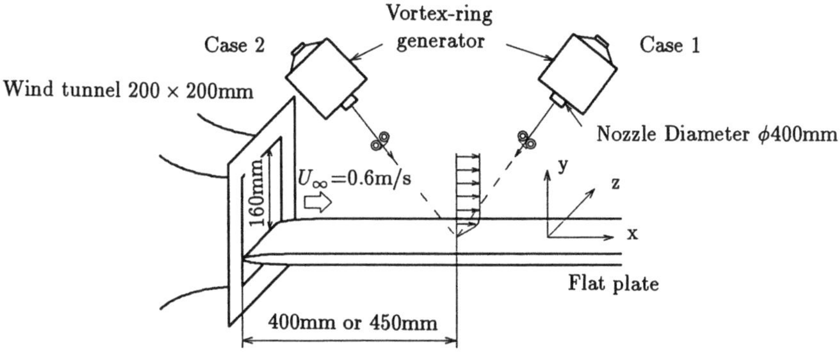

Fig. 1. Experimental Setup.

can see in the picture at t=0.3, the interface between the marching bloblike structure and the free stream oscillates and the inside of the bloblike structure is observed to become complicated. The successive snapshots show that the bloblike structure grows downstream. For Case 2, as we can see in Fig. 3, a part of vortex ring marches downstream. In this case, however, the evolution of the bloblike structure is different from that in Case 1. The bloblike structure is smaller in the streamwise direction than that in Case 1. We also note that the inside of the bloblike structure is visualized by a dense cloud of smoke. The lifting of the bloblike structure is evident in Fig. 3.

3.2.2 End view(not presented in the paper)

The end views show that the vortex ring deforms in the spanwise direction before breakdown in Case 1. After collapse of the vortex ring, fine scale motions were observed. In contrast, flow pictures of end view in Case 2 show that the vortex blob increases in the vertical direction, which is consistent with the observations in the side view.

3.3 Hot wire measurements

3.3.1 Centerline velocity fluctuations

Figures 4(a) and (b) present the time traces of u velocity fluctuation in Case 1 at $x_s = 50mm$ and $150mm$(at the locations of x=500mm and 600mm in Fig. 2), respectively. The streamwise velocity traces at the various y(vertical) locations show the characteristic velocity fluctuations. The fluctuations sustain approximately during 0.5 seconds which corresponds to the streamwise length scale of the visualization picture. The duration of the bloblike structure increases downstream. The u-velocity fluctuation in Case 2 show the sharp peak which represent the arrival of the vortex ring. As observed in the visualization pictures, a part of the vortex ring in Case 2 marches downstream in the laminar boundary layer so that the sharp peak is measured in the boundary layer.

3.3.2 Velocity fluctuations at several spanwise distances

In order to investigate the spreading in the z(spanwise)-direction, velocity fluctuations at six distances from the centerline(z=0,10,20,30,40,50mm) were measured. The measurement station used is at $y/\delta = 0.25$ and $x_s = 0mm$. In this measurement, Reynolds number based on circulation of vortex ring is 2860 and a towing flat plate is employed, because the vortex ring of small circulation

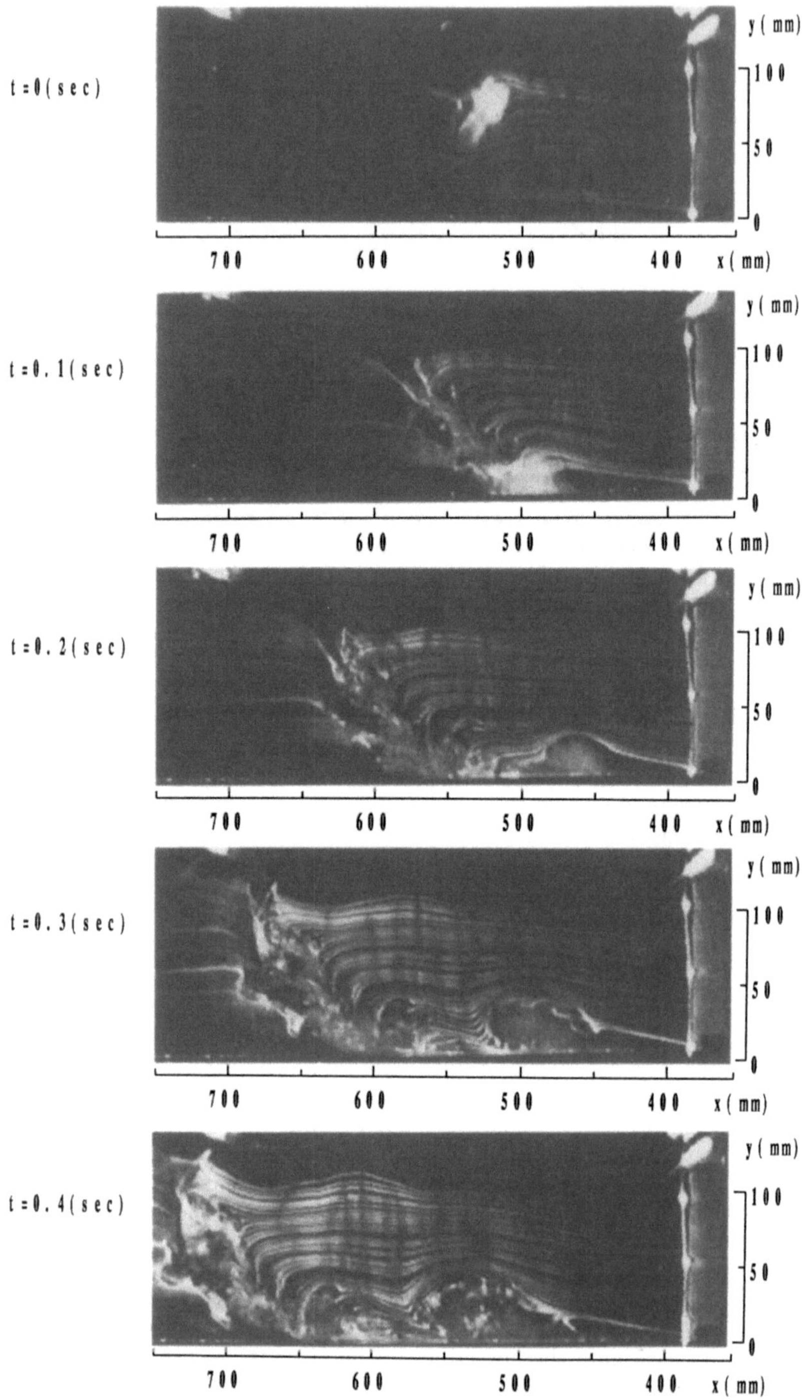

Fig. 2. Visualization Pictures of Side View for Case 1.

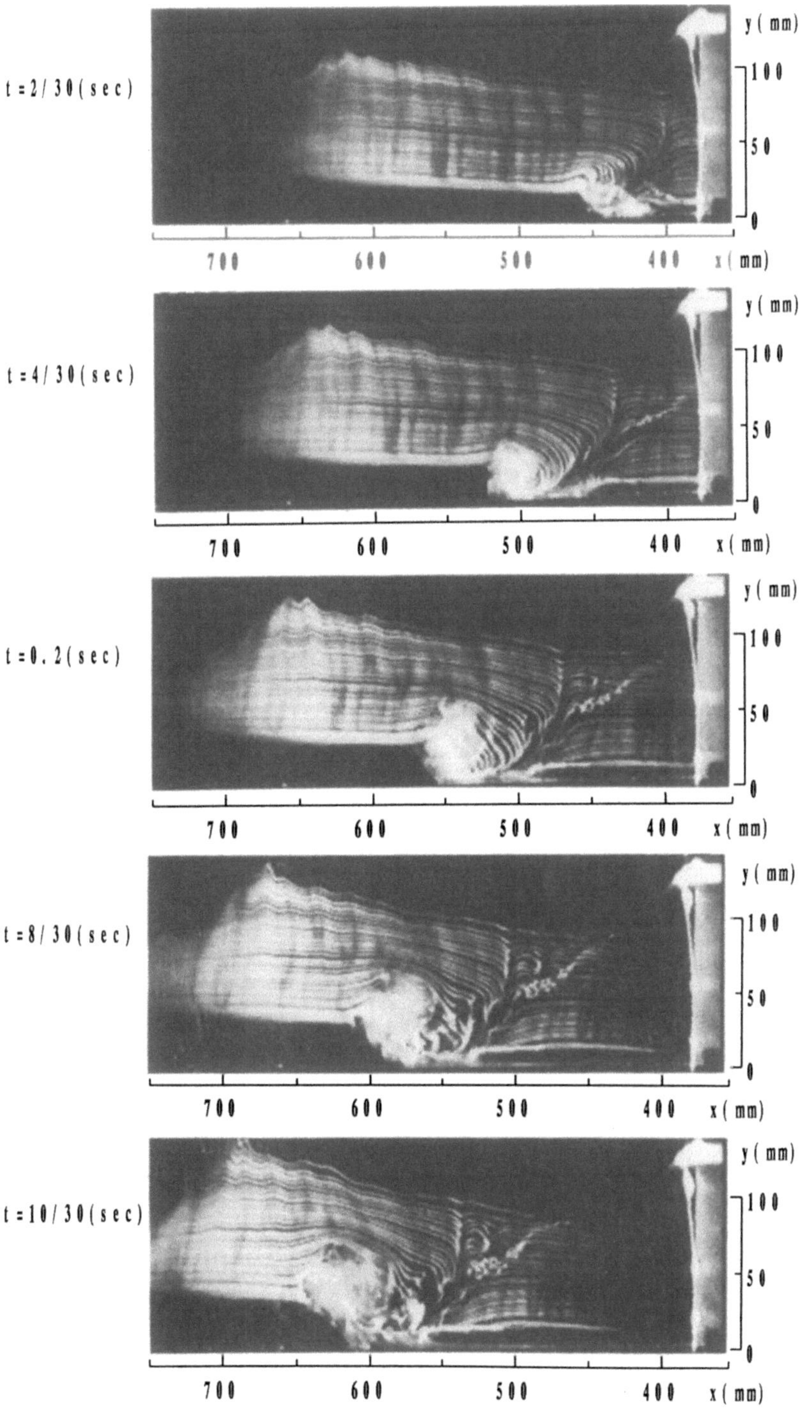

Fig. 3. Visualization Pictures of Side View for Case 2.

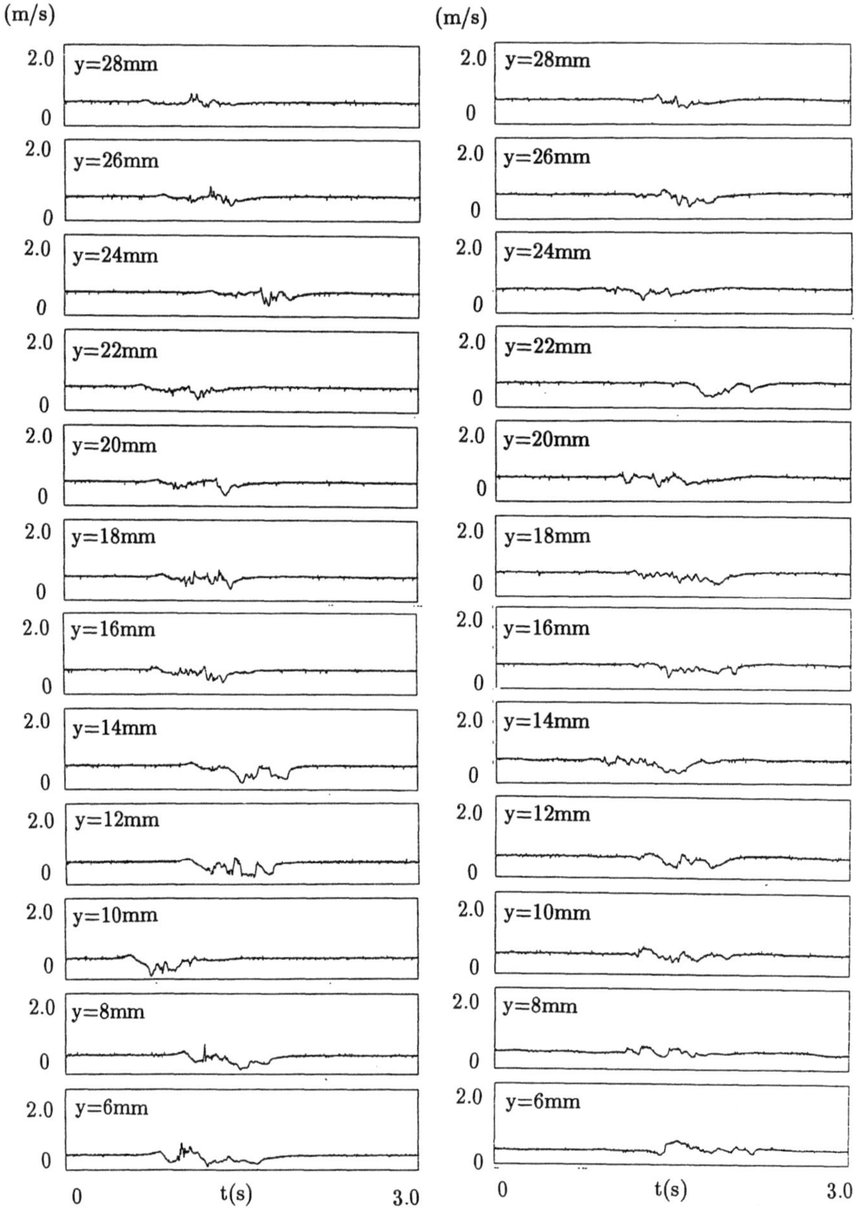

(a) ($x_s = 50mm$) (b) ($x_s = 150mm$)
Fig. 4. Time Traces of u-fluctuation at the centerline for Case 1

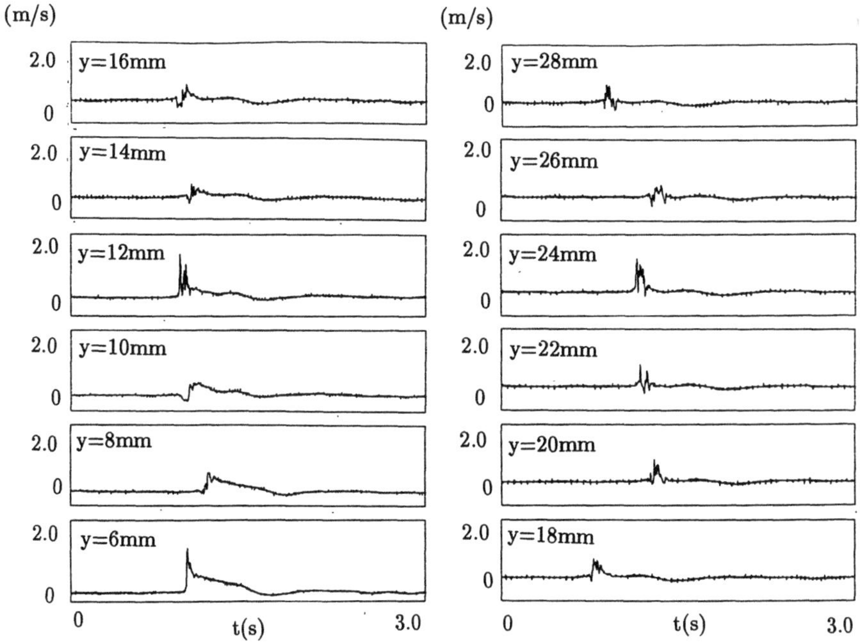

Fig. 5. Time Traces of u-fluctuation at the centerline for Case 2. ($x_s = 150mm$)

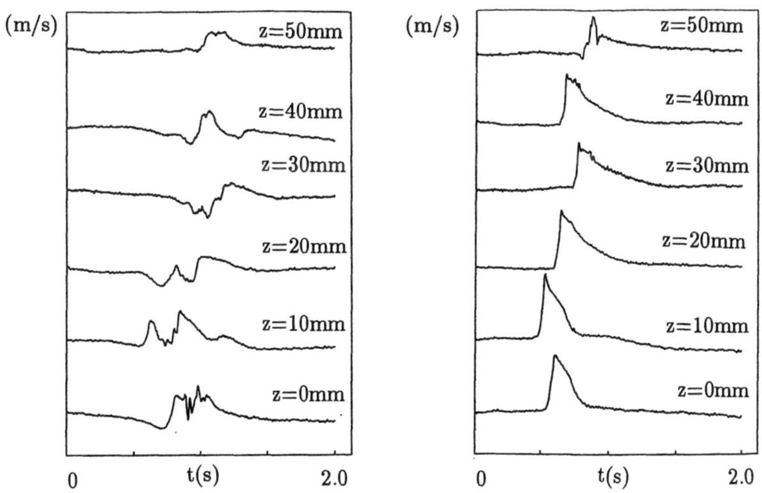

(a) Case 1 (b) Case 2

Fig. 6. Time Traces of u-fluctuation at spanwise distances. ($x_s = 0mm$, $y/\delta = 0.25$)

is hardly collided on the wall boundary in a wind tunnel. In Case 1, velocity fluctuations around the centerline show the dip and recovery. In Case 2, the sharp peaks of the u-fluctuation are spreading in the spanwise direction.

3.3.3 Interface structure

Flow visualization pictures show the clear interfaces between the free stream and the turbulent bloblike structure for both cases. The velocity fluctuations at the interface region in Case 2 show the high frequency fluctuation, similar to that at the interface of the vortex ring. Therefore, we infer that the interface velocity fields are induced by a sharp vorticity within the bloblike structure in Case 2. In contrast, in Case 1, the velocity fluctuations show the dip and recovery, similar to the turbulent spot. The trajectories of the leading and trailing edge were roughly measured using the consecutive flow pictures. In Case 1, the trajectories yield leading- and trailing-edge velocities of $0.8U_\infty$ and $0.4U_\infty$, respectively. These values are smaller than those of turbulent spot by Wignanski et al.[3]. In contrast, the bloblike structure in Case 2 evolves at approximately U_∞.

4. Conclusions

The three-dimensional structures of the turbulent bloblike structure produced by a single vortex ring in a laminar boundary layer were investigated using an experimental method. The disturbances grow downstream after breakdown of the primary vortex ring. A comparison of Case 1 with Case 2 shows that the initial growth of the turbulent structure evolves from the vortex ring depends on the intrinsic vortex dynamics. The inside of the turbulent structure is also strongly dependent on the initial configuration of the vortex ring in the laminar boundary layer.

Further investigations will be made to obtain more detailed insights.

References

[1] Emmons, H. W.:The laminar-Turbulent Transition in a boundary Layer, Part 1. Aeronaut Sci. **18**(17), (1951) 490-498.

[2] Gad-el-Hak, M., Blackwelder, R. F. and Riley, J. J.: On the growth of turbulent regions in a laminar boundary layers, *J. Fluid Mech.*, **110** (1981)73-95.

[3] Wignanski, I., Sokolov, M. and Freidman, D.: On a turbulent spot in a laminar boundary layer, *J. Fluid Mech.*, **78**, (1976), 785-819.

[4] Matsui, T.: Visualization of Turbulent Spots in the Boundary Layer along a Flat Plate in a Water Flow, *Laminar-Turbulent Transition*, (ed. R. Eppler and H. Fasel), (1980), 288-296.

[5] Cantwell, B., Coles, D. and Dimotakis, P.: Structure and Entrainment in the Plane of Symmetry of a Turbulent Spot, *J. Fluid Mech.*, **87**, (1978), 641-672.

[6] Sankaran, R., Sokolov, M. and Antonia, R. A.: Substructures in a Turbulent Spot, *J. Fluid Mech.*, **197**, (1988), 389-414.

[7] Gaster, M. and Grant, I.:An Experimental Investigation of the Formation and Development of a Wave Packet in a Laminar Boundary Layer, *Proc. R. Soc. London* Ser. A. **347** (1975), 253-269.

[8] Maekawa, H. and Nishioka, T.: Control of Mixing Layer Developing Structures Using a Vortex Ring, *Trans. of JSME*, **58**(553), (1992), 2659-2666.

[9] Leonard, A.: Vortex Simulation of Three-Dimensional Spotlike Disturbances in a Laminar Boundary Layer, *Turbulent Shear Flows*, **2**, (ed. L. Bradbury, F. Durst and B. Launder) (1980), 67-77.

[10] Orlandi P.: Vortex Dipole rebound from a wall, *Phys. Fluids.* A. vol. 2(8) (1990) 1429.

Experiments on Secondary Instability of Channel Flow with Body Forces

M. Matsubara and P. H. Alfredsson

Department of Mechanics/Fluid Physics, Royal Institute of Technology, S-10044 Stockholm, SWEDEN.

Summary

Twisting motion, which is one of the secondary instabilities of Dean vortices in a rotating channel, has been studied, triggering it with artificial disturbances. The twisting motion which is an oscillation in the spanwise direction, grows exponentially and induces harmonic modes until it affects the basic flow. The twisting motion has a distinct streamwise wavy structure inclined by the normal shear.

Key words: Coriolis force, Channel flow, Secondary instability, Longitudinal vortex, Dean vortices.

1 Introduction

Laminar-turbulent transition in the presence of streamwise vortices can be studied in wall bounded flows subjected to body forces. Typical flow situations where such forces appear are channel flows subjected to system rotation or streamline curvature. The primary instability in these flows develops in the form of counter-rotating vortices aligned in the streamwise direction. Several different types of secondary instability may appear on top of the primary instability, including merging and splitting of the primary vortex pairs, and streamwise travelling waves. Furthermore, two types of travelling waves are observed, usually denoted as undulating and twisting motions. The twisting motion has a much shorter streamwise wavelength (typically of the same order as the spanwise wavelength of the primary instability) than the undulating motion.

The plane channel flow subjected to spanwise system rotation was studied experimentally by Alfredsson & Persson [1], who visualised the twisting secondary instability as well as the primary instability using reflective flakes in water. Experiments in a flat plate boundary layer subjected to spanwise system rotation have also shown similar travelling secondary instability [2].

The spatial development of the primary as well as naturally occurring twisting and undulating motions in a curved channel was studied by Matsson & Alfredsson [3-4]. The results from such measurements, where the two instabilities have been sorted out through band pass filtering of the velocity fluctuation, indicate that both the twisting and undulating secondary instabilities are related to the low velocity region between a vortex pairs at the concave wall. In this region of the flow strong inflectional velocity profiles exist and may be connected to the instability. For the rotating channel flow three types of secondary instability, splitting and merging of the longitudinal vortices, undulating and twisting motions, were also predicted by numerical simulations [5-6].

In the present paper we will describe experiments in a straight rotating channel where the primary and secondary instabilities are triggered independently. The focus is on the high frequency twisting motion, and its energy growth was determined. A phase averaging technique enabled us to reveal the three-dimensional structure of the twisting motion.

2 Experimental set-up

The experimental set-up is shown in figure 1. A 1 m long channel was mounted on a rotating table, the channel width d is 10 mm, and the spanwise aspect ratio is 29:1. A computer controlled traversing mechanism for hot wires enabled automatic measurements in the spanwise and wall-normal directions. Two 2.5 μm diameter platinum hot-wires with a length of 0.5 mm, separated by 3 mm in the spanwise direction, were sampled simultaneously in order to reduce the data collecting time.

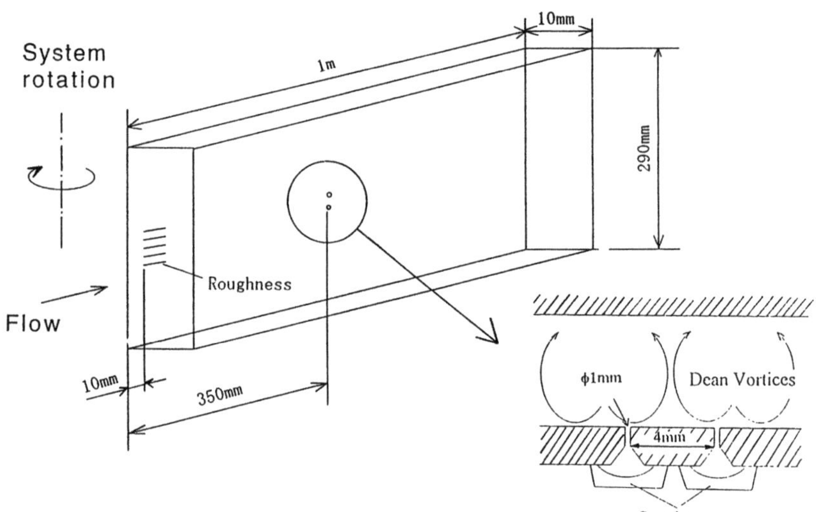

Figure 1: Experimental set-up.

The measurements were made at a mean velocity $U_m = 2.0$ m/s and a rotation rate $\Omega = 5.2$ rad/s, giving $Re = U_m d/\nu = 1300$ and $Ro = \Omega d/U_m = 0.026$, where ν is kinematic viscosity. Roughness elements with 10 mm spanwise spacing were placed at $x = 10$ mm in order to fix the spanwise position of the primary vortices. This roughness induced temporally stable primary vortices with a non-dimensional wave number $\alpha = 2\pi d/\lambda$ of 6.28, where λ is a spanwise wave length of the primary vortices. The secondary instability was triggered through periodic injection and suction from two holes on the unstable wall at $x = 350$ mm. Two small headphone speakers made out-of-phase initial disturbances on both sides of the upflow region as shown in figure 1. The hole diameter was 1 mm and they were separated by 4 mm in the spanwise direction. This distance is comparable with the spanwise size of the upflow region. The intensity of the initial disturbance $v_{rms}/U_m = 1.0$ % was evaluated by the rms of the velocity normal to the wall at the exit of the holes. By varying the frequency it was found that a forcing frequency of 131.6 Hz has the maximum growth rate for the present experimental parameters. The triggering of the secondary instability as well as the sampling of the hot wire signals was computer controlled through a Macintosh II. To obtain the phase averages 300 periods of the signals were averaged.

3 Experimental Results

The hot-wire signals were phase averaged with respect to the speaker signals. The integrated disturbance energy was used as a criterion of its intensity. The total energy of velocity fluctuation, E_{total}, is defined as

$$E_{total} = \frac{1}{d\lambda U_m^2} \int_0^\lambda \int_0^d u_{rms}^2 dydz, \tag{1}$$

where u_{rms} is the rms of the streamwise velocity. $E_{fundamental}$ is energy integration of the forcing frequency component. $E_{harmonic\ modes}$ and E_{random} were estimated as follows,

$$E_{harmonic\ modes} = \frac{1}{d\lambda U_m^2} \int_0^\lambda \int_0^d \langle u \rangle_{rms}^2 dydz - E_{fundamental} \tag{2}$$

$$E_{random} = E_{total} - \frac{1}{d\lambda U_m^2} \int_0^\lambda \int_0^d \langle u \rangle_{rms}^2 dydz, \tag{3}$$

where $\langle\ \rangle$ denotes phase averaging. The fundamental and its harmonic modes dominate the high frequency range of the spectrum, and E_{random}, that comes mainly from the low frequency part, can be regarded as the energy from the undulating motion.

Energy growth of each component with the initial disturbance is shown in figure 2. The induced fundamental mode grows exponentially up to $x = 636$ mm and then begins to decay at $x = 736$ mm. The harmonic modes grow much faster than the fundamental mode or random components. In the non-linear region, the fundamental mode decays with the harmonic modes, and finally random components become dominant further downstream. The total energy without forcing starts to grow at $x = 586$ mm. At $x = 836$ mm it is comparable with the total energy with forcing and that value exceeds E_{random} in the forcing case. This result suggests that the forced twisting motion damps the undulating motion even though it is still growing along the channel.

Figure 3 shows the time averaged disturbance velocity, $\bar{u} - u_{parabolic}$ with and without forcing. At all streamwise positions, the streamwise velocity increases near the unstable wall ($y = 0$ mm) except in a narrow low velocity region around $z = 5$ mm. This low velocity region corresponds to an upflow region. Comparison between figure 4(a) and (b) shows that forcing does not affect the basic flow, indicating that the growth of the twisting motion is linear up to this streamwise position. At the end of linear region (figure 3c, d), however, the forced twisting motion has much larger energy than the natural secondary instabilities, and this distinctly affects the time averaged flow field. The flow field downstream

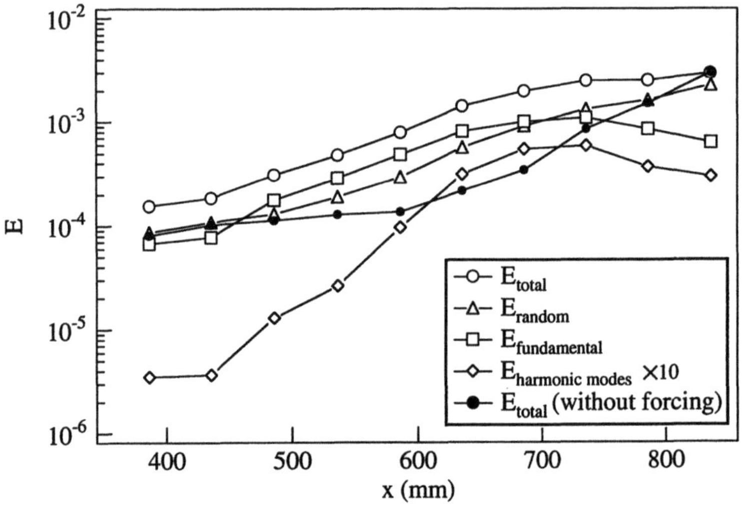

Figure 2: Energy growth with each component.

from this streamwise position is no longer independent of the forced twisting motion. These figures also show that the twisting motion flatten the flow in the spanwise direction, making the spanwise inflectional profile weaker. The velocity field without forcing does not change so much from 200 mm upstream, suggesting that the secondary instabilities are not strong enough to affect the basic flow. Further downstream, the flow with forcing is significantly flattened in the spanwise direction by the twisting motion, but it keeps a symmetric distribution at $z = 5$ mm. It is worth noticing that the flow without forcing is asymmetric suggesting the appearance of another instability that seems to split a vortex pair around $z = 3$ mm. As mentioned above, E_{random} without forcing is larger than with forcing at $x = 836$ mm. These results lead to the conclusion that the forced twisting motion inhibits other secondary instabilities. Sato & Saito [7] found a similar phenomenon in the two-dimensional wake flow that the artificial disturbance inhibits other modes in the inflectional velocity profile, delaying the transition to turbulence.

Figure 4 a, b, c show the rms of streamwise velocity, du/dy and du/dz, respectively. The rms distribution coincides well with $|du/dz|$, while there is no correlation between the rms and du/dy. This coincidence is observed at any streamwise position within the present experimental range including the non-linear region, suggesting that the secondary instabilities are mostly spanwise oscillating disturbances.

The spatial energy distributions in y-z plane are shown in figure 5. The pair of peaks of $E_{fundamental}$ at the beginning of the linear region (figure 5 b) are located further from the unstable wall ($y = 0$ mm) than the peaks of E_{random} (figure 5 a) even though both pairs of peaks are in the low velocity region. This result means that the twisting and undulation motions at first appear at

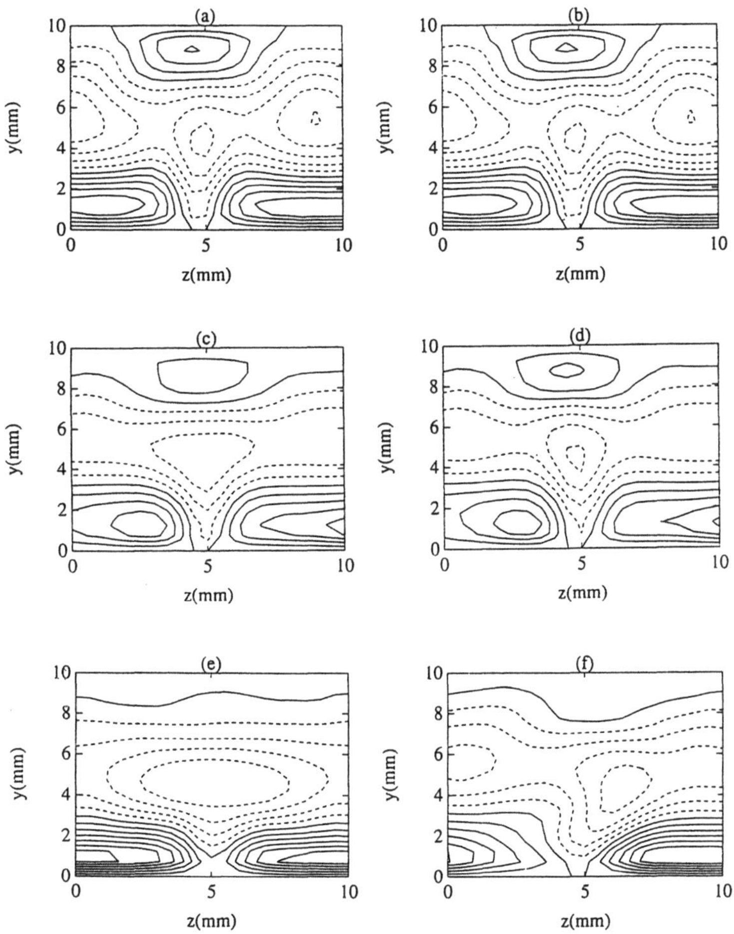

Figure 3: Time averaged disturbance velocity field. (a) $x = 436$ mm with forcing (b) $x = 436$ mm without forcing (c) $x = 636$ mm with forcing (d) $x = 636$ mm without forcing (e) $x = 836$ mm with forcing (f) $x = 836$ mm without forcing. Contour increment is 0.1 U_m and negative contours are dashed.

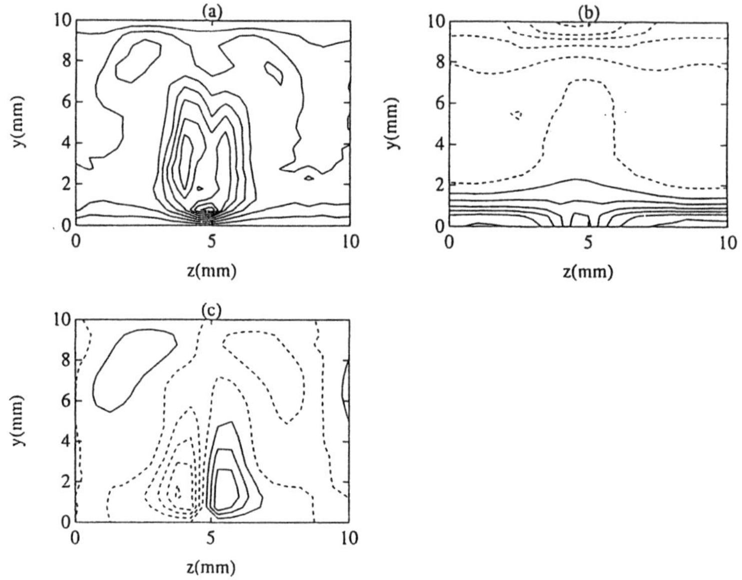

Figure 4: Comparison between fluctuation and velocity gradients with forcing at $x = 436$ mm. (a) rms of streamwise velocity (contour increment is $0.01U_m$). (b) du/dy (contour increment is $2U_m/d$). (c) du/dz. (contour increment is $1U_m/d$).

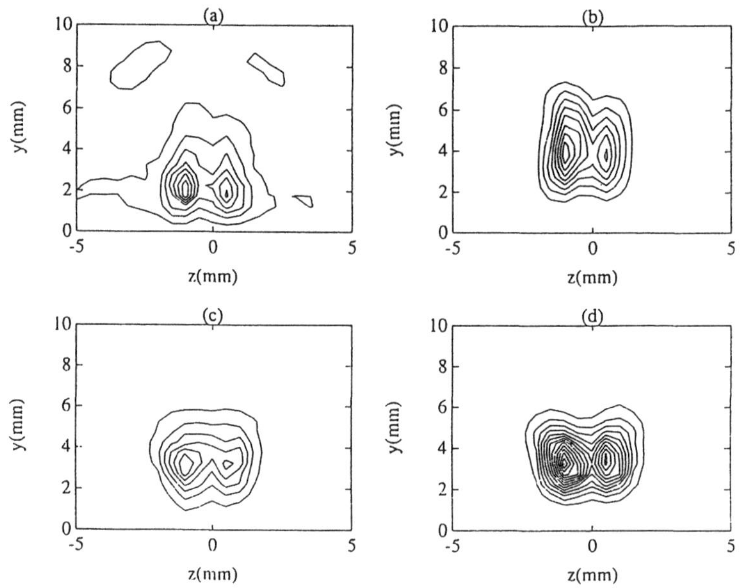

Figure 5: Energy distributions. (a) E_{random} at $x = 436$ mm (b) $E_{fundamental}$ at $x = 436$ mm (c) E_{random} at $x = 636$ mm (d) $E_{fundamental}$ at $x = 636$ mm. Contour increment is $0.00005U_m^2$ for (a) and (b), $0.0005U_m^2$ for (c) and (d).

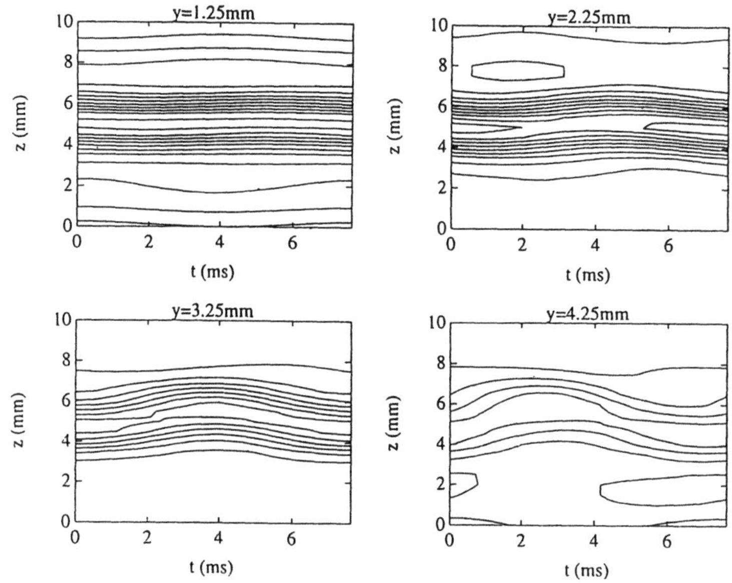

Figure 6: Spanwise wavy structure of the forced twisting motion at $x = 636$ mm. Contour increment is $0.05U_m$.

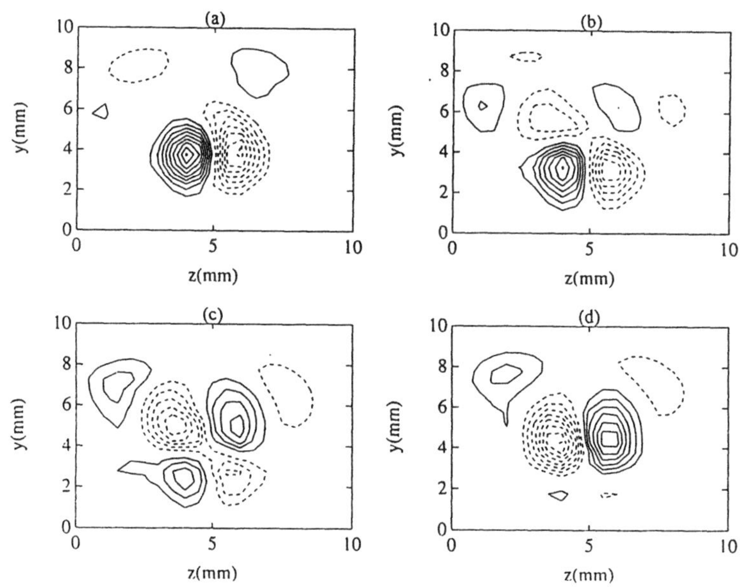

Figure 7: Phase-averaged disturbance velocity with forcing at $x = 636$ mm. (a) 0, (b) $\pi/4$, (c)$\pi/2$,(d) $3\pi/4$. Contour increment is $0.005U_m$.

different y positions. At the end of non-linear region (figure 5 c, d), however, the energy peaks of the fundamental mode move towards the unstable wall, and the peaks of E_{random} move away from the wall, resulting in an overlap of both energy distributions. This tendency was also found in the curved channel flow [4], and seems to be a common feature of the secondary instabilities of Dean vortices.

Contours of $\langle u \rangle$ are shown in figure 6, indicating the spanwise wavy structure of the twisting motion. The area where contour lines are concentrated around $z = 5$ mm is a low velocity region. The phase shifting normal to the wall indicates its inclined structure. The spanwise spatial amplitude is increasing with the distance from the unstable wall up to $y = 4.25$ mm with decreasing disturbance velocity in the low velocity region. Figure 7 shows phase-averaged streamwise velocity $\langle u \rangle - u_{parabolic}$ in the y-z plane. The maximum disturbance positions move towards the unstable wall in the upstream direction, indicating the inclined structure caused by the normal shear.

4 Conclusions

The forced twisting motion grows exponentially until it interacts with and flattens the basic flow in the spanwise direction. In the linear region, the harmonic modes grow much faster than the induced fundamental mode or random components. It decays with the fundamental mode in the non-linear region, however, and finally random components become dominant further downstream, even though the twisting motion inhibits other secondary instabilities. The y-z distribution of the velocity fluctuations coincides with the absolute value of the streamwise velocity gradient in the spanwise direction $|du/dz|$. The twisting motion has a streamwise wavy structure inclined by the normal shear.

The stay of Dr. Matsubara at Royal Institute of Technology was financed through the Swedish Institute, which is gratefully acknowledged.

References

[1] Alfredsson, P. H. and Persson, H.: Instabilities in channel flow with system rotation, *J. Fluid Mech.* **202**, (1989) 543-557.

[2] Masuda, S. and Matsubara, M.: Visual study of boundary layer transition on rotating flat plate, In Laminar-Turbulent Transition (eds. Arnal, D and Michel, R.) (1990), Springer-Verlag 465-474.

[3] Matsson, O. J. E. and Alfredsson, P. H.: Experiments on instabilities in curved channel flow, *Phys. Fluid* **A 4**, (1992) 1666-1676.

[4] Matsson, O. J. E. and Alfredsson, P. H.: The effect of spanwise system rotation on Dean vortices, *J. Fluid Mech.* **274**, (1994) 243-265.

[5] Finlay, W. H.: Transition to oscillatory motion in rotating channel flow, *J. Fluid Mech.* **215**, (1990) 209-227.

[6] Guo, Y. and Finlay, W. H.: Splitting, merging and wavelength selection of vortices in curved and/or rotating channel flow due to Eckhaus instability, *J. Fluid Mech.* **228**, (1991) 661-691.

[7] Sato, H. and Saito, H.: Artificial control of the laminar-turbulent transition of a two-dimensional wake by external sound, *J. Fluid Mech.* **84**, (1978) 657-672.

Numerical Simulation of Boundary Layer Flow over Suction Holes

Hubert L. Meitz and Hermann F. Fasel
Department of Aerospace and Mechanical Engineering
University of Arizona, Tucson, Arizona 85721, USA

Summary

Direct numerical simulations based on the complete Navier-Stokes equations are employed to explore the effects of suction holes on the generation and amplification of Tollmien-Schlichting waves in a flat plate boundary layer. The Navier-Stokes equations are used in vorticity-velocity form. Periodicity is imposed in spanwise direction; all other spatial derivatives are discretized with fourth order compact differences. An explicit fourth order Runge-Kutta scheme is employed for the time-integration of the vorticity transport equations. Suction is applied through a row of holes aligned in spanwise direction. Results are presented for suction holes with two different diameters, and for suction through a slot.

1 Introduction

In laminar flow control (LFC), wall suction may be used to stabilize a laminar boundary layer and delay transition to turbulence. Ideally, one would like to use a permeable wall to obtain a continuous suction over the entire surface. However, permeable materials cannot be used as part of a load carrying structure, e.g. as a wing surface. Therefore, in practical LFC applications suction is applied through slots or holes.

When suction holes are used, the wall normal velocity is zero over most of the surface and varies rapidly over the suction holes. Even though the average suction velocity over the entire surface is very small ($\mathcal{O}(10^{-4}U_\infty)$), the peak suction velocity at the holes can be quite large (up to $\mathcal{O}(10^{-1}U_\infty)$). These variations can cause undesired side effects that counteract the LFC attempts. First, the receptivity of the boundary layer may be increased near the suction holes, i.e. the boundary layer may become more susceptible to external disturbances such as sound waves or freestream turbulence. Second, the streamwise vortices which emerge from the holes can be unstable to small perturbations in the flow. There is now ample evidence that even very weak streamwise vortices can have a strong effect on transition [2, 1]. In fact, experiments [3] and numerical simulations [5] have shown that vortex instability due to strong suction can lead to bypass transition. Third, oscillations generated by the suction pump can feed into the boundary layer and generate disturbances. In addition, pre-existing disturbance waves inside the boundary layer may be scattered by the suction holes.

1.1 Scope of the Present Work

In the present paper emphasis is on the effect of suction holes with suction parameters that are typical of LFC applications, i.e. small holes and weak overall suction flow. We consider a Blasius boundary layer with one row of suction holes aligned along the spanwise direction (see Fig.1). At the inflow boundary a 2-D Tollmien-Schlichting (TS) wave was imposed. The frequency of this wave was chosen such that it was initially damped. In the absence of suction, the TS-wave crossed the lower branch of the neutral stability curve at the x-location of the suction holes and was amplified thereafter. The hole spacing s in spanwise direction z was chosen such that that $s \approx \lambda_{TS}$.

To investigate the effects of the suction holes on the TS-wave we carried out four numerical simulations: As a reference case, we computed the evolution of the TS-wave in the absence of any suction. We then performed two calculations with suction holes of two different diameters. Finally, we performed a simulation with a suction slot in spanwise direction instead of a row of holes (see Fig.1).

2 Numerical Method

The numerical method is the same as one used in [5]. The numerical method is based on the incompressible Navier-Stokes equations in vorticity-velocity form. The spatial derivatives in x, y directions are discretized with 4th order compact differences, while the derivatives in z direction are approximated by a Fourier series. The nonlinear terms are evaluated pseudospectrally. The equations are integrated in time by a 4th order explicit Runge-Kutta method.

2.1 Boundary Conditions

The governing equations are solved inside the computational domain shown schematically in Fig.1. At the **inflow boundary**, a Blasius boundary layer profile is imposed. At the **wall**, no-slip conditions are imposed on u and w. The normal velocity v is zero over the solid part of the wall and is imposed as $v = v_c \cos(\pi r/d)$ over the holes. In addition, $\partial v/\partial y = 0$ is imposed to ensure conservation of mass. At the **freestream boundary**, a decay condition is imposed on v. Near the **outflow boundary**, we use a buffer domain technique similar to the one discussed in [4].

3 Computational Parameters

In all calculations, the integration domain extends from $x = 1.24$ to $x = 2.52$. The suction holes / slot are at $x_c = 1.4$, with a Reynolds number based on the local displacement thickness δ at the location of the holes $Re_\delta = 644$. The suction parameters are chosen such that the total suction flow is the same in all four calculations. The nondimensional frequency of the TS-wave imposed at the inflow boundary is $F = 2\pi\nu/U_\infty^2 \times 10^4 = 4/3$, and its amplitude is $\hat{u}/U_\infty =$

2.16×10^{-6}. At this frequency and Reynolds number, the TS-wavelength is $\lambda_{TS} \approx 25\delta$. As mentioned before, this wave becomes amplified at $Re_\delta = 644$.

Here, velocities are nondimensionalized by the freestream velocity U_∞, and lengths are nondimensionalized by a reference length L such that the global Reynolds number is $Re = U_\infty L/\nu = 10^5$. The parameters used in the four simulations are summarized below:

Suction Through Slot

$d \;= 3.2 \times 10^{-2} \approx 5\delta$ ... width of slot

$v_c \;= 0.8 \times 10^{-3}$ ... maximum suction velocity at slot center

- 641 gridpoints in x direction, $d = 16\ \Delta x$
- 40 gridpoints in y direction, (= 2.5 B.L. thicknesses)
- 2-D computation

Suction Through Large Holes

$s \;= 16 \times 10^{-2} \approx 25\delta$ ... hole spacing

$d \;= 3.2 \times 10^{-2} \approx 5\delta$ ... diameter of holes

$v_c \;= 0.01$ ... maximum suction velocity at hole center

- 641 gridpoints in x direction, $d = 16\ \Delta x$
- 40 gridpoints in y direction, (= 2.5 B.L. thicknesses)
- 21 Fourier modes in z direction, 33 collocation points

Suction Through Small Holes

$s \;= 16 \times 10^{-2} \approx 25\delta$ ... hole spacing

$d \;= 3.2 \times 10^{-2} \approx 2.5\delta$ ... diameter of holes

$v_c \;= 0.04$ ... maximum suction velocity at hole center

- 1281 gridpoints in x direction, $d = 16\ \Delta x$
- 40 gridpoints in y direction, (= 2.5 B.L. thicknesses)
- 41 Fourier modes in z direction, 65 collocation points

4 Results

4.1 Change of the 2-D Meanflow

The primary effect of suction on the amplification/decay of 2-D TS-waves is due to the change of the 2-D meanflow. Figure 2 shows the u-velocity component of the 2-D meanflow distortions induced by a suction slot, large suction holes, and small suction holes, respectively. A close inspection reveals that the change in the meanflow due to the suction slot is slightly stronger than that due to the holes, particularly for the velocity profiles at $x = 1.4$ and $x = 1.44$. However, the differences between the three cases are very small, and vanish further away from the holes / slot.

4.2 Streamwise Vortices Generated by Suction Holes

Each suction hole generates a pair of two counterrotating streamwise vortices that extend downstream. A three-dimensional contour plot of the magnitude of the streamwise vorticity $|\omega_x|$ generated by the large suction holes is shown in Fig.3, as seen at an oblique angle from upstream. For better visibility, the y-coordinate is stretched by a factor of 10 in this plot. The region shown extends from the centerplane of one suction hole at $z = 0$ to the centerplane of the adjacent suction hole to the right at $z = 0.16$. In streamwise direction x, it extends from the inflow boundary at $x = 1.24$ to $x = 2.04$. In wall normal direction, the region extends to $y = 0.02$, or approximately one boundary layer thickness. The streamwise vortices are generated over the holes and decay as they extend downstream. In addition to these vortices, there are streaks of streamwise vorticity along the wall. Near the holes there are also two very small vortex stems above the main vortices.

The vortices generated by the small suction holes are plotted in Fig.4. They are stronger than those generated by the larger holes. The two additional vortices near the top that were barely visible in the case of the large holes are now well developed.

For comparison, the vortices generated by very strong suction are shown in Fig.5 from [5] (this plot is not drawn to the same scale as the previous ones). In this case, the vortices grow in downstream direction and eventually break down and lead to transition.

4.3 Effect of Suction on the 2-D TS-Wave

The amplification rate $\gamma = d/dx \log(u_{amp})$ of the TS-wave is plotted in Fig.7 for four cases: No suction (reference case), suction through slot, suction through large holes, and suction through small holes. There are only minute differences between the curves corresponding to the slot and the holes. In all three cases with suction, the growth rate is reduced relative to the reference case, and the effective reduction diminishes further downstream.

4.4 Generation of 3-D Disturbance Waves

The interaction of the unsteady 2-D disturbance and the steady 3-D flow field caused by suction holes generates 3-D unsteady disturbances. This is in contrast to the calculation with the suction slot, which did not introduce additional 3-D disturbances into the flow. The amplitudes of several 3-D disturbances generated by the suction holes are plotted in Fig.6. The index k indicates the spanwise wavenumber of each disturbance as a multiple of the fundamental spanwise wavenumber $\beta_0 = 2\pi/s = 39.27$. The solid curves correspond to the case of large suction holes, the dashed curves are for the case with small suction holes. In both cases, only the disturbance components with $k = 0$ (i.e. the 2-D TS-wave) and with $k = 1$ (i.e. with $\beta = 39.27$) are amplified at the end of the integration domain, and their respective amplitudes are the same for both large and small suction holes. In contrast, all higher spanwise modes (i.e. $k > 1$) are damped toward the end of the integration domain, and their respective amplitudes are higher in the case of small suction holes. The effects of the suction (both with holes and with a slot) on the 2-D TS-wave are so small that they cannot be distinguished at the scale used in this figure. Also, only disturbances with frequency $F = 4/3$ show sgnificant amplitudes (other frequencies are therefore not shown here).

To better assess the effect of the hole diameter on the generation of 3-D waves, we computed the 2-D Fourier transform of the suction velocity at the holes $F_s(\alpha, k) = \mathcal{F}\{v_s(x, z)\}$. We then divided the amplitudes of the 3-D disturbances by the respective value of $F_s(\alpha, k)$. For α we chose the value α_{TS} of the 2-D TS-wave, however, the result is rather insensitive to the choice of α. The resulting curves are plotted in Fig.8. The scaled curves from the small and large holes are in good agreement for $k = 1, 2, 3$. This indicates that the effects of the suction are still mainly due to linear mechanisms.

5 Conclusions

With the low suction rates considered here, the change of the 2-D meanflow is about the same for the suction holes and the suction slot, with small differences close to the holes / slot. Consequently, the effect of the suction on the 2-D TS wave is very similar for both holes and slot. In contrast, the interaction between the holes and the TS-wave generates 3-D disturbance waves that are absent in the case of the suction slot. Except for the component with spanwise wavenumber $\beta = 2\pi/s$ (where s is the distance between adjacent holes), the amplitudes of these 3-D disturbances are higher in the case of smaller suction holes. This correlates with the observation that small holes generate stronger (steady) vortices. These effects are still linear, however, and the results obtained for different hole diameters scale with the Fourier-transform of the suction velocity over the holes. Most importantly, in the parameter range considered here, there is no significant influence of the hole diameter on the amplitudes of the amplified disturbance waves.

Acknowledgements

This work is supported by the National Aeronautics and Space Administration under NASA-Langley Contract No. NAG-1422.

References

[1] Bertolotti, F., 1991, "Linear and Nonlinear Stability of Boundary Layers with Streamwise Varying Properties," *PhD Dissertation,* Ohio State University.

[2] Butler, K., Farrell, B., 1992, "Three-dimensional optimal perturbations in viscous shear flow," . *Phys. Fluids A,* Vol. 4 (8), pp. 1637–1650.

[3] Goldsmith, J., 1957, "Critical Laminar Suction Parameters for Suction into an Isolated Hole or a Single Row of Holes," *Northrop Aircraft Report* No. BLC-95.

[4] Kloker, M., Konzelmann, U., Fasel, H., 1993, "Outflow boundary conditions for spatial Navier-Stokes simulations of transitional boundary layers," *AIAA Journal,* Vol. 31, No. 4, pp. 620–628.

[5] Meitz, H. L., Fasel, H. F., 1994, "Navier-Stokes Simulations of the Effects of Suction Holes on a Flat Plate Boundary Layer," *Proc. of the AGARD Symposium on DNS and LES, Chania, Greece.*

[6] Schlichting, H., 1987, "Boundary Layer Theory," seventh edition, McGraw-Hill, pp. 506–509

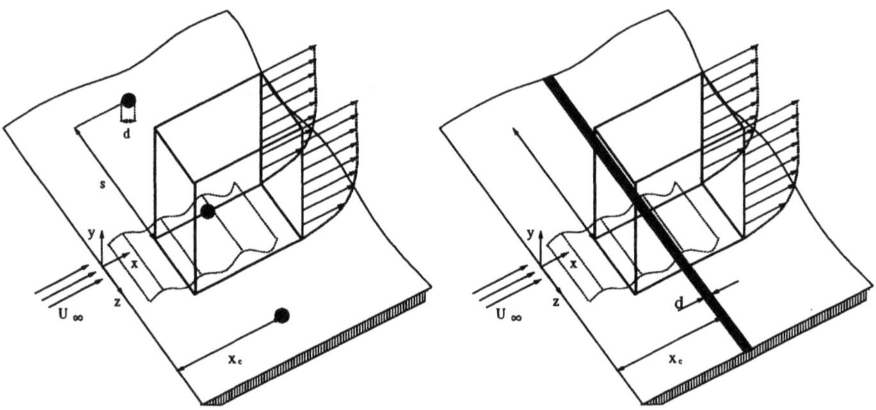

Figure 1: Computational domain with suction holes / slot

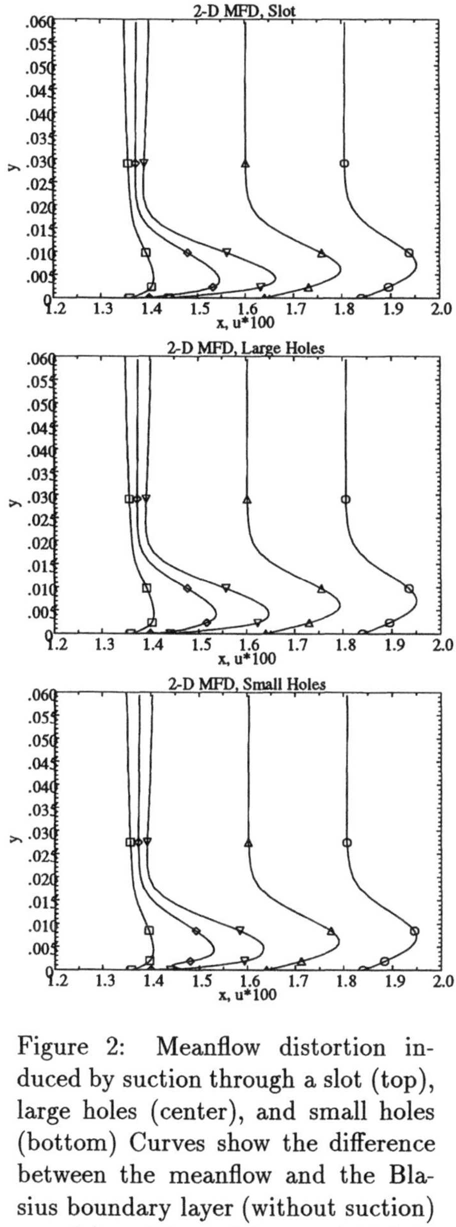

Figure 2: Meanflow distortion induced by suction through a slot (top), large holes (center), and small holes (bottom) Curves show the difference between the meanflow and the Blasius boundary layer (without suction) $u_{mfd}(y) = u(y) - u_{bl}(y)$ at $x = 1.36$ (□), 1.40 (◇), 1.44 (▽), 1.64 (△), and 1.84 (○). One unit on the abscissa corresponds to $u = 0.01$, i.e. the maximum meanflow distortion with a suction slot at $x = 1.44$ is $u_{mfd,max} = 0.22 \times 10^{-2}$.

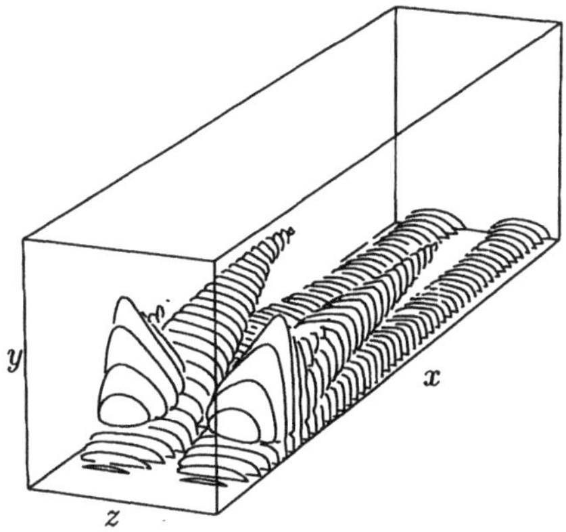

Figure 3: Contours of $|\omega_x| = 0.05$, large holes.

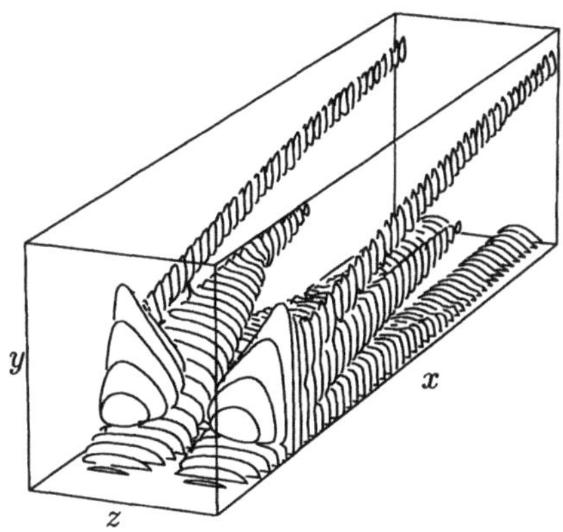

Figure 4: Contours of $|\omega_x| = 0.05$, small holes.

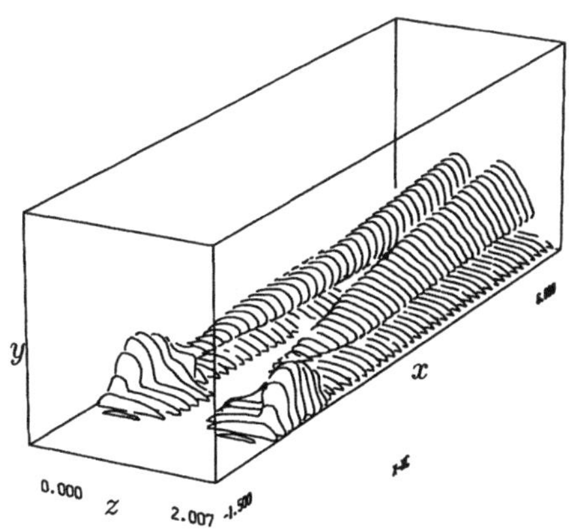

Figure 5: Contours of $|\omega_x| = 30$, strong suction.

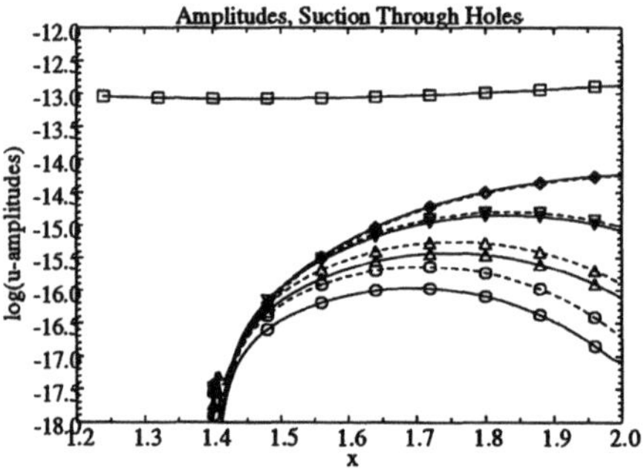

Figure 6: Amplitudes of disturbances with frequency $F=4/3$ and spanwise wavenumber $\beta_k=2\pi k/s$. The curves are $k=0$ ($\Box$), $k=1$ ($\Diamond$), $k=2$ (∇), $k=3$ (Δ), and $k=4$ ($\circ$). Solid curves are for large holes, dashed curves are for small holes.

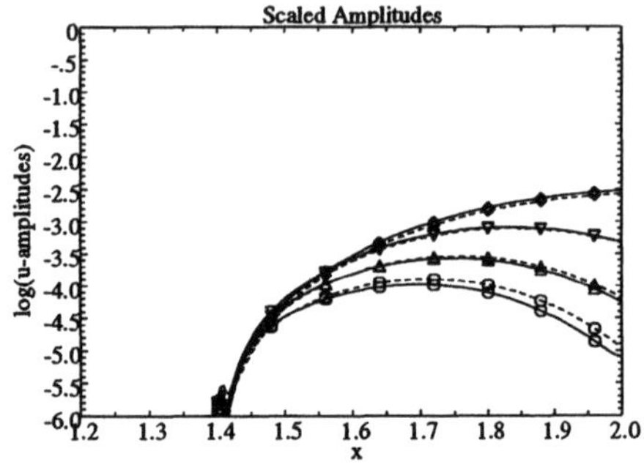

Figure 8: Amplitudes of disturbances with frequency $F=4/3$ and spanwise wavenumber $\beta_k = 2\pi k/s$ scaled by $F_s(\alpha_{TS}, k)$. The curves are $k=1$ ($\Diamond$), $k=2$ (∇), $k=3$ (Δ), and $k=4$ ($\circ$). Solid curves are for large holes, dashed curves are for small holes.

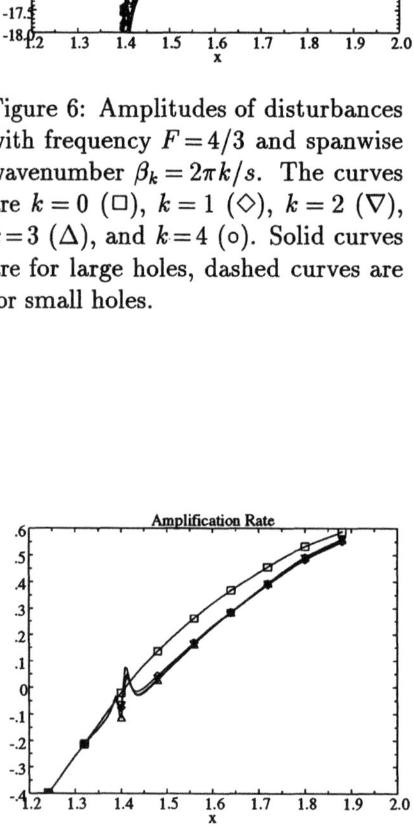

Figure 7: Amplification rate $\gamma(x)$ for reference case without suction ($\Box$), suction through a slot ($\Diamond$), suction through large holes (∇), and suction through small holes (Δ).

Numerical Simulation on Laminar-Turbulent Transition of the Channel Flow with Simulated Wall Roughness

Kiyoshi Yamamoto
National Aerospace Laboratory, Chofu, Tokyo 182, Japan

Summary

Direct numerical simulation on laminar-turbulent transition of the channel flow is conducted to elucidate the effect of distributed wall roughness on the transition. In the numerical simulation, distributed wall roughness is simulated by a periodic suction/blowing (PSB) acted on the channel wall. It is found from the simulation that the PSB simulates well a kind of two-dimensional distributed wall roughness and can strongly destabilize the channel flow. As a result, it induces the transition of the sub-critical channel flow with its Reynolds number equal to 5000 when it has a large amplitude. On the other hand, it accelerates the transition process of the super-critical channel flow with its Reynolds number equal to 10000. The critical amplitude of the PSB under which it has no effect on the transition is found.

Key words : channel flow, transition, wall roughness, numerical simulation

1 Introduction

It is well known that wall roughness affects laminar-turbulent transition of various shear flows, but its dynamical mechanism has been hardly understood yet. Generally speaking, a rough wall destabilizes the boundary layer on it. In the transition process of the channel flow with the smooth wall, it has been observed by the experiments that two-dimensional Tollmien-Schlichting (TS) waves grow at first, and after their amplitudes have grown up a large value, three-dimensional disturbances start to grow. The observation means that the TS wave with the large amplitude produces the distortion of the mean flow profile and induces the secondary instability of the three-dimensional disturbances. What happens, however, in the flow with a rough wall. We are interested in whether the transition process is still held in the case of the rough wall.

So far, wall roughness has been treated two cases, namely, single roughness and distributed one. Klebanoff and Tidstrom [1] studied two-dimensional single roughness in the boundary layer and showed that it destabilizes the disturbances in the flow through the distortion of the mean flow profile. On the other hand, Reshotko [2] studied the effect of distributed roughness in the boundary layer and observed a profound destabilizing effect when its roughness height was large, but its precise mechanism was not explained. Furthermore, Lessen and Gangwani [3] simulated distributed roughness by a wavy wall and studied its effect numerically. Recently, Yamamoto et al [4] and Floryan et al [5] simulated two-dimensional distributed roughness by a periodic suction/blowing

(PSB) acted on the wall and studied numerically its effect on the transition of the channel flow by means of a direct numerical simulation (DNS). Though they studied in the case of sub-critical transition for the Reynolds number equal to 5000, it was found that the PSB can induce the transition of the channel flow.

2 Basic equations and numerical method

Let take the coordinate system (x, y, z) of the channel flow as in Fig. 1 and consider its velocity as $(U, 0, 0) + (u, v, w)$, where $U(= 1 - z^2)$ is the basic channel flow and $\mathbf{u} = (u, v, w)$ the fluctuation velocity. To calculate the Navier-Stokes equation of $\mathbf{u}$, the conditional Fourier spectral method is used [4], [5].

First, $\mathbf{u}$ is expanded by a truncated Fourier series

$$\mathbf{u}(\mathbf{x}, t) = \sum_{k} \hat{\mathbf{u}}(\mathbf{k}) \exp(\, i\, \mathbf{k} \cdot \mathbf{x}) \tag{1}$$

where the wave number $\mathbf{k}$ is defined by $k_\alpha = (2\pi/L_\alpha)n_\alpha$ $(n_\alpha = 0, \pm1, \pm2, \cdots, \pm N_\alpha$; $\alpha = x, y, z)$ and L_α is a length of the calculational domain in the physical space.

The dynamical equation of $\hat{\mathbf{u}}(\mathbf{k})$ is derived from the Navier-Stokes equation as

$$\begin{aligned}\frac{\partial \hat{\mathbf{u}}(\mathbf{k})}{\partial t} = & -ik_x \sum_{k'_z} \hat{\mathbf{u}}(k_x, k_y, k'_z)\hat{U}(k_z - k'_z) - \sum_{k'_z} \hat{w}(k_x, k_y, k'_z)\hat{U}'(k_z - k'_z) \\ & -i \sum_{k'} \hat{\mathbf{u}}(\mathbf{k}')\mathbf{k} \cdot \hat{\mathbf{u}}(\mathbf{k} - \mathbf{k}') - i\, \mathbf{k}\hat{p} - \frac{k^2}{R}\hat{\mathbf{u}}(\mathbf{k})\end{aligned} \tag{2}$$

where $U' = \mathrm{d}U/\mathrm{d}z$, $\hat{p}$ denotes the pressure field and R the Reynolds number based on $U(0)$ and $L_z/2$.

The continuity condition for $\hat{\mathbf{u}}$ is described by

$$\mathbf{k} \cdot \hat{\mathbf{u}}(\mathbf{k}) = 0 \tag{3}$$

The boundary condition of the velocity at the smooth wall imposes the following constraint conditions on the Fourier components.

$$\sum_{n_z} (-1)^{|n_z|} \hat{\mathbf{u}}\,(k_x,\ k_y,\ \frac{2\pi}{L_z} n_z) = 0 \tag{4}$$

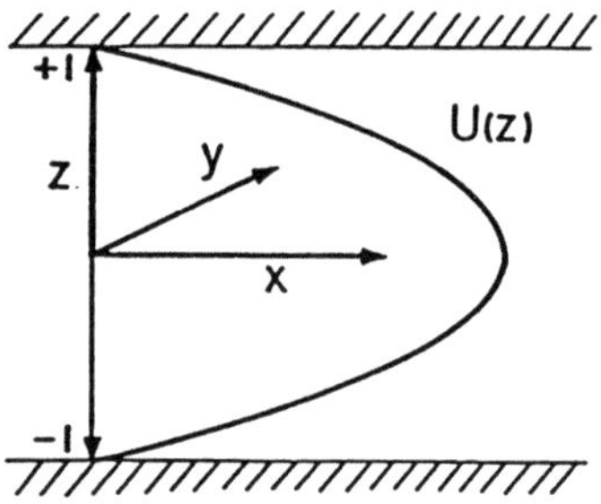

Fig.1 Coordinate system of the channel flow

These two constraint equations, (3) and (4), are calculated together with Eq. (2). Indeed, Eqs. (3) and (4) are satisfied by the orthogonal transform in the spectral space constructed by the Fourier components [6]. For instance, to satisfy Eq. (3), define two unit vectors $\mathbf{e}_1(\mathbf{k})$ and $\mathbf{e}_2(\mathbf{k})$ as $\mathbf{e}_1(\mathbf{k}) = [\ k_x k_z/(kl),\ k_y k_z/(kl),\ -l/k\ ]$, $\mathbf{e}_2(\mathbf{k}) = [\ -k_y/l,\ k_x/l,\ 0\ ]$, which are normal to each other and $\mathbf{k}$, where $l^2 = k_x^2 + k_y^2$ and $k^2 = l^2 + k_z^2$. Then, $\hat{\mathbf{u}}(\mathbf{k})$ is defined as $\hat{\mathbf{u}}(\mathbf{k}) = \eta_1(\mathbf{k})\, \mathbf{e}_1(\mathbf{k}) + \eta_2(\mathbf{k})\, \mathbf{e}_2(\mathbf{k})$, which satisfies Eq. (3). The dynamical equations of the new flow functions $\eta_1(\mathbf{k})$ and $\eta_2(\mathbf{k})$ are derived as

$$\frac{\partial \eta_\mu(\mathbf{k})}{\partial t} = \mathbf{e}_\mu(\mathbf{k}) \cdot \frac{\partial \hat{\mathbf{u}}(\mathbf{k})}{\partial t} \qquad \text{for } \mu = 1,\ 2 \tag{5}$$

Equations of (4) for the various k_x and k_y can be also satisfied by the similar transform [6]. This numerical method is called the conditional Fourier spectral method.

In the calculations, various parameters are chosen as the following values : $N_x = N_y = 10$, $N_z = 41$ and $L_x = L_y = 2\pi$ for R equal to 10000, and for the case of R equal to 5000, $N_x \doteq N_y = 21$, $N_z = 41$ and $L_x = L_y = 4\pi/1.2$ to include the sub-harmonic Fourier components of the TS wave with $k_x = 1.2$ and $k_y = 0$.

For the timewise integration of Eq.(5), the Runge-Kutta-Gill method (the fourth order scheme of Δt) is applied.

When a periodic suction/blowing(PSB), which denoted by w_s here, is applied on the wall, Eq. (4) should be changed for the wall-normal component $\hat{w}$. As for two-dimensional distributed roughness, w_s is given as

$$w_s = A_s \cos(k_s\ x) \tag{6}$$

When the w_s is applied on the wall, the induced flow is still a two-dimensional flow because there exist no three-dimensional components. Figure 2 schematically shows various stream lines of the induced flow when the PSB is applied. It can be observed in this figure that distributed separation ridges are induced by the PSB on the wall, and the stream line through the tops of these ridges is similar to one streaming over two-dimensional distributed roughness. When the amplitude A_s is small, the height of the ridge is equal to $(A_s/k_s)^{1/2}$ and the wave length $2\pi/k_s$. Thus, the PSB can simulates a kind of two-dimensional distributed roughness.

The result of Fig.2 is calculated by the present DNS code. When A_s is too small, the induced flow can be also calculated by the Orr-Sommerfelt equation with a non-homogeneous boundary condition [3]. Figure 3 shows the comparison between the results obtained by an accurate linearized code [5] and by the present DNS code with $A_s = 10^{-5}$. Since both results agree quite well, this provides an examination for the accuracy of the present numerical calculation.

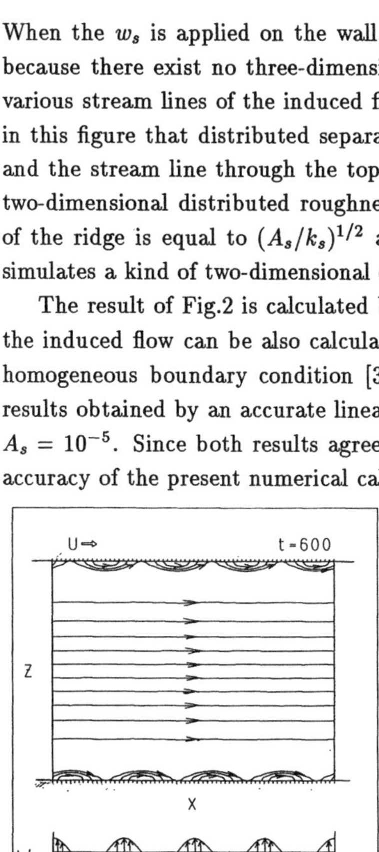

Fig.2 Illustration of the separation ridges induced by a suction/blowing

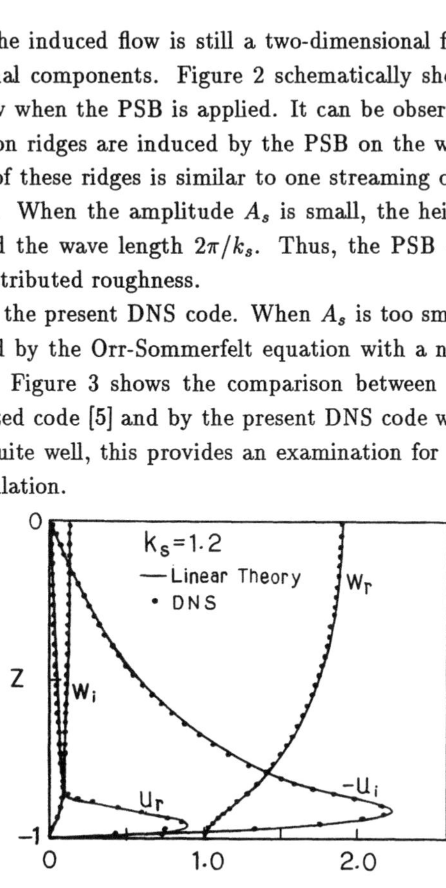

Fig.3 Comparison between the results obtained by the linearized theory [7] and by the present DNS with $A_s = 10^{-5}$

3 Transition of the channel flow with the smooth wall

Laminar-turbulent transition of the channel flow has long been studied by the linear stability theory [7],[8] and the experiments [9]. The critical Reynolds number has been obtained as 5772.22 by the linear stability theory [8]. It is instructive to study the transition process of the channel flow with the smooth wall before the DNS of the flow with wall roughness.

Figure 4 (a) shows the numerical result for the super-critical Reynolds number, $R = 10000$. Solid curves describe the time evolutions of the various Fourier modes, where $E(k_x, k_y)$, twice the energy of the Fourier mode (k_x, k_y), is defined by

$$E(k_x, k_y) \ = \sum_{k_z} |\hat{\mathbf{u}}\ (k_x,\ k_y,\ k_z)|^2 \tag{7}$$

The initial values of $E(k_x, k_y)$ are given as 10^{-10}. Various symbols in the figure distinguish among the wavenumber k_x (in particular, the squares show the modes with $k_x = 0$). During an initial long period, only a TS mode, that is $E(1,0)$, is growing almost linearly in this figure, which means the exponential growth with time, and the slope of this curve agrees quite well with the growth rate derived by the linear stability calculation [8]. As soon as the energy of TS mode reachs to the magnitude about 10^{-4}, other three-dimensional disturbances start to grow abruptly and the transition process breaks down into turbulence at $t_c \sim 2000$, which is called the transition time. During the transition process, E(0, 0) describes a remarkable energy change of the mean flow from the laminar profile to the turbulent one. Thus, this figure describes well the scenario of the super-critical transition of the channel flow.

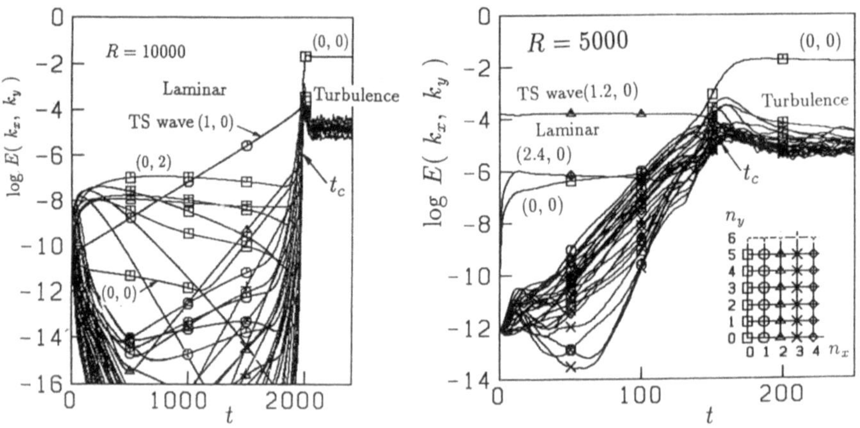

(a) $R = 10000$ (b) $R = 5000$

Fig.4 Time evolution of the dominant Fourier modes in the transition process of the channel flow with the smooth wall.

On the other hand, Fig.4(b) shows the numerical result for the sub-critical Reynolds number, $R = 5000$. Since all the small disturbances are stable in this case, a TS wave should be excited initially at a large amplitude. Indeed, in this figure, $E(1.2, 0)$ is imposed as 1.7×10^{-4} together with other disturbances given as $E(k_x, k_y) = 10^{-12}$. As a result, the disturbances grow almost linearly for an initial period, and as soon as their dominant modes reach to the magnitude about 10^{-5}, the transition process breaks down into turbulence at $t_c \sim 300$. The results of Figs. 4 (a) and (b) indicate the important role of the finite amplitude TS wave, namely, it distorts the mean flow profile and destabilizes the three-dimensional disturbances.

4 Transition of the channel flow with distributed wall roughness

4.1 Numerical result for $R = 10000$

At this Reynolds number, the transition of the flow still occurs even for the smooth wall at $t_c \sim 2000$ (see Fig.4(a)). Thus, we inquire how to affect the presence of distributed wall roughness on the transition. The effect is measured quantitatively by the transition time t_c because the presence of roughness should destabilize the channel flow and accelerate the transition process. Numerical simulations are carried out for the various amplitudes and wavenumbers of the PSB.

Figure 5(a) shows the numerical result in the case of $k_s = 1$ and $A_s = 0.001$. In this case, the PSB directly excites the TS wave (1,0) because the k_s is in the linearly unstable region. Thus, $E(1,0)$ starts from its magnitude nearly equal to the energy of the PSB and grows with time including an oscillation. When its magnitude reachs about 10^{-4}, other three dimensional disturbances grow up and the transition process breaks down into turbulence at $t_c \sim 400$. On the other hand, Fig.5(b) shows the result in the case of $k_s = 2$ and $A_s = 0.001$. The transition process in this case is almost the same as the result for the case of the smooth wall in Fig.4(a). The similar result is obtained in the case of $k_s = 4.8$ and $A_s = 0.001$. Thus, it can be concluded from these results that the PSB with $A_s = 0.001$ has little effect on the transition process except for the case of $k_s = 1$, which is equal to the wavenumber of the TS wave.

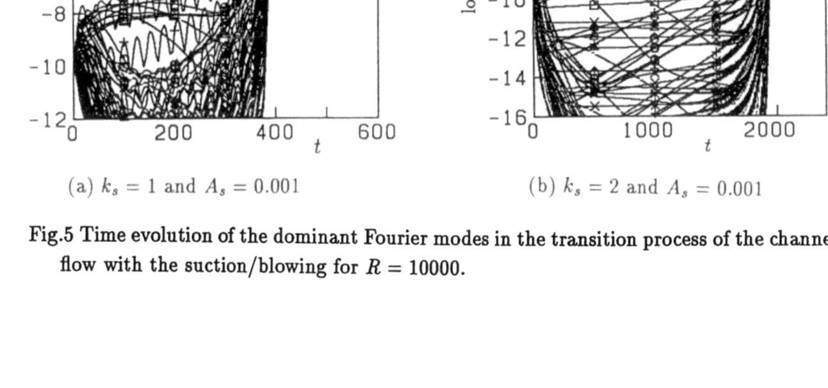

(a) $k_s = 1$ and $A_s = 0.001$

(b) $k_s = 2$ and $A_s = 0.001$

Fig.5 Time evolution of the dominant Fourier modes in the transition process of the channel flow with the suction/blowing for $R = 10000$.

Figure 6(a) shows the result in the case of $k_s = 2$ and $A_s = 0.003$. In this case, the transition process is remarkably enhanced by the PSB, and as a result, it breaks down into turbulence at $t_c \sim 770$. An important feature of this transition process is that there is no dominant TS mode (1, 0), and instead, the most dominant mode is E(0, 5), which is a longitudinal vortex component. Thus, the transition process is different from that in Fig. 4(a) and rather resembles that in Fig.4(b). Furthermore, the result in the case of $k_s = 4$ and $A_s = 0.003$ is shown in Fig.6(b). The transition process is quite similar to that in Fig.6(a) and the transition time is obtained as $t_c \sim 790$.

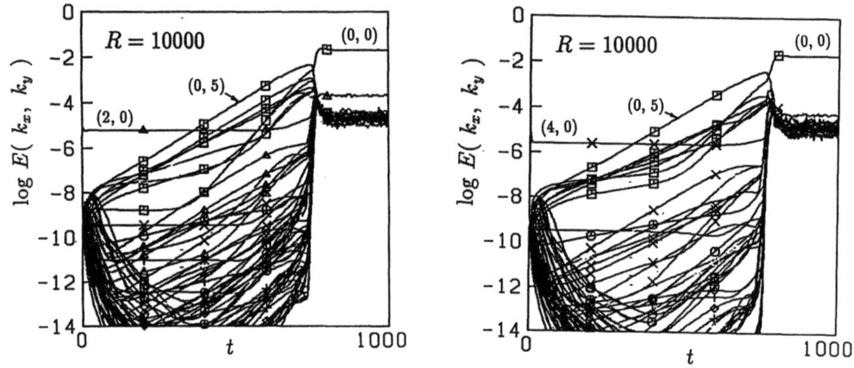

(a) $k_s = 2$ and $A_s = 0.003$ (b) $k_s = 4$ and $A_s = 0.003$

Fig.6 Time evolution of the dominant Fourier modes in the transition process of the channel flow with the suction/blowing for $R = 10000$.

These results indicate that the transition time t_c can be used as the measure of the destabilization by the PSB. Figure 7 shows the numerical results of t_c obtained for the various A_s and k_s of the PSB. It is found from the figure that the results in the cases of $k_s = 2$ and 4 coincide with the solid curve, but the results for $k_s = 1$ are considerably departure from that. Another important result of this figure is that the critical amplitude of the PSB under which the PSB has no effect on the transition process of the flow is obtained as about 0.001 except for the case of k_s equal to 1, the wavenumber of the TS wave.

4.2 Numerical result for $R = 5000$

Numerical simulations on the transition of the channel flow with the PSB for $R =$ 5000 has been carried out by Yamamoto et al [4] and Floryan et al [5]. They treated for the cases of the wavenumbers 0.6, 1.2 and 2.4. Here, to confirm the effect of the wavenumber of the PSB on the transition, the simulation in the case of $k_s = 4.8$ of the PSB is conducted with twice the resolution compared with their simulations.

In this sub-critical Reynolds number case, the transition never occurs in the flow with only the small disturbances. Thus, in Fig.4(b), the finite-amplitude TS wave (1.2, 0) should be imposed over the small disturbances. Instead of the TS wave, here, the

PSB with $k_s = 4.8$ is given at initial time, and its amplitude A_s which can induce the transition is searched. Figure 8 shows the numerical results obtained in the case of the amplitude $A_s = 0.008$. The transition process is similar to that obtained in the cases of k_s which are smaller than 4.8. The transition time t_c is obtained as about 500.

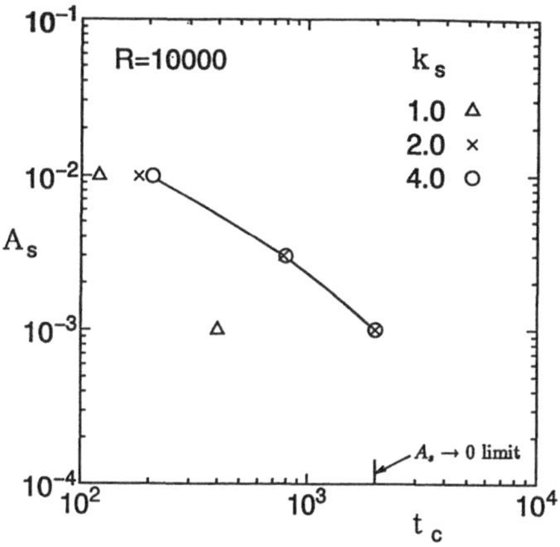

Fig.7 Transition time t_c versus the amplitude A_s of the suction/blowing in the case of super-critical transition for R =10000.

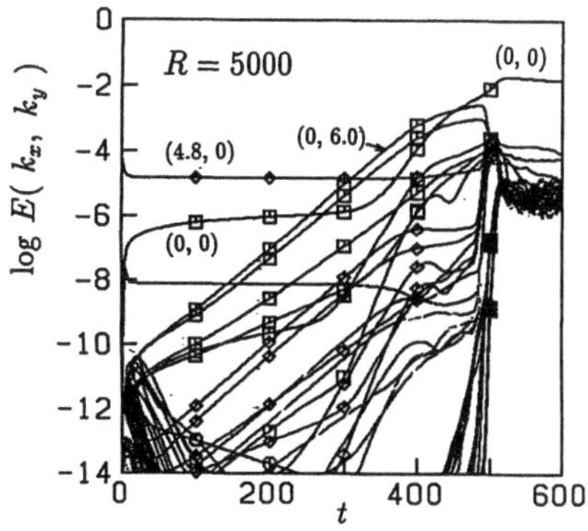

Fig.8 Time evolution of the dominant Fourier modes in the transition process of the channel flow with the suction/blowing for $R = 5000$, $k_s = 4.8$ and $A_s = 0.008$

To see the effect of the amplitude and wavenumber of the PSB on the sub-critical transition, the numerical results of t_c obtained in the cases of $k_s = 0.6, 1.2, 2.4$ [4-5] and 4.8 are shown in Fig. 9. It is found from the figure that the results obtained for $k_s = 1.2, 2.4$ and 4.8 almost coincide with the solid curve, but the results for $k_s = 0.6$ are considerably departure from that, which means the channel flow has a particular reaction on the PSB with wavenumber $k_s = 0.6$, the sub-harmonic wavenumber of the TS wave (1.2, 0). An important result obtained here is that the PSB can destabilize the small disturbances in the channel flow when its amplitude $A_s = 0.002$, which is far smaller than the threshold amplitude of the TS wave which becomes to be unstable for $R = 5000$(see Fig. 4(b))[10], and can induce the sub-critical transition of the flow. Of course, the PSB with the amplitude $A_s = 0.001$ can not induce the transition in this case.

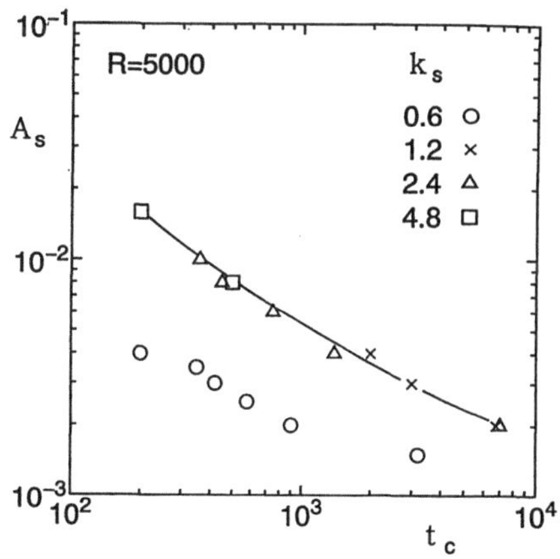

Fig.9 Transition time t_c versus the amplitude A_s of the suction/ blowing in the case of sub-critical transition for $R = 5000$

5 Conclusion

Direct numerical simulations (DNS) on laminar-turbulent transition of the channel flow with a periodic suction/blowing (PSB) are conducted. The PSB simulates well a kind of two-dimensional distributed wall roughness. It is obtained from the DNS that the simulated wall roughness destabilizes the small disturbances in the flow with the sub-critical Reynolds number equal to 5000 and induces the transition into turbulence. Furthermore, it also accelerates the transition process in the flow with the super-critical Reynolds number equal to 10000. To explain the precise mechanism of the transition by the PSB, further simulations are under way.

Acknowledgements

The author thanks to Dr. N. Itoh of NAL who provides the numerical results of the linearized code in Fig. 3.

References

[1] Klebanoff, P. S. and Tidstrom, K. D. : Mechanism by Which a Two-Dimensional Roughness Element Induces Boundary Layer Transition, Phys. Fluids, 15 (1972) 1173-1188.

[2] Reshotko, E. : Disturbances in a Laminar boundary Layer due to Distributed Surface Roughness, Turbulence and Chaotic Phenomena in Fluids, Tstsumi, T. ed., Elsevier Sci. Publ. (1984) 39-46.

[3] Lessen, M. and Gangwani, S. : Effect of small amplitude wall waviness upon the stability of the laminar boundary layer, Phys. Fluids, 19 (1976) 510-513.

[4] Yamamoto, K. , Murase, T. and Floryan, J. M. : Numerical Simulation of Laminar-Turbulent Transition in a Channel Flow with a Periodic Suction and Blowing, Proceedings of the 9th and 10th NAL Workshop on "Investigation and Control of Boudary Layer Transition", NAL Special Publ., SP-18 (1992) 5-9.

[5] Floryan, J. M. , Yamamoto, K. and Murase, T. : Laminar-Turbulent Transition Process in the Presence of Simulated Wall Roughness, Canad. Aero. Space J. 38 (1992) 173-182.

[6] Hosokawa, I. and Yamamoto, K. : On the Direct Numerical Simulation of the Shear Flow Turbulence by the Fourier Spectral Method, J. Phys. Soc. Japan, 55 (1986) 1030-1031.

[7] Itoh, N. : A power Series Method for the Numerical Treatment of the Orr-Sommerfeld Equation, Trans. Japan Soc. Aero. Space Sci. 17 (1974) 65-75.

[8] Orszag, S. A. : Accurate Solution of the Orr-Sommerfeld Stability Equation, J. Fluid Mech. 50 (1971) 689-704.

[9] Nishioka, M., Iida, S. and Ichikawa, Y. : An Experimental Investigation of the Stability of Plane Poiseuille Flow, J. Fluid Mech. 73 (1975) 731-751.

[10] Herbert, T. : Finite Amplitude Stability of Plane Parallel Flow, Laminar-Turbulent Transition, AGARD Conf. Proc. No. 224 (1977) 3.1-3.10.

Session G: Bypass Transition and Turbulent Spot

The Development of a Turbulent Spot in an Adverse Pressure Gradient Boundary Layer

B.F.A. van Hest, D.M. Passchier and J.L. van Ingen
Low Speed Aerodynamics Laboratory
Faculty of Aerospace Engineering
Delft University of Technology, Delft, the Netherlands

Summary

The development of a turbulent spot in a boundary layer with adverse pressure gradient was studied experimentally with a single hotwire probe. Disturbances were triggered in an unstable laminar boundary layer by a small pressure pulse. The turbulent spots developed from these disturbances through a wave packet phase. The leading and trailing edge propagation velocities and the spanwise growth angle were determined from the velocity traces.

In comparison to zero and favourable pressure gradient spots, the present spot was found to propagate with a lower velocity and a larger spanwise growth angle. It was further found that much of the span of the total disturbance was dominated by the different behavioural stages of an amplified wave packet. This wave packet was observed to propagate with the leading edge of the turbulent spot.

Key words : Transition, Turbulent Spot, Adverse Pressure Gradient

1 Introduction

Most experimental investigations on the behaviour of turbulent spots have been performed in zero (ZPG) or favourable pressure gradient (FPG) flow. In ZPG flow [1], leading and trailing edge propagation velocities at the spot centerline are known to scale with the edge velocity U_e as $0.88\,U_e$ and $0.50\,U_e$ respectively. The spanwise growth angle is approximately 10 degrees with respect to the plane of symmetry. In plan view, the spot has an arrowhead shape and often wave packets are observed trailing the spot at the outer span [2]. The spot is followed by a calmed region where the flow is characterised by a relatively full velocity profile, making the flow more stable to disturbances than the surrounding flow. In a FPG [3] the leading and trailing edge propagation velocities have increased and are no longer constant when scaled with U_e, but decrease linearly with streamwise distance. The spanwise growth angle is reduced to typically 5 degrees.

An extensive survey of the development of spots is given by Riley [4].

The present investigation is part of a project concerning both natural and triggered transition in boundary layers developing in an adverse pressure gradient (APG) and with low free stream turbulence. The purpose is to obtain more experimental data on transition for the development and evaluation of numerical models describing the transition region in a boundary layer. Details on the growth of turbulent spots can be used for the statistical description of the intermittent behaviour of the flow in the transition region, to be used in modelling the transition region. Therefore, an experiment was set up to determine the above mentioned elementary characteristics of turbulent spots in a moderate APG. Together with the spot formation rate, it is possible to estimate the extent of the transition region from these parameters.

2 Experiment

2.1 Experimental setup

The measurements were conducted on the flat test wall ($2.0 \times 0.48\,\text{m}^2$) of an open return wind tunnel. The contraction ratio is 16.7 and the entrance of the rectangular test section has a height of 0.08 m. The wall opposite to the test wall is flexible to impose different pressure gradients. Suction was applied on the opposite wall to keep the boundary layer growth within bounds.

The pressure distribution can be measured with 72 pressure orifices fitted in two streamwise rows along the test wall. The pressure distribution was set close to a Hartree $\beta = -0.14$ adverse pressure gradient (figure 1), at a unit Reynolds number of 1.1×10^6. The free stream turbulence was 0.1 %. Natural transition started at 400 mm from the leading edge and was completed at 520 mm.

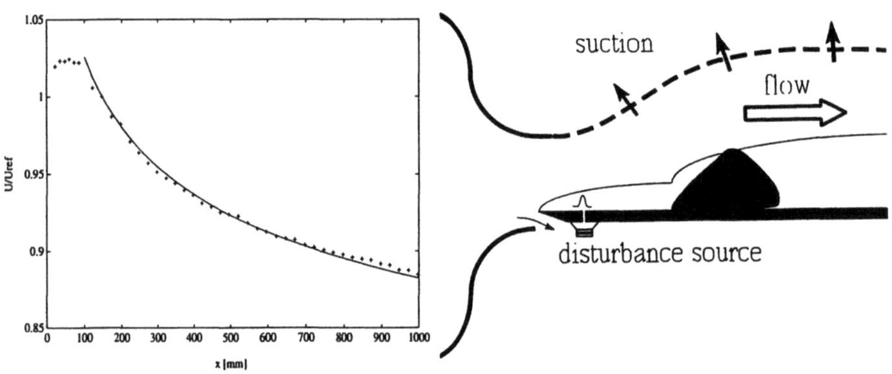

Figure 1: *Experimental set up and pressure distribution.* (—) *Hartree* $\beta = -0.14$, (+) *experiment*

Turbulent spots developed from natural amplification of instabilities, triggered by applying a pressure pulse with a loudspeaker, positioned at 197.5 mm from the nose of the test plate at the tunnel centre line (approximately 120 mm downstream of the point of first instability). The loudspeaker was driven by a differentiated square wave with a period of 0.5 Hz. The pulse width at half height was approximately 0.5 ms. The 0.5 Hz period was chosen to provide sufficient time for the flow to relax after the passage of a spot.

Velocity measurements were performed with a single hotwire at different streamwise, normal and spanwise positions. The tungsten hotwire (1 mm active length, 5 μm diameter) was connected to a constant temperature anemometer operated at an overheat ratio of 0.6. Mean voltages were measured with a digital voltmeter. Fluctuating voltages were bandpass filtered between 0.1 and 2000 Hz, amplified and read with a 12 bits AD-converter. At each position 100 traces of 1000 samples were collected at a sampling rate of 4000 Hz.

The spots were detected from the velocity traces by two different methods: The first method determines the spot edges by distinguishing between laminar and turbulent velocity fluctuations. This is done by comparing a detector function $|d^2u/dt^2|$ with a threshold level. Here the ensemble average of this detector function was taken, which reduced contamination with naturally occurring spots and provided a way to trace the spot until the end of natural transition. The second method is based on the deviation of the velocity from the laminar flow during a spot passage. It determines the spot edges at the point where the velocity perturbation $\tilde{U}(t) = (\langle U(t) \rangle - U_{lam})/U_e$ exeeds a threshold of 0.025. The trailing edge obtained by this method is related to the end of the calmed region instead of that of the spot.

2.2 Experimental results and discussion

The development of the disturbance was determined at five streamwise stations, located between 200 mm and 299 mm behind the loudspeaker. Because most of the measurements were performed in the region where natural transition occurred, contamination with spontaneously developing turbulent spots was possible. We believe that our results are not greatly affected because of their random occurrence; their effect will dissappear by ensemble averaging. The development of the spot can still be affected by the natural spots.

The streamwise development of the turbulent spots is discussed first. Ensemble averaged velocity traces $\langle U \rangle$ and perturbation $\tilde{U}$ contours, obtained in the plane of symmetry of the spot, are presented in figure 2. In all figures the position of the spot LE and TE, as determined with the turbulence detector, are marked by crosses, edges determined from the velocity perturbation are marked by circles. At the first station ($x = 397.5$ mm), most of the disturbance is still a wave packet. In some of the individual velocity traces (not shown) measured near the wall at this station, high frequency oscillations are apparent, indicating that the wave packet is breaking up into turbulence. The frequency of the wave packet is $f \approx 340$ Hz. This value is higher than the frequency of the most amplified Tollmien-Schlichting (TS) wave found from experimental data ($f = 227$ Hz), or by solving the Orr-Sommerfeld equation for the present pressure distribution (f$\approx$ 305 Hz). The wave packet is preceded and followed by TS waves with the most amplified frequency ($f = 227$ Hz).

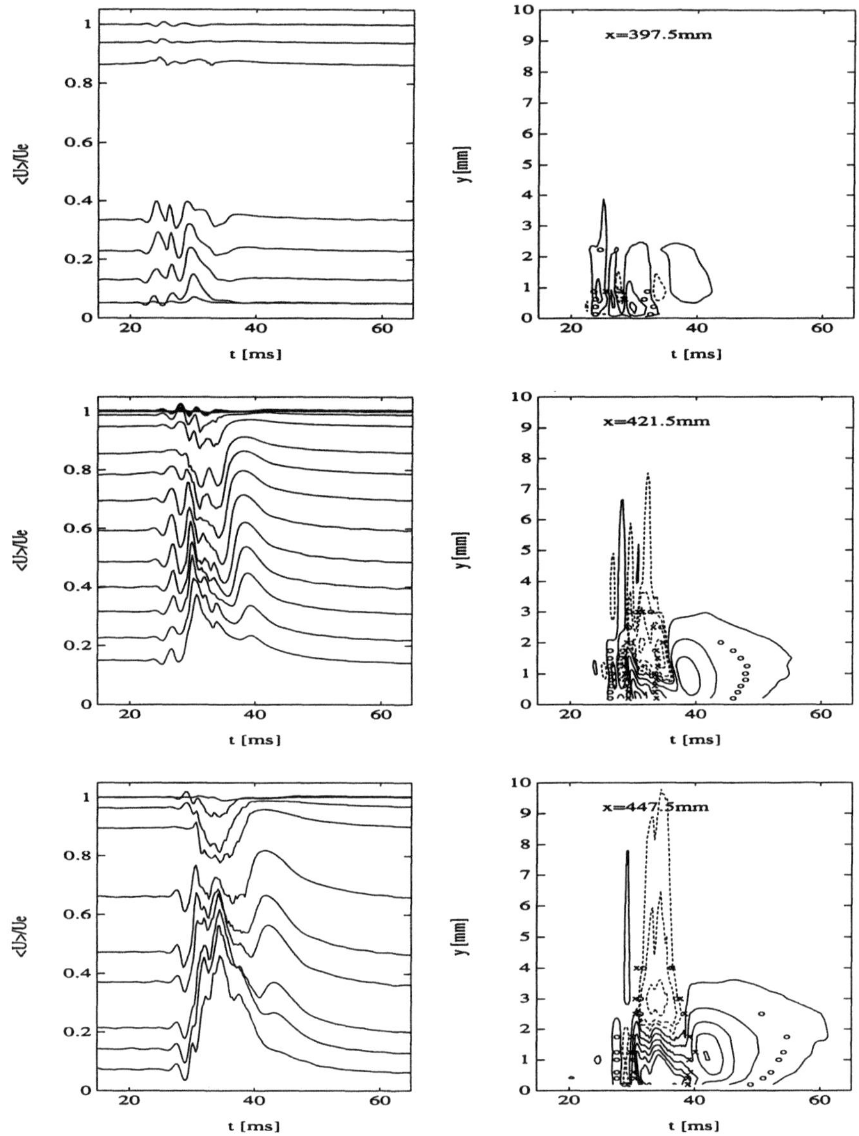

Figure 2: *Ensemble averaged velocity* $\langle U \rangle$, *velocity perturbation* $\tilde{U}$ *contours and spot edge at* $x = 397.5$*mm*, $x = 421.5$ *mm and* $x = 447.5$ *mm.*

Once the wave packet has become a turbulent spot (x = 421.5 mm and x = 447.5 mm), it is preceded by a short oscillation which decreases from approximately two wavelengths at the first station to one wavelength at the last station. The frequency of the oscillation is the same as that of the former wave packet. As turbulence appears first at the center of the wave packet, the oscillation could be the remains of the wave packet which has been overtaken by the turbulent spot. Another explanation could be that the oscillation is the result of the destabilizing effect of the turbulent spot on the surrounding laminar flow, as found by Gad-el-Hak [5]. Inside the spot the velocity fluctuations are larger and of high frequency, indicating turbulent motion. Averaging over 100 realisations is not enough to completely smooth out the turbulent fluctuations.

The spot is followed by a calmed region which is characterised by a fuller velocity profile on which almost no fluctuations are present. TS-waves reappear some time after passage of the spot, indicating that the calmed region is less unstable then the surrounding laminar flow (not shown here). In contrast to ZPG spots, the calmed region is separated from the spot by a velocity dip at intermediate y-values. The velocity at the end of the calmed region differs slightly from the laminar velocity preceding the spot: near the wall the velocity is lower then that of the preceding laminar flow.

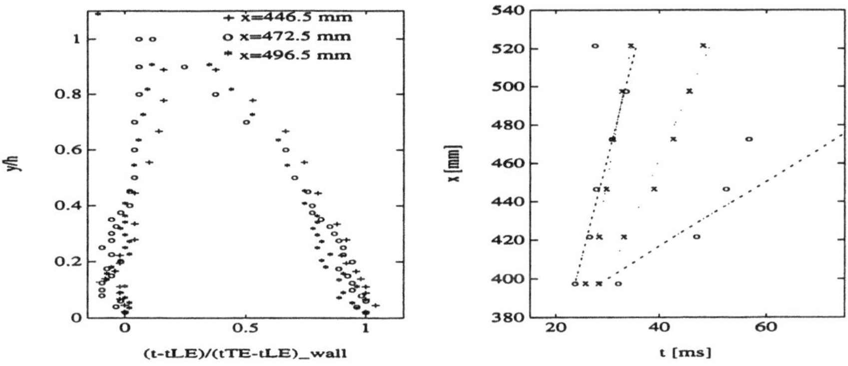

Figure 3: *a) Shape of the turbulent spot, non-dimensionalised with its height and length. b) Average arrival times of the LE and TE edge of the spot.*

The shape of the spot in the streamwise and normal direction at the centerline is shown in figure 3a. The leading edge has a slight overhang at $y/h \approx 0.15$ ($y \approx 1$ mm). All results presented below were determined at this height.

The position of t_{LE} and t_{TE} at six streamwise stations is presented in figure 3b. The propagation velocity of the disturbance was calculated from the slope of a straight line through these points. The propagation velocity was scaled with the edge velocity at x = 397.5 mm. Because there was only a slight decrease in the edge velocity ($\approx$ 1.76 %) over the x-range studied, no definite conclusion can be drawn wether the spot propagation velocities scale with U_e.

It was found that the LE of the turbulent spot, measured at the centerline, propagates with $U_{LE} = 0.92\,U_e$, while the TE velocity was $U_{TE} = 0.38\,U_e$. This speed approximates the phase velocity of the TS-waves. The APG reduces the TE velocity with about 11%. The increased difference between LE and TE velocities results in an increased elongation of the spot in streamwise direction, when compared to the ZPG and FPG spot. The speed of the calmed region is $U_{CR} = 0.10\,U_e$. The turbulent region makes up only a small part of the total disturbance. The calmed region is approximately 1.4 times the duration of the turbulent spot. The ratio of the length of the calmed region to that of the ZPG spot is between 1.6 and 2, while this ratio is about 1 for a FPG spot.

In figure 4 ensemble averaged velocity traces and contours of the velocity perturbations in spanwise direction are presented for $x = 397.5$ mm and $x = 447.5$ mm. Only the positive spanwise positions are plotted. The measurements in the negative spanwise direction were less detailed and were only performed to check the symmetry of the spot.

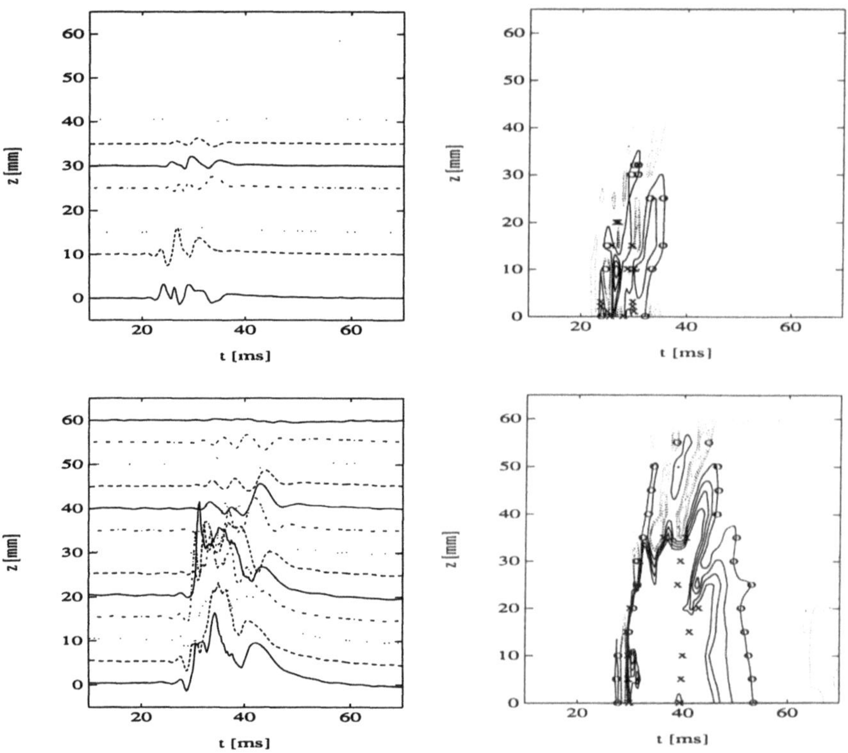

Figure 4: *Ensemble averaged velocity traces $\langle U \rangle$, velocity perturbations $\tilde{U}$ contours and spot edge in spanwise direction at $x = 397.5$mm and $x = 447.5$mm.*

Two regions inside the total disturbance can be distinguished. The central region corresponds to the turbulent spot. It can be observed that the maximum velocity perturbation is located away from the centerline. This may be explained by strong amplification of oblique modes in the non-linear stages of breakdown of the wave packet, which triggers transition first at this spanwise location [6]. This may also explain why the virtual origin of the spot is often located upstream of the trigger position.

The second region shows velocity fluctuations resembling a wave packet. At the outmost spanwise position the wave packet consist of only three oscillations. The amplitude of the oscillations increases for the more inboard positions. The two regions appear to be separated by a small region where the wave packet breaks down through a one- and two spike stage into a turbulent spot. In contrast to the ZPG spot, where a wave packet is usually observed trailing the spot at the outer edges, here the wave packet is observed to propagate along the leading edge of the turbulent spot.

The shape of a turbulent spot under both ZPG and FPG conditions resembles that of an arrowhead. Here the contour of both the total disturbance and the region of the turbulent spot resembles more an ellips.

The two spanwise regions are characterised by different growth angles in spanwise direction. The maximum extent in spanwise direction of the wave packet and the turbulent spot were determined from the deviation in the laminar velocity and with the turbulence detector respectively. The edge of the turbulent spot was defined as the last spanwise position where there were still high frequency fluctuations present in the individual realisations. The edge of the wave packet was defined as the position where $\tilde{U} = 0.025$. The angle was determined from the slope of a straight line through these points. The angle for the turbulent spot was found to be $17^\circ \pm 5^\circ$, for the wave packet the angle is $24^\circ \pm 5^\circ$. The spreading angle in spanwise direction of the turbulent spot under an adverse pressure gradient has increased considerably compared to the value of 10° for a turbulent spot in a ZPG [5], and the value of 5° in a FPG.

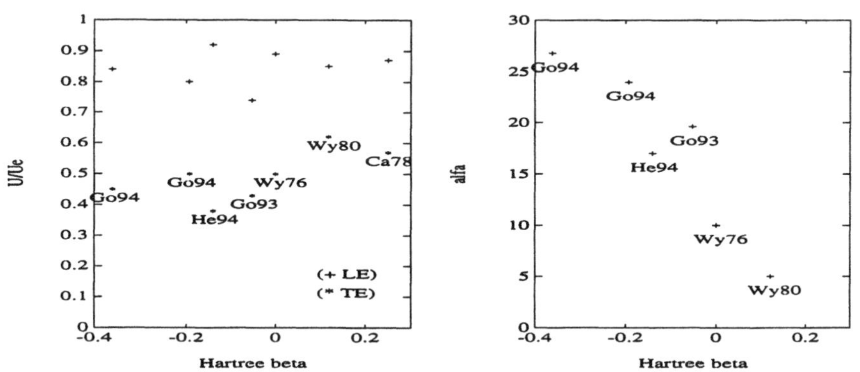

Figure 5: *Spot growth parameters at different pressure gradient*

In figure 5 our results on the spot growth are compared to the results of experiments in other pressure gradients [1] [3][7][8]. Our results agree reasonably well with the general trend.

3 Conclusions

Spot growth in an APG boundary layer has greatly increased in both streamwise and spanwise direction, compared to a ZPG boundary layer. The turbulent spot and the wave packet are part of the same total disturbance, where they reflect different stages of an amplified wave packet. This is in contrast with the ZPG case, were the turbulent spot is trailed by a (separate) wave packet at the outer span position.

The propagation speed of the turbulent spot in a Hartree $\beta = -0.14$ APG is lower than in ZPG and FPG cases, with a leading edge velocity of $0.92\,U_e$ and a trailing edge velocity of $0.38\,U_e$. The growth angle of the disturbance in spanwise direction was found to be 17° for the turbulent spot region and 24° for the wave packet region.

References

[1] Wygnanski, I., Sokolov, M. and Friedman, D. On a Turbulent 'Spot' in a Laminar Boundary Layer, *J. Fluid Mech.* **78** (1976) 785-819

[2] Wygnanski, I., Haritonidis, J.H. and Kaplan, R.E. On a Tollmien-Schlichting wave packet produced by a turbulent spot, *J. Fluid Mech.* **92** (1979) 505-528

[3] Wygnanski, I. The effects of Reynolds number and Pressure Gradient on the Transitional Spot in a Laminar Boundary Layer, in *Lecture notes in Physics: The role of coherent structures in modelling turbulence and mixing layers*, **136**, Springer (1981), pp. 304-332

[4] Riley, J.J. and Gad-el-Hak, M. The Dynamics of Turbulent Spots, in *Frontiers in Fluid Mechanics*, Springer (1985), pp. 123-155

[5] Gad-el-Hak, M., Blackwelder, R.F. and Riley, J.J. On the growth of turbulent regions in laminar boundary layers, *J. Fluid Mech.* **110** (1981) 73-95

[6] Breuer, K.S., Haritonidis, J.H. and Cohen, J. The late stages of transition for a wave packet in a laminar boundary layer, in *Boundary Layer Stability and Transition to Turbulence* FED-**114** ASME (1991), 115-122

[7] Gostelow, J.P., Hong, G., Metwanl, N. and Walker, G.J. Turbulent spot development under a moderate adverse pressure gradient, in proceedings of *International Gas Turbine and Aeroengineering Congress and Exposition*, ASME **93-GT-377** (1993)

[8] Cantwell, B., Coles, D. and Dimotakis, P. Structure and entrainment in the plane of symmetry of a turbulent spot, *J. Fluid Mech.* **87** (1978) 641-675

Spatial Simulations of Bypass Transition in Boundary Layers

D. S. Henningson
FFA, Box 11021, S-16111 Bromma, Sweden
S. Berlin
Dept. Mechanics, KTH, S-10044 Stockholm, Sweden
A. Lundbladh
FFA, Box 11021, S-16111 Bromma, Sweden

Summary

Investigations of the mechanisms underlying bypass transition are presented, and it is shown that a linear growth mechanism is required for energy growth. In shear flows such a mechanism is identified in the rapid transient growth of streaks produced by the three-dimensional lift-up effect. In streak breakdown, oblique transition and receptivity of three-dimensional free-stream disturbances it is shown that this transient growth is utilized as a main mechanism for disturbance growth.

1. Introduction

Disturbances in laminar flows may trigger transition to turbulence if they are strong enough or if the flow is linearly unstable, i.e. an eigenmode of the linearized problem experiences exponential growth. Traditionally, stability analysis has for the most part dealt with the exponential growth and the subsequent secondary instability of the fastest growing eigenmode. In subcritical flows, where all the eigenmodes decay, the linear problem may still have growing solutions although they will eventually decay. This transient growth has received much attention recently and the purpose of this paper is to show the importance of transient growth in various transition scenarios in wall bounded shear flows, particularly the scenario known as oblique transition.

Before we present the spatial simulations of oblique transition we will give some theoretical background and show simulation results from some simplified cases in plane Poiseuille flow in order to put the spatial boundary layer simulations into perspective.

2. Non-linear disturbance equations and the Reynolds-Orr equation

The non-linear disturbance equations for a perturbation velocity u_i on a base flow U_i can be written as

$$\frac{\partial u_i}{\partial t} = -U_j\frac{\partial u_i}{\partial x_j} - u_j\frac{\partial U_i}{\partial x_j} - \frac{\partial p}{\partial x_i} + \frac{1}{R}\nabla^2 u_i - u_j\frac{\partial u_i}{\partial x_j}$$

where $(x_1 = x)$ is the streamwise directions, $(x_2 = y)$ the normal direction and $(x_3 = z)$ the spanwise direction. The Reynolds number will be based on the maximum velocity of the base flow and a typical thickness of the boundary layer, e.g. the displacement thickness or channel half-height. In the streamwise and spanwise directions the domain will be assumed to be either periodic or infinite. These equations, along with the equation for continuity, constitutes an initial value problem for the evolution of an initial disturbance $u_i^0 = u_i(t = 0)$.

We can now derive an equation for the evolution of the kinetic energy of the disturbance,

$$E(t) = \frac{1}{2} \int_V u_i u_i dV,$$

by scalar multiplication of the non-linear disturbance equations by u_i and integration over the domain. We find the Reynolds-Orr equation

$$\frac{dE}{dt} = - \int_V u_i u_j \frac{\partial U_i}{\partial x_j} dV - \frac{1}{R} \int_V \frac{\partial u_i}{\partial x_j} \frac{\partial u_i}{\partial x_j} dV$$

A number of terms have dropped out and the two remaining terms represent the exchange of energy with the mean flow and energy dissipation due to viscosity.

From the Reynolds-Orr equation the important conclusion can be drawn that a linear mechanism is required for energy growth. It is easy to see this by noting that the instantaneous growth rate, $\frac{1}{E}\frac{dE}{dt}$, is independent of the disturbance amplitude. In other words, if a finite amplitude disturbance has a certain instantaneous growth rate, so does the corresponding infinitesimal one. Thus, *there is a need for a growth mechanism in the linearized equations, even for finite amplitude energy growth.* This is a simple consequence of the conservative nature of the non-linear terms in the Navier-Stokes equations.

Morkovin [12] noted, "Apparently, we can bypass the TS-mechanism altogether if we can replace it with another strongly amplifying mechanism". Based on this definition and the conclusion drawn from the Reynolds-Orr equation, we define bypass transition as *transition emanating from linear mechanisms other than exponential instabilities.*

3. The linearized initial value problem and transient growth

Since we have established the requirement of a linear mechanism for growth and transition we now proceed to investigate the linear initial value problem. Based on its solution we define the growth function $G(t)$ as

$$G(t) = \max_{u_i^0} \frac{E(t)}{E(0)}$$

i.e. the energy growth maximized over all initial conditions at each instant of time. For details on the computation of $G(t)$ see Reddy & Henningson [13].

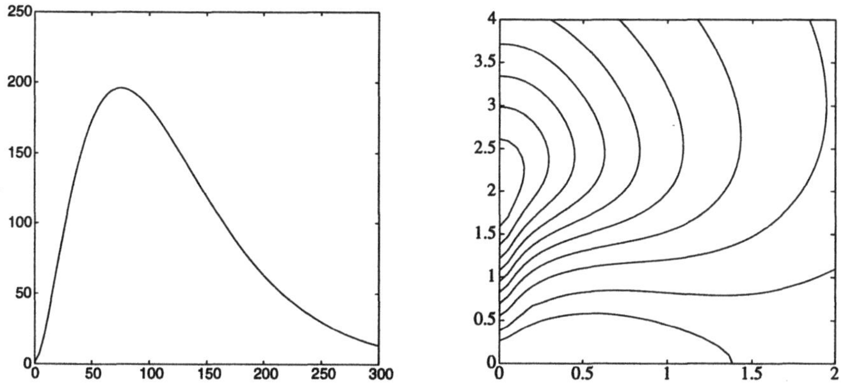

Figure 1: $G(t)$ for plane Poiseuille flow, $R = 1000$. a) $\alpha = 0$, $\beta = 2.04$. b) Contours in the α-β plane. The curves from outer to inner correspond to $\max_t G(t) = 10, 20, 40, ..., 140, 160, 180$. From Reddy & Henningson [13].

Flow	$\max G(t)$ (10^{-3})	t_{max}	α	β
plane Poiseuille [15]	$0.20\ R^2$	$0.076\ R$	0	2.04
plane Couette [15]	$1.18\ R^2$	$0.117\ R$	$35/R$	1.6
circular pipe [14]	$0.07\ R^2$	$0.048\ R$	0	1
Blasius boundary layer [4]	$1.50\ R^2$	$0.778\ R$	0	0.65

Table 1: Reynolds number dependence of $G(t)$ and t_{max} for a number of wall-bounded shear flows.

As an example we now consider the largest growth possible for plane Poiseuille flow at $R = 1000$. This is shown in figure 1a. Note the large transient growth in energy although all eigenfunctions decay.

Since we have assumed a periodic domain in both the streamwise and spanwise directions it is natural to consider the solution in the form of single wave components, $u = \hat{u}(y)e^{i(\alpha x+\beta z)}$ where α is the streamwise wavenumber and β is the spanwise wavenumber. Figure 1b shows contours of the maximum of $G(t)$ as a function of the wavenumbers.

An important property of the growth function is the quadratic dependence of the energy growth on the Reynolds number. In table 1 this dependence, the Reynolds number dependence of the time of the maximum (t_{max}) and the wavenumber combinations for which the maximum occurs, are given for a number of wall-bounded shear flows. Note that in all flows the maximum growth occurs for structures which are essentially independent of the streamwise direction.

The physical mechanisms for the growth may be clarified if we consider the structure of the initial condition giving the optimal growth. Butler & Farrell [4] found that it consists of a decaying streamwise vortex which forces transiently

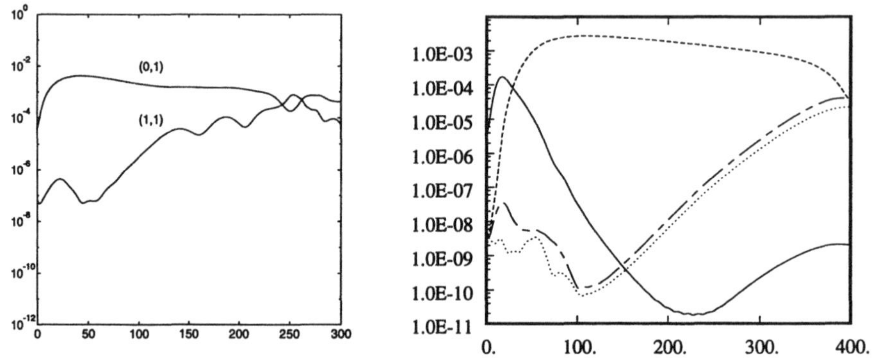

Figure 2: Disturbance energy density evolution in various wave numbers in Poiseuille flow at $R = 1500$. a) Secondary instability of a streak. b) Oblique transition. Solid: (1,1); dashed: (0,2); dotted (1,0); chain-dashed: (1,2). From Lundbladh *et al.* [10].

growing high and low velocity streaks in the streamwise velocity. The horizontal velocity defect associated with the streak is caused by the lift-up of fluid elements that conserve their horizontal momentum. This is referred to as the lift-up effect (Landahl [9]).

The possibility for transient growth has been investigated recently in a number of wall-bounded shear flows, see for example Boberg & Brosa [3], Butler & Farrell [4], Reddy & Henningson [13] and Trefethen, Trefethen, Reddy & Driscoll [15]. In many cases the transient growth has been found to overshadow the exponential growth associated with unstable eigenvalues.

4. Transition and transient growth

4.1 Secondary instability of streaks in plane Poiseuille flow

Let us now consider the non-linear development of an optimal vortex in plane Poiseuille flow. Small amplitude random noise is added to the direct numerical simulations in order to break symmetries and allow instabilities to grow. When the streak produced by the lift-up effect has grown above a threshold, a secondary instability appears in the form of an oscillating motion in the spanwise direction. This secondary instability quickly takes the flow into the turbulent regime.

Due to the periodicity we can expand the solution in a Fourier series in the streamwise and spanwise directions, giving the solution in terms of components of discrete wave numbers in Fourier space, $(\alpha/\alpha_1, \beta/\beta_1)$, where $\alpha_1 = 2\pi$ and $\beta_1 = \pi$ are the fundamental wavenumbers in the streamwise and spanwise directions, respectively. The initial streamwise vortex will then be represented by the (0,1) component, as well as the subsequent growth of the streak, see figure 2a. The secondary instability is seen to emerge from low amplitude random noise in the (1,1) component.

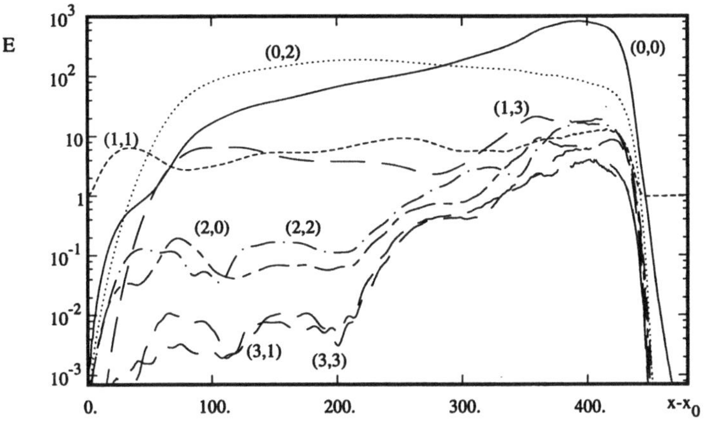

Figure 3: Energy in Fourier components with frequency and spanwise wavenumber $(\omega/\omega_1, \beta/\beta_1)$, where $\omega_1 = 0.08$, $\beta_1 = 0.192$. $R = 400$ at the inflow. The curves are normalized such that the energy of the (1,1) mode at inflow is set to unity.

4.2 Oblique transition in plane Poiseuille flow

Instead of starting with streamwise vortices, it is also possible to generate such disturbances non-linearly, and subsequently obtain the generation of large amplitude streaks. A pair of oblique waves in the $(1, \pm 1)$ components will transfer energy to the streamwise independent mode $(0, 2)$. Here the fundamental wave numbers are $\alpha_1 = \beta_1 = 2\pi$. Once large amplitude streaks have been formed they may become unstable to the same type of secondary disturbances as in the breakdown of the streaks described in the previous subsection. The evolution of energy in selected Fourier components for this *oblique transition* in plane Poiseuille flow can be seen in figure 2b. In this calculation the initial energy in the $(1, \pm 1)$ components are optimal disturbances and thus experience transient growth which enables a stronger excitation of the streamwise vortex. Again, note the exponential growth in the secondary components causing the breakdown of the vortex. Oblique transition in plane Poiseuille flow has also been studied experimentally (Elofsson & Alfredsson, these proceedings).

4.3 Oblique transition in boundary layers

Oblique transition behaves in a similar manner in a spatially developing boundary layer. Streamwise streaks grow to a large amplitude and become unstable to a nonstationary disturbances, resulting in a breakdown to turbulence. Here we present spatial simulations of this transition process in a boundary layer, using the fringe region technique [2] [11].

Figure 3 shows the energy in some of the excited Fourier components during the transition process. Since the development here is spatial the Fourier transform in the streamwise direction is replaced by a transform in time. At $x = 0$ only the $(1, \pm 1)$ components are excited. These components generate $(0, 2)$ $(2, 0)$ $(2, 2)$ by non-linear triad interactions. The $(0, 2)$ mode grows more

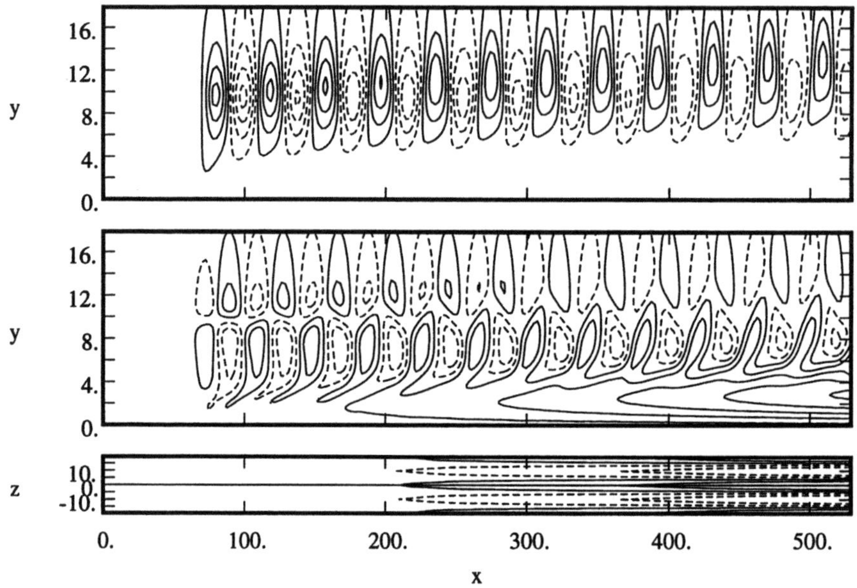

Figure 4: Spatial development of a pair of oblique waves generated in the free-stream. (a) Normal velocity at $z = 0$. (b) Streamwise disturbance velocity at $z = 0$. (c) Streamwise disturbance velocity at $y = 2.03$. $\omega = 0.16$ and $R = 336$ at the inflow. Contour spacing: (a) 0.0025, (b) 0.0025, (c) 0.004.

rapidly than the other modes and continues to grow until about $x = 100$, again due to a linear forcing of the streak from the vortex. A second phase of rapid growth starts for modes with nonzero ω, eventually completing the transition process. This second phase may also be described as a secondary instability on the base flow with a spanwise variation given by the streak. Some more details from this calculation are given in Berlin, Lundbladh & Henningson [1].

Oblique transition has also been investigated by a number of other investigators, for example in the incompressible boundary layer by Joslin, Streett & Chang [8], in compressible boundary layers by Fasel & Thumm [6] and Chang & Malik [5], and in the compressible shear layer by Gathmann, Si-Ameur & Mathey [7]. In these flows the transition scenario remains the same, all of them showing the importance of the $(0,2)$ component. In the compressible boundary case one may also have the added advantage of having the strongest exponential instability in an oblique wave, implying that this scenario may appear naturally from vanishingly small amplitudes.

4.4 Disturbances in the free stream

The oblique transition scenario also has relevance for the receptivity of the boundary layer to free-stream turbulence. Figure 4 shows preliminary calculations where two oblique waves are generated by a normal volume force in the free-stream. In figure 4a the normal velocity in a side view of the boundary layer is shown, where the slowly decaying waves can be seen in the free-stream.

Figure 4b shows the streamwise velocity in the same $x - y$-plane, where a large disturbance inside the boundary layer can be seen. Figure 4c confirms that the disturbance inside the boundary layer are growing streaky structures. Detailed measurements of boundary-layer receptivity to free-stream turbulence have identified the same growing low-frequency disturbances in the streamwise velocity inside the boundary layer, see e.g. [16]. Thus a similar transition scenario as in oblique transition is possible when three-dimensional finite amplitude disturbances exist in the free-stream. However, in the present low amplitude simulation no breakdown to turbulence occurred. This new receptivity mechanism provides an illustration of how free-stream turbulence may enter a laminar boundary layer.

5. Conclusions

Investigations of the mechanisms behind bypass transition have been presented. It was pointed out that a linear growth mechanism is required for energy growth. In shear flows such a mechanism was identified in the rapid transient growth of streaks produced by the three-dimensional lift-up effect. In streak breakdown, oblique transition and receptivity to three-dimensional free-stream disturbances it was shown that this transient growth is utilized as a main mechanism for disturbance growth.

The present work has important implications for transition modeling. When modern transition prediction tools are used, for example based on the parabolized stability equations (PSE), the initial conditions used need not be those associated with traditional secondary instability type disturbances. Instead one could calculate the development and breakdown of streamwise vortices and pairs of oblique waves. Investigations into how this may improve the transition prediction capability in practical applications would be highly desired.

This work has mainly been financed by The Swedish Research Council for the Engineering Sciences (TFR) and the Swedish National Board for Industrial Development (NUTEK).

References

[1] S. Berlin, A. Lundbladh, and D. S. Henningson. Spatial simulations of oblique transition. *Phys. Fluids*, 6:1949–1951, 1994.

[2] F. P. Bertolotti, T. Herbert, and P. R. Spalart. Linear and nonlinear stability of the Blasius boundary layer. *J. Fluid Mech.*, 242:441–474, 1992.

[3] L. Boberg and U. Brosa. Onset of turbulence in a pipe. *Z. Naturforsch.*, 43a:697–726, 1988.

[4] K. M. Butler and B. F. Farrell. Three-dimensional optimal perturbations in viscous shear flow. *Phys. Fluids A*, 4:1637–1650, 1992.

[5] C. L. Chang and M. R. Malik. Non-parallel stability of compressible boundary layers. AIAA Paper 93-2912, 1993.

[6] H. Fasel and A. Thumm. Direct numerical simulation of three-dimensional breakdown in supersonic boundary layer transition. *Bull. Am. Phys. Soc.*, 36:2701, 1991.

[7] R. J. Gathmann, M. Si-Ameur, and F. Mathey. Numerical simulations of three-dimensional natural transition in the compressible confined shear layer. *Phys. Fluids A*, 5:2946–2968, 1993.

[8] R. D. Joslin, C. L. Streett, and C. L. Chang. Spatial direct numerical simulations of boundary-layer transition mechanisms: Validation of PSE theory. *Theoret. Comput. Fluid Dyn.*, 4:271–288, 1993.

[9] M. T. Landahl. Wave breakdown and turbulence. *SIAM J. Appl. Math*, 28:735–756, 1975.

[10] A. Lundbladh, D. S. Henningson, and S. C. Reddy. Threshold amplitudes for transition in channel flows. Proceedings from the 1993 ICASE/NASA Langley Workshop on the Transition to Turbulence, 1993.

[11] A. Lundbladh, P. J. Schmid, S. Berlin, and D. S. Henningson. Simulation of bypass transition in spatially evolving flows. Proceedings of the AGARD symposium on Application of direct and large eddy simulation to transition and turbulence, Chania, Crete, Greece, 18-21 april, 1994, 1994.

[12] M. V. Morkovin. The many faces of transition. In C. S. Wells, editor, *Viscous Drag Reduction*. Plenum Press, 1969.

[13] S. C. Reddy and D. S. Henningson. Energy growth in viscous channel flows. *J. Fluid Mech.*, 252:209–238, 1993.

[14] P. J. Schmid and D. S. Henningson. Optimal energy density growth in Hagen-Poiseuille flow. *J. Fluid Mech.*, 1994. (To Appear).

[15] L. N. Trefethen, A. E. Trefethen, S. C. Reddy, and T. A. Driscoll. Hydrodynamic stability without eigenvalues. *Science*, 261:578–584, 1993.

[16] K. J. A. Westin, A. V. Boiko, B. G. B. Klingmann, V. V. Kozlov, and P. H. Alfredsson. Experiments in a boundary layer subject to free-stream turbulence. Part I: Boundary layer structure and receptivity. *J. Fluid Mech.*, 1994. (To appear).

The Natural Evolution of Turbulent Spots in a Flat Plate Laminar Boundary Layer

F. N. Shaikh and M. Gaster
Cambridge University, U.K.

Summary

Computer generated white noise sequences were used to excite a zero pressure gradient laminar boundary layer. This form of excitation generated modulated disturbance waves in a controlled and deterministic manner. By repeatedly exciting the flow with the same noise sequence, it was possible to examine the spatio-temporal evolution of the disturbance system with a single measurement probe. The development of the resulting flow structures immediately preceding the formation of a turbulent spot was investigated using hot-wire anemometry. The analysis of hot-wire measurements indicated that the resultant highly three-dimensional flow field was associated with the development of a horse-shoe vortex in the flow.

Key Words : Flat plate transition, turbulent spots, white noise excitation.

1.0 INTRODUCTION

The generation of turbulent spots is a stage common to a number of evolutionary routes leading to the creation of turbulence. In general, experimental investigators into this stage of the transition process have used some form of controlled signal to excite the boundary layer in a low turbulence environment. Unfortunately, the majority of these investigations have used a periodic wavemaker to excite the flow. These approaches have ignored the fact that naturally occurring highly modulated waves break down to turbulence through a somewhat different process from that followed by regular periodic wavetrains. The modulated waves breakdown and form turbulent spots at a fraction of the amplitude of the purely periodic disturbance.

[1] used a different form of excitation signal namely, computer generated white noise. It was found that this form of excitation generated disturbances that evolved with streamwise distance in a manner similar to that of naturally excited oscillations. Hot-wire measurements in the region of the flow preceding the appearance of turbulence indicated that the formation of a spot was initiated by the development of a localised large amplitude event in the signals. Here, the three-dimensional structure of this event was explored by repeatedly exciting the flow with a single noise sequence, whilst recording the velocity fluctuations at different points in the flow. The flow field was sampled in a plane normal to the flow direction at a single streamwise location.

The analysis of the results suggested that the formation of a fully turbulent spot was preceded by a highly three-dimensional flow field that could be associated with the development of a horse-shoe vortex.

2.0 ARTIFICIALLY EXCITED DISTURBANCES

2.1 Modulated Disturbances

Naturally excited waves detected by hot-wire anemometers in wind tunnels of moderate turbulence develop a deeply modulated waveform containing bursts of Tollmien-Schlichting (T-S) waves. Wavetrains of this type evolve quite differently from regular harmonically excited waves when the amplitudes are large enough for nonlinear effects to be significant. This type of disturbance therefore warrants further study to gain more insight into the transition of boundary layer flows in both flight and wind tunnel conditions.

The experiment was performed on a flat plate aerofoil mounted vertically in the 3 ft.-square working section of the closed return low turbulence wind tunnel at Cambridge University. A small loudspeaker was buried on the centre-line of the plate at a distance 205 mm from the leading edge. A zero pressure gradient was maintained over the first 1m of the plate by adjusting a two-component flap mounted at the trailing edge. The origin of the co-ordinate system used in this investigation was situated on the centre-line, at the leading edge of the plate: x corresponds to the streamwise ordinate, y the ordinate perpendicular to the wall and z the spanwise ordinate.

Flow measurements were made with a Dantec constant temperature anemometer. The position of the boundary layer probe, data acquisition and the excitation was controlled by a micro-computer. A single time base was used to control both the digital-to-analogue converter feeding the speaker and the signal acquisition analogue-to-digital converter so that the two processes were synchronised. A more detailed description of the apparatus is given by [2].

The free stream velocity was set at 18 m/s for this experiment. The boundary layer that developed on the plate was excited repeatedly with 1 kb of computer generated white noise. The amplitude of the excitation signal was chosen so as to ensure there were no gross distortions of the flow local to the acoustic source and that the boundary layer disturbances were initially of a small amplitude.

The white noise sequence was constructed so as to excite a wide spectral range of disturbance waves. The sequence was produced by specifying the desired Fourier domain characteristics of the signal and then performing an inverse FFT to generate a real time series.

[1] demonstrated that the white noise excitation technique generated the amplitude modulated disturbance waves of interest in a controlled deterministic manner. By repeatedly exciting the boundary layer with the same noise sequence and recording the velocity fluctuations at different points in the flow, it was possible to monitor the spatio-temporal evolution of the wave system. Figure 1 shows a number of ensemble averaged velocity records from different locations in the flow, reproduced from [1].

Mean velocity signals from two non-dimensional wall positions η=0.35 and η=1.1, where $\eta=y\sqrt{Ux/2\upsilon}$, are shown in figure 1a. The ensemble averaging process has removed the background stochastic disturbances that are normally observed in these types of measurements. The remaining signal represents the deterministic response of the boundary layer to the artificial excitation. The figure indicates that, at this position, the disturbance field consists of modulated T-S waves. A comparison of the two velocity traces shows that the characteristic signal shapes found in the near-wall measurements are reproduced very well in the record obtained at η=1.1. The measured velocity fluctuations therefore arose from deterministic structures propagating in the flow.

Similar signals associated with a position further downstream are given in figure 1b. The dramatic increase in the magnitude of the mean velocity signal at η=1.1 can be associated with the appearance of a localised large amplitude event in the record. The remainder of the signal at this position appears quiescent, however, an examination of the measurements from the position nearer the wall indicates that T-S waves are still present in the flow, but are insignificant in amplitude in comparison with the event. Finally, figure 1c, from a position even further downstream, shows that the localised events have developed into turbulent spots.

The development of the localised large amplitude event, observed in the previous velocity records, is a fundamental characteristic of the nonlinear evolution of modulated disturbances and is observed in naturally excited situations, [1], and in the nonlinear evolution of a wavepacket, [3]. [4] examined the breakdown to turbulence of the flow structure associated with the event in the velocity signals. It was found that the breakdown was initiated by the development of high frequency oscillations in the neighbourhood of an event. These oscillations did not correspond to chaotic turbulence, but appeared to consist of a secondary relatively narrow band of frequencies some five or six times that of the basic T-S wave. Figure 2 shows a single realisation of the velocity record obtained from the edge of the boundary layer in this region of the flow field. A high frequency wavepacket can be seen clearly superimposed on the background T-S signal. Slightly further downstream there was also evidence of a tertiary range of frequencies, some twenty times that of the fundamental, prior the formation of a fully turbulent spot. This breakdown process, in which turbulence develops through a cascade of successively higher frequency disturbances was also observed by [3] in the nonlinear development of a wavepacket.

2.2 Examination Of Coherent Flow structures

The evolution of the artificially excited velocity fluctuations was measured at 14 spanwise stations and at 15 locations normal to the wall at a single streamwise position corresponding to $Re_{\delta^*} \cong 1700$. In comparison, the generation of high frequency disturbances was observed to occur at $Re_{\delta^*} \cong 1720$. Figure 3 shows a sequence of ensemble averaged velocity profiles that were obtained on the centre-line of the plate. The profiles cover a portion of the three-dimensional flow field with a time duration of 0.002s and correspond to the passage of a nascent turbulent spot over the hot-wire probe. The development of an inflectional velocity profile and the emergence of a region of high shear are shown clearly. Figure 4 shows a sequence of profiles obtained by [3] from the nonlinear evolution of a wavepacket. The two sets of profiles are remarkably similar.

Figure 5 shows contour plots of the fluctuating streamwise velocity in the plane normal to the flow direction at three time instants. The figure shows the development of a number of concentric regions of alternating sign in the disturbance flow. Although the hot-wire measured only one component of the velocity, it is clear that the circular regions can be associated with longitudinal vortical structures. Finally, figure 6 shows similar plots in the time/span plane at a number of non-dimensional position from the wall. The diagram indicates that the circular regions have a Λ-shaped form in this plane. An examination of the full three-dimensional flow field indicates that the structures identified in the figures correspond to a horse-shoe vortex developing in the flow.

3.0 CONCLUSIONS

The initial development of the horse-shoe structure associated with the nascent turbulent spot can be traced upstream to the emergence of the localised event in the velocity records. [5] have shown that the formation of these events and hence of the turbulent spots is dependent upon the phase dispositions between the spectral components of the excitation signal. Thus the conditions required for the development of an event are determined by the amplitude of the signal, the Reynolds number and the phase relations between the spectral components of the excitation.

It has been suggested, [3], that the modulation of the wavetrains induce Reynolds stresses that are significantly greater than those generated by purely periodic wavetrains. This additional stress would modify the nonlinear development of the wave system significantly and thus provide a ubiquitous route to turbulence.

4.0 REFERENCES

[1] Shaikh, F.N. and Gaster, M.: The generation of turbulent spots in a laminar boundary layer. In preparation (1994).

[2] Shaikh, F.N.: Turbulent spots in a transitional boundary layer. Ph.D Thesis. Cambridge University Engineering Department, U.K. (1994).

[3] Gaster, M.: The origins of turbulence. Proc. conf. on New Approaches and Concepts in Turbulence, Switzerland, (1993). Pub. Birkhauser Verlag Basel.

[4] Shaikh, F.N. and Gaster, M.: Investigation of transition to turbulence using white noise excitation. In preparation (1994).

[5] Shaikh, F.N. and Gaster, M.: The non-linear evolution of modulated waves in a boundary layer. Journal of Eng. Math. 28, (1994) 55-71.

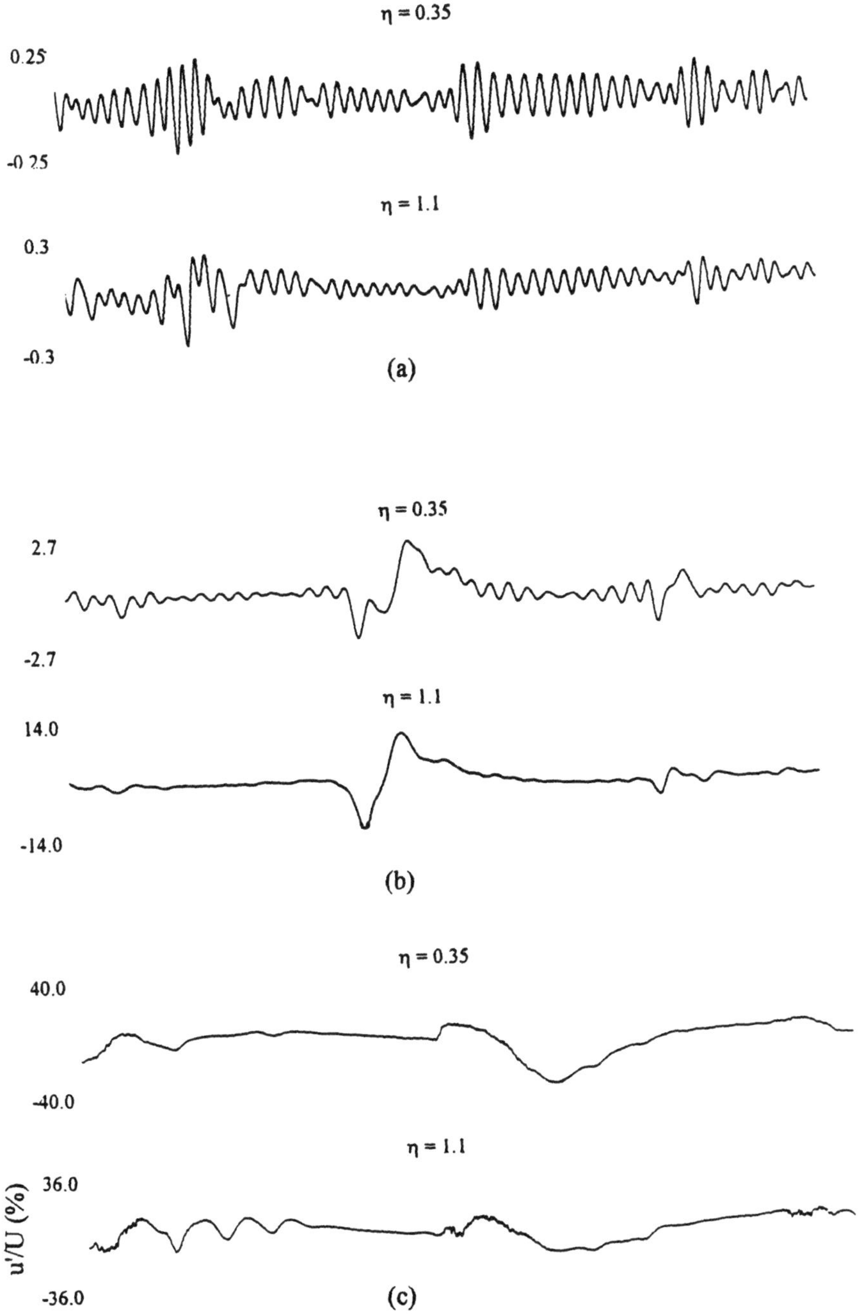

Figure 1 Artificially excited disturbances at three consecutive streamwise locations.

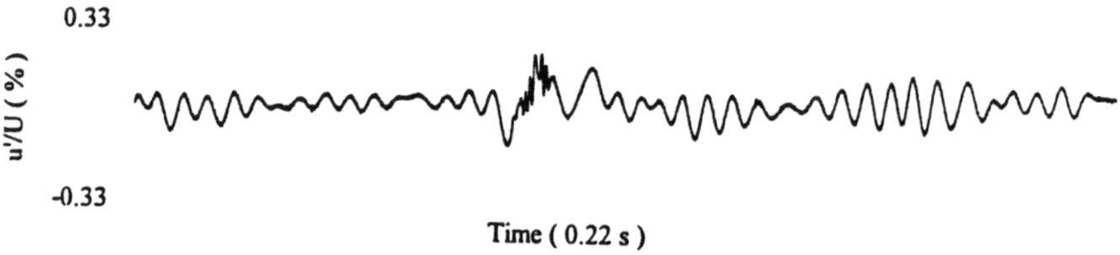

Figure 2 Single realisation of a velocity record showing a high frequency ripple. x=975 mm, η=4.24.

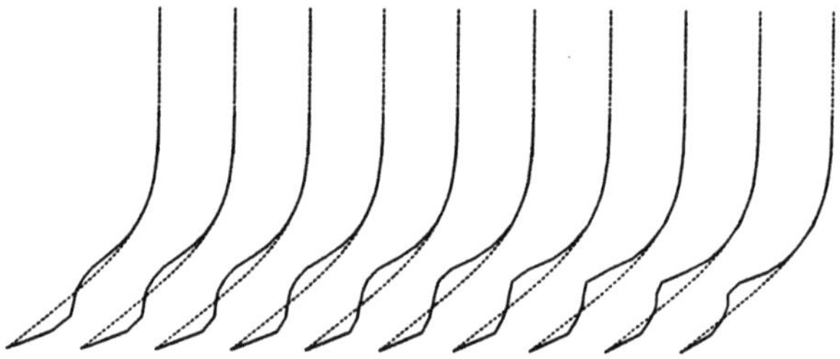

Figure 3 A sequence of velocity profiles centred on the appearance of an event. Time duration=0.002s.

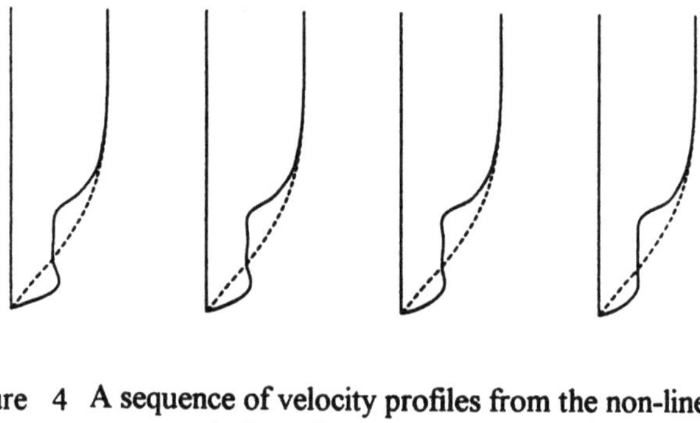

Figure 4 A sequence of velocity profiles from the non-linear evolution of a wavepacket.

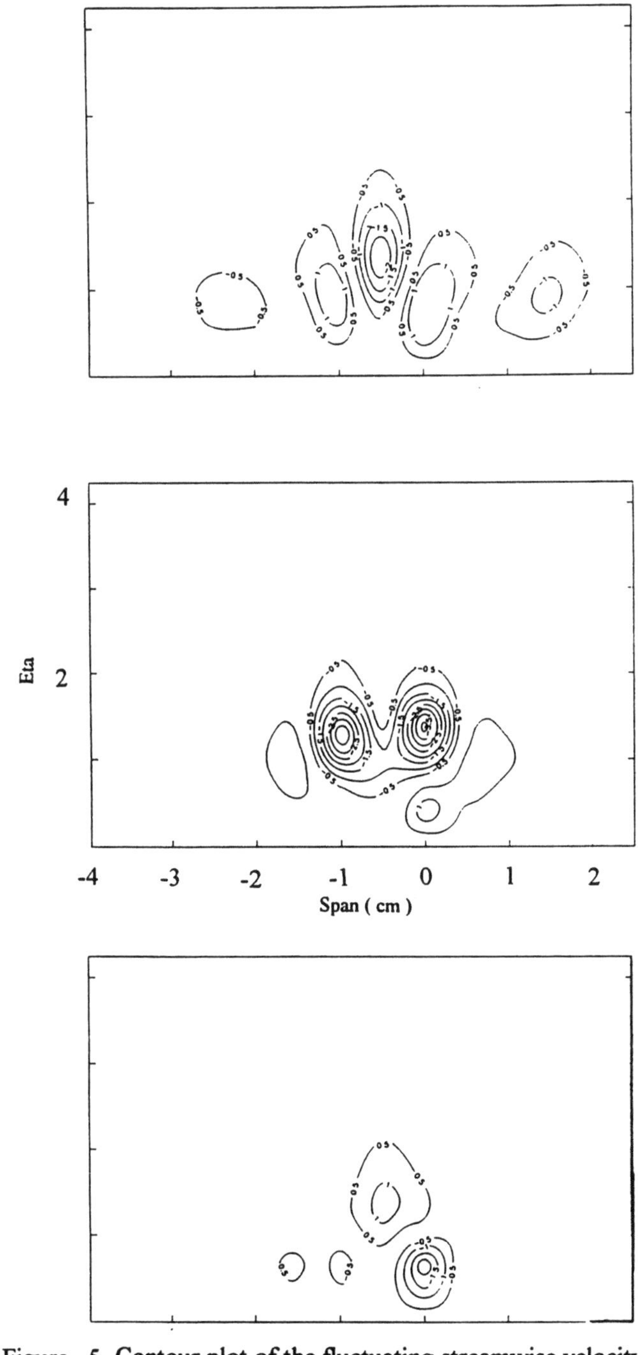

Figure 5 Contour plot of the fluctuating streamwise velocity in the span/normal direction to the wall at three consecutive time instants during the passage of an event.

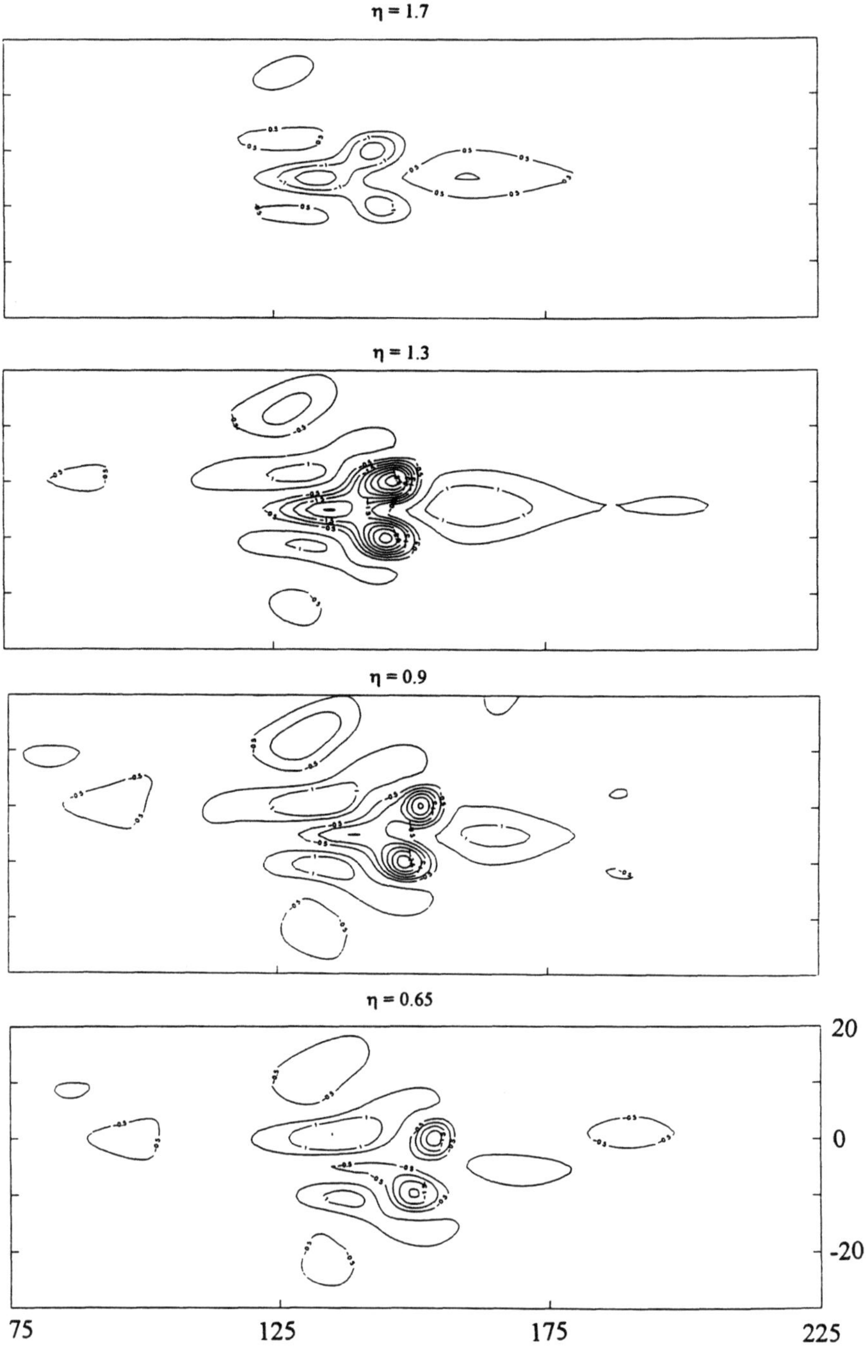

Figure 6 Contour plot of the fluctuating streamwise velocity in the time/span plane during the passage of an event.

Turbulent Spot in 3-D Constant Pressure Flow

M. JAHANMIRI[1], A. PRABHU[1], R. NARASIMHA[2]
([1]Department of Aerospace Engineering,
Indian Institute of Science;
[2]Also: Jawaharlal Nehru Centre for Advanced
Scientfic Research; Bangalore, India)

Summary

We report here on the results of a series of experiments carried out in a distorted duct to study the effects of transverse streamline curvature followed by divergence on the structure of a turbulent spot. It is found that both parameters substantially modify the symmetry of the spot, and in particular significantly alter the angle at which the spot cuts across the local streamline, but otherwise have no great effect on the inner structure of the spot and the eddies therein, or its overall spread rate and celerity.

Key words: 3-D , Transition, Turbulent spot, Streamline curvature, Divergence.

1. Introduction

Many detailed investigations have been carried out in the past few decades on the structure and characteristics of a turbulent spot in zero pressure gradient 2-D flows [1-5]. In a review of recent developments in modelling transitional boundary layers [6], it has been suggested that careful experimental studies of spots subjected to various influences including 3-dimensionality are now badly needed. The few experimental investigations made to-date in 3-D flows [7-9] are generally concerned with bodies of revolution and swept wings; so the understanding of turbulent spots in more general situations remains poor. Although in applications many parameters like pressure gradient, streamline curvature (lateral and longitudinal) and divergence/convergence of the flow all often act simultaneously, it could be of immense value in the understanding of transition if the effect of each of these parameters is studied separately. In this spirit, we report here on the effect of a stretch of transverse streamline curvature followed by divergence (in the absence of any pressure gradient) on the characteristics of an artificially generated turbulent spot. The primary objective is to understand the basic difference from 2-D flows. Preliminary results of the study have been reported elsewhere [10, 11].

2. Experiments

The low turbulence wind tunnel (nominal free-stream turbulence 0.03%) at the Department of Aerospace Engineering has been modified (Fig. 1) to obtain a constant pressure divergent flow. This is achieved by

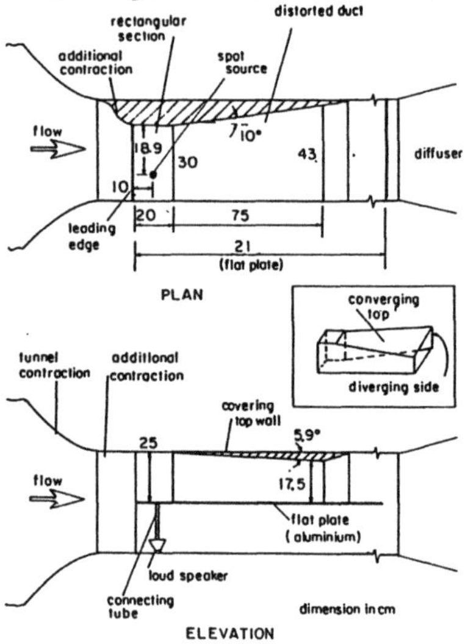

Fig.1 Schematic view of the experimental set-up. Inset shows an isometric view of the distorted duct.

inserting in the tunnel test section a distorted duct that has a converging top-wall and a diverging side-wall, with a flat plate (which is the measuring surface) forming one side of the duct. This avoids longitudinal streamline curvature, but the "bend" in the diverging side-wall provides a short stretch of nearly non-divergent lateral streamline curvature before the flow experiences divergence (with straight streamlines). The distorted duct has a constant cross-sectional area along its length, and thus the pressure gradient is virtually zero.

Flow quality in the distorted duct was verified by surface flow visualization on both flat plate and side wall, and by measurements of (i) the static pressure distribution along the duct and (ii) streamline inclinations on the surface and at the edge of the boundary layer using a three-hole probe. The visualizations showed no flow separation in the duct, either on the side-wall or on the flat plate. The pressure inside the duct was constant to within ±2% of the dynamic head. The velocity in the duct was kept at 10m/s nominal for all the measurements.

An artificial turbulent spot was generated by a loudspeaker exciting the flow at a frequency of 3Hz through one of the 1mm static holes on the plate. Measurements were made using a constant-temperature hot-wire anemometer. A new technique was developed to identify turbulent and non-turbulent zones [12], based on the sensitisation of the signal by squaring its double derivative; this procedure was found to provide excellent results, especially with the incorporation of a variable threshold and a variable hold time proportional to the Kolmogorov time scale.

All overall statistics are generated by averaging over 100 spots. Three different averaging procedures have been adopted in this work: (i) averaging over the turbulent or non-turbulent region based on the discrimination procedure described above, leading to "zonal" averages (e.g. mean flow velocity during the passage of the spot); (ii) averaging with respect to constant phase from the falling edge of the pulse driving the loudspeaker (e.g. average time of arrival of leading edge of the spot at a given point); (iii) averaging with respect to constant phase from the leading edge of the spot (e.g. the average velocity distribution inside a spot).

3. Results and Discussion

Two series of experiments have been conducted. In the "main" series, the spot is generated at x=100mm, in the rectangular section upstream of the entrance to the distorted duct. In the "subsidiary" series the spot is created within the duct in the purely divergent flow, at x=294mm.

The present measurements map (for the first time in turbulent spot studies) the boundaries of the spot in fixed phase time as it propagates downstream; this is done by triggering the spot periodically and taking ensemble averages of type (ii) above over 7 streamwise and 3 spanwise stations as shown in Fig. 2, involving a total of 210 measuring points, and a grid of 144 points at a fixed height of 0.5mm above the surface.

The plan view of the spot at different instants of time, with the superposed local surface streamlines, is shown in Fig. 3 from the main series. It is seen that the spot is quite unsymmetrical: e.g. its nose does not propagate along the axis of the turbulent wedge created by the spot. Some characteristics of the spot are collected in Table 1 and compared with those in 2D flows. It is seen that the overall spread angle of the envelope is 20 deg, and the celerities of the leading and trailing edges are 0.86 U_∞ and 0.56 U_∞ respectively; all nearly the same as in 2-D flow. But it is important to note that, unlike in 2D flow, the spot propagation wedge cuts across streamlines on the inside and outside of the bend at different angles; at x=500mm these angles are respectively 3 deg. and 11 deg. This contradicts the popular assumption in transition zone modelling [7, 13] that the wedge always cuts across the local streamline at the same angle.

Table 1 Characateristics of spots in 2D and in distorted flow.

	2D	Spot A	Spot B
Celerity, L.E.	$0.88U_\infty$	$0.86U_\infty$	$0.87U_\infty$
Celerity, T.E.	$0.5U_\infty$	$0.56U_\infty$	$0.54U_\infty$
Angle beetween locus of spot nose and wedge C.L.	0.0	3.5°	1.2°
Wedge angle with local streamline	11.3°	11°(inner) 3° (outer) (at x=500mm)	11.5° 8.5°
Total spread angle	22.6°	20°	20°

A Spot source located upstream
B Spot source located downstream

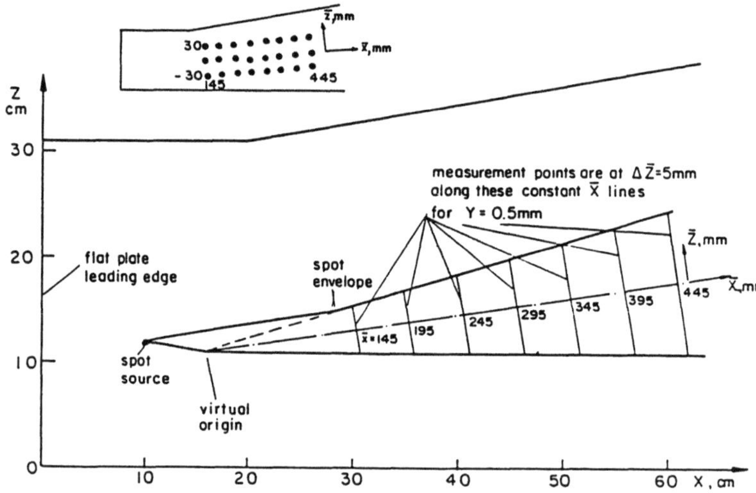

Fig.2 Coordinates and measurement locations. Inset shows stations where data on distribution in y were acquired.

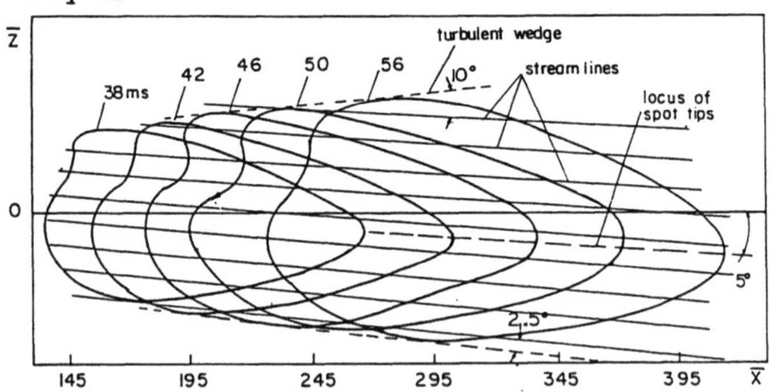

Fig.3 Evolution of turbulent spot in time; plan view at y=0.5mm.

The normalised planform is shown in Fig.4 at all the phase times. Even though there is some scatter in the data shown, a mean shape can be clearly identified and is seen to differ appreciably from the normalised shape in 2-D flow, also shown in Fig. 4 for comparison. This figure reveals that the spot retains a certain universality while propagating downstream. The mean shapes of three sections of the spot, at a time 50ms from the instant of generation, are superposed on the mean plan shape at the same instant and shown in isometric view in Fig. 5 in appropriately normalised coordinates. The effect of distortion on the spot is clearly seen.

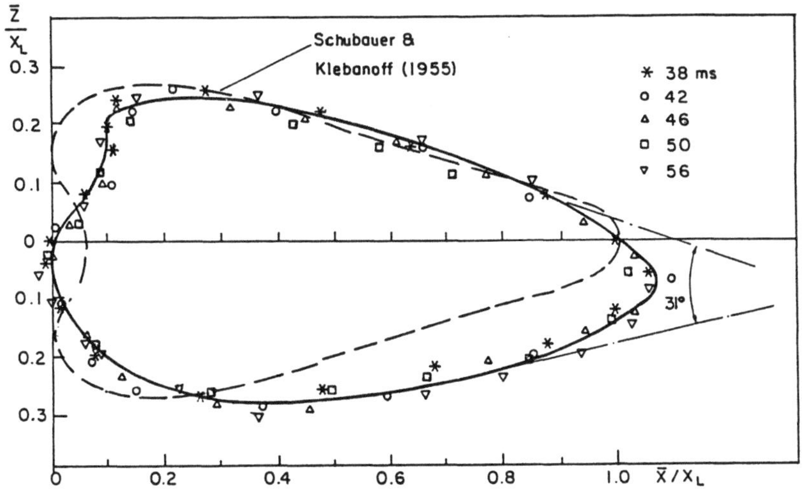

Fig.4 Normalised spot shape: plan view at y=0.5mm, of spot at different times.

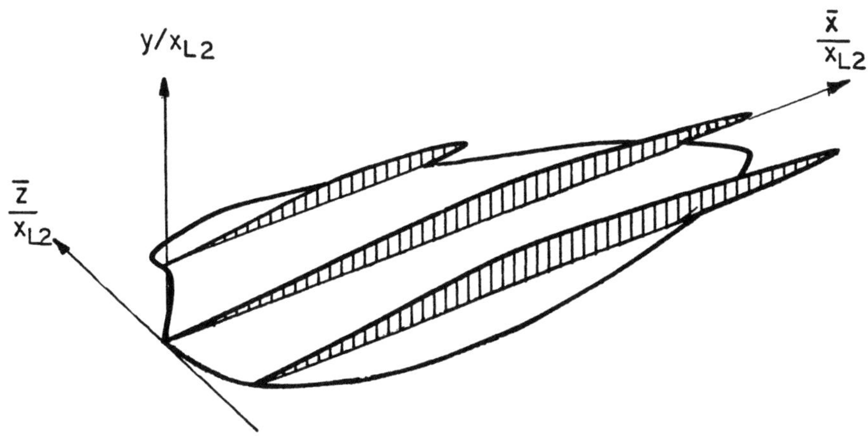

Fig.5 3-D mean shape of turbulent spot at 50ms.

The reasons for the asymmetry of the spot need some consideration, especially as the divergence in the downstream section of the duct is close to that from source flow [14]. One possibility is the effect of the curvature of the streamlines in the corner region of the duct. The supplementary series of measurements made on a spot generated downstream of the bend also revealed asymmetry, although weaker than that seen in Figure 4; some relevant parameters are listed in Table 1. It therefore appears that the observed asymmetry must in part be attributed to the state of the laminar flow into which the spot is injected. It has been demonstrated that, in a divergent flow of the type being studied, the laminar background flow into which the spot propagates has velocity profiles that exhibit local Blasius similarity; however there is a spanwise gradient in boundary layer thickness (higher on the outside of the bend) as seen in Fig. 6. On further examination this is not surprising ; in any cross-flow plane the boundary layer on the outside of the bend has had a longer distance of development compared to that on the inner side of the bend. We must conclude that the spot is extremely sensitive to the asymmetries of the background flow.

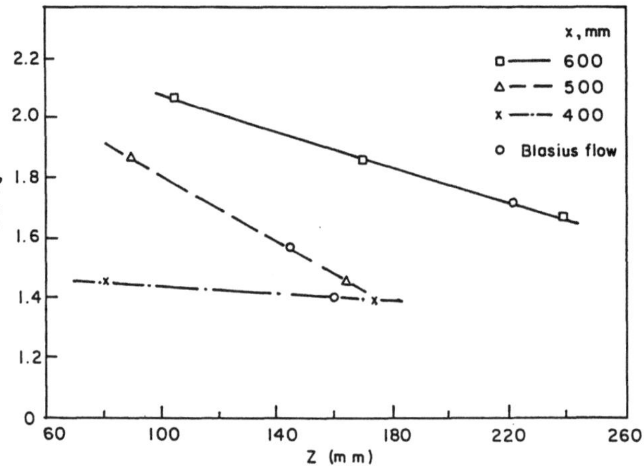

Fig.6 Spanwise variation of boundary layer thickness at different streamwise stations.

We now consider the perturbation flow field of the spot, obtained by subtracting the steady longitudinal velocity of the unperturbed local laminar boundary layer from the ensemble-averaged velocity within the spot. A typical contour plot of the perturbation velocity field at y = 0.5mm is shown in planform view, along with the measured boundary of the turbulent spot, in Fig. 7.The inner structure of the spot shows a small region of velocity excess just downstream of the trailing edge. The presence of peaks and valleys in the central region of the spot, qualitatively similar to those found in 2D spots, may also be seen. These features are consistent with a pair of counter-rotating vortices which entrain fluid from the laminar flow outside the spot. T-S waves may also be seen, but these are much stronger on the outer wing tips than on the inner ones. The asymmetry here is consistent with that in the background laminar flow, as the local boundary layer Reynolds number is higher on the outside of the bend.

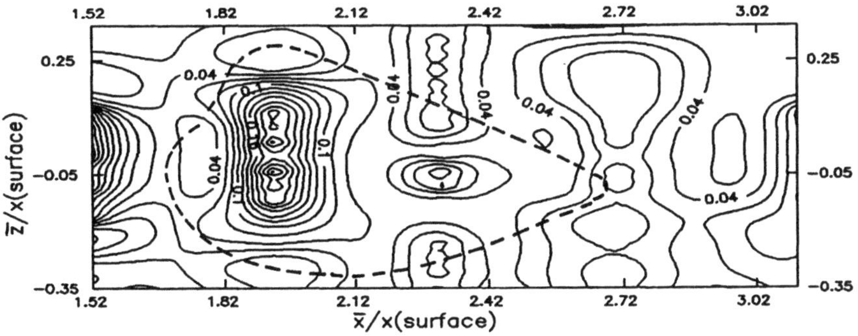

Fig.7 Contour plot of perturbed velocity field in spot at 42ms in normalised $\bar{x}$-$\bar{z}$ plane; y=0.5mm.

4. Concluding remarks

The present experiments reveal that overall spot characteristics like spot envelope (wedge) angle, spread rate, and the celerities of the leading and trailing edges are not significantely altered from their values in 2-D flows under the influence of transverse streamline curvature or divergence. This might help transition modellers in justifiably using certain 2-D flow assumptions in 3-D transition flow. There are some significant differences as well; the angle at which the spot propagates <u>across</u> local streamlines can vary widely although the total spread rate does not. The spot planform can be considerably distorted, and its symmetry is sensitive to that of the background laminar flow into which it propagates; and the TS waves that accompany the spot at its wing tips can be much stronger on one side than the other. The distortion already present in the development of the laminar boundary layer in the background flow thus exerts considerable influence on the symmetry of the spot.

References

[1] Coles, D. and Barker, S.J.: Some Remarks on a Synthetic Turbulent Boundary Layer, in Turbulent Mixing in Non-Reactive and Reactive Flows, Ed. S.N.B. Murthy, Plenum New York (1975) 285-292.

[2] Wygnanski, I., Sokolov, M. and Friedman, D.: On a Turbulent Spot in a Laminar Boundary Layer, J.Fluid Mech. 78 (1976) 785-819.

[3] Cantwell, B., Coles, D. and Dimotakis, D. : Structure and Entrainment in the Plane of Symmetry of a Turbulent Spot, J. Fluid Mech. 87 (1978) 641-672.

[4] Gutmark, E. and Blackwelder, R.F.: On the Structure of a Turbulent Spot in a Heated Laminar Boundary Layer, Expt. in Fluids 5 (1987) 217-229.

[5] Gad-el-Hak, M., Blackwelder, R.F. and Riley, J.J.: On the Growth of the Turbulent Regions in Laminar Boundary Layer, 110 (1981) 73-95.

[6] Narasimha, R.: The Laminar-Turbulent Transition Zone in the Boundary Layer, Prog. Aero. Sci. 22 (1985) 29-80.

[7] Chen, K.K. and Thyson, N.A.: Extension of Emmons Spot Theory to Flows on Blunt Bodies, AIAA Journal 9 (1971) 821-825.

[8] Rao, G.N.V. : Mechanics of Transition in an Axisymmetric Laminar Boundary Layer on a Circular Cylinder, J. Applied Maths. and Physics (ZAMP) 25 (1974) 63-75.

[9] Poll, D.I.A.: The Development of Intermittent Turblence on Swept Attachment Line Including the Effect of Compressibility, C of A Rep 8108 (1981).

[10] Jahanmiri, M., Dey, J., Narasimha, R. and Prabhu, A.: Spot Characteristics in 3-D Flow with Streamline Divergence under Zero Pressure Gradient, Proc. Fifth Asian Congress of Fluid Mechanics, Daejon,Korea (1992) 1026-1030.

[11] Prabhu, A. and Jahanmiri, M.: Developent of a Spot in a Flow with Streamline Divergence, Proc. Third World Conf. on Experimental Heat Transfer, Fluid Mechanics and Thermodynamics, Honolulu, Hawaii, U.S.A. (1993) 316-323.

[12] Jahanmiri, M., Rudra Kumar, S. and Prabhu, A.: A Method for Generating the Turbulent Intermittency Function, Report 91 FM 13, Aero. Engg. Dept., I.I.Sc. (1991).

[13] Emmons, H. W. and Bryson, A. E.: The Laminar -turbulent transition in a boundary layer - Part II. Proc. 1st U.S. Natl. Congr. on Appl. Mech. (1952) 859-868

[14] Ramesh, O. N., Dey, J. and Prabhu, A. : Three-Dimensional Laminar Boundary Layer in a Constant Pressure Diverging Flow - Blasius Equivalent, AIAA Journal 32 (1994) 209-210.

Session H: Swept Wings

Experimental Study of Secondary Instability and Breakdown in a Swept Wing Boundary Layer

A.V.Boiko,V.V.Kozlov,V.V.Syzrantsev,V.A.Scherbakov
Institute of Theoretical and Applied Mechanics,
Novosibirsk, Russia

Summary

In present paper results of experimental simulation of laminar-turbulent transition in a single stationary vortex artificially generated in a swept wing boundary layer are presented. It was observed that two wave packets are formed on the vortex and leads to downstream transition. Also riblet influence on the transition was investigated. It was found that placing the riblets directed along the vortex result in significant prolongation of laminar regime.

Key words : secondary instability, laminar-turbulent transition, boundary layer, riblets

1. Introduction

The most common feature of laminar-turbulent transition in different boundary layers is generation of spatially localized structures or vortices such as Λ–structures on a flat plate, Goertler vortices at curved surfaces or cross flow vortices at swept wings [1], etc. In spite of great variety of mechanisms that lead to generation of the structures the last are characterized by strong elongation in direction parallel to local free stream velocity U_0 vector and can reach amplitudes more then 10% of U_0. The quasistationarity of the vortices in longitudinal direction is a necessary condition for development of their local high frequency instabilities and downstream turbulization.

It was shown [2, 3] that transition to turbulence over swept wing can be observed at subcritical Reynolds numbers. In these cases in transition region high frequency disturbances arise on and travel along the separate stationary vortices. The disturbances grow and turbulize single vortices (saw-tooth visualization patterns) followed by filling of laminar gaps.

The significant role of the initial vortices in laminar-turbulent transition indicates necessity to investigate the transition mechanisms deeper and to find effective control techniques for laminar regime prolongation and drag reduction. It was shown in a number of experiments that special roughnesses called riblets are able to reduce the drag considerably [4]. It is expected that they primarily affect coherent structures appeared in laminar sublayer that are similar to Λ-vortices of laminar flow. Experiments [5] circumstantially confirmed it and pointed out the possibility of riblet application in unusual situation namely for controlling the laminar-turbulent transition.

The aim of the present investigation was to experimentally simulate development of disturbances generated on a single stationary vortex in a swept wing boundary layer and estimate the possibility to control their development by means of riblets mounted in the transition region.

2. Experimental set-up

Experiments were carried out in subsonic low turbulence wind tunnel T-324 of Institute of Theoretical and Applied Mechanics, Novosibirsk, Russia with free stream disturbance level less then 0.04% of U_∞. Experimental set-up is shown in Fig.1 and described below. Wing profile was of C-12 type [6] with swept

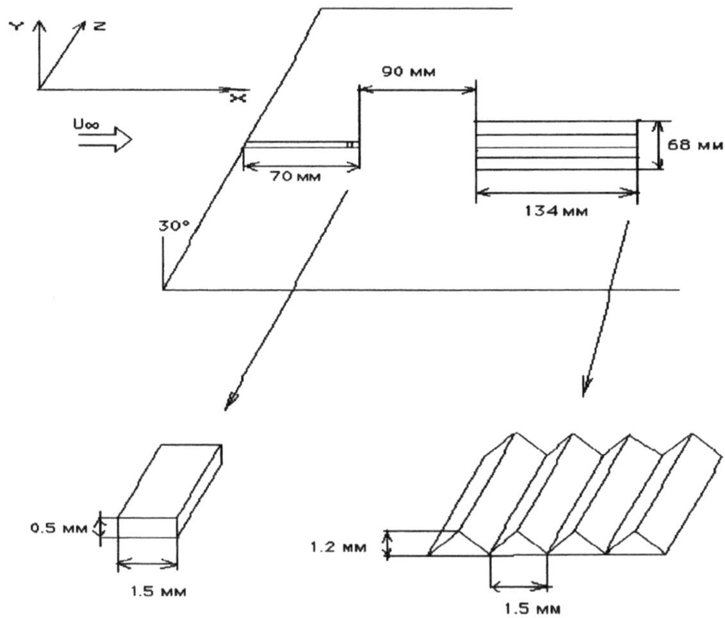

Fig.1 Experimental set-up.

angle of 30^0 and chord length of 500 mm. The profile consists of symmetrical ogive nose 80 mm wide and 5.6 mm long followed by two converging flat plates. The profile was positioned so that its test surface appeared to be at zero angle of incidence. Such an installation significantly simplified data interpretation, hot wire positioning and measurements during the experiments.

A stationary vortex was generated behind an elongated roughness 0.5 mm high and 70 mm long glued to the wing surface beginning from the nose. To prevent local flow separation the roughness was extended to the opposite profile surface. The downstream roughness edge was smoothed for the same purpose.

Coordinate system applied is as follows: X axis is directed from the leading nose along the free stream, Z is parallel to the nose and Y is normal to the wall. Hot wire measurements of streamwise velocity component were carried out in central part of the profile in the range of ΔZ=80 mm at U_∞=7 m/s and X=200-315 mm over gradientless region where secondary flow was negligible. Hot wire data were digitized during the experiments by an analog-to-digital converter and the samples were linearized and processed with a personal computer.

3. Results

Tests showed that mean velocity profiles measured far from the axis of the roughness generated vortex are close to Blasius ones for a flat plate boundary layer with shape factor calculated using experimental data H=2.67±0.11. Boundary layer over the wing was laminar with low disturbance level. Some general shape features of the generated vortex are shown in Fig. 2 as U(Z) distribution measured at X=232 mm at outer border of undisturbed boundary layer. U(Z) shape is typical for a single stationary vortex when local in spanwise direction decrease of the velocity is initialized by low momentum fluid swept to the outer border of boundary layer and increase of velocity is done by high momentum fluid swept to the surface.

The corresponding root mean square (r.m.s.) amplitude distribution u' across vortex axis is presented in Fig. 3. Its maximum is reached in the region of strong mean velocity distortion in Z direction near the vortex core. The same disturbance velocity distribution had been found in [7] at a swept wing for lower frequency instability mode of the two selected there. Similar maximum location relative to U(Z) distribution is typical for sinusoidal instability of Goertler vortices and probably also is connected with appearance of crosswise shear layer.

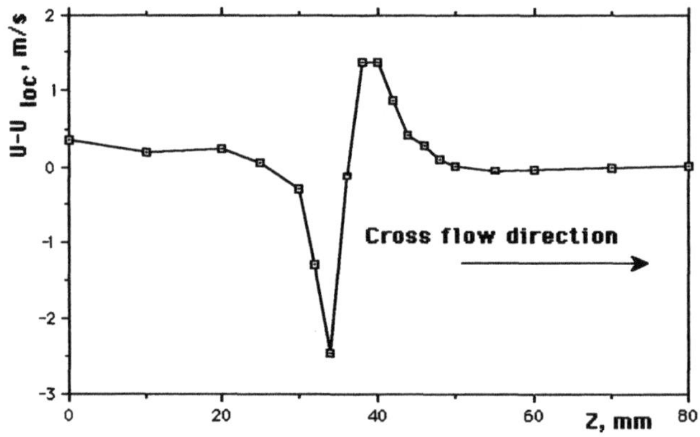

Fig. 2 U(Z) distribution, X=232 mm, "natural" conditions.

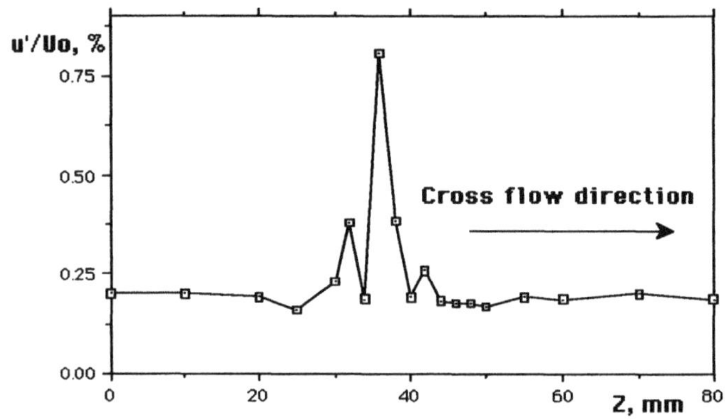

Fig.3 u'(Z) distribution, X=232 mm, "natural" conditions.

It was shown by spectral measurements that two frequency packets are selected inside the vortex with central frequencies close to f=410 and 1800 Hz. Higher frequency packet is amplified downstream and then it is damped whereas the transition to turbulence occurs farther through amplification of lower frequency packet.

The following investigations were carried out under controlled conditions when disturbances were introduced by sound at f=410 and 1800 Hz. Amplification curves of the disturbances measured at their maximums over frequency band of 4 Hz are shown in Fig. 4. The transition here was developed through evolution of the lower frequency oscillations in the region where the higher frequency had not to be found that is similar to results obtained under natural conditions. Maximum increments of lower frequency oscillations were found to be several times larger then for higher ones, whereas their phase velocities were close to each other and ≈ 0.56 of U_0 over almost the whole

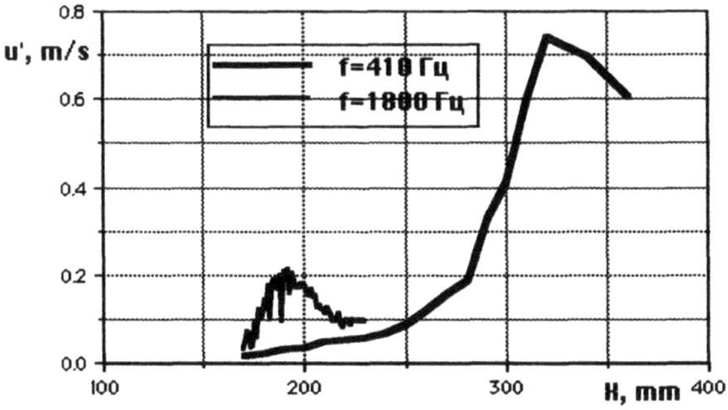

Fig. 4 Disturbance amplification curves.

region of the waves development that provides potential possibility of effective energy exchange between the modes.

Mean velocity and disturbance r.m.s. distributions along Z coordinate indicate that lower frequency wave maximum has origin near inflection point of U(Z) as for integral disturbances under natural conditions whereas high frequency wave maximum changes it location in normal direction. Additionally profiles of U(Y) and $u'_f(Y)$ measured at different Z coordinates for the same X presented in Fig.5 confirm that the lower frequency instability found here is not connected directly with presence of inflection point in U(Y). Oscillations in Fig. 5 are normalized by their maximum at Z = 36 mm $u'_{max} = 2.17$ of U_0.

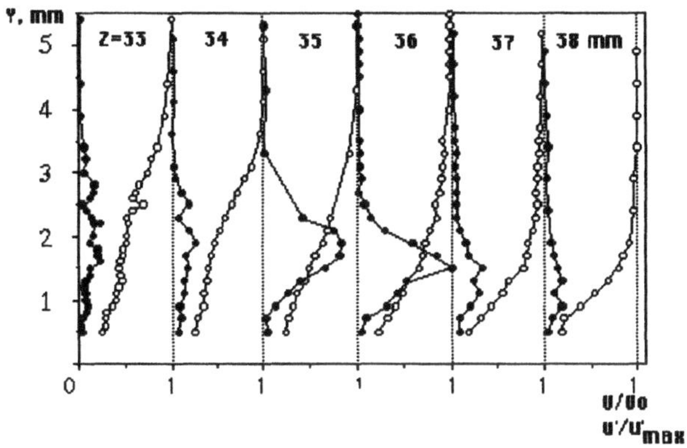

Fig. 5 U(Y) and $u'_f(Y)$ profiles at different Z, X = 232 mm, f=410 Hz.

It is known that additional even rather small disturbance introduction is able to affect mean flow in shear layers [8] modifying the object under investigation. Nevertheless, for lower frequency it was found here that even disturbances with maximum amplitude over 1% of U_0, that as a rule is associated with onset of non-linearity, under experimental accuracy do not change mean velocity shape.

Opposite to the lower frequency oscillations the excitation of the higher ones strongly affect the mean flow. Under their influence change of mean velocity profiles U(Y) is significant, integral disturbance r.m.s. values growing more rapidly, that leads to earlier transition to turbulence then in natural case.

Riblet plate of 134x68 mm was mounted 90 mm downstream the closest roughness edge so that its ribs were parallel to the vortex axis. The shape and size of the ribs were the same as in [5] being of triangular form 1.2 mm height and 1.5 mm base in cross view.

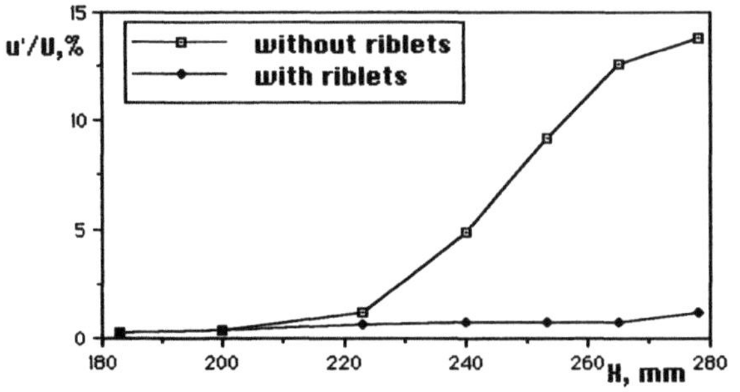

Fig. 6 Riblet influence on disturbance amplification.

To investigate riblet influence on the transition the disturbance maximum r.m.s. growth along X was measured with and without riblets. The comparison is presented in Fig. 6. Beginning from X = 223 mm at the smooth surface the disturbances grow rapidly leading to turbulence whereas at the riblet one their amplitude level is almost unchanged and the flow is laminar. To the end of the measurement region it results in significant difference in amplitude levels.

Taking into account that riblets create a hump over the profile surface special tests with smooth support of effectively the same height and sizes were carried out. In the cases with and without the support similarity of mean velocity profiles and no difference in laminar-turbulent transition were found. The last result indicates that riblets probably do not affect the transition through the breakdown of the initial vortex but rather suppressing instability wave development directly.

4. Conclusions

It was shown that single roughness element placed at the swept wing nose is able to produce stationary vortex in the boundary layer similar to those appeared as a result of crossflow instability in natural case.

Two natural wave packets with central frequencies differ several times were observed on the vortex and their downstream development promoted laminar-turbulent transition in a narrow region of the stationary vortex localization.

The eigen oscillations constituted these wave packets can be significantly forced due to transformation of external acoustic field.

The waves have downstream propagation velocities ≈0.6 of U_∞ and specific u'(Y) distributions that differ them from T-S waves.

It was observed that low frequency packet accelerates the transition due to instability of its waves whereas presence of high frequency disturbance affects primarily mean velocity profiles turning the vortex to more unstable state.

It was found that riblets placed in the transition region can significantly prolong laminar regime suppressing the wave packet development.

5. Acknowledgement

These experiments became possible under the financial support of The Russian Foundation for Fundamental Researches (93-13-17359).

References

[1] Kohama Y.: Some expectation on the mechanism of cross-flow instability in a swept-wing flow, Acta Mech. 66 (1987) 38 p.
[2] Michel R., Arnal D., Coustols E., Juillen J.C.: Experimental and theoretical studies of boundary layer transition on a swept infinite wing, IUTAM Symposium on Laminar-Turbulent Transition, Novosibirsk, 1984, Springer-Verlag, Berlin (1985) 553-562.
[3] Bippes H.: Instability features appearing on swept wing configurations, IUTAM Symposium on Laminar-Turbulent Transition, Zurich, 1989, Springer-Verlag, Berlin, (1990) 419-430.
[4] Kwing-So Choi.: Drag reduction mechanisms and near wall turbulence structure with riblets, IUTAM Symposium on Structure of Turbulence and Drag Reduction, Toulouse, 1989, Springer-Verlag, Berlin, (1990).
[5] Grek G.R., Kozlov V.V., Titarenko S.V.: Investigation of riblets influence on dynamics of isolated wave packet (Λ-vortex) in a laminar boundary layer, Sibirskiy Fiziko-Tekhnicheskiy Zhurnal, 2, (1993). (in Russian).
[6] Zanin B.Yu., Lushin V.N.: Comparing of flowfields over a swept wing in two wind tunnels, Sibirskiy Fiziko-Tekhnicheskiy Zhurnal, 2, (1991) 99–104. (in Russian).
[7] Swearingen J.D., Blackwelder R.F.: The growth and breakdown of streamwise vortices in the presence of the wall, J. Fluid Mech, 182, (1987) 255-290.
[8] Boiko A.V., DovgalA.V., Kozlov V.V.: Nonlinear interactions of disturbances at the transition in a laminar separated boundary layer, Proc. of Siberian Branch USSR Acad. Sci.-Ser. Tech. Sci, 5, No. 18 (1988) 44-49. (in Russian).

Development of Stationary Crossflow Vortices on a Swept Wing

R. H. Radeztsky, Jr.,* M. S. Reibert and W. S. Saric
Mechanical and Aerospace Engineering
Arizona State University, Tempe, AZ 85287-6106, USA

Summary

Stability experiments are conducted in the ASU Unsteady Wind Tunnel on a 45° swept airfoil. The surface of the is polished to 0.25 μm rms. Under these conditions, natural stationary crossflow vortices are not measurable. This state is used to measure roughness-induced stationary crossflow. Spanwise arrays of 70–150 μm roughness elements are introduced near the attachment line. Detailed hot-wire measurements are taken to document the growth of these vortices. The data clearly show that linear stability theory does not accurately predict the growth rates of stationary crossflow waves under these conditions.

Key words: Roughness, Stability, Saturation, Nonlinear

1. Background and motivation

The objective of the present swept-wing research program is to examine the fundamental nature of the crossflow instability which leads to transition in 3-D boundary layers. Reviews of the current literature and problems are given by [1-4]. The DLR efforts in Germany on a swept flat plate are summarized by [5-8]. Another swept-flat-plate experiment is reported by [9]. The CERT/ONERA experiments on swept wings are given by [10-12]. The IFS work in Sendai on cones and spheres is contained in [13-14]. The swept-cylinder experiments are described by [15]. The ASU swept-wing experiments are summarized by [16-18]. This work established the existence of traveling and stationary crossflow vortices, saturation of the stationary crossflow vortex, the nonlinear secondary instability leading to transition, and the sensitivity to freestream disturbances and surface roughness. It results in an Orr-Sommerfeld type solution that can be implemented in a variety of ways ([2-3], [19-21]). See [2-3] for current reviews.

Whereas linear stability theory predicts that the traveling crossflow waves are more amplified than the stationary crossflow waves, many experiments observe stationary waves. Refs. [7-8] describe a series of comparative experiments in a low-turbulence tunnel and a high-turbulence tunnel. Their results show that

---

* Present address: Boeing Commercial Airplane Group, Seattle WA 98124-2207

traveling crossflow waves are observed in the high-turbulence tunnel rich in unsteady freestream disturbances and the dominant structure in a low-turbulence tunnel is a stationary crossflow vortex. One of the important results from the DLR group is the set of data that show early saturation of the disturbance amplitude and the failure of linear theory to predict the growth of the instability. The saturation phenomena of crossflow vortices and the measurements of phase velocity and group velocity directions by [6-8] are important work. They report distorted mean profiles similar to those of [17]. The saturation phenomena are predicted by DNS calculations in the DLR group. This may be part of the Arnal [3] conjecture that e^N methods do not work in crossflow-dominated flows. Another major shortcoming of traditional linear stability theory is the treatment of initial amplitudes. Radeztsky et al. [18] show that the receptivity process for stationary waves is strongly influenced by roughness near the attachment line.

A similarity between the DLR and ASU experiments is the high *N-factors* and high amplitude of the u' distortion. Measurements were taken under conditions of 10%–20% mean-flow distortion. Perhaps then, it is not surprising that linear theory fails. This motivated our recent work, which is an attempt to measure the development of crossflow vortices under conditions of very weak growth.

The present experiment uses the same NLF(2)-0415 airfoil swept at 45° as the previous work [16-18] (Fig. 1). The angle of attack is changed to 0°, significantly reducing crossflow amplification. A new set of contoured wall liners produce an infinite swept-wing flow. Carefully prepared arrays of roughness near the leading edge introduce controlled initial conditions. Specific stationary modes can be identified and followed as they develop over the airfoil. This provides comparisons with single-wavelength stability calculations.

2. Experimental Methods

A new high-resolution, computer-controlled three-dimensional traverse mechanism allows extremely accurate probe placement ($\Delta y = 1$ μm, $\Delta z =$ 1 μm, and $\Delta x =$ 10 μm. Software-corrected scanning techniques produce self-consistent arrays of boundary-layer profiles over a span of 240 mm even in boundary layers less than 1 mm thick. Spanwise scans at constant u/U_e can be maintained over the same spanwise distance. The long measurement times for this experiment require careful attention to hot-wire calibration issues. The temperature response of the hot wires is modeled by a linear correction term [22].

Many experiments which report the variation of total crossflow disturbance amplitude are inaccurate because all stationary modes which are present in the flow are lumped together. If individual crossflow modes can be extracted from the total signal, comparisons with linear theory will be much more meaningful. For the detailed spanwise scans presented here, a span of 240 mm is covered in 256 steps, yielding a wavelength spacing of 0.6 mm at the most-important crossflow wavelength of 12 mm. The lowest resolvable wavelength is 1.9 mm.

The Fast Fourier Transform (FFT) is prone to error when analyzing the small data sets found in these spatial measurements. The Maximum-Entropy Method (MEM) is a transformation that is well-suited for identifying strong spectral features under limited sampling conditions. The drawback of this method is that the user must specify the value of a key parameter in the transformation. Here the results are insensitive to this parameter. Therefore, the MEM is used for all quantitative work, and the FFT is used for qualitative results.

3. Results

3.1 Basic State. The freestream turbulence level is an important parameter in crossflow experiments. Bippes & Müller [6], [7] show that traveling crossflow disturbances dominate the flow in a high-turbulence wind tunnel, and stationary modes are more important in a low-disturbance environment. The ASU Unsteady Wind Tunnel has very low turbulence levels [23]. Most of the disturbance energy is concentrated in the lowest frequencies. With a filter pass band of 1–1000 Hz, the fluctuation level u'/U_∞ is measured to be 0.038%. A pass band of 10–1000 Hz, yields $u'/U_\infty = 0.015\%$. Measurements of the pressure coefficient are very close to that predicted by the theory.

Figure 2 shows a set of 25 boundary-layer profiles obtained without roughness (actual baseline is 0.25 μm rms). These measurements are for direct comparison with later profiles with roughness installed. The profiles cover a span of 24 mm, or two wavelengths of the most-amplified stationary disturbance. The profiles appear as a single line in the plot, indicating a very clean basic state with no measurable stationary crossflow and no spanwise nonuniformities. Extensive spanwise scans verify this.

3.2 Boundary-Layer Scans With Roughness. For this experiment, the goal is to directly influence the *linear* modes of the boundary layer. A spanwise array of 6.2 mm-diameter, 117 μm roughness elements is installed at $x/c = 0.005$, with a spanwise spacing of 12 mm. Figure 3 shows a spanwise array of 25 boundary-layer scans for the same flow conditions as in Fig. 2. With the roughness in place, significant boundary-layer distortions are present, indicating substantial stationary crossflow disturbances. Disturbance contours obtained by subtracting the mean profile are shown in Fig. 4. These shapes clearly indicate a stationary crossflow structure. Velocity contour plots corresponding to Fig. 3 and 4 are shown in Fig. 5 and 6. These plots show the entire velocity field in the (y, z) plane for the u component of velocity. The crossflow disturbances are clearly visible, producing alternating regions of velocity surplus and deficit. A slight asymmetry gives a hint of the "rolling over" motion observed by [17].

Measurements at $x/c = 0.05$, in a boundary layer only 1 mm thick, show a pattern that is more complex due to the presence of shorter-wavelength components. As the vortices develop, the shorter wavelengths decay, and the 12 mm mode corresponding to the roughness spacing begins to dominate. However, in contrast to theory, the amplitude does not increase, but decays.

Figure 7 shows a comparison of linear PSE computations [12] and experimental crossflow modes. In this case, the PSE profiles have been scaled to match the experimental amplitudes, so that the shapes may be compared directly. The remarkable agreement verifies that the experimentally observed disturbances are indeed the same stationary crossflow modes predicted by linear theory. It is important to note that even though linear theory is predicting the correct mode shapes, the growth rates are incorrect. Boundary-layer measurements and mode-shape calculations are performed for many different roughness configurations and flow conditions. The results are similar in all cases, with linear theory severely over-predicting the growth rate.

3.3 Spanwise Scans With Roughness. A more careful comparison with linear theory is possible through the use of spanwise hot-wire scans at constant height in the boundary layer. For the conditions of this experiment, scans at $u/U_e = 0.75$ are close to the maximum of the crossflow disturbance modes. Figure 8 shows a spanwise scan at $x/c = 0.6$ with a full array of roughness installed at $x/c = 0.005$. The dominant 12 mm crossflow mode is clearly visible as a regular variation in the streamwise velocity. At $x/c = 0.3$, as with the boundary-layer scans, shorter wavelengths are present, while the dominant longer-wavelength mode dominates at larger x/c.

Spanwise arrays of roughness elements are applied with a range of spacings in order to produce uniform stationary crossflow waves. Long spanwise scans (240 mm) are used to produce accurate average amplitudes, and also to provide enough resolution for wavelength discrimination. With spanwise wavelengths in the 10–12 mm range, at least 20 wavelengths are covered in a typical scan. The power spectrum is computed for each scan, using both MEM and FFT methods. Figure 9 shows a MEM spectrum for a scan taken with a roughness spacing of 36 mm. Dominant peaks can be observed at 36 mm, 18 mm, 12 mm, and 9 mm. The scans typically show peaks at the roughness spacing and the first few terms in the Fourier spectrum of the roughness. The spectral peaks are roughly fixed for all chord positions, and do not indicate any significant evolution of the dominant frequency. This indicates that initial amplitudes are as important as growth rates in fixing which components will dominate transition.

For configurations with one dominant wavelength, amplitudes can be accurately estimated by simply taking the rms of the disturbance component of the scan. For more complex situations, the high resolution provided by the long scans allows a separate integration of each significant peak. Individual modes can be followed, allowing comparisons with single-mode predictions theory.

The amplitude information extracted from the spanwise scans can be processed to obtain growth rates and N-factors, for direct comparison with linear stability predictions. Figure 10 shows a direct comparison of experimental and linear-theory N-factors for a pure stationary mode with a spanwise wavelength of 12 mm. The Orr-Sommerfeld and linear PSE calculations are performed by [12]. The SALLY results are based on [21]. The MARIA results are based on [20]. The experimental curve shows a behavior similar to that found in the boundary-layer scans. The amplitude decays at first, and then grows very slowly. All of the linear stability calculations show very large growth rates. In order to account for curvature effects, the corrected version of CosCur [19] is used. These results are also shown in Fig. 10. The growth is not as large in this case but the sign is still wrong. Clearly, these linear calculations do not capture the behavior of realistic stationary crossflow waves and a nonlinear calculation is needed.

N-factor comparisons at other wavelengths show similar results. The discrepancy between theory and experiment seems to be largest for the shorter wavelengths. Figure 11 shows N-factors for a spanwise wavelength of 9 mm. A direct comparison of wavelength dependence of growth rates is difficult. An approximate evaluation can be made by comparing final amplitudes at $x/c = 0.6$ with total linear-stability N-factors. Figure 12 shows that linear theory correctly predicts a maximum near $\lambda = 10$ mm.

4. Conclusions

Linear stability theory correctly predicts the expected mode shapes for stationary crossflow. When the computed and measured profiles are scaled and compared, the agreement is excellent. Moreover, linear theory correctly predicts the expected wavelengths for the crossflow modes. Short wavelengths dominate the flow early, and longer wavelengths dominate at larger chord locations. The most-amplified wavelength is correctly predicted.

Linear theory does not correctly predict the growth rates for stationary crossflow waves. The growth curves do not even have the correct sign. Linear theory often predicts strong growth where the amplitude is actually decaying. Previous experiments [5-6] showed the failure of linear theory for extremely large crossflow vortices. This experiment extends these results to include low-amplitude crossflow waves. The inclusion of curvature and nonparallel effects improve the predictions only marginally, and it is conjectured that a nonlinear calculation is required to obtain complete agreement.

Acknowledgments

The work was supported by NASA-Langley Research Center Grants NAG1-937 & 1111 and the Office of Naval Research Fellowship Program. Drs. H. Reed, J.R. Dagenhart, and D. Arnal provided theoretical and computational support.

References

[1] Reed, H.L., Saric, W.S. 1989. Stability of Three-Dimensional Boundary Layers. *Ann. Rev. Fluid Mech.* vol. 21, 235.

[2] Arnal, D. 1992. Boundary-layer transition: Prediction, application to drag reduction. *Special Course on Skin Friction Drag Reduction, AGARD Report 786*, March 1992.

[3] Arnal, D. 1993. Predictions based on linear theory. *Progress in Transition Modelling, AGARD Report 793*, March 1993.

[4] Saric, W.S. 1994. Low-speed boundary-layer transition experiments. *Transition: Experiments, Theory & Computations.* Eds. T.C. Corke, G. Erlebacher, M.Y. Hussaini, Oxford.

[5] Bippes, H. 1990. Instability feature appearing on swept wing configurations. *Laminar-Turbulent Transition, Vol III,* eds. D. Arnal and R. Michel, Springer-Verlag.

[6] Bippes, H. 1991. Experiments on transition in three-dimensional accelerated boundary-layer flows. In *Proc. R.A.S. Boundary-Layer Transition and Control*, Cambridge, U.K.

[7] Bippes, H., Müller, B. 1990. Disturbance growth in an unstable three-dimensional boundary layer. In *Numerical and Physical Aspects of Aerodynamic Flows IV* (ed. T. Cebeci). Springer.

[8] Deyhle, H. Höhler, G., Bippes, H. 1993. Experimental investigation of instability wave-propagation in a 3-D boundary-layer flow. *AIAA J., 31.*

[9] Kachanov, Yu.S., Tararykin, O.I. 1990. The experimental investigation of the travelling waves in a three-dimensional boundary layer. *Laminar-Turbulent Transition, Vol III,* eds. D. Arnal, R. Michel, Springer-Verlag.

[10] Arnal, D., Casalis, G., Juillen, J.C. 1990. Experimental and theoretical analysis of natural transition on infinite swept wing. *Laminar-Turbulent Transition, Vol. III,* eds. D. Arnal and R. Michel, Springer-Verlag.

[11] Arnal, D., Juillen, J.C., Casalis, G. 1991. The effects of wall suction on laminar-turbulent transition in three-dimensional flow. *Boundary Layer Stability and Transition to Turbulence, FED-Vol. 114*, Eds: D.C. Reda, H.L. Reed, R. Kobayashi, ASME.,

[12] Arnal, D., Casalis, G., Copie, M. 1994. *Personal Communication.*

[13] Kobayashi, R., Kohama, Y., Arai, T., Ukaku, M. 1987. The boundary-layer transition on rotating cones in axial flow with freestream turbulence. *JSME Int. J.* **30**(261), 423.

[14] Kohama, Y. 1987a. Some expectation on the mechanism of crossflow instability in a swept-wing flow. *Acta Mech. 66*:21–38.

[15] Poll, D.I.A. 1985. Some observations of the transition process on the windward face of a long yawed cylinder. *J. Fluid Mech.* ***150***, 329

[16] Dagenhart, J.R., Saric, W.S., Mousseux, M.C., Stack, J.P. 1989. Crossflow Vortex Instability and Transition on a 45-Degree Swept Wing. *AIAA Paper No. 89-1892.*

[17] Kohama, Y., Saric, W.S., Hoos, J.A. 1991. A High-Frequency Secondary Instability of Crossflow Vortices that leads to Transition. *Proc. R.A.S. Boundary Layer Transition and Control*, Cambridge UK.

[18] Radeztsky, R.H. Jr., Reibert, M.S., Saric, W.S. 1993a. Effect of Micron-Sized Roughness on Transition in Swept-Wing Flows. *AIAA Paper No. 93-0076.*

[19] Collier, F.S., Malik, M.R. 1988. Curvature effects on stability of three-dimensional laminar boundary layers. *AGARD C-P 438.*

[20] Dagenhart, J.R. 1981. Amplified Crossflow Disturbances in the Laminar Boundary Layer on Swept Wings with Suction. *NASA TP-1902.*

[21] Malik, M. R. 1982. COSAL—A black box compressible stability analysis code for transition prediction in three-dimensional boundary layers. *NASA CR-165952.*

[22] Radeztsky, R.H. Jr., Reibert, M.S., Takagi, S. 1993b. A software solution to temperature-induced hot-wire voltage drift. In *Proc. Third Inter'l Symp. on Thermal Anemometry*, ASME-FED, Washington, DC.

[23] Saric W.S. 1992b. The ASU Transition Research Facility. *AIAA Paper No. 92-3910.*

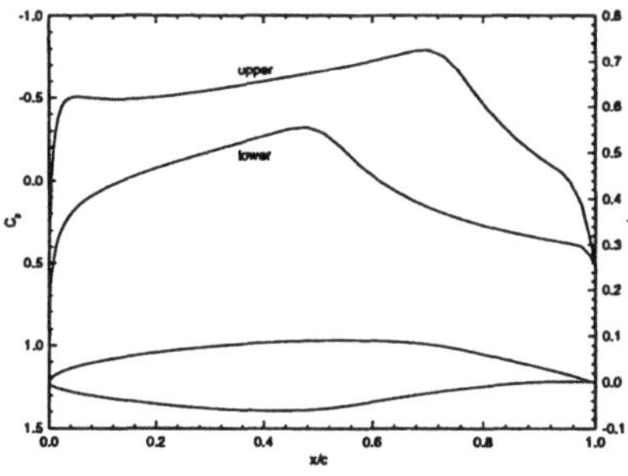

Fig. 1. Unswept NLF(2)-0415 airfoil coordinates and design C_p at $\alpha = 0^\circ$

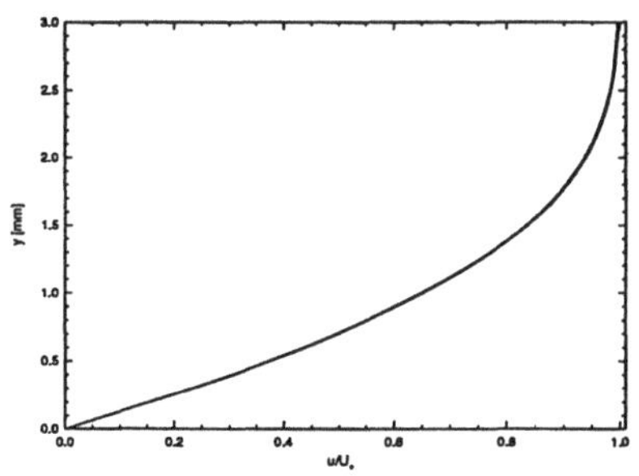

Fig. 2. Spanwise array of boundary-layer scans at $x/c = 0.60$. No roughness. $Re_c = 3.2 \times 10^6$.

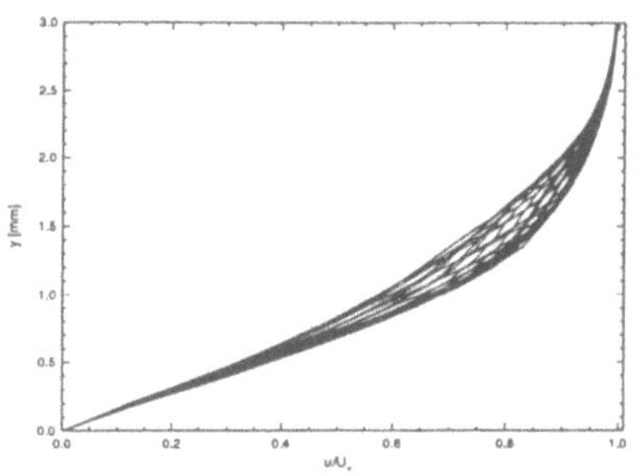

Fig. 3. Spanwise array of boundary-layer scans at x/c = 0.6. A 10-element array of 117 μm roughness with a spacing of 12 mm is at x/c = 0.005, Re_c= 3.2×10^6

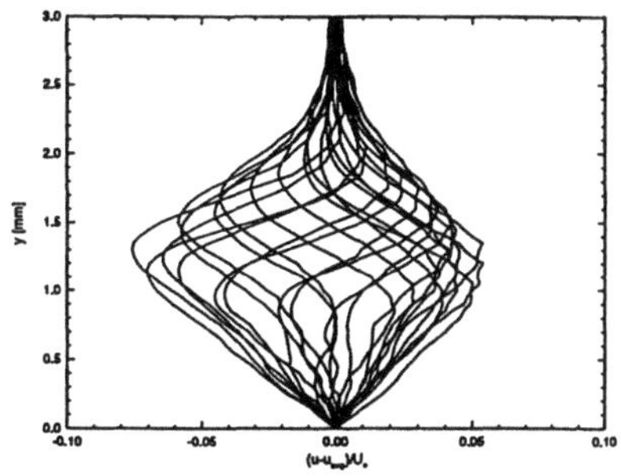

Fig. 4. Disturbance profiles for boundary-layer scans at x/c = 0.60. A 10-element array of 117 μm roughness with a spacing of 12 mm is at $x/c = 0.005$. $Re_c = 3.2 \times 10^6$.

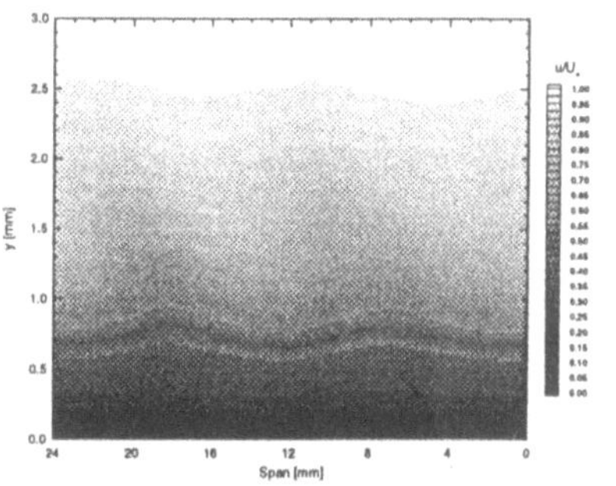

Fig. 5. Velocity contours at $x/c = 0.60$, showing total u component. A 10-element array of 117 μm roughness with a spacing of 12 mm is at $x/c = 0.005$. $Re_c = 3.2 \times 10^6$.

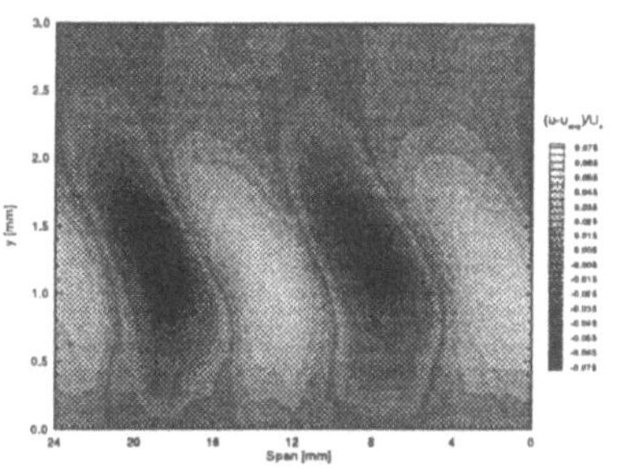

Fig. 6. Disturbance contours for boundary-layer scans at $x/c = 0.60$. Same roughness as Fig. 5.

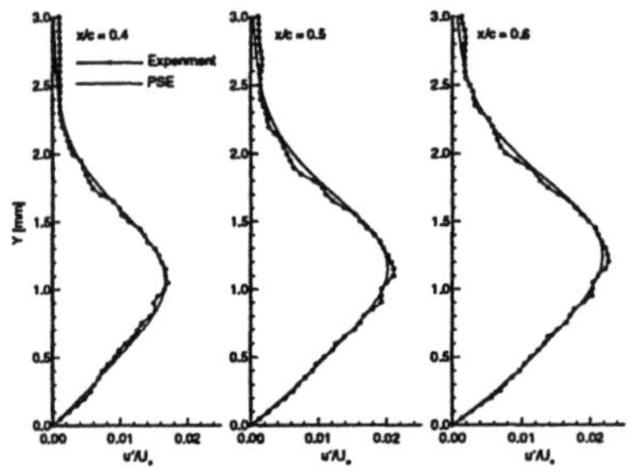

Fig. 7. Comparison of measured and theoretical stationary crossflow mode shapes. Spanwise array of 146 μm roughness at $x/c = 0.005$, 12 mm spacing. $Re_c = 3.0 \times 10^6$.

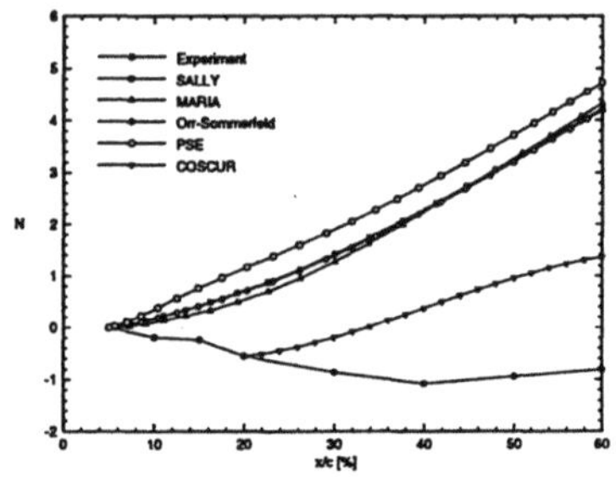

Fig. 10. Measured and theoretical N-factors for $\lambda_s = 12$ mm. Reference point is $x/c = 0.05$. Growing solutions from COSCUR begin at $x/c = 0.20$. $Re_c = 3.0 \times 10^6$.

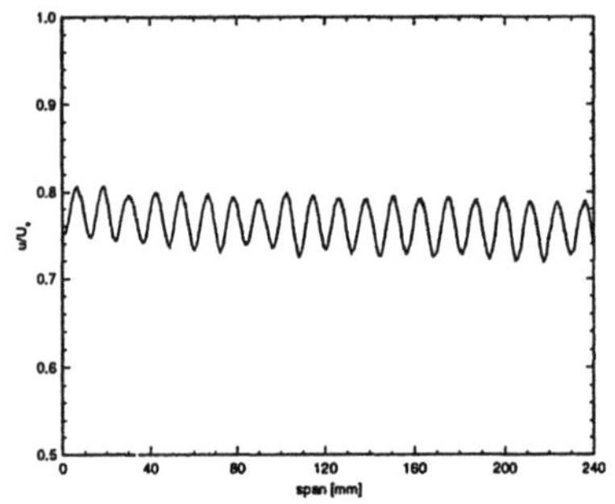

Fig. 8. Spanwise hot-wire scan at $x/c = 0.60$. Spanwise array of 146 μm roughness at $x/c = 0.005$, 12 mm spacing. $Re_c = 3.0 \times 10^6$.

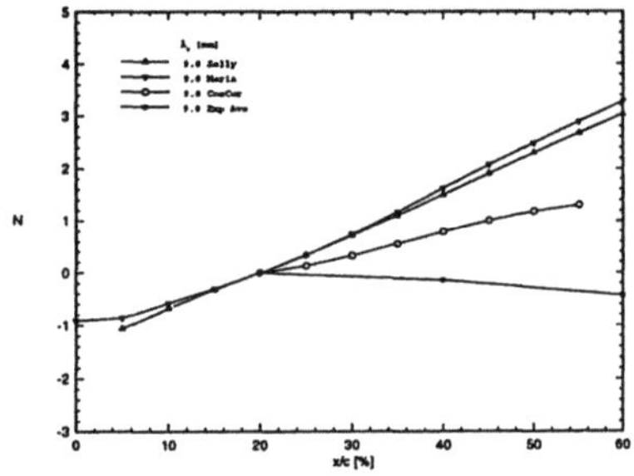

Fig. 11. Measured and theoretical relative N-factors for $\lambda_s = 9$ mm. Reference point is $x/c = 0.20$. $Re_c = 3.0 \times 10^6$.

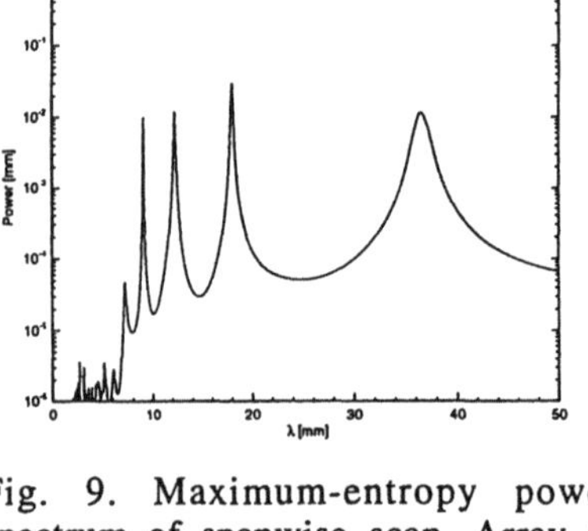

Fig. 9. Maximum-entropy power spectrum of spanwise scan. Array of 146 μm roughness at $x/c = 0.005$, 36mm spacing. $Re_c = 3.0 \times 10^6$.

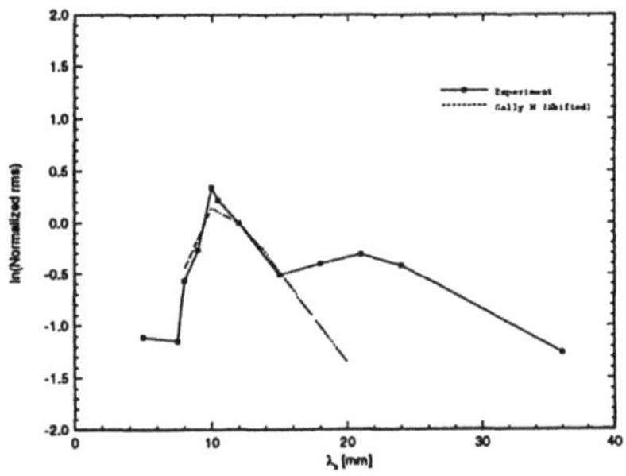

Fig. 12. Comparison of measured roughness sensitivity and SALLY N-factors. $Re_c = 3.0 \times 10^6$.

Session I:
3-D Boundary Layers I

An Experiment on Transition of Three-Dimensional Boundary Layer

Tsuneo Azuma
Department of Mechanical Engineering
Osaka City University, Osaka, Japan

Summary

The laminar-turbulent transition of a three-dimensional boundary layer was studied experimentally using a thin radial liquid-film flow along a stationary and rotating disk, and with a linear stability theory. Because of the very small liquid-film thickness of the flow, a disturbance growing near the wall produces an uneven liquid surface. Hence, the process of transition and the radius of the point of transition could be easily observed through the features of the liquid surface. It was found that two different types of instability, viscous and inflectional, may appear, and that the weak three-dimensionality of the mean flow suppresses the transition due to viscous-type instability, whereas the strong one promotes it due to inflectional-type instability.

Key words : Transition, Boundary layer, Liquid film flow, Rotating disk, Stability

1. Introduction

The transition to turbulence in a three-dimensional boundary layer is the most fundamental problem in fluid mechanics. In order to predict and/or to control the transition, it is indispensable to clarify the effect of the three-dimensionally of the velocity profile on the transition. Although two typical transitions, one in the boundary layer on a flat plate in a stream and the other on a rotating disk in still fluid, have been well understood, various experimental data must be accumulated on the three-dimensional boundary layer to obtain a better understanding of the transition.
A smooth water jet falling vertically onto a horizontal plane spreads out in a thin layer. This liquid film flow could be regarded as a kind of boundary layer [1]. This suggests that the laminar-turbulent transition may occur in the thin layer at high Reynolds number, if the turbulence generated at the moment of jet impingement on the plane is removed.
In the present study, both the transition process and the transition point depending on the three-dimensionally of velocity profile are investigated for a radial liquid film flow on a stationary and rotating disk, along with a linear stability theory.

2. Experiment

Figure 1 shows the test apparatus. Liquid film flow was generated by a water discharge to the atmosphere from the clearance between the lower end of a vertical circular pipe and the horizontal flat surface of a disk concentrically rotating around the pipe axis. The lower end of the vertical

pipe having an inner diameter of 25 mm was formed into a knife edge. The disk of 350 mm in diameter was rotated at n=0, 2, 5, 8 and 10 rps. The clearance between the pipe end and the disk was 0.5 to 1.5 mm.
Since the transition causes an appreciable variation in the features of the liquid surface, the transition process was carefully observed with a stroboscope and also photographed through a high-speed camera.

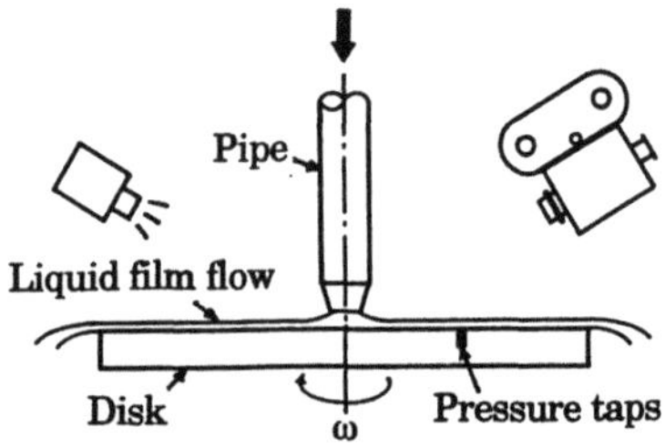

Fig. 1. Experimental apparatus

To disclose the characteristic properties of the disturbance resulting in the transition, the fluctuation of wall pressure was measured with a pressure transducer of a diffusion-type through two pressure taps 0.59 mm apart, each with an inner diameter of 0.15 mm. The pressure taps were made of stainless syringe needles measuring 0.35 mm in external diameter, 0.15 mm in inner diameter and 3.5 mm in length. These needles were embedded vertically in the surface of the disk and soldered onto the disk before the disk surface was machined smoothly. Pressure chambers filled with silicone oil of 5000 cSt were deaired in a vacuum to heighten the frequency response. Signals of the wall pressure were analyzed through a FFT processor.
The experimental flow condition through the clearance is indicated by the discharge Reynolds number, which is defined by $Re=Q/(CDH)^{1/2}\nu$, where Q is the volume flow rate, D the pipe diameter, H the distance of clearance, C the discharge coefficient of the clearance and ν the kinematic viscosity. $(CDH)^{1/2}$ is essentially equal to the liquid jet radius a mentioned below.

3. Linear Stability

3.1 Mean Flow

A liquid jet having a radius a and a speed U_0 impinges onto the disk rotating around its axis with an angular velocity ω. Then a laminar boundary layer grows three-dimensionally from the central stagnation point, and this boundary layer gradually absorbs the whole of the flow until the whole layer is a kind of three-dimensional boundary layer. Since the flow is governed both by the wall friction of the rotating disk and the impinging force, the velocity profile is not similar. Thus, nonlinear differential equations transformed from the boundary layer equations using an appropriate stream function were solved numerically by the finite difference method [2]. Typical examples of velocity profiles changing downstream are shown in Fig. 2, where $Re_0=Q/a\nu$ denotes the jet Reynolds number (Re_0 is equivalent to Re), $m=a\omega/U_0$ the parameter of disk rotation, $\xi=r/a$ the

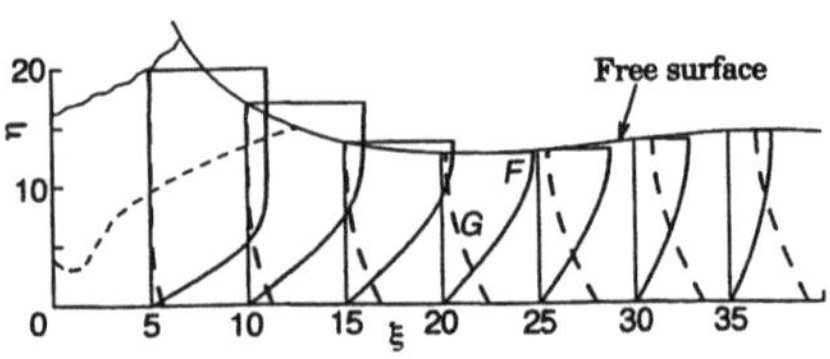

Fig. 2. Velocity profiles ($Re_0=10^5$, $m=0.02$)

dimensionless radius, $\eta = z(2U_0/a\nu)^{1/2}$ the dimensionless distance from the wall, and F and G the radial and the circumferential velocity components divided by U_0, respectively. It was confirmed by LDV measurement [3] that the calculated velocity profiles are in good agreement with those of the laminar liquid film flow used in our experiment.

3.2 Perturbation Equations and Calculated Results

In studying the linear stability of the simplified flow mentioned above, we take the same orthogonal curvilinear coordinate system (Fig. 3) and procedure as adopted by Kobayashi et al. [4]. ε in Fig. 3 denotes the propagation direction of disturbance.

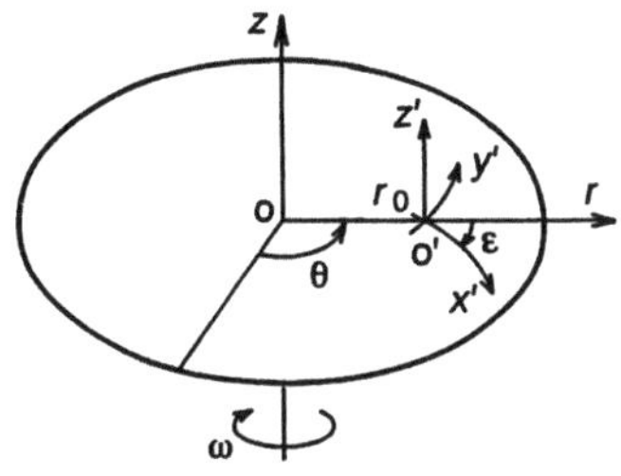

Fig. 3. Coordinate systems

Let $\bar{\mathbf{V}}(\bar{U}, \bar{V}, \bar{W})$, $\bar{P}$ and $\mathbf{v}'(u', v', w')$, p' denote the steady state and the perturbation quantities of velocities in x'-, y'-, and z'-directions and pressure. The perturbation quantities are also assumed to have the form

$$\mathrm{v}' = \hat{\mathrm{v}}(z')\exp\{i(\alpha x' - \beta t)\}, \quad p' = \hat{p}(z')\exp\{i(\alpha x' - \beta t)\}$$

where α is the wave number, and β the complex angular frequency. Furthermore, the following non-dimensional quantities are introduced:

$$z = z'/\delta,\quad u = \hat{u}/F_s,\quad v = \hat{v}/F_s,\quad w = \hat{w}/F_s,\quad U = \bar{U}/F_s,\quad V = \bar{V}/F_s,$$

$$\delta' = \delta/r_0,\quad \Omega = r_0\omega/F_s,\quad Re_\delta = F_s\delta/\nu,\quad c = \beta\delta/\sigma F_s,\quad \sigma = \alpha\delta$$

where δ is the liquid film thickness, F_s the radial surface velocity, ω the angular velocity of the disk, Re_δ the local Reynolds number, c $(=c_r+ic_i)$ the dimensionless complex phase velocity of disturbance, and σ the dimensionless wave number of disturbance.

Consequently, we obtain the following perturbation equations.

$$\begin{aligned}&(U-c)(\frac{d^2w}{dz^2} - \sigma^2 w) - \frac{d^2U}{dz^2}w + \frac{i}{\sigma Re_\delta}(\frac{d^4w}{dz^4} - 2\sigma^2\frac{d^2w}{dz^2} + \sigma^4 w)\\ &+ \delta'[\frac{i}{\sigma}\{(U-c)\cos\varepsilon - V\sin\varepsilon\}\frac{d^2w}{dz^2} + \frac{i}{\sigma}(\frac{dU}{dz}\cos\varepsilon - \frac{dV}{dz}\sin\varepsilon)\frac{dw}{dz}\\ &+ 2\frac{dV}{dz}v\cos\varepsilon + (2V\cos\varepsilon - c\sin\varepsilon - 2\Omega)\frac{dv}{dz}] = 0\end{aligned}$$

$$\begin{aligned}&(U-c)v + \frac{i}{\sigma Re_\delta}(\frac{d^2v}{dz^2} - \sigma^2 v) - \frac{i}{\sigma}\frac{dV}{dz}w\\ &+ \delta'\{-\frac{i}{\sigma}Uv\cos\varepsilon + \frac{1}{\sigma^2}(V\cos\varepsilon - 2U\sin\varepsilon - 2\Omega)\frac{dw}{dz}\} = 0\end{aligned}$$

$$i\sigma u + \frac{dw}{dz} + \delta'(V\sin\varepsilon + \frac{i}{\sigma}\frac{dw}{dz}\cos\varepsilon) = 0$$

Using the radial and the circumferential velocity components F and G, the velocity components U and V in x'- and y'-directions are obtained from

$$U(z) = F(z)\cos\varepsilon - \{G(0) - G(z)\}\sin\varepsilon,\quad V(z) = F(z)\sin\varepsilon + \{G(0) - G(z)\}\cos\varepsilon$$

The boundary conditions at the free surface are obtained from both the absence of shearing stress and the existing balance between the surface tension and the normal stress. Then the boundary conditions, neglecting the terms involving δ', are given by

$$z=0: u=v=w=0$$

$$z=1: \frac{d^2w}{dz^2}+\{\sigma^2-\frac{d^2u}{dz^2}/(U-c)\}w=0, \qquad \frac{dv}{dz}-i\{\frac{d^2V}{dz^2}/\sigma(U-c)\}w=0$$

$$\frac{d^3w}{dz^3}-\{3\sigma^2+i\sigma Re_\delta(U-c)\}\frac{dw}{dz}+i\sigma Re_\delta(\frac{\sigma^2}{We}+\frac{1}{Fr^2})\frac{w}{U-c}=0$$

where *We* denotes the Weber number, and *Fr* the Froude number.

The perturbation equations was solved numerically by the finite difference method, in which both *We* and *Fr* were assumed to be sufficiently large.

The stability characteristics with changing *m* under a typical set of *Re*o and ξ are shown in Figs. 4a-c. The closed curves indicate the neutral curve and x-marked positions indicate points where the growth rate of the disturbance is maximum. It should be noticed that the points of maximum growth rate appear at two different values of ε in Fig. 4b. This means that two different types of instability may occur in the flow. The mean velocity profiles at both ε=4 deg. in Fig. 4a and ε=18 deg. in Fig. 4b have no inflection point, but those at both ε=56 deg. in Fig. 4b and ε=55 deg. in Fig. 4c do. Thus, the instability of Fig. 4a may be considered as a viscous-type instability, and those of Fig. 4c as inflectional-type instability. In other words, the instability changes from viscous to inflectional type, as the three-dimensionality of the flow increases due to the disk rotation.

Figures 5a-c show the stability characteristics with changing ξ under a typical set of *Re*o and *m*. It is found from Figs. 5 that the change from viscous to inflectional instability occurs with increasing ξ, i.e. with increasing

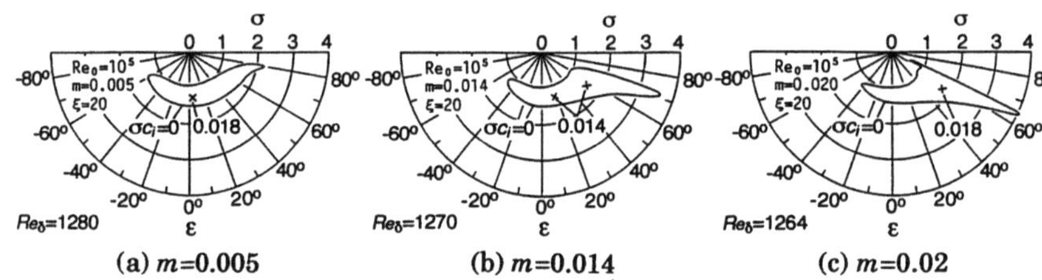

(a) m=0.005 (b) m=0.014 (c) m=0.02

Fig. 4. Stability characteristics under conditions $Re_0=10^5$ and ξ=20

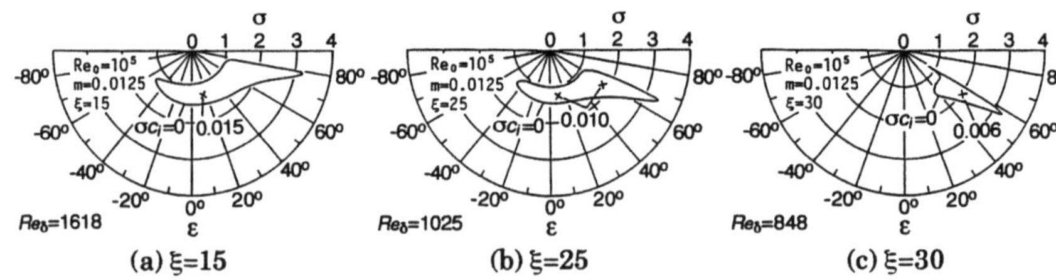

(a) ξ=15 (b) ξ=25 (c) ξ=30

Fig. 5. Stability characteristics under conditions $Re_0=10^5$ and m=0.0125

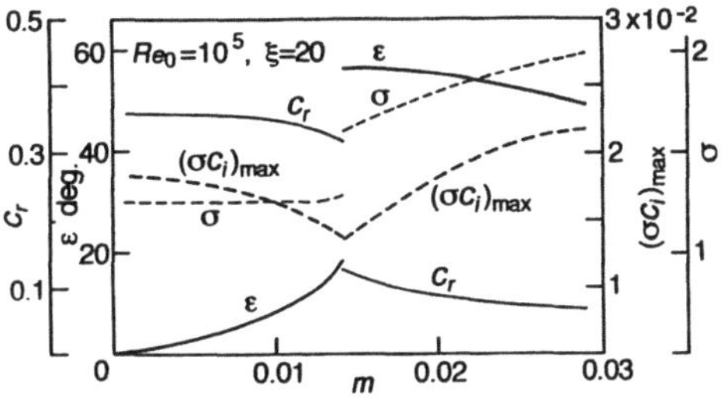

Fig. 6. Characteristics of the most unstable disturbance ($Re_0=10^5$ and $\xi=20$)

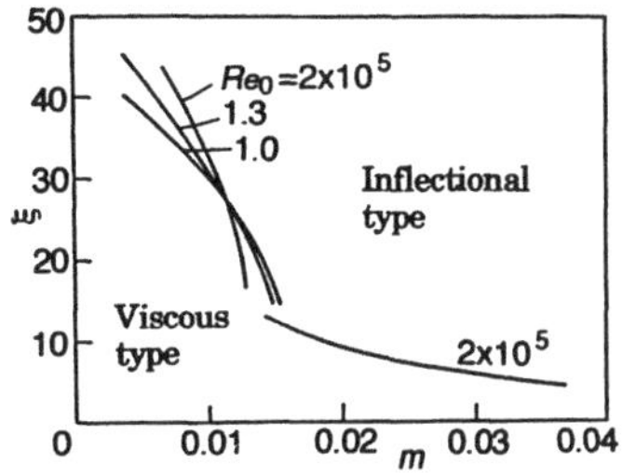

Fig. 7. Boundaries between two types of instability

three-dimensionality in the velocity profile.

Figure 6 shows characteristic properties of the disturbance having its maximum growth rate as a function of m under the conditions $Re_0=10^5$ and $\xi=20$. It is especially interesting, from the experimental point of view discussed later, to see how ε and $(\sigma c_i)_{max}$ change with increasing m; ε monotonously increasing with m in the viscous-type instability region abruptly increases at $m=0.014$, and then decreases slowly in the inflectional-type region. On the other hand, $(\sigma c_i)_{max}$ firstly decreases before $m=0.014$, and after passing this m value, increases gradually with m.

Figure 7 shows the location at which the type of instability for the disturbance having the maximum growth rate changes from viscous to inflectional type as a function of m, indicating that almost whole flow region is subject to inflectional-type instability for $m>0.015$.

4. Experimental Results and Discussions

The radial liquid film on a stationary disk flows as follows. When the discharge Reynolds number Re is relatively small, the water spraeds out in a thin layer with a smooth surface until a circular hydraulic jump takes place. The radius of the jump increases with Re. If Re exceeds $7.4x10^4$, the transition from laminar to turbulent flow occurs at a smaller radius than that of the jump [5].

Figure 8 shows the typical features of the flow in which the transition occurs, indicating that the transition results in a sudden change in the features of the liquid surface.

Figures 9ab, 10ab and 11ab show the transition process effected by the disk rotation. The numerical figures on the scale in each figure indicate the distance from the disk center in cm. Table 1 shows the flow characteristics at a specified point just before the transition shown in Figs. 9-11, where ξ denotes the dimensionless radius, F_s and G_s the radial and circumferential components of the surface

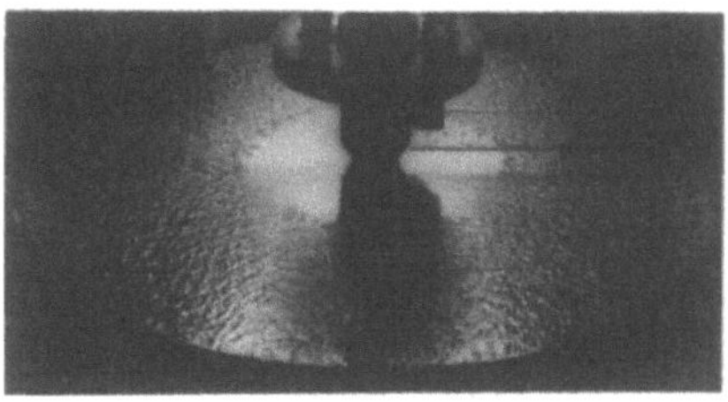

Fig. 8. Transition in radial liquid film flow on a stationary disk under conditions D=25 mm, H=0.5 mm, $Re=9.6x10^4$

velocity, respectively, and G_0 the circumferential speed of disk and h the liquid film thickness at the specified radius r. The parameter of disk rotation m in this table is estimated using the discharge speed from the clearance, the characteristic length of clearance $(CDH)^{1/2}$, and ω $(=2\pi n)$.
It is seen from Fig. 9a that the transition of the flow on the stationary disk proceeds through the following stages: Fine concentrically shaped disturbance waves (denoted as 'D' waves) having a wave-length of about 0.7 mm appear on the liquid surface at r=55 mm (primary instability). They are amplified gradually with an increase in radius and turn into three-dimensional waves ('3D' waves) with a marked peak-valley appearance in the circumferential direction at r=64 mm, i.e., three-dimensional vortex formation (secondary instability). At r=68 mm, '3D' waves change suddenly into granular waves of a very small wave-length appearing as fine eddies ('FE' waves). The fine eddies appear to originate in the valley positions of '3D' waves. (Here, the radius r_t at which 'FE' waves are generated is called the radius of transition). The phase velocity of 'D' waves obtained through high-speed photography was in good agreement with that of the most unstable disturbance obtained by the linear stability theory [3].

Table 1. Flow conditions of Figs. 9-11

H mm	Re x10^{-5}	n s^{-1}	m	r mm	ξ	Fs m/s	Gs m/s	G_0 m/s	h mm
0.5	1.1	0	0	65	22.0	9.5			0.14
		10	0.015	70	24.0	8.4	0.55	4.4	0.14
1.0	1.5	0	0	80	19.4	11.4			0.19
		10	0.021	85	20.7	10.8	0.15	5.3	0.18
1.5	1.2	0	0	105	21.3	6.7			0.23
		10	0.037	85	17.3	7.9	0.03	5.3	0.24

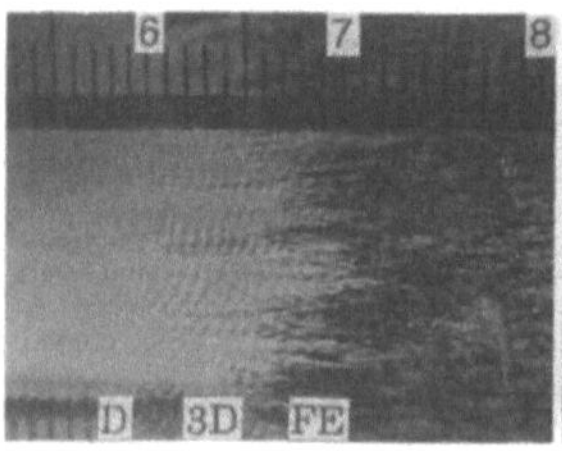

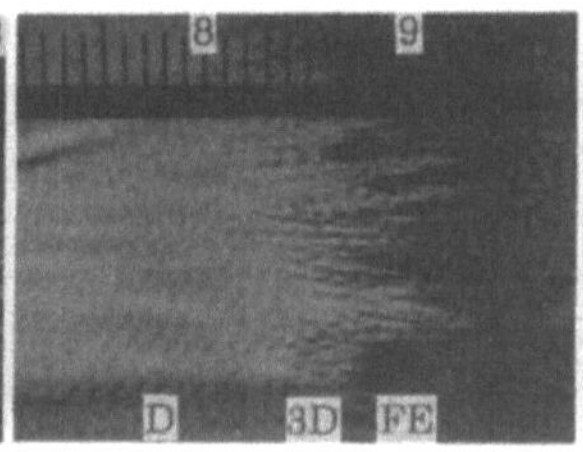

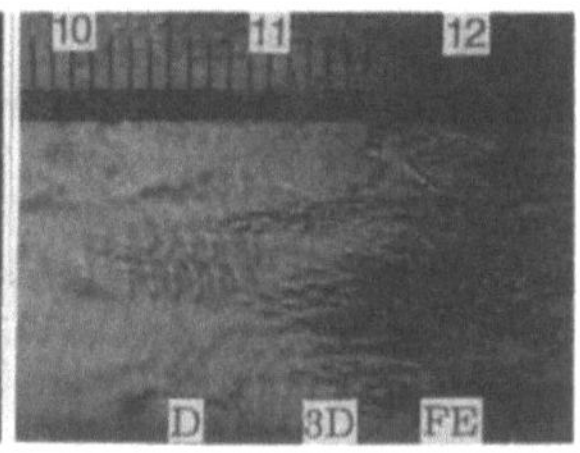

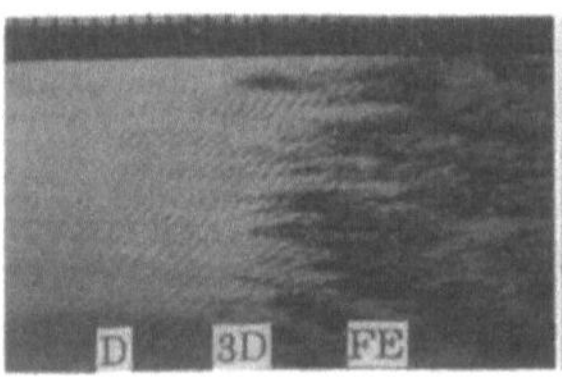

(b) n=10 rps

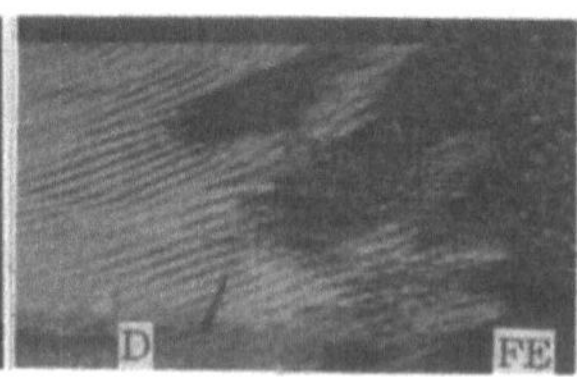

(b) n=10 rps

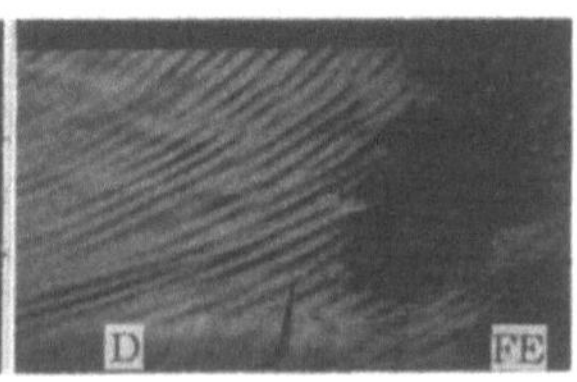

(b) n=10 rps

Fig. 9. Transition process (H=0.5 mm, Re=1.1x10^5)

Fig. 10. Transition process (H=1 mm, Re=1.5x10^5)

Fig. 11. Transition process (H=1.5 mm, Re=1.2x10^5)

When the disk rotates clockwise with n=10 rps, the direction of the crest of 'D' waves, as shown in Fig. 9b, is inclined at about 20 deg. with the tangential direction. However, the circumferential surface velocity G_s is considerably smaller than the radial surface velocity F_s, as shown in Table 1. This means that 'D' waves are generated not by the gas-liquid interaction on the liquid surface, but by vortices existing in the vicinity of the wall. Figure 9b also indicates that the transition proceeds in the same manner as that in Fig. 9a except that the radius of the point of transition in the former is 6 mm larger than that in the latter; the rotation of disk suppresses the transition.

Contrastingly, Fig. 11b indicates that 'D' waves are strongly distorted at about 55 deg. although G_s is considerably smaller than F_s, and greatly amplified to collapse, and that no '3D' waves appear; instead one observes a fine transverse rippling on 'D' waves (see arrows in Fig. 11b). This transition process is clearly different from that in Fig. 9a. Figure 11b also indicates that the radius of transition is smaller than that on the stationary disk in Fig. 11a; the rotation of disk promotes the transition. In comparison with the linear stability theory, it is deduced that the transition in Figs. 9ab, 10a and 11a is due to viscous-type instability, and those in Figs. 10b and 11b is due to inflectional-type instability.

The oscillograms of wall pressure detected simultaneously through the two pressure taps at radii of 73 mm and 73.6 mm under the flow conditions of H=0.5 mm and Re=9.9x10^4, as well as their correlation, are shown in Figs. 12ab. Observing the liquid surface simultaneously with measurement of wall pressure, we verified that concentric 'D' waves were considerably amplified at the radius of 73 mm, but they did not completely turn into '3D' waves. Figure 12a indicates that high frequency disturbance of about 5 kHz exist in the liquid film and they are modulated because of a natural transition. Figure 12b indicates that the two signals in Fig.12a exhibit a good correlation. The phase velocity of the disturbance was estimated by $c_r = l/\Delta\tau$, where l is the distance between two pressure taps. This phase velocity and the frequency in Fig.12a are in good agreement with the linear stability theory [5].

Figure 13 shows examples of disturbance frequencies f detected through wall pressure at r=60 mm. Thick lines give the results calculated using the linear stability theory. Figure 13 indicates that the frequency decreases slightly with an increase of the disk rotational frequency; Data obtained at n=10 rps indicate a notable decrease in the frequency. This suggests that the characteristics of the local instability begin to change from viscous to inflectional type at n=10 rps.

Figure 14 shows the effect of the rotational frequency of the disk on the radius

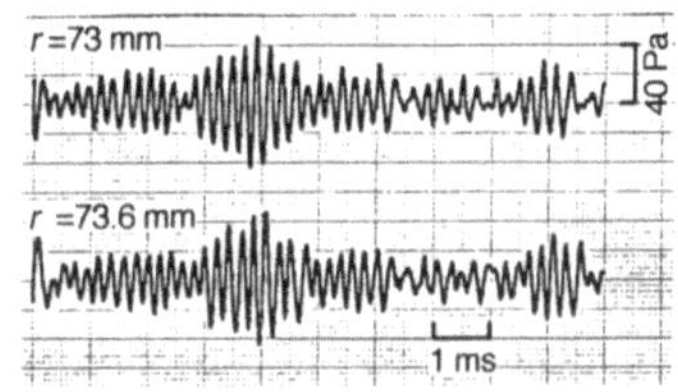

(a) Osscilograms of wall pressure at r=73 mm and r=73.6 mm

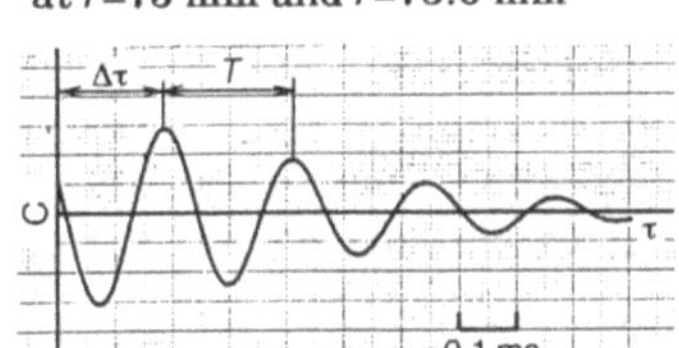

(b) Correlation function of two pressure signals of (a)

Fig. 12. Characteristics of disturbance detected through wall pressure under conditions H=0.5 mm and Re=9.9x10^4

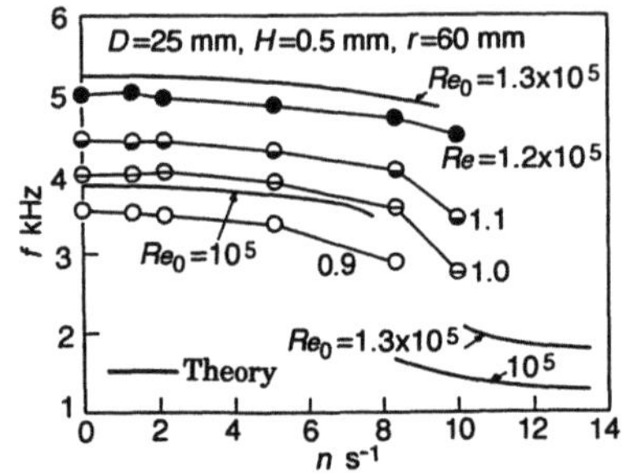

Fig. 13. Disturbance frequency depending on disk rotation

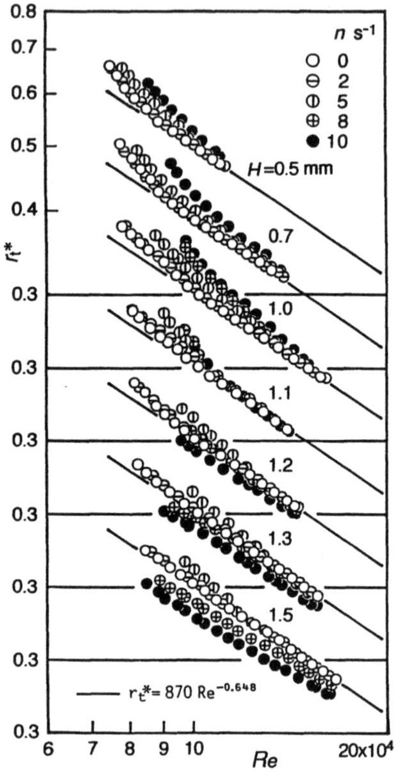

Fig. 14. Point of transition as a function of the discharge Reynolds number

of transition r_t, where r_t^* denotes the dimensionless radius of transition defined as $r_t^*=[r_t/(CDH)^{1/2}]Re^{-1/3}$. Figure 14 indicates that both the radius of transition and the critical discharge Reynolds number (the smallest experimental *Re* indicated in each grouping is the critical discharge Reynolds number) increase with an increase in the ratio of the circumferential velocity to the radial velocity until a maximum value is reached, but that they decrease with a further intensification of three-dimensionality. Consequently, the three-dimensionality of the flow suppresses the transition as long as the latter occurs owing to viscous-type instability; otherwise it promotes the transition.

Conclusions

The laminar-turbulent transition in the liquid film flow along a stationary and rotating disk was investigated both theoretically and experimentally. It was found that two different types (viscous and inflectional) of instability may appear. The findings also revealed that the weak three-dimensionality of the flow suppresses the transition due to viscous-type instability, whereas the strong three-dimensionality promotes the transition because inflectional-type instability becomes dominant.

References

[1] Watson, E. J., J. Fluid Mech., 20(1964), 481-499. [2] Azuma, T. and Nunobe, M., Trans. Japan Soc. Mech. Engrs., 58B(1992), 1005-1010. [3] Azuma, T., Proc. 7th Symp. on Turb. Shear Flows, 1989, 21.4. [4] Kobayashi, R. , Kohama, Y. and Takamadate, C., Acta Mech., 35(1980), 71-82. [5] Azuma, T. and Hoshino, T., Bull. Japan Soc. Mech. Engrs., 27(1984), 2739-2770. [6] Keller, M. D. et al. (Edit.): Exp. Heat Transf., Fluid Mech. & Thermodyn., Elsevier Sci. Pub., Amsterdam (1993), 936-942.

Instabilities Developing in the Three-Dimensional Boundary Layer on Concave and Convex Curved Surfaces

H. Bippes
DLR Institute for Fluid Mechanics, Bunsenstr. 10, 37073 Göttingen, Germany

Summary

The transition process on an infinite swept wing boundary layer subjected to both cross-flow and centrifugal instability is investigated experimentally. The model is a cylindrical wing with concave and convex profile segments. Detailed unsteady flow field measurements allow the analysis of the instability features and the description of the downstream disturbance development. On the concave segment, the quick growth of streamwise vortices gives rise to an early nonlinear development. In contrast to the transition process in the presence of pure cross-flow instability even at low free stream turbulence unsteady modes end up at very high amplitudes. The receptivity to free stream turbulence is increased whereas the receptivity to surface roughness is less pronounced. On the convex segment streamwise vortices are rather damped and the nonlinear development is changed.

Key words: Centrifugal and cross-flow instability, 3-D boundary layer

List of symbols:

c	chord length, 1.4 m
c_p	chordwise pressure coefficient
f	dimensional frequency
$Tu = 100 / \overline{U}_\infty \bullet \sqrt{\left(\overline{\tilde{u}^2} + \overline{\tilde{v}^2} + \overline{\tilde{w}^2}\right) / 3}$	free stream turbulence
	The wall parallel components of:
$\overline{u}_s, \overline{v}_s$	the steady disturbance modes,
$\tilde{u}_s, \tilde{v}_s$	the unsteady disturbance modes,
u_s', v_s'	the r.m.s. values of the unsteady modes
$\overline{U}_s, \overline{V}_s$	the mean flow
$\overline{\overline{U}}_s, \overline{\overline{V}}_s$	the experimental base flow (Sec. 3.1)
$\hat{\overline{u}}_s, \hat{\overline{v}}_s$	amplitudes of the steady modes (Sec. 3.1)
x, y, z	spatial Cartesian coordinates (Fig. 1)
$h^+ = u_\tau h / \rho$	roughness height in wall units
δ_{xs}	99.9% boundary layer thickness along x_s
α, β	streamwise and spanwise wavenumbers
λ	wavelength
Φ	sweep angle

Subscripts:

c	model fixed coordinates, $x_c \perp$ leading edge
e	at the edge of the boundary layer
s	streamline oriented coordinates, $x_s \parallel Q_e$
Tr	at the location of transition
∞	free stream conditions

1. Introduction

Previous experiments on the transition process of three-dimensional boundary-layer flows were performed either on plane or rather weakly curved surfaces with the exception of tests by Poll [1] and Takagi et al [2]. The latter were carried out on swept cylinders of equal geometry, i.e, on convex surfaces. At first sight their results disagree. Whereas Poll [1] established streamwise vortices as well as traveling waves, Takagi et al. [2] did not trace streamwise vortices on the clean surface. However, available experimental results as reviewed by Reed and Saric [3] suggest that this is due to differences in the environmental conditions of the two experiments. Although Malik and Poll [4] made a first attempt to compare the experimental result in [1] with an analysis taking streamline and surface curvature into account, the knowledge on the effect of surface curvature on the disturbance growth seems to be rather scarce especially regarding concave curvature. On the other hand, since the early experiments of M. and F. Clauser [5], we know from two-dimensional flow that concave curvature destabilizes the flow whereas convex curvature has rather a stabilizing effect. Detailed experimental evidence on the influence of centrifugal instability seems still to be missing. For this reason a basic experiment on a highly curved model was performed. The results are compared with those established on a swept flat plate [6] which describe pure cross-flow instability unaffected by centrifugal forces normal to the wall. Special interest is centered on the nonlinear development and on the dependence upon the initial conditions. Concerning cross-flow instability, we know that in low turbulence environments ($Tu \leq 0.15\%$), even on smooth surfaces, streamwise vortices start with considerably larger initial amplitudes than the unsteady modes. They deform the boundary layer flow three-dimensionally to become unstable to secondary instabilities in a very early state of the transition process. On concave surfaces centrifugal forces lead to an unstable stratification normal to the wall. Thus, in addition, the flow becomes highly susceptible to free-stream turbulence. In the current paper some characteristic results are shown.

2. The Experiment

The experiments were performed in a low speed wind tunnel with 3.25 m ×2.8 m cross section. The turbulence level $Tu = 0.08\%$ is measured in the closed test section within the frequency range $2\,Hz \leq f \leq 2\,kHz$. Although the blockage by the model was 8% and the aspect ratio was only 2, for the benefit of lower free stream turbulence a closed test section was chosen.

The model is a cylindrical body with concave and convex profile segments (Fig. 1). The surface is polished to a residual roughness of 5 μm Infinite swept wing conditions are approximated by means of contoured end plates where the contours

are aligned with the streamlines at the edge of the boundary layer. The change of the streamline curvature normal to the surface is neglected. The nose piece is furnished with a 12% Clark-Y profile with the weakly curved side on the upper surface of the model. By this means a suction peak in the nose region with subsequent adverse pressure gradient leading to Tollmien-Schlichting instability is avoided so that the flow first becomes unstable to cross-flow instability. The nose piece, segments of constant concave (radius R = 650 mm) and constant convex curvature (R = 300 mm) are linked by tangential fairings. For the present tests the model is set at an angle of incidence of -4° and a sweep angle of Φ_0 = 30°. The profile of the model and the pressure distribution measured around the nose and all along the upper side are shown in Fig.2. There is a continuous pressure decrease downstream of the stagnation line up to the point of maximum profile thickness. Downstream of this location the pressure increases and Tollmien-Schlichting instability comes into play.
The unsteady flow field is measured by means of hot-wire anemometry using miniaturized V-probes with the two hot-wires adjusted parallel to the wall. For data reduction, the analogue signals are sampled at 3.8 kHz for a period of 6.3 s. With this sample rate the whole band of amplified frequencies is covered. From the instantaneous signals, mean values ($\overline{U}_s, \overline{V}_s$) averaged over the time period of 6.3 s, fluctuations (u_s', v_s') and power spectra, averaged over 11 data windows, were computed.
For the determination of the disturbance motion, a series of velocity profiles along the span, spaced at 1.5 to 2 mm, is measured at the chord positions marked on the profile in Fig.2. Transition is traced with the use of surface hot-films placed at 8 downstream positions also marked in Fig. 2.
The free stream velocity and the model sweep with respect to the tunnel axis have been chosen to be Q_∞ = 15 m/s and Φ_0 = 30°, respectively.

3. Results

3.1 Identification of the Instabilities

In order to provide an insight into the nature of the disturbance motion the velocity distribution at the chord position x_c/c = 0.2421 is given in Fig. 3. It results from a series of profiles spaced along the span at 1.6 mm. The figure displays the lines of constant velocity (isotachs) in a cross section (y_c,z-plane). The upper diagrams show the isotachs of the mean flow for the wall parallel components, $\overline{U}_s, \overline{V}_s$. The spanwise periodic deformation indicates the existence of a stationary disturbance mode of vortical structure. If the spanwise averaged mean flow

$$\overline{\overline{U}}_s(z) = \frac{1}{n}\sum_{i=1}^{n}\overline{U}_s(y_c = i\Delta y_c, z) \qquad \text{and} \qquad \overline{\overline{V}}_s(z) = \frac{1}{n}\sum_{i=1}^{n}\overline{V}_s(y_c = i\Delta y_c, z) \; .$$

is subtracted from the mean flow, the stationary disturbance mode

$$(\overline{u}_s, \overline{v}_s) = \left(\overline{U}_s - \overline{\overline{U}}_s, \; \overline{V}_s - \overline{\overline{V}}_s\right)$$

is obtained. It exhibits the form of counterrotating vortex pairs corresponding to the normal mode solutions of linear theory both for cross-flow and centrifugal instability.

Under natural conditions of transition the wavelength spread is over a wide range as seen in the wavenumber spectra displayed in Fig.4. The result of a test where the streamwise vortices are stimulated artificially is also shown in the figure. The wavenumber chosen was shown to lead to the largest amplitudes at the measuring position x_c/c = 0.3071. It is centered in the range of naturally occurring wavenumbers. The most striking feature is the change of the stimulated wavenumber to smaller values when the pressure decrease accelerates.
Besides of the stationary modes, unsteady disturbances are also traced. The lower diagrams of Fig.3 show the isotachs of the r.m.s. values. The frequency analysis in Fig. 5 reveals that the unsteady disturbance motion comprises a relatively broad band of frequencies as experienced in pure cross-flow instability.

3.2 Downstream growth of the disturbance flow

The growth of the streamwise vortices leads to a considerable distortion of the mean flow (Fig. 6 a). The laminar boundary-layer profiles, upon which linear stability theory is based, no longer exist. Instead, inflexional profiles have developed (Fig. 6b). Hence, the flow becomes susceptible to secondary instabilities. Nonlinear features are observed: The difference between the base flow profiles (U_{th}, V_{th}) and the spanwise averaged experimental profiles $(\overline{\overline{U}}_s, \overline{\overline{V}}_s)$ due to nonlinear interaction between disturbances and base flow [7], the change in the shape of the amplitude functions of the streamwise vortex instabilities $(\hat{\overline{u}}_s, \hat{\overline{v}}_s)$ (Fig. 7), and the r.m.s. values of the traveling modes undergoing a spanwise periodic deformation (Fig. 8) with the wavelength of the steady modes. All these nonlinear characteristics first occur in a very early state of the transition process.
The nonlinear features observed on the swept flat plate (pure cross-flow instability) in the same wind tunnel and for the same level of surface roughness are qualitatively the same. Considering only the disturbance growth, a striking difference becomes evident (Fig. 9). In the case of pure cross-flow instability (Fig. 9a) under the same environmental conditions as in the present experiment, the stationary vortex mode and the traveling modes end up at respective saturation amplitudes of $\approx$ 20% and < 5% of the local velocity outside the boundary layer. On the concave surface (Fig. 9b), however, under natural conditions of transition, the stationary modes are less amplified. The saturation amplitudes are < 15%. Contrarily, the traveling disturbances are much more amplified and grow up to 40 - 50 %.
As the disturbance motion develops in a rather irregular way on the curved model under natural conditions of transition, some additional tests were performed under controlled conditions where the streamwise vortices are stimulated at the most unstable wavelength. In this case the traveling modes end up at even higher values (50 - 60%) whereas in contrast to pure cross-flow instability the end amplitudes of the streamwise vortices are essentially the same. This observation infers that concave curvature influences the transition process significantly. Traveling modes are much more amplified.
On the convex segment the streamwise vortices are rather attenuated. The traveling modes change some of their characteristics but the maximum r.m.s. values integrated over the whole amplified spectrum also decrease rather than further increase.

5. Conclusions

The experiments on the swept flat plate under varied environmental conditions [8] revealed that in low turbulence environments, even on polished surfaces streamwise vortices are the dominant instability. Traveling modes are then rather due to secondary instability [9] and end up at clearly smaller saturation amplitudes. In this case the location of transition can essentially be correlated with surface roughness. At turbulence levels > 0.2% the growth of streamwise vortices becomes reduced and the travelling modes become dominant. On concave surfaces, under the same environmental conditions, the traveling modes grow much faster and end up at very high amplitudes whereas the stationary mode arrives at smaller saturation amplitudes. Furthermore, artificial stimulation of streamwise vortices does not lead to saturation amplitudes as high as on the swept flat plate but rather to smaller amplitudes and above all to higher amplitudes of the traveling modes. Thus, it is suggested that three-dimensional centrifugal instability favors the growth of traveling modes. In the present experiment it could not be made clear whether this is due to linear or nonlinear development.

References

[1] D. I. A. Poll, *J. Fluid Mech.* ***150***, *329-356 (1985).*
[2] S. Takagi and N. Itoh, *Fluid Dynamic Research* ***14*** *(4) pp. 167-189 (1994).*
[3] H. L. Reed and W. S. Saric, *Ann. Rev. Fluid Mech.* ***21***, *235-284 (1989).*
[4] M. R. Malik and D. I. A. Poll, *AIAA Pap. No. 84-1672 (1984).*
[5] M. and F. Clauser, *National Advisory Committee for Aeronautics Techn. Note No. 613, 1937.*
[6] H. Bippes, In: R. Aeron. Soc. Conf. on Boundary Layer Transition and Control, Cambridge 1984
[7] F. Meyer and L. Kleiser, *IUTAM Symp., Toulouse, France, 1989, Ed.: D. Arnal and R. Michel, Springer, 609-619 (1990).*
[8] H. Deyhle, *Thesis, University Hanover, published in: VDI Fortschritt-Bericht Reihe 7 Nr. 226, VDI-Verlag Düsseldorf (Germany), 1993.*
[9] T. M. Fischer and U. Dallmann, *Phys. Fluids A 3, 2378-2391 (1991).*
[10] S. Hein, private communication

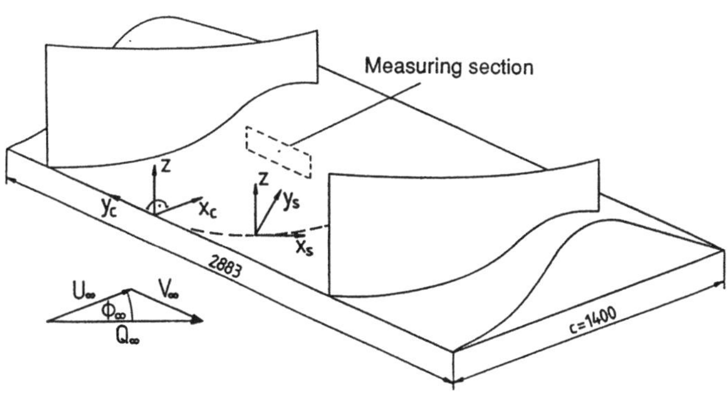

Fig. 1 *Experimental model*

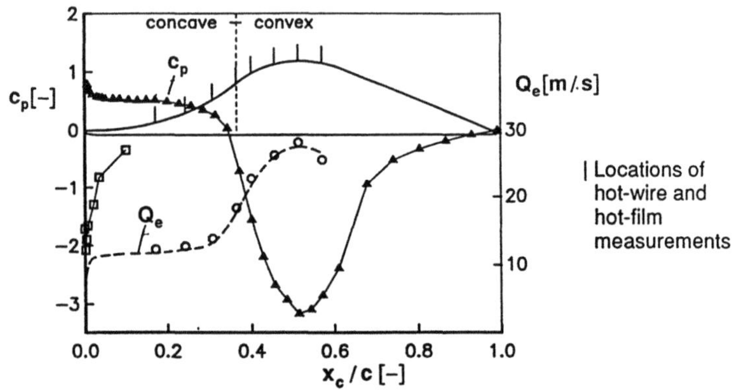

Fig. 2 *Profile of the model, experimental wall pressure* $\mathbf{c_p}$ *(Δ upper side, □ lower side), and inviscid flow velocity* $\mathbf{Q_e}$ *(----calculated from pressure distribution assuming infinite sweep conditions,* o *hot-wire measurements).* $\mathbf{Q_\infty}$ *= 16 m/s.*

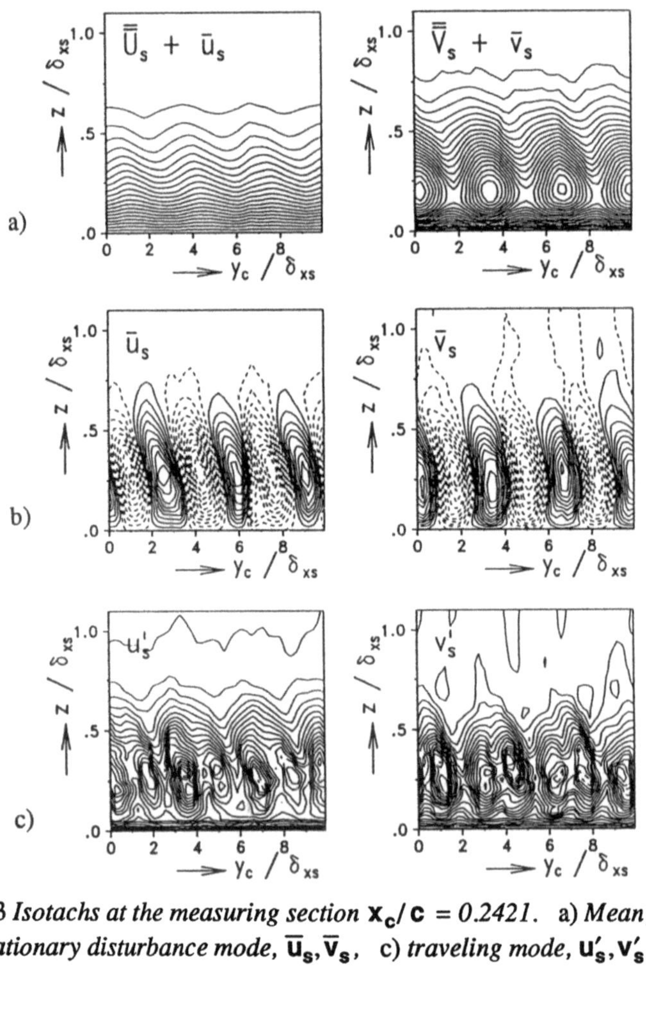

Fig. 3 *Isotachs at the measuring section* $\mathbf{x_c/c}$ *= 0.2421.* a) *Mean flow,* $\mathbf{\overline{U}_s, \overline{V}_s}$, b) *stationary disturbance mode,* $\mathbf{\overline{u}_s, \overline{v}_s}$, c) *traveling mode,* $\mathbf{u'_s, v'_s}$.

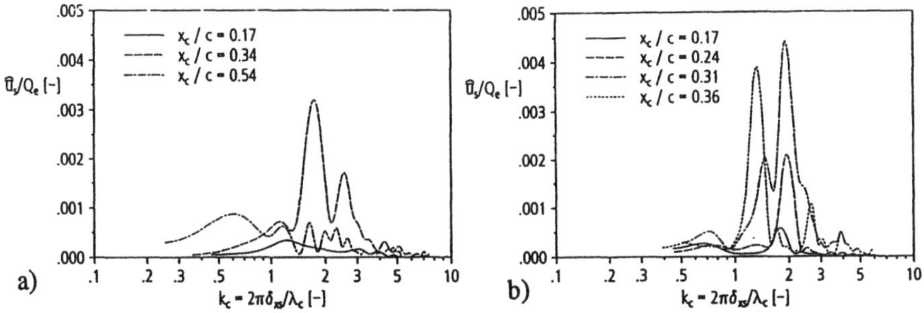

Fig. 4 *Wavenumber spectra measured* a) *under natural conditions of transition,* b) *under controlled stimulation of the stationary disturbance mode.*

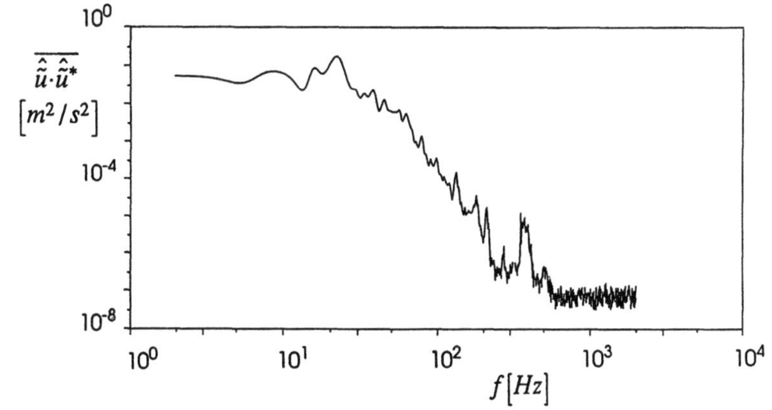

Fig. 5 *Frequency spectrum of the r.m.s. fluctuations measured at* $x_c/c = 0.3071$.

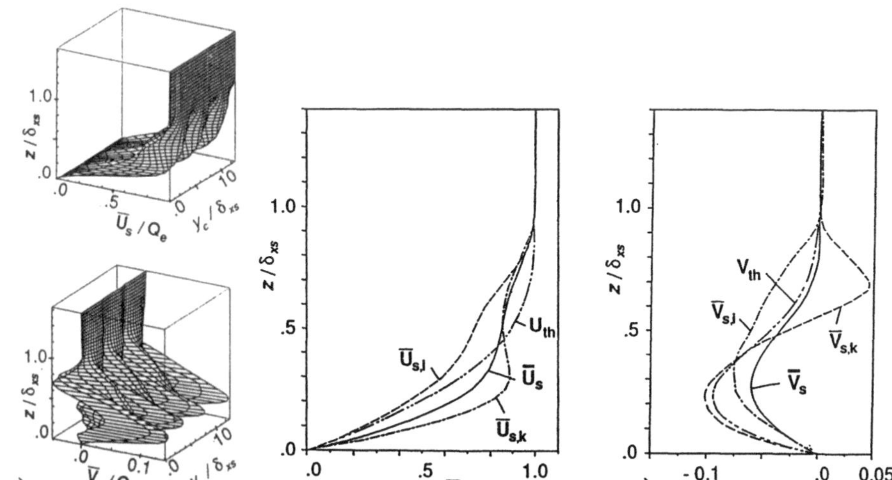

Fig. 6 *Deformation of the base flow under the action of the stationary disturbance mode.* a) *Mean flow profiles,* b) *selected profiles* $(\overline{U}_{s,lk},\overline{V}_{s,lk})$, *the experimental* $(\overline{\overline{U}}_s,\overline{\overline{V}}_s)$, *and the calculated* (U_{th},V_{th}) *base-flow profiles.*

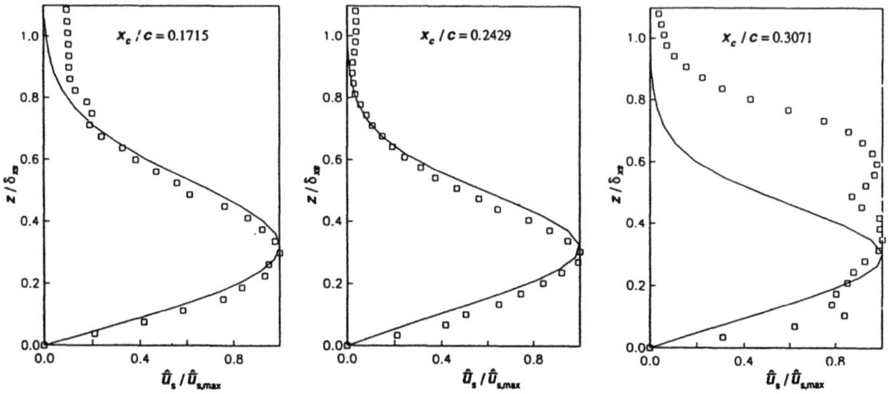

Fig. 7 *Amplitude functions of the stationary disturbance mode measured on the concave profile segment.* □ *experiment,* ___ *local linear analysis by Hein [10].*

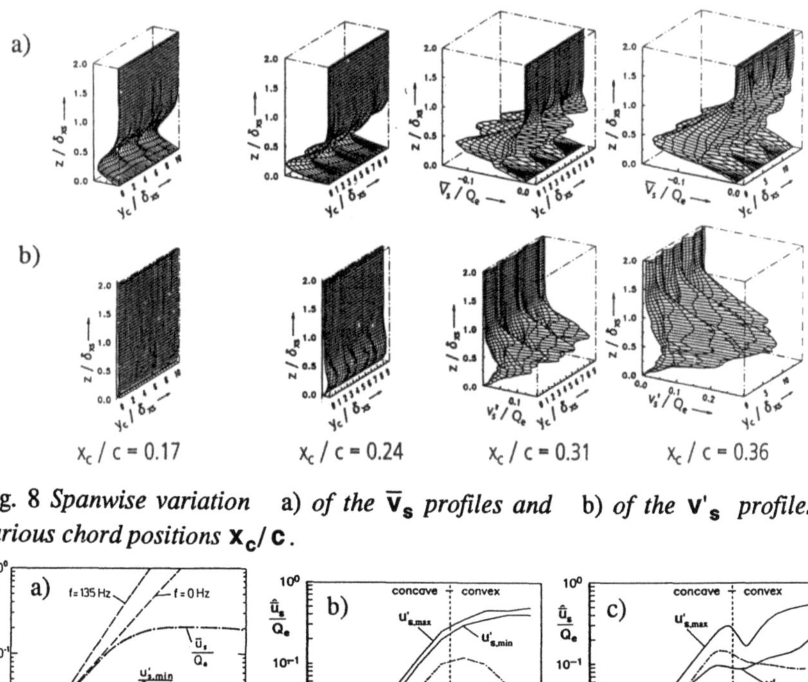

Fig. 8 *Spanwise variation* a) *of the* $\overline{\mathbf{v}}_s$ *profiles and* b) *of the* $\mathbf{v}'_s$ *profiles at various chord positions* $\mathbf{x_c/c}$.

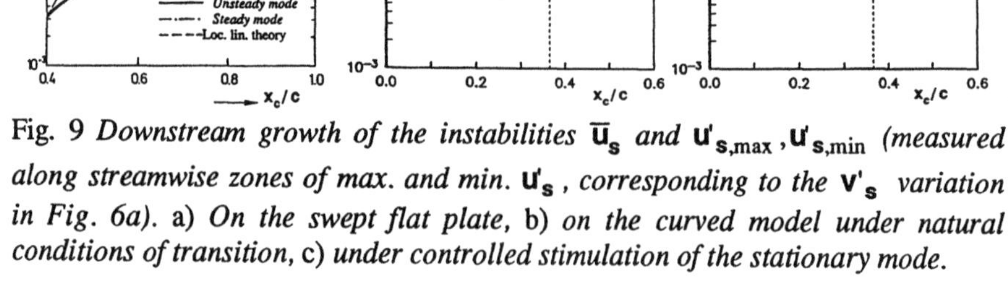

Fig. 9 *Downstream growth of the instabilities* $\overline{\mathbf{u}}_s$ *and* $\mathbf{u}'_{s,max}$, $\mathbf{u}'_{s,min}$ *(measured along streamwise zones of max. and min.* $\mathbf{u}'_s$, *corresponding to the* $\mathbf{v}'_s$ *variation in Fig. 6a).* a) *On the swept flat plate,* b) *on the curved model under natural conditions of transition,* c) *under controlled stimulation of the stationary mode.*

Effects of Wall and Streamline Curvatures on Instability of 3-D Boundary Layers

N. Itoh
National Aerospace Laboratory, Chofu, Tokyo, Japan

Summary

Three-dimensional boundary layers may encounter three possible instabilities; that is, the Tollmien-Schlichting instability, the cross-flow instability and the instability due to curvature of external streamlines. To show competitive appearance of those in the boundary layer on a yawed circular cylinder, local velocity distributions of the boundary layer are approximated with the Falkner-Skan-Cooke velocity family and their stability is examined using a model system of disturbance equations, which include the terms denoting wall and streamline curvatures. Results show that the streamline-curvature instability gives the lowest values of critical Reynolds number near the leading edge.

Key words: 3-D boundary layer, Cross-flow instability, Streamline curvature.

1. Introduction

The boundary layer developing on a swept wing is characterized by curved streamlines of external flow and a cross-flow velocity component inside. Effects of the cross flow on stability characteristics can be investigated with the Orr-Sommerfeld equation, where the smaller-order terms of non-parallelism of boundary layers, wall curvature and streamline curvature are ignored [1,2]. A different system of stability equations is, however, necessary for taking account of the effects of wall and streamline curvatures in stability calculations of three-dimensional boundary layers, as in the problem of centrifugal instability of two-dimensional flows. A previous study of the author [3] dealt with an interesting case of three-dimensional boundary layers along a concave wall and predicted competitive appearance of three kinds of instabilities induced by viscosity, cross flow and centrifugal force. A model system of disturbance equations used there includes the principal curvature term and a part of non-parallel effects, the latter of which has been introduced to remove the well-known defect of the parallel-flow approximation that critical points of the centrifugal instability are not obtained in a finite range of wavenumbers. On appropriate modification of this formulation, the author [4] also showed the possibility of a similar instability of the centrifugal type due to curvature of external streamlines. In the present paper, the modified method, together with an approximate method of boundary-layer calculation [5], is applied to the boundary layer along a yawed circular cylinder to show fundamental effects of wall and streamline curvatures on the stability characteristics.

2. Model equations for linear stability analysis

Orthogonal curvilinear coordinates are used with ξ in the direction of external streamline, η in the direction perpendicular to ξ on the plane parallel to the wall and ζ in the nornal-to-wall direction, and let t denote time, $(U+u^*,V+v^*,W+w^*)$ velocity components, and $P+p^*$ pressure, where all quantities have been made dimensionless with an appropriately defined boundary-layer thickness δ, the external-flow velocity Q_E and the density ρ. Then the local Reynolds number is given by $R=Q_E\delta/\nu$, ν being the kenematic viscosity. The small disturbance superimposed on the basic laminar flow (U,V,W,P) is assumed to be written in the form

$$u^*=\mathrm{Re}[u(\zeta;\xi,\eta)\exp\{i(\alpha\xi+\beta\eta-\omega t)\}], \tag{2.1}$$

together with the same expressions for v^*, w^* and p^*. Here Re denotes the real part, and α and β are real wavenumbers in the ξ and η directions, ω a complex whose real and imaginary parts denote frequency and temporal growth rate, respectively, and the dependence of (u,v,w,p) on ξ and η is assumed to be very weak in comparison with that on ζ. Substituting the above form into the Navier-Stokes and continuity equations, subtracting the basic-flow parts which are assumed to satisfy the equations of motion by themselves and neglecting coupling terms of the small disturbance, we have linear disturbance equations of the partial differential form (cf. Stuart [6]).

In the present paper, we follow the previous work [3,4] and use a simple system of model equations, where the parallel-flow approximation is modified to include the principal terms denoting wall and streamline curvatures and a part of non-parallel effects. The model equations are written in the form

$$i\alpha u + i\beta v + Dw = 0,$$

$$[\frac{1}{R}(D^2-\alpha^2-\beta^2)+i(\omega-\alpha U-\beta V)-\frac{\hat{W}}{R}D]u - U'w - i\alpha p = 0,$$

$$[\frac{1}{R}(D^2-\alpha^2-\beta^2)+i(\omega-\alpha U-\beta V)-\frac{\hat{W}}{R}D]v + 2\kappa_s Uu - V'w - i\beta p = 0,$$

$$[\frac{1}{R}(D^2-\alpha^2-\beta^2)+i(\omega-\alpha U-\beta V)-\frac{\hat{W}}{R}D]w + 2\kappa_w Uu - Dp = 0, \tag{2.2}$$

where the prime denotes differentiation with respect to ζ, κ_w the wall curvature in the direction of external streamlines, κ_s the curvature of streamlines and the non-parallel term $\hat{W}$ is defined by

$$\hat{W}= R\{W-(\delta_\xi\zeta U+\delta_\eta\zeta V)\}, \tag{2.3}$$

subscripts denoting partial derivatives. If the basic flow is a member of the Falkner-Skan-Cooke velocity family, $\hat{W}$ is found to be given by $-(1/2)(1+m)F(\zeta)$, F being the Falkner-Skan function, and to become important outside the boundary layer, because its magnitude increases towards infinity with ζ. For convenience of numerical solution, we assume β not to be zero and eliminate v and p from (2.2) to get the simultaneous equations for u and w as

$$[(\alpha^2+\beta^2)\{\frac{1}{R}(D^2-\alpha^2-\beta^2)+i(\omega-\alpha U-\beta V)-\frac{\hat{W}}{R}D\}-2\alpha\beta\kappa_s U]u$$

$$-[i\alpha\{\frac{1}{R}(D^2-\alpha^2-\beta^2)+i(\omega-\alpha U-\beta V)-\frac{\hat{W}}{R}D\}D+\beta(\beta U'-\alpha V')]w=0,$$

$$[\{\frac{1}{R}(D^2-\alpha^2-\beta^2)+i(\omega-\alpha U-\beta V)-\frac{\hat{W}}{R}D\}(D^2-\alpha^2-\beta^2)+i(\alpha U''+\beta V'')-\frac{\hat{W}'}{R}D^2]w$$
$$-[2(\alpha^2+\beta^2)\kappa_w U+2i\beta\kappa_s(UD+U')]u=0. \qquad (2.4)$$

If we ignore the $\hat{W}$ terms and put $\kappa_w=\kappa_s=0$, the above equations reduce to a three-dimensional version of the Orr-Sommerfeld equation, and if only the κ_w term is retained, we have an equivalent of the Taylor-Görter equations. The latter equations, however, are known to give the critical point located at an infinitesimally small wavenumber for two-dimensional problems, and this inconvenience is removed in the present model equations by including the $\hat{W}$ terms, which have been shown to be very effective for this purpose in the previous work [3]. It should be also noted that the κ_s term bears a certain similarity in its position in (2.2) to the κ_w term, suggesting the possibility of the centrifugal instability due to streamline curvature.

The boundary conditions to be imposed are that disturbance velocities vanish on the wall and far away from the wall. When the outer conditions are replaced with the matching conditions imposed at an appropriate boundary-layer edge ζ_e [3,4], we have the boundary conditions of the form

$$u = w = w' = 0 \quad \text{at } \zeta=0,$$
$$u'+\rho_1 u - \frac{i\alpha}{\rho_2^2}(w''+\rho_1 w')=0,$$
$$w''+(\rho_1+\rho_2)w'+\rho_1\rho_2 w+\frac{2RU_0(i\beta\rho_1\kappa_s-\rho_2^2\kappa_w)}{(2\rho_1+\hat{W}_0)(\rho_1+\rho_2)}u=0,$$
$$w'''+(2\rho_1+\rho_2)w''+\rho_1(\rho_1+2\rho_2)w'+\rho_1^2\rho_2 w=0 \quad \text{at } \zeta=\zeta_e, \qquad (2.5)$$

where $\rho_1=\frac{1}{2}\hat{W}_0+(\frac{1}{4}\hat{W}_0^2+\alpha^2+\beta^2-i\omega R+i\alpha RU_0+i\beta RV_0)^{1/2}$, $\rho_2=(\alpha^2+\beta^2)^{1/2}$, and the subscript $_0$ denotes the value at $\zeta=\zeta_e$. Since these boundary conditions have been derived under an approximate procedure, the boundary-layer edge has to be taken so large that numerical solutions are of an acceptable accuracy. Following the previous study [3], therefore, we take a sufficient value $\zeta_e=10$ in the computations here.

The homogeneous equations (2.4) and boundary conditions (2.5) pose an eigenvalue problem to determine the complex frequency ω as a function of α, β, R, κ_w, κ_s and other flow parameters, and the critical point for a given flow is obtained by equating the imaginary parts of ω and its partial derivatives with respect to α and β to zero. In the present paper, we concentrate our attention on determination of the critical Reynolds number, because it is the simplest and most convenient measure of stability characteristics. Necessary computations are made with the numerical method developed by the auther [7,8].

3. Approximate method of boundary-layer calculation

In the present paper, we assume that general boundary layers on a swept wing with a large aspect ratio can be simulated with the Falkner-Skan-Cooke flow, which represents a two-parameter family of velocity distributions associated with the boundary layer on a yawed infinite wedge placed in a uniform flow [5]. This flow is obtained subject to the external flow with the chordwise velocity U_E proportional to x^m and the spanwise velocity V_E of a constant, where the chordwise and spanwise coordinates on the surface are given by the distance x from the leading edge and the distance y along the leading edge, respectively, while the normal-to-wall coordinate is by z. Introducing the boundary-layer thickness defined by $\delta=\sqrt{\nu x/U_E}$ and the nondimensional variable defined by $\zeta=z/\delta$, we can write the well-known similarity solution of the boundary-layer equations in the form

$$\bar{u}=U_E F'(\zeta), \qquad \bar{v}=V_E G(\zeta), \qquad \bar{w}=\frac{\nu}{\delta}\{\frac{1+m}{2}F(\zeta)-\frac{1-m}{2}\zeta F'(\zeta)\}, \tag{3.1}$$

with the solutions F and G of the ordinary differential equations

$$F''' + \frac{1+m}{2}FF'' + m\{1-(F')^2\}=0, \qquad F(0)=F'(0)=0, \qquad F'(\infty)=1,$$

$$G'' + \frac{1+m}{2}FG' =0, \qquad G(0)=0, \qquad G(\infty)=1. \tag{3.2}$$

The velocities given in (3.1) are transformed into the streamwise component U, the cross-flow component V in dimensionless form and the vertical component $\hat{W}$ defined by (2.3) as

$$U = F'(\zeta;m) + \frac{\gamma^2}{1+\gamma^2}\{G(\zeta;m) - F'(\zeta;m)\},$$

$$V = \frac{\gamma}{1+\gamma^2}\{G(\zeta;m) - F'(\zeta;m)\}, \qquad \hat{W} = -\frac{1+m}{2}F(\zeta;m), \tag{3.3}$$

where the reference velocity is $Q_E=\sqrt{U_E^2+V_E^2}$, $\gamma=V_E/U_E$ represents the angle of external streamline from the x axis and m is a measure of the pressure gradient in the x direction.

Now assuming that any velocity distribution of the boundary layer on a swept wing with a large aspect ratio is expressed with a member of the above family, we determine values of the parameter m as a function of x according to the momentum integral equation. When the velocity in the x direction is given by $\bar{u}=U_E F'(\zeta)$, the wall shear stress τ_0, the displacement thickness δ^*, the momentum thickness θ and the boundary-layer thickness δ are written in the form

$$\frac{\tau_0\delta}{\mu U_E} = F''(0)\equiv\Theta_0(m), \qquad \frac{\delta^*}{\delta} =\int_0^\infty (1-F')d\zeta \equiv\Theta_1(m),$$

$$\frac{\theta}{\delta} =\int_0^\infty F'(1-F')d\zeta \equiv\Theta_2(m), \qquad \delta=\sqrt{\nu m/U_E'}. \tag{3.4}$$

Using these expressions and following the Pohlhausen method [9], we rewrite the momentum integral equation in the x direction into a simple form

$$\frac{dZ}{dx} = \frac{E(K)}{U_E}, \qquad K = U_E' Z, \tag{3.5}$$

where

$$Z \equiv \frac{\theta^2}{\nu}, \qquad E(K) \equiv \frac{2\theta}{\nu}\{\frac{\tau_0}{\rho U_E}-(2\theta+\delta^*)U_E'\}. \tag{3.6}$$

This system may be regarded as nonlinear differential equations for the unknown parameter m and can be integrated numerically from the leading edge (x=0) of the wing subject to appropriate initial values [5].

Once the boundary-layer calculation determines values of γ and m and the curvatures κ_w and κ_s along the surface of a given wing are obtained, we can apply the method of stability calculation presented in the previous section to evaluate the critical Reynolds number for each location on the surface of wing. If the local Reynolds number determined from the boundary-layer calculation exceeds the critical value, the flow is considered to be unstable at this location. This comparison may be done by plotting both the critical values and the local Reynolds numbers for several values of uniform-flow Reynolds number against the distance x. Here the uniform-flow Reynolds number R_Q is defined with the uniform-flow velocity and the chord length of the wing in the uniform-flow direction, as $R_Q = Q_\infty c/(\nu \cos \Lambda_\infty)$, c being the chord length in the x direction and Λ_∞ the sweep angle of the wing. In addition, it is more convenient to rewrite the local Reynolds number into a modified one that is invariant to sweep angle of the wing, such as

$$\hat{R} \equiv \Theta_1 R \frac{\cos \Lambda_E}{\cos \Lambda_\infty} = \hat{U}_E(X)\Theta_1(m)\left(\frac{ms}{\hat{U}_E' c}\right)^{1/2} R_Q^{1/2}, \tag{3.7}$$

where $\Lambda_E = \arctan \gamma$, $\hat{U}_E(X) = U_E/U_\infty$, $U_\infty = Q_\infty \cos \Lambda_\infty$, $\gamma_\infty = \tan \Lambda_\infty$, $X = x/s$ and s is the surface length between the leading and trailing edges. In accordance with this expression, we use non-dimensional representation of the critical Reynolds number R_c and critical values of the dimensional wavelength λ_c, the angle of wave-number vector ϕ_c and the frequency f_c in the form

$$\hat{R}_c \equiv \Theta_1 R_c \frac{\cos \Lambda_E}{\cos \Lambda_\infty} = \Theta_1(m) R_c \left(\frac{1+\gamma_\infty^2}{1+\gamma^2}\right)^{1/2},$$

$$\hat{\lambda}_c \equiv \frac{\lambda_c \cos \Lambda_\infty}{c} R_Q^{1/2} = \frac{2\pi}{(\alpha_c^2+\beta_c^2)^{1/2}}\left(\frac{ms}{\hat{U}_E' c}\right)^{1/2}, \qquad \hat{\phi}_c \equiv \tan^{-1}\frac{\beta_c}{\alpha_c},$$

$$\hat{f}_c \equiv \frac{f_c c}{Q_\infty R_Q^{1/2} \cos \Lambda_\infty} = \frac{\omega_c}{2\pi}\hat{U}_E(X)\left(\frac{\hat{U}_E' c}{ms}\right)^{1/2}\left(\frac{1+\gamma^2}{1+\gamma_\infty^2}\right)^{1/2}. \tag{3.8}$$

4. Application to the flow along a yawed circular cylinder

The external potential flow along a circular cylinder is given by $\hat{U}_E = 2\sin(\pi X)$, while the curvature of external streamlines $1/r_s$ and the wall curvature in the direction along the streamlines $1/r_w$ are written in the non-dimensional form

$$\kappa_s \equiv \frac{\delta}{r_s} = \frac{\gamma m}{R(1+\gamma^2)},$$

$$\kappa_w \equiv \frac{\delta}{r_w} = \frac{\pi \Delta^2 U_E}{R(1+\gamma^2)^{1/2}}, \tag{4.1}$$

where the reference length is the boundary-layer thickness δ and $\Delta = (\delta/s)\sqrt{U_\infty s/\nu}$. Results of the boundary-layer calculation for this potential flow are presented in Fig. 1, where values of δ^*, τ_0, $1/r_w$ and $1/r_s$ in appropriate dimensionless form as well as the flow parameters m and γ are plotted against the

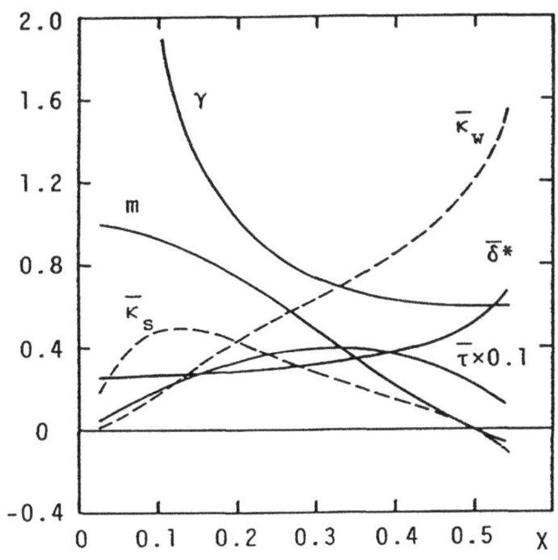

Fig. 1. Results of boundary-layer calculation. ($\bar{\delta}^* = (\delta^*/s)\sqrt{U_\infty s/\nu}$, $\bar{\tau} = (\tau_0/\rho U_\infty)\sqrt{U_\infty s/\nu}$, $\bar{\kappa}_w = (\delta^*/r_w)\sqrt{U_\infty s/\nu}$, $\bar{\kappa}_s = (\delta^*/r_s)\sqrt{U_\infty s/\nu}$.)

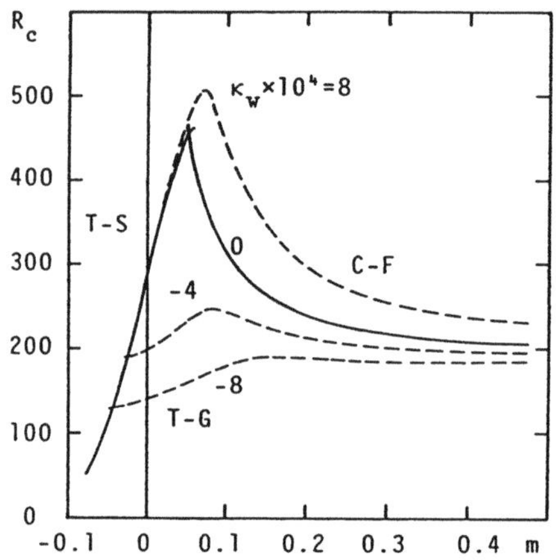

Fig. 2. Effects of wall curvature on the critical Reynolds number of F-S-C flow in the case of $\gamma=1$ [3].

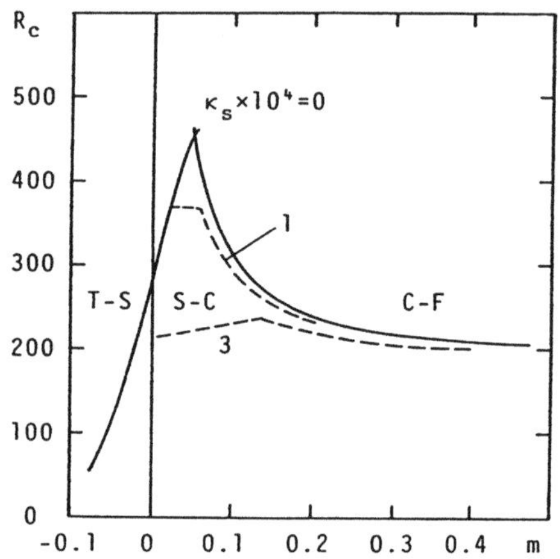

Fig. 3. Effects of streamline curvature on the critical Reynolds number of F-S-C flow in the case of $\gamma=1$ [4].

distance X.

Individual effects of wall and streamline curvatures on stability characteristics of the Falkner-Skan-Cooke flow have been investigated in the previous studies [3,4] and the principal results are shown in Figs. 2 and 3, where the critical Reynolds numbers are plotted against the pressure parameter m for a fixed value of $\gamma=1$. If the curvatures are ignored, the Tollmien-Schlichting (referred to as T-S hereinafter) instability determines the critical value in the range $m<0.05$ and the cross-flow (C-F) instability does it for larger values of m. If the wall curvalure takes a negative value (concave wall), the third critical curve due to Taylor-Görtler (T-G) instability appears between the two old curves, but positive values of κ_w denoting convex walls do not induce the T-G instability and yield only a little shift of the T-S and C-F critical curves, as shown in Fig. 2. On the contrary, addition of the curvature of external streamlines κ_s produces a new instability of the centrifugal type, namely the streamline-curvature (S-C) instability, and the corresponding critical curve first appears between the T-S and C-F critical curves and moves downwards with extending its governing region as κ_s increases, as shown in Fig. 3. It should be noted that the T-S critical curve is very weakly affected by the curvatures but that the C-F critical curve is considerably shifted upwards by a positive value of wall curvature (a convex wall) and downwards by a negative value of wall curvature and also by a streamline curvature.

Now let us examine whether or not the S-C instability can occur in the simple flow around a yawed circular cylinder placed in a uniform flow with the sweep angle $\Lambda_\infty=50°$. Results of the boundary-layer calculation and the stability calculation are presented in Fig. 4, where the fine lines denote variations of the local Reynolds number $\hat{R}$ with the distance X and the thick line denotes the critical curves of the C-F and T-S instabilities, while the broken line indicates the S-C critical curve, which is lower than the C-F one before $X\fallingdotseq0.3$.

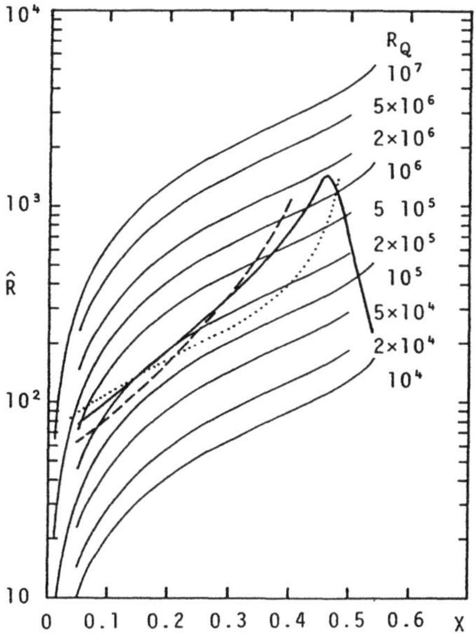

Fig. 4. Variations of R_c along the surface of a yawed circular cylinder with Λ_∞=50° . (Solid line: C-F and T-S instability; broken line: S-C instability; dotted line: C-F one of parallel-flow approximation [5].)

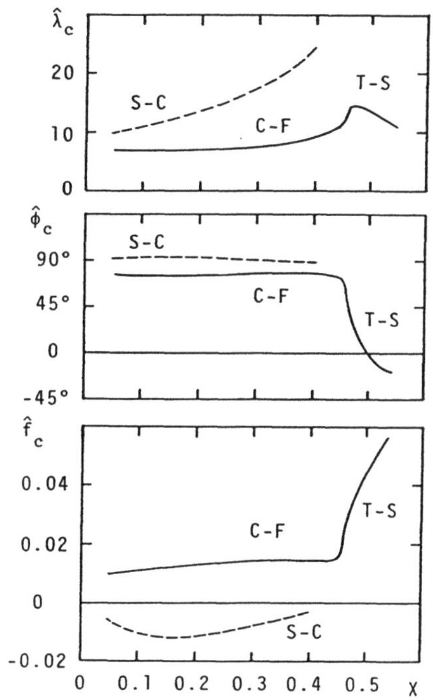

Fig. 5. Variations of wave length, propagation direction and frequency along the critical curves of Fig. 4.

This fact indicates that the S-C instability gives the lowest value of critical Reynolds number in the region near the leading edge of the cylinder. The S-C instability, however, is not yet detected in experiments, probably because the C-F instability occurs soon downstream and the corresponding disturbances may grow faster to a large amplitude so as to veil comparatively small disturbances of the S-C instability. More detailed measurements seem to be necessary for observation of the S-C disturbances.

Fig. 5 shows variations of the wavenumber vector and frequency along the critical curves given in Fig. 4. It can be seen there that wavenumber vectors of S-C disturbances are almost perpendicular to external streamlines, and that frequencies of C-F and S-C disturbances are opposite in sign to each other and are both one order of magnitude smaller than those of T-S disturbances. These features of S-C disturbances are very similar to those of wall-curvature-instability disturbances in three-dimensional boundary layers [3,4].

For comparison of the S-C disturbances with the C-F ones, two points on the critical curves of the C-F and S-C instabilities given in Fig. 4 are taken and amplitude and phase distributions of the corresponding disturbance velocities are shown in Fig. 6, where only the longitudinal component v^* in the direction normal to the wavenumber vector is given, because the other components are much smaller than it. Amplitude distributions of the two disturbances are almost similar and a distinguishable difference is found in phase distributions only.

Disturbances of the C-F instability have a definite advance in phase as the distance from the wall increases, but those of the S-C type show quite weak variations of phase with the distance. This difference in phase distribution of the two kinds of disturbances may be an important key when we attempt to detect the S-C instability in experiments.

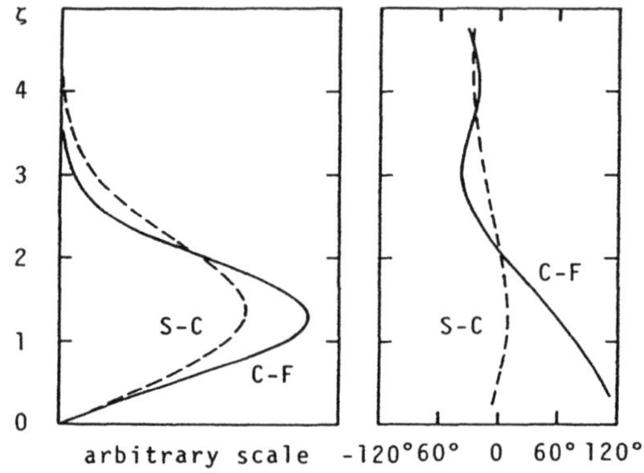

Fig. 6. Comparison in amplitude and phase distributions of C-F and S-C disturbances. (C-F: $\hat{R}$=327.0, $\hat{\lambda}$=7.44, $\hat{\phi}$=71.1°, $\hat{f}$=0.0141; S-C: $\hat{R}$=324.8, $\hat{\lambda}$=17.02, $\hat{\phi}$=90.7°, $\hat{f}$=-0.0082.)

5. Conclusions

A model system of disturbance equations explicitly including the terms of wall and streamline curvatures is applied to stability estimation of the three-dimensional boundary layer along a yawed circular cylinder. Velocity distributions of the boundary layer are approximated with members of the Falkner-Skan-Cooke velocity family, and variation of the Falkner-Skan parameter with the distance from the leading edge has been approximately determined by solution of the momentum integral equation.

Since the surface of a yawed circular cylinder is curved convexly, the wall curvature does not induce its own instability but gives a slightly stabilizing effect on the cross-flow and Tollmien-Schlichting instabilities. On the contrary, the curvature of external streamlines yields a new instability different from the cross-flow and T-S ones. This streamline-curvature instability gives the lowest values of the critical Reynolds number in the region near the leading edge, predicting an earlier appearance of the streamline-curvature disturbances than the cross-flow disturbances observed in current experiments.

References

[1] Mack, L.M.: AGARD Rep. No.709 (1984), 3.1-3.81.
[2] Itoh, N.: Fluid Dyn. Res. 7 (1991), 37-50.
[3] Itoh, N.: Trans. Japan Soc. Aero. Space Sci. 37 (1994), 125-138.
[4] Itoh, N.: Fluid Dyn. Res., in press (1994).
[5] Itoh, N.: J. Japan Soc. Aero. Space Sci. 41 (1993), 213-221 (in Japanese).
[6] Gregory, N., Stuart, J.T. and Walker, W.S.: Phil. Trans. Roy. Soc. London A 248 (1955), 155-199.
[7] Itoh, N.: Trans. Japan Soc. Aero. Space Sci. 17 (1974), 65-75.
[8] Watanuki, T. and Itoh, N.: NAL TR-841 (1984) (in Japanese).
[9] Schlichting, H.: Boundary-Layer Theory, 1968, Chap.10, McGraw-Hill.

Session J:
Poster Session II (Swept Wings, 3-D Boundary Layers, 2-D Boundary Layers and Receptivity)

Stability of Two-Phase Boundary Layer on a Flat Plate

E.S. Asmolov, S.V. Manuilovich
Zhukovsky Central Aero-Hydrodynamics Institute,
Zhukovsky, Moscow region, 140160, Russia

Summary

The linear stability of incompressible boundary-layer flow of dusty gas is considered. Eigenvalue problem for modified Orr-Sommerfeld equation is solved numerically using two approaches: a) directly by orthonormalization method, and b) by perturbation method. Particles most effectively suppress the instability waves when the relaxation length of their velocity is near to the wavelength of Tollmien-Schlicting (TS) wave. In this case 4% of mass particle content may reduce the growth more than one third. There is a region in the stable part of the flow where one more discrete mode of TS wave is added and the region where the eigenvalue problem has no solution.

Key words: Particles, Boundary layer, Relaxation Length of Particle Velocity, Modified Orr-Sommerfeld equation.

1. Governing equation

Particle influence on the stability of two-phase flow is due to two main effects: the change of mean-flow velocity because of the variation of effective density and viscosity and the unsteady momentum transfer between phases. When the particles' content is low, they slightly affect on the fluid mean flow. However, the second mechanism of impact on the instability wave may be significant in this case. In the unsteady perturbed velocity field the particles cannot accommodate rapid acceleration but tend to slip through the gas since the density of their material is much greater than that of the gas. The Stokes drag acts on the disperse phase, and an equal and opposite volume force acts on a gas. For this reason small portion of disperse-phase suppresses the instability wave.

Consider the linear stability problem for the boundary-layer flow of incompressible gas containing the particles of the same diameter a°. Particle phase has a uniform upstream volume density N. Free-stream velocity, gas density and the local width of clean-gas boundary layer are taken as the basic dimension variables so that

$$U^o_\infty = \rho^o = 1 \quad , \qquad \delta^o = \left(\frac{\mu^o X^o}{\rho^o U^o_\infty} \right)^{1/2} = 1 \quad .$$

Parameters of flow disturbances are sought in the form

$$(u \, , \, v \, , \, p \, , \, u_p, \, v_p) = (u^*, \, v^*, \, p^*, \, u_p^*, \, v_p^*) \exp(\, i\alpha x \, - \, i\omega t \,) \, + \, \mathrm{c.c.}$$

Here wavenumber α of TS wave is taken to be the real variable while the frequency ω is the complex one.

Over a wide range of particles' size (up to 1 micrometer) the relaxation length of their velocity is small as compared with characteristic distance from the leading edge of a plate and is of the order of characteristic wavelength of TS wave. For such particles their density in the region under examination is equal to its free-stream value, and their mean-flow velocity equals that of the gas [1]. As a result steady dusty-gas flow considered is equivalent to incompressible boundary-layer flow with the density to be equal to net density of both phases, and one can write

$$U(y) = U_0(\beta y) \, , \tag{1.1}$$

$$\beta = (1+f)^{1/2} \quad , \qquad f = \frac{4\pi}{3} a^{o^3} N \frac{\rho^o_p}{\rho^o} \quad .$$

Hereinafter subscript p characterizes the parameters of particle phase and subscript 0 those of clean-gas flow, f is dimensionless particle density. A dominant mechanism of unsteady momentum transfer is due to small perturbations of velocities is a Stokes drag. Then linearized Navier-Stokes equations accounting for momentum transfer between phases and momentum equations for particle phase can be written as

$$i \, \alpha \, u^* + \frac{dv^*}{dy} = 0 \quad ,$$

$$i \, (\alpha \, U \, - \, \omega) \, u^* + v^* \frac{dU}{dy} + i \, \alpha \, p^* + \mathrm{R} \, \gamma \, f \, (u^* - u_p^*) = \frac{1}{\mathrm{R}} \left(\frac{d^2 u^*}{dy^2} - \alpha^2 u^* \right) \, ,$$

$$i \, (\alpha \, U \, - \, \omega) \, v^* + \frac{dp^*}{dy} + \mathrm{R} \, \gamma \, f \, (v^* - v_p^*) = \frac{1}{\mathrm{R}} \left(\frac{d^2 v^*}{dy^2} - \alpha^2 v^* \right) \, , \tag{1.2}$$

$$i \, (\alpha \, U \, - \, \omega) \, u_p^* + v_p^* \frac{dU}{dy} - \mathrm{R} \, \gamma \, (u^* - u_p^*) = 0 \, ,$$

$$i \, (\alpha \, U \, - \, \omega) \, v_p^* - \mathrm{R} \, \gamma \, (v^* - v_p^*) = 0 \, ,$$

$$u^*(0) = v^*(0) = u^*(\infty) = v^*(\infty) = 0 \, .$$

$$R = \frac{\rho^{o} U^{o}_{\infty} \delta^{o}}{\mu^{o}} \quad , \quad \gamma = \frac{9}{2} \frac{\rho^{o}}{\rho^{o}_{p}} \left(\frac{\rho^{o} U^{o}_{\infty} a^{o}}{\mu^{o}} \right)^{-2} \tag{1.3}$$

Here γ is the dimensionless parameter associated with particle-phase features (particles' size and the density of their material). It remains constant over the entire flow distinct from the Reynolds number R. Particle perturbed velocity can be readily expressed from (1.2) in terms of u^*, v^* as

$$u^*_p = \frac{R\,\gamma}{R\,\gamma + i\,\alpha\,(U-c)}\, u^* - \frac{R\,\gamma}{[\,R\,\gamma + i\,\alpha\,(U-c)\,]^2}\, \frac{dU}{dy}\, v^* \tag{1.4}$$

$$v^*_p = \frac{R\,\gamma}{R\,\gamma + i\,\alpha\,(U-c)}\, v^* \quad , \quad c = \frac{\omega}{\alpha} \, .$$

Navier-Stokes equation can be reduced in the view of (1.4) to Orr-Sommerfeld equation given by

$$L_{O-S}(\omega,\tilde{U})\ \varphi = 0 \quad , \quad \varphi(0) = \frac{d\varphi}{dy}(0) = \varphi(\infty) = 0 \, , \tag{1.5}$$

$$L_{O-S}(\omega,U) = (\,U - c\,)\left(\frac{d^2}{dy^2} - \alpha^2 \right) - \frac{d^2U}{dy^2} - \frac{1}{i\alpha R}\left(\frac{d^2}{dy^2} - \alpha^2 \right)^2 ,$$

$$u^* = \frac{d\varphi}{dy} \quad , \quad v^* = -i\,\alpha\,\varphi \, ,$$

with modified velocity profile

$$\tilde{U} = U + f\, \frac{R\,\gamma\,(U-c)}{R\,\gamma + i\,\alpha\,(U-c)} \, .$$

Equation (1.5) was derived in [2] for plane parallel flow of dusty gas. It differs from ordinary Orr-Sommerfeld equation for clean gas by replacing the fluid velocity by modified velocity profile being the complex function of y. However it should be noted that in [2] mean-flow velocity distributions $U(y)$ for dusty and clean gas were taken to be the same. This is valid, for example, for Poiseuille flow, but in our case, as it seen from (1.1), $U(y)$ for dusty gas differs from that for clean gas.

The first approach that was used to solve eigenvalue problem (1.5) for arbitrary particle density is direct numerical integration of (1.5) using orthonormalization method. The growth rate $Im\,\omega$ and frequency of TS wave were evaluated as the function of wave number, Reynolds number, parameter γ and dimensionless particle density f. Dependencies $Im\,\omega$ vs α calculated for R = 1000, $\gamma = 10^{-4}$ and different f = 0,01; 0,02; 0,04; 0,1 (curves 1-4 correspondingly) are shown in Figure 1. Dashed line represents the TS parameters for clean gas. Dust results in significant reduction of growth rate even for small particle content. For example, for given values R and γ 4% of particle mass content reduces the maximum growth rate more than one third. It is also seen that variation of $Im\,\omega$ is linearly proportional to particle content when this value is small. This suggests the use of second approach at low f - namely, perturbation method.

2. Perturbation method

At small particle content eigenfunction and eigenvalue of modified Orr-Sommerfeld equation is sought in the framework of outlined method as perturbation with respect to those of Orr-Sommerfeld equation for clean gas so that

$$\omega = \omega_0 + f\,\omega_1 + O(f^2)$$

where ω_0 is the solution of ordinary Orr-Sommerfeld equation for clean gas given by

$$L_{0\text{-}S}(\omega_0, U_0)\,\varphi_0 = 0 \quad , \qquad \varphi_0(0) = \frac{d\varphi_0}{dy}(0) = \varphi_0(\infty) = 0 \ . \tag{2.1}$$

Then the perturbation of complex frequency can be determined from the following equation

$$\omega_1 = -\frac{(\,\chi_0,\ L_U\varphi_0\,)}{(\,\chi_0,\ L_\omega\varphi_0\,)} \quad , \qquad (\,\chi\,,\,\varphi\,) = \int_0^\infty \bar{\chi}\,\varphi\,dy \ .$$

$$L_U = U_1\left(\frac{d^2}{dy^2} - \alpha^2\right) - \frac{d^2U_1}{dy^2} \quad , \qquad U_1 = y\,\frac{dU_0}{dy} + \frac{\mathrm{R}\,\gamma\,(U-c)}{\mathrm{R}\,\gamma + i\,\alpha\,(U-c)} \quad ,$$

$$L_\omega = -\frac{1}{\alpha}\left(\frac{d^2}{dy^2} - \alpha^2\right) .$$

Here χ_0 is the solution of adjoint equation defined as

$$L^+_{0\text{-}S}(\omega_0, U_0)\,\chi_0 = 0 \quad , \qquad \chi_0(0) = \frac{d\chi_0}{dy}(0) = \chi_0(\infty) = 0 \ , \tag{2.2}$$

$$L^+_{0\text{-}S}(\omega, U) = (\,U - \bar{c}\,)\left(\frac{d^2}{dy^2} - \alpha^2\right) + 2\,\frac{dU}{dy}\frac{d}{dy} + \frac{1}{i\alpha\mathrm{R}}\left(\frac{d^2}{dy^2} - \alpha^2\right)^2 .$$

Similar approach based on approximate interpolation of eigenfunction φ_0 and adjoint function χ_0 was used early to evaluate neutral stability curves for plane Poiseuille flow of dusty gas [3]. However it yields inaccurate results even at $f = 0,05$.

In present study ω_1 is obtained using exact numerical solution of (2.1), (2.2). The growth rates calculated for $\mathrm{R} = 1000$, $\gamma = 10^{-4}$ using both perturbation and direct integrations method are shown in Figure 2 as the functions of particle mass content (correspondingly dashed and solid lines). Thus a comparison of results of both methods shows that perturbation theory can be successively used up to $f \simeq 0,1$.

3. Optimum damping of instability wave

Now we investigate the effect of dimensionless value γ defined by (1.3) on parameters of TS wave. It characterizes particles' size for given density of their material and varies as $(a^{\circ})^{-2}$. For two limiting cases of fine ($\gamma \to \infty$) and coarse dust ($\gamma \to 0$) solution can be easily expressed through the parameters of TS wave for clean-gas flow. For the first case particle velocities both for the main and perturbed flow equal those of the gas. In other words, particles are frozen into the gas, and stability problem is identical to that for the fluid having the dimensionless density $1 + f$. Hence, we can write

$$\omega (\alpha , R , f , \gamma) = \beta\, \omega_0\left(\frac{\alpha}{\beta} , \beta R \right) \quad \text{as} \quad \gamma \to \infty .$$

For the second limiting case ($\gamma \to 0$) velocities of two phases are equal as well while the perturbed velocity of coarse particles, as it follows from (1.4), is zero since they are not sensitive to the rapid acceleration of the gas. Stokes drag arises in this case but its influence on the TS wave is small since the volume density of the particles N decreases at given mass content linearly with γ or with $(a^{\circ})^{-2}$. Finally we have

$$\omega (\alpha , R , f , \gamma) = \beta\, \omega_0\left(\frac{\alpha}{\beta} , \frac{R}{\beta} \right) \quad \text{as} \quad \gamma \to 0 .$$

Thus both fine and coarse particles slightly effect on the instability wave. The damping influence of dust is most efficient when the relaxation length of particle velocity is of the order of the wavelength of TS or, as the same, $\gamma \sim \alpha/R$. Particles volume density in this case is greatest possible provided that slip velocity and Stokes drag are still sizable. Growth rates evaluated at $R = 1000$, $f = 0{,}01$ and various $\gamma = 10^{-3}$; 10^{-4}; 10^{-5} are drawn in Figure 3 (correspondingly curves 2-4). For comparison line 1 shows $Im\, \omega$ for clean gas. It is seen that dust damps the instability wave in a wide range of γ parameter.

4. Discontinuity of eigenvalues

Distinguishing feature of Orr-Sommerfeld equation with modified velocity profile that was not pointed out early is that it has a singular point y_c in y-variable complex plane. y_c can be determined from the equation

$$R\, \gamma + i\, \alpha\, [\, U(y_c) - c\,] = 0 . \qquad (4.1)$$

When the pole is near a path of integration rapid particle motion in critical layer takes place. Numerical procedure in this case may present some difficulty since $\tilde{U}$ and consequently eigen function rapidly vary in the vicinity of y_c , and for this reason it is required very small steps of integration. This problem is eliminated using the fact that Blasius function and $U(y)$ are the regular functions if y is complex. It makes possible to integrate equation (1.5) along the curved contour in a complex y-plane. We use the contour consisting of segments of real axis and semi-circle bypassing y_c. Semi-circle is located on the upper semi-plane when $Im\ y_c < 0$ and on the lower semi-plane for opposite inequality.

When y_c crosses the real axis, $\omega(\alpha, R)$ has a break. This discontinuity takes place as it follows from (4.1) for $Im\ \omega < 0$, i.e., in the stable part of the flow. As a result there are the region in the $(\alpha$-$R)$ space where one more discrete TS mode arises and the region where the eigenvalue problem has no solution. This conclusion is illustrated in Figure 4.

References

[1] Singleton, R.E.: The compressible gas-solid particle flow over a semi-infinite flat plate, J. Appl. Math. Phys. **16** (1965) 421-449.

[2] Saffman, P.G.: On the stability of laminar flow of a dusty gas, J. Fluid Mech. **13** (1962) 120-128.

[3] Michael, D.H.: The stability of plane Poiseuille flow of a dusty gas, J. Fluid Mech. **18** (1964) 19-32.

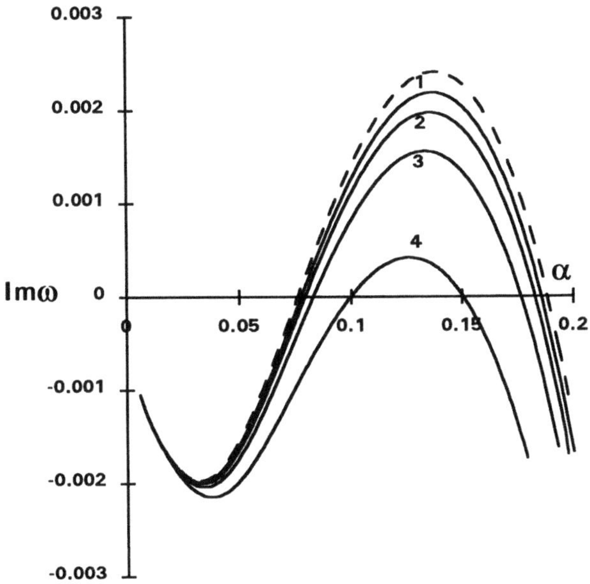

Figure 1. Growth rates at R=1000, γ=1.e-4 and various particle mass content f=0,01; 0,02; 0,04; 0,1 (correspondingly solid curves 1-4). Dashed line is the same value for clean gas.

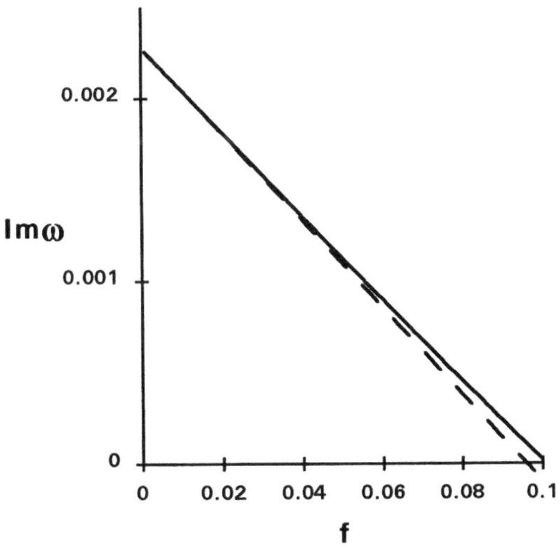

Figure 2. Growth rates at R=1000, γ=1.e-4, α=0,15 evaluated by direct numerical integration (solid line) and by perturbation method (dashed line).

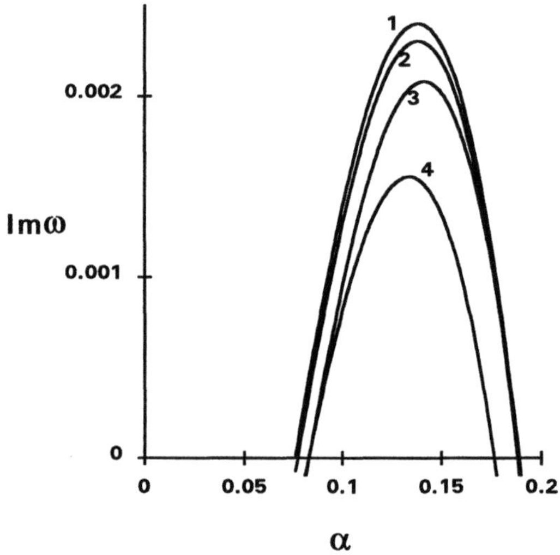

Figure 3. Growth rates at R=1000, f=0,04; and various γ=1.e-3; 1.e-4 ; 1.e-5 (correspondingly curves 2-4). Curve 1 is the same value for clean gas.

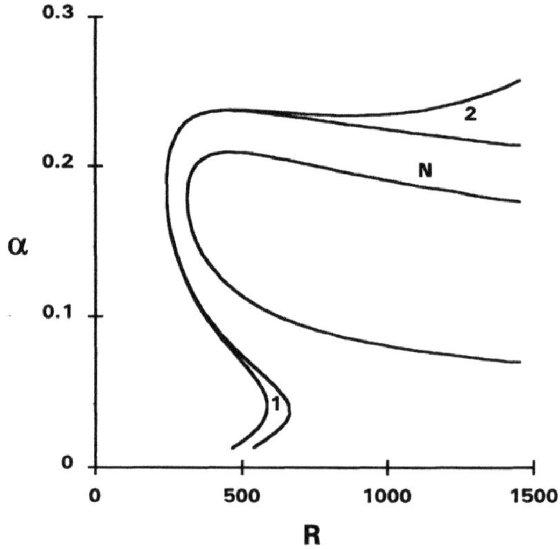

Figure 4. Diagram illustrating the number of TS modes at f=0,04 and γ=5.e-6. Two modes exist in reqion 1 while in region 2 there is no solution of eigenvalue problem. Line N is the neutral-stability curve.

On the Receptivity of a Flat Plate Boundary Layer to Localized Free Stream Disturbances

A.A. Bakchinov[1], K.J.A. Westin[2],V.V. Kozlov[1] & P.H. Alfredsson[2]

[1]Dept. Theoretical and Applied Mechanics, Russian Academy of Sciences, Siberian Branch, 630090 Novosibirsk, Russia.
[2]Dept. Mechanics/Fluid Physics, Royal Institute of Technology, S-10044 Stockholm, Sweden.

Summary
A localized disturbance was introduced into the free stream from a small pipe located upstream of the leading edge of a flat plate. The disturbance penetrated the boundary layer, and a so called boundary layer 'puff' was created, consisting of three longitudinal streaks with the maximum perturbation amplitude approximately in the middle of the boundary layer. The amplitude decayed downstream, however, when a high-frequency Tollmien-Schlichting (TS) wave was generated simultaneously, strong interaction was observed which could lead to transition.

Key words: free stream turbulence, localized disturbance, TS-waves

1 Introduction

High levels of free stream turbulence (FST) can be found in many technical applications, and the corresponding transition problem has faced an increased interest during the last years. However, there are many difficulties connected with experiments at high FST-levels. A laminar boundary layer subjected to FST is characterized by low-frequency fluctuations, with amplitudes of the order of 10-12% before transition occurs (see for instance Westin et al. [1]). To separate and study details in the transitional structures is therefore an intricate task, which need to be simplified as much as possible. Transition at high and moderate levels of FST is often thought of as a result of the evolution of localized, transient disturbances caused by free stream vortices impinging onto the boundary layer. The objective of the present study is to simulate the free stream eddies by disturbances generated in a controlled manner from the free stream, and study their evolution in the boundary layer and their ability to interact with TS-waves.

In studies of localized disturbances in boundary layer flows, the perturbation is usually generated from a disturbance source on the wall, located at supercritical Reynolds-numbers (R) with respect to TS-waves. Gaster & Grant [2] and Cohen, Breuer & Haritonidis [3] used small amplitude disturbances, from which growing wave packets were obtained. The wave packet growth was enhanced by oblique subharmonic waves, which could lead to transition ([3]). A larger initial disturbance (also injected at supercritical R) was used by Amini & Lespinard [4], showing an almost immediate formation of an incipient spot. However, a third type of boundary layer disturbance (beside the wave packet and the incipient spot) was reported by Grek, Kozlov & Ramazanov [5]. They injected a moderate amplitude jet through a hole close to the leading edge (subcritical R), and obtained a localized disturbance, which was named a boundary layer 'puff'. The 'puff' was characterized by narrow spanwise scales, where the downstream lateral growth was almost negligible, and the amplitude was always decaying in their set-up. Also, the amplitude maximum was not found close to the wall, as is the case both for wave-packets and incipient spots. In a later experiment by Grek et al. [6], it was found that the boundary layer puff could interact with artificially generated TS-waves, leading to transition. Similar results have been observed at moderate levels of FST (1.5%), where generated TS-waves could enhance the transition process. (Boiko et al., [7]).

2 Experimental set-up

The experiment was carried out in the MTL-wind tunnel at the Royal Institute of Technology, using a flat plate model with a 2 m long working part. The leading edge has an asymmetric shape and was specially designed to minimize the non-zero pressure gradient region close to the leading edge. The plate and leading edge configuration is extensively described in [1] (there denoted set-up II), but the traversing mechanism is slightly modified. The coordinate system has its origin at the leading edge, where the streamwise, wall-normal and spanwise directions are denoted x, y and z, and the corresponding velocity components are u, v and w. The free stream velocity (U_0) was 6.6 m/s. The measurements were carried out with hot-wire anemometry, where a single wire probe of length 0.5 mm and wire diameter 2.5 μm was used in most cases. For some measurements a cross-wire probe was used, having a measurement cube with side length less than 0.5 mm, and calibrated in the free stream at different angles and flow velocities. The localized disturbance was generated by a loudspeaker connected to a small pipe (inner diameter 1 mm), with its tip positioned 12 mm upstream of the leading edge (see figure 1).

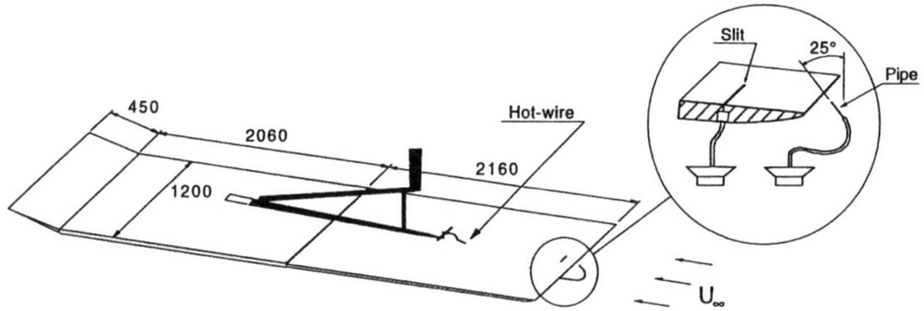

Figure 1: Experimental set-up.

The y-position was carefully adjusted to obtain a disturbance with a trajectory hitting the working side of the plate, and at the same time avoid any influence from the wake of the pipe. To generate the disturbance, a function generator was connected to a pulse generator, giving a short duration pulse at a frequency typically $F_0/64$ (where F_0 is the frequency of the function generator). In the second part of the experiment, TS-waves were generated from a 35 mm long and 0.5 mm wide slit in the plate, positioned at $x = 95$ mm. The slit was connected to a second loudspeaker, driven with frequency F_0 from the same generator, giving a definite phase relation between the TS-wave and the localized disturbance.

3 Receptivity and development of the localized disturbance

3.1 The initial disturbance

An impulsive jet emerging into quiescent air should under ideal conditions generate a vortex ring. In the present case we can expect a more complicated free stream disturbance due to the strong mean flow which deforms the disturbance. Figure 2, shows qualitative measurements of the u- and v-perturbation just upstream of the leading edge. The measurement position is chosen so that the streamline passing the probe will be close to the upper plate surface at the leading edge. The front part of the disturbance consists of a negative u-perturbation, corresponding to a a positive v-component. From a different plot (not shown) contributions of spanwise vorticity could be clearly observed. This negative perturbation in u turns out to be very stable, and is observed far downstream as a velocity deficit convected in the free stream. The rear part of the disturbance is dominated by a positive perturbation in u. The v-component is negative on the centreline, but also a positive region of v can be found on either side of the centreline, corresponding to streamwise vorticity.

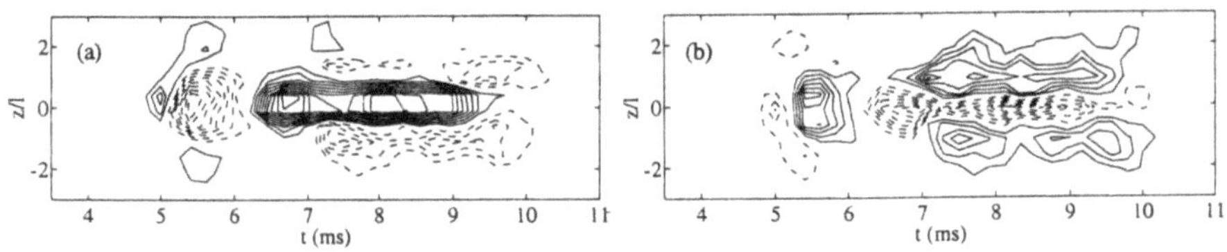

Figure 2: Spanwise distribution of perturbation velocity for the free stream disturbance. (a) u-component, (b) v-component. $x = -2$ mm. z scaled with the pipe diameter (l). Contour spacing in all plots: 1%.

3.2 The boundary layer puff

The first measurement position where the fully developed boundary layer disturbance has been captured is shown in figure 3. In figure 3a, the spanwise distribution shows a positive u-disturbance on the centreline, surrounded by two negative regions. This distribution leads to the assumption that the disturbance mainly consists of two counter rotating streamwise vortices, which is also supported by the two yt-planes shown in figure 3b and c. The plane measured in the negative region extends outside the boundary layer, which is due to the upward motion of low-velocity fluid from the near-wall part of the boundary layer (lift-up effect). The peak values of the perturbation (positive and negative) are located approximately in the middle of the boundary layer. The localized character should also be emphasized. Although the contour spacing is chosen rather small, it is not possible to observe any weak disturbances outside the three streaks in figure 3a. It should be mentioned that the length to width ratio of the disturbance is not correctly described in figure 3, but should be stretched in the x-direction. The duration of 8 ms for the perturbation corresponds to a disturbance length of almost 50 mm at this position.

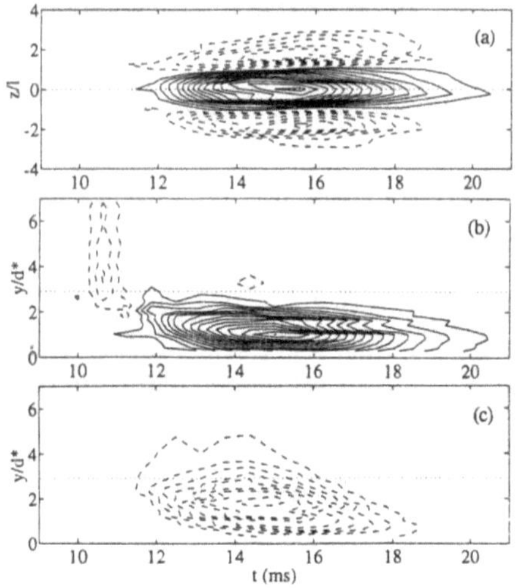

Figure 3: Streamwise perturbation amplitude of the puff at x = 35 mm. (a) Spanwise distribution at $y \approx \delta^*$. Wall-normal distribution at (b) z/l=0 and (c) z/l=–1.75.

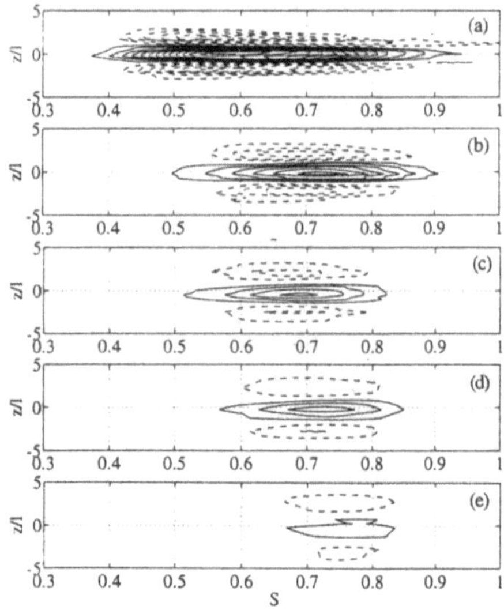

Figure 4: Downstream development of puff (u-component). $S=tU_0/x$. (a) x=35 mm, (b) x=100 mm, (c) x=160 mm, (d) x=200 mm and (e) x=300 mm

In order to find a simple measure of the disturbance amplitude, zt- and yt-planes were measured close to the positive and negative peak values of the disturbance, and from these a maximum peak-to-peak value (u_{pp}) was determined. At x = 7 mm, which is just upstream of the position where measurements can be carried out inside the boundary layer, the disturbance amplitude is $u_{pp} \approx 12\%$ of the free stream velocity. At first sight this value seems to be large, but a sine wave with the same peak-to-peak value results in $u_{rms} \approx 4\%$, which is a typical level of free stream turbulence in many types of fluid machinery. The value of u_{pp} for the boundary layer puff shown in figure 3 is about 25%. The strong response is a consequence of the strong mean velocity gradient in the boundary layer close to the leading edge, where a disturbance entering from the free stream leads to a large perturbation velocity in the u-component.

In the present set-up it was not possible to generate free stream perturbations large enough to obtain growing boundary layer disturbances, and the downstream development of the puff shows a slowly decaying amplitude, without any dramatic changes in the disturbance structure. From figure 4 it can be observed that the spanwise growth downstream is almost negligible in physical units, and the width of the streaks does not scale with the boundary layer thickness. The structure is elongated due to the mean shear in the boundary layer, and attains an inclined shape in the yt-plane. The elongation is not observable in figure 4, which is scaled as $S = tU_0/x$ to show the local propagation velocity. This is close to the local mean velocity, where the front of the disturbance is convected with approximately $0.9U_0$, and the rear part with $0.5U_0$. An attempt to measure the w-component was carried out at $x = 160$ mm, in order to find out if the disturbance consists of two streamwise vortices. The results showed only a weak motion outwards from the centreline on either side, and it was not possible to detect any vortical motion. Apparently the disturbance at this position consists mainly of low and high velocity streaks, where the vortical motion that can be expected close to the leading edge has been damped out.

3.3 Discussion

The boundary layer disturbance that is obtained from the free stream eddy is surprisingly simple, showing just three longitudinal streaks. In previous experiments on the puff, the disturbance was generated from a hole in the plate close to the leading edge. In that case the disturbance consisted of two sets of streaks, one for the upward motion of the loudspeaker and one with opposite signs for the downward motion. When looking at the initial disturbance in the present case it also consists of a positive and a negative v-component, where only the rear part of the disturbance results in a boundary layer perturbation. It can therefore be expected that the sign of the v-component is crucial for whether or not the disturbance can penetrate the boundary layer, and a strong negative v-perturbation should be necessary for the receptivity. However, there are also differences in the vorticity between the front and rear part of the initial disturbance, which could be important for the receptivity process.

An interesting comparison can be done between the puff and the work on optimal disturbances by Butler & Farell [8]. They found that the streamwise vortex is the disturbance with the fastest growth for short times in the Blasius boundary layer. After the initial growth, the amplitude of the disturbance starts to decay, which is first observable in the v-component. The streamwise perturbation also starts to decay, but damping is slower and low- and high-velocity streaks can be observed for long times.

When comparing the puff characteristics with results for boundary layers subjected to free stream turbulence, it is striking that the maximum perturbation amplitude of the puff is found approximately in the middle of the boundary layer, in agreement with the location of the maximum in u_{rms} in the case of FST. Also, the formation of longitudinal streaks can be detected from two-probe correlation measurements in boundary layers subjected to FST. However, the puff amplitude is always decaying downstream, in contrast to the growing u_{rms} that is observed in the FST-case. It is possible that the downstream development of u_{rms} depends to a large extent on the continuous forcing from the FST above the boundary layer, and is not only confined to the leading edge receptivity.

4 Interactions with TS-waves

4.1 Method

The objective of the second part of the study was to generate a controlled TS-wave from the slit, in order to study possible interactions with the boundary layer puff. The frequency was chosen to get as large interaction as possible for a specified forcing amplitude. This was obtained for a non-dimensional frequency F ($F = 2\pi f\nu/U_0^2 \cdot 10^6$) of about 460, which is a highly damped TS-wave in the Blasius boundary layer. However, linear stability calculations for local mean velocity profiles extracted from the high-speed region on the centreline showed that TS-waves can be unstable, and the largest growth was observed for $F = 510$.

In order to study and separate non-linearly induced disturbances, three separate measurements were carried out at each position in space: 1) Generation of localized disturbance (puff), 2) TS-wave generation, 3) Localized disturbance and TS-wave generation simultaneously. The three sets of data were collected sequentially before the probe was moved to the next position. The method provides the possibility to subtract the generated disturbances (puff and TS-wave), giving only the perturbation that is non-linearly induced (figure 5).

4.2 Effect of non-linear interactions

Both the puff and the TS-wave show a downstream decay of the amplitude when they are generated separately. The TS-wave amplitude at $x = 100$ mm (5 mm downstream of the slit) is $u_{rms} = 2.2\%$, but decreases to 0.7% at $x = 200$ mm. However, when generated simultaneously, the total disturbance amplitude is decaying until $x = 160$, after which a strong amplification is observed. With the present choice of initial amplitudes, the majority of disturbances started to decay again downstream of $x = 300$ mm. For this reason the total disturbance amplitude (in average) is reduced at $x = 400$ mm. However, about 15% of the generated disturbances continued to amplify and developed into turbulent spots.

The non-linear part of the total disturbance shows a large increment between $x = 160$ and 200 mm, and from this position and downstream the total disturbance is almost entirely dominated by the non-linear contribution. Consequently, the non-linear interactions between puff and TS-wave must occur between $x = 100$ and 200 mm. The non-linearly induced perturbations measured on the centreline are shown in figure 6, where the puff and TS-wave are subtracted from the total disturbance. Also the position of the puff (when generated separately) is shown as a dotted line. The non-linearity starts within the puff, but at $x = 160$ it is possible to observe that the new perturbation moves with a lower propagation velocity. A new wave length close to the subharmonic of the generated TS-wave is also seen at the rear part of the structure.

The spanwise distributions of the total disturbance structure is shown in figure 7 for the downstream positions. Initially, the non-linear part of the perturbation shows streamwise peaks according to the fundamental wave and a spanwise modulation following the streaks of the puff (not shown). At $x = 160$ the previously mentioned subharmonic modulation can be observed at the rear, and at $x = 200$ mm (fig. 7a) there is a tendency for the formation of oblique structures. At the two last positions, a new structure has formed which consists of several longitudinal streaks. In the new structure there are regions with strong negative perturbations (see for instance fig. 6d), which results in local inflectional mean velocity profiles.

5 Conclusions

The present work can be summarized as follows:

i) It has been possible to verify the existence of the previously observed boundary layer 'puff' in the present experimental set-up. The puff is characterized by longitudinal high and low velocity streaks with the maximum perturbation amplitude approximately in the middle of the boundary layer. The lateral growth of the structure is negligible, and the perturbation amplitude is decaying downstream.

ii) The disturbance structure shows some similarities with perturbations observed in boundary layers subjected to free stream turbulence, although the downstream decay of the amplitude is in contrast to the FST-case.

iii) The puff can interact non-linearly with TS-waves, which can lead to transition. The strongest interaction is observed for high-frequency waves, which are strongly damped in the unperturbed Blasius boundary layer.

At present time it is not clear which are the mechanisms behind the interaction. Both wave interactions and inflectional instabilities are possible, and furher analysis as well as new experiments are therefore necessary.

This work was supported by the Swedish National Board for Industrial and Technical Development (NUTEK). The visit of Prof. V.V. Kozlov and A.A. Bakchinov to Stockholm was made possible through the scientific exchange program sponsored by the Royal Swedish Academy of Sciences (KVA) in cooperation with the Russian Academy of Sciences.

References

[1] K.J.A. Westin, A.V. Boiko, B.G.B. Klingmann, V.V. Kozlov and P.H. Alfredsson (1994) Experiments in a boundary layer subjected to free stream turbulence. Part I: Boundary layer structure and receptivity. *J. Fluid Mech.* (in press)

[2] Gaster, M. & Grant, I. 1975 An experimental investigation of the formation and development of a wave packet in a laminar boundary layer. *Proc. R. Soc. Lond.* A **347**, 253–269.

[3] Cohen, J., Breuer, K.S. & Haritonidis, J.H 1991 On the evolution of a wave packet in a laminar boundary layer. *J. Fluid Mech.* **225**, 575–606.

[4] Amini, J. & Lespinard, G. 1982 Experimental study of an "incipient spot" in a transitional boundary layer. *Phys. Fluids* **25**(10), 1743–1750.

[5] Grek, H.R., Kozlov, V.V. & Ramazanov, M.P. 1985 Three types of disturbances from the point source in the boundary layer. In *Laminar-Turbulent Transition 2* (ed. V.V. Kozlov), pp. 267–272. Springer.

[6] Grek, H.R., Dey, J., Kozlov, V.V., Ramazanov, M.P. & Tuchto, O.N. 1991 Experimental analysis of the process of the formation of turbulence in the boundary layer at higher degree of turbulence of windstream. *Rep. 91-FM-2, Indian Inst. Science, Bangalore, 560012, India.*

[7] A.V. Boiko, K.J.A. Westin, B.G.B. Klingmann, V.V. Kozlov and P.H. Alfredsson (1994) Experiments in a boundary layer subjected to free stream turbulence. Part II: The role of TS-waves in the transition process. *J. Fluid Mech.* (in press)

[8] Butler, K.M. & Farrell, B.F. 1992 Three-dimensional optimal perturbations in viscous shear flow. *Phys. Fluids A* **4**(8), 1637–1650.

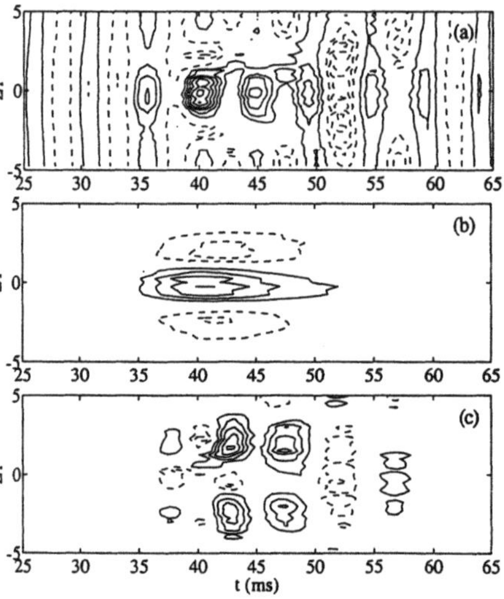

Figure 5: Streamwise perturbation at $x = 160$ mm. (a) Total disturbance, (b) puff and (c) total disturbance with puff and TS-wave subtracted.

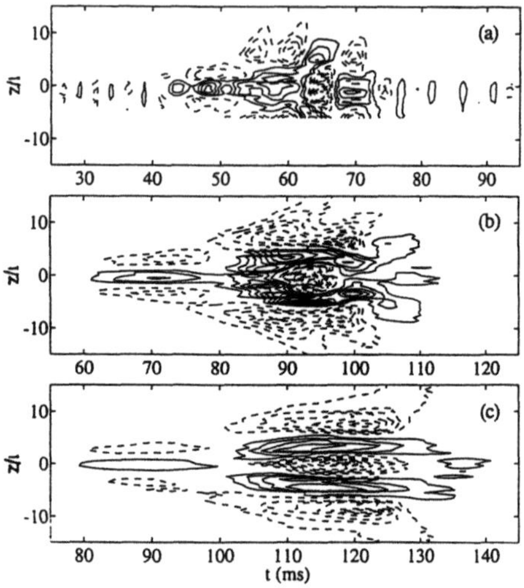

Figure 7: Streamwise perturbation for the total disturbance at $y \approx 1.2\delta^*$ (a) x=200 mm, (b) x=300 mm and (c) x=400 mm.

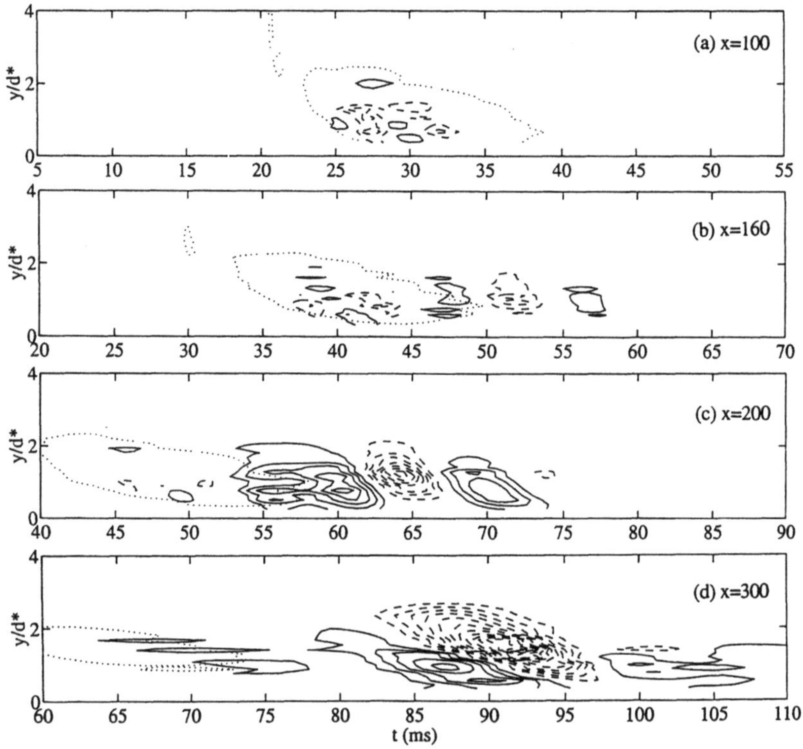

Figure 6: Non-linearly induced u-perturbation (puff and TS-wave subtracted). $z/l = 0$. Dotted line shows position of the puff.

Comparison of Asymptotic Prediction Methods for Boundary-Layer Receptivity

P.W. Hammerton
Department of Applied Mathematics & Theoretical Physics,
University of Cambridge, Cambridge CB3 9EW, UK

M. Gaster
Department of Engineering,
University of Cambridge, Cambridge UK

Summary

Orr-Sommerfeld theory and triple deck theory are compared for receptivity problems. Predictions for the perturbed velocity field are obtained and compared with experimental results for one of the simplest cases of receptivity, namely an isolated oscillating roughness element on a flat-plate.

Key Words : Receptivity, Orr-Sommerfeld, Triple-Deck

1 Introduction

When studying the phenomena of boundary-layer receptivity, three approaches are possible– experimental, numerical solution of the full set of equations, or asymptotic analysis based on large Reynolds number. Careful quantitative comparison of the results obtained by the different methods appears vital in unifying the various approaches to boundary-layer transition. While most previous calculations of boundary-layer flows have tended to focus on wall pressure or wall shear, the most convenient measurement taken during wind tunnel tests is the horizontal velocity perturbation, measured by hot-wire probes.

In this paper we discuss two quite different asymptotic methods and compare the results with experimental measurements. The receptivity problem considered is the simplest possible, namely the generation of boundary-layer disturbances by an oscillating two-dimensional hump/ribbon on an otherwise flat plate, which is closely related to the problem of the interaction of free-stream disturbances with a steady hump. The second case is of more interest for practical purposes, but the first case is more suitable for experimental investigation since the source of excitation can be controlled more easily.

Triple-deck theory is an asymptotic analysis of boundary-layer structure based on large Reynolds number [1, 2]. For the scalings associated with lower-branch stability, the standard boundary layer contains an inner layer close to the wall, while a third outer layer interacts with the main layer. Triple-deck has the advantage of being rigorously formulated in the large-Re limit, and hence can be extended, at least formally, to whatever order may be required. Effects such as non-parallel base flow and non-linearity due to surface perturbation can then be incorporated easily into the theory. The major disadvantage of this approach for practical purposes is that the underlying asymptotic structure must be fully understood before predictions of the perturbation flow can be interpreted. Moreover, the small parameter of triple-deck theory is $(Re_\delta)^{-1/4}$, where Re_δ is the Reynolds number based on displacement thickness, suggesting that the Reynolds number must be very large for leading-order triple-deck results to be a valid approximation.

An alternative approach utilises numerical integration of the Orr-Sommerfeld equation. Orr-Sommerfeld theory is based on the flow being parallel, and one weakness of such an approach is that inclusion of higher order effects in a consistent way is still a contentious issue. However, this prediction method is widely used for practical applications. It can be applied much more easily than triple-deck theory and both the upper- and lower-branches of the stability curve can be calculated from the same formulation.

2 Triple deck theory for a two-dimensional boundary layer

The triple-deck structure for flow over an unsteady hump is well described elsewhere [3]. Here we present only the details necessary for illustrating the results of this paper. Defining a small parameter $\epsilon = (Re_L)^{-1/8}$, where $Re_L = U_\infty L/\nu$ is the Reynolds number based on distance of the hump from the leading edge, interaction of the lower deck with the outer deck occurs when the streamwise coordinate scales on ϵ^3. The lower, main and outer decks of the boundary layer then scale on ϵ^5, ϵ^4 and ϵ^3, respectively, with the scaled non-dimensional frequency given by $\Omega = \epsilon^2 2\pi f L/U_\infty$. Determining whether such a three-deck structure has any meaning for the moderate values of Re_L typical of wind-tunnel tests is one aim of the present paper. When the hump height is small compared with the lower deck, the lower deck equations can be linearised about the Blasius solution. In this case the equations, supplemented by matching conditions, can be solved in Fourier space, to give

$$U(X,Y) = -\frac{1}{2\pi}\int_{-\infty}^{\infty} \frac{w^4\{1-3I(wY+z_0)\} - 3i\hat{\Lambda}\mathrm{Ai}'(z_0)}{w^4\{1-3I(z_0)\} - 3i\hat{\Lambda}\mathrm{Ai}'(z_0)} F_f(k) e^{ikX}\, dk,$$

where $w = (ik)^{1/3}$, $z_0 = i\Omega/w^2$, $I(z) = \int_0^z \mathrm{Ai}(z')dz'$ and $\hat{\Lambda} = \pm 1$ according to whether $k > 0$ or $k < 0$. Here $F_f(k)$ is the transform of the hump height $F(X)$. In the main deck of the boundary layer, streamlines are displaced upwards in phase with the hump and the velocity perturbation is readily calculated in terms of a displacement function. Although this gives a closed form solution for the whole disturbance field, inversion of this Fourier transform requires care since in the complex plane the integrand has branch cuts along both the positive and negative imaginary axes. In addition there is an infinite set of poles corresponding to roots of the dispersion relation $D(k,\omega) = 0$. Hence, the solution can be re-expressed as the sum of a local solution given by the contribution from the branch cut, together with a set of normal modes, either growing or damped. For small Ω, all the poles of the integrand lie in the upper half plane and hence the corresponding modes are damped. As Ω increases, one of these roots approaches the real axis, and for $\Omega > 0.44$ this root lies in the lower half plane, corresponding to a mode which grows downstream [3].

One of the aims of the present study was to develop a robust scheme for efficiently calculating both the near- and the far-field disturbance. Fourier inversion by contour deformation requires calculation of all the roots of the dispersion relation, together with the corresponding residues, which will be different for each inversion required. Direct numerical inversion breaks down as the neutral frequency is approached and the disturbances extend a long distance downstream. The approach taken here combines the results of contour deformation and numerical inversion. Writing

$$\Phi = \frac{1}{2\pi}\int_{-\infty}^{\infty} \frac{N(k)}{D(k)} e^{ikX}\, dk,$$

where $\Phi(X)$ represents either pressure, displacement or lower deck velocity, this can be expressed in the form of a convolution,

$$\Phi = e^{ik_1X}\int_{-\infty}^{X} \phi(y)e^{-ik_1y}dy, \qquad \phi(X) = \frac{1}{2\pi}\int_{-\infty}^{\infty} \frac{N(k)}{D(k)}(k-k_1)e^{ikX}\, dk,$$

where k_1 is the one root of $D(k) = 0$ that may lie in the lower half plane. In this way, all problems with numerical inversion are avoided, since the inversion to obtain $\phi(X)$ has no poles close to the real axis, and none in the lower half plane. The next correction term to the leading-order disturbance field has order-two poles arising in the inversion integral, but the inversion can again be expressed in the form of a convolution.

In Figure 1, the velocity perturbation field obtained using this convolution method is plotted. This corresponds to a two-dimensional oscillator, forced at a frequency above the neutral frequency, situated 40 cm downstream of the leading-edge, where the triple-deck parameter $\epsilon = 0.18$. The wall normal co-ordinate is the usual boundary layer variable $\eta = y\sqrt{U_\infty/2x\nu}$, and the lower deck solution is patched to the main deck at $\eta = 0.7$. It is seen that this method satisfactorily gives the local field as well as the growing disturbance downstream.

For the two-dimensional case described here, evaluation by contour deformation is possible, but in the three-dimensional case briefly described in §5, the approach described here seems the only practical method.

3 Orr-Sommerfeld Theory in two dimensions

An alternative approach to solving boundary layer flows is to assume that the mean flow is parallel, perform a linear perturbation about the mean flow and then look for solutions proportional to $\exp(i\alpha x - i\omega t)$, which yields a fourth-order ODE in the boundary normal coordinate. Solution of this equation subject to boundary conditions then yields numerical eigenvalues for α. For the boundary layer over a flat plate, the base flow is not strictly parallel and hence it is assumed that Re_δ is asymptotically large to allow the quasi-parallel assumption. In contrast the derivation of the ODE requires the assumption that Re_δ is finite. Thus formally the Orr-Sommerfeld equation can be considered an *ad hoc* approximation [4]. However stability calculations using OS theory are widely used and can yield good agreement with experiment. Implementation of OS theory to obtain full flow fields is described elsewhere [5]. In the next section we compare the results from OS calculations with results from triple-deck theory and with experimental measurements.

4 Experimental Results and comparison with theory

Experiments were conducted on a flat plate in the Low Turbulence Wind Tunnel in the Engineering Department of the University of Cambridge. A disturbance was generated in the boundary layer using an oscillating blade, encased in a rubber mould, which was inserted into a slot machined into the plate 400mm downstream of the leading-edge. The slot measured 145mm in the spanwise direction and 5mm in the streamwise direction. Experimental measurements together with numerical investigation of three-dimensional disturbance fields demonstrated that the disturbance field along the centreline of the oscillator was essentially two-dimensional over the downstream ranges considered. The base flow was set to conform as closely as possible to the Blasius boundary layer profile with zero pressure gradient. The resulting oscillating flow field was then surveyed with a single hot-wire. Experiments were carried out at nominal speeds between 17.5 and 18.3 m/sec to keep the Reynolds number based on displacement $Re_\delta = 1224$ constant. This gives a triple-deck parameter $\epsilon = 0.18$, and a non-dimensional hump width of 1.8 on a triple-deck scaling. The maximum height of the hump was approximately 0.01mm, corresponding to a height relative to the inner layer of 0.184, which suggests that linearised theory should be sufficient. Initially the surface perturbation was forced at a frequency of 2Hz, corresponding to the quasi-steady case since $\Omega = 0.01$. In Figure 2 the horizontal perturbation velocity is plotted, with the streamwise coordinate measured in centimetres from the leading edge, and the wall normal distance expressed in terms of the dimensionless boundary layer coordinate, $\eta = y\sqrt{U_\infty/2x\nu}$. Throughout most of the boundary layer, the velocity perturbation is in the opposite direction to the mean flow and assumes a double lobed structure over the wall hump. This structure is related to the appearance of a small region of positive velocity perturbation directly above the surface roughness element, further away from the wall but still in the main part of the boundary layer.

The perturbation velocity calculated using Orr-Sommerfeld theory for the wind tunnel parameter values is plotted in Figure 3. Clearly there is very good agreement with the experimental measurements. The minor quantitative difference in velocity levels is almost certainly attributable to the difficulty in measuring accurately the amplitude of the oscillating wall element. This good agreement is hardly surprising since the significant assumptions of Orr-Sommerfeld theory are satisfied for the experimental conditions under consideration, because non-parallel and non-linear effects are both small. Comparison with triple-deck predictions is more interesting. The velocity field from considering only the leading-order term, u_0, is shown in Figure 4, while the effect of including the first correction term (i.e. $u_0 + \epsilon u_1$) is shown in Figure 5. Here the lower layer has been patched to the main layer at $\eta = 0.8$. Clearly for the parameter range under consideration ($\epsilon = 0.18$), agreement with experiment is considerably improved with the inclusion of the first correction term. In particular, the region of positive velocity perturbation in the main part of the boundary layer appears as part of the first correction term, being absent from the leading-order solution.

5 Discussion

Comparison of triple-deck solutions with Orr-Sommerfeld theory and experimental measurements shows that triple-deck theory can be successfully used for quantitative predictions of flow fields, rather than merely as an indication of the scalings significant for a particular physical problem. Having demonstrated the applicability of asymptotic methods for the simple case considered here, asymptotic results for situations of more physical interest may possibly be used with renewed confidence, without the need for such rigorous comparison with experiment (*eg* the analysis of free-stream interaction with a fixed surface roughness, which is closely related to the present problem [6]).

A robust method is described for dealing with Fourier inversions involving singularities, which arise when solving the linearised triple-deck equations. As previously illustrated in Figure 1, results are obtained using this method for a forcing frequency above the neutral frequency. The results show good qualitative agreement with preliminary wind-tunnel measurements, though careful experimental investigation has been hampered by problems of plate vibration at the higher forcing frequencies. The same method for obtaining flow-field predictions from triple-deck theory has been used for three-dimensional disturbances. The lower-deck equations are solved in Fourier space and at each span-wise wave number, l, the pole in the integrand $k(l, \Omega)$ is calculated and then eliminated using convolutions. Results can then be compared with asymptotic far-field results [7] and with measurements from wind tunnel tests currently in progress.

Acknowledgements

The authors wish to thank Mr. S. Bake and Dr. Shaikh for providing preliminary results from experiments currently in progress.

References

[1] Stewartson, K., 1969, "On the flow near the trailing edge of a flat plate," *Mathematika* **16**, 106–121.

[2] Smith, F.T., 1979, "On the non-parallel flow stability of the Blasius boundary layer," *Proc. R. Soc. Lond.* A**366**, 91–109.

[3] Duck, P.W., 1985, "Laminar-flow over unsteady humps–the formation of waves," *J. Fluid Mech.* **160**, 465–498.

[4] Cowley, S.J. & Wu, X. , 1994, "Asymptotic approaches to transition modelling," *AGARD Report.*

[5] Gaster, M. & Sen Gupta, T.K., 1993, "The generation of disturbances in a boundary layer by wall perturbations: the vibrating ribbon revisited once more," In *Instabilities and Turbulence in Engineering Flows* ed. D.E. Ashpis *et al.* Kluwer, 1993.

[6] Goldstein M.E. 1985 "Scattering of acoustic waves into Tollmien-Schlichting waves by small streamwise variations in surface geometry," *J. Fluid Mech.* **154**, 509–529.

[7] Choudhari, M. & Kerschen, E.J., 1990, "Instability wave patterns generated by interaction of sound waves with three-dimensional wall suction or roughness," AIAA–90–0119.

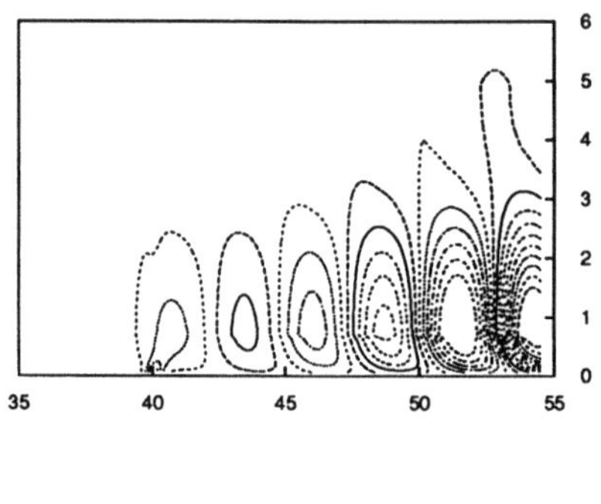

Figure 1: Horizontal perturbation velocity field predicted by the first two terms of triple-deck theory for a two-dimensional hump oscillating at a frequency above the neutral frequency.

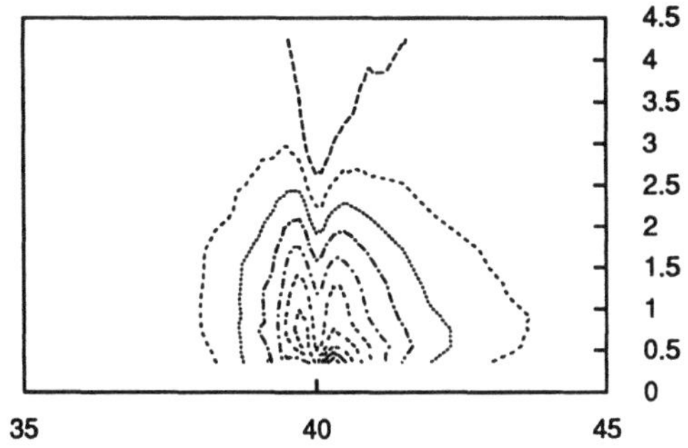

Figure 2: Experimental measurements of horizontal perturbation velocities for a 2Hz oscillating hump. Contours of u'/U_∞ are plotted at levels of -20×10^{-4} to 2×10^{-4} at increments of 2×10^{-4}.

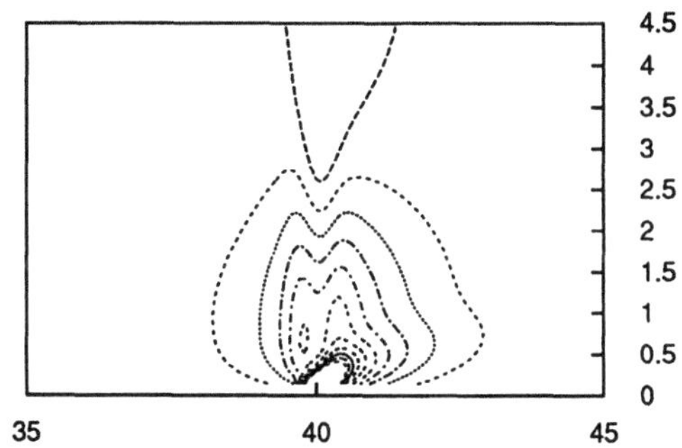

Figure 3: Orr-Sommerfeld calculations corresponding to the experimental measurements illustrated in Figure 2.

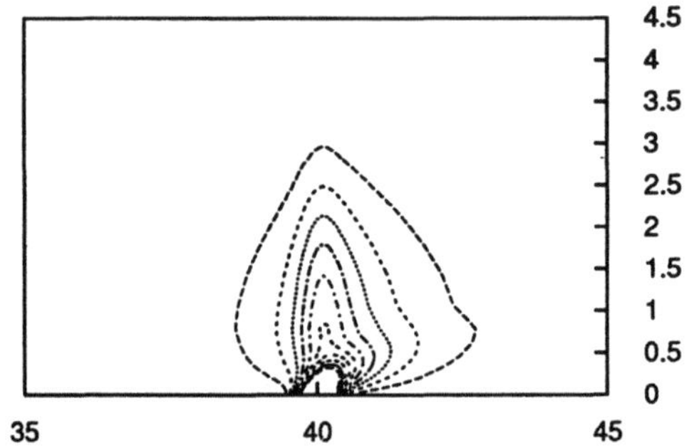

Figure 4: Leading-order triple-deck calculations corresponding to the experimental measurements illustrated in Figure 2. For details see the caption to Figure 2.

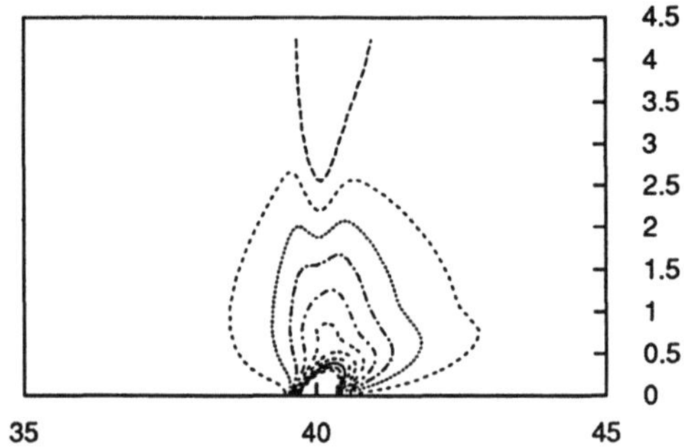

Figure 5: Effect of including the first correction to leading-order triple-deck.

On Excitation of Unstable Boundary Layer Waves by Freestream Sound

E.S. Holm
Department of Mechanical Engineering, University of Wyoming
Laramie, WY, USA

V.S. Kosorygin
Institute of Theoretical and Applied Mechanics SB RAS
Novosibirsk 630090, Russia

Summary

The response of boundary layers under the global favourable, "zero", and unfavorable longitudinal pressure gradients as well as the local variations in the pressure distribution over a flat plate model with respect to an external acoustic disturbance is investigated. The forcing frequencies is chosen to provoke the highest response. The initial Tollmien-Schlichting wave amplitudes in the region of their appearance are directly measured. The boundary layer domain having real influence on a formation of the initial amplitude is detected.

Key words : Laminar boundary layer, Pressure gradient, Sound, Receptivity, Direct measurements

1. Introduction

It is well known at low subsonic velocity the three types disturbances in natural environment affect on a location of laminar-turbulent boundary layer transition: (1)free-stream sound, (2)free-stream turbulence; and (3)vibration of model. The first two kinds of disturbances are most interesting for understanding of physics of transition process and practical application, third one has probably some effect on the location of transition at relatively large vibration amplitudes especially when the unfavorable pressure gradients exist near to the leading edge or the vibrations are localized close to the lower branch of neutral curve. The appearance of most unstable Tollmien-Schlichting (TS) waves responsible for the natural laminar-turbulent transition and which have very low frequency parameters (F= 30-40*10^{-6}) is determined by the receptivity of boundary layers with respect to the external disturbances. Receptivity mechanisms cause environmental disturbances to enter the boundary layer and create the initial amplitudes of the unstable waves. Receptivity has many different paths through which to introduce the disturbances into the boundary layer. Any one or a combination of these may lead to the unstable waves.

In case of acoustic environment the coupling between the long-wavelength sound wave and short-wavelength TS wave occurs when the boundary layer is required to adjust locally [1-6,9]. For the simplest situation in the flat plate experiments this can occur at four positions: the leading edge, the discontinuity in the surface curvature occurring at the flat plate and leading edge junction, the presence of pressure gradients in initial region of boundary layer development, and any surface non-homogeneities. Other routes of the T-S wave generation are connected with non-uniformity of boundary layer and boundary conditions [3,7,8]. At present, there is a possibility to calculate the receptivity coefficients for some particular cases [4,6,9,10] but there is no quite clear which route is most dangerous for laminar-turbulent transition.

The present experiments focus on the qualitative study of so-called "leading edge" receptivity of the smooth flat plate boundary layer with respect to free-stream sound. As a rule, there are simultaneously a few sources of receptivity in the wind tunnel experiments. Therefore a comparison of theoretical and experimental data is a difficult problem. A separation of main receptivity process is a complicated task due to the same reason. It seems the best way in experimental work is an attempt to create a situation when some of sources are turned off. Thereupon it is needed to modify the situation by an appropriate way and to investigate the turned on source of receptivity. One of the objectives of this experiment is to realize the scheme in a practice. Another goal of this study is to measure the extremely small initial amplitudes of unstable wave in boundary layer to detect a real location of its appearance. The measuring of these small initial values could lead to the direct experimental estimations of receptivity coefficients. The main objective of this experiment is to study the leading edge and pressure gradients effects in sound receptivity of laminar boundary layer.

2. Experimental equipment

The experiments are conducted in the Wind Tunnel T-324 at the Institute of Theoretical and Applied Mechanics of the Russian Academy of Sciences (Novosibirsk). This facility is a low-turbulence wind tunnel well suited for stability and receptivity experiments. The T-324 ITAM wind tunnel is a closed-loop with a heavy 1m x 1m x 4m steel test section. The flat plate model used in experiments is 0.996 m wide and 2.0 m long. The plate is manufactured from 6 mm aluminum alloy sheet. The leading edge is machined directly on the flat plate model. Its shape is a semicylinder with radius r = 1.5 mm. The thickness of leading edge is constant along 60 mm from the origin of model and equal to 3 mm. This design of model permits to move upstream a position of curvature discontinuity at the semicylinder/flat plate juncture and the region of pressure recovery upstream as far from the neutral curve as possible. The flat plate is horizontally mounted inside the test tunnel part. It contains one row of 25 static pressure orifices with 0.35 mm in diameter that is parallel to the center line of the plate and has a shift in the span direction. The static pressure measured by means of Pitot tube located at x = -800 mm is applied as reference pressure. The tail part of model is supplied by a flap driving outside of test section to adjust the position of stagnation line and pressure distribution in a vicinity of leading edge during the tests.

The acoustical disturbances are introduced in the stream by means of loudspeaker that have been removed downstream from the flat plate model. An 1"-inch microphone is employed for measuring of sound level created in the stream. The microphone has an anti-wind fairing and is set at the same, as the Pitot tube, reference cross section.

A 3D traversing system is used. The mechanical pieces of system is mounted inside the wind tunnel. The y-traverse unit rested on the flat plate surface to reduce a false signal due to the vibrations of a hot-wire probe in the flow with the velocity gradients. The y-traverse unit is attached by means of variable length stick to the x,z-traverse mechanism mounted downstream from the plate model.

Velocity measurements are conducted using a single hot-wire probe in-house made with 5 μm x 1.25 mm tungsten wire having a platinum coating. Both mean flow velocity and root-mean-squared (RMS) velocity oscillations are measured in the free stream and boundary layers placing the probe at different y-locations. The probe is connected to a Dantec 55M01 Main Unit which has a 55M10 Constant Temperature Anemometer (CTA). The output CTA signal is processed by a Dantec 55D10 Linearizer. The RMS velocity amplitudes are measured by passing the linearized signal through a Rohde & Schwarz frequency narrow band analyzer FAT1 with bandpass $\Delta f = 4$ Hz. The instantaneous output signal of analyzer is monitored by a two-beams oscilloscope to measure the phases. Second input and synchronization of the oscilloscope is supplied by the sine wave oscillator which also gave the sound frequency under investigation.

3. Scheme of experiments

The initial amplitudes of TS wave excited by the external disturbances should be determined when the problem of receptivity is being solved. Usually, the initial amplitude is attributed to lower branch of neutral stability curve. If the TS wave due to some receptivity mechanism appears upstream from the neutral stability curve an experimentalist or theorist could detect and watch for it up to the neutral curve to determine the desirable initial amplitude. In experimental study it is possible also to use the data got far downstream from the lower neutral stability curve where the TS-wave amplitude has large enough magnitude due to a linear amplification. Then this data can be extrapolated to the neutral curve employing the theoretical or found in experiment amplification factor. However in this case the peculiarities of TS wave rising can not be reveal. This is a shortcoming of similar kind of receptivity study. To find the real cause and region of TS wave appearance in some experimental environment the direct measurements are demanded. The direct measuring of initial TS amplitude near to its localization is difficult task because the amplitude is usually extremely small. Previous experiments show that its magnitude should be about a few millionth parts of free stream mean velocity or less.

In present experiments we shall conduct the direct recording of initial TS wave amplitudes and apply the next experimental scheme. First of all, some measurements will be taken in the boundary layer of smooth flat plate under "zero" pressure gradients and acoustic excitation of unstable wave. These results will give us the reference data for the following comparisons. Then the definite changes of global pressure distribution over the same flat plate will be only made. The same kinds of measurements will be repeated. A comparison of these sets of data should lead to some conclusions about a significance of global pressure gradients for receptivity process.

If the TS wave have not been detected anywhere upstream from the neutral curve during the first part of investigation, the boundary layer with the same global pressure distribution will be modified by the local pressure gradients having an appropriate scale. Due to the local gradients the acoustic external wave will give rise to a new *trial* TS wave. A place of trial wave rising can be easy changed. The trial wave will superimpose to the natural TS wave arising in the smooth flat plate boundary layer. Both waves are coherent. The probe placed far downstream should record some constructive or destructive changes, if any, [11,12]. Varying an onset position of trial TS wave along x-coordinate it is possible to detect a region of boundary layer that can exert an influence for a transformation of long wave acoustic disturbances into the short wave unstable TS waves.

4. Results

There are three means to modify a global pressure distribution over the flat plate. First one is a variation of its installation angle of attack. The pressure in some neighbourhood of leading edge can be adjusted by means of the flap. The utilizing of a false ceiling is third one. Each of them is applied in present experiments. The basic distributions of the pressure coefficients used in the study are collected in Fig.1. The distributions #1 and #3 correspond to a "zero" pressure gradient case. The distribution #2 and #4 are accordingly the favourable and adverse pressure gradient cases. Data of Fig.1 show that a strong pressure gradients is localized in the small vicinity of leading edge at $x < 50$ mm. It should be noted the initial pressure distribution, $x < 50$ mm, in the adverse gradient case is exactly the same as in the case of "zero" pressure gradients.

In present experiments a rectangular domain under study is limited by a front boundary AB that has the coordinates $x = 28$ mm and $y = 40$ mm, a back one CD with coordinates at $x = 1460$ mm and $y = 40$ mm, and an upper boundary BC is set at $28 \leq x \leq 1460$ mm and $y = 40$ mm. A bottom limit AD is the flat plate surface at $28 \leq x \leq 1460$ mm and $y = 0$ mm. The experiments have been conducted at the free stream velocity 18 m/s. A mean longitudinal velocity U as well as its oscillations u' are measured within this domain. The measuring of the mean velocity profiles in the boundary layers have shown the pressure gradients #2 and #4 produce the detectable alterations of basic flow in comparison with "zero" pressure gradient case. To select the most unstable frequencies the hot-wire probe is placed at x=1400 mm and y-coordinate corresponding to $U(y)/U_b = 0.4$ where U_b is the mean velocity at the boundary BC. The acoustic excitation with variable frequency is utilized to reveal the frequencies having the highest response. The frequencies are equal to 109 Hz (pressure distribution #3) and 110 Hz (p.d. #1).

When the unstable wave have been excited by a free stream sound wave with a frequency under consideration a total hot-wire signal is a combination of the

"Stokes" wave and the TS wave component. Turbulence and vibration of wire in a flow with velocity gradient create the additional contributions. If a traverse system has low rigidity and is not connected with the surface of model the vibration can produce significant contribution into the total signal. Scanning the boundary layer along x-coordinate it is easy to detect an appearance of TS wave. The Stokes wave has a wavelength of the same order as free stream acoustic wave (for f=110 Hz λ_a= 3.14 m) whereas TS wave has very short wavelength λ_{ts} (about 54 mm in present experiments). Short-wave oscillations of total signal RMS amplitude at some place of boundary layer testify to the TS wave rising.

Linear stability theory predicts the first neutral point position in Blasius boundary layer at Reynolds numbers R = 755 for frequency parameter F = $34.6*10^{-6}$ (f = 110 Hz, p.d.#1) and R = 769 for F = $33.6*10^{-6}$ (f = 109 Hz, p.d.#3). Taking into account experimental conditions these values result accordingly in the coordinates of neutral points for physical plane at x = 0.512 m and x = 0.514 m.

Experiment #1. In this trail the hot-wire probe is moved inside the smooth flat plate boundary layer at constant relative mean velocity $U(y)/U_b$ = 0.4 along x-coordinate. The data obtained is plotted in Fig.2. RMS amplitude of total hot-wire signal begins to oscillate downstream from the coordinate x = 460 mm (f = 109 Hz) and x = 480 mm (f = 110 Hz) approximately. A spatial scale of oscillations in boundary layer is much less than at upper limit of domain under study. The distance between nearest minima or maxima gives an estimation for the TS wavelength and a half of fluctuation swing can be applied to get an evaluation for RMS amplitude of TS wave. The data in Fig.2 show that the region of initial amplitude formation is placed near to the lower branch of neutral curve. Upstream from this region we could not detect any systematic spatial variations in the total hot-wire signal.

A magnitude of initial TS wave amplitude have been obtained using the first spatial oscillation of total signal. As preliminary result the RMS amplitudes u'/U_b are equal to $8*10^{-6}$ for f = 109 Hz and L_a= 96 dB and $5*10^{-6}$ for f = 110 Hz and L_a= 96 dB, where L_a is a pressure sound level at the reference point. If receptivity coefficient is determined as a ratio of the initial TS wave amplitude to the free stream oscillations amplitude at the upper limit of domain at the same x-coordinate we shall obtain its values 0.048 (f = 109 Hz) and 0.034 (f = 110 Hz). The wavelength of TS wave is about 54 mm. The same kind of experiment have been repeated with favourable and adverse pressure gradients. The data obtained lead to the receptivity coefficients 0.024 (f = 110 Hz, p.d.#2) and 0.074 (f = 109 Hz, p.d.#4). The TS rising position shifts slightly towards the leading edge for adverse gradients and downstream in opposite case.

Experiment #2. In this trial we shall try to determine a significance of TS wave generated in boundary layer between the leading edge and the neutral curve for the receptivity process. The hot-wire probe is located within boundary layer at fixed position. The 2D roughness strip manufactured from mylar is used as a new source of receptivity. The strip has a width b = 25 mm and height h = 20 μm. Inasmuch as the trial TS wave mentioned above and the natural TS wave in the smooth flat plate boundary layer is excited by the same acoustic field they will be coherent. A probe placed anywhere downstream from the lower neutral curve should indicate the constructive/destructive wave superposition. Moving the roughness from the neutral curve towards the leading edge we shall observe some changes in boundary layer. The results of this trial are shown in Fig.3. The dark circles in this chart correspond to the measurements when the roughness is set on the flat plate at different position x_h. The open square is the reference magnitude of oscillations in the same point of boundary layer without roughness. It follows from the data the influence of 2D roughness element decrease when the distance from the neutral curve is increased. Starting from some distance the influence disappears. What does it mean? It means that the trial wave, if any, must strongly decay near to the leading edge and does not make any contribution into the initial TS wave amplitude. To argue this statement it is demanded to get an evidence of TS wave existing near to the roughness. Desirable measurements have been done, Fig.4a,b. These data prove the presence of TS wave. It can be seen the less distance from the leading edge the larger TS wave amplitude and stronger decay. This trail have been repeated under adverse gradients (p.d.#4). Results point on the sprawling of region of probable influence towards the leading edge.

Conclusions

1. The region of TS wave appearance under acoustic excitation is localized near to the lower neutral curve in the "zero" pressure gradients boundary layer. The direct measuring of TS wave amplitudes in area of TS wave appearance is possible. A variation in definite range of global pressure gradients results to a weak shift of location of TS wave rising. The adverse pressure gradients lead to the amplification of initial TS wave amplitude and the favourable pressure gradients attenuate it.

2. The boundary layer domain which could affect on the formation of initial TS wave amplitude is detected. This domain locates on some distance downstream from the leading edge. It is also correct when the boundary layer is subjected to the global adverse pressure gradients starting directly from the leading edge. Despite the unfavorable pressure gradients the TS wave could not be detected in nearest vicinity of smooth flat plate leading edge but due to them the domain of probable influence expands towards the leading edge.

Acknowledgments

The authors are indebted to Messrs. I.P. Nosyrev and N.I. Rudenko for their efforts during the preparation of equipment and conducting of these experiments. This work is supported by Boeing Operations International, Inc. of Agreement #106 between Boeing and Institute of the Theoretical and Applied Mechanics SB RAS.

References

[1] Aizin, L.B. and Polyakov, N.Ph.: Acoustic Generation of Tollmien-Schlichting Waves over Local Unevenness of Surface Immersed in Stream,

Preprint 17, Acad. Nauk USSR Sib. Div., Inst. Theor.&Appl. Mech., Novosibirsk (1979) (in Russian).
[2] Kachanov, Yu.S., Kozlov, V.V. and Levchenko, B.Ya.: Beginning of Turbulence in Boundary Layer, Nauka Publ., Novosibirsk (1982) (in Russian).
[3] Nishioka, M. and Morkovin, M.V.: Boundary Layer Receptivity to Unsteady Pressure Gradients: Experiments and Overview, J. Fluid Mech. 171 (1986) 219-261.
[4] Heinrich, R.A., Chrouhary, M. and Kerschen E.J.: A Comparison of Boundary Layer Receptivity Mechanisms, AIAA Paper 88-3758.
[5] Kerschen, E.J.: Boundary Layer Rreceptivity, Apll.Mech. Rev. 43 (1990) 5 2 S152-S157.
[6] Crouch, J.D.: Localized Receptivity of Boundary Layer, Phys. Fluids A4(7) (1992) 1408-1414.
[7] Jigulev, V.N. and Tumin, A.M.: Beginning of Turbulence. Dynamics Theory of Excitation and Development of Instabilities in Boundary Layer, Nauka Publ., Novosibirsk (1987) (in Russian).
[8] Crouch, J.D.: Non-localized Receptivity of Boundary Layers, J. Fluid Mech. 244 (1992) 567-581.
[9] Goldstein, M.E.: Scattering of Acoustic Waves into Tollmien-Schlichting Waves by Small Streamwise Variation in Surface Geometry, J. Fluid Mech. 154 (1986) 509-529.
[10] Lin, N., Buter, T.A. and Reed, H.L.: Nonlinear Instability of Nonparallel Flow, IUTAM Symp. Abstracts., Clarkson University, Potsdam, NY USA (1993) 26-28.
[11] Kosorygin, V.S. and Polyakov, N.Ph.: Auto Destruction of Unstable Boundary Layer Waves, Preprint 11-90, Acad. Nauk USSR Sib. Div., Inst. Theor.&Appl. Mech., Novosibirsk (1990) (in Russian). See also: Bulletin of Amer. Phys. Soc. 35 (1990) 10 2291.
[12] Kosorygin, V.S., Radeztsky, R.H.,Jr. and Saric, W.S.: Boundary Layer Receptivity and Control: Free-stream Sound, 2-D Roughness, Preprint 19-92, Russian Acad. Nauk Sib. Div., Inst. Theor.&Appl. Mech., Novosibirsk (1992) (in Russian).

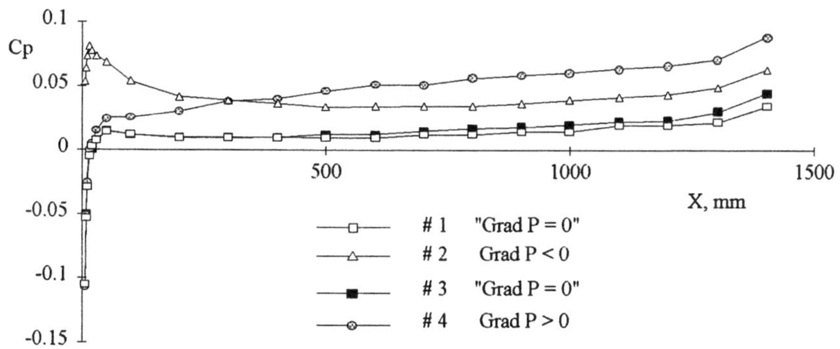

Fig.1 Distributions of pressure coefficients

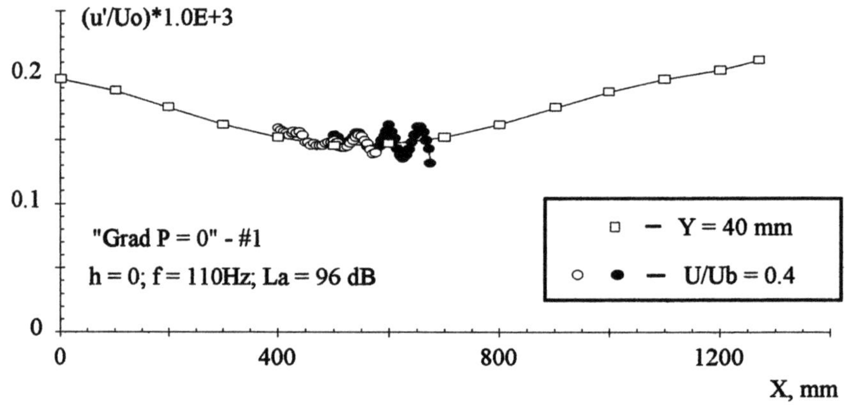

Fig.2. Variations of total amplitude of longitudinal velocity oscillations

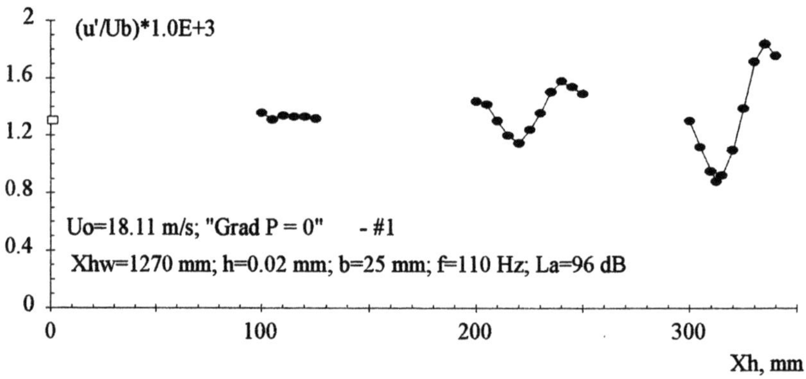

Fig.3. Velocity oscillations in the boundary layer with roughness. $U/U_b = 0.4$

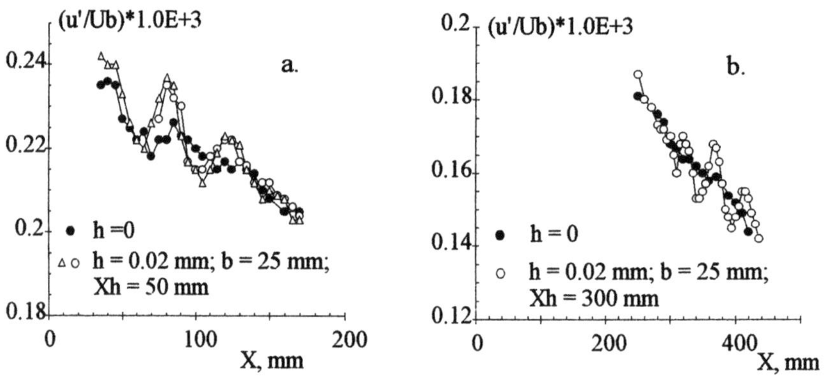

Fig.4. The velocity oscillations with and without roughness: U/U_b=0.4; f=110 Hz; L_a = 96 dB; "Grad P = 0" - #1.

The Resonant-Triad Interaction in Boundary-Layer Transition

M.J. Jennings and P.A.Stewart,
Department of Applied Mathematics and Theoretical Physics
University of Cambridge, Silver Street, Cambridge CB3 9EW, UK

Xuesong Wu
Department of Mathematics, Imperial College,
180 Queens Gate, London SW7 2BZ, UK

Summary

The growth of Tollmien-Schlichting (T-S) waves in a Blasius boundary layer is studied using high-Reynolds-number asymptotic methods. Attention is focused on weakly nonlinear *resonant-triad* interactions. Most previous work has concentrated on the interaction of a two-dimensional wave with a pair of oblique subharmonics. Here, we consider non-aligned triads in which all three modes may be oblique. Moreover, we allow the waves to be weakly modulated in time and in the streamwise and spanwise directions, so that localised wave-packets can be described. Analytic and numerical solutions of the governing amplitude equations indicate that the two most oblique modes may grow super-exponentially for a time, as a result of quadratic interaction with the third mode, in both the aligned and non-aligned cases. Subsequently, however, cubic ('wave-vortex') interactions may become significant, and this stage of development may be quite different in the two cases.

Key words: Resonant triad, Wave-vortex interaction, Super-exponential growth, Transition, Turbulence.

1. Introduction

Laminar-turbulent transition in a boundary layer is often initiated by instability. The initial disturbance may be a two-dimensional mode, and be described by linear instability theory. Further downstream, however, three-dimensional disturbances grow faster than predicted by the linear theory. Various non-linear mechanisms have been suggested to account for this. For example, the *resonant-triad interaction* theory ([1], [2]) predicts the dominant three-dimensional modes to be subharmonics of the two-dimensional mode. This is observed in some experiments where the initial disturbance amplitude is very small, e.g. [3], [4]. Other proposed mechanisms (e.g. [5]) focus on the presence of *streamwise vortex* structures, and their effect on the development of the instabilities.

An asymptotic description of the resonant-triad mechanism, based on weakly nonlinear theory in conjunction with a high-Reynolds-number assumption, has recently been given by [6], [7] and [8]. These studies considered the interaction between a two-dimensional mode and a pair of oblique subharmonics propagating at equal and opposite angles to the flow. The waves were assumed to be introduced with fixed frequency, and to be modulated in the streamwise direction only. The amplitudes were fixed so that (weakly) nonlinear effects become significant in the 'upper-branch' scaling régime. It was found that the oblique modes can grow *super-exponentially*, as a result of interaction with the planar mode. However, nonlinear interaction between the oblique modes generates a streamwise vortex, which acts to reduce the growth-rate of the instability.

Here, we extend the aforementioned analyses to the more general resonant triad identified in [9], in which all three modes may be oblique. Attention is principally focused, however, on the case when one mode is *nearly* two-dimensional. We also allow for the effects of slow spanwise and/or temporal modulation.

2. Resonant Triads

We take a Blasius boundary layer as the basic flow, and consider a pressure perturbation, P, of the form

$$\frac{P}{\rho U_\infty^2} \sim \epsilon\, P_0(X,Z,T)\, E_0 + \delta_1\, P_1(X,Z,T)\, E_1 + \delta_2\, P_2(X,Z,T)\, E_2 \quad (2.1)$$

$$+ \quad \text{complex conjugate} \quad + \quad \text{higher-order terms}\ , \quad (2.2)$$

where the E_i represent the rapid variation of the T-S waves,

$$E_i = \exp(i\alpha_i x + i\beta_i z - i\omega_i t)\ , \quad (2.3)$$

and X, Z and T represent slow 'modulation' length- and time-scales.[1] Linear theory gives the dispersion relation

$$\omega_i = \alpha_i(\alpha_i^2 + \beta_i^2)^{\frac{1}{2}}\ . \quad (2.4)$$

A weakly nonlinear *resonant-triad* interaction can take place if

$$E_1 E_2 = E_0\ , \quad (2.5)$$

i.e. if

$$\alpha_1 + \alpha_2 = \alpha_0\ , \quad \beta_1 + \beta_2 = \beta_0\ , \quad \omega_1 + \omega_2 = \omega_0\ . \quad (2.6)$$

This is obtained for

$$\alpha_0 = c\cos\phi\,, \qquad \beta_0 = c\sin\phi\,, \qquad \omega_0 = c^2\cos\phi\,,$$

[1]The independent variables have been scaled as appropriate for the upper-branch régime ([10]).

$$\alpha_1 = c\cos(\phi - \frac{\pi}{3}), \qquad \beta_1 = c\sin(\phi - \frac{\pi}{3}), \qquad \omega_1 = c^2\cos(\phi - \frac{\pi}{3}),$$
$$\alpha_2 = c\cos(\phi + \frac{\pi}{3}), \qquad \beta_2 = c\sin(\phi + \frac{\pi}{3}), \qquad \omega_2 = c^2\cos(\phi + \frac{\pi}{3}),$$

where c and ϕ are arbitrary [9]. The three modes have the same phase speed

$$\frac{\omega_i}{\alpha_i} = c\,, \tag{2.7}$$

and hence a common *critical layer.*

The dominant nonlinear interactions take place in the critical layer, and can significantly affect the evolution of the disturbance when the amplitude of the 'central' mode scales as

$$\epsilon \sim Re^{-\frac{11}{20}}\,, \tag{2.8}$$

where Re is the Reynolds number based on distance from the leading edge ([8]). We assume for the moment that δ_1 and δ_2 are infinitesimal. The amplitude equations then take the form

$$\left(\frac{\partial}{\partial T} + v_{gx}^{(0)}\frac{\partial}{\partial X} + v_{gz}^{(0)}\frac{\partial}{\partial Z}\right) P_0 = g_0 P_0\,, \tag{2.9}$$
$$\left(\frac{\partial}{\partial T} + v_{gx}^{(1)}\frac{\partial}{\partial X} + v_{gz}^{(1)}\frac{\partial}{\partial Z}\right) P_1 = g_1 P_1 + ih_1 P_0 P_2^*\,, \tag{2.10}$$
$$\left(\frac{\partial}{\partial T} + v_{gx}^{(2)}\frac{\partial}{\partial X} + v_{gz}^{(2)}\frac{\partial}{\partial Z}\right) P_2 = g_2 P_2 + ih_2 P_0 P_1^*\,. \tag{2.11}$$

The coefficients h_1, h_2 are real and positive, as are the group velocities $v_{gx}^{(i)}$.

In the case where the P_i depend on X only, an explicit analytic solution can be obtained (cf. [11]). If

$$P_0 = A_0 \exp\left(\frac{g_o}{v_{gx}^{(0)}} X\right)\,, \tag{2.12}$$

then at large X,

$$P_{1,2} \sim \exp\left(se^{GX} + \ldots\right)\,, \quad G = \frac{Re(g_0)}{v_{gx}^{(0)}}\,, \quad s = \frac{h_1 h_2 |A_0|^2}{v_{gx}^{(1)} v_{gx.}^{(2)}}\,; \tag{2.13}$$

i.e. the 'side' modes grow *super-exponentially.* This is also observed in numerical solutions with both X- and Z-dependence present.

3. Wave-Vortex Interactions

When the outside modes become nonlinear, the effect on the solution to the problem depends on the magnitude of the skewing angle ϕ. We restrict attention to *small* ϕ, i.e. a nearly two-dimensional mode interacting with a pair of near-subharmonics propagating at angles of approximately $\pm\pi/3$ to the flow. This is possibly the case of most relevance to experiments.

A distinguished scaling has $\phi = O(Re^{-\frac{1}{10}})$. Thus, we consider a triad of modes with phase terms

$$E_0 = \exp\left(2i\alpha(x-ct)\right)\exp\left(iRe^{-\frac{1}{10}}.\frac{2Kz}{\sqrt{3}}\right), \tag{3.14}$$

$$E_1 = \exp\left(i\alpha(x-\sqrt{3}z-ct)\right)\exp\left(iRe^{-\frac{1}{10}}K(x+\frac{z}{\sqrt{3}}-ct)\right), \tag{3.15}$$

$$E_2 = \exp\left(i\alpha(x+\sqrt{3}z-ct)\right)\exp\left(-iRe^{-\frac{1}{10}}K(x-\frac{z}{\sqrt{3}}-ct)\right), \tag{3.16}$$

where the skewness parameter K may be taken to be positive. The oblique modes E_1, E_2 generate a *streamwise vortex* proportional to

$$E_1E_2^* \equiv \exp\left(2\sqrt{3}i\alpha z\right)\exp\left(2iRe^{-\frac{1}{10}}K(x-ct)\right), \tag{3.17}$$

and this leads to a coupling between the oblique modes *via* the relations

$$(E_1E_2^*)E_2 = E_1\ , \quad (E_1E_2^*)^*E_1 = E_2\ . \tag{3.18}$$

These nonlinear 'wave-vortex' interactions occur predominantly in the neighbourhood of the critical layer, and become significant when

$$\delta_1 \sim \delta_2 \sim Re^{-\frac{9}{20}} \tag{3.19}$$

([8]). After a scaling, the amplitude equations can then be written as

$$\left(\frac{1}{2}\frac{\partial}{\partial T}+\frac{\partial}{\partial X}\right)P_0 = \frac{(1-i)}{\sqrt{2}}P_0\ , \tag{3.20}$$

$$\left(\frac{\partial}{\partial T}+\frac{\partial}{\partial X}+\frac{\partial}{\partial Z}\right)P_1 = (1-i)P_1+iP_0P_2^*+iP_2Q\ , \tag{3.21}$$

$$\left(\frac{\partial}{\partial T}+\frac{\partial}{\partial X}-\frac{\partial}{\partial Z}\right)P_2 = (1-i)P_2+iP_0P_1^*+iP_1Q^*+iKP_1R^*\ , \tag{3.22}$$

$$\left(\frac{5}{4}\frac{\partial}{\partial T}+\frac{\partial}{\partial X}\right)Q = \left(\frac{5}{4}\frac{\partial}{\partial T}+\frac{\partial}{\partial X}\right)^5 R = P_1P_2^*\ , \tag{3.23}$$

where the terms Q and R represent the streamwise vortex.

4. Numerical solutions

Numerical solutions of (2.9)-(2.11) and (3.20)-(3.23) have been obtained with dependence on X and Z but not T. In each case we allow initially separated wavepackets, each containing disturbances of a *single* mode, to pass through each other and interact non-linearly as they do so.

4.1 Interactions in the unskewed ($K = 0$) case

When $K = 0$, and there is no two-dimensional mode present (i.e. $P_0 = 0$), it can be shown that

$$\frac{\partial}{\partial X}\int_{-\infty}^{\infty}\left(|P_1|^2 + |P_1|^2\right) dZ = 2\int_{-\infty}^{\infty}\left(|P_1|^2 + |P_1|^2\right) dZ \; , \tag{4.24}$$

i.e. the total 'energy' grows as in linear theory. Numerical solutions (see Fig. 1) show the development of relatively short-scaled oscillations in both X and Z. When the central two dimensional mode is also present (see Fig. 2), there is a period of super-exponential growth arising from the resonant-triad interaction. This is suppressed, however, when the wave-vortex interaction comes into play. The subsequent growth of the oblique modes is merely exponential, with growth-rate as in linear theory, and accompanied by rapid oscillations.

4.2 Interactions in the skewed ($K \neq 0$) case

In the case in which the modes are slightly skewed, there is the possibility of faster 'energy' growth than occurs in the unskewed case. This can happen even when the quasi-two-dimensional 'central' mode is absent , as shown in Fig. 3. When the central mode is present (Fig. 4), the resonant-triad interaction once again gives rise to superexponential growth of the oblique modes , and this does not appear to revert to linear growth as the oblique modes become nonlinear.

5. Conclusions

We have extended the resonant-triad analysis of [6], [7] , [8] to include the effects of

1. spanwise (and temporal) modulation, and
2. the triad not being aligned with the flow.

The former does not give rise, it would seem, to qualitatively new behaviour. On the other hand, after the wave-vortex interaction comes into play, a slightly skewed triad may show quite different development from the exactly aligned triad. Even when the (quasi-)2-D mode is absent, a pair of *skewed* oblique modes may exhibit faster-than-exponential growth.

We have only presented results for cases when the resonance condition (2.6) is exactly satisfied. A consequence of this is that the modes share a common critical layer, and thus a particularly efficient nonlinear interaction can take place. It is also possible to consider *detuning* from exact resonance. In that case, the the resonant-triad interaction may still give rise to superexponential growth, but with reduced growth-rate. Likewise, the skewed wave-vortex interaction appears to have a less powerful effect when the oblique modes have (slightly) different phase speeds, and so their critical layers do not coincide. We believe, therefore, that there may be particular interest in the study of interactions between modes which are 'phase-locked', in the sense that they have the same phase speed (see also the paper by Wu & Stewart in these proceedings).

References

[1] Raetz, G.S. 1959 A new theory of the cause of transition in fluid flows. *Norair Rep.* **NOR-59-383**, Hawthorne, California.

[2] Craik, A.D.D. 1971 Non-linear resonant instability in boundary layers. *J. Fluid Mech.* **50**, 393–413.

[3] Kachanov, Yu.S. & Levchenko, V.Ya. 1984 The resonant interaction of disturbances at laminar-turbulent transition in a boundary layer. *J. Fluid Mech.* **138**, 209–247.

[4] Saric, W.S. & Thomas, A.S.W. 1984 Experiments on subharmonic route to turbulence in boundary layers. In *Turbulence and Chaotic Phenomena in Fluids* Ed. Tatsumi, T. 117-22, Amsterdam: North-Holland

[5] Jang, P.S., Benney, D.J. & Gran, R.L. 1986 On the origin of streamwise vortices in a turbulent boundary layer. *J. Fluid Mech.* **169**, 109–123.

[6] Goldstein, M.E. & Lee, S.S. 1992 Fully coupled resonant-triad interaction in an adverse-pressure-gradient boundary layer. *J. Fluid Mech.* **245**, 523–551.

[7] Wu, X, 1993 On critical-layer and diffusion-layer nonlinearity in the three-dimensional stage of boundary-layer transition. *Proc. Roy. Soc. Lond.* **A443**, 95–106.

[8] Mankbadi, R.R., Wu, X. & Lee, S.S. 1993 A critical-layer analysis of the resonant triad in Blasius boundary-layer transition: nonlinear interactions *J. Fluid Mech.* **256**, 85–106.

[9] Khokhlov, A.P. 1991 The temporal evolution of Tollmien-Schlichting waves near nonlinear resonance. Preprint.

[10] Bodonyi, R.J. & Smith, F.T. 1981 The upper-branch stability of the Blasius boundary layer, including non-parallel flow effects. *Proc. Roy. Soc. Lond.* **A375**, 65–92.

[11] Wundrow, D.W., Hultgren, L.S. & Goldstein, M.E. 1994 Interaction of oblique stability waves with a nonlinear planar wave. *J. Fluid Mech.* **262**, 343.

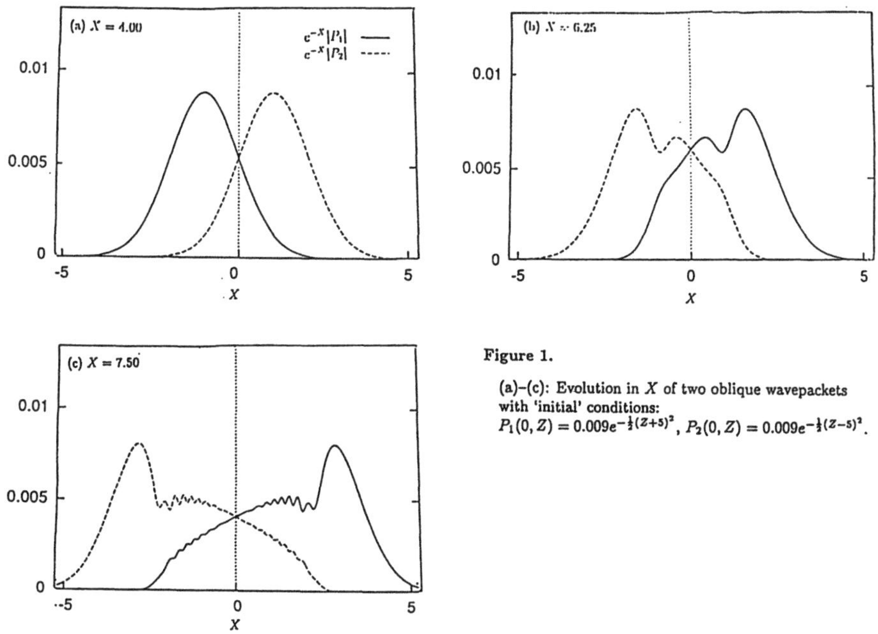

Figure 1.

(a)–(c): Evolution in X of two oblique wavepackets with 'initial' conditions:
$P_1(0, Z) = 0.009e^{-\frac{1}{2}(Z+5)^2}$, $P_2(0, Z) = 0.009e^{-\frac{1}{2}(Z-5)^2}$.

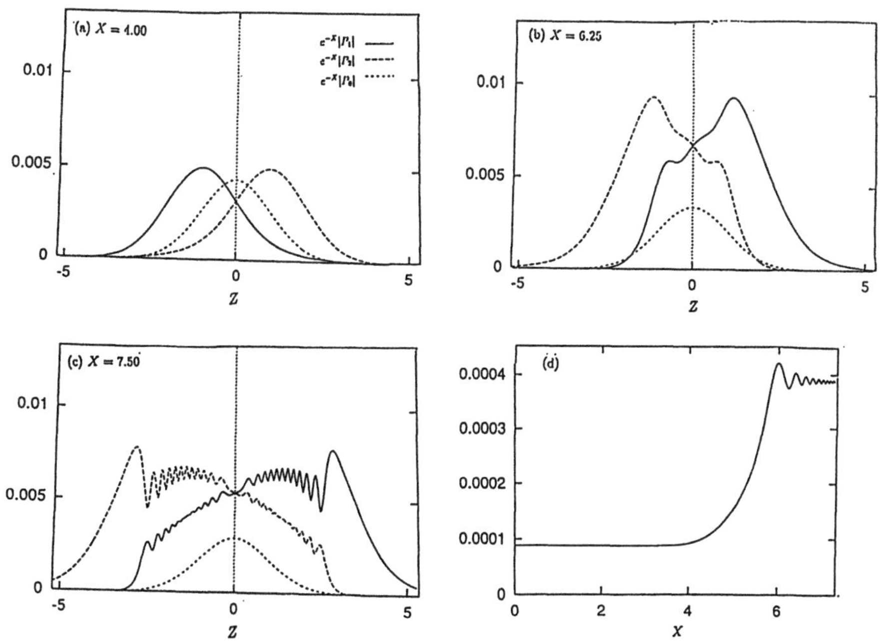

Figure 2. (a)–(c): Evolution in X of wavepackets with 'initial' conditions:
$P_1(0, Z) = 0.005e^{-\frac{1}{2}(Z+5)^2}$, $P_2(0, Z) = 0.005e^{-\frac{1}{2}(Z-5)^2}$, $P_0(0, Z) = 0.007e^{-\frac{1}{2}Z^2}$, for unskewed resonant triad.

(d) Evolution of $e^{-2X} \int_{-\infty}^{\infty} (|P_1|^2 + |P_2|^2)\, dZ$ for the same case.

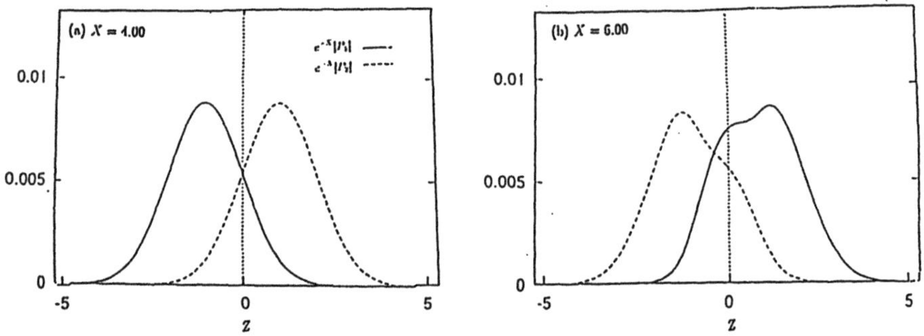

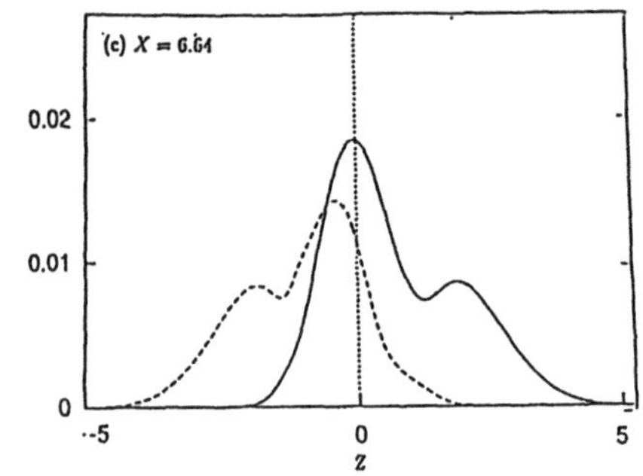

Figure 3.

(a)–(c): Evolution in X of two oblique wavepackets with skewing parameter $K = 0.6$ with 'initial' conditions: $P_1(0, Z) = 0.009e^{-\frac{1}{2}(Z+5)^2}$, $P_2(0, Z) = 0.009e^{-\frac{1}{2}(Z-5)^2}$.

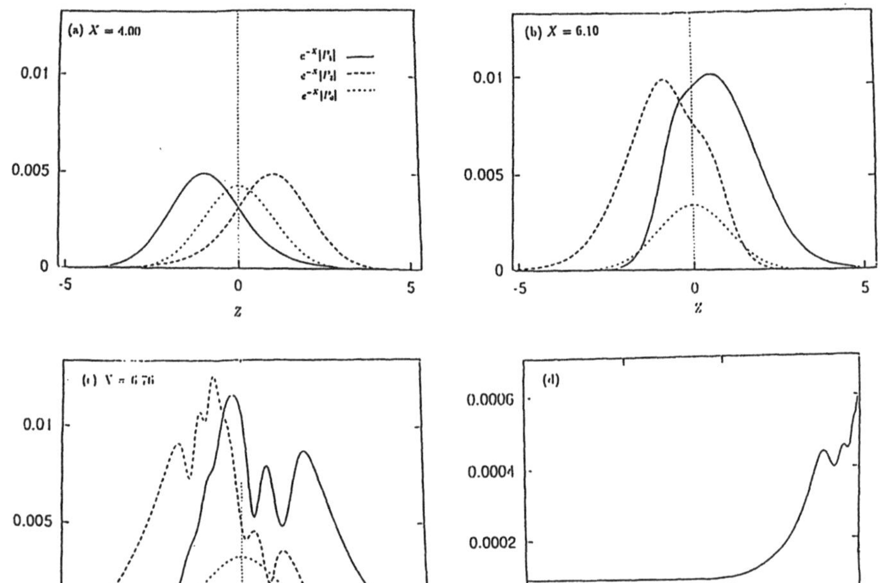

Figure 4. (a)–(c): Evolution in X of wavepackets with 'initial' conditions: $P_1(0, Z) = 0.005e^{-\frac{1}{2}(Z+5)^2}$, $P_2(0, Z) = 0.005e^{-\frac{1}{2}(Z-5)^2}$, $P_0(0, Z) = 0.007e^{-\frac{1}{2}Z^2}$, for skewed resonant-triad with $K = 0.6$.

(d) Evolution of $e^{-2X} \int_{-\infty}^{\infty} (|P_1|^2 + |P_2|^2)\, dZ$ for the same case.

Experimental Study of Cross-Flow Instability of a Swept-Wing Boundary Layer with Respect to Traveling Waves

V.R. Gaponenko, A.V. Ivanov, Y.S. Kachanov
Institute of Theoretical & Applied Mechanics, Novosibirsk, 630090, Russia

Summary

The paper is devoted to experimental study of the linear stability a 3D boundary layer on a model of a swept wing. Namely, the development of spatial wave packets (wave trains) generated by localized disturbance source is investigated. Decomposition of these wave trains to oblique harmonic (in time and space) modes is carried out. These wave packets are shown to consist mainly of the cross-flow instability modes. A complete set of stability and dispersion characteristics of the flow with respect to traveling waves inclined at various angles to the flow direction is obtained.

Key words : Swept wing, 3D laminar boundary layer, Cross-flow instability.

1. Introduction

The subject of the present study is the boundary layer on a swept wing in the region of the flow acceleration. One of the most important kind of instability in such a flow is the cross-flow instability, which is the main object of the present experiments. At present there are only several experimental investigations where some quantitative results on the linear stability of the swept-wing boundary layer to traveling waves were obtained. These experiments were conducted under so-called "natural" conditions when the initial disturbance spectrum is uncontrolled. In such experiments it is possible to obtain only information about the development of some most intensive fluctuations which are attributed usually to the most unstable modes. However, comparison of such kind of data with the properties of the most unstable modes predicted by the linear stability theory demonstrates sometimes a significant disagreement, in particular in the disturbance amplification rates.

In the present paper the problem of the 3D boundary-layer stability to traveling waves is studied at controlled disturbance conditions using a generator of the 3D instability modes. This method, together with a special procedure of Fourier decomposition of the generated wave-trains to harmonic (in time and space) oblique modes, gives possibility to obtain a complete set of stability characteristics of the flow with respect to each mode of the frequency-wavenumber spectrum.

2. Experimental procedure

To extract in a clean, refined state the cross-flow instability phenomenon only, and to obtain in an experiment the data which could be directly compared with the linear stability theory, we carried out an experimental simulation of both the mean swept-wing flow and the initial disturbance field. The mean flow on an initial section of a swept wing was simulated by means of a flow over a model of a swept plate with the pressure gradient induced by a contoured wall bump putted onto the test-section wall just above the plate. Both the plate and the bump had the same sweep angle χ = 25.0°. See [1] for more detail. The experiments were conducted at a low-turbulence wind-tunnel of the Institute of Theoretical and Applied Mechanics. The experimental model was mounted in the test section (1 x 1 x 4 m) at a zero attack angle. The turbulence level at the experimental conditions was about 0.065 % in the frequency range higher than 1 Hz. The mean-flow structure and the velocity disturbances were measured by means of a hot-wire anemometer. The main coordinate systems which were used in the present paper are the following. The system (x,z) is connected with the free stream direction with the vector $\boldsymbol{U}_\infty$ parallel to the x-coordinate. The axis z' is parallel to the leading edge of the model. The coordinate system (x^*,z^*) is local, such as the x^*-axis is directed along the vector $\boldsymbol{U}_0$ of the potential flow near the external edge of the boundary layer.

The disturbances were harmonic in time and introduced into the flow by means of an electro-magnetic generator which simulated localized surface vibrations. The source (36.0 mm in diameter) was mounted flush with the model surface at a distance x = 446 mm from the leading edge (see [2] for more detail). There were studied several disturbance frequencies f = 8.3, 12.15, 18.0, 25.0, 35.0 and 42.0 Hz. These frequencies correspond to nondimensional frequency parameters: $F \cdot 10^6$ = $2\pi f \nu / U_o^2 \cdot 10^6$ = 16.2, 23.8, 35.4, 48.8, 68.6, and 82.4, for x' = 528.8 mm (i.e. in the middle of the region of the stability measurements).

3. Mean flow structure

The characteristics of the 3D mean flow field over this model were measured carefully in both the potential flow and the boundary layer using, in particular, the X- and V-shaped hot-wire probes [1].

Potential flow. It was found that the characteristics of the potential flow over this model are independent, in practice, of the spanwise z'-coordinate in a region of the all main measurements (about z ± 100 mm from the central line). The experiments [1] showed also an independence (within experimental accuracy) of the potential flow structure from the free stream velocity U_∞ in the region (at least) from $U_\infty \approx 3$ to 11 m/s. The local sweep angle in the center of the region of main stability measurements (x' = 528.8 mm) was 22.85 degrees.

In the present experiments the free-stream velocity was U_∞ = 6.22 m/s. Within the region of the measurements (x' = 510.7 to 547.0 mm) the potential flow velocity $U_o(x)$ grows with x' in linear way and has the value U_o= 7.00 m/s at x' = 528.8 mm. The value of Hartree parameter β_H in this point is equal to 0.48.

Boundary layer. The 3D structure of the mean flow inside the boundary layer on the present model was also studied in detail in [1]. The normal-to-wall distributions of the U^*- and the W^*-components of the mean velocity vector and the yaw angle γ^* of the flow were measured and compared with calculations of A.V. Fyodorov for the swept-wing flow using experimental distribution of $U_o(x)$ only. A very good agreement is observed between the theory and experiment. Thus, we can conclude that the structure of the mean flow over the model used is almost completely adequate to a swept-wing flow in the region of the flow acceleration.

In the present studies a set of U-velocity profiles was also measured in the chosen flow regime. The profiles taken at various positions on the (x',z')-plane (Fig 1) demonstrate an independence of their shape from the spatial coordinates within the region of the stability measurements. It was not also found significant downstream and spanwise (along z' axis) dependences of the boundary-layer displacement thickness δ_1, momentum thickness δ_2 and the shape factor $H = \delta_1/\delta_2$. Their averaged values (within the region of the stability measurements) were: $\delta_1 = 1.19 \pm 0.04$ mm, $\delta_2 = 0.534 \pm 0.02$ mm, and $H = 2.24 \pm 0.06$. The local Reynolds number $Re = U_o\delta_1/\nu$ grew slowly downstream (mainly due to the growth of U_o) and was equal in average to $Re = 547$ (at $x' = 528.8$ mm).

Thus, we can conclude that within the region of the main stability measurements the flow over the experimental model was: (i) self-similar, (ii) uniform along the z'-coordinate, and (iii) almost uniform along the x'-coordinate.

4. Properties of wave trains

The levels of the disturbance excitation have been chosen very small to provide the conditions when the stability problem is linear and the normal-to-wall, streamwise, and spanwise distributions of the disturbance phase and normalized (by maxima) amplitudes are independent of the initial amplitude.

Normal-to-wall profiles. A set of y-profiles of the disturbance amplitude and phase within the wave train, measured at three downstream positions along the central line ($z = 0$), is shown in Fig. 2. Similar profiles were obtained at fixed x' for different spanwise positions and at fixed spatial position for different disturbance frequencies. It was found that the form of the disturbance profiles is almost self-similar. The positions of the amplitude maxima are almost independent of downstream and spanwise coordinate and depend very weakly on the frequency. The disturbance maxima are concentrated near $y/\delta_1 \approx 1.0$. This corresponds to the nondimensional mean flow velocity $U/U_o = 0.60$. The behaviour of the disturbance phases is found to be rather conservative near the amplitude maxima for all streamwise and spanwise positions. The form of the profiles, even before decomposition of the wave trains to oblique modes (see below), is very similar to eigenfunctions of the cross-flow instability waves predicted by theory [3]. These results give us possibility to choose the distance from the wall for all main stability measurements (for except a special study of the eigenfunctions of instability modes) to be equal to $y/\delta_1 \approx 1.0$, that corresponds to $U/U_o = 0.60$.

Downstream evolution. The disturbance amplitudes measured along the x-coordinate at $z = 0$ (Fig. 3a) have their local maxima near the source center and then develop downstream. The disturbance phases (Fig. 3b) are constant over the source and then start to grow according to almost linear laws. Their growth gates depend on the disturbance frequency.

Fig. 4 shows an example of the spanwise distributions of the disturbance amplitude and phase within the wave train, obtained for three downstream positions for f=18.0 Hz. In the central parts of these distributions the disturbance amplitude behaviour is rather complicated but the "tails" were found to attenuate exponentially. At the same time, the phase distributions are seen to behave themselves in a very conservative way. For all downstream positions and frequencies the disturbance phase grows along the z'-coordinate almost in a monotonous way, close to the linear one (for except a region lying approximately at $z' < -20$ mm). This behaviour can be interpreted as a propagation of a most intensive mode (contained within the wave train) along the z'-axis.

5. Properties of spectral modes and stability characteristics

Wavenumber spectra. An example of spanwise wavenumber spectra, obtained for f=18.0 Hz at three downstream positions, are shown in Fig. 5. It is seen that during the downstream development the two distinct peaks appear in the amplitude part of the spectra. They are localized around $|\beta^*\delta_1| \approx 0.13$ to 0.18. The same spectra, but shown as functions of the propagation angle θ^*, are presented in Fig. 6. The peaks, mentioned above, proved to be localized near rather large propagation angles with $|\theta^*|$ between 50 and 80° (these typical angles decrease with frequency and increase with downstream coordinate). The disturbance energy is mainly contained within these peaks, especially at late stages of development. However, these most intensive disturbances, as shown below, do not coincide with the most unstable modes.

Eigenfunctions. To determine them a special large set of measurements was carried out. This set consisted of an array of 12 spanwise distributions of the disturbance amplitude and phase measured at fixed frequency and x'-coordinate ($x' = 510.7$ mm) but at different distances from the wall. After spatial Fourier decomposition a set of eigenfunctions was obtained for various wave propagation angles. Some of them are shown in Fig. 7. It is seen that the position of the disturbance amplitude maxima is almost independent of the propagation angle θ^* and is localized near $y/\delta_1 = 1.0$. This observation is in agreement with stability theory and confirms that the choice of the distance from the wall for all main stability measurements was correct (see section 4).

Streamwise wavenumbers. Almost all results presented below were obtained in two sets of the measurements conducted in September 1992 (set 1) and in April 1993 (set 2). Dependences of streamwise wavenumber $\alpha_r^*\delta_1$ from $\beta^*\delta_1$ were obtained for all the frequencies studied. Five of them are shown in Fig. 8. It is seen that in the coordinate system (x^*,z^*) these dependences are almost symmetric and very smoothed for low frequencies. At the same time, at high frequencies there is a peak near $\beta^*\delta_1$ close to zero. The typical values of $\alpha_r^*\delta_1$ are seen to be by an order less than those of $\beta^*\delta_1$ for all frequencies.

Amplification rates. A set of disturbance spatial (along the x'-coordinate) amplification rates $-\alpha'_i\delta_1$ versus the spanwise wavenumber $\beta^*\delta_1$ is presented in Fig. 9. Two distinct maxima localized around $|\beta^*\delta_1| \approx 0.4$ - 0.6 are observed for all disturbance frequencies which, most probably, correspond to two different instability modes. The shape of the dependences agrees qualitatively with the theoretical one calculated, for example, in [4] (for $\beta^* > 0$). It is interesting to note that the amplification maxima are localized near very large values of the disturbance propagation angles θ^* (Fig. 10). The most unstable disturbances propagate at the angles with $|\theta^*|$ between 80 and 90°. This is very typical property of the cross-flow instability modes predicted by all theories (see for example [3]). At the same time, note that the propagation angles of the most unstable modes differ significantly from the angles of the most intensive disturbances observed in the spanwise-wavenumber spectra (Fig. 6). Dependences presented in Figs. 9, 10 show that the typical wavenumbers of the most unstable modes increase with frequency. The difference in the growth rates for modes with different spanwise wavenumber also increase with frequency due to a faster attenuation of both quasi 2D modes (with $\beta^*\delta_1$ close to zero) and strongly 3D ones (with high values of $\beta^*\delta_1$). At the same time the amplification rates of the most unstable modes depend very slowly (if any) from the frequency in the region of parameters studied. The waves with positive wavenumbers (the first instability mode) are, as a role, a little bit more unstable then those with negative wave numbers (the second instability mode). The presence of a shelf at $\beta^*\delta_1$ close to zero in Fig. 9f is, probably, explained by an admixture of the Tollmien-Schlichting waves which have relatively higher values of $-\alpha_i$ for higher frequencies.

Phase velocities. C_k are the velocities along the wavevectors. They are presented in Fig. 11(a) as functions of the spanwise wavenumber for all the frequencies studied. It is seen that quasi-2D modes (with β^* close to zero) propagate with velocities around 0.3 - 0.6 of U_o . At the same time, the most unstable modes (with $|\beta^*\delta_1| \approx$ 0.4 - 0.6) have phase velocities C_k around $0.05U_o$. These values are typical for the cross-flow instability waves predicted by theory. Dependences of C_k from the wave propagation angle (Fig. 11(b)) demonstrate that they decrease rapidly with θ^* for $|\theta^*| > 60°$ for all frequencies, but there is a "shelf" in the region $|\theta^*| < 60°$ for high-frequency modes with $C_k/U_o \approx 0.30$ - 0.35. For strongly 3D modes close to the most unstable (with $|\beta^*\delta_1| \approx 0.4$ - 0.6) the values of C_k/U_o tend to zero when the frequency goes to zero; the same behaviour is observed for the most intensive fluctuations. However the waves with small propagation angles ($|\theta^*| < 50°$) have phase velocities which depend slowly on the disturbance frequency and do not tend to zero at small frequencies. It is interesting that the downstream (along x'-coordinate) phase velocities C_x^* of the almost all instability modes for all frequencies are rather close to each other and have typical values about 0.3 - 0.5 for not very high propagation angles ($|\theta^*| < 60°$) and about 0.6 - 0.8 for the most unstable modes ($|\theta^*| \approx 84$ - $87°$).

Group velocities. The group velocities were determined for the most unstable disturbances with the help of the formulas $C_{gx}^* = d\omega/d\alpha_r^*$ and $C_{gz}^* = d\omega/d\beta^*$

using the distributions shown in Fig. 8 which were approximated to reduce an error of differentiation. The streamwise component C_{gx}* of the group velocity vector for both instability modes (with β* >0 and with β* <0) are found to be rather close to a half of the free-stream velocity (mainly between 0.4 and 0.7). (This is very similar to the behaviour of the streamwise phase velocity C_x*.) Meanwhile, the values of the spanwise component of the group velocity vector C_{gz}* were found to have very small values around $0.01U_o$ to $0.03U_o$ (that is close to the phase velocity C_k) The inclination angles ψ_g* of the group velocity vectors are very close to zero too (about 2°-5°) and were found to be positive for the first instability mode (with β* >0) and negative for the second mode (with β* <0). (Compare them with the propagation angles of these modes close to ± 85°!) For high-frequency modes and $|\beta^*\delta_1| < 0.2$ (i.e. for $|\theta^*| < 50 - 60°$) the values of $|\psi_g^*|$ increase significantly with frequency and approach 20 - 30 degrees (although ψ_g* and C_{gz}* are very small for the modes with $\beta^*\delta_1$ exactly equal to zero). The maximum values of $|\psi_g^*|$ are observed for f = 38.5 Hz and $\beta^*\delta_1 \approx 0.05$.

Note in the end, that the most of the features of instability modes discussed in section 5 are in a good qualitative (and quantitative) agreement with the properties of the cross-flow instability modes predicted by linear stability theory.

6. Conclusions

1. A complete set of stability and dispersion characteristics of a swept-wing boundary layer with respect to traveling waves (with various frequencies, inclined at various angles to the flow direction) is obtained in the present experimental study .

2. All the results obtained can be directly quantitatively compared with the linear stability theory without any additional assumptions and hypothesis. These results are independent of initial disturbance spectrum and can be considered as an experimental base for validation of 3D stability theories and computer codes.

The authors express their gratitude to the Russian Foundation of Basic Research for support of this investigation (Grant No 94-01-00062-a).

References

[1] Kachanov, Y.S., Tararykin, O.I. : The experimental investigation of stability and receptivity of a swept-wing flow. In: *Laminar-Turbulent Transition* (eds. D.Arnal, R.Michel) Springer-Verlag, 1990, p. 499-509.

[2] Ivanov, A.V., Kachanov, Y.S.: A method of study of the stability of 3D boundary layers using a new disturbance generator. In: *Proc. Intl. Conference on Methods of Aerophysical Research. Part 1.* Novosibirsk: Inst. Theor. & Appl. Mech., Siberian Div. Russian Acad. Sci., 1994, p. 125-130.

[3] Dallmann, U., Bieler, H. : Analysis and simplified prediction of primary instability of three-dimensional boundary layer flows. *AIAA -87-1337*, 1987.

[4] Meyer, F. : *Numerische Simulation der Transition in dreidimensionalen Grenzschichten.* PhD thesis,DLR Inst. für Theoretische Strömungsmechanik, Göttingen, 1989, 101 p.

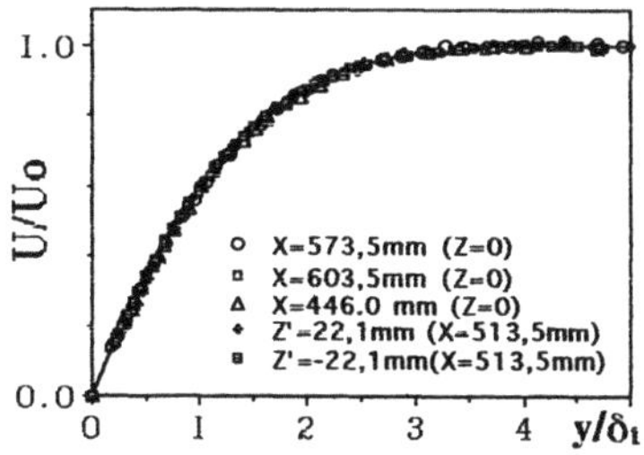

Fig.1 Mean velocity profiles

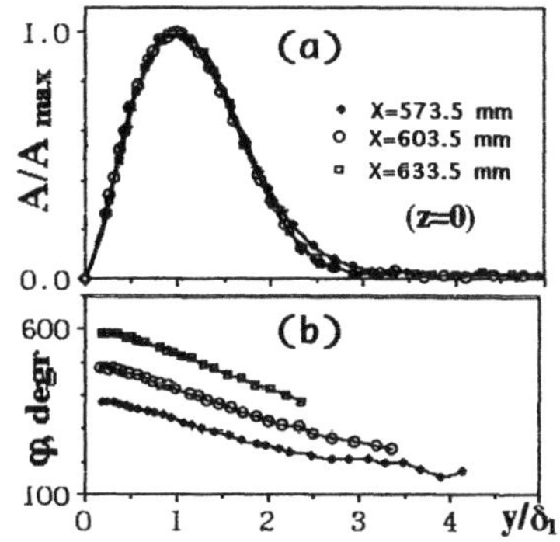

Fig.2 Disturbance profiles within the wave-trane

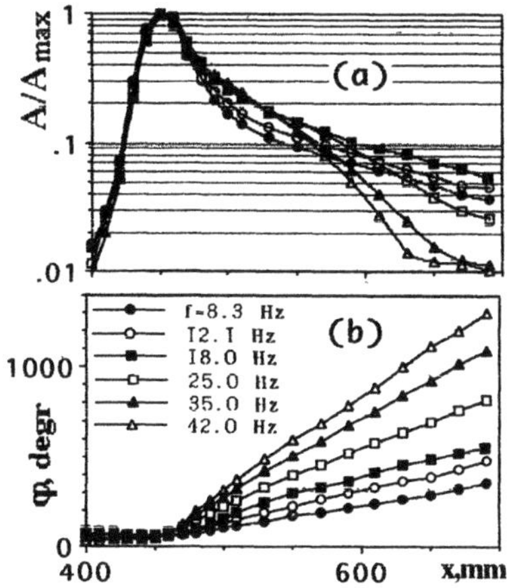

Fig.3 Downstream evolution of disturbances along z=0

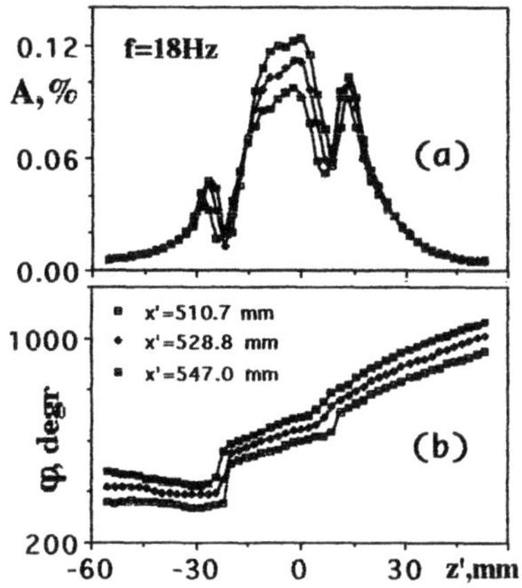

Fig.4 Evolution of spanwise disturbance distributions

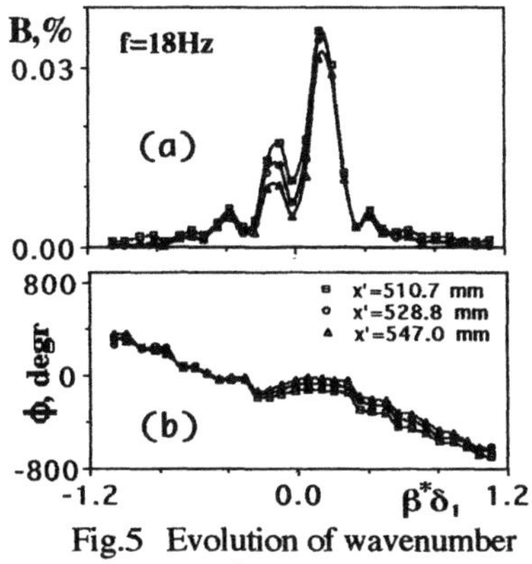

Fig.5 Evolution of wavenumber spectra

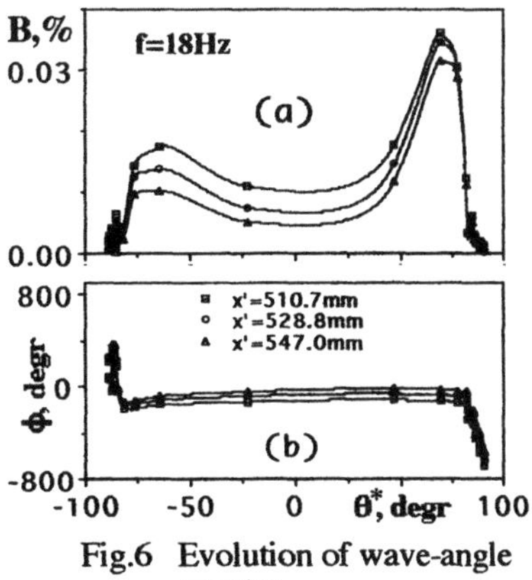

Fig.6 Evolution of wave-angle spectra

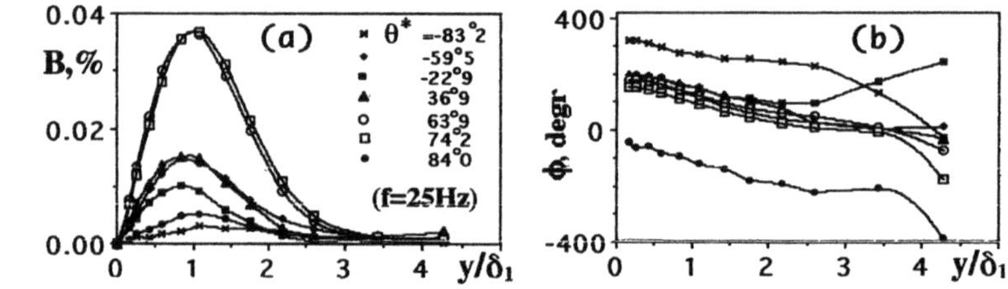

Fig.7 Eigenfunctions

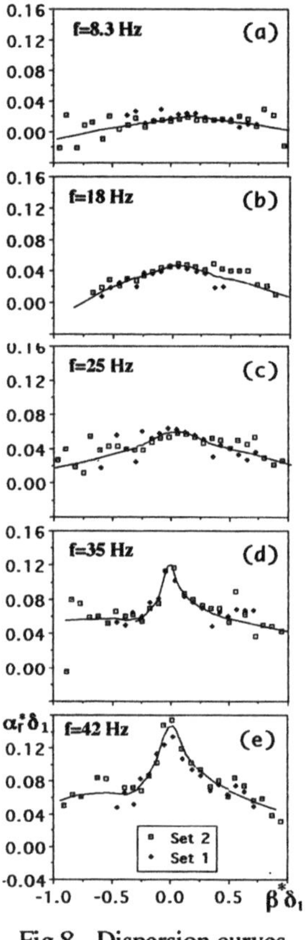

Fig.8 Dispersion curves

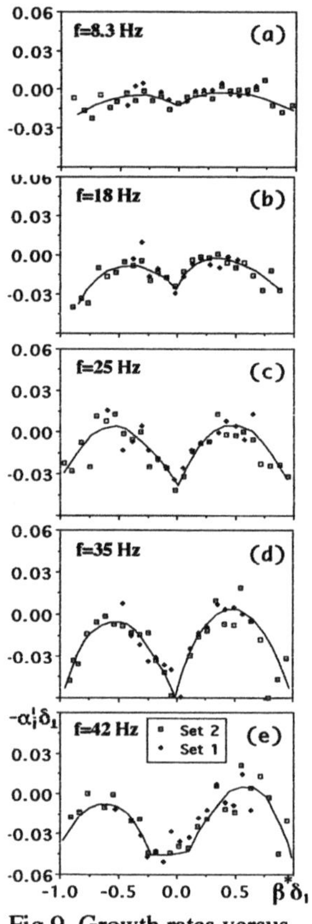

Fig.9 Growth rates versus spanwise wavenumber

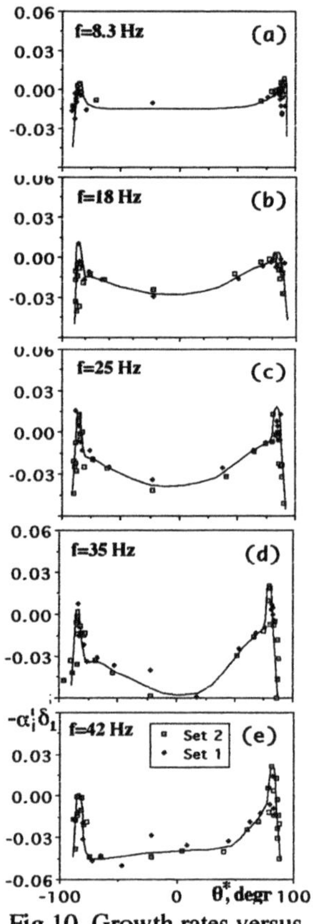

Fig.10 Growth rates versus propagation angle

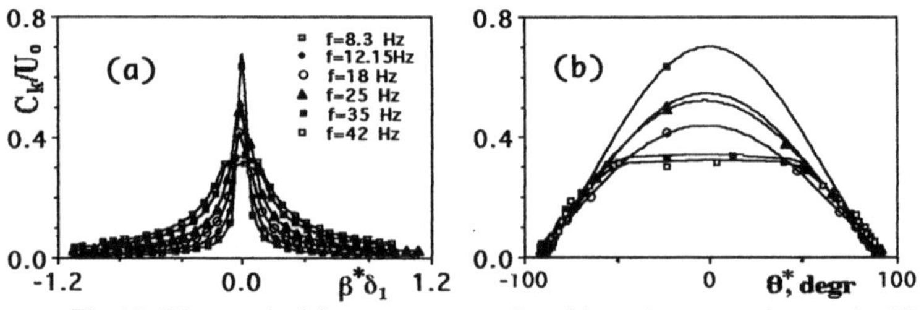

Fig.11 Phase velocities vs. wavenumber (a), and propagation angle (b)

Stability Characteristics of 3-D Boundary Layer on a Yawed Cone

A. Hanifi and A. Dahlkild

Department of Mechanics/Fluid Physics, Royal Institute of Technology, S-10044 Stockholm, SWEDEN.

Summary

The effects of yaw angle on the stability characteristics of the 3-D supersonic boundary layer are investigated by means of linear stability analysis. A simple method for calculation of the basic flow around a yawed cone is developed. Results of the stability analysis show that cross flow caused by a small yaw angle has a significant destabilizing effect on the first-mode disturbances, while it only slightly affects the second-mode growth rate. In the presence of the cross flow the most unstable second-mode disturbances was oblique.

Keyword : Supersonic, boundary layer stability, cross flow, yawed cone.

1 Introduction

Theoretical studies on stability and transition to turbulence of the supersonic boundary layer on a cone has mainly been focused on the axisymmetric case. Since Battin & Lin's [2] transformation of the flat plate linear stability characteristics to the axisymmetric boundary layer on a cone, several approaches have tried to improve and refine the linear theory. Additional curvature effects, not accounted for by the Battin & Lin transformation, have been dealt with by e.g. Gasperas [3] and Malik & Spall [12] who included curvature for the disturbances. The latter authors also used an extended formulation for a truly non-similar basic boundary layer flow. Refined methods for the mean flow have been tried by Malik, Zang & Bushnell [11] who used PNS and Simen & Dallman [16] who used "Thin layer Navier-Stokes" solutions. Generally, comparisons with experiments show that these refinements of the linear theory are necessary for a more quantitative agreement.

For a cone at an angle of attack there are few theoretical results presented in the literature. Experiments at supersonic or hypersonic speed by e.g. Krogmann [8], Stetson et al. [17] and King [6] show that the transition line moves downstream on the windward side and upstream on the leeward side of the cone for increasing angle of attack. As pointed out by Moore [14] this can qualitatively be understood from the changes of the effective Reynolds number, based on boundary layer thickness, since for increasing angle of attack the boundary layer decreases in thickness on the windward and increases on the leeward side of the cone. Simen & Dallmann [16] used "Thin layer Navier-Stokes" solutions for a 7° half-angle cone with 2° angle of attack at freestream Mach number 8 to compare the theoretical results with data from experiments of Stetson et al.

[17]. The predicted growth rates of second-mode disturbances, on both leeward and windward meridians, were reported to be in close agreement with the experimental data. At the leeward meridian even predicted first-mode growth rates were close to the experimental ones. Hanifi & Dahlkild [4] used non-similar boundary layer solutions on the extreme windward and leeward sides of the cone to compare their theoretical results with the experiments of Krogmann [8] for a 5° half-angle cone at Mach number 5. N-factor calculations were performed which partly recovered the movement of the transition point for various yaw angles. At angles of attack cross flow on the sides of the cone changes the structure of the boundary layer and affects the stability characteristics. Longitudinal patterns have been observed by Mc Devitt & Mellenthin [13] which could be associated with cross flow vortices. Simen & Dallman [16] obtained growth rates at the middle meridian for a 7° cone at 2° angle of attack at Mach 8. They found the stationary vortex line to be approximately parallel to the local stream line of the inviscid flow. Balakumar & Reed [1] investigated stability of three-dimensional supersonic boundary layers by studying a rotating cone. They found the effects of the cross flow to be destablizing. They also found that the most unstable second-mode wave was oblique.

The fully three-dimensional boundary layer around a yawed cone of course requires an extra computational effort for the mean flow as compared to the axisymmetric case. In the present paper we mimic the behaviour of a perturbation solution for small angles of attack, originally by Moore [14], to obtain a parametric representation of the boundary layer for various meridians just from solutions of the boundary layer equations at the extreme windward and leeward meridians. Non-similar mean flow profiles can then be obtained with a minimum amount of computational work at any position of the cone.

2 Basic laminar flow

We restrict the analysis to the flow past a sharp cone, see figure 1, with zero pressure gradient in the streamwise direction. Thus it is assumed that no interactions with the inviscid flow, nor with the bow shock wave, take place that affect the boundary layer. In order to derive the boundary layer equations we assume, in addition to the usual 2-D boundary layer approximations, that the ϕ-component of the velocity, w, is of the same order of magnitude as u. Also, we keep all terms whose relative magnitude is of order 1, δ/L and transverse curvature terms with relative magnitude of order δ^2/L^2, where δ is the boundary layer thickness and L is the length along the cone surface. The resulting boundary layer equations are given by Hanifi & Dahlkild [4]. The velocity components as well as normall derivative of the temperature vanish at the surface of the cone. The boundary layer edge conditions are taken from a perturbation solution for small angles of attack of the outer inviscid flow according to Stone [18].

$$\begin{cases} u \to 1 + \varepsilon \cos\phi U_\alpha, \\ w \to \varepsilon \sin\phi W_\alpha, \\ T \to 1 - (\gamma - 1)M^2 \varepsilon \cos\phi U_\alpha \end{cases} \qquad \text{as } \eta \to \infty, \tag{1}$$

where M is the edge Mach number for zero yaw angle. The values of U_α and W_α are obtained by solving the small disturbance equations in the *Tables of Supersonic Flow Around Yawing Cones* published at M.I.T. under the direction of Kopal [7] and corrected according to Moore [14]. In the calculations presented here all thermodynamic variables are assumed to be functions of temperature. The first coefficient of viscosity, μ, is calculated using two part Sutherland's formula and the thermal conductivity, κ, is calculated in a similar way using Keye's formula (see Mack [9]). The specific heat, c_p is obtained by a fifth order polynomial fit to the experimental data of air at atmospheric pressure. The data is taken from the *Tables of Thermal Properties of Gases* [5]. Furthermore, in our calculations the ratio of specific heats is 1.4 and assumed to be constant. The symmetry of the flow in a plane normal to the flow direction provides

$$w = 0, \frac{\partial}{\partial \phi}(u, T, V, \rho) = 0 \text{ at } \phi = 0, \pi. \tag{2}$$

These symmetry properties can be used to solve the boundary layer equations at the windward and leeward meridians separately. By inserting the equalities (2) in the equations one recovers the x-momentum and energy equations for axisymmetric flow. Also, by expanding w in a Taylor series around $\phi = 0$ and $\phi = \pi$

$$w(\xi, \eta, \phi) = \Delta\phi \frac{\partial}{\partial \phi} w(\xi, \eta, \phi = 0, \pi) + O(\Delta\phi^3), \tag{3}$$

and keeping terms to lowest order in the ϕ-component of the momentum equation completes the system of equations. Observe that we do not make any restriction on the value of ε, except from the requirement that no boundary layer separation occurs on the cone. It should be mentioned though that the edge values of the variables used in our calculations are the same as for the perturbation series (1), i.e. the boundary conditions are in fact not the exact ones. To obtain then the boundary layer profiles at the meridians with polar angles $0 < \phi < \pi$, we make the approximate ansatz

$$u(\xi, \eta, \phi; \varepsilon) = u^0(\xi, \eta; \varepsilon) + u^I(\xi, \eta; \varepsilon) \cos\phi + u^{II}(\xi, \phi; \varepsilon) \cos 2\phi, \tag{4}$$

similarly for $v(\xi, \eta, \phi; \varepsilon)$ and $T(\xi, \eta, \phi; \varepsilon)$ and finally

$$w(\xi, \eta, \phi; \varepsilon) = w^I(\xi, \eta; \varepsilon) \sin\phi + w^{II}(\xi, \eta; \varepsilon) \sin 2\phi. \tag{5}$$

This is in formal agreement with the first terms in a regular perturbation expansion for small angles of attack, ε, which at the boundary layer edge must approach the corresponding series of the inviscid outer flow as given by Munson [15]. Here, we approximate u^0, v^0 and T^0 with the corresponding profiles at zero angle of attack whereas the other two functions for each variable are obtained by equating the ansatz with the exact solutions on the windward and leeward meridians. The polar coordinate ϕ can then be used as a parameter to obtain the boundary layer profiles at any meridian of the cone. (One may note that to obtain complete formal agreement between the ansatzes (4), (5) and the inviscid flow, the next term in the expansions (1) should be included. These terms were omitted here since indeed they are not as easly obtained as the first terms of the expansions.)

Boundary layer profiles for a 10° cone and 1° yaw angle at various meridians are shown in figure 2. As this figure shows the boundary layer thickness increases from the windward meridian, $\phi = 0$, to the leeward meridian, $\phi = \pi$. It can also be observed that the maximum azimuthal velocity, w, occurs at $\phi = \pi/2$.

3 Linear stability analysis

In order to derive the stability equations all flow variables are made non-dimensional using their edge value at zero angle of attack as reference and the length scale used is $\sqrt{\nu_e^0 x / u_e^0}$. Furthermore, all flow quantities are decomposed into a mean and a fluctuating part:

$$\mathbf{q} = \bar{\mathbf{q}} + \tilde{\mathbf{q}} \tag{6}$$

where bar and tilde refer to the mean and fluctuating part of the quantities respectively. The assumption of a locally parallel flow yields

$$\bar{\mathbf{q}} = (u(y), 0, w(y), T(y), p(y), \rho(y)) . \tag{7}$$

The fluctuating parts are assumed to be harmonic waves of the form

$$\tilde{\mathbf{q}} = \left(\hat{u}(y), \hat{v}(y), \hat{w}(y), \hat{T}(y), \hat{p}(y), \hat{\rho}(y)\right) e^{i(\alpha x + n\phi - \omega t)}, \tag{8}$$

where α is the wavenumber along the body, n the number of waves in the azimutal direction and ω the disturbance frequency. Inserting (7)–(8) in the linearized Navier-Stokes equations gives an eigenvalue problem for the determination of a dispersion relation of the type

$$F(\alpha, n, \omega, R) = 0. \tag{9}$$

If we assume α to be complex and ω to be real, then the amplitude growth rate is $-\alpha_i$. The magnitude of the wavenumber vector $\mathbf{k}$ is

$$k = \left(\alpha_r^2 + \beta^2\right)^{\frac{1}{2}}, \qquad \beta = \frac{n}{r}. \tag{10}$$

To solve the stability equations, a fourth order shooting method (see Mack [9]) with asymptotic boundary conditions was used. In all calculations presented here, the stagnation temperature was taken to be 600 K and all material properties were assumed to be temperature dependent. All curvature and body divergence terms were included in the stability equations. Furthermore, for the simplicity the spatial amplification rate was assumed to be in the direction of the meridians.

4 Results

The boundary-layer profiles obtained from the method described in section 2 were used to study the effects of cross flow on the stability characteristics of a supersonic boundary layer on a yawed cone. Figure 3 shows the maximum growth rate of first- and second-mode disturbances as a function of reduced frequency $F = \omega / R$. The maximum growth rate of second mode waves grows and moves to the lower frequency as the azimuthal position is moved from the windward meridian, $\phi = 0$, to the leeward meridian, $\phi = \pi$. Since the wavelength of the second-mode disturbances scales with the boundary-layer thickness, the shifting of the unstable frequencies can be explained by the thickening of the boundary layer from the windward to the leeward side. The behaviour of the first-mode disturbances is different from those of the second-mode. These waves are most unstable at $\phi = -\pi/2$ where w is largest. Since w vanishes at $\phi = 0, \pi$ no effect of cross flow can be expected at these meridians, and the differences between growth rates of the unstable modes at the leeward and the windward meridians are only caused by changes of the boundary layer thickness and its shape. In contrast to the second-mode waves there is no shifting of the unstable frequency band.

In order to demonstrate the effects of the cross flow, the growth rate of the most unstable waves are calculated when the cross flow component of the basic flow (velocity component in the direction prependicular to the inviscid streamline) is subtracted from the basic flow. The maximum growth rates of the first-and second-mode disturbances for a 10° cone with $M = 5$ (edge Mach number for zero yaw angle) and $\varepsilon = 1^o$, with and without cross flow component of the basic flow, are plotted in figure 5. As is shown in this figure the second-mode disturbances are hardly affected by the absence of the cross flow component, while the maximum first-mode growth rate decreases by 44%. The maximum cross flow, in this case, is about 3% of u only. This large effect of cross flow on the first-mode disturbances was also found by Balakumar & Reed [1]. The most unstable wave angle is also affected by the presence of the cross flow (see figure 6). The most unstable second-mode waves are not 2-D waves, as in the axisymmetric case. It was also observed that the maximum amplification of the second mode did not occur along the inviscid-flow direction. At $\phi = -\pi/2$ the most unstable second-mode disturbance is inclined at about 3.5^o to the inviscid flow direction.

In order to investigate the effect of the cross flow for the different Mach numbers a series of calculations for a 10° cone at 1° angle of attack was performed. The results can be found in figure 4. As this figure demonstrates, the effect of the cross flow on the maximum growth rates of the first-mode increases with increasing Mach number. At $\phi = -\pi/2$, in absence of a cross flow component, the relative decrease of the maximum growth rates is 31%, 37% and 43% for Mach 1.6, 2.5 and 5 respectively. Figure 7 shows the maximum growth rates of a 10° cone at different yaw angle and $M = 5$. As a result of increasing cross flow, the maximum first-mode growth rates increase with increasing yaw angle. On the windward side of the cone, $\phi = -\pi/4$, the maximum growth rates of the second-mode disturbances slightly decrease. This behaviour may be explained by the fact that the effect of cross flow on the growth rates of second-mode disturbances is small. On the leeward side of the cone, $\phi = -3\pi/4$, the maximum second-mode amplification rates increase with increasing angle of attack. At $\phi = -\pi/2$, the maximum second-mode growth rates decrease for $\varepsilon = .5^o$ and increase for $\varepsilon = 1^o$. This may be due to the competing effects of cross flow and slightly decreasing boundary layer thickness. In all these cases, the changing boundary layer thickness causes a shift of the unstable second-mode frequency band.

5 Conclusions

The basic laminar supersonic flow around a yawed cone was obtaind using a simplyfied boundary-layer method. Destabilizing effects of cross flow on the first-mode disturbances were substantial even for small yaw angles. The main effect on the second-mode was to change the most amplified waves from 2-D to oblique, whereas the growth rate was less sensitive. In the presence of the cross flow the most unstable second-mode disturbances were not aligned with the inviscid streamline. Also, the effects of cross flow increased with increasing Mach number.

Acknowledgements

This work was sponsored by the Aeronautical Research Institute of Sweden (FFA) through the Swedish Space Corporation on behalf of the Swedish National Space Board and was a part of Swedish participation within Sänger program.

References

[1] Balakumar, P. and Reed, H.L. 1990 Stability of Three-Dimensional Supersonic Boundary Layers. Phys. Fluids A **3** (4), pp. 617–632.

[2] Battin, R.H. and Lin, C.C. 1950 On the Stability of the Boundary Layer Over a Cone. Journal of the Aeronautical Sciences, Vol. 17, p. 453.

[3] Gasperas, G. 1985 Instability of compressible laminar boundary layers. AIAA Paper No. 85-1699.

[4] Hanifi, A. and Dahlkild, A.A. 1993 Some Stability Characteristics of the Boundary Layer on a Yawed Cone. AIAA Paper No. 93-3048.

[5] Hilsenrath, J. et al. 1955 Tables of Thermal Properties of Gases. U.S. Department of Commerce, National Bureau of Standards Circular 564.

[6] King, R.A. 1991 Mach 3.5 Boundary Layer Transition on a Cone at Angle of Attack. AIAA Paper No. 91-1804.

[7] Kopal, Z. 1947 Tables of supersonic flow around cones. Technical report No. 1–3, Dept. Elec. Eng., M.I.T.

[8] Krogmann, P. 1977 An experimental study of boundary layer transition on a slender cone at Mach 5. AGARD-CPP 224, Laminar-Turbulent Transition, Lyngby, Denmark.

[9] Mack, L.M. 1965 Computation of the Stability of the Laminar Boundary Layer. Methods Computational Physics, Vol. 4, pp. 247–299, Academic, N.Y.

[10] Mack, L.M. 1987 Stability of Axisymmetric Boundary Layers on Sharp Cones at Hypersonic Mach Numbers. AIAA Paper No. 87-1413.

[11] Malik, M., Zang, T. and Bushnell, D. 1990 Boundary Layer Transition in Hypersonic Flows. AIAA paper No. 90-5232.

[12] Malik, M.R. and Spall, R.E. 1991 On the stability of compressible flow past axisymmetric bodies. J. Fluid Mech., 228, pp. 443–463.

[13] Mc Devitt, J.B. and Mellenthin, J.A. 1969 Upwind Patterns on Ablating and Nonablating Cones at Hypersonic Speeds. NASA TN D-5346.

[14] Moore, F.K. 1951 Laminar boundary layer on a circular cone in supersonic flow at a small angle of attack. NACA TN 2521.

[15] Munson, A.G. 1964 The vortical layer on an inclined cone. J. Fluid Mech., 20, pp. 625–643.

[16] Simen, M. and Dallman, U. 1992 On the Instability of Hypersonic Flow Past a Pointed Cone - Comparison of Theoretical and Experimental Results at Mach 8. Deutscher Luft- und Raumfahrtkongre /DGLR-Jahrestagung 1992.

[17] Stetson, K.F., Thompson, E.R., Donaldson, J.C. and Siler, L.G., 1985 Laminar Boundary Layer Stability Experiments on a Cone at Mach 8. Part 3. Sharp Cone at Angle of Attack. AIAA Paper No. 85-0492.

[18] Stone, A.H. 1948 On supersonic flow past a slightly yawing cone. Journal of Math. and Phys., vol. XXVII, no. 1, April 1948, pp. 67–81.

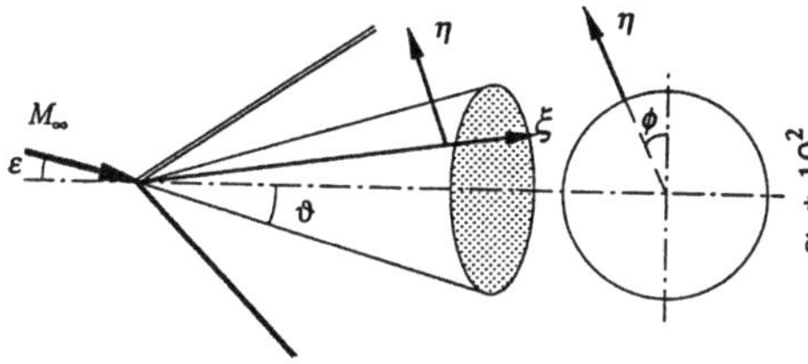

Figure 1: Geometry and coordinate system

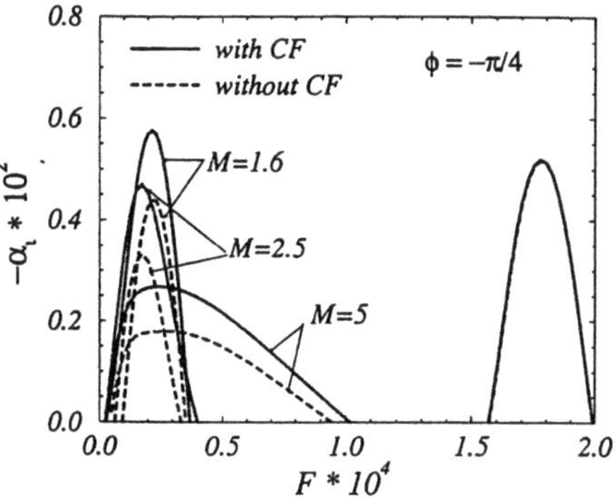

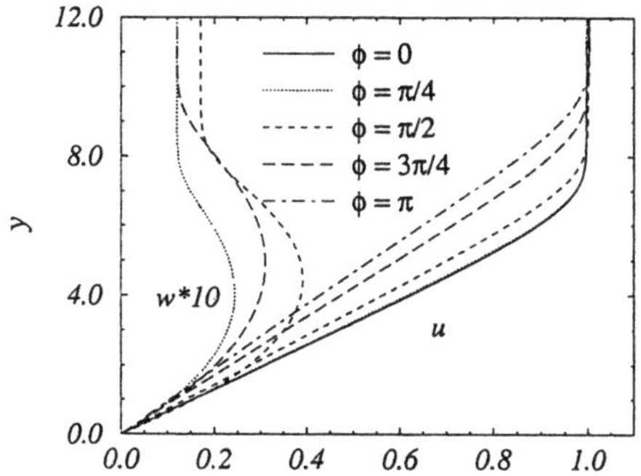

Figure 2: Boundary layer profiles. $M = 5$, $\vartheta = 10^\circ \varepsilon = 1^\circ$ and $R = 2000$.

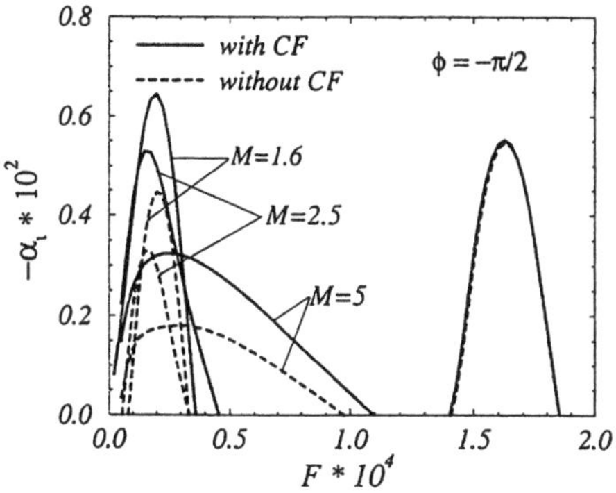

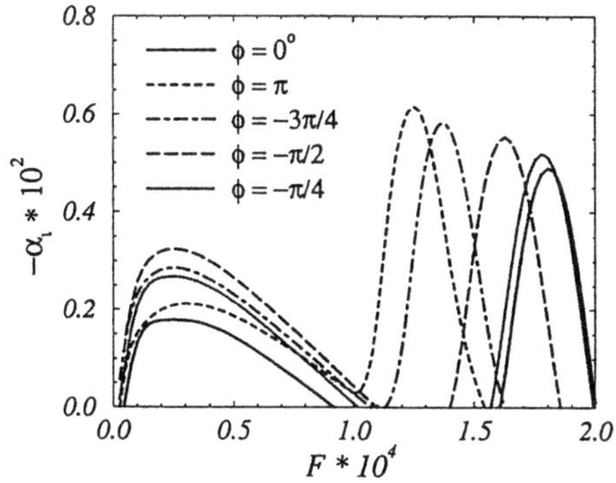

Figure 3: Maximum growth rate, $M = 5$, $\vartheta = 10^\circ$, $\varepsilon = 1^\circ$ and $R = 2000$.

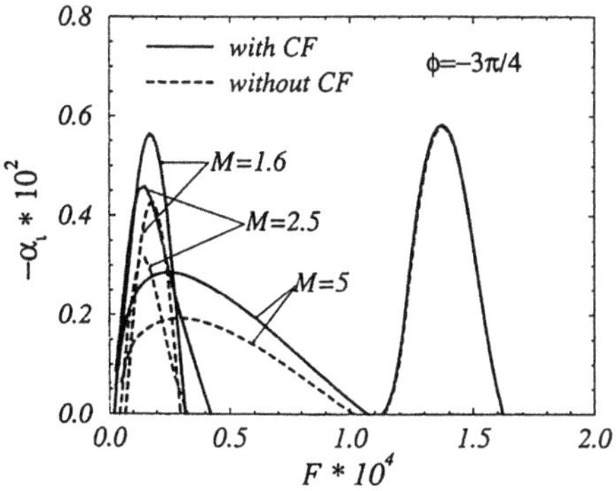

Figure 4: Maximum growth rate for $M = 1.6, 2.5, 5$. $\vartheta = 10^\circ$, $\varepsilon = 1^\circ$ and $R = 2000$.

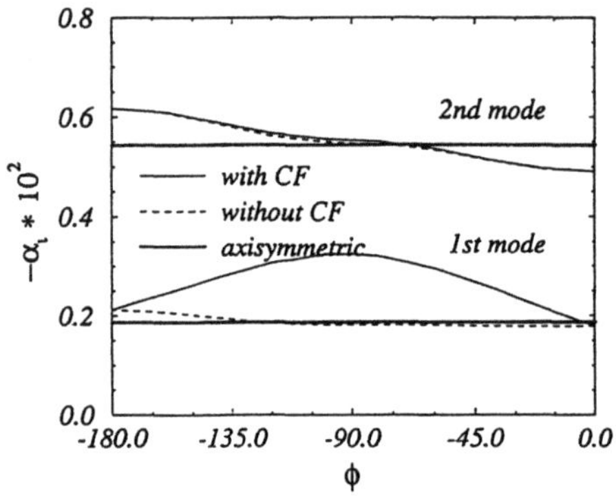

Figure 5: Maximum growth rate. $M = 5$, $\vartheta = 10^o$ and $\varepsilon = 1^o$ and $R = 2000$.

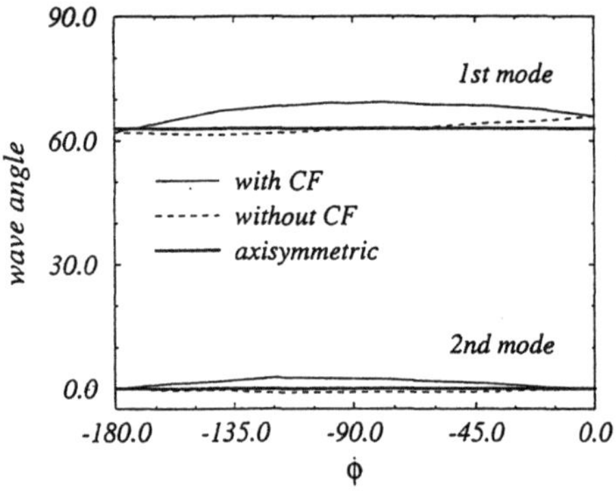

Figure 6: Angle of the most unstable waves, $M = 5$, $\vartheta = 10^o$, $\varepsilon = 1^o$ and $R = 2000$.

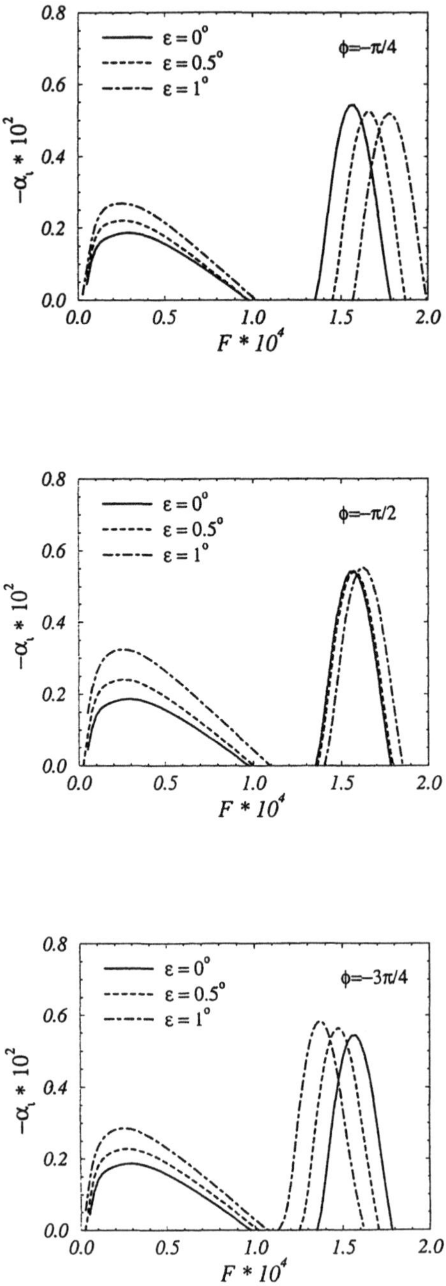

Figure 7: Effects of yaw angle on the maximum growth rate. $M = 5$, $\vartheta = 10^o$, $\varepsilon = 1^o$ and $R = 2000$.

Absolute Instability of the Rotating-Disk Boundary Layer

R.J. Lingwood
University of Cambridge, Department of Engineering
Trumpington Street, Cambridge CB2 1PZ, UK

Summary

This paper is concerned with the theoretical behaviour of the boundary-layer flow over a disk rotating in otherwise still fluid. The flow is excited impulsively at a certain radius at time $t = 0$. It is found that above a critical Reynolds number *absolute* instability occurs and a disturbance grows in time at every fixed point in space. Below the critical Reynolds number, the flow is convectively unstable. The absolute or convective nature of the instabilities is determined by examining the branch-point singularities of the dispersion relation. It is suggested that this absolute instability may trigger abrupt transition from laminar to turbulent flow.

Key words : Rotating disk, boundary layer, absolute instability, pinch-point.

1. Introduction

The response of the flow to impulsive forcing shows whether it is convectively or absolutely unstable. If the response to the transient disturbance grows with time at a fixed location in space, then the flow is absolutely unstable. In contrast, if the impulse response is swept away as it grows and leaves the point of disturbance unperturbed, then the flow is convectively unstable. In terms of figure 1, the non-dimensional radius is r, the radial location of the source of the impulse is given by r_s and the disturbances are growing within the dashed lines. Figure 1 shows a stable case, a convectively unstable case and an absolutely unstable case. Huerre & Monkewitz [1] give several examples of a change in behaviour from convectively to absolutely unstable.

Previous work on this problem has concentrated on the disturbances created by roughness elements fixed to the disk. It has been suggested [2] that the well-known stationary vortex lines, observed in the laminar region prior to transition, emanate from such point sources, and that the observed pattern results from the superposition of the entire circumferential wavenumber spectrum of the equal phase, zero-frequency normal modes. There is convincing experimental confirmation of this theory [3]. Here, to study transition, linear stability of travelling waves (non-zero frequency) is considered. Viscous effects as well as Coriolis and streamline curvature effects are included. The initial disturbance is provided by an impulsive circumferential line forcing, $\delta(r-r_s)\delta(t)e^{i\beta\theta}$, where

$\delta(r - r_s)$ and $\delta(t)$ are the Dirac delta functions at r_s and at $t = 0$, respectively, β is the prescribed integer circumferential wavenumber and θ is the circumferential co-ordinate. The response to point forcing can be obtained by summing over all values of β.

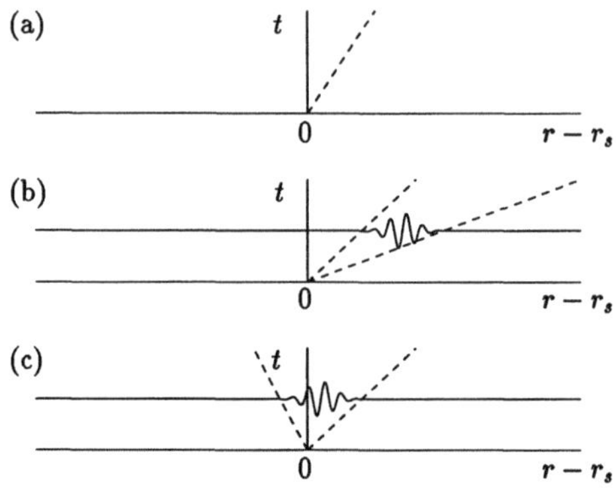

Figure 1. Sketches of typical impulse responses: (a) stable, (b) convectively unstable, (c) absolutely unstable.

2. Formulation of the problem

The disk is modelled as an infinite planar disk rotating at constant angular frequency, Ω^*, about the vertical axis z^* (asterisks indicate dimensional quantities). The mean flow *relative* to the disk is given by von Kármán's [4] exact similarity solution to the Navier-Stokes equations. The dimensionless similarity variables of the solution are defined by

$$U(z) = \frac{U^*(z^*)}{r^*\Omega^*}, \quad V(z) = \frac{V^*(z^*)}{r^*\Omega^*}, \quad W(z) = \frac{W^*(z^*)}{\sqrt{(\nu^*\Omega^*)}}, \quad P(z) = \frac{P^*(z^*)}{\rho^*\nu^*\Omega^*}, \tag{1}$$

where U, V, W, are the radial, circumferential and axial base flow velocities, respectively, P is the pressure, ρ^* and ν^* are the density and kinematic viscosity, whilst $z = z^*/L^*$ is the non-dimensional axial co-ordinate, with $L^* = \sqrt{(\nu^*/\Omega^*)}$ as the non-dimensionalizing length-scale.

The stability analysis is applied at say $r_a = r_a^*/L^* = r_a^*\Omega^*L^*/\nu^* = R$, the local Reynolds number. Infinitesimally small disturbances are imposed on the mean flow. The dimensionless Navier-Stokes equations are linearized with respect to the small perturbation quantities. Terms $O(R^{-2})$ are neglected and the variable r, which appears in coefficients of the linearized equations, is replaced by R. The latter is equivalent to the *parallel-flow* approach and means that the equations are separable in r, θ and t but this does introduce some inconsistencies in to the analysis, as discussed by Lingwood [5,6]. The perturbation quantities may now be assumed to have normal-mode form, e.g. $u = \hat{u}(z; \alpha, \omega; \beta, R)e^{i(\alpha r + \beta\theta - \omega t)}$ (v, w and p are defined similarly), ω is the frequency of the disturbance in the rotating frame and α is the radial wavenumber.

Naturally, the value of β is equal to that imposed by the forcing.

The equations may be written as a set of six first-order ordinary differential equations in the following transformed variables

$$z_1(z;\alpha,\omega;\beta,R) = (\alpha - \frac{i}{R})\hat{u} + \bar{\beta}\hat{v}, \quad z_2(z;\alpha,\omega;\beta,R) = (\alpha - \frac{i}{R})D\hat{u} + \bar{\beta}D\hat{v},$$

$$z_3(z;\alpha,\omega;\beta,R) = \hat{w}, \quad z_4(z;\alpha,\omega;\beta,R) = \hat{p},$$

$$z_5(z;\alpha,\omega;\beta,R) = (\alpha - \frac{i}{R})\hat{v} - \bar{\beta}\hat{u}, \quad z_6(z;\alpha,\omega;\beta,R) = (\alpha - \frac{i}{R})D\hat{v} - \bar{\beta}D\hat{u}, \quad (2)$$

where $\bar{\beta} = \beta/R$ and D represents differentiation with respect to z. The solutions to the equations at the outer boundary are found analytically from their asymptotic forms, assuming all perturbations decay exponentially as $z \to \infty$. The three independent solutions are summed, with appropriate weightings to satisfy the boundary conditions at the wall, to give Z_i $(i = 1,2,3,..,6)$. For the unforced problem the boundary conditions at $z = 0$ are

$$Z_1(0;\alpha,\omega;\beta,R) = Z_3(0;\alpha,\omega;\beta,R) = Z_5(0;\alpha,\omega;\beta,R) = 0. \quad (3)$$

The eigenfunctions Z_i $(i = 1,2,3,..,6)$ exist only if the eigenvalues, α and ω, satisfy the dispersion relation $d(\alpha,\omega;\beta,R) = 0$. The eigenvalues are found using a fourth-order Runge-Kutta integrator and Newton-Raphson linear search procedure. Gram-Schmidt orthonormalization is applied to maintain the linear independence of the three independent solution vectors prior to summing.

Now, the boundary condition on $Z_3(0;\alpha,\omega;\beta,R_s)$ given in (3), where R_s is the Reynolds number at the source of the impulsive disturbance, needs to be reconsidered. The generated waves propagate at finite velocities and are therefore confined to a finite region at any finite time. Therefore, Fourier transform with respect to r at fixed t can be taken with α real. However, due to the suspected absolute instability the transform with respect to t may not exist for real frequencies. However, the Causality Principle states that there is no response prior to the forcing and therefore the transform may be taken provided the frequency has positive imaginary part ω_i greater than any of the temporal growth rates given by real α. Applying these constraints, the double Fourier transform of the forcing, after extracting the factor $e^{i\beta\theta}$, is

$$\int_{-\infty}^{+\infty}\int_{-\infty}^{+\infty} \delta(r - r_s)\delta(t)e^{-i(\alpha r - \omega t)}dr\,dt = e^{-i\alpha r_s}. \quad (4)$$

Following Gaster [7], the the forcing is introduced by making the boundary conditions inhomogeneous. The response is given by a form of the homogeneous solution, weighted in such a way as to give the appropriate boundary value at $z = 0$, namely $\Gamma(\alpha,\omega;\beta,R_s)Z_3(z;\alpha,\omega;\beta,R_s)$, where Z_3 is the homogeneous solution as above and Γ the weighting function. At the wall $\Gamma(\alpha,\omega;\beta,R_s)Z_3(0;\alpha,\omega;\beta,R_s) = e^{-i\alpha r_s}$. Hence, the axial perturbation velocity $w(r,\theta,t,z)$ is given by the inverse Fourier transforms

$$w(r,\theta,t,z) = \frac{1}{(2\pi)^2}\int_F\int_A \frac{Z_3(z;\alpha,\omega;\beta,R_s)}{Z_3(0;\alpha,\omega;\beta,R_s)} e^{i(\alpha(r-r_s)+\beta\theta-\omega t)}d\alpha d\omega. \quad (5)$$

The A-contour is taken along the real α-axis, giving temporal branches of the dispersion relation in the ω-plane, whilst the horizontal F-contour must lie above all the temporal branches and gives spatial branches in the α-plane. These contours ensure convergence of the transforms and satisfy causality. The spatial branches cannot cross the A-contour therefore they are damped. It follows that branches lying in the upper half α-plane correspond to the region $r > r_s$ and branches lying in the lower half α-plane correspond to the region $r < r_s$.

The α-integral can be evaluated by closing the A-contour in the upper and lower half-planes for $r > r_s$ and $r < r_s$, respectively, which then gives contributions due to zeros of the dispersion relation (where $Z_3(0;\alpha,\omega;\beta,R_s) = 0$) and due to the branch-cuts. The zeros of the dispersion relation give discrete modes and the branch-cuts give the continuous spectra. It is assumed that the contributions from the discrete modes are dominant. The response is now

$$w(r,\theta,t,z) = \frac{ie^{i\beta\theta}}{2\pi}\int_F H(r-r_s)\sum_u \frac{Z_3(z;\alpha_u(\omega),\omega;\beta,R_s)e^{i(\alpha_u(\omega;\beta,R_s)(r-r_s)-\omega t)}}{\left(\frac{\partial Z_3(0;\alpha_u(\omega),\omega;\beta,R_s)}{\partial\alpha}\right)_{\alpha_u}} -$$

$$H(r_s-r)\sum_l \frac{Z_3(z;\alpha_l(\omega),\omega;\beta,R_s)e^{i(\alpha_l(\omega;\beta,R_s)(r-r_s)-\omega t)}}{\left(\frac{\partial Z_3(0;\alpha_l(\omega),\omega;\beta,R_s)}{\partial\alpha}\right)_{\alpha_l}}d\omega, \tag{6}$$

where H is a unit-step function and the subscripts u and l refer to spatial branches in the upper and lower half α-planes, respectively. The residues, which then form the ω-integrand, are assumed to be first-order poles in the α-plane.

In order to distinguish between convectively and absolutely unstable time-asymptotic responses, Briggs [8] used analytic continuation to bring the original F-contour towards the real ω-axis; if the flow is absolutely unstable then the time-asymptotic response will tend to infinity for all points in space and it will not be possible to move the original F-contour on to the real ω-axis. If the F-contour is lowered below the most amplified temporal branch causality would be violated and there would be a loss of analyticity on the F-contour. If, however, the A-contour is first deformed off the real α-axis, such that the contour still passes on the correct side of the spatial branches, causing the temporal branch to lie lower in the ω-plane then causality and analyticity are satisfied. However, if two branches from different sides of the A-contour move towards each other and the A-contour becomes pinched between them, then the F-contour cannot be lowered any further to give the contribution of each real frequency. At the pinch-point $\partial\omega/\partial\alpha$ is identically zero, as is $\partial Z_3(0;\alpha(\omega),\omega;\beta,R_s)/\partial\alpha$, and the ω-integrand appearing in equation (6) exhibits a singularity in ω at ω^o and $\alpha(\omega^o) = \alpha^o$, say. A point where $\partial\omega/\partial\alpha = 0$ causes a singularity in the ω-integrand only if it also a pinch-point. Then, as α_u and α_l tend to α^o equation (6) can be approximated, for all r, by

$$w(r,\theta,t,z) \simeq \frac{e^{i\beta\theta}}{2\pi}\int_F \frac{Z_3(z)e^{i(\alpha^o(r-r_s)-\omega t)}}{\left[2\frac{\partial Z_3(0)}{\partial\omega}\frac{\partial^2 Z_3(0)}{\partial\alpha^2}\right]^{\frac{1}{2}}_{\alpha^o,\omega^o}}\frac{1}{(\omega-\omega^o)^{\frac{1}{2}}}d\omega. \tag{7}$$

If the two coalescing spatial branches are both from the same half α-plane then both give residue contributions of the same type in equation (6) and these cancel in the limit of $\alpha \to \alpha^o$ and there is no singularity in the ω-integrand. This leaves only a finite residue contribution from the second-order pole. This causes a short period of algebraic growth, which may be important if the second-order pole is near neutral. This idea will not be persued here but has been studied by Benney & Gustavsson [9] and Koch [10].

For the case where the coalescing spatial branches meet at a pinch-point, the integral over ω can now be evaluated along the deformed F-contour. The time-asymptotic response for all r is given by

$$w(r,\theta,t,z) \simeq \frac{Z_3(z;\alpha^o,\omega^o,\beta,R_s)e^{i(\alpha^o(r-r_s)+\beta\theta-\omega^o t)}}{\left[2\frac{\partial Z_3(0)}{\partial\omega}\frac{\partial^2 Z_3(0)}{\partial\alpha^2}t\right]^{\frac{1}{2}}_{\alpha^o,\omega^o}}. \quad (8)$$

If imaginary part of the frequency ω_i^o is positive, then the response given by equation (8) grows in time and indicates absolute instability; otherwise it is at worst convectively unstable and the disturbances are swept away from the source. Moreover, because the response now applies for all r and is continuous through r_s, there is no need for the source at r_s and an absolute instability can be thought of as a normal mode, which can be initiated by noise or any other excitation.

It will be shown that, in distinct parameter regions, branch-coalescing of both types occurs and, therefore, the flow instability changes from convective to absolute.

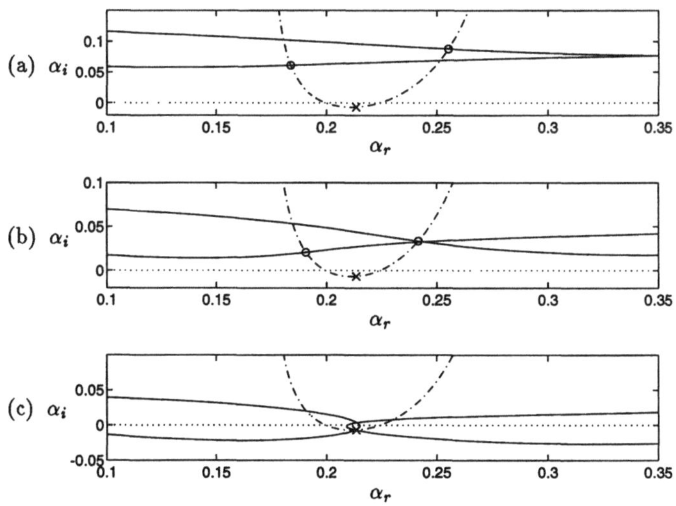

Figure 2. Families 1 and 2 (_), in the α-plane, for $R_s = 515$ and $\beta = 6$ given by F-contours: (a) $\omega_i = 0.0100$, (b) $\omega_i = 0.00400$, (c) $\omega_i = \omega_i^o \simeq 0$. The branch-point is marked by $\times$. The broken line (_ .) traces the path of the branches as ω_i is reduced. The circles (o) in (a) and (b) mark the points on the broken line where $\omega_i = 0.0100$ and $\omega_i = 0.00400$, respectively.

3. Results

Mack [2] calculated two spatial branches, or *families* for $\omega = 0$ and real β. Both originate in the upper half α-plane when ω_i is large and positive. Here, branch-point singularities have been found between these two families. Figure 2 shows family 1 and 2 in the α-plane as ω_r traverses the horizontal F-contour, at $R_s = 515$ and $\beta = 6$. Figures 2(a) and 2(b) show the spatial branches when the F-contour is at $\omega_i = 0.0100$ and $\omega_i = 0.00400$, respectively. Figure 2(c) corresponds to the F-contour lying along the real ω-axis. The branch-point, $\omega \simeq 0.0243 + i0$, lies at $\alpha^o \simeq 0.214 - i0.00701$. The broken line (_ .) traces the path of α as ω_i is decreased from infinity to zero, with ω_r, β and R_s kept constant (ω_r at the branch-point value). At such branch-points $\partial\omega/\partial\alpha$ is zero. However, both branches emanate from the upper half α-plane and therefore *do not* constitute an absolute instability. For Reynolds numbers greater than about 460 these points indicate a convective instability; the branches then cross into the unstable, lower half α-plane when the F-contour is lowered to the real ω axis. Below about 460 these branch-points are stable.

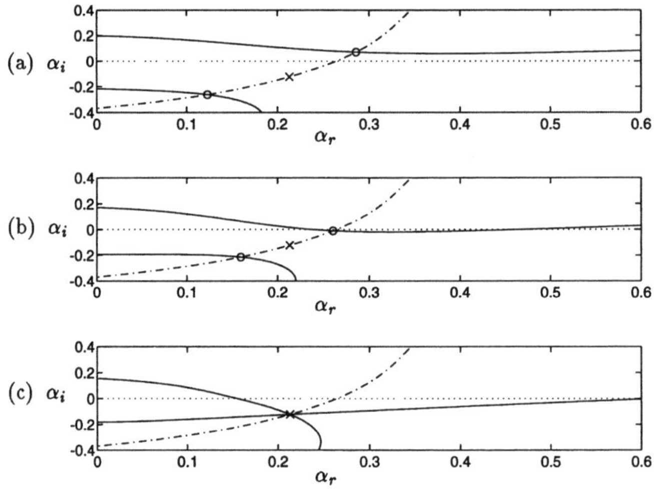

Figure 3. Families 1 and 3 (_), in the α-plane, for $R_s = 530$ and $\beta = 67$ given by F-contours: (a) $\omega_i = 0.0100$, (b) $\omega_i = 0.00400$, (c) $\omega_i = \omega_i^o \simeq 0.000289$. The branch-point is marked by $\times$. The broken line (_ .) traces the path of the branches as ω_i is reduced, on which the circles (o) in (a) and (b) mark the points where $\omega_i = 0.0100$ and $\omega_i = 0.00400$, respectively.

It has been thought that the onset of transition, which is observed experimentally at Reynolds numbers ranging from about 500 [11], to 513 – 526 [12], is caused by the radial, convective growth of disturbances to amplitudes large enough to cause non-linearities. A certain amount of variation in the reported transition Reynolds numbers is due to different methods of determination. However, we offer a quite different explanation for transition. There is a third spatial branch that Mack [2] mentions briefly. This family resides in the lower half α-

plane for large positive ω_i and therefore corresponds to a damped response in the region $r < r_s$. Here, pinch-points between this branch, henceforth called *family 3*, and the family 1 have been found. At large positive ω_i these branches lie in the distinct half α-planes and, for certain values of β, above a critical Reynolds number of 510.625, ω_i^o at the pinch-point is positive. As described in §2, these points correspond to absolute instabilities. Such an example is shown in figure 3 for $R_s = 530$ and $\beta = 67$. Figure 3(a) and 3(b) show the branches when the F-contour has values of $\omega_i = 0.0100$ and $\omega_i = 0.00400$, respectively. Figure 3(c) shows the branch-point, $\omega^o \simeq -0.0311 + i0.000289$, at $\alpha^o \simeq 0.213 - i0.123$ and where the wave angle $\varepsilon = \tan^{-1}(\bar{\beta}/\alpha_r) \simeq 30.7°$. Again, the broken line (_ .) shows the path in the α-plane as ω_i is decreased from infinity to the positive imaginary pinching frequency ω_i^o, with R_s and β constant (ω_r at the branch-point value). Below $R_s = 510.625$, ω_i^o can be reduced to zero before pinching occurs and the flow is then only convectively unstable.

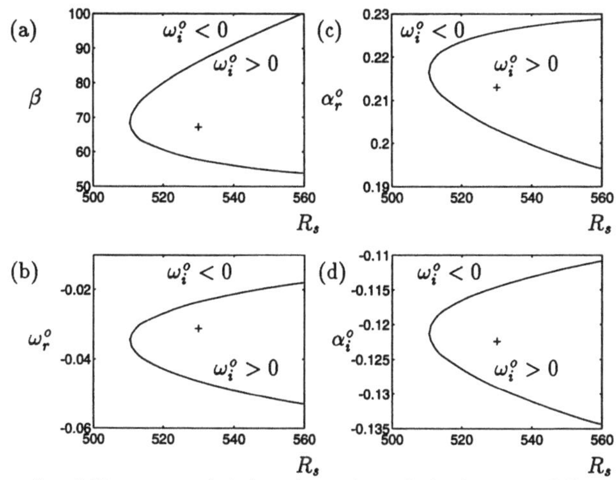

Figure 4. Neutral stability curves defining the region of absolute instability; absolutely unstable where $\omega_i^o > 0$ and convectively unstable where $\omega_i^o < 0$. The cross (+) marks the pinch-point depicted in figure 3.

Figure 4 shows the neutral stability curves for absolute instability; inside the stability curves ω_i^o is positive and represents absolutely instability. The parameter region over which the flow is absolutely unstable is shown to expand at Reynolds numbers above the critical value.

4. Conclusions

Absolute instability is identified by looking at the branch-point singularities in the dispersion relation and applying Briggs criterion [8]. A third branch of the dispersion relation has been found, that is generally heavily damped, but does meet the spatial branch termed family 1 by Mack [2] at pinch-points with positive ω_i. These points indicate absolute instability for $R_s > 510.625$ and suitable β. Below the critical Reynolds number, the flow is convectively unstable or stable depending on the parameter values. The onset of absolute instability is consistent with experimental observations of the critical Reynolds number for the onset of transition and may, therefore, be a better explanation for transition than the convective radial growth of disturbances, leading to non-linearity and may have implications for transition on highly swept wings.

Acknowledgements

This work has benefitted from discussions with Professor M. Gaster and Professor D.G. Crighton and I am grateful to Dr L.M. Mack for advice on the Gram-Schmidt orthonormalization technique. This work was supported by the Engineering and Physical Sciences Research Council and the Boeing Commercial Airplane Group.

5. References

[1] Huerre, P. and Monkewitz, P.A.: Local and global instabilities in spatially developing flows, Ann. Rev. Fluid Mech. 22 (1990) 473-537.

[2] Mack, L.M.: The wave pattern produced by point source on a rotating disk, AIAA 85-0490 (1985).

[3] Wilkinson, S.P. and Malik, M.R.: Stability experiments in the flow over a rotating disk, AIAA J. 21 (1985) 588-595.

[4] Kármán, Th. von: Über laminare und turbulente Reibung, Zeitschrift für angewandte Mathematik und Mechanik 1 (1921) 233-252.

[5] Lingwood, R.J.: Absolute instability of the boundary layer on a rotating disk. Part I: Viscous stability study, J. Fluid Mech. (1994) (submitted).

[6] Lingwood, R.J.: Absolute instability of the boundary layer on a rotating disk. Part II: Inviscid stability study, J. Fluid Mech. (1994) (submitted).

[7] Gaster, M.: On the generation of spatially growing waves in a boundary layer, J. Fluid Mech. 22 (1965) 433-441.

[8] Briggs, R.J.: Electron-stream interaction with plasmas, MIT Press, Cambridge, Mass. (1964).

[9] Benney, D.J. nd Gustavsson, L.H.: A new mechanism for linear and nonlinear hydrodynamic instability, Studies in Appl. Math. 64 (1981) 185-209.

[10] Koch, W.: Direct resonance in Orr-Sommerfeld problems, Acta Mechanica 58 (1986) 11-29.

[11] Kobayashi, R., Kohama, Y. and Takamadate, Ch.: Spiral vortices in boundary layer transition regime on a rotating disk, Acta Mechanica 35 (1980) 71-82.

[12] Malik, M.R., Wilkinson, S.P. and Orszag, S.A.: Instability and transition in rotating disk flow, AIAA J. 19 (1981) 1131-1138.

Spatial Direct Numerical Simulation of Transition in a Three-Dimensional Boundary Layer

W. Müller*, H. Bestek*, H. Fasel‡

*Institut für Aerodynamik und Gasdynamik, University of Stuttgart, Pfaffenwaldring 21, 70550 Stuttgart, Germany,

‡Aerospace and Mechanical Engineering Department, University of Arizona, Tucson, AZ 85721, U.S.A.

Summary

The spatial development of artificially excited crossflow vortices and superposed travelling waves in a three-dimensional boundary layer is investigated by direct numerical simulation. For this investigation the pressure distribution of the transition experiment at DLR Göttingen is used. The simulations are based on a well tested finite difference/spectral method [1], which allows the numerical investigation of an incompressible 3-D boundary layer flow over an infinite swept wing where effects of surface curvature are neglected. All other effects, such as of nonparallel base flow, streamline curvature and nonlinearity are included in the numerical model. Two simulations under conditions close to the experimental situation of the DLR swept-wing experiment [2, 3, 4] were carried out. In one simulation, a steady vortex was generated, which developes very similar to the vortex dominated experiment. For a simulation with a vortex-wave interaction the steady vortices saturate at a significant lower level.

Keywords: 3-D boundary layer, crossflow instability, direct numerical simulation.

1 Introduction

To understand the key mechanisms that lead to transition in a three-dimensional boundary layer flow is still one of the great challenges in transition research. A few experiments exists that allow for the separate investigation of the main instabilities at swept wing configurations [5, 6]. Together with well-documented experiments that mostly focussed on 'natural' transition, spatial direct simulations are an useful tool to contribute towards understanding of the relevant phenomena.

In the natural transition experiment by Bippes and coworkers [3] the mechanisms of crossflow instability in a three-dimensional boundary layer are investigated. The experimental set-up consists of a flat plate with 45° geometric sweep angle (Figure 1). As reported by Müller [4], the effective sweep is 42.5°, which is due to a slight deflection of the oncoming flow. With the flat plate geometry surface curvature effects are excluded and therefore it is easier to compare experimental and theoretical results. For comparison with the direct numerical simulations the measurements obtained in a low turbulence windtunnel with $Tu = 0.05\%$ and a plate with low surface roughness (approx. $6\mu m$, [7]) were used as a database. It was found, that under these conditions the steady disturbances (crossflow vortices) dominate over the unsteady disturbances. The steady disturbances saturate at a high amplitude level. The crossflow vortices have a nearly constant spanwise wavelength of $\lambda_z \approx 12mm$ [8]. For unsteady waves, peak amplitudes in the spectra were found for frequencies of approximately 70 Hz and 135 Hz but at amplitude levels significant lower than those of the crossflow vortices. According to linear stability theory calculations for this flow, the 135 Hz-disturbances have the strongest amplification.

2 Numerical Model

The numerical method is based on the three-dimensional Navier-Stokes equations for incompressible flow in vorticity-velocity formulation (see for example [9]). The "spatial" model used here, allows for simulations of spatially evolving, three-dimensional disturbances in a growing boundary layer. All variables are nondimensional, relating to the corresponding dimensional variables (denoted by overbars) as

$$x = \bar{x}/\bar{L}, \quad y = \bar{y}/\bar{L}\sqrt{Re}, \quad z = \bar{z}/\bar{L}, \quad t = \bar{t}\bar{U}_\infty/\bar{L},$$
$$u = \bar{u}/\bar{U}_\infty, \quad v = \bar{v}/\bar{U}_\infty\sqrt{Re}, \quad w = \bar{w}/\bar{U}_\infty, \quad Re = \bar{U}_\infty\bar{L}/\bar{\nu}.$$

where u, v and w are the velocity components in x-, y-, z-direction, respectively, $\bar{U}_\infty$ is a reference velocity, $\bar{L}$ is a fixed reference length and $\bar{\nu}$ is the kinematic viscosity. For the results presented in this paper, the Reynolds number is $Re = 92000$ ($\bar{U}_\infty = 14m/s, \bar{L} = 0.1m, \bar{\nu} = 1.5217 * 10^{-5}m^2/s$). The basic equations are split into a set of equations for the three-dimensional steady base flow (denoted by subscript B) and the three-dimensional disturbance flow (denoted by a prime), i.e. for all variables we have $f = f_B + f'$. The integration domain (Fig. 2) is a rectangular box located downstream of the leading edge. It extends from $x = x_0$ to $x = x_N$, with x_0 being the distance from the leading edge. In the y-direction, the height of the domain is chosen such that it covers approximately 1.7 boundary layer thicknesses at $x = 5.0$. In the spanwise z-direction, the domain extends over one spanwise wavelength of the expected crossflow vortices. For comparison with the experimental results, the flow quantities are considered in a local, flow orientated coordinate system denoted by index s, which is aligned with the curved external streamline (Fig. 2).

The pressure distribution of the transition experiment [3] is implemented into the numerical model by prescribing the corresponding streamwise velocity distribution of the external flow at the freestream boundary. Due to the infinite swept wing assumption the spanwise velocity component $w_B(y)$ outside the boundary layer is constant along the x- and z-axis. At the inflow boundary, the velocity components $u_B(y)$ and $w_B(y)$ can be taken from the Falkner-Skan-Cooke profiles that correspond to the local pressure gradient. Then, the $v_B(y)$ profile can be obtained by integration of the continuity equation. At the outflow boundary, all equations are solved by neglecting the terms with second derivatives with respect to x. At the wall no-slip conditions are used.

For the disturbance flow, we prescribe the normal velocity component at the wall as a function of x, z and t. This allows to introduce controlled disturbances within a narrow disturbance strip. Outside the disturbance strip the normal velocity component is zero. At the freestream boundary, vanishing vorticity fluctuations and an exponential decay of the velocity disturbances are assumed. At the inflow boundary, all velocity and vorticity components are set to zero. For the treatment of the outflow boundary, an artificial suppression of the disturbance vorticity components is enforced within a region upstream of the outflow boundary. With this measure the disturbance level at the outflow boundary is substantially reduced (see Kloker *et al.* [10] for details). Together with a decay-condition for the velocity disturbances employed directly at the outflow boundary reflections of disturbances through passing the outflow boundary are minimized. A detailed description of the boundary conditions used for the calculations of the three-dimensional disturbance flow is given by Fasel *et al.* [9].

For the numerical solution of both, the base flow and the disturbance flow equations, fourth-order accurate finite-difference approximations are employed in x- and y- direction, respectively. For the discretisation in spanwise direction of the disturbance flow a spectral ansatz is used exploiting the periodicity with respect to z. For the base flow, the vorticity transport equation is solved by a semi-explicit Euler scheme using artificial time stepping. For the disturbance flow the integration in time is performed using an explicit Runge-Kutta scheme of fourth-order accuracy.

3 Numerical Results and Discussion

In order to investigate the nonlinear disturbance development, two extensive simulations with identical resolution were carried out matching the parameters and conditions of the natural transition experiment as closely as possible. In the first simulation case, called VORTEX, a steady crossflow vortex is generated at the disturbance strip, with an amplitude level comparable to that of the experiment [2]. In the second simulation case, called WAVE, the same steady vortex is excited at the disturbance strip, but now superposed is a single 135 Hz travelling wave with same spanwise wavelength and initial amplitude as the vortex. In this second case, the vortex/wave interaction is then studied by comparing time-averaged and rms-amplitudes with results of a double Fourier decomposition (frequency, spanwise wavenumber) of the disturbance flow.

3.1 Base Flow and Linear Theory

For the calculation of the baseflow a constant pressure gradient was approximated from the experimental data as it was done by Meyer & Kleiser in [8]. The integration domain begins at 20% chord ($x_0 = 1.0$) and extends to $x = 6.0$. A regular grid of 121 $\times$ 3834 points (normal, streamwise) and ten spanwise Fourier modes (after de-aliasing) are used. Thus, together with the spanwise periodicity 80% of the plate's surface are covered by the simulation. At the inflow boundary the local Reynolds number was $Re_{\delta_1,s} = 570$ and the local flow direction was about 62.5° (Fig. 2). A comparison of the local boundary-layer profiles at 60% chord ($Re_{\delta_1,s} = 804$) in Fig. 3 shows very good agreement with experiment.

For velocity profiles at $x = 2.0$ a complete stability diagram was calculated in order to predict the disturbance development of steady vortices and travelling waves, see Fig. 4. Here, the amplification rate α_i is obtained directly from spatial linear stability theory (Orr-Sommerfeld) as a function of the frequency β ($\beta = \bar{\beta}\bar{L}/\bar{U}_\infty$) and the spanwise wavenumber γ ($\gamma = \bar{\gamma}\bar{L} = 2\pi/\lambda_z$). The upper half (positive γ) is related to the more unstable waves, for which the wavenumber vectors are usually directed towards the leading edge. Due to the weaker amplification, the linear waves corresponding to the lower half seem to play no important role during the transition process.

As can be seen from this diagram, linear crossflow vortices with $\gamma \approx 48$ are most unstable. The most unstable travelling waves with a similar spanwise wavenumber are at a frequency $\beta \approx 7$ ($\bar{f} \approx 156Hz$), and have a 50 % stronger amplification rate than the crossflow vortex. Due to the slow variation of the amplification rates in streamwise direction these results are within the bandwidth of the experimentally observed values for the most amplified spanwise wavelength of the steady disturbances and the frequency of the travelling waves [3].

As indicated in Fig. 4 (see marks), for the initially disturbed modes of the numerical simulations discussed in the following paragraphs, a spanwise wavenumber $\gamma = 52.4$ and a frequency $\beta = 6$ was used.

3.2 Evolution of a Crossflow Vortex

For the simulation case VORTEX a single crossflow vortex was generated at the disturbance strip. Its initial amplitude was chosen such that the computed and measured [2] crossflow vortex amplitudes at $x = 2.0$ were approximately equal. In Figure 5, the amplitudes from the time-averaged data $\langle u'_{3D} \rangle$ at $y/\delta = 0.24$ from simulation (solid line) and experiment [2] (dashed line, symbols) show the same amplification until about $x = 3.5$. This is in satisfactory agreement with the amplitude growth as calculated from linear theory (dashed-dotted), although the linear amplification rate is slightly smaller [1]. Downstream of this location, an amplitude saturation at an amplitude level of $\langle u'_{3D} \rangle = 20\%$ can be observed. The spanwise Fourier decomposition of the data into modes (0,k), where k are the integer multiples of the fundamental wavenumber leads to the following situation (Fig. 6): The first spanwise harmonic (mode (0,1)) has a growth rate that is nearly equal to the time-averaged data in Figure 5, i.e. it contains most of the information of the crossflow vortex. The second and zeroth harmonic show almost identical amplitudes until mode (0,1) reaches its saturation level. Then, the amplitudes of these two modes start to deviate and the mean flow distortion (mode (0,0)) passes the amplitude of mode (0,1), while mode (0,2) maintains its amplitude downstream of $x = 3.5$. The higher harmonics have a similar behaviour up to mode (0,2), however at altogether lower amplitudes.

3.3 Vortex-Wave-Interaction

Superposing the crossflow vortex of case VORTEX with a travelling wave (case WAVE), the disturbance development changes completely. This is obvious for example from Figure 7, where the u'_{rms}- and $\langle u'_{3D} \rangle$-amplitudes are plotted. In this figure an additional curve for the amplitude growth of the steady vortex as obtained from linear stability theory is also shown for comparison. Due to its stronger linear growth, the travelling wave reaches its saturation amplitude of $u'_{rms} \approx 12\%$ much further upstream than the vortex in the former case. Therefore, the travelling wave dominantes the flow behaviour. As known from experiments [2], maximal and minimal rms-fluctuations were found at spanwise positions of minimal and maximal $\langle u'_{3D} \rangle(z)$- amplitudes, respectively. The growth rate of the steady vortex deviates from linear theory prediction even at early downstream positions. Further downstream, the steady vortex reaches an amplitude level of only 1/10 of that of case VORTEX in Fig. 5.

In order to isolate the main mechanisms of the vortex-wave-interaction, the flow was analysed by a double Fourier decomposition into modes (h,k) with h being integer multiples of the basic frequency β. The amplitude development of the modes (h,k) with h = k is shown in Figure 8. Similar to case VORTEX, the mean-flow distortion mode (0,0) and the higher harmonic to the fundamental wave (mode (2,2)) have nearly equal amplitudes until mode (1,1) reaches its saturation amplitude. Then, mode (0,0) catches up with the fundamental wave, while mode (2,2) roughly maintains its amplitude. Now even mode (7,7) reaches an amplitude of 0.1%. The 0-Hz modes of case WAVE are plotted in Figure 9 together with the fundamental wave. While the amplitude of the generated vortex mode (0,1) saturates at about 2%, the higher spanwise harmonics (0,k) grow further. Comparing Fig. 6 and Fig. 9 it appears, that the mean-flow distortion in case WAVE is generated by the travelling wave which has therefore a completely different growth rate than mode (0,2).

From the double Fourier decomposition one may conjecture, that even modes with high frequencies contribute considerably to the nonlinear transition process. This can be expressed schematically in the frequency-spanwise wavenumber spectra in Figure 10. The dominant fundamental wave (1,1) generates the mean-flow distortion (0,0) and its higher harmonics (h,k) (with h = k). The nonlinear interaction of the crossflow vortex mode (0,1) with the fundamental wave and its higher harmonics generates modes (h,k) on the lines parallel to the diagonal $k = h \pm 1$, (see Fig. 11). This nonlinear mechanism is present even at lowest amplitudes. The generation of high frequency harmonics might also be an explanation for the different frequency ranges reported in [3].

References

[1] Müller, W., Bestek, H., Fasel, H.: Numerical Simulation of the Spatial Disturbance Development in Transitional Boundary Layers along a Swept Plate, *Proc. Int. Conf. on Near-Wall Turbulent Flows*, Tempe, Arizona, March 15-17, 1993.

[2] Müller, B., Bippes, H.: Experimental Study of Instability Modes in a Three-Dimensional Boundary Layer, AGARD-CP-438, 1989, pp. 13.1 - 13.15.

[3] Müller, B.: Experimental Study of the Travelling Waves in a Three-Dimensional Boundary Layer, *Proc. 3rd IUTAM Symposium on Laminar-Turbulent Transition*, Toulouse, France, Springer 1990.

[4] Müller, B.: Experimentelle Untersuchung der Querströmungsinstabilität im linearen und nichtlinearen Bereich des Transitionsgebietes, DLR-FB 90-09, Göttingen, Germany, 1990.

[5] Bippes, H.: Instability Features Appearing on Swept Wing Configurations. *Proc. 3rd IUTAM Symposium on Laminar-Turbulent Transition*, Toulouse, France, Springer 1990.

[6] Radeztsky Jr., R.H., Reibert, M.S., Saric, W.S.: Development of Stationary Crossflow Vortices on a Swept Wing. AIAA 94-2373, 1994.

[7] Deyhle, H.: Einfluß der äußeren Strömungsbedingungen auf den Transitionsprozeß einer dreidimensionalen Grenzschicht, Ph.D. thesis, Fortsch.-Ber. VDI Reihe 7 Nr. 226. VDI-Verlag, Düsseldorf, 1993.

[8] Meyer, F., Kleiser, L.: Numerical Investigation of Transition in 3D Boundary Layers, AGARD-CP-438, 1989, pp. 16.1 - 16.17.

[9] Fasel, H., Rist, U., Konzelmann, U.: Numerical Investigation of the Three-Dimensional Development in Boundary-Layer Transition, AIAA J. 28, 1990, pp. 29-37.

[10] Kloker, M., Konzelmann, U., Fasel, H.: Outflow Boundary Condition for Spatial Navier-Stokes Simulations of Laminar-Turbulent Transition, AIAA J. 31, 1993, pp. 620-628.

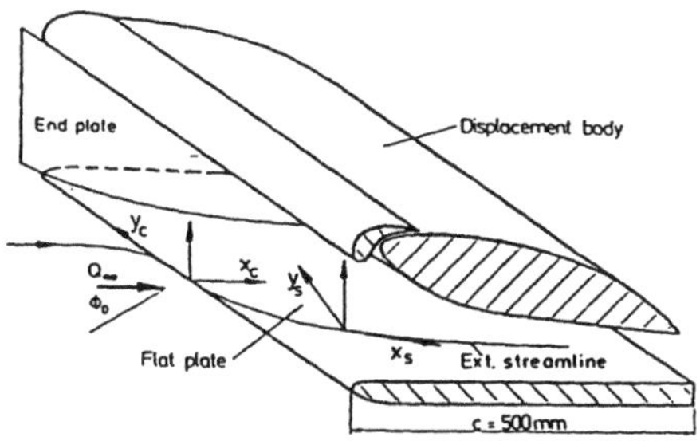

Figure 1: Set-up of the DLR transition experiment 'crossflow instability' [3]

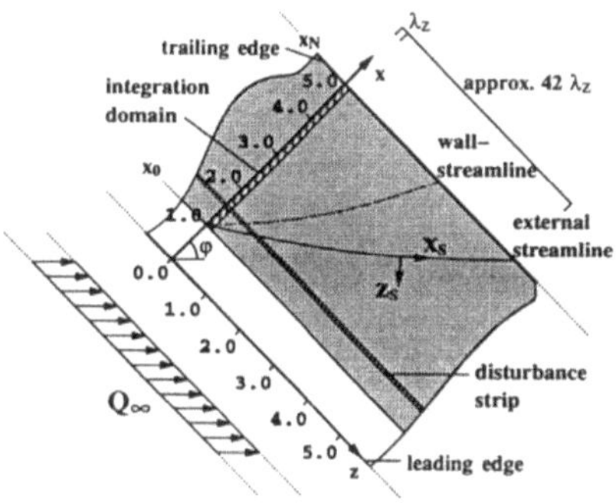

Figure 2: Top view of the integration domain (in scale)

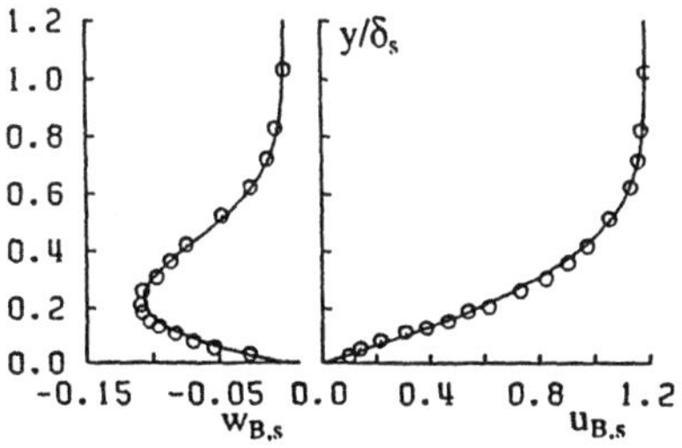

Figure 3: Comparison of boundary layer profiles between experiment (symbols, [3]) and simulation (solid line), x = 3.0

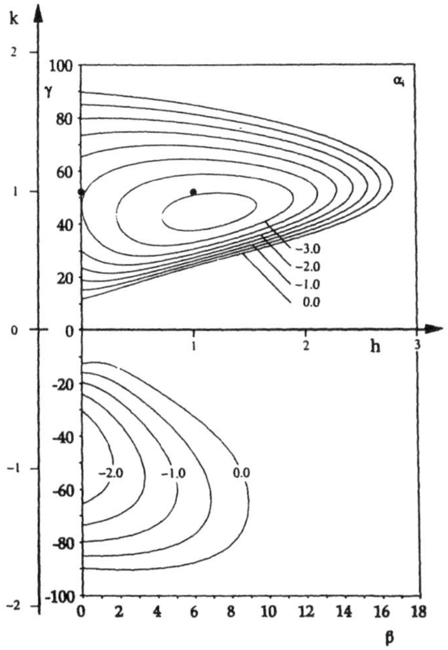

Figure 4: Stability diagram for the DLR transition experiment at x=2.0. • Generated modes for case VORTEX (0,1) and WAVE (0,1);(1,1)

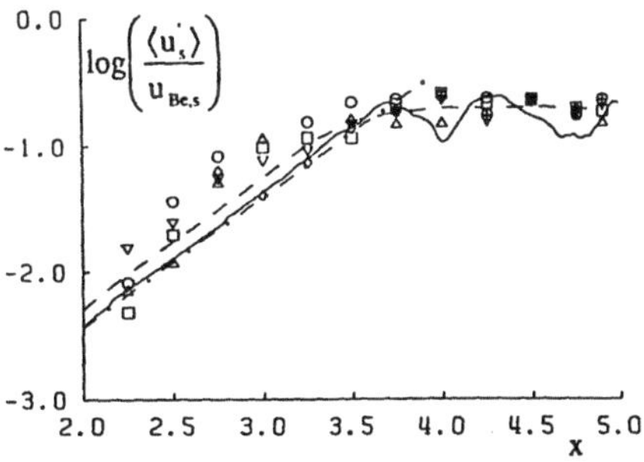

Figure 5: Comparison of time- averaged amplitudes from experiment [2] (symbols, dashed line), simulation (solid) and lin. theory (dashed-dotted). Case VORTEX

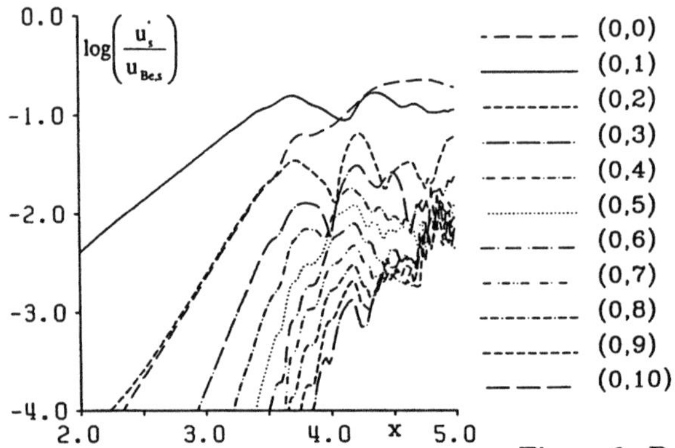

Figure 6: Development of modes (0,k). Case VORTEX

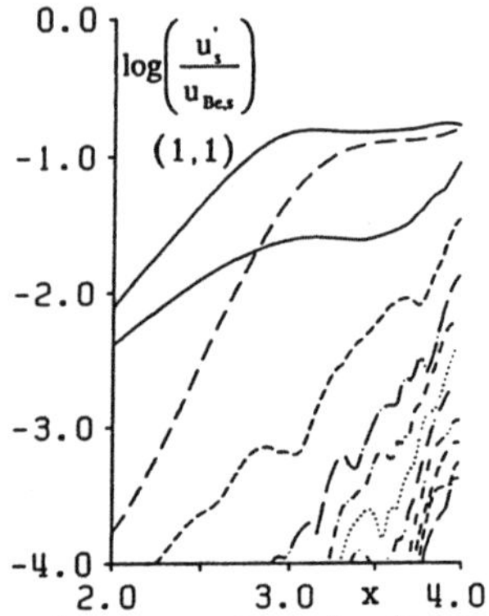

Figure 9: Development of modes (0,k). Case WAVE

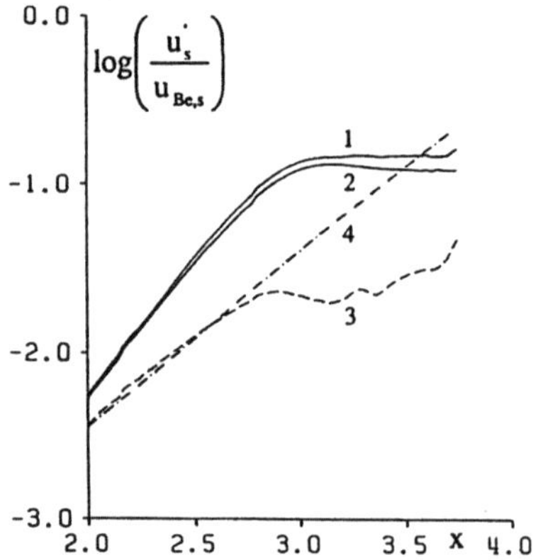

Figure 7: Development of max. and min. rms- amplitudes (curve 1,2), stat. disturbances (3) and lin. theory for CF-vortex (4). Case WAVE.

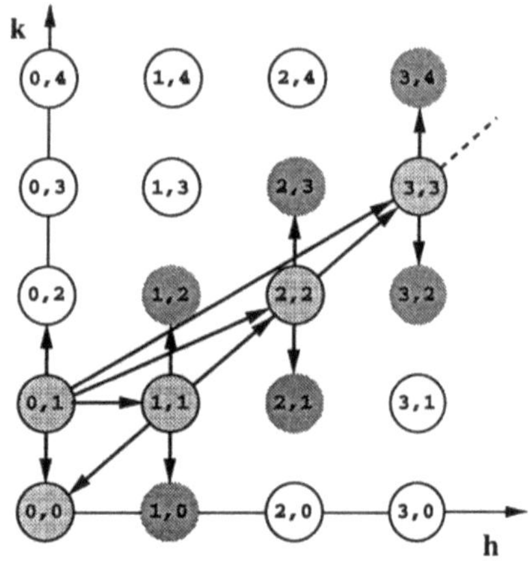

Figure 10: Leading nonlinear vortex-wave-interactions in frequency- spanwise wavenumber plane

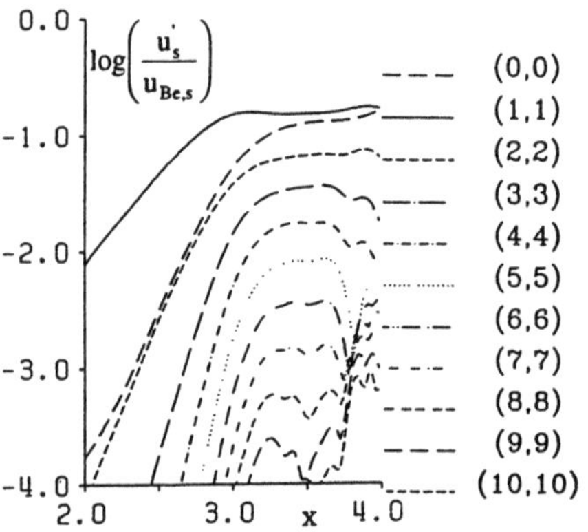

Figure 8: Development of modes (h,k), h=k. Case WAVE

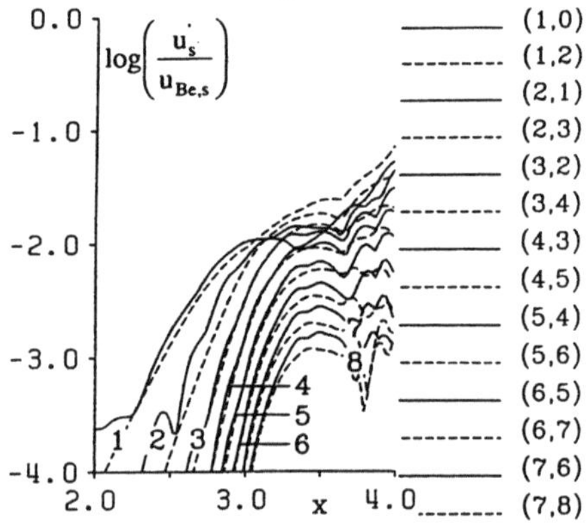

Figure 11: Development of modes (h,k), k=h±1. Case WAVE

Numerical and Experimental Investigation of the K-Regime of Boundary-Layer Transition

U. Rist
Institut für Aerodynamik und Gasdynamik, University of Stuttgart,
70550 Stuttgart, Germany
Y. S. Kachanov
Institute of Theoretical and Applied Mechanics, 630090 Novosibirsk, Russia

Summary

One particular case of K-type transition has been investigated using hot-wire measurements and spatial direct numerical simulation (*DNS*). Detailed quantitative comparisons of the results of both approaches showed very good agreement of the spatial disturbance development, the disturbance spectra, the instantaneous velocity traces, and the local frequency–spanwise-wave-number spectra. Indications for a direct generation of three-dimensional modes as higher harmonics of the fundamental modes were found. A closer look at the phase speeds of these modes, however, revealed that weak-nonlinear interactions are only initially appropriate to describe the flow, they fail when local events dominate, like, for example the formation of small-scale vortices in the boundary layer. The investigation of the later stages showed that the hot-wire 'spike'-signals are connected with small ring-like vortices.

1. Introduction

Since its discovery more than thirty years ago [1], the K-regime of boundary-layer transition has repeatedly attracted scientific attention due to its complicated nonlinear nature. Different, partly controversal, theoretical models to describe the initial disturbance development appeared during this period (cf. discussions in [1] – [5], for example). A verified numerical/experimental data base could help to evaluate such models and to more clearly identify the relevant mechanisms. Additional difficulties in exploring K-type transition are connected with the investigation of the 'late stage' which is dominated by the nonlinear, three-dimensional, and unsteady character of the disturbances. At present, there are only two feasible approaches for the investigation of this regime: the laboratory experiment, and the direct numerical simulation using the complete Navier-Stokes equations.

First comparisons of the experimental data obtained in [6, 7] with results of DNS carried out in [8, 9] proved to be very fruitful and showed very good quantitative agreement for the initial stage leading to the appearance of high-frequency fluctuations called 'spikes'. Here we continue our combined theoretical and experimental studies of the K-breakdown with the purpose of verification of each others data and providing a more detailed data base for identification of the main physical mechanisms of this type of transition.

2. Experimental and numerical data base

The experiments were performed in Novosibirsk on a flat plate mounted in the ITAM low-turbulence wind tunnel running at a speed of $U_0 = 9.18m/s$. Disturbances were introduced by a vibrating ribbon driven at $f_1 = 96.4Hz$ ($F = 2\pi f_1 \nu / U_0^2 = 1.1 \cdot 10^{-4}$) and spanwise spacers with a periodicity of $\lambda_z = 25$mm in spanwise direction (z) placed at $x = 250$mm from the leading edge of the flat plate ($R = (U_0 \delta_1 / \nu)^{1/2} = 670$, where δ_1 is the displacement thickness). Covering the region from $x = 300$mm to $x = 700$mm, signals of the downstream velocity component u were registered using a single hot-wire probe. All time traces of the probe were recorded relative to the forcing current and the instantaneous 3-D flow field could thus be reconstructed from these recordings. The amplitude of the vibrating ribbon was adjusted so that the first spikes in the disturbance velocity traces occured at $x \approx 430$mm.

Several numerical simulations of these experiments have been performed in Stuttgart using the DNS-scheme originally developed by Fasel et al. [10], improved by Kloker et al. [11], and more recently by Kloker [12]. The scheme is based on spectral approximations of the spanwise z-coordinate by a truncated Fourier series ($-K \leq k \leq K$), and on fourth-order accurate finite-differences in downstream x- and wall-normal y-directions. Time integration of the discretized Navier-Stokes equations was performed by a fourth-order accurate Runge-Kutta scheme. Results of two such simulations are presented here: one on a 2674×121 grid using $K = 15$, and one using 1394×273 grid points and $K = 64$. The first integration domain extended over 10 δ_1 in y-direction[1], and from $x_0 = 188$mm to $x_N = 572$mm in downstream direction, the second extended over 11.4 δ_1 at $x_1 = 396$mm and until $x_N = 767$mm, with the 'artificial relaminarization zones'[11] in both simulations starting at $x = 502$mm, and $x = 663$mm, respectively.

Disturbances consisting of a large-amplitude 2-D TS-wave with nondimensional frequency $F = 1.1 \cdot 10^{-4}$ and a steady spanwise modulation with period $\lambda_z = 25$mm were introduced in the first simulation at $x = 250$mm using suction and blowing at the wall [10]. The disturbance amplitudes of the 2-D periodic forcing and of the 3-D steady mode at $x = 250$mm were chosen to closely match the u'-disturbance amplitudes at $x = 300$mm in the experiment. The second simulation was started using data and instantaneous boundary conditions at x_1 obtained from an earlier simulation.

3. Comparison of results

Detailed quantitative comparisons of the numerical results with the experimental measurements of the u-velocity component are presented both in physi-

[1] at inflow

cal space (section 3.1) and in spectral space after Fourier decomposition into frequency–spanwise-wave-number harmonics (section 3.2). A comparison of the 'late-stage' structures is presented in section 3.3. Data of the first simulation are shown, except were noted.

3.1 Comparison of disturbance amplitudes and phases

Amplitude and phase distributions of the u-velocity component versus x and y are directly compared in Figure 1. In addition to the *rms*-amplitudes of the disturbance signals, their harmonic content (after Fourier analysis) is also shown. A constant distance of y = 4mm from the flat plate was chosen in Figure 1a in order to demonstrate the good quantitative correlation of the data in the 'initial' and 'late-stage' of K-breakdown. The sudden increase of all amplitudes at the spanwise peak station quite clearly separates the two regimes. At the valley station, nothing similar is observed. A small branching of different curves in the DNS-data which is not observed in the experiments, indicates that the DNS-results using more spanwise modes (129 instead of 31) must be used for comparisons downstream of x = 450mm.

Amplitude and phase profiles for two x-stations, one before, and one after the sudden increase of the disturbance amplitudes are shown in Figures 1b and 1c, respectively. Although some differences appear in the comparison of the mean flow[2], especially in the outer part of the boundary layer, the mean-flow influence does not appear to be crucial for the disturbances, since their profiles agree qualitatively very well. Due to the rapid downstream amplification of the initial disturbances one could expect that a small raise of the upstream forcing amplitudes would lead to a quite perfect quantitative agreement of the numerical results with the experimental measurements.

The phase profiles in Figure 1b and c show the characteristic phase synchronization originally observed by Kachanov et al. [6] which is due to the 'spikes'. The phases are plotted relative to the instant of the passage of the 'spike'. Using this normalization, a small but throughout the boundary layer approximately constant phase shift of the numerical data relative to the measurements may be observed at x = 450mm. The shift amounts to $\phi_n \approx n \cdot 20^o$, where $n = f_n/f_0$, indicating that the 'spike' appears at a slightly different time (6% of the disturbance cycle). Such a good agreement of the numerical results with the measurements for x = 450mm was originally not expected from a simulation using 31 spanwise Fourier modes only.

The effect of increasing the number of spanwise modes in the computations is shown in the next figure together with a comparison of instantaneous 'peak'-disturbance profiles at x = 450mm. The results of the two data sets described in section 2 are shown in Figure 2a and b. Both agree quantitatively very well with the measurements in Figure 2c. As could be expected already from Figure 1, the agreement for the data using fewer spanwise modes is already rather good but it is improved using more spectral modes. We will nevertheless continue using

[2] At x = 400mm: $\delta_{1_{DNS}}$ = 1.52mm, $\delta_{1_{Exp}}$ = 1.66mm and at x = 450mm: $\delta_{1_{DNS}}$ = 1.66mm, $\delta_{1_{Exp}}$ = 1.92mm.

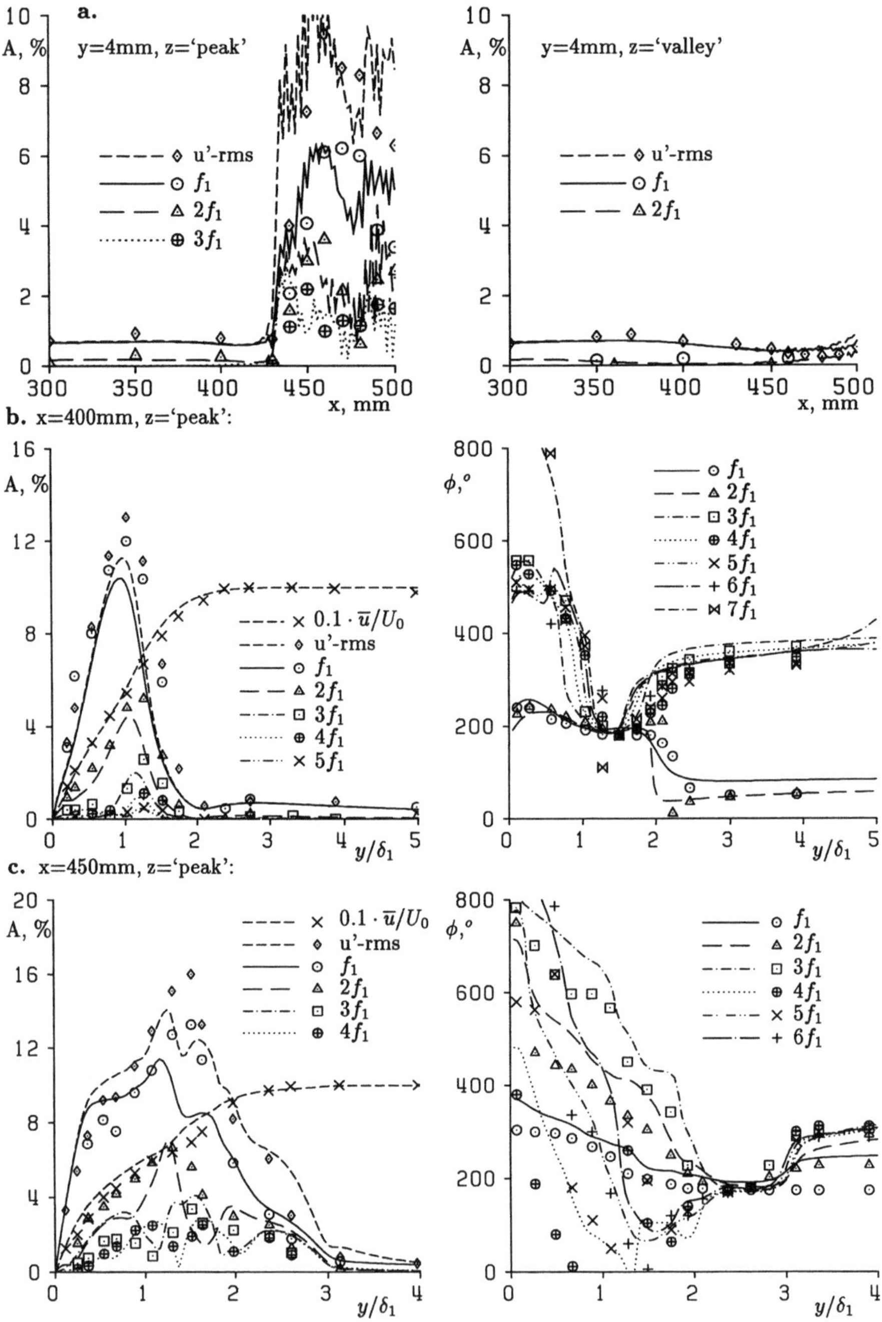

Figure 1: Comparison of mean flow ($\overline{u}$), u'-disturbance amplitudes (A), and phase profiles (ϕ). Lines = DNS, symbols = experimental measurements, f_1 = disturbance frequency.

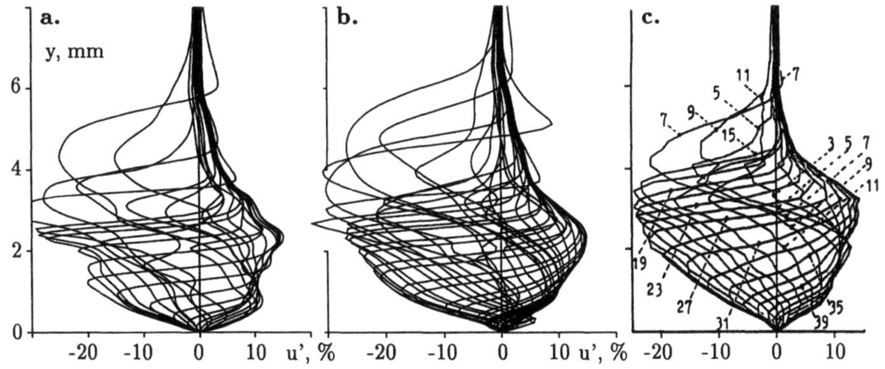

Figure 2: Comparison of instantaneous u'-amplitudes at x = 450mm, z ='peak'. **a.** Simulation using 31 spanwise modes (22 instants/period shown), **b.** simulation using 129 modes (28 instants/period), **c.** experiment (48 instants/period).

the first data set here, except for the comparison of the 'late-stage' structures in section 3.3.

3.2 Comparison of disturbance spectra

The study of the behavior of modes in frequency and frequency-wave-number-spectra as performed in [6, 7], for example gives information about possible weak-nonlinear disturbance interactions, as well as about the geometrical (shape) characteristics of the coherent structures observed at late stages of K-breakdown. The detailed comparison of the spectral amplitudes B defined by $u(x,y,z,t) = \sum_n \sum_k B_{n,k}(x,y) \cdot e^{i(nf_1 t + k\beta_0 z)} + c.c.$ at y = 1mm in Figure 3a illustrates the formation of spanwise and frequency higher harmonics with increasing x in very good quantitative agreement.

In a first step to isolate possible weak-nonlinear interactions between different (n,k)-modes, the approximate amplification rates $\kappa = B(x_1)/B(x_0)$ have been computed from Figure 3a and also compared in Figure 3b. Numerical and experimental results agree well for different frequencies and moderate values of the spanwise wave numbers. However, in contrast to the experimental data, the disturbances propagating at very high angles of inclination with respect to the flow direction grow much faster in the numerical simulation. One possible cause for this discrepancy might be due to less accurate amplitudes $B(x_0)$ in the experimental data when computing κ.

The strong increase of the amplification rates with the spanwise wave number indicates that the large-β modes are most probably just higher-harmonic disturbance components of the fundamental 3-D modes. Additional numerical computations showed that the large amplification rates just observed cannot be explained by parametric resonance of the 3-D modes with higher-harmonic 2-D modes, which is an idea originally stated in [7]. Test calculations using different initial amplitudes for these higher harmonics indicated that their downstream development (after an initial transient) is independent of the initial amplitude.

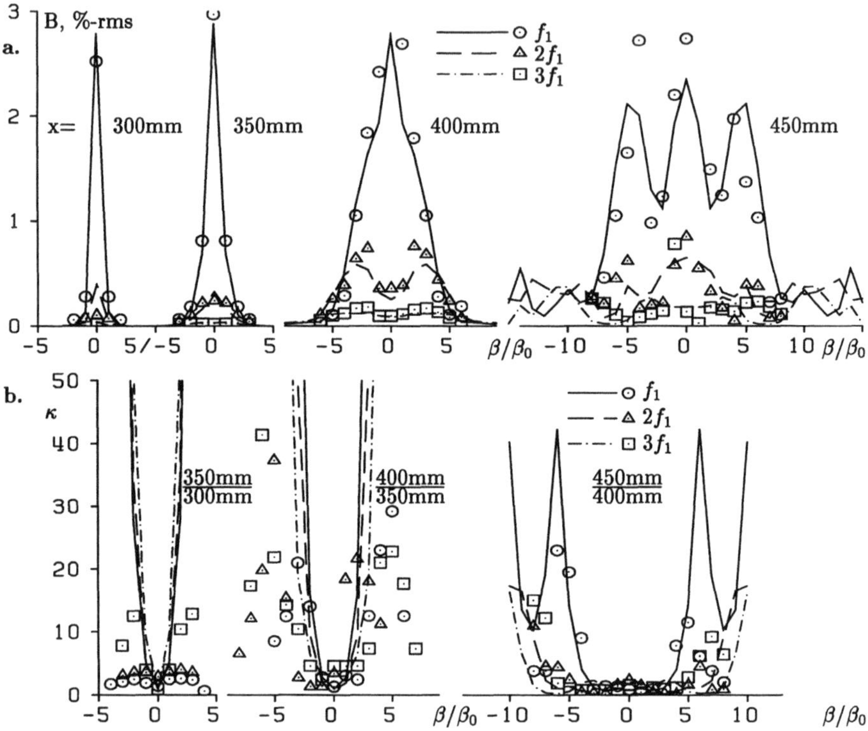

Figure 3: Comparison of spectral amplitudes B (**a.**) and amplification factors $\kappa = B(x_1)/B(x_0)$ (**b.**) vs. spanwise wave number at $y = 1$mm. Lines = DNS, symbols = experiments, spanwise wave number $\beta_0 = 2\pi/\lambda_z$, disturbance frequency f_1.

This also supports our interpretation.

Additional interesting spectral features are observed in the phase speeds of the (n,k)-modes in Figure 4. Initially (i.e., for small x), all modes have a 'wave-like' behavior, i.e., their phase speeds vs. y are constant and approximately equal, indicating a 'phase-lock' of higher harmonic modes with the fundamental modes. Also, one may correlate the amplitudes of single modes with the critical layer (y-position where $c_r = \overline{u}$): it appears that the nonlinearly generated higher harmonics do not necessarily have a maximum at y_{crit}. For larger x, however, c_r vs. y is no longer constant and each mode has a different phase speed outside of the boundary layer. In addition, the critical layer is at a y-position where no particularity is observed in the amplitude profiles. Around $y \approx 2$mm, instead, a new region appears where all modes have identical phase speed and rather large local disturbance amplitude of the higher harmonics. This is exactly the position where the 'spike' occurs. The 'spike' apparently belongs to a coherent structure that travels with its own speed. Weak-nonlinear wave interactions are most probably no longer appropriate to describe this behavior; local dynamics dominate, instead.

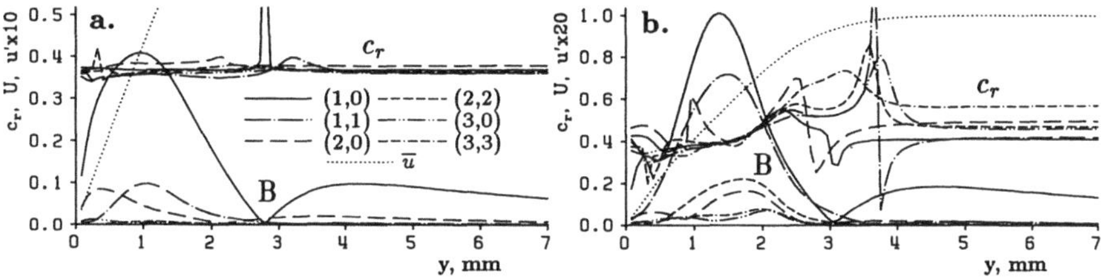

Figure 4: Comparison of phase speeds c_r with mean-velocity $\overline{u}$ and spectral amplitudes B for various frequency–spanwise-wave-number modes (n, k). **a.** $x = 350$mm, **b.** $x = 410$mm. (Simulation)

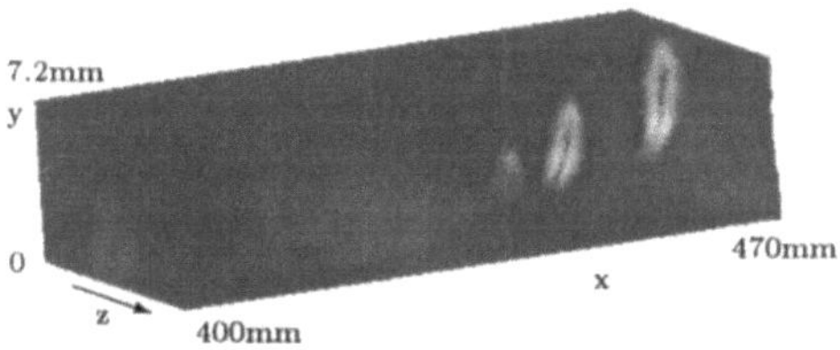

Figure 5: Volumetric perspective rendering of computed instantaneous 3-D pressure field. The grey scale used is proportional to $-p$, low-pressure regions therefore appear as light structures.

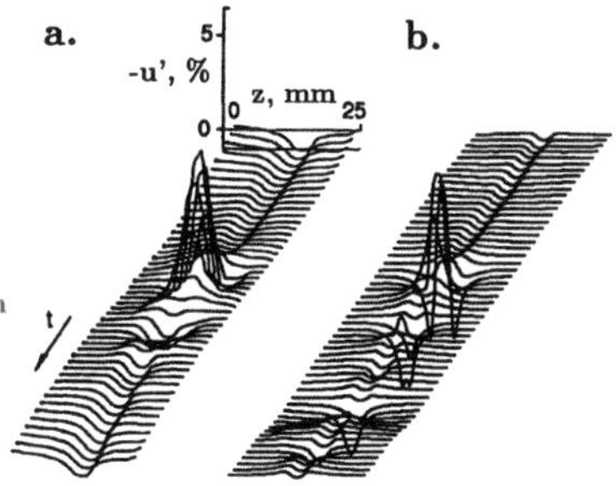

Figure 6: Comparison of 'spike'-signals vs. z and t at $x = 500$mm, $y = 8$mm. **a.** Experiment, **b.** Simulation.

3.3 Late-stage structures

Regarding the further development of spikes, including their multiplication and downstream convection in a soliton-like manner (for a discussion of these aspects see, e.g., [5, 13]), the comparison with the DNS gives also new insights: The spikes are connected with ring-like vortices snatching off from the downstream ends of Λ-vortices. This behavior can be seen in flow visualizations using particles, velocity, vorticity and pressure (cf. Figure 5). The typical scale of these new structures is of the same order as the boundary layer displacement thickness, and their shape, orientation and position within the boundary layer is very similar to those of so-called 'typical eddies' found in experiments by Falco [14] in the external part of developed turbulent flow. This indicates that the study of K-type transition is also important for the understanding of turbulent flows.

Using data of the latest finely resolved simulation, a detailed investigation of the final breakdown to turbulence is now possible. A first direct comparison of the 'spike'-amplitudes in the 'late stage' of K-breakdown is shown in Figure 6. Again, a very good quantitative agreement is observed using these new data.

4. Conclusions

By detailed quantitative comparisons of experimental measurements and results of direct numerical simulations, a validated data base has been generated that may be used for the validation of different theories, e.g., weakly nonlinear and strongly nonlinear wave interactions, local inflectional instability, wave kinematics, soliton theory, etc. Evaluation of the 'late-stage' data has just begun, interesting new results may be expected.

Acknowledgements

The contributions of V.I. Borodulin, Novosibirsk to the experiments, and of M. Kloker, Stuttgart to the simulations are gratefully acknowledged.

References

[1] Klebanoff, P.S., Tidstrom, K.D. and Sargent, L.M.: The three-dimensional nature of boundary-layer instability, J. Fluid Mech. 12 (1962) 1-34.

[2] Craik, A.D.D.: Nonlinear evolution and breakdown in unstable boundary layers, J. Fluid Mech. 99 (1980) 247-265.

[3] Nishioka, M., Asai, M. and Iida S.: An experimental investigation of the secondary instability, *Laminar-Turbulent Transition* (eds. R. Eppler and H. Fasel), Springer, Berlin, Heidelberg (1980) 37-46.

[4] Borodulin, V.I. and Kachanov, Y.S.: Role of the mechanism of local secondary instability in K-breakdown of boundary layer. Izv. Sib. Otd. Akad. Nauk SSSR, Ser. Tekh. Nauk. 18 (1988) 65-77 (in Russian). (Transl. Soviet J. Appl. Phys. 3(2) (1989) 70-81.)

[5] Kachanov, Y.S.: Physical Mechanisms of laminar-boundary-layer transition, Ann. Rev. Fluid Mech. 26 (1994), 411-482.

[6] Kachanov, Y.S., Kozlov, V.V., Levchenko, V.Y. and Ramazanov, M.P.: On nature of K-breakdown of a laminar boundary-layer; new experimental data. *Laminar-Turbulent Transition* (ed. V.V. Kozlov), Springer, New York (1986) 61-73.

[7] Kachanov, Y.S.: On the resonant nature of the breakdown of a laminar boundary layer. J. Fluid Mech. 184 (1987) 43-74.

[8] Rist, U.: Numerische Untersuchung der räumlichen, dreidimensionalen Störungsentwicklung beim Grenzschichtumschlag. Dissertation Universität Stuttgart (1990).

[9] Rist, U. and Fasel, H.: Direct numerical simulation of controlled transition in a flat-plate boundary layer, J. Fluid Mech. (1985).

[10] Fasel, H.F., Rist, U. and Konzelmann, U.: Numerical investigation of the three-dimensional development in boundary-layer transition, AIAA J. 28 (1990) 29-37.

[11] Kloker, M., Konzelmann, U. and Fasel, H.: Outflow boundary conditions for spatial Navier-Stokes simulations of transitional boundary layers, AIAA J. 31 (1993) 620-628.

[12] Kloker, M.: Direkte Numerische Simulation des laminar-turbulenten Strömungsumschlages in einer stark verzögerten Grenzschicht, Dissertation Universität Stuttgart (1993).

[13] Kachanov, Y.S., Ryzhov, O.S. and Smith, F.T.: Formation of solitons in transitional boundary layers: theory and experiments. J. Fluid Mech. 251 (1993) 273-297.

[14] Falco, R.E.: Coherent motions in the outer region of turbulent boundary layers. Phys. Fluids Suppl. 20(10) (1977) S124-132.

Effect of Leading Edge on Sound Receptivity

William S. Saric, Wei Wei, and Brenda K. Rasmussen*
Mechanical and Aerospace Engineering
Arizona State University, Tempe, AZ 85287-6106, USA

Summary

Experiments are conducted to determine the acoustic receptivity in the leading-edge region of a Blasius boundary layer. Receptivity coefficients are determined for two different leading edges under different frequency and speed conditions. It is shown that only a rather narrow band of frequencies generate measurable unstable wave amplitudes. Comparisons are made with existing theory.

Key words: Sound Receptivity, Leading Edge, T-S Waves

1. Introduction

The process by which the disturbances are introduced into the boundary layer is called *receptivity*, a term first used by Morkovin [1]. Disturbances in the freestream, such as sound or vorticity, enter the boundary layer as small fluctuations of the basic state and excite unstable modes. The receptivity stage of the transition process is the least understood, but is extremely important because it provides the initial condition on the disturbance amplitude.

The second stage is the initial growth of these boundary-layer disturbances which can be described by *linear stability theory*; that is, the linearized, unsteady, Navier-Stokes equations. If the initial amplitudes of the disturbances are small, they will tend to excite the linear normal modes of the boundary layer. These normal modes in the case of a Blasius boundary layer are of the Tollmien-Schlichting (T-S) type. Their growth is weak, occurs over a viscous length scale, and can be modified, for example, by pressure gradients, surface mass transfer, and temperature gradients.

Since receptivity deals with the generation rather than the evolution of instability waves in a boundary layer, neither departures from the linear-mode scenario, called *bypass phenomena* [2,3], nor details of the transition process itself [4] are discussed here. To generate instability waves, energy from the long-wavelength external disturbances must be transferred to the shorter wavelength T-S waves through some kind of *local* [5-7] or *global* [8] flow adjustment. Reviews of earlier receptivity research are given in [3, 9-13]. The reader is referred to these reviews and their references since space is limited here.

Here we consider disturbances consisting of only plane acoustic waves. We ignore freestream turbulence and artificial disturbances within the boundary layer. Moreover, the main objective of this experiment is to isolate the influence of the leading edge on the initial amplitudes of T-S waves. These experiments then limit receptivity to the leading-edge region.

2. Experimental Procedures

2.1 Basic State

This experimental investigation is conducted in the ASU Unsteady Wind Tunnel whose details are given in [14]. The tunnel is a low-turbulence, closed-return wind

---

* Present address: MSD-4839B, NASA-Dryden Flight Research Center, Edwards, CA 93523-0203

tunnel. Several features that contribute to the receptivity experiments are pointed out here. Both the motor housing and the test section are independently mounted on isolated concrete foundations and are connected to the rest of the tunnel by flexible couplings. In addition, a sound insulated wall separates the test section and the motor and the plenum is lined with sound reducing foam. Previous experiments have established ideal flow conditions for transition experiments.

The computer system at the Unsteady Wind Tunnel consists of a 50 MHz 80486DX PC compatible, a DECstation 5000/200, a Macintosh Quadra 650, two AT-compatible PCs, and two VT100 terminals. They are all connected to a local Ethernet. The tunnel has automatic speed control for either constant speed or constant Reynolds number. The hot wires are automatically temperature compensated [15].

A new 3-D traverse is mounted exterior to the test section. A 45° forward-swept, carbon-composite sting with a thickness of 3 mm is passed through the test-side window. The entire traverse system is enclosed in a pressure vessel in order to avoid mass flow through the sting-access slot in the Plexiglas window. Hot-wire probes on the sting are used for boundary-layer and freestream measurements. The range and step sizes are: streamwise $x < 1250$ mm, $\Delta x = 12\ \mu$m; spanwise $z < 100$ mm, $\Delta z = 2\ \mu$m; and wall normal $y < 100$ mm, $\Delta y = 1\ \mu$m.

The flat plate is Aluminum-Nickel alloy, 9.53 mm thick, 4000 mm long, and 1370 mm wide. It is ground flat, polished to 0.2 μm rms and mounted vertically in the test-section approximately 782 mm from the contraction. Each end of the plate is machined with a unique leading edge. One is a 20:1 modified super ellipse (MSE) and the other is a 40:1 MSE. These shapes were used by [16] in a DNS of the receptivity problem. The geometry is described by:

$$\left(\frac{y}{b}\right)^2 + \left(\frac{a-x}{a}\right)^m = 1, \quad m = 2 + \left(\frac{x}{a}\right)^2, \tag{1}$$

where a is the major axis, b is the minor axis, and the coordinate origin is at the stagnation line. This shape eliminates the curvature discontinuity at the ellipse/flat-plate juncture and moves the minimum pressure region toward the leading edge. The ellipse is machined directly on the flat plate to ensure a smooth transition at the juncture. Without roughness, receptivity is limited to the leading edge.

A 500 mm flap with a sharp trailing edge is attached at the end of the plate and is used to control the location of the stagnation line. The flat plate has two rows of 25 static pressure ports on either side of the centerline. The pressure distribution along the plate is used in the initial stage to align the plate for Blasius flow. These measurements are supplemented by velocity-profile measurements that yield the shape factor, $H = \delta^*/\theta$, which is a more accurate means for establishing Blasius flow [17]. The flat plate is mounted in the test section at a slight angle of attack to compensate for the blockage due to the boundary-layer growth. The plate undergoes a series of fine adjustment until $H = 2.59 \pm 0.7\%$ everywhere on the plate past Branch I of the neutral stability point. Comparisons are done with the theoretical Blasius flow and the virtual leading edge is determined. For these experiments, the virtual origin was 110 mm from the leading edge.

On the back side of the plate, "Velcro" strips are pasted near the pressure recovery region of the leading edge in order to fix the transition-to-turbulence location. To achieve symmetric flow, the stagnation line is moved to the leading edge by adjusting the flap while measuring the pressure differential of two Preston tubes placed on either side of the plate, 75 mm from the leading edge. The leading edge is then firmly secured in order to prevent acoustically-excited, leading-edge vibrations.

2.2 Disturbance State

The acoustic disturbances are introduced by a set of 9 McCauley Sound Inc. 250 mm diameter, 300 Watt speakers located on the plenum wall upstream of the honeycomb section. The sound pressure level in the test section is limited to 95 dB to avoid excessive forcing [9]. The incidence angle and amplitude of the sound wave are measured by 20 microphones placed in the walls at the leading-edge location. The relative phase measurements show an essentially plane sound wave at zero incidence. Hot-wire measurements at the same stream location calibrated the microphones.

At a given speed, a dimensionless frequency, F, is chosen and the physical frequency f is calculated. Hot-wire position and flow-speed provide the Reynolds number. In this case,

$$F = 2\pi f \nu / U_\infty^2 \quad \text{and} \quad R = \sqrt{U_\infty \hat{x} / \nu} \tag{2}$$

where ν is the kinematic viscosity, U_∞ is the freestream speed, $\hat{x}$ is the distance from the virtual origin, and the appropriate length scale is $L = \sqrt{\nu \hat{x} / U_\infty}$.

2.3 Hot-Wire Signal Processing

The typical hot-wire signal at the driving frequency, contains the T-S wave, the Stokes wave within the acoustic boundary layer, and the probe vibrations. The hot-wire signals, measured directly with a narrow band-pass, produce something that does not resemble a T-S amplitude profile. Prior to [9], this was a typical measurement in the literature. These signals are separated using a technique [9, 10, 17] that is briefly described.

Since the acoustic wavelength is two orders of magnitude longer than the T-S wavelength, a streamwise scan that covers one T-S wavelength at constant height from the plate, is essentially one of constant phase for both the Stokes wave and the sting vibrations. A plot of the amplitude and phase of the total signal in the complex plane is a spiral. The radius of the spiral is the T-S amplitude and the vector pointing from the origin to the center of the spiral represents the contamination of the Stokes wave and the sting vibration. Figure 1 is an example where the T-S amplitude is much greater than the Stokes wave and Fig. 2 is an example where the signal to noise ratio is less than one. In each case the T-S amplitude can be recovered. This procedure is used at different heights in the boundary layer until the full T-S profile is obtained. The result is shown and compared with theory in Fig. 3.

3. Results

3.1 Receptivity Coefficients

The receptivity coefficient, K_S, is the ratio of the T-S amplitude and the acoustic amplitude is defined by Eq. (3).

$$K_S = |u'_{TS}|_I / |u'_{AC}|_{LE} \tag{3}$$

where the T-S wave amplitude is referenced to Branch I and the acoustic wave is measured at the leading edge. This particular definition is convenient for the experiments because it is almost impossible to separate and measure the embryonic T-S wave near the leading edge. However, by using linear stability theory, the result can be scaled to the leading edge and compared with receptivity theory [5,6].

All T-S amplitudes are measured near Branch *II* using the streamwise-scan technique for different F and R combinations. The data are generally presented in the polar plane as shown in Fig. 2. Because the T-S amplitudes at Branch I are of the order of $10^{-6}U_\infty$ the linear growth from Branch *I* to near Branch *II* is used to amplify the signal. The amplitude near Branch *II* is of the order of $10^{-4}U_\infty$ and is then scaled to Branch *I* using the Orr-Sommerfeld equation. The freestream acoustic disturbances are measured at the leading edge with a hot wire. The receptivity coefficient is calculated from

$$K_S = \frac{|u'_{TS}|_R e^{-N}}{|u'_{AC}|_{LE}}. \tag{4}$$

where N is the integrated growth rate from Branch *I* to the Reynolds number of the measurement.

Figure 4 is a summary of the results for two leading edges and five different speeds. The first surprising feature is the fact that only a narrow band of frequencies are excited. The typical band of amplified T-S waves is much broader, The narrow frequency-band characteristic is the same for both leading edges, even thought the peak frequencies shift slightly. In fact, the frequency band is so narrow that the first-order effect can be observed directly on an oscilloscope. The disturbance amplitude (measured with a hot wire near the maximum of the T-S profile) drops off quickly as the forcing frequency is detuned from the peak frequency.

A comparison with receptivity theory [5,6] is done for the case of the 20:1 MSE and $F = 84 \times 10^{-6}$. Based on the scaling of the asymptotic theory, we assume that the Orr-Sommerfeld region begins around $\hat{x} \approx \lambda_{TS}/2$. The amplitude-ratio decay from this point to Branch *I* is 0.154 based on stability calculations of the basic state determined from DNS [19]. With K_S= 0.036, the receptivity coefficient, scaled to the leading edge, is 0.24 which compares with the theoretical calculations of 0.95 by [6].

3.2 T-S Wavelength

The second surprising feature is that the T-S wavelengths corresponding to the peaks of Fig. 4 are all near 51 mm. Again, the 20:1 and 40:1 leading edges show almost identical results, despite the significant differences in pressure distribution, minimum pressure location, and nose radius. The convective length scale, defined as $L_c = U_\infty/2\pi f$, remains approximately 25 mm. This same length scale is used in the asymptotic analysis [5,6]. Table 1 summarizes the results.

In order to test the conclusion that the T-S wavelength and the convective length scale are only affected slightly by the pressure distribution at the leading edge, the stagnation line was moved by changing the flap angle. When the stagnation line is placed at different locations aft of the symmetry line, the convective length scale and the T-S wavelength only change slightly.

Table 1. Convective wavelengths with acoustic forcing; 20:1 ellipse

U_∞ [m/s]	F	R	ν [m^2/s]	L_c [mm]	λ_{TS} [mm]
8.0	85	788	16.8×10^{-6}	24.1	51.0
8.5	85	805	16.9×10^{-6}	23.6	50.3
12.0	53	1101	16.3×10^{-6}	26.7	50.9
15.0	44	1228	16.8×10^{-6}	25.3	50.9
18.0	37	1327	17.9×10^{-6}	26.1	50.6

A speaker is mounted downstream of the turning vanes, between the honeycomb and the first screen, in order to determine whether the turning vanes play a role in the focusing.. Streamwise scans at both 12 m/s and 15 m/s show the same 51 mm T-S wavelength and 25 mm convective length scale.

It is found that the receptivity for the 20:1 leading edge is strong enough that at high speeds the freestream disturbances can generate detectable instability waves. Assuming that these waves are T-S waves, they seem to have a typical low-frequency modulation that is absent in the signals obtained by acoustic forcing and they have more of a wave-packet-like structure. The frequency content of the signals are analyzed using FFT. Both hot wires and microphones are used to acquire the signal to ensure that sting vibration is not causing the oscillations. The spectrum obtained from both methods are essentially the same and Table 2 summarizes the results. One should note that the convective length scale and the T-S wavelength are both slightly shorter. This result is also observed by [19] using an entirely different leading edge and flat plate.

Table 2. Convective wavelengths without acoustic forcing

U_∞ [m/s]	F	ν [m^2/s]	L_c [mm]	λ_{TS} [mm]
12.0	65	17.2×10^{-6}	22.3	48.3
15.0	51	17.2×10^{-6}	22.7	47.1
18.0	43	17.3×10^{-6}	22.6	45.8
19.5	37	16.7×10^{-6}	23.0	45.8
21.0	35	17.1×10^{-6}	23.0	45.5

4. Conclusions

Receptivity in the leading edge region of a Blasius boundary layer is examined. The leading edges are either a 20:1 or 40:1 modified super ellipse. The new microphone system confirms that the wavefront of the incident sound is normal to the leading edge. The receptivity coefficients for the 20:1 modified super ellipse are greater than those for the 40:1 leading edge. With both leading edges, a wavelength focusing and a narrow frequency band phenomena are observed. At each freestream velocity, the T-S wavelength corresponding to the maximum receptivity coefficient is approximately 51 mm; the convective length scale is approximately 25 mm. Movement of the stagnation line does not change this result.

Due to greater receptivity with the 20:1 leading edge, it is possible to observe the instability waves produced by freestream disturbances. The T-S wavelength remains constant for all speeds at approximately 45 mm instead of 51 mm.

Acknowledgments

The help of Mr. D. Clevenger and Dr. R.H. Radeztsky, Jr. is greatly appreciated. This work is supported by the Air Force Office of Scientific Research Grant, *AFOSR Grant Number F49620-94-2-002-1*. The acoustic speakers were donated by McCauley Sound Inc., Payallup, Washington.

References

[1] Morkovin, M.V. On the many faces of transition. *Viscous Drag Reduction* ed.: C.S. Wells, Plenum (1969).

[2] Morkovin, M.V. Bypass-transition research: Issues and philosophy. *Instabilities and Turbulence in Engineering Flows*, Eds: D.E. Ashpis, T.B. Gatski, R. Hirsh, Kluwer Academic (1993).

[3] Reshotko, E. Boundary layer instability, transition, and control. *AIAA Paper 94-0001* (1994).

[4] Morkovin, M.V. Panoramic view of changes in vorticity distribution in transition instabilities and turbulence. *Boundary Layer Stability and Transition to Turbulence, FED-Vol. 114*, pp 1-12. Eds: D.C. Reda, H.L. Reed, R. Kobayashi, ASME (1991).

[5] Goldstein, M.E. and Hultgren, L.S. Boundary-layer receptivity to long-wave disturbances. *Ann. Rev. Fluid Mech., 21*, 137 (1989).

[6] Kerschen, E.J., Choudhari, M. and Heinrich, R.A. Generation of boundary instability waves by acoustic and vortical freestream disturbances. *Laminar-Turbulent Transition vol III*, Eds. D. Arnal and R. Michel, Springer-Verlag (1990).

[7] Crouch, J.D. Localized receptivity of boundary layers. *Phys. Fluids A 4*, 1408 (1992).

[8] Crouch, J.D. Non-localized receptivity of boundary layers. *J. Fluid Mech., 244*, 567 (1992).

[9] Nishioka, M. and Morkovin, M.V.: Boundary-Layer Receptivity to Unsteady Pressure Gradients: Experiments and Overview, *J. Fluid Mech. 171* (1986).

[10] Wlezien, R.W. Measurement of acoustic receptivity. *AIAA Paper 94-2221*.

[11] Choudhari, M. and Streett, C. Theoretical predictions of boundary layer receptivity. *AIAA Paper 94-2223* .

[12] Crouch, J.D. Receptivity of boundary layers. *AIAA Paper 94-2224*.

[13] Saric, W.S., Reed, H.L. and Kerschen, E.J. Leading-edge receptivity to sound: Experiments, DNS, theory. *AIAA Paper No. 94-2222* (1994).

[14] Saric, W.S. The ASU transition research facility. *AIAA Paper 92-3910*.

[15] Radeztsky, R.H. Jr., Reibert, M.S. and Takagi, S. A software solution to temperature-induced hot-wire voltage drift. *Proc. 3rd International Symposium on Thermal Anemometry*, ASME-FED, June 1993.

[16] Lin, N., Reed, H.L. and Saric, W.S. Effect of leading edge geometry on boundary-layer receptivity to freestream sound. *Stability, Transition and Turbulence*, eds. M.Y. Hussaini, A. Kumar, C.L. Streett Springer-Verlag, New York (1992).

[17] Saric, W.S. Low-speed boundary-layer transition experiments. *Transition: Experiments, Theory and Computations*. Eds. T.C. Corke, G. Erlebacher, M.Y. Hussaini, Oxford (1994).

[18] Kosorygin, V.S., Radeztsky, R.H., Jr., and Saric, W.S. Laminar boundary layer sound receptivity and control. Laminar-*Turbulent Transition, Vol. IV*, Ed. R. Kobayashi, Springer, 1995.

[19] Fuciarelli, D.A. and Reed, H.L. *Private Communication,* Sept. 1994.

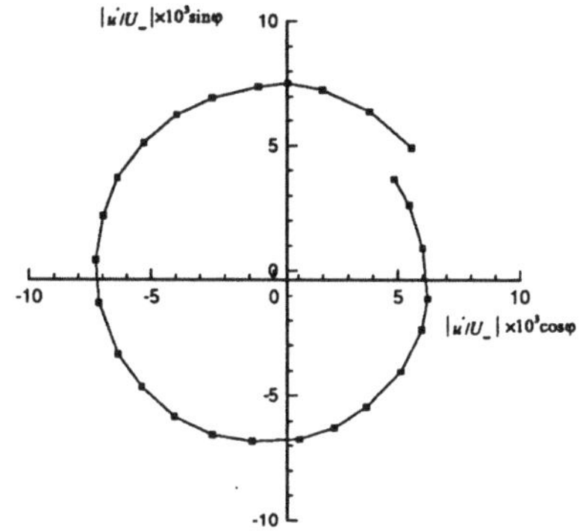

Figure 1. Complex polar plot resulting from a streamwise scan with $U_\infty = 18.0$ m/s and $F = 37 \times 10^{-6}$. Measurements taken at $u(y)/U_\infty = 0.12$ in the range $1700\,\text{mm} \leq \hat{x} \leq 1750\,\text{mm}$

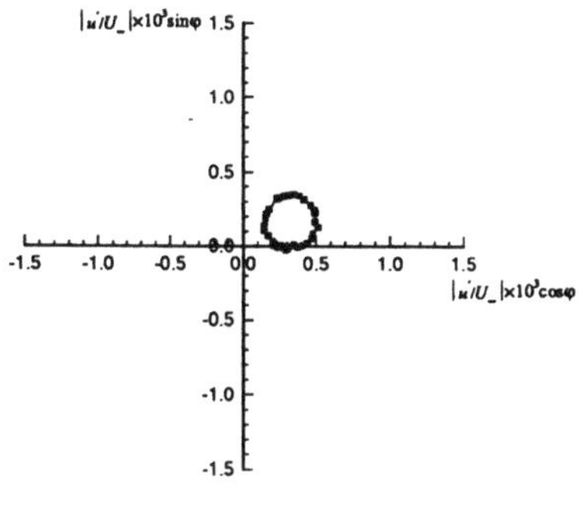

Figure 2. Complex polar plot resulting from a streamwise scan with $U_\infty = 12.0$ m/s and $F = 54 \times 10^{-6}$. Measurements taken at $\hat{y} = 11$ mm (outside the boundary layer) in the range $1700\,\text{mm} \leq \hat{x} \leq 1750\,\text{mm}$

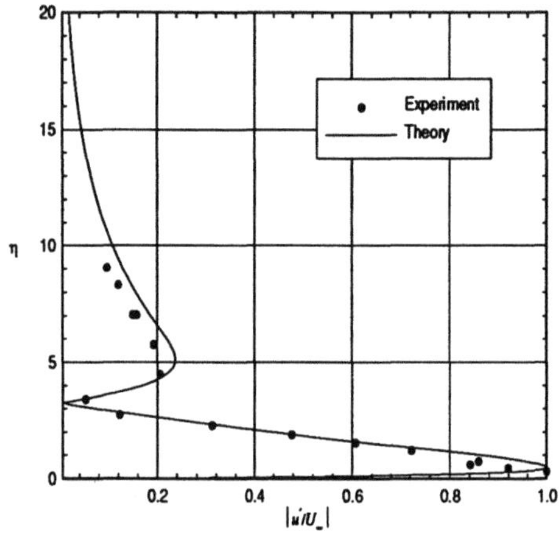

Figure 3. RMS amplitude of T-S wave after signal separation for U_∞= 12.0 m/s and $F = 54 \times 10^{-6}$.

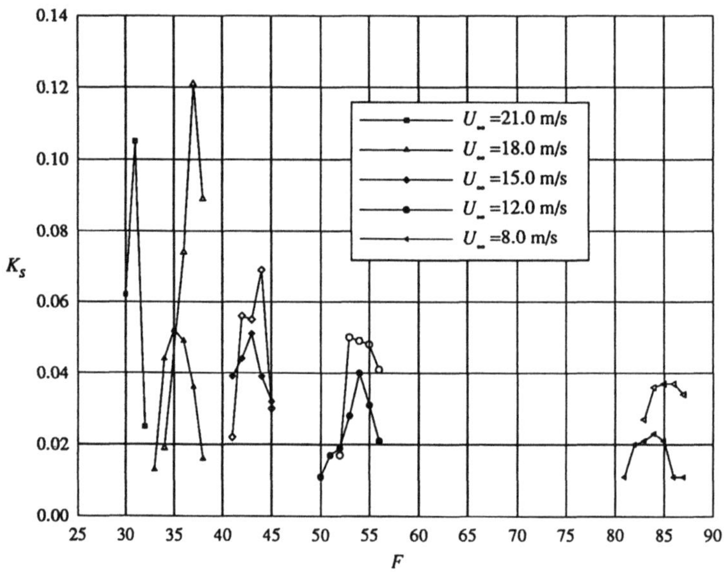

Figure 4. Comparisons of receptivity coefficients as a function of frequency at different speeds. K_S defined in Eq. (3). Solid symbols are for the 20:1 MSE and the open symbols are for the 40:1 MSE.

On Similarity Law of the Crossflow Instability in Three-Dimensional Boundary Layers on Yawed Cylinders

Shohei Takagi and Nobutake Itoh
National Aerospace Laboratory
Chofu, Tokyo 182, JAPAN

Summary
Three-dimensional boundary layers are unstable to both stationary and traveling crossflow disturbances. Early experiments show the center frequency of traveling disturbances to shift according to experimental conditions such as freestream turbulence and surface roughness. A natural speculation is that the observed shift of the center frequency of traveling disturbances is due mainly to the presence of stationary vortices, which surely deform the basic flow; no frequency shift should be observed in the flow with no stationary vortices. On this speculation, special cares are paid to establish traveling-mode dominated flows free from the stationary mode on yawed-circular cylinders. Then, a series of parametric studies focusing on the center frequency of traveling disturbances is conducted in combination of three yawed cylinders with different size and three wind tunnels with different turbulence levels under a constant Reynolds number. Experimental results show that the nondimensional center frequencies in totally six cases tested are almost independent of freestream turbulence levels and their spectra, and of the size of models with different surface conditions, providing a strong support to the above speculation.

Key words : 3-D boundary layer, Crossflow instability, Traveling waves, Stationary vortices.

1. Introduction

Four decades passed since it had been found that the sweep angle of main wings of aircraft has a destabilizing effect on the 3-D boundary layer. It is well-known that this effect is caused by crossflow, which has an inflection point in the velocity distribution normal to the wing surface. This inflection-type instability yields regularly spaced stationary vortices approximately aligned to local flow. Besides the stationary vortices, Poll [1] observed time-dependent unsteady disturbances in the boundary layer on the windward region of yawed circular cylinders. Subsequent experiments have endorsed that both stationary and traveling disturbances are amplified in 3-D boundary layers. A common feature seen in these experiments is that the stationary mode is dominant [e.g.2,3] in contradiction to linear stability theory, which predicts unsteady mode to be more unstable than stationary mode. In these stationary-mode dominated flows, experimental results show that the center frequency of traveling disturbances shifts with different levels of freestream turbulence and surface roughness, preventing us from unique determination of the center frequency as a function of Reynolds number. On the other hand, the authors' recent experiments [4] succeeded in obtaining traveling-mode dominated flows free from stationary vortices and presented the center frequency of traveling disturbances in quite good agreement with prediction of linear stability theory. These contradictory facts naturally lead to a speculation that the frequency shift observed in current experiments may be related to the presence of stationary vortices, which will distort the mean flow and result in a slight modification of the stability characteristics. If this is the cause, no frequency shift should be observed in the flow where stationary vortices are absent.

The present study is undertaken to find an experimental evidence in favor of the above speculation. For this purpose, traveling-mode dominated flows are first established free from stationary mode on yawed-circular cylinders, which simulate the leading-edge region of swept wings in a similar manner with the previous experiment [4]. Then, a series of parametric studies to determine the center frequency of unsteady disturbances are conducted in combination of three kinds of cylinder models with different size but similar shape, and three kinds of wind tunnels with different levels and spectra of freestream turbulence when the Reynolds number based on the model chordlength and the uniform velocity is held fixed.

2. Linear stability calculations

Our understanding of a physical phenomenon is attained when experimental observations are fully described by theoretical descriptions. In the present study, therefore, experimental results are compared with linear-stability calculations. Boundary-layer calculations are based upon momentum integral equations, where the basic flow is approximated with a family of the Falkner-Skan-Cooke velocity profiles. If the local Reynolds number is defined by $R^*= \delta^* Q_e /\nu$, where δ^* is the displacement thickness, Q_e the local velocity at the edge of boundary layer,and ν the kinematic viscosity, then $\hat{R} = R^* cos\Lambda / cos\Lambda_\infty$ becomes invariant to sweep angle and is determined as a function of the dimensionless distance from the leading edge and the wing Reynolds number defined by $R_Q = Q_\infty L/\nu$, where Λ_∞ is the sweep angle of wing, Λ is the local swept angle of external streamline, Q_∞ is the freestream velocity and L is the chord length in the streamwise direction.

Once the 3-D velocity distribution at each location along the wing surface is obtained from the boundary-layer calculations, linear stability of the flow is evaluated by solution of the modified Orr-Sommerfeld equation, which includes two terms of wall and streamline curvatures and a part of non-parallelism [5]. Results are shown in Fig.1. In Fig.1(a), a thin line denotes the development of local Reynolds number with a constant value of $R_Q = 7.5x10^5$ corresponding to the present experimental conditions. In addition, the two critical curves denoted by C-F and T-S are also presented for a circular-cylinder flow with the sweep angle $\Lambda_\infty = 50°$, indicating that two different instabilites occur in this flow. Curve C-F represents the critical Reynolds number of the crossflow instability in a forward-facing region of the cylinder. Curve T-S is due to Tollmien-Schlichting instability. The first point of intersection between the curve of local Reynolds number $\hat{R}$ for $R_Q = 7.5x10^5$ and the curve C-F or T-S gives the initiation of the instability. The local Reynolds number $\hat{R}$ exceeds the critical value at point A, where the boundary layer is unstable to the C-F instability. Then, the boundary layer becomes stable again downstream of point B and immediately unstable in T-S mode at point C.

The critical dimensionless frequency $\hat{f} = f\,L/(Q_\infty\sqrt{R}_Q)$ of the disturbances along the solid lines given in Fig. 1(a) are plottted in Fig. 1(b). The frequency with respect to the crossflow instability tends to slightly increase downstream, while the T-S frequency abruptly increases. The more precise description is given in this volume [5].

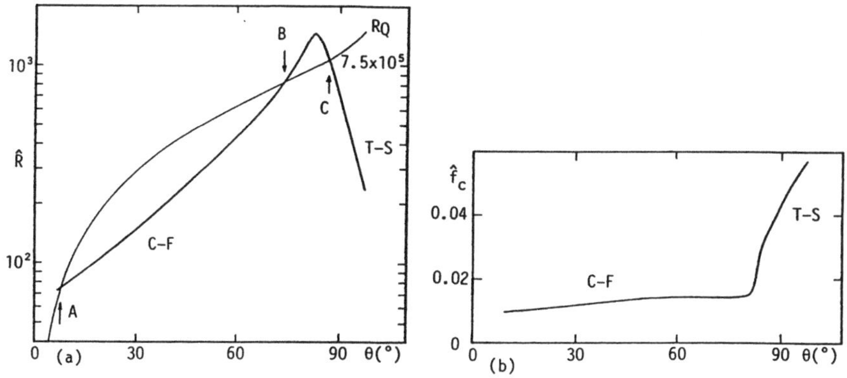

Fig.1. Critical Reynolds number $\hat{R}$ plotted against the circumferential distance θ from the attachment line with respect to the wing Reynolds number R_Q=7.5x10^5. (a) The solid lines represent neutral positions for crossflow instability and for Tollmien-Schlichting instability. The points A, B and C are intersections of the solid lines for C-F instability and for T-S instability, respectively. (b) The dimensionless frequency of crossflow disturbances along the critical curves in Fig.1(a). $\hat{f}$ is dimensionless frequency $fL/(Q_\infty\sqrt{R}_Q)$.

3. Experimental setup

3.1 Cylinder models

For the parametric studies, three kinds of yawed-circular cylinders have been used. The basic shape of the models is kept similar with a constant sweep angle 50° as illustrated in Fig.2 and the model dimensions are compared in Table 1. The first two models labeled M_1 and M_2 are made of steel and acrylic pipes, respectively, and are the same ones that have been used in the experiment of Takagi and Itoh [4]. The third model M_3 was newly constructed so as to have an excellent surface smoothness (denoted by κ in rms value in Table 1) in comparison with M_1 and M_2. Straight

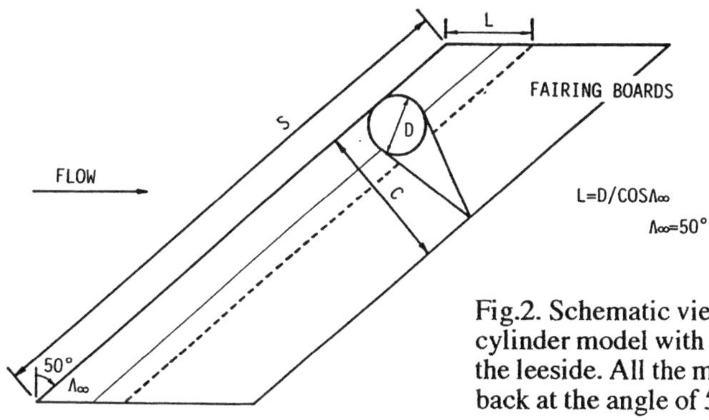

Fig.2. Schematic view of the yawed cylinder model with fairing boards on the leeside. All the model are swept back at the angle of 50°.

Table 1. Comparison of the model dimensions labeled M_1, M_2 and M_3.

MODEL	D(mm)	S(mm)	S/D	C(mm)	C/D	κ (μm)
M_1	500	3004	6.01	1240	2.48	1.3
M_2	300	1642	5.47	624	2.03	0.33
M_3	138	686	4.96	287	2.08	0.07

fairing boards were attached on the leeward of all the cylinders to decrease mechanical vibration owing to vortex shedding from the cylinder.

A part, which is called drum, of the models M_1 and M_3 is rotatable around the cylinder axis and is itself used for traverse in the circumferential direction. Of course, a careful adjustment has been made for the drum surface in order to keep flush to the cylinder surface. Inside the drum, a traversing mechanism for the direction normal to the surface is installed. These traversing systems including the drum are used for hot-wire measurements of fluctuating velocity component due to unsteady mode and mean velocity profile along the circumferential direction. The velocity profile is used to check as to whether or not stationary mode is amplified.

All the models were set up in order for measuring positions to be located around the center of the test section of each wind tunnel. For this arrangement, a vertical riser with a diameter of 500mm and a height of 2m was used as the basis of each holder of M_2 and M_3 in the Lowspeed Wind Tunnel.

The aspect ratio S/D is a key parameter for infinite yawed wing conditions in case of having no end plates. The experiment of Takagi and Itoh [4] showed that the static pressure measurements along the circumferential direction at the midspan of M_2 was in good agreement with the potential theory. Thus, it can be concluded that the flow on the cylinder having an aspect ratio of about 5 and more is a very close approximation to the infinite conditions.

3.2 Wind tunnels

The Lowspeed Wind Tunnel (abbreviated LWT), the Gust Wind Tunnel (GWT) and the Calibration Wind Tunnel (CWT) at the NAL were used throughout the series of experiments. These three tunnels are closed-circuit. General wind-tunnel specifications such as overall streamwise turbulence levels in the corresponding test section and acoustic sound levels outside the test section are compared in Table 2. General performance of the GWT and CWT is found to be good, while the LWT may belong to the category of high-turbulence wind tunnel. However, spectral analyses of freestream turbulence showed that hot-wire anemometer outputs in the LWT and CWT were contaminated by acoutic sound fluctuations coming from the motor-fan blades. It is obvious, therefore, that the actual vortical fluctuation level is generally less than the nominal level listed in Table 2. For example, the level of the CWT at a freestream of 30m/s is reduced to 0.022% of the freestream velocity [6]. Typical spectra in the LWT, GWT and CWT test sections are shown in Fig.3. It should be noted that in the LWT freestream fluctuations in the higher frequency range than about 500Hz are sufficiently large compared to those of the other two wind tunnels.

3.3 Hot-wire and hot-film sensors

A single hot-wire, which consists of 5μm dia. tungsten wire with copper-plated sleeves, on the traversing mechanism was fabricated in house, while a hot-film sensor

(Dantec Model 55R47) was adhered at the location θ=80° from the attachment line. Although the hot-wire anemometer with a constant temperature mode responds to more than 10kHz, the hot-film in operation with low overheat ratio is found to respond to about 4~5kHz according to the conventional square-wave test. This frequency response of the film covers the center frequency of unsteady disturbances as shown later.

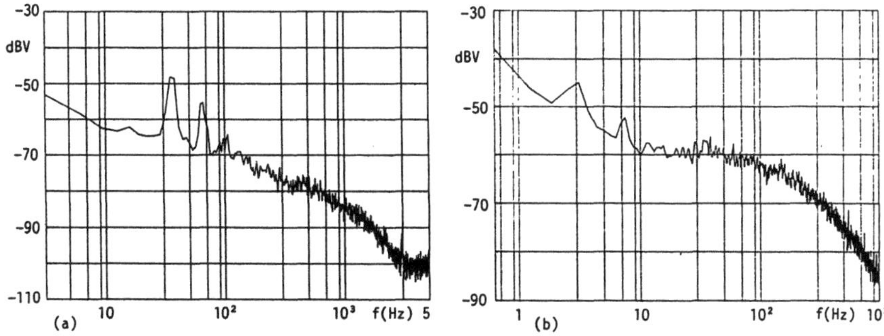

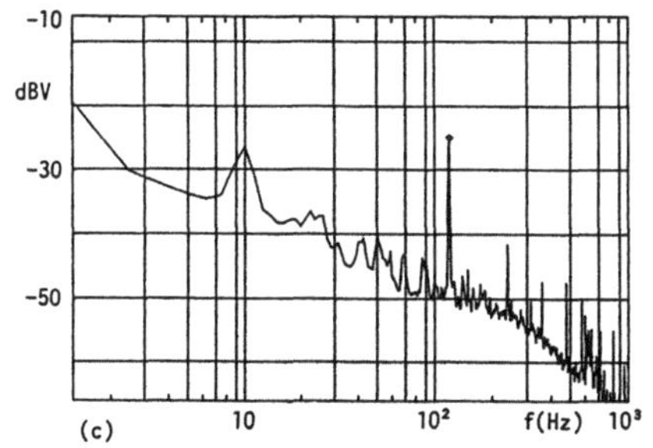

Fig.3. Freestream turbulence spectra.
(a) Lowspeed Wind Tunnel at 15m/s,
(b) Gust Wind Tunnel at 30m/s,
(c) Calibration Wind Tunnel at 30m/s.

Table 2. Comparison of windtunnel specifications.

	Turbulence level(%)	Acoustic sound(dB)	Test section	Contraction ratio
LWT	0.18 at 15m/s	98	5.5m×6.5m 9m in length	5.3:1
GWT	0.046 at 30m/s	81	2m×2m 4m in length	9:1
CWT	0.05 at 30m/s	82	0.55m×0.65m 1.5m in length	9:1

4. Experimental results and discussion

4.1 Appearance of stationary vortices

It is generally accepted that stationary vortices in crossflow dominated flow tend to appear in low-turbulence environment [2-4]. Takagi and Itoh [7] also observed altenatively modulated structures in the mean velocity profiles along the circumferential direction of the model M_3 in the CWT. Such structures are attributed to action of a row of stationary vortices, but their amplitude in the velocity profile is drastically weakened on the smooth surface. It was finally confirmed that these structures disappear on the specially cleaned model surface in the CWT. This result suggests that even invisible motes or finger prints favor stationary vortices in low-turbulence flow. In the present experiment, same special care was taken of model surface. In consequence, traveling mode instead of stationary mode dominates the transition process. Such observations are in reasonable agreement with theory as mentioned before. On the other hand, besides traveling modes no stationary disturbances were observed on the M_1 model in the LWT, whose freestream turbulence level is the highest in the three wind tunnels used in the present experiment. This observation is consistent with a series of the previous experiments [2] and [4]. Furthermore, a new observation by Takagi and Itoh [7] indicates that roughness elements give rise to traveling disturbances as well as stationary vortices in the wavepacket development.

4.2 Observation of unsteady mode

Three spectra of traveling disturbances detected on M_3 model in the three wind tunnels are compared in Fig4. Figs. 4(a)-(c) were observed by the use of hot-film and hot-wire sensors at the circumferential position θ=80° from the attachment line, respectively. Several peaks at lower frequency range in Fig.4(b) are due to aerodynamical vibration of hot-wire probe on the traversing mechanism, but there are no such undesirable peaks in Fig.4(a) and (c) by means of surface sensors. A distinctive peak at about 3kHz due to crossflow instability is seen in Figs. 4(b) and (c) in the GWT and CWT, while the corresponding frequency in Fig.4(a) is about 2.4 kHz. Consequently, it can be concluded that the center frequency of the traveling waves in higher freestream turbulence flow shifts down. Such a frequency shift is also observed by Müller and Bippes. It is noticeable that there is no valley structure at about 1kHz in Fig.4(a) in contrast with Figs.4(b) and (c). Freestream fluctuation at this frequency range in the LWT is large compared to those of the GWT and CWT as already pointed out in Fig.3. Therefore, the center frequency of a broad hump on the slope of spectrum like Fig.4(a) may apparently shift down to some extent, but its shift is not so large as the difference between Fig.4(a) and Figs.4(b)-(c). For the main cause of such a frequency shift, however, further investigations are required.

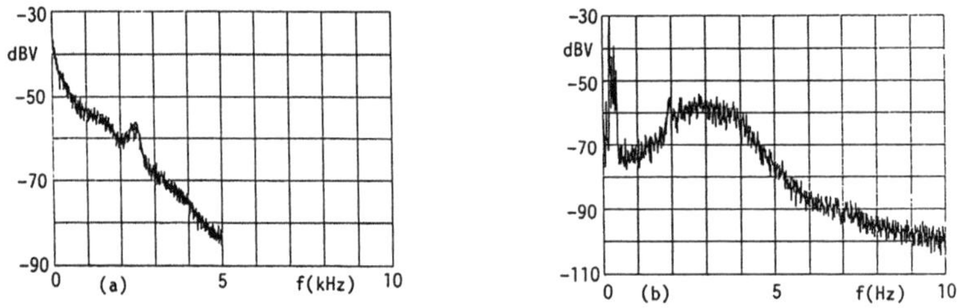

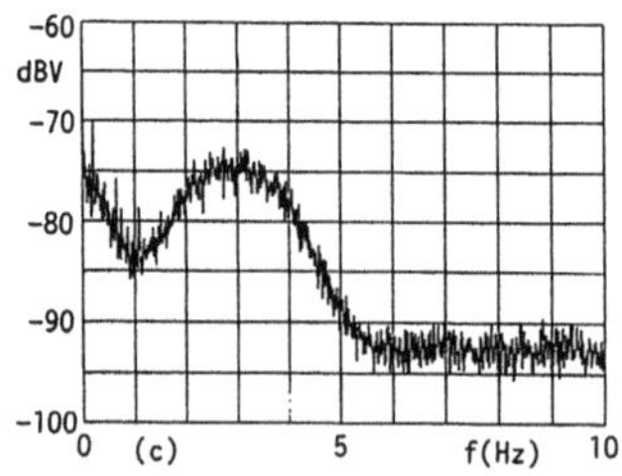

Fig.4. Spectra of crossflow traveling waves on M_3 model. (a) Lowspeed Wind Tunnel, (b) Gust Wind Tunnel, (c) Calibration Wind Tunnel.

Fig.5 shows two spectra of traveling disturbances detected with the aid of hot-flm adhered at θ=80° on M_2 model in the LWT and GWT. The center frequency of traveling waves is about 600Hz in both cases. The fact that the amplitude of the traveling waves in the LWT seems to be lower than that in GWT may be attributed to the different level of freestream turbulence as aforementioned.

Detection of the traveling waves using the M_1 model was performed only in the LWT. Fig.6 shows a spectrum of traveling waves, whose center frequency is very ambiguous, because of obvious contamination of freestream turbulence, while an instantaneous waveform seems to be fairly periodic as inset in Fig.6. To avoid misleading of the center-frequency determination of traveling disturbances, spectral analyses (not shown here) were carefully repeated changing the freestream velocity. In fact, the same procedure has been applied for the spectral measurements in Fig.4(a).

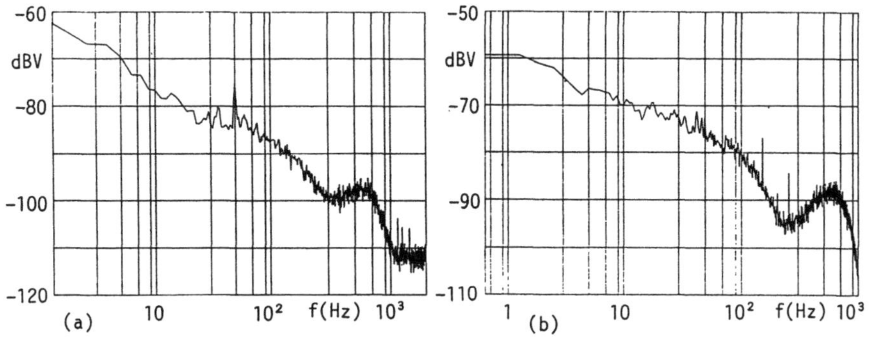

Fig.5. Spectra of crossflow traveling waves on M_2 model. (a) Lowspeed Wind Tunnel, (b) Gust Wind Tunnel.

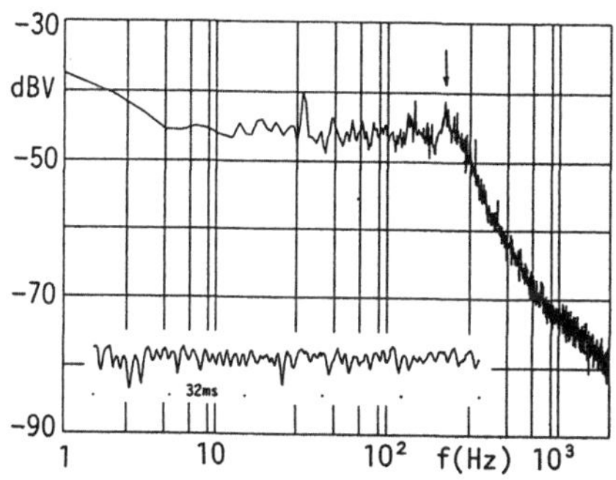

Fig.6. Spectrum of crossflow traveling waves on M_1 model in Lowspeed Wind Tunnel.

4.3 Comparison between experiment and prediction

In sequence of the present experiment focusing on the center frequency of traveling disturbances, totally six cases in combination of three models and three wind tunnels have already been conducted under a constant Reynolds number based on the chordwise length and the uniform velocity. The experimental results in dimensional and nondimensional forms are summarized in Table 3. The dimensionless frequencies $\hat{f}$ lie between 0.0112 and 0.014 and seem to be slightly low in high-turbulence environment. According to Fig.1(b) obtained from linear stability calculations, $\hat{f}$ is about 0.01 at the critical point of crossflow instability and 0.014 at the circumferential location θ=80°. The observed center frequency seems to be in very good agreement with the calculations. Generally, experiments observe the frequency of the spatially growing disturbances with maximum amplitude at the observed location. It can be conjectured, therefore, that the critical disturbances originating at point A in Fig.1(a) maintain the large amplitude sufficiently downstream of the critical point. Thus, the observed frequency should be compared with the critical frequency rather than that of locally most unstable disturbances at the measuring location [4]. From this point of view, the observed frequency is slightly higher than the predicted value.

Table 3. Dimensional and nondimensional center frequency of crossflow traveling waves.

Model	LWT		GWT		CWT	
	f(Hz)	f	f(Hz)	f	f(Hz)	f
M_3	2.4k	0.0112	3.0k	0.014	3.0k	0.014
M_2	600	0.0129	600	0.0129	N.A	
M_1	225	0.0126	N.A		N.A	

5. Conclusions

Three-dimensional boundary layers are unstable to both stationary and traveling crossflow disturbances. Early experiments have observed that the stationary mode is dominant rather than the unsteady mode and that the center frequency of traveling disturbances shifts with different levels of freestream turbulence even though the identical model is used. In the present paper, the main reason of such a frequency shift is attributed to the presence of stationary vortices, and experimental observations for an evidence of no frequency shift in the flow free from stationary vortices have been made in the well-controlled flows on yawed circular cylinders of three different sizes in the three windtunnels with different turbulence levels.

Hot-wire surveys showed that even unmeasurable surface roughness such as finger prints favored stationary vortices especially in low-turbulence freestream. If the surface was carefully cleaned, however, the stationary mode disappeared in all six cases tested. Detailed observations in this flow condition have shown that the nondimensional center frequencies in all cases are almost independent of freestream turbulence levels and their spectra, and of the size of models that have different surface-roughness levels. This result may be considered to support our crude conjecture that the frequency shift observed in current experiments occurs through spatial distortion of the basic laminar flow due to the presence of stationary vortices with considerably large amplitude. It should be noted however that comparison of the observed frequency with the critical frequency predicted by local linear stability theory including two essential terms of streamline and wall curvatures shows still minor disagreement.

References

[1] Poll, D.I.A.: J. Fluid Mech.150 (1985) 329-356.
[2] Müller, B. and Bippes, H.: AGARD Conference Proceedings No. 438, 13.1.
[3] Radeztsky, R.H., Reibert Jr., M.S., Saric, W.S. and Takagi, S.: AIAA Pap. No. 93-0076.
[4] Takagi S. and Itoh, N.: Fluid Dynamics Research, 14(4) (1994) 167-189.
[5] Itoh, N.: in this volume.
[6] Takagi, S.: NAL TR-1054 (1990) (in Japanese).
[7] Takagi, S. and Itoh, N.: Proc. of Turbulence Symposium (1993) (in Japanese).
[8] Deyhle, H., Höhler, G and Bippes, H.: AIAA J. 31(4) (1993) 637-645.

Numerical and Experimental Investigation of Secondary Instability in Leading Edge Boundary Layer Flow

V. THEOFILIS

Department of Applied Mathematics, University of Twente, P.O. Box 217, 7500 AE Enschede, the Netherlands

and

D. I. A. POLL

Department of Aeronautical Engineering, University of Manchester, Manchester M13 9PL, United Kingdom

Abstract.

Subharmonic growth of secondary disturbances is considered in boundary layer flow in the infinite swept attachment-line at high Reynolds number. Results are obtained in the neighbourhood of Branch I, where instability waves are observed experimentally. The amplified range of wavenumbers is identified, as is the dependence of the secondary growth rate on (a) the Reynolds number and (b) the amplitude of the superimposed primary disturbance. The spatial structure of the secondary disturbance at conditions of maximum growth is made available for comparisons with experiment.

1. Introduction

Secondary stability theory has proven an invaluable tool in analysing the late stages of shear flow transition from laminar to turbulent state (Herbert, 1983, 1988; Kachanov and Levchenko, 1984; Orszag and Patera, 1983). In archetypal flows (e.g. incompressible flat plate boundary layer) it has been demonstrated that a basic state comprising the laminar profile perturbed by a finite amplitude viscous primary disturbance is susceptible to convective growth of three-dimensional disturbances. Secondary theory describes the subsequent development of this flow with notable success in predicting the critical Reynolds number as well as the flow structures observed in experiments.

In incompressible flow in an infinite swept attachment-line boundary layer, on the other hand, experimental evidence suggests that sharply defined frequencies, distinct from that of the Tollmien-Schlichting mode, appear in the spectrum during the late stages of transition. Although a clear precursor of onset of nonlinear flow behaviour, the origin of these frequencies is presently unknown. We commence presently an investigation of secondary instability mechanisms in attachment-line boundary layer flow, in an attempt to discern between parametric excitation of the primary unstable wave on one hand and nonlinearity on the other.

Our first target is high Reynolds number flow. This Reynolds number range was chosen because the system governing primary stability may be drastically simplified, albeit non-rationally, and still deliver accurate predictions for the primary destabilisation mechanism. Moreover, we currently restrict ourselves to subharmonic growth; the other two identified secondary mechanisms, fundamental and detuned resonance, are also under investigation and results will be reported elsewhere. The systems of equations governing the base flow, primary and subharmonic secondary stability are discussed in the next section. The numerical techniques for their solution, as well as validation results are presented in §3. Results on the secondary stability of the attachment-line boundary layer may be found in §4 and concluding remarks are offered in §5.

2. Governing Equations

The generalised Hiemenz flow is taken to be the model describing the incompressible base flow velocity components in the infinite swept attachment-line boundary layer. The temporal linear stability of this flow is described by the system (henceforth referred to as AL) derived and solved by Hall *et al.* (1984). While the neutral loops yielded by the linear analysis agree favourably with experiments (Pfenninger and Bacon, 1969; Poll, 1979) the equations governing primary stability may be simplified to yield the Orr-Sommerfeld (OS) equation (Poll, 1978). The latter simplification results in overprediction of the linear critical Reynolds number; however, at large Reynolds numbers the stability problem described by the attachment-line system and the Orr-Sommerfeld equation yield equivalent results (Arnal, 1993; Theofilis, 1994a).

While work is underway on the full system, in the present paper we restrict ourselves to high Reynolds number results delivered by a secondary analysis of the primary stability problem described by the Orr-Sommerfeld equation. Some comments on validation of this choice are provided in the next section. Moreover, of the possible resonances, we currently focus on subharmonic (H-type) growth. One of the possible scalings of the primary disturbance, chosen here because of its direct analogy to the Blasius problem, is

$$\max_{0\le y\le y_\infty} 2\,|\,\hat{v}'(y)\,|^2 = 1, \tag{1}$$

prime denoting derivative w.r.t. the normal coordinate y of the normal perturbation velocity component $\hat{v}$. We subsequently build the secondary problem by constructing a spanwise periodic flow (in the direction along the attachment line) using the shape assumption. This flow may be susceptible to three-dimensional streamwise disturbances (in the direction normal to the attachment-line and to the body surface), the simplest non-trivial form of which is, in the subharmonic framework,

$$\mathbf{u}_3 = 2\,e^{\sigma t}\,e^{i\alpha x}\left[\mathbf{u}^+(y)\cos\hat{\beta}z - \mathbf{u}^-(y)\sin\hat{\beta}z\right], \tag{2}$$

with $\hat{\beta} = \beta/2$ (Herbert, 1984). Here x and z denote coordinates along the streamwise and spanwise directions with respective wavenumbers α and β and t denotes time. The objective becomes determination of the secondary growth rate σ, in general complex, and the secondary eigenfunctions $\mathbf{u}^{\pm}$. Using a global method, such as that presented in the sequel, one obtains the full spectrum of the secondary eigenvalues σ; of particular interest of these is that with $\sigma_i = 0$ which, according to Herbert (1984), provides the conditions for perfect synchronisation between the (primary) Tollmien-Schlichting wave and the three-dimensional disturbance.

3. Numerical Methods - Validation

Spectral collocation has been used throughout for the determination of the base flow, as well as the primary and the secondary stability problems. The base flow (of the Falkner-Skan class) was solved in a manner similar to that discussed by Streett *et al.* (1984); in contrast to these authors who used preconditioned Richardson iteration, however, Newton-Kantorowitz iteration (Boyd,

1989) is used. While a straightforward shooting approach might have been used to determine the base flow, collocation was chosen in view of at least two reasons. Firstly, the number of nodes utilised is minimised for a given level of accuracy when one uses collocation; the convergence history of our solution scheme, presented in Table I, suggests that ten significant places of the converged result may be obtained with a mere $N = 32$ collocation points. Secondly, one may choose the grid on which the base flow is solved to be that on which the stability problems are subsequently set up; this eliminates interpolation-induced error in the stability calculations.

The Orr-Sommerfeld equation is solved by spectral collocation based on Chebyshev polynomials (Theofilis, 1994a). Some validation may be found in Table II, where the convergence history is presented of the temporal eigenvalue $c_r + i c_i$ in Blasius flow at the set of parameters presented by Herbert (1984), $R = 550$, $\alpha = 0.16869$. Our numerical techniques are similar to those of the latter author and one may observe that both calculations converge at a similar level of number of collocation nodes; excellent agreement is demonstrated between the converged results.

The range of Reynolds numbers in which the Orr-Sommerfeld equation is expected to yield predictions relevant to the attachment-line linear stability problem is established next. In Table III we present a comparison of the results delivered by (spectral collocation) numerical solution of the temporal AL problem (Theofilis, 1994b) and the OS equation, for linear perturbations developing about the generalised Hiemenz flow in the AL- and its spanwise component in the OS case. In both cases a fixed value of the spanwise wavenumber $\beta = 0.2$ is taken, $N = 64$ collocation nodes are used and the Reynolds number is allowed to take values in the range that a reasonable agreement between the AL and OS approaches is expected. One notices that on a qualitative level the error between the frequency results delivered by the OS equation and the AL system is much lower than that of the respective growth rates. However, as the Reynolds number is increased, one sees that the two approaches yield results that may be considered to be equivalent (cf. Arnal, 1993). Of interest for the secondary problem are the eigenfunctions of the primary perturbation; the results for the normal perturbation velocity $\hat{v}$ are presented in figure 1. The real parts, $\hat{v}_r$, of the unstable mode predicted by the AL and the OS systems may be seen to be within line-thickness at all Reynolds numbers presented. In contrast, it may be noticed that the imaginary parts, $\hat{v}_i$, of the eigenvectors yielded by the two approaches are different at $y \approx 2$, with the component resulting from the Orr-Sommerfeld equation approaching that of the attachment-line system as the Reynolds number increases. Guided by these results we conclude that the secondary stability results to be obtained will be of increasing relevance to the attachment-line problem as the Reynolds number increases.

The subharmonic secondary stability problem is solved next. In the incompressible regime considered here, the system presented by Herbert (1984) is relevant. This system has been set up and solved using a numerical approach similar in spirit to that used for the primary stability problem. Guided by the convergence history of the Fourier modes presented in later work of Herbert and co-workers (f.e. Herbert *et al.* 1985) we only consider here the minimum non-trivial number of two Fourier modes in the x direction. Again, validation of our numerical scheme is provided by the classical Blasius flow. Herbert (1984) has presented the dependence of the secondary growth rate σ_r on the spanwise wavenumber at varying amplitudes A of the primary disturbance in Blasius flow at Branch II neutral conditions (Kachanov and Levchenko, 1984). Our result may be found in figure 2. At $A = 0.01$, Branch II neutral conditions ($R = 606$, streamwise wavenumber $\alpha = 0.20335$, spanwise wavenumber β obtained by the relation $10^3\beta/R = 0.33$), Herbert *et al.* (1985) and Ng and Erlebacher (1992) have calculated the secondary growth rate value to be $\sigma_r \approx 8.12 \times 10^{-3}$; in our calculations we find $\sigma_r \approx 8.09 \times 10^{-3}$. This result does not change with an increasing number of Chebyshev nodes or extent of the integration domain in the normal direction provided, of course, that the respective primary stability results have converged. However, our primary stability result at $R = 606$, $\alpha = 0.20335$ suggests that the eigenvalue of the Tollmien-Schlichting wave at these conditions is $c = c_r + i c_i \approx 0.369558 + 0.000015\, i$, i.e. *not* machine-precision neutral. We are therefore inclined to attribute this .4% discrepancy in secondary growth rate to difference in the parameters used for the primary problem.

4. Results

Having determined the Reynolds number range, we are next faced with the task of choosing wavenumbers for the calculations that may be relevant to the experimentally observed waves. The comparison of linear stability and experimental results reveals that naturally occurring waves may be observed in the neighbourhood of Branch I (Hall *et al.* 1984). We thus restrict ourselves to a streamwise wavenumber $\beta = 0.2$ and present results in the Reynolds number range $R \in [1000, 2000]$. Throughout our secondary stability calculations we used $N = 64$ collocation points that were found to be sufficient for the Blasius problem discussed earlier.

We first present in figure 3 the secondary growth rate σ_r dependence upon the streamwise wavenumber parameter $a = 10^3\alpha/R$ at three Reynolds numbers $R = 1200, 1600$ and $R = 2000$. An increase of the amplitude of the fundamental is considered from a low value $A = 0.4\%$ up to a value that secondary theory suggests as an approximate upper limit for which the shape assumption still holds, $A = 1\%$. The first result of significance that may be observed is that, with increasing Reynolds number, larger values of secondary growth rate are obtained at the same amplitude of the fundamental A. This may be linked to the fact that the respective primary wave considered is amplified stronger as β is kept constant and the Reynolds number increases. In a manner qualitatively similar to Blasius flow, a sharp cutoff wavenumber is observed at the low end of the range of secondary amplified modes at all Reynolds numbers considered. At constant Reynolds number an increase of the amplitude of the primary wave results in a wider range of streamwise wavenumbers being potentially dangerous. Further, with increasing Reynolds number, an ever wider band of modes that may be amplified due to secondary interactions is calculated at the same level of primary excitation.

The condition $|\omega_i| = |\beta c_i| \ll |\sigma_r|$ is inherent in the assumptions of the secondary stability theory. As an a-posteriori check we observe that this condition is satisfied in the results obtained, the secondary growth rate being always larger to that of the primary wave by a factor between two and three at $A = 1\%$. However, we note that a fixed value of the spanwise wavenumber $\beta = 0.2$ results in an increasingly large value of the primary growth rate ω_i as R increases, i.e. departure from the near neutral conditions considered at $R = 1000$.

Turning to the dependence of the wavenumbers of the amplified secondary modes on Reynolds number we predict that, in the Reynolds number range considered, the wavenumber of the maximally amplified mode lies in the narrow band $a \in (0.14, 0.20)$, corresponding to streamwise wavenumbers $\alpha \in (0.14, 0.40)$. This range is, in fact, comparable to the range of spanwise wavenumbers amplified according to linear theory.

Finally, while possibly least relevant to the attachment-line problem, the lowest Reynolds number value considered, $R = 1000$, is that for which it is most feasible to obtain experimental results. For comparison with experiment, when available, we present in figure 4 the spatial structure of the maximally amplified secondary mode at $A = 0.01$; this occurs at $a \approx 0.185$ and has a growth rate of $\sigma_r \approx 3.2 \times 10^{-3}$.

5. Discussion

Subharmonic stability results have been obtained at high Reynolds number, where the primary stability problem based on the Orr-Sommerfeld equation is a good approximation to the attachment-line system. The range of wavenumbers favouring secondary growth has been identified and the dependence of the secondary growth rate on the streamwise wavenumber, the Reynolds number and the amplitude of primary excitation has been obtained. The spatial structure of the respective (secondary) eigenvectors has also been made available for comparison with experiment. Work is currently underway to assess the relevance of the predictions delivered by our model to the experimental results.

As has been discussed, the problem discussed presently is expected to be the large Reynolds number limit of the secondary stability problem in incompressible flow in the infinite-swept attachment-line boundary layer. Work is also underway on the study of the full AL secondary problem. In this respect it is of interest to compare the predictions of the secondary AL system to those delivered by the present model at a substantially lower computing effort. Results of this work will be reported elsewhere.

References

Arnal, D., (1993). AGARD-FDP-VKI LS-06: *Progress in Transition Modelling,* VKI, Brussels, Belgium, March 29 - April 1, 1993.

Boyd, J. P., (1989). *Chebyshev and Fourier Spectral Methods,* Lecture Notes in Engineering **49**, Springer.

Hall, P., Malik, M. R. and Poll, D. I. A., (1984). On the Stability of an Infinite Swept Attachment-Line Boundary Layer. *Proc. Roy. Soc.* **A 395**, p.229.

Herbert, Th., (1983). Secondary instability of plane channel flow to subharmonic three-dimensional disturbances. *Phys. Fluids* **26** (4), p.871.

Herbert, Th., (1984). Analysis of the subharmonic route to transition in boundary layers. *AIAA*, Paper 84-0009.

Herbert, Th., Bertolotti, F. P., and Santos, G. R., (1985). Floquet analysis of secondary instability in shear flows. in *Stability of time dependent and spatially varying flows,* D. L. Dwoyer and M. Y. Hussaini (eds.), Springer, p.43.

Herbert, Th., (1988). Secondary instability of boundary layers. *Ann. Rev. Fluid Mech.*, **20**, p.487.

Kachanov, Yu. S., and Levchenko, V. Ya., (1984). The resonant interaction of disturbances at laminar-turbulent transition in a boundary layer. *J. Fluid Mech.* **138**, p.209.

Ng, L. L., and Erlebacher, G., (1992). Secondary instabilities in compressible boundary layers. *Phys. Fluids A* **4** (4), p. 710.

Orszag, S. A., and Patera, A. T., (1983). Secondary instability of wall-bounded shear flows. *J. Fluid Mech.* **128**, p. 347.

Pfenninger, W., and Bacon, J. W., (1969). Amplified laminar boundary layer oscillations and transition at the front of a 45° swept flat-nosed wing with and without boundary layer suction. in *Viscous Drag Reduction,* (C. S. Wells ed.), Plenum Press, p.85.

Poll, D. I. A., (1978). *College of Aeronautics* Rep. **7805**.

Poll, D. I. A., (1979). Transition in the infinite swept attachment line boundary layer. *The Aeronautical Quart.* **30**, p.607.

Streett, C. L., Zang, T. A. and Hussaini, M. Y., (1984). Spectral methods for solution of the boundary layer equations. *AIAA Pap. No. 84-0170.*

Theofilis, V., (1994a). The discrete temporal eigenvalue spectrum of the generalised Hiemenz flow as solution of the Orr-Sommerfeld equation. *J. Eng. Math.* (to appear in Vol. **28**)

Theofilis, V., (1994b). On subcritical instability of the attachment-line boundary layer. in review in *J. Fluid Mech.*

# of Nodes	$v''(0)$	$v'(\infty)$	CPU (secs)
8	-1.240338101715	-1.133332778613	0.074797
16	-1.232572595378	-0.999434577839	0.216552
24	-1.232587607999	-0.999994296376	0.358365
32	-1.232587656865	-1.000000008974	1.032070
48	-1.232587656820	-0.999999999953	2.102004
64	-1.232587656820	-0.999999999900	2.885904

TABLE I
Convergence history of collocation solution for the base flow normal velocity component v against the number of (mapped) Chebyshev nodes utilised N. Also shown the respective CPU time on a typical workstation.

# of Nodes	c_r	c_i
16	0.3657138	0.0078122
32	0.3675448	0.0088640
48	0.3675377	0.0088558
64	0.3675377	0.0088559

TABLE II
Convergence history of collocation solution for the temporal primary stability problem in Blasius flow at $R = 500, \alpha = 0.16869$.

	AL		OS	
R	c_r	c_i	c_r	c_i
1000	0.331329	0.005653	0.331948	0.003336
1200	0.323049	0.009650	0.323438	0.007794
1400	0.316111	0.012467	0.316351	0.010929
1600	0.310171	0.014502	0.310308	0.013196
1800	0.304998	0.015993	0.305062	0.014863
2000	0.300431	0.017091	0.300442	0.016098

TABLE III
Frequency and growth rate delivered by spectral collocation solution of the temporal linear stability generalised eigenvalue problem in incompressible infinite swept attachment-line boundary layer flow (Theofilis, 1994b) against those delivered by numerical solution of the Orr-Sommerfeld approximation to the linear stability problem (Poll, 1978; Theofilis, 1994a).

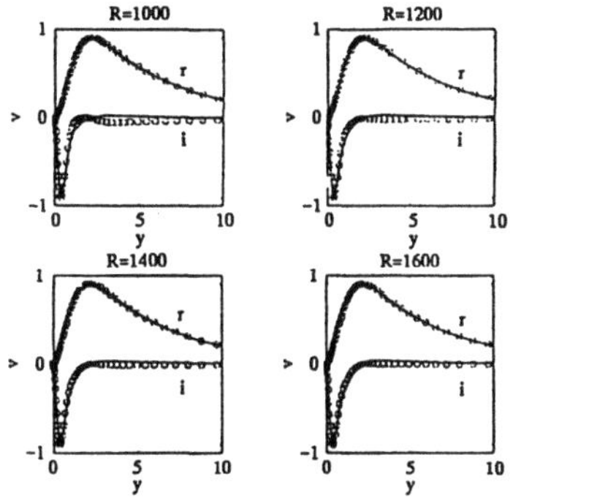

Fig. 1. Spatial structure of the normal perturbation velocity $\hat{v}$ delivered by the AL system (solid line) and the OS equation (symbols) at $R = 1000(200)1600$.

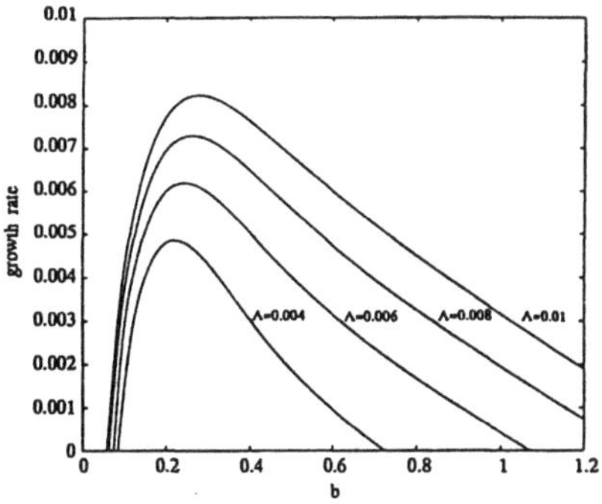

Fig. 2. Growth rate σ_r as a function of wavenumber b in Blasius flow at Branch II (Herbert, 1984).

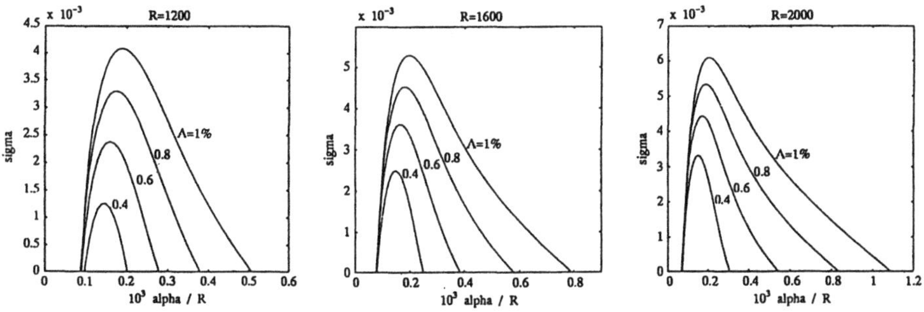

Fig. 3. Secondary growth rate σ_r against spanwise wavenumber parameter $a = 10^3\alpha/R$ at $R = 1200, 1600$ and $R = 2000$ and primary disturbance amplitude $A = 0.004(0.002)0.010$.

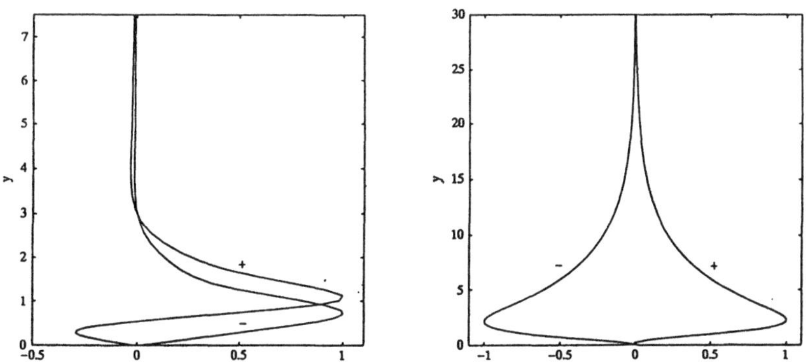

Fig. 4. Secondary eigenfunctions $w^{\pm}$ and $v^{\pm}$ at $R = 1000, a = 0.185$. Calculated growth rate $\sigma = \sigma_r \approx 0.003218$.

Comparison between the Orr-Sommerfeld and Asymptotic Theories for the Flat-Plate Boundary Layer

J.J. Healey
Department of Engineering
Cambridge University
Trumpington Street
Cambridge, UK

Summary

The linear stability of the flat-plate boundary layer has been investigated using the Orr–Sommerfeld (OS) theory and compared with consistent asymptotic expansions. In the conventional picture, OS theory provides better agreement with experiment, but the asymptotic theory gives greater physical insight and can be extended to include nonlinear and nonparallel effects. The nonlinear asymptotic theory has concentrated mainly on the upper branch of the neutral curve. In this paper results are presented that suggest that the upper-branch asymptotic theory is not applicable to boundary-layer transition.

Key words: Linear stability, 2-D boundary layers, growth rates.

1. Introduction

The transition of a boundary layer from a laminar to a turbulent state is not properly understood and this is a major obstacle to the accurate prediction of drag on aircraft wings. The simplest case to study in an experiment is a flat plate mounted at zero incidence to the freestream. If care is taken with the leading edge design, then an essentially zero pressure gradient can be established over a large part of the plate and many experimental investigations into the linear stability of the resulting Blasius velocity profile have been performed.

Although incompressible two-dimensional zero pressure-gradient boundary layers are not of direct interest in commercial applications, the same transition scenario can be seen in many boundary layer flows. The first phase is the growth of travelling instability waves (Tollmien–Schlichting waves) followed by nonlinear interactions with oblique modes (resonant triads, secondary instabilities, streamwise vortices) which lead to the intermittant formation of turbulent spots and their coalescence to form a turbulent boundary layer. For this reason the flat-plate boundary layer has been used as starting point for developing nonlinear models for transition by many authors and is the case studied here.

The nonlinear models are determined by the types of nonlinear interactions being investigated and this in turn is determined by the linear stability characteristics of the flow. The use of the parabolized stability equations will not be considered here, but the two other most widely used stability methods are the Orr–Sommerfeld (OS) equation and consistent asympotic expansions in inverse powers of the Reynolds number. The OS equation is commonly used by experimentalists since the phase speed and growth rates of waves predicted by this theory are in close agreement with those measured in experiment. The disadvantage of the OS theory is that it is not clear how nonparallelism, which is arbitrarily excluded in the derivation of the OS equation, can be consistently re-introduced at higher order. This also makes the consistent introduction of nonlinearity problematic.

Alternatively, the asymptotic theory does provide a framework for the systematic inclusion of nonparallelism and nonlinearity. It also identifies physical structures within the boundary layer enabling further nonlinear types of behaviour, like critical layers, to be studied. The asymptotic theory can be subdivided into two main parts; one for the lower branch neutral curve, the other for the upper branch neutral curve. Each branch corresponds to a different distinguished Reynolds number scaling and a different physical structure. For disturbances near the lower branch, the boundary layer has a triple-deck structure. Nearest the wall is a layer where the viscous terms are of the same order as the unsteady and convective terms, this region is called the viscous wall layer (or lower deck) and the critical layer (where phase speed of the wave matches the local mean flow speed) lies in this layer too. In this region the mean velocity profile is essentially linear. Outside this layer is a region of inviscid but rotational flow (the main deck), which is the main part of the boundary layer, and whose height is of the order of the displacement thickness. Above this deck is the outer deck where the flow is inviscid and irrotational, i.e. potential flow.

For the upper branch there are five decks because now the critical layer is well separated from the viscous layer. The viscosity is now confined to two layers, the wall layer and the critical layer and there is a phase jump in the disturbance velocity across the critical layer. This phase jump is stabilising or destabilising depending on the sign of the second derivitive of the mean profile at the critical layer. Pressure gradients affect the curvature of the profile and this is why the stability of the flow near the upper branch is sensitive to the pressure gradient, but that near the lower branch the stability is much less sensitive to the pressure gradient.

The strongest growth rates are found on the lower-branch scaling and so weakly nonlinear theory, which can only be applied when the growth rates are small, has a limited range of applicability on this scaling. The upper-branch scaling has small growth rates (good for weakly nonlinear theory) and applies over a much larger part of the plate compared to the lower-branch region. In fact the net growth over the upper branch is of the same order as that for the lower branch despite the smaller growth rates. Resonant triad calculations can be performed on the upper branch but not the lower branch. In addition, there is nonlinear critical layer behaviour for the upper branch. All these reasons make the upper-branch scaling more attractive for doing nonlinear asymptotic calculations than the lower branch.

However, the upper-branch asymptotics have a much worse quantitative agreement with experiment than the lower-branch asymptotics (or the OS calculations). In this paper we investigate the possibilty that on the upper-branch, the critical layer, which is separate from the viscous wall layer at large Reynolds numbers, may submerge into the wall layer at some finite Reynolds number; the five-deck theory would then not be applicable below this Reynolds number. This has been done by studying the eigenvalues and eigenfunctions of the OS equation at larger Reynolds numbers than those used in other investigations (normally the eigenvalues are only calculated for Reynolds numbers relevant to experiment). It turns out that structures found in the OS eigenfunctions can be related to the decks of the upper- and lower-branch asymptotics, and that the upper-branch structure only applies above a certain finite Reynolds number (below which both branches of the neutral curve have a triple-deck structure). This Reynolds number is so high that five-deck asymptotics can play no role in boundary layer transition. In this paper, only an outline of the main results will be presented, the full details are given in [1].

2. Results

Length, time, velocity and pressure are nondimensionalised with respect to δ, δ/U_0, U_0 and ρU_0^2 respectively, where δ is the displacement thickness given by

$$\delta = 1.7208\sqrt{\frac{x\nu}{U_0}}, \tag{1}$$

U_0 is the freestream velocity, x is the streamwise co-ordinate, ρ is the density and ν is the kinematic viscosity. The Reynolds number is then given by

$$R_\delta = \frac{\delta U_0}{\nu}. \tag{2}$$

The leading order term of the lower-branch asymptotic expansion for the neutral curve is

$$\omega \sim 0.995 R_\delta^{-1/2}, \tag{3}$$

see [2], and for the upper branch it is

$$\omega \sim 0.2027 R_\delta^{-1/5}, \tag{4}$$

see [3], where ω is the nondimensional neutral frequency.

Fig. 1 shows these asymptotic results for the neutral curves on the same graph as the OS neutral curve and a set of experimental results from [4]. (More recent experimental results, see [5], show that better agreement with the OS neutral curve can be obtained if greater effort is taken to achieve a zero pressure gradient.) These results are usually plotted on linear scales for $R_\delta < 2500$ and for that range there is a big quantitative difference between the upper branch

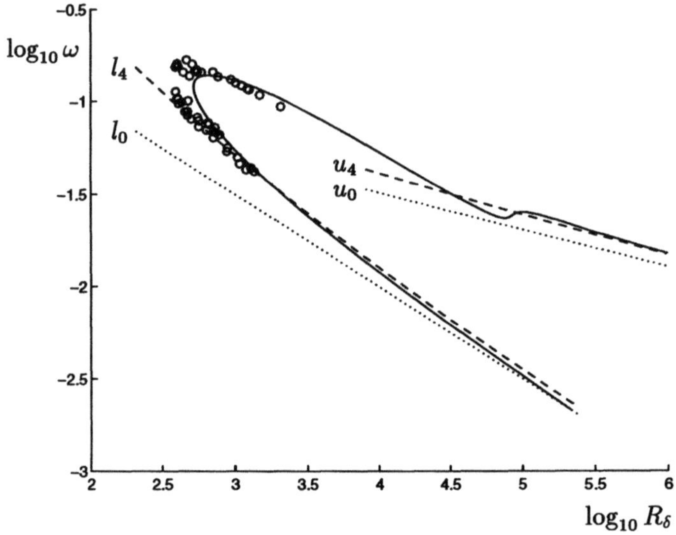

FIGURE 1. Log-log plot of neutral stability curves. The dotted lines l_0 and u_0 are (3-4). The dashed lines l_4 and u_4 are the four term (but still parallel) expansions given in [2-3], see also [1]. The circles are experimental data from [4] and the solid line is the OS curve.

results and the others. However, fig. 1 shows that at $R_\delta \approx 10^5$ there is a relatively sharp kink in the OS neutral curve and that for $R_\delta > 10^5$ there is good agreement with the upper branch asymptotics. In [1] it is shown that this kink can be related to the close proximity of a square-root branch point singularity in the Tietjens function, and thus marks a region in the (R_δ, ω) plane where analytic expansions for the neutral curve will become inaccurate. After failing to negotiate the kink, the upper-branch asymptotics are inaccurate for $R_\delta < 10^5$.

As well as the sharp change in the eigenvalues near the kink, the eigenfunctions also show a change in structure in this region. According to the upper-branch asymptotic theory, the phase jump, $\Delta\phi$, of the streamwise component of the fluctuating velocity, across the critical layer is given by

$$\Delta\phi = \left(\frac{\omega}{2\lambda\alpha}\right)^2 \left(\frac{R_\delta}{1.7208}\right)^{\frac{1}{5}} \pi P_0 \tag{5}$$

where α is the wavenumber, P_0 is the pressure perturbation at the wall and $\lambda = 0.33206$ (the gradient of the mean profile at the wall), see [3]. This predicted phase jump has been compared with the phase jump found in the OS eigenfunctions and is shown in fig. 2. The emergence of a new layer at the height of the critical layer and with a phase jump in close agreement with the asymptotic theory is clearly visible between figs. 2(b) and (c). In fig. 2(d) there is a region of almost constant phase above and below the critical layer and also

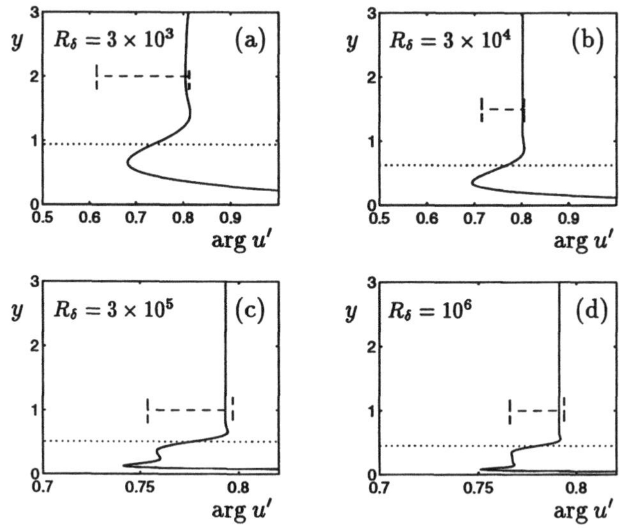

FIGURE 2. Variation of phase through the boundary layer for four different Reynolds numbers, all on the upper-branch OS neutral curve. y is the nondimensional distance from the wall, u' is the streamwise perturbation velocity, the dotted line shows the height of the critical layer and the dashed bars correspond to the phase jump (5).

a region of rapid phase variation close to the wall which can be associated with the viscous wall layer.

The turning points in the phase are a good indicator of the structure of the eigenfunction, e.g. in fig. 2(d) the critical layer is bracketed by turning points in the phase on either side. The heights of the turning points have been plotted as a function of Reynolds number for both the upper and lower branches of the OS neutral curve in fig. 3. On the lower branch, fig. 3(b), the critical layer lies below two turning points in the phase of u' (the streamwise velocity fluctuation), and both of these turning points and the critical layer scale at $R_\delta^{-1/4}$ which is the triple-deck asymptotic scaling for the height of the viscous wall layer. On the upper branch, for $R_\delta < 10^5$, the turning points also scale by the triple-deck result for the height of the viscous wall layer. The critical layer now lies between these two turning points and gently rises up towards the upper turning point as R_δ increases. As R_δ increases through 10^5 the critical layer breaks free from the viscous wall layer, and for $R_\delta > 10^5$ there is a new structure for the eigenfunctions. In this region the viscous wall layer scales as $R_\delta^{-2/5}$ which is in agreement with the five-deck asymptotic scaling, and the height of the critical layer scales as $R_\delta^{-1/10}$ which is also in agreement with the five-deck asymptotic scaling.

These results show that, for the neutral curve at least, the five-deck asymptotics shouldn't be used for $R_\delta < 10^5$. In order to investigate the applicability of triple-deck and five-deck asymptotics for the growing modes (which are much

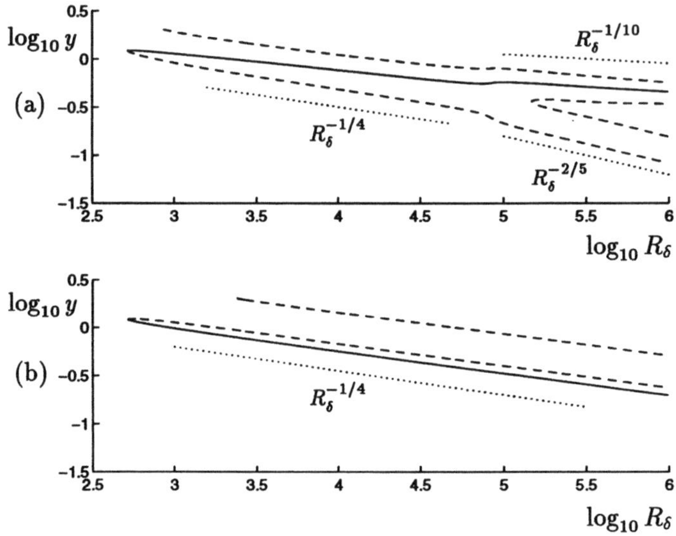

FIGURE 3. Log-log plots of the heights of turning points of arg u' (dashed lines) against Reynolds numbers for (a) OS upper branch and (b) OS lower branch. The solid line is the height of the critical layer. The dotted lines show triple-deck and five-deck asymptotic scalings (see text).

more important for transition), contours of nonzero growth rates have also been calculated using the OS theory.

There are three different ways of defining the nondimensional growth rate of a wave: $\delta\alpha_i^*$, $\nu\alpha_i^*/U_0$ or α_i^*/α_r^* where α_i^* is the dimensional growth rate. Of these, the first is most commonly used in OS calculations since all lengths are nondimensionalised with displacement thickness anyway. A draw back, however, is that δ varies over the plate and this makes it difficult to make comparisons of growth rates at different Reynolds numbers. The second choice gives a growth rate independent of Reynolds number and is the natural choice if the net growth in amplitude of a disturbance down the plate is required. Third choice is the relevant quanitity for weakly nonlinear analyses and has the advantage of being independent of the nondimensionalizing length scale. One consequence of the different definitions is that each has a different Reynolds number at which the maximum growth rate occurs. For the first choice the maximum growth rate is at $R_\delta \approx 4930$, for the second $R_\delta \approx 1050$ and for the third the growth rate approaches a maximum, which is $O(1)$, as $R_\delta \to \infty$ and lies on the triple-deck asymptotic scaling.

In fig. 4 the contours of constant α_i/α_r are plotted on a logarithimic scale over a large range of Reynolds numbers. The strongest growth rates are found in a band that lies approximately parallel to the lower branch and this region can be associated with the triple-deck structure. At large ω, the contours for damped waves are almost horizontal which is characteristic of damped Rayleigh

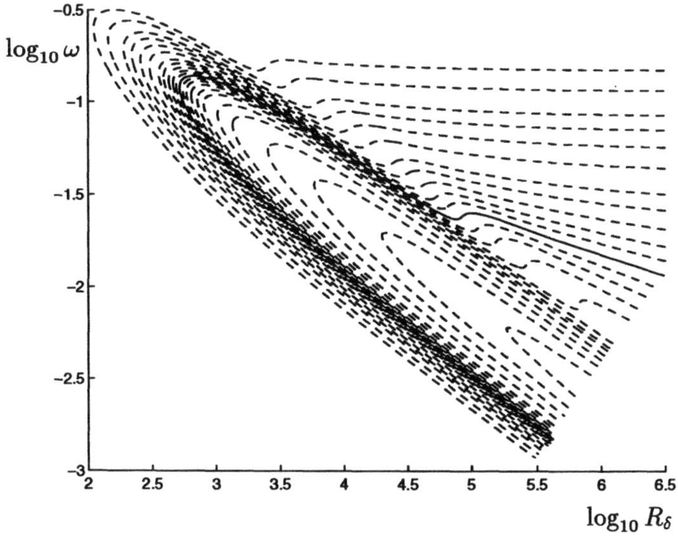

FIGURE 4. Log-log plot of contours of constant growth rate. The solid line is the neutral curve of fig. 1. The dased curves are unequally spaced as follows: 10%, 8%, 6%, 5%, 4%, 3%, 2%, 1.5%, 1%, 0.6%, 0.3%, -0.3%, -0.6%, -1%, -1.5%, -2%, -3%, -4%, -6%, -8%, -10%, -12%, where negative values are unstable.

(inviscid) modes. In [6], spatial solutions of the Rayleigh equation have been calculated and the OS and Rayleigh eigenvalues agree very well in this region. At $R_\delta > 10^5$ there is a range of weakly unstable modes that covers a larger part of the plate than the strongly growing triple-deck modes (when $\log_{10} R_\delta$ increases by 0.5, the downstream distance increases by a factor of 10). This weakly growing area corresponds to the five-deck region of the plate and forms a wedge between the strongly growing triple-deck modes and the damped Rayleigh modes. The apex of this wedge, which gives the lowest Reynolds number at which the five-deck asymptotics can be applied is close to $R_\delta = 10^5$. At lower R_δ the triple-deck modes match directly onto the Rayleigh modes along a 'fault line' where all the contours show kinks. The kinks are sharper at higher Reynolds numbers.

Calculations have also been performed for Falkner–Skan profiles with weak favourable and adverse pressure gradients, see [6]. When the constant growth rate contours are plotted for favourable and adverse pressure gradients, the graphs are qualitatively very similar to that shown in fig. 4. The main difference lies in the position of the neutral curve relative to the other contours. For the favourable pressure gradient the kink in the neutral curve moves to higher R_δ and beyond this point the neutral curve lies on a new asymptotic scaling. For the adverse pressure gradient the kink in the neutral curve moves to a lower R_δ and then the neutral curve follows a Rayleigh scaling.

3. Conclusions

A great deal of the theory of the nonlinear stages of boundary layer transition has been concentrated on the asymptotic structure of the upper branch of the neutral stability curve. This is because the growth rates are relatively weak, and nonlinear critical layer behaviour and resonant triads can all be studied consistently on the scalings of the upper branch. In this paper, the eigenvalues and eigenfunctions of the OS equation at very large Reynolds number have been studied. The results show that the five-deck asymptotic structure only exists for $R_\delta > 10^5$.

In a wind tunnel running at 20ms^{-1} this Reynolds number is only reached some 2km downstream of the leading edge of the plate. For flight speeds of say 250ms^{-1} this distance is of order 180m (a Boeing 747 is about 70m long). Since the flow would always be expected to be turbulent at such Reynolds numbers, the five-deck asymptotic structure is not relevant to boundary layer transition.

References

[1] Healey, J.J.: On the neutral curve of the flat-plate boundary layer: Comparison between experiment, Orr–Sommerfeld theory and asymptotic theory, J. Fluid Mech. (1995) accepted, subject to revision.

[2] Smith, F.T.: On the non-parallel flow stability of the Blasius boundary layer, Proc. R. Soc. Lond. A 366 (1979) 91–109.

[3] Bodonyi, R.J. and Smith, F.T.: Proc. R. Soc. Lond. A 375 (1981) 65–92.

[4] Ross, J.A., Barnes, F.H., Burnes, J.G. & Ross, M.A.S.: The flat plate boundary layer. Part 3. Comparison of theory with experiment, J. Fluid Mech. 43 (1970) 819–832.

[5] Klingmann, B.G.B., Boiko, A.V., Westin, K.J.A., Kozlov, V.V. and Alfredsson, P.H.: Experiments on the stability of Tollmien Schlichting waves, Eur. J. Mech. B 12 (1993) 493–514.

[6] Healey, J.J.: A comparison between Orr-Sommerfeld theory and asymptotic theories for damped and growing boundary layer modes, J. Fluid Mech. (1995) submitted.

Session K:
3-D Boundary Layers II

Analysis of the Instability of the Boundary Layer over a Rotating Disk

S. Jarre, P. Le Gal, M.P. Chauve

Institut de Mécanique Statistique de la Turbulence
U. M. 0033, Université d'Aix-Marseille II, C.N.R.S.
12, Avenue Général Leclerc
13003 Marseille

Summary

The instability of the three-dimensional flow over a rotating disk is experimentally investigated by hot-wire measurements. The first part of this paper presents the experimental marginal stability curve and dispersion curves (linking frequencies to wave vector components). It is then observed that non-linear effects play a major role very early in the boundary layer. The second part is thus devoted to the study of a well controlled wave pattern excited by a roughness element placed under linear threshold. The system exhibits a weak non-linear behaviour which leads to the appearance of super harmonics modes and to a shift of the dispersion curves as compared to their linear values. Finally, we present an original method in order to infer the group velocity components from the dispersion curves associated with the propagation angle of the wave packet. In particular, the results demonstrates the convective nature of the instability.

Key words: Rotating disk flow instability, Marginal stability, Dispersion relation, Group velocity.

1. Introduction

This paper presents an experimental analysis of the instability of the flow over a rotating disk. Our device consists of a rotating stainless-steel disk immersed in a cylindrical container filled with water. When the disk rotates in still water, an Ekman boundary layer of thickness $\delta = \sqrt{\nu/\Omega}$ develops over the entire disk (in the following, the rotation speed Ω is fixed to 1 revolution per second and $\delta = 0.4$ mm). The control parameter associated with the flow is the Reynolds number using the local radius R: $Re = R/\delta$. One of the laminar velocity profiles inside this boundary layer is of peculiar type: the centrifugally induced radial velocity profile posesses an inflexion point. This situation is known in hydrodynamics to be highly unstable, and leads to the so called "cross-flow" instability.

The system amplifies infinitesimal perturbations creating a periodic wave pattern at the surface of the disk corresponding to an averaged value of 30 spiral vortices first visualized by Gregory et al .(1955) [1]. Further along the radius, these vortices transit to turbulence. Thus, this experiment is able to reproduce the complete scenario of the transition to turbulence of a three-dimensional boundary layer flow. In particular, this "cross-flow" instability is the same as that which takes place on airplane wings. It is well known that the control of the transition to turbulence in such flows is of primary importance in reducing the drag on airplanes.

2. Linear analysis of the instability

By an inviscid linear analysis, Stuart et al. (1955) [1] determined the existence of more than one hundred stationary spiral vortices in the boundary layer of a rotating disk. This value, four times greater than the experimental one, was corrected later by Brown (1959) [2] and Cebecci & Stewartson (1980) [3], by taking the viscous effects into account. But, their calculated critical Reynolds numbers were smaller than the observed value. Kobayashi et al. (1979) [4] and later, Malik et al.. (1981) [5], showed that the Coriolis force and stream line curvature cannot be neglected. Their calculations predicted marginal stability curves which separate domain of stable waves from that of unstable waves. These parabola-shaped curves are calculated in the (Re,k) plane, where k is the wavenumber, and require experimental verification. This section is devoted to the experimental lirear stability analysis of the flow, leading to the experimental marginal stability curve (Jarre et al, 1991 [6]).
We measure azimuthal velocity fluctuations by means of a hot film probe located inside the boundary layer, at a distance of 0.6 mm from the disk (the experimental boundary layer is about 1 mm thick). The wire is aligned with the disk radius in order to be sensitive to azimuthal velocity fluctuations. Figure 1 presents some aspects of the signals recorded at two different Reynolds numbers.

(Re=393)

(Re=488)

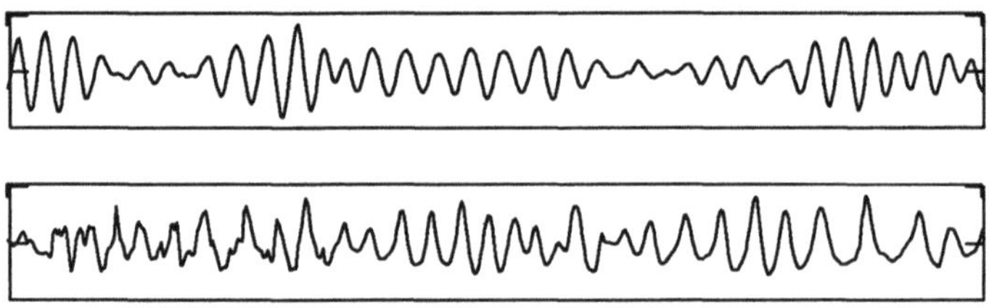

Figure 1: Velocity signals during 1 disk revolution at two Reynolds numbers.

Fast Fourier Transforms of these signals are then computed and averaged over 60 disk revolutions Figure 2, shows the evolution of the Fourier spectrum versus Reynolds number and reveals the merging of a wave packet centered on 30 Hertz. As this packet grows, a low frequency band is viscously damped till the 30 Hertz band spreads out and invades the whole spectrum. This large bandwidth spectrum is a characteristic feature of turbulence, although in our case, the moderate Reynods numbers do not permit the appearance of developed turbulence.

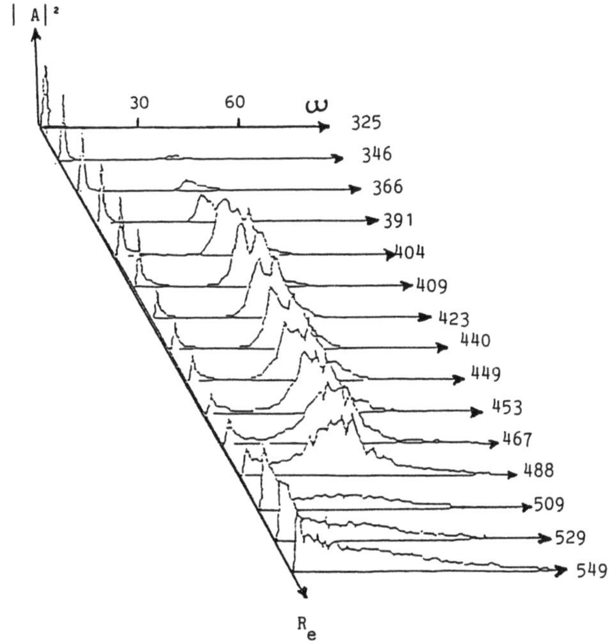

Figure 2 : Evolution of the Fourier spectrum with the Reynolds number.

Plotting the amplitude of a given frequency ($\omega = \frac{\omega^*}{\Omega}$) versus the Reynolds number, yields the global amplification diagram of each Fourier mode. Figure 3 shows such a diagram. In deriving this curve, we obtain the evolution of the local amplification rate. If it is negative, the wave is damped, where as a positive value corresponds to an exponential growth of the amplitude. The critical threshold of the mode is obtained for a zero value of this local rate which permit the experimental determination of the marginal stability curve in the (ω,Re) plane. In order to compare our experimental results with numerical ones, we used a zero-phase velocity assumption in order to transform frequencies into azimuthal wave numbers ($\beta = \frac{2\pi}{\lambda_\theta / R}$). This curve is given in Figure 4 together with numerical ones. It can be seen that there is very good agreement between the experimental ($Re_C = 264$) and the numerical determinations of the critical Reynolds number. But we notice a shift of our experimental curve towards higher wave numbers. In fact, there is nothing to indicate that the 30-Hertz frecluency corresponds to 30 vortices rotating rigidly with the disk. Indeed, a 30-hertz periodic signal can also result from fewer (more) than 30 vortices rotating with a positive (respectively negative) phase velocity. In order to verify this assumption, we measure dispersion relations which associate frequencies to wave vectors. The next section presents double probe measurements that lead to such relations.

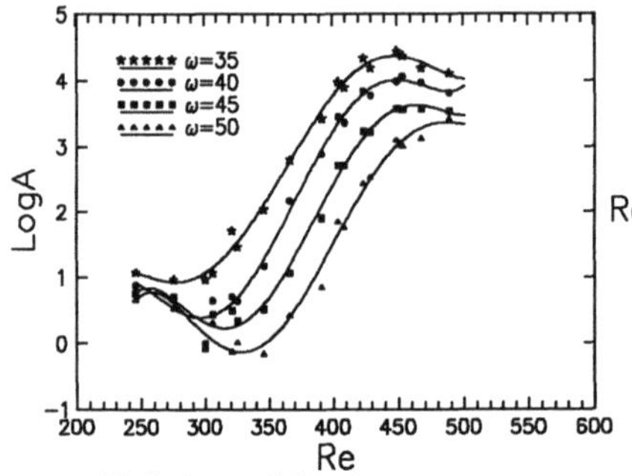

Figure 3 : Global amplification rate versus Reynolds number for selected frequencies.

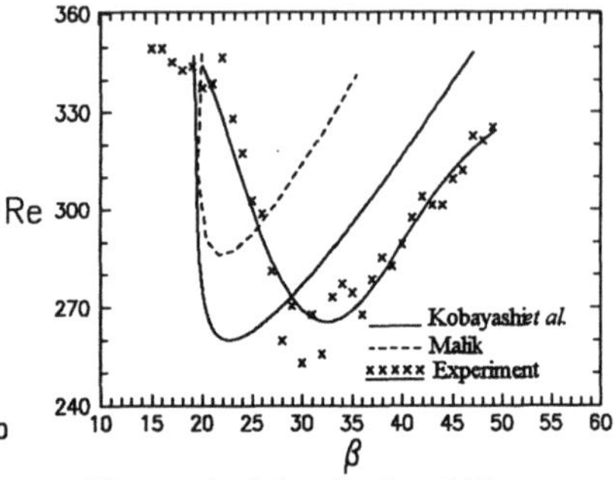

Figure 4 : Marginal stability curve of the rotating disk flow instability.

3. Dispersion relation measurement

When a wave of frequency ω and wave number k travels in a medium, the phase difference between two positions is directly linked to k. If the distance between the two probes is Δx, the phase difference is simply: $\Delta\psi=\Delta x\ k$. In fact our problem is more complicated because of the two-dimensional aspect of the wave propagation on the disk. Indeed, we need two sets of measurements in order to obtain the two components (α for the radial direction and β for the azimuthal direction) Of the wave vector k. Figure 5 shows two different dispersion relations, ω=ω(α) and ω=ω(β), averaged over 60 disk revolutions, and measured in the laboratory frame.

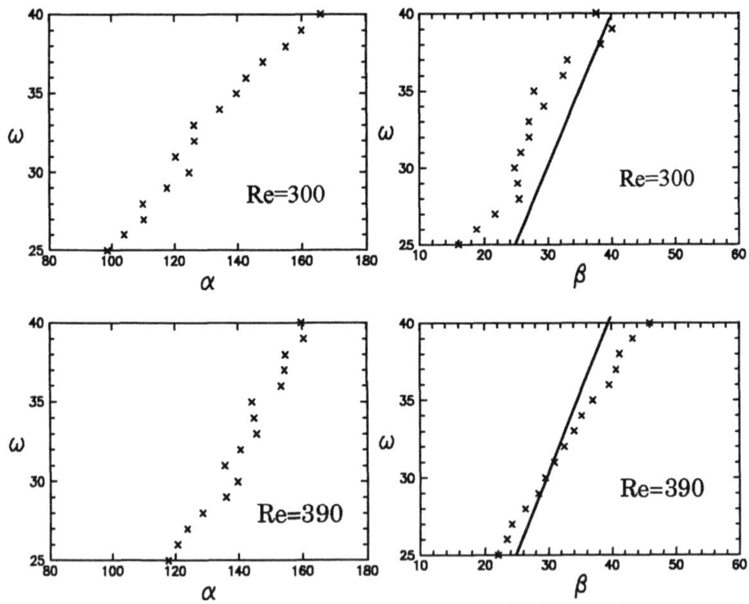

Figure 5 : Dispersion relation curves for two typical Reynolds numbers.

On the (β,ω) diagrams, the line (sw) denotes the zero-phase velocity wave dispersion relation. At low Reynolds numbers, the (β,ω) curves are shifted with respect to the stationary wave line, indicating that the unstable waves are travelling. Note that the total derivatives $d\omega/d\alpha$ and $d\omega/d\beta$, (the slopes of the experimental curves) are not the components of the group velocity as it would be the case in a one-dimensionnal space. Indeed, the dispersion relation is a two-dimensionnal surface $\omega=\omega(\alpha,\beta)$ and its partial derivatives are needed to determine the group velocity components. We will come back later on the determination of this velocity $\vec{V}_g$. When the Reynolds number is increased, the experimental data points are nearly aligned on the (sw) line. Thus the unstable waves can be considered as stationary on the disk. At higher Reynolds numbers, the dispersion curve tilts away from the stationary wave line. This effect is linked to the dispersion of the waves leading to the spreading of wave packets; some of the modes travel faster than the disk and others with a lower phase speed.

4. Non linear effects

When the amplitude of the waves is not small enough, non-linearity can affect the characteristics of their propagation. In this section, we study large amplitude waves generated by a roughness element located on the disk surface. This roughness element consists of a 0.5 mm diameter and 0.6 mm long cylinder and is placed at Re=250, just before the linear threshold. Figure 6 presents the wave pattern generated by the roughness element which is similar to Wilkinson & Malik (1983)'s result [7]. Because of wave dispersion, the wave pattern spreads out in the manner of a "ship wave pattern".

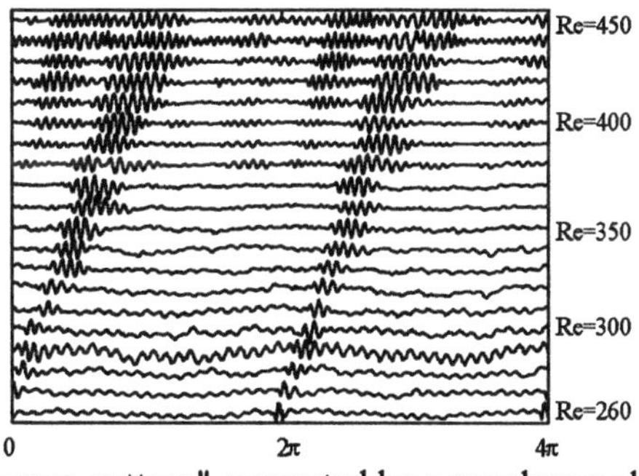

Figure 6 : The "ship wave pattern" generated by a roughness element over two disk revolutions.

The marginal stability curve (figure 7) has been derived from the previous signals, and plotted in the (ω,Re) plane. The curve has a double-parabola shape; the first is centered on the most unstable mode (ω_c=30) and is similar to the curve obtained in the case of natural excitation The second is centered around the super harmonic critical mode of frequency $2\omega_c$= 60. We confirm that this effect is produced by the non linearities, in plotting the amplitude of a 60 Hertz-centered band versus the squared amplitude of the corresponding fundamental band, in a log-log diagram. As expected, figure 8 shows a line of slope equal to one in accordance with the quadratic non linear term of the Navier-Stokes equation.

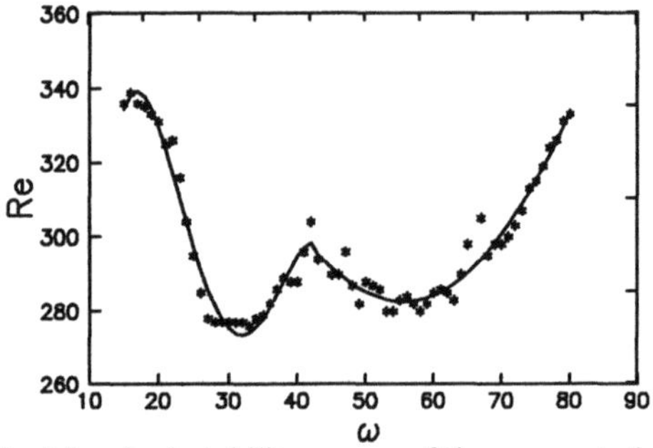

Figure 7 : Marginal stability curve of the generated wave pattern.

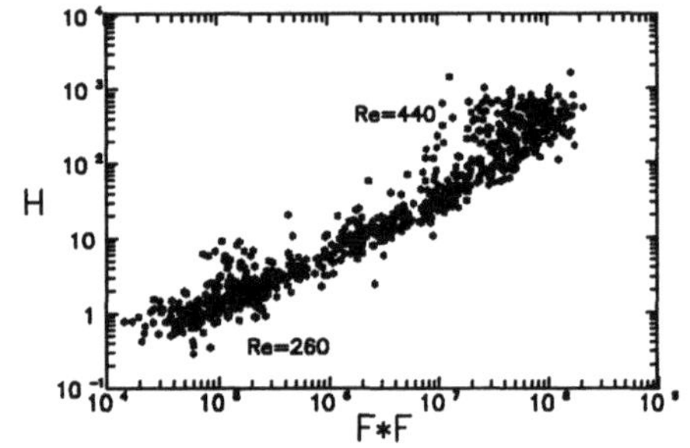

Figure 8 : Amplitude of the harmonic band versus the squared amplitude of the fundamental band.

In order to observe effects of non linearities on dispersion curves, two-probe measurements have also been peformed. Results are presented in figure 9, and show a shift of the dispersion relations towards higher wave numbers. In this case, travelling waves generated by the roughness element propagate with a negative phase velocity in the rotating frame. Besides these dispersion relations, the generation of a well-controlled wave packet enables its group velocity to be determined.

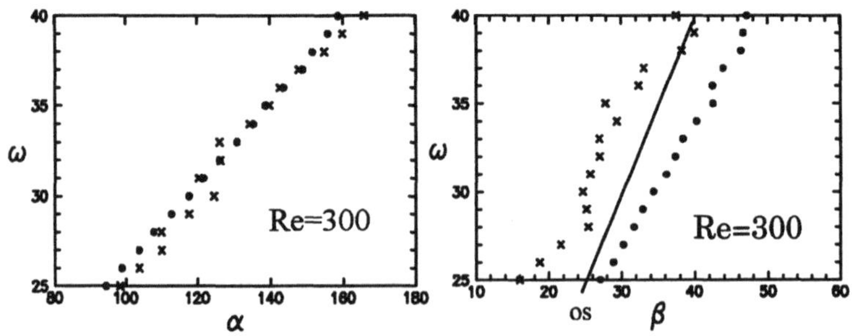

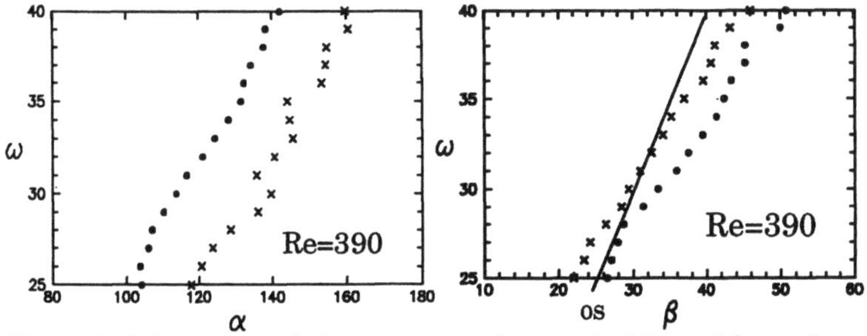

Figure 9 : Dispersion relation curves at three typical Reynolds numbers (xxxx Natural flow; .... Forced flow).

5. Group velocity measurement

Figure 6 shows the wave pattern issued from the roughness element on the disk. γ is the angle between the direction of the group velocity vector and the radius. The experimental determination of γ (-80° in accordance with Wilkinson & Malik experiment) leeds to the calculation of the ratio of the azimuthal ($vg\theta$) to the radial (vgr) components of the group velocity vector: By definition, the group velocity components write:

$$\begin{cases} v_{gr} = \dfrac{\partial \omega_d}{\partial \alpha} = \dfrac{\partial(\omega - \beta)}{\partial \alpha} = \dfrac{\partial \omega}{\partial \alpha} \\ v_{g\theta} = \dfrac{\partial \omega_d}{\partial \beta} = \dfrac{\partial(\omega - \beta)}{\partial \beta} = \dfrac{\partial \omega}{\partial \beta} - 1 \end{cases}$$

where the subscript d denotes the rotating frame. The partial derivatives can be linked to the total derivatives of the frequency and leed to the explicit formulation of the group velocity components versus the total derivatives:

$$v_{gr} = \frac{\dfrac{d\omega}{d\alpha} - \dfrac{d\beta}{d\alpha}}{1 + tg\gamma \dfrac{d\beta}{d\alpha}} \quad \text{and} \quad v_{g\theta} = \frac{\dfrac{d\omega}{d\beta} - 1}{1 + \cot g\gamma \dfrac{d\alpha}{d\beta}}$$

Therefore, we obtain the experimental values of the radial and azimuthal group velocity components versus the Reynolds number (Fig. 11) using the slopes of the dispersion relations given in Fig. 9. Figure 10 shows that the radial component is closed to $0.1V_d$ and the azimuthal component is closed to $-0.6V_d$ (where V_d is the local linear disk velocity). These values are very closed to the mean flow velocity components. Moreover, note that the positive value of the radial group velocity component demonstrates the convective nature of the instability and shows that the flow can be considered as a noise-amplifier.

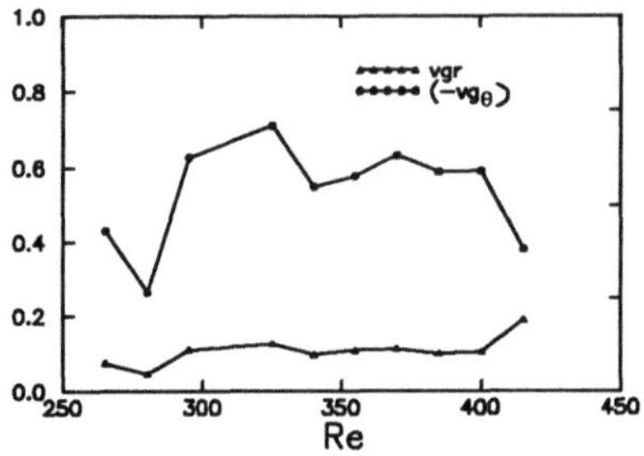

Figure 10 Group velocity components in the rotating frame.

6. Conclusion

An experimental description of the linear and non linear unstable processes has been obtained. The marginal stability curve of the natural transitional flow has been measured; the critical Reynolds number is in good agreement with its computed values, but a shift of the total unstable band towards high wavenumbers is clearly observed. Dispersion relation curves have also been obtained by means of two-probe measurements. They prove that the unstable waves are not stationary at linear threshold and become stationary as the Reynolds number increases. At higher Reynolds number, they clearly show a dispersive behaviour linked to the spreading of wave packets. By means of a roughness element, we excited a well-controlled wave pattern of finite amplitude. Weak non-linear effects have been noticed. In particular,. we observed the appearance of super-harmonic modes, and a shift of the dispersion curves as compared to their linear values. We think that this study will lead to a better understanding of linear and non linear wave propagation in three dimensionnal boundary layer flows. It will certainly increase our knowledge of the transition to turbulence and may contribute to the elaboration. of new transition criteria improving those usually used in aeronautics.

VI - REFERENCES :

[1] Gregory N., Stuart J.T., and Walker W.S. (1955) Proc. Roy. Soc. London Ser. A, 248, 155
[2] Brown W. E. (1959), Northrop Rep. Nal-59-5.
[3] Cebecci T. and Stewartson K. (1980), AIAA Journal, 18, 398.
[4] Kobayashi R., Kohama Y., and Takadamate Ch. (1980), Acta mechanica, 35, 71.
[5] Malik M.R. (1 986) J. Fluid Mech., 164, 275 .
[6] Jarre S., Le Gal P., and Chauve M.P. (1991), Europhys. Lett., 14, 649
[7] Wikinson S. P. and Malik M. R. (1983), AIAA Pap. N° 83-1760.

Randomization Process in Crossflow Instability Dominant Three-Dimensional Boundary-Layer Transitions

Y. Kohama, Y. Kodashima, and H. Watanabe

Institute of Fluid Science
Tohoku University, 2-1-1 Katahira, Sendai 980, Japan

Key words: Randomization, Unsteady instabilities, High frequency disturbances, Nonlinear, Cascade process,

Summary

Present investigation is concerned with the clarification of the randomization process in crossflow instability dominant boundary-layer transition. A swept wing model and a spinning disk, both of which produce crossflow dominant three-dimensional (3-D) boundary layer transition, are investigated experimentally in detail. Using accurate 3-D traversing system, hot wire is scanned in the transition region of a swept wing boundary layer, and also disk wall fixed hot wire system is newly developed in order to measure unsteady disturbances in a spinning disk boundary layer. Results showed that two different unsteady disturbances, frequencies of which are different in about one order, are detected in the transition regions of two different crossflow dominant experimental models. Such unsteady disturbances are also detected in a swept cylinder boundary layer transition region by Poll[1]. In the present paper, low frequency disturbance is called as f_1, while high frequency one is as f_2. Using rotatable parallel hot wire system, nature of those unsteady disturbances are considerably clarified. Results showed that origin of the low frequency disturbance f_1 is sitting very close to the wall in the boundary layer, where as that for the secondary instability is at the edge of the boundary layer. Numerical results obtained by Malik[2] showed the similar tendency with present results. One of the possible explanation for f_1 will be the crossflow primary instability unsteady mode. However, traveling direction is almost 180° different from that of measurement and calculation in different crossflow experimental models[3, 4]. Nature of the instabilities such as phase velocities and traveling directions are successfully detected, and discussion concerning the detailed randomization process are included in the present paper.

1. Introduction

How the high frequency unsteadiness, or randomization is inherited to 3-D laminar boundary layer in general crossflow instability dominant transition process? Streamwise crossflow vortices itself do not possess time dependent nature. This question has long been one of the major target to work with in aircraft drag reduction research. Recent investigations both numerically[5] and experimentally[6,7] have started to answer the questions.

Namely, Primary Instability, unsteady disturbance is detected at the early stage of the transition region, which has rather low frequency unsteadiness. Then at the later stage of the transition region, inflectional inviscid instability is induced locally by crossflow vortices (Primary Instability, steady mode), and high frequency Secondary Instability appears in the later stage of the transition region. However, in order to develop LFC technology for such complicated boundary layer transition, much more quantitative information about the transition structure should be obtained. Therefore, for the purpose of obtaining more information about such transition process, systematic experiment is performed using low turbulence wind tunnel[8] equipped at the Institute of Fluid Science, Tohoku University together with high accuracy 3-D traversing system.

2. Description of the Experiment

Figure 1 (a)～(c) show experimental setups. Figure (a) shows swept wing measurement condition in an open jet type working section of the Low Turbulence Wind Tunnel. Sweep angle Λ and attack angle α of the wing model is arbitrary changeable. In the present experiment, these are fixed to $\Lambda = 45°$, $\alpha = -20°$. Since purely crossflow instability dominant transition is expected in the suction side of the wing, such condition is selected for the swept wing measurement. Figure (b) shows the configuration of the F-4 ($C = 0.9$ m) wing and its pressure distribution on suction surface at $\alpha = 0°$. Figure (c) shows on the other hand, spinning disk ($D = 0.4$ m) model. Spin direction is clockwise, and as it is seen, two smaller rotatable disks are flush mounted on the disk surface so that a hot wire probe mounted on the outer edge of a smallest disk can be traversed at arbitrary radius positions ($r = 62$～182mm) by rotating two mounted disks. Signal from a hot wire on a spinning disk is taken out through rotary connector equipped at the end of disk shaft to the hot wire bridge system equipped on the laboratory floor.

3. Results and discussions

Before going into detailed measurements, flow visualizations are performed to both experimental models. Figure 2 (a), (b) is a wall visualization by Naphthalene method while (b) is a 3-D smoke ($TiCl_4$) visualization. Naphthalene streaks almost aligned to outer streamline are visible in figure (a), while spiral smoke streaks are visible in figure (b). Flow conditions are shown in each captions. Owing to crossflow instability, stationary crossflow vortices are generated in the transition regions (Primary Instability, steady mode). When a hot wire is put in the boundary layer and traversed along the span, clear sinusoidal velocity fluctuations u(y) are visible as shown in Figure 3 (a), (b). Signal in figure (b) is taken by a hot wire probe fixed to the laboratory. Typical time signals for both models are shown in Figure 4 (a), (b). Important sequence is that as proceed to downstream in the transition region, two different disturbances, frequencies of which are different in about one order, are appearing at the same time. From here after, two different frequency disturbances are named as f_1 (low frequency), f_2 (high frequency). Previously, such condition is observed by Poll [1] in the case of a yawed cylinder without expressing detailed explanation. Appearance of f_2 is somewhat intermittent as seen in the signals, while f_1 appears rather stable. Such appearance might strongly dependent on the transition mechanism. Figure 5 (a), (b) show FFT analysis of these hot wire signals. Two peaks associated with f_1, f_2 are observable. In the case of a swept wing ($U_\infty = 26.6$ m/s, $x/C =$

0.57), f_1 = 200, 300 Hz, f_2 = 3.5 KHz, while in the case of a spinning disk (N = 1,200 rpm, $Re = 2.14 \times 10^5$), f_1 = 150, 300 Hz, f_2 = 3.5 KHz. Low frequency disturbance tends to have also two frequencies. Fundamental and its higher harmonics. In the case of Figure 5 (b) for a spinning disk (N= 1,200rpm), they are 150 and 300 Hz. This might be "wave doubling" explained by Kohama [9] which is non-linear vortex deformation in a shear flow. Similar situation is observed in Ref. [10]. In their investigations, f_1 is said to be the higher harmonics of crossflow instability unsteady mode. It was decided from the fact that fundamental frequency calculated was the same with measured ones. However, more detailed information like phase velocity and traveling direction is needed in order to settle the kind of disturbances. Figure 6 is the turbulence intensity variations along the wing chord. Three profiles are plotted. Namely, overall, band pass 200-500Hz (f_1), and band pass 2-5KHz (f_2). Critical points f_{1c}, f_{2c} for f_1 and f_2 are also indicated. From this figure, it can be said that low frequency instability f_1 appears before f_2, and growth rate of f_2 seems to be larger than f_1 at the final stage of the transition.

Fine mesh measurement using a single hot wire with 3-D traversing system is performed in the case of a swept wing as shown in Figure 7 (a)~(c). Mesh points are shown in figure (a). Namely in y-direction, 150 points, and in z-direction, 94 points are taken. Mesh sizes are Δy = 0.2, Δz = 0.04 respectively. Total measured area is 29.7mm x 5mm. Some of the obtained results are shown in Figures (b), (c). From above, figures are, velocity contour, turbulence intensity (over all), turbulence intensity (200-500 Hz band pass), and turbulence intensity (2-5 KHz band pass) respectively. Namely, 200-500 Hz band pass results tells information about f_2 disturbance. From figure (b), one can see that f_1 starts to appear first very close to the wall surface, where high velocity flow is squeezing into low velocity flow near the wall. As proceed to more downstream as shown in figure (c), which is at x/C = 0.6, f_2 starts to appear near the outer edge region of the boundary layer, and the place where low velocity flow is pushed up by high velocity flow motion. Therefore, at those positions, local high shear layer seems to exist. Figure 8 (a), (b) show u(z) profiles obtained at four different span positions (a)~(d) (indicated in figure 6 (b) in the top figure). With reference to one crossflow vortex structure, those positions are chosen. Velocity profiles changes greatly from place to place at very narrow spacings. At position (a), where high momentum flow is coming into the boundary layer (see Figure 6 (c)), $u(z)$ profile is full and stable one. However, at position (b), which is only 3.4mm left from position (a), profile becomes much thicker. atposition (c), which is again 3.4mm further left, u(z) profile becomes most unstable profile with two points of inflection in its profile. Position (d) is another unstable profile with two points of inflection at 3.4mm further left. From these results, it can be seen that such a complicated fine structure is created by streamwise vortices (crossflow vortices) in the boundary layer. From Figure 7 (c), it is also clear that low frequency disturbance f_1 is originating at very low in the boundary layer, and high frequency disturbance f_2 is near at the edge of the boundary layer where point of inflection exists in $u(z)$ profile.

It is quite interesting to know more details about those disturbances. Using parallel probe and rotatable hot wire system, phase velocity Up and direction ϕ of those disturbances are measured on both test models. In Figure 9 (a), (b), measured directions of f_1, f_2 are indicated by arrows in the case of a swept wing. f_1 travels nearly in the direction to crossflow velocity profile, while f_2 is nearly to uniform flow direction. Phase velocity of f_1 is very low, it is about 7% of the uniform flow, while that of f_2 is about 35%.

In the case of a spinning disk, nature of the low frequency disturbance f_1 was not able to measure with present measuring procedure. Figure 10 (a), (b) show schematic sketch of the parallel probe mounted on a spinning disk, and the obtained directions of f_2 disturbance together with schematic sketch of the visualized structure refered from Ref.[11] in the final stage of the transition region where another fine systematic structure is visible on a spinning disk. One of the possible reason will be the small circular dip (diameter $d \fallingdotseq 6$mm, depth $l \fallingdotseq 5$mm) on the disk surface in the parallel hot wire sensor area as shown in Figure 10 (a). Since f_1 is found to be originating very close to the wall surface, this small dip might have disturbed local flow close to the wall.
Variation in the directions obtained by present procedure is reasonable while the structure of the Secondary Instability is complicated and as a result, direction changes from place to place depending on the height in the boundary layer as shown in the sketch.
Therefore, it can be said that f_2 detected in the spinning disk boundary layer is the Secondary Instability which is visible in the smoke visualization in Ref. [11] , and same as detected in the present swept wing model.

4. Concluding remarks

Through systematic experimental investigations using two different test models, nature of unsteady disturbances appearing in the transition region of the crossflow instability field is examined. Obtained major results are listed in the Table 1. Considering from these data, it can be said that in crossflow dominant instability field, low frequency disturbance f_1 first appears after the appearance of stationary instability (crossflow vortices). direction of the disturbance is in the maximum crossflow velocity, and the phase velocity is less than 10 % of the free stream velocity. It maybe unsteady crossflow instability (Primary Instability). However, traveling direction is almost 180° different from calculated results[3] and also from measured results in different crossflow models[3,4]. Therefore, f_1 might be another instability like streamline curvature instability which is the suggestion of Dr. N. Ito. Direction of the disturbance is in the measured direction in the case of the streamline curvature instability. Further investigation should be done for the clarification of the low frequency instability region. After the appearance of f_1, about one order high frequency f_2 disturbance start to appear in the outer edge of the boundary layer. direction is nearly equal to outer streamline, and phase velocity is around 35 % of the uniform flow velocity. Since measured direction of f_2 disturbance coincides with both in hot wire and smoke visualization measurements, it can be said that f_2 disturbance is the high frequency Secondary Instability[7] appeared in the streamwise velocity profiles.
Since these highly unstable velocity profile is generated by streamwise vortex motion, it can be said that high frequency unsteady nature is produced through the cascade lake disturbance generating process. Namely, steady streamwise crossflow vortex first appears, and this disturbance deforms the base flow field into locally unstable conditions for the low and high frequency instabilities. Through such cascade process energy of large scale and steady instability is therefore transferred to smaller scales and high frequency unsteady instabilities. Once such high frequency instability appears, almost abrupt break-up of turbulence production process occurs and attains to fully turbulent state. Such systematic turbulence production mechanism, or randomization process seems to be the general transition process in the case of crossflow dominant 3-D boundary layers.

Reference

[1] Poll, D. I. A.: Some observations of the transition process on the windward face of a long yawed cylinder, J. Fluid Mech. 150 (1985), 329-356.

[2] Malik, M. R., Li, F.: Transition studies for swept wing flows using PSE, AIAA Pap. No.93-0077, (1993).

[3] Deyhle, H., Holer, G., Bippes, H.: Experimental investigation of instability wave propagation in a three-dimensional boundary-layer flow, AIAA J., 31 (1993), 637-645.

[4] Kohama, Y., Motegi, D.: Traveling disturbance appearing in boundary layer transition in a yawed cylinder, Experimental Therapy and Fluid Science, 8 (1994), 273-278.

[5] Balakumar, P., Malik, M. R.: On the receptivity and nonparallel stability of traveling disturbances in rotating-disk flow, Theoret. Comput. Fluid Dynamics, Springer Verlag, (1991), 125-140.

[6] Radeztsky, R. H. Jr., Reibert, M. S., Saric, W. S.: Effect of micron-sized roughness on transition in swept-wing flows, AIAA Pap. No.93-0076, (1993).

[7] Kohama, Y., Saric, W. S., Hoos, J. A.: A high-frequency secondary instability of crossflow vortices that leads to transition, Proc. Roy. Aeron. Soc. Boundary Layer Transition and Control, Cambridge, UK, (1991).

[8] Ito, H., Kobayashi, R., Kohama, Y.: The low-turbulence wind tunnel at Tohoku University, Aeron. J. of the Roy. Aeron. Soc., (1992), 141-151.

[9] Kohama, Y.: Some expectation on the mechanism of crossflow instability in a swept wing flow, Acta Mech., 66(1987), 21-38.

[10] Saric, W. S., Yeates, L. G.: Experiments on the stability of crossflow vortices in swept-wing flows, AIAA Pap. No.85-0493, (1985).

[11] Kohama, Y., Suda, K.: Crossflow instability in a spinning disk boundary layer, AIAA J., 31, (1993), 212-214.

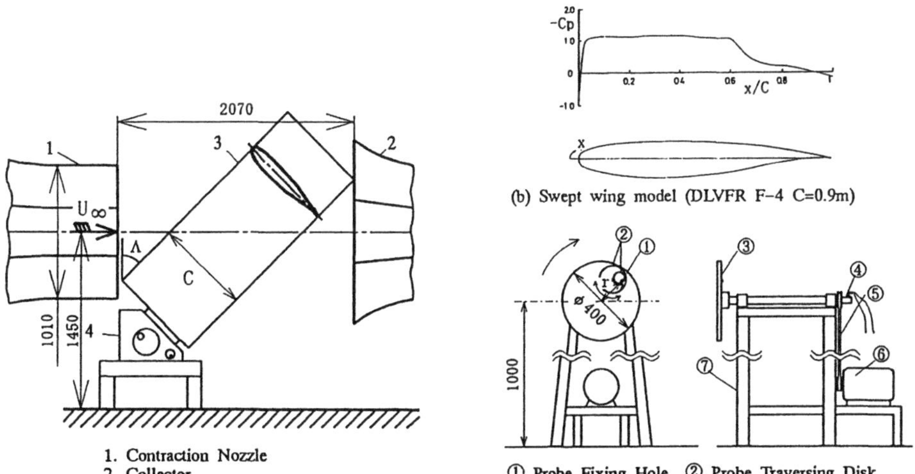

1. Contraction Nozzle
2. Collector
3. F–4 model (C=0.9m)
4. Tilting table (α, Λ variable)

(a) Swept wing model in a low-turbulence wind tunnel

(b) Swept wing model (DLVFR F–4 C=0.9m)

① Probe Fixing Hole ② Probe Traversing Disk ③ Spinning Disk ④ Rotating Connector ⑤ V–Belt ⑥ DC Motor ⑦ Supporting Table

(c) Rotating disk model with wall fixed hot wire system

Fig.1 Experimental setups

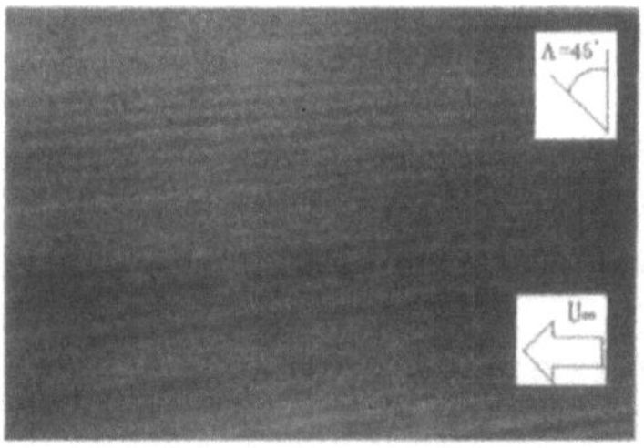

(a) Naphthalene visualization on a swept wing ($\Lambda=45°$, $U_\infty=24$m/s, $\alpha=-20°$)

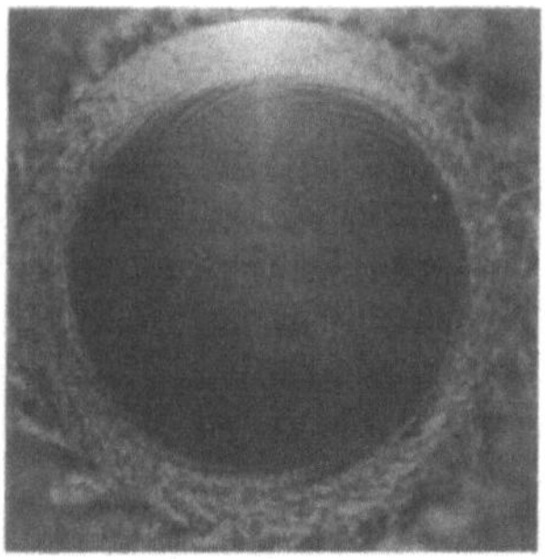

(b) $TiCl_4$ smoke a spinning disk visualization on ($D=0.4$m, $N=1{,}500$rpm, anti-clockwise spin)

Fig.2 Flow visualizations on both models

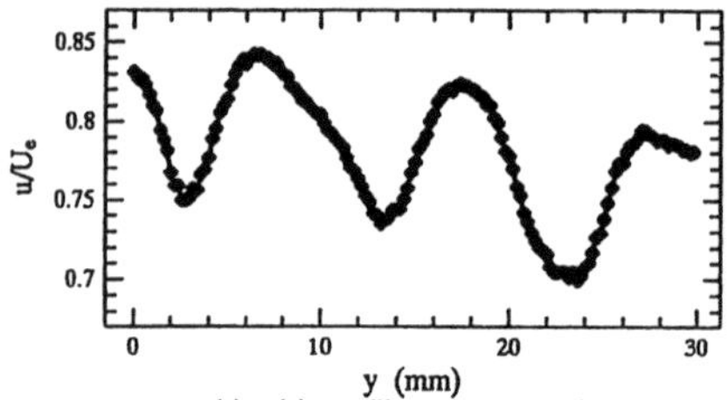

(a) u(y) profile on swept wing ($\Lambda=40°$, $\alpha=-20°$, $U_\infty=30$m/s, $x/C=0.3$, $z=0.5$mm)

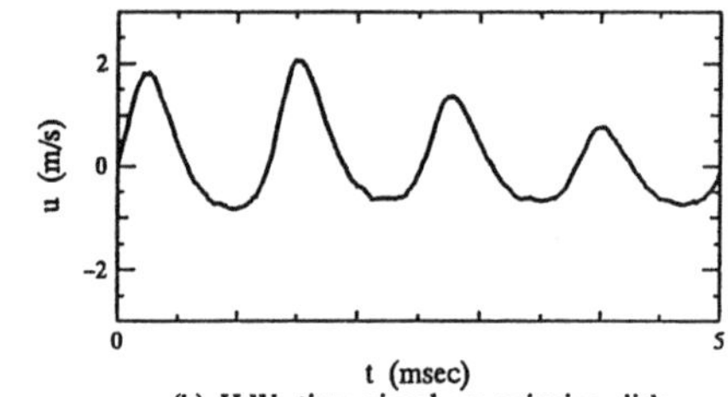

(b) H.W. time signal on spinning disk ($D=0.4$m, $N=1{,}500$rpm, measured from outside probe)

Fig.3 Velocity distribution in the transitional boundary layer

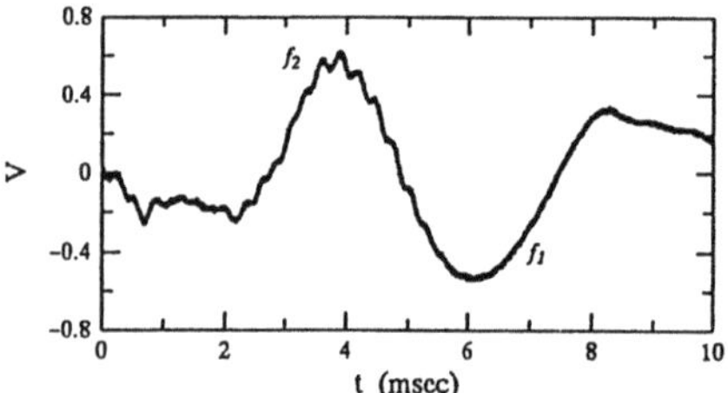

(a) Hot wire time signal on swept wing ($\Lambda=45°$, $\alpha=-20°$, $U_\infty=26.6$m/s, $x/C=0.57$, $z=0.7$mm)

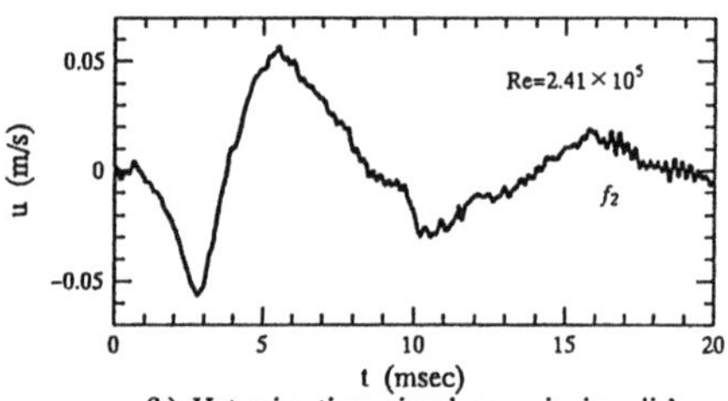

(b) Hot wire time signal on spinning disk ($D=0.4$m, $N=1350$rpm, measured on a disk fixed probe)

Fig.4 Hot wire time signals where unsteady disturbances are appearing

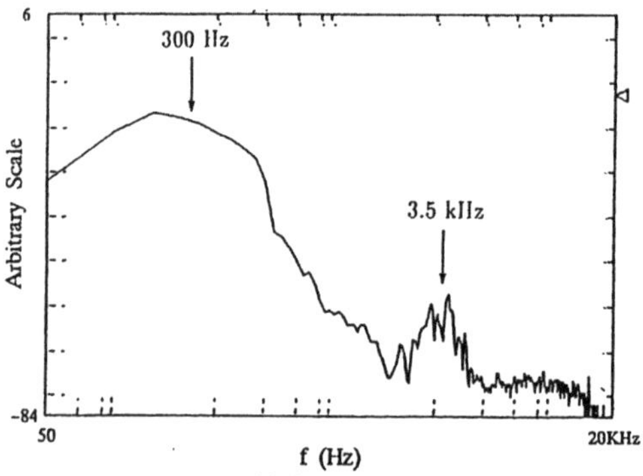

(a) Swept wing ($\Lambda=45°$, $\alpha=-20°$, $U_\infty=26.6$m/s, $x/C=0.57$, $z=0.7$mm)

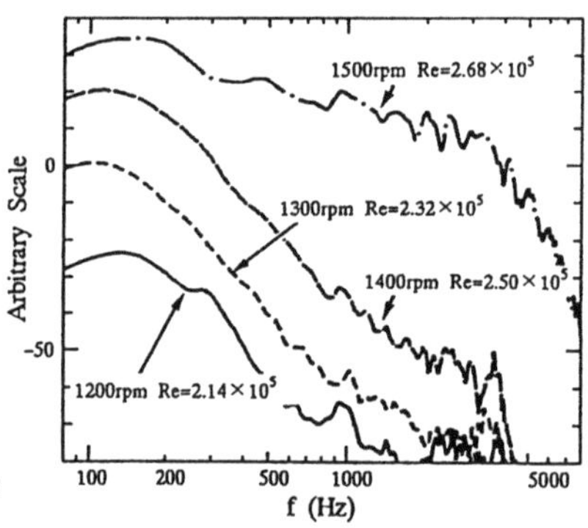

(b) Spinning disk ($D=0.4$m, measured on a disk fixed probe)

Fig.5 FFT analysis of the hot wire signals

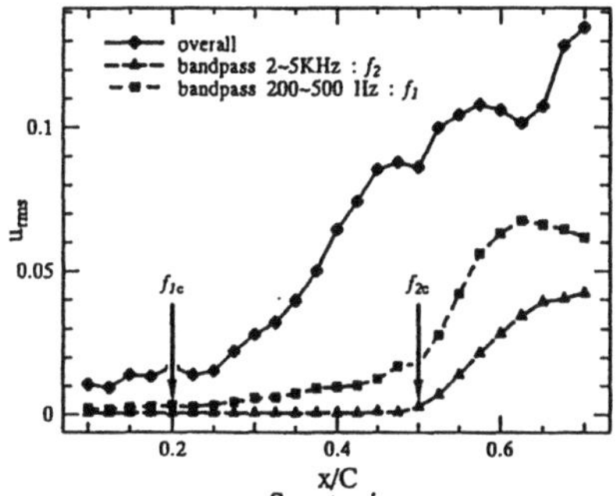

Swept wing
(Λ=45°, α=-20°, U_∞=26.6m/s, z=0.7mm)

Fig.6 Chordwise turbulence intensity variations

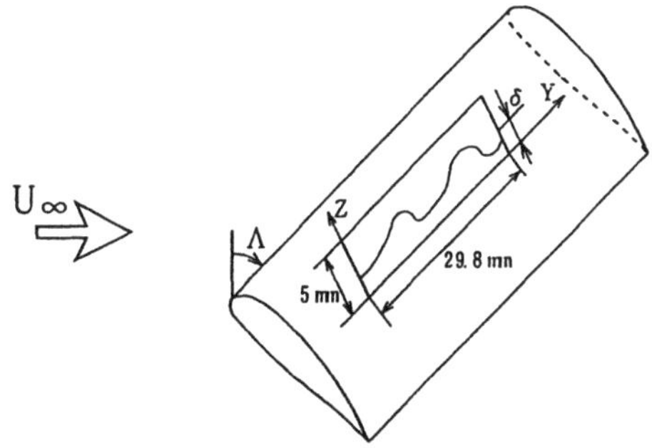

(a) Schematic sketch of the measured mesh points
(Δy=0.2mm, Δz=0.04mm, 150×94points)

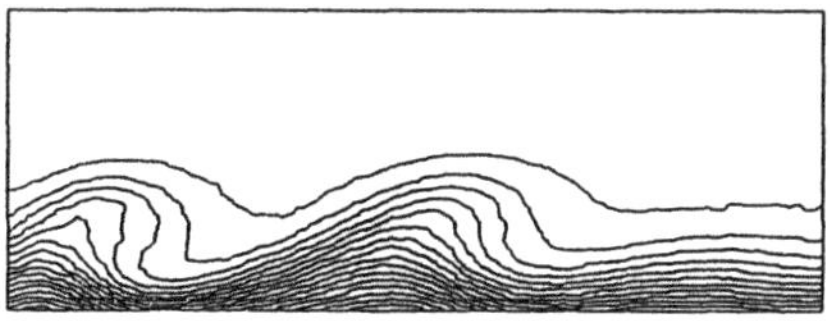

velocity contour

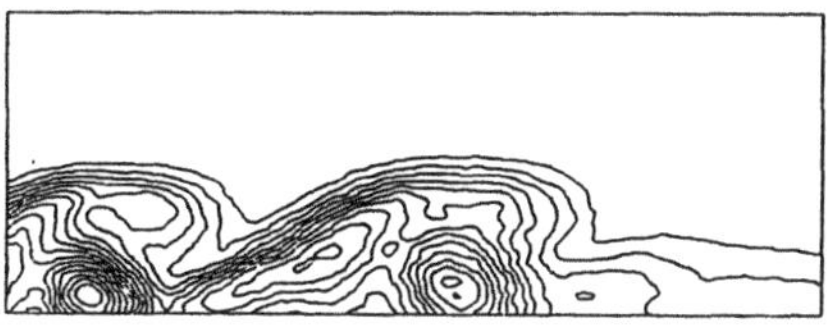

turbulence intensity contour (over all)

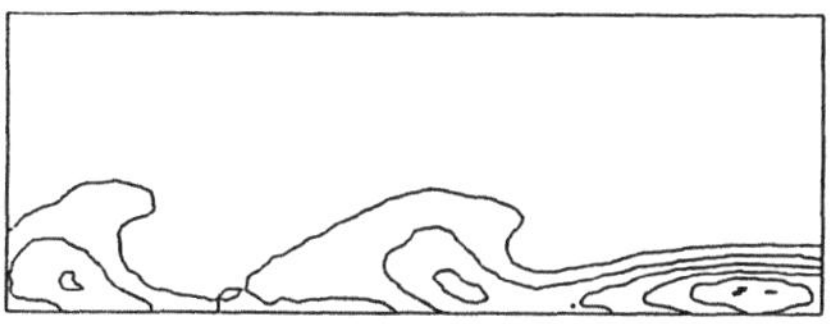

turbulence intensity contour (200~500 Hz : f_1)

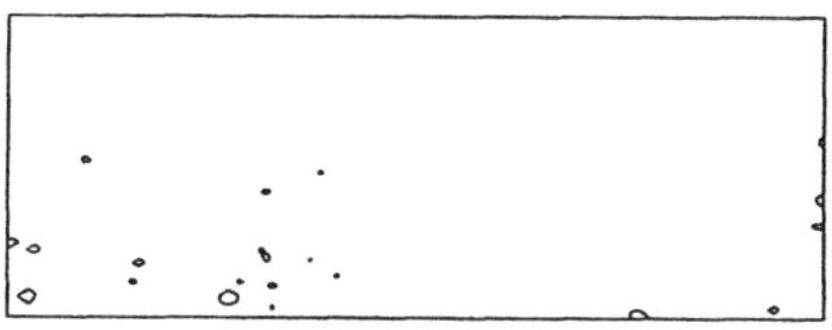

turbulence intensity contour (2~5 KHz : f_2)

(b) Contour of velocity and turbulence intensity
(Λ=45°, α=-20°, U_∞=26.6m/s, x/C=0.5)

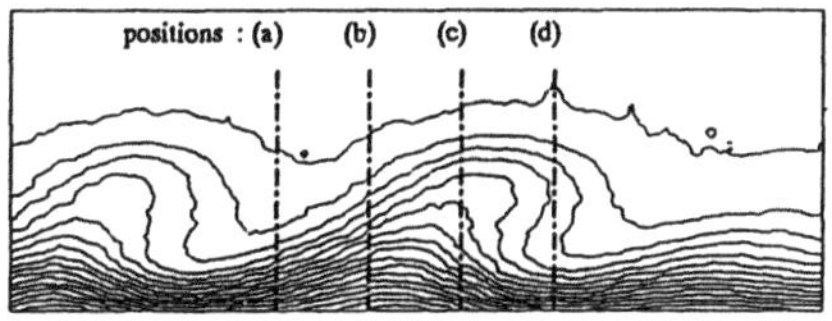

velocity contour

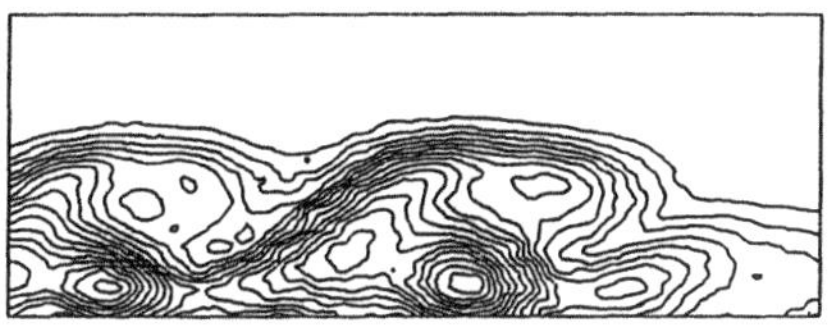

turbulence intensity contour (over all)

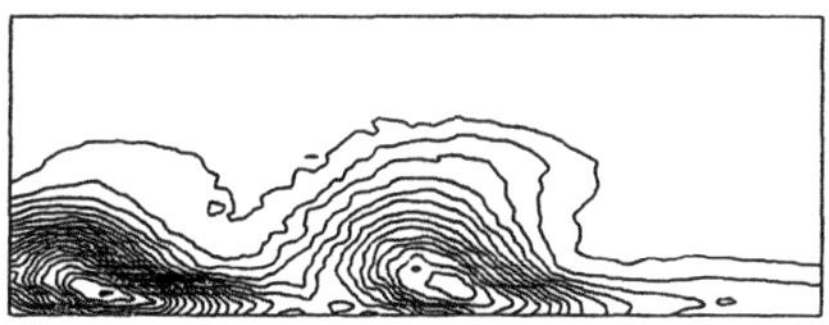

turbulence intensity contour (200~500 Hz : f_1)

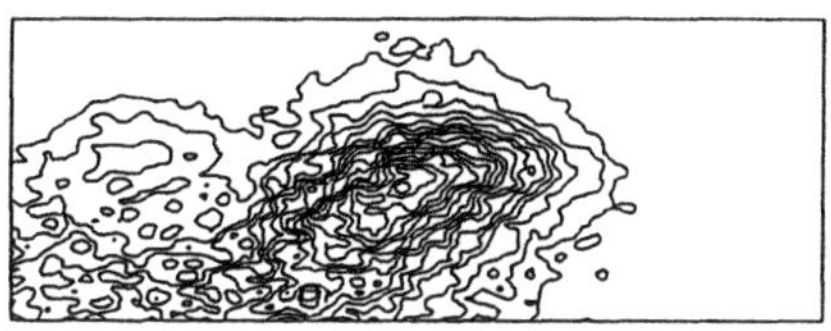

turbulence intensity contour (2~5 KHz : f_2)

(c) Contour of velocity and turbulence intensity
(Λ=45°, α=-20°, U_∞=26.6m/s, x/C=0.6)

Fig.7 Cross sectional flow structure in a swept wing boundary layer

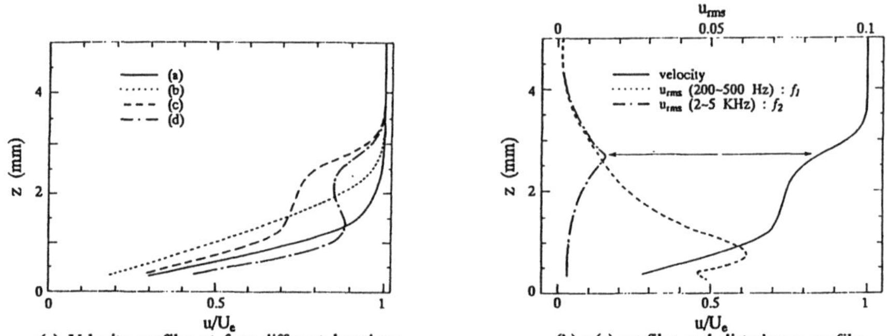

(a) Velocity profiles at four different locations (b) u(z) profiles and disturbance profiles

Fig.8 Velocity profiles on swept wing boundary layer

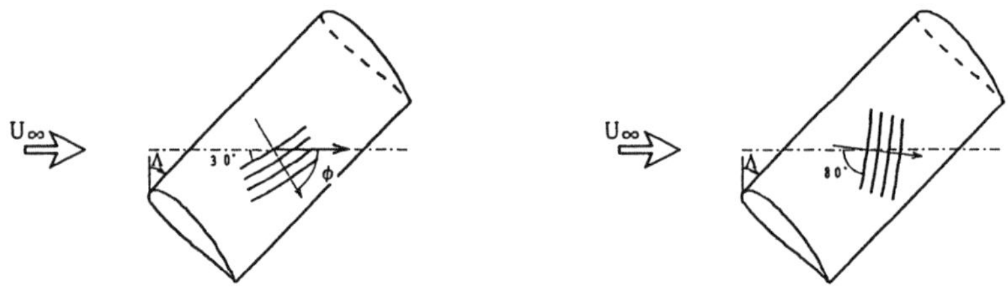

(a) Obtained direction for f_1 (b) Obtained direction for f_2

(Λ=45°, α=−20°, U_∞=26.6m/s, x/C=0.6)

Fig.9 Traveling direction and phase velocity measurements on a swept wing boundary layer

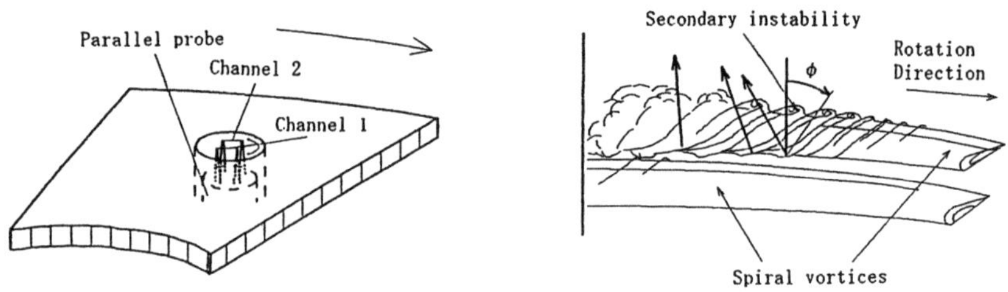

(a) Schematic sketch of the measurement technique (b) Obtained direction for f_2

Fig.10 Phase velocity measurements on a spinning disk boundary layer

Experimental models	Nature of unsteady disturbances	Swept wing (U_∞=26.6 m/s, Λ=45°, α=20°, x/C=0.6)	Spinning disk (D=400 mm, N=1,350 rpm)
f_1 disturbance	f	200, 300 Hz	200 Hz
	ϕ	60°	not measurable
	U_p/U_∞	7.2 × 10^2	not measurable
f_2 disturbance	f	3.5 KHz	4 KHz
	ϕ	10°	−5° ∼ −30°
	U_p/U_∞	0.34	0.2 ∼ 0.4

Table 1. Unsteady disturbance nature obtained in crossfrow instability flow field

Session L:
2-D Boundary Layers

Experiments on Nonlinear Interaction between Oblique Tollmien-Schlichting Waves

P.A. Elofsson and P.H. Alfredsson

Department of Mechanics/Fluid Physics, Royal Institute of Technology, S-10044 Stockholm, SWEDEN.

Summary
Results from an experimental investigation on the interaction between a pair of oblique waves in plane Poiseuille flow are presented. Non-linear interaction resulted in a streaky flow structure. For high initial wave amplitude higher harmonics occurred, which finally developed into a turbulent-like state. Comparisons were made with results from earlier spatial simulations and close agreement was found except for differences in symmetry properties that were observed in cross-stream planes.
Key words: Stability, Tollmien-Schlichting waves, Non-linear interaction

1 Introduction

Subcritical transition is usually denoted as by-pass transition, although it is not a single scenario but can take various forms. Experimental studies of by pass transition often utilises a controlled localised disturbance introduced at the wall, for instance by a short duration, small diameter jet (Gaster & Grant [1], Klingmann [2]), deflection of the wall through a membrane (Breuer & Haritonidis [3]) or an electrical discharge (Wygnanski, Haritonidis & Kaplan [4]). With a strong disturbance a turbulent spot is created at the disturbance itself, however at smaller amplitudes streaky streamwise structures and/or wave packets are produced which may or may not grow depending on the R and on the initial disturbance strength. Unfortunately, the disturbance velocities and the size of the disturbance cannot be exactly determined, and dissipative effects will also be strong because of the localised nature of the disturbance. In numerical simulations of by pass transition in plane channel flow two pairs of streamwise counter rotating vortices have been used (Henningson, Lundbladh & Johansson [5]). The vortices produce a localised disturbance which partly resembles the injection of a jet pulse at the wall.

Recently a transition scenario that resembles what is seen for a localised disturbance was investigated through a temporal numerical simulation by Schmid & Henningson [6]. The initial disturbance consisted of two finite amplitude, oblique waves, which interacted non-linearly to form longitudinal streaks. These streaks were susceptible to secondary instabilities. Schmid, Lundbladh & Henningson [7] later extended this study through a spatial simulation, where the initial waves were given as $(\omega, \pm\beta)$, where ω is the TS-wave frequency. The first generation of non-linear interaction between the waves may result in the following components (0,0), (2ω,0), ($2\omega, \pm 2\beta$) and ($0, \pm 2\beta$). The first corresponds to a mean flow distortion, the second to a travelling two-dimensional wave with the double streamwise wave number, the third to oblique travelling waves with the same inclination as the primary but with half the wave length, whereas finally the fourth corresponds to a stationary streamwise disturbance. It was found that the highest energy growth was obtained in the ($0, \pm 2\beta$)-mode which gives a streaky structure in the streamwise velocity. In fact this possibility was already pointed out by Corke & Mangano [8] for interacting oblique waves in a boundary layer. Due to extreme demands on computer power the simulations have not been taken to the breakdown stage in more than a few cases, one which was reported for a boundary layer flow by Berlin, Lundbladh & Henningson [9].

The present study is the first experimental investigation of the non-linear interaction of oblique waves in plane Poiseuille flow. The waves were generated by vibrating ribbons, one at each wall, which made it possible to generate two equally strong TS-waves and follow their subsequent interaction. Two amplitudes of the initial disturbance were used, a low amplitude where only the streaky structure survives and a high amplitude where second generation interactions were observed. Most of the results of the numerical simulation were confirmed, although the experiments did not show the perfect symmetry with respect to the channel centreplane which was obtained in the simulations.

2 Experimental set-up

The experiments were carried out in an air-flow channel consisting of two parallel glass plates separated by 8.2 mm (see Fig. 1). The spanwise aspect-ratio was 101. The air, driven by a variable speed fan, passed through three damping screens and a contraction with a ratio of 40:1 before entering the channel. Wave disturbances were generated with the aid of vibrating ribbons positioned one meter downstream of the channel entrance. The ribbons were made of a phosphor bronze alloy and they had a width, thickness and working length of 3

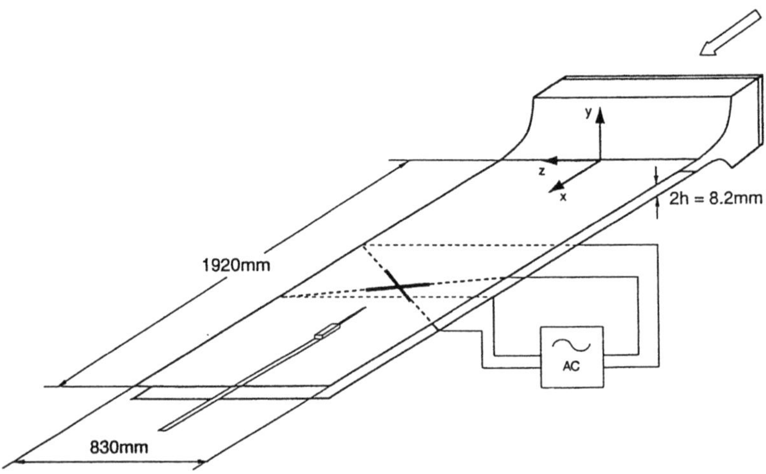

Figure 1: Experimental set-up.

mm, 0.05 mm and 320 mm, respectively. The ribbons were positioned in a magnetic field from five horseshoe type permanent magnets evenly distributed along each ribbon and mounted outside the channel walls. A sinusoidal signal provided by a function generator and amplified by a stereo amplifier excited the ribbons. The streamwise velocity component was measured with a constant-temperature anemometer using a 2.5 μm diameter platinum hot-wire with a length of 0.6 mm. The hot-wire probe was traversed automatically using stepper motors in the spanwise and wall-normal directions and manually in the streamwise direction. The hot-wire was calibrated against the basic Poiseuille profile and a relative error less than $\pm 0.7\%$ was obtained for all the calibration points. The hot-wire was recalibrated if the temperature during measurements deviated more than 0.5 degrees from the calibration temperature. A Macintosh computer and a MacADIOS adio were used for sampling the hot-wire and generator signals and controlling the stepper motors.

3 Basic flow

In the following all lengths are scaled with the channel half-height, h and the velocities with the centreline velocity, U_{CL}. Measurements were carried out at two different subcritical Reynolds numbers, R=1600 and R=2000 where R is defined as $R=U_{CL}h/\nu$ (ν is the kinematic viscosity). In order to document the basic flow initial measurements were carried out with and without ribbons mounted inside the channel. The hot-wire probe was traversed both in the wall-normal and spanwise directions at different streamwise positions. The background disturbance level was found to be less than 0.16% of the centreline velocity for the worst case with two ribbons mounted inside the channel at R=2000. At a streamwise position five half-channel heights downstream of the ribbons (x-x_r=5), the mean velocity profile was slightly influenced by the ribbons but further downstream (at x-x_r>35) the measured velocity profiles fall upon the parabolic Poiseuille profile. The measured profile at x-x_r=5 deviated from the parabolic profile at y-positions near the height where the ribbons were mounted but closer to the centreline no effect could be observed. As mentioned earlier the hot-wire probe was also traversed in the spanwise direction at different y-positions in order to detect any spanwise periodicity. The spanwise variation in U_{CL} was $\pm 0.3\%$ at R=2000 and no periodicity could be found.

4 One ribbon measurements

Before starting the investigation of the two oblique waves, measurements were carried out with only one ribbon placed inside the channel. TS-waves with low initial amplitudes were generated at the lower wall and the experimental results obtained could then be compared with results from linear theory. The ribbons spanned the whole channel width and the tension of the ribbons could be adjusted with mechanisms placed outside the sidewalls of the channel. The tension was adjusted to give a resonant frequency of approximately 220 Hz and the part of the ribbons that were not moving were attached to the walls using a thin tape. In order to ascertain that the ribbons were working in a proper way the spanwise variation of the TS-wave maximum was measured and wall-normal amplitude profiles and spectra were also obtained. If the results indicated any errors the ribbons were remounted. The data were obtained by simultaneous sampling of the anemometer signal and the input signal to the ribbons, followed by phase-alignment of the hot-wire signal to the ribbons input signal and Gauss-filtering in a 4 Hz band centred at the generator frequency. The phase-alignment was achieved with a method similar to the one described by Boiko *et al.* [10]. Fig. 2a & b show the amplitude and phase distributions of a two-dimensional TS-wave at R=1600 measured at x-x_r=40. Also shown in the figure are the theoretical distributions for the least stable mode which have been calculated with two-dimensional linear spatial theory. The experimental amplitude maximum in the upper part of the channel (y>0) is 10% lower than the maximum in the lower part of the channel and the phase shift near the walls start closer to the centre of the channel in the experiment as compared with the theoretical results. Overall, the experimentally determined distributions can be viewed to be in good agreement with the theoretical ones. Fig. 2c & d display the streamwise amplitude and phase evolution measured at R=1600 and ω=0.42 (f=95 Hz) and lines corresponding to linear theory are also shown for comparison. The damping rate from the measurements is slightly larger

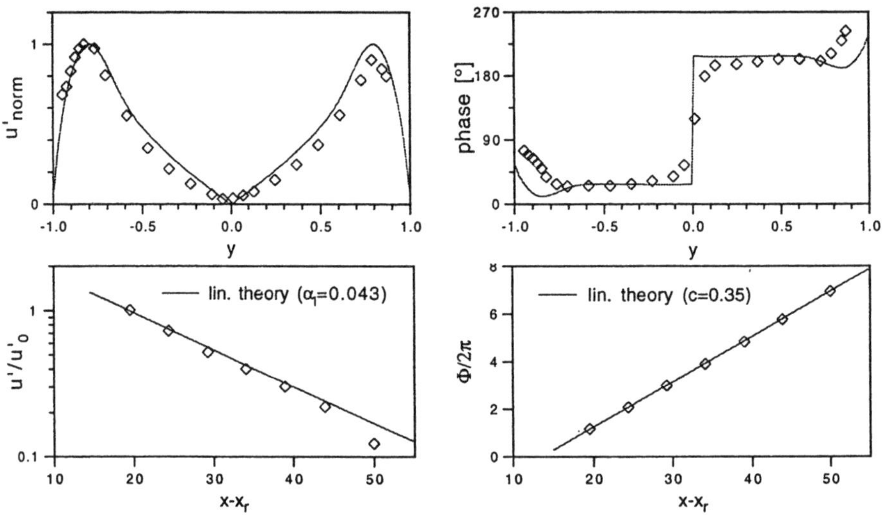

Figure 2: Characteristics of 2D TS-wave at R=1600 and ω=0.42. Amplitude (top, left), phase (top, right), damping rate (bottom, left) and phase evolution.

than the one given by linear theory (α_i=0.043) but no deviation between the measured phase evolution and the theoretical phase velocity (c=0.35) was seen.

Fig. 3 shows the amplitude and phase distributions for an oblique wave with a low initial amplitude at R=2000 and ω=0.33 (f=94 Hz). The profiles were measured at z=0 and at a streamwise position 35h downstream from the centre of the ribbon, which in this case is mounted at the lower wall with an angle ϕ=45°. ϕ is defined as the angle between the wave vector and the streamwise direction. Also shown in the figure are the distributions obtained from three-dimensional linear theory and as in the two-dimensional case the experimental data were found to be in good agreement with the data from the calculations. However, the difference between the experimentally determined

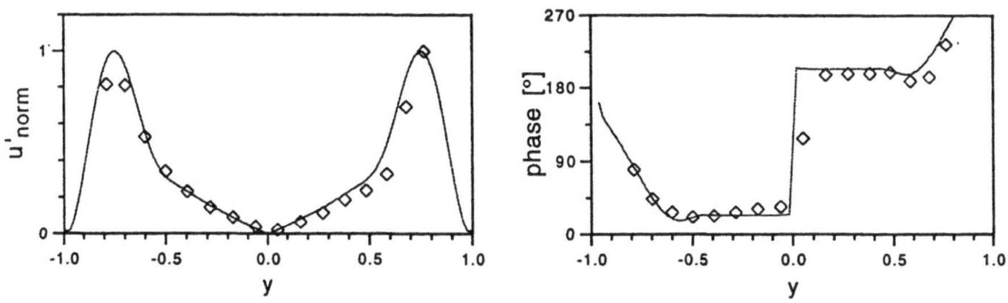

Figure 3: Amplitude and phase distributions for 3D TS-wave at R=2000 and ω=0.33.

damping rate and the theoretical one was larger for the oblique TS-wave than for the two-dimensional TS-wave. For the oblique wave the experimentally obtained damping rate was α_i= 0.051, compared to α_i=0.044 for three-dimensional linear theory. The agreement between the experiment and linear theory was better for the phase velocity where the theoretical value was c=0.37 and the result from the experiment was c=0.38.

5 Two ribbon measurements

5.1 Structure of flow field

Measurements on the interaction between a pair of oblique waves were carried out at R=2000 with two different angles of the ribbons (ϕ=34° and ϕ=45°), and for two different initial wave amplitudes. The initial amplitudes were measured at a fixed position $12h$ downstream of the centre of the ribbons. The maximal amplitude for each ribbon differed slightly between the lower and the upper part of the channel but the current to the ribbons were individually adjusted to give the same total maximal amplitude in the lower and upper channel halves. Fig. 4 shows contours of $(U - U_{ave})$ measured in a xz-plane at y=0.75 with ϕ=34°, where U_{ave} is the spanwise average at each streamwise position. Data were taken at 10 streamwise and 64 spanwise positions and the velocity at each position was the average from 25 sets consisting of 1024 points sampled with a frequency of 2880 Hz.

For both the low and high amplitude cases the interaction between the two oblique waves gave rise to a spanwise mean velocity structure of large amplitude. In the low amplitude case this structure was preserved far downstream of the ribbons despite a decay of the fundamental, and for the high amplitude both the mean velocity structure and the fundamental was growing downstream until a turbulent-like state was reached. Also, in spectra obtained for the high amplitude a strong growth in higher harmonics was observed starting at $x - x_r \approx 40$, and further downstream the amplitude in the spectra was evenly distributed over all frequencies except for a peak at the fundamental frequency. The mean velocity structure, with zero frequency and half the original spanwise wavelength, was a result of non-linear interactions between the two oblique waves that were initially generated with the vibrating ribbons. These interactions can be described using (ω, β)-modes where the initial waves with fundamental frequency and spanwise wave number $(\omega, \pm\beta)$ interact non-linearly to produce (ω_m, β_n)-modes which are multiples of the fundamental frequency and spanwise

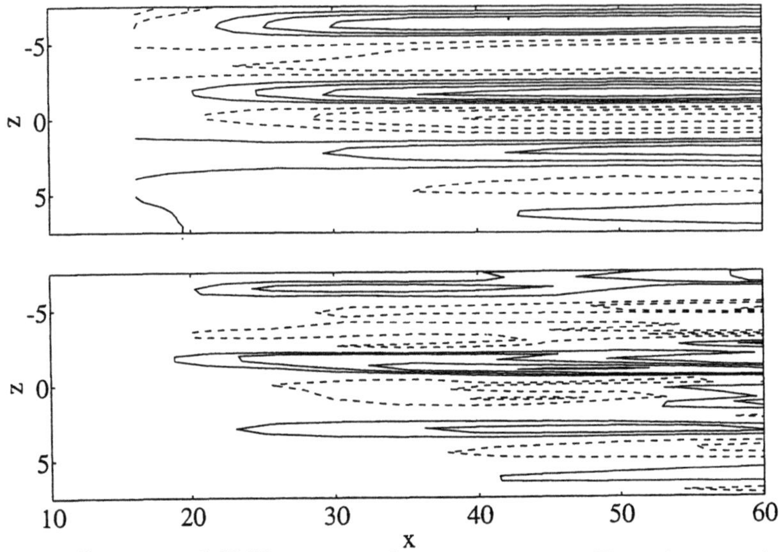

Figure 4: Contours of U-U_{ave} in xz-planes at y=0.75. Top: low amplitude (spacing=1%). Bottom: high amplitude (spacing=6%).

wave number, respectively. The $(0, \pm 2\beta)$-mode dominated the flow field in the low amplitude case but other modes generated through the non-linear interactions were also observed. Measurements were carried out at two different wave angles (ϕ=34° and ϕ=45°) and the $(0, \pm 2\beta)$-mode dominated in both cases. An expression for the spanwise distance (d) between consecutive low or high velocity regions can be given as $d = \lambda_x / 2tan\phi$, where λ_x is the streamwise wavelength. Fig. 5 shows contour plots of $(U - U_{ave})$ in yz-planes at three different streamwise positions. The data was measured at a low amplitude and with ϕ=34°. The streaks were not developed at the first streamwise position, x-x_r=15, but further downstream high and low speed streamwise streaks were formed with a spanwise wavelength half of that of the fundamental. The streaks in the lower and upper half of the channel were shifted in the spanwise direction and formed a staggered pattern in the cross-stream plane, but the individual streaks were almost perfectly straight in the downstream direction. In the spatial simulations of ref. [8] symmetry was imposed and in this case the streaks appeared fully anti-symmetric with respect to the centreplane (y=0).

5.2 Evolution of energy

Fig. 6 displays the streamwise evolution of energy in different (ω, β)-modes with ϕ=45° at R=2000. 25 sets of data each consisting of 1024 points were acquired at a sampling frequency of 2880 Hz from 9 streamwise and 64 spanwise positions. The data was obtained by traversing the hot-wire probe in the spanwise direction at a fixed wall-normal position (y=0.75). Fourier transforms of the time series were computed on-line and the obtained Fourier components were stored on file for later processing. After completion of the measurements a spanwise Fourier transform was made and a Gaussian filtering was applied in 4 Hz bands around f=0, the fundamental frequency and the first 10 harmonics.

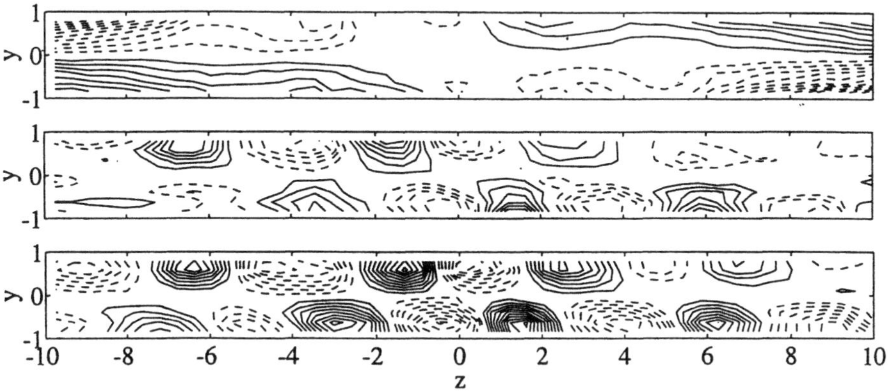

Figure 5: Contours of U-U_{ave} in yz-planes, spacing=0.5%. Top: x-x_r=15. Centre: x-x_r=30. Bottom: x-x_r=60.

As a measure of the energy the following expression was used, $E = |\hat{u}(\omega_m, \beta_n)|^2$ where $\hat{u}(\omega_m, \beta_n)$ is the Fourier component for ω=ω_m and β=β_n. When the initial wave amplitude was low all the modes decayed for x-x_r>28 except for the $(0, \pm 2\beta)$-mode which continued to grow before it finally levelled off at the last measurement positions. For the high amplitude case the growth scenario was different. The behaviour of the $(0, \pm 2\beta)$-mode was similar to the low amplitude case although the energy level was higher, but for this case also other modes grew. Initially the $(0, 2\beta)$ and $(\omega, 3\beta)$ modes experienced the largest growth but further downstream the modes that were multiples of the fundamental frequency grew the most. In Fig. 6 the strong growth of the $(2\omega, 2\beta)$, $(3\omega, \beta)$ and $(3\omega, 3\beta)$ modes can be observed but also higher modes had a large growth. The development in the low amplitude case was similar to what is observed in the simulations by Schmid *et al.* [8]. They show the spatial evolution of (ω, β)-modes with a low initial amplitude and after a strong initial growth all the modes except the $(0, 2\beta)$ and the $(\omega, 3\beta)$ mode decay. Beyond $x \approx 20$ the $(\omega, 3\beta)$-mode starts to decay while the $(0, 2\beta)$-mode is still growing.

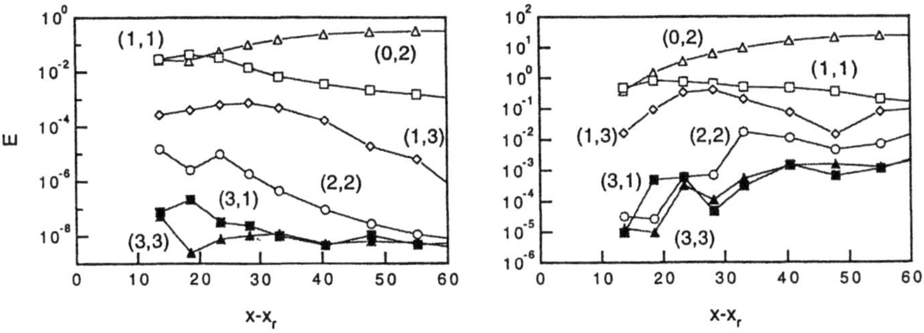

Figure 6: Energy in (ω, β)-modes. Left: low amplitude. Right: high amplitude.

6 Conclusions

Results from experiments on the interaction between a pair of oblique waves generated with two vibrating ribbons mounted on opposite sides of a Poiseuille flow channel wall were presented. The non-linear interaction resulted in a spanwise streaky structure. Depending on the initial wave amplitude the structure either persisted far downstream of the ribbons or higher harmonics were generated which eventually broke down into a turbulent-like state. The low and the high velocity regions appeared in a staggered pattern in a cross-stream plane indicating a mixture between a symmetric and an anti-symmetric mode. The experimental results have also been compared with results from simulations [8] and it was found that the spatial evolution of energy in (ω, β)-modes seen in the experiments are in close agreement with the simulation results, although the symmetry properties of the spanwise mean velocity structure was different. Also, results from initial measurements that were carried out with only one vibrating ribbon mounted inside the channel were found to be in good agreement with linear spatial theory.

This work was supported by the Swedish Research Council for the Engineering Sciences (TFR).

References

[1] Gaster, M. & Grant, I. 1975 An experimental investigation of the formation and development of a wave packet in a laminar boundary layer. *Proc. R. Soc. Lond.* A **347**, 253-269.

[2] Klingmann, B.G.B. 1992 On transition due to three-dimensional disturbances in plane Poiseuille flow. *J. Fluid Mech.* **240**, 167-195.

[3] Breuer, K.S. & Haritonidis, J.H. 1990 The evolution of a localized disturbance in a laminar boundary layer. Part I. Weak disturbances. *J. Fluid Mech.* **220**, 569-594.

[4] Wygnanski, I., Haritonidis, J.H. & Kaplan, R.E. 1979 On a Tollmien-Schlichting wave packet produced by a turbulent spot. *J. Fluid Mech.* **92**, 505-528.

[5] Henningson, D.S., Lundbladh, A. & Johansson, A.V. 1993 A mechanism for bypass transition from localized disturbances in wall bounded shear flows. *J. Fluid Mech.* **250**, 169-207.

[6] Schmid, P.J. & Henningson, D.S. 1992 A new mechanism for rapid transition involving a pair of oblique waves. *Phys. Fluids* A**4**, 1986-1989.

[7] Schmid, P.J., Lundbladh, A. & Henningson, D.S. 1993 Spatial evolution of disturbances in plane Poiseuille flow. Proceedings from the 1993 ICASE/NASA Langley Workshop on the Transition to Turbulence.

[8] Corke, T.C. & Mangano, R.A. 1989 Resonant growth of three-dimensional modes in transitioning Blasius boundary layers. *J. Fluid Mech.* **209**, 93-150.

[9] Berlin, S., Lundbladh, A. & Henningson, D.S. 1994 Spatial simulations of oblique transition in a boundary layer. *Phys. Fluids* **6**, 1949-1951.

[10] Boiko, A.V., Westin, K.J.A., Klingmann, B.G.B., Kozlov, V.V. & Alfredsson, P.H. Experiments in a boundary layer subjected to free stream turbulence. Part II: The role of TS-waves in the transition process. *J. Fluid Mech.* (in press).

Three-Dimensional Instability of the Blasius Boundary Layer

Y.S. Kachanov*, A. Michalke**
*) Institute of Theoretical & Applied Mechanics, Russian Academy of Sciences, Novosibirsk, Russia
**) Hermann-Föttinger-Institut für Thermo- und Fluiddynamik, Technische Universitat Berlin, F.R.G.

Summary

The 3D instability of the flat-plate boundary layer is investigated. Experimental results obtained in [2] for the Blasius flow excited by a point source were compared with theoretical solutions of the Orr-Sommerfeld equation based on a parallel-flow approximation. Very good agreement was found for the dispersion characteristics of the 3D instability waves as functions of the propagation angle, frequency parameter and Reynolds number. The growth rates showed mainly a qualitative agreement, but a rather good one for not too high propagation angles.

Key words : Blasius boundary layer, 3D linear stability, Quantitative comparison of theory and experiment.

1. Introduction

The linear hydrodynamic stability theory is undoubtedly the most developed branch in the description of the process of turbulence onset. At the same time, the most of experimental studies of 2D flows were devoted to their 2D instability, even in the canonical case of the flat-plate boundary layer. First successful experimental investigation of the 3D instability waves in this flow was conducted in [1]. Later a complete set of stability characteristics of the flat-plate boundary layer was obtained experimentally in [2]. However a quantitative comparison of these characteristics with those predicted by the linear stability theory was not performed.

In the present paper, the experimental data obtained in [2] are used for additional processing and for a quantitative comparison with the results of calculations performed within the framework of the parallel linear stability theory for the conditions of those experiments. Such a comparison seems to be an inherent step in validation of the computer codes for investigation of the 3D instability.

2. Methods of studies

Experimental procedure. A detailed description of the experimental conditions and the procedure of data processing is presented in [2] (see also [3]). The experiments were conducted at a low-turbulence wind-tunnel of the Institute of Theoretical and Applied Mechanics (Novosibirsk), at a free-stream velocity U_o = 9.18 m/s with the turbulence level less than 0.02 % (measured at frequencies higher than 1 Hz). A flat plate having a chord length of 1.0 m, a span 1.0 m, a thickness of 10 mm, and elliptic leading edge was used. The plate was mounted in the test section under zero attack angle. A wave-trains of instability waves (sinusoidal in time) were excited within the laminar boundary layer by means of a "point" source similar to that used in [1] (blowing-suction through a small hole) mounted at a distance $x = 250$ mm from the leading edge of the plate. All main measurements discussed in the present paper were conducted by means of a hot-wire anemometer in the region from $x = 350$ to 700 mm. The spanwise distributions of the disturbance amplitude and phase, used for determination of the stability characteristics, were measured mainly at constant non-dimensional distance from the wall $U/U_o = 0.416$ ($y/\delta_1 = 0.74$, $\eta = 1.7208(y/\delta_1) = 1.273$, $y/\delta \approx 0.25$) which corresponds approximately to the position of the inner maxima in the amplitude of the u-profiles of both 2D and 3D instability waves (see [2,4]). A part of data was obtained at a distance y = 4.5 mm from the wall that approximately corresponds to the position of the external (outer) maximum. The procedure of decomposition of a wave-train into a fan of oblique waves, used in the experiments, is described in [2,3].

Besides of the point source, a vibrating ribbon technique was also used for obtaining some additional information about the development of both 2D and 3D waves. In the 3D case, the ribbon was inclined to the flow direction by the angles 20 and 42°.

Method of calculations. The calculations are based on the conventional assumptions used in the linear stability theory: The incompressible basic flow is assumed to be parallel with only an x-component U, depending only on the normal-to-wall coordinate y. The disturbances are assumed to be small compared with the basic flow, so that nonlinear terms in the disturbances can be neglected. A normal-mode decomposition is used in that all disturbance quantities are proportional to $\exp[i(\alpha x + \beta z - \omega t)]$, where $\alpha = \alpha_r + i\alpha_i$ is the complex eigenvalue to be determined, α_r is the streamwise wavenumber and $-\alpha_i > 0$ is the spatial growth rate of the spatially growing disturbances; β is the real spanwise wavenumber and $\omega = 2\pi f$ is the cyclic frequency. (All quantities used *in this section* are assumed to be normalized by the free-stream velocity U_o of the basic flow and by the local boundary layer displacement thickness δ_1.) The frequency parameter is $F = 2\pi f \nu / U_o$ and $Re = U_o \delta_1 / \nu$ is the Reynolds number based on δ_1, where ν is the kinematic viscosity of the fluid.

If in the linearized Navier-Stokes equation all quantities, except for the normal-to-wall velocity component, are eliminated, we obtain for its amplitude $v(y)$ the well-known Orr-Sommerfeld equation:

(1) $$(v'' - \mu^2 v)'' - [\mu^2 + iRe(\alpha U - \omega)](v'' - \mu^2 v) + i\alpha ReU''v = 0$$

where

(2) $$\mu^2 = \alpha^2 + \beta^2$$

and $(\ )' = d(\ )/dy$. The boundary conditions to be satisfied by $v(y)$ are the no-penetration and no-slip condition at the wall ($y = 0$):

(3) $$v(0) = v'(0) = 0$$

and the vanishing of the disturbance for $y \rightarrow \infty$:

(4) $$v(\infty) = v'(\infty) = 0.$$

For $y \to \infty$ we have $U(y) \to 1$ and $U''(y) \to 0$. Then, asymptotically, equation (1) has two solutions satisfying (4): The "inviscid" solution v_i and the "viscous" solution v_v which are of the type:

(5) $$v_i(y) = \exp(-\mu y); \qquad v_v(y) = \exp(-\lambda y)$$

where

(6) $$\lambda = [\mu^2 + iRe(\alpha - \omega)]^{1/2}$$

and where the part of μ and λ are assumed as being positive. By starting with the asymptotic solutions (5), two linear independent solutions have been obtained by numerical integration of (1), simultaneously, with a Runge-Kutta-Fehlberg procedure. At the wall ($y = 0$) the combined solution $v = v_i + c_1 v_v$ has to satisfy (3) leading to:

(7) $$c_1 = -v_i'(0)/v_v'(0); \quad v(0) = v_i(0) + c_1 v_v(0) = 0.$$

The latter condition constitutes an eigenvalue equation for the complex eigenvalue $\alpha(\omega, Re, \beta)$ which has been solved by a parabolic zero-search procedure. For more detail of the method of calculations see [3].

3. Mean flow in the experiment

The y-profiles of the mean flow velocity have been measured in the experiment [2] at positions $x = 300$ - 700 mm with a step of 50 mm, as well as for a number of various spanwise positions (from -42 to + 62 mm) at some distances from the leading edge. All of them proved to be in excellent agreement with the Blasius profile (Fig. 1). Maximum deviation of the experimental points from the theoretical curve is not higher than ±0.5 %. The comparison of experimental downstream dependence of the boundary-layer displacement thickness δ_1 with that for the Blasius flow and for a shifted (upstream by $\Delta x = 123$ mm) Blasius dependence (Fig. 2) demonstrates an essential influence of the finite thickness of the plate in experiments on the boundary layer. This influence results in a shift of an effective position of the leading edge by Δx. The shifted theoretical curve is in a very good agreement with experiment. Note, that this shift did not influence the results presented below, since starting with x = 350 mm the boundary layer develops according to the theory and the values of the local Reynolds numbers used for comparison with theory were determined using the experimental values of δ_1. Note also, that the average experimental value of the shape parameter $H = \delta_1/\delta_2$ (where δ_2 is the momentum thickness) differs from its theoretical value only by +0.3 %.

Thus, the structure of the mean flow, realized at the experimental conditions, corresponds locally to the Blasius boundary layer with high accuracy.

4. Dispersion characteristics of 3D instability waves

To obtain experimentally the stability and dispersion characteristics of the "normal modes" (harmonic in time and space instability waves inclined at various angles to the flow direction) a procedure of a complete decomposition of the wave-trains, developing within the boundary layer, to oblique modes was carried out in [2].

Streamwise wavenumbers. The dependences of the streamwise wavenumber $\alpha_r\delta_1$ from the propagation angle σ ($\sigma = \arctan[\beta/\alpha_r(\beta)]$) for three different frequency parameters, shown in Fig. 3, demonstrate a good quantitative agreement between the theory and experiment. A small deviation of the experimental points from the theory has a character of a systematic shift and does not exceed 2 - 3 % for σ from 0 to 50° and 5 % for other angles. Fig. 3a shows that the streamwise wavenumbers decrease slowly with σ for each constant frequency and increase with frequency parameter F for each fixed σ (Fig. 3b). The data obtained from disturbances excited by the point source (with a subsequent Fourier decomposition) agree very well with the data obtained by means of an inclined (or flat) vibrating ribbon; the corresponding points for $\sigma = 0$ merge, in practice, in Fig. 3a (as well as in Figs. 4, 6 and 7). All the data from the experiments [2] are in a very good agreement with the results [5] obtained with the help of a vibrating ribbon for 2D waves (Fig. 3b).

Dependences of $\alpha_r\delta_1$ from the Reynolds number Re is shown in Fig. 4, where again a good quantitative agreement between the theory and experiment is observed for both methods of the disturbance excitation. In the experiment the dimensional values of α_r were found to be independent of the x-coordinate (i.e. of Re) for each fixed frequency parameter F. The dependence $\alpha_r\delta_1(x)$ is conditioned by that of δ_1 of x. This means that the streamwise wavenumbers of the 3D modes normalized by (for example) corresponding values for the 2D waves ($\alpha_{r0}\delta_1$) have to be independent of Re in both the theory and experiment. This fact is really observed and shown in Fig. 5, where a set of experimental points for $F = 94.1 \cdot 10^{-6}$ corresponds to a region of Re = 827 to 1226. Very similar result have been obtained in [5] for 2D instability waves.

Phase velocities. Downstream phase velocities ($c_x/U_o = \omega/\alpha_r$), shown in Fig. 6, again demonstrate very good agreement between the theory and experiments, especially for the angles less than 50°. The small difference has again a systematic character and does not exceed of 3 % for σ from 0 to 50°, and 5 % for other compared angles. Fig. 7 demonstrates an independence of c_x/U_o on the Reynolds number, found both in the theory and experiment, and a very good agreement of the present results with experimental data obtained in [6] under absolutely other experimental conditions (in a water flow, without spanwise Fourier transform, from the investigation of 2D and oblique waves, harmonic in time and space, introduced into the flow by means of a set of heating elements). Fig. 8 shows that the phase velocities of 3D modes, normalized by those of 2D waves, are almost completely independent of Re and have a very week dependence (if any) from F.

Thus, the data presented in Figures 3 - 8 demonstrate that the parallel-flow linear stability theory describes very well the dispersion characteristics of the 3D instability waves found in the experiment.

5. Amplification rates of 3D instability modes

The comparison of the growth rates represents a much more complicated problem. Good enough quantitative agreement for α_i between the experimental and theoretical values have never been obtained for the flat-plate boundary layer even for 2D modes, especially not for the 3D ones. Some typical problems, which arise in this field, are discussed for the 2D instability waves in [4,7,8]. These problems are mainly connected with the downstream deformation of the disturbance amplitude *y*-profiles that results in a situation when there is no a unique stability criterion. The first theoretical study of the non-parallel effects for 3D modes [4] showed that in the Blasius flow these effects are significantly more important for 3D waves than for the 2D ones. In view of this, the main purpose of the present work, concerning the amplification of 3D modes, was to carry out only a first step in this field and to undertake the first attempt of direct quantitative comparison of the growth rates observed in the experiment with results of the parallel stability theory.

Dependences of nondimensional amplification rates $-\alpha_i\delta_1$ from the propagation angle σ, obtained in both theory and experiment for the frequency parameter $F = F_1 = 94.1\cdot10^{-6}$ and Reynolds number $Re = Re_1 = 1089$, are shown in Fig. 9. It is seen that theory and experiment demonstrate a very similar general behaviour of the growth rates with propagation angle, and even a rather good agreement is observed for not too large values of σ. At the same time the experimental and theoretical data deviate from each other for large values of the propagation angle. This is in consistence with theory [4], where it was found, in particular, that the parallel-flow approach gives too low growth rates, especially for high angles σ. Note that the experimental data obtained at $y = 4.5$ mm display better agreement with the parallel-flow theory than the points measured along $y/\delta_1 = 0.74$. [It is interesting that the growth rates calculated for somewhat smaller value of Re (namely for $Re'_1 = 0.87Re_1 = 963$) demonstrate very good agreement with the experimental points.] Very similar results are observed at another value of Reynolds number ($Re = Re_2 = 1129$) and for the same frequency parameter (Fig. 10). In both figures the growth rates measured directly (for 2D waves generated by the vibrating ribbon) demonstrate very good agreement with the data obtained by means of the Fourier decomposition of the wave-trains.

For the frequency parameters $F = F_2 = 63.6\cdot10^{-6}$ and $F = F_3 = 138.7\cdot10^{-6}$, the measurements have been conducted only near the inner maximum, i.e. at $y/\delta_1 = 0.74$. For $F = F_2$ and $Re = Re_3 = 1071$ (Fig. 11), the theory predicts somewhat higher growth rates for the unstable waves ($\sigma < 60°$) and lower growth rates for the stable modes. At the same time, the general behaviour of $-\alpha_i\delta_1$ with σ and the position of the neutral point are predicted quite well. For $F = F_3$ and $Re = Re_3 = 1071$ (Fig. 12), the disagreement between the theoretical curve and the experimental points is the largest, although a general behaviour of the growth rates with σ is the same. In Figs. 11 and 12 again the theoretical curves calculated for a 13 % lower Reynolds number ($Re = 948$) demonstrate significantly better agreement with the experiment.

6. Conclusions

1. The mean flow over the flat plate in experiments [2] is found to be in a very good agreement with the Blasius boundary layer.

2. Very good quantitative agreement of the dispersion characteristics of 3D waves obtained in experiments [2] with predictions of the parallel stability theory is found.

3. No differences are found between the experimental data obtained using the Fourier decomposition and those measured using a vibrating ribbon. (For oblique vibrating ribbon only dispersion characteristics have been measured and compared.)

4. The values of the streamwise wavenumbers of the 2D waves are found to be in a very good agreement with those obtained in the previous experiments [5].

5. The values of the phase velocities of 3D modes are compared with experimental data [6] obtained in the water flow without Fourier decomposition for harmonic in space oblique waves. Very good quantitative agreement is found.

6. As expected, a qualitative, mainly, agreement is found for the growth rates of 3D modes. However, for some values of the frequency parameters and Reynolds numbers rather good accordance is observed for the waves inclined at not too large angles to the flow direction. For high propagation angles the parallel-flow theory predicts smaller growth rates than those found in the experiment. This observation is in consistence with the theoretical results obtained in [4].

Thus, we can conclude that the parallel-flow linear stability theory predicts very well the dispersion characteristics of 3D waves in the Blasius boundary layer, but their amplification is predicted mainly qualitatively. The growth rates and the stability criteria for 3D modes need more precise investigation in both theory and experiment.

Acknowledgment

One of the authors (Y.K.) wishes to express his gratitude to Prof. H.H. Fernholz for the support of the idea of this paper and to the Technische Universität Berlin for the financial support of his visit to TUB. The paper has been prepared during this visit.

References

[1] Gaster, M., Grant, T. : An experimental investigation of the formation and development of a wave packet in a laminar boundary layer. *Proc. Roy. Soc. London.* A 347 (1975) 253 - 269.

[2] Gilyov, V.M., Kachanov, Y.S., Kozlov, V.V. : Development of a spatial wave packet in a boundary layer. *Izv. Sib. Otd. Akad. Nauk SSSR, Ser. Tekh. Nauk.* 13 (1983) 27 - 37 (in Russian). (See also Kachanov, Y.S. In: *Laminar-Turbulent Transition* (ed. V.V.Kozlov) Springer, 1985, p. 115 - 123.)

[3] Kachanov, Y.S., Michalke, A. : 3D instability of flat-plate boundary layer. Theory and experiment. *Eur.J. Mech., B/Fluids.* 1994 (accepted for publication).

[4] Bertolotti, F.P. : *Linear and nonlinear stability of boundary layers with streamwise varying properties.* PhD thesis. The Ohio State Univ., USA, 1991.

[5] Ross, J.A., Barnes, F.H., Burns, J.G., Ross, M.A.S. : The flat plate boundary layer. Part 3. Comparison of theory with experiment. *J. Fluid Mech.* 43 (1970) 819 - 832

[6] Schneider, S.P. : *Effects of controlled three-dimensional perturbations on boundary layer transition.* PhD thesis. Calif. Inst of Tech., Pasadena, 1989.

[7] Fasel, H, Konzelmann, U. : Non-parallel stability of a flat plate boundary layer using the complete Navier-Stokes equations. *J.Fluid Mech.* 221(1990) 311-347.

[8] Bertolotti, F.P., Herbert T., Spalart, P.R. : Linear and nonlinear stability of the Blasius boundary layer. *J. Fluid Mech.* 242 (1992) 441 - 474.

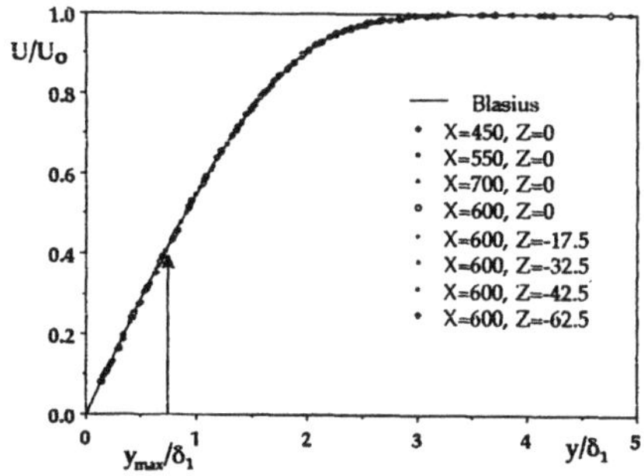

Fig.1. Mean velocity profiles. Comparison with Blasius.

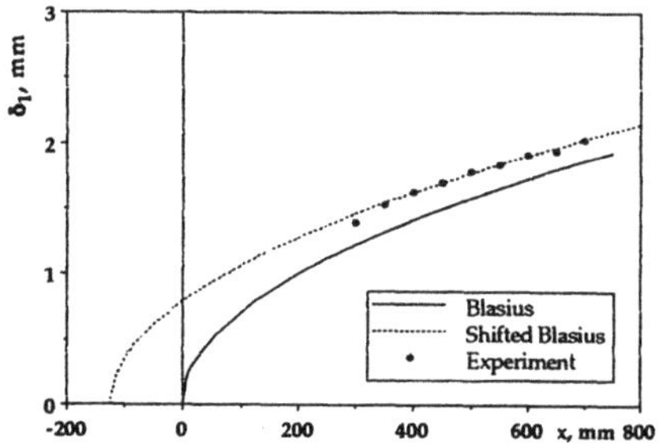

Fig.2. Downstream growth of boundary layer displacement thickness.

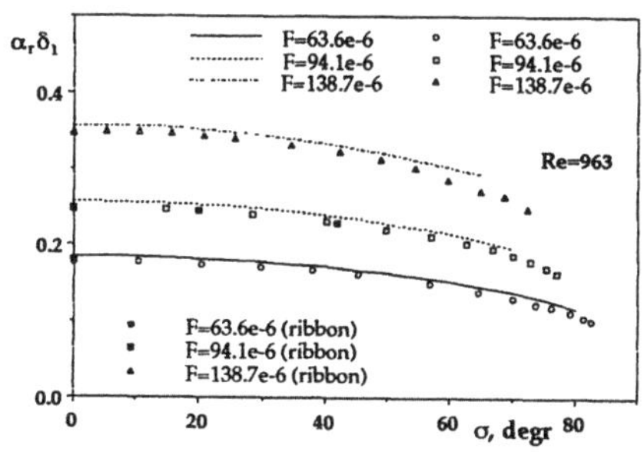

Fid.3a. Streamwise wavenumber vs. propagation angle. Dependence on frequency.

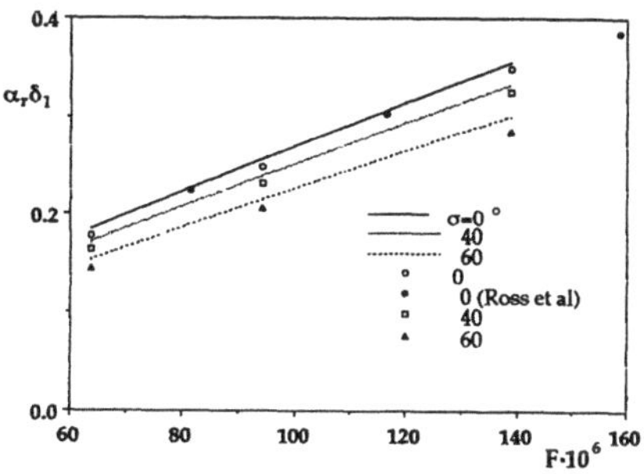

Fig.3b. Streamwise wavenumber vs. frequency parameter for three wave angles.

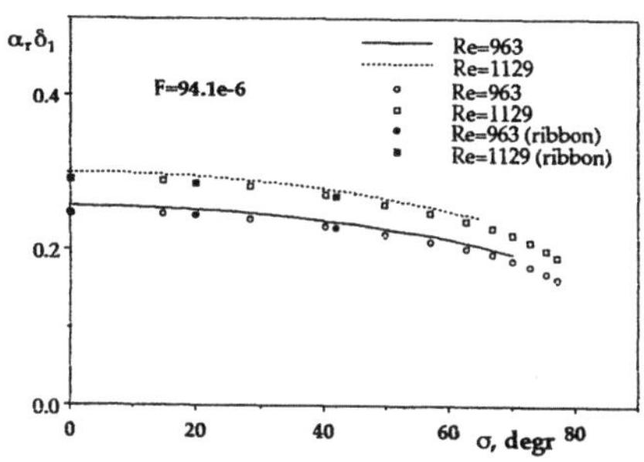

Fig.4. Streamwise wavenumber vs. propagation angle. Dependence on Reynolds number.

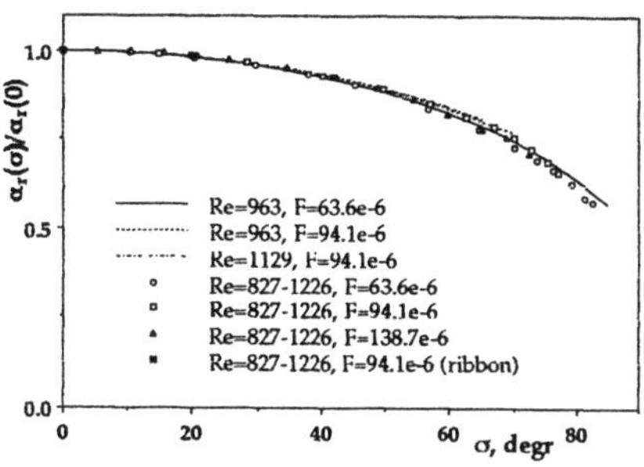

Fig.5. Normalized streamwise wavenumber vs. propagation angle.

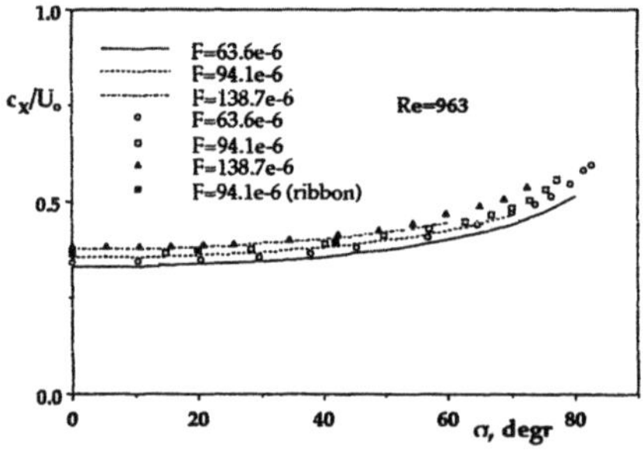

Fig.6. Phase velocities for different frequencies.

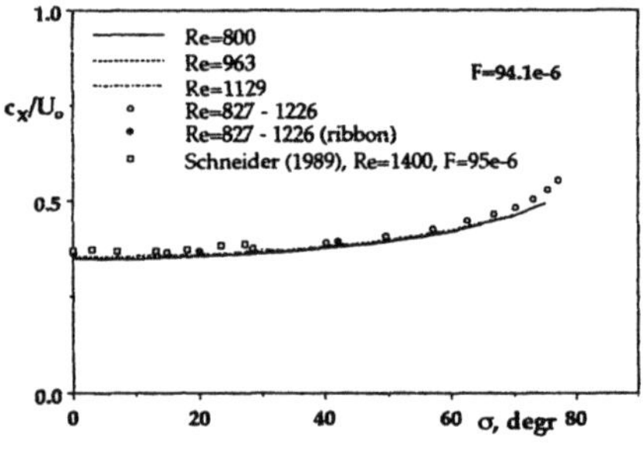

Fig.7. Phase velocities for different *Re*.

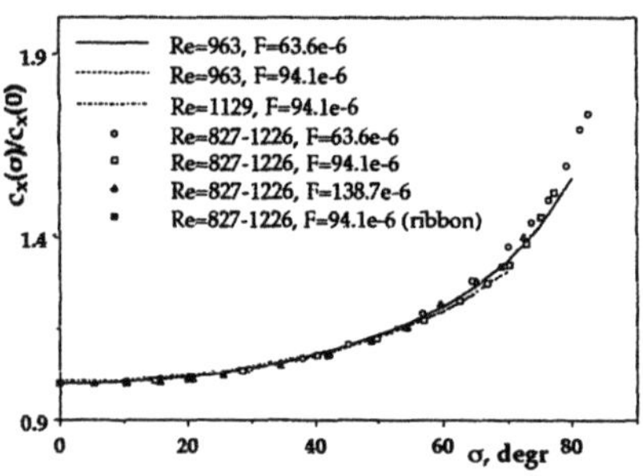

Fig.8. Normalized phase velocities.

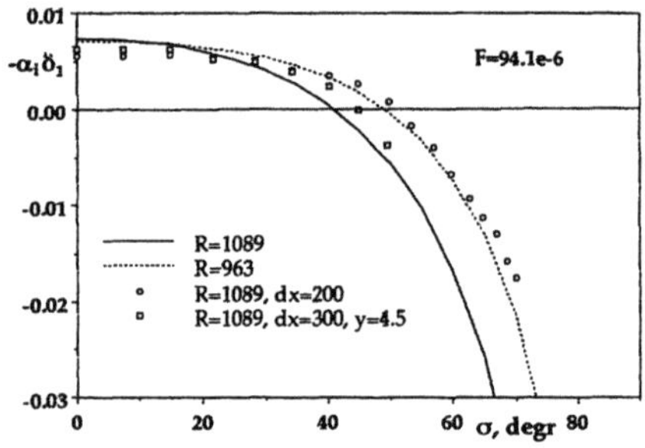

Fig.9. Growth rates for middle frequency.

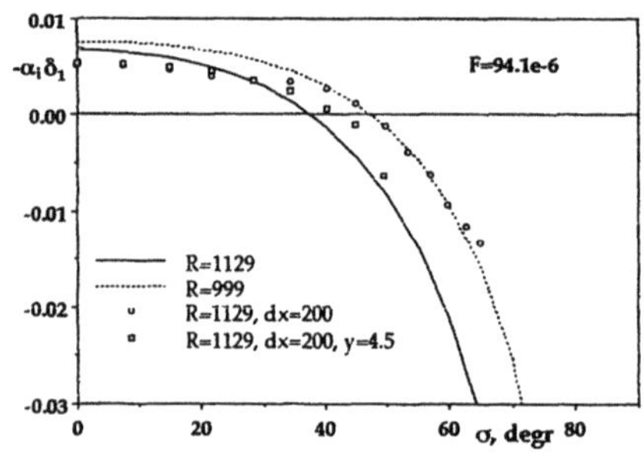

Fig.10. Growth rates for middle frequency.

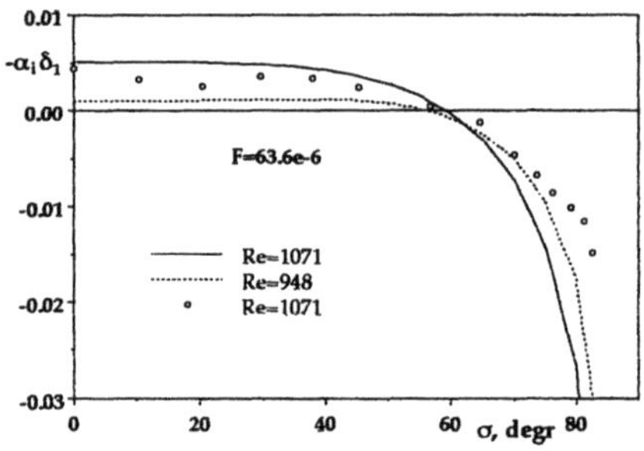

Fig.11. Growth rates for low frequency.

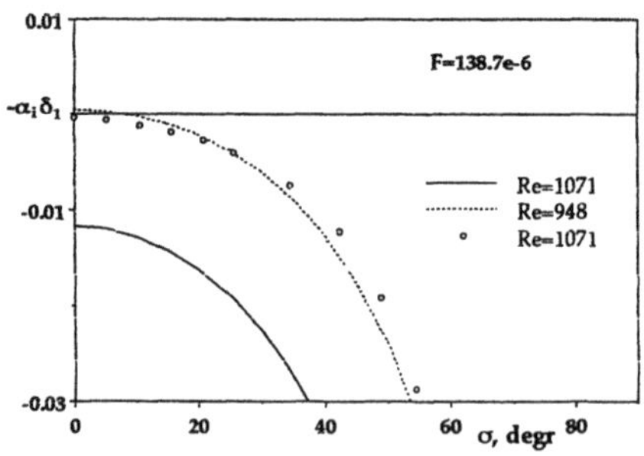

Fig.12. Growth rates for high frequency.

Direct Numerical Simulation of Boundary-Layer Transition with Strong Adverse Pressure Gradient

M. Kloker
Institut für Aerodynamik und Gasdynamik, Universität Stuttgart
D-70550 Stuttgart, FRG
H. Fasel
Aerospace and Mechanical Engineering Dept., University of Arizona
Tucson, AZ 85721, USA

Summary

Fundamental breakdown of a strongly decelerated Falkner-Skan-type boundary layer (Hartree-Parameter β_H=-0.18) is investigated up to late stages by direct numerical simulation using the complete Navier-Stokes equations for three-dimensional incompressible flow and the so-called spatial model. Transition is initiated by a timewise-periodic small amplitude 2-D wave and a pair of oblique 3-D waves with identical frequency. It is observed that the resulting breakdown process under adverse pressure gradient is dramatically more complex than the well-known K-breakdown in the Blasius flow: In addition to the (upper) high-shear layer on top of the lambda (Λ-) vortex, a (lower) characteristic high-shear layer is formed simultaneously in between neighbouring Λ-vortices. This shear layer, induced by a secondary vortex system close to the wall, precipitates ultimate breakdown to turbulence.

Keywords : Fundamental breakdown, High-shear layers, 3-D Structures

1. Introduction

Transition in boundary layers is a spatially evolving complicated process influenced by many parameters. In order to reduce the number of these parameters, transition research has focussed on "controlled" transition in simple model flows, like flat-plate boundary layers. In classical experiments (see [1], eg), a vibrating ribbon is used to periodically excite disturbance waves. The streamwise evolution of the disturbances eventually causing turbulent flow is then observed using flow visualization and hot-wire measurements.

In addition to experimental (and theoretical) work, direct numerical simulations based on the solution of the complete three-dimensional Navier-Stokes equations have played an increasingly important role in transition research in the past decade. The basic approach is similar to that of controlled experiments. Some regular perturbations are introduced into the integration domain, and their subsequent unstable, nonlinear development is calculated numerically. Two basically different kinds of numerical models have been used until the beginning of the '90s, the "temporal model" and the "spatial model" (for a thorough discussion of respective advantages and shortcomings see Fasel [2], and Kleiser & Zang [3]). Nowadays, the computationally less demanding temporal model with its underlying unphysical assumptions has been widely supplanted by the *spatial* model, which has been used by the authors of this article for many years of research.

In the spatial model the streamwise evolution of disturbances is simulated in a fixed integration domain extending over a large downstream distance. With this model, realistic simulations of controlled experiments even in flows with large streamwise gradients and/or feedback, or local flow reversal are possible, like flows with strong adverse pressure gradient, as presented in this paper. However, to devise a properly working numerical method for this model is a difficult task, particularly due to the outflow conditions required and for reasons of numerical stability.

The capability of the numerical methods developed at the University of Stuttgart [2-7] for the spatial model has been convincingly demonstrated by simulations of two different transition experiments. Rist et al. ([4], [2], see also the article in this book) performed a simulation of the vibrating-ribbon experiments by Kachanov et al. (see [1]) for the fundamental breakdown in a Blasius boundary layer, and Konzelmann et al. (in the same proceedings as [4]) simulated the nonlinear development of a three-dimensional wave packet according to the experiments by Gaster & Grant. Both simulations yielded excellent *quantitative* agreement with the experimental measurements.

The effects of a (strong) adverse streamwise pressure gradient – known to boost laminar breakdown in practical flows – on the nonlinear evolution and structure formation in the transition process are largely unknown. Controlled experiments are scarce due to the great difficulties arising from the proneness to separation and the strong disturbance growth in these flows – the instability mechanism is now both of viscous and inviscid nature.

2. Numerical method

The numerical method used in this paper is discussed in detail by Kloker et al. [6, 7]. It is based on the complete three-dimensional, unsteady Navier-Stokes equations in vorticity-velocity formulation for incompressible flow. The vorticity components are denoted by ω_x, ω_y and ω_z, and u, v, and w are the velocity components in the x, y and z directions. All variables are nondimensionalized with a reference length L and the free stream velocity U_∞, relating to the dimensional variables (denoted by bars) as

$$x = \bar{x}/L, \quad y = \bar{y}/L \cdot \sqrt{Re}, \quad z = \bar{z}/L, \quad t = \bar{t}\,U_\infty/L,$$
$$u = \bar{u}/U_\infty, \quad v = \bar{v}/U_\infty \cdot \sqrt{Re}, \quad w = \bar{w}/U_\infty; \quad Re = U_\infty L/\bar{\nu} = 10^5.$$

The equations are solved in the domain shown in Fig. 1. In the spanwise direction, the flow is assumed to be periodic with the domain extending from z=0 to $z=\lambda_z$ (λ_z-basic wavelength in z-direction). The flow variables (f) are decomposed into those of the two-dimensional steady base flow (f_B) and of the three-dimensional unsteady disturbance flow (f') with $f=f_B+f'$. Consequently, for the base flow one vorticity (ω_{zB}) transport equation and two Poisson equations for u_B and v_B have to be solved, and for the disturbance flow three vorticity transport equations (in conservative formulation) and three Poisson equations for the velocity vector have to be solved. A mean-flow change is thus represented by a non-zero time-average of the disturbance flow.

Boundary Conditions.—For the base flow, a streamwise pressure gradient is imposed by prescribing the streamwise velocity of the external flow $u_e(x)$ at the upper boundary, assuming inviscid flow. For the simulations discussed in this paper, $u_e(x)$ is chosen according to the Fálkner-Skan ansatz $u_e(x)=u_0 \cdot x^m$, $m=\beta_H/(2-\beta_H)$, $u_0=x_0^{-m}$, β_H-Hartree-Parameter. For the Blasius boundary layer we have $\beta_H=0$, and for the strongly decelerated flow $\beta_H=-0.18$ was chosen. At the inflow boundary, corresponding Falkner-Skan profiles are specified. At the outflow boundary, all equations are solved neglecting the terms containing the factor 1/Re (second x-derivatives); at the wall, the velocity vector is zero. For the disturbance flow, the velocity vector is also zero at the wall, except within the disturbance strip, where the v'-component can be prescribed as a function of x (see f_v in Fig. 1), z and t in order to introduce 2-D/3-D disturbance waves by localized blowing and suction. At the freestream boundary, vanishing vorticity fluctuations and an exponential decay of the v'-velocity are assumed. At the inflow boundary, all disturbances are set to zero. As for the outflow boundary, the implementation of appropriate, non-reflecting outflow conditions is of crucial importance for the feasibility of numerically stable and reliable simulations up to late breakdown stages or developed turbulence. Here we used the method of artificial relaminarization (see Kloker et al. [6]). A gradual artificial decay of disturbances is enforced in a region somewhat upstream of the outflow boundary ($x_3 \le x \le x_N$) to substantially (orders of magnitude) reduce the disturbance level at this boundary. This is achieved by direct suppression of the disturbance vorticity vector. Thus, possible undue reflections caused by large amplitude, broad-band disturbances passing the outflow boundary are prevented.

Solution Method.— The equation set is discretized on a uniform grid employing 4th-order accurate (mainly central) finite differences in streamwise direction and normal to the wall. For the disturbance flow a spectral Fourier ansatz is employed in spanwise direction to exploit the z-periodicity. The nonlinear terms in the vorticity transport equations (in conservative form) are evaluated pseudospectrally, i.e. in physical space, including de-aliasing.

After the calculation of the laminar steady base flow using a temporally dissipative time-stepping scheme, the unsteady disturbance flow with active disturbance strip is simulated. The integration in time is performed with a 4th-order Runge-Kutta scheme. The four stages per time step are coupled with a centered-upwind-downwind (and vice versa)-centered discretisation for the x-convection terms to ensure numerical stability at *no* additional costs by effective damping of badly calculated high wavenumber (>approx. $\pi/(2\Delta x)$) components. Since this damping is proportional to the square of the convection velocity, no problems are encountered at flow reversal. The v'-Poisson equation is solved with a multigrid method, using a SOR line iteration technique.

3. Numerical Results

The goal of this study was to explore the effects of a strong streamwise adverse pressure gradient on the evolution of the transition process (for preliminary studies see [5]). Here, we will concentrate on comparing the formation of flow structures with the transitional Blasius flow as obtained from a simulation of Kachanov's experiment (this simulation was performed with much finer spatial resolution than the original simulation by Rist [2, 4]). The simulation results for the Kachanov case are subsequently denoted by the acronym KACH. In both cases, the Blasius and strongly decelerated flow, the fundamental breakdown is triggered by the input at the disturbance strip. The following table summarizes the relevant parameters:

Table 1 Case	β_H=0 (KACH)	β_H=-0.18
Ref. velocity / length	U_∞=9.09m/s, L=165mm	
x_0; x_1–x_2	1.14; 1.35–1.56	1.59; 2.08–2.29
$Re_{\delta_1}(x_2)$ (δ_1-displac. ths.)	679	1303
Start of relam. x_3; $u_e(x_3)$	3.04; 1.00	5.38; 0.90
Width of domain λ_z	0.148	0.205
Dist. frequency $\beta=\bar{\beta}L/U_\infty$	11	10.8
Dist. amplitude 2-D; 3-D	3.0%; 0.2% (freq. β=0)	0.001%; 0.002%
Δx, Δz_{min} (both$*10^3$); Δy	0.873, 3.083; 0.153	2.64, 1.60; 0.348
Max. grid size ($N_x*N_y*N_z$)	4500*121*48	1850*121*128
Central memory req.	1.8 gbyte	1 gbyte
CPU CRAY-2 (based on 1 cpu)	400 hrs.	150 hrs.
Taped data	13 gbyte	6.7 gbyte

Reference case KACH.—In the experiment by Kachanov et al. [1], the boundary layer flow over a flat plate without pressure gradient was periodically disturbed by a vibrating ribbon at a distance of 250mm from the flat-plate leading edge to introduce a 2-D disturbance wave. A spanwise row of small pieces of tape was attached to the plate underneath the ribbon, in order to induce a 3-D disturbance with a fixed spanwise wavelength λ_z. In fact, such a disturbance generates a system of small-amplitude longitudinal vortices. In direct succession, this system interacts nonlinearly with the large amplitude 2-D wave producing a pair of oblique 3-D waves with the frequency of the 2-D wave and spanwise wavelengths $\pm\lambda_z$. This phase-locked combination of fundamental disturbance waves grows while propagating downstream and generates higher harmonics in the frequency – spanwise-wavenumber spectrum. The disturbance amplitudes in physical space exhibit "peaks" and "valleys" in spanwise direction, half a spanwise wavelength $\lambda_z/2$ apart (see Fig. 2b). An overall view of the dynamic processes involved in the K-breakdown process can be obtained from Fig. 3, where the instantaneous ω_z-vorticity contours at the spanwise peak and valley location are shown. At peak position (Fig. 3a), the formation and subsequent breakdown of the instantaneous high-shear layer into, at first, deterministic smaller vorticity concentrations can be observed which, further downstream, lead to a chaotic motion. In contrast, the disturbance development at the valley position (Fig. 3b) shows, at this stage, no pronounced characteristic structures.

Adverse pressure-gradient case.— The deceleration of the external flow according to the Falkner-Skan law with β_H=-0.18 corresponds to a decreasingly strong adverse pressure gradient ($dp_e/dx \sim x^{-1.165}$) in the flow. The adverse pressure causes a shift of the inflexion point in the $u_B(y)$-profile, corresponding to the maximum of the base-flow vorticity, away from the wall, thus giving rise to inviscid instability. The flow is near separation (note the u_B-profile sketched in Fig. 4a). Transition was initiated by introducing a small-amplitude disturbance combination (see Table 1), consisting of a 2-D wave and a pair of oblique 3-D waves with identical frequency. At first, both modes grow according to linear instability, followed by the onset of fundamental resonant growth of the 3-D mode and the simultaneous, nonlinear generation of the so-called longitudinal vortex mode (see [5, 7]). It should be noted that, in the presence of a sufficiently large 2-D wave, the existence of a longitudinal vortex mode is inherently connected with the presence of a pair of oblique 3-D waves, and vice versa. Thus, if one mode is part of the disturbance input, the other is nonlinearly generated. Therefore, characteristic differences in the fundamental breakdown between the two cases investigated in this paper are *not* due to the disturbance input.

The final breakdown stages, visualized in the same manner as in Fig. 3 for the zero pressure-gradient case, are shown in Fig. 4. At the peak location, a qualitatively similar formation of a high-shear layer can be observed, albeit more intense due to larger amplitudes (cf. Fig. 2), and undergoing a more rapid breakdown. Drastic differences, however, appear in the plane half a spanwise wavelength apart, which is the valley plane in the case KACH. Now, in addition to the "upper" high-shear layer at the peak position, a "lower inverted" high-shear layer evolves close to the wall at this "co-peak" position, where the largest amplitudes occur except for those at the peak position (Fig. 2a). The inverted layer, generated by a downstream travelling, instantaneous separation zone[1] (see the streamline plot inserted in Fig. 4b and note $w \equiv 0$ in this x-y plane) then merges with a second high-shear layer, and subsequently this extended layer crosses the entire boundary-layer flow. Thus the shear-layer separates a large region of very slow, partly recirculating fluid from the energetic motion above. The break-up of this very unstable layer into smaller structures, that represent longitudinal cuts through the heads of small hairpin vortices, occurs obviously *earlier*, i.e further upstream, than the high-shear-layer decay at the peak station.

A cut through the instantaneous flow field at $x=4.6$ (Fig. 5, ω_z-, ω_x-contours plus velocity vectors) reveals that the high-shear layer at the co-peak position is induced by a *secondary vortex system* under it (close to the wall, Fig. 5b). This system transports very slow fluid from the wall upwards at the co-peak station, as do the legs of the higher, larger Λ-vortex (almond-shaped ω_x-concentrations) at the peak station.

As the numerical simulation yields data of the entire flow field, we are able to visualize the dominant, three-dimensional flow structures of the transition process. In Fig. 6, iso-surfaces $|\omega_x|$=const. and ω_z=const. visualize (in perspective view) the longitudinal vortical structures and instantaneous high-shear layers at the same time level as for the previous figures. Now, the size and position of the secondary vortex system between two neighbouring

[1] Such a region of flow separation has also been observed in the recent experiments by Kosorygin [8] ("microseparation lenses").

Λ-vortices becomes evident (ω_x-plot); for the case without pressure gradient, no such corresponding structure can be detected. The ω_z iso-surface shows the delta-wing topology of the high-shear layer above the Λ-vortex, with the downstream head marking the tip of the Λ, and the tongue-like lower layer, with its upstream head located near the wall between the upstream ends of two Λ-vortices. We note that the small, tube-like ω_z-structures near the valley position downstream of $x \approx 2.6$ in the KACH case are unphysical, as they are caused by insufficient spanwise resolution.

Finally, in Fig. 7 the flow structures are visualized by numerically generated particle (time-) lines like they can be generated experimentally by the hydrogen bubble technique in water flow using a horizontal wire placed near the y-position of the critical layer. The particles are plotted in perspective view like the iso-surfaces in Fig. 6. For the KACH case, the particle-lines develop into the well-known Λ-vortex structures that are connected by straight lines, while in the decelerated flow, the formation of smooth, inverted Λ-shaped vortices is observed at corresponding spanwise positions. The rapid spatial redistribution of particles in the latter case documents the violent transition process. More details of this transition case are discussed in [7].

4. Acknowledgements

This paper benefited from many years of research sponsored by different agencies, mainly the Deutsche Forschungsgemeinschaft, FRG; the Stiftung Volkswagenwerk, FRG; and the Office of Naval Research, USA.

5. References

[1] Kachanov, Y.S.: On the Resonant Nature of the Breakdown of a Laminar Boundary Layer. *J. Fluid Mech.* 184 (1987), pp. 43-74.

[2] Fasel, H.: Numerical Simulation of Instability and Transition in Boundary—Layer Flows. In: Arnal, D., Michel, R. (eds.) *Laminar-Turbulent Transition.* IUTAM-Sym., Toulouse, France, 1989, Springer, pp. 587-598 (1990).

[3] Kleiser, L.; Zang, T.: Numerical Simulation of Transition in Wall-Bounded Shear Flows. *Ann. Rev. Fluid Mech.* 23, pp. 495-537 (1991).

[4] Rist, U.; Fasel, H.: Spatial Three-Dimensional Numerical Simulation of Laminar-Turbulent Transition in a Flat-Plate Boundary Layer. *Proc. Conference on Boundary Layer Transition and Control*, April 8-12, 1991, Cambridge, U.K., RAeS, London (1991).

[5] Kloker, M., Fasel, H.: Numerical Simulation of 2-D and 3-D Instability Waves in 2-D Boundary Layers with Streamwise Pressure Gradient. See [2].

[6] Kloker, M.; Konzelmann, U.; Fasel, H.: Outflow Boundary Conditions for Spatial Navier-Stokes Simulations of Laminar-Turbulent Transition in Boundary Layers. *AIAA J.* 31, No. 4 (1993).

[7] Kloker, M.: *Direkte numerische Simulation des laminar-turbulenten Strömungsumschlages in einer stark verzögerten Grenzschicht.* Dissertation, Universität Stuttgart (1993).

[8] Kosorygin, V.S.: Stability and Transition to Turbulence in 2-D Boundary Layer under the Influence of Adverse Pressure Gradients. In: *Nonlinear Instability of Nonparallel Flows.* IUTAM-Sym., Potsdam, USA, 1993 (1994).

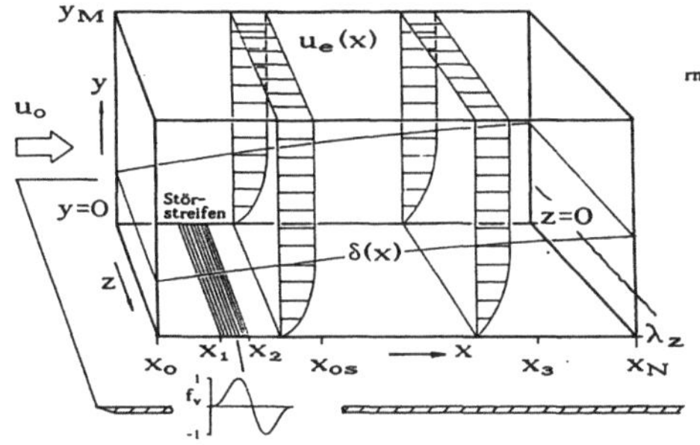

Fig. 1 Integration domain

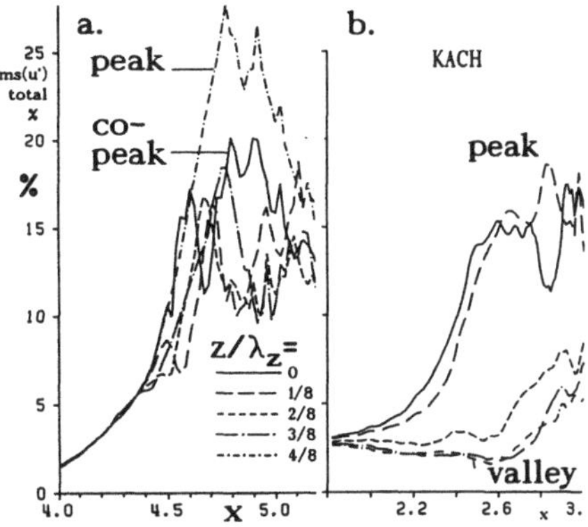

Fig. 2 (top right) Total rms amplitudes of u' (maximum over y) versus x at different z-positions: a) strong adverse pressure-gradient case, b) reference case KACH without pressure gradient.

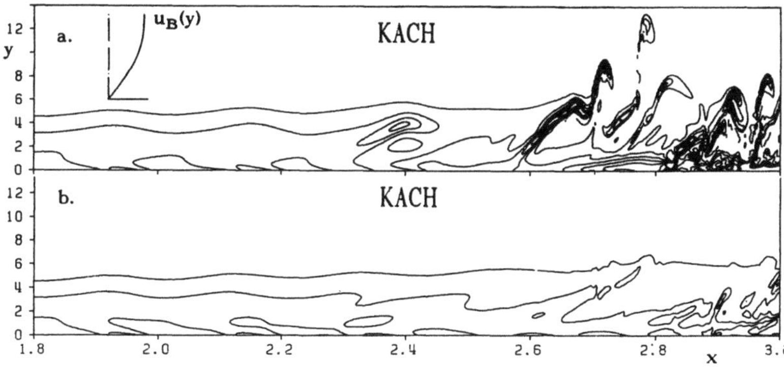

Fig. 3 Longitudinal sections of the flow field, showing instantaneous spanwise vorticity (ω_z) contours at spanwise peak (a) and valley (b) position for reference case KACH

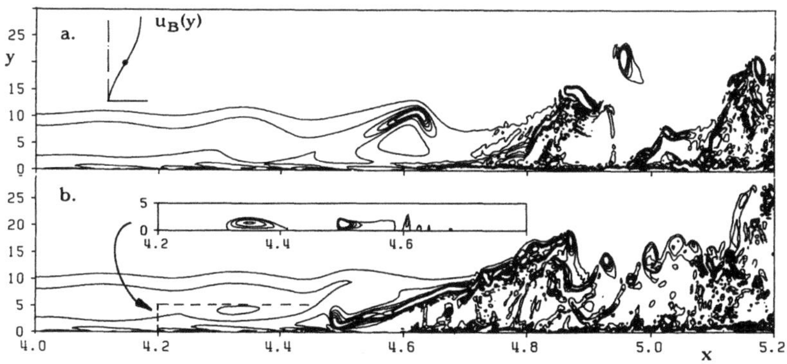

Fig. 4 Same as Fig. 3, now for strong adverse pressure-gradient case ($t=t_1$): (a) peak station ($z/\lambda_z=0.5$) (b) co-peak station (z=0). Inserted in (b): instantaneous streamlines showing separation zones.

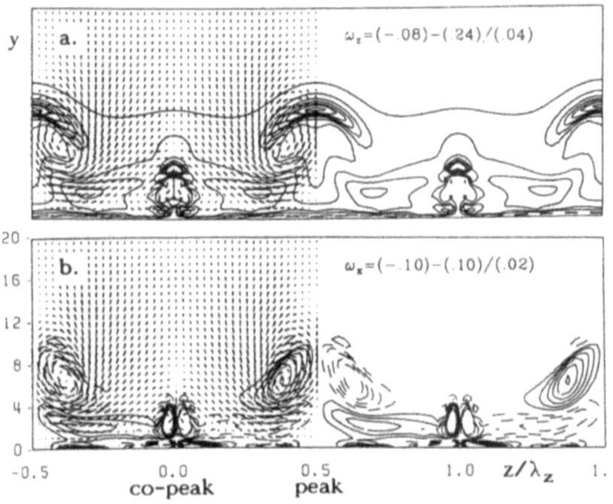

Fig. 5
Cross-sections of the flow field at x=4.6: instantaneous ($t=t_1$) spanwise (a) and streamwise (b) vorticity contours (ω_z, ω_x) with velocity vectors

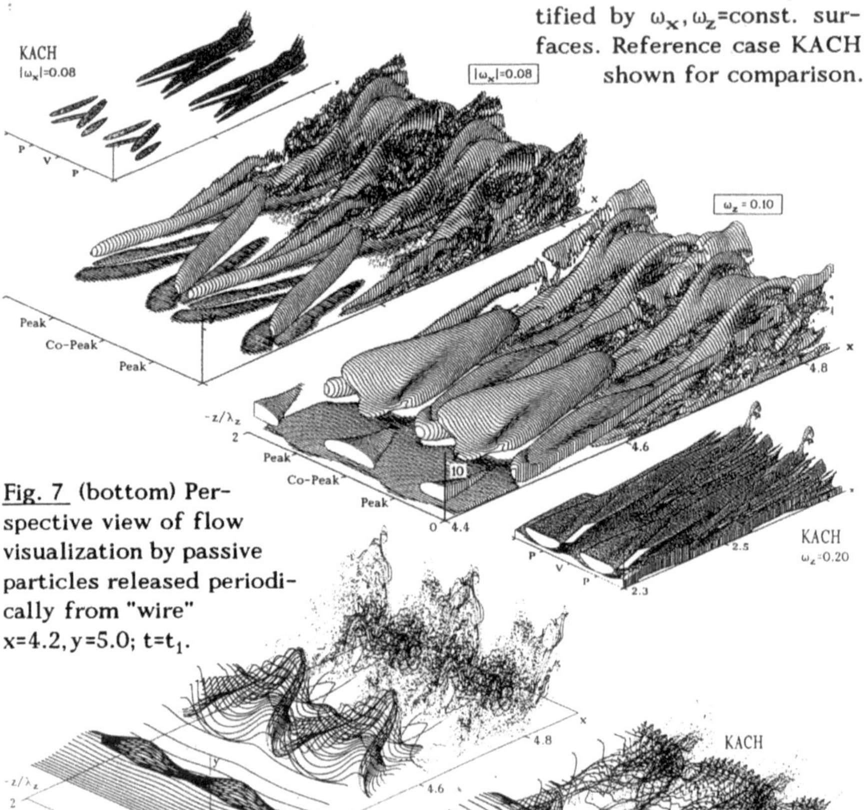

Fig. 6 (below) Perspective view of the dominant structures at $t=t_1$, identified by ω_x, ω_z=const. surfaces. Reference case KACH shown for comparison.

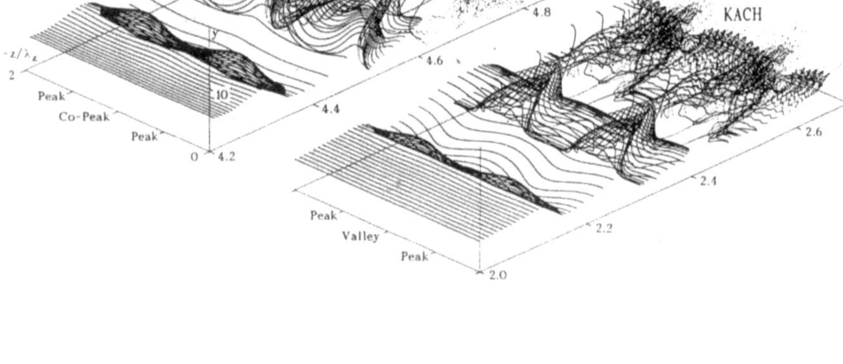

Fig. 7 (bottom) Perspective view of flow visualization by passive particles released periodically from "wire" x=4.2, y=5.0; $t=t_1$.

Session M: Receptivity I

A Partial Simulation of Receptivity and Transition in 3-D Boundary Layers

F.P. Bertolotti
Institut für Strömungsmechanik
DLR, 37073 Göttingen, Germany

Summary: A model for the generation of steady disturbances by localized and distributed surface irregularities in three-dimensional boundary layers is presented. The subsequent nonlinear development of the flow is computed with the parabolized stability equations. Comparisons with experimental results suggest that the disturbance generation phenomena is modeled correctly.

Key words: Cross-Flow Vortex, Receptivity, PSE.

1 Introduction

This work is part of the current effort at DLR to study and understand the laminar-turbulent transition over wings of commercial civil aircraft. The motivation comes, in part, from the interest of the aerospace industry in the potential fuel savings realizable with the establishment of laminar flow over part or all of the upper and lower wing surfaces.

In three-dimensional boundary layers, such as those over swept wings, experimental measurements have revealed a more complex dependence of transition on environmental conditions than the one displayed in two-dimensional boundary layes. A longer transition region has been observed in which disturbances approach a quasi-saturated state. The saturation levels have been correlated with the free-stream turbulence, but not with free stream sound level [6], while the seeding of the steady disturbances by very small surface roughness has been confirmed in a number of experiments [4],[7].

These phenomena render difficult the application of the current standard transition-prediction technique (e.g. e^N method) because the technique relies solely on the amplification rates of eigenmodes derived from the analysis of infinitesimally small disturbances. We employ a new, "extended," local procedure to model the creation of disturbances by distributed or localized surface roughness, and the parabolized stability equations [3] (PSE) to study the subsequent nonlinear development. Currently, a traveling eigenmode of pre-selected amplitude is specified at the initial marching location to simulate the effect of free-stream turbulence. The effect of distributed roughness on a Görtler vortex is presented in [1], but is not included herein due to lack of space. Instead, we focus entirely on the transition over a swept smooth plate with localized roughness.

2 The experiment

An experiment has been performed by Bippes and colleagues at DLR [4] to elucidate the basic properties of transition in 3D boundary-layers, and to aid the development of analytical models. Compressibility and surface curvature have been omitted in order to reduce the number of independent parameters, and simplify the analysis.

The measurements are made over a flat plate with a chord of $c = 500$ mm and 45 degrees of sweep. A wing-like body with circulation is placed above the plate to create a uniformly accelerated flow over the plate. Figure 1a displays a perspective view of the geometry, and figure 1b displays the sweep angle of the leading edge, as well as the coordinate system used in our mathematical work. This system has x normal to the leading edge, z parallel to the leading edge, and y normal to the plate.

Two experimental campaigns were performed, one to observe natural transition over a smooth surface (polished to 1.8 micrometers), the other to observed the transition in the presence of cross-flow vortices generated by an array of cylinders of 3.3 mm diameter and 10 ± 1 micron height, placed parallel to the leading edge at 7% chord and spaced 12 mm apart, matching the spanwise wavelength of the most amplified steady mode.

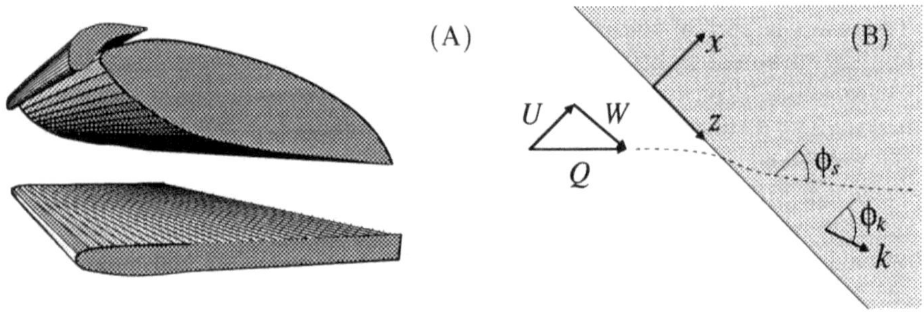

Figure 1: (A) Perspective view of geometry. (B) Leading edge sweep, and coordinate system employed. The dashed line is (roughly) the path of the edge streamline. **k** denotes the disturbance's wavevector.

3 Governing equations

The reference velocity and length used to nondimensionalize all quantities are,

$$Q_\infty = 20.5 \quad (\text{m/s}) \qquad , \qquad \delta_r = \sqrt{\nu c / Q_\infty} = 0.5814 \quad (\text{mm}). \qquad (1)$$

The nondimensional frequency F is defined as $F = 2\pi f \nu / Q_\infty^2 \times 10^6$, where f is the frequency in hertz. The Reynolds number is $R_o = 851.3$. We denote the velocities along x, y and z as u, v and w, and the velocities in a coordinate system rotated along the local free-stream direction as u_s and w_s.

The laminar flow, denoted $\mathbf{Q}_B$, is calculated using the boundary-layer equations and the experimentally measured edge velocity, U_e. The solution is discretized in x using first-order finite differences, and approximated in y with a multidomain Chebyshev-polynomial expansion.

The disturbance field is denoted by $\mathbf{q} = \{\mathbf{v}, p\} = \{u, v, w, p\}$, and satisfies,

$$\nabla \cdot \mathbf{v} = 0 \tag{2}$$

$$\partial \mathbf{v}/\partial t + (\mathbf{v} \cdot \nabla)\mathbf{V}_B + (\mathbf{V}_B \cdot \nabla)\mathbf{v} + \nabla p - 1/R_o \nabla^2 \mathbf{v} = -(\mathbf{v} \cdot \nabla)\mathbf{v} \tag{3}$$

Homogeneous Dirichlet boundary conditions far from the plate are satisfied by u, v, w and p of the time-varying disturbances as well as by u, w and p of the steady modes. The v velocity component of the steady modes satisfies homogeneous Neumann boundary conditions to allow changes in the boundary layer's displacement thickness. In the absence of distributed roughness, all velocity components satisfy homogeneous Dirichlet conditions at the wall.

The PSE equations [3] are a set of equation that model the evolution of infinitesimal, as well as finite amplitude, disturbances in flows with non-parallel streamlines. The PSE are related to the WKB method used, for example, in acoustic wave-propagation studies, although the PSE formulation allows non-linear solutions, and can be extended to handle wave transmission and reflection [2]. The disturbance field is expanded in a Fourier series in time and z using fixed wavenumbers ω_0 and β_0, respectively, and in x using a variable streamwise wavenumber $\alpha_0(x)$,

$$\mathbf{q}(x, y, z, t) = \sum_{l=-L}^{L} \sum_{n=-N}^{N} \sum_{k=-K}^{K} \mathbf{q}_p(x, y)\chi_p(x, z, t) \tag{4}$$

$$\chi_p(x, z, t) = \exp[\Theta_p(x) + ik\beta z - il\omega t] \tag{5}$$

where the growth rate, $\gamma(x)$, and streamwise wave-number, $\alpha \approx n\alpha_0$, of mode $\mathbf{p} \equiv (l, n, k)$ is given by the real and imaginary part of $d\Theta_p/dx$, respectively.

Introducing these expansions into the governing equations (2,3) and collecting terms multiplied by the same exponential function χ_p yields a coupled set of partial differential equations. Neglecting the second order streamwise derivative $\partial^2/\partial x^2$ of the functions $\mathbf{q}_p$ produces a parabolic system of equations,

$$\mathbf{L}\,[\mathbf{q_p}] \;+\; \mathbf{M}\left[\frac{\partial \mathbf{q_p}}{\partial x}\right] \;+\; \frac{d^2\Theta_\mathbf{p}}{dx^2}\mathbf{N}\,[\hat{q}_\mathbf{p}] \;=\; \sum_{\mathbf{r}} \mathbf{Q}\,[\mathbf{q_r}\,,\,\mathbf{q_{p-r}}]\,, \tag{6}$$

where $\mathbf{L}, \mathbf{M}, \mathbf{N}, \mathbf{Q}$ are operators containing derivatives in y only. Lastly, adding an auxiliary condition for each mode $\mathbf{p}$ defines a unique value of the function $\Theta_\mathbf{p}$, and renders the problem solvable,

$$\int_0^\infty \hat{q}_\mathbf{p}^\dagger \cdot \frac{\partial \hat{q}_\mathbf{p}}{\partial x}\, dy \;= 0\,. \tag{7}$$

4 The localized receptivity problem

If receptivity is neglected, one may (usually) obtain the initial conditions for the PSE equations from either the eigenfunction of the Orr-Sommerfeld equation, or the eigenfunction of the local procedure [3] obtained by extending the Orr-Sommerfeld equation to incorporate the local divergence of the mean-flow.

Herein we present an expanded "local procedure" for modeling the generation of cross-flow vortices by either a wall undulation of constant maximum height in the streamwise direction (i.e. ribblets), or a localized disturbance (i.e a hump). The procedure rests on the expansion of the geometry and the velocity field in Chebyshev polynomials $T_n(x)$, yielding a coupled set of ordinary differential equations. The first two of these equations, namely at order T_0 and T_1, are similar to those of the standard local procedure mentioned above, while equations at higher order provide additional accuracy. A converged solution satisfies the full disturbance equations in the neighborhood $\Omega = [x_0 - \Delta, x_0 + \Delta]$. We map $x \in \Omega$ into $\xi \in [-1, 1]$ via the linear mapping $\xi = (x - x_0)/\Delta$.

The boundary conditions at the wall incorporate the information of the wall geometry. In the current work, the surface of the plate is steady, and allowed to vary from the plane $y = 0$ by a small amount, $h(x, z) << 1$. Typical peak values for h are between 0.005 and 0.02 (e.g. 3 to 12 micrometers) and, thus, we transfer the no-slip boundary condition on the surface to the plane $y = 0$ via a Taylor series in $h(x, z)$, keeping only the first two terms,

$$u(x,0,z) = -h\,U_{B,y} \quad , \quad v(x,0,z) = 0 \quad , \quad w(x,0,z) = -h\,W_{B,y}\,. \tag{13}$$

The surface geometry is represented by

$$h(x,z) = \mathcal{F}(\xi(x)) \sum_{n=0}^{N} \mathcal{W}_n e^{in(\alpha_0 x + \alpha_1 x^2/2 + \beta z)} + c.c. \tag{14}$$

where $\alpha_0 x + \alpha_1 x^2/2$ matches $\alpha(x)$ of the eigenmode in Ω. The constants $\mathcal{W}_n$ and the function $\mathcal{F}$ prescribe the surface shape in the direction normal and parallel to $\alpha x + \alpha_1 x^2/2 + \beta z = const$, respectively. We consider three cases,

surface A)	$\mathcal{F}(\xi) = 1$	$\Delta = 40$
surface B)	$\mathcal{F}(\xi) = \tanh(2\xi + 2)/\tanh(4)$	$\Delta = 20$
surface C)	$\mathcal{F}(\xi) = (\exp(-3\xi^2) - \exp(-3))/(1 - \exp(-3))$	$\Delta = 10$

The function $\mathcal{F}$ for the three different cases is plotted in 2-b, and a perspective view of the hump geometry for surface C is shown in 2-a.

If the extent of the domain Ω is not too large, the variation of the mean-flow, $\mathbf{Q}_B$, and of the streamwise wavenumber $\alpha(x)$ is small, and we can expand these two quantities using only the first two Chebyshev polynomials. On the other hand, the surface variation $h(x, z)$ can be rapid, and, consequently, so can the disturbance flow field $\mathbf{q}$; thus, we approximate these two quantities with N Chebyshev polynomials. We obtain,

$$\mathbf{Q}_B(\xi, y) = \mathbf{Q}_o(y) + T_1(\xi)\mathbf{Q}_1(y) \tag{15}$$

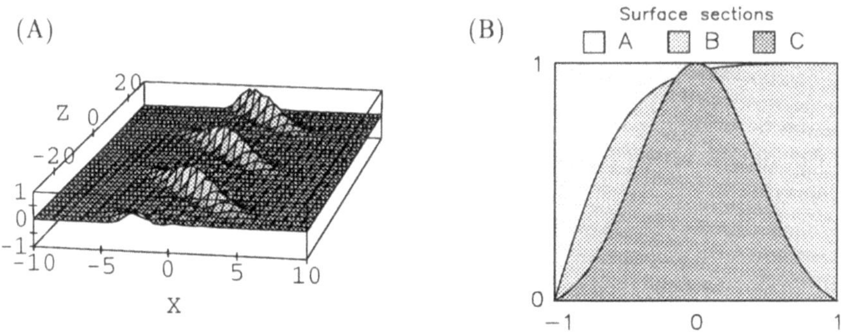

Figure 2: (A) Perspective view of surface hump. (B) Variation of surface height along the line $\alpha_0 x + \alpha_1 x^2/2 + \beta z = 0$, for surfaces A, B, and C.

$$\chi(\xi, z) = \exp[T_1(\xi)(i\alpha_0\Delta) + T_2(\xi)(i\alpha_1\Delta^2/4) + i\beta z] \tag{16}$$

$$h(\xi, z) = \sum_0^N \mathcal{W}_n T_n(\xi)\,\chi(\xi, z) + \text{compl. conj.} \tag{17}$$

$$\mathbf{q}(\xi, y, z) = \sum_0^N \mathbf{q}_n(y) T_n(\xi)\,\chi(\xi, z) + \text{compl. conj.} \tag{18}$$

Inserting these expansions into the linearized disturbance equations, (2,3) and boundary conditions (13), and collecting terms multiplied by the same Chebyshev polynomial yields N coupled ordinary differential equations in y. The continuity, and the u, v, w momentum equations are (summation on k):

$$\begin{aligned}
i\alpha_0 u_n + v_n' + i\beta w_n + i\alpha_1\Delta[P1]_n^k u_k + [D1]_n^k u_k &= 0,\\
(\mathcal{A} - i\omega)u_n + \mathcal{B}_n^k u_k + V_0 u_n' + U_0' v_n + 1/\Delta U_1 u_n + V_1[P1]_n^k u_k' + U_1'[P1]_n^k v_k &+\\
i\alpha_0 p_n + (i\alpha_1\Delta[P1]_n^k + [D1]_n^k)p_k - 1/R_0(u_n'' + c_1 u_n + \mathcal{C}_n^k u_k) &= 0,\\
(\mathcal{A} - i\omega)v_n + \mathcal{B}_n^k v_k + V_0 v_n' + V_0' v_n + 1/\Delta V_1 u_n + V_1[P1]_n^k v_k' &+\\
V_1'[P1]_n^k v_k + p_n' - 1/R_0(v_n'' + c_1 v_n + \mathcal{C}_n^k v_k) &= 0,\\
(\mathcal{A} - i\omega)v_n + \mathcal{B}_n^k w_k + V_0 w_n' + W_0' w_n + 1/\Delta W_1 u_n + V_1[P1]_n^k w_k' &+\\
W_1'[P1]_n^k v_k + i\beta p_n - 1/R_0(w_n'' + c_1 w_n + \mathcal{C}_n^k w_k) &= 0
\end{aligned} \tag{19a}$$

where,

$$\begin{aligned}
\mathcal{A} &= i\alpha_0 U_0 + i\alpha_1\Delta U_1/2 + i\beta W_0\\
\mathcal{B}_n^k &= (i\alpha_0 U_1 + i\alpha_1\Delta U_0 + i\beta W_1)[P1]_n^k + U_0[D1]_n^k + U_1[Q1]_n^k + i\alpha_1\Delta U_1[P2]_n^k\\
\mathcal{C}_n^k &= c_2[P1]_n^k + c_3[P2]_n^k + c_4[D1]_n^k + c_5[Q1]_n^k + c_6[D2]_n^k
\end{aligned}$$

and the matrices $[P1]_n^k$, etc, are defined as,

$$T_1 T_n = [P1]_n^k T_k \qquad dT_n/d\xi = [D1]_n^k T_k \qquad T_1\, dT_n/d\xi = [Q1]_n^k T_k$$

$$T_2 T_n = [P2]_n^k T_k \qquad d^2T_n/d\xi^2 = [D2]_n^k T_k \qquad T_2\, dT_n/d\xi = [Q2]_n^k T_k$$

The boundary conditions at the wall (13) become,

$$u_n = -U_0'\mathcal{W}_n - U_1'[P1]_n^k\mathcal{W}_k \quad , \quad v_n = 0 \quad , \quad w_n = -W_0'\mathcal{W}_n - W_1'[P1]_n^k\mathcal{W}_k \quad (19b)$$

and the boundary conditions at infinity are $u_n = 0$, $v_n' = 0$; $w_n = 0$ and $p_n = 0$.

Optionally, initial conditions can be specified. For the case of a localized hump lying inside Ω, we require vanishing u and w velocities at the upstream-end of the domain Ω. (Note that since u_x is non-zero at $\xi = -1$, the v component is also non-zero). Thus, we drop the u and w momentum equations at order N, and impose, instead, zero disturbance at the inflow,

$$\sum_0^N u_n(y)\,(-1)^n = 0 \quad , \quad \sum_0^N w_n(y)\,(-1)^n = 0 \qquad (20)$$

For the case of ribblets that extend beyond the region Ω (e.g. surface A), the velocity components are finite at the inflow plane of Ω, and initial conditions are not imposed. The freedom to impose or neglect initial conditions is a characteristic that distinguishes the local procedure from the standard Galerkin method.

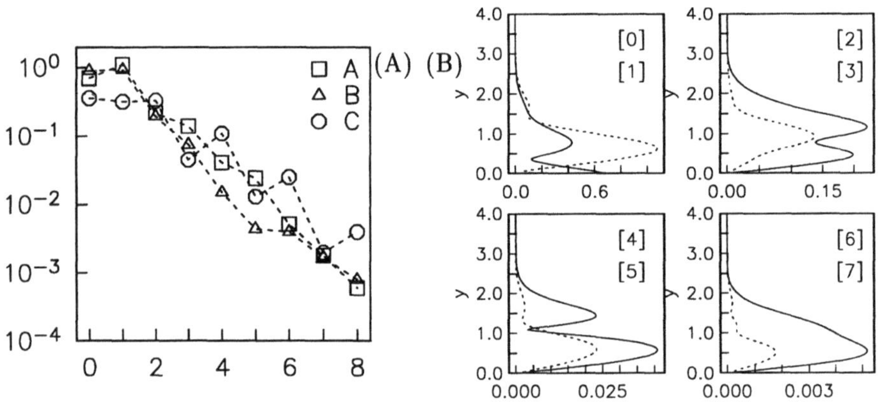

Figure 3: (A) Convergence rate: $|(u_n)_s|_{max}$ vs Chebyshev polynomial n, for the three surface types. (B) Functions $|(u_n)_s|$ for even (solid line) and odd (dashed) Chebyshev polynomials n, surface A.

We solve equation (19) using nine Chebyshev polynomials in x, and seventy Chebyshev polynomials in y, equally distributed in four subdomains spanning $y = 0$ to $y = 10$. The entire problem is solved in under four minutes on a Sun Spark-10 workstation. The point $\xi = 0$ is mapped to 7% chord and the values of $\alpha_0 = -0.608$, $\alpha_1 = 0.002$ are selected to match those of the eigenmode with zero frequency and $\beta = 0.3$ (the fastest growing steady eigenmode).

The convergence rate of expansion (18) is shown in Figure 3-a, for the three surface types. $(u_n)_s$ denotes the velocity component of $T_n(\xi)$ in the coordinate system locally aligned with the boundary-layer edge streamline. All three solutions converge to sufficient accuracy. This fact suggests strongly that the solution satisfies, up to a small error, the linearized Navier-Stokes equations.

The profiles $(u_n)_s$ are shown in 3-b, for the case of surface A. Mode 0 is similar to the solution obtained with the parallel-flow approximation. Mode 1 has an amplitude of equal order of magnitude, and, more importantly, is closer to the shape of the eigenmode. Modes 3 and 5 also have this property, but are of lower amplitude. These results suggest that an extended local procedure using at least 5 Chebyshev polynomials should be used to resolve the disturbance field generated by both distributed and localized surface roughness. We note in passing that the local procedure can accommodate a time-dependent surface roughness, hence it can model the harmonic point-source used in the generation of disturbances in experiments of wave-packets.

4.1 Receptivity results

According to Bippes, the experimental measurements downstream of the surface roughness display the presence of nonlinear effects at the location where the steady disturbances become of measurable size (i.e. 1 to 2 %). Consequently, the numerical results must model both the nonlinear and the receptivity stages in order to match quantitatively the results with experiment. The experimentally measured peak amplitudes of the u_s component of the steady velocity field, with the spanwise mean subtracted, are shown as symbols in figure 4-a.

The initial conditions for the steady modes were obtained for surface C, with a maximum height of 0.009, which translates to 5 microns. The curve labeled (1) contains no additional modes in the I.C. (i.e. a purely steady calculation). Curve (2) has an eigenmode of frequency $F = 40$, $\beta = 0.3$, and amplitude 0.0035% r.m.s. added to the I.C., while curve (3) has the eigenmode amplitude raised to 0.007% r.m.s. From Crouch's model [5], we can infer that the amplitude of the eigenmode will be of order $\mathcal{Q}(\epsilon\delta)$, where ϵ is the amplitude of the a free-stream turbulence, and δ is the height of the roughness. Thus, curve (2) corresponds to an $\epsilon = 0.2\%$, and curve (b) to $\epsilon = 0.4\%$. The experimental value for the turbulence level is 0.15%. These calculations employed a total of 325 modes, which is sufficient to display the lowering of the steady-mode's saturation amplitude with increased amplitude of the traveling modes, but is probably insufficient to resolve accurately the flow. The damping effect of traveling modes on the amplitude of the steady modes agrees with results from direct numerical simulations [8]. Figure 4-b displays the amplitude, based on the integral of u for the cross-flow vortex mode with $\beta = 0.3$, i.e. equal to the roughness spacing (solid line), and of the traveling mode at $F = 40$ (dashed line), for the three runs made. The saturation amplitude for the traveling mode is almost equal in runs 2 and 3, but the vortex's ampliude saturates at different levels. This indicates that the stabilizing influence of the traveling mode comes through a distortion of the mean-flow.

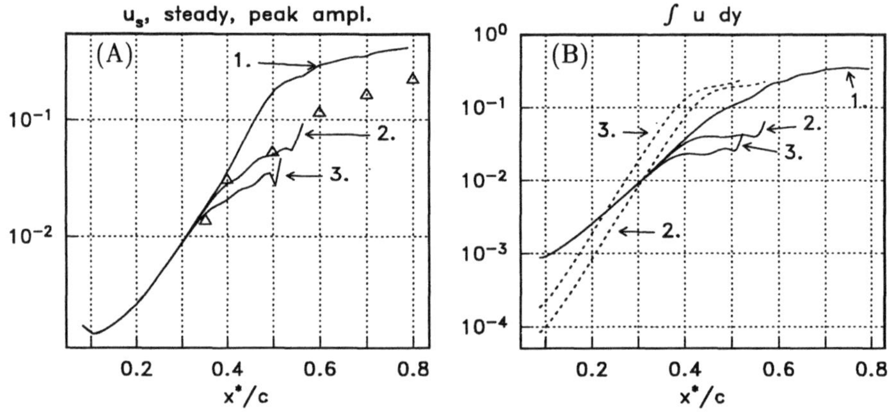

Figure 4: (A) Experimental (symbol) and computed maximum values of u_s. (B) Amplitudes of steady mode (solid) and traveling mode (dashed) for runs 1. 2. and 3.

References

[1] BERTOLOTTI, F. P.. Vortex generation and wave-vortex interaction over a concave plate with roughness and suction. ICASE Report 93-101, Nasa Langley, Hampton Va., 1993. Submitted to *Theo. Comp. Fluid Dyn.*

[2] BERTOLOTTI, F. P., An introduction to the parabolized stability equations, in *Transition: Experiment, Theory, and Computation*, (eds. T. C. Corke, G. Erlebacher, M. Y. Hussaini). Oxford University Press, 1994.

[3] BERTOLOTTI, F. P., HERBERT, TH., SPALART, P. R.. Linear and nonlinear stability of the blasius boundary layer. *J. Fluid Mech.*, 242: 441–474, 1992.

[4] BIPPES, H.. Experiments on transition in three-dimensional accelerated boundary-layer flows. In *Proc. R.A.S. Conf. "Boundary Layer Transition and Control"*, Cambride, England, 1991.

[5] CROUCH, J. D.. Distributed excitation of tollmien-schlicting waves by vortical free-stream disturbances. *Phys. Fluids*, 6 (1): 217–223, 1994.

[6] DEYHLE, H.. Einfluss der äusseren Strömungsbedingungen auf den Transitionsprozess einer dreidimensionalen Grenzschicht. Fortschrittberichte Reihe 7: Strömungstechnik, Nr 226, VDI, 1993.

[7] RADEZTSKY JR., R. H., REIBERT, M. S. ANDSARIC, W. S., TAGAKI, S.. Effect of micron-sized roughness on transition in swept-wing flows. AIAA Paper 93-0076, 1993.

[8] WAGNER, M., KLEISER, L.. Nonlinear disturbance development in three-dimensional boundary layer transition. In *Proceedings of 1st Euro. Fluid Mech. Conf.*, Cambridge, UK, 1991.

Distributed Excitation of Cross-Flow Vortices in Three-Dimensional Boundary Layers

J.D. Crouch
Boeing Commercial Airplane Group, Seattle, WA, U.S.A.

Summary

The receptivity of swept Hiemenz flow to stationary cross-flow vortices is analyzed using a generalization of the classical stability theory. The vortices are excited by surface perturbations, either localized or nonlocalized in their chordwise extent. The analysis provides the receptivity amplitude as a function of the perturbation characteristics. Results are presented for roughness and nonuniform-suction perturbations. Nonlocalized receptivity occurs for a broad band of surface wave numbers. The results for an array of surface perturbations show that the receptivity is 3 to 7 times larger than for a single spanwise row. The results are in good agreement with earlier results from direct numerical simulations.

1. Introduction

In quiet flow conditions, the transition to turbulence in boundary layers is linked to the growth and nonlinear interaction of instabilities. The details of the nonlinear development depend on the initial amplitudes of these instabilities. Receptivity theory provides the link between 'environmental inputs' (e.g. sound, free-stream turbulence, surface vibration, surface roughness) and the resulting instability amplitudes. A range of mechanisms for exciting the instabilities have been identified; they can be grouped into 'leading-edge' receptivity, localized receptivity, and nonlocalized receptivity.

Nonlocalized receptivity involves a distributed excitation of boundary-layer instabilities due to a near-resonant forcing in the neighborhood of branch I. Earlier analyses have considered the nonlocalized receptivity to Tollmien-Schlichting (TS) waves in a flat-plate boundary layer [1, 2, 3]. A recent experiment of Wiegel & Wlezien [4] shows good agreement with earlier theoretical results of Crouch [2] and Choudhari [3]. Calculations with the parabolized stability equations [5] and with direct numerical simulations [6] also show good agreement with the theory. Results show large receptivity amplitudes over a modest band of surface wave numbers. Meanwhile, the theory has been extended to consider pressure gradients [7] and surface disturbances with wavelengths typical of roughness experiments [8]. When operative, nonlocalized receptivity can dominate other mechanisms for generating TS waves in boundary layers.

Most of the work on receptivity to cross-flow vortices has focused on localized mechanisms. Fedorov [9] and Manuilovich [10] considered the excitation of vortices due to localized surface variations. Localized roughness with a height of $\mathcal{O}(.01\delta^*)$ was shown to produce a disturbance of amplitude $\mathcal{O}(.1\%)$. Crouch [11] considered the relative strength of steady and traveling cross-flow vortices in the presence of acoustic free-stream disturbances. For acoustic amplitudes typical of TS receptivity experiments, the steady vortices have initial amplitudes approximately 100 times greater than the unsteady vortices. Even though the unsteady vortices have higher growth rates, they are unlikely to be significant unless the free-stream disturbance levels are very high. This is consistent with the experiments of Müller & Bippes [12] and Takagi *et al.* [13]. The analysis of Choudhari [14] also shows a significant bias towards stationary modes in low-disturbance environments. Thus, for aerodynamic applications, steady vortices will be the most critical form of cross-flow instability. Choudhari [14] has also briefly considered the nonlocalized receptivity to cross-flow vortices due to roughness; this study suggests that a distributed array of roughness elements can produce stationary vortices approximately three times larger than a single row of roughness elements. Spalart [15] considered both localized and nonlocalized receptivity to steady cross-flow vortices due to suction using DNS. Nonlocalized receptivity resulted in vortex amplitudes approximately five times larger than those resulting from localized receptivity.

In this paper, we analyze the nonlocalized receptivity to steady cross-flow vortices resulting from both roughness and suction. The problem formulation is presented in § 2. Sections 3 and 4 present results for localized and nonlocalized receptivity, respectively. These results are applied to specific surface variations for distributed suction in § 5. Conclusions are presented in § 6.

2. Formulation

We consider the nonlocalized receptivity to steady cross-flow vortices. The vortices are excited by distributed variations in surface roughness or suction (as for laminar-flow-control applications). The basic state is given by the swept Hiemenz flow with chordwise edge velocity $U_e(x) = Sx$, and constant spanwise edge velocity $W_e = W_\infty$. The variables x, y, z represent the chordwise, surface normal, and spanwise directions, respectively, in a right-hand reference frame. Flow quantities are nondimensionalized using W_∞ and $\delta_l = [\nu/S]^{1/2}$; this leads to the global Reynolds number $\bar{R} = W_\infty \delta_l/\nu$. Results are presented for $\bar{R} = 400$ – the value considered in the DNS of Spalart [15]. Nonparallel effects on the receptivity should be small for $\bar{R} = 400$. Earlier studies for swept Hiemenz flow have shown that nonparallel effects on the stability are relatively small for $\bar{R} = 500$, although they are more significant for $\bar{R} = 250$ [16, 17]. Nonparallel effects on the receptivity amplitude are usually smaller than the effects on the instability growth rate.

Since the basic flow is independent of z, we consider the spanwise wave number β to be real and independent of x for a given mode. The receptivity is analyzed under the quasi-parallel-flow assumption. Disturbances are locally governed by a nonhomogeneous Orr-Sommerfeld problem. The local solutions are used to construct an O.D.E. for the chordwise variation of the instability amplitude. The surface variations are represented by the boundary conditions

$$u = v = w = 0 \quad at \quad y = hH(x, z) \tag{1}$$

for roughness, or

$$u = w = 0 \ , \ v = -v_s H(x,z) \ \ at \ \ y = 0 \qquad (2)$$

for suction. The quantities h and v_s characterize the roughness height and suction amplitude, respectively. We represent the surface variation $H(x,z)$ by the convolution of a localized perturbation shape $G(x,z)$ and a function describing the distribution $F(x,z)$ of the localized perturbations

$$H(x,z) = \int_{-\infty}^{\infty} \int_{-\infty}^{\infty} G(x - \hat{x}, z - \hat{z}) \ F(\hat{x}, \hat{z}) \ d\hat{x} d\hat{z} \ . \qquad (3)$$

Two distributions of surface variation are considered in the results: a single row at x^*, and a distributed array over the entire surface. This leads to the Fourier transform $\tilde{H}_\beta(\alpha) = \tilde{\tilde{G}}(\alpha,\beta)/\lambda_z$ for localized receptivity to a single row of surface perturbations. The Fourier coefficient for nonlocalized receptivity is given by $H_{\alpha\beta} = \tilde{\tilde{G}}(\alpha,\beta)/(\lambda_x\lambda_z)$. The two-dimensional Fourier transform of $G(x,z)$ is obtained from

$$\tilde{\tilde{G}}(\alpha,\beta) = \int_{-\infty}^{\infty} \int_{-\infty}^{\infty} G(x,z) \ e^{-i\alpha x} \ e^{-i\beta z} dx dz \ . \qquad (4)$$

The velocity field resulting from the small-amplitude surface variation can be represented by

$$\mathbf{v}(x,y,z) = \mathbf{v}_0(y;x) \ + \ \epsilon \mathbf{v}_\epsilon(x,y,z) \ . \qquad (5)$$

The component $\mathbf{v}_0$ is the swept Hiemenz flow in the absence of the surface variation, and the component $\mathbf{v}_\epsilon$ is the modification to the basic flow resulting from the surface variation. The small parameter ϵ is defined as $\epsilon = h \ll 1$ for roughness and $\epsilon = v_s \ll 1$ for suction. The receptivity is given by the magnitudes of the eigenmodes at $\mathcal{O}(\epsilon)$ as a function of the surface perturbation.

3. Localized Receptivity

For comparison purposes, we first consider the receptivity resulting from a surface variation which is localized in the chordwise direction and centered at x^*. Downstream of the perturbation, the cross-flow instability amplitude (for a particular spanwise wave number β) is given by the residue associated with the corresponding eigenmode with chordwise wave number $\alpha_e = \alpha_{er} + i\alpha_{ei}$. The amplitude of the eigenmode at the location of the surface perturbation is

$$A_\beta(x^*) = \epsilon \ |K_\beta \ \tilde{H}_\beta(\alpha_e)| \ , \qquad (6)$$

where A_β is defined as the maximum (over y) of the chordwise velocity component (see [11] for more details). K_β is the complex residue from the forced response at α_e. The geometry factor $\tilde{H}_\beta$ is the Fourier transform of the surface variation evaluated at α_e – discussed in more detail in §5.

The response residue K_β provides a measure of the localized receptivity independent of the details of the surface geometry. Figure 1 shows the chordwise variation of $|K_\beta|$ for different spanwise wave numbers. Branch I locations are

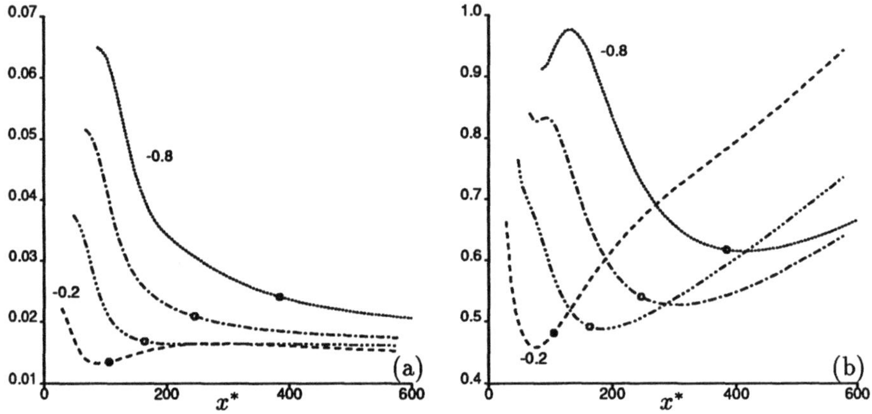

Figure 1: Chordwise variation of the response residue $|K_\beta|$ for $\beta = -0.2, -0.4, -0.6, -0.8$; (a) surface roughness, (b) surface suction.

shown by the symbols (o). The results for roughness show a general decrease in $|K_\beta|$ with increasing chordwise position, upstream of branch I. The residue magnitude shows little variation downstream of branch I. The results for suction follow a similar trend upstream of branch I, but show an increase downstream of branch I. The overall magnitudes for the suction case (with $\epsilon = v_s$) are significantly larger than for the roughness (with $\epsilon = h$).

4. Nonlocalized Receptivity

Nonlocalized receptivity results from a distributed energy transfer to the eigenmode due to a near-resonant forcing. The resonance can be exploited to provide a nonhomogeneous amplitude equation for the spatial growth of the eigenmode. This is the approach used by Crouch [2] to study the nonlocalized receptivity of TS waves. An alternative approach, based on a Green's function, is to treat the distributed forcing as a collection of local sources; the nonlocalized receptivity is then given by an integral over x of the local source contributions. This approach was used to study the nonlocalized receptivity to TS waves in a Blasius boundary layer [3] and in Falkner-Skan flow [7]. Results from the Green's function approach are in good agreement with the resonance approach [4, 7]. When focusing only on the downstream amplitude of the eigenmode (i.e. the total receptivity magnitude), the Green's function approach is relatively more efficient.

Following this approach, the receptivity resulting from a surface variation characterized by the Fourier mode $\exp(i\alpha x + i\beta z)$ is given by

$$A(x)e^{i\psi(x)} = \exp\Big[i\alpha x_0 + i\int_{x_0}^{x} \alpha_{er}(s)ds\Big] \times$$

$$\int_{x_0}^{x} iK_\beta(\hat{x}) \exp\Big[i\alpha(\hat{x} - x_0) - i\int_{x_0}^{\hat{x}} \alpha_{er}(s)ds + \int_{\hat{x}}^{x} \alpha_{ei}(s)ds\Big] d\hat{x} \ , \quad (7)$$

where x_0 is a reference location upstream of the receptivity region [2], and K_β is the response residue discussed above. After integrating eq.(7) over the region of

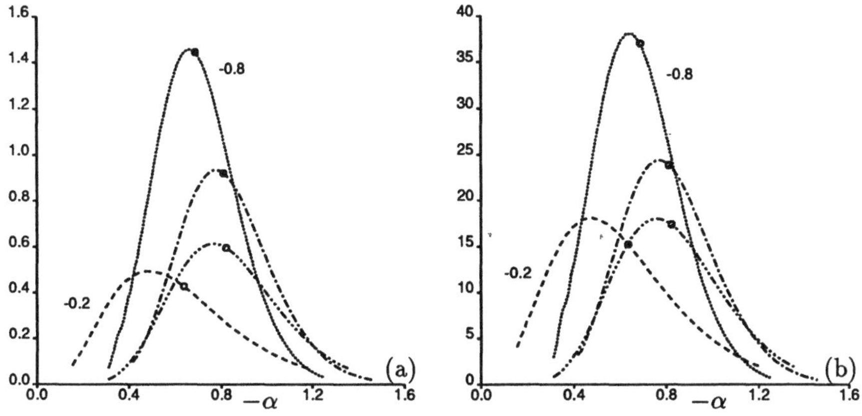

Figure 2: Variation of the receptivity amplitude $A_{I\beta}$ with chordwise wave number α for $\beta = -0.2, -0.4, -0.6, -0.8$; (a) surface roughness, (b) surface suction.

receptivity, the downstream amplitude can be projected back to branch I using the integrated linear growth. This provides an effective branch I amplitude, $A_{I\beta}(\alpha)$, that would produce an equivalent disturbance downstream. The recent roughness work of Choudhari [14] follows a similar approach, but focuses on an approximate solution for $A_{I\beta}(\alpha)$, based on the method of steepest descent.

For a given surface perturbation, the effective instability amplitude at branch I for each spanwise mode β is given by

$$\text{eff.}A_\beta(x_I) = \epsilon\ A_{I\beta}(\alpha)\ H_{\alpha\beta} \quad . \tag{8}$$

$H_{\alpha\beta}$ is the discrete Fourier coefficient of the surface variation at the wave numbers α, β. A measure of the nonlocalized receptivity at these wave numbers is given by the receptivity amplitude $A_{I\beta}(\alpha)$. Figure 2 shows the variation of $A_{I\beta}$ with α for different values of β. The symbols (o) denote the eigenmode wave number at branch I. The results for distributed roughness and suction are qualitatively similar. The larger magnitudes for the suction could be anticipated from the localized results of figure 1. In general, the peak receptivity amplitude increases with spanwise wave number $|\beta|$. Along with the increase in peak receptivity amplitude, there is a reduction in the α bandwidth over which significant receptivity occurs. For $\beta = -0.2$, the half bandwidth of the $A_{I\beta}$ curves is approximately 125%; this is reduced to approximately 65% for $\beta = -0.8$. The receptivity to cross-flow vortices is relatively broad banded compared to TS waves in Falkner-Skan boundary layers where the half bandwidths are approximately 10 to 35% [7].

The variation in the magnitude and bandwidth of the receptivity amplitude stems from the differences in detuning for the different spanwise modes. Stability results show that the rate of variation of α_{er}, at branch I, decreases with increasing $|\beta|$. Thus, over the window in which the detuning due to α_{ei} is small, α_{er} varies more rapidly for $\beta = -0.2$ compared to $\beta = -0.8$. This results in a larger band of α wave numbers which can lead to significant receptivity. However, the increased variation of α_e also means that the window over which α_e is tuned to α becomes smaller; this results in a reduced receptivity amplitude.

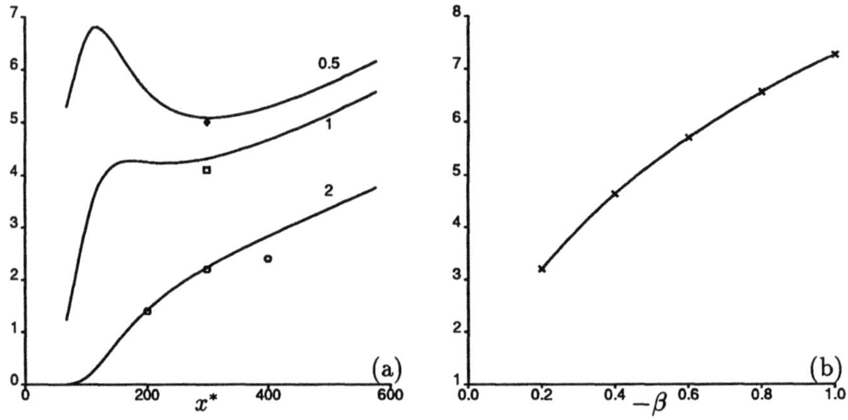

Figure 3: (a) Variation of the receptivity amplitude $A_\beta(x^*)/\epsilon$ with the source location x^* for $\sigma = 0.5, 1, 2$. Symbols show the DNS results of Spalart [15]. (b) Ratio of receptivity amplitudes eff.$A_\beta(x_I)/A_\beta(x^* = x_I)$ as a function of spanwise wave number.

5. Distributed Suction

We now consider the application of the receptivity results of sections 3 and 4 to specific surface variations. This permits a detailed quantitative comparison between the localized and nonlocalized results. We focus on surface perturbations created by a distribution of suction holes (i.e. as used for laminar-flow-control applications). The local perturbation shape is approximated by a Gaussian function

$$G(x,z) = \frac{\lambda_x \lambda_z}{\pi \sigma^2} \exp\left[-\left(\frac{x^2+z^2}{\sigma^2}\right)\right] . \tag{9}$$

Using this description, the nondimensional suction velocity v_s represents the average suction per unit area and λ_x, λ_z are the spacing between holes. The size of the holes is represented by σ. v_s is the quantity which characterizes the level of stabilization for laminar-flow control. Holding v_s constant and increasing λ_x or λ_z, or decreasing σ, results in an increase in the peak flow velocity through the holes.

The localized receptivity resulting from a single row of holes is given by eq.(6) with $\epsilon = v_s$, K_β as given in figure 1b, and $\tilde{H}_\beta(\alpha) = \lambda_x \exp[-\sigma^2(\alpha^2+\beta^2)/4]$, (with $\lambda_x = \lambda_z$). Figure 3a shows the variation of the receptivity amplitude $A_\beta(x^*)/\epsilon$ with x^* for different values of σ. The spanwise spacing is fixed at $\lambda_z = 10$, which gives $\beta = -0.6283$; this is the spanwise wave number considered by Spalart [15]. The symbols show Spalart's DNS results. For $\sigma = 2$, the receptivity increases continuously with x^*. At smaller values of σ, the receptivity shows a local maximum between $x^* = 100$ and $x^* = 200$ followed by an increase with x^*. The results show very good agreement with the DNS for $x^* < 400$. We attribute the differences for $x^* = 400$ to local transients which persist downstream and affect the DNS results; the DNS results were obtained by measuring the amplitude downstream and then projecting the value upstream using the linear growth curve. The variation of the receptivity with σ, at $x^* = 300$, shows very good agreement between the theory and simulations.

The total receptivity resulting from the distributed excitation of cross-flow vortices is described by eq.(8). Values for the receptivity amplitude $A_{I\beta}$ are given in figure 2b, and the Fourier coefficient for an array of holes is given by $H_{\alpha\beta} = \exp[-\sigma^2(\alpha^2+\beta^2)/4]$. For a given hole pattern, the values of α, β, and $A_{I\beta}$ are fixed and the effective branch I amplitude depends only on σ. As the hole size becomes small ($\sigma/\lambda \ll 1$, where $\lambda = [\lambda_x^2+\lambda_z^2]^{1/2}$), eff.$A_\beta(x_I)$ approaches the limiting value of $\epsilon A_{I\beta}$. The effective branch I amplitude for the conditions $\lambda_x = \lambda_z = 10$ and $\sigma = 2$ is eff.$A_\beta(x_I)/\epsilon = 9.2$; DNS results of Spalart [15] give a value of 8.4. The smaller value of the DNS can be attributed to the damping effect of the suction on the amplitude growth – not included in the theory. As the hole size is reduced, the effective branch I amplitude is increased to 16.6 for $\sigma = 1$ and to 19.2 for $\sigma = 0.5$. The corresponding DNS results show an increase to approximately 13 for both values of σ. Part of this disparity appears to stem from transient and damping effects in the DNS.

Finally, figure 3b shows the ratio of the nonlocalized-receptivity amplitude from an array of perturbations to the localized-receptivity amplitude from a single row placed at branch I. The chordwise spacing of the array is chosen to match the eigenmode wavelength at branch I. This result is valid for any perturbation shape $G(x,z)$, for both suction and roughness. The ratio varies from approximately 3 to 7 over the range of spanwise wave numbers considered.

6. Conclusions

Results are presented for the excitation of stationary cross-flow vortices due to surface roughness or nonuniform suction. Earlier studies have shown the stationary vortices to be dominant over traveling vortices in low-disturbance environments. Nonlocalized receptivity from an array of surface perturbations produces disturbance amplitudes 3 to 7 times larger than a single spanwise row of the same perturbation. Results for the receptivity amplitude are in good agreement with direct numerical simulations of Spalart [15]. The nonlocalized receptivity occurs over a relatively broad band of surface wave numbers. For smaller spanwise wave numbers, the half bandwidth exceeds 100%. The results suggest that the nonlocalized receptivity will be broad banded in the attachment-line region of a swept wing due to the rapid variation of the boundary layer and the associated eigenmodes.

Acknowledgements

P.R. Spalart and L.L. Ng provided a number of discussions. The manuscript was reviewed by P.R. Spalart and I.W.M. Crouch.

References

[1] Zavol'skii, N.A., Reutov, V.P. & Ryboushkina, G.V. (1983) "Excitation of Tollmien-Schlichting waves by acoustic and vortex disturbance scattering in boundary layer on a wavy surface," *J. Appl. Mech. Tech. Phys.* **24**, 355.

[2] Crouch, J.D. (1992) "Non-localized receptivity of boundary layers," *J. Fluid Mech.*, **244**, 567-581.

[3] Choudhari, M. (1993) "Boundary-layer receptivity due to distributed surface imperfections of a deterministic or random nature," *Theoret. Comput. Fluid Dyn.* **4**, 101.

[4] Wiegel, M. and Wlezien R.W. (1993) "Acoustic receptivity of laminar boundary layers over wavy walls," AIAA Paper No. 93-3280.

[5] Crouch, J.D. and Bertolotti, F.P. (1992) "Nonlocalized receptivity of boundary layers to three-dimensional disturbances," AIAA Paper No. 92-0740.

[6] Crouch, J.D. and Spalart, P.R. (1992) "Analytical and numerical studies of boundary-layer receptivity," *Bull. Am. Phys. Soc.*, **37**, 1720.

[7] Crouch, J.D. (1994) "Influence of pressure gradient on the nonlocalized receptivity of boundary layers," in: *Nonlinear Instability of Nonparallel Flows*, IUTAM Symposium, Springer-Verlag, 106-116.

[8] Crouch, J.D. (1993) "Receptivity and the evolution of boundary-layer instabilities over short-scale waviness," *Phys. Fluids A*, **5**, 561-567.

[9] Fedorov, A.V. (1989) "Excitation of waves of instability of the secondary flow in the boundary layer on a swept wing," *J. App. Mech. Tech. Phys.*, **29**, 643-648.

[10] Manuilovich, S.V. (1990) "Disturbances of a three-dimensional boundary layer generated by surface roughness," *Fluid Dyn.*, **24**, 764-769.

[11] Crouch, J.D. (1993) "Receptivity of three-dimensional boundary layers," AIAA Paper No. 93-0074.

[12] Müller, B. and Bippes, H. (1988) "Experimental study of instability modes in a three-dimensional boundary layer," in: *AGARD Symp. on Fluid Dynamics of Three-Dimensional Turbulent Shear Flows and Transition*, AGARD-CP-438.

[13] Takagi, S., Saric, W.S., Radeztsky, R.H., Spencer, S.A. and Orr, D.J. (1991) "Effect of sound and micro-sized roughness on crossflow dominated transition," *Bull. Am. Phys. Soc.*, **36**, 2630.

[14] Choudhari, M. (1994) "Roughness-induced generation of crossflow vortices in three-dimensional boundary layers," *Theoret. Comput. Fluid Dyn.* **6**, 1-30.

[15] Spalart, P.R. (1993) "Numerical study of transition induced by suction devices," in: *Near-Wall Turbulent Flows*, Elsevier Science Pub. B.V., 849-858.

[16] Malik, M.R. and Li, F. (1993) "Transition studies for swept wing flows using PSE," AIAA Paper No. 93-0077.

[17] Herbert, Th. and Lin, N. (1993) "Studies of boundary-layer receptivity with parabolized stability equations," AIAA Paper No. 93-3053.

The Receptivity of Flat-Plate Boundary-Layers with Two-Dimensional Roughness Elements to Freestream Sound and Its Control

R. Kobayashi, Y. Fukunishi, T. Nishikawa and T. Kato
Department of Machine Intelligence and Systems Engineering
Tohoku University, Sendai, Japan

Summary

The receptivity of a flat-plate boundary layer to freestream sound when one or two two-dimensional roughness elements are attached to the surface is experimentally investigated. It is shown that a thin two-dimensional roughness element together with sound can generate T-S wave and its amplitude can be controlled by adding the second roughness element and changing the distance between the two elements. This effect is explained by the superposition of the waves generated by each roughness element. It is also shown that in the case of single roughness element, the amplitude of the generated T-S wave is a function of roughness-width. It is revealed that the T-S wave generated at the leading edge of the roughness element and that generated at the trailing edge are 180 degrees out of phase and they either cancel or amplify depending on the distance between the edges.

Key words : Receptivity, Two-Dimensional Roughness, Sound, Control.

1. Introduction

Receptivity to freestream disturbances, which is the initial stage of the whole transition process, has been the subject of intensive studies[1-2]. It has been found that even a very thin roughness element can dramatically increase the receptivity of sound[3].

The objective of this research is to find out the feasibility of controlling the receptivity of a flat-plate boundary layer to freestream sound by simply attaching one or two thin two-dimensional roughness elements on the surface of the plate. The work is a part of our research project on receptivity[4], which includes the study by numerical simulations which we have already reported[5]. The present study focuses on the control of the receptivity by changing the configuration of the two-dimensional roughness element.

2. Experimental Setup

The low turbulence wind-tunnel[6] of Institute of Fluid Science, Tohoku University was used in the experiment. The working section of the wind-tunnel has an octagonal cross section whose distance is 1010mm from wall to wall. Freestream velocity $U_\infty$was 20m/s for most of the experiments. Turbulence level of the freestream at that wind velocity was $0.00016U_\infty$. The velocity was raised to 30m/s for the experiment determining the transition points in order to obtain higher Reynolds numbers. A loud speaker was located upstream of the settling chamber. The distance from the loud speaker to the measuring section was around 12m. It was checked out that the presence of the loud speaker would not increase the turbulence level at the measuring area. The wave front of the sound wave introduced by the loud speaker was perpendicular to the main flow. The intensity of sound was 90.2dB at the measuring section. A smooth flat aluminum-alloy plate, 10mm thick and 3.2m long, was placed in the working section dividing the section into halves. A 36:1 modified super ellipse[7] was used as the leading edge of the flat plate in order to minimize the sound receptivity at the leading edge.

Two-dimensional roughness element which actually was two layers of 50μm thick adhesive tape (Scotch tape) was attached on the surface at x=480mm where x=0 is the leading edge. This is the location where the lower branch of the neutral stability curve obtained from the linear stability theory crosses the frequency of the naturally developing T-S wave, 133.3Hz. The height of the roughness element was 100μm which was roughly 10% of the displacement thickness δ^*=1.03mm. The width of the roughness element was 12mm except for the experiment changing the element widths. Hot-wires were used for the measurement.

3. Results and Discussions

3.1 Receptivity of Sound with Single Two-Dimensional Roughness Element

Velocity fluctuation $\sqrt{\overline{u'^2}}/U_\infty$ along y-axis perpendicular to the wall is shown in Fig. 1. This velocity fluctuation was found to be a T-S wave with a phase speed approximately one third of the freestream velocity. It can be noticed that the combination of sound and roughness element creates a very strong velocity fluctuation inside the boundary layer. Low level of fluctuations when sound is present and roughness element is not attached is an indication of the effectiveness of the modified super ellipse leading edge[6] in reducing receptivity there.

3.2 Receptivity Control by Two Two-Dimensional Roughness Elements

In Fig. 2, two cases of growths of velocity fluctuations along the x-axis when two elements are present are compared to the case with only one element. Velocity fluctuation was measured at y=0.84mm which is close to the location where velocity fluctuation along the y-axis reaches maximum (see Fig. 1). It is shown that the T-S wave generated by two elements can be stronger or weaker compared to the single element case depending on the location of the second roughness element.

Figure 3 shows variations of the maximum amplitude of velocity fluctuations when the second roughness element is placed at various distances d from the first element. It can be noticed that the maximum amplitude of the T-S wave changes periodically with the increase of distance between the first and second roughness elements. The amplitude is found to be large when the distance is an integer multiple of the wavelength of the T-S wave λ_{TS}(=48 mm in this case). This phenomenon can be explained by the superposition of the wave generated at the upstream roughness element and that of the downstream element. When the sources of the waves are separated by a multiple of the T-S wavelength, the phases of the two waves will match and the waves will overlap and amplify each other creating a wave of twice the amplitude of the original wave. With a distance between the wave sources is 0.5, 1.5, ... times the wavelength of the T-S wave, the superposition of two waves will cause cancellation.

Velocity fluctuations along x-axis are shown in Fig. 4. The transition points determined from the shape factor are indicated in the figure. It should be noticed that the transition point can be shifted downstream by adding the second roughness element.

3.3 Receptivity Control by Width of the Single Roughness Element

Velocity fluctuation profiles along the y-axis are compared for single roughness elements of different streamwise widths D in Fig. 5.
Figure 6 gives variations of the maximum amplitude of the T-S wave depending on the roughness element D. The amplitude of the T-S wave changes periodically with the width D of the roughness element in a same manner as shown in Fig. 3 where the distance between the two roughness elements has been changed. However, unlike the case shown in Fig. 3, the amplitude is found to be smallest when the distance between the leading and trailing edges is a multiple of the wavelength of the T-S wave.

In order to understand the phenomena, the following experiment was carried out to find out the nature of the wave generated at the leading edge and at the trailing edge. The leading edge can be regarded as a tiny forward facing step while the trailing edge is a backward facing step. However, it is extremely difficult to create an isolated forward facing step or a backward one whose height is only 100μm on the surface of the flat plate in the wind tunnel. So the roughness element configurations shown in Fig. 7, which simulate the forward and backward facing steps, were designed. They are designed based on the idea that: Waves generated at edges α' and α", or β' and β", should be of the same phase with the same wavelength λ_{TS}. Therefore, superposition of these waves will result in cancellation of these waves, while the wave generated at α or β will remain.

Figure 8 shows the velocity fluctuations observed at locations of the same distance 1350 mm downstream from either edge α or β where the wave generation supposedly takes place. It can be found that the wave generated at the leading edge and that of the trailing edge are 180 degrees out of phase. So, the results of Fig. 6 can be explained in the same way as in Fig. 3, that is, the amplification or the cancellation takes place as a result of the superposition of the waves generated at two different locations. Only difference is that in Fig. 6 case, the waves are 180 degrees out of phase at its origin.

4. Conclusions

The experimental study on receptivity of a flat-plate boundary layer to freestream sound has revealed that the receptivity can be controlled by the configuration of the roughness elements. The receptivity can be enhanced or reduced by adding the second two-dimensional roughness element, or changing the width of the element. It was shown that these controls utilize the superposition of waves generated at two different locations.

Acknowledgements

This investigation is a part of the research project initiated while Profs. H. L. Reed and W. S. Saric were visiting Tohoku University from September 1991 through May 1992. The authors would like to acknowledge their encouragements and advises.

References

[1] Morkovin, M.V. : Critical Evaluation of Transition from Laminar to Turbulent Shear Layers with Emphasis on Hypersonically Traveling Bodies, AFFDL-TR-68-149, Air Force Flight Dynamics Lab., Wright-Patterson AFB, Ohio (1980).

[2] Nishioka, M. and Morkovin, M.V. : Boundary-Layer Receptivity to Unsteady Pressure Gradients : Experiments and Overview, J. Fluid Mech. 171 (1986) 219-261.

[3] Saric, W.S., Hoos and J.A., Radeztsky, R.H. : Boundary-Layer Receptivity of Sound with Roughness, ASME FED 114 (1991) 17-22.

[4] Kobayashi, R., Fukunishi, Y., Nishikawa, T., Ishii, T., Kohama, Y., Saric, W.S. and Reed, H.L. : The Receptivity of Two-Dimensional Roughness Element in Flat-Plate Boundary-Layer to Sound, Proc. 25th Turbulent Flow Symp. - Sendai (1993) 96-99 (in Japanese).

[5] Kobayashi, R., Fukunishi, Y., Ishii, T. and Reed, H.L. : The Receptivity of Flat-Plate Boundary-Layers to Freestream Sound, Proc. 5th Int. Symp. on Computational Fluid Dynamics - Sendai 2 (1993) 65-69.

[6] Ito H., Kobayashi, R. and Kohama, Y.: The Low-Turbulence Wind Tunnel at Tohoku University, Aero. J. Royal Aero. Soc. (1992) 141-151.

[7] Lin, N., Reed, H.L. and Saric, W.S. : Effect of Leading Edge Geometry on Boundary-Layer Receptivity to Freestream Sound, "Instability, Transition, and Turbulence", ed. M.Y. Hussaini, A. Kumar, and C.L. Streett, Springer-Verlag (1992) 421-440.

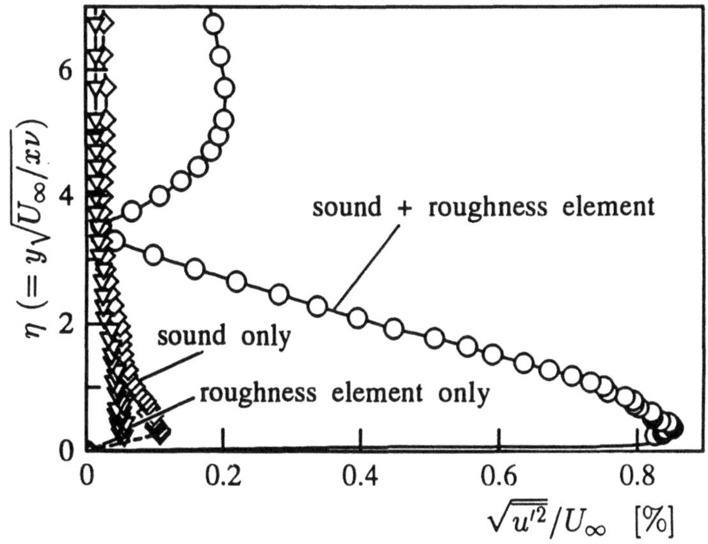

Fig. 1 Velocity fluctuation $\sqrt{\overline{u'^2}}/U_\infty$ along y-axis.

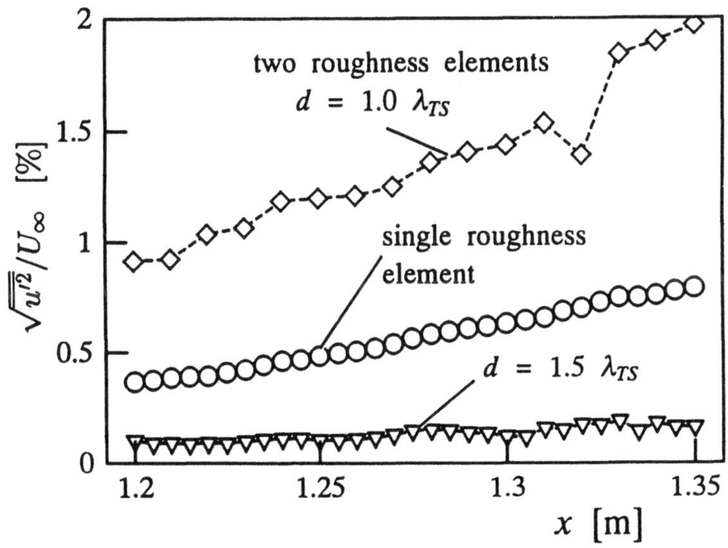

Fig. 2 Effect of distance d between the two roughness elements on velocity fluctuation $\sqrt{\overline{u'^2}}/U_\infty$ along x-axis.
d denotes a distance between the leading edges of two elements and λ_{TS} is the wavelenght of the T-S wave.

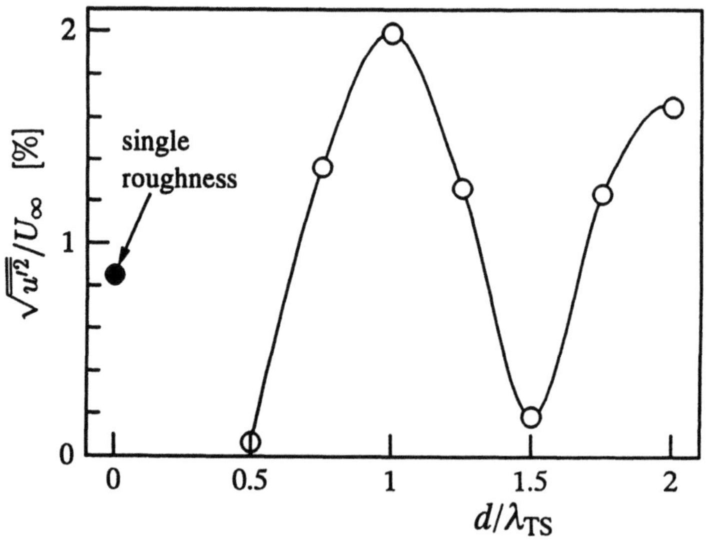

Fig. 3 Maximum amplitude of velocity fluctuations $\sqrt{\overline{u'^2}}/U_\infty$ when distance d between two roughness elements are changed.

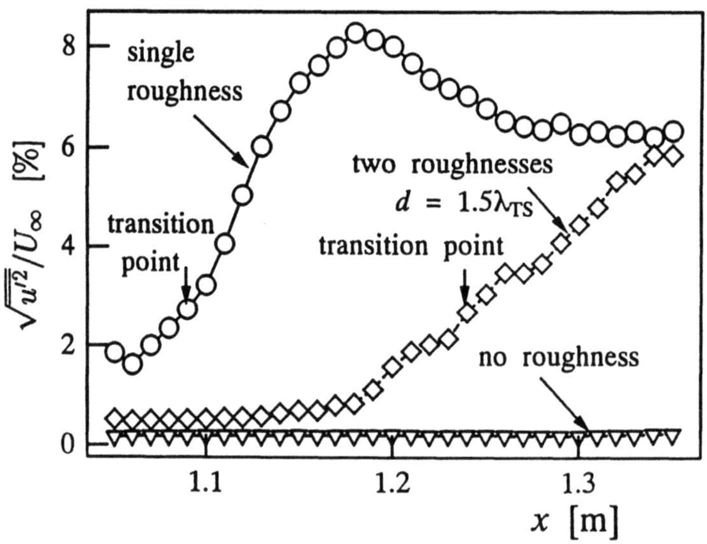

Fig. 4 Effect of adding the second roughness element to the transition point. (U_o =30.0m/s, y=0.79mm)

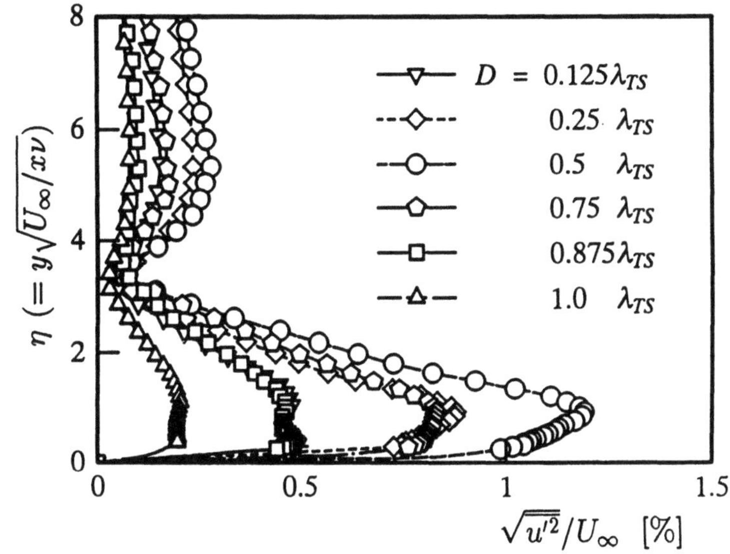

Fig. 5 Effect of distance D between the edges of the roughness element to the velocity fluctuation $\sqrt{u'^2}/U_\infty$ along y-axis.

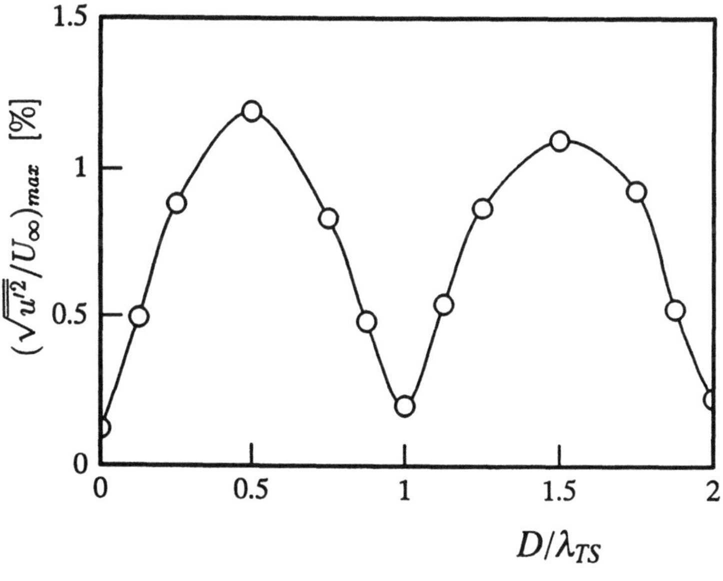

Fig. 6 Maximum amplitude of velocity fluctuation $\sqrt{u'^2}/U_\infty$ with various roughness element widths D.

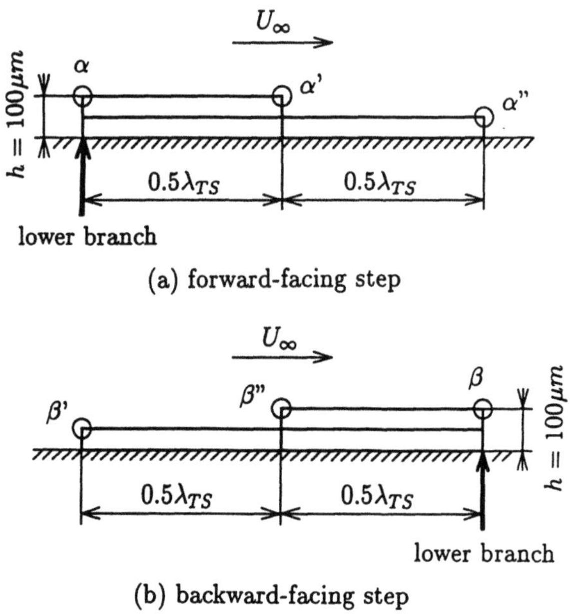

Fig. 7 Roughness element configurations simulating the forward and backward facing steps.

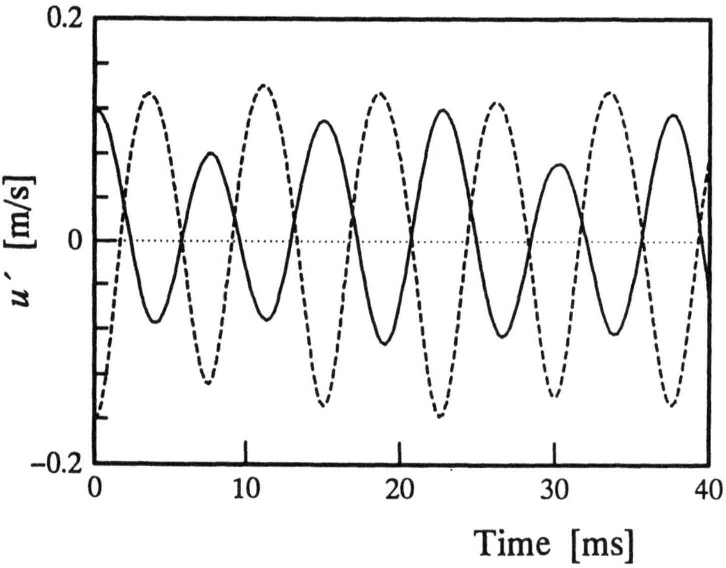

Fig. 8 Velocity fluctuation waves generated by the forward facing step (solid line) and backward facing step (broken line). (x = 1350 mm, y = 1.0 mm ($y\sqrt{U_\infty/x\nu}$=1.0))

Session N: Receptivity II

Laminar Boundary-Layer, Sound Receptivity and Control

V. S. Kosorygin
Institute of Theoretical and Applied Mechanics SB RAS
630090, Novosibirsk, Russia

R. H. Radeztsky, Jr.* and W. S. Saric
Mechanical and Aerospace Engineering
Arizona State University, Tempe, AZ 85287-6106, USA

Summary

The receptivity of a laminar boundary layer on a flat plate is experimentally studied in the presence of 2-D roughness and freestream sound. The work is carried out in the ASU Unsteady Wind Tunnel. It is shown that the T-S wave amplitude can be less than that of the smooth-surface case downstream from the roughness due to a superposition of unstable waves. Thus, receptivity studies of localized sources must consider constructive-destructive interference.

Key words: Sound Receptivity, 2-D Roughness, T-S waves, Control

1. Introduction

This work addresses the combination of leading-edge receptivity and receptivity of two-dimensional (2-D) roughness to freestream sound. These two mechanisms create independent Tollmien-Schlichting (T-S) waves that can undergo destructive or constructive interference with each other. Thus the opportunity for control of T-S waves can be investigated. Recent reviews of boundary-layer receptivity are given by [1-3]. Since the present paper is restricted to experiments on the role of acoustic receptivity, one should consult these reviews for a general treatment of the problem.

When considering a boundary layer over the smooth surface, it is observed that the presence of freestream sound gives rise to unstable waves due to rapid changes in the flow near the leading edge and around non-uniformities of the boundary conditions [1-3], [4-9]. When a roughness element is placed on the surface, it generates additional regions with strong local changes of the flow which in turn are generators of unstable waves [10-11]. A coupling between the different waves initiated by these mechanisms takes place in the vicinity of the roughness. Since the disturbance amplitudes are typically small, the coupling is in the form of linear superposition which gives rise to destructive or constructive interference [12-13]. The Novosibirsk experiments, [12-13], had a high leading-edge receptivity with a blunt leading edge and showed the effects of

---

* Present address: Boeing Commercial Airplane Group, Seattle WA 98124-2207

wave interference. On the other hand, experiments with 2-D roughness [10-11] typically had sharp leading edges and ignored the role of the leading edge.

The goals of this work are to examine the results of [12-13] in a different facility in the case of a sharp leading edge. Thus the basic phenomenon will be confirmed and the value of earlier 2-D roughness experiments [10-11] will be assessed. The basic unknowns of receptivity are the receptivity coefficients which link the initial amplitude of unstable waves with the amplitude of the external disturbances and the identification of the dominant receptivity mechanism. Realization of these unknowns could promote an effective means of laminar-turbulent transition prediction and control in boundary layers.

2. Experimental Facility

These experiments are carried out in the Arizona State University Unsteady Wind Tunnel which is well suited for stability experiments. The flow characteristics of the tunnel are documented in [14]. The computer-based, data-acquisition system of real-time processing, three-dimensional (3-D) traversing mechanism and speaker system developed previously at ASU [13] are employed. The loudspeaker is mounted on the back wall of the plenum, upstream of the honeycomb and screens. An essentially planer sound wave of 0° incidence moves over the plate. The detailed descriptions are found in [14-15].

The flat-plate model, used in [13], has a 1372 mm span, 20.6 mm thickness, and 3700 mm chord. The leading edge has a 340 mm chord and an asymmetric elliptical profile with aspect ratio of 67:1. Surface defects and scratches are hand polished until the *rms* surface roughness is 0.2 μm. The leading-edge junction gap is filled using a polyester patching compound. A trailing-edge flap (280 mm chord) is used to position the stagnation line for symmetric flow. We use boundary-layer shape factors to align the plate for zero pressure gradient.

The 3-D traverse is mounted exterior to the test section. A 45° forward-swept, carbon composite sting with a thickness of 3 mm is passed through the test-side window. Hot-wire probes are mounted on the sting for both boundary-layer and freestream measurements. A special support is mounted on the free end of the sting to eliminate any probe vibrations. The range and steps for the system are: streamwise $x < 1200$ mm, $\Delta x = 200$ μm; spanwise $z < 90$ mm, $\Delta z = 100$ μm; and wall normal $y < 100$ mm, $\Delta y = 12$ μm. Details are in [14-15].

3. Results

A direct measurement of T-S wave amplitudes near the lower branch (Branch I) of the neutral stability curve is impractical because in this region, the amplitudes are extremely small, $O(10^{-6})$, [11]. As a result, we use measurements near the upper branch (Branch II) of the neutral stability curve and linear-stability calculations to determine the initial T-S wave amplitudes at Branch I. The disturbances are at a low level where linear stability has been verified [9], [11-13], [15-16]. In this case, measurements are taken far enough downstream, where the unstable waves are large, and the data are extrapolated to the location of the known source of receptivity.

3.1 Natural disturbances without surface roughness

It is necessary to show the level of T-S waves on the smooth surface with *natural* background disturbances. The basic-state characteristics correspond to the Blasius flow. The spectra of *rms* streamwise fluctuations, $|u'|$, outside (y = 200 mm) and inside boundary layer ($u/U_\infty = 0.3$) for freestream velocity U_∞ = 18 m/s show that inside the boundary layer there is a dominant peak frequency, f =110 Hz, that is not present in the freestream. Similar data are obtained for different freestream velocities at the same location within the boundary layer. The most amplified frequencies are shown in Fig. 1 and are connected by means of a straight line. The other frequencies showing growth are also presented in this plot (solid symbols). The same data are presented in Fig. 2 in the dimensionless form to show their position with respect to the neutral curve. The frequencies increase with the freestream velocity. Other discrete frequencies observed in the spectra do not change with speed and are vibrations of the probe-sting-surface system. Thus, the boundary layer over the smooth flat plate, having a large-aspect-ratio leading edge, under the natural conditions, exhibits a narrow band of frequencies that lie within the region of the unstable T-S waves.

3.2 Acoustic disturbances without roughness

To study the boundary-layer response with respect to the external acoustic disturbances, the hot wire is set at x =1800 mm from the leading edge and a constant distance from the plate surface y that corresponds to the relative dimensionless mean velocity $u/U_\infty = 0.25$ at the freestream velocity U_∞ =18 m/s. The forcing frequency is varied with discrete steps at three different freestream speeds. The results show that the most amplified frequencies correlate well with the data found under natural disturbances.

The conditions are fixed at U_∞ =18 m/s and f = 112.3 Hz ($F = 37\times10^{-6}$). It should be noted this frequency almost coincides with the dominant frequency observed in the Novosibirsk experiments [6], [12-13] which were conducted on a semi-cylindrical leading edge. The parallel linear stability theory predicts the positions of the Branch I and Branch II neutral points at x = 504 mm (R_I = 727) and x = 2130 mm (R_{II} = 1494), respectively. In this range of Reynolds numbers, the measurements inside the boundary layer show a spatially growing T-S wave. The results of boundary-layer scans along the x-coordinate are shown in Fig. 3. The experimental estimations for the T-S wavelength and amplitude can be extracted from these data. The T-S wavelength, determined as a slope of phase function, is $\lambda_{TS} \approx$ 53 mm. Stability theory predicts λ_{TS}= 51 mm. In as much as the total hot-wire signal within the boundary layer is greater than the freestream one ($|u'_{ac}| = 6.8\times10^{-5}U_\infty$) one half the sum of the values at the nearest minima and maxima produces the desirable average magnitudes. This average is used as an estimate for the receptivity coefficient. The data are extrapolated to Branch I using the N-factor calculated from linear theory. The receptivity coefficient is defined by

$$K_s = |u'_{TS}|_I / |u'_{ac}|_I \tag{1}$$

and is found to be $K_s \approx 0.031$, where $|u'_{TS}|$ is the initial *rms* T-S amplitude and $|u'_{ac}|$ is the *rms* acoustic amplitude in the freestream at the same x-coordinate. The linearity of transformation of the external acoustic disturbances into the T-S waves has been verified up to the sound pressure level in 95.6 dB.

Thus, in the boundary layer over the smooth flat plate, having an elliptic, large-aspect-ratio leading edge, the freestream acoustic wave excites unstable T-S waves. The conversion is linear and the receptivity coefficient estimate is close to the value ($K_s \approx 0.042$) found in the blunt-nose experiments [6], [12]. In our case and in the Novosibirsk experiments, this receptivity coefficient is the combination of leading-edge receptivity and curvature-discontinuity receptivity which may have the same interference as discussed in the next section.

3.3 Destructive wave superposition: 2-D roughness

Taking into account the results described above and previous data of the experiments [11-13] in which the linearity of roughness induced sound receptivity has been verified, it is possible to demonstrate the constructive-destructive interference effects. The preliminary measurements with a rectangular strip of 25 mm width and 30 μm height showed that there are some distances of roughness from the leading edge, x_h, for which the velocity oscillations taken far downstream (x =1750 mm) in the boundary layer are less than in the smooth surface case. However, the destructive effect was not so obvious when the boundary-layer scans along the x-coordinate were attempted. The experiments were repeated with a roughness strip* having a height of $h \approx 15\mu$m and located at x_h = 641 mm. Two different cases are shown in Fig. 4.

The same roughness is used to verify the effect under the natural conditions and with "white" noise disturbances. The hot wire is placed at x = 1775 mm. The *rms* amplitudes measured inside and outside boundary layer are collected in a table of $|u'|$. It follows from these data the destructive interference takes place for the natural and "white" noise disturbances in the tunnel.

Table 1.: *rms* disturbance amplitudes $|u'|/U_\infty \times 10^5$ at f = 112.3 Hz

Type:	*mono*		*natural*		*"white" noise*	
$h\backslash\ u/U_\infty$	0.226	1.0	0.226	1.0	0.226	1.0
0	116.3	8.67	7.03	0.68	13.56	0.84
0	113.3	9.12	7.6	0.44	12.55	0.96
0			8.1	0.58	10.4	0.83
0			7.05	0.96	11.9	0.92
15 μm	56.1	8.55	6.4	0.55	8.71	0.89
15 μm	54.1	9.12	7.6	0.49	7.5	0.92
15 μm			6.3	1.05		

* To obtain the small roughness heights of these experiments and those of [10], one unwraps the aluminum foil from heavy-duty transmitter capacitors. The foil is secured to the plate by a very thin coating of light machine oil.

Thus, the destructive/constructive superposition of unstable boundary layer waves on the surfaces with the roughness have been confirmed in a new experimental environment with a much different leading edge.

3.4 Effect of sound level and roughness height

Taking into account the importance of verification of the theoretical receptivity models reviewed in [3] there is a question regarding the influence of the constructive-destructive interference on receptivity when the measurements are conducted away from the receptivity source. It appears as though measurements downstream from a source of receptivity is a result of the superposition of *at least* two unstable waves. The experimental estimations of receptivity will be subject to the alterations due to the small variations of the source location, $\Delta x_h = O(\lambda_{TS}/2)$. In this connection we have studied a response of boundary layer with respect to freestream sound amplitude and roughness height.

Conditions are set at U_∞ =13.3 m/s and f = 75.8 Hz ($F = 45.8\times10^{-6}$, λ_{TS} = 55 mm) using 25 mm strips. The boundary-layer probe is placed at x = 1800 mm and u/U_∞ = 0.304. The $|u'|$ amplitudes versus roughness location are shown in Fig. 5 for a few roughness heights. The height variation in the roughness does not change the position of characteristic minimum in these curves. This implies that the phase relationship of the roughness-induced and smooth-flat-plate T-S waves does not change over this height range. The data obtained with larger heights are presented in Fig. 6 for two freestream sound intensities. The open symbols correspond to the constructive interference and the solid ones to the destructive interference. The horizontal line is the $|u'|$ amplitude in the smooth flat plate case. The intersection of the smooth data and the lower curves give a limit of roughness heights beyond which the $|u'|$ amplitude is higher than the smooth case. We estimate that the limit is about h/δ^* = 0.025 and δ^* is the displacement thickness at the roughness position.

The increase of height influences the local changes to the flow near the roughness element. As a result, the phase and amplitude correlations are changed and the interference pattern measured downstream is transformed. An increase in freestream sound pressure level gives similar effects.

The experimental appraisal of localized source receptivity depends on the destructive-constructive effect. Figure 7 shows the amplitude of the total T-S wave, normalized by the acoustic amplitude, as a function of roughness height for two roughness position x_h. The range of heights, where a linearity of function exists, coincides with the range found in [13] (h < 120 μm). But the slope of functions is different and is determined by the constructive or destructive interference. The correct experimental estimations of localized source receptivity must take into account the described phenomenon.

Acknowledgments

The authors are indebted to Prof. H. L. Reed for stability calculations. The help of Messrs. D. Clevenger and M. C. Mousseux during the experiments is greatly appreciated. This work is supported by AFOSR Grant *90-NA-234*, NASA-Ames *NCC-2-659* and Arizona State University.

References

[1] Nishioka, M. and Morkovin, M.V.: Boundary-Layer Receptivity to Unsteady Pressure Gradients: Experiments and Overview, *J. Fluid Mech.* ***171*** (1986).

[2] Reshotko, E. Boundary layer instability, transition, and control. *AIAA Paper 94-0001* (1994).

[3] Saric, W.S., Reed, H.L. and Kerschen, E.J. Leading-edge receptivity to sound: Experiments, DNS, theory. *AIAA Paper No. 94-2222* (1994).

[4] Leehey, P. and Shapiro, P.J.: Leading Edge Effect in laminar Boundary Layer Excitation by Sound, *Laminar-Turbulent Transition, **I***, Eds. Eppler & Fasel, Springer-Verlag (1980).

[5] Kachanov, Yu.S., Kozlov, V.V. and Levchenko, V.Ya.: Beginning of Turbulence in Boundary Layer *Nauka Sib. Div. Pub.* Novosibirsk (1982)

[6] Kosorygin, V.S., Levchenko, V.Ya., Polyakov, N.Ph.: On Generation and Evolution of Waves in a Laminar Boundary Layer, *Laminar-Turbulent Transision, **II**,* Ed. V.V. Kozlov, Springer-Verlag (1985).

[7] Wlezien, R.W. Measurement of boundary layer receptivity at suction surfaces. *AIAA Paper 89-1006* (1989).

[8] Wlezien, R.W., Parekh, D.E. and Island, T.C. Measurement of acoustic receptivity at leading edges. *Appl. Mech. Rev., **43***, (1990).

[9] Saric, W.S., Wei, W., and Rasmussen, B.K. Effect of leading edge on sound receptivity. *Laminar-Turbulent Transition, IV,* Ed. R. Kobayashi, Springer-Verlag (1995).

[10] Aizin, L.B. and Polyakov, N.Ph.: Acoustic Generation of Tollmien-Schlichting Waves over Local Unevenness of Surface Immersed in Stream, *Preprint 17 ITPM*, Acad. Nauk USSR, Siberian Div. (1979) (in Russian) (see also Nishioka and Morkovin, 1986).

[11] Saric, W.S., Hoos, J.A., Radeztsky, R.H. Jr. Boundary-layer receptivity of sound with roughness. *Boundary Layer Stability and Transition, FED-Vol. 114*, Eds: D.C. Reda, H.L. Reed, R. Kobayashi, ASME (1991).

[12] Kosorygin, V.S.: Experimental Investigation of Flat Plate Laminar Boundary Layer under Low Natural and Acoustic Disturbances, *Ph.D.Thesis, ITAM SB RAS*, Novosibirsk (1986) 202 p.

[13] Kosorygin, V.S., and Polyakov, N.Ph.: Auto Destruction of Unstable Laminar Boundary Layer Waves, *Preprint 10-90, ITAM SB RAS*, (1990).

[14] Saric, W.S. The ASU transition research facility. *AIAA Paper 92-3910.*

[15] Kosorygin, V.S., Radeztsky, R.H., Jr., Saric, W.S.: Laminar Boundary Layer Receptivity and Control: Freestream sound disturbances, 2D roughness, *Preprint 19-92, ITPM, Rus. Acad. Nauk, Sib. Otd.* (1992).

[16] Saric, W.S. Low-speed boundary-layer transition experiments. *Transition: Experiments, Theory and Computations*. Eds. T.C. Corke, G. Erlebacher, M.Y. Hussaini, Oxford (1994).

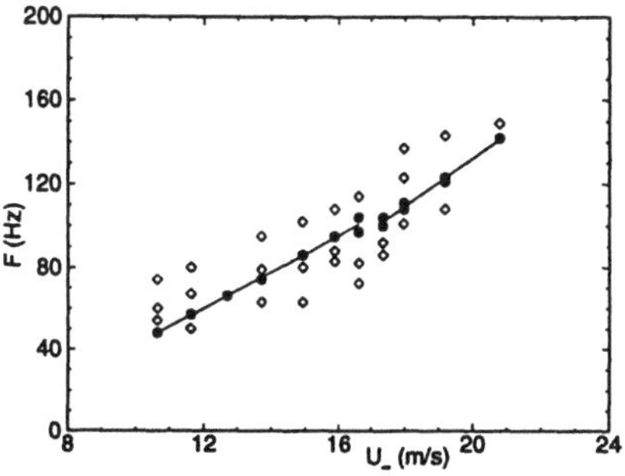

Fig. 1 Locus of most amplified frequencies under *natural* conditions at different speeds. • denotes principal peak, ○ denotes minor peaks.

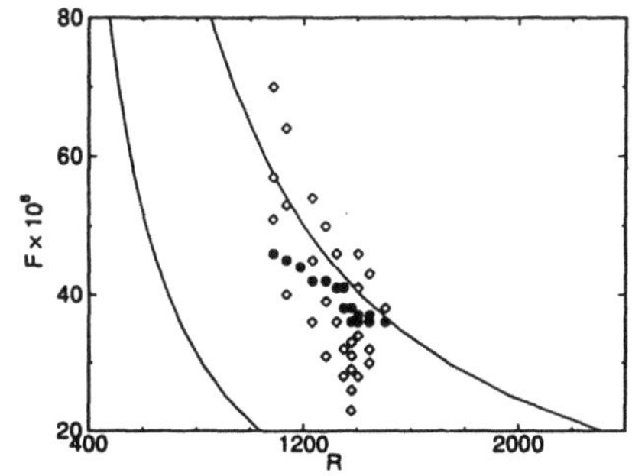

Fig. 2 Dimensionless display of data of Fig. 1, locus of most amplified frequencies under *natural* conditions at different speeds.

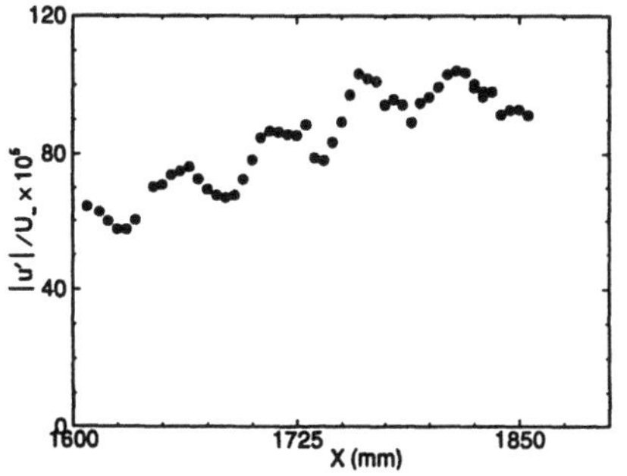

Fig. 3a rms u' amplitude within the boundary layer (u/U_∞ = 0.252) as a function of chord position. U_∞ = 18 m/s, F = 37.2 $\times 10^{-6}$, (f = 112.3 Hz), $|u'_{ac}|$=6.8$\times 10^{-5} U_\infty$.

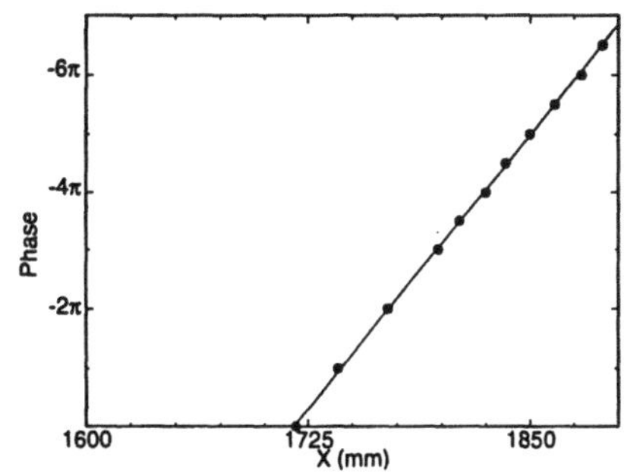

Fig. 3b Phase within the boundary layer (u/U_∞ = 0.252) as a function of chord position. U_∞ = 18 m/s, F = 37.2 $\times 10^{-6}$, (f = 112.3 Hz), $|u'_{ac}|$=6.8$\times 10^{-5} U_\infty$.

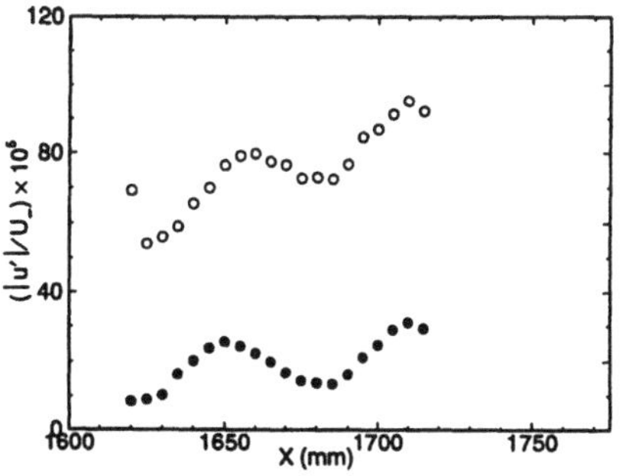

Fig. 4a rms u' amplitude within the boundary layer (u/U_∞ = 0.257) vs x. U_∞ = 18 m/s, F = 37.2 $\times 10^{-6}$, $|u'_{ac}|$=7.6$\times 10^{-5} U_\infty$. ○- no roughness, •- 15 μm rougness at x_h = 641 mm.

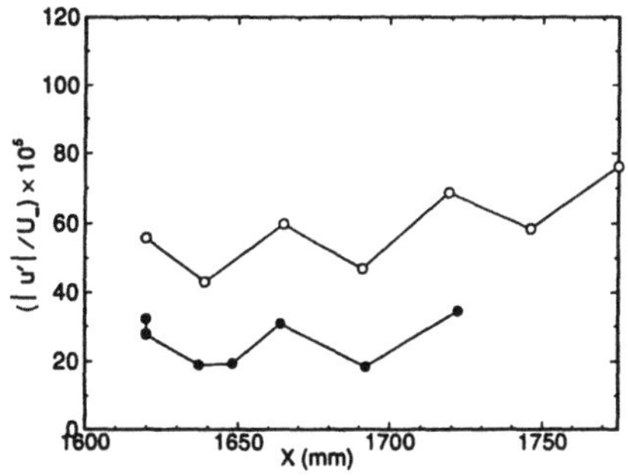

Fig. 4b rms u' amplitude within the boundary layer as a function of chord position. Same as Fig. 4a except u/U_∞ = 0.19 and $|u'_{ac}|$=9.2$\times 10^{-5} U_\infty$

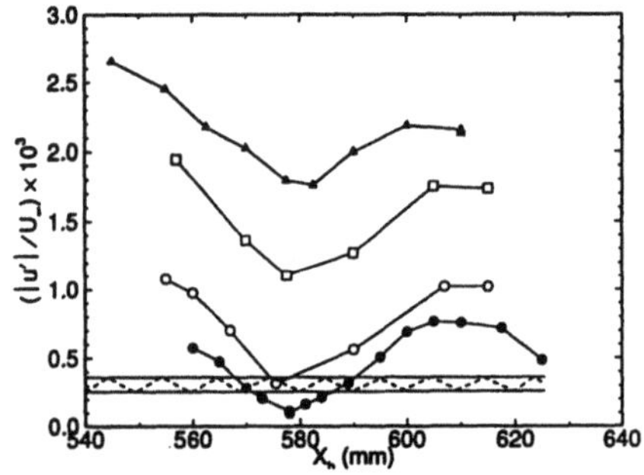

Fig. 5 rms u' amplitude within the boundary layer (u/U_∞ = 0.304) as a function of roughness position x_h. U_∞= 13.3 m/s, F= 45.8 $\times10^{-6}$, (f= 75.8 Hz), $|u'_{ac}|$=7.0$\times10^{-5}U_\infty$. Shaded area is no roughness, •- h= 30μm, ○- 45μm, □- 95μm, ▲- 135μm.

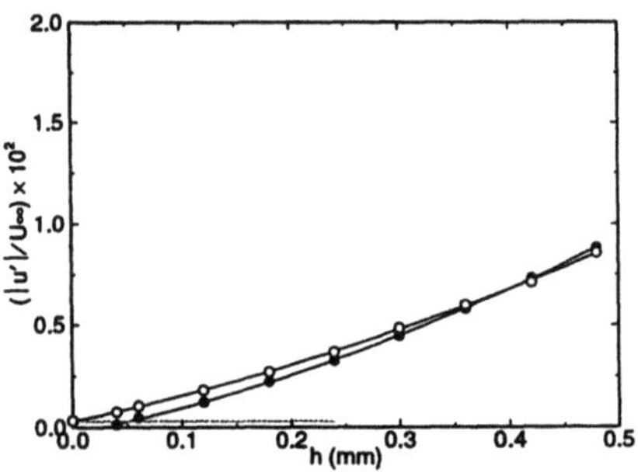

Fig.6a rms u' amplitude as a function of roughness height. Flow conditions of Fig. 5 except $|u'_{ac}|$=7.0$\times10^{-5}U_\infty$. •- x_h= 586 mm, ○- x_h= 611 mm.

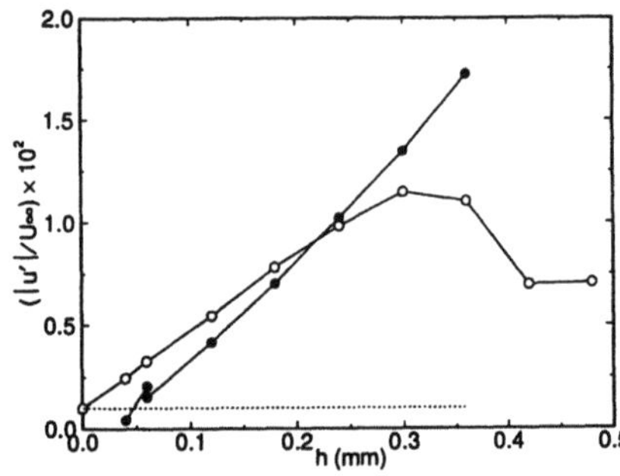

Fig.6b rms u' amplitude as a function of roughness height. Flow conditions of Fig. 5 except $|u'_{ac}|$=11.2$\times10^{-5}U_\infty$. •- x_h= 586 mm, ○- x_h= 611 mm.

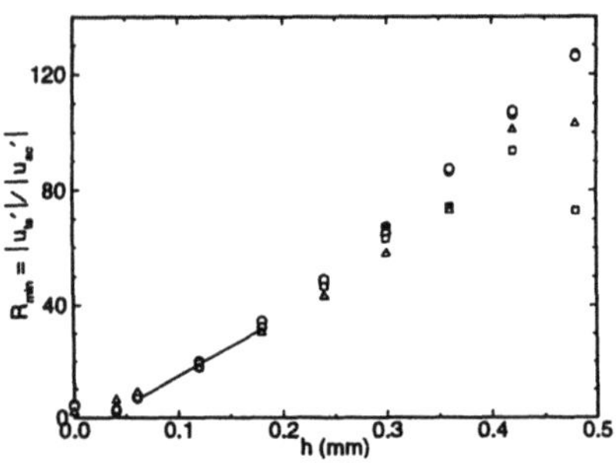

Fig. 7a Minimum T-S amplitude at x = 1800 mm, normalized with freestream sound, as a function of roughness height at different sound levels. Conditions of Fig. 5.

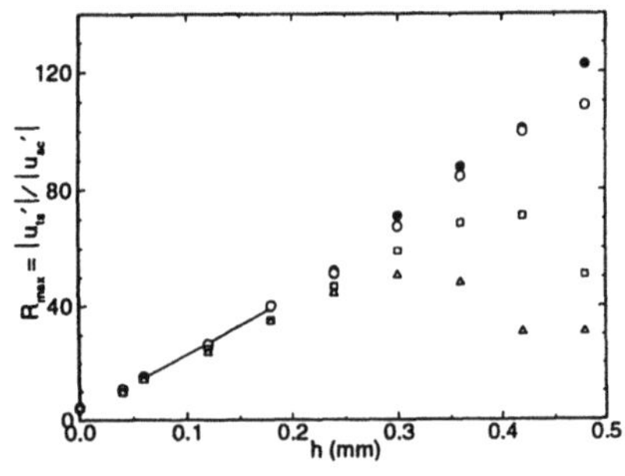

Fig.7b Maximum normalized T-S amplitude. Δ- $|u'_{ac}|$=6.9$\times10^{-5}U_\infty$, □- $|u'_{ac}|$=11.2$\times10^{-5}U_\infty$, ○- $|u'_{ac}|$=16.3$\times10^{-5}U_\infty$ •- $|u'_{ac}|$=23.1$\times10^{-5}U_\infty$

On 3-D Boundary Layer Receptivity

V.Ya.Levchenko, V.A.Scherbakov
Institute of Theoretical and Applied Mechanics,
SB RAS, Novosibirsk, Russia

Summary

Experimental data on receptivity of three-dimensional boundary layer to free-stream turbulence and sound are presented. The model was a flat plate with relatively sharp 45^0 swept leading edge. The flow above the plate had a negative pressure gradient. Traveling waves were object of investigation. The region of their generation was a neighbourhood of the leading edge. Vibration of the leading edge under action of outer turbulence and sound was a generator of traveling waves.

Key words. Laminar boundary layer, receptivity, sound, turbulence, experiment.

INTRODUCTION

It is well-known that the receptivity, i.e. the transformation of background disturbances (turbulence, acoustic, surface vibration etc.) into boundary layer eigenoscillations, is one of the important stages of laminar-turbulent transition process in boundary layers. The solution of receptivity problem has for an object to determine initial amplitudes of instability waves which are developing downstream according to linear stability theory. There is a definite understanding of receptivity process for 2D boundary layers /1-5/. The transformation of background disturbances into T-S waves proved to occur on non-iniformities both the flow and the surface. In particular, the leading-edge region /1/ and small single roughness of the surface /2/ are most sensitive to background disturbances. At present, there is a possibility to calculate receptivity coefficients for some particular cases /4-5/

The problem is much more complex for 3D case, particularly in the region of negative pressure gradients. In this case, additional modes of instability waves occur besides well-known T-S waves /6/. They connected with the cross-flow in a boundary layer and can be both station ary vortices and amplified traveling waves. The relative significance of these instability modes in the transition process is not clear /6/. Obviously, it depends strongly on initial amplitudes of different modes. Therefore, it is necessary to investigate the process of transformation of different background disturbances into different instability modes, i.e. to solve the receptivity problem

The problem of receptivity to steady cross-flow vortices was investigated both theoretically /7/,/8/, and experimentally /9/, /10/ recently and there is a definite understanding of this process. Experimental investigations of receptivity to traveling waves for 3D boundary layers are at very beginning /11/,/12/ and there are only preliminary results and ideas.

Experimental set-up

The experiments were carried out in ITAM subsonic wing tunnel T-324.

The model was a metal flat plate 2m long, 1m wide and 7mm thick. The 45^0 swept leading edge had radius 1.5mm. The plate was mounted horizontally. A flexible wall for creation of negative streamwise pressure gradients was placed above the plate. The measurements were traditional for transition experiments. Hot-wire anemometer was used for measurements of mean flow and disturbances velocities. The traverse mechanism had steps 1mm in streamwise direction (x-axis), 1mm in perpendicular direction (z-axis) and 0.05mm in direction perpendicular to the surface (y-axis). The data were processed with help of Macintosh Classic II computer. The naphthalene-acetone spray was used for visualization of surface flow pattern.

Results of measurements

First series experiments

The first series of experiments was conducted at conditions of "natural" transition. The second one was with excitation of sound.

A typical visualization picture is shown in fig.1. It indicates on presence of stationary vortices. The measurements of their development were not made. Streamwise variation of velocity U/U_0, $U_0 = 21.5$m/s, is shown in fig.2. Measurements were made at distance from surface y=11 mm for -250mm£x£900mm, x=0 corresponds to the leading edge. The measurements of mean velocity and disturbances profiles were made at x=50,-17,-10,-5,-3, 2, 6, 10, 20, 30, 50, 70, 100, 200, 300, 400, 500, 600, 700, 800 and 900 mm. Spectra were measured at points of maxima in profiles of RMS values of disturbances amplitudes u, and out of boundary layer. At some sections spectra were measured in dependence on y-coordinate.

Fig.3a shows mean velocity profile in the neihbourhood of leading edge. U is velocity out of boundary layer. Typical boundary layer profiles are formed between x=6 mm and 10 mm. Fig.3 shows also typical profile of R.M.S. disturbances amplitudes.

Fig.4 shows typical spectrum before the plate. Fig.5 shows the same above the plate. While spectra in fig.4 as well as on the edge of boundary layer are continuons ones, at x=6mm and more, a strong wave packet with frequency range of 20-80 Hz is separated out inside the boundary layer. Also is separated out a more slow wave packet with central frequency ≈280 Hz. It is possible to see separating out of other wave packets downstream.

Fig.6 shows downstream development of u'_{max} for different wave packets as well as for whole spectra. The packet of 20-80 Hz prevails over others. R.M.S. values of amplitudes of "natural" disturbances and 20-80 Hz packet reach several percents while the turbulence level out of boundary layer is approximately 0.3

The low frequency packet of 0-10 Hz proved to be the reason of high turbulence level. It prevails out of the boundary layer. Strong periodical oscillations with the frequency equal to 4Hz were revealed. The analogical oscilloscope traces were taken in second series of experiments at free-stream velocity equal to 12.6m/s. They are presented in fig.7. A low frequency packet is effect of model. It is presented in any experiments and as a rule is filtered at measurements. We decided do not filter it in order to look for transformation and development of periodical vortex disturbance of low frequency. It is seen that the amplification of this wave packet is very week. However, this low frequency packet seems to play a great role in transition process. The transition in these experiments took place from x≈600mm. This process demands special investigation.

Second series of experiments

Because of separations out some wave packets in "natural" conditions, some disturbances of discrete frequencies were excited artificially at more low free-stream speed U =12.6m/s. A loudspeaker was placed several meters downstream. A microphone was used for crude control of sound amplitude level. Two kinds of two-dimensional roughnesses 50mk thick were placed on the surface parallel to the leading edge.one, 33mm wide, began from x=6mm, another, 2mm wide, began from x=0. Two frequencies were chosen which had largest response in the boundary layers: f=62 Hz and f=280 Hz. Some measurements were made without roughness. An analysis showed that roughness and kind of roughness influenced on the date very week. Below we give some results for roughness 2 mm wide and f=62 Hz. In order to check the linearity of the process we made measurements at two different amplitudes of sound levels in microphone location : 115 dB and 105 dB. Corresponding disturbances profiles and ones without sound are given in fig8.

Downstream development of maxima in disturbances profiles for 2 levels of sound are given in fig.9a in nondimensional form where U_1 is corresponding first measured point. We can see an amplification near the leading edge with farther decreasing of amplitude like data of first series of experiments. However, in this case we have amplification with following decreasing at x≥300mm what looks like a neutral curve points. It should be noted that in this series of experiments we had no amplifica tion of disturbances up to x=900mm. Disturbances were periodical in the whole measurements range. The same date in enlarged scale near the leading edge are given in fig.9b.

Because a roughness had not influence on receptivity process, a vibrometer was placed on some point of surface. Spectra of vibration amplitudes in the same arbitrary linear scales of ordinates axis are given in fig.10a-e. Fig.10a-c show spectra without sound (a), for sound levels of 105 dB (b) and 115 dB (c) for frequency of 62 Hz. Figs.10d-e show the sane for frequency of 280 Hz. R.M.S. values of vibrations velocities were 0.22 mm/s, 0.5 mm/s, 1.25 mm/s in figs.10a-c and 0.3 mm/s, 0.52 mm/s in fig.10 d,e. Spectra of vibration velocities also in an the same arbitrary scales of ordinate are given in figs.11a-e. Figs.11a-c show spectra without sound (a), for sound levels of 105 dB (b), 115 dB (c) for frequency of 62 Hz. The same for frequency of 280 Hz is given in figs.11d,e. RMS values of vibration amplitudes were 2.5 mk, 4.3 mk, 9mk correspondingly in figs.11a-c and 2.5mk, 2.8 mk in figs.11d,e. It is seen very clear a strong excitation of the plate vibrations by a sound.

Conclusion

A large experimental material was obtained in series of experiments on receptivity of a 3-D boundary layer to free-stream turbulence and sound.

As a results of these series of experiments we obtained a good idea on receptivity process to traveling wave and their downstream behavior on swept flat plate with sharp leading edge under negative streamwise pressure gradients. The generation of traveling waves takes place near the leading edge both at "natural" conditions and in an acoustic field.

Vibration of the plate seems to be generator of disturbances in the boundary layer under studied conditions

Acknowledgments

This work was supported by Boeing Operations International, Inc. under Agreement 106 with ITAM /1994/ and by the Russian Fundamental Research Foundation /GRANT NUMBER 93-013-173258/

References

[1] Kachanov Yu.S., Kozlov V.V. & Levchenko V.Ya. (1982). Begining of Turbulnce in Boundary Layer. Nauka Publ., 152.(in Russian).

[2]. Aizin L.B. & Polyakov N.Ph. (1979). Acoustic Generation of Tollmien-Schlichting Waves Over Local Unevenness of Surface in Streams. Preprint 17,USSR,Akad,Siberian Division,Inst. Appl.Mech., Novosibirsk.

[3]. Nishioka M. & Morkovin M.V. (1986). Boundary Layer Receptivity to Unsteady Pressure Gradients. Experiments and Overview.J.Fl.Mech. 171.

[4]. Kershen E.J., Chondhary M. & Heinrich R.A. Generation of Boundary Layer Instability Waves by Acoustic and Vortical Free-Stream Disturbances. (1990). In "Laminar-Turbulent Transition" (D.Arnal R.Michel eds), 477-488, Springer Verlag, Berlin et al.

[5]. Crouch J.D. (1992). Non-Localized Receptivity of Boundary Layers. J.Fl.Mech., 244, 567-581.

[6] Reed H.L. & Saric W.S. Stability of three-dimensional boundary layer. Ann.Rev.Fluid. Mech. (1989). 21, 235-284

[7] Crouch J.D. (1993). Receptivity of Three-Dimensional Boundary Layer. AIAA Paper N93-0074.

[8]. Crouch J.D. Distributed Excitation of Cross-flow Vortices in Three-Dimensional Boundary Layers. (1994). IUTAM Symposium on LaminarTurbulent Transition. Abstracts. Sendai, Japan.

[9]. Kachanov Yu.S. & Tararykin O.I. (1990). The Experimental Investigation of Stability and Receptivity of a Swept Wing Flow. Laminar-Turbulent Transition.Springer-Verlag.Berlin,Heidelberg.

[10]. Radezky R.H.,Jr., Reibert M.S. & Saric W.S. Development of Stationary Crossflow Vortices on a Swept Wing. Abstract. IUTAM Symposium on Laminar-Turbulent Transition. Sept.5-9, (1994), Sendai, Japan.

[11]. Bippes H. Experiments of Transition in Three-Dimensional Accelerated Boundary Layer Flows. (1991). Preprint for Int.Conf. on Boundary Layer Transition and Control, organized by Royal Aeronautical Society, UK

[12] Levchenko V.Ya. Report on IUTAM Symposium on Non-Linear Stability of Non-Parallel Flows. (1993). Potsdam, USA.

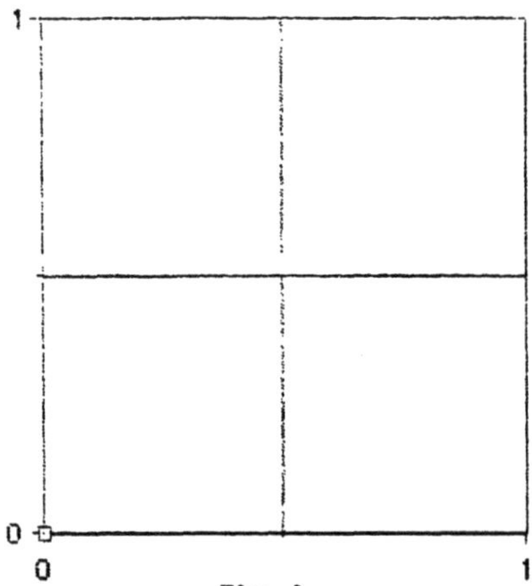

Fig. 1

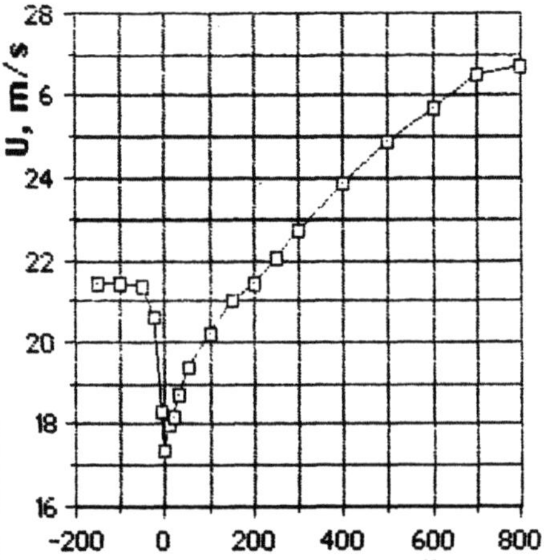

Fig. 2

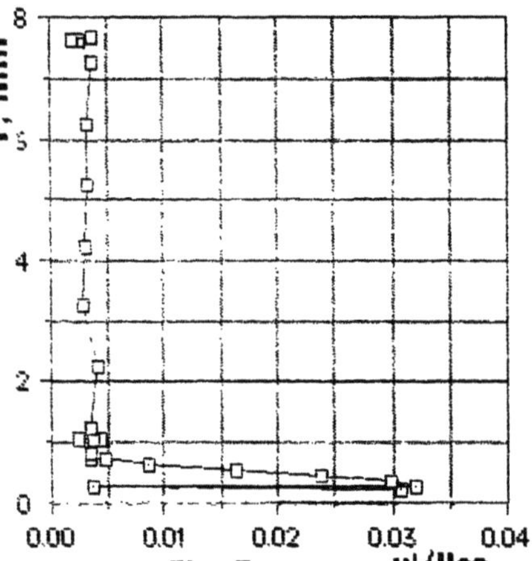

Fig. 3a

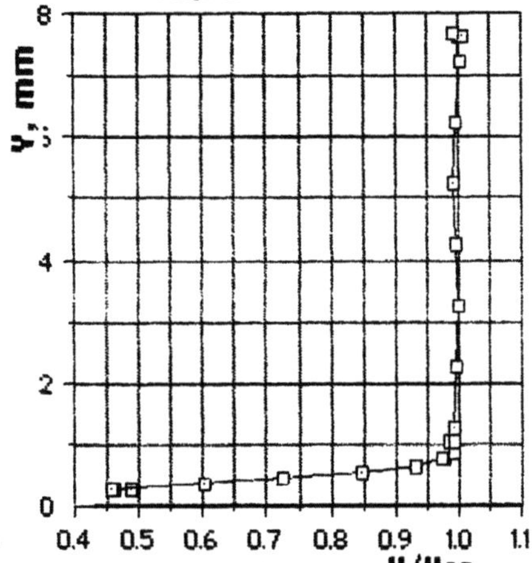

Fig. 3b

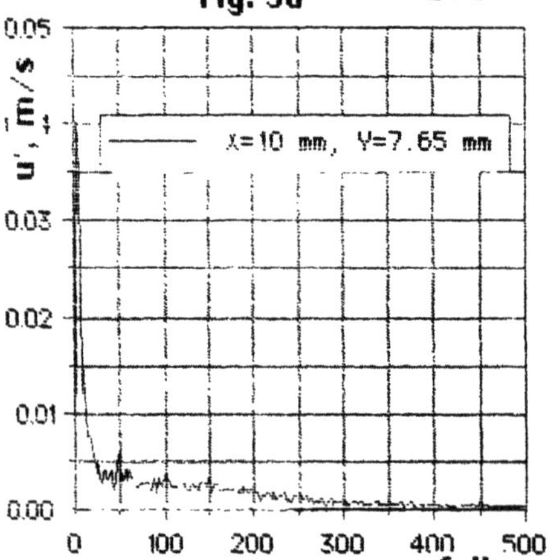

Fig. 5a

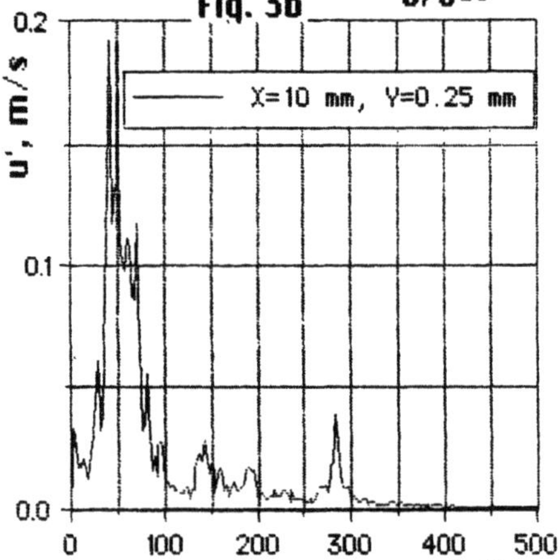

Fig. 5b

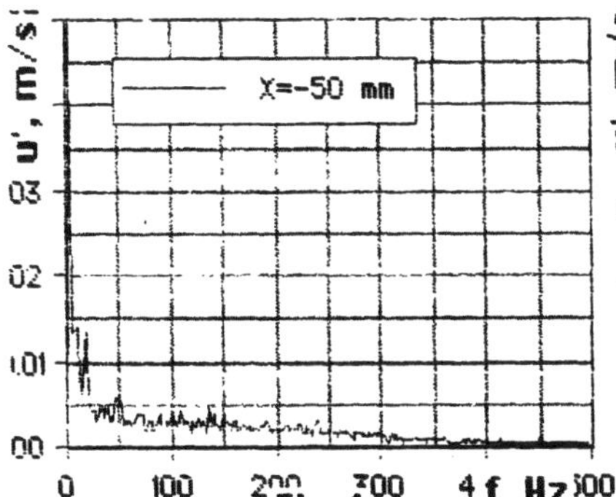

Fig.4

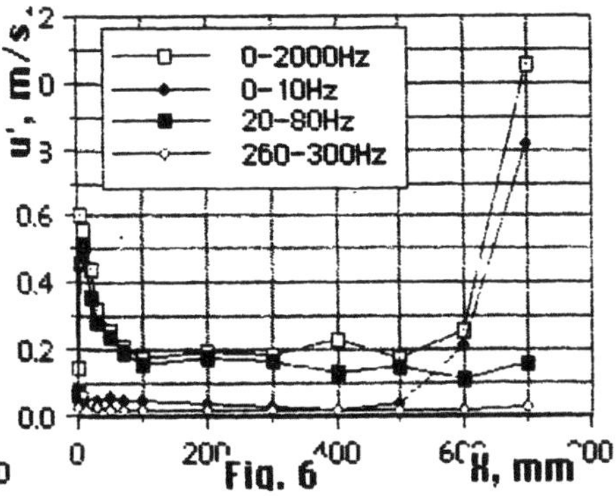

Fig. 6

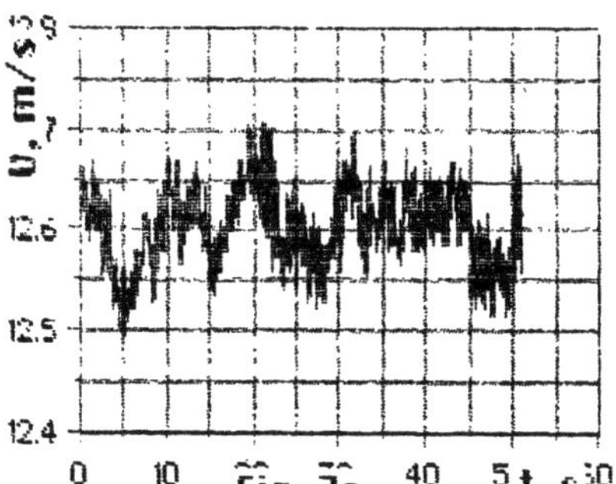

Fig. 7a

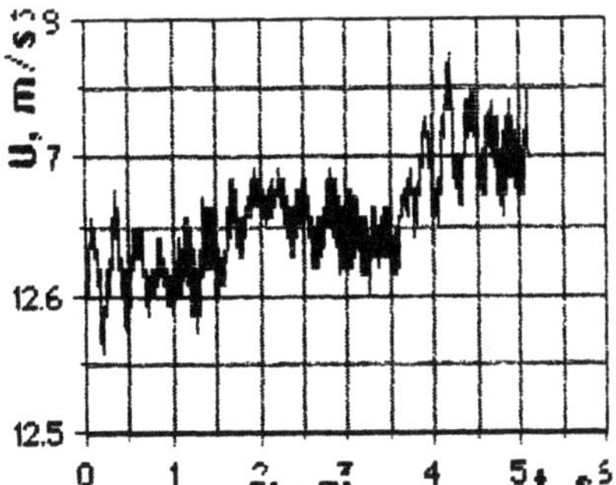

Fig.7b

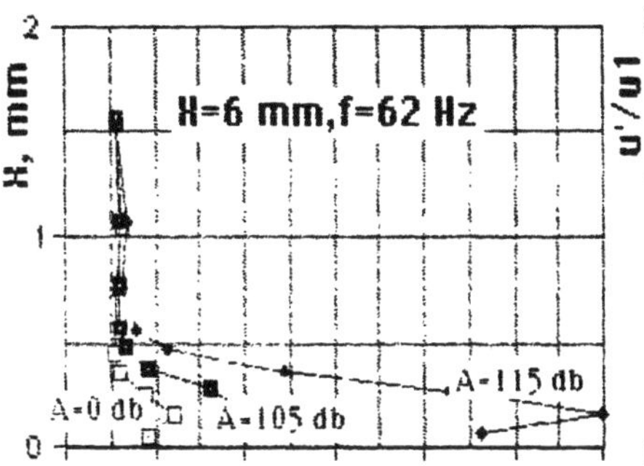

Fig. 8

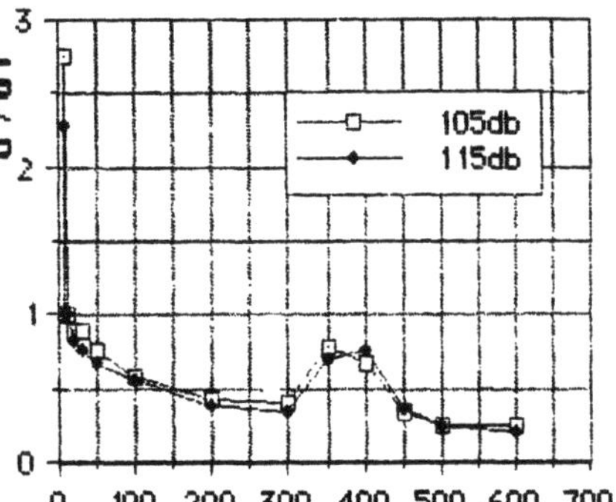

Fig. 9a

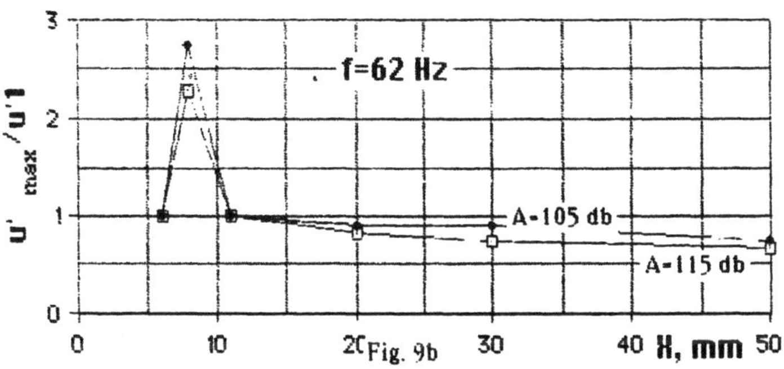

Fig. 9b

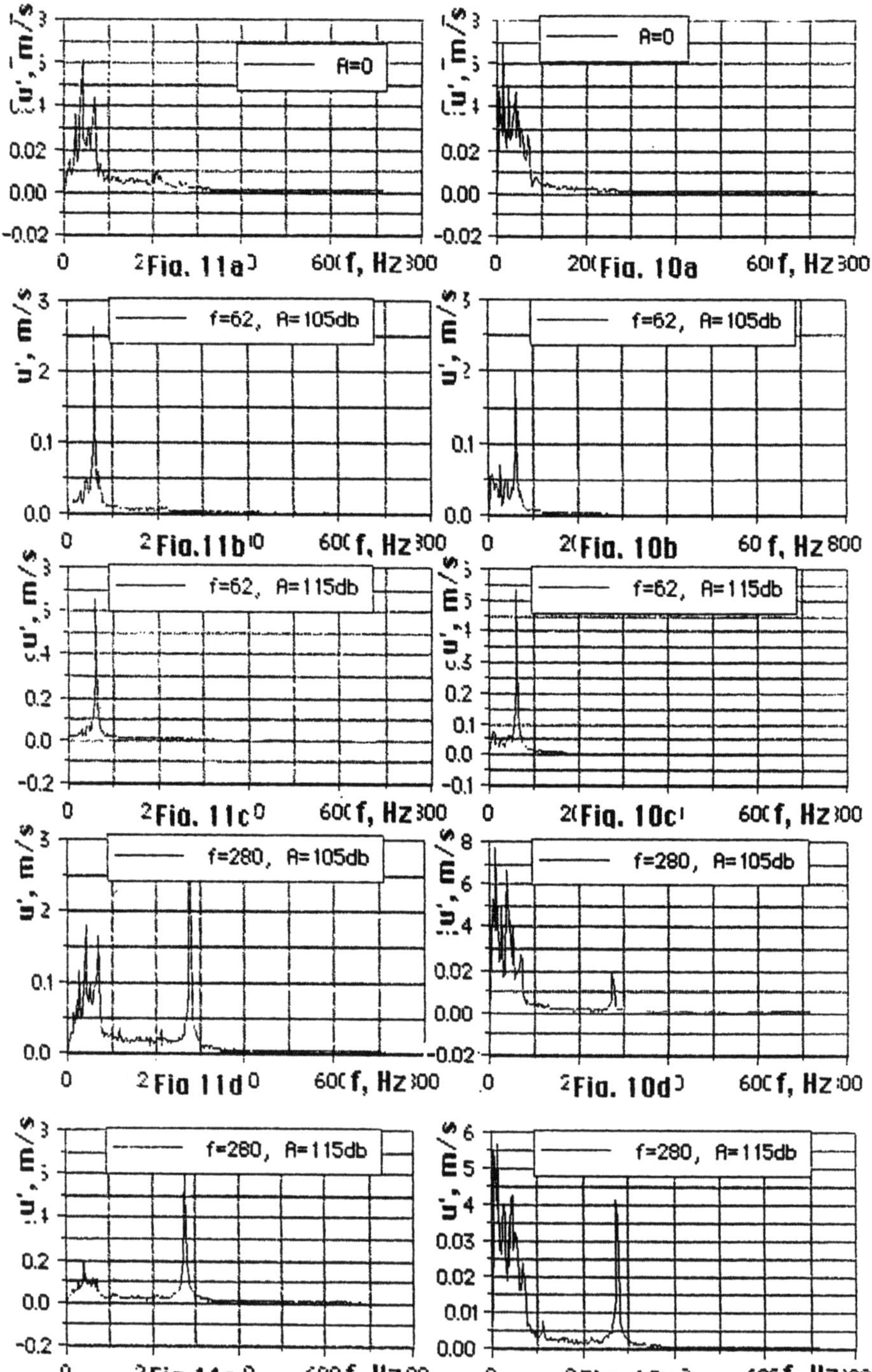

Printing: Mercedesdruck, Berlin
Binding: Buchbinderei Lüderitz & Bauer, Berlin

MIX
Papier aus verantwortungsvollen Quellen
Paper from responsible sources
FSC® C105338

If you have any concerns about our products,
you can contact us on
ProductSafety@springernature.com

In case Publisher is established outside the EU,
the EU authorized representative is:
Springer Nature Customer Service Center GmbH
Europaplatz 3, 69115 Heidelberg, Germany

Printed by Libri Plureos GmbH
in Hamburg, Germany